The Geologic Time Scale

Time Units of the Geologic Time Scale					Distinctive Plants and Animals
Eon	*Era*	*Period*	*Epoch*		
Phanerozoic Eon	Cenozoic Era	Quaternary	Recent or Holocene		First hominids
			Pleistocene		
		Tertiary	Pliocene	2 / 5	Mammals develop and become dominant
			Miocene	24	
			Oligocene	37	
			Eocene	58	
			Paleocene		
	Mesozoic Era	Cretaceous		66	Extinction of many species. First flowering plants, climax of dinosaurs
		Jurassic		144	First birds and mammals, abundant dinosaurs
		Triassic		213	First dinosaurs
	Paleozoic Era	Permian		248	Extinction of trilobites and many other marine animals
		Carboniferous — Pennsylvanian		286	Great coal forests; abundant insects, first reptiles
		Carboniferous — Mississippian		320	Large primitive trees
		Devonian		360	First amphibians
		Silurian		408	First land plant fossils
		Ordovician		438	First fish
		Cambrian		505 / 590	First organisms with shells, trilobites dominant
Proterozoic Eon		} Sometimes collectively called Precambrian			First multicelled organisms
Archean Eon				2500	First one-celled organisms
Hadean Eon				3800	Approximate age of oldest rocks
				4600 ±	Origin of the Earth

Time is given in millions of years (for example, 1000 stands for 1000 million, which is one billion). The table is *not* drawn to scale. We know relatively little about events that occurred during the early part of the Earth's history. Therefore, the first four billion years are given relatively little space on this chart, while the more recent Phanerozoic Eon, which only spans 570 million years, receives proportionally more space.

Zoology

Zoology

Robert L. Dorit

Harvard University

Warren F. Walker, Jr.

Emeritus Professor, Oberlin College

Robert D. Barnes

Gettysburg College

S A U N D E R S   C O L L E G E   P U B L I S H I N G

PHILADELPHIA   FT. WORTH   CHICAGO   SAN FRANCISCO   MONTREAL   TORONTO   LONDON   SYDNEY   TOKYO

Text Typeface: ITC Garamond Light
Compositor: General Graphic Services
Senior Acquisitions Editor: Julie Levin Alexander
Developmental Editor: Christine Connelly
Managing Editor: Carol Field
Project Editor: Martha Brown
Developmental and Copy Editor: Joanne Fraser
Manager of Art and Design: Carol Bleistine
Art Director: Christine Schueler
Art and Design Coordinator: Doris Bruey
Text Designer and Layout Artist: Tracy Baldwin
Cover Designer: Lawrence R. Didona
Text Artwork: J&R Art Services, Inc.
Photo Editor: Robin Bonner
Director of EDP: Tim Frelick
Production Manager: Charlene Squibb
Marketing Manager: Marjorie Waldron

Front cover: Alaskan Brown Bears. Thomas Mangelsen/Images of Nature
Back cover: Alaskan Brown Bears. John Shaw/Tom Stack & Associates
Half-title page: Sally light-foot crabs (*Grapsus grapsus*) on lava at tide line. Frans Lanting/Minden Pictures
Frontispiece: Sally light-foot crab (*Grapsus grapsus*). Frans Lanting/Minden Pictures

Printed in the United States of America

ZOOLOGY

ISBN 0-03-030504-7

Library of Congress Catalog Card Number: 90-050861

1234   032   98765432

PREFACE

Modern zoology is a dynamic branch of science. The proper study of animals draws from every subdiscipline in biology—from the study of genes and DNA sequences to the analysis of complex ecosystems and of the biosphere. If we are to understand animal structure and function, behavior and social organization, diversity and community complexity, we cannot be content with simply describing living animals. We need to understand the biological past, the evolutionary history of the living world.

APPROACH

Throughout this book, we have emphasized the comparative approach to the study of animals, contrasting living species to their extinct ancestors, and tracing the similarities among organisms that bear clear witness to their common ancestry. Evolution—the notion that all animals are part of a single genealogical network, and that descent with modification is a sufficient explanation for the shape of the living world—forms the conceptual core of this book. Without the golden thread of evolutionary theory, the animal world cannot be truly understood.

The four primary goals of *Zoology* are (1) to introduce students to the diversity and interrelationships among animals; (2) to explain the basic molecular, physiological, and developmental mechanisms that underlie and unite all animal life; (3) to explain the forces that have shaped and continue to shape animal form and function, thus enabling students to understand relationships and the predictability of animal design; and (4) to convey some of the excitement of science by examining its methods: the interplay of ideas, experiments, and observations that lead to the formulation of theories and hypotheses, and the profound influence of social context on the work of scientists.

ORGANIZATION

Part I, *The Conceptual Basis of Zoology*, begins with a discussion of evolution as an example of what science is and is not, and of what it can and cannot explain. We emphasize that biology as a science seeks to explain both the unity and diversity of the living world. Biology utilizes the methods and concepts of other sciences, but biology differs from other sciences in having a strong historical component. Past events in the history of life greatly influence the diversity and character of modern organisms and their structure and function. Biology is unique among the natural sciences in its emphasis on hierarchy: the living world reveals order at many levels of organization, from the structure of DNA to the flow of energy in the biosphere. We examine how these phenomena have been explained, culminating with Charles Darwin's theory of evolution.

The foundations for discussing life and evolution are laid in Part II, *Life and Its Continuity*. Since the chemical background of students is so diverse, ranging from secondary school chemistry through one or more college courses, we review the basic chemical and physical concepts that students will need to know in Appendix A. The material is available to those who need it, but its placement in the appendix preserves the flow we intend as we develop biological concepts. We begin Part II by discussing the unique molecules of life, and go on to examine hypotheses for the origins of life and of cells. A consideration of Mendelian and molecular genetics lays the foundation for discussing the mechanisms of evolutionary change.

How animals obtain the materials they need, eliminate wastes, integrate their activities, reproduce, and meet the challenges of survival are the issues dealt with in Part III, *Animal Form and Function*. Cell physiology is integrated with discussions of the structure and physiology of the organ systems. Although we include numerous examples drawn from human physiology, our coverage is broadly comparative. The themes of diversity and contingency, the evolution in different groups of alternate solutions to different problems, are explained.

Evolution has resulted in vast animal diversity and we explore this issue in Part IV, *Animal Diversity*. By building upon the unifying principles developed in Part III, we emphasize the distinctive aspects of the biology of the various groups, their adaptive diversity, and their evolutionary interrelationships.

We conclude in Part V, *Interactions of Organisms and Their Environment,* by considering the complex interrelationships that have evolved between animals and

the totality of the biologic and abiotic environments in which they live. We must elucidate these relationships in order to understand how the world works, how dependent we are on other organisms, and how our activities affect the natural world and, ultimately, our own well being.

TEACHING OPTIONS

Modern zoology is a complex subject, and we have tried to build the fundamentals gradually so as to make the material accessible to all students. We believe, however, that higher education is best served if students are challenged. Our expectations of the level of students' reading and understanding are high. Although ideally suited for a two-semester freshman or sophomore zoology course, this book is sufficiently flexible to be used in a one-semester course, or in a biology course with a strong zoological component. A zoology course that follows a course in cell and molecular biology and evolutionary principles can omit Parts I and II. Some instructors may wish to emphasize the unifying physiological principles underlying zoology and will rely on Part III (Animal Form and Function), which precedes and is not dependent on the survey of the animal kingdom. The animal examples used in Part III will be familiar to most students. Part IV, Animal Diversity, may be omitted or abbreviated by selecting groups of particular interest. Other courses may wish to emphasize animal diversity and classification and put the emphasis on Part IV. Many cross-references in Part IV will lead the students back to fuller discussions of anatomy and physiology presented in Part III.

FEATURES

Many features of this book are designed to enhance the learning process. The use of color throughout the book has enabled us to render drawings and diagrams clearly, and to show students what animals really look like. *Important terms* are set in boldface type when they first appear, and their classical *derivation* is given in the text when this would be helpful in remembering the term. In many chapters we *Focus* on certain timely topics not discussed in the main body of the text. In some chapters we have included short essays, *Connections*, that cut across animal groups and are not limited to the group under consideration. A *Summary* at the end of each chapter emphasizes the most important material that students should have gained. An annotated list of *References and Selected Readings* serves as a starting point for students or instructors who wish to explore the subject of the chapter in greater depth. These reading lists have been selected to emphasize recent material along with particularly important classic works. An abbreviated phylogenetic tree or cladogram at the beginning of each chapter on animal groups highlights the relationship of the group under consideration to other groups, and a list of *Characteristics* defines the group. *Classifications* at the end of these chapters list the subgroups, the derivation of their names, and their distinctive features. Important terms and concepts are defined in a *Glossary*. Page range entries in the *Index* will lead students directly to the pages where the topic is discussed most fully, and single page entries to other important places. Many general topics can be located either through the *Contents Overview* or the *Contents*. The latter includes a list of the primary and secondary topics in the chapters.

ANCILLARY AIDS

Many ancillary aids are available to help users of this textbook. These include an *Instructor's Manual with Test Bank* prepared by Vicky McMillan of Colgate University and Peter Dalby of Clarion University of Pennsylvania, an IBM and a MacIntosh *Computerized Test Bank*, and a set of 100 *Overhead Transparency Acetates*. In selecting figures for the transparencies, we have emphasized ones that cannot be easily and quickly drawn by the instructor in class. Also, a new edition of Boolootian/Heyneman's *An Illustrated Laboratory Text in Zoology* will be available for the fall semester to further enhance the student's understanding of zoology.

ACKNOWLEDGMENTS

Writing, illustrating, and producing a book of this scope is a major task that the authors could not have carried out without the help and commitment of many others.

We are very grateful to Edward Murphy, former Senior Acquisitions Editor of Saunders College Publishing for launching this book, and to our current Senior Acquisitions Editor, Julie Levin Alexander, for her continuing support. The guidance of Christine Connelly, our Managing Developmental Editor, was invaluable as we prepared and revised the manuscript. We also owe much to Joanne Fraser, our Developmental and Copy Editor, for the many hours she spent critically reading our material, helping us smooth out rough spots, and ensuring that the treatment of material was consistent throughout the book. Photo Editor Robin Bonner knew just where to find the outstanding photographs that add so much to our text. We owe much to Laszlo Meszoly for the initial preparation of new art. Christine Schueler, Art Director, supervised the preparation of the drawings and skillfully put the art program and design together. Carol Field, Managing Editor, and Martha Brown, Project Editor, supervised the production of the book, bringing all parts together at the right time, and keeping a complex and tight production schedule running smoothly. One of our wives, Tensy Walker, helped with the proofreading. We are also much indebted to Janis Diehl of Gettysburg College for help in preparing the glossary and index and for other manuscript and production tasks. We also wish to thank Stephen Hart for his help in revising the final draft of the last two chapters on ecology.

All parts of the manuscript were carefully read and in some cases reread by teachers and experts in various subjects. We wish to thank all of our reviewers for taking time from busy schedules to help us in this way. Their comments and suggestions have helped us to improve the quality and readability of the book, but the authors, of course, take responsibility for any errors or shortcomings that remain. We hope users of the book will call these to our or to the publisher's attention. A list of our reviewers follows:

S. Kris Ballal, Tennessee Technological University
Earl B. Barnawell, University of Nebraska at Lincoln
Malcolm R. Braid, University of Montevallo
Judy Brown, San Jacinto College-North Campus
Gary J. Brusca, Humboldt State University
Alan Brush, University of Connecticut

Frank J. Bulow, Tennessee Technological University
Warren Burggren, University of Massachusetts at Amherst
Bryan Burrage, College of the Desert
Ann Cali, Rutgers University
David B. Campbell, Rider College
Albert G. Canaris, University of Texas at El Paso
Greg Capelli, College of William and Mary
David J. Cotter, Georgia College
David L. Cox, Illinois Central College
Peter Dalby, Clarion University
George Delahunty, Goucher College
Peter K. Ducey, University of Texas at Arlington
Kay Etheridge, Gettysburg College
Linda S. Eyster, Tufts University
Arthur J. Goven, University of North Texas
Karen Hart, Peninsula College
Sherman Hendrix, Gettysburg College
Janann V. Jenner, New York University
Ira Jones, California State University at Long Beach
Valerie M. Kish, Hobart and William Smith Colleges
Stanley E. Lewis, St. Cloud State University
Steele Lunt, University of Nebraska at Omaha
Robert F. McMahon, University of Texas at Arlington
John S. Mecham, Texas Technological University
David Nunnally, Vanderbilt University
Larry R. Petersen, San Jacinto College Central Campus
Raleigh K. Pettegrew, Denison University
Edwin C. Powell, Iowa State University
Jonathan L. Richardson, Franklin and Marshall College
William Sheppard, California State University
  at Sacramento
Ralph Sorensen, Gettysburg College
Jerry P. Suits, Brazosport College
Elliot J. Tramer, University of Toledo
Stuart L. Warter, California State University at Long Beach
Olivia White, University of North Texas
Edward J. Zalisko, Blackburn College

Robert L. Dorit

Warren F. Walker, Jr.

Robert D. Barnes

February 1991

CONTENTS OVERVIEW

PART I

The Conceptual Basis of Zoology   1

1 The Nature of Science and the Science of Nature   3

2 The Core of Modern Biology: Evolution   15

PART II

Life and Its Continuity   33

3 The Molecules of Life   35

4 The Origin of Life   53

5 The Eukaryotic Cell: Structure and Evolution   63

6 Genetics I: The Fundamental Observations of Heredity   87

7 Genetics II: The Molecular Basis of Inheritance   109

8 Evolution I: The Origin of Evolutionary Novelty   135

9 Evolution II: Populations, Species, and Macroevolution   153

PART III

Animal Form and Function   183

10 Symmetry, Form, and Life Style   185

11 Body Covering, Support, and Movement   199

12 Digestion and Nutrition   235

13 Cellular Respiration, Gas Exchange, and Metabolic Rates   261

14 Internal Transport   295

15 The Immune System   323

16 Excretion and Water Balance   347

17 The Nervous System   371

18 Receptors and Sense Organs   401

19 The Endocrine System   433

20 Behavior   457

21 Reproduction   483

22 Embryonic Development   509

PART IV

Animal Diversity 539

23 Taxonomy, Phylogeny, and the Origin and Evolution of the Animal Kingdom 541

24 Protozoa 561

25 Sponges 585

26 Cnidarians 595

27 Flatworms 621

28 Aschelminths 641

29 Molluscs 657

30 Annelids 693

31 Arthropods I: Chelicerates and Crustaceans 715

32 Arthropods II: Uniramians 749

33 Bryozoans 769

34 Echinoderms 777

35 Hemichordates and Chordates 797

36 Fishes 815

37 Amphibians 843

38 Reptiles 861

39 Birds 881

40 Mammals 909

41 Primates and Human Evolution 935

PART V

Interactions of Organisms and Their Environment 957

42 Population Ecology 959

43 Communities and Ecosystems 979

Appendix A The Chemical and Physical Basis for Zoology A-1

Appendix B Scientific Units and Quantities A-11

Glossary G-1

Index I-1

CONTENTS

Preface   v

I   THE CONCEPTUAL BASIS OF ZOOLOGY   1

1   The Nature of Science and the Science of Nature   3

The Domain of Science   4
What Do Creationists Want?   5
Science and the Material World   6
   The Characteristics of Science   7
   The Terms of Science   8
   The Fallacy of Creation-Science   9
The Nature of Biological Explanation   10
Summary   13
References and Selected Readings   13

2   The Core of Modern Biology: Evolution   15

The Major Features of the Living World: What Must We Explain?   16
   Unity and Diversity   16
   The Nature of Biological Adaptation   17
   History and Contingency   18
Darwin's World: Biological Thought in the Early 19th Century   19
   The Facts Are Already in Place   19
   Natural Theology   21
Darwin and the Structure of *On the Origin of Species*   21
   Establishing the Fact of Evolution   22
   The Agent of Evolutionary Change: Natural Selection   27
■ *Focus 2.1   Alfred Russel Wallace or "Why Do Great Minds Think Alike?"*   30
Summary   31
References and Selected Readings   31

II   LIFE AND ITS CONTINUITY   33

3   The Molecules of Life   35

Organic Compounds   36
Carbohydrates   37
Lipids   39
Proteins   41
   Protein Structure   41
   Enzymes   44
Nucleic Acids   46
   Nitrogenous Bases as Energy and Signaling Molecules   49
■ *Focus 3.1   The Study of Proteins*   50
Summary   50
References and Selected Readings   51

4   The Origin of Life   53

The Evidence: The Fossil Record of Early Life   54
Making Life Possible: The Prebiotic World   56
   A Single Origin of Life   56
   The Origin of Complex Molecules   56
Life and Nonlife: Where Is the Boundary?   58
   The Essential Characteristics of Life   59
   Cutting the Gordian Knot: Catalytic RNA   59
Summary   61
References and Selected Readings   61

5 The Eukaryotic Cell: Structure and Evolution   63

Organizing Complexity   *64*
   Defying the Second Law   *64*
   The Cell and the Second Law   *64*
The Cell Theory   *65*
Two Underlying Cell Plans: Prokaryotes and Eukaryotes   *66*
The Basic Features of the Eukaryotic Cell   *66*
   The Cell Membrane   *67*
   Transport of Materials Across the Membrane   *69*
   Cytoplasmic Organelles   *73*
   The Nucleus   *78*
   The Cytoskeleton   *80*
The Evolutionary Origin of the Eukaryotic Cell   *80*
   The Endosymbiotic Theory   *81*
■ *Focus 5.1   Revealing Cell Ultrastructure   82*
   The Direct Filiation Theory   *83*
Summary   *84*
References and Selected Readings   *84*

6 Genetics I: The Fundamental Observations of Heredity   87

The Fundamental Observation: Like Begets Like   *88*
Agriculture and Domestication: The Earliest Genetics   *88*
Mendel and the Laws of Inheritance   *88*
   Mendel's Insight: Quantify Variation   *88*
   The Basic Vocabulary of Genetics   *91*
   The Law of Equal Segregation   *91*
   The Law of Independent Assortment   *94*
The Material Basis of Mendelian Genetics   *95*
   Why Was Mendel Ignored?   *95*
   The Rediscovery of Mendel   *95*
■ *Focus 6.1   Gene Interaction   97*
The Somatic Cell Cycle and Mitotic Division   *98*
   The Cell Cycle   *98*
   The Events of Mitosis   *98*
Meiosis: The Production of Gametes   *101*
   Meiosis I   *101*
   Meiosis II   *102*
The Chromosomes Carry Mendel's "Particulate Factors"   *103*
Sex Linkage   *104*
Summary   *107*
References and Selected Readings   *107*

7 Genetics II: The Molecular Basis of Inheritance 109

The Nature of the Gene *110*
 The Importance of the Nucleus *110*
 The Transforming Principle *112*
 Nucleic Acid Is the Hereditary Material *113*
 The One Gene–One Enzyme
  Hypothesis *115*
The Structure of DNA: The Double
 Helix *117*
The Replication of DNA: Semiconservative
 Copying *119*
Interpreting DNA: The Genetic Code *122*
 How Long Are the Words? *122*
 The Grammar of DNA *122*
 Cracking the Wobbly Code *122*
The Central Dogma: Protein Synthesis *124*
 Transcription *125*
 Translation *127*
 Assembling the Protein *128*
Summary *132*
References and Selected Readings *132*

8 Evolution I: The Origin of Evolutionary Novelty 135

The Origin of Novelty *136*
 Point Mutations *136*
 Frame Shift Mutations *137*
 Transposable Elements *138*
 Chromosomal Changes *140*
■*Focus 8.1 The Molecular Basis of
Mendel's Wrinkled Seed Trait 140*
The Shuffling of Existing Variation:
 Recombination *144*
 The Physical Basis of Recombination:
  Crossing-Over *144*
 The Genetic Consequences of
  Recombination *144*
 Chromosome Mapping *145*
 Lucky Mendel *146*
The Consequences of Mutation *147*
 Mutations Are Overwhelmingly
  Deleterious *147*
 The Effect of Mutations Is Context
  Dependent *148*
 The Direction and Rate of Mutation *149*
 The Random Nature of Mutation *150*
Summary *151*
References and Selected Readings *151*

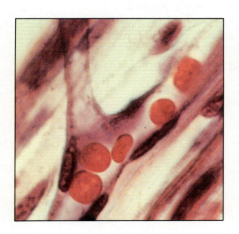

9 Evolution II: Populations, Species, and Macroevolution 153

The Evolving Population *154*
    Evolution Depends on Variation *154*
    The Hardy-Weinberg-Castle Law *154*
    Microevolutionary Forces *160*
◼ *Focus 9.1 Selection 162*
◼ *Focus 9.2 How Is Genetic Variation Maintained?: The Cause of Sickle Cell Anemia 164*
The Origin of New Species *168*
    Darwin and the Problem of Diversity *169*
    The Biological Species Concept *170*
    The Maintenance of Reproductive Isolation *170*
    The Evolution of Reproductive Isolation *172*
    Allopatric Speciation *172*
◼ *Focus 9.3 Continental Drift 174*
Macroevolution: Large-Scale Patterns in the History of Life *175*
    Pattern and Tempo in Evolution *175*
Summary *181*
References and Selected Readings *181*

III ANIMAL FORM AND FUNCTION 183

10 Symmetry, Form, and Life Style 185

Motility and Symmetry *186*
◼ *Focus 10.1 Anatomical Planes 186*
Animal Architecture *189*
Size and Shape *190*
Colonial Organization *192*
Symbiotic Organization *192*
Predictability in Animal Design *193*
Animal Groups *194*
Summary *196*
References and Selected Readings *196*

11 Body Covering, Support, and Movement 199

Body Covering *200*
    Epithelial and Connective Tissues *201*
    The Integument of Invertebrates *202*
    The Vertebrate Integument *203*
    Skin Coloration *204*
Support *204*
    Functions of Skeletons *204*
    Skeletal Materials and Types *205*
    Histology of Cartilage and Bone *207*
    Development and Growth of Cartilage and Bone *209*
    Support and Architecture in Appendicular Skeletons *210*
    The Vertebrate Skeleton *210*
Movement *214*
    Muscle *214*
    Biochemistry of Muscle Contraction *216*
    Physiology of Muscle Contraction *218*
    The Vertebrate Muscular System *221*
    Nonmuscular Movement *223*
Locomotion *225*
    Ciliary Propulsion *226*
    Propulsion by Changes in Body Shape *226*
    Propulsion by Body Undulations *228*
    Appendicular Propulsion *228*
◼ *Focus 11.1 Some Physics of Appendicular Movement 230*
Summary *232*
References and Selected Readings *233*

12 Digestion and Nutrition 235

Kinds of Diets *236*
Steps in Food Processing *236*
    Digestion *237*
    Absorption *239*
Evolution of the Animal Gut *241*
Feeding Mechanisms and Gut
    Specializations *243*
    Sucking and Cropping Herbivores *243*
    Raptorial Feeding *244*
    Parasitism *245*
    Suspension Feeding *245*
    Deposit Feeding *246*
The Vertebrate Pattern *246*
    Mouth *246*
    Pharynx and Esophagus *248*
    Stomach *249*
    Liver and Pancreas *249*
    Intestine *251*
    Regulation of Digestive Secretions *254*
Nutrition *254*
    Water and Minerals *254*
    Organic Molecules *254*
    Energy Needs *255*
    Vitamins *256*
Summary *258*
References and Selected Readings *259*

13 Cellular Respiration, Gas Exchange, and Metabolic Rates 261

Cellular Respiration *262*
    Reactions of Cellular Respiration *263*
    Glycolysis *263*
    Fermentation *263*
    Aerobic Respiration *265*
    ATP Yield *269*
    Cellular Respiration Controls *269*
    Utilization of Other Fuels *270*
Balancing Food Supply with Cellular
    Demand *271*
    Storage *271*
    Interconversion of Food Compounds *271*
Gas Exchange *272*
    Availability of Oxygen *272*
    Steps in Gas Exchange *273*
    Environmental Gas Exchange in Aquatic
        Animals *273*
    Environmental Gas Exchange in Terrestrial
        Animals *276*
■*Focus 13.1 Adaptations for Diving 281*
    Gas Transport *284*
Fuel Utilization: Metabolic Rates and
    Energy *288*
    Temperature *288*
    Body Size *290*
Summary *291*
References and Selected Readings *292*

14 Internal Transport 295

Functions of Transport Systems *296*
Types of Transport Systems *296*
Blood and Interstitial Fluid *297*
   Plasma *297*
■ *Focus 14.1 The Circulation of Blood 298*
   Red Blood Cells *299*
   White Blood Cells *300*
Hemostasis *300*
Invertebrate Circulatory Patterns *302*
Vertebrate Circulatory Patterns *304*
   Primitive Fishes *304*
   Evolution of the Double Circulation *305*
   Amphibians and Reptiles *306*
   Birds and Mammals *307*
   Fetal and Neonatal Circulations *308*
The Propulsion of Blood and Hemolymph *311*
   Hearts *311*
   The Mammalian Heart *311*
   Heartbeat and Its Integration *312*
   Cardiac Output *315*
   Cardiac Control *315*
The Peripheral Flow of Blood and Hemolymph *315*
   The Flow of Liquids in Pipes and Blood Vessels *315*
   Regulation of Peripheral Flow *316*
   Capillary Exchange *318*
   Venous and Lymphatic Return *319*
Summary *320*
References and Selected Readings *321*

15 The Immune System 323

Types of Immune Response *324*
   Nonspecific Mechanisms *324*
   The Humoral Immune Response *325*
   The Cell-Mediated Immune Response *335*
   Self-tolerance and Autoimmune Disease *337*
   Hypersensitivity: Allergies and Anaphylaxis *338*
   Blood Groups *340*
The Evolution of the Immune Response *341*
   Invertebrate Immune Responses *341*
   The Vertebrate Immune System *342*
   The Immunoglobulin Superfamily: Variations on a Theme *342*
Evading the Immune System: A Target's-Eye View *343*
   The Moving Target: Accelerated Antigenic Variation *343*
   Weakening the System: The Immunosuppression Strategy *344*
   The Camouflage Strategy *344*
Summary *344*
References and Selected Readings *345*

16 Excretion and Water Balance 347

Nitrogenous Wastes *348*
  Protein Metabolism *348*
  Forms of Nitrogenous Wastes *349*
Excretory Organs *350*
  Nephridia *350*
  Green Glands or Antennal Glands *351*
  Malpighian Tubules *351*
The Vertebrate Kidney *353*
  Nephron Structure *355*
  Urine Formation *356*
■ *Focus 16.1 Dialysis 360*
  Control of Kidney Function *361*
Osmotic Regulation in Marine Animals *361*
  Osmoconformers *361*
  Osmoregulators *362*
  The Vertebrate Problem *362*
Osmoregulation in Freshwater Animals *363*
Osmoregulation in Terrestrial Animals *364*
  Adaptive Strategies *365*
  Adaptations of Animal Groups *366*
Summary *368*
References and Selected Readings *369*

17 The Nervous System 371

The Neuron and Surrounding Cells *372*
The Nerve Impulse *374*
  The Resting Potential *374*
  The Action Potential *376*
  Impulse Propagation and Neuron
    Recovery *377*
  Velocity of Nerve Impulses *377*
Synaptic Transmission *378*
  Electrical Synapses *378*
  Chemical Synapses *378*
  Excitation and Inhibition *379*
Organization of the Nervous System *379*
  Types of Neurons and Their
    Arrangement *380*
  Reflexes *381*
  Pathways between Cord and Brain *381*
  Neuron Pools *381*

Evolutionary Trends in the Nervous
  System *381*
The Vertebrate Nervous System *385*
  The Peripheral Nervous System *386*
  The Development and Evolution of the
    Brain *390*
  The Mammalian Cerebrum *394*
  Neurotransmitters in the Brain *397*
  Brain Nutrition *397*
Summary *398*
References and Selected Readings *399*

18 Receptors and Sense Organs 401

Receptor Mechanisms *402*
  Reception and Transduction of Stimuli *402*
  Sensory Coding and Sensation *403*
  Physiological Adaptation *403*
Chemoreceptors *404*
  Olfactory Receptors *404*
  Gustatory Receptors *406*
Mechanoreceptors *406*
  Touch and Pressure *407*
  Motion Detectors *408*
  Gravitational Detectors *408*
■ *Focus 18.1 Electroreception by a
  Shark 411*
  Sound and Its Detection *413*
■ *Focus 18.2 The Evolution of
  Mammalian Auditory Ossicles 416*
  Proprioceptors *417*
Photoreceptors *419*
  Light-Orienting Eyes *420*
  Image-Forming Eyes *420*
  Compound Eyes *422*
  The Vertebrate Eye *424*
Thermoreceptors *429*
Summary *430*
References and Selected Readings *430*

19 The Endocrine System 433

Endocrine and Nervous Integration *435*
The Nature of Hormones and Hormonal
  action *436*
■ *Focus 19.1 Hormonal Regulation of*
  Genes 439
Regulation of Hormone Levels *440*
Invertebrate Hormones: Arthropod
  Molting *440*
Vertebrate Hormones *442*
  The Hypothalamus and Pituitary Gland *442*
  Pineal Gland *445*
  Hormones Related to Growth and
    Maturation *446*
  Hormones Related to Metabolism and
    Homeostasis *448*
  Other Hormone-like Substances *454*
Summary *454*
References and Selected Readings *455*

20 Behavior 457

The Nature and Development of
  Ethology *458*
Causation *459*
  Elements of Behavior *459*
  Physiology of Behavior *460*
  Timing of Behavior *462*
Development *464*
  Instincts *465*
  Restricted Learning *465*
  Flexible Learning *467*
Genetics and the Evolution of
  Behavior *468*
Behavioral Ecology *468*
  Foraging Strategies *468*
  Defense *470*
  Agonistic Behavior and Dominance *471*
  Territoriality *472*
  Communication *473*
  Sexual Behavior and Reproduction *475*
  Social Behavior *476*
Summary *479*
References and Selected Readings *480*

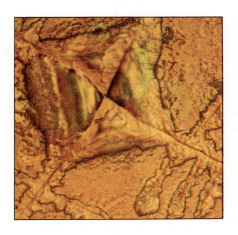

21 Reproduction 483

Modes of Reproduction *484*
   Asexual Reproduction *484*
   Sexual Reproduction *485*
   Parthenogenesis *485*
   Advantages and Disadvantages of Asexual and Sexual Reproduction *485*
Gametogenesis *486*
   Spermatogenesis *486*
   Oogenesis *489*
Fertilization *490*
Adaptations for Fertilization *491*
   Reproductive Synchrony *491*
   Hermaphroditism *495*
Egg Deposition *496*
Predictability in Reproductive Tract Design *496*
Vertebrate Reproductive Patterns *496*
   Male Reproductive Tracts *496*
   Female Reproductive Tracts *499*
   Mammalian Fertilization *499*
The Hormones of Reproduction and Pregnancy *501*
   Male Reproductive Hormones *502*
   Hormones of the Ovarian and Menstrual Cycles *502*
   The Hormones of Fertilization, Pregnancy, and Lactation *505*
Summary *506*
References and Selected Readings *507*

22 Embryonic Development 509

Egg Types *510*
Cleavage *510*
   Major Types of Holoblastic Cleavage *511*
   Types of Meroblastic Cleavage *512*
Gastrulation and Coelom Formation *513*
   Gastrulation in Isolecithal Eggs *514*
   Gastrulation in Amphibians *515*
   Gastrulation in Reptiles, Birds, and Mammals *516*
Organogenesis *517*
   Ectoderm Differentiation *517*
   Mesoderm Differentiation *520*
   Endoderm Differentiation *522*
Development, Environment, and Life Style *522*
   Planktonic Development and Larvae *522*
   Cleidoic Eggs and Extraembryonic Membranes *523*
   Brooding *524*
   Mammalian Placentation *524*
Human Development *526*
   The Development of Body Form *526*
   Birth *526*
■ *Focus 22.1 Twinning 528*
   Postnatal Development *528*
Control of Development *529*
   Mosaic and Regulative Development *529*
   Induction in Regulative Development *530*
   Differential Gene Expression *530*
   Morphogenesis: Cell Movements and Interactions *532*
   Pattern Formation *532*
Sex Determination and Sex Differentiation *535*
Summary *536*
References and Selected Readings *537*

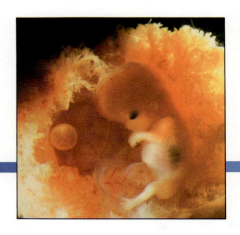

IV ANIMAL DIVERSITY 539

23 Taxonomy, Phylogeny, and the Origin and Evolution of the Animal Kingdom 541

Taxonomy 542
Phylogeny and Systematics 542
  Evidence for Evolutionary Relationships 544
  Approaches to Systematics 545
Origin of the Animal Kingdom 547
  The Fossil Record 548
  Evidence from Comparative Anatomy 549
Grouping of Phyla 552
  Symmetry 552
  Presence or Absence of Body Cavity 553
  Developmental Patterns 553
Adaptive Diversity 554
How to Study the Animal Phyla 554
Summary 558
References and Selected Readings 558

24 Protozoa 561

Phylum Sarcomastigophora 562
  Flagellates 562
  Phytoflagellates 562
  Reproduction 566
  Sarcodines 567
Sporozoans 571
Phylum Ciliophora 573
  Locomotion 576
■ *Connections 24.1 The Spatial World of Microorganisms* 576
  Nutrition 577
  Water Balance 580
  Reproduction 580
  Classification of Protozoa 581
Summary 582
References and Selected Readings 583

25 Sponges 585

Structure and Function 586
  Water Flow and Feeding 587
  Grades of Sponge Structure 588
Regeneration and Reproduction 590
■ *Connections 25.1 Marine Borers* 591
Evolutionary Relationships 592
  Classification of Phylum Porifera 592
Another Primitive Phylum 593
Summary 593
References and Selected Readings 593

26 Cnidarians 595

Structure and Function 596
  Movement 596
  Nutrition and Nematocysts 596
  Nervous System 598
  Regeneration and Reproduction 599
Class Hydrozoa 599
  Colony Formation 599
  Skeleton Formation 600
  Polymorphism 601
  Medusae 602
  Life Cycles 603
Classes Scyphozoa and Cubozoa 605
Class Anthozoa 608
  Sea Anemones 609
  Scleractinian Corals 611
  Octocorals 612
■ *Connections 26.1 Algal Symbiosis* 612
Coral Reefs 614
Evolutionary Relationships 616
Another Gelatinous Phylum 616
  Classification of Phylum Cnidaria 617
Summary 619
References and Selected Readings 619

27 Flatworms 621

■ *Connections 27.1 The Interstitial Fauna 622*
Class Turbellaria 623
   Locomotion 624
   Nutrition 625
   Gas Exchange, Internal Transport, and Water Balance 626
   Nervous System and Sense Organs 626
   Regeneration and Reproduction 626
■ *Connections 27.2 Parasitism 628*
Parasitic Flatworms 628
   Flukes (Class Trematoda) 630
   Class Monogenea 633
   Tapeworms (Class Cestoda) 634
Evolutionary Relationships 634
Related Phyla 636
   Phylum Nemertea or Rhynchocoela 636
   Phylum Mesozoa 637
   Classification of Phylum Platyhelminthes 637
Summary 638
References and Selected Readings 639

28 Aschelminths 641

Phylum Rotifera 643
■ *Connections 28.1 Cryptobiosis: True "Suspended Animation" 645*
Phylum Gastrotricha 646
Phylum Nematoda 646
   Parasitic Nematodes 649
Evolutionary Relationships 652
A Related Phylum? 653
   Phylum Gnathostomulida 653
   Classification of Aschelminths 653
Summary 655
References and Selected Readings 655

29 Molluscs 657

The Molluscan Body Plan 658
Classes Monoplacophora and Polyplacophora 661
Class Gastropoda 662
   Evolution of Gastropods 663
   Shell 665
   Locomotion 666
   Evolution of Water Circulation and Gas Exchange 666
■ *Connections 29.1 Homing in Invertebrates 667*
   Circulation, Excretion, and Water Balance 669
   Nutrition 670
   Sense Organs 672
   Reproduction and Development 672
Class Bivalvia 673
   Mantle and Shell 673
   Nutrition 675
■ *Connections 29.2 Records in Skeletons and Shells 676*
   Adaptive Groups of Bivalves 679
   Reproduction and Development 682
Class Cephalopoda 682
   Shell 682
   Locomotion 685
   Feeding 685
   Internal Structure and Physiology 686
   Reproduction and Development 686
■ *Connections 29.3 Bioluminescence 687*
Evolutionary Relationships 687
   Classification of Phylum Mollusca 688
Summary 689
References and Selected Readings 691

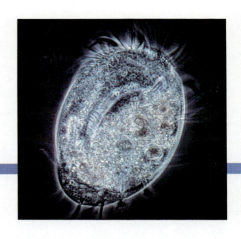

30 Annelids 693

Metamerism and Locomotion *694*
Class Polychaeta *695*
  Adaptive Groups of Polychaetes *695*
  Internal Transport, Gas Exchange, and
    Excretion *701*
    Sense Organs and Nervous System *701*
    Reproduction and Development *701*
  Class Oligochaeta *702*
    Nutrition *703*
    Internal Transport, Excretion, and Nervous
      System *704*
    Reproduction and Development *705*
  Class Hirudinea *707*
    Locomotion *707*
    Nutrition *707*
    Internal Transport, Excretion, and Sense
      Organs *709*
    Reproduction *709*
Evolutionary Relationships *709*
Related Phyla *709*
■ *Connections 30.1 Symbiosis between*
  Animals and Chemosynthetic
  Bacteria 710
    Classification of Phylum Annelida *711*
Summary *712*
References and Selected Readings *713*

**31 Arthropods I: Chelicerates and
Crustaceans 715**

The Arthropod Design *716*
  Annelidan Features *716*
  The Arthropod Skeleton *716*
  Locomotion *717*
  Reduction of Metamerism *717*
  Digestive System *718*
  Internal Transport, Gas Exchange, and
    Excretion *718*
  Nervous System and Sense Organs *719*
  Reproduction *719*
Subphylum Chelicerata *719*
  Class Merostomata *719*
  Class Arachnida *720*
■ *Connections 31.1 Leaf Mold*
  Fauna 722
  A Related Class *728*
Subphylum Crustacea *729*
  Class Malacostraca *731*
  Class Branchiopoda *737*
  Class Copepoda *739*
  Class Ostracoda *740*
  Class Cirripedia *741*
  Class Remipedia *743*
  Crustacean Radiation *744*
  Classification of Chelicerates and
    Crustaceans *745*
Summary *746*
References and Selected Readings *747*

32 Arthropods II: Uniramians   749

Myriapods   *750*
   Millipedes   *751*
   Centipedes   *751*
Insects   *752*
■ *Connections 32.1   Life in the Tree*
  Tops   752
   The Insect Ground Plan   *753*
   The Radiation of Insects   *754*
   Nutrition   *758*
   Insect-Plant Interactions   *758*
   Parasitism   *759*
   Communication   *760*
   Social Insects   *761*
   Reproduction   *763*
Evolutionary Relationships   *764*
Related Phyla   *764*
   Classification of Uniramians   *765*
Summary   *766*
References and Selected Readings   *767*

33 Bryozoans   769

Structure of a Bryozoan Individual   *770*
Organization of Colonies   *771*
Reproduction   *773*
Other Lophophorate Phyla   *774*
Evolutionary Relationships   *775*
Summary   *775*
References and Selected Readings   *775*

34 Echinoderms   777

Class Stelleroidea   *779*
   Asteroids   *779*
   Ophiuroids   *783*
Class Echinoidea   *785*
   Sea Urchins   *785*
   Sand Dollars and Heart Urchins   *786*
   Reproduction and Development   *787*
Class Holothuroidea   *788*
Class Crinoidea   *790*
Class Concentricycloidea   *792*
Evolutionary Relationships   *792*
Another Deuterostome Phylum   *794*
   Classification of Phylum Echinodermata   *794*
Summary   *795*
References and Selected Readings   *795*

35 Hemichordates and Chordates   797

Phylum Hemichordata   *798*
   Class Enteropneusta   *798*
   Class Pterobranchia   *800*
Phylum Chordata   *800*
   Chordate Characteristics   *800*
   Subphylum Urochordata (Tunicata)   *802*
   Chordate Metamerism   *804*
   Subphylum Cephalochordata   *804*
   Subphylum Vertebrata   *807*
   Vertebrate Evolution   *809*
Evolutionary Relationships of Deuterostomes
  and Chordates   *809*
   Classification of the Phyla Hemichordata and
    Chordata   *811*
Summary   *811*
References and Selected Readings   *812*

36 Fishes 815

Adaptations of Fishes *816*
   Locomotion and Buoyancy *816*
   Sense Organs *816*
   Metabolism *818*
   Excretion and Water Balance *819*
   Reproduction and Development *819*
Class Agnatha *819*
   The Ostracoderms *819*
   Living Jawless Vertebrates *820*
Early Jawed Vertebrates *822*
   Class Acanthodii *823*
   Class Placodermi *824*
Class Chondrichthyes *824*
   Basic Features of Cartilaginous Fishes *825*
   Adaptive Radiation of Cartilaginous
      Fishes *826*
Class Osteichthyes *828*
   Basic Features of Bony Fishes *828*
   Actinopterygians *830*
   Sarcopterygians *837*
   Classification of Fishes *838*
Summary *839*
References and Selected Readings *840*

37 Amphibians 843

The Transition from Water to Land *844*
Adaptations of Amphibians *845*
   Integument *845*
   Skeletomuscular System *846*
   Sense Organs and Nervous System *846*
   Metabolism *847*
   Excretion and Water Balance *848*
   Reproduction and Development *849*
Evolution of Amphibians *849*
Salamanders *850*
   Adaptations of Salamanders *850*
   Reproduction and Life Cycles *853*

Frogs and Toads *854*
   Adaptations of Frogs and Toads *854*
   Reproduction and Development *857*
   Classification of Amphibians *858*
Caecilians *858*
Summary *859*
References and Selected Readings *859*

38 Reptiles 861

Adaptations of Reptiles *863*
   Integument *863*
   Skeletomuscular System, Locomotion, and
      Nervous System *863*
   Metabolism *863*
   Digestion *864*
   Respiration *865*
   Circulation *865*
   Excretion *865*
   Reproduction and Development *865*
Evolution and Adaptive Radiation of
   Reptiles *865*
   Early Reptiles *865*
   Turtles *866*
   Primitive Diapsids: Lepidosaurs *868*
■*Focus 38.1 How Does a Snake
  Move? 872*
   Blind Alleys: Extinct Marine Reptiles *873*
   Advanced Diapsids: Archosaurs *874*
■*Focus 38.2 Were the Dinosaurs Warm
  Blooded? 876*
   Classification of Reptiles *877*
Summary *878*
References and Selected Readings *878*

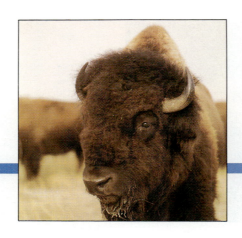

39 Birds 881

Bird Flight *882*
■ *Focus 39.1 Flight 885*
Adaptations of Birds *885*
    Endothermy *885*
    Feathers *886*
    Skeleton *888*
    Muscles *888*
    Sense Organs and Nervous System *889*
    Digestion *890*
    Gas Exchange *892*
    Circulation *892*
    Excretion and Water Balance *892*
    Reproduction and Development *894*
Migration and Navigation *895*
Evolution and Adaptive Radiation of
    Birds *897*
    The Origin of Birds *897*
    Cretaceous Toothed Birds *897*
    Ratites *898*
    Land Birds *899*
    Shore and Water Birds *902*
    Classification of Birds *904*
Summary *906*
References and Selected Readings *906*

40 Mammals 909

Adaptations of Mammals *910*
    Endothermy and Temperature
        Regulation *910*
    Locomotion and Coordination *912*
    Digestion *912*
    Respiration *912*
    Circulation *913*
    Excretion and Water Balance *914*
    Reproduction and Care of the Young *914*
The Origin of Mammals *914*

Primitive Mammals: Prototherians *915*
    Early Mammalian Evolution *915*
    Monotremes *916*
Therians *917*
    Reproductive Strategies *917*
    Adaptive Radiation of Marsupials *918*
    Adaptive Radiation of Eutherians *919*
    Classification of Mammals *930*
Summary *931*
References and Selected Readings *932*

41 Primates and Human Evolution 935

Primate Adaptations *936*
The Groups of Primates *936*
    Prosimians *936*
    Anthropoids *938*
Human Characteristics *942*
Early Evolution of Apes and Hominids *943*
The Australopithecines *944*
    Evolution of the Australopithecines *946*
Early *Homo* *946*
    Homo habilis *946*
    Homo erectus *947*
■ *Focus 41.1 Larger Primates, Larger
    Brains 948*
Homo sapiens *950*
Human Potential and Responsibilities *951*
    Classification of Primates *951*
■ *Focus 41.2 A Molecular Look at* Homo
    sapiens *952*
Summary *954*
References and Selected Readings *954*

V   INTERACTIONS OF ORGANISMS AND THEIR ENVIRONMENT   957

42   Population Ecology   959

The Spatial Distribution of Populations   *960*
   Geographical Range   *960*
   Dispersal Patterns   *962*
   Density   *964*
Dynamics of Populations   *964*
   Birth, Death, Emigration, and
     Immigration   *964*
   Survivorship Curves   *965*
   Age and Sex Distribution   *965*
Population Growth   *966*
   Intrinsic Rate of Growth   *967*
   Exponential Growth   *967*
   Logistic Growth   *968*
   Population Growth in Nature   *969*
   K-Selection and *r*-Selection: Extremes in a
     Continuum   *972*
■ *Focus 42.1   Fire   973*
Human Populations   *974*
■ *Focus 42.2   Doubling Time   975*
Summary   *975*
References and Selected Readings   *977*

43   Communities and Ecosystems   979

Communities   *980*
   Dominance   *980*
   Succession   *980*
   Niche   *984*
   Diversity: Why so Many Species?   *984*
   Resource Partitioning   *986*
The Earth's Major Ecosystems   *987*
   Biomes   *988*
   Aquatic Ecosystems   *992*
Energy Flow Through Food Webs   *995*
   Trophic Levels   *995*
■ *Focus 43.1   Phytoplankton Blooms and Busts   996*
   Production   *998*
   Energy Transfer Efficiency   *998*
Solar Radiation: Where the Chain
   Begins   *1001*
   The Ozone Layer   *1001*
   The Greenhouse Effect   *1001*
■ *Focus 43.2   Food Chains and Human Welfare   1002*
Biogeochemical Cycles   *1004*
   Hydrological Cycle   *1004*
   Carbon Cycle   *1005*
   Nitrogen Cycle   *1005*
   Phosphorus and Sulfur Cycles   *1006*
Epilogue: Global Change   *1007*
■ *Focus 43.3   Biogeochemical Cycles and Acid Rain   1007*
Summary   *1008*
References and Selected Readings   *1009*

Appendix A   The Chemical and Physical Basis for Zoology   A-1

Appendix B   Scientific Units and Quantities   A-11

Glossary   G-1

Index   I-1

I

The Conceptual Basis of Zoology

1 The Nature of Science and the Science of Nature

2 The Core of Modern Biology: Evolution

(Left) The goldenrod crab spider (Misumenta vatia) *has a coloration that resembles the flowers where it lies in wait for prey. (Above) The goldenrod crab spider stalking a hover fly.*

1

The Nature of Science and the Science of Nature

(Left) The Greater flamingo. The inverted, scooping bill enables flamingos to filter small invertebrates and algae from the water. (Above) A flock of Greater flamingos (Phoenicopterus ruber), *photographed in one of the soda lakes of Kenya.*

Zoology has been written with four goals in mind. First, we intend to introduce you to the diversity of living and extinct animals that have populated this Earth. Second, we discuss the laws of form and function, the basic biochemical, physiological and developmental mechanisms that underlie all animal life. Third, we hope to make you aware of the physical and biological forces that have shaped, and continue to shape, animal form and diversity. Finally, we wish to convey the methods of science—the interaction of ideas, facts, and observations that lead to an understanding of pattern and process in the natural world.

One central theme echoes throughout this book: the study of animals, or **zoology**, a subdivision of **biology**, the study of all life, can only be understood in the context of **evolution**, the theory that all organisms are related by genealogy—all ultimately descended from a common ancestor. The geneticist Theodosius Dobzhansky put it succinctly when he declared that "nothing in biology makes sense except in the light of evolution." Without the power of evolutionary explanation, we could never move beyond simply describing the living world; we would certainly not comprehend it. Evolution is the fundamental unifying concept in the study of life. So too in this book, the reality of evolution, the process Charles Darwin was to call "descent with modification," will be the golden thread running throughout all the chapters.

Unfortunately, the teaching of evolution is under attack from certain quarters. A small group of religious fundamentalists has repeatedly sought to block teachers from discussing evolutionary theory and evidence in their classrooms. More recently, creationists have insisted that the accounts of Creation found in the Bible be taught in science classes. These attacks are not to be taken lightly: they challenge not only the intellectual core of biology, but also the very nature of the scientific enterprise. We therefore begin our exploration of zoology with a detailed look at the nature of science, using the controversy surrounding "creation-science" as an illustration. Having developed an appropriate general definition of science itself, we end this chapter by detailing certain features that make biology a unique and characteristic form of human inquiry.

THE DOMAIN OF SCIENCE

On January 5th, 1982, Judge William Overton struck down Arkansas Law 590, euphemistically known as "the balanced-treatment law." Despite its harmless title, Law 590—and similar laws in Louisiana and Tennessee—represented a major obstacle to the proper teaching of science in public schools. Supporters of this legislation, first enacted in 1981, argued that evolution was "just a theory" and therefore on the same footing as other "theories" about the natural world and the origin of modern humans, *Homo sapiens*. In particular, Law 590 required

Figure 1.1
This illustration by William Blake, "The Ancient of Days Striking the First Circle of the Earth" (1794), is one of the many dramatic interpretations of the Genesis account of the creation of the Earth. The compass reflects the growing appreciation by Blake and his contemporaries of the many symmetries and regularities governing the movements of the heavens.

that creationist accounts be taught in biology classes. After all, creationists argued, "In the beginning, God created the Heaven and the Earth" is also a theory (Fig. 1.1). What could be fairer, more in keeping with the American tradition of free expression and open debate, than to teach these two accounts side by side, and let the students decide?

This flawed appeal to equal time for "creation-science" masked a much more profound rejection of modern science. The individuals and institutions that challenged Law 590 realized that "creation-science" was nothing more than an attempt to impose a particular set of religious beliefs on the schoolchildren of Arkansas. Only by abandoning the methods and practices of science could the biblical account of the origins of the universe, the Earth, and life be taught as science in biology classrooms.

There is no culture on Earth that has not fashioned some account of its own creation (Fig. 1.2). Many of these accounts, like the opening verses of Genesis, are stirring poetry that inspire poets and writers, painters and sculptors, religious leaders, and even scientists. But to argue that the account in Genesis represents the factual chronicle of the forces that have shaped us and our world requires dismissing the last 300 years of scientific achievement.

There need not be a contradiction between science and religion. Each of these forms of human thought has its domain, a set of questions that it is uniquely qualified to address. In our view, for example, the answers to questions of free will, of morality, and of human ethics are not to be found through scientific inquiry. For some, the answers to these questions may come from religious inquiry.

Conversely, if we seek to understand the material basis of life—the forces and causes and effects that have given rise to past and present organisms—it is to science that we must turn. That is the task we now set for ourselves.

WHAT DO CREATIONISTS WANT?

In the summer of 1925, high-school biology teacher John Scopes was convicted of violating the Butler Act, a Tennessee statute that explicitly prohibited the teaching of evolution. The statute was, if anything, direct. It stated that

> . . . it shall be unlawful for any teacher in any of the Universities, Normals and all other public schools of the state . . . to teach any theory that denies the story of the Divine Creation of man as taught in the Bible, and to teach instead that man has descended from a lower order of animals.

This statute so clearly violated the First Amendment of the Constitution, mandating the separation of church and state, that opponents of the Butler Act were certain that it would be declared unconstitutional by the Supreme Court. In order to get to the Supreme Court, however, someone would have to be convicted of violating the Butler Act. The American Civil Liberties Union (ACLU) had advertised in the local papers for someone willing to face charges for violating the Butler Act. They had finally found a high-school teacher in Dayton, Tennessee, who was willing to be tried: John Scopes (Fig. 1.3). Scopes was an athletics coach who for a brief period had been the substitute teacher for "general science." During that period, he had assigned the chapter in the biology textbook dealing with evolution and hence was technically guilty of violating the Butler Act.

The trial that ensued was a true media circus, and both sides would summon formidable orators to plead their case (Fig. 1.4). William Jennings Bryan, three-time presidential candidate and former Secretary of State, would argue the prosecution's case, against the teaching of evolution. Clarence Darrow, a gifted defense lawyer who had argued (and won) many important civil liberties cases, would defend John Scopes. As portrayed in the movie *Inherit the Wind* (1960), great drama and oratory took place in that Dayton courtroom. In fact, fine speeches were made, but the outcome of the trial was not much in doubt. The ACLU wanted a quick conviction in order to set their appeal in motion. Scopes was convicted and fined $100.00 by the judge. Here, Darrow and his associates made one serious mistake in failing to realize that the judge, under

Figure 1.2

(a) A rendition of the Hindu creation myth. A turtle carries the 21 worlds composing the physical and spiritual universe upon its back. The turtle, in turn, is surrounded by a cobra consuming its own tail. (b) A Chinese creation myth: The sculptor P'an Ka chisels out the universe and its inhabitants.

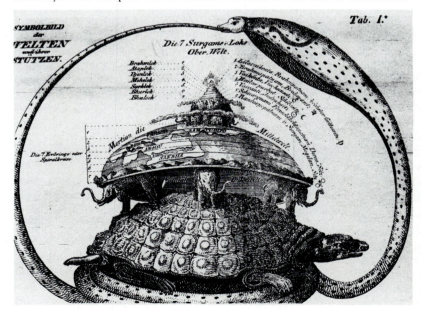

(a)

(b)

Figure 1.3
John Thomas Scopes, the coach and general science instructor of the Dayton, Tennessee high school. By assigning the material on evolution in the science textbook, Scopes would set in motion the events leading to one of the best-known trials in America, dubbed "the Scopes Monkey Trial."

Figure 1.4
The two famed orators of the Scopes trial: William Jennings Bryan for the prosecution; Clarence Darrow for the defense.

Tennessee law, could not impose any fine above $50.00 without the consent of a jury. It was solely on that technicality that Scopes's conviction was overturned by the Tennessee Supreme Court; the constitutionality of the Butler Act was never addressed. That act, never again enforced, remained on the books until 1967.

The Scopes trial would be little more than a quaint episode in American legal history (and the context for one of Spencer Tracy's best movie performances) were it not for the fact that 56 years later, in 1981, another trial involving disturbingly similar issues took place in U.S. District Court in Little Rock, Arkansas. This time, the issue was the constitutionality of Arkansas Act 590, which stated that "Public schools within this state shall give balanced treatment to creation-science and evolution-science."

This is how Act 590 defined "creation-science":

"Creation-science" includes the scientific evidences and related inferences that indicate: (1) Sudden creation of the universe, energy and life from nothing; (2) The insufficiency of mutation and natural selection in bringing about development of all living kinds from a single organism; (3) Changes only within fixed limits of originally created kinds of plants and animals; (4) Separate ancestry for man and apes; (5) Explanation of the earth's geology by catastrophism, including the occurrence of a worldwide flood; and (6) A relatively recent inception of the earth and somewhat later of life.

On the surface, this creationist challenge seemed far more sophisticated than its predecessors that had sought to ban the teaching of evolution outright. Yet despite the ostensibly scientific tone of Act 590, it only represented a change in the strategy of creationists. Act 590 is still religious belief in science clothing. "Creation-science," as we will soon discover, is no science at all.

SCIENCE AND THE MATERIAL WORLD

Science cannot address the realms of the imaginary, the metaphysical, or the spiritual. The aim of science is to classify, understand, and unify the objects and phenomena of the material world. By using a mixture of observation and experimentation, logic and intuition, scientists seek to understand the rules and laws that govern the universe (Fig. 1.5). Physicists, for example, study the fundamental particles of matter and the laws that govern them. Astronomers strive to reconstruct the origin of stars, planets and moons, and the "big bang" that gave rise to the universe.

The task of biology is to understand the living world at all levels of the biological hierarchy—from the molecular basis of inheritance to the complex ecological interactions that shape the Amazon rainforest. Each of these sciences focuses on a different aspect of the material world, but their methods and approaches are remarkably similar.

The Characteristics of Science

Philosophers and historians of science, and scientists themselves, have honest disagreements about the details of what constitutes proper science. Nonetheless, broad agreement about the main characteristics of a true science exists. These characteristics include:

   1. **Materialism**: scientific explanations must be grounded in material cause, and must not violate natural law.

   2. **Testability**: science makes predictions that can be tested against the material world.

   3. **Falsifiability**: scientific explanations must be stated in a manner that makes it possible, in principle (given new evidence or insight), to refute the explanation. This feature lies at the root of the self-correcting character of science.

   Each of these key features of science carries with it important implications. The first of these, *materialism*, defines the content of explanations that can properly be considered scientific. Consider the problem of accounting for the fact that the sun always rises in the east. The fact that you wake every morning to sunlight coming from the east can be accounted for by our current understanding of celestial mechanics: from where you stand, the Earth rotates around its axis in a counterclockwise direction. In

Figure 1.5

A biologist checks the growth of human cells in culture. These cells are being grown in order to isolate interferon, a protein with strong antiviral action produced by cells of the immune system.

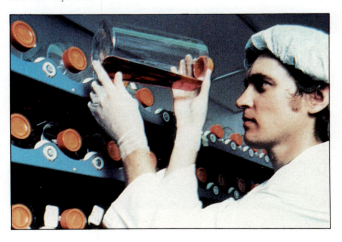

Figure 1.6

Helios, the greek sun god. His chariot, drawn by four white horses, would travel from East to West across the heavens every day.

this simple example, we appeal to a material cause (the direction of the Earth's rotation) to explain an observation about the material world (sunrise is always from the east).

   The ancient Greeks, too, knew the sun rose in the east, and they also had an explanation for its regular motion. The sun was Helios's chariot, drawn across the heavens by a team of horses, arriving at the western edge of the Earth by dusk (Fig. 1.6). By night, a boat that sailed the ocean encircling the earth carried the chariot back east. This account, though lyrical, would fail our first test for scientific explanation. Helios and his chariot are mythological figures, whose existence cannot be independently confirmed and whose presence has no other consequence except to account for the fact that the sun rises from the east.

   The second characteristic, *testability*, defines the way in which scientific explanations are constructed. In the example above, we cannot test for the existence of Helios. In contrast, the rotation of the Earth is a material explanation for the direction and regularity of the sunrise, an explanation firmly grounded in the laws of physics. Furthermore, a large body of evidence supports the notion that the Earth rotates on its axis, and a vast number of consequences (in addition to the direction of the sunrise) flow from it. Again, there is nothing wrong with postulating the existence of Helios, but Helios is not a material explanation that can be confirmed or falsified. It is simply not a scientific account.

   The explanations we set forth in this book are best viewed as hypotheses—arguments put forth by scientists to explain the workings of the living world. The value of an explanation will depend directly on its ability to withstand challenges to its validity. By the same token, any explanation formulated in a way that cannot be tested, and, as we will see below, disproven, is not a *scientific* explanation. To return once again to our Helios example, there is no way to test this explanation. No other fact, pattern, or observation about the material world (other than the sunrise) depends on the presence of Helios.

Finally, the third condition, *falsifiability*, goes to the heart of what makes science a unique form of human inquiry. The nature of science is such that its conclusions are always subject to correction or revision as additional tools, observations, or theories emerge. On the surface, this appears to contradict the usual perception that science is a collection of clear, established facts about the world. But this emphasis on "facts" reveals a profound misconception about the nature of science, one that emphasizes the end product of the enterprise rather than the process itself. The enormous power of science derives from the approaches and methods we use to understand the world, not from the individual "facts" that are the result of this process. All scientific knowledge is *by definition* provisional and subject to revision or rejection. This does not, of course, mean that all interpretations of material phenomena are equally valid—there are clear standards of proof and evidence by which we judge the validity of scientific propositions. Certain scientific claims—that the Earth is round, for example, and that it orbits around the Sun—are supported by so much evidence that we cannot

imagine anything tentative or provisional about those propositions (Fig. 1.7). Naturally, it is not likely that our views on the shape of the Earth will be overturned anytime soon. Nevertheless, these claims are testable and can *in principle* be proven false. Keep in mind that only a few centuries ago, scientists argued strongly that the Earth was flat and that the Sun revolved around the Earth.

The Terms of Science

It is appropriate at this point to clarify the meaning of certain crucial terms we use in discussing the scientific enterprise, the manner in which scientists pursue knowledge. We begin with the concept of **theory**. As noted earlier, creationists have sought to dismiss the concept of evolution as "just a theory." Presumably, they wish to imply that it is "not a fact." But a theory is not a second-rate fact. Instead, a theory is a system of ideas and concepts assembled to order and make sense out of data or observations. Theories are the intellectual core of science, the overarching structures we create in order to make sense of the world.

The specific, testable claims that arise from scientific theories are referred to as **hypotheses**. More limited and specific in their scope, they are the scaffolding of science, the ideas for (or against) which individual experiments and observations are carried out.

Figure 1.7
A view of our Earth, taken from space by the Apollo astronauts.

Finally, the **data** or facts of science come from a variety of sources: observation, manipulation, experimentation, and comparison. A common misconception about the scientific enterprise holds that good scientists simply collect data about the world with no preconceived notion, letting the data speak directly to them. Nothing could be farther from the truth. Data are always collected with a particular hypothesis or theory in mind. The very choice about what information is "relevant" suggests that the scientist's mind is more than a blank slate.

Naturally, there is a constant interplay between theory and data, and good scientists modify their theories when the facts compel them to do so. The theories, hypotheses, and even the "facts" of science are always subject to revision or rejection. Does the emphasis on the provisional nature of scientific knowledge somehow weaken the value of the scientific enterprise? Does it somehow suggest that we can never really "know" anything, that there are no truths about the world of which we can be certain? In our view, the provisional character of scientific knowledge reflects the strength of the scientific enterprise, because it places the emphasis on the methods by which we acquire knowledge, rather than on the results alone. Scientific knowledge undergoes constant revision and occasionally radical alteration as scientists constantly probe the nature of the world. Flaws and errors are found and corrected, previously contradictory results are harmonized under a new theory, neglected facts suddenly gain importance. Science is a dynamic process taking place simultaneously on many fronts. Hypotheses are framed, experiments and tests are devised, theories are supported or rejected, and no conclusion is beyond revision if new observations or results warrant the change. It is from this constant intellectual ferment that the real strength and excitement of science springs.

The Fallacy of Creation-Science

As we will see below, so-called "creation-science" fails all of the conditions we have set forth for a true science (materialism, testability, and falsifiability).

Not surprisingly, given that the Bible is the source of "creation-science," creationists frequently appeal to forces and events that cannot be considered material law. Claims about the creation of the universe "from nothing" (as the result of divine intervention) or appeals to the "great flood" at the center of the Biblical story of Noah cannot be considered scientific explanations. Instead, they are appeals to unique, untestable phenomena that require the suspension of all natural laws.

Secondly, because creationists consider the opening chapters of Genesis to be the literal account of the origin of the Earth and of life, the core of "creation-science" does not need to be tested, since the complete and correct

Figure 1.8
The creation of Adam, as depicted on the ceiling of the Sistine Chapel in Rome by Michelangelo Buonarotti.

answer is already known in advance. The literal Genesis account cannot in fact be independently tested because it has no independent consequences. The claims of creation-science—the "insufficiency of mutation," the "fixed limits" of change, the "relatively recent" origin of the earth and life—are deliberately vague, making their testing or refutation impossible.

Finally, "creation-science" cannot be falsified. There is no amount of evidence that will persuade creationists to abandon their conception of the origin and fixity of species. Unlike true science, with its emphasis on the *process* whereby we come to understand the world, "creation-science" only emphasizes the result—the story of Genesis (Fig. 1.8). Evidence that contradicts the literal interpretation of Genesis is either ignored or dismissed as irrelevant, since creationists assert that the account *cannot* be other than the literal truth. Creation "scientists" have been careful to make few specific, positive statements that can actually be falsified. The rare specific claims of creationism (such as the separate ancestry of man and apes, and the coexistence of dinosaurs and humans) have been thoroughly and overwhelmingly contradicted by available evidence; these claims have yet to be abandoned by "creation-science."

Creationism is not science and cannot be taught as such. In declaring Act 590 to be unconstitutional, Judge Overton would write:

> . . . the evidence is overwhelming that both the purpose and effect of Act 590 is the advancement of religion in the public schools. . . . The Court would never criticize or discredit any person's testimony based on his or her beliefs. While anybody is free to approach a scientific inquiry in any fashion they choose, they cannot properly describe the method used as scientific, if they start with a conclusion and refuse to change it regardless of the evidence developed during the course of the investigation.

THE NATURE OF BIOLOGICAL EXPLANATION

Having described the general conditions under which science operates, we can now begin to focus more specifically on the nature of the biological sciences. As you read this book, it will become clear to you that modern zoology combines methods and information from a vast array of scientific fields. We will, for example, use chemical principles when we explain the intricacies of vertebrate metabolism or the transmission of nerve impulses. The evolution of vertebrate limbs, the size and shape of animals, and the structure of the mammalian ear cannot be accounted for without using the laws of physics. The distribution of animals on different continents and the nature of the fossil record are best explained using geological information. In short, modern zoology is a highly synthetic science, a field of inquiry that lies at the intersection of several other disciplines.

But zoology is much more than just a collection of carefully selected bits of information developed in other disciplines. The aim of zoology is to understand animal life: its origin, its evolution, its function, and its organization. As we discussed in the previous section, all sciences adhere to the same methods and practices. In certain fundamental ways, however, zoology and the other branches of biology are unlike most other sciences. Two features of the living world—historicity and hierarchy—uniquely shape the character of biological inquiry.

The first of these features, **historicity**, alludes to the profound importance of history in shaping the living world. We refer of course to biological history (not human history): past events in the history of life have influenced the biological picture we see today. Mass extinctions, for example, leave a deep imprint, reshaping the evolution of biological diversity. Our very presence as a species may well have depended on a mass extinction: the diversification of mammals was only made possible by the extinction of dinosaurs 65 million years ago. Had this extinction not happened, our species, *Homo sapiens*, might have never evolved, and the world would have been a very different place. The entire history of life is replete with such forks in the road: the world we see today is only one of many possible outcomes. The role of history is also evident in every biochemical, physiological, and morphological feature of organisms—evolution operates by modifying what is already available and does not build organisms from scratch. Thus, for example, the human fetus has a tail for a few weeks in early development (Fig. 1.9). It is a useless feature that will eventually be lost, but it is also a clear reminder of our biological past.

Biological explanations must acknowledge the central, crucial role played by history in shaping the living world. Most other sciences seek to uncover the universal, absolute, and unchanging laws that govern the world.

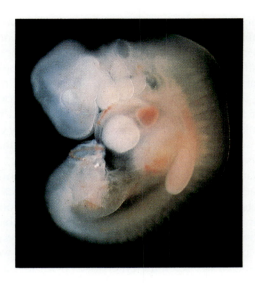

Figure 1.9
A photograph of a 29-day-old human embryo. The embryo is approximately 7 mm long. Note the gill clefts and the slender tail, two features that reflect the evolutionary history of our species. Both of these traits will disappear at a later stage of development.

Ideally these laws represent "the great uniformities": laws that are true and constant anywhere in time or space. For example, Newton's third law states that "for any action there is an opposite and equal reaction." As you know, this law applies whether you are trying to push your car in the snow or seeking to explain the irregular motions of the planets. It was true a billion years ago, and we expect it to be true a billion years hence. A simple, elegant statement that can be expressed mathematically explains a multitude of observations in the world of physics.

Are we likely to discover such laws in biology? Do such laws in fact exist when we are trying to explain the living world? This is a difficult question, but we would argue that biologists are unlikely to come up with a limited set of universal laws that explain all of the living world. There are, to be sure, laws and regularities in biology, but they generally apply only to a very circumscribed domain, or else are so broad as to preclude specific and detailed prediction or explanation. We can, for example, calculate the maximum weight at which an animal may still fly. Using our knowledge of the laws of physics and aerodynamics, we can develop a law—organisms weighing more than 25 kilograms would require such enormous wingspans (15 meters) that they are extremely unlikely ever to evolve. While this might be a fascinating law about the limits of biological design, it does not help us to understand the actual diversity of birds and bats. If we wish to understand the flying fauna, we need to go beyond the simple laws of physics and explore the history that first allowed organisms to take to the air. Naturally, the living world does not exist in violation of the uniformities of physics or chemistry: rheas and ostriches cannot fly, and Kori bustards do so with difficulty (Fig. 1.10). Organisms, like all other matter in the universe, must obey gravity. No object in this world can escape physical law. But biology is not physics, and organisms are not just a collection of atoms. The challenge of biology is precisely to sail a middle course between ahistorical explanations (which seek to reduce all the diversity and complexity of the living world to a small set of constant laws) and simple descriptions (which do not seek out the underlying regularities and causes of biological phenomena). Because history and unique, contingent events are so important in shaping the history of life, the future course of evolution cannot be predicted.

Hierarchy is the second characteristic of the living world that makes biology qualitatively different from most of the other sciences. We see order and organization at many different levels: from the regular, elegant structure of the DNA molecule, which contains genetic information that is passed from generation to generation, to the complex ecosystem of the African savannah. Each of these levels has its own players, its own field, and its own ground rules. To be sure, there are important links between levels: the fate of individual organisms does in part depend on the genetic information contained in the organism, and the interactions among species are affected by the reproductive success of individual organisms. But unlike many other sciences, which seek to reduce complexity to the

(a)

(b)

Figure 1.10

(a) The largest of living birds, the African ostrich (*Struthio camelus*). Weighing approximately 70 kgs., ostriches are very fast runners (up to 50 km/hr), but they cannot fly. (b) The Kori bustard (*Ardeotis kori*), at 16 kg., the largest bird still capable of flight (although only for short distances).

(a)

(b)

(c)

(d)

(e)

Figure 1.11

The many levels of the biological hierarchy (a) A computer-generated illustration of the deoxyribonucleic acid (DNA) double-helix. This structure, finally elucidated in 1956, is responsible for the transmission of genetic information from one generation to the next. (b) The cell, the fundamental unit of all living tissues. Shown here are typical eukaryotic cells in culture. (c) The organism, primary focus of natural selection. These king penguins have evolved in the rigorous conditions of the Antarctic. (d) The population, in this case a king penguin colony, is a constantly evolving entity composed of interacting and interbreeding organisms. (e) The ecosystem includes all interacting organisms, as well as the physical characteristics of the environment in which these interactions take place. Shown here is a colony of king penguins on South Georgia Island in the Southern Atlantic Ocean.

simplest level, all biological phenomena cannot be explained or accounted for at the molecular level. Each of the levels of biological organization—such as the gene, the chromosome, the cell, the organism, the population, the species, and the ecosystem—has irreducible features (Fig. 1.11). Biological explanations, therefore, must be geared to the level they seek to explain. The **reductionist approach** to the living world, which tries to explain all phenomena at the molecular level, has been immensely successful over the past 25 years, but the hierarchy of biological organization is real and must be accounted for.

The living world has followed only one of many possible paths as it has evolved over the past four billion years. From a primordial broth containing a few organic molecules, life has diversified and organized itself into increasingly complex structures and systems, eventually giving rise to a biosphere, a world of living organisms, composed of as many as 30 million species. It is to the humble beginnings of life on Earth and to the forces and conditions that shaped them that we now turn.

Summary

1. In 1925 John Scopes, a high school teacher, was brought to trial and convicted by the state of Tennessee for violating a statute that declared the teaching of evolution to be unlawful. The constitutional issues involved were never tested in federal courts because a higher Tennessee court reversed the conviction on a technicality.

2. More recently, religious fundamentalists in several states have sought to pass legislation requiring that a balanced treatment be given to the teaching of "creation-science" along with evolution. In 1981 an Arkansas law to this effect was found to be unconstitutional by a U.S. District Court.

3. The controversy surrounding "creation-science" illustrates a profound misconception in the minds of many about the nature of science. Scientific explanations must be (1) grounded in material cause, (2) make predictions that can be tested, and (3) arrive at conclusions that, in principle, are capable of being proven false. Scientific knowledge undergoes constant revision, and flaws and errors are found and corrected. The postulates of "creation-science" fail to meet all of these conditions for a true science.

4. The aim of zoology and other biological sciences is to understand life: its origin, evolution, function, and organization. Biological sciences make use of the concepts and methods of chemistry, physics, and other sciences, but are unique in acknowledging the roles of history and hierarchy. The history of life itself has played a central role in shaping the living world. The history of life is full of chance events and roads not taken; life easily could have turned out differently.

References and Selected Readings

Futuyma, D. J. *Science on Trial: The Case for Evolution.* New York: Pantheon Books, 1983. A book addressed to the general reader explaining the evidence for evolution and the scientific fallacies in the creationist arguments, and placing the controversy in a larger scientific and social context.

Gould, S. J. Justice Scalia's misunderstanding. *Natural History* (Oct. 1987):14–21. An analysis of the misconceptions of the nature of science in Justice Scalia's dissent from the U.S. Supreme Court's ruling in 1987 declaring that the Louisiana statute requiring a balanced treatment of creation-science and evolution-science is unconstitutional.

Hanson, R. W., ed. *Science and Creation: Geological,* *Theological and Educational Perspectives.* American Association for the Advancement of Science, Issues in Science and Technology Series. New York: Macmillan, 1986. A collection of essays on the nature of science with a focus on the evolution-creation controversy. The full texts of the Arkansas Creation-Science Statute (Act 590) and Judge Overton's decision in *McLean v. Arkansas Board of Education* are given in appendices.

Kitcher, P. *Abusing Science: The Case Against Creationism.* Cambridge: MIT Press, 1982. A compelling and clear exploration of the fallacies underlying creationism and of the philosophical structure of evolutionary biology.

2

The Core of Modern Biology: Evolution

(Left) Fernandina Island, one of a group of volcanic islands, the Galapagos. Darwin visited the Galapagos in September 1835, aboard the H.M.S. Beagle. (Above) In isolation from the South American mainland, penguins, iguanas, and many other species of plants and animals have evolved unusual features for life on the Galapagos. Sea lions are frequent visitors to these islands.

THE MAJOR FEATURES OF THE LIVING WORLD: WHAT MUST WE EXPLAIN?

If we were to define the task of biology as a science, it would be to understand the nature of the living world. But that is not a very specific definition. There are four overarching features of the biological world that any successful and complete theory must account for:

1. The *unity* of all life, visible in the significant similarities that are shared by all organisms. All life, for example, relies on nucleic acids (DNA or RNA) for the transmission of information across generations.

2. The striking *diversity* of the living world. Current estimates suggest that up to 30 million species populate the Earth, the majority of which have yet to be studied or even formally described.

3. The existence of *biological adaptations*, compelling examples of a fit between organisms and their environments.

4. The deep, dynamic *history* of life on Earth. The history of life on this planet extends back at least 3.4 billion years; it is a history of constant change. New forms appear, diversify, and go extinct, leaving behind traces both in the fossil record and in living organisms. From these traces, we seek to reconstruct and understand what we did not witness.

In this chapter, we examine these four central features of the biological world. As we will see, it is only in the last 130 years, beginning with the publication in 1859 of Charles Darwin's *On the Origin of Species*, that a single coherent theory, capable of accounting for all four features, begins to take shape. We will take a brief look at biological thought around Darwin's time, and a close look at the principal arguments of *The Origin*.

Unity and Diversity

The first two general features, unity and diversity, appear initially to contradict each other. How can a single theory encompass these two aspects of the living world? The paradox, however, is only superficial. The biological world is exceedingly diverse, composed of millions of different species of organisms, each with its own characteristic appearance, ecological requirements, physiology, behavior, and social structure. Each species is truly unique, the product of its own particular history, distinct and distinguishable from all other species. Yet underlying this apparent chaos and variety are profound similarities in the ways in which all species go about their business. The oak tree (*Quercus lobata*), the malarial protozoan parasite (*Plasmodium falciparum*), and a human (*Homo sapiens*) all use the same chemical structures, chains of nucleic acids (DNA or RNA), to pass genetic information from one generation to the next (Fig. 2.1). The arrangement of the units is different, but the alphabet is the same. Similarly, all organisms make the proteins needed for their survival out of the same 20 amino acids. The coat protein that envelops the HIV particle (the virus responsible for AIDS) and the protein that brings oxygen to your brain while you read this page are made of the same stuff. Only the number and order of the amino acids are different. Deep unities among all organisms also exist at the metabolic level—all aerobic (oxygen-utilizing) life uses one of two compounds, adeno-

Figure 2.1

(a) An oak tree (genus *Quercus*) in evening light. (b) The parasite *Plasmodium falciparum*, responsible for the most common human infectious disease, malaria. (c) *Homo sapiens*, in its early years.

(a)

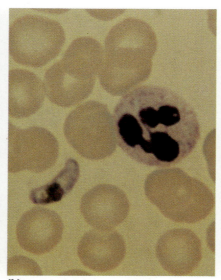

(b)

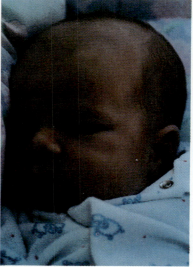

(c)

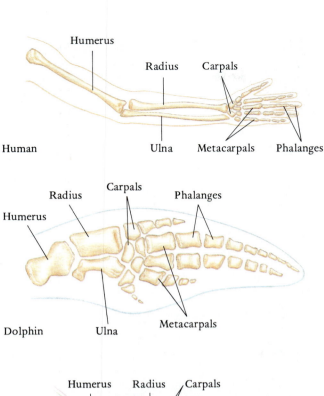

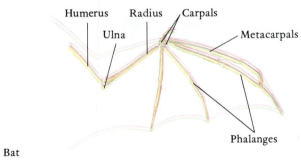

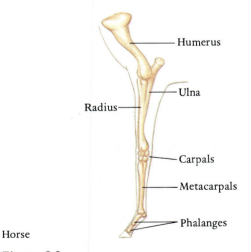

Figure 2.2
Although specialized for different tasks, the detailed underlying similarities in the bones of the homologous structures shown above are compelling evidence for the shared common ancestry of these species.

sine triphosphate (ATP) and guanosine triphosphate (GTP), as their means of energy exchange.

Other biological similarities are not shared by all living organisms, but nonetheless unite major groups of apparently disparate organisms. Why, for example, do all terrestrial vertebrates go through a stage before birth where pharyngeal pouches are present in the developing embryo? Why are bones in the human forelimb, the porpoise flipper, the bat's wing, and the horse's leg all organized in similar fashion: a long bone (the humerus) articulating to two thinner bones (radius and ulna), followed by several very small cuboidal bones (carpals), terminating in five (or fewer) digits (Fig. 2.2)? We will be looking in more detail at such similarities and assessing their significance, when we begin our consideration of different animal groups. For now, keep in mind that animal groups that appear very different on the surface often share important underlying biological similarities.

The Nature of Biological Adaptation

In our examination of the animal kingdom, we will see many **adaptations**, features of organisms that seem specifically designed to meet the needs of their bearers. Examples can be drawn from any group of organisms, and occur at every level of biological organization. Certain adaptations concern appearance: insects frequently avoid detection by looking like twigs, leaves, thorns, and even bird droppings (Fig. 2.3). Other adaptations are clearly functional, such as the powerful jaw muscles and short, stout jaw of the hyena, which allow it to crush the bones of the carcasses it scavenges (Fig. 2.4). Yet other adaptations are physiological. During a dive, whales, seals, and other diving mammals benefit from a complex series of adaptations that allow them to cope with the absence of oxygen,

Figure 2.3
A katydid from Peru displays protective coloration, mimicking the partly decomposed leaves on which it is frequently found. Note the detailed similarity between the wings and the venation pattern of the leaf.

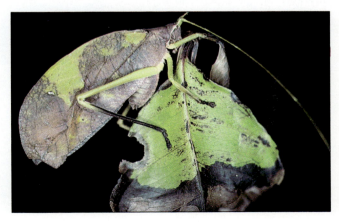

the cold water temperatures, and the risk of "the bends" (Fig. 2.5; see also Focus 13.1: Adaptations for Diving). Diving triggers a "diving reflex" in these animals—a set of physiological changes that reduce heart rate and shunt blood primarily to the heart and brain.

Finally, biological adaptations can be found at the molecular level. Diseases caused by pathogenic bacteria have, since World War II, been treated primarily by the widespread use of antibiotics. Over the last 25 years, a large number of antibiotic-resistant bacterial strains have appeared: the bacteria have developed a variety of biochemical mechanisms that enable them to inactivate or elude previously lethal antibiotics.

Each example has been extensively studied, described, and analyzed. A coherent biological theory, however, seeks not only to describe, but also to explain how these and other examples of the close, elegant fit between an organism and its environment have come about.

History and Contingency

As we discussed in the first chapter, biology must take into account the role of history in shaping life as we now see it. The fossil record bears witness to the constantly changing nature of the living world and to the emergence and inevitable extinction of species throughout the history of life. In addition to this constant turnover of species, the fossil record also preserves major, cataclysmic events—the mass extinctions—in which huge, diversified, previously successful groups of organisms disappear, never to be seen again (Fig. 2.6). Thus, 230 million years ago, as many as 90% of the species of marine organisms disappeared forever. There have been many such mass extinctions throughout the history of life; although we do not fully understand the causes of major extinctions, there is little doubt that they fundamentally alter the history of life,

Figure 2.4

An efficient hunter and scavenger, the spotted hyena (*Crocuta crocuta*) has evolved a compact set of jaws and a powerful set of jaw muscles that enable it to tear flesh and crush bone. Despite their reputation, hyenas are among the most social predatory mammals.

Figure 2.5

A humpback whale, *Megaptera novaeanglie*, beginning a dive. These mammals have evolved a set of anatomical, physiological, and biochemical features that enable them to remain submerged for extended periods of time.

Figure 2.6
Fossil ammonites from the lower Jurassic (appox. 195 million years ago). The evolutionary history of this group is profoundly affected by mass extinctions. Once an extremely diverse component of marine faunas, the ammonites survived three major extinctions before finally disappearing during the Cretaceous extinction, 65 million years ago.

making way for new species to arise from the survivors of these debacles. Mass extinctions are but one reminder of the contingent nature of the history of life: geological, climatic, and biological events have redirected the course of biological evolution. The current number, diversity, and kinds of organisms that we see today are the result of only one of many possible paths that life on Earth could have taken.

The history of life has unfolded over time scales that are incomprehensibly vast relative to our own lifetimes. *Homo sapiens*, the species that seeks to understand this biological past, is but a very recent newcomer, present only for the last two-thousandths of 1% of the story. If we are to understand the evolution of the living world, we need to develop methods to reconstruct the events that we did not witness. Fortunately for us, living organisms carry within them—in their genes, in their structures, and in their behavior—the traces of their past. Our task as biologists is to tap into those traces in order to reconstruct the past.

DARWIN'S WORLD: BIOLOGICAL THOUGHT IN THE EARLY 19TH CENTURY

The Facts Are Already in Place

Many accounts of Charles Darwin (Fig. 2.7) paint him as a brilliant man who, after years of patient, careful observation, managed to break the shackles of superstition and ignorance that prevented all of his predecessors and con-

temporaries from seeing "the facts." Armed now with the truth, Mr. Darwin waged battle against religious zealots and eventually triumphed.

Superficially satisfying as that account may be, it is an oversimplification. The notion that good scientists have no ideas or preconceived notions, that they simply let nature and the facts speak directly to them, distorts the nature of scientific achievement. In reality, scientists do not operate in an intellectual or cultural vacuum. The ideas, insights, and preconceptions of their time do exert an influence on the way scientists think (see Focus 2.1). Even today, despite our best efforts, we will make mistakes and misinterpret the evidence according to our own preconceptions. With luck, future scientists will correct them.

Our preconceptions are a lens through which we perceive not only the natural world but also the achievements of past scientists. As a result, we judge those scientists that predicted what we accept today as good and able, and those whose arguments have been refuted, modified, or discarded as simple-minded, short-sighted, and blind to the truths of nature. But we learn nothing about the history of science by judging from the perspective of what we know today. We can only really grasp the magnitude of Darwin's achievement, and the radical nature of his insights, if we place him in the intellectual context of his time.

The remarkable way in which organisms appear to be built to perform the tasks that confront them had caught the attention of natural historians from the time of the Greeks. Aristotle, for example, described the specialized arm of the male octopus that it uses to transfer packages of sperm to the female. Similarly, the notion that the living

Figure 2.7
Portrait of Charles Darwin painted following his return from the five year circumglobal voyage of exploration on the H.M.S. *Beagle*.

Carolina Biological Supply Co.

(a)

(b)

Figure 2.8

(a) The zebra, *Equus burchelli*, a close relative of the horse. Its conspicuous stripes, whose function is not understood, occasionally appear (in muted form) in certain domestic horse breeds. The similarities in appearance, skull and limb anatomy, social structure, and behavior between zebras and horses reflect their common ancestry. (b) The horse, *Equus caballus*, is placed in the same genus as the zebra.

world was ordered and not chaotic was practically commonplace by the beginning of the 18th century. The idea that resemblances existed among groups of organisms—dogs and wolves, for example, or horses and zebras (Fig. 2.8)—was not only acknowledged, but formed the basis for the discipline of classification, or **taxonomy**. In the mid-18th century (100 years before Darwin) Carolus Linnaeus, a Swedish botanist, developed a complete system of classification for plants and animals (Fig. 2.9). To this day, we still classify organisms using the Linnaean hierarchy. Similar individuals that interbreed, or are potentially capable of doing so, form a *species*. Different species that resemble one another closely are grouped into *genera*, similar genera into *families*, and so on through progressively inclusive *orders*, *classes*, *phyla*, and *kingdoms* (see Chapter 23).

The fossil record, with its preserved evidence of extinct animals and plants, was also well explored by Darwin's time. The discipline of **paleontology**, the study of fossils, was a strong and vibrant enterprise in England, France, and Germany. The publication of Nicolaus Steno's *Prodromus of a dissertation on solid bodies which are naturally contained in other solid bodies* (1667), in which he argued that the **glossopetrae** (tongue stones) found in sedimentary rocks in Italy were in fact fossil shark teeth, is the first clear acknowledgment that fossils were the remains of once-living animals and plants (Fig. 2.10). Since then, extensive analyses of fossil deposits had been carried out, and the similarities that existed between certain living animals and the remains in the fossil record were appreci-

ated by most practicing scientists. Even the vastness of geological time was clear to Darwin's immediate predecessors and contemporaries. Although considerable debate raged about the exact age of the Earth, geologists had fully debunked the notion that a few thousand years would have sufficed to fashion the Earth as they saw it.

Figure 2.9

Carolus Linnaeus, the Swedish botanist whose 1758 *Systema Naturae* sought to order and classify all living organisms into a taxonomic hierarchy.

Figure 2.10
Fossil shark teeth removed from the Devonian sandstone matrix in which they were embedded.

Natural Theology

What were the scientists of the time making of these observations and ideas? There is no single answer to this question, since every European country had its own intellectual traditions and styles, and the United States was a young country with only a fledgling scientific practice. Nevertheless, most of the natural historians of the time saw the biological world as the embodiment of God's design. This approach came to be known as **natural theology**. For natural theologians, there could be no nobler purpose than to study nature in hopes of glimpsing the intent and design of the creator. After all, the countless adaptations and elegant interactions in nature could only have been fashioned by a designer. This was the popular "argument for design," which stated that the thoughts of God were encoded in Nature, and the task of natural history was to decode these thoughts.

It is easy to view these scientists today as biblical apologists who betrayed the cause of science. But to do so is to miss the point entirely. Some of the finest work of the time in botany, zoology, and geology was carried out under the framework of natural theology. Careful and detailed experiments and observations were made by scientists seeking this deeper message. Ironically, this melding of science and theology focused the considerable talents of individuals educated by and for the church on important problems in natural history. To be sure, certain scientists and philosophers had begun to question the validity of this unholy alliance between the methods of science and the message of theology. Darwin would undercut the notion that *design* need imply a *designer* by showing that the operation of material forces could produce order and design. Nevertheless, natural theology had cast the foundation upon which the intellectual revolution in biology was to take place.

DARWIN AND THE STRUCTURE OF ON THE ORIGIN OF SPECIES

What then, exactly, was the nature of Darwin's revolutionary genius? He did not discover a whole new class of "facts." As we have just discussed, the fundamental empirical observations that Darwin would use were already in place: the vast expanse of geological time, the fossil record, similarities among organisms. Nor, as we noted before, was Darwin a "great empiricist," able to "do science" without any preconceived ideas or notions in his head. In fact, Darwin would write in 1861 about the importance of information gathered with a hypothesis in mind:

> About thirty years ago there was much talk that geologists ought only to observe and not to theorize; and I well remember someone saying that at this rate a man might as well go into a gravel-pit and count the pebbles and describe the colors. How odd it is that anyone should not see that all observation must be for or against a point of view if it is to be of any service!

What Charles Darwin did, with the publication on November 24, 1859, of *On the Origin of Species by Means of Natural Selection, or the Preservation of Favoured Races in the Struggle for Life*, would forever change the way we saw the world (Fig. 2.11). With the publication of this 490-

Figure 2.11
The title page of the first edition of Charles Darwin's *On the Origin of Species*. This work, whose first edition of 1250 copies sold out the day it appeared, revolutionized our understanding of the living world.

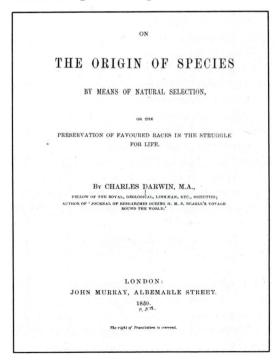

ON

THE ORIGIN OF SPECIES

BY MEANS OF NATURAL SELECTION,

OR THE

PRESERVATION OF FAVOURED RACES IN THE STRUGGLE
FOR LIFE.

By CHARLES DARWIN, M.A.,
FELLOW OF THE ROYAL, GEOLOGICAL, LINNÆAN, ETC., SOCIETIES;
AUTHOR OF ' JOURNAL OF RESEARCHES DURING H. M. S. BEAGLE'S VOYAGE
ROUND THE WORLD.'

LONDON:
JOHN MURRAY, ALBEMARLE STREET.
1859.

The right of Translation is reserved.

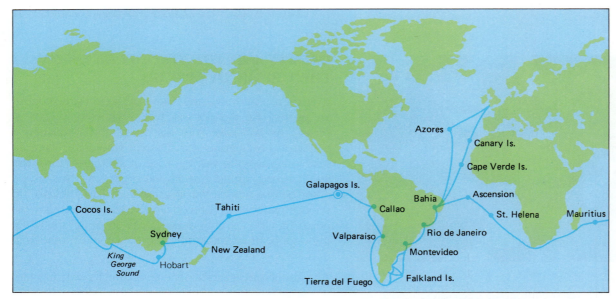

(a)

page book, whose first edition of 1250 copies sold out the same day it appeared, the natural world would cease to be the consequence of supernatural forces or divine intervention. Henceforth, only material cause could be used to explain the phenomena of natural science. Darwin had also succeeded in synthesizing what appeared to be disparate observations from various fields—geology, paleontology, taxonomy, and animal husbandry—into a single coherent theoretical framework. The concept of evolution, and the idea that adaptation arose from the operation of material forces—specifically natural selection—revolutionized the study of organisms.

Darwin would put forth two separate but interrelated arguments in *On the Origin of Species*. First, he would establish that change, *evolution*, was a characteristic of the living world. He would argue that the fossil record and the distribution and appearance of living animals and plants could not be coherently explained unless (1) all organisms are descended from a common ancestor, and hence all life forms a single genealogical network, and (2) a process of evolution, or descent with modification, is constantly taking place from generation to generation.

Darwin's second argument concerned natural selection, which he claimed to be the mechanism propelling evolutionary change. He would argue that the "preservation of favourable variations and the rejection of injurious variations," the consequence of the struggle for existence being waged by every individual organism, was a sufficient explanation for the order, design, and apparent harmony of the natural world.

How did Darwin go about establishing the two claims? What kind of evidence formed the core of his arguments? How, in short, did he hope to persuade both his colleagues and the educated lay reader of the validity of his radical claims?

(b)

Figure 2.12

(a) The itinerary of the surveying voyage undertaken from 1831 to 1836 by the H.M.S. *Beagle*. (b) A replica of the H.M.S. *Beagle*, built for an English television recreation of Darwin's voyage.

Establishing the Fact of Evolution

The Voyage of H.M.S. *Beagle*

By age 21, the young Charles Darwin had already undertaken to become a medical doctor, at Edinburgh University, and then a country clergyman at Cambridge University. These vocations held little appeal for Darwin, who had relatively undistinguished academic performances in both. In contrast, Darwin was fascinated by entomology, natural history, geology, hunting, and taxidermy, and would purse those interests aggressively. Through his studies and hobbies, he came to know many of the illustrious geologists and naturalists at Cambridge and Edinburgh Universities. Darwin spent the summer of 1831 in North Wales, examining geological formations with the geologist Adam Sedgewick. Upon his return to Cambridge, a letter would alter the course of Darwin's life: Sedgewick had recommended Darwin for the position of captain's companion aboard the H.M.S. *Beagle*. This surveying vessel, commanded by Captain Robert FitzRoy, was to undertake a surveying voyage around the coast of South America. Darwin's father, already frustrated by his son's indecision and underachievement, tried hard to discourage Charles from accepting the invitation. But Charles' naturalist friends and other members of his family urged him to agree to the offer. On December 27, 1831, the *Beagle* sailed from Devonport harbor, with Charles Darwin aboard (Fig. 2.12).

Darwin did not take kindly to the sea. Near the beginning of the trip, he would write to his father, "I was unspeakably miserable from sea-sickness"; three years later, to his sister, "I positively suffer more from sea-sickness now than three years ago." Yet despite his discomfort and frequent altercations with FitzRoy, Darwin was indefatigable in his collections of specimens, in his observations of natural history and geology, and in his readings of philosophy and geology. The *Beagle* did not return to England for 5 years. During that time, it called at ports and islands throughout the Southern Hemisphere (Fig. 2.12a). At each of these stops, Darwin would disembark and collect specimens of plants, animals, birds, and insects. Frequently, he would undertake inland voyages, observing and collecting, and keeping a detailed journal of the voyage. Curiously, Darwin's journals and notebooks while on the H.M.S. *Beagle* do not contain any mention of the theory of evolution by natural selection and only foreshadow his concept of evolution. The importance of the observations made while on the *Beagle* (Fig. 2.13) would only become apparent to Darwin several years after his return (see Focus 2.1 on page 30).

(a)

(b)

(c)

Figure 2.13
The fauna and flora of the Galapagos Islands. (a) A saddleback tortoise from Hood Island (*Geochelone elephantopus hoodensis*). The giant tortoises on each of the islands in the Galapagos archipelago exhibit characteristic markings and patterns on their carapaces. (b) The booby (*Sula sula*), whose webbed feet enable it to swim effectively, is also able to perch on tree limbs. (c) A cactus in the genus *Opuntia*. Despite their superficial resemblance to trees, these cacti, widely distributed in the New World, are not trees.

The Mutability of Species

There is little doubt that what Darwin saw firsthand during his voyage had much to do with the development of his ideas. But he had not come across some brand new "kind" of fact or observation, that in a single, blinding "Eureka!" led to the theory of evolution. Darwin's most profound breakthrough was conceptual and synthetic. He did not so much see new things, as old things anew.

Historians and philosophers of science identify two central notions that Darwin directly challenged. The first, that of separate origins of fixed species or "kinds," held that the different kinds of organisms that existed had been separately and independently created, and had not given rise to other kinds. Secondly, Darwin challenged the essentialist or **typological view** of the natural world. Essentialists, basing their arguments on the writings of Plato, argued that every species had an "essence," an unchanging, ideal form. The similarity between animals in a given species (all wolves, for example) reflected their shared essence. The variation and differences that exist between one wolf and another were seen as the result of the imperfect expression of the ideal, distracting noise to be ignored by naturalists in search of the essence. Again, essences were considered eternal and invariant, reaffirming the notion that species were constant and unchanging as well.

In one stroke, Darwin would undercut this entire view of the world. For him, the similarities between organisms arose not out of a shared essence, but out of common descent. This was true at all levels of the biological hierarchy: wolves resembled other wolves because of common ancestry. Wolves and dogs were similar because they too had shared a common ancestor (Fig. 2.14). Wolves, dogs, foxes, and coyotes had also shared a common ancestor, albeit much earlier in evolutionary time. Picture every species of animal that had ever lived as each sitting on a branch tip. We could work our way back from the tips and find that the twigs joined into small branches, the small branches into larger ones, and so on, until we could trace all life back to a single trunk and root.

Darwin would supply a great deal of evidence for his assertion that species were not fixed. It is here that the material collected and the observations made during his *Beagle* voyage would prove so useful. The fossils that Darwin had seen embedded in the sedimentary layers of South America had convinced him that species, far from being immutable and immortal, were in fact in constant flux, appearing, changing, and going extinct. He had witnessed the fossil remains of enormous and now extinct South American mammals (Fig. 2.15), of snails and bivalve mollusks, and he would emphasize in the *Origin* that all showed affinities and similarities to living organisms. In some cases, the extinct forms appeared as intermediates between now distinct groups of organisms. Ideally, Darwin argued, a perfect fossil record should yield a complete series of intermediate forms that would connect all extant organisms into a vast, branching network. He was, of course, well aware of the scarcity of intermediate

Figure 2.14

(a) The timber wolf (*Canis lupus*), the closest relative and likely ancestor of the domestic dog. (b) The Siberian husky, one of more than 300 breeds of domestic dog. Despite the enormous differences in appearance between different breeds, they are all part of the same species and are (at least potentially) capable of interbreeding.

(a)

(b)

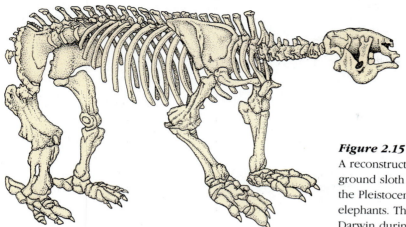

Figure 2.15
A reconstruction of the skeleton of *Megatherium*, the giant ground sloth of South America. These animals lived during the Pleistocene and could grow to the size of modern elephants. The first fossil of *Megatherium* was discovered by Darwin during the voyage of the H.M.S. *Beagle*.

forms in the fossil record, a situation he would attribute to the numerous imperfections of the fossil record. Ironically, one of the best examples of an intermediate fossil form, *Archaeopteryx* (Fig. 2.16), was only discovered two years after the publication of the *Origin*. In the 6th edition of the *Origin*, Darwin would seize upon this example with noticeable relief, since *Archaeopteryx* was an excellent example of intermediacy, combining features of both birds and reptiles, as though it lay at the branchpoint of what now were very distinct lineages.

The theory of evolution as Darwin conceived of it assumed the slow and steady transformation of species into new forms. Of this transformation, however, only occasional snapshots were provided by the fossil record.

Figure 2.16
Fossil of *Archaeopteryx*, a bird-like reptile from the middle Jurassic. This important fossil is a mosaic of reptile and bird features, illustrating evolutionary intermediacy between two important groups.

Darwin would go so far as to declare that "he who rejects this view of the imperfection of the fossil record will rightly reject the whole theory." As we will see in Chapter 9, Darwin may have saddled himself too heavily with the assumption that evolution can *only* proceed by imperceptibly slow, small steps that would leave behind a complete trail of intermediate forms. In certain cases, the absence of intermediates may reflect not only the imperfection of the fossil record, but also the more episodic and rapid changes that take place in evolution.

Darwin had succeeded in accounting for fossil forms by the same arguments used to explain the diversity of living forms: fossils were part of the same branching tree and fossil forms resembled their living counterparts more and more as one moved closer to the present. In short, fossil forms simply reflected the operation of evolution in times past.

The Importance of Geographical Variation

In support of the reality of evolution, Darwin would also marshall an entirely different set of observations regarding the geographical distribution of animals and plants, and the existence of endemic (unique to a particular locality) faunas. Having used an argument on the reality of organic change over time (as seen in the fossil record), he would now use the argument of variation over space (geographical variation) to support the theory of evolution. In his journal, he would write ". . . opened first notebook on 'Transmutation of Species'—Had been greatly struck from about month of previous March [1836] on character of South American fossils—and species on Galapagos Archipelago. These facts (especially latter) origin of all my views."

The fictional "Eureka" tale of Darwin's insight would have him arriving at the Galapagos, where, glimpsing the mockingbirds and finches of those islands, he would immediately understand evolution. In fact, it was only years after his return that Darwin would realize the central role that geographical variation would play in his theory.

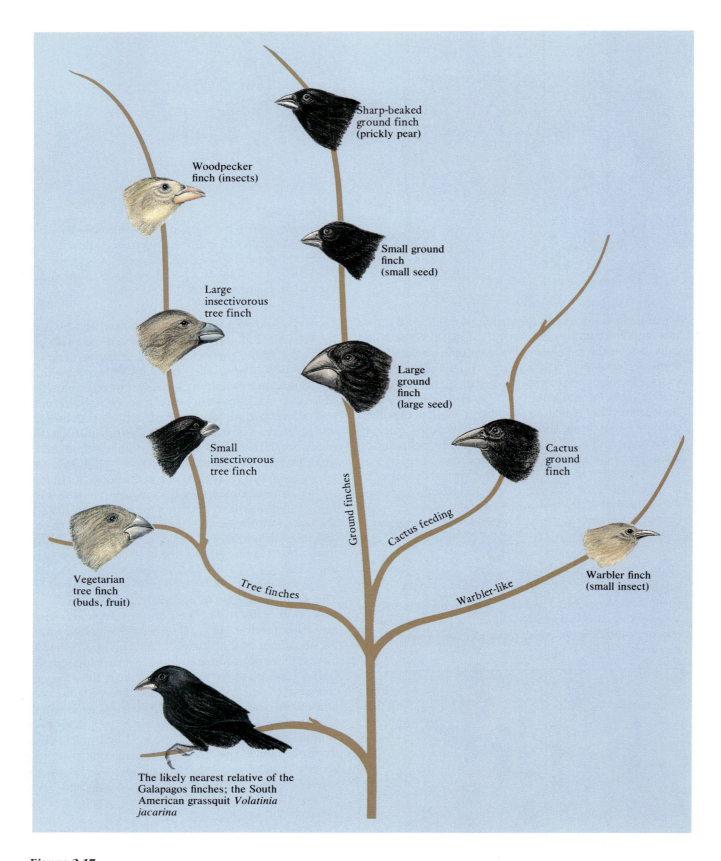

Figure 2.17

Galapagos finches. The fourteen species of Galapagos finches evolved from a small population of a single ancestral species blown over to the Galapagos Islands from South America. In the course of their evolutionary radiation on the Galapagos Islands different species became adapted for a variety of feeding habits (in parentheses) reflected in their beak forms.

Darwin had carried out extensive collections of birds in the Galapagos (Fig. 2.17), but had failed to note the specific island at which each specimen had been collected. Apparently, at the time Darwin was collecting, the variation that occurred from island to island seemed to him irrelevant, and on many of the collection tags he would simply note "Galapagos." Two years after his return to England, Darwin would eventually depend on the substantial patience and assistance of John Gould, a leading British ornithologist, to reconstruct the locality information so crucial to the theory of natural selection.

In retrospect, three aspects of island faunas would strike Darwin as particularly significant. The first of these, as we have already mentioned, was the endemicity of island forms. The birds, tortoises, and plants of the Galapagos were all obviously related to forms living on the South American mainland, but they were somehow slightly different. Clearly, the original colonists of these islands had come from the mainland. Now, isolated from the source and subjected to the new conditions of the volcanic Galapagos, these faunas had changed and adapted to the new circumstances. The governor of the Galapagos had pointed out to Darwin that the shells of the land tortoises differed from island to island. Years later, Darwin would still recall the significance of that comment. The bird faunas of the Galapagos were even more striking in their variety and endemicity. Each island appeared to harbor its own variety of mockingbird and one or more species of finch. Although these birds could fly, they appeared to have evolved in isolation not just from the mainland, but also from island to island. Darwin would correctly point out that the currents between islands were swift, and the gale winds rare. In consequence, "the islands are far more effectually separated from each other than they appear on a map."

A second feature of island faunas would also catch Darwin's eye: entire classes of organisms were frequently absent from island faunas. Terrestrial mammals and frogs, with their limited powers of dispersal, would frequently fail to colonize oceanic islands. Despite the presence of appropriate climatic conditions, the absence of these organisms argued against a theory of separate creation.

One final set of observations about island faunas would prove important to Darwin's theory: the faunas of the Cape Verde islands, in the North Atlantic, did not resemble those of the Galapagos (Fig. 2.18). Instead, the Cape Verde animals and birds clearly resembled those of the nearby West African mainland. In contrast, the Galapagos fauna had clear affinities with South American species. Geologically, climatically, and topographically, the Galapagos and the Cape Verdes were quite similar. If island organisms had been separately and especially created to inhabit islands, ought we not to expect greater similarity between the two island faunas? For Darwin, the following conclusions were inescapable:

1. Island faunas did not arise as the result of special creation, but as a consequence of the colonization by species from neighboring regions. The imprint of common descent (for the mainland and corresponding island forms) was therefore stamped on island faunas and floras.

2. Dispersal and the ability to successfully compete in and colonize a new island accounted for the composition of island biotas.

3. In many instances, although the affinities to mainland biotas could be readily seen, the species inhabiting oceanic islands had changed. Many birds, for example, would become flightless when isolated on islands. The island forms were readily distinguishable; evolution, descent with modification, was taking place.

The Agent of Evolutionary Change: Natural Selection

The true power of any scientific theory resides in part in its ability to account for patterns or observations that previously appeared unrelated. Darwin had succeeded in bringing together observations from all corners of natural history. The exquisite design of organisms, the distribution of animals on different continents, the extinct forms in the fossil record, the behavior of animals, the place of human beings in the great scheme of nature—he had woven all these together as evidence for the reality of evolution. But the question of mechanism still remained. If the scientific community and lay establishment were to accept evolution, a convincing agent of evolution had to be put forth

Figure 2.18

Galapagos tortoises on Isabela Island. The realization that the tortoises on each of the different islands exhibited a different shell pattern would play an important part in the development of Darwin's theory.

and defended. The motor of evolutionary change, Darwin argued, was the principle of **natural selection**.

The Logic of Natural Selection

Let us begin by analyzing the logic behind the theory of natural selection. In its purest form, natural selection is a set of deductions that flows from three basic facts about the natural world.

First, any population of organisms is potentially capable of exponential growth over time, becoming so numerous that it would soon overwhelm resources, which, at best, can only grow arithmetically. This is known as the Malthusian principle of **superfecundity**: Darwin had certainly read and been greatly impressed by Thomas Malthus's Essay *On Population* (1798).

Second, in all sexually reproducing species, individual organisms always differ. This is the principle of **individual variability**, the notion that variation is always present in natural populations.

Third, some proportion of the differences between individuals in a population has a genetic component. This is the principle of **heritability**, the basis of all of genetics.

From these three basic facts, the theory of natural selection would flow:

1. Given that resources cannot accommodate the potential fecundity of populations, some competition for resources must ensue (the "struggle for existence").

2. Success in the struggle for existence is in part determined by the characteristics of the individuals involved, and these characteristics are passed on to the offspring of the successful individuals. The offspring of organisms possessing characteristics that increase survival and reproduction thus form a larger proportion of the next generation.

3. The consequence of this process is a change in the overall composition and character of the population from generation to generation. As the result of the process of natural selection, certain features of organisms become more (or less) prevalent. Evolution is taking place.

Before proceeding, it is useful to clear up certain misconceptions about natural selection. The first of these concerns the concept of "struggle for existence." Many would come to see this phrase as a metaphor for brutal, bloody competition between individuals, and would decry this vision of nature. But Darwin meant nothing of the sort. He would write:

> I should premise that I use the term Struggle for Existence in a large and metaphorical way, including dependence of one being on another, and including (which is more important) not only the life of an individual, but success in leaving progeny. Two canine animals in a time of dearth, may be truly said to struggle with each other which shall get food and live. But a plant on the edge of a desert is said to struggle for life against the drought.

Ironically, the phrase "survival of the fittest" was not Darwin's. Instead, it was coined by Herbert Spencer following the publication of the *Origin*. Well aware of the image it evoked, Darwin would not use it until the 6th edition.

Second, keep in mind that although natural selection targets individual organisms, what actually evolves are **populations**, groups of organisms of the same species. The theory of evolution is a theory about change from generation to generation, eventually telescoped into deep time. But although it is individuals that are better able to survive and reproduce, what is changing is the composition of the next generation.

Third, evolution is not a theory about increasing perfection. The fact that, on average, individuals possessing characteristics that give them the slightest edge in survival or reproduction leave a greater proportion of offspring does lead to adaptation. Features that confer advantage are retained and combined at the expense of other available alternatives. But it is only *adaptation to local circumstance*. Natural selection is a force with no foresight: it can select only from among the available variants, and only for the features that are advantageous to local circumstances.

The Defense of Natural Selection

While the Victorian age knew that some kind of "selection" operated in nature—offspring born with severe defects often died, the smallest of the litter often failed to reach adulthood—this was a view of selection as a primarily cleansing force, destroying the unfit and preserving the type. Darwin's argument turned selection into a *creative* force, capable of producing change and novelty over evolutionary time. Many of his contemporaries would not take kindly to this proposition (Fig. 2.19). The *Origin* is, in some sense, a legal brief in defense of the theory of evolution by natural selection ("one long argument," Darwin would call it). Like any good lawyer, Darwin began with an example his audience was familiar with and would have little trouble accepting: **artificial selection**, the managed breeding of certain domestic animals, selected for their desirable features (appearance, milk yield, herding ability, coat color, and so on.) His readers were quite familiar with the numerous varieties of livestock, dogs, and pigeons that had been produced through the breeding of particular members of a herd or flock (Fig. 2.20). Artificial selection eventually produced animals that differed dramatically from the original stock from which they had come. These breeds and varieties illustrated two important premises of the theory of natural selection: the enormous variability of natural populations (from which the breeder could then choose) and the relative ease and speed with which the appearance of animals could be changed.

Having managed to get his foot in the door by reminding his audience of what they already knew, Darwin

Figure 2.19
One of many political cartoons that appeared after the publication of the *Origin*. This one shows Darwin and Bergh, the founder of the Society for the Prevention of Cruelty to Animals. The caption reads: "The defrauded gorilla: 'That man wants to claim my pedigree. He says he is one of my descendants.' Mr. Bergh: 'Now, Mr. Darwin, how could you insult him?'"

would use subsequent chapters to draw a parallel between artificial and natural selection. To be sure, the differences were important; for one thing, natural selection was not the result of a conscious act by an agent of selection. Natural selection was in fact far subtler, distinguishing between slightly different variants. Natural selection, finally, was a probabilistic force, selecting those features that *on average* would confer greater survival and reproduction. The order of nature, however, flowed entirely from this simple process operating between individual organisms within a species. Natural selection was, in a profound way, the invisible hand of nature. Simply stated, there was nothing more to the theory than the notion that organisms differ in their ability to survive and reproduce; the playing out of those differences caused populations to change and evolve over time. In Darwin's words:

> . . . if variations useful to any organic being do occur, assuredly individuals thus characterised will have the best chance of being preserved in the struggle for life; and from the strong principle of inheritance they will tend to produce offspring similarly characterised. This principle of preservation I have called, for the sake of brevity, Natural Selection.

We shall have more to say about evolution and about the concept of natural selection as the mechanism of organic change. What you must keep in mind is that they are separate strands running through *On the Origin of Species*. As in all healthy sciences, there continues to be much debate about the exact mechanisms responsible for evolution, about the importance of natural selection, and about the possibility that other material forces may be responsible for evolutionary change. That evolution occurs, however, is no longer under question.

Figure 2.20
A few of the many different breeds of cattle that have resulted from the operation of artificial selection and selective breeding by pastoralists and cattle breeders.

In 1858, Darwin was finally hard at work on his "Big Book." Twenty years had elapsed since he had first conceived of the theory of evolution by natural selection. Since then, he had completed his reports on the *Beagle* voyage, and an extensive monograph on the Cirripedia, the barnacles. Finally, in 1856, encouraged and pressured by his friend Charles Lyell, Darwin had begun his book, *Transmutation of Species*, in which he intended to lay out the evidence for evolution and to propose the theory of natural selection. In June 1858, he had drafted the first ten chapters, when a package arrived in the mail from the East Indies. It had been sent by Alfred Russel Wallace, and it contained a manuscript entitled "On the Tendency of Varieties to Depart Indefinitely from the Original Type." Also included was a letter from Wallace, asking Darwin to read the enclosed manuscript, and, should he find it of interest, to forward it to Charles Lyell for publication. Darwin was crushed. Wallace had independently arrived at the same conclusions as Darwin. The manuscript was a concise statement of both the constantly changing character of the species (as illustrated by endemic island species) and of the "struggle for existence" that was responsible for evolutionary change. Darwin had little choice but to forward the manuscript to Lyell, accompanied by this sad note: "I never saw a more striking coincidence: if Wallace had my manuscript sketch written out in 1842, he could not have made a better short abstract! . . . All my originality, whatever it may amount to, will be smashed." Fortunately for Darwin, Lyell decided to present Wallace's paper, together with portions of an essay written by Darwin in 1844, at a meeting of the Linnean Society of London, on July 1st, 1858.

Wallace, himself, was an interesting character. Having quit school at age 13, he was an avid reader who had become an accomplished naturalist. He made his living dangerously, by collecting tropical birds, insects, and butterflies for European museums and private collections. In 1848 he had set out for the Amazon to collect specimens and to try to find evidence for how new species arose. Four years later, the ship in which he was returning to England would catch on fire and sink, taking with it all of Wallace's collections and journals. Still, 2 years later, in 1854, he would set out again, this time for the Malay Archipelago, convinced that islands and isolated land masses were the place to find the necessary evidence for evolution. In 1855, he would write his first famous paper, "On the Law which has Regulated the Introduction of New Species," in which he sketched out a theory of organic change and a model of speciation. Yet missing from that paper was a mechanism, a material cause for the changes he knew were taking place over geological time.

That insight would come to him during a malarial episode in 1858. Wallace wrote: "At that time I was suffering of intermittent fever at Ternate in the Moluccas, and one day, while lying on my bed during the cold fit, wrapped in blankets . . . something led me to think of the 'positive checks' described by Malthus in his *Essay on Population*, a work I had read several years before and which had made a deep and permanent impression on my mind."

The credit for the idea of evolution, propelled by natural selection, properly belongs to both Darwin and Wallace. But what are we to make of the coincidence that led two naturalists who had never met or spoken, who lived thousands of miles apart, to arrive at the same detailed conclusions and mechanisms concerning the nature of organic change? Historians of science suggest that these are not mere coincidences. Instead, they reflect once again on the importance of the social and intellectual context in which science is done. Darwin and Wallace had read many of the same books and had both been impressed by Malthus, Lyell, and Chamber's *Vestiges of the Natural History of Creation*. Both were products of the British Victorian age, a time when many different ideas about change and evolution were "in the air." It is important to remember that both the scientists and their ideas were products of their time. Nevertheless, it still required individual synthetic genius for them to see and articulate a complete material theory of evolution.

Summary

1. Any theory of the biological world must account for four major features of this world. (1) Organisms are exceedingly diverse; up to 30 million species currently inhabit the Earth. (2) Despite this diversity, there are profound biological similarities among all organisms, e.g., in the ways species transmit information from generation to generation, construct proteins, and utilize energy. (3) Organisms display an elegant fit between their structural and functional features and the environments in which they live. (4) The history of life has been contingent; its course has been profoundly affected by geological and climactic events and by the history of life itself.

2. Before Darwin proposed his theory of evolution, scientists already recognized that organisms were remarkably adapted to their environment, that the living world was ordered and species could be grouped according to their resemblances, that fossils are remains of once-living plants and animals, and that the earth was far older than a few thousand years. These observations were interpreted in the 18th and early 19th centuries as evidence of the existence of a divine designer.

3. The publication of Darwin's book *On the Origin of Species* (1859) exploded the notion that design and order in the natural world implied a designer and showed that material causes could explain the phenomena of natural sciences.

4. Darwin made two separate but interrelated arguments in the *Origin*: first, that evolution had occurred and second, that natural selection was the mechanism of evolution. In arguing for the reality of evolution, Darwin challenged the Platonic view that species were fixed and had an unchanging essence or ideal form. To Darwin, similarities between species arose not from a shared essence but from common descent. Variation was the raw material for evolutionary change and not the result of an imperfect expression of an ideal essence. Darwin argued that the fossil record, although imperfect and incomplete, was evidence that species were in constant flux, appearing, changing, and going extinct. Observations regarding geographical variation, the distribution of plants and animals, and the peculiarities of island faunas constituted further evidence for the reality of evolution.

5. Convincing as the evidence for evolution was, Darwin felt that the acceptance of his ideas required a mechanism of evolutionary change. His theory of natural selection as the agent of organic change rests on three uncontested facts: (1) populations of organisms have the potential for exponential growth, (2) individuals in populations vary, and (3) some of this variation is heritable. Since resources cannot support an ever-growing population, some competition between individuals for resources must ensue. Success in this struggle for existence is determined in part by the inherited characteristics of individuals. Individuals that are most successful pass on their characteristics to the next generation to a greater extent than do the less successful. The consequence is a gradual change, generation by generation, in the composition of populations. Darwin used the example of artificial selection practiced for centuries by plant and animal breeders as an illustration of the creative power of selection.

References and Selected Readings

Darwin, C. *On the Origin of Species by Means of Natural Selection, or the Preservation of Favoured Races in the Struggle for Life.* London: John Murray, 1859. Darwin's own arguments, required reading.

Gingerich, O., ed. *Scientific Genius and Creativity.* New York: W.H. Freeman, 1952–1987. Essays from *Scientific American* on the nature of scientific discovery and short biographies of scientists who have built our scientific edifice. A biography of Charles Darwin, written by Loren Eiseley, is included.

Eiseley, L.C. *Darwin's Century: Evolution and the Men who Discovered it.* Garden City, N.Y.: Doubleday, 1958. A history of the Darwinian era, published to commemorate the 100th anniversary of *On the Origin of Species.*

Gould, S.J. *Wonderful Life: The Burgess Shale and the Nature of History.* New York: W.W. Norton, 1989. A fascinating account of the contingent nature of evolution, and how different the outcome would have been had different organisms survived the great extinctions of the past.

Irving, W. *Apes, Angels, and Victorians.* New York: McGraw-Hill, 1955. An outstanding analysis of the life and work of Charles Darwin; Thomas Henry Huxley, the leading advocate of Darwin's views; and the Victorian society in which they lived and worked.

Mayr, E. *The Growth of Biological Thought: Diversity, Evolution, and Inheritance.* Cambridge: Harvard University Press, 1982. A comprehensive history of evolutionary biology written by one of the major figures in the development of modern evolutionary thought.

II

Life and Its Continuity

3 *The Molecules of Life*

4 *The Origin of Life*

5 *The Eukaryotic Cell: Structure and Evolution*

6 *Genetics I: The Fundamental Observations of Heredity*

7 *Genetics II: The Molecular Basis of Inheritance*

8 *Evolution I: The Origin of Evolutionary Novelty*

9 *Evolution II: Populations, Species, and Macroevolution*

(Left) A lioness (Panthera leo) *with her cub, photographed at dawn in the Masai Mara of Kenya. (Above) A male lion, yawning.*

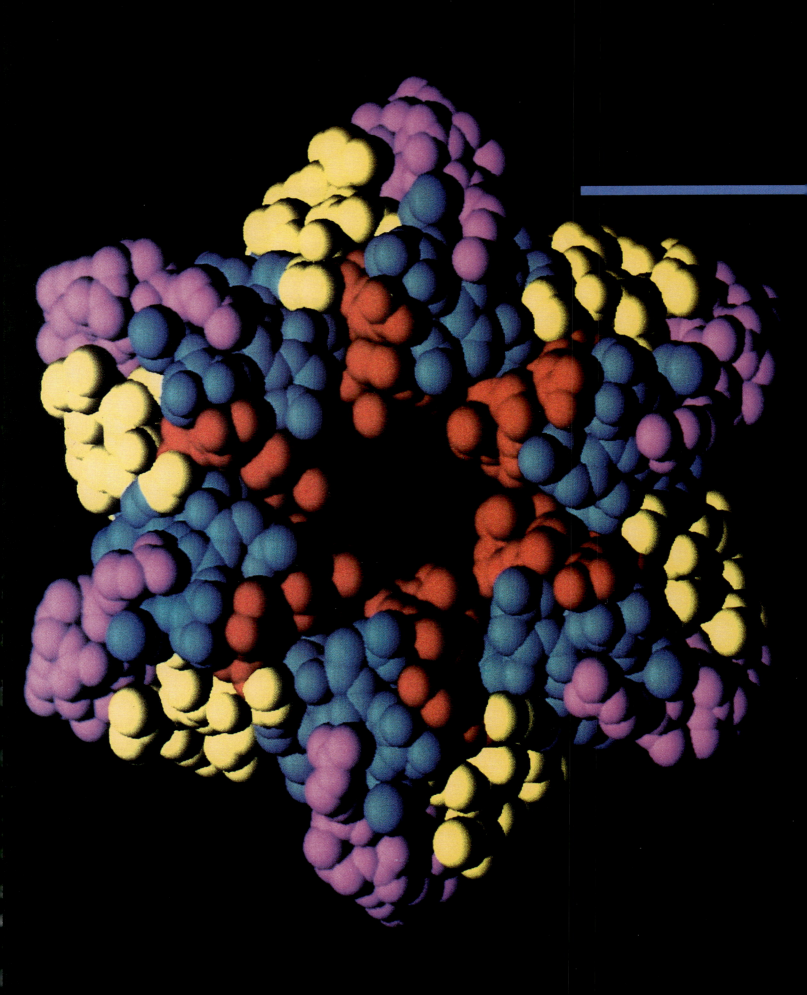

3

The Molecules of Life

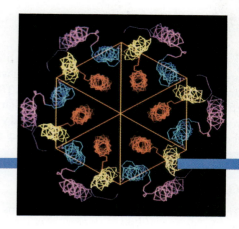

(Left) A computer-generated space-filling molecular model of the 16K protein, a protein that traverses the cell membrane, forming a pore through which ceratin ions and molecules may flow. (Above) A skeletal model of the same protein, showing the structural backbone of the different units.

This chapter is meant to introduce you to the fundamental organic molecules that make up the living world: carbohydrates, lipids, proteins, and nucleic acids. Throughout this chapter, we will assume that you are familiar with the basic vocabulary and concepts of chemistry and with the laws of thermodynamics. If you wish to review that material, it can be found in Appendix A.

ORGANIC COMPOUNDS

The complex chemistry of living organisms is mediated by a wide variety of organic compounds. Life on this planet is carbon-based: the majority of organic molecules are skeletons of carbon atoms to which hydrogen, oxygen, and other atoms are frequently attached. Curiously, the designation of these molecules as "organic" reflects the obsolete notion that these compounds could *only* be synthesized by living organisms. We now know that the chemistry of living systems obeys the same laws as the chemistry of the test tube. Nucleic acids, amino acid chains, and carbohydrates are now routinely synthesized in the laboratory.

Many of the molecules we will discuss are complex structures, frequently composed of repeated simpler units. These subunits are known as **monomers** (Gr. *monos*, single + *meros*, part). When many monomers are linked together, they give rise to long-chain molecules, collectively known as **polymers** (Gr. *polys*, many).

Organic compounds are often shown by structural formulas in which each atom is designated by its chemical symbol and the covalent bonds between them as lines. Some common carbon skeletons are shown in Figure 3.1. Notice in these formulas that each carbon atom, which has four electrons in its outer shell and needs four more to complete it, always makes four covalent bonds with other atoms. Similarly, oxygen, with six electrons in its outer shell, will make two bonds; nitrogen, three bonds; and hydrogen, one bond.

Organic molecules frequently display **functional groups,** clusters of atoms that give each molecule its specific identity and chemical reactivity. Some common functional groups are shown in Figure 3.2. The functional group in water is the hydroxyl group. Many organic acids contain a carboxyl group that releases hydrogen ions, as do all acids. Amino acids have a carboxyl group on one side and an amino group on the other. Alcohol, aldehyde, and keto groups occur in many sugars and related compounds. Methyl groups occur in fatty acids. Phosphate groups are found in nucleic acids and in many other energy-rich compounds.

Figure 3.1

The carbon skeletons of some simple organic compounds. The carbon skeletons (shown in blue) may be chains of different lengths or form double bonds, branched chains, rings, or a combination of these configurations.

(a) Carbon atoms can form chains of varying length.

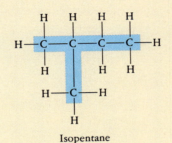

(b) Carbon atoms may form double bonds with one another.

(c) Carbon atoms can form branched chains.

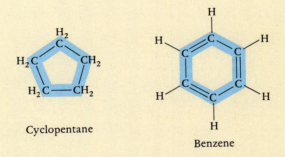

(d) Carbon atoms can be joined to form rings.

Cyclopentane

Benzene

Histidine (an amino acid)

(e) Rings and chains may be joined.

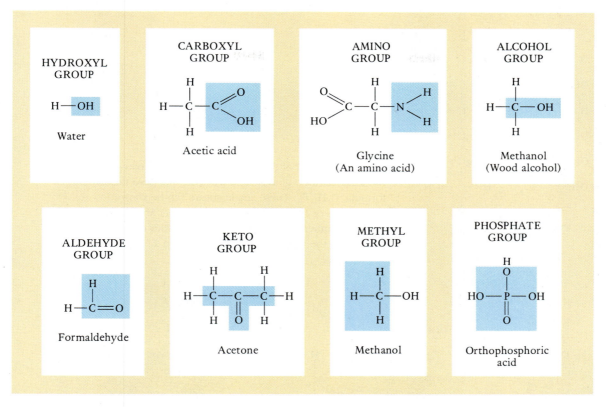

Figure 3.2
The identity and chemical reactivity of molecules depend on their functional groups, shown in blue.

CARBOHYDRATES

Carbohydrates include sugars, starches, cellulose, and related compounds in which carbon, hydrogen, and oxygen have a ratio of approximately 1:2:1. They are called carbohydrates because the hydrogen and oxygen have the same ratio as in water. Carbohydrates are the primary source of energy used in **metabolism**, the integrated set of biochemical reactions in living organisms. More complex carbohydrates also perform structural roles in animal and plant cells.

Among the simplest carbohydrate monomers are the **monosaccharides** (Gr. *monos*, single + *sakcharon*, sugar) such as glucose and fructose, which differ only in the arrangement of their atoms (Fig. 3.3). When in aqueous solution, the straight-carbon chains of these sugars assume a ring configuration. The carbons are numbered clockwise, with the 3 o'clock position in the ring being number 1. Notice that the rings for glucose and fructose are slightly different, with glucose having six facets and fructose five. Glucose also can assume two different forms, alpha or beta, depending on whether the hydroxyl group on carbon 1 is above or below the ring.

Monosaccharides are classed according to the number of carbons present. This number ranges from three to seven carbons. Glucose and fructose, each with six carbons, are known as **hexoses**. The sugar fractions of deoxyribonucleic acid (DNA) and ribonucleic acid (RNA) are five-carbon sugars, or **pentoses**.

Two monosaccharides may be covalently joined to form a **disaccharide** (Gr. *dis*, twice), or double sugar, in an energy-requiring reaction that removes a hydrogen from one sugar and a hydroxyl ion from the other. These liberated ions subsequently unite to form water: this type of reaction is called a **dehydration** or **condensation reaction**. The remainder of the sugar monomers, called **residues**, now share the oxygen atom between them. Two glucose molecules can combine to form malt sugar, or maltose (Fig. 3.4). Similarly, a glucose and fructose combine to form table sugar, or sucrose; a glucose and galactose, milk sugar, or lactose.

Monosaccharides can be assembled into polymers, called **polysaccharides**, some of which contain hundreds of residues. Excess circulating glucose in animals (derived from the digestion of foodstuffs) is stored as animal starch, or **glycogen**. Glycogen is composed of many alpha glucose residues joined by oxygen bridges connecting carbons 1 and 4 (also known as alpha 1-4 linkages). Branching occurs every two to four residues, this time through a 1-6 linkage. The storage molecule in plants is **starch**, also a chain of

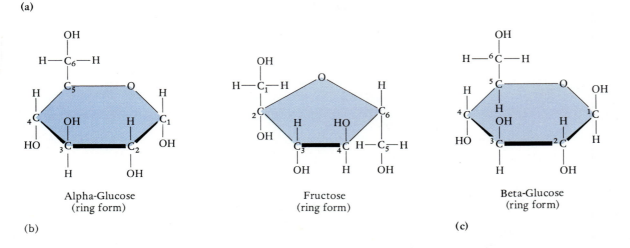

Figure 3.3

(a) A comparison of the structural formulas of two monosaccharides, glucose and fructose, that have the same formula, $C_6H_{12}O_6$. These single sugars can be shown in linear form or (b) in ring configuration. (c) An alternate form, or isomer, of glucose (β-glucose). Note the different position of the OH group bound to carbon 1.

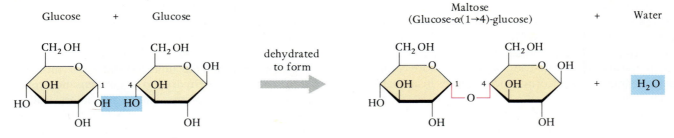

Figure 3.4

Two glucose molecules unite to form maltose or, in chemical terms, glucose alpha 1-4 glucose. (The 1-4 indicates the carbons that are bonded by an oxygen.) In this figure, carbon rings are shown in their simplified form.

alpha glucoses connected by 1-4 linkages, but branching less often (Fig. 3.5). Plant glucose may also occur in the beta glucose form. Long, unbranched chains of beta glucose, connected by oxygen bridges, form **cellulose**. This molecule plays a structural rather than a storage role. It is the most abundant organic molecule on the planet, for it is the primary component of plant cell walls. Half of the organic carbon on Earth is in the form of cellulose molecules. Adjacent cellulose molecules, linked by hydrogen bonds, form cellulose fibers (such as those in cotton wool and wood).

Chitin is another polysaccharide made up of modified glucose molecules in which a hydroxyl group has been replaced by an acetyl group. These **acetyl-**

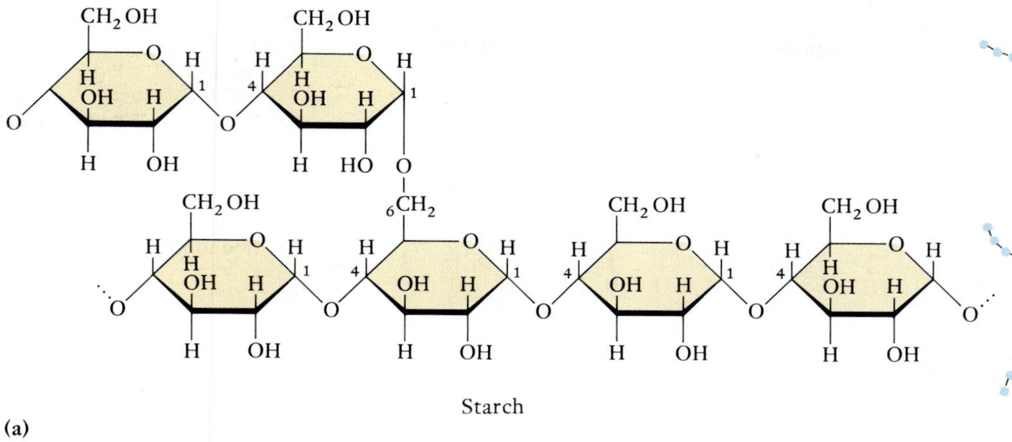

Starch

(a)

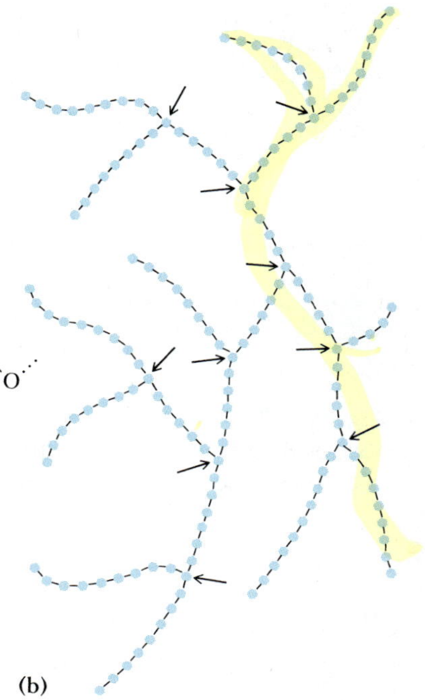

(b)

Figure 3.5

Glucose monomers can be linked to produce starch, the main form of carbohydrate storage in plants. (a) Glucose monomers form 1-4 linkages, but also branch by way of 1-6 linkages. (b) Overall view of a starch molecule. Arrows indicate branching 1-6 linkages.

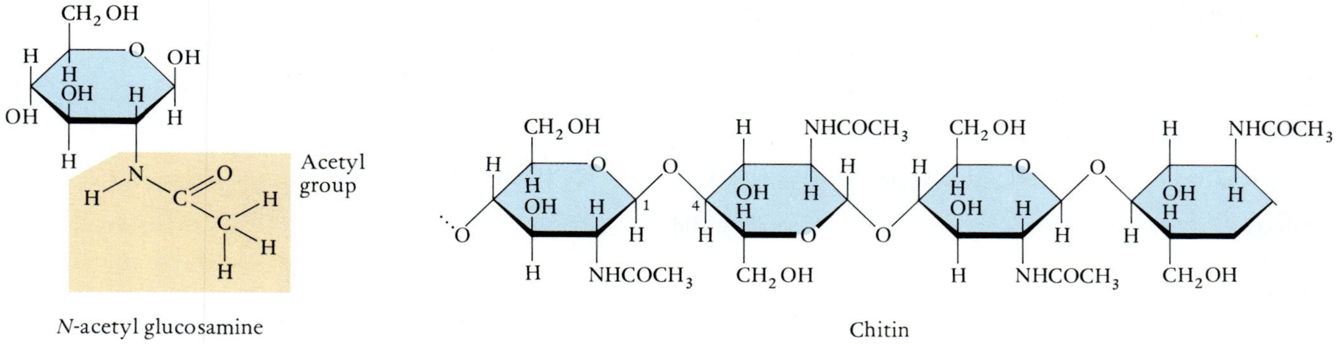

Acetyl group

N-acetyl glucosamine

(a)

Chitin

(b)

Figure 3.6

(a) Acetylglucosamine, a modified glucose molecule in which an acetyl group replaces a hydroxyl group. (b) Chitin, the main constituent of insect exoskeletons, is a polymer of acetylglucosamine units in 1-4 linkages.

glucosamines (Fig. 3.6) are linked together in beta 1-4 linkages. An important component of the exoskeleton of arthropods and some other invertebrates (Ch. 31), chitin is extremely flexible and resistant to degradation. Chitin is the second most abundant organic molecule on Earth.

LIPIDS

The **fatty acids** and the **lipids** are a heterogeneous group of fats and fatlike substances that are insoluble in water but do dissolve in benzene, ether, and other organic solvents. **Fatty acids** are long-chain hydrocarbons composed of an even number of carbon atoms to which hydrogen is

attached (Fig. 3.7). When each carbon within the chain bears two hydrogen atoms, the fatty acid is said to be **saturated**. When one pair of carbons is double bonded, the fatty acid carries fewer hydrogens than possible and is described as **monounsaturated**; when more than one pair are double bonded, it is **polyunsaturated**.

Fatty acids can also combine into more complex molecules. **Triacylglycerols** are formed by dehydration reactions that link three molecules of fatty acids with one of an alcohol known as glycerol (Fig. 3.8). The bonds between them are called ester bonds because they link an acid and an alcohol. Triacylglycerols are the most efficient form of energy storage: a gram of triacylglycerol releases

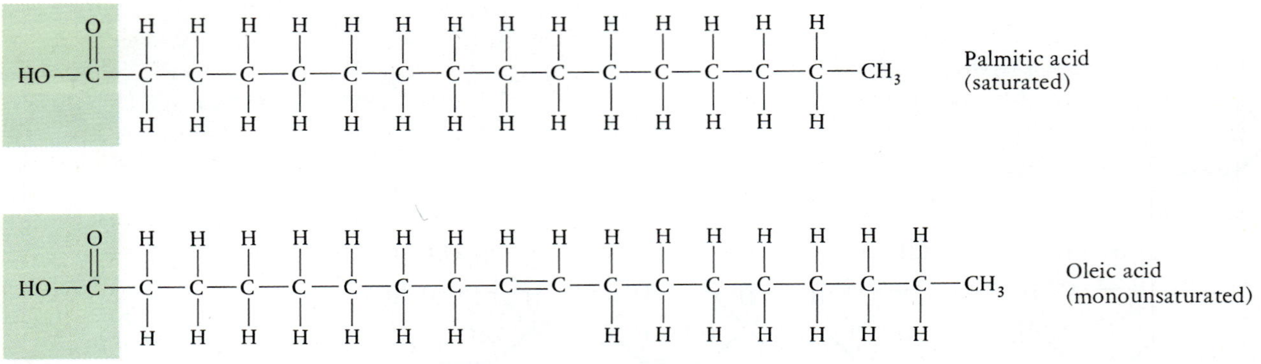

Figure 3.7

Fatty acids: palmitic acid is a saturated fatty acid; oleic acid, a monounsaturated fatty acid where one carbon pair is linked by a double bond.

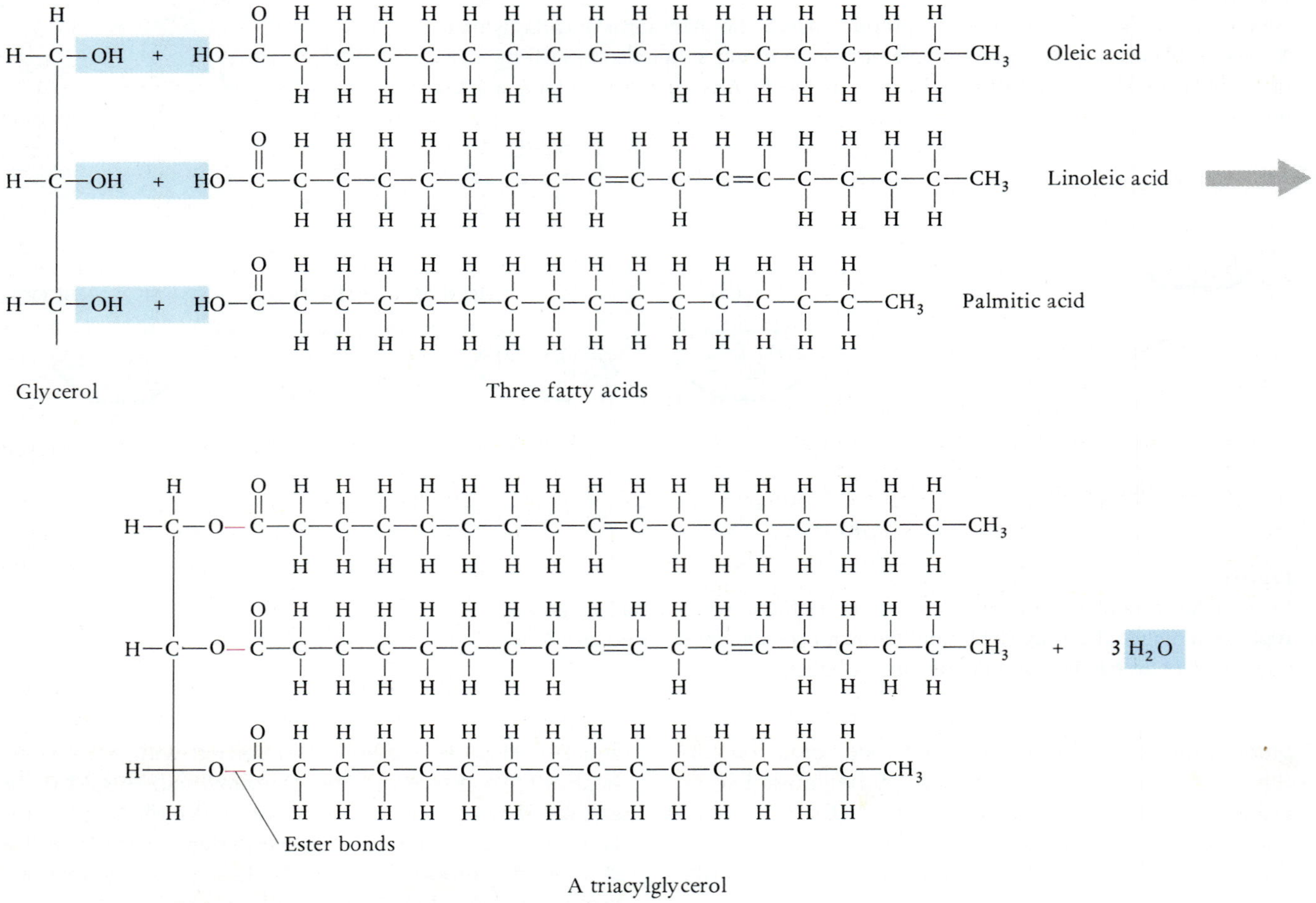

Figure 3.8

Three fatty acids are joined by dehydration reactions (in which molecules of water are removed) with a molecule of glycerol to form a triacylglycerol and three molecules of water. The bonds linking the fatty acids to the glycerol backbone are known as ester bonds.

2.25 times as much energy as a gram of carbohydrate. Excess calories eventually accumulate as triacylglycerols, primarily in the adipose or fat cells of the body.

A pair of fatty acid molecules attached to a glycerol molecule and modified by the addition of a phosphate group forms a **phospholipid**. If a sugar is added instead of the phosphate group, the resultant molecule is known as a **glycolipid**. These different lipids play several roles in animal metabolism: as energy stores, as hormones or other signaling molecules, and as the main component of cell membranes (see p. 67).

Waxes contain a single very-long-chain fatty acid attached to an alcohol other than glycerol. The long fatty acid has a high melting point, making waxes quite solid at high temperatures. Waxes occur on the surface of many organisms and act to reduce water loss.

Steroids differ from other lipids in being composed of four interlocking carbon rings, three of which contain six carbons, and one of which contains five carbons (Fig. 3.9). Cholesterol is the most abundant of the steroids. Although an excess of cholesterol can be harmful, it is an integral element of cell membranes and a major component of the myelin insulation around many nerve cells. Other steroids include hormones produced by parts of the adrenal gland, ovary, and testis (see Ch. 21).

PROTEINS

Proteins constitute over 50% of the dry weight of most cells. Their importance is reflected in their name (Greek *protos*, first). Like carbohydrates and lipids, they are composed of carbon, oxygen, and hydrogen. In addition, all contain nitrogen and often sulfur and phosphorus as well.

Proteins are the most chemically versatile of organic molecules in living systems. They perform a wide range of structural and functional tasks. Certain proteins, such as actin and myosin in muscle, keratin in skin, and collagen in connective tissues, contribute to the structure, stability, or

Cholesterol

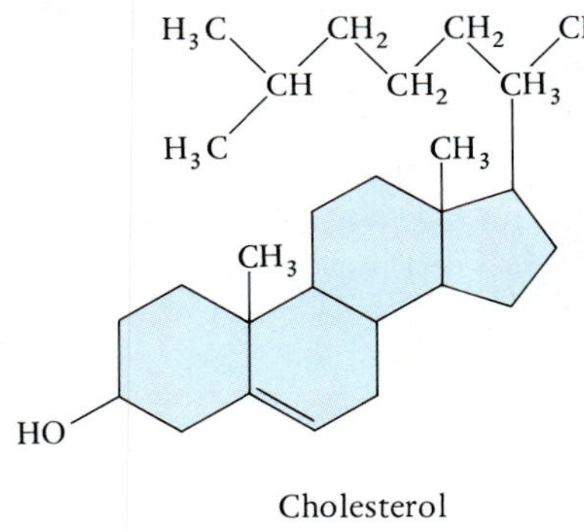

Figure 3.9
The structure of an important steroid, cholesterol. Other steroids exhibit different side chains but share the basic four-ring structure.

cohesion of the cell. Proteins are an important component of the cell membrane and regulate the passage of materials into and out of the cell. Other proteins perform important transport functions; hemoglobin, for example, carries oxygen to all animal tissues. Proteins also serve as long-distance messengers (hormones). Finally, proteins perform myriad tasks as **enzymes**, chemical catalysts that facilitate chemical reactions by bringing the reacting molecules together and lowering the amount of energy needed to initiate the reaction (activation energy). With the exception of a few nucleic acids (RNAs) recently discovered to have enzymatic activity, all enzymes are proteins.

Protein Structure

The building blocks of proteins are the **amino acids**, all of which consist of central or alpha carbon to which a hydrogen, an amino group (NH_2), and a carboxyl group (COOH) are attached (Fig. 3.10). Amino acids differ from each other in the functional side group attached at the remaining position on the alpha carbon. Twenty common amino acids are used in the synthesis of the proteins of all organisms. Two additional modified amino acids occur in a small subset of specialized proteins. Plants have the necessary machinery to synthesize all 20 amino acids, but most animals appear to have lost the ability to synthesize 9 of these amino acids (referred to as the **essential amino acids**) and must obtain them from food sources.

Amino acids are covalently linked into a polymer, known as a **polypeptide**, in an order carefully specified in the instructions carried by genes. The amino acids are forged together in a dehydration reaction that forms a **peptide bond** between the carboxyl group of one amino acid and the amino group of the next (Fig. 3.11).

Proteins vary in length from 124 to 1800 amino acids. Proteins also differ in the sequence of their amino acids. Thousands of different proteins are known, for like the letters of the alphabet, the limited array of amino acids can be assembled in a practically infinite number of combinations. Many proteins, such as the enzymes involved in cellular respiration (Chap. 13), occur in all cells, but some are restricted to cells with specialized functions. Hemoglobin, for example, is found in red blood cells and certain muscle cells, and not normally in others. Similar proteins in different species also vary. All vertebrates and some invertebrates have hemoglobin, but that of each species is unique in the number of amino acids it contains or in the sequences in some segments of the chain.

The amino acid sequence of a protein is referred to as its **primary structure**. But a protein is not simply a linear chain of amino acids: it takes on a three-dimensional structure, on which its function depends. The three-dimensional conformation of a protein is entirely dictated by its primary structure, but the rules of this folding are as yet not well understood. Hydrogen bonds forming between

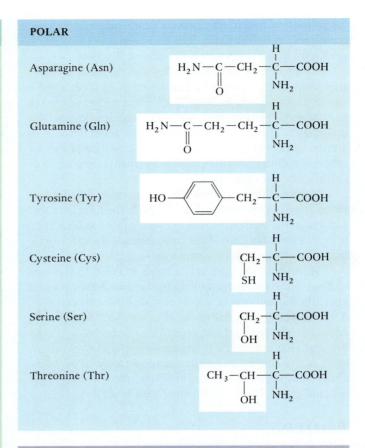

Figure 3.10
Structural formulas of 20 commonly occurring amino acids. They can be sorted into polar, nonpolar, and electrically charged groups, based on the features of their functional groups, shown in white boxes.

R group   Carboxyl   Amino   R group
             group   group

H   H   O   H   CH₃ O   Enzyme
 \ | ‖   \ | ‖
  N—C—C   +   N—C—C
 / | \   / | \
H H   (OH H)   H H   OH

Glycine   Alanine

                                          Peptide bond
                    H   H   O   CH₃ O
                     \ | ‖ | ‖
                      N—C—C—N—C—C   + H₂O
                     / | | \
                    H H   H   OH

                    Glycylalanine (a dipeptide)

H   H   O   CH₃ O   H   CH₂SH O   Enzyme
 \ | ‖ | ‖   \ | ‖
  N—C—C—N—C—C   +   N—C—C
 / | | \   / | \
H H   H   (OH H)   H H   OH

Glycylalanine   Cysteine

                    H   H   O   CH₃ O   CH₂SH O
                     \ | ‖ | ‖ | ‖
                      N—C—C—N—C—C—N—C—C   + H₂O
                     / | | | \
                    H H   H   H   OH

                    Glycylalanylcysteine (a tripeptide)

Figure 3.11

Amino acids are joined by dehydration reactions. Two joined amino acids form a dipeptide; the addition of a third forms a tripeptide. Polypeptide chains are formed by the joining of several amino acids. The sequence of amino acids in a polypeptide is known as the primary structure of a protein.

nearby residues (separated by three or four amino acids) give rise to a very common structure, the alpha helix (Fig. 3.12a). Hydrogen bonds between two different (antiparallel) chains give rise to another common structure, the beta-pleated sheet (Fig. 3.12b). Other parts of the polypeptide may act as hinges, connecting these sheets or helices. This fundamental level of local three-dimensional organization is known as the **secondary structure** of the protein.

More distant amino acids may further interact, forming hydrogen bonds (and occasionally covalent bonds) within the protein and further specifying the three-dimensional structure of the protein. Reactions between the side groups in the amino acid residues probably cause the already coiled or pleated polypeptide to assume its **tertiary structure**, which may be globular or fibrous (Fig. 3.13). As the protein takes on its tertiary configuration, residues containing hydrophobic groups orient towards the center of the molecule away from surrounding water, creating an unusual microenvironment at the center of the molecule. This is frequently the reactive core of many enzymes.

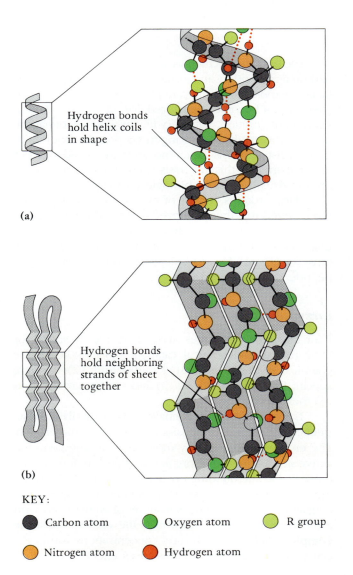

(a) Hydrogen bonds hold helix coils in shape

(b) Hydrogen bonds hold neighboring strands of sheet together

KEY:

● Carbon atom   ● Oxygen atom   ● R group

● Nitrogen atom   ● Hydrogen atom

Figure 3.12

The secondary structure of a protein. Bonds between parts of a polypeptide chain cause the chain to assume the structure of an alpha helix (a) or beta-pleated sheet (b), depending on the sequence of amino acids in the protein.

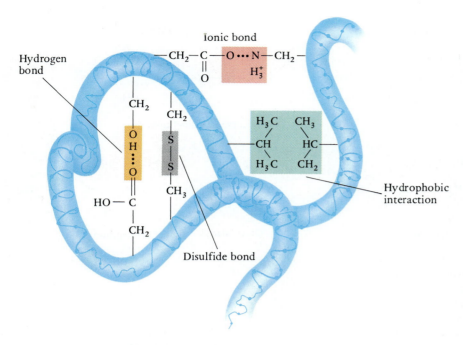

Ionic bond

Hydrogen bond

Hydrophobic interaction

Disulfide bond

Figure 3.13
The tertiary structure of a protein results from the folding of the secondary structure into a globular, fibrous, or other shape. The alpha helix (secondary structure) is formed by bonds between certain repeating amino acids. In this diagram, the alpha helix is further folded into a tertiary structure, shown here as a loop, as a result of ionic, disulfide, or hydrogen bonds and hydrophobic interactions.

The biological activity of any protein depends in large part on its secondary and tertiary structures. If a protein is heated, exposed to a different pH, or treated with certain chemicals, it becomes denatured, losing its three-dimensional configuration. With very few exceptions, denaturation is irreversible, suggesting that a protein acquires its characteristic form as the amino acids are joined together rather than after this synthesis is complete. A denatured protein usually loses its biological activity.

Many proteins consist of two or more polypeptide chains fitted together in a precise way. The interactions between the different subunits are described as the **quaternary structure** of a protein. The hemoglobin of most vertebrates, for example, is composed of four polypeptide chains (Fig. 3.14).

Enzymes

Four features of enzymes make them ideally suited for their role as biochemical catalysts: (1) they accelerate chemical reactions at least a millionfold, (2) they are specific for single reactions, (3) they can be recycled, and (4) they can be regulated.

The catalytic ability of enzymes is primarily a consequence of the three-dimensional structure of the protein. The enzyme creates a tiny molecular environment, the **active site**, where conditions are frequently different from those of the surrounding environment (Fig. 3.15). Here, the reactants, or **substrates**, are brought into close physical contact within the enzyme. Substrates often react and form a transient association, known as the **enzyme-substrate complex**, with very high local concentrations within the active site. Under these circumstances, chemical reactions

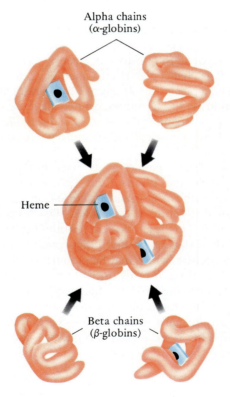

Alpha chains (α-globins)

Heme

Beta chains (β-globins)

Figure 3.14
When a protein consists of two or more polypeptide strands, as hemoglobin does, it is said to have a quaternary structure. Each of the four polypeptides in hemoglobin is joined to an iron-containing molecule, the heme, but only two of the hemes can be seen in this view of the molecule.

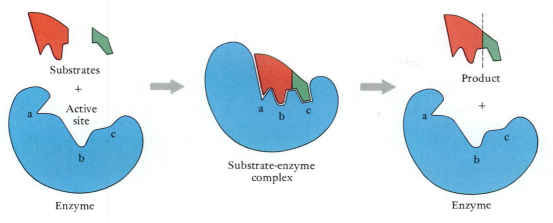

Figure 3.15

An enzyme-catalyzed reaction. An enzyme may change shape slightly as it binds with its substrates. The exact geometric fit between enzyme and substrates is responsible for the specificity and speed of enzyme action. As the reaction proceeds, the substrates combine into a single product, and the enzyme is released and recycled. Some reactions proceed in the opposite direction, splitting a substrate into two smaller product molecules.

are far more likely to occur, as well as to occur faster. The consequence of the interaction between substrates and the enzyme is a decrease in the **activation energy**, the energy "hump" that must be overcome for any chemical reaction to proceed (Fig. 3.16). As the reaction proceeds, the enzyme and product of the reaction are released, and the enzyme is recycled and can be utilized repeatedly to catalyze the same reaction.

The specificity of enzyme action also depends on the geometry of the active site, which will only allow the entry of the correct substrates. Further control is exerted when substrates are bound to the active site, a step that depends on the close contact between substrate molecules and the key amino acid residues in the active site.

Enzymes are regulated by a variety of molecules that interact with the enzyme and alter its three-dimensional structure. Certain enzymes require the presence of a small organic compound known as a **coenzyme** to be fully active. The importance of many vitamins (Ch. 12) stems from their role as coenzymes. Some other enzymes need a **cofactor** before the reaction can proceed. Metallic ions including magnesium (Mg^{2+}), copper (Cu^{2+}), zinc (Zn^{2+}), and iron (Fe^{2+}) often act as cofactors.

Enzymes frequently work in teams in a series of chemical reactions that constitute a **metabolic pathway**. For example, the breakdown of glucose in the cell to carbon dioxide and water with the release of energy requires nearly two dozen separate enzyme-mediated steps (see Ch. 13). Certain of the products of one reaction serve as the substrates for the next reaction; this pattern continues down the entire pathway.

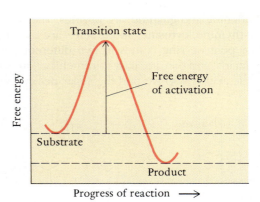

(a)

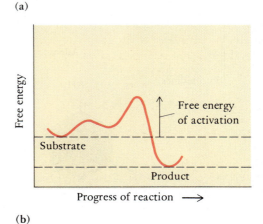

(b)

Figure 3.16

(a) Any chemical reaction that transforms substrate into product must proceed through a transition state. The energy required to reach this state is known as the free energy of activation. (b) A catalyst, such as an enzyme, acts principally by lowering the free energy of activation of the reaction, thus facilitating the reaction. Note that the energy states of the substrate and product are not altered by the catalyst.

NUCLEIC ACIDS

In all organisms on Earth, genetic information is transmitted from generation to generation in one of two forms: **deoxyribonucleic acid (DNA)** or **ribonucleic acid (RNA)**. These long chain-like molecules are present in practically every cell of every organism. The discovery, properties, and features of nucleic acids are discussed in Chapter 7. Here, we will primarily be concerned with the chemical identity of DNA and RNA.

The building blocks of nucleic acids are the **nucleotides**. Each nucleotide is composed of three distinguishable parts: a nitrogen-containing ring structure (the **nitrogenous base**), a five-carbon ring sugar (the **pentose ring**), and a **phosphate group** attached to the pentose.

Two basic types of bases are readily distinguishable: a "double-ring" base (adjoining six-carbon and five-carbon rings), known as a **purine**; and a smaller six-carbon single-ring base, known as a **pyrimidine** (Fig. 3.17). Two of the bases, adenine and guanine, are purines. The remaining bases—**thymine**, **cytosine**, and **uracil**—are pyrimidines.

The pentose ring comes in two different forms in nucleic acids, and they are easy to remember: in deoxyribonucleic acid (DNA) we find the sugar **deoxyribose**; in ribonucleic acid (RNA) we find the sugar **ribose** (Fig. 3.17). The difference between the two sugars is slight: a single hydroxyl (OH) group attached to carbon 2 in ribose. The base and sugar are joined by a covalent bond between carbon 1 in the sugar and nitrogen 1 in pyrimidines or nitrogen 9 in purines (Fig. 3.18). The generic name for this combination of a nitrogenous base and a sugar is a **nucleoside**.

One final portion completes the picture: the phosphate group. It is usually attached to the carbon 5 position of the sugar, generally called the "five prime" (5') position. This phosphate entity may include one, two, or three phosphates. The bonds connecting the first phosphate (the alpha phosphate) to the second phosphate (the beta phosphate), and the second to the third (the gamma phosphate) are extremely energy rich.

How are these nucleotides assembled into DNA or RNA? The precursors of DNA and RNA are the triphosphate forms of the nucleosides. During DNA synthesis, these triphosphates are added one by one to the growing nucleic acid chain by the formation of a covalent bond that connects the 5' position of the nucleotide (the carbon 5 position of ribose) with the 3' position (the carbon 3 position of ribose) of the last nucleotide in the chain, with

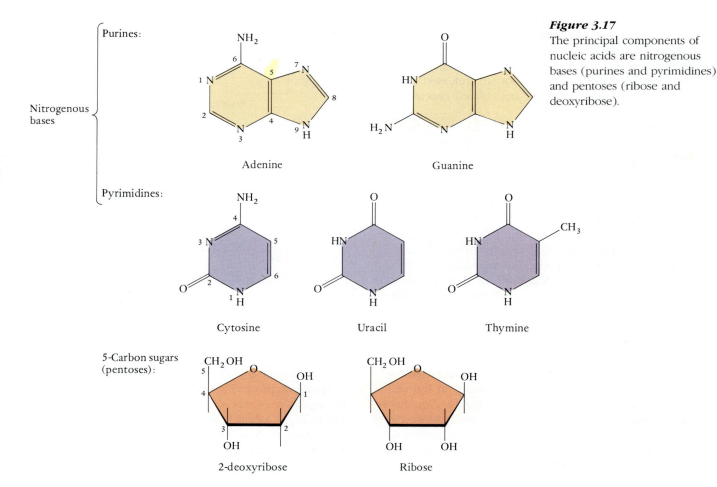

Figure 3.17

The principal components of nucleic acids are nitrogenous bases (purines and pyrimidines) and pentoses (ribose and deoxyribose).

Purines:

Adenine

Guanine

Pyrimidines:

Cytosine

Uracil

Thymine

5-Carbon sugars (pentoses):

2-deoxyribose

Ribose

Nitrogenous bases

the resultant release of the alpha and beta phosphates from the original triphosphate. The resulting DNA structure therefore displays a ribose-phosphate-ribose "backbone," with the purine or pyrimidine fractions jutting out like ribs (Fig. 3.18). It is this three-dimensional structure that makes the pairing of nucleic acid strands possible.

The structure of DNA is now clearly understood, characterized, and recently photographed. The two nucleic acid strands are oriented in an antiparallel fashion—pointing in opposite directions, if you will (Fig. 3.19). The bases point inward, perpendicular to the phosphate backbone. Think of a ladder: the phosphate-ribose backbones correspond to the long sidepieces; the bases, to the rungs. The bases on either strand pair by the formation of two or three hydrogen bonds between them. As in a ladder, the width between the two strands is constant (2 nm), reflecting the invariant pairing rule: purine with pyrimidine. A purine-purine pairing would create a bulge in the width; conversely, a pyrimidine-pyrimidine pair would constrict the width. In fact, adenine will always pair with thymine; guanine will always pair with cytosine. To continue our metaphor, picture the ladder as made of clay. Hold the top and bottom of the ladder and gently twist it clockwise, such that it makes a full turn every 10 "rungs" (nucleotides) or 3.4 nm. This begins to capture the three-dimensional structure of the DNA helix: the coiled ladder of inherited information (Fig. 3.20).

The vast complexity of genetic information is specified using only four basic nucleotide building blocks. Adenylic acid, cytidylic acid, guanylic acid, and thymidylic acid are the sole components of DNA; uridylic acid replaces thymidylic acid in RNA. It is the specific linear combination of these nucleotides, assembled into the DNA (or RNA) chain, that conveys biological information. While it may seem at first that an alphabet of only four letters severely limits the complexity of information that DNA can

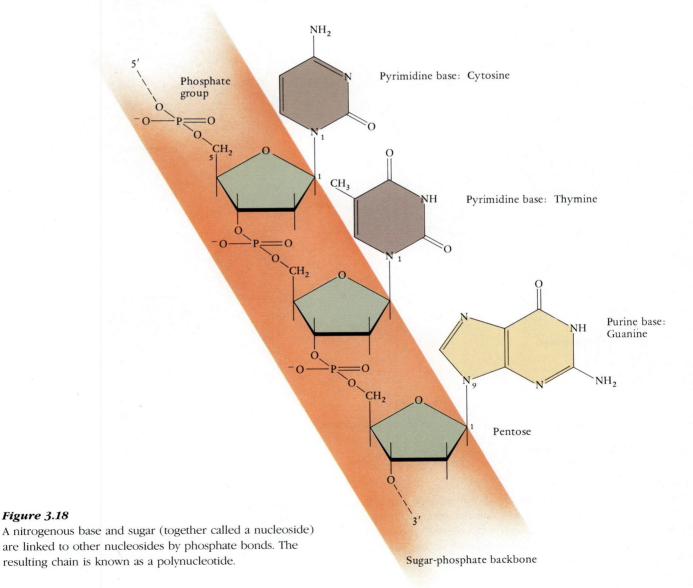

Figure 3.18
A nitrogenous base and sugar (together called a nucleoside) are linked to other nucleosides by phosphate bonds. The resulting chain is known as a polynucleotide.

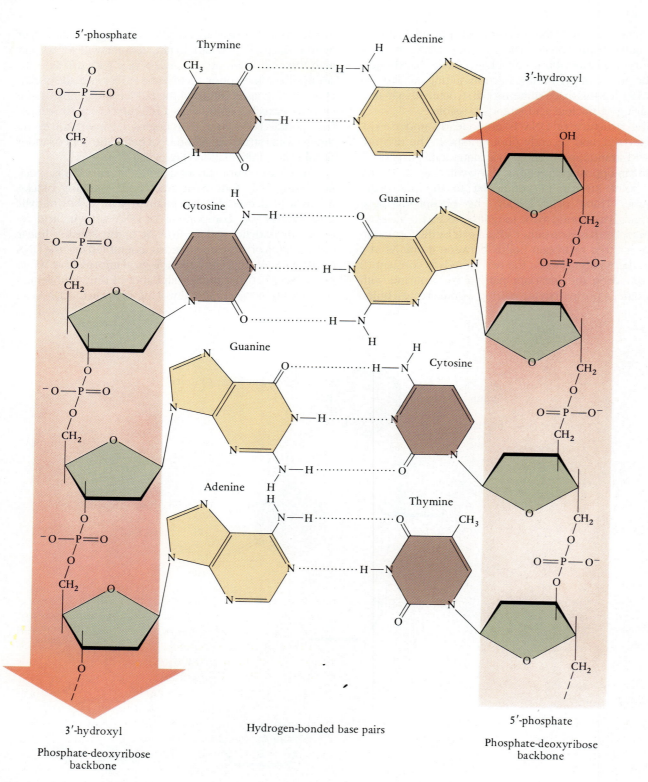

Figure 3.19

DNA consists of two complementary polynucleotide chains linked by hydrogen bonds. Note that the pyrimidine in one chain pairs with a purine in the other strand, thus maintaining a constant width between the two chains. The faithful transmission of genetic information depends on this pairing between complementary bases.

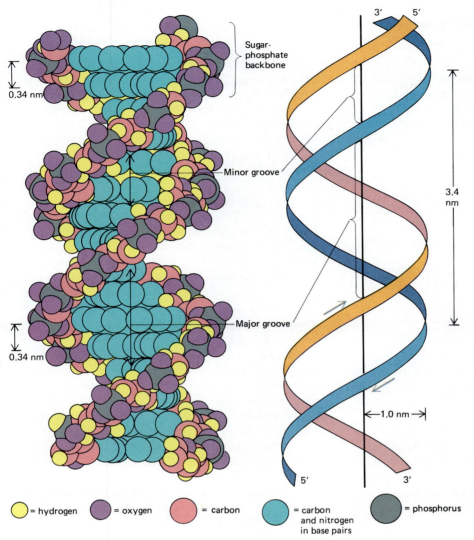

Sugar-
phosphate
backbone

0.34 nm

Minor groove

Major groove

0.34 nm

3′ 5′

3.4
nm

1.0 nm

5′ 3′

○ = hydrogen   ○ = oxygen   ○ = carbon   ○ = carbon
and nitrogen
in base pairs   ○ = phosphorus

Figure 3.20
The two strands of DNA are twisted
into a double helix (right). A space-
filling model of the double helix is
shown on the left.

convey, keep in mind that a linear chain 50 nucleotides long can be assembled in 4^{50} different ways. The limited size of the molecular alphabet in no way constrains the information content of DNA.

Nitrogenous Bases as Energy and Signaling Molecules

Finally, the nucleosides adenosine and, on occasion, guanosine may play more than one crucial role in living systems. As we have seen above, these nucleosides are components of DNA. But the triphosphate forms of adenosine and guanosine, **adenosine triphosphate (ATP) and guanosine triphosphate (GTP)**, are also the energy molecules of the cell. Precisely because the bonds connecting phosphates are so energy rich, they can be synthesized to store the energy of the exergonic (energy-producing)

reactions of metabolism. Conversely, they can be hydrolyzed to release energy to power endergonic (energy-requiring) metabolic reactions. Cells take in sugars and other high-energy compounds, but the energy is not available to power the metabolic activity of the cell until it is changed to ATP or GTP (Chap. 13).

And there are yet more functions: the monophosphate form of adenosine, **adenosine monophosphate (AMP)**, undergoes a slight chemical modification to become a major "signal molecule" in the cell, mediating the action of several important hormones (Ch. 19). Finally, nucleosides can also form part of larger molecules, coenzymes, that facilitate the catalytic action of enzymes. These multiple roles reflect a common evolutionary phenomenon, in which a molecule that may have been present early in cellular evolution has been drafted into a variety of different roles.

Carbohydrates, lipids, nucleic acids, and proteins are the essential molecules in living systems. While all perform crucial functions, it is proteins that are most versatile, and carry out the widest range of functions.

Modern molecular techniques make it possible to isolate, characterize, and determine the amino acid sequence of a particular protein. These methods usually involve the following basic steps:

1) The protein of interest must be separated from the many other proteins present in the cells or tissues of the organism.

2) The protein must be specifically identified or visualized.

3) If required, the overall amino acid composition, and the precise amino acid sequence of the protein, can be determined.

1. **Separation.** When studying the proteins of living systems, biologists generally begin with intact cells, tissues, or even whole organisms. In order to release the proteins from the cells, the tissues are first homogenized, using a mortar and pestle, ultrasonic vibrations, or a blender, or by chemical means, using detergents or enzymes that rupture cell membranes. The resultant slurry is then centrifuged at high speeds, a procedure that causes larger (and denser) particles to settle at the bottom of the tube. By controlling the speed and duration of the centrifugation (differential centrifugation), specific cell components can be recovered. (Thus, for example, a slow spin will cause the larger nuclei to aggregate at the bottom of the tube, but will not pellet the smaller mitochondria.)

Once the fraction containing the protein of interest is recovered, further isolation steps take advantage of three basic properties unique to every protein: size, solubility, and charge.

Electrophoretic techniques exploit the differences in size and amino acid composition among different proteins. A sample containing a mixture of proteins is passed through a gel matrix (usually a slab of agarose or acrylamide) to which a current is applied. Small proteins will make their way through the pores of the gel more quickly than will large proteins; similarly, proteins composed of many charged amino acids will move more rapidly in response to the electric field. Proteins thus end up arrayed throughout the length of the gel. **Chromatographic techniques** separate proteins on the basis of their solubility (which in turn depends on the proportion of polar to nonpolar amino acids) and their size. The sample in this case is applied to specially treated paper, gel, or other matrix. A particular organic solvent is then placed in contact with the matrix. As the solvent migrates upward, nonpolar proteins (soluble in the solvent, but not in water) migrate with it. Again, the net result is the separation of different proteins.

Summary

1. Carbohydrates include sugars, starches, cellulose, and related compounds in which carbon, hydrogen, and oxygen have a ratio of approximately 1:2:1. They are composed of single sugars, often containing five or six carbon atoms, that are linked together by dehydration reactions to form long polymers. Many carbohydrates are energy sources, but a few, such as cellulose and chitin, are structural components of cells.

2. Fatty acids and lipids are a heterogeneous group of fats and fatlike compounds that are insoluble in water but do dissolve in organic solvents. Three fatty acids can be linked in dehydration reactions with glycerol to form a fat, triacylglycerol. Fats are long-term energy stores. Other lipids include phospholipids, glycolipids, waxes, and steroids.

3. Proteins constitute over 50% of the dry weight of cells. Their building blocks are amino acids, which are characterized by an amino group (NH_2) at one end of the molecule and a carboxyl group (COOH) at the other end. Twenty common amino acids are used by all organisms in the synthesis of their proteins. Amino acids are linked in dehydration reactions to form long polypeptide chains. The sequence of the amino acids in the chain is the primary structure of the protein. This determines the pattern of twisting or folding (alpha helix or beta-pleated sheet) that forms the secondary structure of a protein. Additional folding into globular or fibrous configurations results in tertiary structure. The biological activity of proteins depends in large part on their secondary and tertiary structures.

4. Some proteins are important structural components of cells, some control the passage of materials through membranes, some transport materials through the body, some are long-distance messengers, and a great many proteins act as enzymes. Enzymes are catalytic mediators of chemical reactions in cells, bringing molecules together in sufficient concentration for reactions to proceed. They lower the activation energy needed to start a

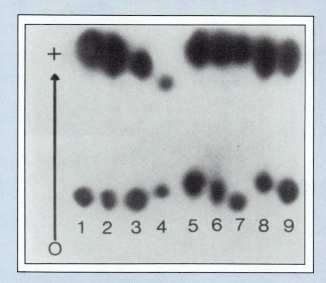

The patterns produced by two different proteins (top and bottom rows) separated by electrophoresis and specifically stained. Nine different *Drosophilia* lines are shown; the variability within each row illustrates the genetic differences among the different lines.

changes color when catalysis occurs is attached to the substrate. When the enzyme catalyzes the reaction, a color band appears at the place in the gel where the protein is found. A second method of identification uses antibodies designed specifically to bind to a portion of the protein. Once bound, these antibodies can then be easily visualized. The protein again appears as a specific band (or spot) on the gel or chromatograph.

In many cases, no further characterization of the proteins may be necessary. We may, for example, be interested in the **genetic polymorphism**, or **allelic frequencies** within a population—the number and relative frequency of different versions of the same protein present in a given sample. For such questions, electrophoretic techniques provide an initial answer. Similarly, **diagnostic alleles**, versions of a protein that occur only in a single population or species, can be sought by electrophoretic methods.

2. **Identification.** Once proteins have been separated, the individual protein must be identified within the gel or chromatograph. One commonly used method for visualizing enzymes involves supplying the enzyme with its specific substrate. A compound (chromatophore) that

3. **Sequence Determination.** For certain biological problems, the specific amino acid sequence of a protein must be determined. Direct methods consist of cleaving the protein with enzymes or reagents that specifically cleave at certain peptide bonds. By examining the resulting fragments, and by treating the fragments further in order to remove the terminal amino acid (which can then be identified by chromatography), the full amino acid sequence of the protein can be obtained. More recently, the sequence of many proteins has been decoded by isolating the gene for the particular protein, determining its DNA sequence, and then using the genetic code to translate DNA sequence into amino acid sequence.

reaction and accelerate the reaction at least a millionfold, they are specific for single reactions, they can be regulated, and they can be recycled.

5. Genetic information is transmitted from generation to generation in all organisms by either deoxyribonucleic acid (DNA) or ribonucleic acid (RNA). These nucleic acids are composed of four nitrogenous bases and pentose sugars (either deoxyribose or ribose) that are linked together by phosphate groups to form long polynucleotide chains. The genetic information is conveyed by the sequence of the four bases, two of which are purines (adenine and guanine) and two, pyrimidines (thymine and

cytosine). Uracil replaces thymine in RNA. DNA consists of two paired strands in which adenine always pairs with thymine and guanine with cytosine. The two strands are twisted into a double helix.

6. The synthesis of the nucleosides adenosine triphosphate (ATP) and guanosine triphosphate (GTP) requires a great deal of energy. This is stored in certain of the phosphate bonds and can be released as needed to power the metabolic machinery of the cell. The monophosphate form of adenosine (AMP) is an important signal molecule, and some nucleotides are constituents of coenzymes.

References and Selected Readings

Atkins, P. W. *Molecules.* New York: Scientific American Library, 1987. A beautifully illustrated and very readable account of the molecules we encounter in our daily lives. Many of the organic molecules considered in this chapter are diagrammed and discussed.

The molecules of life. *Scientific American* 253 (Oct. 1985). A special issue devoted to the molecules of life. Includes articles on DNA, RNA, proteins, molecules of the cell membrane, and various aspects of cell biology and evolution.

4

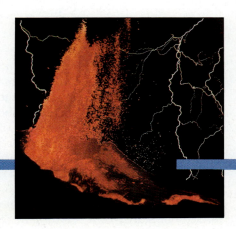

The Origin of Life

*(Left) The energy required for the synthesis of
prebiotic compounds may have been
supplied by the constant lightning storms
that lighted the skies 4 billion years ago.
(Above) The early Earth was an inhospitable
planet with no atmosphere, no oxygen, and
constant volcanic activity.*

In Chapter 3, we examined the basic molecules that make life possible. The majority of them are complicated organic molecules that are synthesized and degraded in multistep metabolic pathways. But these metabolic pathways are themselves the products of hundreds of millions of years of evolution, and cannot account for the origin of the organic molecules we have been discussing. We must take one step back, and ask how the earliest organic molecules might have arisen in a **prebiotic** Earth, an Earth without organisms, without metabolism, without life. We must retrace our story to the very origins of our planet.

The Earth, as best can be ascertained, is approximately 4.6 billion years old. Created by the coalescence of gases and debris from exploding supernovas, the Earth began as a molten core of iron and nickel surrounded by silicate rock. It was a barren planet, enveloped by an atmosphere of toxic, reducing gases—ammonia, methane, and hydrogen (Fig. 4.1). There was nothing hospitable about this young planet: little freestanding water, no oxygen, and a steady bombardment of cosmic and ultraviolet radiation.

How could life possibly evolve in this context? What are the events that permit (or precipitate) the appearance of living organisms? How and when does this tiny rocky outpost at the edge of the Milky Way come to be populated by organisms? These questions seem at first to be the province of science fiction, but when properly phrased they are central issues in the biological sciences. In this chapter, we will focus on the problem of the origin and early history of life on Earth by asking three separate but interrelated questions:

1. What evidence still remains, either in living organisms or in the fossil record, that allows us to reconstruct the early history of life?

2. What are the necessary chemical and biochemical events that must have preceded the earliest forms of life?

3. Where is the boundary between "life" and "nonlife"?

THE EVIDENCE: THE FOSSIL RECORD OF EARLY LIFE

Though the telescopes and radiotransmitters of the SETI (Search for Extra-Terrestrial Intelligence) project scan a wide range of frequencies in the sky, 24 hours a day, and have done so for more than a decade, there is as yet no evidence of life—at least in a form we are likely to recognize—anywhere else but on Earth (Fig. 4.2). There are tantalizing clues that suggest that the conditions that ex-

Figure 4.1

An artist's rendition of the early Earth. The first steps in the evolution of life took place in an environment far different from today's Earth. There was no oxygen, intense solar and ultraviolet radiation, violent thunderstorms, and constant volcanic activity.

Figure 4.2
The radiotelescope antennae of the SETI (Search for Extra-Terrestrial Intelligence) project. These antennae scan and monitor deep space for any structured or repetitive pattern of radio waves that might indicate the presence of life elsewhere in the universe.

back, we will reach the trunk, and eventually the root. There is certainly a beginning to the history of life; our task is to reconstruct it and to determine the forces that have propelled it to its present state.

The oldest rocks ever found on Earth come from the Isua Formation, on the western edge of Greenland, and are approximately 3.7 billion years old. The Earth was already 900 million years old when the sediments that form the Isua formation were deposited. Over the next 3.7 billion years, the explosive heat and constant pressure of the Earth's geological activity transformed, compressed, and baked these sediments into a different kind of rock: metamorphic rock. No fossil can possibly survive the process that creates metamorphic rock; we will never know if any evidence of life existed in these earliest sediments when they were first deposited.

The oldest sedimentary rocks that have not undergone extensive metamorphosis, and might therefore conceivably contain fossils, are found in southern Africa and in western Australia. Startlingly, these sedimentary deposits, the Fig Tree group in Africa and the Warrawoona group in Australia, both contain evidence of early life. Embedded in the 3.4- to 3.5-billion-year-old sedimentary rocks of these formations, tiny spheres (2.5 to 8 µm in diameter) and sheathed rods represent the earliest record of life on Earth (Fig. 4.3). Three billion four hundred million years ago,

isted on the early Earth may exist in other planets or moons: standing water, volcanic activity, winds, and an atmosphere. But we have so far discovered only one rocky body in the universe that is clearly alive: the Earth. Though we do not often stop to reflect upon it, the mere existence of life anywhere in the universe, including here on Earth, seems practically unbelievable. Yet this planet is most certainly alive, and it is up to biologists and paleontologists to reconstruct its earliest history.

The 18th century geologist James Hutton sought to encompass the vast and complex history of the Earth when he declared it to have "no vestige of a beginning, no prospect of an end." Hutton and his contemporaries had begun to grasp the real age of the Earth, and had come to understand the repeated cycles of sedimentation, upheaval, and erosion that shape the geological record. Hutton had further extended his vision of an endlessly cyclical geological history to encompass biological history as well—species arising, thriving, going extinct in a repeating spiral with no retrievable beginning.

Powerful and compelling as Hutton's geological metaphor is, it is not a correct account of biological history. The history of life is not so much a spiral as a complex, ever-branching tree. If we carefully follow the branches

Figure 4.3
Photomicrograph of fossils of the earliest known life on Earth, spherical cells trapped in the cherts of the Fig Tree group of southern Africa. Some of the fossils appear to show binary fission. Arrows point to individual cells, which to some extent resemble modern cyanobacteria. (Magnification 1600 ×.)

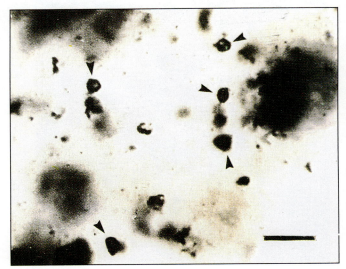

these earliest cells were trapped in a gelatinous, silica-rich matrix that eventually became chert, a finely crystalline form of quartz. Readily identifiable biological processes have been captured in these sediments: cell division, for example, appears to be taking place. The much younger 2 billion-year-old sediments of the Canadian Gunflint Formation already contain bacterial mats (stromatolites) remarkably similar to modern blue-green algal mats (Fig. 4.4).

The fact that we find fossil evidence of life in the oldest possible rocks that might contain it may be the single most remarkable observation about early life. Bear in mind that the Fig Tree cells are already complex cells, capable of dividing; the earliest protocell, which gave rise to the cells preserved in the Fig Tree formation, must be older still. The evidence is clear: life begins on this planet very soon after the Earth is formed and has had time to cool. What are we to make of this observation? Does it suggest that given the conditions of the early Earth, it was inevitable that life would evolve? Or are the chances of life arising vanishingly small, and this planet simply exceedingly fortunate? We cannot answer this question based on a single sample. Its resolution will depend on finding other planets or moons with conditions similar to those of the early Earth, and on probing them for the existence of life. If such

Figure 4.4

Stromatolites form Shark's Bay, in Western Australia. These stromatolites are in fact 2000-year-old, blue-green algal mats, in which silt and sediment have been trapped. Fossil stromatolites, approximately 2,000,000,000 years old, have been found in the Gunflint Formation in Ontario, Canada. These early blue-green algae may have been the first producers of atmospheric oxygen on the planet, and they are virtually identical in appearance to modern stromatolites.

planets exist, we will learn much about ourselves. But likely or improbable, life has arisen on Earth. We must now focus our attention on the events and circumstances that must have preceeded (and permitted) life to arise. How were the molecules necessary to build a living cell formed? Where did the energy required for life come from?

MAKING LIFE POSSIBLE: THE PREBIOTIC WORLD

A Single Origin of Life

Life on Earth is **monophyletic**—(Gr. *monos* = single, + *phyle* = tribe)—the vast and complex tree of life arises from a single root. A dazzling diversity of organisms, both living and extinct, have populated the Earth, all of them ultimately descended from a single ancestral stock.

Perhaps the strongest evidence of this common ancestry can be found at the molecular level, where a large part of the metabolic machinery of life is shared by all organisms. The pathway for the synthesis of ATP, for example, is the same for all aerobic organisms. Similarly, all organisms transmit heritable information from generation to generation using nucleic acids and catalyze chemical reactions using enzymes. These biochemical similarities are detailed, complex, and compelling. They are neither coincidental nor the result of the separate origin of identical metabolic pathways in plants, animals, and bacteria. Instead, these fundamental similarities are the clearest evidence of the monophyletic nature of life. Furthermore, these shared traits were most likely already present in the common ancestor of all organisms; it, too, must have used nucleic acids to encode information.

The Origin of Complex Molecules

In order to fully understand the early history of life, we need to ask an even more fundamental question: how did the organic molecules that make life possible arise on the early Earth?

Two general theories have been put forth to account for the origin of the complex organic molecules that underlie all living processes. The first of these theories suggests that complex organic molecules were generated throughout the solar system, during the early stages of its formation, as a result of the violent physical and chemical processes accompanying the origin of planets. In this model, stellar bodies other than the Earth are thought to contain these precursor molecules, but only on Earth do we know that these molecules coalesced and organized to give rise to life.

Perhaps the most interesting evidence compatible with this theory comes from the analysis of a class of meteorites called carbonaceous chondrites (or chondritic meteorites). These meteorites, thought to have formed early in the history of the solar system, 4.6 billion years ago,

Figure 4.5
The Leonid meteorite shower. The Earth is bombarded by a constant shower of particles, ranging in size from tiny dust specks to larger rocky meteorites. Certain important organic molecules have been found in the carbon-containing chondritic meteorites, suggesting a very early abiotic origin for these chemical compounds.

constitute 1 to 2% of the meteorites found on Earth. Composed of small spheres (chondrules) of the simplest silica minerals and containing a significant amount of carbon, these meteorites differ from the more usual iron-core meteorites, which contain practically no carbon. The chondritic meteorites have been carefully analyzed for the presence of biologically important molecules. Much to our surprise, they contain small but detectable amounts of amino acids, sugars, guanine (a component of DNA), long-chain fatty acids, and several other potentially significant organic molecules. These results suggest that several of the precursor molecules necessary for the evolution of life may have formed early in the history of our solar system as a result of nonbiological processes. Some scientists have suggested that the early Earth may have been "seeded" by the bombardment of these meteorites (Fig. 4.5). This hypothesis seems improbable, however, because these organic chemicals are unlikely to have survived either the levels of ultraviolet radiation or the oxygen-free environment of the early Earth. These same meteorites would likely have impacted on the Moon and on Mars, two bodies from which rock samples have been analyzed for the presence of organic carbon, and in neither case have any organic molecules been found. Tantalizing as the existence of these organic compounds in meteorites actually is, it most likely does not point to the source of the molecules that give rise to life on our planet.

A second set of theories argues that the crucial prebiotic compounds arose only on Earth as a result of the particular geological, atmospheric, and chemical conditions prevailing 4 billion years ago. We cannot, of course, hope to recover direct evidence of these original organic molecules, since they leave no trace or fossil. Support for this theory must come from experiments that seek to replicate in the laboratory the conditions of the primitive Earth. Ever since the 1920s, beginning with the work of the Russian scientist A. I. Oparin, scientists have tried to re-create a synthetic "primordial soup" and have sought to analyze its composition. These experiments in prebiotic chemistry share the same basic design, which includes:

1. A solution (soup) frequently containing some mixture of the organic and inorganic compounds thought to have been present on the Earth.
2. An "atmosphere," generally an oxygen-free (reducing) atmosphere.
3. A source of energy for synthesis, either in the form of electrical sparks (resembling lightning), heat, or ultraviolet radiation.

The most interesting of the early attempts were carried out in the mid-1950s by Stanley M. Miller and Harold Urey. They prepared a sealed glass flask containing water and a reducing atmosphere of methane, ammonia, and hydrogen (Fig. 4.6). Strong electrical discharges into this sealed flask resulted in the abiotic synthesis of a tarry organic soup, which contained various biological precursors including the amino acids glycine and alanine. Surprisingly, these experiments did not result in some random mixture of all sorts of organic molecules. Instead, substantial yields of a small, reproducible set of biologically important compounds were routinely obtained. When hydrogen sulfide, a gas generated by a variety of geochemical processes, was added to the original atmosphere, Miller also recovered significant amounts of two other amino acids, cysteine and methionine. Subsequent experiments, varying the concentrations and pressures of the gases composing the original atmosphere, yielded measurable amounts of 13 amino acids.

The basic components of nucleic acids—purines and pyrimidines—are not quite as easily obtained in the "early atmosphere"–type laboratory experiments. To date, the most remarkable yields were obtained by allowing a concentrated aqueous solution of hydrogen cyanide to remain undisturbed for several days. In this fashion, appreciable amounts of adenine, guanine, and other purines appeared in the solution. The synthesis of these compounds thus appeared to occur spontaneously, but only in very concentrated solutions of hydrogen cyanide. Would such concentrations have been reached in any of the environments of the early Earth? There is debate on this matter, but hydrogen cyanide was likely present in ponds and other small bodies of water, and the freezing of these ponds would

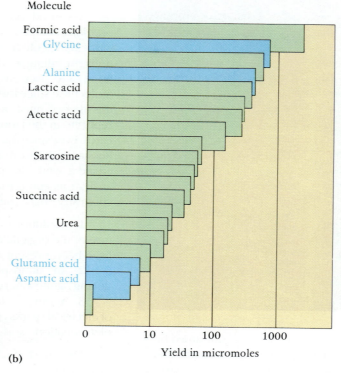

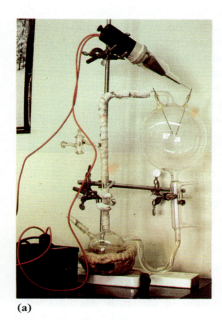

Figure 4.6

(a) A replica of the Miller-Urey experimental apparatus. A mixture of different gases and molecules, meant to simulate the conditions of the early Earth, is introduced into the device, which is then sealed off. An electrical spark in the glass bulb (*right*) provides the energy for chemical reactions to occur. The products and remaining reagents are collected in the flask (*left*) and subsequently analyzed. (b) This graph shows the yield of various important organic molecules generated in a series of Miller-Urey prebiotic experiments. Note that several amino acids, the building blocks of proteins, are produced in appreciable amounts.

have gradually increased the concentrations of hydrogen cyanide. This model for the prebiotic origin of purines and pyrimidines is thus plausible, but requires a very specific set of conditions. We cannot be sure at present that this is the pathway that gave rise to these compounds.

Curiously, sugars have proven the most difficult molecules to synthesize in these experiments. In particular, ribose—a basic component of all nucleotides—is rarely found in these experiments. Only concentrated solutions of formaldehyde give rise to detectable concentrations of ribose. While there is little doubt that formaldehyde may have been present in the primordial atmosphere, the necessary concentrations are exceedingly high, and therefore very unlikely to have occurred.

The successful synthesis of crucial compounds in these experiments in early chemical evolution is encouraging, but only begins to answer the questions surrounding the origin of life. As is the case for all chemical reactions, there is an equilibrium between the synthesis and the degradation of a given compound; the overwhelming majority of chemical reactions are reversible. The concentrations of amino acids, nucleic acids, and lipids

produced by the types of reactions described above would have remained extremely low, too low for life to evolve. Our next challenge, therefore, is to examine the early mechanisms that bring together and concentrate these critical molecules. How do the chemical reactions necessary to life become organized? How does that organization get transmitted over time? We now turn our attention to the origins of life and focus in on the earliest protocell.

LIFE AND NONLIFE: WHERE IS THE BOUNDARY?

The fact that biologically important molecules—carbohydrates, nucleic acids, lipids, and amino acids—can be generated is in itself astounding, but it is only a partial solution to the scientific problem of life's origins. After all, exciting as the results of these early atmosphere experiments are, they generate organic chemical compounds, not living organisms. The experiments certainly suggest that the necessary building blocks for life could arise from the operation of chemical and physical forces. But blocks do not a building make.

When in time can we say that life has begun? As we mentioned before, the fossils of the Fig Tree group appear already to be complex, dividing cells. Much of the early history of life had already taken place before the cells of the Fig Tree formation arose. They are unquestionably the earliest fossils we have so far encountered, but they are not likely to have been the earliest life on Earth.

The problem we have yet to address is the following: at what point (and under what circumstances) do these molecules and compounds come together in a sufficiently organized manner to refer to the result as "living"? Are there certain fundamental characteristics or criteria that must be met before we consider life to have evolved?

The Essential Characteristics of Life

There can be endless debate on the question of what constitutes life, but we would like to suggest that there are three fundamental features that characterize all life on this planet:

1. Organisms, as the name suggests, exhibit some degree of *organization* and are not merely random collections of molecules.
2. Organisms contain, interpret, and transmit *information* from one generation to the next.
3. Organisms are capable of *function*: they can catalyze and regulate metabolic activities.

It is easy to identify and detail these three characteristics when we examine life today. The cells and tissues of modern organisms represent extraordinary examples of organization, carrying out thousands of different metabolic tasks in a coordinated way. Similarly, the complex architecture of DNA, with its unique and precise sequence of nucleotides, embodies the information necessary to accurately synthesize the many proteins the organism requires. And it is this DNA chain that is transmitted from parent to offspring—the phenomenon of inheritance. Finally, the proteins of the cell catalyze and regulate complex metabolic functions, synthesizing, transporting, and destroying the compounds necessary to the survival of the cell and consequently of the organism.

From an evolutionary standpoint, a reconstruction of the initial events that led to life on this planet must therefore account for the origin of organization, information, and function. We are faced with the following problem: if we reconstruct how proteins came about, we have accounted for the origin of function. But proteins are not the stuff of heritable information; they cannot be directly transmitted from one "generation" to the next. Conversely, we could plausibly reconstruct the evolutionary origins of DNA (or some other nucleic acid chain) and in so doing explain the inception of biological information. But DNA cannot by itself carry out function; its value as information in the modern cell depends on protein en-

zymes capable of interpreting and carrying out the instructions encoded in DNA. A coherent theory of life's origins therefore requires the simultaneous, independent evolution and bringing together of informational molecules (e.g., DNA) and functional molecules (e.g., proteins). The conjunction of two complex systems, each of little value without the other, evolving independently and fortuitously brought together seems exceedingly improbable. Francis Crick, one of the discoverers of the structure of DNA, would write in 1981:

> An honest man, armed with all the knowledge available to us now, could only state that in some sense, the origin of life appears at the moment to be almost a miracle, so many are the conditions which would have had to have been satisfied to get it going.

Cutting the Gordian Knot: Catalytic RNA

As is often the case in science, a possible resolution to this complex problem came from a completely unexpected discovery. Beginning in the early 1980s, Tom Cech, Sidney Altman, and other researchers reported that certain biochemical reactions, thought to be possible only when catalyzed by specific proteins (enzymes), could in fact take place in the complete absence of any proteins. Even more exciting was the realization that these reactions were indeed being catalyzed (they were not just occurring spontaneously) and that the catalyst was RNA, a nucleic acid. The implications of this finding were immediate and profound: a single molecule of RNA (in one case only 31 bases long) was capable of *both* containing information and carrying out function. These tiny **ribozymes**, as they were soon dubbed, could serve as an information template for their own precise replication and could catalyze specific biochemical reactions. In 1989, Cech and Altman were awarded the Nobel Prize in Chemistry for their discovery of catalytic RNA.

It was as though we had suddenly caught a glimpse of the very early stages of biological evolution, and it had forced us to look at the problem in a brand new way. These ribozymes suggested that information and function need not have arisen independently, but may instead have had a joint origin in a single type of molecule. What appeared almost impossible now seemed far more plausible. Proteins may have been late arrivals onto the biological scene, with the evolutionary game well under way by the time they appeared. Proteins eventually proved far more efficient and versatile catalysts than RNA, and it is proteins that now mediate the chemical reactions necessary for life to exist. In modern cells, RNA is now a molecule dedicated entirely to the transmission of information. In all but a few viruses, RNA serves as a messenger, shuttling information from the DNA to the machinery required to synthesize proteins. Much then has changed from those early stages of life on Earth. RNA, once a molecule doing double duty as

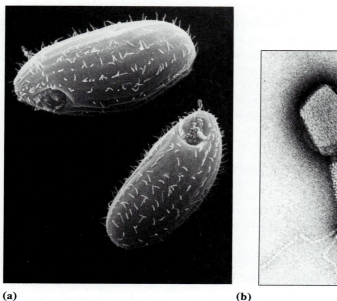

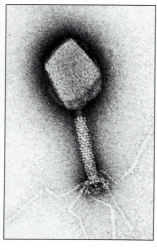

(a)   (b)

Figure 4.7

(a) The unicellular ciliated eukaryote, *Tetrahymena*, in which ribozymes were first discovered. (Magnification 3000 ×). (b) T₄ bacteriophage, a polyhedral virus that infects bacterial cells. (Magnification approximately 275,000 ×.)

source of information and chemical catalyst, now serves to shuttle information between a newer information molecule (DNA) and the new functional molecules (proteins). Yet, embedded in the complex, highly evolved machinery of modern cells is the track of history. These ribozymes still survive in the genetic material of the ciliated protozoan *Tetrahymena pyriformis* and in the tiny virus that infects bacteria, the T4 phage, where they are still able to carry out their double task (Fig. 4.7). And they have made us think quite differently about the way in which life may have first evolved. In fact, they force us to rethink what we mean by "life." These ribozymes, after all, carry specific information that can be faithfully copied and transmitted through time. They can carry out complex enzymatic function, catalyzing a difficult reaction with great precision. Are these ribozymes the first organisms? Has life at this point already begun?

The realization that certain ribozymes can serve both as information and as functional molecules has led a number of investigators to suggest that the earliest life was RNA based—the so called **RNA world**. But the evolution of ribozymes may itself have been a long process, and we have yet to account for the third fundamental feature of life: organization. Even the simplest of contemporary cells is surrounded by a membrane. This membrane allows materials to be concentrated inside the cell, and results in an *internal environment* that differs from the surrounding world. Cell organization has been well investigated over the past century, as we will see in the next chapter. Only now, however, are we beginning to understand some

aspects of its evolution, and the earliest steps in the development of cellular organization remain something of a mystery. Some tantalizing clues are emerging from the work of Sidney Fox and other investigators who are experimenting with *protobionts*. These are small microspheres, some only 1 to 2 μm in diameter (Fig. 4.8) that are formed

Figure 4.8

Proteinoid microspheres, 1 to 2 μm in diameter. Certain investigators suggest that these microspheres resemble the earliest protocells that evolved some 3.8 to 4 billion years ago.

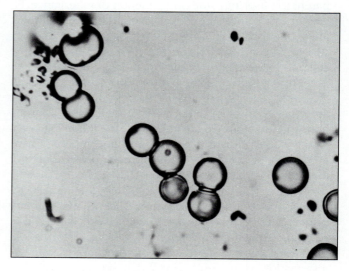

by heating dry amino acids and other organic molecules and then adding water to the mixture. Some of these microspheres show amazing life-like properties: they can differentially absorb certain molecules from the surrounding environment, they can develop an electric potential across their surface (although they lack a cell membrane as such), they can increase in size ("grow"), and some seem to divide. The earliest cells might have begun in a similar way, but we cannot address this question until we understand how the cell is organized. It is therefore to the structure, organization, and evolution of cells that we now turn.

Summary

1. The earth is approximately 4.6 billion years old and began as a barren and inhospitable planet. Life arose surprisingly early, for the earliest known sedimentary rocks that might contain fossils, the 3.4-billion-year-old Fig Tree deposits from southern Africa, do appear to contain fossilized cells. The earliest protocells that gave rise to these must be older still.

2. We know that all living organisms arose from a single ancestral stock because all organisms share a fundamental metabolic machinery and transmit information from generation to generation by nucleic acids. It is unlikely that such similar and complex biochemical mechanisms would have evolved more than once.

3. One theory of the origin of complex organic molecules argues that these molecules arose only on Earth as a result of the particular geological, atmospheric, and chemical conditions that prevailed 4 billion years ago. Scientists have been able to synthesize many of these compounds in an oxygen-free (reducing) atmosphere by exposing a "soup" of inorganic compounds thought to have been present on the early Earth to sources of energy such as electric sparks (simulating lightning), heat, and ultraviolet radiation. Although the necessary compounds for life could have arisen in this fashion from the operation of physical and chemical forces, their concentration would at first have been too low for life to have evolved. They must have been concentrated in some way in order for life to evolve.

4. Living organisms are not random collections of molecules, but exhibit some degree of organization; they also contain, interpret, and transmit information from generation to generation; and they catalyze and regulate the metabolic functions of cells.

5. A theory of life's origin requires the simultaneous and independent evolution and bringing together of informational molecules (the nucleic acids) and functional molecules (the proteins). The evolution of this conjunction seemed to be highly improbable until the discovery in the early 1980s that certain reactions could be catalyzed not by a protein but by an informational molecule, ribonucleic acid (RNA). Information and function may have had a joint origin in a single type of nucleic acid molecule. Would aggregations of such molecules have been the first life? Proteins may have been later arrivals on the biological scene. With the advent of proteins, which are more efficient catalysts, a division of labor occurred. In all but a few viruses and protozoans, RNA now serves only to shuttle information, and proteins serve to catalyze the metabolic functions of cells.

References and Selected Readings

Cech, T. R. RNA as an enzyme. *Scientific American* 255 (Nov. 1986):64–75. RNA is a functional as well as an informational molecule.

Margulis, L. *Early Life*. Boston: Scientific Books International, 1982. An account of the early development of life.

Schopf, J. W. The evolution of the earliest cells. *Scientific American* 239 (Sept. 1978). Evidence for the early history of life.

Waldrop, M. M. Did life really start out in an RNA world? *Science* 246 (Dec. 1989):1248–1249. A brief statement of remaining problems in hypotheses about the origin of life.

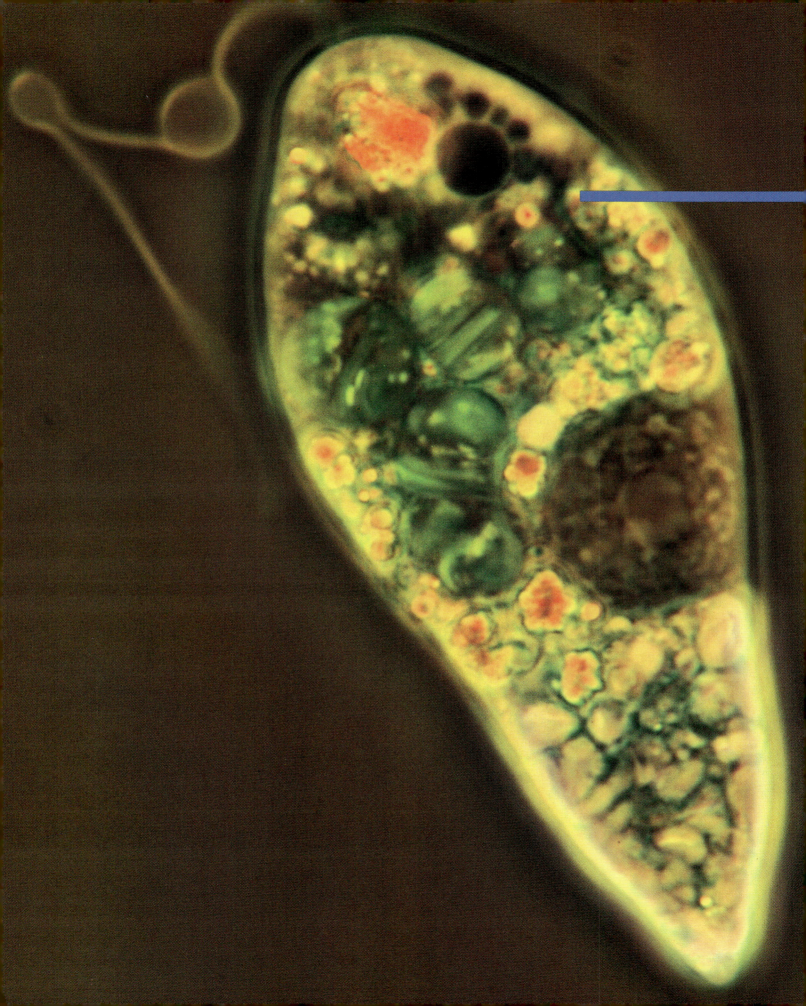

5

The Eukaryotic Cell: Structure and Evolution

(Left) The single-celled eukaryote Euglena gracilis *is a protoctistan that acquires energy via photosynthesis. The nucleus (large dark spot) and the chloroplasts (green structures) are clearly visible, as is the flagellar tail used for propulsion. (Above) Two representatives of the phylum* Euglenophyta, Euglena *and* Trachelomonas, *are simple eukaryotes capable of photosynthesis.*

In Chapter 4, we discussed the origin of two key aspects of living systems: information and function. Our discussion focused on the molecules that make life possible and on how those molecules may have first arisen on a lifeless Earth. In this chapter, we address the final general feature of living systems: organization. We will examine the **cell**, the basic unit of biological structure. In so doing, we continue our ascent up the biological hierarchy, from the molecules of living systems to the architecture of the animal cell.

ORGANIZING COMPLEXITY

Defying the Second Law

The starting point for our discussion of the cell is the second law of thermodynamics, which states that all closed systems—chemical, physical, and biological—tend over time to become increasingly disordered, or **entropic**. (The laws of thermodynamics are reviewed in Appendix A.)

At first glance, living systems appear to violate the second law, also known as the principle of increasing entropy. The one overwhelming fact of life at the biochemical level is the exquisite order and coordination of metabolic reactions. If you are to remain alive until the end of this page (and beyond), thousands of chemical reactions must take place, with precision and delicate timing. Are you somehow violating the second law with every word you read?

In order to answer this question, let us take a look at an example of the second law in action. If you add a spoonful of sugar to your morning coffee, the sugar will immediately go to the bottom of the cup. If at that point we were to measure the concentration of sugar at the various levels in your cup, we would initially find a very high concentration at the bottom and a very low concentration at the top. But if you leave your coffee cup undisturbed for some time, the sugar will not remain at the bottom. Instead, it will go into solution; eventually the concentration of sugar throughout the whole cup will be equal, even if you did not stir your coffee. From a thermodynamic standpoint, the initial situation (with the sugar at the bottom) is a highly ordered state, with a gradient of sugar concentration extending from the bottom to the top of the cup. As the second law predicts, the system (in this case, the coffee cup) will tend to increase in entropy. The final state, with sugar molecules spread throughout the coffee, is a more disordered, entropic state.

The behavior of amino acids, nucleic acids, and other important molecules present in the primordial pools or oceans of the early Earth would have been no different. These crucial molecules would have been present in infinitesimally small amounts; driven by the second law, they would have diffused throughout the pools where they were formed. Under those circumstances, the concentra-

tions of these biologically important molecules would have remained so low as to make the evolution of life virtually impossible.

The Cell and the Second Law

One simplistic way to think of the cell is essentially as life's answer to the second law of thermodynamics; the earliest function of a cell would have been to concentrate and retain the necessary molecules for life to proceed. The cell, in its simplest form, is an enclosed space surrounded by a membrane that delays (or in some cases entirely prevents) certain molecules from entering or exiting. If we were to examine a variety of molecules and ions both inside and outside a cell, we would find their relative concentrations to be different. Does this mean that living cells somehow violate the second law of thermodynamics? Not at all. The very improbable, low-entropy situation we describe comes at a very dear price: the cell must constantly expend energy to maintain this orderly state. If the cell in our above experiment were to die, and we were then to measure the concentrations of molecules inside and outside the cell, they would soon be equal, just like the sugar in our coffee cup.

Why is it important that the cell maintain higher (or lower) concentrations of particular molecules than can be found in its surroundings? To answer this fully, we must introduce the chemical concept of **reaction kinetics**: the factors that regulate the likelihood and velocity of a chemical reaction.

All chemical reactions are, at least in theory, reversible. Thus, a reaction combining molecules A and B (known as the "substrates" in most biological reactions) to give rise to a novel compound C (the "product") can be run in either direction and is written as such:

$$A + B \rightleftharpoons C$$

Let us imagine that the organism, in order to survive, requires the constant production of compound C. Yet the reaction not only produces C, it may also metabolize C into compounds A and B. If the reaction is equally likely to occur in either direction, an equilibrium would eventually be reached between the synthesis and the destruction of C. But the cell prevents this equilibrium from being reached by *driving* the reaction: if there is a constant excess of A and B, the reaction will most often proceed towards C. To do this, the cell must be able to control the local concentrations of all three molecules. It cannot simply passively accept whatever concentrations of A, B, and C are present in its surroundings. The cell exercises control over concentrations in one of three ways:

1. By using product C as a substrate for a subsequent reaction, as happens in most metabolic pathways involving a chain of reactions.

2. By actively pumping substrates A and B into the cell from the surrounding environment, or by pumping product C out of the cell.

3. By sequestering product C of a reaction into a compartment within the cell.

The ability to create, control, and regulate an *internal environment* makes life possible; it is the cell that defines and separates the internal living world from the outside.

THE CELL THEORY

It is difficult for us to imagine, in this age of satellites and computers, that there was ever a time when we did not suspect the existence of a world smaller than what the naked eye could see. Yet it was not until the early 17th century that a rudimentary microscope was first assembled and used to examine the living world. In 1665, Robert Hooke, observing the fine structure of cork under a simple microscope, was the first to see a lattice of cell walls (Fig. 5.1). He later noted a similar arrangement in a living leaf and used the term "cell" to describe the repeating elements he had observed. Anton van Leeuwenhoek, a contemporary of Hooke, would be the first to glimpse many other types of cells: single-celled algae and protozoans in pond water, sperm cells, and red blood cells. He referred to these structures by the generic term "animalcules."

These initial depictions of cells show little more than the thick cell walls of plant tissue. It would take almost 200 years of technical improvements in the design of microscopes and lenses before Theodor Schwann, Matthias Schleiden, and Rudolf Virchow stated the principle now known as the **cell theory**: that all tissues are composed of cells or of the products of cells. The cell theory established the cell as the fundamental unit of the living organism; life could now be studied at the cellular level. The cell theory further underscored the unity of all life, and became a central pillar in the rising discipline of biology. The corollary realization, that cells must come from other cells and therefore do not arise spontaneously, was clearly demonstrated in 1861 by Louis Pasteur in a classic set of experiments (Fig. 5.2) that stand as an important milestone in the search for material explanations in biology.

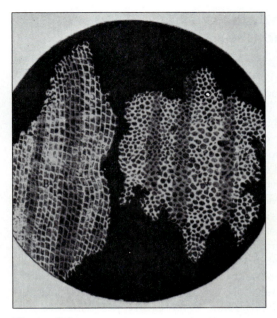

Figure 5.1
Robert Hooke's diagram of cells (actually, plant cell walls) in a thin slice of cork as seen through his primitive microscope.

Figure 5.2
Pasteur's experiment. (a) A broth exposed to the air was soon contaminated with bacteria. (b) Pasteur placed the broth in a flask with a long **S**-shaped neck open to the air. Bacteria in the air were trapped in the low part of the neck and the medium remained uncontaminated. (c) If the neck was removed, the medium quickly became contaminated.

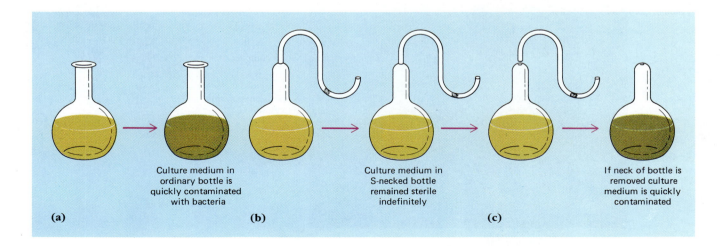

(a) Culture medium in ordinary bottle is quickly contaminated with bacteria

(b) Culture medium in S-necked bottle remained sterile indefinitely

(c) If neck of bottle is removed culture medium is quickly contaminated

TWO UNDERLYING CELL PLANS: PROKARYOTES AND EUKARYOTES

While all living organisms are composed of cells, and all cells share certain fundamental structural and biochemical features, there are many different kinds of cells. The most important distinction made at this level divides all cells into two basic categories: the prokaryotic cell and the eukaryotic cell. The **prokaryotic cell** (Gr. *pro*, before + *karyon*, nucleus) does not have a nucleus; its genetic material is confined to a region of the cell, the **nucleoid**. The **eukaryotic cell** (Gr. *eu*, true) has a clearly discernible nucleus, with a nuclear membrane that encloses the genetic material (Fig. 5.3). Broadly speaking, the prokaryotic cell is less complex than its eukaryotic counterpart. Prokaryotic cells are much smaller than eukaryotes (ranging in size from 1 to 5 μm), and they lack internal membrane-bound organelles (structures devoted to a subset of metabolic tasks). All available evidence suggests that prokaryote-like cells evolve long before the eukaryotes first appear. The earliest fossils in the record, dating back some 3.4 billion years, are prokaryote-sized and lack a nucleus. The Prokaryotae, which includes both the Eubacteria (true bacteria) and the Archaebacteria (ancient bacteria), are an extremely successful and diverse group. They now occupy practically all terrestrial and aquatic environments, from the permafrost layer of the Arctic to the boiling sulfurous hot springs of Yellowstone. The enormous evolutionary versatility of the prokaryotes comes about, at least in part, because of their short generation times (as little as 20 min/generation for certain species, under ideal circumstances), which allow for rapid evolution.

With the exception of the viruses, the rest of the living world—protoctistans (protozoans, algae, and slime molds), fungi, plants, and animals—are all **eukaryotes**. Whether single-celled or multicellular, they are all built using the same basic unit, the eukaryotic cell. This exquisitely complex structure, which we will discuss in further detail below, is unlikely to have evolved more than once in the history of life; as a result, we consider the eukaryotes to be a monophyletic group: they are all descendants of a single common ancestor.

THE BASIC FEATURES OF THE EUKARYOTIC CELL

Eukaryotes come in many shapes and sizes, ranging from small, single-celled organisms, such as the protozoan *Paramecium* (Fig. 5.4), to complex, multicellular organisms such as ourselves. Our bodies, and those of all multicellular animals, contain a large variety of different

Figure 5.3

An electron micrograph of a portion of a eukaryotic cell from the human pancreas. Note the presence of an organized nucleus and cytoplasmic organelles.

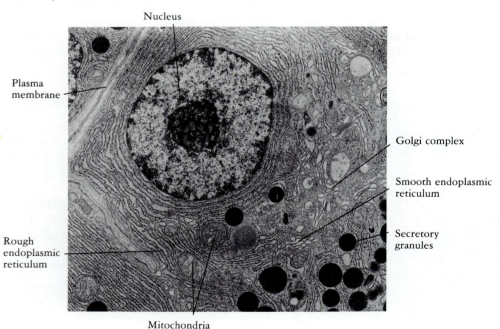

Nucleus

Plasma membrane

Golgi complex

Smooth endoplasmic reticulum

Secretory granules

Rough endoplasmic reticulum

Mitochondria

Figure 5.4

A stained photograph of a protozoan, *Paramecium caudatum*, seen through an optical microscope.

specialized cells, including muscle cells, nerve cells, and red blood cells. Yet underlying this vast array of specializations is the basic eukaryotic cell.

What are the contents of this cell? Thanks to the further improvement of optical microscopes and the invention of the electron microscope, our knowledge of the eukaryotic cell has increased dramatically. With the early microscopes, the nucleus was barely discernible, and there was little possibility of seeing any of the smaller components or compartments of cells. Cells were known to contain a viscous or sticky substance, the **protoplasm**, which was further subdivided into the nucleus and the cytoplasm, and upon which biologists conferred all sorts of mysterious and vital powers. As with the discovery of cells themselves, it was technical innovation that eventually demystified the nature of protoplasm. The dramatic increase in resolution brought about by increasingly sophisticated light microscopes, and eventually by electron microscopes, permitted scientists to reconstruct the complex furnishings of the cell (see Focus 5.1). Far from a homogeneous fluid, the eukaryotic cell is instead a complex arrangement of compartments and subdivisions. The entire cell is spanned by an organized network of filaments that permit the cell to change shape and to actually move along certain surfaces. The genetic information is packed into chromosomes, themselves enclosed in the nucleus. The eukaryotic cell is a highly organized, dynamic entity;

molecules and information flow within and between cells in a carefully orchestrated way. If we wish to understand this flow, we must now focus on the architecture of the intracellular world. The principal components of the eukaryotic cell are shown in the transparency insert between pages 70 and 71.

The Cell Membrane

The cell membrane marks the boundary between the internal and external environments; it is the border separating the minuscule ordered world of the cell from the disorder of its surrounding environment.

The Phospholipid Bilayer

The earliest cells may have been no more than packages of chemicals surrounded by some sort of membrane, a barrier that would have prevented the free diffusion of important molecules into the environment. These earliest membranes had a simple and passive role to play, serving only to enclose and concentrate some of the crucial molecules we have already described. These membranes were most likely composed of phospholipids. Recall that phospholipids are composed of two fatty acid chains connected to a glycerol molecule and modified by the addition of a phosphate group (see Ch. 3). The result is a molecule with a somewhat schizophrenic character. The fatty acids are strongly **hydrophobic**—they do not mix readily with water. In contrast, the phosphate-glycerol part of the molecule is electrically charged, or polar, making it **hydrophilic**—it is attracted to water. A phospholipid molecule is thus said to be **amphipathic** ("tolerant of both"). What then will happen if we mix amphipathic molecules with water? The tails of these molecules will arrange themselves in such a way as to minimize contact with water and maximize contact with other hydrophobic tails; the heads will seek to maximize their contact with the aqueous environment. How can both of these goals be met simultaneously? Two basic geometrical configurations can arise when phospholipid molecules are mixed in water:

1. **Micelles**, spheres of phospholipids in which all the polar heads face outward and all the hydrophobic tails face inward, creating a strongly hydrophobic internal environment (Fig. 5.5a).

2. Phospholipid **bilayers**, linear sheets of double thickness in which the polar heads face outward (on either side of the bilayer), thus creating an internal hydrophobic layer where both sets of tails face each other (Fig. 5.5b,d). Under certain conditions, phospholipid bilayers may form spheres, known as **liposomes**, which contain an internal aqueous core (Fig. 5.5c).

The capacity of phospholipids to form organized bilayer structures (spheres or sheets) in aqueous environ-

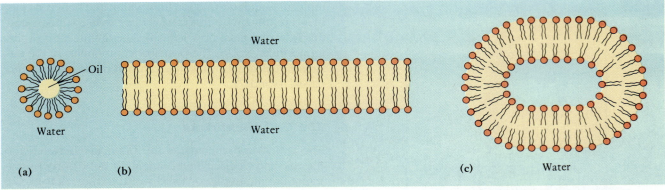

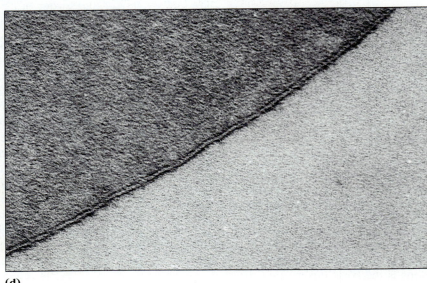

Figure 5.5
Geometric configurations of phospholipid layers. (a) A micelle, a spherical monolipid layer surrounding an oil or fat droplet. (b) A bilipid membrane, or bilayer, with an aqueous medium on each side of it, as occurs at the surface of cells. (c) A liposome, a bilipid layer with an aqueous core. The earliest protocells may have resembled liposomes. (d) An electron micrograph of the plasma membrane of a cell. The two parallel dark lines represent the hydrophilic heads of the lipid molecules.

ments suggests a model for the earliest protocells, in which tiny droplets of water containing RNA, carbohydrates, amino acids, and other molecules would be enclosed in phospholipid liposomes. It is from these humble beginnings, tiny oily droplets enclosing a brew of chemicals, that the complex cell likely evolved. The cell has come a long way from the early days, 3.8 billion years ago. Nonetheless, the evidence of the past remains: all eukaryotic cell membranes studied thus far are composed of a bilayer of amphipathic lipids (phospholipids or glycolipids).

Membrane Proteins

A large number of proteins are associated with the phospholipid cell membrane (Fig. 5.6). **Peripheral proteins** are attached to either the outer or the inner (cytoplasmic) surface of the membrane, and **transmembrane proteins** extend through it. The cell membrane, along with associated proteins and other molecules such as cholesterol, is not a static structure. The proteins move like small rafts in a sea of phospholipids. The cell membrane should be thought of as a "fluid mosaic," a dynamic entity that is constantly changing.

Many peripheral proteins on the inside of the membrane serve as anchors for the cytoskeleton, a complex of various proteins that maintains cell shape and allows directional movement. Peripheral proteins on the outside of the membrane and some transmembrane proteins, both of which may bind with distinctive carbohydrate chains (Fig. 5.6), give cells unique identities. Many of these proteins are receptors to which specific chemical messengers, or hormones, bind, initiating a cascade of events that affect the biochemistry of the cell (Chap. 19). Other surface proteins are signature proteins that identify the cell to the outside world. Such proteins, present on the surface of red blood cells, determine blood types of individuals (A, B, Rh). The ability of the immune system to distinguish between "self" and "nonself" is also based almost entirely on membrane-associated proteins (Chap. 15). Other signature proteins enable cells to recognize each other during development and allow cells to migrate to appropriate parts of the embryo (Chap. 22).

A great many transmembrane proteins are transport proteins that regulate the movement of molecules into and out of cells. The cell membrane is the marketplace of the

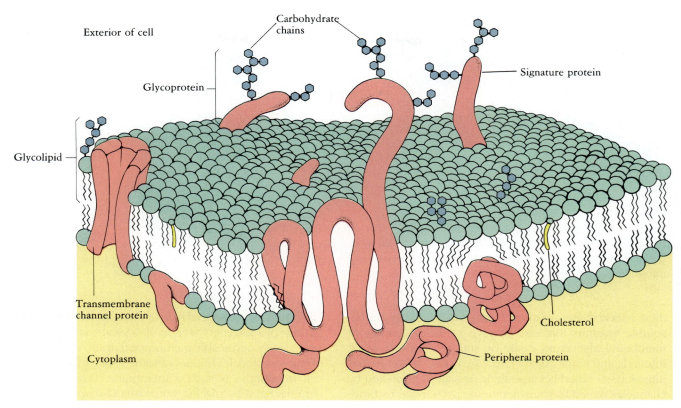

Figure 5.6
A diagram of the plasma membrane of the eukaryotic cell showing some of the
types of proteins associated with the membrane.

cell where materials are constantly exchanged, traded, imported, and exported. The transport proteins are the clearing agents for this traffic. A simple phospholipid bilayer would not permit much of this commerce since any charged molecule, such as a sodium ion, might get past the heads of the phospholipids but would have great difficulty crossing the hydrophobic layer formed by the fatty acid tails of the phospholipids.

Transport of Materials Across the Membrane

Diffusion and Osmosis

Several quite different processes are involved in the movements of materials across the cell membrane. **Diffusion**, the *net* movement of materials from an area of high concentration of the material to an area of lower concentration, is responsible for the movement of some types of molecules across cell membranes as well as for their dispersion within the cell. Diffusion results from the kinetic energy of molecules, their constant random motion (see Appendix A). Individual molecules move in straight lines, but they cannot go far before colliding with other molecules and bounding off in a different direction. When a spoonful of sugar is placed in a cup of coffee, the sugar molecules will at first be very concentrated where the spoon is placed (Fig. 5.7), but the kinetic energy of the molecules gradually disperses the sugar molecules through the coffee (and the coffee through the sugar). Eventually the concentration of the sugar becomes equal throughout the system, even if the coffee is not stirred. Molecular motion does not cease when equality is reached, but there is no longer a net movement in one direction.

Diffusion is a slow process, so it is an efficient means of transport only over short distances. Diffusion rates depend not only on the difference between high and low concentrations (the concentration gradient) but on many other factors. As temperature increases, molecular motion and diffusion rates increase. If the material that is diffusing is subject to an increased pressure, it diffuses faster. The electrical charge on a molecule, molecular size, and molecular shape also have effects. Each type of molecule in a solution will diffuse independently of any others.

Certain small, uncharged molecules such as oxygen (O_2), nitrogen (N_2), carbon dioxide (CO_2), and urea can cross the cell membrane by **simple diffusion**, following

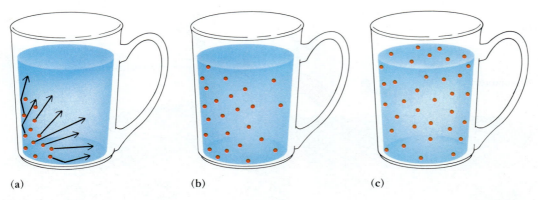

(a)   (b)   (c)

Figure 5.7

Diffusion. (a) A spoonful of sugar (*red dots*) is placed in a cup of coffee. A net movement of sugar molecules will occur (*arrows*) from their region of high concentration to low concentration (b) until their concentration is equal throughout the solution (c).

their concentration gradients. Fatty acids and certain fat-soluble vitamins that can dissolve in the phospholipid membrane also cross easily by simple diffusion. Many minerals and some sugars (fructose) cross by **facilitated diffusion**, so called because their passage is facilitated by channels created by transmembrane proteins or by binding with transport proteins in the membrane. These channels and transport proteins are hydrophilic passages through an otherwise hydrophobic phospholipid bilayer.

Osmosis is a special type of diffusion that involves the passage of a solvent (water, in biological systems) through a selectively permeable membrane that allows the passage of water molecules but not of some solute molecules. The cell membrane is such a membrane. Osmosis can be demonstrated by placing water in a U-shaped tube (Fig. 5.8) that is divided by a membrane permeable to water but not to a molecule such as sucrose. Sucrose is added to the left side of the tube. Because of the presence of the sugar molecules, there are fewer water molecules on the left side

Figure 5.8

Osmosis. (a) A sugar solution in a U-shaped tube is separated from pure water by a membrane that is permeable to water but not to sugar. A net movement of water molecules occurs by osmosis (*arrows*) from their area of high concentration on the right to the area of lower concentration on the left. (b) Water rises in the left side of the tube. Its height creates a hydrostatic pressure that drives water to the right and equals the tendency of water to move to the left by osmosis. This hydrostatic pressure is a measure of the osmotic pressure of the sugar solution.

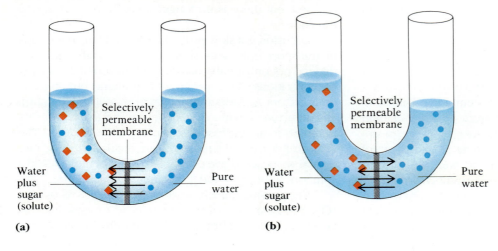

(a)   (b)

than on the right side. Water molecules will move in both directions, but because they are more concentrated on the right, there will be a net movement toward the left. As water accumulates on the left side, the water colum will rise on this side of the tube. Since the solute molecules cannot pass through the membrane, the concentration of water on the two sides can never become equal. However, the weight of the higher column of water on the left exerts a hydrostatic pressure that tends to push water back to the right. The height of the column of water is a measure of the **osmotic pressure** of the sugar solution. An equilibrium will be reached when the tendency for water to move to the left by osmosis will be balanced by the hydrostatic pressure on the left that drives water molecules back. Osmotic effects, as we shall see, are very important in controlling the water and solute balances of cells (Chap. 16).

Active Transport

The passing of materials through the cell membrane by diffusion and osmosis are passive processes that do not require a special energy expenditure by the cell. But the list of readily diffusible compounds is small. Many molecules need to be moved into (or out of) the cell against their concentration gradients. Such "uphill" transport is accomplished by certain transport proteins that expend energy (in the form of ATP) to drive the pump. This process is known as **active transport**. One well-studied example of active transport is the **sodium-potassium pump**, a crucial element in the transmission of nerve impulses in neurons (see Ch. 17). This membrane protein is actually a **cotransport system** that moves one type of ion (sodium) out of the cell, while simultaneously pumping a different type of ion (potassium) in, using one molecule of ATP to accomplish the task (Fig. 5.9). The active transport

Figure 5.9

A model of the sodium-potassium pump in the plasma membrane. When a sodium ion (*1*) and a phosphate ion derived from the breakdown of ATP (*2*) bind to the proteins of the pump, the proteins change their shape and release the sodium ion to the outside (*3*). A potassium ion from the outside now binds to the proteins (*4*), and causes them to release the phosphate ion (*5*). The proteins of the pump return to their original shape and release the potassium into the cell (*6*). The result is a net movement of Na$^+$ ions out of the cell, and an increase in the intracellular K$^+$ concentration.

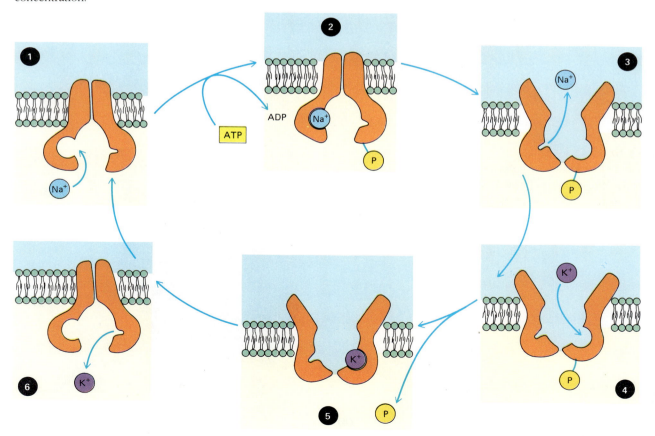

of important molecules, such as glucose, into the cell is frequently coupled to the activity of the sodium-potassium pump.

The ability of membrane proteins to move ions against their concentration gradients has many important physiological consequences. The acidic environment of the stomach, for example, is generated by the continuous pumping of H^+ ions into the stomach lining, with a consequent drop in pH. More generally, ion pumps generate an asymmetrical distribution of ions on either side of the membrane, creating an electrical gradient across the membrane. In general, the outside of a cell is positively charged, and the inside has a slight negative charge. The cell's ability to transport substances against concentration gradients is crucial to the survival of the cell. In the absence of such active transport (or if the action of such pumps is experimentally blocked), the ion concentration inside and outside the cell begins to equilibrate, and the cell dies.

Endocytosis and Exocytosis

On occasion, certain macromolecules or larger particles that cannot fit into the channels created by transmembrane proteins nor bind with the transport proteins must be transported across the cell membrane. In addition, cells may also produce certain proteins exclusively for export—proteins whose activity may be potentially harmful to the cell that produced them. In such circumstances, the cell uses small vesicles for transport into and out of the cell. Macromolecules, such as proteins and large polysaccharides, are transported into the cell by **endocytosis** (Fig. 5.10a). This process involves the formation of a small cavity that eventually surrounds the material being ingested. This

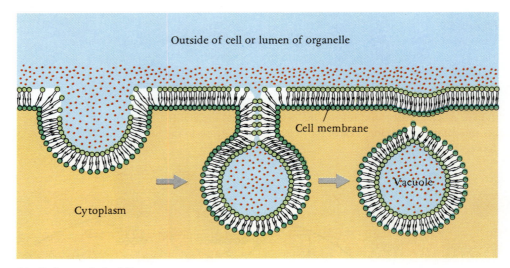

(a) Endocytosis, budding

Figure 5.10

(a) A model of molecules or particles entering the cell by endocytosis. The vacuole is formed by the invagination and eventual pinching off of a portion of the cell membrane. (b) Electron micrograph of a white blood cell phagocytosing a dividing bacterial cell. The bacterium will be digested within the cell.

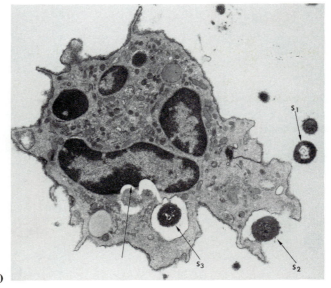

(b)

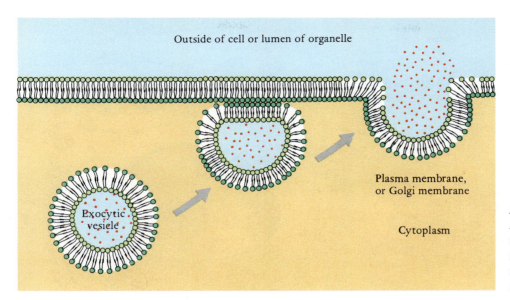

Outside of cell or lumen of organelle

Exocytic vesicle

Plasma membrane, or Golgi membrane

Cytoplasm

Figure 5.11
A model of exocytosis. Once the contents of the vesicle are released, the vesicular membrane fuses with the cell membrane.

cavity continues to invaginate, eventually forming a small spherical vesicle or **vacuole**, surrounded by a phospholipid bilayer derived from the cell membrane. These sealed vesicles are then transported to specific locations within the cell. They will eventually release their contents by fusing with the membrane of the target organelle.

Specialized cells within certain blood vessels, the **macrophages**, are capable of trapping and endocytosing large particles, including viruses and bacteria (Fig. 5.10b). These trapped particles are quickly digested. Such ingestion and digestion of materials is known as **phagocytosis** (Gr. *phagein*, to eat + *kytos*, cell). Not surprisingly, macrophages are an important component of the vertebrate immune response, protecting organisms from a variety of infectious agents.

The reverse process, in which molecules synthesized within the cell are released to the external environment, is called **exocytosis** (Fig. 5.11). Insulin, for example, a protein involved in the regulation of glucose metabolism (see Chap. 19), is produced in vast amounts by certain cells in the pancreas. These cells release insulin into the bloodstream by the process of exocytosis. In this case, compartments within the cell (where the export protein is manufactured) form small vesicles that then migrate towards the cell membrane. Eventually, these exocytic vesicles fuse with the cell membrane, emptying their contents into the external environment.

Cytoplasmic Organelles

If we take a look inside the eukaryotic cell, what we find is far from a homogeneous cytoplasmic fluid. Instead, the cell is subdivided and highly compartmentalized (Fig. 5.12). These compartments, the **organelles**, are clearly defined structures within the cell. The nucleus and many others are surrounded by a bilayer membrane. Both the mitochondrion and the nucleus are enclosed by double phospholipid bilayer membranes.

These organelles perform a variety of specialized metabolic tasks: protein synthesis, ATP synthesis, cell maintenance, and the organization and distribution of genetic information during cell division. As we will see below, the different organelles frequently act in concert. Their coordinated action is responsible for cell function. The evolution of organelles enables the cell to compartmentalize the many reactions of metabolism. This has one important consequence: different and often incompatible metabolic reactions can proceed simultaneously in separate compartments within the cell.

The Endoplasmic Reticulum

The most extensive of the organelles is the **endoplasmic reticulum** (L. *rete*, net), which appears as a collection of membranes spread throughout the cell (Fig. 5.12). In fact, the endoplasmic reticulum is a set of connecting thin closed vesicles and is the site of synthesis of some of the major cellular components, including proteins, lipids and steroids (such as the hormones testosterone and estrogen). When viewed under the microscope, two types of endoplasmic reticulum are readily discerned.

1. The **smooth endoplasmic reticulum (ER)**, as the name indicates, shows a smooth appearance (Fig. 5.13a). The smooth ER is the membrane factory of the cell. It is here that the fatty acids and phospholipids that form all the membranes in the eukaryotic cell are synthesized. The

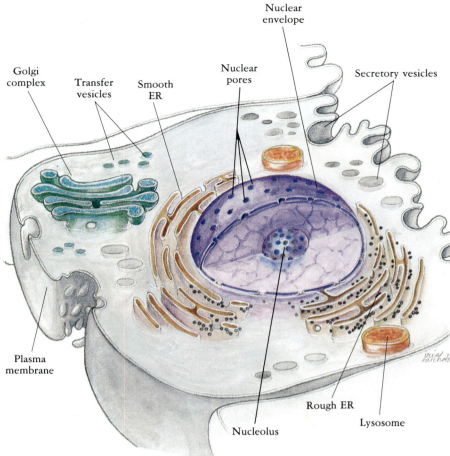

Nuclear envelope

Golgi complex

Transfer vesicles

Smooth ER

Nuclear pores

Secretory vesicles

Plasma membrane

Rough ER

Nucleolus

Lysosome

Figure 5.12

Diagram of the eukaryotic cell, showing the nucleus and many of the cytoplasmic organelles. Note the interconnected membrane systems between the nucleus and the endoplasmic reticulum. Vesicular traffic among organelles further strengthens the interconnections among the various components of the cell.

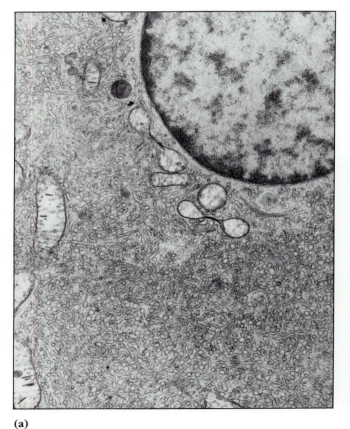

(a)

Figure 5.13

(a) An electron micrograph of the smooth ER, which is composed of branching and connecting networks. The nucleus occupies the upper right corner. (b) An electron micrograph of the rough ER, which is composed of parallel arrays of broad, flat sacs. Their outer (cytoplasmic) surface is studded with small black dots, the ribosomes. Free ribosomes are in the cytoplasm between the rough ER.

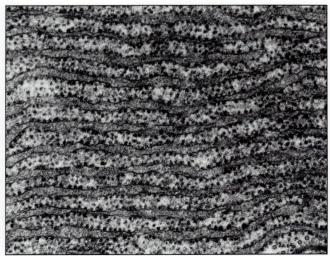

(b)

smooth ER also appears to be the site where many drugs, as well as several toxic and carcinogenic chemicals that enter the body, are detoxified. This detoxification generally occurs in the cells of the liver.

2. The studded appearance of the **rough endoplasmic reticulum** is caused by the associated ribosomes, key structures in protein synthesis (Fig. 5.13b). The rough ER is the site of active protein synthesis, particularly of proteins that are produced by certain cell types for later export. As we discuss in Chapter 7, there are many ribosomes that are not associated with the endoplasmic reticulum. These free ribosomes synthesize proteins in the cytoplasm for internal use by the cell. Export proteins, however, must somehow be packaged in a manner that permits them to cross the cell membrane, and that packaging process begins in the rough ER. The completed proteins are often modified by the addition of a carbohydrate molecule and then packaged for export into transfer exocytotic vesicles. These transfer vesicles then bud off from the endoplasmic reticulum, and most of them will proceed to the Golgi apparatus.

The Golgi Apparatus

The **Golgi apparatus**, looking not unlike a stack of pancakes, is responsible for the final sorting, packaging, and export of the proteins synthesized in the ER (Figs. 5.12 and 5.14a,b). The small transfer vesicles from the rough ER fuse with the membranes of the Golgi apparatus, dumping their contents into the Golgi. Once there, proteins are modified: they may be phosphorylated (by the addition of phosphate groups), turned into glycoproteins (by the addition of carbohydrates), or cleaved by specific enzymes (proteases). These modifications frequently act as luggage tags, directing particular proteins to their final destinations. The Golgi also forms small secretory vesicles that will shuttle proteins out of the Golgi and towards the cell membrane. These secretory vesicles are surrounded by a bilayer much like the cell membrane. In most cases, the vesicle will fuse with the cell membrane when it comes into contact with it, and the contents of the vesicle will be discharged from the cell by exocytosis. Not surprisingly, this is also how membrane-associated proteins are synthesized and eventually embedded in the cell membrane.

Figure 5.14

(a) Diagram of the Golgi apparatus. Note that the individual Golgi sacs are interconnected, allowing for the movement of materials within the apparatus. Proteins synthesized in the ER are brought into the Golgi by vesicular traffic. After further modification, proteins are exported to other organelles or out of the cell by secretory vesicles budding off the upper face of the Golgi. (b) An electron micrograph of the Golgi apparatus.

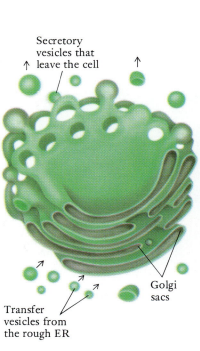

Secretory vesicles that leave the cell

Transfer vesicles from the rough ER

Golgi sacs

(a)

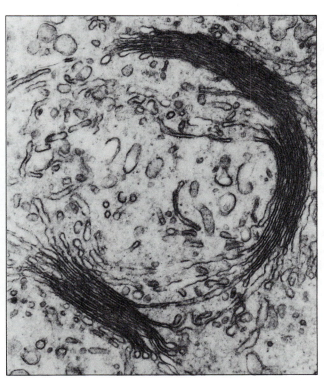

(b)

The Microbodies: Lysosomes and Peroxisomes

Lysosomes can be thought of as the janitors of the cell—their role is to find and dispose of a large variety of different molecules that enter the cell, as well as to recycle certain cell components. Lysosomes are specialized vesicles that form by budding off the Golgi apparatus (Figs. 5.12 and 5.15). Lysosomes contain a broad range of enzymes that can digest proteins (proteases), lipids (lipases), nucleic acids (nucleases), and can remove phosphate groups (phosphatases). Obviously, if all these enzymes were to operate freely within the cell, the cell would immediately be destroyed from within. The cell, however, has evolved two mechanisms to prevent self-destruction. First, these enzymes are contained in specific organelles (lysosomes). Second, these lysosome enzymes are only active at an acidic pH of 4.8. An acid environment occurs within the lysosomes because their membrane contains a proton pump that actively pumps H^+ ions (protons) into the lysosome, lowering the pH and activating the enzymes. If a lysosome should rupture and release its contents into the cytoplasm, where the pH is around 7.2, the enzymes would be immediately inactivated.

Lysosomes must deal with any foreign substance brought into the cell. A large variety of molecules that the cell ingests through endocytosis are degraded by lysosomal enzymes, and the debris expelled by exocytosis. The role of lysosomes is particularly obvious in macrophages, cells of the immune response that actively patrol the body and ingest bacteria when they are encountered (Fig. 5.15, *right side*). Lysosomes also perform important housekeeping tasks: as organelles and macromolecules in the cell age and break down, the lysosomes capture and destory them, and recycle their basic components (Fig. 5.15, *left side*).

Peroxisomes are formed by budding from the smooth ER. These small vesicles are primarily responsible for the degradation of hydrogen peroxide, a strong oxidizing agent formed as a byproduct of many metabolic reactions. Peroxisomes contain large amounts of an enzyme, catalase, which breaks hydrogen peroxide down into water and oxygen.

Mitochondria

Mitochondria are round to rod-shaped organelles that are easily recognized under the microscope (Figs. 5.3, 5.16a, and transparency insert). They contain the enzymes that perform an absolutely critical function, that of cellular respiration. This process, discussed more fully in Chapter 13, results in the net release of energy. This energy is captured in the mitochondria and applied to the synthesis of the cells' main energy molecule, adenosine triphosphate (ATP). Ten to 100 or more mitochondria may be present in a single cell. Metabolically active cells (such as muscle cells and sperm cells), with their higher energy demands, contain larger numbers of mitochondria.

Unlike the other organelles we have discussed, mitochondria have two membranes (Fig. 5.16b). The outer membrane simply envelops the organelle; the inner

Figure 5.15

A diagram of lysosomal action. On the right, a bacterium is being digested. The process of endocytosis brings the bacteria into the cell; the endocytotic vacuole fuses with several lysosomes, and the bacterium is digested. The left side of the diagram shows the cell carrying out its housecleaning functions. Organelles, membranes, or other cellular structures that have aged or are damaged are engulfed by autophagic vacuoles. These vacuoles fuse with lysosomes, and the contents of the vacuole are digested. Many of the products of this digestion will be recycled by the cells.

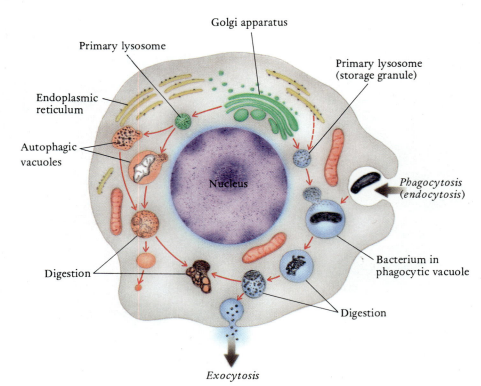

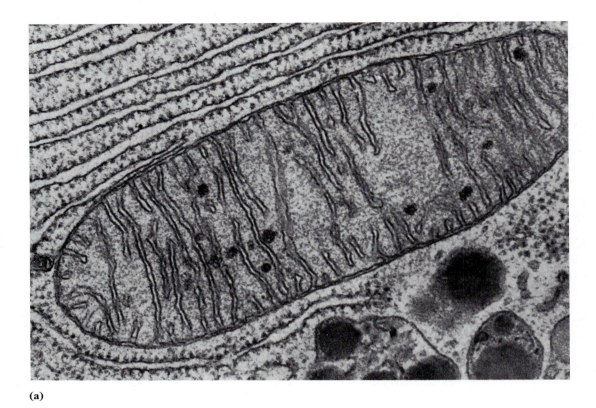

(a)

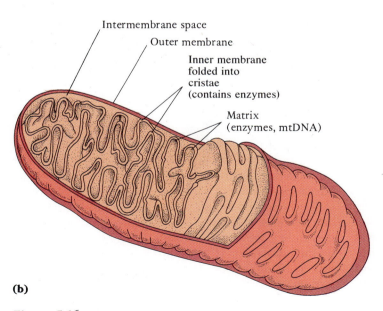

Intermembrane space

Outer membrane

Inner membrane folded into cristae (contains enzymes)

Matrix (enzymes, mtDNA)

(b)

Figure 5.16

The mitochondria are the powerhouses of the cell. (a) An electron micrograph of a mitochondrion, showing the outer surrounding bilayer membrane and the internal bilayer, invaginated to form the characteristic cristae. (b) Diagram of the mitochondrion showing the two membrane system. The reactions resulting in ATP synthesis take place on the surface of the inner membrane.

membrane forms a maze of projections (**cristae**) to the interior of the mitochondria. These cristae may simply represent additional surface area upon which the complex set of reactions involved in synthesizing ATP may occur.

But perhaps the most surprising feature of mitochondria is the presence of a circle of DNA, called **mitochondrial DNA (mtDNA)**, that directs the synthesis of several mitochondrial proteins. Why would an organelle be carrying its own DNA, transmitting it from generation to generation independently of the DNA in the cell nucleus? In fact, in the majority of organisms studied, mtDNA is maternally inherited—passed on from females to their offspring in the fertilized egg. In the overwhelming majority of species, male mtDNA is not passed on to the next generation, probably because paternal mitochondria in the sperm either do not enter the egg upon fertilization or are selectively destroyed if they do.

The presence of DNA in the mitochondrion became even more difficult to comprehend when it was discovered that many of the proteins necessary to the mitochondrion are encoded in the nuclear DNA. The mitochondrion, therefore, is not even self-sufficient—it requires help from the nucleus. How can we explain the presence of mtDNA? We will return to this question in the final part of this chapter, when we consider the origin of the eukaryotic cell.

The Nucleus

We come now to the feature originally used to define the eukaryotes, a group to which all animals belong. The nucleus is perhaps the most prominent feature of the cell when viewed under a microscope (Fig. 5.3 and transparency insert). Generally, it is a somewhat flattened sphere, located roughly at the center of the cell. It represents anywhere from 10 to 90% of the total cell volume.

The nucleus is surrounded by a **nuclear envelope** consisting of a pair of membranes (Fig. 5.17). The inner membrane circumscribes the nucleus; the outer membrane surrounds the nucleus, but with frequent gaps. Detailed photographs of the nuclear surface show that it is studded with tiny nuclear pores, where the outer and inner nuclear membranes may fuse. These pores are the only site of contact and material exchange between the nucleus and the cytoplasm.

With the exception of small fragments or circles of DNA present in mitochondria and in some other organelles, the DNA of the cell is contained entirely within the nucleus. When viewed through a light microscope, the nuclear DNA in a cell that is not dividing appears as a network of **chromatin** (Fig. 5.17). Prior to a cell division, the chromatin shortens and coils into a fixed number of long, continuous chains, the **chromosomes** (Fig. 5.18).

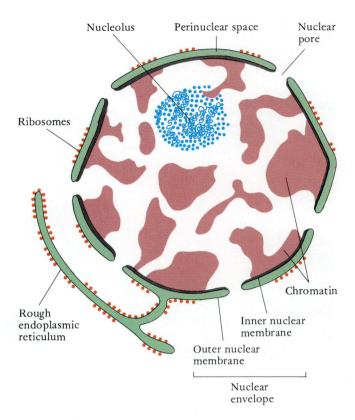

Figure 5.17

A diagram of the nucleus. Note that the outer nuclear membrane is continuous with the rough ER. The chromatin contains DNA and associated proteins; at certain stages of the cell cycle, this material condenses into chromosomes.

Each chromosome is a single, continuous double helix of DNA. Human chromosomes, by no means the largest ones found in the animal kingdom, contain on average from 100 to 300 million base pairs each. The packaging of such a long and complex thread of DNA (which, if stretched out, would measure 5 cm) into regular, discrete, and well-behaved chromosomes is only possible through the concerted action of a number of proteins and enzymes. In effect, the genetic material of the eukaryotic cell is organized around a scaffold of structural proteins, the **histones**. The amino acid sequence of these histones has undergone very little change throughout the evolution of the eukaryotes. Histones are small proteins (approximately 100 amino acids long) containing a large number of positively charged amino acids. These amino acids cause the histones to bind to the negatively charged DNA double helix. The DNA wraps twice around a complex formed by several histones, giving rise to a bead-like structure known as the **nucleosome**. Several nucleosomes in turn will associate into regular **chromatin fiber structures** (Fig. 5.18b). These fibers condense even further, eventually

giving rise to the characteristic packing of the eukaryotic chromosome (Fig. 5.18a). The chromosomes are stored and copied and their information transcribed entirely in the nucleus. The two thin membranes that enclose the nucleus are walls that the DNA never breaches. Like a captive for life, the chromosomal DNA's only contact with the outside world is through messages sent, appropriately enough, in the form of messenger RNA (mRNA). The nucleus is therefore primarily involved with the organization and maintenance of DNA, and with the appropriate and timely transformation of genetic instructions into messenger RNA. The ways in which these feats are accomplished will be described in detail in Chapters 6 and 7.

A look inside the nucleus reveals, in addition to the chromosomes, a discrete dark region, the **nucleolus** (see Fig. 5.17). This structure is the site of ribosome synthesis. Ribosomes are composed of a mixture of proteins and a special type of RNA (ribosomal RNA); they are made in parts in the nucleus and are then exported and assembled into their final form in the cytoplasm. The nucleolus is composed of small regions of one or more chromosomes, the **nucleolar organizer**, where the instructions and materials for ribosome assembly are contained.

Figure 5.18

(a) A scanning electron micrograph of a single chromosome of a hamster cell just before cell division, when the threads of DNA are tightly coiled. (b) A diagram showing the packaging of the genetic material in a chromosome (as it appears during cell division) under increasing magnifications.

(a)

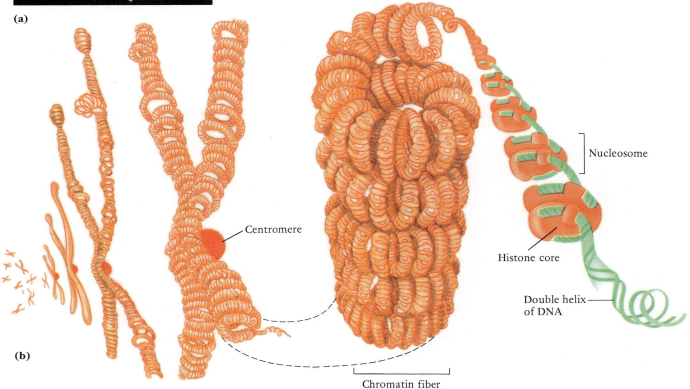

(b)

Nucleosome

Centromere

Histone core

Double helix of DNA

Chromatin fiber

The Cytoskeleton

We have, up to now, paid little attention to the behavior of cells. But cells are capable of movement. Think of the concerted swimming of a single-celled eukaryote (such as *Paramecium*) or of the migration of cells within the embryo that takes place during development (Chap. 20).

The movement of cells, as well as the transport of certain structures (chromosomes, organelles, and vesicles) within a cell, is made possible by a protein lattice that extends throughout the cell: the **cytoskeleton** (Fig. 5.19). This lattice, composed of three basic types of protein fibers (or filaments) makes it possible for cells to move across a surface, change shape, and divide.

The three types of filaments comprising the cytoskeleton can be distinguished by their appearance, as well as by their chemical composition. All of them are formed by the polymerization of different protein monomers. The filaments themselves are dynamic entities: they polymerize and dissociate under appropriate physiological conditions, and in response to the needs of the cell. The filaments that comprise the eukaryotic cytoskeleton are the **microtubules**, the **actin filaments**, and the **intermediate filaments**. Below, we briefly describe the role and basic characteristics of each filament type.

Microtubules

These filaments are involved in the beating of cilia, projections on the surface of the cells that create orderly fluid flow over the surface of the cell, and flagella, the tail-like structures present in sperm cells and many unicellular protozoans (see Fig. 11.30). Each individual microtubule is formed by an interlocking set of 13 tubulin molecules that

Figure 5.19
Elements of the cytoskeleton can be seen in this micrograph of a fibroblast cell. Microfilaments are stained in red and microtubules are stained in green.

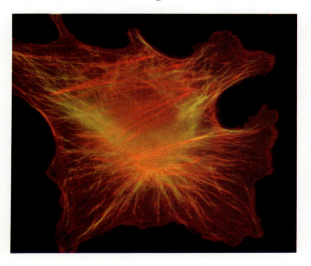

aggregate to form an elastic braid. When seen in cross section, a cilium is in fact an intricate arrangement of pairs of microtubules, forming a well-known "9 + 2" pattern. The individual elements in the 9 + 2 arrangement can slide relative to each other, resulting in the organized beating of the entire structure.

Microtubules are also the main component of **centrioles**, organelles involved in the formation of the mitotic spindle along which chromosomes travel during mitosis (see Ch. 6). Curiously, the centrioles in a cell almost always arise by duplication from a previous pair of centrioles. Coupled with the presence of DNA in the centriole itself, this mode of "reproduction" by duplication has led to the suggestion that centrioles may be semiautonomous organelles (like the mitochondria) trapped inside the eukaryotic cell.

Finally, microtubules exist as cytoplasmic filaments, extending throughout the cell. These filaments act as anchors for certain structures (Golgi apparatus and the ER). They are also involved in the rapid movement of mitochondria and vesicles within the cell.

Actin Filaments

These filaments are formed by the polymerization of **actin** monomers. Individual filaments are connected by cross-linking proteins into a vast network stretching throughout the entire cell, directly underneath the cell membrane. This complex, flexible web is actually anchored to the cell membrane by a series of transmembrane proteins, and can thus induce a change in the shape of the cell. As in the case of muscle contraction, the movement of actin filaments involves interactions with a different protein, myosin (see Ch. 11). The actin network in the cell is primarily responsible for the movement of cells along a substrate, for the mixing of components within the cytoplasm ("cytoplasmic streaming"), and for the changes in cell shape in phagocytosis.

Intermediate Filaments

These extremely strong filaments are formed by the braiding together of four distinct kinds of intermediate filament proteins. Within the cell, the network of intermediate filaments appears to be cradling the nucleus, and sends out slight projections to the cell membrane. Interestingly, the nucleus itself appears to have a "nucleoskeleton" of its own, composed mainly of a specific class of intermediate filaments. Unlike the actin and microtubule networks, which are involved in specific alterations of cell shape, the intermediate filaments serve primarily as structural support for the cell and the nucleus.

THE EVOLUTIONARY ORIGIN OF THE EUKARYOTIC CELL

Complex biological features, such as the eukaryotic cell, pose a particular challenge to evolutionary biology. The cell, with its complex and delicate orchestrating of biochemical reactions, provides clear advantages to its owner. Although multicellular prokaryotic organisms do exist, the most important corollary of multicellular life—the specialization of particular cell types—is only fully possible in the eukaryotic world. This ability to specialize probably arises from the very complexity of the eukaryotic cell and makes possible the evolution of tissues and organs, leading to the diversity of eukaryotic organisms.

But how might the eukaryotic cell have first evolved? What, if any, would have been the evolutionary advantages of a protoeukaryote? In short, how can a structure as complex as the eukaryotic cell evolve by natural selection?

The modern eukaryotic cell still preserves traces of its evolutionary past. Perhaps the most interesting feature in this respect is the mitochondrion. Here, inside the cell, is an organelle about the size and shape of a bacterium. It contains its own DNA (mtDNA) and its own ribosomes, which, unlike the ribosomes in the cell cytoplasm (but like the ribosomes of many bacteria), are sensitive to certain antibiotics. Its double membrane is unlike any other membrane system in the animal cell. In many ways, the mitochondrion appears like a bacterium caught inside a cell.

The Endosymbiotic Theory

The notion that the mitochondrion might have once been a free-living prokaryote forms the basis of the **endosymbiotic theory** for the origin of the eukaryotic cell (Fig. 5.20). In its simplest form, this theory argues that many features of the eukaryotic cell are the result of a permanent partnership, or symbiosis, between two or

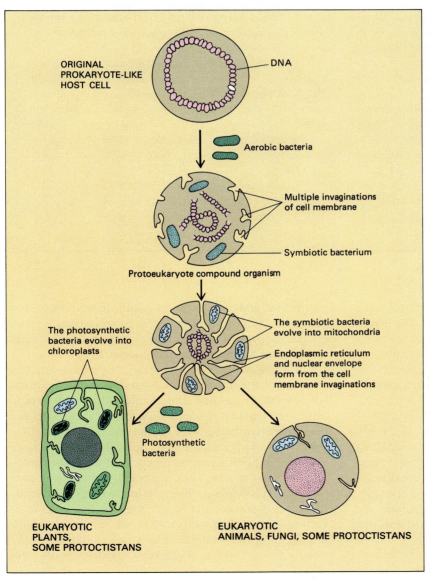

Figure 5.20
A diagram for the origin of the eukaryotic cell. The mitochondria, and the chloroplasts of plant cells, as well as several other organelles (not shown) are seen to arise endosymbiotically. In contrast, other features of the eukaryotic cell, such as the nucleus and the ER, evolve for the first time in the early proto-eukaryotic cell.

ORIGINAL PROKARYOTE-LIKE HOST CELL

DNA

Aerobic bacteria

Multiple invaginations of cell membrane

Symbiotic bacterium

Protoeukaryote compound organism

The photosynthetic bacteria evolve into chloroplasts

The symbiotic bacteria evolve into mitochondria

Endoplasmic reticulum and nuclear envelope form from the cell membrane invaginations

Photosynthetic bacteria

EUKARYOTIC PLANTS, SOME PROTOCTISTANS

EUKARYOTIC ANIMALS, FUNGI, SOME PROTOCTISTANS

more prokaryotic (or at least non-eukaryotic) organisms. There are many examples of symbiosis in the natural world: your healthy digestion depends on the colonies of bacteria living permanently in your intestines. You provide them with a safe haven and a constant food supply; they help you digest, and they synthesize several critical components (including vitamin B_{12}).

The eukaryotic cell may be the product of several such successful partnerships. The evidence for the endosymbiotic origin of mitochondria is the most compelling: the protomitochondrion was some kind of aerobic bacterium, capable of oxidizing glucose to carbon dioxide and water. Certain modern bacteria are thought to be good models for the protomitochondrion: *Paracoccus* and *Bdellovibrio* are two possibilities. The partnership may originally have been beneficial to both partners, but not essential. Later, the association evolved to become more

permanent, and the protomitochondrion would have lost the cell wall characteristic of most bacteria; certain genes encoding crucial mitochondrial proteins would be transferred from the mtDNA to the nuclear DNA.

There is some evidence for the endosymbiotic origin of other eukaryotic organelles. Flagella and cilia, the small, whiplike structures that allow single-celled eukaryotes to move (see Fig. 11.30), are thought to have evolved from a permanent symbiosis with spiral-shaped bacteria called spirochetes. In plants, the chloroplast—the main organelle involved in the photosynthetic capture of sunlight—also contains its own DNA, suggesting a free-living past.

The endosymbiotic theory is an exciting evolutionary idea. If it is correct, the theory suggests that even in its earliest form, the protoeukaryotic cell was already a functioning entity that would have been selected by the evolutionary process. The endosymbiotic model proposes that

(a) Modern scanning electron microscope.

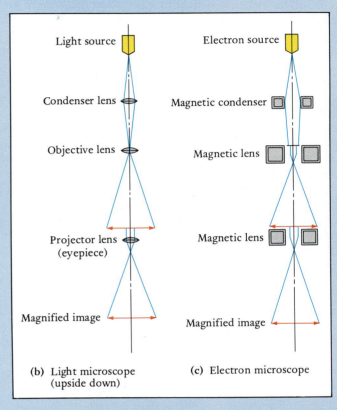

Light source

Condenser lens

Objective lens

Projector lens
(eyepiece)

Magnified image

(b) Light microscope
(upside down)

Electron source

Magnetic condenser

Magnetic lens

Magnetic lens

Magnified image

(c) Electron microscope

the early eukaryotic cell was not so much a single cell as a colony of associated prokaryotes, permanent partners in the most successful evolutionary venture seen thus far.

The Direct Filiation Theory

We do not want to leave you with the impression that all biologists accept this account of eukaryotic origins nor that all features of the eukaryotic cell arose by endosymbiosis. The main alternative is known as the **direct filiation (descent) theory**: it argues that the main features of the eukaryotic cell derive from the differentiation and compartmentalization of the cytoplasm in the ancestor of eukaryotes. In other words, an ancestral prokaryote gave rise to the first eukaryote, which contained few or no organelles. The organelles developed gradually in this early protoeukaryote, culminating eventually in the modern eukaryotic cell.

The eukaryotic nucleus is the one feature that has not been satisfactorily accounted for by the endosymbiotic hypothesis, which suggests that both endosymbiotic and direct filiation theories may be necessary to account for the evolution of the eukaryotic cell. According to the direct filiation theory, the nuclear material came to be surrounded by membrane that invaginated from the cell surface (see Fig. 5.20).

The nucleus represents a unique, novel feature of eukaryotes. In it is stored the very material that insures the continuity of life—DNA, that unbroken chain of information that unites all living and extinct organisms on Earth. It is to the features of the chain, and to the process of heredity—the transmission of genetic information from parents to offspring—that we now turn.

Summary

1. The cell enables organisms to concentrate and retain the molecules necessary for life to proceed. The cell exercises control over concentrations by (1) pumping molecules in and out, (2) using the product of one reaction as the substrate of the next in a metabolic pathway, and (c) sequestering the products of reactions into compartments within the cell.

2. Cells were discovered in the early 17th century when primitive microscopes came into use. Nearly 200 years later biologists recognized that all organisms are composed of cells or cell products and that all cells come from preexisting cells. This concept is known as the cell theory.

3. Prokaryotic cells differ from eukaryotic cells in being smaller and lacking a definite nucleus and internal membrane-bound organelles. Such cells are not capable of the types of specialization that occur among eukaryotic cells. Algae, protozoans, fungi, plants, and animals share the eukaryotic cell.

4. The eukaryotic cell is enclosed in a phospholipid bilayer that is studded with many types of protein that perform many functions: anchoring the cytoskeleton, recognizing and binding with specific hormones, identifying the cell to other cells, and assisting in the transport of materials into and out of cells.

5. Some materials move in and out of cells by diffusion, which is the net movement of a substance from an area of high concentration to an area of low concentration. Osmosis is the diffusion of water through selectively permeable membranes that allow the passage of water molecules, but not of some solute molecules. Other materials can be moved into and out of the cell against their concentration gradients by active transport. The sodium-potassium pump is an example of a mechanism of active transport. Large molecules can be transported across the cell membrane by endocytosis and exocytosis. In these processes, the molecules become enclosed in a vesicle that unites with the cell membrane and then discharges on the other side.

6. Many organelles lie in the cytoplasm. The smooth endoplasmic reticulum synthesizes the phospholipids that make up the membrane. The rough endoplasmic reticulum and its associated ribosomes are sites of protein synthesis. The Golgi apparatus sorts and packages proteins for export, or for transport to other parts of the cell. It is particularly well developed in secretory cells. Lysosomes contain enzymes that digest foreign substances brought into the cell and destroy aged and broken down organelles. Peroxisomes degrade hydrogen peroxide, a toxic material formed as a byproduct of many intracellular reactions. Mitochondria contain the enzymes essential for the oxidative phase of cellular respiration, when most of the energy in food molecules is trapped as ATP. Mitochondria contain some of their own DNA (mtDNA) that is passed on from generation to generation only through the mitochondria in the egg cell.

7. Apart from mitochondrial DNA, all of the DNA in the cell is contained within chromosomes in the nucleus. The nucleus organizes and maintains DNA and controls the timely transfer of genetic information to the cytoplasm, in the form of messenger RNA (mRNA).

8. Eukaryotic cells contain a cytoskeleton composed of three distinct classes of protein filaments: microtubules, actin filaments, and intermediate filaments. These filaments hold the nucleus and organelles in place, are responsible for the transport of materials within the cell, and permit cell movement and migration.

9. The eukaryotic cell evolved from a prokaryote-like precursor. Many of its organelles, including mitochondria, cilia and flagella, and (in plant cells) chloroplasts, may have evolved when a protoeukaryotic cell captured and established a symbiotic relationship with other prokaryotic cells that had the features represented by the organelles. The eukaryotic nucleus cannot be explained in this way, and it may have evolved by the differentiation and compartmentalization of the cytoplasm in the ancestor of eukaryotes.

References and Selected Readings

Bretscher, M.S. The molecules of the cell membrane. *Scientific American* 253 (Oct. 1985):100−109. An excellent review of the structure and many functions of the cell membrane.

Darnell, J., H. Lodish, and D. Baltimore. *Molecular Cell Biology.* New York: W.H. Freeman and Co., 1986. A comprehensive and well-illustrated text on cell biology.

de Duve, C. *A Guided Tour of the Living Cell.* New York: Scientific American Library, 1984. A superbly illustrated tour through the living cell.

Fawcett, D.W. *Bloom and Fawcett: A Textbook of Histology,* 11th ed. Philadelphia: W. B. Saunders, 1986. The first chapter of this standard histology text has a detailed account of the structure of the eukaryotic cell.

Margulis, L., and K.V. Schwartz. *The Five Kingdoms: An Illustrated Guide to the Phyla of Life on Earth,* 2nd ed. New York: W.H. Freeman and Co., 1988. An excellent review of the differences between prokaryotes and eukaryotes and of the evolution of the five major groups of organisms.

Rothman, J.E. The compartmental organization of the Golgi apparatus. *Scientific American* 253 (Sept. 1985):74–89. A good discussion of the structure and function of the Golgi apparatus.

Sagan, D., and L. Margulis. Bacterial bedfellows. *Natural History* March 3 (1987):26–33. A review of the endosymbiotic origin of the eukaryotic cell.

6

Genetics I: The Fundamental Observations of Heredity

(Left) Albinism, a trait controlled by a single gene, occurs frequently in vertebrates. Here, an albino racoon (Procyon lotor) *is shown with its normally-colored parent. (Above) A close-up of an albino racoon. This phenotype arises from a heritable defect in the synthesis of pigment.*

In the following four chapters, we explore the principles of genetics, examining the process of heredity from many different vantage points and at many different scales. This chapter will deal with basic observations about the nature of heredity. We will take a look at the historical development of the science of genetics, beginning with the work of Gregor Mendel. We will also examine the material basis of inheritance—the chromosomes—and their behavior in cell division. In Chapter 7, we will focus on the modern face of genetics by turning to the molecular aspects of heredity. We will consider the nature of genes and how they control the synthesis of proteins. In Chapter 8, we will discuss the origin of inherited variation by mutation and recombination. Finally, in Chapter 9, we examine the dynamics of genes in populations—the factors that produce change in the genetic composition of populations, and the forces responsible for long-term evolutionary change.

THE FUNDAMENTAL OBSERVATION: LIKE BEGETS LIKE

Nature, in some ways, is remarkably orderly and consistent. Japanese beetles give rise to other Japanese beetles, oak trees to new oak trees, people to new people. Over the time scales that we can observe, the boundaries that separate one species from another are never breached.

If we look within a single species, we again see certain regularities: offspring, for example, resemble their parents more than they resemble other unrelated individuals. But the resemblance is often incomplete—in the majority of animal species, offspring are not identical copies of either parent. On occasion, offspring may show traits present in neither parent, but shared perhaps with a grandparent. Siblings, too, resemble one another, but only in the case of monozygotic (single egg) twins are they truly identical. As biologists, we must always ask the same question: what is the material basis of these similarities between parents and offspring? What are the causes of variation among individuals? How can both similarity and variation arise from a single underlying mechanism?

Let us begin our analysis by making a fundamental point: similarities between parents and offspring arise from a variety of sources. Similarity *per se* is not proof of genetic inheritance. There is, for example, a very strong correlation between the religious affiliation of parents and that of their offspring. Yet no one, we trust, would argue that religious preference is genetically inherited. Shared environments and cultural norms often produce very strong similarities between human parents and children. Similarly, in the case of other animals, we must set strong standards of proof before we declare a particular trait to be "genetic." The distinction between familial correlation and genetic inheritance should be kept firmly in mind.

AGRICULTURE AND DOMESTICATION: THE EARLIEST GENETICS

Long before the physical basis of genetics came to be understood, our species was already engaged in several experiments in applied genetics. Agriculture, thought to originate in the crescent between the Tigris and Euphrates Rivers some 12,000 years ago, began when humans first selected those grasses that seemed to yield edible grains year after year. Similarly, we have long managed certain animals, first by domesticating them and then by carefully breeding only certain individuals who displayed characteristics we valued, in the process known as artificial selection (see p. 29, Fig. 2.20). The earliest practitioners of artificial selection grasped some of the fundamental aspects of genetics and understood that future generations could be shaped by arranging the parental matings. They also realized that different traits of organisms could be combined by manipulating the crosses that took place. Thus, for example, a rapid-growing strain of corn could be crossed with a relatively drought-resistant strain of corn, and at least some of the offspring would prove to be both fast growing and drought resistant. The whole of human history has been profoundly shaped by our ability to successfully domesticate and manipulate other species. Beginning 12,000 years ago, long before the discovery of DNA, farmers, pastoralists, and nomads understood something fundamental about the processes of inheritance.

MENDEL AND THE LAWS OF INHERITANCE

Mendel's Insight: Quantify Variation

The earliest thread in the complex tapestry of modern genetics traces back some 135 years, to a small monastery in what is now Brno, Czechoslovakia. There, Gregor Mendel, a priest who had also been trained as a high-school physics teacher, took a strong interest in the problem of how new varieties, and eventually new species, arise (Fig. 6.1). He was particularly interested in testing the results of crosses between different "varieties," or strains, of garden peas. He had chosen garden peas because a large number of different varieties existed and were readily available. Pea varieties appeared to breed "true"—the offspring of a mating between two plants of the same variety always resembled the parents. Finally, peas are capable of self-pollinating, but can also be pollinated experimentally using pollen from a different variety. Mendel painstakingly carried out an extensive series of crosses between varieties, carefully noting the characteristics of each offspring produced by a cross. Partly by design (and partly by good fortune) Mendel focused on a small suite of features (characters) that could be easily scored: (1) flower color, (2) seed coat color, (3) seed shape, (4) pod shape, (5) pod color, (6) flower position, and (7) stem length (Fig. 6.2).

Figure 6.1
Gregor Mendel (1822–1884) was the first to quantify the consequences of particular genetic crosses.

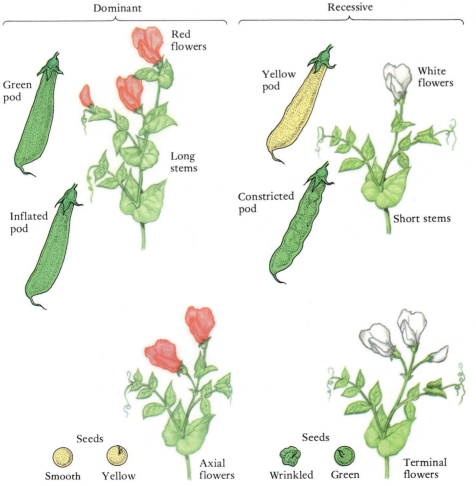

Figure 6.2
The characteristics of garden pea plants (*Pisum sativum*) studied by Mendel. Each trait exists in two alternative versions: a dominant and a recessive phenotype. A phenotype is a reflection of the underlying genetic basis of the traits.

The results that Mendel obtained form the basis of modern genetics. In his first set of experiments, Mendel performed **monohybrid crosses**—crosses between two varieties that differed only in a single character. In one such experiment, depicted in Figure 6.3, he crossed two varieties differing only in the shape of their seeds (smooth or wrinkled). Each of the offspring in the first filial (F_1) generation was scored for seed shape, and, surprisingly, all the seeds were smooth. The plants were then allowed to self-pollinate, letting the pollen of each flower fall on the stigma of the same flower. All 7324 offspring from those self-pollinated crosses, known as the second filial (F_2) generation, were scored for seed shape: 5474 were smooth, and 1850 were wrinkled. Two aspects of these results are worth noting. First, a feature that seemed to have disappeared (wrinkled shape) in the F_1 generation reappeared in the F_2 generation. Second, the ratio of smooth to wrinkled seeds was very close to 3:1 (2.96:1, to be exact). Mendel carried out six similar experiments, each time crossing two lines that differed by just one character—stem height, flower color, and so on. In every experiment, the F_1 generation was homogeneous, showing only one of the two character states (Table 6.1). In every experiment, the absent character state reappeared in plants in the F_2 generation, always in the mysterious 3:1 ratio.

Even before Mendel began to interpret the results of his experiments, he had already achieved a major breakthrough in the nascent science of genetics: *he had observed and quantified variation.* Mendel had set up a series of experiments that considered every individual specimen in the P (parental), F_1, and F_2 generations. His focus was not on essences or unchanging types, but on the actual appearance of individual plants.

Clearly, despite the disappearance in the F_1 generation of one of the two alternate character states, the "factor" allowing plants to produce wrinkled seeds had not disappeared, but was simply not *expressed* in F_1 plants. The smooth factor masked, or dominated, the wrinkled seed factor—in genetic terms, we refer to the factor producing smooth seeds as **dominant**, and to the factor producing wrinkled seeds as **recessive**. Equally importantly, Mendel had disproven the theory of blending inheritance, which

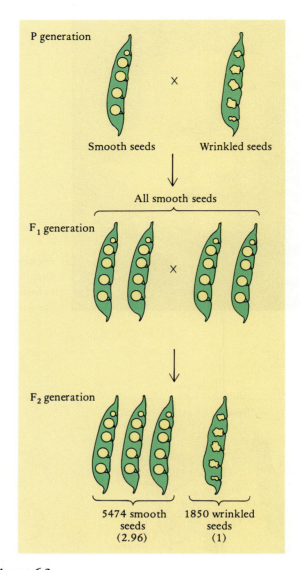

Figure 6.3

A monohybrid cross in which Mendel bred pea plants differing only in the shape of the seeds. The F_1 generation showed only smooth-seeded plants. When F_1 individuals are crossed, the recessive phenotype (wrinkled) reappears in the F_2 generation.

TABLE 6.1			
An Abstract of the Data Obtained by Mendel from His Breeding Experiments with Garden Peas			
Parental Characters	**F_1 Generation**	**F_2 Generation**	**Ratios**
Yellow seeds × green seeds	All yellow	6022 yellow : 2001 green	3.01 : 1
Smooth seeds × wrinkled seeds	All smooth	5474 smooth : 1850 wrinkled	2.96 : 1
Green pods × yellow pods	All green	428 green : 152 yellow	2.82 : 1
Long stems × short stems	All long	787 long : 277 short	2.84 : 1
Axial flowers × terminal flowers	All axial	651 axial : 207 terminal	3.14 : 1
Inflated pods × constricted pods	All inflated	882 inflated : 299 constricted	2.95 : 1
Red flowers × white flowers	All red	705 red : 224 white	3.15 : 1

argued that offspring were the result of a blending or mixing of fluids from either parent. Crossing a purple-flowered plant with a white-flowered plant did not result in offspring with all pink flowers. Instead, flower color or seed shape appeared to be determined by *particulate factors* that remained intact from one generation to the next.

The 3:1 ratio in the F_2 generation was harder to explain. Mendel carried out further crosses, allowing the F_2 plants to self-pollinate. He found that wrinkled seeds gave rise only to wrinkled-seed F_3 plants. In contrast, smooth seeds gave rise in some cases to only smooth-seeded plants, but in other cases to a mixture of smooth and wrinkled seeds, again in a 3:1 ratio. All of the wrinkled seeds bred "true," and $\frac{1}{3}$ of the smooth seeds also bred "true." In summary, the 3:1 ratio of smooth to wrinkled seeds could actually be broken down further: $\frac{1}{4}$ true-breeding smooth seeds, $\frac{2}{4}$ mixed-breeding smooth seeds, and $\frac{1}{4}$ true-breeding wrinkled seeds, or a 1:2:1 ratio.

In all of the monohybrid character crosses carried out by Mendel, he found that the 3:1 ratio in the F_2 generation could be broken down to a 1:2:1 ratio. The conclusions that Mendel drew seem almost obvious to us now, but they represented a startling revolution at the time of their discovery. Mendel argued that the reappearance of the "masked" character in the F_2 and F_3 generations must imply that each individual plant carries two copies of the "factor." In the cases where the plants bred "true," the two copies are identical; in the cases where a cross gives rise to mixed-character offspring, one (or both) parents must be carrying two different copies of the "factor" (one dominant copy and one recessive copy). Furthermore, Mendel argued that the parents produce both female and male **gametes** (reproductive or germ cells) each of which contains one of the pair of factors present in the parent. The first cell of the next generation is formed by the random union of two gametes, restoring two copies of every factor to every offspring.

The Basic Vocabulary of Genetics

We can now review and illustrate Mendel's conclusions, but we will use more modern terminology. The "factors" of which we have been speaking are referred to as **genes**; alternate versions of the same gene are known as **alleles**. In our original example, "smooth seed" and "wrinkled seed" are alleles of the gene controlling seed appearance. Alleles of the same gene occupy the same position, or **locus**, on the chromosomes. All cells in the pea plant (except for the gametes) contain two alleles of any given gene—we refer to such cells as **diploid** (Gr. *diploos*, double + *eidos*, form) cells. The gametes contain only a single copy of each gene, and are referred to as **haploid** (Gr. *haploos*, single) cells. The same situation exists in the vast majority of animal species: all cells in the body, also known as **somatic cells**, are diploid; the gametes (sperm and egg) are haploid.

A standard way of depicting the alleles of a particular gene is to use a capital letter—"*S*" for seed shape, for example—to represent the dominant allele (in this case, smooth coat) and a lower-case letter ("*s*") to represent the recessive allele (in this case, wrinkled coat). A diploid individual could therefore contain one of three possible combinations of alleles at the seed shape locus: *SS*, *Ss*, and *ss*. Plants that have two copies of the *same* allele are called **homozygotes** (Gr. *homos*, same). Plants that contain two different alleles for a given gene are said to be **heterozygotes** (Gr. *heteros*, different).

At this point, we need to focus our attention on the difference between the **genotype**, or the underlying genetic composition of an individual, and the **phenotype**, or the actual appearance of an individual. When we say a given plant is "*ss*" (or homozygous for "*s*"), we are referring to its genotype. The plant, of course, does not have its genotype tatooed on the leaves; we must infer it from the plant's appearance. When we say a plant has wrinkled seeds or purple flowers, we are describing its phenotype, which we can observe directly. As we will see in Chapter 22, the way in which genotypes are transformed into phenotypes is through the complex process of development.

For certain phenotypic traits, such as seed shape, the relationship between phenotype and genotype is fairly straightforward. Seed shape appears to be controlled by a single gene, and we can cautiously infer the genotype of an individual by looking at the phenotype and by carrying out certain crosses (see Fig. 6.3). As we mentioned earlier, when wrinkled seed plants are allowed to self-fertilize or when wrinkled seed plants are crossed with other wrinkled seed plants, the offspring always show wrinkled seeds. The diploid genotype of these plants is *ss*, and they can only produce one type of haploid gamete: *s*. When they self-pollinate, the diploid zygote (fertilized egg) they produce can only be *ss*, and their progeny will therefore only exhibit a single phenotype: wrinkled seeds. Note that even though we said that the *s* allele was recessive, the issue does not arise in this particular cross, since no *S* alleles are present. Dominance and recessivity are relative and can only be determined when more than one type of allele is present.

In certain cases, two alleles of a given gene may give rise to a heterozygote that is intermediate in appearance (Fig. 6.4). This suggests that neither of the alleles is dominant over the other; they are considered **codominant**. Despite the appearance of blending inheritance in the F_1 generation, subsequent crosses again confirm the particulate nature of the alleles.

The Law of Equal Segregation

If we allow all the smooth-seeded F_2 individuals to self-pollinate (or if we cross two smooth-seeded plants from the F_2 generation), they produce both smooth-seeded and

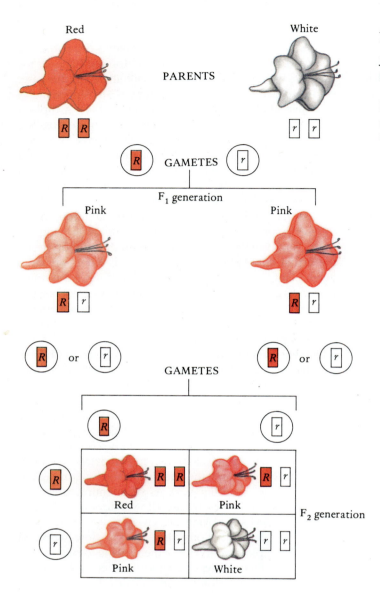

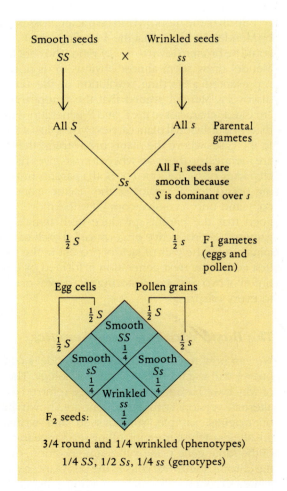

Figure 6.4

The alleles for red and white flowers are codominant in Japanese four-o'clock flowers; so heterozygous plants have flowers of intermediate color. The F_2 generation have a 1:2:1 phenotypic ratio: heterozygote flowers are pink, homozygous flowers breed true (red or white). The matrix showing the genotypes of the F_2 plants is known as the Punnett square, after the geneticist R. C. Punnett.

Figure 6.5

A monohybrid cross carried out over two generations, showing the phenotype, genotype, and gamete types produced.

wrinkled-seed plants (see Fig. 6.3). Why? Clearly, if all the smooth-seeded plants had the *SS* genotype, they would only give rise to smooth-seeded progeny. Some of the smooth-seeded parents, however, are heterozygotes, and have the *Ss* genotype (Fig. 6.5). Phenotypically, they cannot be distinguished from the *SS* plants—all of them are smooth seeded. But the heterozygotes can produce *two* kinds of gametes in equal proportions. On average, half of their gametes will carry the *s* allele; the other half, the *S* allele. When fertilization occurs, three kinds of new zygotes can be produced: *SS* (smooth seeded), *Ss* (smooth seeded) and *ss* (wrinkled seeded). Even though only two phenotypic states appear for this character, three different genotypic states underlie it.

The argument that the two alleles for a given gene separate when the gametes are formed, with one or the other allele going into each gamete, is known as Mendel's first law, the **law of equal segregation**. With very rare exceptions, $\frac{1}{2}$ of the gametes produced by the organism contain one of the alleles, and $\frac{1}{2}$ contain the other.

This law also allows us to make quantitative predictions about the outcome of particular crosses. For example, a cross between two *Ss* heterozygotes (two F$_1$ plants in Mendel's experiments, for example) can be analyzed as shown in Figure 6.5. A heterozygous plant produces two sorts of gametes in equal proportion: *s*-containing gametes and *S*-containing gametes. During reproduction, two gametes unite, and they do so at random with respect to the allele they contain—an *s* gamete will not preferentially unite with another *s* gamete. The probability of a particular union of gametes is simply a function of the relative proportion of each of the gametes present. In the case of a mating between heterozygotes, we expect that $\frac{1}{4}$ of the zygotes will have an *SS* genotype (smooth-seed phenotype), $\frac{1}{2}$ will be *Ss* heterozygotes (smooth-seed phenotype, because *S* is dominant), and $\frac{1}{4}$ will be *ss* homozygotes (wrinkled-seed phenotype). This should look familiar: it is the 3:1 phenotype ratio of smooth to wrinkled seeds that Mendel observed.

Mendel had succeeded in demonstrating that the plants in the F$_2$ generation that showed the dominant phenotype were in fact a mixture of two different genotypes: *Ss* and *SS*. This discovery was possible because pea plants can self-pollinate. The majority of organisms, however, cannot be crossed with themselves. In such cases, the genotypic basis of dominant phenotypes is investigated using a **test cross**, also called a **backcross**, in which individuals showing the dominant phenotype are crossed to homozygous recessive individuals (Fig. 6.6). If the individual with the dominant phenotype is a homozygous dominant (*BB*), the test cross with *bb* individuals will yield only heterozygous (*Bb*) progeny, all of which show the dominant phenotype (Fig. 6.6, *left*).

On the other hand, if the individual with the dominant phenotype is in fact a heterozygote (*Bb*), the test cross with *bb* will yield two types of progeny: homozygous recessives (*bb*), which show the recessive phenotype, and heterozygotes (*Bb*), which show the dominant phenotype. As you can see from Figure 6.6 (*right*), the two types of progeny will occur in equal numbers (a 1:1 ratio). Test crosses are now widely used to investigate the genetic basis of particular phenotypic traits.

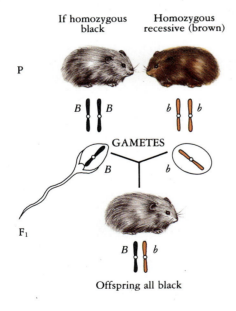

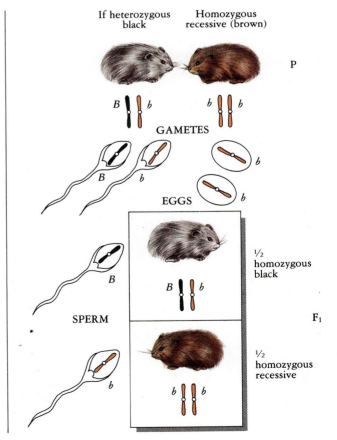

Figure 6.6
In this test cross, a dominant black guinea pig of unknown genotype is crossed with a homozygous recessive brown guinea pig. The outcome of the cross indicates whether the black guinea pig is homozygous, resulting in the outcome on the left, or heterozygous, with the outcome on the right.

The Law of Independent Assortment

The next series of experiments that Mendel undertook were more complex versions of his single-character crosses. He wondered what the outcome would be if he crossed plants that differed in two characters, such as seed coat color and seed shape. Such experiments are known as **dihybrid crosses**. Once again, Mendel began with lines that bred "true" for both characters (Fig. 6.7). We now know that means that the parent plants were homozygous for both of the characters. Mendel knew from previous experiments that the allele (S) for smooth seeds was dominant over the allele (s) for wrinkled seeds, and the allele (Y) for yellow seeds was dominant over the allele (y) for green seeds. Thus, one of the parent lines had an $SSYY$ genotype, the other line had an $ssyy$ genotype. The F_1 generation, as you might have predicted, was made up entirely of plants that produced smooth-yellow seeds; their genotype was heterozygous for both genes: $SsYy$. When the F_1 plants were crossed, they produced a set of off-

Figure 6.7

A dihybrid cross where parental plants differ in two traits: seed color and seed shape. Note the gamete types produced, the independence of the two traits, and the characteristic 9:3:3:1 phenotypic ratios obtained in the F_2 generation.

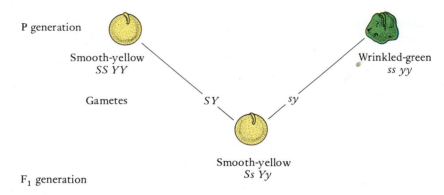

P generation

Smooth-yellow
$SS\ YY$

Wrinkled-green
$ss\ yy$

Gametes SY sy

Smooth-yellow
$Ss\ Yy$

F_1 generation

Gametes

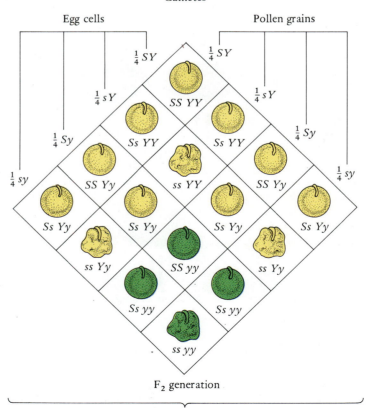

Egg cells   Pollen grains

$\frac{1}{4}SY$ $\frac{1}{4}SY$

$\frac{1}{4}sY$ $\frac{1}{4}sY$

$\frac{1}{4}Sy$ $\frac{1}{4}Sy$

$\frac{1}{4}sy$ $\frac{1}{4}sy$

$SS\ YY$

$Ss\ YY$ $Ss\ YY$

$Ss\ YY$ $ss\ YY$ $SS\ Yy$

$Ss\ Yy$ $Ss\ Yy$ $Ss\ Yy$ $Ss\ Yy$

$ss\ Yy$ $SS\ yy$ $ss\ Yy$

$Ss\ yy$ $Ss\ yy$

$ss\ yy$

F_2 generation

$1/16(SS\ YY) + 2/16(Ss\ YY) + 2/16(SS\ Yy) + 4/16(Ss\ Yy) = $ 9/16 smooth-yellow seeds
$1/16(SS\ yy)\ + 2/16(Ss\ yy) = $ 3/16 smooth-green seeds
$1/16(ss\ YY)\ + 2/16(ss\ Yy) = $ 3/16 wrinkled-yellow seeds
$1/16(ss\ yy) = $ 1/16 wrinkled-green seeds

spring that occurred in a characteristic phenotypic ratio: $\frac{9}{16}$ of the F_2 plants had smooth-yellow seeds, $\frac{3}{16}$ had wrinkled-yellow seeds, $\frac{3}{16}$ had smooth-green seeds, and $\frac{1}{16}$ had wrinkled-green seeds. Can we account for this more complex 9:3:3:1 phenotypic ratio?

Once again, to understand phenotypic ratios, we must examine the underlying genotypic ratios. Mendel argued that the two characters continued to obey the first law, but in addition the two characters independently assort (segregate) into gametes. Let us examine the gametes that can be produced by an *SsYy* "double heterozygote." Each gamete will receive one of the two alleles for each of the two characters. The possible gametes are therefore *SY, sY, Sy,* and *sy*. As seen in Figure 6.7, these gametes can combine into a variety of zygotes: *SSYY; SsYY; SSYy; SSyy; SsYy; Ssyy; ssYY; ssYy,* and *ssyy*. Because certain alleles are dominant, only four phenotypes appear:

1. Smooth-yellow seeds (from genotypes *SSYY; SSYy; SsYY,* and *SsYy*),
2. Wrinkled-yellow seeds (from genotypes *ssYY* and *ssYy*),
3. Smooth-green seeds (from genotypes *SSyy* and *Ssyy*),
4. Wrinkled-green seeds (from genotype *ssyy*).

The fact that each of the characters behaves independently of the other makes each of the four combinations of seed shape and color possible. This is Mendel's second law, the **law of independent assortment**: during the formation of gametes, the segregation of alleles of a gene pair is *independent* of the segregation of all other allelic pairs. This fundamental genetic insight came about solely because Mendel was lucky enough to select characters that are truly independent of each other. As we discuss in Chapter 8, the actual physical organization of genes sometimes results in violations of Mendel's second law.

By continually testing the consequences of different crosses, making predictions about the outcomes, chronicling the results, and comparing them to his predictions, Mendel had founded the science of genetics. Without using the actual words, he had developed the concepts of phenotype and genotype, and had inferred that independent, particulate factors were passed on from one generation to the next. This process of inheritance was complex but orderly (see Focus 6.1). Mendel's genius lay in his ability to grasp and quantify the rules of inheritance.

THE MATERIAL BASIS OF MENDELIAN GENETICS

Why Was Mendel Ignored?

Two poignant historical ironies will always accompany any chronicle of Mendel's achievement. The first of these involves Charles Darwin. From the publication of *On the Origin of Species* until the time of his death, Darwin wres-

tled, unsuccessfully, with the genetic mechanisms that generated and maintained variation, making the process of natural selection possible. The prevailing explanations for inheritance all assumed some form of blending inheritance. Under such models, it was difficult to understand how a favorable genetic variation that arose in a single individual would ever spread through a population rather than simply being diluted away by the blending process. Darwin had no useful theory of inheritance to put forth, and it remained a serious stumbling block to the acceptance of the theory of evolution. Mendel's paper appeared in 1865, six years after the 1st edition of *The Origin*. It now appears that Darwin owned a copy of Mendel's paper, but nowhere in his extensive journals or diaries does he mention having read it—or if he read it, it did not make an impression. Right there, in Darwin's library, lay the key to the genetic underpinnings of the evolutionary process. And there it would remain, unread or unappreciated by Darwin.

The second irony concerns the total and complete oblivion into which Mendel's laws immediately sank. Historians of science have tried to account for this fate in numerous ways: the results were published in an exceedingly obscure journal, scientific communication was poor, France and England were the centers of biological inquiry at the time and scientists there paid little attention to work carried out in other countries. But perhaps the most compelling explanation involves the absence of an actual mechanism or observable agent that would have explained Mendel's hypotheses. What, after all, were these "factors" of which Mendel spoke? Why had no one seen them? In the absence of any physical manifestation of these particulate factors, the scientific community was reluctant to embrace these new and radical notions about the nature of heredity.

The Rediscovery of Mendel

It was not until the early 1900s that the stage was set for the rediscovery of Mendel. The cell theory was well in place by the 1880s, and biologists recognized that some component, likely present in all cells, must be the physical basis of heredity. Aware that the gametes were the cell type that began a new generation, much of the attention focused on gamete structure. One of the first observations concerned the different sizes of male and female gametes (Fig. 6.8). The female gamete—the egg—contained a nucleus, as well as a large amount of cytoplasm. In contrast, the male gamete—the sperm—was little more than a nucleus with a tail, with very little surrounding cytoplasm. Biologists realized that both gametes made an equal contribution to the new zygote, and therefore concluded that the stuff of heredity was unlikely to reside in the cytoplasm. In contrast, the nucleus of both sperm and egg were of roughly equal size, making it the most likely place of residence of any hereditary factors.

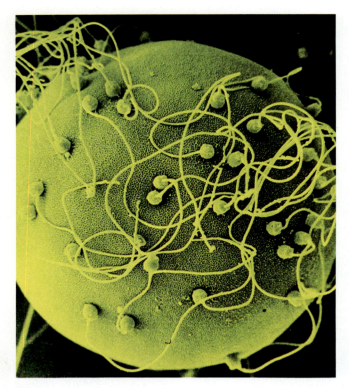

Figure 6.8
Scanning electron micrograph (2100×) of sperm swarming around a surf clam egg. Note the dramatic difference in size between the two gametes. Only a single sperm will successfully penetrate the egg and fertilize it.

Much of this early work involved careful microscopic observations and experimentation with a variety of dyes that stained specific components of the cell. One feature that attracted notice was a set of fibers in the nucleus that could be seen under the light microscope. These structures could not be clearly resolved in all cells at all times. They appeared at certain stages in the life of the cell, later to diffuse into a web of dark material. Most important, the same number of these structures, later to be called **chromosomes** (Gr. *chroma*, color + *soma*, body), were present in all somatic cells of an individual and in all individuals of a given species. The number of chromosomes could vary from one species to the next (Fig. 6.9), but seldom varied within species.

All cells appeared to go through a stage in which a copy of every chromosome was made, followed by the

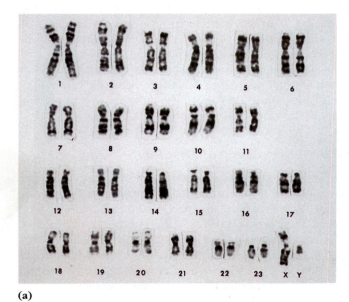

(a)

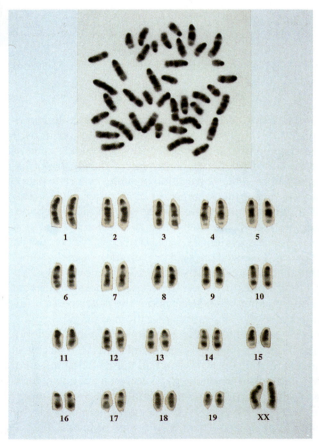

(b)

Figure 6.9
(a) The chromosomal complement, or karyotype, of a male chimpanzee (*Pan troglodytes*). Despite being our nearest living relative, this species has one more pair of chromosomes (23 autosomal pairs + 1 pair of sex chromosomes) than does *Homo sapiens* (22 autosomal pairs + 1 pair of sex chromosomes). Nonetheless, there is probably less than 1% overall difference between the genetic material of chimps and humans. (b) The karotype of a female house mouse (*Mus musculus*). The upper panel shows the chromosomes as they would appear in the cell nucleus, after staining. The lower panel shows the paired homologous maternal and paternal chromosomes, conventionally displayed in order of decreasing size, with the sex chromosomes (XX) displayed last. The number and shape of the chromosomes frequently vary from one mouse species to the next, but will seldom vary within a species.

Focus 6.1   *Gene Interaction*

The dihybrid crosses we examined earlier in this chapter involved two different, independent pairs of genes, each determining a separate phenotypic trait. The majority of phenotypic traits, however, are not controlled by single loci. Instead, they arise through the action and interaction of multiple genes, and are consequently referred to as **polygenic** traits. We will examine the genetics of polygenic traits, such as height in Chapter 9. For now, we direct our attention to some possible patterns of interaction that may occur among two (or more) genes acting upon the same phenotypic trait.

In a simple case, two genes may have an **additive effect** upon a phenotype. An example of this interaction involves the determination of kernel color in wheat, controlled by two different genes (Fig. a). The dominant allele at each gene provides for the same amount of red pigment in the kernel, and the intensity of the red color is directly proportional to the number of dominant alleles present. Thus, a plant that is homozygous dominant at both loci will have red kernels; conversely, a plant that is homozygous recessive (at both loci) will have white kernels. When a plant with red kernels (*AABB*) is crossed with a plant with white kernels (*aabb*), the F₁ offspring, which receive two dominant alleles (*AaBb*), all have pink kernels. When two plants with pink kernels are crossed (Fig. a), the F₂ plants fall into five color groups resulting from the presence of 4 (red), 3 (light red), 2 (pink), 1 (pale pink), and 0 (white) dominant alleles. As with other dihybrid crosses where the loci are not closely linked, the F₂ plants show the characteristic phenotypic ratio: 1 red:4 light red:6 pink:4 pale pink:1 white.

In certain cases, the interaction among loci controlling a polygenic trait is more complex. One such case involves the determination of comb shape in fowl (Fig. b). The shape of the fleshy red comb located above the beak is determined by two unlinked genes, *P* and *R*. Dominant and recessive alleles exist at both of these loci. The homozygous recessive condition (*pprr*) results in a "single" comb shape; the homozygous dominant condition (*PPRR*), or the simultaneous presence of a dominant allele at each of the two loci (*PpRr*), results in a "walnut" comb shape. A single dominant *P* allele (*Pprr*) produces a "pea-shaped" comb; a single *R* allele (*ppRr*) produces a "rose-shaped" comb. Crosses among individuals with different comb shapes reveal that although the loci are segregating in appropriate Mendelian fashion, their effect on the phenotype involves a more complex interaction. When a walnut-combed bird (*PPRR*) is crossed with one having a single comb (*pprr*), the resulting F₁ will be

heterozygous at both loci (*PpRr*), and all will have walnut combs. Two such double heterozygotes, when crossed, produce the phenotypic ratios shown in Fig. b: 9 walnut (*P__ R __*):3 pea (*P__ rr*):3 rose (*ppR__*):1 single (*pprr*). Such non-additive interactions among two or more loci are frequently known as **epistatic** interactions.

The physiological and developmental basis of gene interaction is still unclear, since we do not yet fully understand the relationship between genotypes and phenotypes. In cases like this, the phenotype may result from a multistep process, with each locus controlling a single step in the pathway. But in most cases the mechanisms of gene interaction cannot be so simply explained. The existence of interaction, however, underscores again the complexity of biological systems.

Parents (pink-kerneled)
AaBb × AaBb

Gametes	AB	Ab	aB	ab
AB	AABB	AABb	AaBB	AaBb
Ab	AABb	AAbb	AaBb	Aabb
aB	AaBB	AaBb	aaBB	aaBb
ab	AaBb	Aabb	aaBb	aabb

(a)

P__ R__ Walnut-shaped comb	P__ rr Pea-shaped comb	ppR__ Rose-shaped comb	pp rr Single comb

Dihybrid walnut parents
PpRr × PpRr

Gametes	PR	Pr	pR	pr	
PR	PPRR	PPRr	PpRR	PpRr	——Walnut
Pr	PPRr	PPrr	PpRr	Pprr	——Pea
pR	PpRR	PpRr	ppRR	ppRr	——Rose
pr	PpRr	Pprr	ppRr	pprr	——Single

(b)

exact distribution of identical sets of chromosomes to each of two descendant (or daughter) cells. Every time a cell divided, each of the two daughter cells ended up with a full set (and the same number) of chromosomes. We now understand the process responsible for the faithful transmission of complete genetic information from a parent somatic cell to its two descendants during cell division; it is the process of **mitosis** (Gr. *mitos*, thread).

The transformation from a single cell (the fertilized egg) into a complete organism involves thousands of cell divisions. Cell division is also constantly taking place to replace damaged, aged, or diseased cells. As you read this, cells from your intestinal lining are being sloughed off, replaced by new cells. Similarly, cells from the surface layer of your skin are shed, and new cells divide to replace them. With a few important exceptions (most notably the neurons, or nerve cells), the majority of specialized cells in the body undergo mitotic division to replace cells that are lost.

THE SOMATIC CELL CYCLE AND MITOTIC DIVISION

The Cell Cycle

Cells, like the organisms they compose, have finite lifespans. In this section, we will focus on the lifetime of a single somatic cell, beginning with its birth as a daughter cell of a previous cell division, through its maturation, and culminating in a new round of cell division.

Every cell in the body of a living organism travels through the **cell cycle**. A host of complex events take place throughout the cell cycle. Nevertheless, we can divide the cell cycle into four discrete periods (Fig. 6.10), each of which is defined by characteristic molecular and morphological events:

1. The **M** or **mitotic period**, when mitosis actually occurs. In the majority of eukaryotic cells, this phase takes between 45 and 90 minutes to complete.

2. The **G_1** or **gap one period**, a somewhat misleading name (because a lot is actually going on) for the interval between the *M* period and *S* phase. No net synthesis of DNA occurs during this gap, but the cell prepares for the active period of chromosome synthesis to follow. This stage can last anywhere from 90 minutes to several days, depending on the organism, the cell type, and the growth conditions.

3. The **S** or **synthesis period**, when copies of the chromosomes are made. This stage, which generally lasts from 5 to 7 hours, is the crucial phase during which each chromosome is copied, resulting in the two sets of chromosomes (known as **chromosome complements**) that will eventually be distributed during cell division to the two daughter cells.

4. The **G_2** or **gap two period**, lasting 60 to 90 minutes in most cells, precedes the mitosis phase, during which the cell will actually divide.

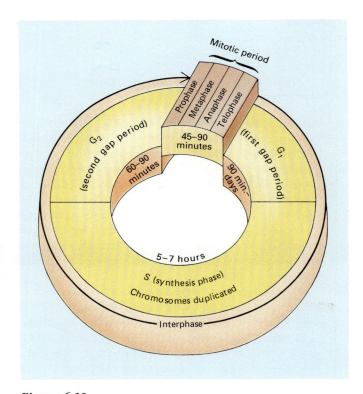

Figure 6.10

A diagram of the cell cycle of a representative cell. The G_1 period, S phase, and G_2 period together constitute the interphase between cell divisions.

The Events of Mitosis

The G_1, S, and G_2 periods are jointly known as **interphase**. The actual division period (M) is further subdivided because several crucial and visually dramatic events are taking place. At mitosis, the cell must carry out its complex genetic bookkeeping, sorting out the individual copies of each of the chromosomes and insuring that complete chromosomal complements go to each of the daughter cells. Based mainly on the observable differences that occur in the nucleus, mitosis is divided into four substages: **prophase**, **metaphase**, **anaphase**, and **telophase**. Certain investigators separate the actual splitting of the two daughter cells following telophase into yet another substage, **cytokinesis**.

Prophase

The chromosomes, previously appearing as thin threads dispersed throughout the nucleus (Fig. 6.11 and 6.12a), begin to take on their characteristic rod shapes, in a process known as **chromosome condensation**. Picture a spring (or a "Slinky") as you compress it—the coils draw closer, or condense, giving you a denser, more compact structure. Keep in mind that each chromosome has already been copied in the previous S phase. At the end of prophase, each chromosome appears as a pair of identical

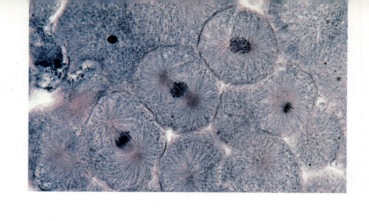

Figure 6.11
Mitotic cell divisions, occurring in the whitefish blastula (320×). The dark staining chromosomes and the mitotic spindle, stretched between the star-shaped centrioles, are clearly visible.

Figure 6.12
A diagram of mitosis in a eukaryotic cell. For simplicity, only a single chromosome pair is shown. The paternal chromosome is shown in red; the maternal chromosome is shown in green.

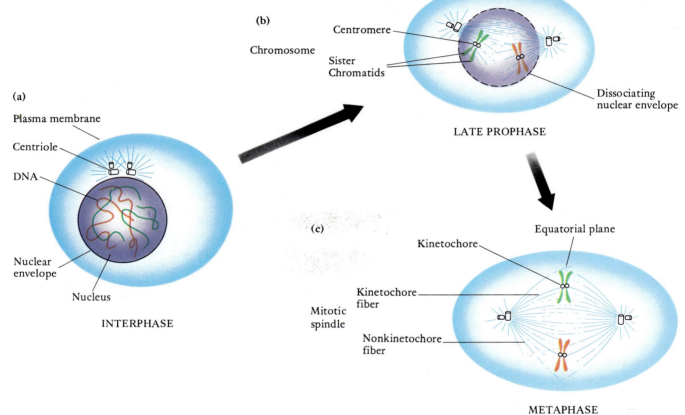

(a)

Plasma membrane

Centriole

DNA

Nuclear envelope

Nucleus

INTERPHASE

(b)

Chromosome

Centromere

Sister Chromatids

Dissociating nuclear envelope

LATE PROPHASE

(c)

Mitotic spindle

Kinetochore

Equatorial plane

Kinetochore fiber

Nonkinetochore fiber

METAPHASE

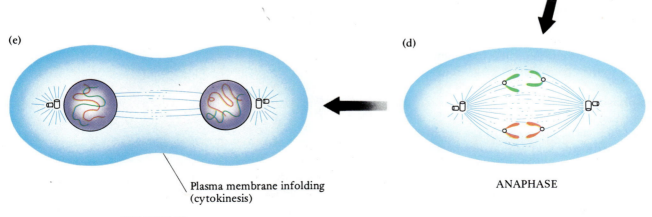

(e)

Plasma membrane infolding (cytokinesis)

TELOPHASE

(d)

ANAPHASE

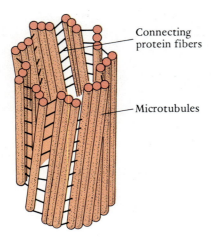

Figure 6.13
A schematic view of the centriole, showing the 9 sets of microtubule triplets, or "spokes," connected by protein crossfibers. Two additional triplets often occur in the center of this arrangement.

strands, joined like Siamese twins at a single point, the **centromere**. Each of the twin strands is called a **sister chromatid** (Fig. 6.12b).

The nucleoli, dark structures normally visible inside the nucleus (see p. 78), begin to dissolve. Two pairs of tiny structures, the **centrioles**, which look like a pair of barrels lying at right angles to one another, begin to move to opposite ends of the cell. The centrioles are considered cellular organelles, composed of an organized array of microtubules (Fig. 6.13). They act as a "seed" in assembling a group of microtubules that consists primarily of polymerized protein subunits. These grow out toward the cell center (Fig. 6.12b). A symmetrical web of microtubules stretches between the two centriole pairs. Two types of microtubules develop. Nonkinetochore fibers extend toward and sometimes across the equator of the cell but do not attach to the chromosomes. Kinetochore fibers extend from the centrioles to the chromosomes to which they will attach. The **mitotic spindle** formed by these fibers will help organize the chromosomes and guide their movement. The end of prophase is marked by the relatively rapid disappearance of the nuclear envelope, at which time the kinetochore fibers attach to a protein-rich region of the chromosome, the **kinetochore**, that lies adjacent to the centromere.

Metaphase

The chromosomes begin to move along the spindle, aided by the kinetochore fibers (Fig. 6.12c), until all of the centromeres are aligned along a single equatorial plane at the midpoint of the spindle. The chromosome bookkeeping is now clearly visible: the chromosomes are fully condensed, visible, and aligned. The chromosomes are held perpendicular to the fibers of the spindle, held in place by a cradle

of kinetochore and nonkinetochore fibers. In certain cell types, this substage of mitosis may last for several hours.

Anaphase

Abruptly, the centromere is torn apart, and each of the sister chromatids, which are identical copies of each other, moves very quickly in opposite directions towards the centrioles (Fig. 6.12d). This is the stage in which each pair of sister chromatids is split up, resulting in identical chromosome sets at either end of the mitotic spindle.

At least two factors are involved in the movement of the chromosomes toward the poles of the cell. The kinetochore fibers shorten, apparently because their subunits are disassembled and removed by the kinetochore. Second, the nonkinetochore fibers lengthen by the addition of subunits to the ends near the cell center. As they lengthen, the centrioles at the poles are pushed apart; this pulls the chromosomes along, for the kinetochore fibers remain attached to the chromosomes. Other factors, including sliding movements between microtubules, may also be involved.

Telophase

The nuclear envelope begins to re-form around each of the chromatid sets, eventually giving rise to nuclei in each of the two daughter cells (Fig. 6.12e). The spindle fibers disappear, the chromosomes begin to unravel and become far less compact, and the nucleoli reappear.

Finally, during **cytokinesis** the cell membrane begins to invaginate perpendicularly to itself, along the equatorial plane, forming the cleavage furrow (Fig. 6.14). Eventually, the two cleavage furrows meet, the membrane is pinched

Figure 6.14
Scanning electron micrograph of a dividing frog egg (30×). The furrow developing in the plasma membrane is caused by the contraction of the cytoskeleton, which is anchored to the surface of the cell by membrane proteins.

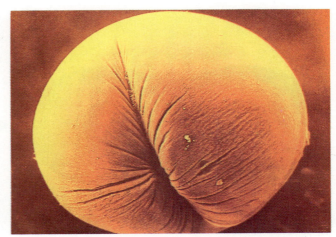

off, and two new, independent, daughter cells have come into being. Keep in mind that at this point, the two daughter cells have a set or complement of chromosomes virtually identical to that of the parent cell: a diploid somatic cell has given rise to two diploid descendant cells.

MEIOSIS: THE PRODUCTION OF GAMETES

Sexual reproduction involves the coming together of two gametes, one from each parent. At first glance, this would seem to raise an arithmetic problem: how is the chromosome number kept constant if two cells come together? After all, every diploid cell has two copies of every chromosome, suggesting that the fertilized egg of a diploid animal would then have four copies of each chromosome. But in fact the fertilized egg is itself a diploid cell. This observation led to the prediction that fertilization is the result of the union of two haploid cells—gametes that contain only one copy of each chromosome. This suggested that a special mechanism must exist for the production of gametes, resulting in cells having only one copy of each chromosome. The specialized process of cell division that gives rise to gametes is known as **meiosis**.

Gametes connect one generation of organisms to the next. The continuity of all sexually reproducing life depends on the fusion of two cells at fertilization. In a curious way, organisms carry within them a completely separate apparatus for the maintenance and production of these intergenerational modules. There is no mingling of somatic cells and germ cells: gametes are produced and stored in specific organs (ovaries and testicles, for example), using the separate rules of meiosis. As far as we can tell, the integrity and independence of the germ line are never breached; from the earliest stages of development, the cells that will give rise to the specialized reproductive tissues are sequestered away from the rest of the organism.

The process of meiosis can immediately be divided into two main segments, meiosis I and meiosis II. Within each of these segments, there are stages similar to those of mitosis: prophase, metaphase, anaphase, and telophase (Fig. 6.15).

Meiosis I

Prophase I

The diploid gamete-producing cells, called **meiocytes** or **gametocytes**, contain one pair of each kind of chromosome. One member of the pair originally came from the maternal parent and one from the paternal parent. Both members of a chromosome pair contain the same genes (although the alleles of these genes may differ), so the members of the pair are described as **homologous chromosomes**. As the chromosomes condense and appear during prophase I of meiosis, they behave differently than they do during mitosis. Instead of moving toward the equatorial plane independently, as they do in mitosis, the members of a pair align intimately side by side, a process termed **synapsis**. A human gametocyte at this stage would show 23 of these pairings, corresponding to the 23 different chromosome pairs that form the human complement. Each homologous pair consists of two homologous chromosomes; since each chromosome has been duplicated prior to meiosis, each pair appears as four chromatids or a **tetrad** (Fig. 6.15). A homologous pair thus consists of two different pairs of sister chromatids.

As synapsis continues, **chiasmata** (Gr. *chiasma*, cross) appear between two nonsister chromatids within a homologous pair (Fig. 6.15). These chiasmata are exactly what they appear to be: the result of the breakage and subsequent rejoining of two similar *but not sister* chromatids. The breaks and sutures occur at precisely the same point in the two chromatids, ensuring that no genetic information is either lost or duplicated. This process is known as **crossing-over**. During crossing-over, genetic information is shuffled as two versions of a chromosome (containing different alleles of many loci) combine into a third, novel combination. This is an extremely important phenomenon from a genetic standpoint, and we will return to it in Chapter 8.

By the end of prophase I, the nuclear envelope has dissolved.

Metaphase I

The homologous chromosome pairs now align along the equatorial plane of the cell. As in mitosis, the chromosomes lie in a spindle of fibers attached to the centromere. Remember that each homologous chromosome pair contains two centromeres. The two centromeres in a homologous pair will eventually move to opposite poles of the cell, dragging their chromosome with them.

Anaphase I

The homologs now migrate to opposite poles of the cell. A full complement of chromosomes thus ends up at each of the poles, but because of crossing-over, the chromosomes are slightly different from those of the original gametocyte.

Telophase I

This stage marks the end of the Meiosis I segment. The two chromosome complements are surrounded by new nuclear envelopes. These two daughter cells are effectively haploid cells, each having a single copy of each chromosome (each containing two sister chromatids). We began Meiosis I with a cell that was diploid and have now produced two haploid cells. This process is consequently referred to as a **reduction division**, because the diploid number of chromosomes is reduced to a haploid complement in each of the two descendant cells.

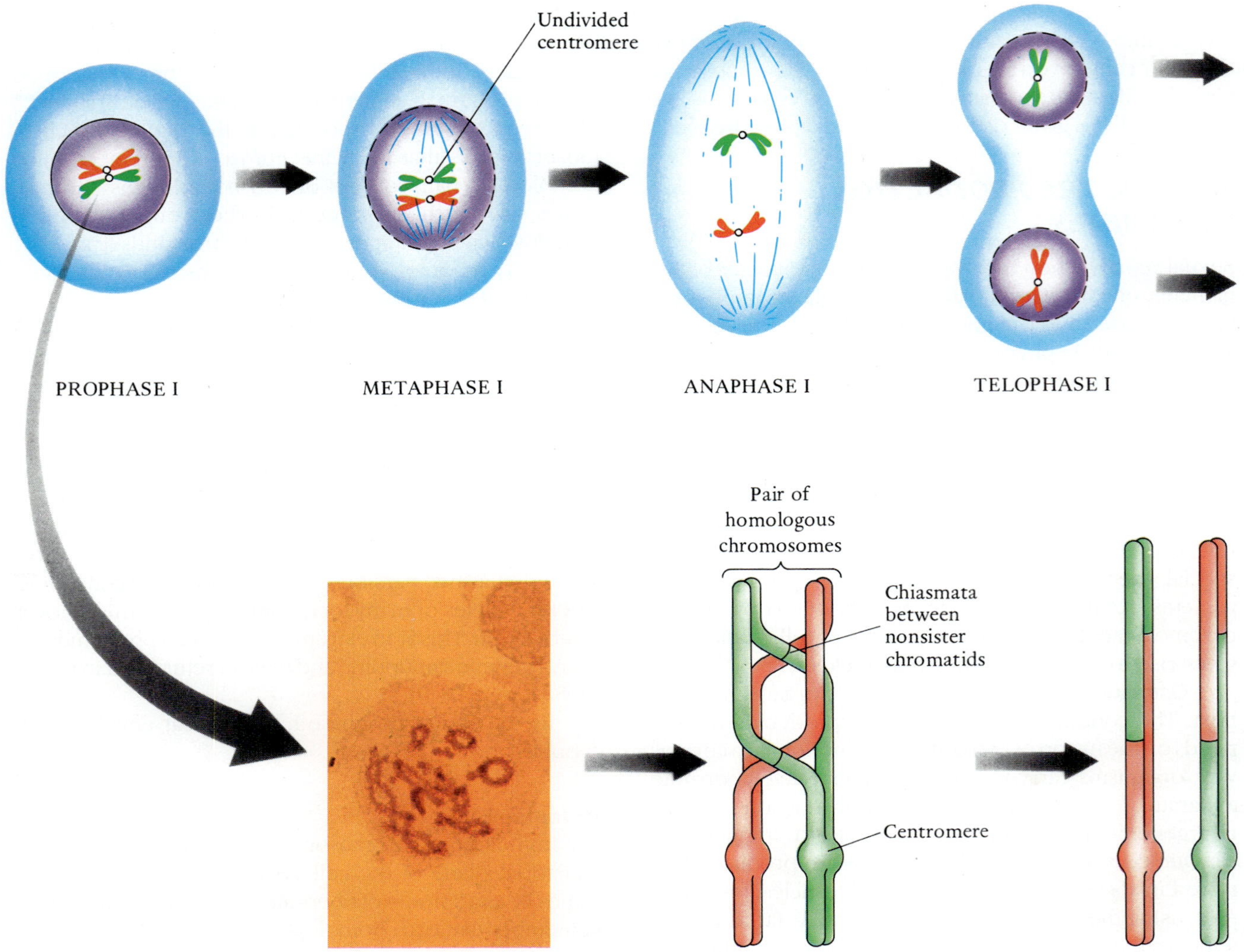

Undivided centromere

PROPHASE I   METAPHASE I   ANAPHASE I   TELOPHASE I

Pair of homologous chromosomes

Chiasmata between nonsister chromatids

Centromere

Meiosis II

No further DNA synthesis will occur from this point. In meiosis II, each of the two haploid daughter cells produced at meiosis I will in turn give rise to two haploid daughter cells, this time by a process of **equational division**.

Prophase II

The chromosomes once again condense and compact, much as they did during mitotic prophase. Keep in mind, however, that in mitosis, two individual chromatids of each of the homologous chromosomes are present in the cell. At this point in meiosis, homologous chromosomes have separated and are in different cells. Each of these cells contains a single homolog of each chromosome, and each chromosome is represented by two joined sister chromatids.

Metaphase II

The centromeres align along the equatorial spindle plane, dragging the chromosomes to a position perpendicular to the spindle.

Anaphase II

In this meiotic stage (unlike at anaphase I), the centromeres do split, and the two sister chromatids migrate to opposite poles along the spindle fibers.

Telophase II

The movement of the chromosomes is completed, leaving a haploid chromosome complement (in the form of a single chromatid of every chromosome) at each of the

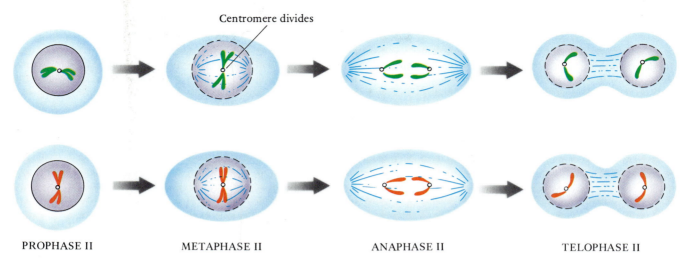

Centromere divides

PROPHASE II   METAPHASE II   ANAPHASE II   TELOPHASE II

Figure 6.15

Crossing-over and the formation of novel gametes during meiosis. The upper portion of the figure illustrates the stages of meiosis, using only a single chromosome pair. At Prophase I, the paternal and maternal versions of the chromosome (homologs), each composed of two identical sister chromatids, line up side by side. At Metaphase I, the nuclear membrane dissolves, and the two chromosomes begin their migration to opposite poles of the cell. By Telophase I, the two daughter cells each contain a single copy of the chromosome. In the subsequent stages of meiosis, the centromere splits and the two sister chromatids separate. The net result of meiosis is the formation of four haploid cells, each containing a single chromatid version of the chromosome. The lower left portion of the figure details the process of crossing-over. The photograph shows the paired chromosomes in the testis cells of the grasshopper *Romalia microptera*, prior to the onset of Metaphase I. As diagrammed in the lower part of the figure, the tight alignment and pairing of homologs that occur at this stage make it possible for crossing-over to occur. Chiasmata form between non-sister chromatids, resulting in the formation of a novel chromatid containing material derived from both the maternal and paternal versions of the chromosome. This process of chromatid breakage and rejoining is extremely precise; no net gain or loss of genetic material occurs. Instead the result is a shuffling or *recombination* of the two slightly different versions of chromosome present in every diploid cell. The figure shows two chiasmata formed between non-sister chromatids. This "double cross-over" generates four novel chromatids, thus ensuring that the four haploid cells that will eventually be produced will differ genetically from one another.

poles of the cell. During cytokinesis, the spindle dissolves and the nuclear envelope re-forms around the chromosomes.

Through the events of meiosis, a single, diploid meiocyte has given rise to four haploid cells. In many organisms, not all four cells will become functional gametes (see Fig. 21.4).

The rest of the meiotic cell cycle is, by comparison, relatively uneventful. Before meiosis begins, the cell has undergone an S stage, in which copies of the chromosomes are synthesized. Following meiosis, the daughter cells will not undergo any subsequent meiotic or mitotic events—they are now terminal haploid gametes that will combine with haploid gametes of the opposite sex at fertilization or, failing that, will die. If fertilization does take place, the two haploid nuclei will fuse, reconstituting a

diploid cell, the **zygote**, that eventually develops into a complete organism. We return to the nature of the gametes and fertilization in Chapter 21.

THE CHROMOSOMES CARRY MENDEL'S "PARTICULATE FACTORS"

Having examined in some detail the processes of mitosis and meiosis, we now return to Mendel's observations. Mendel, after all, had described the consequences of inheritance. By carefully counting the ratios of offspring obtained from particular crosses, he had uncovered certain regularities, rules that eventually allowed him to make a prediction about the ratios to expect. Furthermore, Mendel had succeeded in demonstrating that the ratios were independent of the particular character he analyzed. Thus, in a dihybrid cross, the F_2 generation showed a 3:1 ratio both of smooth to wrinkled seeds, and of yellow to green seeds.

We now consider Mendel's work a cornerstone of modern biology. Yet, as we mentioned previously, it remained obscure for close to 40 years following its publication. Mendel's results depended on a material process that had not been observed. He had correctly argued that inheritance was particulate, but he did not know what these particles were or where they could be found.

Now that we have examined the behavior of chromosomes, we can reconstruct the logic that led, in the early 1900s, to the **chromosome theory of inheritance**, the hypothesis that the hereditary factors are located on the chromosomes. Careful observations of the chromosomal mechanics of meiosis suggested an immediate parallel with the behavior of Mendel's factors. Several investigators noticed the correlation between Mendel's observations and the emerging picture of chromosomal behavior, and independently proposed that the factors were located on chromosomes. They argued that Mendel's rules could be rephrased in terms of chromosome movement. Furthermore, the particulate nature of Mendel's factors was echoed by the integrity of the chromosomes—although they were more or less visible throughout the cell cycle, chromosomes did not blend or dissolve. The fact that offspring appeared to receive a copy of a given factor from each parent corresponded to the observation that chromosomes were segregating (at anaphase I) to give rise to haploid cells (gametes). Finally, the independence that Mendel had observed when looking at two or more characters (seed shape and color, for example) could be explained by postulating that the two factors resided on different chromosomes and hence were physically independent from one another.

SEX LINKAGE

The chromosome theory provided a clear explanation for the phenotypic ratios that Mendel had observed in his crossing experiments. However, not all crosses carried out by breeders and geneticists obeyed Mendel's ratios. Among the first to notice this discrepancy was the geneticist Thomas H. Morgan, who observed crosses in the fruit fly, *Drosophila*, in which the pattern of inheritance was affected by the sex of the individual. It was known from earlier studies that the sex of *Drosophila* and many other animals is determined by the distribution of the **sex chromosomes**. Members of all other chromosome pairs, known as the **autosomes**, are identical, but this is not always the case for the sex chromosomes. Female *Drosophila* have two X chromosomes and are the **homogametic** sex (all eggs contain a single X chromosome). Male *Drosophila* have a single X chromosome and a differently shaped Y chromosome, and are the **heterogametic** sex (one-half of the sperm contain an X chromosome and one-half contain a Y chromosome). The fertilization of an egg by a Y-containing sperm results in a male individual (XY); fertilization by an X-containing sperm results in a female (XX).

The conspicuous difference in the chromosomal complements of males and females exists in practically every species with separate sexes. In *Drosophila*, and in all placental mammals (including humans), females are homogametic and carry two X chromosomes, while the males are heterogametic and carry an XY pair of sex chromosomes. The situation is reversed in birds, with the female being heterogametic (carrying the so-called ZW pair) and the males homogametic (ZZ). The sex chromosomes are thought to have evolved from an original autosomal pair. Despite their differences, X and Y chromosomes still pair along part of their length (the pseudoautosomal region) during male meiosis. We will return to the various mechanisms of sex determination in Chapter 21.

The inheritance of eye color in *Drosophila* was one of the first cases in which the sex of the individual was found to be critical in the expression of the gene. In one experiment, Morgan crossed mutant white-eyed males to females with the wild-type, or normal, red eye color (Fig. 6.16). All the resulting F_1 progeny had red eyes, as expected in any cross involving recessive and dominant traits (white eye = w, red eye = W). When the F_1 males and females were crossed, Morgan obtained an F_2 generation with the expected 3:1 Mendelian ratio of red-eyed flies to white-eyed flies. He soon realized, however, that something unusual was going on: all the white-eyed flies were male ($\frac{1}{3}$ of the red-eyed flies were also male). When heterozygous red-eyed F_2 females were backcrossed to white-eyed males, an equal proportion of red-eyed males, red-eyed females, white-eyed males, and white-eyed females resulted. What was going on?

Morgan would resolve this riddle by proposing that the gene for red (or white) eye color resided on the X chromosome. White-eyed males could therefore only arise from the joining of an egg bearing the recessive allele (w) with a sperm bearing no X chromosome, and therefore no eye color allele. As Figure 6.16 shows, this hypothesis explained the results fully. If the allele was recessive and carried on the X chromosome, we should expect no white-eyed males (or females) in the F_1 generation. In the F_2 generation, 50% of the males should have white eyes. Far from contradicting the chromosome theory, the discovery of **sex-linked traits** would provide additional support for it. In these cases, the phenotypic ratios obtained in the crosses mirrored the behavior of the sex chromosomes.

Several important human sex-linked diseases have now been identified, including a form of hemophilia (Fig. 6.17), red-green color blindness, and Duchenne muscular dystrophy. Because the defective genes that produce these conditions in humans are located on the X chromosome, the frequency of these conditions is far higher in males. Duchenne muscular dystrophy, for example, occurs in 1

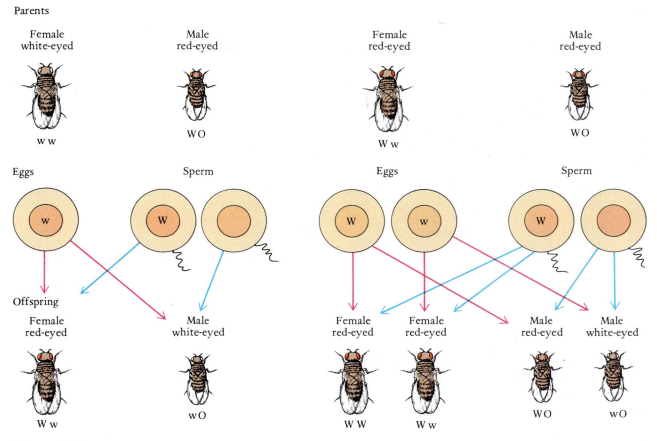

Figure 6.16

The inheritance of sex-linked traits. The gene determining red or white eye in *Drosophila* is found on the X-chromosome. Consequently females carry two copies of this gene and may be homozygous-dominant (red-eye), heterozygous (red-eye), or homozygous-recessive (white-eye). In contrast, the males are *hemizygous* and carry only a single copy of this gene. As can be seen in this diagram, sex-linked traits result in an association between the sex of the offspring and the phenotypic characteristic (red or white eye).

out of 7000 males, but is never seen in females. In effect, males, with only a single X chromosome, cannot carry a defective copy of this gene without suffering from the disease, because it is the only copy they carry. In contrast, females can be heterozygous carriers at this locus and express few or no symptoms. (The situation is made somewhat more complicated by the fact that in females one of the two copies of the X chromosome is always inactivated.) The chromosomal location of these genes ensures that there is no father to son transmission of these conditions: sons can only inherit the defective allele from their mothers. Daughters must be homozygous recessives to exhibit the condition, and must therefore be the offspring of a mating between an afflicted male and a carrier (or afflicted) female. The only X-linked condition that occurs at an appreciable frequency in human females is

red-green color blindness, since this benign condition occurs at high frequency in Caucasian males (8%) and does not interfere with survival or reproduction.

Most sex-linked traits reside on the X chromosome. Interestingly, very few genes have been found on the human Y chromosome.

Even in circumstances where the genes are not on the sex chromosomes, the law of independent assortment sometimes appears to fail, and characters do not segregate independently. This phenomenon of **linkage** occurs when two (or more) characters occur in close proximity on the same chromosome. Such linked characters in effect travel together at meiosis and are not likely to be separated by recombination events. We will return to the issue of linkage when we examine the origin of evolutionary novelty in Chapter 8.

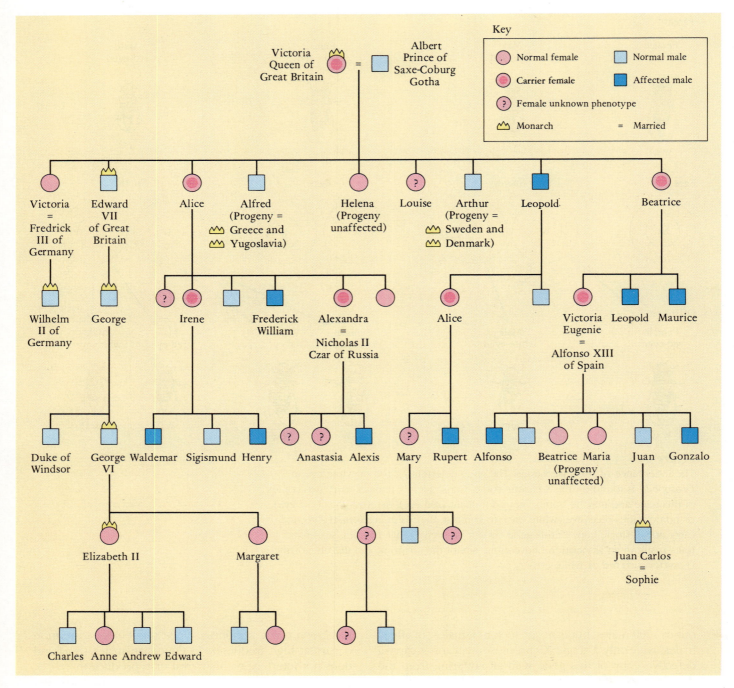

Figure 6.17

Hemophilia is a sex-linked blood-clotting disorder. If untreated, affected individuals show a tendency to bleed excessively following minor trauma. One form of this disease affected a number of individuals in the European royal families, as shown in this pedigree. The defect appears to have arisen in the germ cells of Queen Victoria; it was soon introduced into the Russian and Austrian royal families.

Summary

1. Human beings have long known that many characters of organisms are passed on from generation to generation, and have applied this knowledge to improve crops and domestic animals. In the early 19th century, Gregor Mendel was the first to undertake a successful analysis of the mechanisms of inheritance. He selected characters that could be easily identified and he kept careful quantitative records of his results.

2. Working with crosses of garden peas that differed in a single character (monohybrid crosses), Mendel concluded that the inheritance of the characters he studied depended on two parental factors that separated during the production of gametes and reunited at fertilization. This principle is known as Mendel's law of equal segregation. Inheritance depended on the transmission of particulate factors and not on a blending of characteristics.

3. We now call Mendel's factors genes, and alternate versions are alleles. The diploid parents contain two versions of each allele and the haploid gametes only one. If one parent is homozygous for a dominant trait, and the other homozygous for the recessive trait, the F_1 offspring will all be heterozygous because they contain one allele from each parent. The recessive allele is not expressed in the F_1 but reemerges in the F_2 in the phenotypic ratio of 3 dominant to 1 recessive. The genotypic ratio is 1:2:1 because one third of the phenotypic dominants is homozygous and two thirds are heterozygous.

4. Mendel continued his experiments by crossing peas that differed in two characters (dihybrid crosses). He discovered that the two factors behaved independently of each other and resulted in a F_2 phenotypic ratio of 9:3:3:1. Mendel's law of independent assortment states that the segregation of alleles of a gene pair occurs independently of all other allele pairs. Mendel's work was overlooked until the early 20th century when biologists recognized that the behavior of chromosomes paralleled that of Mendel's factors and that chromosomes were doubtlessly the bearers of the genes.

5. All somatic cells go through a cell cycle. During mitotic division (the M period), the chromosomes, which were previously duplicated, are evenly distributed to the daughter cells. Each cell receives one of each type of chromosome. Mitosis is followed by a G_1 period, an S period when the chromosomes are duplicated, and a G_2 period that precedes the next division. Chromosomes become more tightly coiled and microscopically visible in the prophase of mitosis, and the duplicates become visible as sister chromatids. The chromosomes line up on the equatorial plane at metaphase, and the sister chromatids (now considered to be chromosomes) are drawn apart during anaphase. The mitotic spindle plays an active role in chromosome movements. The nuclei re-form in the telophase, and the cytoplasm divides during cytokinesis.

6. Haploid gametes are produced during two meiotic divisions. In prophase I, homologous chromosomes intimately pair to form a tetrad of four chromatids. At this time chiasmata form between nonsister chromatids and they exchange segments. During the rest of the first meiotic division, each chromosome pair separates and the chromosomes move to opposite poles of the cell, but sister chromatids are held together by the centromere. No further duplication occurs before the second meiotic division. During this division the centromere divides and sister chromatids move apart. Meiosis results in four haploid cells that contain one of each type of chromosome.

7. Sex is determined in many animals by an unequal distribution of the sex chromosomes. In most species the male is the heterogametic sex (XY), and the female is homogametic (XX). In birds and some other species the female is heterogametic (WZ) and the male is homogametic (ZZ).

8. Genetic traits that are carried on the X chromosome are sex linked. Males inherit such characters only from their mothers.

References and Selected Readings

Ayala, F.J., and J.A. Kiger, Jr. *Modern Genetics*, 2nd ed. Menlo Park, Ca.: Benjamin/Cummings Publishing Co., 1984. A good textbook covering all aspects of genetics.

McIntosh, J.R., and K.L. McDonald. The mitotic spindle. *Scientific American* 261 (Oct. 1989):48–56. A review of recent research on the role of the spindle in cell division.

Miller, J.A. Mendel's peas: a matter of genius or guile? *Science News* 125 (Feb. 1984):108. Did Mendel fudge his data?

Stern, C., and E.R. Sherwood, eds. *The Origin of Genetics*. San Francisco: W.H. Freeman Co., 1966. A reprint and discussion of classic papers in genetics, including Mendel's overlooked work.

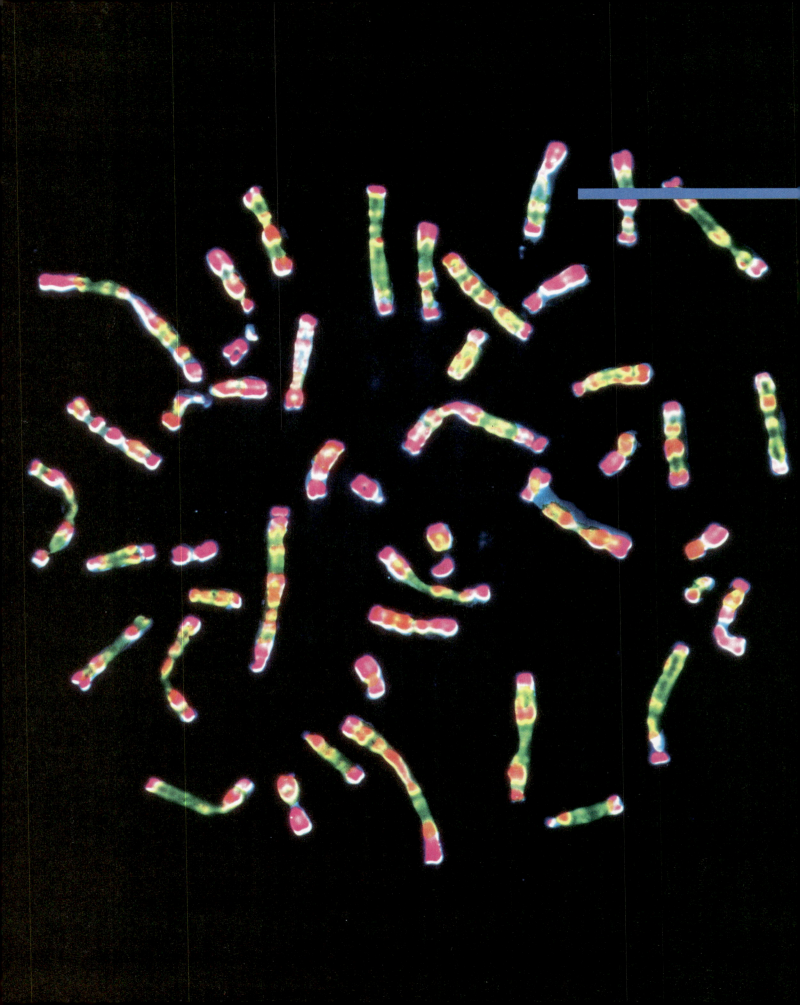

7

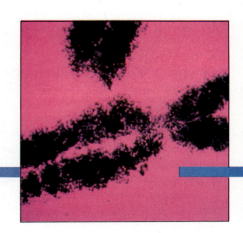

Genetics II: The Molecular Basis of Inheritance

(Left) The human chromosomal complement, or karotype, shown in a false-color photograph. Twenty-three chromosome pairs, each composed of one maternal and one paternal version, contain all the information inherited from generation to generation (magnification × 500). (Above) A single human chromosome, composed of two identical sister chromatids. The chromosome is actually a single, continuous molecule of DNA, carefully wrapped around a protein lattice.

In Chapter 6, we explored the fundamental phenotypic observations that would eventually lead to the basic rules of genetics. Observations made under the microscope had given rise to the chromosome theory, which established the chromosomes as the physical entities that carried the genes. But the quest for the material basis of inheritance was far from over: the chemical identity and organization of genetic material still remained to be determined.

Scientists suspected that an understanding of the chemical basis of inheritance would also reveal *how* heredity worked. At least in one respect, nobody anticipated just how right that prediction would turn out to be. The elucidation of the structure of DNA, the "stuff" of heredity, immediately resolved the question of genetic transmission. In all of its elegant simplicity, the organization of DNA would disclose just how a cell made faithful copies of its genetic information and would help explain how variation, the essence of evolutionary change, arose.

In this chapter, we will examine the nature of inheritance at the molecular level. We will also review the ways in which genetic information is read and interpreted by the cell in order to make specific proteins when and where they are needed. As you will see, the quest for the chemical nature of the genetic material is one of the most exciting episodes in the history of biology.

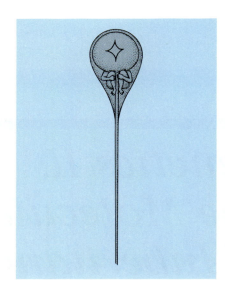

Figure 7.1
Many scientists of the 17th and 18th centuries considered heredity to be the consequence of growth in the womb of a miniature, fully formed individual contained in the sperm. This diagram illustrates such a *homonculus*, purportedly as seen under a microscope.

THE NATURE OF THE GENE

Although we have already introduced the concept of a gene, we have treated genes as "factors," idealized constructs that help us account for the results of crosses. But genes are more than just convenient terms. Following the acceptance of the chromosome theory in the early part of this century, the search was on for the nature and physical reality of the gene. What, investigators asked, are genes made of? What do they actually do?

In this molecular age, it is difficult to picture that 50 years ago, scientists had only a faint notion of what "genetic instructions" were instructions for. After all, it was one thing to say that chromosomes contained the information that resulted in the resemblances between parents and offspring, but it was quite another to explain how that was done. Early theories of preformation (Fig. 7.1) were disproven, but no clear alternative had arisen in their place. Important clues, however, were already available.

The Importance of the Nucleus

Support for the chromosome theory came indirectly from experiments that sought to pinpoint the cellular location of genetic information. One of the earliest experiments was carried out in 1943 by the Danish biologist Joachim Hammerling. He used one of the largest single-celled organisms known, the green alga *Acetabularia*. The different species in this genus are readily distinguished by the shape of their "caps." Hammerling used two very different species, *A. crenulata*, with a disc-shaped cap, and *A. mediterranea*, with a tulip-shaped cap (Fig. 7.2a). He sought to determine where the information that dictated the shape of the cap resided. In order to do so, he carried out a set of **reciprocal transplantation** experiments, taking the middle section ("stem") from one species and grafting it onto the other, and then allowing the alga to regrow a new cap. Would the shape of the new cap be dictated by the transplanted stem or by the original base? The results are depicted in Figure. 7.2b. In both cases, the cap shape corresponded to the identity of the base and not to the stem. Hammerling concluded that the information that determined the cap shape resided in the base structure, where the nucleus of the alga is found. Hammerling's experiments also demonstrated that the information present in the nucleus was sufficient to regenerate the entire alga, with all of its specialized structures. This suggested, at least in principle, that the nucleus retained the capacity to direct the complete development of *Acetabularia*.

The definitive proof that genetic information resided in the nucleus came from the **nuclear transplantation** experiments carried out in 1952 by Thomas King and

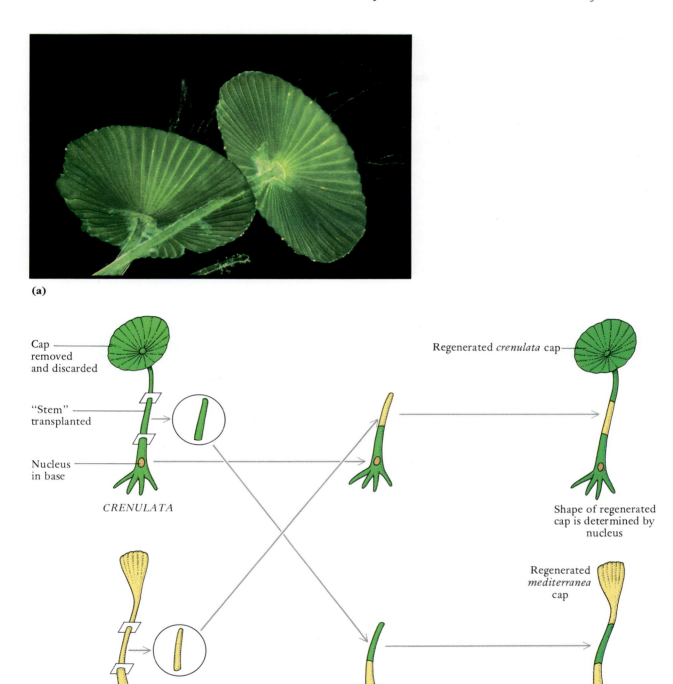

(a)

Cap removed and discarded

"Stem" transplanted

Nucleus in base

CRENULATA

Regenerated *crenulata* cap

Shape of regenerated cap is determined by nucleus

Regenerated *mediterranea* cap

(b)   *MEDITERRANEA*

Figure 7.2

(a) *Acetabularia crenulata,* a large single celled alga used in the reciprocal transplant experiments that extablished the role of the nucleus in heredity. (b) By grafting the stem of two distinct *Acetabularia* species onto the reciprocal base (containing the nucleus), Hammerling showed that the shape of the regenerated cap was dictated by the identity of the base.

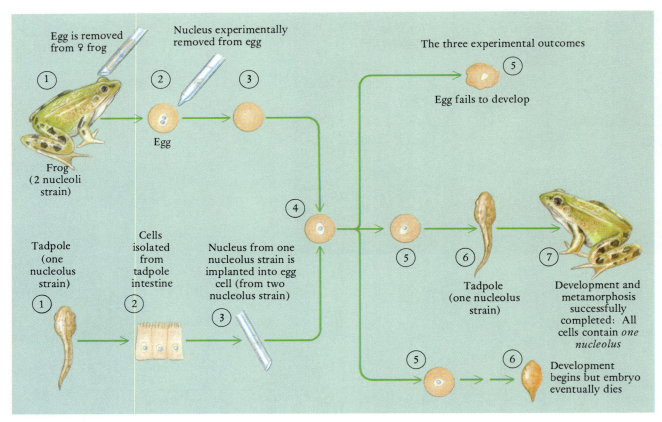

Figure 7.3

The reciprocal nuclear transplant experiments of King and Briggs demonstrated that the information necessary for the development of a viable frog was carried entirely in the nucleus. Although the manipulations resulted in a number of inviable embryos, the frogs that developed always showed the nuclear phenotype (one nucleolus) of the donor nucleus, not of the recipient egg.

Robert Briggs (Fig. 7.3). They showed that a nucleus taken from the intestine of a tadpole could be implanted into a fertilized frog egg whose own nucleus had been experimentally destroyed, or enucleated. These experimental manipulations frequently resulted in the abnormal development or death of the embryos. But several of the experiments did yield fully developed, healthy frogs. King and Briggs had cleverly designed their experiment such that the tadpoles and the enucleated eggs came from slightly different frog strains, whose cells contained different numbers of nucleoli. In the successful nuclear transplantation experiments, the frog that developed was of the same strain as the nuclear-donor frog. Once again, these results suggested that the nucleus was the site of hereditary information and that cell nuclei taken from a specialized tissue (the lining of the intestine) nonetheless contained all of the information necessary to direct the complete development of a healthy individual. A host of subsequent

experiments have confirmed this startling observation: with very few exceptions, all cells in a multicellular eukaryote contain the same complete genetic information. The process of differentiation that gives rise to specific, specialized cells does not result in the loss of any genetic information. Thus, the nuclei of your liver cells, brain cells, and skin cells contain the same information and could, in principle, direct the development of a new you. Specialization and differentiation, therefore, involve the selective reading and use of genetic information by the cell and not the erasing of unnecessary genetic material.

The Transforming Principle

Other investigators were pursuing an entirely different set of leads in the search for the nature of genetic material. In 1944, Oswald Avery, Maclyn McCarty, and Colin MacLeod were attempting to elaborate on a set of experiments carried out almost 20 years before by Frederick Griffith. Griffith worked on bacteria of the genus *Pneumococcus*. Certain strains of this bacteria were pathogenic (disease-producing), causing pneumonia and blood poisoning. Mice infected with these pathogenic strains would eventually die. Other strains of this bacteria, however, were nonpathogenic and produced no symptoms or ill effects in

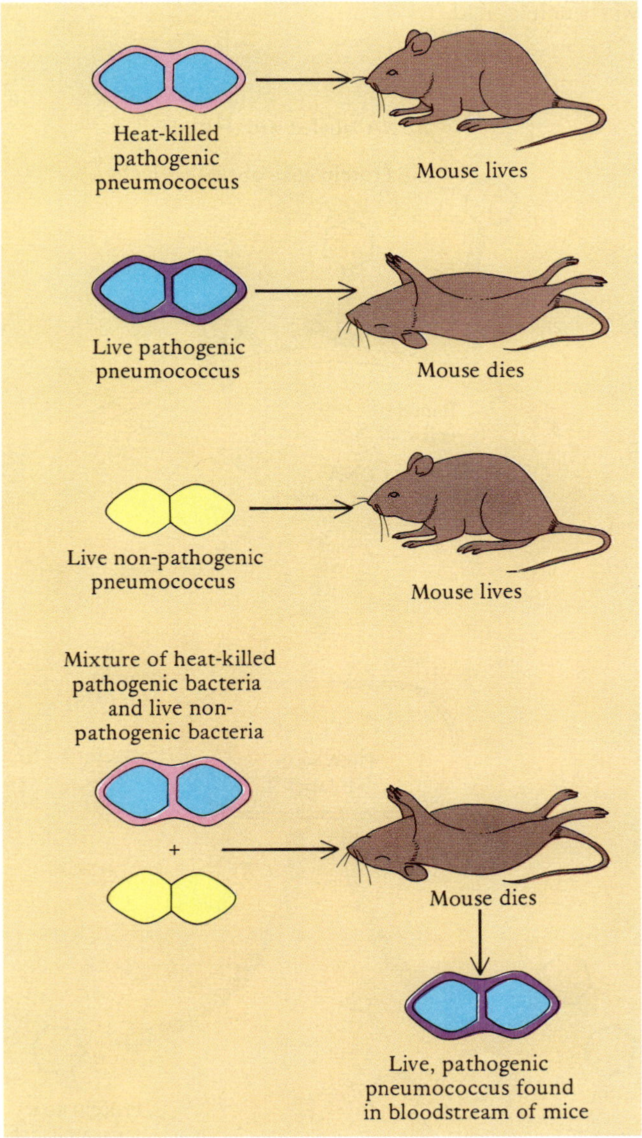

Figure 7.4
Avery and MacLeod, basing their work on previous experiments by Griffith, demonstrated that heat-inactivated bacteria were still capable of transforming non-pathogenic *Pneumococcus* into a pathogenic strain. This transforming principle would later be shown to be composed of nucleic acid.

the mice. Previous work indicated that one of the differences between the pathogenic and nonpathogenic strains was the presence of a polysaccharide cell wall, or coat, in the pathogenic strains.

In a set of experiments designed to confirm the importance of the polysaccharide coat, Avery killed the pathogenic bacteria by exposing them to high temperatures (Fig. 7.4). When the heat-treated bacteria were injected into mice, the mice, as expected, developed no symptoms. Avery had also confirmed that bacteria with no coats were indeed nonpathogenic. The most surprising result came when Avery made a mixture of killed pathogenic bacteria and live nonpathogenic bacteria, and injected that mixture into mice. A large number of the mice died, and autopsies revealed that they harbored large amounts of live, polysaccharide-coated bacteria circulating in their blood. Either the live bacteria had somehow stolen the coats from their dead relatives, or the information necessary to make a coat had been transferred from the dead bacteria to the live ones. Avery would refer to this transferred information as the **transforming principle**.

Avery and his colleagues sought to establish the identity of this transforming substance. They had finally narrowed down the nature of this inducing stimulus: it appeared to be a nucleic acid, specifically deoxyribonucleic acid (DNA). This substance continued to transform nonpathogenic strains into pathogenic ones by directing the synthesis of a polysaccharide coat, even after the mixture was treated with proteinases, enzymes that digest and destroy proteins. Proteins had long been thought to be the likely repository of genetic information, but the outcome of the proteinase treatments appeared to rule proteins out. If, on the other hand, the mixture was treated with nucleases, enzymes that digest nucleic acids, no transformation of nonpathogenic bacteria occurred, and no pathogenic bacteria appeared in the Griffith mixture. The trail was heating up: the genetic material appeared indeed to be DNA.

Nucleic Acid Is the Hereditary Material

Despite these elegant experiments, scientists were unwilling to abandon the notion that proteins somehow were responsible for the transmission of hereditary information. The reasons for this reluctance are, in retrospect, understandable. After all, the information required to direct the development of any organism was likely to be extremely complex. Since nucleic acids are composed of nucleotides containing only four different bases—adenine, guanine, cytosine, and thymine—they did not appear sufficiently complex to carry the necessary information. In contrast, proteins appeared much more numerous and varied, occurring in many different sizes and shapes. Proteins still seemed to be the only biological molecule capable of carrying so much information.

This predilection for proteins would prevail until the important set of experiments carried out in 1952 and 1953 by Martha Chase and Alfred Hershey. They used a different set of hosts and pathogens: this time, the intestinal bacteria *Escherichia coli* was the host and a small bacterial virus, bacteriophage T_2, was the pathogen. The bacteriophage was known to infect the bacterial cell, taking over the cell's biochemical machinery in the process and forcing it to make multiple copies of the bacteriophage. This process of infection clearly indicated that instructions for making bacteriophage T_2 were somehow being injected into the bacteria when infection occurred. Microphotographs showed that the bacteriophage was not actually entering the bacterial cell, but instead remained at the cell surface, finding a way to introduce its own instructions into the bacterial cell (Fig. 7.5a). In order to determine the nature of the injected instructions, Chase and Hershey labeled the protein coat of the bacteriophage with one radioactive tag (^{35}S) and the nucleic acid core of the bacteriophage with another (^{32}P). They then asked the following question: Which of the two labels do we find inside the *E. coli* cell once infection has occurred? In their experiments, they placed *E. coli* and the bacteriophage together in an appropriate medium and allowed the bacteriophage to adhere

to the bacterial cell wall. After some time, they shook the cells violently, dislodging the bacteriophage particles from their point of attachment. When they examined the infected cells, they could find practically no trace of the ^{35}S-labeled material, but large amounts of ^{32}P label were present inside the *E. coli* (Fig. 7.5b). Clearly, the bacteriophage was injecting nucleic acid, and not protein, into the host cell. Furthermore, Chase and Hershey found traces of the ^{32}P-labeled substance in the newly synthesized bacteriophages, confirming once again that the genetic material was nucleic acid.

Figure 7.5

(a) Electron micrograph of bacteriophages attached to the wall of the bacterium *E. coli*. Note the characteristic hexagonal shape of the viral capsule, and the slender tail through which the viral DNA is injected into the host.
(b) Diagram of the Chase and Hershey experiment. Bacteriophage were prepared in the presence of radioactive phosphorus (^{32}P) and sulfur (^{35}S), resulting in distinctively labeled DNA and protein coats. The phage viruses were permitted to infect bacterial cells; the cells were then agitated vigorously, stripping the phage coats off their surface. The bacterial cells contained only ^{32}P, and no ^{35}S, confirming that it was DNA, and not protein, being injected into the host cell. Furthermore, the new generation of phages contained residual amounts of labeled DNA.

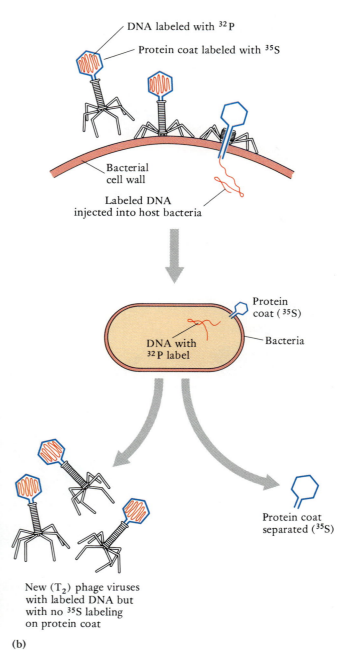

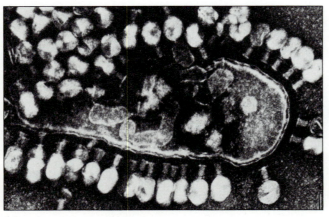

(a)

(b)

The One Gene–One Enzyme Hypothesis

A third strand of evidence about the nature of genetic information was also being woven into this developing tapestry. In 1902, Alfred Garrod, a British doctor, had noted that a certain rare disease, **alkaptonuria**, occurred very frequently in certain families. When he examined the family histories of these afflicted individuals, he found that the disease was inherited (Fig. 7.6). The pedigrees of affected families suggested that the condition might be caused by a recessive allele. It tended to appear in alternate generations, in the same way that certain characters in Mendel's experiments disappeared in the F_1 generation, only to reappear in the F_2 generation. One of the symptoms of alkaptonuria is that the urine of afflicted individuals turns black soon after being excreted. The underlying cause of this symptom is the presence in the urine of homogentisic acid, a substance that oxidizes readily and turns black. Garrod suggested that in alkaptonurics some normal metabolic pathway in the body was interrupted at the step that would ordinarily break down homogentisic acid (Fig. 7.7). In afflicted individuals, the acid was building up and being expelled in the urine. Garrod concluded that alkaptonuria was an **inborn error of metabolism**, a disease caused by the absence of a specific enzyme, in this case, the enzyme responsible for the breakdown of homogentisic acid. This metabolic defect was caused by a genetic (heritable) error that had arisen in one of the ancestors of the afflicted family. The descendants of that ancestor (homozygous for the recessive defect) therefore produced an inoperative enzyme, causing a breakdown of the normal metabolic pathway. Garrod's work would inspire physicians to examine family histories as part of the diagnostic evaluations. Since Garrod's studies, a large number

Figure 7.6

A pedigree showing the incidence of alkaptonuria in a Lebanese family. Note that the very first generation already consists of two related individuals; and a number of second cousin and uncle–niece marriages have also taken place in this family. This level of consanguineous matings results in an increased incidence of homozygous recessives exhibiting the alkaptonuric phenotype.

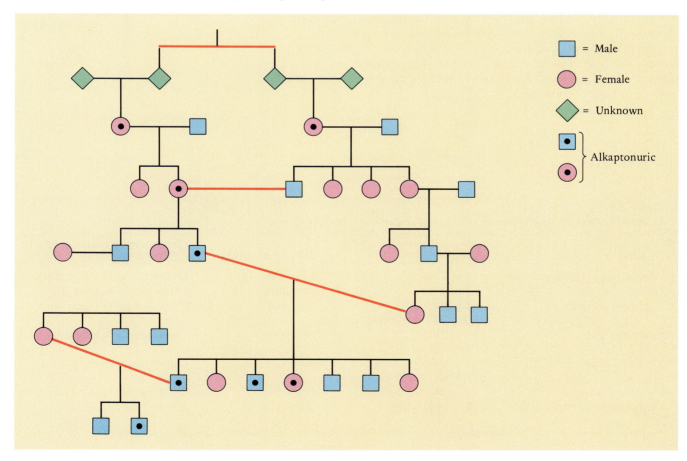

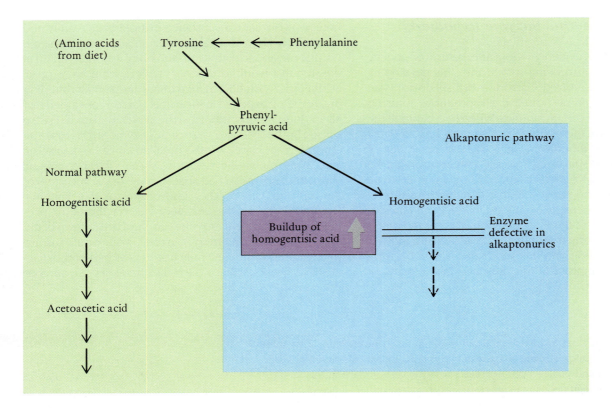

Figure 7.7

The defect in the metabolic pathway of tyrosine that results in alkaptonuria. Alkaptonurics exhibit a defect in one of the enzymes in the pathway that transforms homogentisic acid to acetoacetic acid, resulting in a buildup of homogentisic acid. Homogentistic acid oxidizes on contact with air, turning the urine of afflicted individuals black. Alkaptonuria was the first heritable human single gene defect we identified.

Table 7.1
Certain Common Single-Gene Diseases (Frequencies/Live Births)

Disease	Frequency	Type of Mutation	Chromosomal Location of Affected Gene
Adult polycystic kidney disease	1 in 1250	Dominant	Chromosome 16 (short arm)
Huntington's chorea	1 in 2500	Dominant	Chromosome 4 (short arm)
Sickle-cell anemia	1 in 655 (U.S. African-Americans)	Recessive	Chromosome 11 (short arm)
Cystic fibrosis	1 in 2500 (Caucasian)	Recessive	Chromosome 7 (long arm)
Tay-Sachs disease	1 in 3000 (Ashkenazi Jews)	Recessive	Chromosome 15 (long arm)
Phenylketonuria	1 in 12,000	Recessive	Chromosome 12 (long arm)
Duchenne muscular dystrophy	1 in 7000 males	X-linked	X chromosome

Data from Scriver, C.R., *et al*; *The Metabolic Basis of Inherited Disease*. New York, McGraw-Hill Information Services (1989).

of heritable diseases have been identified through the use of pedigree analysis. In the last ten years, the actual gene defects responsible for certain genetic diseases have been identified (see Table 7.1).

Garrod's hypothesis about the causes of alkaptonuria was the first statement of what would come to be known as the **one gene–one enzyme** hypothesis of gene action. But it would be nearly a half century before the one-to-one relationship between a gene and an enzyme (or more generally, a polypeptide chain) would be established. The final proof was provided in 1941 by George Beadle and Edward Tatum. They experimentally induced **mutations** (a change in the structure of DNA) in the fungus *Neurospora* by bombarding the spores with radiation. They then established the metabolic consequences of the mutations they had produced. Some of these *Neurospora* mutants lacked the ability to synthesize certain amino acids that they usually produced. Beadle and Tatum then determined the position on the chromosome where each mutation had occurred (Fig. 7.8). In short, the gene for a particular enzyme had been repeatedly hit by X-rays, producing mutants that interrupted the normal metabolic pathway in the synthesis of an amino acid. Different clusters of mutations affected different enzymes in the metabolic pathway. Beadle and Tatum concluded that each enzyme in the pathway was produced by one (and only one) gene on the chromosome.

THE STRUCTURE OF DNA: THE DOUBLE HELIX

Once it had become clear that genetic information was carried by nucleic acids, a feverish search began for the mechanisms and methods that made possible the transmission of information from one generation to the next.

Investigators knew that there were four different bases in DNA: adenine, thymine, cytosine, and guanine (A, T, C, and G, respectively). In analyzing the nucleic acids of a variety of organisms, several investigators had concluded that the relative proportions of each of these bases varied considerably from one species to another, with no apparent rhyme or reason. However, by 1948, Edwin Chargaff had noticed that the total amounts of A and G (A + G) always equaled the total amount of T and C (T + C), regardless of the species being examined. He also noted that the amounts of A and T were always equal, as were the amounts of G and C (Table 7.2). These were clearly interesting observations, but it was difficult to incorporate them into the prevailing models about the structure of DNA.

The work of Rosalind Franklin, in 1952 and 1953, would provide the missing piece of the puzzle. She used a technique known as X-ray crystallography, in which crystals (of DNA, in this case) are bombarded with X-rays, and the pattern produced by the diffraction of these rays along

Figure 7.8
Beadle and Tatum showed that the experimentally induced mutations that interfered with the synthesis of arginine in *Neurospora* were in fact affecting enzymes along that metabolic pathway. Furthermore, they showed that these mutations were, in fact, clustered in specific regions of the chromosome that corresponded to the location of the genes for the enzymes in the arginine pathway.

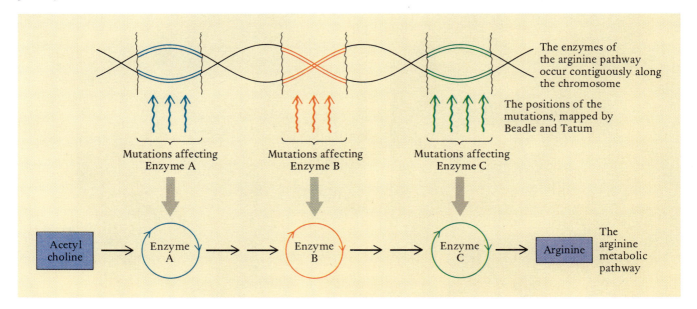

Table 7-2
A Comparison of the Base Composition of DNA from Several Different Organisms

	Base Composition of DNA			
Source of DNA	Adenine	Thymine	Guanine	Cytosine
Micrococcus lysodeiticus (a bacterium)	0.15	0.15	0.35	0.35
Aerobacter aerogenes (a bacterium)	0.22	0.22	0.28	0.28
Escherichia coli	0.25	0.25	0.25	0.25
Calf thymus	0.29	0.28	0.21	0.22
Phage T₂	0.32	0.32	0.18	0.18

the crystal faces is examined. She concluded from these patterns that the molecule was probably an extended, stringlike structure. It appeared to be composed of two similar, parallel strands. Within it, two regularly repeating patterns appeared: one every 2 nm and one every 3.4 nm (Fig. 7.9).

The credit for finally resolving the structure of DNA belongs to two young biochemists: James Watson and Francis Crick. They chose to tackle the issue strictly as a problem in three-dimensional geometry. They knew that the structure had to accommodate Franklin's diffraction patterns and could not violate the rules that Chargaff had discovered. Equally importantly, they knew that the structure would have to explain both how DNA could contain the enormous complexity of genetic information *and* how this information could be faithfully replicated at every

Figure 7.9
X-ray diffraction patterns of DNA. Different preparations of DNA are bombarded with X-rays, and the resulting reflections and shadows are photographed. Pictures like these, obtained by Rosalind Franklin, were crucial to resolving the structure of DNA. The two diagonals at the center (X-shape) in (a) and (b) suggested a double-helical pattern of "rungs"; the dark shadows at the top and bottom of the picture suggested that the bases were stacked parallel to the axis of the double helix.

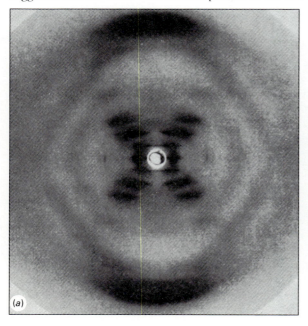
(a)

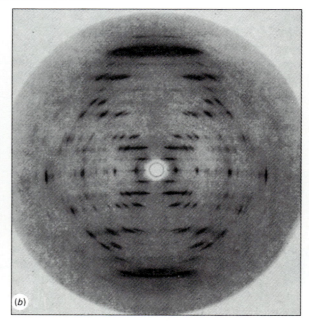
(b)

event of cell division. They arrived at the solution in 1953 by making cardboard scale models of each of the four nucleotides and then fitting them together in a structure that would meet all of the above criteria.

The structure—the double helix—is one of the icons of the 20th century (Fig. 7.10; see also p. 48, Fig. 3.19). Each of the strands of the helix consists of a linear sucession of nucleotides, covalently bound along their sugar portions. This characteristic part of each nucleotide—the base— points towards the inside of the helix. The two strands of the helix interact in a very specific fashion, each of the bases pairing only with its **complement**. The **base-pairing rules** soon became apparent: adenine can only pair with a thymine from the other strand; guanine can only pair with a cytosine. This pairing is mediated by hydrogen bonds, a molecular interaction strong enough to confer stability to the double helix, yet weak enough to be broken without an enormous expenditure of energy.

The elegant double helix structure immediately suggested to Watson and Crick how faithful copies of DNA might be produced. The strands, or parts of them, unwound and separated. Then the base-pairing rules—A with T, C with G—meant that each of the strands could act as a **template**, or model, dictating the synthesis of a new strand by ensuring that only the appropriate nucleotide would be incorporated into the newly synthesized strand. This meant that DNA did not actually make direct copies of itself, but rather **complementary** copies, which in turn could direct the synthesis of the original DNA nucleotide arrangement. Each of the two strands in the double helix acted as a template, giving rise to exact replicas of the original double helix.

Figure 7.10

A computer-generated image of the DNA double helix, viewed on end (like a kaleidoscope). (Graphic kindly prepared by L. Schoenbach)

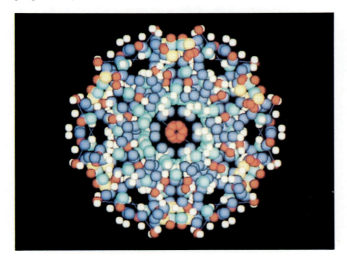

THE REPLICATION OF DNA: SEMICONSERVATIVE COPYING

Although the Watson-Crick double helix suggested the outlines of a model for DNA replication, important questions remained unanswered. What, for example, was the nature of the association between the original template and the newly synthesized copy? In 1958, Matthew Meselson and Franklin Stahl would answer that question with the experiment shown in Figure 7.11. They showed that DNA replication was **semiconservative**. Initially, the two strands of the original double helix are separated by severing the hydrogen bonds that link the complementary bases. Each of the two old strands then serves as a template, and eventually as the partner, of a newly synthesized strand. The resultant DNA molecule is therefore composed of one completely new strand, paired with one old strand. In the subsequent cycle of replication, the old strand once again serves as a template, and ends up paired with a brand new strand. The "middle-aged" strands also serve as templates, and also end up paired with new strands. In effect, the double helix, with its two strands oriented in opposite directions (antiparallel), provides two templates upon which new DNA strands can be synthesized. The elegant, simple rules of base pairing give stability to the double helix. But more importantly, the base-pairing rules account for the accurate replication, cell division after cell division, generation after generation, of the genetic information.

Elegant work carried out over the past twenty years has elucidated the details of DNA replication. This complex process requires the action of more than a dozen different enzymes. In brief, DNA replication is **bidirectional**, with each of the original strands of the double helix acting as templates for the synthesis of new strands. The DNA is first unwound from its tightly coiled arrangement by the enzyme **DNA topoisomerase**. Once the DNA is accessible, replication begins with the attachment of **DNA polymerase** to a specific point in the sequence, the **origin of replication**. DNA polymerases act by adding single nucleotides to the 3′ end of a pre-existing chain of nucleotides (synthesis is said to be 5′→3′). DNA polymerases cannot begin DNA synthesis "from scratch," but instead require a **primer**, a short nucleic chain that acts as an initiator. Curiously, the primers are short RNA molecules to which the new deoxynucleotides are added. The primers are later removed. As you can see from Figure 7.12, one of the new strands (the **leading strand**) is synthesized continuously, in the 5′ to 3′ direction. In contrast, the other strand cannot be synthesized continuously, since no DNA polymerase is capable of synthesis in a 3′ to 5′ direction. The synthesis of this **lagging strand** is carried out in small discontinuous fragments, which are later stitched together into a single nucleic acid chain by the enzyme **DNA ligase**.

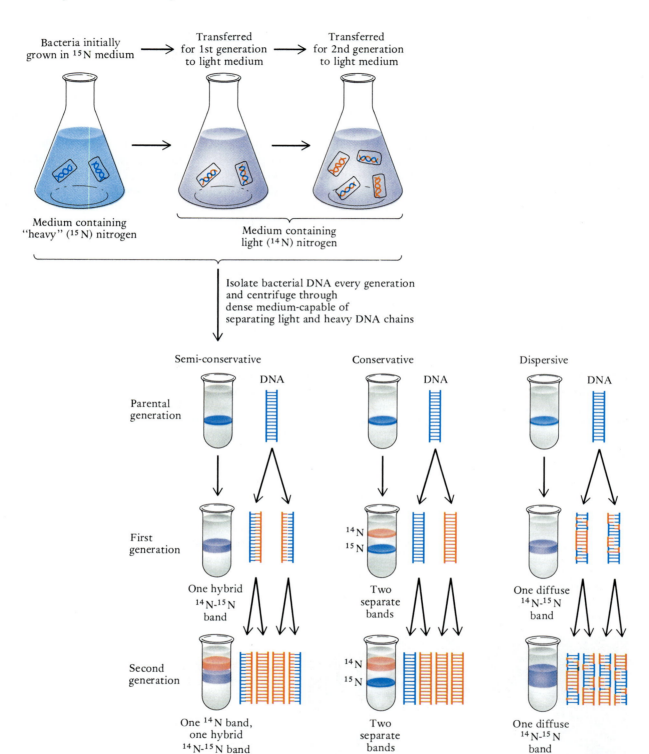

Bacteria initially grown in ¹⁵N medium → Transferred for 1st generation to light medium → Transferred for 2nd generation to light medium

Medium containing "heavy" (¹⁵N) nitrogen

Medium containing light (¹⁴N) nitrogen

Isolate bacterial DNA every generation and centrifuge through dense medium-capable of separating light and heavy DNA chains

Semi-conservative

Conservative

Dispersive

DNA

DNA

DNA

Parental generation

First generation

One hybrid ¹⁴N-¹⁵N band

Two separate bands

One diffuse ¹⁴N-¹⁵N band

Second generation

One ¹⁴N band, one hybrid ¹⁴N-¹⁵N band

Two separate bands

One diffuse ¹⁴N-¹⁵N band

The results of the Meselson-Stahl experiment

The two alternative models of DNA replication disproven by Meselson and Stahl

Figure 7.11

Diagram of the Meselson–Stahl experiment. In order to determine the exact pattern of DNA replication, bacterial cells were first grown in medium containing a heavy isotope of nitrogen ^{15}N. This isotope is incorporated into the DNA of the dividing cells. The cells are then transferred for the next two generations into medium containing a light isotope of nitrogen ^{14}N. "Heavy" and "light" DNA can be told apart by differential centrifugation through a dense medium; the results obtained by Meselson and Stahl, shown on the left, confirmed that DNA replication was semiconservative. The competing dispersive and conservative models of DNA replication made specific predictions about the outcome of this experiment that were never seen.

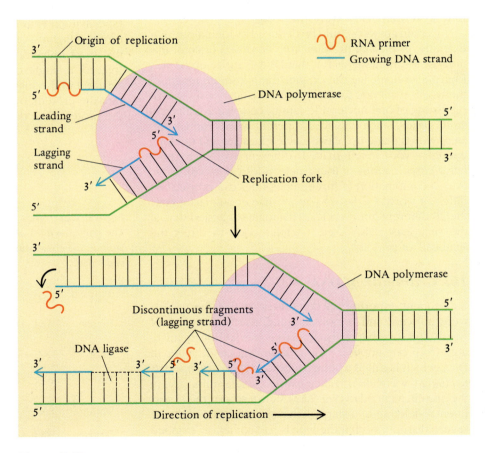

Figure 7.12

The replication of DNA involves the simultaneous synthesis of two new strands, using both of the old strands as templates. DNA, however, can only be synthesized by the addition of new nucleotides at the 3′ end of a growing molecule. The DNA polymerase can thus make the leading strand as a continuous 5′→ 3′ chain, but the lagging strand must be synthesized in the opposite direction as a set of short discontinuous fragments that are then stitched together by DNA ligase. The synthesis of both strands begins at a particular point in the sequence, the **origin of replication**, and uses short RNA primers to initiate the formation of new DNA molecules.

INTERPRETING DNA: THE GENETIC CODE

How Long Are the Words?

The experiments we have been discussing (and many we do not have room to describe) confirmed beyond reasonable doubt that DNA was the chemical molecule of genetic information. The cell was taking information in the form of a long string of covalently bound nucleotides (DNA) and using it to build proteins, long chains of covalently bound amino acids. The question, of course, was "how?"

The structure and composition of DNA suggested that combinations of nucleotides might form molecular "words" that the cell was somehow interpreting. While this was quickly accepted as the correct metaphor, the details needed to be worked out. What were these "words?" How long were they? How were they read?

Certain answers could be deduced by logic alone. For example, given that proteins are composed of combinations of 20 amino acids but that DNA is made up of only four nucleotides, we can deduce the minimum length of the words that code for an amino acid. Obviously, words consisting of one nucleotide cannot code for 20 different amino acids, since only four different words can be created. Words made up of two nucleotides are still not long enough—you can only make 16 different words using two-letter combinations of four nucleotides ($4^2 = 16$). However, a combination of three nucleotides can produce 64 different words ($4^3 = 64$)—more than enough to specify the 20 different amino acids. This logic suggested that, at the very least, DNA used a **triplet code** to direct the synthesis of proteins.

The Grammar of DNA

Subsequent experiments demonstrated that the triplets were adjacent and nonoverlapping, much like a set of words run together without spaces, but still distinguishable from one another. The change of a single letter in DNA, which is called a **point mutation** (Ch. 8), (L. *mutare*, to change) was shown to alter at most a single amino acid in the protein it encoded (Fig. 7.13). Furthermore, the experimental deletion of a single letter in a triplet caused very severe disruption, as if the words had somehow stopped making sense. But perhaps the most surprising finding in this set of experiments concerned the results of experiments that removed three adjacent nucleotides (a full word, if you will). In these experiments, a protein was still made, albeit a slightly different one, since it was missing a single amino acid. The grammar of DNA was beginning to emerge: the **codons** (the "words" we have been discussing) are three nucleotides long and they are read consecutively, without overlap, and in a single direction. Only the meaning of the words remained a mystery: what amino acid did each of the triplets encode?

Cracking the Wobbly Code

The realization that every codon in DNA was three nucleotides long presented biologists with a glaring paradox. The four letters of DNA—A, G, C, and T—could give rise to 64 different triplet codons. But only 20 amino acids existed in living systems; why use a 64-word molecular vocabulary to designate 20 objects? To be sure, a two-nucleotide codon could only specify 16 different amino acids, but that is an insufficient answer. The real question concerned the translation of 64 different codons into 20 amino acids.

Perhaps, some speculated, only 20 of the 64 possible codons actually encode for amino acids, and the remaining 44 are "stop" or "nonsense" codons. If that were the case, however, changing a single base within a codon would most frequently (44 out of 64 times) transform a useful codon (specifying an amino acid) into a nonsense codon (specifying no amino acid), and thus result in the premature termination of the protein. Yet experiments showed that most mutations did not result in early termination. Instead, a single change in one base in a triplet often resulted in the incorporation of an incorrect amino acid into the protein, but seldom interrupted the synthesis of the protein. The data suggested that the triplet code was

Figure 7.13

The consequences of point and frame-shift mutations on the amino acid chain translated from a hypothetical DNA sequence.

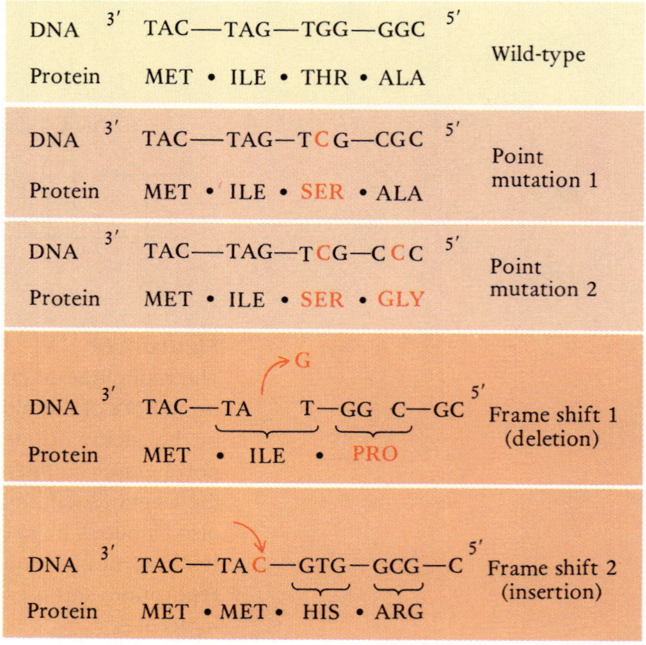

more like a thesaurus than like a dictionary: different codons were clearly specifying the same amino acid. The triplet code was said to be a **degenerate code**, in which more than one codon can specify the same amino acid.

The actual meaning of each triplet still remained to be worked out. H. Ghobind Khorana, Marshall Nirenberg, and their colleagues aggressively and ingeniously tackled this problem using a variety of methods. A short stretch of messenger RNA (mRNA), one of the intermediate molecules used by the cell to convert the information in DNA into proteins (discussed below), could be synthesized *in vitro*. When a message composed solely of the ribonucleotide uridylic acid (U) is placed in a system containing all of the necessary components for protein synthesis, the result is a protein composed entirely of a string of phenylalanines. The triplet UUU thus coded for the amino acid phenylalanine. This experimental approach, however, could only resolve the meaning of four of the 64 possible codons. A complementary set of experiments involved the chemical synthesis of three-nucleotide-long mRNA molecules. These short messages directed the assembly of a complex that included the synthetic mRNA, a ribosome, and the appropriate tRNA (transfer RNA) with its specific amino acid (see below, p. 128). Each mRNA triplet thus ended up in a complex with a single, specific amino acid, whose identity could be readily determined. In this manner, the meaning of the remaining codons was quickly elucidated. The result of these and other experiments is the **genetic code**, shown in Table 7.3. This is the Rosetta stone of molecular biology, the key to deciphering every codon in DNA.

Certain features of the genetic code deserve special mention:

1. By convention, Table 7.3 lists the mRNA triplet and its amino acid translation. Given the base-pairing rules, you should easily be able to determine the corresponding nucleotide triplet in DNA. For example, the tryptophan codon UGG will appear in the DNA strand as ACC (and as TGG in the other strand of the DNA double helix).

2. Synonymous triplets are grouped together in the table. With very few exceptions, the synonymous codons differ only in their third positions, a feature known as **wobble**. Serine is the only amino acid encoded by codons that differ at their first or second positions (UCU and AGU, for example). The third codon position is not always insignificant: the two codons for phenylalanine (UUU and UUC) permit either pyrimidine (U or C) to occur at the third position. If that position is occupied by a purine (A or G), the meaning of the codon changes and now encodes for leucine. Finally, the amino acids methionine and tryptophan are only encoded by a single triplet.

3. Three of the triplets (UAA, UAG, and UGA) do not code for any amino acid, acting instead as **termination** or **stop codons**. These triplets interact directly with a small set of proteins, the **termination factors**, that cause the breakup of the protein synthesis complex. Whenever these codons appear, either at the end of a coding sequence or within a coding sequence as a result of mutation, the consequences are the same: protein synthesis is terminated. Bear in mind that mutations in the DNA can transform a stop codon into an amino acid encoding triplet. It is perhaps to guard

Table 7.3
The Genetic Code

First position	Second position				Third position
	U	C	A	G	
U	Phenylalanine	Serine	Tyrosine	Cysteine	U
	Phenylalanine	Serine	Tyrosine	Cysteine	C
	Leucine	Serine	Stop	Stop	A
	Leucine	Serine	Stop	Tryptophan	G
C	Leucine	Proline	Histidine	Arginine	U
	Leucine	Proline	Histidine	Arginine	C
	Leucine	Proline	Glutamine	Arginine	A
	Leucine	Proline	Glutamine	Arginine	G
A	Isoleucine	Threonine	Asparagine	Serine	U
	Isoleucine	Threonine	Asparagine	Serine	C
	Isoleucine	Threonine	Lysine	Arginine	A
	Methionine	Threonine	Lysine	Arginine	G
G	Valine	Alanine	Aspartic acid	Glycine	U
	Valine	Alanine	Aspartic acid	Glycine	C
	Valine	Alanine	Glutamic acid	Glycine	A
	Valine	Alanine	Glutamic acid	Glycine	G

against such run-on protein synthesis that most mRNAs contain two contiguous stop codons at the end of the coding sequence.

THE CENTRAL DOGMA: PROTEIN SYNTHESIS

Although we have been speaking of DNA "encoding" the information for the synthesis of proteins, we have up to now said little about the actual machinery involved in protein synthesis. The molecular interpretation of DNA consists of two basic steps:

1. **Transcription.** DNA is transcribed into a molecule that will carry the genetic information out of the nucleus and into the cytoplasm, where the machinery for protein synthesis in eukaryotes is found (see Chap. 5). This molecule is an RNA chain, **messenger RNA (mRNA)**.

2. **Translation.** In the second phase of the process, the information carried by mRNA is translated into an amino acid chain, a polypeptide. Translation is carried out by a complex machinery centering on tiny cellular particles, the ribosomes, and involving another class of RNA, **transfer RNA (tRNA)**.

The flow of biological information from DNA to mRNA and finally to protein is known as the **central dogma** of molecular biology (Fig. 7.14). In the words of Francis Crick, one of the discoverers of the structure of DNA,

> The central dogma states that once "information" has passed into protein it cannot get out again. The transfer of information from nucleic acid to nucleic acid, or from nucleic acid to protein, may be possible, but transfer from protein to protein, or from protein to nucleic acid, is impossible.

To date, the central dogma has held up well. No case has yet been found where information in the form of a protein directs the synthesis of a specific nucleic acid sequence.

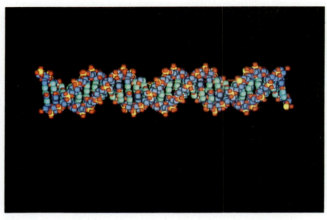

(a)

Transcription and translation

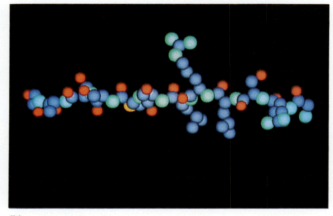

(b)

Protein folding

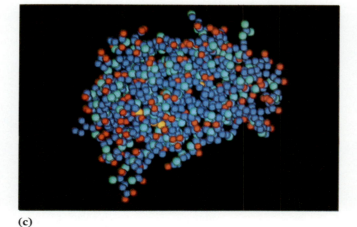

(c)

Figure 7.14
The central dogma: information flow in living systems. The inherited information necessary to synthesize proteins is encoded in nucleic acids (in this case, DNA). The DNA (a) is transcribed into mRNA and translated into a polypeptide (b), a linear series of amino acids. In turn this linear chain adopts a particular three-dimensional shape (c) necessary to protein function. While the rules governing the first step (DNA→peptide) are well understood, the second step (amino acid sequence→three-dimensional protein structure) is not yet fully elucidated.

Transcription

Synthesizing the Primary Transcript

Although DNA is double-stranded, only one of the two chains, the **coding strand**, actually carries the information necessary to direct the synthesis of a protein. It is this coding strand that serves as the template for the synthesis of the mRNA. mRNA is a single-stranded molecule, synthesized in strict adherence to the base-pairing rules. As a result, the sequence of the mRNA molecule is identical to that of the complementary non-coding strand of the double helix, except that mRNA uses ribose instead of deoxyribose as the sugar, and uracil (U) takes the place of thymine (T). The synthesis of mRNA is catalyzed by a family of enzymes, the **RNA polymerases**. RNA polymerase binds to a specific triplet sequence, the **promoter**, located upstream from the initiation (first) codon of the gene being transcribed, "melts" or separates the double helix, exposing single strands, and begins the synthesis of the mRNA **primary transcript**, complementary to the coding strand of DNA (Fig. 7.15). The presence of promoter sequences, which the RNA polymerase complex can recognize, distinguishes the coding strand from the complementary non-coding strand.

Editing and Exporting the Transcript

Surprisingly, the coding regions of most eukaryotic genes are interrupted by **introns**, long stretches of non-coding DNA. The full set of DNA triplets necessary to make a complete protein is therefore frequently divided into blocks of coding sequence, the **exons**, separated by these introns (Fig. 7.16). Picture a page from this book in which the actual text was occasionally interrupted by paragraphs of random words. If you were editing the book, you would have to somehow remove the nonsense paragraphs and join together the actual text into a coherent page. The eukaryotic cell faces much the same task when the primary mRNA transcript is first synthesized. This transcript is not yet suitable for translation into protein and must be edited by removing the intron sequences and joining together the exon messages into a single continuous message containing only the coding sequence. This process, known as **splicing**, involves a complex of proteins and RNAs that form a small particle, the **spliceosome**, which precisely recognizes the boundaries between exon and intron, cleaves the mRNA at the junctions, and rejoins the two ends of the exons into a continuous coding sequence (Fig. 7.16). This fully edited mRNA is then exported from the nucleus

Figure 7.15

Transcription involves the melting of the DNA double helix and the binding of RNA polymerase to a promoter sequence on the coding strand of the DNA. The synthesis of the mRNA transcript begins at the first start codon, and proceeds in a 5′ to 3′ direction until a stop codon is reached.

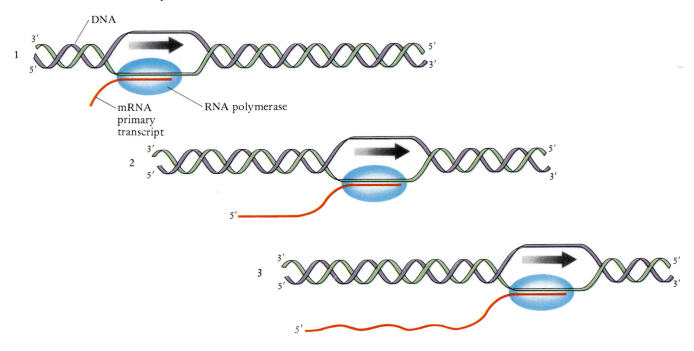

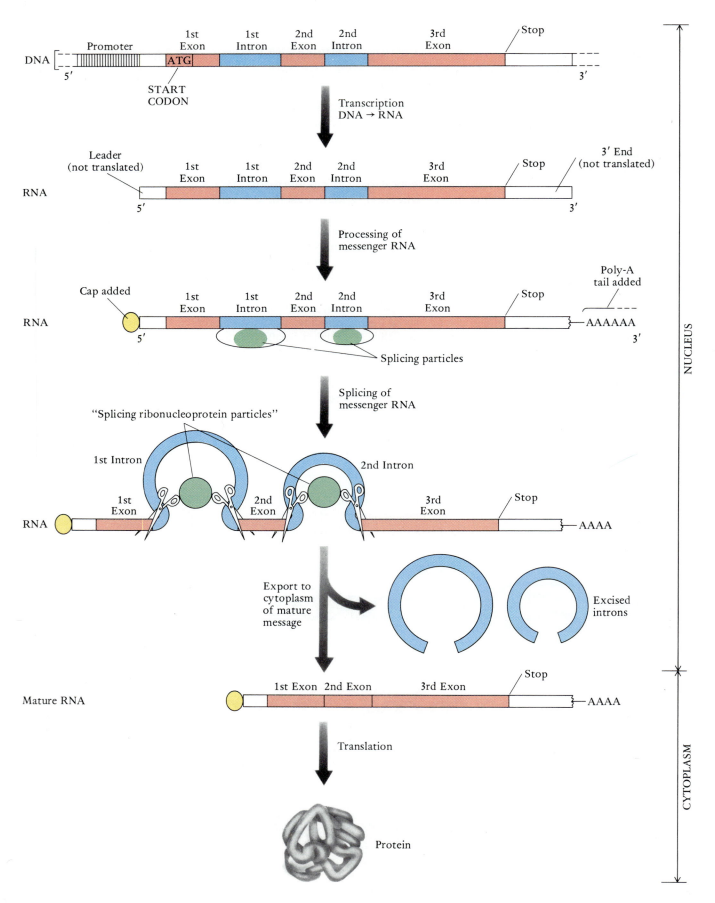

Figure 7.16

Most eukaryotic genes are discontinuous: portions of the coding sequence (exons) are interrupted by stretches of non-coding DNA (introns). The making of a functional mRNA molecule involves a series of steps, beginning with the synthesis of an RNA strand, using the DNA coding strand as a template. The mRNA is then modified by the addition of a "cap" and a "tail," sequences that facilitate translation, and by the removal of the introns. In this splicing step, the introns are precisely cut out from the message by spliceosomes, complexes of RNA and protein (intron size is exaggerated to show splicing particles). The mature mRNA, now encoding a set of continuous instructions for making a peptide, is exported into the cytoplasm, where it is translated into a complete protein.

to the cytoplasm through pores in the nuclear envelope. This message is now ready for translation into a polypeptide.

Translation

The Ribosome

A battery of molecular interpreters must now undertake the task of translating the mRNA into a specific amino acid chain. The primary agent in this process is a small particle, the ribosome, which consists of a "small" and a "large" subunit (Fig. 7.17). The subunits themselves are made up of a third and extremely abundant type of RNA, **ribosomal RNA (rRNA)**, and together contain up to 82 specific proteins assembled in a precise sequence. The ribosome resembles an interlocking Chinese puzzle, whose constituents must be put together in an extremely precise position and sequence. Ribosomes are very similar in structure

Figure 7.17

(a) Diagram showing the fit between the small and large subunits of the ribosome. When assembled, the ribosome exhibits a groove between the two clefts where the active P and A sites for translation are found. (b) Electron micrograph of the rough endoplasmic reticulum of a mouse liver cell ($100,000 \times$), showing the abundant ribosomes bound to the E.R. membrane. This cell is actively synthesizing proteins.

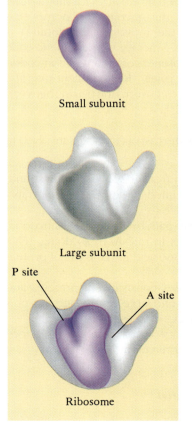

Small subunit

Large subunit

P site

A site

Ribosome

(a)

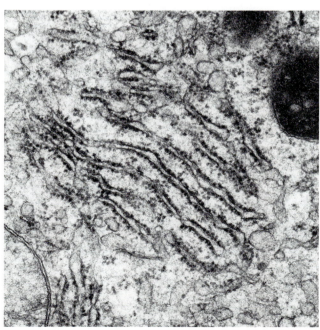

(b)

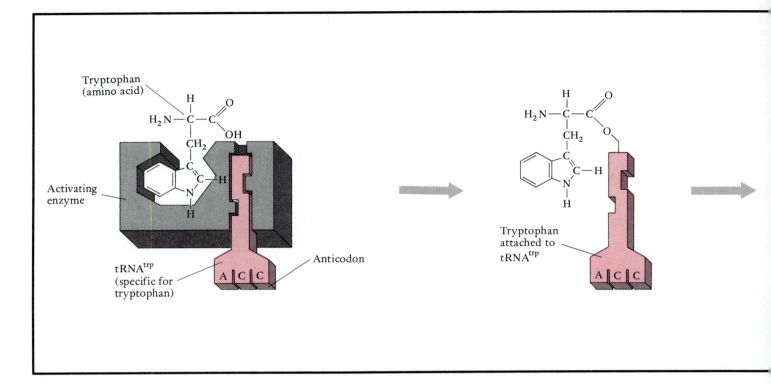

and organization in all organisms: the basic assembly instructions for the ribosome have evolved little since they appeared early in the history of life.

The assembled ribosome displays a series of small grooves, tunnels, and platforms, where the action of protein synthesis occurs. These are the **active sites**, each of them dedicated to one of the tasks required for the translation of mRNA into protein.

Transfer RNA (tRNA)

As noted previously, tRNA is the actual "translator" molecule. The genetic instructions for the synthesis of tRNAs are contained in the genome of all eukaryotes. As with rRNAs, multiple copies of tRNA "genes" are present; these genes are distributed throughout the chomosomes. Two critical events in protein synthesis involve tRNA: **charging** and **pairing**. The charging of a tRNA molecule refers to the addition of a specific amino acid to the appropriate tRNA. This critical and delicate step in protein synthesis is mediated by an **activating enzyme**; a different activating enzyme exists for each of the 20 different tRNAs (Fig. 7.18). Any error in the charging of tRNA will result in the incorporation of an incorrect amino acid during protein synthesis. Activating enzymes have therefore evolved to insure great accuracy.

The pairing between mRNA and tRNA takes place at the **anticodon**, a set of three ribonucleotides found in the anticodon loop of the tRNA molecule. The anticodon pairs with its complementary codon in the mRNA molecule (Fig. 7.19). A tRNA molecule (with its appropriate anticodon and amino acid) exists for every codon in the genetic code

(except the stop codons). Thus, for example, the AUG codon in mRNA is recognized by the tRNA for methionine (abbreviated tRNA^{met}). As this recognition occurs, the amino acid specified in the message (methionine) is brought into place. The degeneracy of the genetic code (p. 123) arises from the looseness of the anticodon-codon pairing. A set of "wobble rules," less stringent than the traditional base-pairing rules, governs the pairing between mRNA and tRNA.

Let us now assemble the cast of characters involved in protein synthesis:

1. mRNA, transcribed from the DNA coding strand, spliced and otherwise modified in the nucleus, and transported into the cytoplasm.

2. tRNAs, each able to pair, by virtue of its anticodon, to a complementary triplet in mRNA. In so doing, tRNA brings the appropriate amino acid into the site of protein synthesis.

3. Ribosomes, particles consisting of proteins and rRNA. Two subunits (small and large) come together, providing the stage for protein synthesis.

4. A variety of accessory proteins—initiation factors, elongation factors, and termination factors.

Assembling the Protein

The process of protein synthesis in eukaryotes is depicted in Figure 7.19. It begins with the capture of tRNA^{met} by an initiation factor, which binds to a small ribosomal subunit.

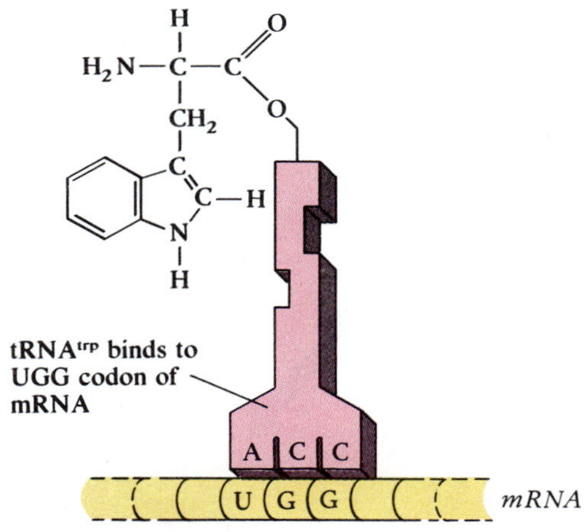

Figure 7.18
A diagram of a tRNA molecule as it is charged with the appropriate amino acid. An activating enzyme that is specific for each amino acid binds with the amino acid (in this case, tryptophan) and also with a specific tRNA (one specific for tryptophan). The amino acid and tRNA are bound together, and the enzyme separates. The tRNA has a specific anticodon at one end that recognizes and binds with the complementary codon on mRNA.

This initiation complex recognizes and binds to the 5' end of an mRNA molecule and slides down to the **initiation codon**, always an AUG.

The large subunit of the ribosome now joins the complex; the tRNA^{met} occupies one of the active sites in the ribosome, the P (protein) site. A second tRNA is now brought into the ribosome by the **elongation factor**. If the anticodon of this tRNA pairs with the next codon of the message, the tRNA occupies the A (acceptor) site on the ribosome, thus positioning the second amino acid adjacent to the initiation methionine. An enzyme, **peptidyl transferase** (part of the large ribosomal subunit) mediates the separation of the first amino acid from its tRNA and the formation of a peptide bond between the first and second amino acids. The nascent chain, bound to the second tRNA, occupies the A site. The P site is now occupied by an uncharged tRNA molecule.

The ribosome will now move down the mRNA by one codon, a process known as **translocation**. This movement shifts the growing polypeptide chain to the P position, and results in an empty A site, where a new charged tRNA can enter and pair. The uncharged tRNA that previously occupied the P site is booted out of the ribosome and will be recharged and recycled by the cell. Like all motion, the translocation step requires energy, provided by the hydrolysis of a guanosine triphosphate (GTP) molecule (see p. 49). The process continues along the length of the mRNA, until the first stop codon is encountered. At that point, the action of a **termination factor** releases the completed protein from the last tRNA, and the ribosome dissociates into its component parts.

As we saw in Chapter 5, proteins being synthesized for export out of the cell are made by ribosomes attached to the rough endoplasmic reticulum. In contrast, proteins for use by the cell are generally made in the cytoplasm by free ribosomes. Several of these free ribosomes may attach to a single mRNA molecule, giving rise to a characteristic structure, the **polyribosome** or **polysome** (Fig. 7.20). A single message is thus read several times, and multiple polypeptide molecules can be synthesized simultaneously.

Not all genes in the eukaryotic genome are transcribed to the same degree or in all tissues. Thus, for example, mammalian fetuses make large amounts of a red blood cell protein, fetal hemoglobin. Shortly after birth, the gene for fetal hemoglobin is switched off, and the gene for adult hemoglobin switched on. The synthesis of certain proteins is also **tissue-** or **cell-specific**: immunoglobulins, for example, are synthesized only by specialized cells of the immune system (see Ch. 15), albumin is produced only by liver cells. The control of gene expression lies at the root of the molecular understanding of development. The complex mechanisms of transcription and translation provide multiple control points. The expression of a gene can therefore be regulated at several stages: transcription, message splicing and editing, message transport, ribosome assembly, and translation. The complex process of development—a process in which a set of instructions for the synthesis of individual proteins gives rise to an actual organism—is made possible by the coordinated spatial and temporal control of gene expression. We will return to the molecular events of early development in Chapter 22.

INITIATION OF PROTEIN SYNTHESIS

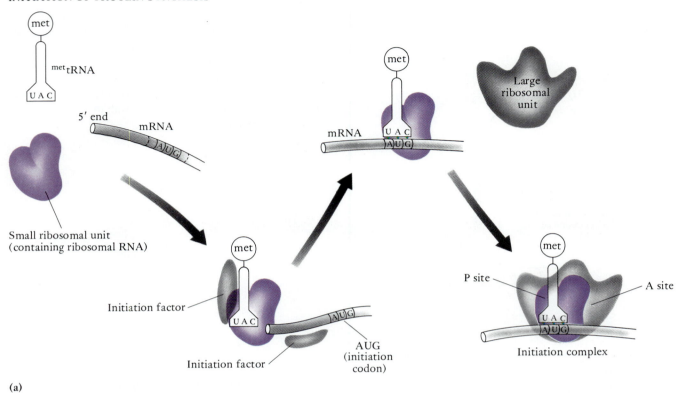

(a)

ELONGATION STEP OF PROTEIN SYNTHESIS

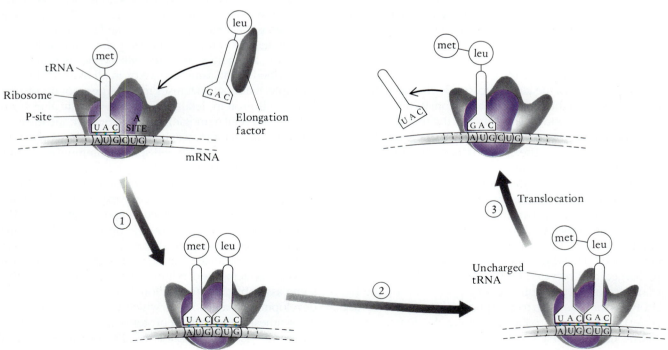

(b)

TERMINATION OF PROTEIN SYNTHESIS

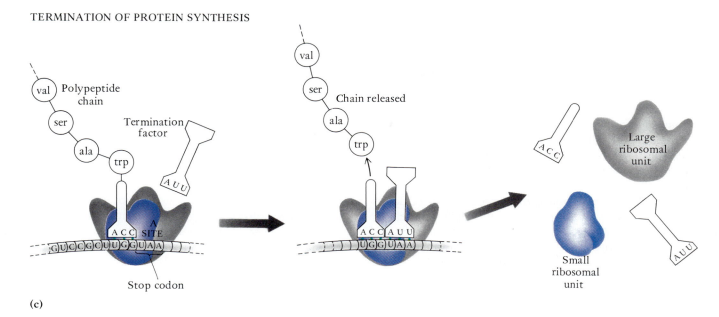

(c)

Figure 7.19

Translation begins with the assembly of an initiation complex including a ^(met)tRNA (paired with the AUG start codon of the mRNA), the two subunits of the ribosome, and a complex of proteins (initiation factors). The ^(met)tRNA occupies the P site of the ribosome. During elongation, an elongation factor brings a new charged tRNA into the A (acceptor) site of the ribosome. If the anticodon of this tRNA is complementary to the next mRNA codon, the ribosome then forges a peptide bond between the initial methionine and the new amino acid (step 2). The growing peptide, attached to the tRNA, is then translocated (step 3) from the A to the P (peptide site) by the movement of the ribosome along the mRNA, and the uncharged tRNA is released. This process will continue until the first stop codon (UAA) in the mRNA is reached. At that point, a termination factor will occupy the A site, the bond holding the completed peptide to the tRNA is severed, and the entire translation apparatus dissociates into its component parts, which are then recycled.

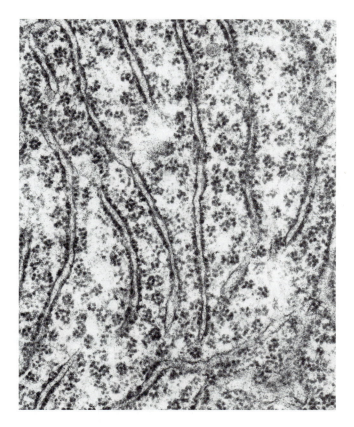

Figure 7.20

Electron micrograph of rabbit nerve cell (102,600 ×), showing ribosomes not bound to the E.R. These polysomes frequently adopt a rosette shape; they bind to a single mRMA molecule and permit the simultaneous translation of several copies of a protein from a single mRNA.

Summary

1. Many experiments pinpointed the nucleus as the site of genetic information, but experiments by Avery and his colleagues were the first to show that the deoxyribonucleic acid (DNA) of the chromosomes was the repository of genetic information. This work was confirmed by the experiments of Chase and Hershey, which showed that it was the DNA core of the T$_2$ bacteriophage and not its protein coat that entered bacteria, usurped their metabolic machinery, and led to the formation of more phage.

2. That genes control the synthesis of specific enzymes (proteins) was demonstrated early in this century by genetic and biochemical studies of alkaptonuria and other human diseases that are caused by inborn errors of metabolism. The one-to-one relationship between a gene and a protein product was confirmed by Beadle and Tatum in experiments with mutants of the mold *Neurospora*.

3. Watson and Crick, building upon the discoveries of others, showed that DNA consists of two nucleotide strands arranged in a double helix. Each strand is the complement of the other, for the base adenine on one strand only pairs with the base thymine on the other, and guanine on one strand only pairs with cytosine on the other. The strands are held together by weak hydrogen bonds that, on breaking, permit the strands to separate. Each strand acts as a template for the copying of, and eventually as the partner of, a new complementary strand. The resulting DNA molecule is composed of one old and one new strand. This process is known as semi-conservative copying. The successful replication of DNA involves the concerted action of several enzymes. DNA synthesis is bidirectional, beginning at the origin of replication, and is carried out by DNA polymerase. Synthesis always proceeds in a 5′ to 3′ direction. As a result, the leading strand is synthesized continuously, but the lagging strand must be synthesized in short fragments that are subsequently joined together by DNA ligase.

4. The sequence of nucleotides in the strands, form a series of molecular "words" that code for the sequence of amino acids in proteins. The code is a set of nonoverlapping nucleotide triplets, which enables it to code for 64 different "words," or codons. Since 64 codons are more than enough to code for the 20 amino acids used in protein synthesis, the code is termed "degenerate" in the sense that many codons, which differ in their third position, code for the same amino acid. Three of the codons are stop codons and cause termination of protein synthesis. The code has been determined in terms of the sequence of nucleotides in messenger RNA (mRNA).

5. During protein synthesis, information flows from DNA through mRNA to protein. This one-way flow of information is termed the central dogma of molecular biology. Information may flow from one nucleic acid to another, as it does in certain viruses, but information in protein never directs the synthesis of nucleic acids.

6. The flow of information begins with the transcription within the nucleus of information on one of the DNA strands (the coding strand) into a complementary mRNA strand. mRNA differs from DNA in having a different sugar (ribose) and in having the base uracil (U) instead of thymine. The first mRNA formed is the primary transcript, which must be "edited" in the nucleus to remove noncoding sequences of nucleotides (the introns) and splice meaningful coding sequences (the exons) together. Edited mRNA leaves the nucleus to enter the cytoplasm, where it is translated into a specific amino acid sequence.

7. When mRNA molecules leave the nucleus, their 5′ ends attach to small cytoplasmic particles known as ribosomes. Ribosomes consist of ribosomal RNA (rRNA) and protein, and they are the sites where the code is translated and protein synthesized. Other small nucleic acids, the transfer RNAs (tRNA), attach to free amino acids in the cytoplasm. Each tRNA molecule is specifically and uniquely charged with a single amino acid. An anticodon loop within the tRNA molecule pairs with a codon on mRNA and brings the appropriate amino acid to the ribosome. As the ribosomes move down the mRNA strand, the ribosomes read the code, accept the appropriate tRNA molecules, form peptide bonds between adjacent amino acids as they come into place, and then release the tRNA and eventually the completed polypeptide chain. All of these reactions are catalyzed by enzymes and involve the action of initiation factors, elongation factors, and termination factors. The energy needed for protein synthesis is provided by the breakdown of guanosine triphosphate (GTP).

References and Selected Readings

Crick, F.H. The genetic code (III). Scientific American 215 (Apr. 1966): 55–62. A classic description of the experiments that eventually broke the genetic code.

Darnell, J.E., Jr. RNA. *Scientific American* 253 (Oct. 1985):68–88. The role of RNA in transcribing and translating DNA into protein, and the role of RNA in molecular evolution.

Felsenfeld, G. DNA. *Scientific American* 253 (Oct.

1985):58–67. The structure and organization of DNA.

Judson, H.F. *The Eighth Day of Creation: Makers of the Revolution in Biology.* New York: Simon and Schuster, 1979. A fascinating and thorough chronicle of the birth and development of molecular biology. Judson manages to effectively convey both the science and the scientists. The personalities, pressures, conflicts and motivations that drive any scientific field are apparent in this account.

Kornberg, A. *DNA Replication.* San Francisco: W.H. Freeman, 1980. A very detailed but clearly written textbook emphasizing the complex machinery responsible for the copying and transmission of genetic information.

Lake, J.A. The ribosome. Scientific American 245 (Feb. 1981): 84–97. Describes how electron micrographs and biochemical evidence are combined to give a full picture of the structure and function of ribosomes.

McCarty, M. *The Transforming Principle: Discovering that Genes Are Made of DNA.* New York: W.W. Norton, 1986. A history of the discovery of the chemical nature of the genes.

Peters, J.A. *Classic Papers in Genetics.* New York: Prentice Hall, 1959. This collection includes all the milestones in the field of genetics, from Mendel to the discovery of the structure of DNA. Illustrates not only how much we have learned in the last century, but also how differently we now go about understanding genetic mechanisms.

Watson, J.D. *The Double Helix.* New York: Atheneum, 1968. Watson's lively story of his and Crick's discovery of the structure of DNA.

Watson, J.D., N.H. Hopkins, and J.A. Steitz. *Molecular Biology of the Gene*, 4th ed. Menlo Park, Ca.: Benjamin Cummings, 1987. A textbook on the structure and action of the genetic material; advanced, but a valuable reference.

8

Evolution I: The Origin of Evolutionary Novelty

(Left) A white-footed mouse (Peromyscus leucopus) *feeding on an ear of corn* (Zea mays)*. (Above) A single recessive mutation is responsible for the* Nude *phenotype in laboratory mice. These mice are not only hairless, but lack a thymus gland, have inefficient immune systems, and reduced fertility and longevity. The gene affected by the mutation has not yet been identified.*

Having sketched out the two fundamental pillars upon which all of modern biology rests—the Darwinian theory of evolution and the Mendelian theory of genetics—it is now time to build an arch between them. In the next two chapters, our task will be to unite the historical perspective of evolutionary biology with the functional perspective of modern genetics. A successful and complete theory of biological change needs to synthesize these two approaches: we wish to know both *what* has happened in the history of life as well as *how* it has happened.

This chapter will deal principally with the origin of genetic novelty. We will detail the mechanisms involved in the production of new DNA sequences, new gene arrangements, and new chromosome combinations. The majority of these changes have negative consequences on the organism—they are mistakes for which the organism pays dearly. But occasionally these mistakes are actually advantageous. Propelled by natural selection, they will spread in the population. Such novelties are the raw material of evolutionary change.

THE ORIGIN OF NOVELTY

The discovery of the architecture of DNA, and of the elegant molecular machinery involved in its replication, made it clear how faithful, error-free copies of the genetic instructions could be made generation after generation. Yet the existence of such complex mechanisms to ensure the fidelity of genetic transmission presented biologists with a new problem: how did genetic variation ever arise?

The problem was not simply academic. As you recall, Darwin's theory is a *variational* theory. Natural selection does not transform entire populations in a single generation. Instead, certain individuals, by virtue of possessing heritable traits advantageous in the local environment, are more likely to survive and reproduce, and hence to achieve greater representation in subsequent generations. Heritable variation among individuals becomes a central concept in evolutionary theory; without it, there can be no evolutionary change. In this section we take a closer look at the mechanisms responsible for the generation of variation in natural populations.

Point Mutations

DNA is the material of heredity, the substance linking one generation to the next. Elaborate mechanisms have evolved to ensure its successful, letter-perfect copying. Thus, the enzymes responsible for the copying of DNA (the DNA polymerases, p. 119) are usually also capable of "proofreading" the strand that they are synthesizing. In bacteria, a DNA polymerase verifies that the correct nucleotides are always added to the DNA strand being synthesized. If it detects a mistake in base pairing—e.g., a thymine, for example, has been added as a complement to a guanine instead of the normal cytosine—the polymerase is capable of removing the mispaired base. This enzyme takes one step forward (synthesis), checks what it has just done (proofreading), and, if necessary, takes one step back to correct the mistake (excision).

In eukaryotes, a similar but more complex **excision repair system** serves the same function. The enzymes involved in excision repair survey newly synthesized double-stranded DNA helices. If a mispairing is detected, enzymes remove a whole section of the damaged strand and synthesize a new section to replace it (Fig. 8.1).

Despite the existence of this active proofreading machinery, mistakes do occur and may go undetected. Perhaps the simplest mutation is the chemical change that transforms one base in DNA to another. Such **point mutations** may come about because of a mistake in base pairing during the replication of DNA (Fig. 7.13) or as a result of the modification or damage of a given base in the DNA chain. Organisms are under constant attack from various physical forces, including ultraviolet (UV) rays from sunlight and X-rays, and from a variety of chemicals. These agents, known as **mutagens**, are capable of inducing a change in a DNA base. Thus, even if the replication system were totally foolproof, **spontaneous mutations** in the DNA would still arise. Of course, the rate of mutation is greatly increased by the presence of mutagens; much of our concern with toxic wastes and with the depletion of the ozone layer (which increases the incident amount of UV rays) centers on the possible increase in the induced mutation rate.

Any base pair in any DNA strand is susceptible to mutation. In principle, any nucleotide can mutate to any other, although **transitions**, in which a purine is replaced by another purine (e.g., adenine to guanine or vice versa) or a pyrimidine by another pyrimidine, are far more common than **transversions**, in which a purine replaces a pyrimidine (or vice versa). Mutations that occur in cells other than sperm or eggs are known as **somatic mutations**. Many cancers, for example, arise as a result of somatic mutations that activate a class of genes, the **oncogenes** (Gr. *onkos*, mass), that would normally remain inactive. While such somatic mutations affect all the daughter cells derived from the cell in which the mutation originally arose, these changes are not passed on from one generation to the next. Mutations that occur in the DNA of sperm or egg cells can potentially be transmitted to offspring: these are termed **germ-line mutations**. Our attention will focus primarily on these heritable mutations.

What are the potential consequences of a single point mutation? The answer in part depends on the exact location in the DNA chain where the change occurs. For the moment, we will concentrate on a point mutation that occurs in a stretch of DNA that directs the synthesis of a particular protein. Recall that the DNA code is degenerate: a single amino acid can be encoded by as many as six

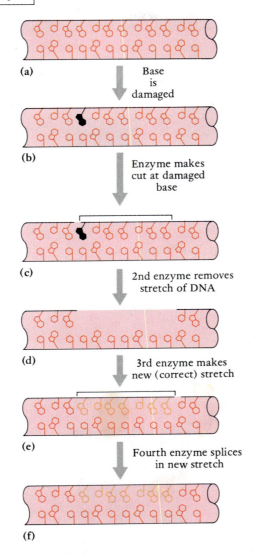

Excision—repair

(a)

Base
is
damaged

(b)

Enzyme makes
cut at damaged
base

(c)

2nd enzyme removes
stretch of DNA

(d)

3rd enzyme makes
new (correct) stretch

(e)

Fourth enzyme splices
in new stretch

(f)

Figure 8.1
A diagram of the eukaryotic excision-repair system. One base in a region of double-stranded DNA is damaged (b). One enzyme cuts the damaged strand at the damaged base (c), and a second enzyme removes a segment of DNA that includes the damaged base (d). A third enzyme makes a new and correct segment of DNA and inserts it into the strand (e), and a fourth enzyme completes the splicing (f).

different DNA codons (p. 123, Table 7.3). As a general rule, synonymous codons generally share the first and second positions, but differ in their third position. Thus, for example, the triplets CCU, CCC, CCA, and CCG all code for the amino acid proline. A point mutation at the third position of this codon, one that might transform a C to a G and thus change the codon from CCC to CCG, would have no real effect on the protein being synthesized. In contrast, if the same point mutation occurred in the second posi-

tion, it would change the codon from CCC to CGC. This new codon has a different meaning and would direct the incorporation of the amino acid arginine in the place of proline.

The effects of a single amino acid change on the function of a protein range from the imperceptible to the very dramatic. Many important heritable diseases, including sickle-cell anemia and cystic fibrosis, result from a single point mutation in an important protein (Fig. 8.2). The most dramatic point mutations are those that transform an amino acid–encoding codon into a stop codon. The codons for cysteine (UGU and UGC) are but a point mutation away from the stop codon UGA. The mutation of a C to an A at the third position of this codon would likely have disastrous consequences, leading to the premature termination of protein synthesis and the production of a truncated and most likely nonfunctional protein.

Frame Shift Mutations

A more complex modification of the DNA involves the addition (or deletion) of one or more nucleotides. As you might predict, these events are often far more disruptive than point mutations. Let us consider the case of a single base insertion into a coding region. As you remember, DNA is read as a linear chain of contiguous codons. The "information" in DNA derives from the precise linear order of these three-letter words. Thus, by analogy, the following sentence can be thought of as a stretch of DNA:

THEFATBATANDTHEMATRATSAT

If you stare at this for a bit, you'll realize that there is a set of three-letter words, which, when properly divided, do have (some) meaning:

THE FAT BAT AND THE MAT RAT SAT

Note that if you divide the triplets differently, the meaning is hard to see. As with DNA, the appropriate "reading frame" is necessary. Now imagine that a single base insertion takes place. The words are still three letters long, but the insertion completely disrupts the meaning, as in the following:

THE FAT BAT sAN DTH EMA TRA TSA T

The insertion of a single "s" results in a **frame shift**. Such frame shifts inevitably alter the amino acid sequence being directed by that stretch of DNA and often result in the downstream appearance of one or more stop codons.

The role of insertions and deletions as agents of genetic change has attracted increasing attention. For a long time, the majority of genetic changes were thought to involve point mutations. The theoretical possibility of insertions and deletions was recognized, and some experimental evidence suggested that certain errors during rep-

NORMAL

| VALINE | HISTIDINE | LEUCINE | THREONINE | PROLINE | GLUTAMIC ACID | GLUTAMIC ACID | - - - |

SICKLE CELL
ANEMIA

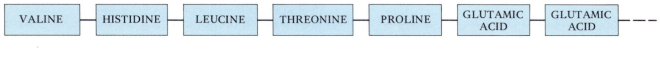

| VALINE | HISTIDINE | LEUCINE | THREONINE | PROLINE | VALINE | GLUTAMIC ACID | - - - |

(a)

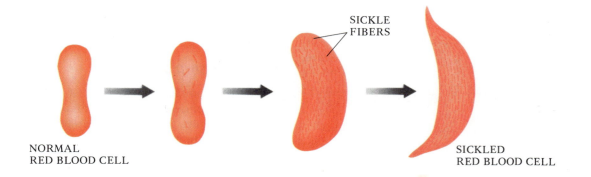

SICKLE
FIBERS

NORMAL
RED BLOOD CELL

SICKLED
RED BLOOD CELL

(b)

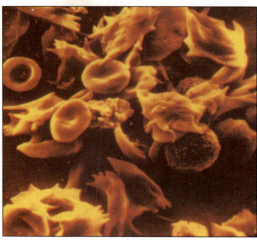

(c)

Figure 8.2

(a) The first seven amino acids in normal and sickle-cell hemoglobin. Glutamic acid, which is normally present at the 6th amino acid position, is replaced by valine. (b) The formation of a sickled red blood cell. The abnormal binding of valine causes a deformation in the normally globular structure of hemoglobin. Hemoglobin molecules aggregate to form chains that distort the shape of the red blood cell, giving it a characteristic sickle shape and reducing its oxygen-carrying ability. (c) Red blood cells from a patient with sickle cell anemia showing characteristic sickle shape ($\sim 4000 \times$).

lication might result in the gain or loss of nucleotides. However, the frequency of such events was thought to be very low. Recently, however, the discovery of **transposable elements**, a large class of DNA sequences that appear capable of moving from one location on the chromosome to another, suggests that insertions and deletions may occur far more frequently than we expected.

Transposable Elements

As early as the 1950s, Barbara McClintock, working on the genetics of corn, had found a gene that frequently caused chromosomes to break. When she tried to map the location of that gene, using the standard recombination methods (see below, p. 145), she obtained a puzzling result.

When looking at different plants that were presumably genetically identical, the same gene was in a different chromosomal position in every plant. Furthermore, this gene did not remain in the same position on the chromosome from one generation to the next. The gene was obviously capable of moving from one place to another on the chromosome, and even from one chromosome to another. These **transposable elements**, or "jumping genes," as they are sometimes called, contradicted the prevailing (and largely correct) notion that the linear arrangement of DNA did not change from generation to generation. It would take 20 years for McClintock's work finally to be embraced by geneticists. Since then, more and

more examples of transposable elements have been discovered, both in prokaryotes and in eukaryotes.

One of the most dramatic examples of moving genetic elements involves **bacterial transposons**. These elements have attracted a good deal of attention because they frequently carry within them genes that confer antibiotic resistance. Disease-causing bacteria carrying such genes are frequently very difficult to combat. To make matters worse, these transposons frequently lodge in **plasmids**, small circular molecules of DNA within the bacterium that exist (and replicate) independently of the bacterial chromosome. The first of these plasmids containing a resistance transposon was identified in the 1950s, when resistant strains of the bacteria that causes dysentery began spreading rapidly in Japan. Such plasmids are easily transmitted from one bacterium to another (and even between bacterial strains) during the process of bacterial conjugation ("bacterial sex") (Fig. 8.3). Resistance to antibiotics thus spreads quickly through a bacterial population. In many Third World countries, where antibiotics are frequently overprescribed, such resistant bacterial strains present significant public health problems.

Other transposable elements such as the **ALU sequences** in human DNA or the **P-elements** in *Drosophila* are more like small invaders that may exist in thousands, or even hundreds of thousands, of copies within the genome of the organism. These transposable elements carry only

(a)

Figure 8.3

(a) An electron micrograph of a conjugation bridge between two bacterial cells. (b) A diagram of the transfer of a resistance plasmid from one cell to another. Plasmids are circular elements of double-stranded DNA (1). A conjugation bridge forms between two bacterial cells (2). One strand of the DNA in the donor crosses the conjugation bridge to the recipient (3 and 4). The strand remaining in the donor and the strand entering the recipient act as templates for the synthesis of new complementary strands (4).

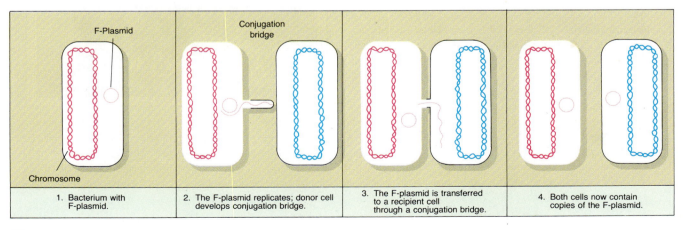

F-Plasmid	Conjugation bridge		
Chromosome			
1. Bacterium with F-plasmid.	2. The F-plasmid replicates; donor cell develops conjugation bridge.	3. The F-plasmid is transferred to a recipient cell through a conjugation bridge.	4. Both cells now contain copies of the F-plasmid.

(b)

the absolute minimum genetic information required for their movement: frequently, these are short characteristic DNA sequences at the ends of the element that appear to facilitate insertion. Certain elements also carry the instructions directing the synthesis of a enzyme (transposase) that enables these elements to insert into the host genome. Transposable elements vary in size from a few hundred bases to 5000 to 6000 bases. Transposable elements may frequently insert within the coding region of a gene. Predictably, such an insertion completely inactivates the gene. (See Focus 8.1.)

Chromosomal Changes

Infrequently, variation within a population is produced by a major change in the structure or number of chromosomes, a far more dramatic alteration of the genetic material of an organism. Such changes are often directly observable under the microscope, particularly in the giant salivary gland chromosomes of certain insects, such as in the fruit fly *Drosophila*. Among the most readily visible changes are chromosomal **inversions**, in which a large segment of a chromosome is flipped around, or inverted (Fig. 8.4). No genetic material is actually lost, but the order

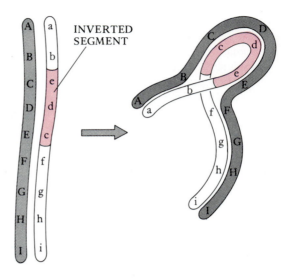

Figure 8.4

A change in the gene sequence resulting from an inversion has occurred in one chromosome, but not in its homologue. The close pairing, gene for gene, that takes place between homologous chromosomes during meiosis can occur only if the inverted segment forms a loop within its homologue.

Focus 8.1   The Molecular Basis of Mendel's Wrinkled Seed Trait

125 years after Mendel first published the results of his crossing experiments, we finally understand the molecular basis of one of the characters he used. In January of 1990, M.K. Bhattacharyya and his colleagues published a description of the gene and the mutation that underlies the wrinkled seed phenotype that Mendel described.

For some time, the biochemical basis of the difference between smooth and wrinkled seeds was thought to involve starch metabolism. The *ss* double recessive seeds (in contrast to the *Ss* heterozygotes or the *SS* dominant homozygotes) contain more free sucrose as they develop. This results in higher osmotic pressure, and more water enters the developing seed by osmosis (see Ch. 5). When the seed finally matures, a substantial amount of water is lost in *ss* seeds. The skin of the seed cannot actually shrink and wrinkles instead.

Using the same European pea strains that Mendel used, these scientists have shown that sucrose accumulates in *ss* seeds because of a defect in an enzyme that catalyzes branching and so sucrose molecules do not polymerize to starch (see p. 37). They also show that the recessive defect in

the starch-branching enzyme (or SBEI) is due to an 800 base pair insertion in the gene encoding SBEI. This insertion probably represents a transposable element. As a result of this insertion, the SBEI gene is inactivated. In *ss* seeds, this defect in the synthesis of starch results in an accumulation of the carbohydrate monomer, sucrose, and the eventual appearance of wrinkled seeds.

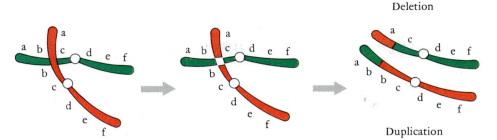

Deletion

Duplication

Figure 8.5

A diagram of how a deletion and duplication can occur during meiosis by an unequal exchange of segments between two homologous chromosomes.

of genes along the chromosome is altered in the region spanning the inversion. These inversions generally result from the breakage and inappropriate repair of chromosomes during crossing-over (p. 101). A number of mutagens may also induce chromosomal breakage, leading to inversions.

Chromosome structure may also be radically altered by the translocation, deletion, or duplication of a region of a chromosome. In chromosomal **translocation**, a segment of one chromosome may break off and attach to the end of a different (nonhomologous) chromosome. On occasion, an entire chromosome arm may be involved in translocation. In the case of **deletions**, significant amounts of genetic material may be lost (Fig. 8.5). Individuals homozygous for such deletions seldom survive beyond the early stages of development. Individuals heterozygous for deletions (individuals with one complete chromosome and one chromosome with a missing section) also frequently fail to develop. Most organisms cannot tolerate the loss of large amounts of genetic material. A number of human heritable

diseases appear to arise from translocation and/or deletion events (Table 8.1). In the cri du chat syndrome, for example, a small segment of the 5th chromosome is translocated to the 13th chromosome (Fig. 8.6). Carriers of this translocation are phenotypically normal, but some of their offspring may lack the translocated segment on chromosome 5 and they will display the syndrome. Afflicted babies have a characteristic catlike cry, abnormally small heads, growth abnormalities, and severe mental retardation. Most die in infancy or early childhood.

Duplications, on the other hand, are seldom lethal. On the contrary, the duplication of a segment of a chromosome may provide an exciting evolutionary opportunity by creating "spare copies" of certain genes. One of the copies retains its original function, but the duplicated material is in some sense free to acquire a new function. Such duplications are a kind of evolutionary testing ground where new protein functions may originate. The newly duplicated material will, in general, not be necessary for the survival of the organism and can therefore accumulate

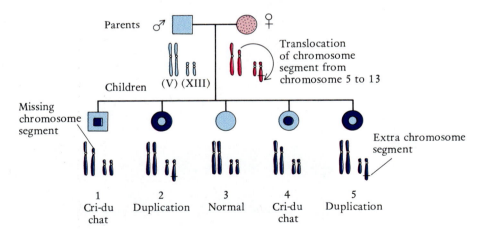

Parents

Children

Missing chromosome segment

Translocation of chromosome segment from chromosome 5 to 13

(V) (XIII)

Extra chromosome segment

| 1 | 2 | 3 | 4 | 5 |
| Cri-du chat | Duplication | Normal | Cri-du chat | Duplication |

Figure 8.6

The pedigree of cri-du chat syndrome. The parents are phenotypically normal, but a segment of one of the mother's two 5th chromosomes has been translocated to one of her 13th chromosomes. These parents have five children, one son (1) and four daughters (2–5). The boy (1) and one of the girls (4) have one abnormal 5th chromosome. Thus they are completely missing a chromosome segment and have the syndrome. The remaining children are phenotypically normal, but two of them (2 and 5) have a duplication and thus carry an extra segment of the 5th chromosome on their 13th chromosome.

Table 8.1
Some Human Chromosomal Abnormalities

Abnormality	Genetic Features	Clinical Aspects
Turner's syndrome	XO (absence of 2nd X chromosome)	Short stature, undifferentiated ovary, juvenile female genitalia, poorly developed breasts
Klinefelter's syndrome	XXY	Gynecomastia, small testes
Triple X females	XXX	Two "Barr bodies" present, fairly normal females, but secondary sex characteristics may be poorly developed
Down's syndrome	Trisomy 21	Epicanthal folds, protruding tongue, hypotonia, mental retardation
Trisomy 18 syndrome	Trisomy 18	Mental retardation, multiple congenital malformations
Trisomy 13 syndrome	Trisomy 13	Mental retardation, severe multiple anomalies, cleft palate, polydactyly, central nervous system defects, eye defects
Translocation Down's syndrome	15/21, 21/22 translocation	Clinically similar to Down's syndrome
Philadelphia chromosome	Deletion of one arm of chromosome 22, usually translocated to chromosome 9	Chronic granulocytic leukemia
Orofaciodigital syndrome	Translocation of part of chromosome 6 to 1	Defects of upper lip, palate, and mouth; stubby toes with short nails
Cri du chat syndrome	Deletion of short arm of chromosome 5, translocated to chromosome 13	Mental retardation, facial anomalies

genetic changes which may end up generating new gene functions.

Finally, the number of chromosomes may change, as the result of the appearance (or disappearance) of entire chromosomes and, on occasion, of full chromosome sets. One well-known case of a change in chromosome number involves the human condition known as **Down's syndrome**. When we examine the chromosome pattern (karyotype) of most affected individuals, an additional "mini-chromosome" is present (Fig. 8.7a). This additional

Figure 8.7

(a) The karyotype of a patient with Down's syndrome. (b) Down's syndrome. (c) Two pathways leading to the production of gametes involved in Down's syndrome. The pathway on the left illustrates the normal events of meiosis. In the middle pathway a failure of chromosomal segregation at Anaphase I results in four abnormal gametes, two of them containing an extra copy of Chromosome 21. In the pathway on the right nondisjunction of sister chromatids during Meosis II produces two haploid gametes with normal chromosome complements, one gamete lacking Chromosome 21 and one with an extra copy. Zygotes with an extra copy of Chromosome 21 are responsible for Down's syndrome; those lacking it die before birth.

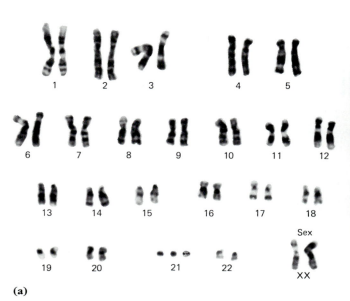

(a)

(b)

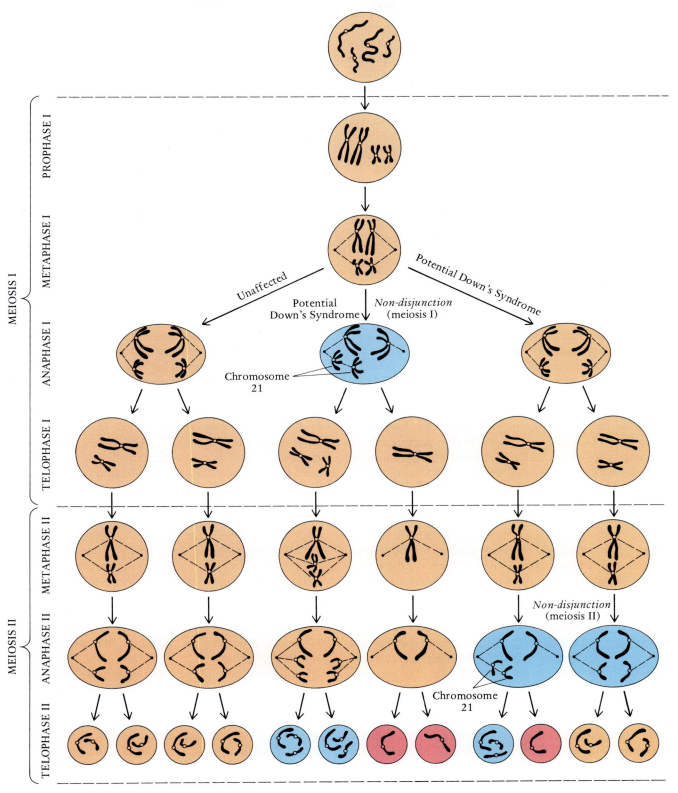

(c)

chromosome is in fact an extra copy of the long arm of chromosome 21. **Trisomy**—the presence of an additional chromosome—arises when paired chromosomes fail to segregate (nondisjunction) during female meiosis (Fig. 8.7c). The presence of a third copy of an entire chromosomal arm brings about severe consequences. Down's syndrome includes mild to severe mental retardation, reduced life span (often 11 years or less), unusual facial features, altered neuronal pathways in the visual system, and obesity (Fig. 8.7b). We do not yet understand the exact mechanisms responsible for producing these characteristics in Down's syndrome individuals. Certain other human conditions result from the inappropriate disjunction of sex chromosomes at meiosis (Table 8.1).

THE SHUFFLING OF EXISTING VARIATION: RECOMBINATION

The Physical Basis of Recombination: Crossing-Over

Thus far, we have focused our attention on the processes of mutation. These spontaneous events, occurring initially in single individuals, introduce genuinely novel variation into a population. But there is a second class of mechanisms that generate variation. In **recombination**, genetic material is exchanged between two homologous (but not identical) chromosomes. Here, although no new genetic material is generated, existing variation is shuffled and arranged into new combinations.

To understand recombination, we must draw back a bit from the gene-by-gene perspective we have been adopting so far and focus instead on the organization of chromosomes. Arrayed along the length of a chromosome are several thousand genes, stretches of DNA that direct the synthesis of proteins. At the prophase I stage of meiosis, homologous chromosomes—each composed of two sister chromatids—pair. The result is a four-stranded complex composed of two pairs of sister chromatids. It is at this stage that a remarkable exchange takes place, in which a stretch of genetic material from one chromosome is precisely exchanged with the equivalent material from its homologue. Strictly speaking, no material is either lost or gained during this process of crossing-over (Fig. 6.15). However, a shuffling of genetic information has taken place: recombination produces new chromosomes that contain a mixture of material from both parental chromosomes. In the absence of recombination, the chromosomes derived from each of the parents would be duplicated, undergo a reduction division, and be parcelled out into the new gametes; the new gametes would contain either the paternal or the maternal version of each chromosome (Fig. 8.8). But the process of recombination makes meiosis an important source of evolutionary novelty, one that acts to *shuffle and recombine* the variation present in each of the two homologous chromosomes.

Recombination results in gametes that contain chromosomes that differ from either of the two parental versions, combining instead some of both. Note that, unlike the process of chromosomal inversion, the linear order of genes on the chromosome is not normally altered by recombination, nor is a single base pair gained or lost. Instead, the subtly different alleles that may be present in the maternal and paternal chromosomes are now combined into a new recombinant chromosome.

The Genetic Consequences of Recombination

How was recombination actually discovered? To answer this question, we need to return to Mendel's original rules. You'll recall that Mendel's laws allow specific predictions about the expected offspring ratios. When Mendel's work was rediscovered, many crosses were undertaken to test his hypotheses. Early in this century, Thomas Hunt Morgan discovered that the law of independent assortment was frequently violated. He crossed two lines of fruit flies, *Drosophila*, that differed in wing shape and body color. One line was homozygous for normal wings and gray bodies (*VVBB*) and one was homozygous for small nonfunctional (vestigial) wings and black body (*vvbb*). As expected the F_1 generation had normal wings and gray bodies (*VvBb*).

He made a testcross or backcross between these heterozygous flies and homozygous recessive flies (Fig. 8.9). If the genes for wing shape and body color were assorting independently, the heterozygous flies would produce four types of gametes (*Vb, vb, vB,* and *VB*) in equal numbers. Their union with the single type of gamete produced by the homozygous recessive flies (*vb*) should produce gray normal (*VvBb*), black vestigial (*vvbb*), gray vestigial (*vvBb*), and black normal (*Vvbb*) flies in equal proportion. Much to Morgan's surprise, most of the flies resembled one parent or the other (Fig. 8.9), and only a few gray vestigial and black normal flies were produced. Despite being controlled by separate loci, the traits for wing size and body color tended to stay together, in apparent violation of Mendel's law of independent assortment. The reason, as we now understand, can be found in the physical location of the genes controlling body color and wing shape. They lie on the same chromosome only a short distance apart, hence they are said to be **linked** (p. 105) and tend to be inherited together.

However, a small number flies were obtained in which the association between gray body and normal wings (or black body and vestigial wings) was broken, giving rise to flies that did not resemble *either* parent. Furthermore, the two alternative combinations (flies with gray bodies and vestigial wings and flies with black bodies and normal wings) occur in equal numbers. These flies could only arise from a combination of a *vB* or a *Vb* gamete with the double recessive gamete, *vb*. These novel gametes

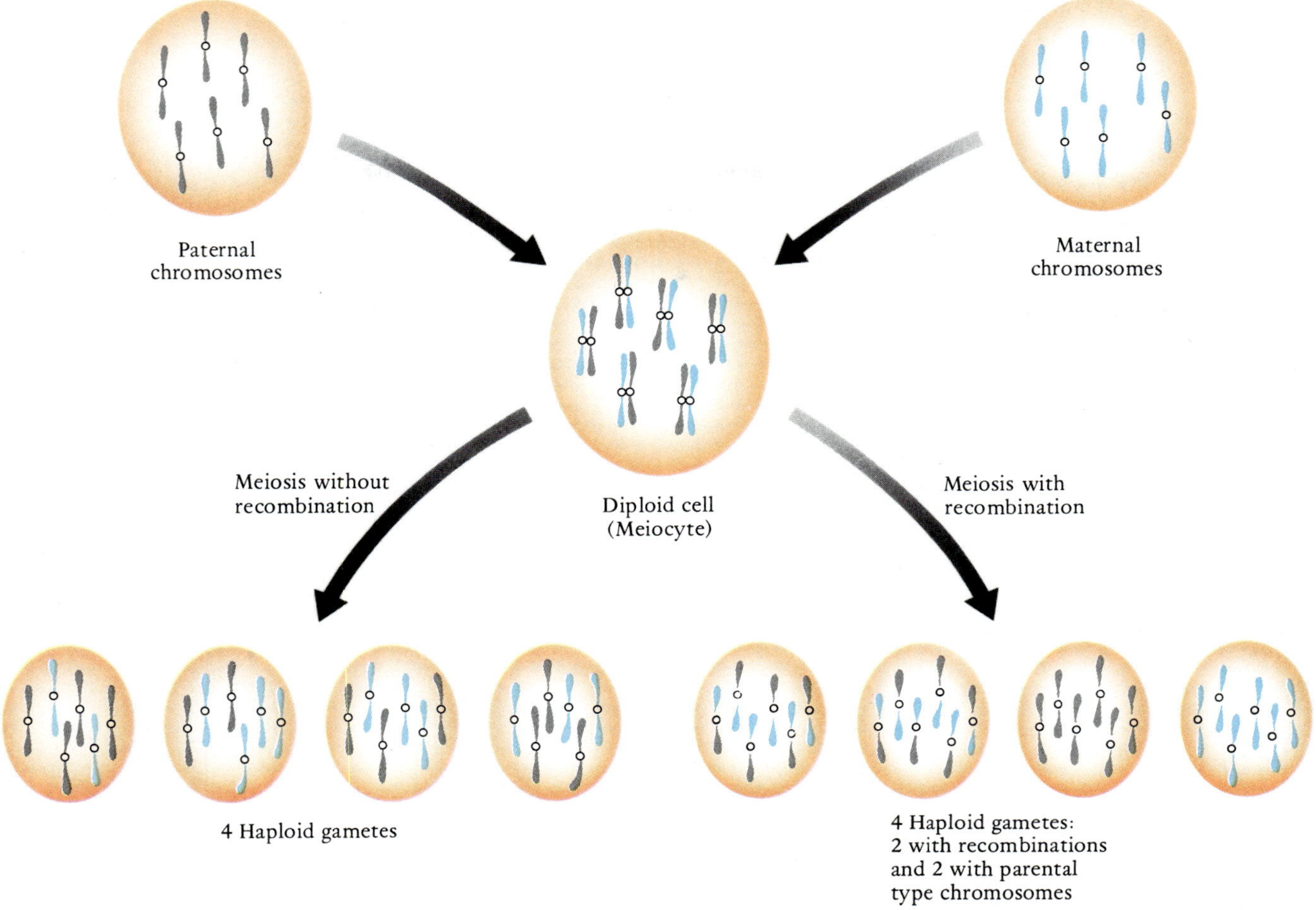
Paternal chromosomes
Maternal chromosomes
Diploid cell (Meiocyte)
Meiosis without recombination
Meiosis with recombination
4 Haploid gametes
4 Haploid gametes: 2 with recombinations and 2 with parental type chromosomes

Figure 8.8
The meiocytes, diploid precursors of all haploid gametes, contain two versions of every chromosome: one derived from the male, the other from the female. During meiosis, the cells undergo reduction division, resulting in gametes that contain only one copy of each chromosome. If recombination does not occur (left), the gametes contain a mixture of intact paternal and maternal chromosomes. If recombination takes place (right), some of the gametes contain novel chromosomes that combine genetic material derived from both the maternal and paternal versions of the chromosome.

could in turn only arise as a result of recombination between the two homologous chromosomes present during meiosis in the F_1 generation. We now know that Morgan's results came about because recombination had managed to separate two genes that were on the same chromosome, breaking up the linkage between the allele for normal wings at the wing shape locus and the allele for gray body at the body color locus (Fig. 8.10). Curiously, recombination does not take place during male meiosis in *Drosophila*. The novel recombinant gametes in Morgan's experiment were all recombinant eggs that had arisen during female meiosis.

Chromosome Mapping

It soon became clear that the probability of recombination between two genes on the same chromosome was a function of the physical distance between them. The farther apart two loci were, the higher the probability that they would be split up as a result of a recombination event. Conversely, two loci that are closely linked, lying close or adjacent to each other on a chromosome, have a relatively small chance of being split up by recombination. This observation forms the basis of a very important genetic technique, **chromosome mapping**, which enables us to assign genes to particular chromosomes and to determine the relative distance between the genes. In effect, the relative positions of gene pairs on a chromosome can be determined by looking at the proportion of recombinants produced in a backcross. The distance between a pair of genes is measured in **centimorgans** (named after Morgan), a unit defined as the distance that results in a 1% frequency of recombination. If we return to Morgan's example (Fig. 8.9), the frequency of recombinants is 10.7%. The genes controlling body color and wing shape are approximately 10.7 centimorgans (or "map units") apart. Using this same

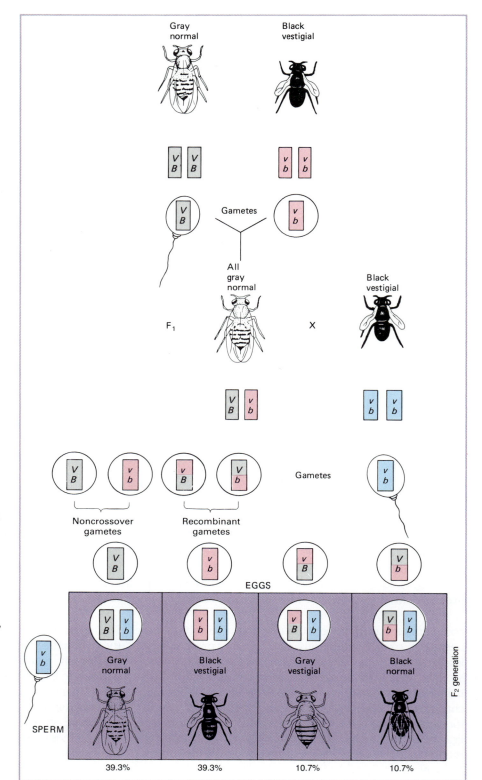

Figure 8.9

A diagram of a cross between a fly with normal wings and gray body and one with vestigial wings and black body. Since these factors are linked, all of the F₁ flies have normal wings and gray bodies. An F₁ fly is backcrossed with a homozygous recessive individual. If the features remained linked (gametes *VB* and *vb*), 50% of the flies should have normal wings and gray bodies, and 50% should have vestigial wings and black bodies. The appearance of 10% of the flies with normal wings and black bodies and 10% with vestigial wings and gray bodies indicates that crossing over and recombination occurred. Some gametes must have been vB and Vb.

logic, it is possible to construct a more complete linkage map, showing the relative positions of many genes (Fig. 8.11). Such linkage maps provide a crucial entry point into the complex genomes of many eukaryotes, including that of human beings.

Lucky Mendel

We end our discussion of recombination by analyzing what was learned, almost 100 years later, about the chromosomal positions of the characters Mendel used. The pea plant, *Pisum*, has seven chromosomes. The loci determin-

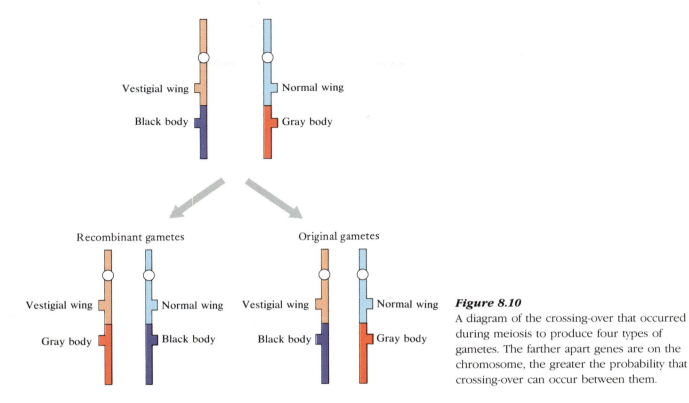

Figure 8.10

A diagram of the crossing-over that occurred during meiosis to produce four types of gametes. The farther apart genes are on the chromosome, the greater the probability that crossing-over can occur between them.

ing flower color and seed color are found on chromosome 1; plant height, flower position, and pod shape are located on chromosome 4; pod color on chromosome 5; and seed shape on chromosome 7 (Fig. 8.12). A paradox should strike you immediately: Mendel had used these characters to illustrate independent assortment, yet several of these characters are located on the same chromosome and hence physically linked. The resolution to the paradox is provided by recombination itself. Characters that are sufficiently distant from one another on the same chromosome are so frequently broken up by recombination that they are effectively unlinked. Thus, independent assortment of two characters comes about either because the characters are on different chromosomes or because the process of recombination effectively breaks up the partnership between two distant loci on the same chromosome. Mendel was either very lucky or very clever. Had he chosen to study different character pairs that might have been closely linked on the same chromosome, he might not have been able to show the law of independent assortment for which he is so justly famous. One final irony: as you can see from Figure 8.12, two of the characters, pod shape and plant height, are controlled by loci that are very close to one another on chromosome 4. As far as we can ascertain, however, Mendel never chose to focus simultaneously on that particular pair of traits.

THE CONSEQUENCES OF MUTATION
Mutations Are Overwhelmingly Deleterious

Regardless of how they occur, the overwhelming majority of mutations are harmful, or **deleterious**, to the organism; very often they are lethal. The genetic instructions of every living organism have been crafted and refined by millions of years of evolution. Any change or error at the DNA level is likely to disrupt the carefully orchestrated set of events involved in successful development.

The deleterious nature of most mutations is illustrated by a rare inherited human disease, xeroderma pigmentosum (XP). XP is caused by a recessive defect in one enzyme of the DNA excision repair system that patrols the double helix of DNA in search of defects or mispairings (see Fig. 8.1). Patients afflicted by XP suffer a number of symptoms—the most dramatic being the appearance of a large number of small, cancerous growths on the arms and back. These lesions are caused by the damage that exposure to the sun's UV rays produces in the DNA of skin cells. In individuals homozygous for the XP defect, the DNA is not repaired and cancerous growth ensues. In most of us, exposure to sunlight also produces DNA defects, but these are routinely repaired and produce no undesirable effects. XP is one illustration of the effects of unrepaired mutations, in this case somatic mutations.

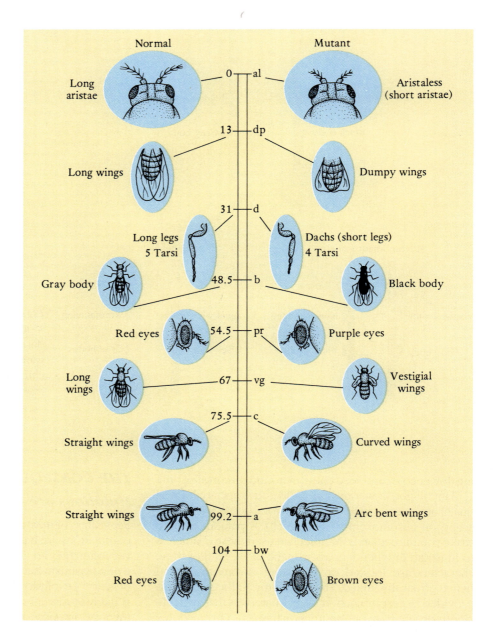

Figure 8.11
A map of a portion of chromosome 2 of *Drosophila* showing the location of genes controlling nine characters. The positions were determined using crossing-over frequencies. The numbers are map units, centimorgans.

A different line of evidence is provided by studies on the mutagenic effects of radiation. These studies had been initiated in the late 1920s by Hermann J. Muller, but had acquired greater urgency after nuclear weapons were dropped on the Japanese cities of Hiroshima and Nagasaki during World War II. The delayed effects of the bombings—including a dramatic increase in the incidence of cancers and birth defects in the inhabitants of those cities—made it clear that the damage inflicted by the bombs did not end after the explosion.

Experiments carried out on the fruit fly *Drosophila* subsequently showed that a direct relationship exists between the exposure to X-rays and the proportions of mutations that appear. More important, the majority of the detected mutations induced by radiation were **recessive lethals**, mutations that cause death in the homozygous condition.

The Effect of Mutations Is Context Dependent

Keep in mind that the consequences of mutation are not always absolute and independent of the context in which the organism exists. For example, a large class of **conditional mutants** have been isolated in various organisms. Temperature-sensitive lethals have been found in *Drosophila*—alleles that, as homozygotes, have no effect if the flies are raised at 25°C but become lethal at 28°C. This temperature dependence may reflect the stability of the protein affected by the mutation: at or below 25°C, the

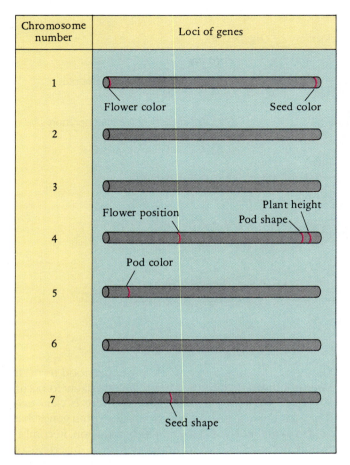

Chromosome number	Loci of genes

Figure 8.12
A diagram of the chromosomes of the garden pea (*Pisum sativum*), showing the location of the seven factors that Mendel studied.

variant protein is stable and functions as well as the wild type protein (the name used to describe the allele most commonly found in nature). Only at higher temperatures is the difference between the mutant and wild type protein revealed, with the mutant protein losing its function entirely.

A more subtle example of the context dependence involves a class of mutations known as **auxotrophic mutations**. In order for an organism carrying an auxotrophic mutation to grow, a particular nutrient—an amino acid, for example, or a vitamin—must be supplied to the organism (referred to as an auxotroph mutant). Under normal circumstances, the organism can synthesize the particular nutrient, but the mutation affects a key biosynthetic pathway and makes the mutant dependent on an external source. As long as the nutrient is made available, the mutation is not really deleterious (and may even be advantageous, since the organism need not expend

energy to synthesize the nutrient). But in the absence of the nutrient, auxotrophic mutants are at a severe disadvantage.

Certain mutations are only deleterious in the presence of certain external conditions. One such case involves the human heritable disease phenylketonuria (PKU), which is caused by a single point mutation affecting the metabolic degradation of the amino acid phenylalanine (the same metabolic pathway affected by alkaptonuria (see Fig. 7.7). If untreated, PKU can cause severe neurological damage, mental retardation, and early death. However, all of these symptoms can be avoided with a diet low in the amino acid phenylalanine. Under those circumstances, the deleterious nature of this mutation is practically invisible. Next time you drink a can of diet soda sweetened with aspartame, read the can carefully: you will see an attempt to change the environment in which PKU is severely deleterious.

Finally, certain mutations may result in amino acid changes that neither enhance nor impede the performance of the protein. The mutant protein is essentially equivalent to the wild type. Such equivalent or **neutral mutations** undoubtedly occur, but there is substantial debate about their relative frequency.

The Direction and Rate of Mutation

We appear to have painted ourselves into a corner. We began the chapter by emphasizing the importance of variation as the raw material of all evolutionary change. We then proceeded to detail the many mechanisms by which heritable change arises, emphasizing that most mutations disrupt a fine-tuned balance achieved by millions of years of evolution and are therefore overwhelmingly deleterious. What is left, therefore, for evolution to act upon? There is no contradiction here: occasionally, a mutation arises that is not deleterious and may even be advantageous in the environment in which it first appears. It is on these *rare events* that evolution depends and on these slight advantages that natural selection trades. You must keep in mind the vast numbers of genes in the genome, of organisms in a species, and of generations in a species' lifetime on Earth. The mutation rate in humans is estimated to be around 10^{-5}/gene/generation. The human genome is composed of roughly 100,000 genes, so on average 2 mutations arise in every generation in every human being. With 5 billion inhabitants on this planet, we are talking about 10 billion mutations arising in *every generation*. Thus, even if only a minuscule fraction of these mutations is not deleterious, the playground of evolution is still immense. The mutation rates for genes in several species are shown in Table 8.2. Again, they suggest that significant numbers of mutations arise in a population in every generation.

You will notice from Table 8.2 that some point mutations are reversible. The allele *his⁻* of the bacterium *Escherichia coli* cannot synthesize the amino acid histi-

Table 8.2
Some Representative Spontaneous Mutation Rates in a Variety of Organisms

Organism	Type of Mutation	Mutation Rate
E. coli (bacteria)	HIS⁻ → HIS⁺ (requires external source of amino acid histidine)	4×10^{-8} mutant bacteria/generation
	HIS⁺ → H1S (able to synthesize histidine)	2×10^{-6} mutant bacteria/generation
Drosophila melanogaster (fruit fly)	W → w (changes eye pigment from red to white)	4×10^{-5} mutations/gamete
Human dominant autosomal mutations	Huntington's chorea (Woody Guthrie's disease)	1×10^{-6} mutations/gamete
	Achondroplasia (dwarfism)	7×10^{-5} mutations/gamete
	Neurofibromatosis (tumors of the nervous system)	10×10^{-5} mutations/gamete
X-linked recessive mutations	Duchenne muscular dystrophy	6×10^{-5} mutations/gamete

(Source: Sager, R. and F.J. Ryan, *Heredity*; NY: Wiley (1961))

dine, which is needed for growth, so bacteria that have this allele can only grow on a medium containing histidine. *His⁻* mutates to *his⁺*, an allele that can synthesize histidine, at a rate of 4×10^{-8}, and *his⁺* mutates back to *his⁻* at a rate of 4×10^{-6}.

An allele at a gene locus may also mutate in more than one direction, giving rise to **multiple alleles**. Three or more alleles may be present at a given locus in a population, although an individual (in a diploid species) will have only two alleles and gametes will contain only one. An example is the human ABO system of blood groups, which depends on the type of glycoprotein present on the surface of the red blood cells. Three alleles are present in the human gene pool: the allele I^A synthesizes the A protein, I^B synthesizes the B protein, and i^O synthesizes neither. I^A and I^B are codominant, and i^O is recessive to both. Thus the blood types (the phenotypes) and their genotypes are:

Phenotype	*Genotype*
A	$= I^A I^A$ or $I^A i^O$
B	$= I^B I^B$ or $I^B i^O$
AB	$= I^A I^B$
O	$= i^O i^O.$

These glycoproteins have important immunological consequences if incompatible bloods are mixed in transfusion (see Ch. 15). Many multiple allele series are known for a variety of loci, and they add an important component to the genetic variability of a population.

The Random Nature of Mutation

The fact that the overwhelming majority of mutations are deleterious underscores the random nature of mutation. We must be careful to analyze just exactly what is meant by "random," for it is a frequently misunderstood term. It does not mean that mutations are equally likely to occur anywhere within the genome. In fact, there is clear evidence for the existence of **hotspots**, sites within genes that are far more likely to be the targets of mutation. Recombination hotspots and even hotspots for the insertion of transposable elements have been identified, although the features of the DNA that make it a more or less likely target are as yet unclear. The randomness of mutations does not mean that all types of changes are equally likely; as we have seen, transitions are far more common than transversions.

The random nature of mutations really refers to the *effects* of the mutation on the organism. In other words, a mutation that is beneficial or advantageous to the organism is not more likely to arise than one that is deleterious or lethal (in fact, it is far less likely to arise). As we will see in the next chapter, the fate of a particular mutation does depend on its effects, as natural selection retains or eliminates particular variants. Despite some puzzling results that claim to demonstrate directed mutation, there is no evidence that specific variants are generated in response to the needs of the organism. However, there is good experimental evidence, particularly in bacteria, that suggests that the *rate* of mutation is turned up in certain organisms under stressful conditions.

Mutation and recombination provide the fuel for the engine of evolution. The ability of populations to adapt in a constantly changing environment depends on the existence of variation. In the next chapter, we will take a closer look at the processes and forces that permit a single mutation, arising in a single individual, to spread through a population. In doing so, we move to the actual evolving entities of the living world: populations.

Summary

1. Enzymatic mechanisms have evolved to ensure the faithful replication of DNA. Errors in base pairing are detected and excised, and correct replacement segments inserted. But despite these safeguards, mistakes sometimes go undetected and result in a change, or mutation, in the genetic material. Evolution depends on the occurrence of mutations, for they introduce heritable variation into populations upon which natural selection and other forces can work.

2. Point mutations result from a change in a single base and can occur during DNA replication or be induced by mutagens such as ultraviolet radiation, X-rays, and certain chemicals in the environment. Because of the degeneracy of the genetic code, some point mutations (especially in the third position of a codon) may have no effect on the protein synthesized by a particular gene. A point mutation in the first or second position in a codon is more likely to lead to the incorporation of a different amino acid. A point mutation that transforms an amino acid–encoding codon into a stop codon can completely disrupt the synthesis of a protein.

3. The insertion (or deletion) of one or more nucleotides into a DNA strand disrupts the sequential reading of the three-letter codons. These frame shift mutations have serious consequences and often result in the downstream appearance of one or more stop codons.

4. Transposable elements are segments of DNA that move from place to place in the genome. Resistance plasmids in bacteria are one example of moveable genetic elements that are transmitted from one bacterium to another, independently of the chromosome. They confer antibiotic resistance to the recipient bacterium and its descendants. Other transposable elements may also insert into the coding region of a gene, consequently inactivating it.

5. Other mutations result from changes in the sequence of genes on chromosomes (inversions), the translocation of part of a chromosome to a nonhomologous chromosome, the deletion of a chromosome segment, and the duplication of a segment. Duplications produce redundant genetic material that may acquire new functions. Changes in the number of individual chromosomes present can also occur.

6. The crossing-over and exchange of segments between nonsister (but homologous) chromatids that occurs during the prophase I stage of meiosis leads to the recombination of existing genetic variation. Most gametes retain the parental arrangement of alleles, but the recombinant gametes have a mixture of paternal and maternal alleles. Since recombination is more likely to occur between distant loci on a chromosome than between ones that are close together, or linked, the frequency of recombinant gametes is a measure of the distance between loci. Recombination frequencies provide the necessary information for mapping the position of loci on a chromosome.

7. Because the genome of any organism has been crafted and refined by millions of years of evolution, random changes (mutations) tend to be deleterious. But mutations must be evaluated in a particular context. Sometimes a mutation that is deleterious in the normal environment may have no effect or be beneficial in a different environment.

8. Occasionally a mutation is beneficial, and it is on these rare events that evolution depends. Even though mutation rates are low (10^{-5}/gene/generation in humans), there are so many genes, and so many individuals in a species, that there is an abundant supply of mutations on which evolutionary forces can act. Mutations can give rise to many different alleles at a given locus. Although a diploid individual can only carry a maximum of two alleles at a given locus, multiple alleles have been documented for many loci in natural populations.

9. Mutations are termed random relative to their effects upon an organism. Advantageous mutations are not generated in response to the needs of an organism. In certain rare cases, the overall rate of mutation may increase in response to stressful conditions.

References and Selected Readings

Ayala, F.J., and J.A. Kiger, Jr. *Modern Genetics*, 2nd ed. Menlo Park, Ca.: Benjamin/Cummings Publishing Co., 1984. Two chapters in this textbook on genetics deal with mutations.

Croce, C.M., and G. Klein. Chromosome translocations and human cancer. *Scientific American* 254 (Mar. 1986):54–60. Discusses the activation of cancer-causing oncogenes.

Futuyma, D.J. *Evolutionary Biology*, 2nd ed. Sunderland, Mass.: Sinauer Associates, Inc., 1986. Contains a very good chapter on mutations.

Stahl, F.W. Genetic recombination. *Scientific American* 256 (Feb. 1987): 91–101. The molecular mechanisms of recombination are being discovered from studies in bacteria.

Suzuki, D.T., A.J.F. Griffins, J.H. Miller, and R.C. Lewontin. *An Introduction to Genetic Analysis*, 4th ed. New York: W.H. Freeman and Co., 1990. A superb, lucid textbook covering all aspects of modern genetics.

9

Evolution II: Populations, Species, and Macroevolution

(Left) Naturally occurring genetic variability in the shell banding patterns of the snail Cepaea nemoralis. Banding pattern in this species is controlled by several loci. (Above) Variation in the shell coloration and pattern of the sea scallop.

Up to now, we have been focusing on the individual organism and on the molecular and chromosomal mechanisms that generate genetic variation. It is now time to turn to the actual evolving entities of the living world: populations and species. In the first section of this chapter, we examine the microevolutionary forces that transform populations over evolutionary time by altering their genetic composition. In the second section, we will tackle the question of organismal diversity, focusing on the origin of species. Finally, we scrutinize the macroevolutionary processes that may be responsible for the broad patterns and trends seen in the history of life.

THE EVOLVING POPULATION
Evolution Depends on Variation

Without heritable variation, there can be no evolution. In effect, a population composed of genetically identical organisms is an evolutionary dead end. Without variation, natural selection has nothing to select, and adaptation to local circumstances cannot take place. An interesting set of experiments carried out in the 1960s graphically illustrates the link between variation and the ability of populations to adapt to local circumstance. Two populations of *Drosophila serrata* were placed in population cages, small wire mesh boxes where the flies compete for food and space to lay eggs. One population was made up from a single strain (type) and hence contained little genetic variation. The other population was derived from a mixture of several different strains and thus was far more genetically variable at the outset. The researchers then monitored the number of flies produced in every generation, over 25 generations (500 days). The experiment assumed that population growth was a good *indirect* measure of how well the flies were adapting to the conditions in the population cage. The results were clear: the mixed population produced more flies, and the rate of population increase was also higher for the mixed strain (Fig. 9.1).

The experiment showed that the presence of increased genetic variation resulted in more successful adaptation to the stressful, crowded conditions of the population cage.

Although variation is clearly the raw material of evolution, we need to understand the fate of this raw material. How do advantageous mutations spread through a population? How are deleterious mutations eliminated? How is genetic variability maintained in populations? These questions are the domain of **population genetics**, the discipline that examines the fate of genes in populations over evolutionary time. In the following section, we will briefly examine the fundamentals of this important branch of biology.

The Hardy-Weinberg-Castle Law

At first glance, accounting for the spread through a population of a new allele (one that has arisen as the result of mutation) seems a simple task. After all, if the organism carrying the new allele survives and reproduces, the allele is passed along to its offspring. If these offspring survive and reproduce, the allele continues to spread. But we cannot be content with such a simple verbal description. What we seek is a more formal *quantitative* answer, one that tracks the change in the frequencies of particular alleles in a population over time. To do so, we must step up one level in the biological hierarchy and begin to consider the genetics of populations.

Even the simplest population genetic model we develop must keep track of three separate but interrelated quantities:

1. The frequency of phenotypes in a population (e.g., how many fruit flies have vestigial wings, how many have normal wings).

2. The frequency of genotypes in a population (e.g., how many organisms in the population have the genotype *AA*, *Aa*, or *aa*).

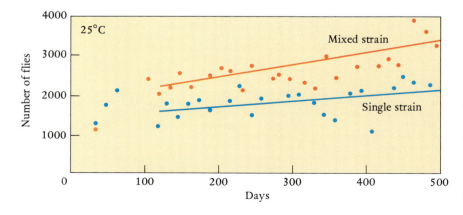

Figure 9.1
The growth of two populations of *Drosophila serrata* in population cages. The population derived from mixed strains (*red*) grew more rapidly than the population derived from a single strain (blue). The more rapid growth of the mixed strain reflected better adaptation to the crowded conditions of the cages.

3. The frequency of alternative alleles at a locus in a population (frequency of allele *A*, frequency of allele *a*). This quantity is usually referred to as the gene frequency, but should more properly be called the allele frequency.

The simplest model we can imagine will consider only a single locus, with two alternative alleles present at that locus. Remember that the independent segregation of genes makes it reasonable, at least initially, to think of genes as independent, individual entities. As we shall see, this simple model explains a great deal about naturally occurring genetic variation.

Let us imagine a population of diploid, sexually reproducing organisms (such as fruit flies, *Drosophila melanogaster*). In our experiment, the population is composed of 900 normal-winged flies that are homozygous for the dominant allele (*WW*). Into the population cage we now introduce 600 vestigial-winged flies that are homozygous for the recessive allele (*ww*) and allow the flies to mate and produce offspring. Can we estimate the phenotype, genotype, and allele frequencies of the F_1 (and subsequent) generations?

In order to answer this question, we begin by focusing on the kind of haploid gametes that can be generated by the parental generation. The normal-winged flies can only produce gametes containing the *W* allele (since they are *WW* homozygotes). The vestigial-winged flies can only produce gametes containing *w* alleles. Picture all of the gametes from all of the parents going into a single **gene pool**, containing all of the haploid eggs and sperm produced. The composition of this gene pool—the *proportion* of *W* gametes and *w* gametes—can be calculated from the proportions of each genotype (*WW*) or (*ww*) in the parental population. In our example, 900 of 1500 flies (or 0.6) of the population are *WW*, and 600 of 1500 flies (or 0.4) are *ww*. The proportions of *W* and *w* gametes they will produce reflect the parental genotypes: 0.6 of the gametes will be *W* gametes, 0.4 will be *w* gametes (Fig. 9.2).

To continue our analogy, we can conceive of fertilization as the random coming together of two gametes within this gene pool. We can now ask about the *probability* of obtaining each of the three possible diploid zygotes: *WW*, *Ww*, and *ww*. Clearly, these probabilities depend on the frequency of *W* and *w* gametes in the gene pool. The probability of generating a *WW* zygote is thus the product of the probability that the first gamete is *W* × the probability that the second gamete is *W*. In this example, the probability of forming *WW* zygotes is 0.6 × 0.6 = 0.36. The probability of generating a *ww* zygote is, by the same logic, 0.4 × 0.4 = 0.16. The probability of generating a heterozygous zygote is somewhat higher, since there are two ways of doing it: we could pick a *W* egg and then a *w* sperm, or pick a *w* egg and then a *W* sperm. That probability is therefore (0.6 × 0.4) + (0.4 × 0.6) = 0.48.

Population geneticists express the frequency of the

dominant allele as *p* and that of the recessive allele as *q*. The frequencies of the three genotypes generated in this example are shown in Figure 9.2. If you look at the matrix in this figure, you will see that we have arrived at the expected genotype frequencies by squaring the sum of the known allele frequencies:

$$(p + q)^2 = p^2 \text{ (freq. of } WW) + 2\,pq \text{ (freq. of } Ww)$$
$$+ q^2 \text{ (freq. of } ww) = 1.$$

This equation, which links allele frequencies with genotype frequencies, is known as the **Hardy-Weinberg-Castle (HWC) law**, after its codiscoverers: H. Castle in 1903, and G. H. Hardy and W. Weinberg in 1908. It is the basic theorem of population genetics.

We can now estimate the allele frequencies in this new F_1 generation. Allele *W* can be found in the *WW* homozygotes and in the *Ww* heterozygotes. All parental *WW* homozygotes, and $1/2$ of the *Ww* heterozygotes contribute *W* alleles. Similarly, *w* alleles are contributed by *ww* homozygotes and by *Ww* heterozygotes. Note that the allele frequencies sum to 1 (0.4 + 0.6 = 1), as expected, since only *w* and *W* alleles exist in our example. Using the terms of the HWC law

$$p = p^2 + 2pq/2$$
$$q = q^2 + 2pq/2.$$

The frequency of the alleles can also be computed from the frequency of the homozygous recessive F_1 individuals, provided that they can be distinguished phenotypically from all other genotypes. The recessive homozygotes are represented by q^2 in the equation, so $q = \sqrt{q^2}$. Since $p + q = 1$, then $p = 1 - q$.

If we look back at our example in Figure 9.2, three important conclusions emerge:

1. The allele frequencies have not changed from one generation to the next. The frequency of *W* remains 0.6, that of *w* remains 0.4.

2. The genotype frequencies did change from the parental generation (*WW* genotype = 0.6, *Ww* genotype = 0; *ww* genotype = 0.4) to the F_1 generation (*WW* = 0.36, *Ww* = 0.48; *ww* = 0.16). If you continue our exercise for another generation, however, you will find that the genotype frequencies do not change between the F_1 and F_2 generations. The population reaches **equilibrium genotype frequencies** in a single generation.

3. The phenotype frequencies change from the parental generation (0.6 normal-winged, 0.4 vestigial-winged) to the F_1 generation (0.84 normal-winged, 0.16 vestigial-winged). Like the genotype frequencies, however, the equilibrium phenotype frequencies will not change after the first generation.

In addition to its mathematical simplicity, the HWC law is rich in its implications. It underscores the profoundly conservative nature of heredity by proving that

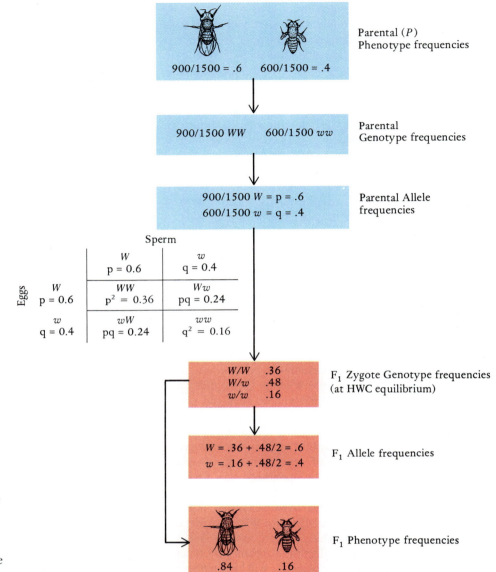

Figure 9.2
The Hardy-Weinberg-Castle law.
This flow chart illustrates the
procedure used by geneticists to
predict the genotype and
phenotype frequencies of a
population in HWC equilibrium.
The parental (*P*) phenotype,
genotype, and allele frequencies
are shown in the blue boxes; the
corresponding frequencies for
the F₁ generation are shown in
red. The matrix at the center of
the chart illustrates the use of the
Hardy-Weinberg-Castle equation.

allele frequencies do not change as the result of the process of heredity. If the allele frequencies are *p* and *q* in the first generation, they remain *p* and *q* in subsequent generations. Furthermore, genotype frequencies reach an equilibrium after one generation of random mating. In subsequent generations, the genotype frequencies remain p^2, $2pq$, and q^2. The implications of this conclusion are subtle but profound: the process of heredity (in the absence of other evolutionary forces, discussed below) preserves genetic variation by maintaining the allele frequencies constant. This is the answer to one of the most vexing challenges raised during Darwin's lifetime to the theory of evolution by natural selection: why are rare, advantageous mutants not simply diluted out as the result of mating with

the existing "types"? Under the notion of blending inheritance that prevailed during Darwin's lifetime, it was a fair objection: rare mutants, no matter how selectively advantageous, would eventually be "diluted out." The Hardy-Weinberg-Castle law, and the particulate character of genes, answer this challenge: no matter how rare an allele is, it is neither destroyed nor diluted by heredity or by the random mating that takes place in natural populations.

The Forces of Evolutionary Change
The HWC law applies specifically to diploid, sexually reproducing organisms. A population at HWC equilibrium experiences no change in genotype, phenotype, or allele frequencies from one generation to the next. But the HWC

law makes a number of assumptions about the forces of evolutionary change, which are best made explicit. They are:

1. *No mutation is taking place*. The HWC law assumes that the allele *W* will not mutate to the *w* allele (or to any other allele). In order to incorporate the effects of mutation, we would need to add an additional term to our equations, *m*, the mutation rate. We will not do so here, but keep in mind that mutation rates are relatively low (p. 149) and mutations arise in single individuals. Consequently, mutation by itself does not radically change allele frequencies from one generation to the next. Calculations have shown that if allele *A* steadily mutates to *a* at a rate of 10^{-5}/nucleotide/generation, it would take about 70,000 generations for the frequency of *A* to drop from 1.0 to 0.5. Assuming a human generation time of 20 years, this would take 1.4 million years, far longer than our species, *Homo sapiens*, has existed.

2. *Mating in the population takes place at random*. While this assumption may conjure up a variety of images, it has a specific and rather dry meaning. Geneticists speak of **random mating** with respect to the particular phenotypic trait being analyzed. In our example, we assume that normal-winged flies are not mating preferentially with (nor avoiding) other normal-winged flies. In effect, the frequency of normal-winged to normal-winged matings depends solely on the frequency of the normal-winged phenotype in the population. In human beings, there is random mating within populations with respect to a number of phenotypes. One well-studied example involves MN blood groups (Table 9.1). One of the criteria that potential sexual partners tend *not* to use in making their decision is the blood group of their partner.

In contrast, nonrandom or **assortative mating** takes place in human populations for a variety of biological, cultural, and social traits, including geographic origin, height, socioeconomic status, and religion (Table 9.2). Nevertheless, for the majority of heritable traits controlled by a single locus, mating in human populations is effectively random. The consequences of nonrandom mating are often quite dramatic. Although the allele frequencies are not affected, nonrandom mating does alter the geno-

Table 9.1
The Genotype and Allele Frequencies of MN Blood Groups in Various Human Populations*

Population	Genotype MM	Genotype MN	Genotype NN	Allele Frequencies p(M)	Allele Frequencies q(N)
Eskimos	0.835	0.156	0.009	0.913	0.087
Australian aborigines	0.024	0.304	0.672	0.176	0.824
Egyptians	0.278	0.489	0.233	0.523	0.477
Germans	0.297	0.507	0.196	0.550	0.450
Chinese	0.332	0.486	0.182	0.575	0.425
Nigerians	0.301	0.495	0.204	0.548	0.452

*From W. C. Boyd, *Genetics and the Races of Man*. Boston: D.C.Heath, 1950. These blood groups are determined by a set of two codominant alleles. The genotype frequencies are at HWC equilibrium and reflect random mating (with respect to MN phenotype) within each of these populations.

Table 9.2
Frequencies of Marriages Based on Geographical Origin*

Marriage Class†	Observed Number of Marriages	Expected Number of Marriages	Expected/ Observed
E × E	141	110.231	0.782
E × S	148	115.481	0.780
E × I	100	82.895	0.829
S × S	50	30.245	0.605
S × I	41	43.421	1.059
I × I	26	15.584	0.599

*Data from Cavalli-Sforza and Bodmer, *The Genetics of Human Populations*.
†Geographical origin was determined by surname (E = English; I = Irish; S = Scottish).

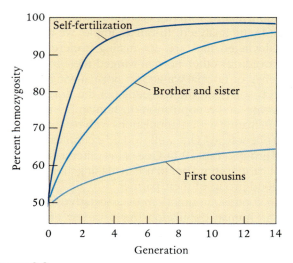

Figure 9.3
Inbreeding between close relatives results in an increase in homozygosity; the closer the relationship, the more rapid and greater the increase. Only a few species are capable of self-fertilization.

type frequencies in a population, since the gametes will not be paired at random. **Inbreeding**, an extreme form of assortative mating in which closely related individuals mate preferentially, generally results in a strong decrease of heterozygotes, with a consequent increase in the frequency of the two homozygotes (Fig. 9.3). A highly inbred population becomes even more susceptible to the effects

of genetic drift and selection (discussed below), as recessive alleles are exposed in homozygous condition. Frequently, these forces result in the disappearance of one or another of the alleles and a consequent reduction in the genetic variability of the population.

3. *The population is relatively large.* As a general rule, large populations are far less vulnerable to chance events. Chance, in this context, takes the form of a process known as **random genetic drift**. An analogy helps illustrate this phenomenon. Imagine we have a bowl with 100 marbles in it: 50 red ones and 50 blue ones (Fig. 9.4). If you pick 100 marbles out of the bowl, you will end up with a collection that reflects the actual frequencies of the marbles: 50 red, 50 blue. Now imagine instead taking 10 marbles out of the bowl. By chance alone, you may end up with something other than 5 red marbles and 5 blue ones. The same is true if we think of allele frequencies in a small population derived from a larger one. By sampling error alone, this small population may not contain the same allelic frequencies as the overall species gene pool.

Small populations may also come about through demographic (population size) fluctuations. All natural populations undergo some fluctuation in their population size from generation to generation, and not all individuals in a population will reproduce equally every generation (by chance alone). In a small population, a rare allele may be represented in only a single individual. If that individual fails to reproduce, the allele may be lost forever. Even if an allele is present in multiple copies, small fluctuations in the reproductive success of individuals from generation to generation will cause changes in the allele frequencies. In fact, it can be shown mathematically (we won't do it here)

Figure 9.4
A example demonstrating that a small sample can by chance alone lead to a significant deviation in allele frequencies. Such sampling errors inevitably lead to the loss of one of the alleles from the population.

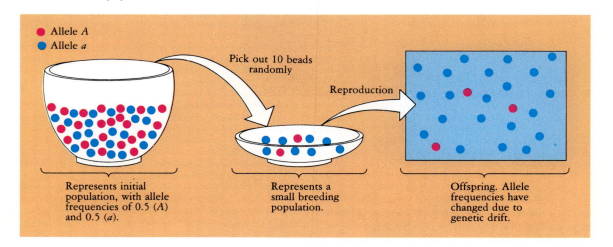

that the process of genetic drift eventually leads to the fixation of one allele and the loss of the other. Genetic drift is exacerbated in small populations because of the inbreeding that may occur; heterozygotes, normally a "hiding place" for the rare alleles, become increasingly less frequent, and the allele is thus even more likely to disappear (Fig. 9.5). Chance fluctuations and sampling errors also occur in large populations, but their effect is proportionately smaller and can frequently be ignored.

4. *No migration into or out of the population is taking place.* This condition is particularly relevant in species that are geographically arrayed as a set of small, partially isolated populations, as many species are. As the result of genetic drift and selection, these populations will generally exhibit different allele frequencies, if not different alleles altogether. Under those circumstances, you can easily picture how migration between two such populations would alter allele and genotype frequencies away from the expectations of the HWC law (Fig. 9.6a). The genetic manifestation of migration (assuming the migrants survive and reproduce at their new locality) is termed **gene flow**, and it is a powerful homogenizing force in population genetics (Fig. 9.6b).

5. *No natural selection is occurring.* The Hardy-Weinberg-Castle law assumes that there is no selection operating to change the relative frequencies of genotypes. Any difference between the relative survival or reproductive success of the different phenotypes will naturally result in a change in the allele frequency of the gene pool and will lead to departures from the Hardy-Weinberg-Castle equilibrium.

The Utility of the Hardy-Weinberg-Castle Law

Given the numerous restrictions and assumptions underlying the HWC law, one might wonder about the possible significance or applicability of this law. The answer can be briefly stated: the HWC law acts as a baseline against which actual allele and genotype frequencies can be compared. The HWC law assumes that no directional or accidental evolutionary forces are operating on the locus under study: no selection, no mutation, no drift. If we find a set of genotypes that are *not* at HWC equilibrium, we know that one or more of the assumptions of the law have been violated. However, a departure from the expectations of the HWC law will not tell us which of the assumptions does not hold true.

Figure 9.5
Results of computer simulation showing the effect of population size (*N*) on gene frequencies. Each of the three populations had initial allele frequencies of $p = q = 0.5$. Notice how the frequency drifted to 1.00 (fixation) in the small population ($N = 25$) and that effects of genetic drift were less pronounced as population size increased.

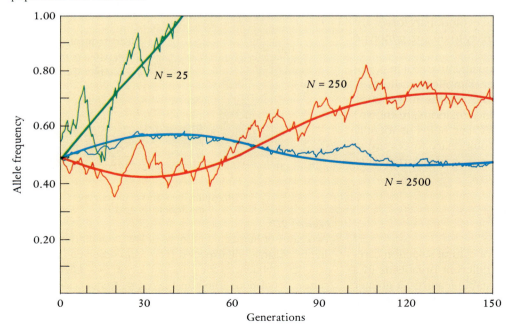

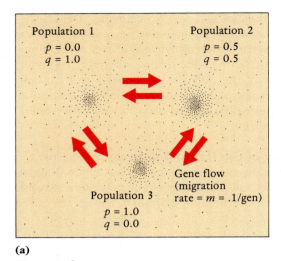

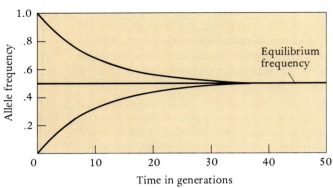

(a)   **(b)**

Figure 9.6

The effect of gene flow between populations. (a) Individuals, represented by dots, are frequently clustered in local populations with a scattering of individuals between them. The three populations depicted initially differ in their allele frequencies. They now begin exchanging migrants at a rate m. This gene flow, shown by the arrows, homogenizes the gene frequencies of the three populations. (b) The allele frequencies in all three populations converge on a value of 0.5. Conversely, the cessation of gene flow leads to the differentiation of isolated populations.

Given the long list of assumptions made by the HWC law, geneticists were surprised by the number of loci in natural populations that are at HWC equilibrium. Selected examples in human populations are shown in Figure 9.7. In such cases, we can conclude that evolutionary forces are not operating with sufficient intensity to disturb genotype frequencies.

The HWC law is a powerful quantitative tool in population genetics. Despite its apparent simplicity, it allows us to move between the three basic entities in genetics: alleles, genotypes, and (if dominance relations are known) phenotypes. In addition, the HWC law helps us identify those cases where important evolutionary forces are at work. It is to these evolutionary forces, and in particular to natural selection, that we now turn.

Microevolutionary Forces

Natural Selection

Darwin had framed the concept of natural selection carefully, emphasizing the fact that different organisms in a population had, by virtue of heritable features, a greater chance of survival and reproduction within their local environment. As a consequence, the offspring of these

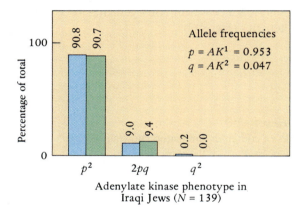

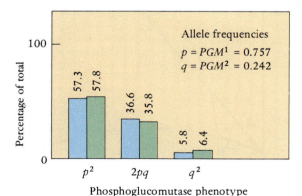

Figure 9.7

Two cases of human populations at HWC equilibrium. The graphs illustrate the expected (*blue*) and observed (*green*) genotype frequencies for particular loci in two different human populations. The actual genotype frequencies correspond to the predictions (p^2, $2pq$, and q^2) of the HWC law.

individuals formed a larger proportion of the next generation. This process, generation after generation, extended over geological time, caused populations to change and eventually gave rise to new species.

Natural selection is a powerful force that allows a population to respond to changing environmental conditions. Keep in mind, however, that natural selection acts principally on individual organisms. More specifically, selection operates by making distinctions between the different phenotypes present in the population. Stated simply, selection can only act on what it can "see"—organisms exhibit phenotypic traits (molecular, physiological, morphological, or behavioral) that confer increased survivorship and reproduction in a given environment. The phenotype of the organism is thus the object of natural selection. Individual organisms, however, do not evolve: they simply are born, survive, reproduce, and die. The actual entity that evolves is the population, as a consequence of changes in its genetic makeup from generation to generation. It is here that the relationship between genotype and phenotype becomes important, since evolution will only occur if the phenotypic differences among individual organisms have an underlying genotypic basis.

The efficacy of natural selection has been demonstrated in many laboratory and field experiments. Perhaps the best studied example of the action of natural selection involves the British peppered moth, *Biston betularia*. These moths are active at night; by day, they rest on lichen-covered tree trunks, camouflaged from bird predators by their protective coloration: white background with small dark patches (Fig. 9.8a). Beginning with the industrial revolution in the 1850s, factories around Manchester, England, began generating large amounts of coal soot and other pollutants, resulting in a darkening of the tree trunks

in the area. Against this darker background, the moths were suddenly very conspicuous to their predators. Examination of moth collections made in the area since the mid-19th century showed a dramatic increase in a dark or melanic phenotype (black with small white spots), a pattern known as **industrial melanism**. Beginning in the early 1950's, H. Bernard Kettlewell and his students showed that environmental changes brought about by industrial activity (darkening tree trunks) and the consequent increase in bird predation acted as selective pressures and favored the darker phenotype of *Biston betularia* (Fig. 9.8b). Now that air pollution standards are in effect in this region, the trees are once again lighter in color, and the frequency of the lighter-colored moths is on the increase. While a number of other evolutionary forces, including drift, assortative mating, and migration act on the genetic composition of *Biston betularia*, this is nevertheless a compelling example of evolutionary change produced by natural selection. Not surprisingly, up to 100 different butterfly and moth species in this region exhibit this pattern of industrial melanism.

The Concept of Fitness

It is now time for us to define the notion of natural selection in more quantitative terms that emphasize the differential success of certain genetic variants (at the expense of the alternatives). To do so, we consider a population of organisms that are genetically identical except at a single locus, where two alleles, A and a, and therefore three genotypes, are present in the population. In this simple case, we will assume that the three genotypes AA, Aa, and aa result in three slightly different phenotypes. We then carry out an experiment to determine the average number of eggs produced by females of each phenotype.

Figure 9.8

(a) In a rural area where tree trunks are covered with lichen, the light-colored form of the British peppered moth, *Biston betularia*, is concealed and the melanic form is conspicuous. (b) The situation is reversed in an industrial area where the lichens have been killed by pollution and the tree trunks are covered with coal soot.

(a)   (b)

Our experiment reveals that *AA* flies produce 26 eggs/fly/day, *Aa* flies produce 26 eggs/fly/day, and *aa* flies produce 13 eggs/fly/day. The genotypes clearly differ in their reproductive rate, and we capture the difference between them with a parameter known as **relative fitness** or **Darwinian fitness** (represented as *W*). Because natural selection acts by choosing among available alternatives, we are interested in the relative Darwinian fitness of each genotype. We therefore *normalize* to the highest reproductive output, assigning a value of 1 to the genotype with the highest reproductive output and expressing the fitness of the other genotypes relative to that value. In our example, the fitnesses (*W*) of the three genotypes are:

$$W(AA): 26/26 = 1.0$$
$$W(Aa): 26/26 = 1.0$$
$$W(aa): 13/26 = 0.5$$

We can easily calculate a reciprocal measure, the **selective coefficient** (*s*), defined as $s = 1 - W$. We can use the fitnesses or the selective coefficients to predict the genotype frequencies in the next generation. To do so, we simply modify the HWC equation to account for the differential fitnesses of the three genotypes (see Focus 9.1). The consequences are not surprising. The frequency of *aa* homozygotes decreases at the expense of the other two genotypes; the frequency of allele *a* therefore decreases every generation and that of *A* increases. Keep in mind that the rate of change, the speed at which a particular allele spreads through the population, depends both on the intensity of selection (*s* or *W*) as well as on the frequency of the allele (Fig. 9.9).

What exactly is this quantity "fitness"? It certainly has nothing to do with the vernacular meaning of the word (healthy, or in good shape). To a biologist, it represents the

Focus 9.1   **Selection**

The effect of natural selection on genotypes with different relative fitnesses can be shown in a simple arithmetic model of natural selection acting over one generation. We assume that the starting allele frequencies are $p = 0.7, q = 0.3$. The outcome is tabulated below (some numbers have been rounded).

Keep in mind that the genotype frequencies after selection must be normalized, so that the sum of the frequencies

equals 1. This is done by dividing the new genotype frequencies by the mean fitness of the population, defined as:

$$\overline{W} = p^2W_{AA} + (2pq)W_{Aa} + q^2W_{aa}$$

Note how the allele and resulting genotype frequencies have changed as a consequence of this simple selection against the recessive homozygote. This same procedure is used to calculate these values for subsequent generations.

Genotypes	AA	Aa	aa	Total
Initial frequency	p^2 0.49	$2pq$ 0.42	q^2 0.09	1.00
Relative fitness (*w*)	1.00	1.00	0.50	
Frequencies after selection (initial frequency × *w*)	.49	.42	0.045	
Mean fitness of the population	$\overline{W}$ = .49 + .42 + 0.045 = 0.955			
Normalized genotype frequency (after selection)	.49/.955 = .513	.42/.955 = .440	.045/.955 = .047	1.00
New allele frequencies (after selection)	$p = .513 + .440/2 = .733$	$q = .047 + .440/2 = .267$		
Genotype frequencies (next generation, before selection)	0.537	0.392	0.071	1.00

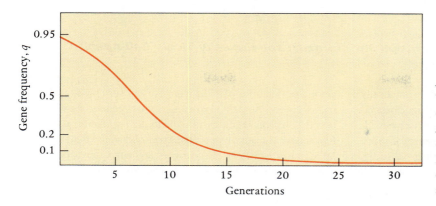

Figure 9.9
The effect of allele frequency on the outcome of constant selection against a recessive allele. As the frequency of the recessive allele declines, the number of recessive homozygotes also decreases. Natural selection consequently becomes less and less efficient at eliminating the recessive allele.

net result of several biological features: the reproductive success of a particular phenotype, its longevity, its competitive ability relative to other phenotypes in a particular environment and context. In effect, it is a shorthand way of expressing the average chance of survival and the reproductive rate of a given phenotype. But it is important to keep in mind that the fitness of a given genotype is never an absolute and invariant property of that genotype. A particular genotype is seldom "best" (or "fittest") under all circumstances and environments.

The phenomenon of **developmental plasticity** frequently allows organisms with identical genotypes to develop into quite different phenotypes under the influence of different environmental conditions. Keep in mind as well that the fitness of a particular genotype is always context dependent. Even in those cases where genotype differences translate into clear phenotype differences under all environmental circumstances, the particular environment in which the "struggle for existence" is being waged determines the relative fitnesses of the competing genotypes (Fig. 9.10). Variation in relative fitness with respect to environment also helps explain the maintenance of genetic variation in a population (see Focus 9.2).

Selection on Complex Traits
Up to this point, we have analyzed fitness and selection in an exceedingly simple situation: one locus, two alleles. Most phenotypic features of organisms, however, are not encoded by a single locus with two alleles. Furthermore, most phenotypic features of organisms show continuous variation in a population. The height of humans in a population cannot be classed into three categories, e.g., short, medium, and tall. Instead, it shows a quantitative continuous variation from very short to very tall. The same is true of milk yield in cows, the size of ears of corn, the weight of chickens, and so on. Variation of this type results from the interaction of genetic and environmental factors. The genetic component of human height depends on the

action and interaction of alleles at many different loci; height is thus a **polygenic** (Gr. *poly*, many + *genos*, descent) **trait**.

Assume, for example, that the expression of a particular phenotypic trait is controlled by three independently segregating loci: A, B, and C. Two alleles—one dominant and one recessive—are present at each locus. The cross between *AABBCC* and *aabbcc* produces a heterozygous F$_1$

Figure 9.10
The fitness of genotypes depends on the environment. The graph depicts the relative fitness of four different strains of *Drosophila pseudoobscura*, homozygous for their 4th chromosome. Note that no one genotype is superior at all the temperatures.

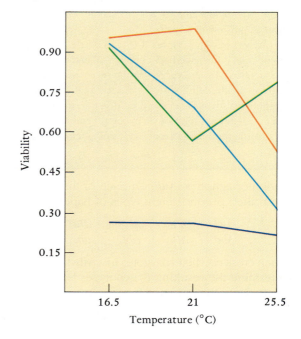

The extensive level of genetic variation present in practically all natural populations is one of the most surprising observations of modern biology. Specifically, geneticists have been able to examine allelic variation—the presence of more than one version of a particular gene product—directly, using the techniques of gel electrophoresis (see Focus 3.1). Proteins move through this matrix as a function of their charge (proteins composed of more charged amino acids move faster), as well as of their size (small portions move faster than large ones). In this manner, two proteins that differ even slightly in composition can be told apart. The extensive surveys carried out on a large number of vertebrate species reveal that, on average, 25% of the genes examined are *polymorphic*, having two (or more) alleles present in the population. In invertebrates, the proportion of polymorphic loci is almost 50%.

The presence of so much genetic variation immediately raises questions about the causes of this variation. After all, if natural selection picks out the best of the available variants, one would expect to find only a single allele at every locus (or only a transient polymorphism if selection has not yet completed its job). Did the data suggest that all these alleles were equivalent, as far as natural selection was concerned? Or instead, was selection picking out different alleles in different environments? Were elevated mutation rates perhaps to blame? The debate on this matter continues, but there is at least one example, involving the red blood cell protein *hemoglobin*, where selection can and does maintain two alleles in the human population. Even more surprisingly, one of the alleles is responsible for an inherited disease, sickle-cell anemia.

As we saw in Chapter 8 (Fig. 8.2), there is only one amino acid difference between normal and sickling hemoglobin (*HbS*). Homozygotes for the recessive sickling allele develop sickle cell anemia, a disease that is usually fatal. Despite this strong selective pressure, the *HbS* allele occurs at a frequency as high as 16% in certain parts of West Africa. As a result, the incidence of sickle cell anemia, particularly among populations of African ancestry, is very high (1/650 births among African-Americans). What, geneticists wondered, could be keeping the allele in the population at that frequency? The answer was first proposed by the geneticist J.B.S. Haldane in 1949. He noticed that the frequency of the sickling allele was highest in the regions where the most aggressive form of malaria, that caused by the protozoan *Plasmodium falciparum*, was present. He proposed that heterozygotes for the sickling allele not only didn't develop full-blown sickle cell anemia, but they also somehow exhibited increased resistance to *falciparum* malaria. In effect, there is a *positive* selection for the allele in the heterozygote that counters the negative selection against the allele in recessive homozygotes. The consequences of this *heterozygote advantage* (also known as heterosis) are easy to picture: both the normal allele and the sickling allele are kept in a *balanced polymorphism* in the population.

A great deal of evidence has accumulated to support Haldane's suggestion. Heterozygotes show reduced levels of infection and milder symptoms when they are infected with the malarial parasite, and heterozygotes survive severe infections of malaria far more successfully (in one study, only 1 of 100 children who actually died of malaria turned out to be a heterozygote, even though the frequency of heterozygotes in the population was approximately 23%). More importantly, the introduction of mosquito control and antimalarial drugs has led to a reduction in the incidence of *falciparum* malaria in the last 40 years. This reduction has been accompanied by a decrease in the frequency of the *HbS* allele in all malarial areas. The mechanism of malarial resistance in heterozygotes is now understood and appears to depend on the reduced level of oxygen that mildly sickling red blood cells can carry (relative to normal red blood cells).

Very few cases of allelic polymorphism in any species are understood in such detail. Nevertheless, the fundamental observation remains: variation at all levels is the hallmark of natural populations.

---

generation of genotype *AaBbCc*. Each of these heterozygotes can then produce eight types of haploid gametes:

$$ABC \quad ABc \quad AbC \quad aBC$$
$$Abc \quad aBc \quad abC \quad abc$$

If we then cross two heterozygotes in a trihybrid cross, we can determine the resulting F_2 genotypes by setting up a matrix similar to the one used in monohybrid or dihybrid crosses (p. 94). Given eight haploid gamete types, 8^2 or 64 possible diploid zygotes can be generated in the following proportions:

$1/_{64}$ have the genotype *aabbcc*
$6/_{64}$ have one dominant allele (e.g., Aabbcc)
$15/_{64}$ have two dominant alleles (e.g., AaBbcc)
$20/_{64}$ have three dominant alleles (e.g., AaBbCc)
$15/_{64}$ have four dominant alleles (e.g., AaBBCc)
$6/_{64}$ have five dominant alleles (e.g., AABbCC)
$1/_{64}$ have the genotype AABBCC

If we assume that each dominant allele makes a contribution towards increased height, the result of this genotype distribution is a **continuous phenotypic distri-**

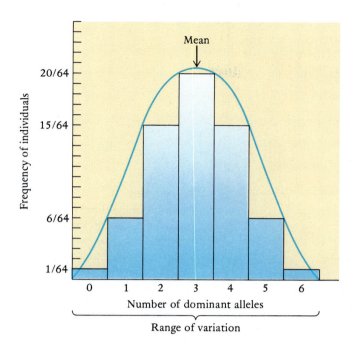

Figure 9.11

The normal curve of variation that results from the interaction of two independently assorting alleles at three loci. As the number of loci increases, the number of categories increases and the gradations become smaller.

bution. Graphing this data produces a bell-shaped or normal curve, with a characteristic range of variation and an average (the mean) (Fig. 9.11). The majority of the individuals cluster near the mean, with few individuals at either extreme.

The expression of polygenic characters is often affected by the environment in which an organism develops. Thus, although body height has a genetic component, the final height of an individual will also be influenced by nutritional factors operating during growth. It is not a simple matter to separate the relative contributions of genetic and environmental factors or to evaluate their interaction. For practical purposes, however, the degree to which selection can modify the phenotypic distribution of a particular trait in a given population is known as the *heritability* of that trait. Figure 9.12 shows the relative heritabilities of various economically important traits in domesticated animals.

Despite the increased complexity of polygenic inheritance, we can still identify and categorize the effects of natural selection on the heritability component, for selection can operate to modify both the mean and the variation within that population. Three basic types of selection can be clearly identified by examining their effect on the phenotypic distribution of a selected population: stabilizing selection, directional selection, and disruptive selection.

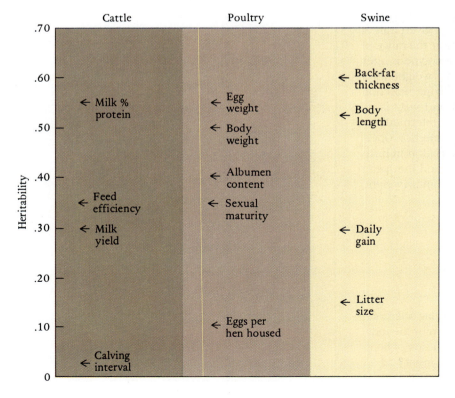

Figure 9.12

Relative heritabilities of different polygenic traits in domestic animals. The arrows indicate the relative extent to which particular traits can be altered by selection. Low heritabilities indicate either the limited involvement of genetic factors in determining the trait and/or the absence of genetic variation for that trait in the population.

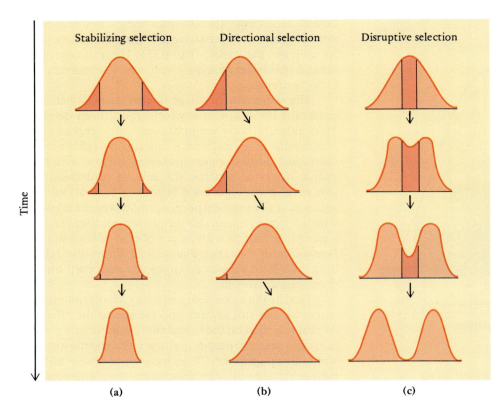

Figure 9.13

Diagrams of the three types of selection that act on variation in a polygenic character. The part of the population selected against is shaded. (a) Stabilizing selection clusters the population near the mean. (b) Directional selection shifts the mean and the distribution in one direction. (c) Disruptive selection divides the population into two groups.

Stabilizing Selection. **Stabilizing selection** acts to stabilize both the mean and the distribution of a given trait in the population. This is conservative selection: the range of variation is narrowed and the average phenotype is fitter than either extreme (Fig. 9.13a). Human birth weights are a particularly compelling example of stabilizing selection (Fig. 9.14). The mortality of very small and very large babies is substantially higher than that of intermediate-weight babies. The consequence is clear: selection operates against the extremes. The range of variation in birth weight in human populations is reduced, and the mean birth weight is very close to the predicted optimum weight.

Figure 9.14

Human birth weights are an example of stabilizing selection. The distribution of infant body weights is shown by the red line; the percent mortality of newborns as a function of body weight, by the blue line. Note that the mean of infant body weights is close to the weight exhibiting lowest mortality.

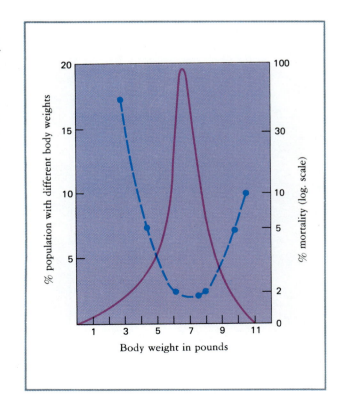

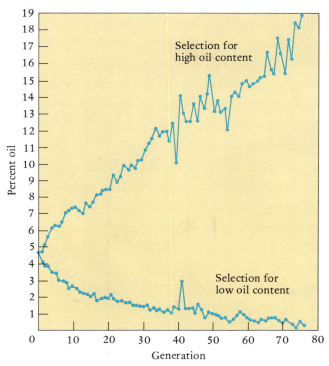

Figure 9.15
Results of long-term experiments on directional selection for high and low oil content in corn seeds. Begun in 1896, the experiments continue today at the University of Illinois. Since it is possible to select for both high and low oil content, the primary underlying genetic variation must have been in the original population rather than being introduced by mutation. Oil content has continued to increase in the high line, presumably because previously hidden genetic variation is exposed. The rate of oil decrease in the low line obviously declines as the oil content approaches zero.

Directional Selection. Under **directional selection**, individuals at one side of the mean of the phenotypic distribution have higher fitness. As a result, the mean of the population distribution shifts in the direction of the fitter phenotype (Fig. 9.13b). Much of the artificial selection carried out by plant and animal breeders is of this sort. As a general rule, breeders select to increase economically important phenotypic traits such as milk production, egg size, or oil content in seeds (Fig. 9.15). To do so, breeders select the organisms that exhibit the highest score for the trait of interest. The mean will shift upwards *as long as genetic variation for that trait exists in that population*. Eventually, the response to artificial selection will slow, either because the remaining variation in the population is largely environmental or because the constant directional selection brings with it unwanted, and often unexpected, negative biological consequences, frequently as a result of decreased genetic variability within the selected population.

Disruptive Selection. In **disruptive selection**, intermediate phenotypes have the lowest fitness, and selection simultaneously favors both extremes of the phenotypic distribution (Fig. 9.13c). Disruptive selection is involved in certain cases of Batesian mimicry, in which species that are palatable to predators (mimics) evolve to resemble species that are toxic or distasteful to the predators (models) (see Ch. 20). Disruptive selection plays a role when a single palatable species is broken up into phenotypically distinct populations, each resembling the particular noxious species present in the area (Fig. 9.16). As you might imagine, populations of the palatable species that look like one of the toxic species are avoided by predators and are thus at some selective advantage. In contrast, individuals that are intermediate in their phenotype—combining features of more than one toxic species—do not remind the predator of anything unpalatable. These intermediate forms experi-

Figure 9.16
Disruptive selection can lead to a type of Batesian mimicry in which a species in broken up into two or more forms that mimic different noxious species living in their geographical area. Females of the African swallowtail butterfly, *Papilio dardanus* (*top row*), gain protection from bird predators by mimicking several different noxious species: *Amarurus niavius*, *Dermatistes poggei*, and *Dermatistes tellus* (*bottom row, left to right*).

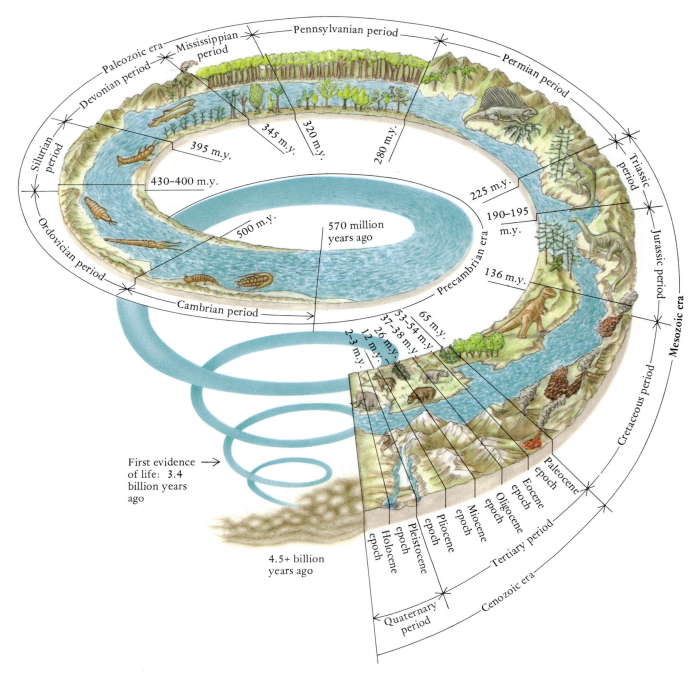

Figure 9.17

An overview of the history of life on Earth. Geologists, students of the Earth's history, divide the distant past into eras and periods based on changes in the Earth's crust associated with mountain formation and erosion, and on the accompanying changes in the fossil record. The two most recent periods are further subdivided into epochs. Dates are determined by using certain isotopes that change over time (see Appendix A). Indications of unicellular organisms have been found in rocks 3.4 billion years old, but fossils do not become abundant until the Cambrian period.

ence far higher predation. In such cases, selection operates against the intermediate phenotypes, favoring the extremes instead.

THE ORIGIN OF NEW SPECIES

The living world is not a continuous smear of overlapping forms. Instead, it is divided into discrete units, each with its own appearance, behavior, and ecology. Every organism on Earth can be classed into one and only one of these discrete units, termed **species**. According to the latest estimates, there may be as many as 30 million different species of living organisms on this planet, the majority of which have yet to be studied, described, or named. How has all this diversity arisen? What are the processes that result in the appearance of new species?

As you can see in Figure 9.17, the history of diversity is a relatively peculiar one. While the earliest record of life is contained in rocks that are 3.4 billion years old (p. 55), it is only in the last 570 million years, beginning with the Cambrian explosion, that diversity began its exponential increase. In this section, we focus primarily on the mechanisms of species formation, or **speciation**, that have given rise to the biological diversity we see today.

Darwin and the Problem of Diversity

Despite its title, *On the Origin of Species* is not really about speciation at all. In an intellectual context that emphasized separate creation and the constancy of species, the title was meant to emphasize the mutability of living forms, their common ancestry, and their constant transformation by the hand of natural selection. Darwin's great work is, in effect, about the origin of adaptations and the creative power of natural selection.

Curiously, Darwin paid little attention to the issue of how new species arose. In part, the neglect stemmed from his conviction that "species" were simply our attempts to make order out of a constantly changing world. In effect, lineages were undergoing constant change over geological time (Fig. 9.18). Our decisions about what to call a new species were entirely arbitrary and reflected our sense that enough change had taken place within a single lineage to justify a new name. Naturally, Darwin realized that this process of change within lineages, or **anagenesis** (Gr. *ana*, again + *genesis*, origin), would not result in increased diversity. An additional process of lineage splitting or

Figure 9.18

The increase in biological diversity throughout the history of life. The graph shows the dramatic increase in overall diversity that has taken place from the Precambrian era to the present. Note that the time scale on the right axis is logarithmic. Despite mass extinctions and a constant, steady rate of extinction, the overall diversity of life on this planet is constantly increasing.

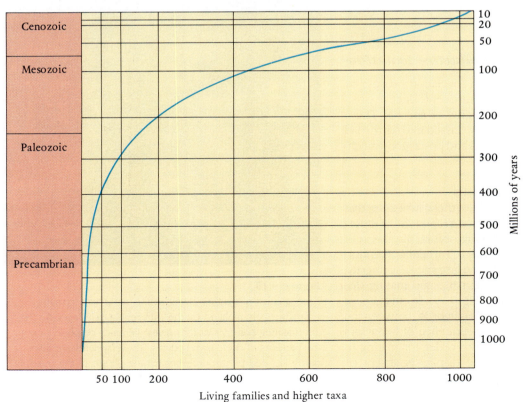

cladogenesis (Gr. *klados*, branch) had to be at work. Although Darwin would comment briefly on the forces responsible for lineage splitting, it is not until 90 years later that the problem of speciation would be tackled head on.

The Biological Species Concept

A significant breakthrough in the study of speciation involved the realization that species were real biological entities and not simply arbitrary names imposed by biologists. As we will see, species were the product of evolution and would pass through a number of intermediate stages before becoming full species. Nonetheless, the species was a meaningful evolutionary unit. The reality of species is best captured in the **biological species concept**, first proposed in 1937 by the geneticist Theodosius Dobzhansky and fully developed in 1942 by the zoologist Ernst Mayr. **Species** are defined as "groups of actually or potentially interbreeding natural populations, which are reproductively isolated from other such groups."

This definition captures many important aspects of speciation. It defines reproductive isolation as a criterion for species status and emphasizes the population as the object of study. The importance of reproductive isolation can be best understood at the genetic level. A species is in effect a closed genetic system—individuals within the system can exchange genetic material through sexual reproduction; individuals belonging to different systems (different species) cannot. The genetic coherence of a species, and its isolation from other species, confer upon it its evolutionary independence.

The Maintenance of Reproductive Isolation

We begin our examination of the speciation process by looking at the mechanisms that ensure reproductive isolation and hence maintain the integrity of species. These isolation mechanisms are generally classed into two broad categories: prezygotic mechanisms and postzygotic mechanisms (Table 9.3).

Prezygotic Isolating Mechanisms

Prezygotic isolating mechanisms are those processes that prevent mating or copulation between members of different species, or ensure that if mating occurs, the union of sperm and eggs does not take place. Particular attention has focused on the behaviors that ensure species integrity. Investigators have found that courtship and mating displays are always species-specific (Fig. 9.19). In effect, these courtship rituals act as a series of "passwords" between the sexes. In most species, a precise sequence of behaviors and responses must be followed in order for mating to occur. If a single element of the display is absent or aberrant, the encounter will usually not proceed. The reasons for the evolution of mating displays will be dis-

Table 9.3
Reproductive Isolating Mechanisms

Prezygotic Isolating Mechanisms

Ethological isolation
Ecological isolation
    Habitat isolation
    Seasonal and temporal isolation
Mechanical isolation
Immunological isolation
    Sperm die in female reproductive tract
    Sperm cannot penetrate egg

Postzygotic Isolating Mechanisms

Hybrid mortality
Reduced hybrid viability
Hybrid sterility
Distorted sex ratios

Figure 9.19
Nine species of fireflies living in the same area remain reproductively isolated by the different species-specific flight and flashing patterns of the males. Females respond only to the patterns of their own species. These species are also partly isolated by the different habitats (high in the trees, in shrubs, or close to the ground) where flight and mating occur.

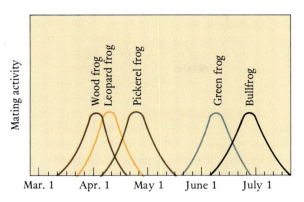

Figure 9.20

The five species of frogs (genus *Rana*), all occurring in the vicinity of Ithaca, N.Y., reach their peak mating activity at different times of the year. These differences in timing (referred to as allochronic isolation) are one of the isolating mechanisms that prevent gene flow between the different species.

cussed in more detail in Chapter 20. For our purposes, we note that the different courtship behaviors of different species ensure that interspecific mismatings will not take place, a process known as **ethological isolation**. The role of behavior as a barrier to gene flow between species is particularly clear in cases where several closely related species live in the same geographical area. In such situations, the species-specific mating displays or calls become more exaggerated, possibly in response to the increased risk of mismating.

Ecological isolation takes place when two or more closely related species occupy different habitats at a particular location or when the reproductively mature adults emerge or are active at different times of day or at different seasons (Fig. 9.20). Again, such processes serve to ensure species integrity under circumstances where mismatings could occur.

In species that do not engage in complex courtship behavior, or in cases where ethological isolation is breached, mating between two individuals from different species may be attempted. Reproductive isolation may still be ensured by **mechanical barriers**, such as an inappropriate fit between male and female copulatory organs of different species (Fig. 9.21). Finally, even when actual sperm transfer takes place, **immunological barriers** can result in the destruction of foreign sperm in the female, prior to fertilization, or in the inability of the sperm to penetrate the egg. Such **insemination reactions** have been reported between different species of *Drosophila*.

Postzygotic Isolating Mechanism

Postzygotic isolating mechanisms operate in cases where fertilization occurs between two species. These mechanisms act by reducing the fitness or the viability of interspecific offspring (or zygotes). The free passage of genes from one species to another is thus prevented. Crosses between different species often result in increased hybrid mortality, as the hybrid zygote fails to develop to term. In effect, species differ in the genetic instructions that transform a fertilized zygote into a complete organism. The interspecific zygote, faced with two different sets of developmental instructions, will frequently be aborted, usually quite early in development.

If the zygote develops normally, the hybrid individual may show **reduced viability** relative to the offspring of homogametic matings. This reduced viability may take different forms. In ecologically specialized species, for example, hybrid offspring may not be able to exploit food resources as efficiently as either of the parental species. In cryptically colored or camouflaged species, hybrids may be at increased risk of predation.

Hybrid individuals may be fully viable but partially or completely sterile. Mules, for example, are the product of a cross between a horse and an ass, and they are fully sterile. Although the hybrid survives and even shows exceptional vigor, it is in effect an evolutionary dead end, since it cannot reproduce. In many of the interspecific crosses that have been studied, the broods produced frequently show **distorted sex ratios** and may even lack one sex entirely.

Species are frequently reproductively isolated by a combination of pre- and postzygotic mechanisms. Such

Figure 9.21

Mechanical isolation in millipedes (genus *Brachioria*). The figure shows the male gonopods of six different species of millipedes. The structural differences in the male copulatory organs inhibit or prevent the successful copulation of males with females of a different species.

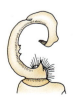

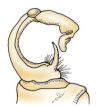

mechanisms, while efficient, are not foolproof and may on occasion be breached. Furthermore, these mechanisms operate most efficiently under natural conditions. Species barriers are frequently breached in captivity: zoos, for example, report crosses between tigers and lions, between gibbons and siamangs, and between wolves and coyotes. Such instances emphasize the finely tuned nature of many of these isolation mechanisms.

The Evolution of Reproductive Isolation

We are now in a position to ask a more dynamic question: how does reproductive isolation evolve? Unlike many other features that evolve by the spread of a favorable mutation through the population, reproductive isolation is not so readily explained. A genetic change that confers reproductive isolation—by altering the mating display of the organism, for example—will simply result in a lonely and unmated individual. Such a change will not spread through the population, since its bearer will more than likely leave no offspring. Yet populations do become reproductively isolated. How can this paradox be resolved? The evolution of reproductive isolation does not begin as a direct result of natural selection. Rather, it arises initially as the *byproduct* of the evolutionary divergence of populations. As differences between two populations accumulate as the result of chance and natural selection, the genetic composition of the two populations becomes increasingly

dissimilar. Natural selection is most certainly involved in the maintenance of reproductive cohesion within a species, since the costs of mismatings are high. Under certain circumstances, selection will favor "choosiness" and act to maintain and even reinforce prezygotic isolation. But the initial steps in the evolution of reproductive isolation are primarily the incidental byproduct of genetic differences between populations. Therefore, if we wish to understand the origins of new species, we must focus on the circumstances that result in the genetic divergence of populations.

Allopatric Speciation

The crucial role of *geographical isolation* is the cornerstone of the most commonly accepted model of speciation, **allopatric** (Gr. *allos*, other + *patra*, native land) **speciation.** This scenario was first proposed in the early 1940s by Ernst Mayr, who argued that an extrinsic (external) barrier to gene flow between two populations was the critical first step in the genetic differentiation of populations. The argument acknowledged the powerful homogenizing effect of gene flow (p. 159). Unless a geological, physical, or ecological barrier interrupted gene flow, or unless a small group of individuals (a **propagule**) dispersed and became isolated from the main population, divergence could not take place. Mountain ranges, glaciers, deserts, and even the changing course of a river

Figure 9.22
Model of vicariant speciation. (a) Gene flow between populations maintains the integrity of a species. (b) An extrinsic barrier, such as a continental glacier, divides the original species into two isolated groups of populations. Genetic differences between the populations develop in isolation, as symbolized by the different colors. (c) The extrinsic barrier may eventually break down; when the two groups meet, gene flow is restricted because of incipient pre- or post-zygotic isolation. (d) Under certain circumstances, selection may reinforce reproductive isolation. The two species overlap geographically, but do not interbreed.

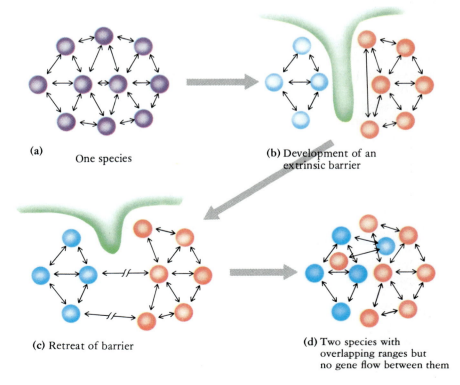

(a) One species

(b) Development of an extrinsic barrier

(c) Retreat of barrier

(d) Two species with overlapping ranges but no gene flow between them could all act as the initial extrinsic barrier. The colonization by mainland organisms of previously sterile volcanic islands (such as Hawaii) supported the idea that rare dispersal over long distances can and does take place.

Two versions of allopatric speciation are thought to be important in the generation of organismal diversity. The first model, known as **vicariant speciation**, suggests that the range of a single species is interrupted by the development of a major extrinsic barrier, dividing the species into two large populations separated by the barrier (Fig. 9.22). Over evolutionary time, different mutations will arise in the two populations. The two populations will also be subjected to different selective pressures, since climate and other environmental factors are unlikely to be exactly the same on opposite sides of the barrier. Evolution will thus guide the two populations in different directions. Naturally, the extrinsic barrier does not automatically mean that the two populations are different species; many species are composed of populations that are physically separated from one another. The test of species status comes when (and if) these two previously isolated populations reestablish contact, in a process known as **secondary sympatry**. If the populations have diversed sufficiently, some degree of pre- or postzygotic isolation is probably present: the offspring of hybrid matings may be less fit, or the courtship and mating displays of the two populations may have diverged to the extent that they are no longer compatible. In such cases, there will be no flow of genes from one population to the other and speciation will have indeed taken place.

The advance of continental glaciers and the concomitant climactic changes during the Pleistocene epoch in North America acted as an extrinsic barrier dividing up once contiguous species (Fig. 9.23a). As the glacier retreated the ranges of new species expanded. Some closely related species pairs are still physically separated (Fig. 9.23b), but others, including some butterfly and bird

Figure 9.23

(a) The advance of the last Pleistocene ice sheet about 60,000 years ago broke up many species into two or more isolated parts. (b) An example of one effect of isolation. (1) Five species (a tree, an insect, a frog, a snake, and a bird) are at first distributed over a wide geographical area. (2) Climactic changes brought about by glaciation form four refuges in which certain of these species become isolated and differentiate genetically, as symbolized by the dark color of some of them. (3) When the glacier retreats and climates again become favorable, ranges of the species expand, and the previously isolated populations meet and overlap in a suture zone. A limited amount of interbreeding or hybridization does occur in the suture zone, as symbolized by the horizontal stripes. But genetic differences between the dark and light populations are sufficiently great that gene flow between them is limited to the narrow suture zone.

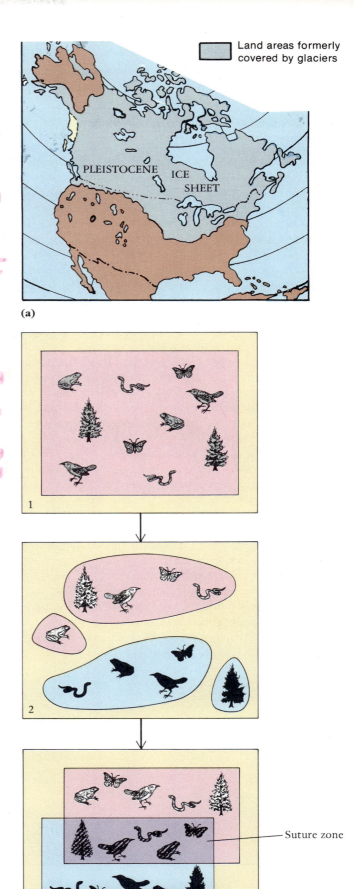

Land areas formerly covered by glaciers

PLEISTOCENE ICE SHEET

(a)

Suture zone

(b)

The continents and ocean basins rest on plates of the Earth's crust that move on deeper semimolten material. The present outlines of these plates and their names are shown in the Focus Figure d. Molten rock is now upwelling in the ocean basins at the boundaries of many of these plates and pushing them farther apart, in a process called continental drift. Mountains form where plates collide; for example, the mountains of western North America have developed where the Pacific Plate has collided with and pushed under the American Plate. Earthquakes also occur frequently along these plate margins.

By studying the outlines of the continents, the types of rocks on the boundaries of continents, and the distribution of extinct groups of animals, geologists have been able to reconstruct their past history. In the early Jurassic period (a), the continents formed a single gigantic landmass known as Pangaea. By the late Jurassic, crustal movements caused Pangaea to break up into a northern landmass (Laurasia) and southern landmass (Gondwana) (b), which became further divided in the late Mesozoic and early Tertiary periods (c). Notice that South America, Antarctica, and Australia remained united for a relatively long time and have only recently separated, geologically speaking. The separation of North America and Eurasia and the connection of North and South America are also relatively recent geological events.

Oceanic barriers to the movement of land animals arise as continents drift apart, and corridors for their movements are formed when continents are joined, as in the case of the Isthmus of Panama, now connecting North and South America. Continental drift has had a profound effect on speciation and on the present distribution of groups of organisms, as we shall see when we discuss the evolution of mammals (Chap. 40).

(a) 200 million years ago, Early Jurassic

(b) 135 million years ago, Late Jurassic

(c) 65 million years ago, Early Tertiary

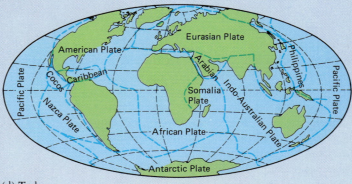

(d) Today

species, have met along "suture zones" and do not interbreed freely, forming instead narrow hybrid zones where limited gene flow takes place (Fig. 9.23b).

The drifting of continents as the result of plate tectonics has also resulted in the breakup of previously continuous species ranges (see Focus 9.3). The fossil and living marsupials of Australia and South America, and the unusual lemur fauna of Madagascar, attest to the effects of vicariant speciation and isolation.

A second model of allopatric speciation, thought to be far more common, is known as **peripatric speciation** (Gr. *peri*, around + *patra*, native land). Here, a small propagule (sometimes no more than a single gravid female) will either disperse or be carried by storms, floating mats of vegetation, or other chance factors across water or inhospitable habitat. The propagules are deposited onto a new habitat far from the original species range, or end up isolated at the periphery of the species range. The majority of such propagules most likely become extinct soon after arrival. But those that manage to gain a foothold are now in a completely new ecological context, geographically isolated from the parent species. Furthermore, as discussed previously (p. 158), such small populations are subjected to genetic drift, where particular alleles can be accidentally lost (or fixed) in a population. Finally, the **founder effect** contributes to the early genetic differentiation between the propagule and the parent species, since the propagule is unlikely to reflect the exact genetic composition of the parent population. This genetic differentiation makes the onset of reproductive isolation between the isolate and the parent species far more probable.

The propagule population, if successfully established, will begin to expand its own range and may once again contact the parent population in a process known as **secondary sympatry**. As in the case of vicariant speciation, the fate of the two previously isolated populations will depend on the outcome of this secondary contact. If prezygotic or postzygotic barriers have already emerged, the two populations will maintain their integrity. Conversely, if the barriers are weak or easily breached, gene flow will take place between the two populations, and their distinct identities will be lost. It is difficult to estimate how often populations are homogenized in the course of secondary sympatry. Nevertheless, incipient reproductive isolation can and frequently does develop between a parent species and its isolated descendant population. The high levels of endemism (species occurring in only one location) and the unusual composition of most island biotas provide strong support for the model of peripatric speciation (Table 9.4).

MACROEVOLUTION: LARGE-SCALE PATTERNS IN THE HISTORY OF LIFE

Pattern and Tempo in Evolution

From the time of Darwin, the process of evolution was seen as slow, steady, and inexorable. Natural selection produced only imperceptible alterations from one generation to the next. But, as Darwin himself would argue, these minute alterations in the genetic makeup of a population would add up to major changes over the vast span of geological time. Since biologists were unlikely to witness major changes during their lifetime, questions about the tempo of evolutionary change could only be answered by looking at the fossil record. The origin of new species, their steady transformation, and their inevitable extinction would all be preserved in the sedimentary layers of the Earth's geological record. It was up to paleontologists to supply confirmation for the gradual nature of evolutionary change (Fig. 9.24).

By 1850, European paleontologists had unearthed, analyzed, and classified a vast number of invertebrate and vertebrate fossils and had already discovered a number of patterns in the fossil record. Darwin was well acquainted with the paleontological literature. He was clearly aware that the fossil record did not show a imperceptibly slow, gradual progression of forms. Instead, the pattern consisted of the sudden appearance (in geological time) of new species, relatively long periods of stasis where the species showed little or no morphological change, and the sudden disappearance of species, frequently coinciding with the arrival of yet another recognizably different form.

Table 9.4
Endemism in Oceanic Islands: The Case of the Hawaiian Archipelago

	Total Number of Genera	Endemic Genera (%)	Total Number of Species or Subspecies	Endemic (Sub)Species (%)
Birds	41	39	71	99
Insects	377	53	3750	>99
Land molluscs	37	51	1064	>99

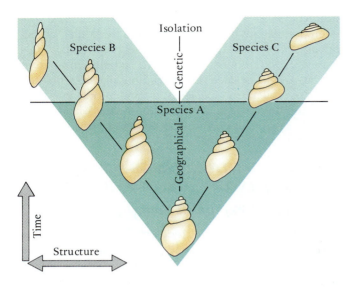

Figure 9.24
Species A, an ancestral snail species, becomes geographically split into two populations. A continued, gradual divergence of the two populations eventually results in sufficient genetic differentiation that they become reproductively isolated and form distinct species, B and C. Both gradual divergence (anagenesis) and branching (cladogenesis) are shown in this hypothetical example.

This pattern placed Darwin and his theories in a difficult situation. The argument for natural selection, after all, depended on the improved survival and reproduction of slight variants within a population. Extreme, dramatic variants were likely to be universally selected against. As Darwin saw it, evolution proceeded in minute steps, not in great leaps. Yet the fossil record did not record long series of gradual transformations. Where was the evidence for imperceptible steps building over evolutionary time? Darwin would confront this discrepancy by arguing that the fossil record was imperfect and that a complete fossil record, with every organism preserved, would indeed bear witness to the gradualistic character of evolutionary change. But the fossil record was far from ideal. In a metaphor borrowed from his mentor, Charles Lyell, Darwin would write of the fossil record:

> "... I look at the geological record as a history of the world imperfectly kept . . .; of this history we possess the last volume alone . . . Of this volume, only here and there has a chapter been preserved; and of each page, only here and there a few lines. . . ."

The actual temporal pattern of change in the fossil record—sudden appearance, stasis, and abrupt replacement—was therefore to be seen as an illusion created by the imperfect character of the fossil record.

In one sense, Darwin was absolutely right—the fossil record is incomplete. A host of important geological forces, including erosion, tectonic and volcanic activity, and restrictive conditions for fossilization, make the record far from perfect. If all evolutionary change proceeds by the steady transformation of lineages, we still could not expect to find every intermediate form preserved.

In 1972, two young paleontologists, Niles Eldredge and Stephen Jay Gould, suggested a different interpretation of the pattern in the fossil record. They argued that

while the fossil record was indeed imperfect, the empirical pattern of evolutionary change—abrupt appearance, stasis, and replacement—nonetheless reflected the *actual* pattern and tempo of evolutionary change over geological time. In contrast to the prevailing gradualist model, their theory of **punctuated equilibrium** suggested that change really was sudden and episodic (Fig. 9.25). The fossils spoke imperfectly, but they nonetheless spoke the truth.

The punctuated equilibrium model consists of two separate but interrelated claims:

1. Stasis: the majority of a species' lifetime is characterized by little or no net morphological change. In contrast to the expectations of phyletic gradualism, species do not undergo continuous directional change in morphology.

2. Abrupt appearance: the morphological changes that accompany (and in fact make it possible to identify) new species occur abruptly in geological time. Consequently, the details of the transition from an ancestral to a descendant species are rarely preserved in the fossil record. Instead, most species appear suddenly in the record and remain in morphological stasis following their appearance. In most cases, the new "daughter" species coexists, at least for some time, with the older "parent" species. This pattern would never be seen if a new species arose simply by the anagenetic transformation of an ancestral species.

If the pattern of stasis and punctuation seen in the fossil record of many species is real, what evolutionary forces could have produced it? One possibility is that most morphological change takes place during speciation, specifically in the small peripheral isolates that play a central role in the allopatric model (p. 175). The events in the peripheral isolate are unlikely to be preserved in the fossil record. If reproductive isolation develops and the new species is successful, it will begin to expand its numbers

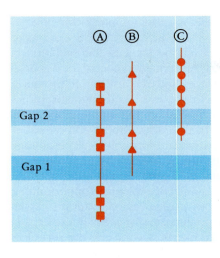

(a) Data collected in the
fossil record

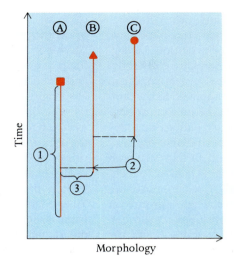

(b) Pattern of evolutionary change
inferred from the data

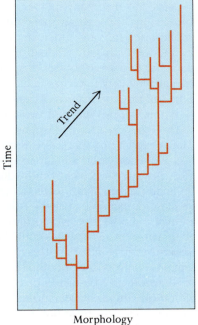

(c) Trends in the fossil record

Figure 9.25
Punctuated equilibrium. (a) The diagram depicts a hypothetical set of fossil beds
from which samples (shown as symbols) are collected. Two gaps in the record are
also shown. (b) The reconstruction of evolutionary change under the model of
punctuated equilibrium. The fossil samples are used to establish stasis (1), the
abrupt appearance of new species (2), and the coexistence of ancestral and
descendant species (3). The horizontal lines connecting the three lineages are
inferred from the data, but are not directly observed. In one case (from A to B),
gap 1 in the record includes the speciation event. Although the speciation event
giving rise to C from B could in theory be preserved in the record, such
intermediate or transitional forms are not likely to be found. (c) Trends in the
fossil record arise because certain lineages speciate more frequently or are more
resistant to extinction. In this diagram, no single species shows phyletic change in
the direction of the trend.

and its range. Only following this expansion are we likely
to see it, as it makes its sudden appearance in the record. In
addition, the development of reproductive isolation in an
incipient species is, from a geological standpoint, a rapid
event. Conversely, for both genetic and ecological reasons,
large populations are relatively buffered against change:
once established, such populations do not usually
undergo major changes in their morphology.

The theory of punctuated equilibrium, though by no
means universally subscribed to, has generated interesting
and fruitful debate. Paleontologists have looked in detail at
the record of morphological change in a variety of organisms. Many marine invertebrate groups, and several vertebrate lineages, appear to show a punctuated pattern of
change. In other cases, including many benthic fo-
raminifera (marine shelled protozoans, Ch. 24), morphological changes do seem to occur gradually and continuously.

The pattern of stasis and sudden appearance in no way
contradicted the Darwinian notion of evolution by natural
selection. Proponents of punctuated equilibrium emphasized that rapid change was far more likely to take place in
small, isolated populations, where drift and selection
could effect major changes, and where these changes
would be "protected" by the development of reproductive
isolation from the parent species. In effect, punctuated
equilibrium was a statement about the consequences of
allopatric speciation over geological time.

Other scientists argue that there is no necessary
correlation between morphological change and speciation, and suggest that stasis is the result of stabilizing
selection operating over long periods. Occasionally, radical shifts in the environment bring about new selective
regimes. These regimes result in rapid changes and hence
the geologically sudden appearance of new morphol-
ogies.

The fact that many species show little or no discernible change throughout their "life span" has important implications for our understanding of evolution. Under the model of phyletic gradualism, lineages are constantly changing, and we arbitrarily assign a new species name to a lineage when enough change has accumulated to warrant it. In contrast, the pattern of punctuated equilibrium suggests that species are real evolutionary entities that are "born" (speciate), have life spans (the period of stasis), and "die" (become extinct). This argument further suggests that a phenomenon similar to natural selection (which takes place at the level of individual organisms) may be occurring at the species level. Certain species or groups of species, for example, may owe their evolutionary success to their ability to speciate more readily, survive longer, or better resist extinction. This could lead to long-term evolutionary trends (Fig. 9.25c). This notion of **species selection** remains controversial, but the suggestion that some form of selection may be taking place at several levels of the biological hierarchy (including genes, organisms, and species) deserves serious consideration.

The idea that species are meaningful biological units has focused attention on macroevolutionary patterns in the fossil record. **Macroevolution** is the study of the large-scale processes and events that have taken place throughout the history of life on this planet. Examples of macroevolutionary phenomena include:

1. Trends in the fossil record, such as the general increase in size in horses (genus *Equus*) from the Eocene to the present (Fig. 9.26), or the overall increase in the cranial capacity of the hominids from the Pliocene to the present (Ch. 41). Traditionally, many of these trends have been seen as the result of gradual change within lineages, possibly as the result of directional selection. More recent reanalysis of trends suggests that within-lineage change is frequently small, and that the trend results from the sorting that occurs among the lineages. Thus, the size trend in horses comes about because the earliest equuids were very small, and the sole surviving lineage (*Equus*) is relatively large. In the interim, many species, both large and small, have appeared and become extinct.

2. Radiations of particular higher taxa (families, orders, and so on) with the consequent change in the overall composition of the living biota. One well known—and from our standpoint, exceedingly important—example is the enormous diversification of mammals beginning in the Late Cretaceous (Fig. 40.12). Of particular interest are so-called **adaptive radiations**, bursts of species diversification that are frequently associated with the evolution of a particular feature or "key innovation". Biologists have put forth a number of potential key inovations to help explain the sudden diversification and evolutionary success of a particular group. Examples include the evolution of a hard calcium carbonate skeleton (actually, interlocking plates) in echinoderms (Ch. 34), which allows echinoderms to survive in a variety of marine habitats; the evolution of flight in birds, with the subsequent potential for dispersal and predator avoidance; and the evolution of shock-absorbing ankle bone in the even-toed ungulates (artiodactyls, Ch. 40), which enables fast running, and thus escape from carnivorous predators. In all of these cases, the appearance of a particular morphological, physiological, or molecular feature in some sense unleashes the evolutionary potential of a given group, enabling it to exploit previously unutilized ecological opportunities. This interplay between ecological opportunity, organismal design, and the potential to speciate may also explain the radiations that frequently accompany the arrival of certain groups onto previously barren habitats. Thus, for example, over 450 endemic species of drosophilid fruit flies have evolved on the Hawaiian archipelago, and now exploit almost every conceivable ecological niche.

3. Mass extinctions, in which entire groups may disappear, leaving behind no descendants (Fig. 9.27). The disappearance of the ammonites, a very diverse group of coiled molluscs, at the Permian-Triassic boundary, profoundly altered the composition of the marine fauna. Similarly, without the extinction of the dinosaurs at the Cretaceous-Tertiary boundary, mammals might have remained a group of rat-sized creatures in a world dominated by huge reptilian forms.

4. Rates of evolution vary widely from group to group. Certain species have remained practically unchanged over millions of years—the horseshoe crab (*Limulus*) and the coelacanth fish (*Latimeria*) have remained virtually unchanged since their first appearance in the fossil record (earning them the mistaken label of "living fossils"). Conversely, other groups, such as the primates, have diversified and differentiated rapidly, and now occupy a wide range of environments. The factors controlling evolutionary rates are the focus of macroevolutionary studies.

A detailed discussion of macroevolution is beyond the scope of this book, but we wish to emphasize the hierarchy of patterns that evolutionary biologists must explain. For some, these patterns can be accounted for by the operation of natural selection on individual organisms. Trends are seen simply as the consequence of directional selection over geological time, species extinction as the result of the inability of organisms to adapt to changing circumstances. In contrast, certain investigators argue that the hierarchy of patterns results from a hierarchy of processes—selection and chance operating at many levels of biological organization. Regardless of the eventual resolution of this debate, its very existence is evidence of the vitality and excitement of evolutionary biology.

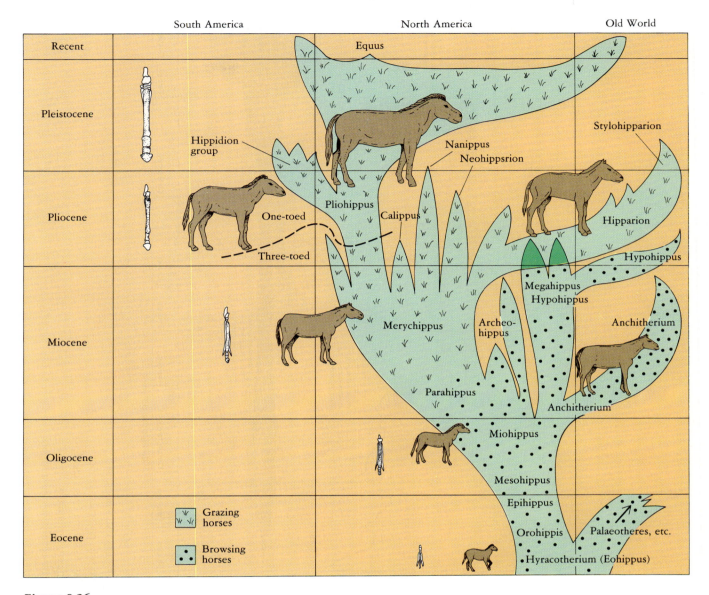

Figure 9.26

Horse evolution was centered in North America and as climates changed during the Tertiary period, horses shifted from browsing in woodlands to grazing on grasslands. Among the many morphological changes were an increase in body size, the reduction of toes from three to one, and the development of grinding teeth. These trends, however, did not come about as the result of directional selection on a single lineage. A large number of speciation and extinction events have taken place in this group since the Eocene; the trends arise as a result of these events.

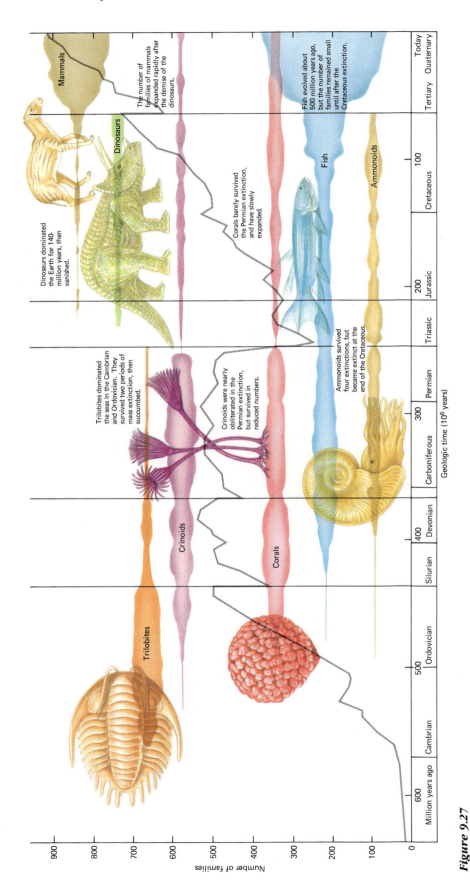

Number of families

Geologic time (10⁶ years)

Mammals

The number of families of mammals expanded rapidly after the demise of the dinosaurs.

Dinosaurs dominated the Earth for 140-million years, then vanished.

Dinosaurs

Fish evolved about 500 million years ago, but the number of families remained small until after the Cretaceous extinction.

Fish

Corals barely survived the Permian extinction, and have slowly expanded.

Ammonoids

Trilobites dominated the seas in the Cambrian and Ordovician. They survived two periods of mass extinction, then succumbed.

Crinoids were nearly obliterated in the Permian extinction, but survived in reduced numbers.

Ammonoids survived four extinctions, but became extinct at the end of the Cretaceous.

Crinoids

Corals

Trilobites

Today | Quaternary | Tertiary | Cretaceous | Jurassic | Triassic | Permian | Carboniferous | Devonian | Silurian | Ordovician | Cambrian | Million years ago

100 | 200 | 300 | 400 | 500 | 600

900 | 800 | 700 | 600 | 500 | 400 | 300 | 200 | 100 | 0

Figure 9.27

The history of life has been characterized by a number of mass extinctions, in which large numbers of species disappear forever. Conversely, particular groups have undergone radiations, dramatic increases in diversity, following episodes of mass extinction. The causes and consequences of these macroevolutionary phenomena are only now being investigated, but they have clearly exerted a major influence on the evolution of life on this planet.

Summary

1. Heritable variation is the raw material for evolutionary change. In diploid, sexually reproducing species, allele frequencies in a population remain in Hardy-Weinberg-Castle equilibrium provided there is no mutation, mating is random, population size is large, migration does not occur between populations, and natural selection is not operating. If a population is found to be out of equilibrium, one or more of these factors is operating.

2. Inbreeding, an extreme form of nonrandom mating, leads to increased homozygosity. Recessive alleles become exposed to selection. Allele frequencies are subject to random changes in small populations, a process called genetic drift. Certain alleles may be lost or become fixed (frequency = 1.0) by chance alone. Migration or gene flow between populations has a homogenizing effect.

3. Natural selection filters the variation in a population to the demands of environment and leads to adaptation. Although individual phenotypes are the target of selection, evolutionary changes occur not in individuals but in the genetic makeup of the populations of which they are a part. The effects of natural selection can be quantified by determining the relative fitness of the different phenotypes and their underlying genotypes. Fitness is not absolute and may change with circumstances.

4. Most traits of organisms, such as body height, are controlled by the interaction of alleles at many different loci (polygenes). The phenotypic variation is continuous and usually distributed along a normal curve. Environmental factors frequently affect this type of variation. Natural selection works in three ways on the heritable component of continuous variation. Stabilizing selection favors individuals close to the mean, and the range of variation of the population is reduced. Directional selection favors individuals at one end of the normal curve and the mean shifts in this direction. Disruptive selection favors individuals at each end of the curve and tends to split the population into two groups.

5. A great increase in biological diversity has occurred over geologic time and there may now be as many as 30 million species. Changes have occurred within lineages (anagenesis) and by the splitting of lineages (cladogenesis).

6. Species are groups of actually or potentially interbreeding populations that are reproductively isolated from other such groups. They are reproductively isolated by a combination of prezygotic mechanisms, such as courtship behavior, and postzygotic mechanisms, such as the reduced viability of hybrids, that restrict gene flow if successful mating occurs.

7. Most species appear to arise by allopatric speciation in which an extrinsic barrier, such as a continental glacier, breaks up a continuous species into two or more parts, or small numbers of individuals succeed in establishing a colony beyond the periphery of their normal range. Genetic differences that accumulate in these separated populations during this period of geographic isolation may lead to some degree of genetic disharmony (postzygotic isolation) if the two groups subsequently meet and hybridize. Differentiation also results in the evolution of prezygotic isolating mechanisms.

8. Darwin believed that small changes in populations that accumulated over geologic time resulted in long-term macroevolutionary changes, a process known as phyletic gradualism. The division of a fossil lineage into species was arbitrary and could be done because of the imperfection of the fossil record. The fossil record, however, often shows little evidence of gradualism. Most new species appear suddenly, persist over long periods of time without change, and then become extinct. This observation has led to the theory of punctuated equilibrium suggesting that bursts of morphological change are correlated speciation.

9. Long-term trends, extensive radiations or branchings, mass extinctions, and rates of evolution are some of the topics addressed by macroevolutionary studies. Some authorities believe that these can be explained by chance and natural selection operating on individuals; others believe that these forces also operate on species and higher levels of biological organization.

References and Selected Readings

Endler, J.A. *Natural Selection in the Wild*. Princeton, N.J.: Princeton University Press, 1986. A lucid account of natural selection theory with an analysis of many field examples.

Futuyma, D.J. *Evolutionary Biology*, 2nd ed. Sunderland, Mass.: Sinauer Associates, 1986. A modern textbook on all aspects of evolution. Includes chapters on the origin of evolutionary thought, human evolution, and social issues.

Grant, P.R. *Ecology and Evolution of Darwin's Finches*. Princeton, N.J.: Princeton Unversity Press, 1986. Over a decade of field analysis of allopatric speciation of the finches of the Galapagos Islands.

Grene, M. Hierarchies in biology. *American Scientists* 75 (1987):504−510. Natural selection and other evolutionary mechanisms extend beyond the gene and organism to the species and possibly higher levels of organization.

Levinton, J. *Genetics, Paleontology, and Macroevolution*. New York: Cambridge University Press, 1987. An advanced text emphasizing processes above the level of populations that affect macroevolutionary changes, and reconciling them with natural selection as the principal cause of organic evolution.

III

Animal Form and Function

10 *Symmetry, Form, and Life Style*

11 *Body Covering, Support, and Movement*

12 *Digestion and Nutrition*

13 *Cellular Respiration, Gas Exchange, and Metabolic Rates*

14 *Internal Transport*

15 *The Immune System*

16 *Excretion and Water Balance*

17 *The Nervous System*

18 *Receptors and Sense Organs*

19 *The Endocrine System*

20 *Behavior*

21 *Reproduction*

22 *Embryonic Development*

(Left) A parakeet, Perruche ondulee, *in motion. (Above) A parakeet, perched.*

10

Symmetry, Form, and Life Style

(Left) Two starfish. (Above) A sun star from New Zealand.

Animal life is exceedingly diverse, and it exhibits an almost endless variety of structural and functional modifications. Animals, of which over a million species have been described to date, are distributed almost everywhere over the Earth's surface and have adapted for many different habitats and life styles. Yet throughout this diversity a number of common morphological features can be distinguished.* All animals ultimately share a common evolutionary origin; different groups have diverged at different points along evolutionary lines and thus still share many features that developed in earlier ancestral forms. In addition, many unrelated groups have invaded and occupied similar environments and habitats and have thus been subjected to similar selection pressures. It is not surprising then that similar solutions to common problems (convergent evolution) appear frequently in animal evolution.

Chapters 10 to 22 explore some of the problems of animal existence and some of the solutions that metazoan (multicellular) animals have evolved during the three quarters of a billion years they have occupied this planet. This chapter will introduce the general architecture of animals and its relationship to animal life styles.

*The term **morphology** refers to the study of form and structure as compared to **anatomy**, which refers to the study of structure and the relationship of parts. However, the two terms are commonly used interchangeably.

MOTILITY AND SYMMETRY

Certainly the most distinguishing feature of animals is their ability to move. Some forms live attached to the substratum, but the ancestral metazoan was probably motile (p. 552), so these attached species evolved from motile ancestors. Attached species typically retain motile larval stages and can still move parts of their body.

Most animals exhibit **bilateral symmetry**, a type of symmetry in which the body can be divided through only one plane to produce two mirror-image halves (Focus 10.1). Bilateral symmetry is a natural consequence of animal motility. The front or **anterior** end of the body is directed forward and meets the environment first. It is thus different from the rear or **posterior** end. The **ventral** body surface facing the ground or bottom (**substratum**) is different from the **dorsal** surface directed upward. The two **lateral** body surfaces contact the environment in much the same way and are similar.

Correlated with motility and bilateral symmetry is the development of a head, termed **cephalization** (Gr. *kephale,* head). Sense organs and associated nervous tissues tend to be more highly concentrated at the forward-moving anterior end, and the mouth is also most frequently located in this region of the body.

In contrast to bilaterally symmetrical motile animals, most attached, or **sessile**, organisms show some degree of

---

Focus 10.1   Anatomical Planes

The plane dividing the body of a bilaterally symmetrical animal into two portions is parallel to the anterior-posterior axis of the body and passes from dorsal to ventral; it is called the **sagittal** plane (or section when cut). The plane is said to be **midsagittal** if it passes through the midline of the body, dividing the body into mirror-image halves, and **parasagittal** if it passes lateral to either side of the midline. A **transverse** plane is perpendicular to the anterior-posterior axis and passes between lateral surfaces from dorsal to ventral. A **frontal** plane passes from anterior to posterior and from left to right sides. It is perpendicular to both the transverse and the sagittal planes. Proximal and distal are also commonly used anatomical terms of reference. **Proximal** means close to the base or to some point of reference; **distal** means away from the base or from some point of reference.

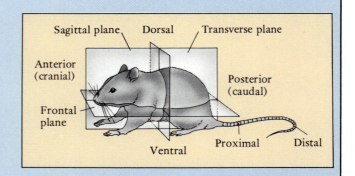

Central axis   Radial section

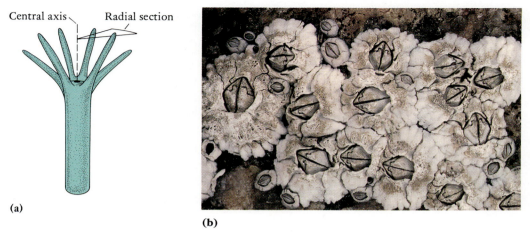

(a)

(b)

Figure 10.1
Radial symmetry. (a) Hydra, an animal with primary radial symmetry. (b) Sessile barnacles, bilateral animals with some secondary radially symmetrical features, such as the circular ring of wall plates.

radial symmetry. Radial symmetry is an arrangement of similar parts around a central body axis, as in a pie, wheel, or column (Fig. 10.1a). Such a symmetry is adaptive for a sessile life style, for the organism can meet its environment equally from all directions. Plants (which are usually attached to the substratum) typically exhibit a radial body plan (think of the perfect Christmas tree).

The ability of most animals to move from place to place has made possible the invasion of a vast number of different environments and hence the evolution of an enormous range of adaptations. Adaptations involving diets, feeding modes, and locomotion are only some of the most conspicuous. Sessile animals, on the other hand, are more limited in their range of life styles. They depend upon food that floats or swims close by or they must create a water current to bring in food. Sessile animals also cannot seek mates. For fertilization most rely upon the chance meeting of sperm and eggs shed freely into the water or upon sperm brought to the body in water currents.

Many zoologists believe that the first multicellular animals probably were radially symmetrical though motile. (We explore the origins of multicellularity in Chapter 23, p. 549). Such an ancestral radial form would likely have moved through the water with the same end or pole forward (Fig. 10.2a) and would thus have possessed anterior and posterior ends but no dorsal and ventral surfaces. From such ancestors the sessile and radial hydras, sea anemones, and corals may have evolved. If so, their radial symmetry would be **primary** (primitive), i.e., like that of

Figure 10.2
Hypothetical ancestral multicellular animals. (a) Free-swimming, radially symmetrical form. (b) Creeping, bilaterally symmetrical form. Arrows indicate direction of forward motion.

(a)

(b)

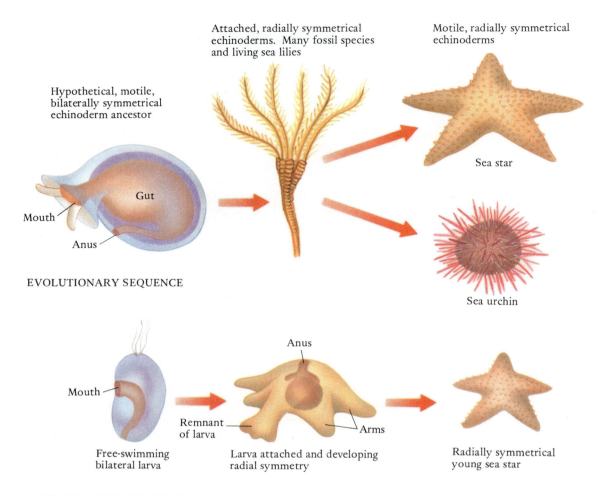

Hypothetical, motile, bilaterally symmetrical echinoderm ancestor

Attached, radially symmetrical echinoderms. Many fossil species and living sea lilies

Motile, radially symmetrical echinoderms

Mouth

Gut

Anus

Sea star

EVOLUTIONARY SEQUENCE

Sea urchin

Anus

Mouth

Remnant of larva

Arms

Free-swimming bilateral larva

Larva attached and developing radial symmetry

Radially symmetrical young sea star

DEVELOPMENTAL SEQUENCE

Figure 10.3
Evolution of symmetry in echinoderms. The evolutionary sequence is paralleled in the developmental sequence of many living sea stars. The free-swimming larva that develops from the egg is bilaterally symmetrical. It settles to the bottom, becomes attached, and then changes to the radially symmetrical adult form. On completion of the transformation, the little sea star becomes free of its attachment and crawls away.

their immediate ancestors. In contrast, the radial symmetry of the sea stars and sea urchins appears to be **secondary** (derived); because the free-swimming larvae of living sea stars and sea urchins are bilateral, the ancestors of these animals are thought to have been free-swimming and bilaterally symmetrical. In the evolutionary history of this group, some ancestral form may have become attached, leading to the evolution of radial symmetry (Fig. 10.3). This idea is supported by the many sessile fossil echinoderms (which lived attached to the bottom) that are found in

Paleozoic deposits. Later echinoderms became motile again but retained the radially symmetrical body plan.*

Aside from echinoderms, most sessile animals with bilateral ancestors show only a slight tendency toward

*The concepts of primary and secondary origins are not limited to symmetry. They may be applied to any structure or condition. For example, the anal opening has disappeared in some animals, and the absence of an anus in such animals is said to be secondary. The absence of an anus in hydras and flatworms, however, is said to be primary because we believe they evolved from ancestors that never possessed an anus.

radial symmetry. Barnacles, for example, are basically bilateral, but the outer protective plates of stalkless forms are arranged in a ring around the body (Fig. 10.1b).

If the first animals were radial and motile, what forces led to the evolution of bilaterality? Perhaps some early population of the ancestral radial form took up an existence close to the bottom, creeping or swimming over the surface (Fig. 10.2b). They would already have been elongate, with an anterior and posterior end. With the assumption of a bottom-dwelling habit, one surface became directed toward the substratum. Ventral and dorsal sides could then have differentiated, and the animal would have become bilateral.

ANIMAL ARCHITECTURE

Most bilateral animals have a relatively uniform basic and consistent design. The mouth is typically located at the anterior end, and the gut tube runs through the body, parallel to the anterior-posterior axis.

The animal's outer casing, the **body wall**, is composed of the exterior integument, or skin, plus underlying muscles. The integument consists of one or more layers of epithelial cells that may also form such specialized structures as hairs, spines, feathers, and nails. In a number of animal groups, such as crabs and clams, all or part of the integumentary epithelium secretes an outer, protective, nonliving shell or a cuticular covering.

The integument is the principal barrier between the external environment and the inner environment of the animal. In addition to an obviously important role in protection, it may also be involved in various regulatory functions that will be discussed in later chapters.

Beneath the integument the body wall is composed of one to several muscle layers. An outer layer of circular muscles and an inner layer of longitudinal muscles, as in earthworms, is a common arrangement, but other arrangements occur. The vertebrate body wall is composed of distinct muscles that are oriented in various ways in different regions of the body (Fig. 10.4).

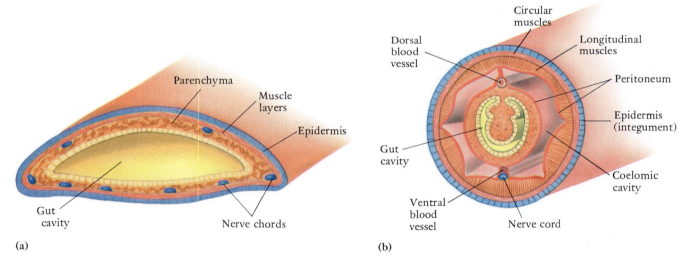

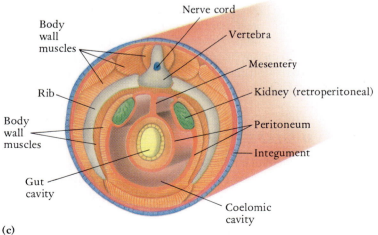

Figure 10.4

Three-dimensional cross sections through the middle of the body of three bilateral metazoans: (a) flatworm, which has a solid type of body structure; (b) earthworm and (c) vertebrate, both of which have a coelomate type of structure. A fluid-filled coelom, or body cavity, fills the interior of the body.

In many animals there is a large, continuous, fluid-filled space, or **body cavity**, located between the body wall and the central gut tube and other internal organs. The body cavity serves different functions in different animals. The fluid may function as one or more of the following: a hydroskeleton, a medium for internal transport, a reservoir for waste deposition, or a site for sperm and egg maturation, and the space may facilitate growth and movement of internal organs. These functions will be discussed in later chapters.

Body cavities are defined on the basis of the way they form in embryonic development. The most widespread type of body cavity is a **coelom** (Gr. *koiloma*, hollow cavity) that arises as a space within the middle tissue layer (mesoderm) of an embryo (Fig. 22.6; see also p. 515). The cavity comes to be lined by a layer of epithelial cells, called **peritoneum** (Gr. *peri*, around + *teinein*, to stretch) (Fig. 10.4b). Various internal organs may bulge partway into the coelom, pushing the peritoneum ahead (Fig. 10.4c). Such organs thus always lie behind the peritoneum and are said to be **retroperitoneal** (Gr. *retro*, behind). Other organs may extend so far into the coelom that they are suspended by a fold of peritoneum that is then called a **mesentery** (Gr. *mesos*, middle + *enteron*, gut) (Fig. 10.4c).

Many animals possess only a small body cavity or lack one altogether. Those having such a solid type of body structure include insects and crabs, which are believed to have lost the body cavity in the course of their evolutionary history, and flatworms, which may never have possessed one (Fig. 10.4a).

The structural design of many animals involves the division of the body into a linear series of similar parts or **segments** (Fig. 10.5). The segmental repetition of parts, called **metamerism** (Gr. *meta*, after + *meros*, part), includes both internal and external structures, such as excretory organs, lateral blood vessels, lateral nerves, and so on. The anterior part of the body bearing the sense organs and brain is not a segment nor is the terminal section bearing the anus. During embryonic and larval development, new segments are formed at the rear, in front of the terminal section (Fig. 10.5). Metamerism is believed to have evolved twice in the animal kingdom, once in the ancestors of the segmented worms (annelids) and arthropods and once in the ancestors of the vertebrates. In both it is believed to have represented an adaptation for locomotion (p. 226).

SIZE AND SHAPE

The size of multicellular animals varies over a wide range. At one extreme are microscopic forms, such as rotifers and certain flatworms; at the other are animals of vast bulk, such as elephants and whales. Many structural and functional features of animals correlate with their size, because

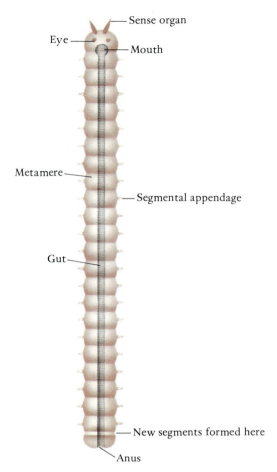

Figure 10.5

Diagram of a metameric animal, such as an annelid worm. In this animal the unsegmented anterior part of the body (acron) composes most of the head, but in many metameric animals a varying number of anterior segments combine with the acron to form the head.

as size increases, volume or weight increases by the cube, whereas surface area increases by the square of the linear dimension. For example, the volume of a sphere = $(4/3)\pi r^3$ (where r = the radius); its surface area = $4\pi r^2$. As a sphere increases in size, its volume increases relatively faster than its surface area so the ratio of surface area to volume decreases, i.e., large animals have less surface per unit of volume than do small ones (Fig. 10.6). The same factors operate in reverse, with a smaller sphere having more surface area relative to its volume than a larger one. Animals are not spheres, but the same principle applies. This means that a significant change in body size, unless compensated for by other factors, will disrupt structures and functions that are dependent on a particular suface to volume relationship. For example, if a sparrow grew to the size of a vulture and maintained the same size wings relative to the body, there would not be enough wing

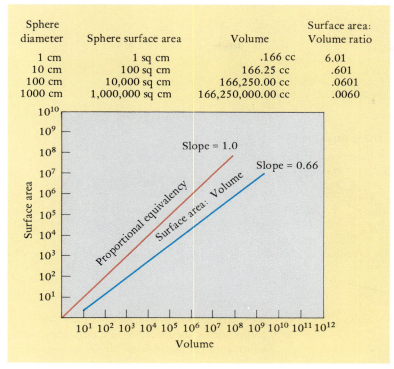

Sphere diameter	Sphere surface area	Volume	Surface area: Volume ratio
1 cm	1 sq cm	.166 cc	6.01
10 cm	100 sq cm	166.25 cc	.601
100 cm	10,000 sq cm	166,250.00 cc	.0601
1000 cm	1,000,000 sq cm	166,250,000.00 cc	.0060

Figure 10.6

Relationship of surface area to volume in spheres of increasing size. When plotted logarithmically the slope is 0.66, as opposed to a slope of 1.0 if their increase were equivalent.

surface area to carry the increased body weight. A microscopic roundworm that lives within algae at the bottom of a pond and lacks gills or internal transport system would have insufficient surface area for gas exchange if it were the size of a fish. *Therefore, change in size usually demands some change in design.* The design change may be a disproportionate increase in the size of different parts of the body. For example, a vulture must have relatively larger wings than a sparrow and the cross-sectional area of the legs of elephants must be disproportionately greater than that of other mammals in order to hold up the elephant's vast bulk. Alternatively, a design change may involve the evolution of new structures to compensate for altered functional demands. Gills, for example, make their appearance when the general body surface area of aquatic animals is inadequate for the exchange of gases with the internal volume of tissue.

There is a common misconception that small animals are always primitive and large animals are specialized. Small size, or miniaturization, has been an important part of the adaptive evolution of a number of animal groups, such as mites, bryozoans, hummingbirds, and shrews. However, the large amount of surface area relative to

volume of very small animals poses particular problems. They may be subject, for example, to much greater heat and water loss than are larger animals.

When one part of the body grows faster or slower than another part, the growth rates are said to be **allometric** to each other (as opposed to isometric growth, in which the rates are the same). Allometric growth is an important mechanism in the development of different shapes in animals. For example, the long shape of the shells of razor clams comes about because the rate of growth at the posterior margin of the shell is much faster than at the ventral and anterior margins (Fig. 10.7). The flatter faces of humans and Persian cats as compared to other primates and cats results from the lower growth rate of the muzzle as compared to the rest of the head. Thus both humans and Persian cats keep their "baby faces" throughout life.

Allometric growth can be an important mechanism through which the evolutionary degeneration of organs takes place. For example, the eyes of many cave fishes are small and nonfunctional. In a completely dark environment, the development of fully functional eyes may represent an unnecessary waste of energy and is likely selected against. Thus arthropods, fishes, and mammals living in

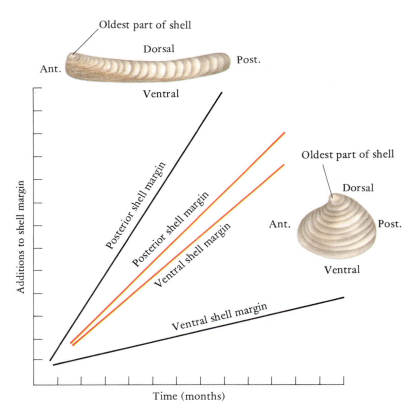

Figure 10.7
Allometric growth rates (*black*) of the ventral and posterior shell margins of a razor clam, which result in the elongate shape of the shell, compared to the nearly isometric growth rates (*red*) of the same areas of a disk-shaped clam.

caves are generally eyeless. Natural selection may be operating on the *rate* of growth of eye structures, slowing it relative to the rest of the body structures and eventually resulting in an eye that is only rudimentary in form, i.e., degenerate.

COLONIAL ORGANIZATION

Most animals are solitary, that is, individuals of a given species live independently of each other except for sexual pairing and sometimes the rearing of young. But individuals of certain animal species cannot exist except in close association; such forms are said to be **colonial**. In many colonial animals, including the familiar corals, the individuals within the colony are connected anatomically (Fig.10.8a and b). All individuals in the colony arise by budding except the first, which develops from the fertilized egg. In the colonies of ants, bees, termites, baboons, and humans, the individuals are not anatomically connected but live together and are functionally or behaviorally interdependent (Fig. 10.8c and d). Such groups of animals are usually said to be **social** colonies.

Morphological polymorphism (Gr. *poly*, many + *morphe*, form), the specialization of forms for the division of labor, characterizes many colonial species. Individuals have become specialized for functions such as feeding, defense, or reproduction. The worker and soldier castes of ants are familiar examples of morphological polymorphism in social colonies, but structurally united colonies may also exhibit relatively complex polymorphism. Not all colonial animals are polymorphic, but morphological polymorphism occurs only in species with colonial organization. Sexual dimorphism, which refers to various differences of size, color, and structure between males and females of a given species, is not morphological polymorphism. Many noncolonial animals are sexually dimorphic.

SYMBIOTIC ORGANIZATION

Many organisms depend for their existence upon an intimate physical relationship with an organism of a different species, a relationship known as **symbiosis** (Gr. *symbiosis*, living together). A symbiotic life style commonly involves structural modifications in one or both partners of the relationship. The larger partner is termed the **host** and the smaller, the **symbiont**. A symbiotic relationship is usually **obligate** for the symbiont, meaning that the symbiont cannot live without the host. Where the symbiont can exist either with or without the host, the relationship is said to be **facultative**.

The symbiont always derives some benefit from the relationship, but the consequences to the host vary. When the host benefits, the relationship is called **mutualism**. The dinoflagellates (zooxanthellae) living within certain cells

Figure 10.8

Four examples of colonial animals. The individuals of (a) coral and (b) sea squirt colonies are attached together. Individuals of (c) bee and (d) baboon colonies live together and are dependent upon each other through their social organization.

of corals and many other cnidarians are a good example of mutualism. In this instance the metabolism of each species is enhanced by the presence of the symbiotic partner.

In **commensalism**, the host neither benefits nor is harmed. The symbiont, or commensal, is protected by the host or utilizes excess food of the host. The clown fish living among the tentacles of sea anemones are commensals. The sea anemones can exist without their symbionts, but the fish cannot exist alone.

In **parasitism**, the symbiont (parasite) benefits at the expense of the host. Later chapters will provide many examples of symbiosis and further explanations of the various types of symbiotic relationships.

PREDICTABILITY IN ANIMAL DESIGN

The sea is the ancestral environment of animals. From the sea there have been numerous invasions to fresh water and land. Of these three major environments—sea, fresh water, and land—the sea is in many ways the most uniform and least stressful. Oxygen is usually adequate, there is usually no danger of desiccation (drying up), and the salt content of the sea is somewhat similar to that of body fluids that bathe the cells.

Other environments are much more stressful. The salt content of fresh water is much less than that of tissue fluids. Pools, ponds, and even streams may periodically dry up, streams fluctuate in velocity and turbidity, and the oxygen content may be very low in swamps and at the bottom of some lakes. On land oxygen is plentiful, but desiccation is an ever-present danger. Each of these three environments poses different problems of existence, and animals have evolved a vast repertoire of structural and functional adaptations to cope with them.

Many structural and functional adaptive themes appear over and over again in animal design, for they are often related to size, degree of motility, or environment. A knowledge of these relationships can have great predictive value. Thus if the size of an animal, whether it is sessile or sedentary, and something of its life style and environment are known, it is possible to predict, or make educated guesses about, many aspects of its structure and physiology. Chapters 11 to 22 will provide a basis of understanding for making many such predictions.

ANIMAL GROUPS

In the following chapters on animal form and function we will refer to different animal groups, although these groups are not described in detail until Section IV. A brief synopsis that will acquaint you with the major divisions (phyla and classes) of animals is therefore provided below. Only one or two of the most conspicuous characteristics are indicated, along with depictions of some of the representative forms.

Phylum Porifera. Sponges. Sessile animals constructed around a system of canals providing for a flow of water through the body.

Phylum Cnidaria. Hydras, jellyfish, sea anemones, and corals. Sessile or free swimming. Mouth surrounded by a ring of tentacles at one end of the radially symmetrical body.

Phylum Platyhelminthes. Flatworms. Free-living and parasitic worms (flukes and tapeworms) having dorsoventrally flattened bodies.

Phylum Nematoda. Roundworms. Free-living and parasitic worms having elongate, cylindrical bodies tapered at each end.

Phylum Rotifera. Rotifers. Microscopic, creeping and swimming, mostly freshwater animals having a crown of cilia at the anterior end.

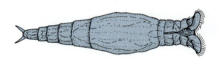

Phylum Mollusca. Molluscs. Marine, freshwater, and terrestrial animals having a muscular, locomotor, ventral surface (foot) and a region of the integument (mantle) that usually secretes a calcium carbonate shell.
  Class Gastropoda. Snails, conchs, and slugs.
  Class Bivalvia. Clams, mussels, oysters, and scallops.
  Class Cephalopoda. Squids, cuttlefish, and octopods.

Phylum Annelida. Segmented worms—polychaete worms, earthworms, and leeches. Marine, freshwater, and terrestrial worms having metameric (segmented) bodies.

Phylum Arthropoda. Arthropods. Marine, freshwater, and terrestrial animals having a body encased within a chitinous exoskeleton and jointed appendages.

    Subphylum Chelicerata. Horseshoe crabs and arachnids (spiders, mites, scorpions, and false scorpions).
    Subphylum Crustacea. Water fleas, copepods, barnacles, shrimps, lobsters, and crabs.
    Subphylum Uniramia. Centipedes, millipedes, and insects.

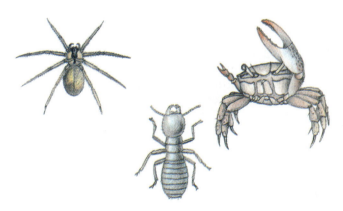

Phylum Bryozoa. Bryozoans. Mostly marine, sessile, colonial animals in which the individuals are less than 1 mm in length.

Phylum Echinodermata. Sea stars, brittle stars, feather stars, sea lilies, sea urchins, sand dollars, and sea cucumbers. Marine animals having a five-part radial symmetry and an internal calcium carbonate skeleton. Body commonly bears spines.

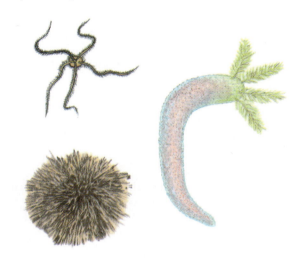

Phylum Chordata. Chordates. Marine, freshwater, and terrestrial animals having pharyngeal pouches (gill slits), a notochord (skeletal rod), and a dorsal hollow nerve cord at some time in their life cycle.

    Subphylum Urochordata. Tunicates, or sea squirts.
    Subphylum Cephalochordata. Amphioxus.
    Subphylum Vertebrata. Sharks, bony fish, amphibians, reptiles, birds, and mammals.

Summary

1. Most animals are bilaterally symmetrical; the body can be cut along the midsagittal plane to produce two mirror-image halves. Most animals also exhibit some degree of cephalization, the concentration of sense organs and nervous tissue at the anterior end of the body. Bilateral symmetry and cephalization are correlated with animal motility, in which one end of the animal's body meets the environment first.

2. Sessile animals are commonly radially symmetrical or exhibit some tendency toward radial symmetry; the body is composed of similar parts arranged around a central axis. Radial symmetry is of advantage in sessility, for the animal can meet the environment equally from all directions.

3. Most animals possess a body design in which there is an outer casing called the body wall, composed of the skin (integument) and one or more layers of muscle. A gut tube generally extends through the middle of the body from the mouth near the anterior end to the anus toward the posterior end.

4. In many animals a fluid-filled space, called a body cavity, occupies much of the interior of the body and partially or completely surrounds many internal organs, such as the gut. A common type of body cavity is a coelom, which is lined by peritoneum. Body cavities serve a variety of functions depending upon the group of animals in which they occur.

5. Three groups of animals—annelids, arthropods, and vertebrates—exhibit a segmental organization, called metamerism, in which the body is divided into a linear series of similar parts (segments). Metamerism probably evolved as an adaptation for locomotion.

6. The structural evolution of animals has been greatly influenced by the limitations of size, for as animals become larger or smaller the volume increases or decreases by the cube of the linear dimension, but the surface area increases or decreases by the square of the linear dimension. Thus the ratio of surface area to volume changes as the size of the animal changes.

7. Allometric growth, in which one part of the body grows at a different rate than another, is an important mechanism in determining the shape of animals.

8. Most animals are solitary in that they are not dependent upon other individuals of the same species except for purposes of reproduction. Some, however, exhibit colonial organization, in which the individuals of a colony are structurally united together (e.g., corals) or live in behaviorally dependent associations, called social colonies (e.g., bees and ants). The colonies of many species are composed of structurally and functionally different members, a condition known as morphological polymorphism.

9. Many animals are partners in symbiotic relationships, a close physical relationship with another species of organism. If both the host and the symbiont are benefited, the relationship is called mutualism; if the host neither benefits nor is harmed, it is called commensalism; and if the symbiont exists at the expense of the host, it is called parasitism.

References and Selected Readings

Most of the following general references contain a great deal of information on the anatomy and physiology of organ systems.

Ahmadjian, V., and S. Paracer. *Symbiosis: An Introduction to Biological Associations.* Hanover, N.H.: University Press of New England, 1986. A broad coverage of symbiotic relationships.

Alexander, R.M. *Animal Mechanics.* 2nd ed. Cambridge, Mass.: Blackwell Scientific Publications, 1983. The biophysics of vertebrate support and movement.

———. *The Chordates.* 2nd ed. Cambridge, England: Cambridge University Press, 1981. An excellent analysis in functional terms of the structure of selected chordates.

———. *The Invertebrates.* Cambridge, England: Cambridge University Press, 1979. A somewhat quantitative approach to the study of invertebrates.

Barnes, R.D. *Invertebrate Zoology,* 5th ed. Philadelphia: Saunders College Publishing, 1987. A textbook and reference on the anatomy, physiology, ecology, and classification of invertebrates.

Bloom, W., and D.W. Fawcett. *A Textbook of Histology,* 11th ed. Philadelphia: W.B. Saunders Co., 1986. The microscopic structure and ultrastructure of the cells and tissues of mammalian organ systems are thoroughly described.

Eckert, R. *Animal Physiology: Mechanisms and Adaptations,* 3rd ed. New York: W.H. Freeman, 1988. A comparative physiology of animals, emphasizing principles and illustrated with a wide variety of animal groups.

Guyton, A.C. *A Textbook of Medical Physiology,* 7th ed. Philadelphia: W.B. Saunders Co., 1986. A detailed consideration of mammalian physiology is presented in this standard textbook.

Hildebrand, M. *Analysis of Vertebrate Structure,* 3rd ed. New York: John Wiley & Sons, 1988. A textbook of comparative anatomy.

Rhoades, R., and R. Pflanzer. *Human Physiology.* Philadelphia: Saunders College Publishing, 1989. A general physiology of the human.

Romer, A.S. and T.S. Parsons: *The Vertebrate Body,* 6th ed. Philadelphia: W.B. Saunders Co., 1986. The morphological aspects of the evolution of vertebrates are thoroughly considered in this widely used textbook of comparative anatomy.

Schmidt-Nielsen, K. *Animal Physiology: Adaptation and Environment,* 4th ed. Cambridge, England: Cambridge University Press, 1990. A physiological account of the interrelationships between animals and their environments, with emphasis upon oxygen, food, temperature, water, movement, and integration.

———. *Scaling: Why Is Animal Size So Important?* Cambridge, England: Cambridge University Press, 1984. The relationship of physiology to animal size.

Wainwright, S.A. *Axis and Circumference: The Cylindrical Shape of Plants and Animals.* Cambridge, Mass.: Harvard University Press, 1988. A good short analysis of biomechanics as it relates to symmetry and form in animals and plants.

Walker, W.F., Jr. *Functional Anatomy of the Vertebrates: An Evolutionary Prospective.* Philadelphia: Saunders College Publishing, 1987. A functional comparative anatomy of vertebrates and its evolutionary setting.

11

Body Covering, Support, and Movement

(Left) The African cheetah is the fastest running mammal, reaching a speed of 112 kilometers per hour over a short distance when chasing an antelope. (Above) A cheetah family in Kenya, Africa.

BODY COVERING

Animals, like all organisms, have a body covering separating their internal environment from the outside world. This **integument** (L. *integumentum*, a covering) may be a single layer of cells or a multilayered complex of tissues like the skin of vertebrates. Its complexity is related in part to the degree of chemical and physical difference between internal and external environments and to the size and mode of life of the animal. The body covering may have many functions. It may protect against abrasion, bacterial penetration, ultraviolet radiation, and other assaults from the external world. In addition to meeting these needs, the body covering may also become hard enough to serve as the main support or skeleton of the animal, e.g., the shell of a snail or the exoskeleton of a crayfish. Although the body covering protects the organism, it does not isolate it from the external environment. It has a regulatory role in many animals, and exchanges of gases, ions, and water take place

Figure 11.1

Epithelial tissues. (a) A typical epithelial cell with microvilli on its exposed surface. (b) Types of junctions between epithelial cells. At tight junctions apposing cell membranes are bound together by surface proteins occluding the intercellular space. At spot desmosomes the cells are united by microfilaments that emerge from a plaque-like thickening on the cytoplasmic side of each cell. At gap junctions the apposing cell membranes are close together and the narrowed intercellular space is bridged by similar membrane channels. (c) Squamous epithelium. (d) Cuboidal epithelium. (e) Columnar epithelium.

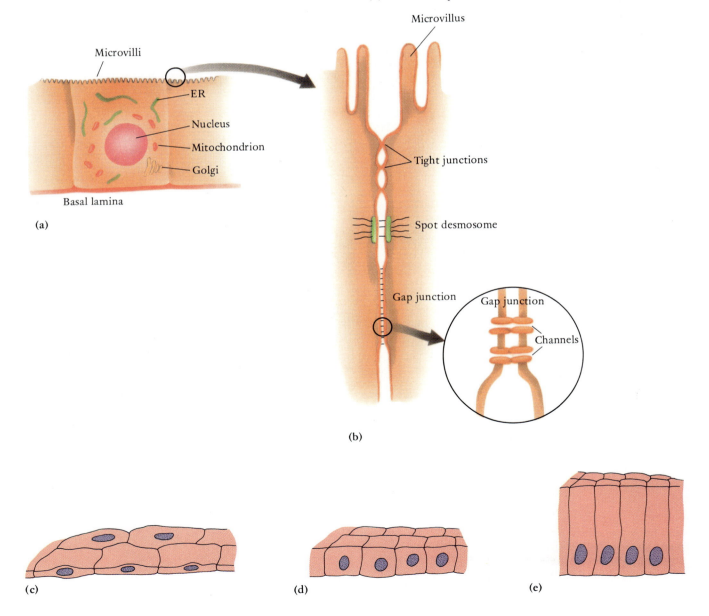

across it. It also contains many receptor cells that detect changes in the external environment. Glands, scales, claws, feathers, and hair develop from the integument and serve a variety of functions. Pigment cells come to lie within it. In birds and mammals, the skin helps to control the loss of body heat and is important in thermoregulation.

Epithelial and Connective Tissues

The integument of most animals is composed of two types of tissues: epithelial tissue and connective tissue. **Epithelium** (Gr. *epi*, on + *thele*, nipple) is the first tissue formed in the developing embryo and gives rise directly or indirectly to all others. It consists of closely placed cells that have little intercellular space between them (Fig. 11.1). Tissues of this sort cover the body surface and line internal cavities, such as the gut, coelom, and vertebrate blood vessels. They form the bulk of the tissues composing glands and secretory organs such as the liver and pancreas. Epithelial cells rest upon a **basal lamina** (**basement membrane**), a thin, fibrous, extracellular sheet. The free surface of an epithelial layer that borders an internal cavity or the external environment often bears microvilli or cilia (Fig. 11.1a). Epithelial cells are typically united by tight junctions, gap junctions, desmosomes, and other intercellular connections, which control the passage of substances across and between them (Fig. 11.1b).

There are three types of epithelial tissue differentiated on the basis of cell shape (Fig. 11.1c–e). **Squamous epithelium** is composed of flattened cells with bulging nuclei. The cells of **cuboidal epithelium** are more or less cube shaped, as their name implies. **Columnar epithelium** is composed of cells shaped like columns, with the nucleus located near the base of the cell. Single layers of epithelium are called **simple epithelium**. Where there are two or more layers, the epithelium is said to be **stratified** and is classified by the shape of its surface cells. For example, the name *stratified squamous epithelium* indicates that the outer layer of cells is of the squamous type.

Connective tissue is characterized by widely separated cells (Fig. 11.2). The large amount of intercellular space is filled with a viscous solution (ground substance) and fibers secreted by the connective tissue cell; together these constitute the **matrix**. The specific nature of the matrix differentiates the four major types of connective tissue: **loose connective tissue, dense connective tissue, cartilage,** and **bone**.

Loose and dense connective tissues are binding tissues and are widespread throughout the body of animals. In loose connective tissue the matrix is composed of irregularly directed fibers and a large amount of ground substance containing macromolecules known as **proteoglycans**. These molecules consist of a core protein to which are attached strings of disaccharides. Different core proteins, and different types and numbers of disac-

charides, are responsible for many of the distinctive properties of connective tissues and enable them to adapt to different functions.

Connective tissue fibers are of two main types: **collagen** and **elastic** (Fig. 11.2a). Collagen fibers are the most abundant and are composed of collagen, a unique animal protein. Each fiber is in turn composed of many parallel fibrils, an arrangement that contributes greatly to the tensile strength of connective tissue. In vertebrates as much as 30% of the body protein is collagen. Elastic fibers are composed of a different protein, elastin, and possess an elasticity lacking in collagen.

The fibers are secreted by **fibroblasts**, star- or spindle-shaped cells anchored by long cytoplasmic strands. The cell secretes the basic units of the fiber, which then polymerize in the matrix to form the actual fiber. Two other types of cells that may be found in loose connective tissue

Figure 11.2

Connective tissue. (a) Loose connective tissue, showing the widely dispersed cells and collagen and elastic fibers within the matrix. (b) Dense connective tissue of a tendon, in which the fibroblasts are in rows between many parallel intercellular fibers.

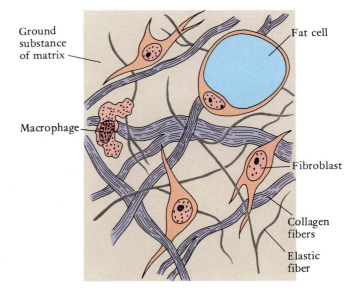

Ground substance of matrix

Fat cell

Macrophage

Fibroblast

Collagen fibers

Elastic fiber

(a)

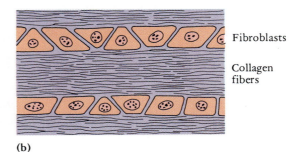

Fibroblasts

Collagen fibers

(b)

are macrophages and fat cells. **Macrophages** (Fig. 11.2a) are wandering, phagocytic cells that engulf dead or damaged cells and foreign particles. Their defensive function is described in Chapter 15 (p. 333). Fat cells are occupied by a large fat-filled vacuole (Fig. 11.2a). Aggregations of these cells, called **adipose tissue**, are the fat depots of the body.

Dense connective tissue differs from the loose type only in having a much greater quantity of fibers, which may be irregularly arranged or directed parallel to each other, as in ligaments and tendons (Fig. 11.2b).

The Integument of Invertebrates

The integument of invertebrate metazoans usually consists of a single layer of columnar epithelial cells known as the **epidermis** (Gr. *epi*, upon + *derma*, skin) that rests upon a basement membrane (Fig. 11.3). When the epidermis is exposed to the surface, many of its cells may be ciliated, and others are glandular. In some cases the glands secrete an overlying, noncellular cuticle or shell, in which case the

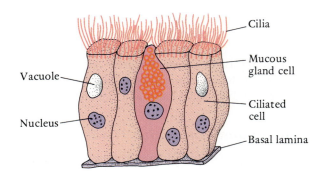

Figure 11.3
The integument of free-living flatworms and many other invertebrates is a simple epidermis consisting of a single layer of columnar epithelial cells resting on a basal lamina.

Figure 11.4
Diagrammatic vertical section through human skin.

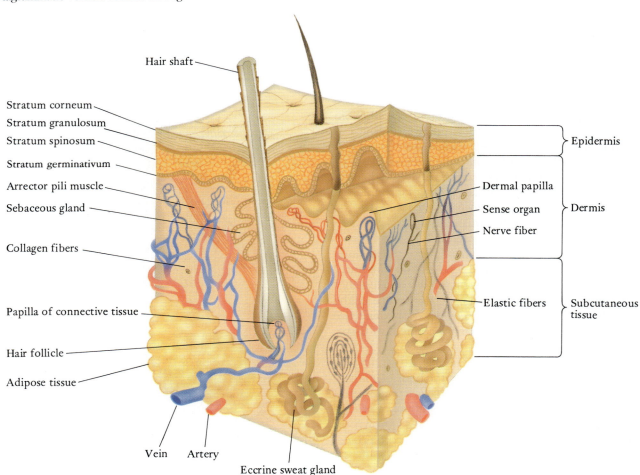

epidermal layer is generally nonciliated. Beneath the basement membrane of the epidermis is a layer of connective tissue fibers and cells, but in most invertebrates this layer is so thin that it is not easily seen.

The Vertebrate Integument

The integument of all vertebrates consists of an epidermis of stratified squamous epithelium overlaying a **dermis** composed of a thick layer of dense connective tissue made up mostly of collagen fibers (Fig. 11.4). When these fibers in cattle hides are reinforced with phenolic cross bonds (in a process called tanning), the dermis becomes leather. The basal cells of the epidermis that form the **stratum germinativum** are cuboidal to columnar in shape and divide mitotically to form new cells, many of which move to the surface, where they finally slough off. This process takes 2 to 4 weeks in humans. As they shift outward, the cells change in shape, eventually becoming flattened or squamous. In some vertebrates, the epithelial cells synthesize and accumulate a horny, water-insoluble protein called **keratin** (Gr. *keras*, horn), and die in the process. The dermis contains blood vessels, nerve fibers, and receptors that may come close to the surface in dermal papillae that project into the base of the epidermis. Bone may form in the dermis, and glands may bulge into it from the epidermis. Often fat is deposited in the deeper part of the dermis and just beneath it.

The epidermis of fishes is relatively thin and contains no keratin. Mucous cells and multicellular glands within it cover the surface with mucus that reduces friction and prevents exchanges of water with the environment and the attachment of many parasites. The skin often forms overlapping folds, and **bony scales** develop within the dermis of these folds (Fig. 11.5a). The superficial parts of the scales may become exposed as the epidermis wears off. Many spines are modified scales. Amphibian skin is similar to fish skin except for the absence of bony scales and the synthesis of some keratin.

As vertebrates adapted to terrestrial conditions, the amount of keratin in the epidermis increased, and the outer layers of cells formed a well-defined horny, protective layer, the **stratum corneum** (L. *corneus*, horny), that greatly reduces evaporative water loss. This layer is particularly thick in reptiles and forms **horny scales** and **plates** (Fig. 11.5b). The **feathers** of birds are essentially elongated and frayed horny scales.

Mammalian **hair** is another type of epidermal derivative, one that emerges from a follicle of epithelial cells that has grown down into the dermis (Fig. 11.4). Feathers and hair entrap a layer of still air and provide insulation against heat loss. The relatively hairless condition of humans varies slightly between races and between sexes, but the variations result not from differences in numbers of hair follicles but in the type of hair produced by the follicle (coarse or fine). This capacity is under both genetic and hormonal control.

Glands, another type of epidermal derivative, are virtually absent from the dry, horny skin of reptiles. Contrary to popular opinion, snakes are not the least bit slimy. Only a few **scent glands** producing odoriferous secretions used in sexual recognition remain. Bird skin is also nearly

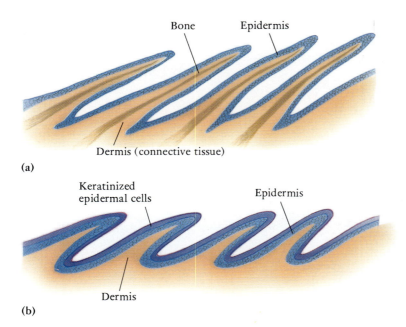

Bone   Epidermis

Dermis (connective tissue)

(a)

Keratinized epidermal cells

Epidermis

Dermis

(b)

Figure 11.5
Vertical sections through the scales of a bony fish (a) and through those of a reptile (b).

aglandular, but many glands have newly evolved in mammalian skin. All develop as ingrowths from the epidermis into the dermis (Fig. 11.4). **Sebaceous glands** (L. *sebum*, tallow) are sac-shaped structures whose cells break down and release an oily secretion into the hair follicles. They are particularly abundant in our scalp. **Eccrine sweat glands** are coiled, tubular glands that secrete a watery solution onto the skin surface. Evaporation of this solution helps cool the body. Similar **apocrine sweat glands**, found in the armpit and pubic areas, produce a more odoriferous secretion. In many mammals these secretions are chemical signals involved in courtship, the marking of territories, defense, and other functions (p. 474).

Skin Coloration

Pigments within the skin and the degree of vascularization produce integumentary coloration. In some animals, the presence of refractive granules or surface striations interact to produce iridescent colors. Some color patterns conceal the animal from predators or enable an animal to lie unseen as it awaits its prey, but others are used for species recognition or as warning signals to frighten another animal or establish territory. Squids, shrimp, crabs, and lower vertebrates, such as the frog, may have several kinds of pigment contained in starshaped cells known as **chromatophores** (Gr. *chroma*, color + *phoros*, bearing). Changes in general color tone are effected by the expansion and contraction of chromatophores or by the movement of pigment within these cells. The skin of a crab

or frog darkens when **melanin** (Gr. *melas*, black) pigment streams into the processes of the cells; it lightens when the pigment concentrates near the center of the cell (Fig. 11.6). In mammals, melanin is present both within and between basal cells of the epidermis. Some melanin is present in the skin of all human beings, except albinos, but it is especially abundant in the skin of blacks. However, the number of pigment-producing cells is about the same in all humans; it is pigment production that varies.

SUPPORT

Functions of Skeletons

All animals have some sort of skeleton, within or on the outside of their bodies, that maintains body shape and provides support. In its simplest form, this consists of the body wall of the animal acting against the noncompressible fluid (a **hydroskeleton**) located between the mesh of internal cells, similar to a bag filled with water. This continues to be the primary support system for many small, aquatic animals. However, large animals and those that live on land where the buoyant effect of water is lacking, require additional means of support.

Support is not the only function of the skeleton. Protection is frequently an important skeletal function. Sessile animals, which cannot retreat to shelter if the need arises, commonly possess a protective skeleton. Skeletons may store calcium, phosphorus, and other important minerals, and may provide a site for the manufacture of blood cells. Of great importance is the interaction of the skeleton

Figure 11.6

Chromatophores of a crab, each composed of a number of branching cells. (a) Pigment dispersed throughout chromatophore, producing a darker or more fully colored integument. (b) Pigment concentrated within the center of chromatophore, producing a paler or less fully colored integument.

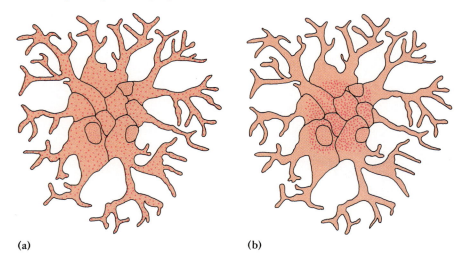

(a)   (b)

with muscles to bring about movement, such as biting and locomotion. Thus the great variation in the development of skeletal systems across the animal kingdom is correlated with (1) the size of the animal, (2) the environment in which the animal lives, and the extent to which the skeleton functions in (3) protection and (4) movement.

Skeletal Materials and Types

The skeletons of many animals are composed of hard parts. Calcium compounds are the most common component of mineral skeletons, present in groups ranging from protozoans to vertebrates. Calcium occurs as calcium carbonate in most invertebrate skeletons and as calcium phosphate in vertebrate skeletons. Siliceous compounds (e.g., silicon dioxide, a component of glass) form skeletons in some protozoans and sponges. Skeletons may be composed of specialized organic compounds, such as chitin or collagen, or complexes of organic and inorganic materials. The latter are particularly strong, for the organic components resist tension and provide elasticity, and the inorganic crystals resist compression.

Hard materials may be secreted on the outside of the body by the underlying integument to form **exoskeletons**, such as those of corals, snails, and crabs. In contrast, many animals, such as sponges, sea stars, and vertebrates, have endoskeletons located within the body wall or deeper body tissues. Not all skeletons are composed of hard materials. A fluid endoskeleton or hydroskeleton in the body cavity, blood vessels, or other body spaces can provide support or induce movement.

A few representative skeletons are considered here to illustrate their variety, advantages, and limitations; others will be described in the survey of the animal kingdom.

Exoskeletons

Snails and clams are covered by a shell secreted by areas of the integument (the mantle). The shell is composed of layers of calcium carbonate prisms or plates deposited within an organic framework. **Conchiolin**, a horny protein, covers the outer surface of the shell (p. 658).

Arthropods, including crabs, insects, and spiders, have an exoskeleton, or **cuticle** (L. *cuticula*, diminutive of *cutis*, skin), composed of the polysaccharide chitin and protein bound together to form a complex glycoprotein. In addition, calcium carbonate is incorporated into the exoskeletons of crabs and other crustaceans. The exoskeleton covers the entire body surface and extends inward to line much of the digestive tract, as well as the air-transporting tubes (tracheae) of insects.

The exoskeleton of arthropods is secreted by the underlying epidermis (Fig. 11.7). It is composed of a thin,

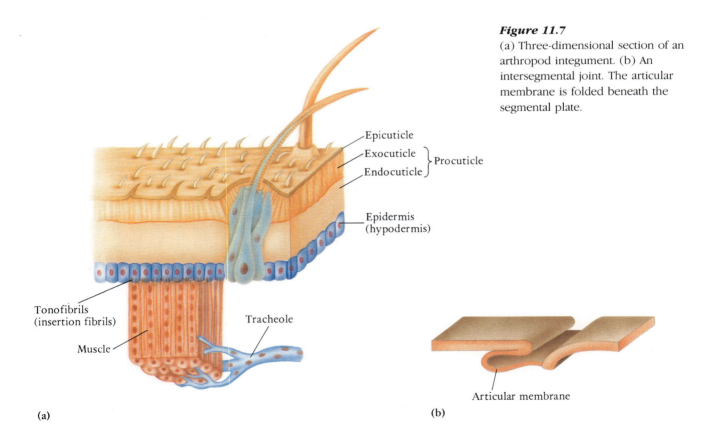

Figure 11.7
(a) Three-dimensional section of an arthropod integument. (b) An intersegmental joint. The articular membrane is folded beneath the segmental plate.

Epicuticle
Exocuticle ⎫
Endocuticle ⎬ Procuticle
Epidermis (hypodermis)
Tonofibrils (insertion fibrils)
Tracheole
Muscle
Articular membrane

(a)

(b)

surface **epicuticle** and a much thicker **procuticle** that is subdivided into an **exocuticle** and an **endocuticle**. The epicuticle, especially of terrestrial species, contains waxes; when these are absent, the exoskeleton is relatively permeable to water and gases. The exocuticle is stronger than other parts, for it has been tanned; its molecular structure has been stabilized by reactions with phenols and by the formation of additional cross linkages in the protein molecules. Sensory processes are borne on the exoskeleton, and it frequently is perforated by pores through which secretions are discharged.

The arthropod exoskeleton is divided into plates on the head and trunk and into a series of strong, tubelike segments on the appendages. These are connected to each other by thinner, folded **articular membranes**, which allow movement (Fig. 11.7b). A major disadvantage of the arthropod exoskeleton is its restriction upon growth. This limitation is circumvented by periodic shedding, or molting, of the old skeleton. During the process of molting, or **ecdysis** (Gr. *ekdysis*, shedding), the epidermis detaches from the exoskeleton and secretes a new epicuticle. Enzymes are secreted into a **molting fluid** that accumulates beneath the old skeleton. The untanned endocuticle is gradually digested while a new skeleton is secreted beneath the old one (Fig. 11.8). Eventually, the old exoskeleton ruptures along lines where the exocuticle is very thin, and the animal pulls out of its old encasement. The new

skeleton is soft and pliable for a number of hours and is stretched by internal fluid pressure to accommodate the increased size of the animal. Animals that have recently molted are very vulnerable to predators, for they are soft and their weak skeletons do not transmit muscle forces well.

Endoskeletons

Endoskeletons occur in animals of many different phyla. Sponges are supported and their shape maintained by internal needle-like mineral spicules or a network of organic spongin fibers, or both, that are secreted by special cells (see Fig. 25.3). Depending upon the class of sponges, the spicules are composed of calcium carbonate or silicon dioxide. Spicules make the body of the sponge stiff, reducing deformation by water currents.

The dermis of the body wall of starfish, sea urchins, and most other echinoderms is supported by small internal calcareous pieces, **ossicles**, that are interconnected in a lattice-like array in the wall of sea stars (see Fig. 25.2) but form a nearly solid internal shell, or **test**, in sea urchins. The test is perforated by many small openings for various organs.

The skeleton of fishes, frogs, cats, and other vertebrates is an endoskeleton composed of many individual bones and cartilages joined to each other to provide a strong but movable framework.

Figure 11.8

Molting in an arthropod. (a) The fully formed exoskeleton and underlying epidermis between molts. (b) Separation of the epidermis and secretion of molting fluid and the new epicuticle. (c) Digestion of the old endocuticle and secretion of the new procuticle. (d) The animal just before molting, encased within both new and old skeletons.

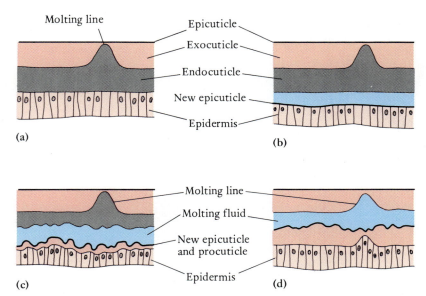

Hydroskeletons

Hydroskeletons, also called **hydrostatic skeletons**, may provide the primary skeletal support of some animals. The coelomic fluid of an earthworm is an example of a primary hydroskeleton, as is the blood flowing into the foot of a clam and pushing the foot into the sand or mud. Jumping spiders have only flexor muscles that can bend their legs. The forceful extension of the legs during a jump comes from the rush of body fluids into the leg.

Buoyancy

Much of the support for aquatic animals, even for those with a skeleton, comes from the surrounding water. If the **density** (mass per unit volume) of the animal is the same as that of the water, the animal has **neutral buoyancy** and neither rises nor sinks. Many aquatic organisms have a density near that of water, and expend little energy to keep afloat. Others have tissues denser than water. In these organisms the tendency to sink may be mitigated by including within their mass some buoyant material of lower density than water. Examples are the numerous gas-filled chambers in the shell of *Nautilus*, a molluscan relative of the octopus, and the **swim bladder** of most bony fishes. Gas, largely oxygen, usually is secreted into or reabsorbed from the fish's swim bladder by means of special glands associated with the circulatory system. The amount of gas in the bladder is regulated as the animal sinks or rises in the water column; the fish can thus maintain a neutral buoyancy over a wide range of depths.

Other aquatic organisms may have a great deal of oil in their tissues. This is true for many small planktonic organisms. Sharks do not have swim bladders but have very large livers with a high oil content. The oil, of course, serves as an important energy reserve and reduces the density of the animal to some extent. Still other marine animals, such as many jellyfish, achieve near neutral buoyancy by regulating their ion content, increasing the lighter ions and reducing certain heavy ones, such as sulfate ions. Some deep-water squids substitute ammonium ions for heavier ones.

Histology of Cartilage and Bone

Vertebrate skeletons are composed primarily of **cartilage** and **bone**. These are connective tissues in which the material forming the matrix between the cells is dense and relatively rigid. Compared with bone, cartilage exhibits substantial flexibility. Embedded within the matrix of cartilage are proteoglycans and collagen fibers, and the cartilage cells, or **chondrocytes** (Gr. *chondros*, cartilage + *kytos*, hollow vessel), are located in spaces called **lacunae** (Fig. 11.9). Considerable water is contained within the matrix, some free and some bound with the proteoglycans. Many of the properties of cartilage, including its resiliency (ability to return to original shape), resistance to compres-

sion, flexibility, and smoothness, derive from the water within it. Capillaries are absent in cartilage and nutritive substances reach the cartilage cells by diffusion through the matrix.

Cartilage is enveloped in a layer of dense connective tissue, the **perichondrium** (Gr. *peri*, on). Cartilage grows internally by division of the chondrocytes and peripherally by the transformation of fibroblasts in the perichondrium into chondrocytes (Fig. 11.9). Because of the ease with which it grows, cartilage forms most of the skeleton of embryonic vertebrates. Some is retained in adults, especially where support with some degree of flexibility and smoothness are required.

There are several types of cartilage, differentiated by the amount and kind of fibers in the matrix. **Hyaline cartilage** is somewhat translucent because of the small amount of fibers in the matrix. It is found in various places, such as the rings of the trachea (windpipe) and the articulating surfaces of joints. It is the principal type of cartilage composing the skeletons of adult sharks and rays, which lack bone. **Elastic cartilage**, in contrast, contains a large number of elastic fibers in the matrix and forms flexible skeletons, such as that of the ear. **Fibrous cartilage** contains a large amount of collagen fibers and composes the discs between the vertebrae.

Figure 11.9

Section through a piece of hyaline cartilage adjacent to the connective tissue perichondrium. Cartilage of this type would be found in joints and the rings of the trachea (windpipe).

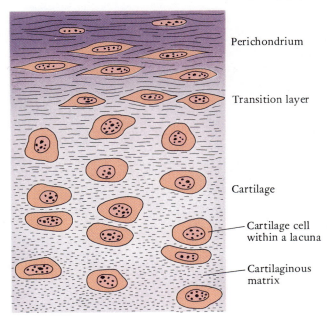

Perichondrium

Transition layer

Cartilage

Cartilage cell within a lacuna

Cartilaginous matrix

Although the matrix of bone contains proteoglycans, it is largely composed of collagen fibers and crystals of calcium triphosphate (plus a small amount of calcium

Figure 11.10
Compact bone, showing haversian systems. (a) Photograph of a section of dried compact bone. Lacunae appear black because of air trapped within them. (b) Structure of compact bone in cross section.

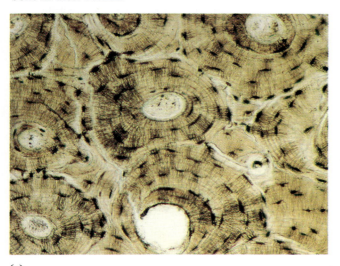

(a)

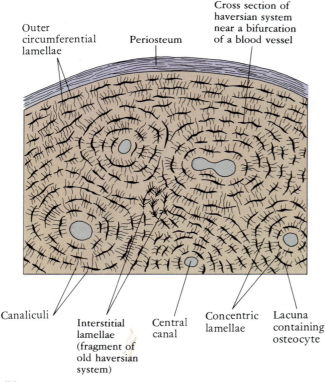

Outer circumferential lamellae   Periosteum   Cross section of haversian system near a bifurcation of a blood vessel

Canaliculi   Interstitial lamellae (fragment of old haversian system)   Central canal   Concentric lamellae   Lacuna containing osteocyte

(b)

carbonate) bound to them. The density is too great to permit diffusion, but tiny canals, or **canaliculi**, connect one lacuna with another (Fig. 11.10). Within the canaliculi, fine extensions of the bone cell, or **osteocyte** (Gr. *osteon*, bone), make contact with similar extensions from adjacent cells, and no bone cell is far removed from a blood supply. The spacing of bone cells is such that the matrix is secreted in sheets, or **lamellae**, with the bone cells separating one lamella from another.

In **compact bone** the lamellae are arranged in concentric rings around a central **haversian canal** containing one or two blood vessels, usually capillaries. The canal and associated rings of lamellae constitute a **haversian system**, which is in effect a microscopic column of bone (Fig. 11.10). Large numbers of such columns, running parallel to each other, compose the visible mass of bone tissue. The space between adjacent, cylindrical haversian systems is filled with **interstitial lamellae**, the remains of old systems, for there is continual turnover of bone material, with old haversian systems being removed and new ones formed. The haversian canals interconnect and accommodate the branching blood vessels.

In **cancellous** or **spongy bone**, which is found, for example, in the ends of long bones, there are no haversian systems, and the lamellae form interconnecting bars or plates that separate large spaces (Fig. 11.11). These spaces are filled with tissue that produces blood cells and constitutes the **red bone marrow**.

Most bones contain both compact and cancellous bone tissue but vary in the amount of each. Flat bones, such as those forming the skull, are composed of spongy bone between two outer layers of compact bone. Like cartilage, the entire bone is surrounded by a layer of connective tissue called the **periosteum**.

A typical long bone, such as we find in the legs of mammals, consists of a central **shaft** and a knoblike **epiphysis** (Gr. *epi*, on + *physis*, growth) at each end (Fig. 11.11). The shaft is composed of compact bone surrounding a large, central **yellow bone marrow** cavity that stores fat. The interior of the epiphysis is occupied by spongy bone tissue and its associated red bone marrow. The periosteum that surrounds a long bone provides the means by which **ligaments** binding bone to bone and **tendons** binding muscle to bone are anchored (Fig. 11.12). The fibers of both ligaments and tendons become interwoven with those of the periosteum, and some penetrate the bone.

The ends of a long bone articulate with adjacent bones, and the articulating surface is capped with a layer of cartilage (Figs. 11.11 and 11.12). The surrounding ligaments enclose the articulating area as a capsule. The joint cavity within the capsule contains a lubricating **synovial fluid** (Fig. 11.12). The configurations of the articulating protuberances and sockets are related to the type of movement that can be performed.

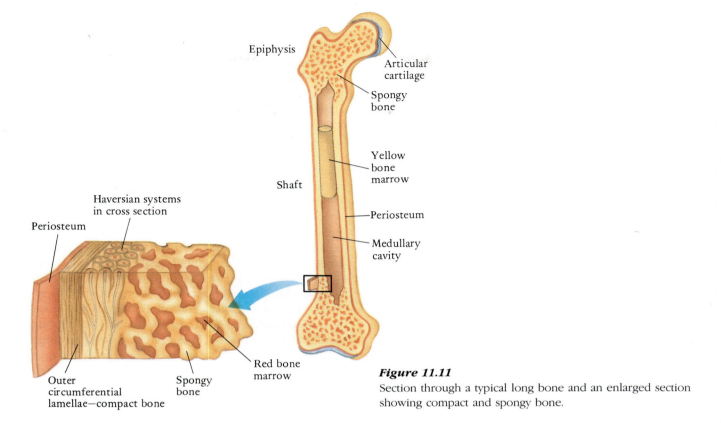

Epiphysis

Articular cartilage

Spongy bone

Shaft

Yellow bone marrow

Periosteum

Medullary cavity

Periosteum

Haversian systems in cross section

Outer circumferential lamellae—compact bone

Spongy bone

Red bone marrow

Figure 11.11
Section through a typical long bone and an enlarged section showing compact and spongy bone.

Development and Growth of Cartilage and Bone

During embryonic development most of the vertebrate skeleton is composed of cartilage, but in most adults the cartilage is replaced by bone. This bone is called **cartilage replacement bone** to distinguish it from the **dermal bone** that develops within or just beneath the skin without any cartilaginous precursor. These two types of bone differ only in their mode of development; they are similar histologically.

Bones increase in size peripherally by the conversion of connective tissue fibroblasts in the periosteum into osteoblasts. In addition, a long bone has an **epiphyseal plate** of cartilage within each end, which provides for increase in length (Fig. 11.12). On either side of the epiphyseal plate chondrocytes are continually being destroyed and replaced by bone cells and bone matrix. The epiphyseal plate is maintained by mitotic division of the cartilage cells, but at maturity in birds and mammals, when growth hormones decrease and sex hormones increase (p. 445), the chondrocytes cease to grow and soon are completely replaced by bone cells. Peripheral growth also ceases. Although we tend to think of bone as being relatively static, it does exhibit some plasticity and is molded in part by the loads it bears and by the pull of ligaments and tendons.

Figure 11.12
Diagram of a joint between two long bones with epiphyseal plates.

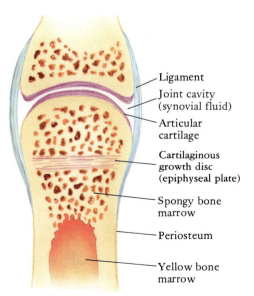

Ligament

Joint cavity (synovial fluid)

Articular cartilage

Cartilaginous growth disc (epiphyseal plate)

Spongy bone marrow

Periosteum

Yellow bone marrow

Support and Architecture in Appendicular Skeletons

The most common means of movement in animals is by appendages. Both arthropods and vertebrates employ appendages, and their appendicular skeletons are similar in many ways. Although arthropods have an exoskeleton with muscles attached to the inner side and vertebrates have an endoskeleton with the muscles attached to the outer side (Fig. 11.13a), both are "jointed" and, as we will describe later, the skeleton and associated muscles function together as lever systems in both groups. The skeletons must resist loads that bend, twist, and shear (skeletal materials slide past each other). For good architectural reasons both the arthropod appendicular skeleton and the vertebrate long bones are hollow cylinders. Imagine instead that both were solid rods. Much of the load would be carried parallel to the long axis, but frequently these rods would be subjected to bending loads from the side. When the rod is bent, the supporting material on the concave side is stressed by compression and that on the convex side is stressed by tension (stretched) (Fig. 11.13b). If the skeletal material were in the center, it would be of little use because this region is stress neutral, i.e., the tissue here is neither stretched nor compressed. The presence of a stress-free region is taken into account in I-beams, girders used in construction. The material is divided into two parallel bars placed at the periphery and connected by a flange that prevents shearing. I-beams work well as supports for roofs and floor slabs of buildings because the bearing load is coming from one direction. Appendages can be subjected to loads from all directions. Thus, all of the skeletal material is placed peripherally but in a ring, which forms a cylinder. Cylinders are resistant to bending, and the greater the diameter, the greater the resistance. However, although they may not bend readily, the wall will buckle easily if too thin. For example, when a tin can is kicked, the impact does not bend the can but it collapses the wall.

We are now in a position to understand the advantages and disadvantages of the two types of appendicular skeletons. The arthropod exoskeleton provides both support and protection, and offers a large surface area for muscle attachments. As long as the body is small, the cylindrical skeleton will resist bending loads and a relatively small amount of skeletal material need be expended to keep the wall from buckling. But this design imposes limits on body size. A really big terrestrial arthropod would require a large amount of skeletal material and would still collapse following a molt, when the skeleton was soft. The endoskeleton of vertebrates, on the other hand, does not provide the protection of the arthropod skeleton, but when the animal is large, the central position of the skeletal cylinder within the appendage does not demand a large diameter. A smaller amount of skeletal material can thus be concentrated to form a thick wall that will not easily buckle under heavy loads.

The Vertebrate Skeleton

The skeleton can be divided into a **somatic skeleton** (Gr. *soma*, body), located in the body wall and appendages, and a deeper **visceral skeleton** (L. *viscera*, bowels), associated primitively with the pharyngeal wall and gills. The somatic skeleton can be further subdivided into an **axial skeleton** (vertebral column, ribs, sternum, and most of the skull) and an **appendicular skeleton** (girdles and limb bones).

Figure 11.13

(a) Diagrammatic sections of the appendages of an arthropod and a vertebrate comparing the external and internal positions of the skeletons and muscles. (b) Regions of stress when a skeletal rod is bent.

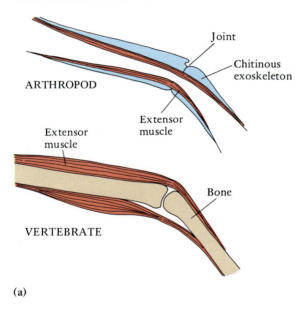

ARTHROPOD

Joint

Chitinous exoskeleton

Extensor muscle

Extensor muscle

Bone

VERTEBRATE

(a)

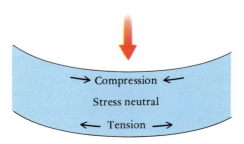

→ Compression ←

Stress neutral

← Tension →

(b)

The Fish Skeleton

The parts of the skeleton can be seen more clearly in a dogfish (a small shark) than in terrestrial vertebrates (Fig. 11.14). The **vertebral column** is composed of **vertebrae**, each of which has a biconcave **centrum** that develops around and largely replaces the embryonic notochord. Between each vertebra and enclosed within the vertebral concavities is an **intervertebral disc**, composed of a large amount of connective tissue. When the vertebral column bends, the intervertebral discs function to distribute force evenly from one vertebra to the next. Dorsal to each vertebral centrum is a **vertebral arch** surrounding the **vertebral canal**, in which the spinal cord lies. Short **ribs** attach to the vertebrae, a sternum is absent, and the individual vertebrae are rather loosely held together. Strong vertebral support is not required in the aquatic environment.

The skull of the dogfish is an odd-shaped box of cartilage, the **chondrocranium** (Gr. *chondros*, cartilage + *kranion*, braincase), that encases the brain, nose, eyes, and inner ear. It forms the core of the skull of all vertebrates. Other components of a vertebrate skull include the anterior arches of the visceral skeleton and dermal bones that encase the chondrocranium and anterior visceral arches. These dermal bones have been lost during the evolution of cartilaginous fishes but were present in the fishes ancestral to terrestrial vertebrates, or tetrapods.

The visceral skeleton consists of seven pairs of >-shaped visceral arches hinged at the apex of the >. They are interconnected ventrally but are free dorsally. Each arch lies in the wall of the pharynx and supports gills in very primitive vertebrates. In jawed vertebrates the first or **mandibular arch** becomes enlarged and, together with

associated dermal bones, forms the upper and lower jaws. The entire jaw in the dogfish is derived from the mandibular arch; there are no surrounding dermal bones. The second or **hyoid arch** of the dogfish helps to support the jaws. Its dorsal portion (hyomandibular) extends as a prop from the **otic capsule** (Gr. *otikos*, ear), the part of the chondrocranium housing the inner ear, to the angle of the jaw. The third to seventh visceral arches, the **branchial arches** (Gr. *branchia*, gills), support the gills; gill slits lie between them.

The appendicular skeleton is very simple in the dogfish. A U-shaped bar of cartilage, the **pectoral girdle**, in the body wall posterior to the gill region supports the **pectoral fins**. The **pelvic girdle** is a transverse bar of cartilage in the ventral body wall anterior to the termination of the digestive tract. It supports the **pelvic fins** but is not connected to the vertebral column.

The Tetrapod Skeleton

The invasion of the land by primitive tetrapods brought about radical changes in the vertebrate skeleton. No longer buoyed by water, terrestrial vertebrates evolved skeletons capable of providing support and increased activity in the new environments. The human skeleton, although adapted to upright posture, illustrates the main features (Figs. 11.15–11.17). The vertebral column in all tetrapods is thoroughly ossified, and the individual vertebrae are strongly united by overlapping **articular processes**, borne on the vertebral arches. A well-developed **vertebral spine**, extending dorsally from the arch, and lateral **transverse processes** serve for the attachment of ligaments and muscles. Ribs articulate on the centrum and transverse processes.

Figure 11.14
A lateral view of the skeleton of a dogfish.

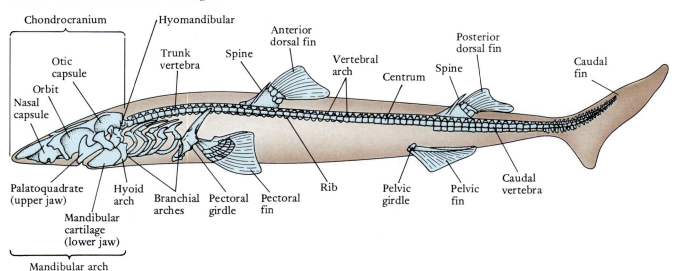

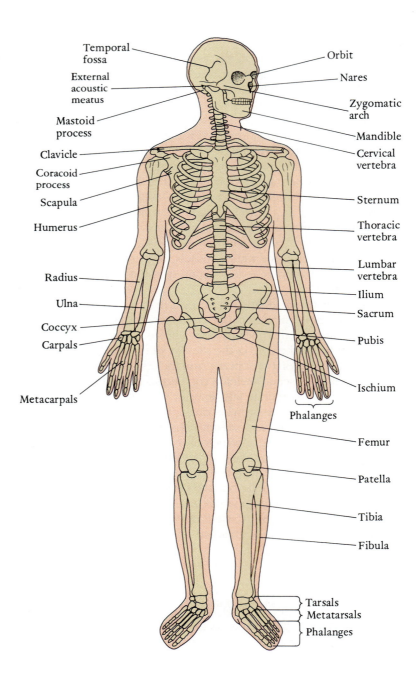

Figure 11.15
The human skeleton.

Correlated with changes in the method of locomotion and the independent movement of various parts of the body, there is extensive regional differentiation of the vertebral column. Mammals have 7 **cervical vertebrae** in the neck region, of which the first two, the **atlas** (Gr. *Atlas*, the Greek god who supported the heavens upon his shoulders) and **axis**, are modified to permit extensive movement of the head (Fig. 11.16). The skull rocks up and down at the joint between the skull and atlas; turning movements occur between the atlas and axis. The vertebral column of the mammalian trunk is differentiated into thoracic and lumbar regions. We have 12 **thoracic** and 6 **lumbar vertebrae**. Only the thoracic vertebrae bear distinct ribs, most of which connect via the costal cartilages with the ventral breast bone, or **sternum**. Rudimentary ribs, present in the other regions during embryonic development, fuse onto the transverse processes. The 5 **sacral vertebrae**, which are fused to form the **sacrum** (L. *sacred*, because this region was used in sacrifices), transfer weight from the trunk to the pelvic girdle and hind legs. Humans have no external tail, but 4 **caudal vertebrae** remain. They are fused together to form an internal **coccyx** (Gr. *kokkyx*,

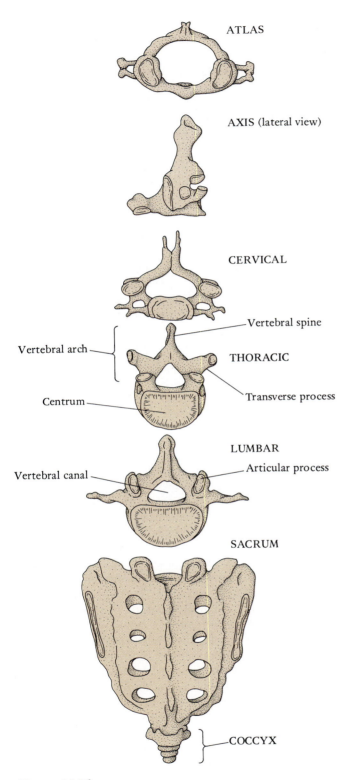

ATLAS

AXIS (lateral view)

CERVICAL

Vertebral spine

Vertebral arch

THORACIC

Centrum

Transverse process

LUMBAR

Vertebral canal

Articular process

SACRUM

COCCYX

Figure 11.16
Types of mammalian vertebrae as represented by those of a
human being. The axis is seen in lateral view; the sacrum
and coccyx, in dorsal view; the others are viewed from the
anterior end.

bone shaped like a cuckoo's bill) to which certain anal
muscles attach. The basic pattern of vertebrae is much the
same in other terrestrial vertebrates, although there are
variations in the number of vertebrae and in their regional
differentiation, depending upon methods of locomotion
and support.

The skull (Fig. 11.17) can be divided into a **cranium**
that houses the brain and encases the inner ear and a **facial
skeleton** that comprises the bones forming the jaws and
encasing the eyes and nose. There are many foramina
(openings) for blood vessels and nerves; most conspicuous is the large **foramen magnum** through which the spinal
cord enters the skull. Many of the individual bones forming
the skull of the human are shown in Figure 11.17.

The jaws of most vertebrates are formed of dermal
bones that encase the cartilaginous mandibular arch (mandibular and palatoquadrate cartilages), but the jaw joint

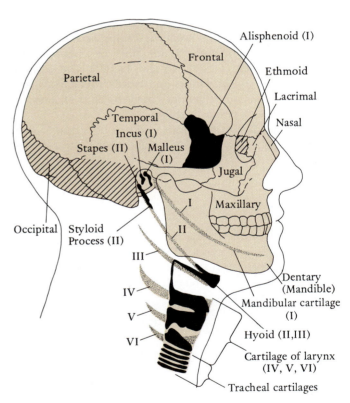

Figure 11.17
Components of the human skull. Dermal bones have been
left plain, chondrocranial derivatives are hatched, those parts
of the embryonic visceral skeleton that disappear are
stippled, and parts of the visceral skeleton that persist are
shown in black. Roman numerals refer to visceral arches and
their derivatives.

lies between the posterior ends of these cartilages. In most tetrapods part of the second visceral arch has been transformed into the **stapes** (L. *stirrup*), a slender rod of bone extending from the eardrum to the part of the skull housing the inner ear, but it still retains a connection with the jaw (first visceral arch). During the evolution of mammals the jaw mechanism became stronger and the bite more powerful. A new joint evolved between certain of the dermal jaw bones. The parts of the mandibular and palatoquadrate cartilages that bore the jaw joint had become quite small and were overspread by the eardrum and incorporated in the middle ear as the outer two auditory ossicles, **malleus** and **incus**, respectively (Fig. 11.17; see also p. 414). The incus of mammals connects with the stapes. In mammals these three auditory ossicles form a pressure-amplifying leverage system (Chap. 18).

The ventral part of the second visceral arch, together with the third arch, forms the hyoid bone, which provides support for the tongue. Parts of the remaining arches contribute to the cartilages of the larynx.

The appendicular skeleton of tetrapods is derived from the pectoral and pelvic fin skeleton of ancestral crossopterygian fish (Fig. 11.18). The appendicular skeleton is composed of the **humerus**, **radius**, and **ulna** of the forelimb or arm and the **femur**, **tibia**, and **fibula** of the hindlimb or leg (Figs. 11.15 and 11.18b). The **carpals**, **metacarpals**, and **phalanges** and **tarsals**, **metatarsals**, and **phalanges** constitute the bones of the forefoot (hand) and hindfoot (foot), respectively.

The girdles of tetrapods need to be stronger than those of fish because body weight must be transferred through them to the limbs and ground. The pectoral girdle is bound onto the body by muscles, but the pelvic girdle extends dorsally and is firmly attached to the vertebral column. A **pubis**, **ischium**, and **ilium** are present on each side of the pelvic girdle, but are fused together in the adult (Fig. 11.15). Our pectoral girdle consists of a **scapula** and a **clavicle**. A projection of the scapula, the **coracoid process**, is a distinct bone in most lower tetrapods. The clavicle is the only remnant of a series of dermal bones that are primitively associated with the pectoral girdle. All other girdle bones are cartilage replacement bones.

MOVEMENT

Movement in animals results from the beating of cytoplasmic processes (cilia) or the contraction of muscle cells. Although some types of coelomic and blood cells move by amoeboid flow (p. 223), metazoan animals do not move in this manner. Ciliary locomotion is widespread among small aquatic animals and sometimes plays a role in moving substances over or within the bodies of larger animals. However, most movement in large animals is powered by muscle contraction.

Muscle

Muscle is a biocontractile system in which cells, or parts of them, are elongated and specialized to develop tension along their axis of elongation. Contractile stalks are found in some protozoans, and relatively unspecialized contractile cells occur in sponges around the opening through which water is discharged from the central cavity. As animals and their movements become more complex, simple contractile cells are replaced by well-differentiated muscular tissue. Simple contractile cells persist in certain higher animals; the myoepithelial cells that aid in the discharge of secretions from mammalian sweat and mammary glands are an example. But most animal muscle cells are part of organized **smooth** or **striated muscle tissues**. The elongated cells, or fibers, of these tissues contain parallel, contractile **myofibrils**. Smooth muscle tissue, which in vertebrates is associated with the gut, blood vessels and other internal organs, consists of long, spindle-shaped cells with a centrally located nucleus (Fig. 11.19a and b). The tissue is called smooth because the myofibrils appear to be homogeneous under a light microscope. Striated muscle tissue composes the skeletal muscles of vertebrates and the muscles of arthropods and is found in many other animal groups as well. The very large cells can reach several centimeters in length and are generally multinucleated. Nuclei are located at the periphery beneath the plasma membrane or **sarcolemma** (Fig. 11.19c). The contractile units, or **sarcomeres**, of the myofibrils are so arranged that the muscle fibers have a cross-striated appearance when seen under a light microscope.

Figure 11.18
Lateral views of the pectoral appendicular skeleton of an ancestral crossopterygian fish (a) and amphibian (b) to show the changes that occurred in the transition from water to land. Dermal bones have been left plain; cartilage replacement bones are stippled.

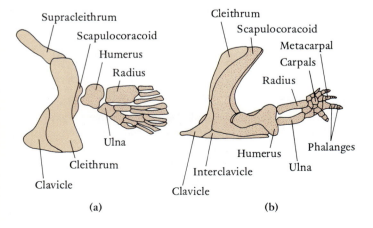

Supracleithrum
Scapulocoracoid
Humerus
Radius
Ulna
Cleithrum
Clavicle

(a)

Cleithrum
Scapulocoracoid
Metacarpal
Carpals
Radius
Humerus
Phalanges
Interclavicle
Ulna
Clavicle

(b)

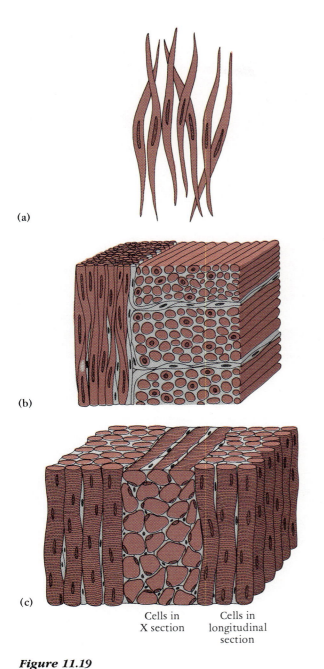

(a)

(b)

(c)

Cells in
X section

Cells in
longitudinal
section

Figure 11.19
Types of muscle tissue. (a) Teased smooth muscle cells. (b)
Section of smooth muscle tissue in the intestinal wall. (c)
Striated or skeletal muscle tissue.

The ultrastructure of these units will be described later (p.
311). The heart muscle of vertebrates (cardiac muscle) is a
specialized type of striated muscle and is discussed in
Chapter 14.

In most animals, groups of muscle fibers are bound
together and invested by connective tissue (epimysium
and perimysium) through which nerves and blood vessels
run. Cuts of meat are tender or tough because of varying

amounts of connective tissue present. However, some of
the very small muscles of invertebrates consist of only a
few muscle fibers, with little connective tissue between
them.

Muscles perform work only by contracting and must
work against **antagonistic** (opposing) forces. The adductor
muscle that holds the shell of a clam closed works against
the force of an elastic spring ligament in the shell hinge.
When the muscle relaxes, the ligament opens the valves of
the shell and stretches the muscle to its original length.
Most muscles are arranged in antagonistic sets, such as the
biceps and triceps of the arm (Fig. 11.20). Contraction of
the biceps causes **flexion** of the forearm and contraction of
the triceps results in **extension** of the forearm. In each case
the antagonistic muscle is stretched. Note that one of the
attachment ends of the muscle generates most of the movement. This end is called the **insertion**; the opposite, less
movable end, is the **origin** (Fig. 11.20).

Many muscles attach to parts of the skeleton, but
others do not and form a mass of interlacing muscle fibers,
as in the foot of a snail, or form layers in the walls of hollow
organs, such as the intestine and blood vessels (see Fig.
12.16). These fibers do not have definable origins and
insertions but pull upon each other and squeeze upon

Figure 11.20
Antagonistic skeletal muscles attached to bones in the human
arm.

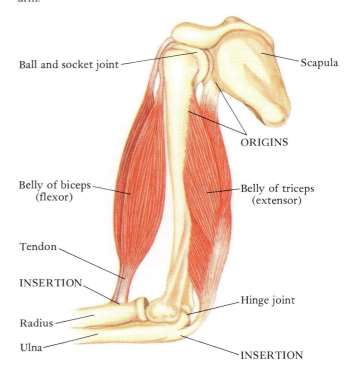

Ball and socket joint

Scapula

ORIGINS

Belly of biceps
(flexor)

Belly of triceps
(extensor)

Tendon

INSERTION

Hinge joint

Radius

Ulna

INSERTION

fluids within the organ so as to change the size and shape of the organ of which they are a part. The original length of the fibers is restored by the pull of fibers oriented in a different direction or by forces generated in the fluids.

Biochemistry of Muscle Contraction

Striated Muscle Contraction

The mechanism of muscle contraction is best understood from studies of vertebrate skeletal muscle. The many myofibrils that pack the cytoplasm of a striated muscle fiber (one cell) can be seen with a light microscope, and the repeating bands (striations) of the myofibrils had been observed long before their ultrastructure was understood (Figs. 11.19c and 11.21a). The repeating units are composed of a dark **A band** bounded on either side by a light **I band** (Fig. 11.21). Another light band, the **H band**, divides the A band, and a **Z line** divides the I band. The Z lines demarcate the repeating units, or **sarcomeres**.

Electron microscopy and biochemical analysis show that each myofibril is composed, in turn, of many longitudinal protein myofilaments of two types, thick and thin. The **thin myofilaments** are composed of globular **actin** monomers with which **tropomyosin** and **troponin** complexes are associated (Fig. 11.21a). The **thick myofilaments** are composed of the tails of elongated **myosin** molecules. Heads of the myosin molecules protrude toward the thin filaments as potential **cross-bridges** that can link the filaments. Each head and tail are joined by a compliant hinge. Thick and thin filaments are packed in such a way that each thick filament is surrounded by six thin ones. Since the myosin heads spiral around the thick filament, cross-bridges can be formed with all of the surrounding thin filaments (Figs. 11.21a and 11.22). The denser, thick filaments are aligned with each other and account for the darker A bands. The thin filaments, also in alignment, account for the lighter I bands. The two types of filament only partially overlap, which explains the somewhat less dense H zone portions of the A bands (Fig. 11.21a). Z lines, where the thin filaments of one sarcomere attach to those of the next sarcomere, cross the entire myofibril in the center of the I bands.

When striated muscle contracts, the I bands narrow and the H zones may disappear as the Z lines move closer together (Fig. 11.21a). The degree of narrowing of the I bands is a function of the extent of the contraction. Such observations in the late 1950s led H.E. Huxley and Jean Hanson of Cambridge University and King's College, London, to propose the **sliding filament hypothesis** of muscle contraction, a theory that has since been elaborated by many investigators.

No interaction occurs between the filaments at rest because the active sites on the actin filaments to which myosin heads can attach are blocked by tropomyosin. When a muscle fiber is stimulated, changes occur in its

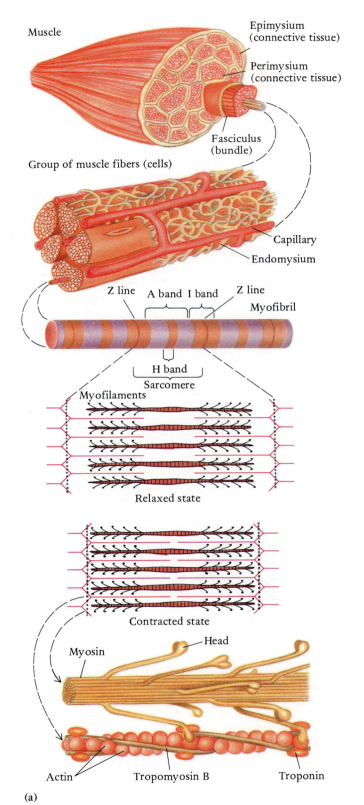

Muscle
Epimysium (connective tissue)
Perimysium (connective tissue)
Fasciculus (bundle)
Group of muscle fibers (cells)
Capillary
Endomysium
Z line   A band I band   Z line
Myofibril
H band
Sarcomere
Myofilaments
Relaxed state
Contracted state
Myosin
Head
Actin
Tropomyosin B
Troponin

(a)

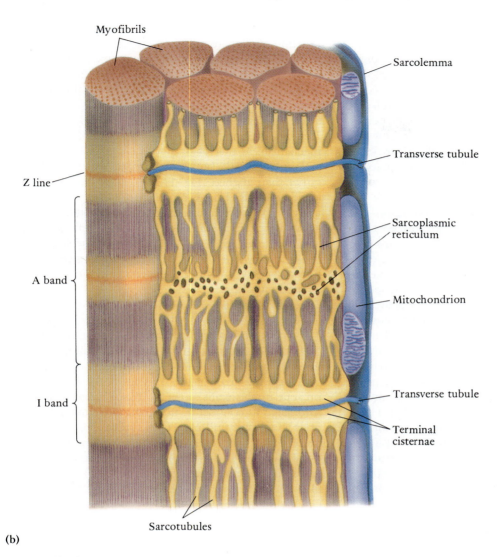

(b)

Figure 11.21

(a) Structure of a striated muscle and its components at successive levels of magnification. (b) A drawing of a portion of a sarcomere of frog striated muscle to show the interrelations of the sarcolemma, sarcoplasmic reticulum, and myofibrils.

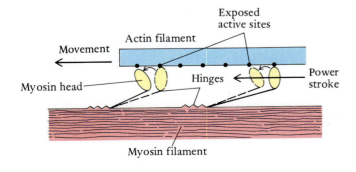

Figure 11.22

Diagram of the ratchet mechanism for muscle contraction. As shown here, some myosin heads are attached and undergoing a configurational change that exerts a power stroke, while other ones are detached and preparing to reattach.

sarcolemma that lead to the unblocking of the active sites. At rest, the sarcolemma has an electrical charge, known as the resting potential, that derives from a large concentration of Na$^+$ ions on the outside. When a nerve impulse reaches the **motor end plate** at the end of a nerve cell, a transmitter substance (usually **acetylcholine**) is released by the neuronal ending and causes a change in the permeability of the sarcolemma such that Na$^+$ ions quickly flow inward. This momentarily reverses the electrical charge and gives rise to the **action potential**. The action potential is self-propagating and spreads quickly along the sarcolemma and deep into the fiber, where the sarcolemma is infolded as **transverse tubules** (Fig. 11.21b). In many muscles these tubules are adjacent to the Z lines. Ca^{2+} ions in adjacent storage sacs of the sarcoplasmic reticulum, the **terminal cisternae**, are then released and combine with troponin (one of the actin proteins), changing its configuration in such a way that the troponin-tropomyosin complex is pulled away from the active sites on the actin. Myosin heads now bind to these sites.

How this results in the sliding of the myofilaments past each other is not entirely clear. Probably the myosin heads, which previously have been energetically charged, or cocked, use their stored energy to bond in sequence to several sites on the actin filament, causing the myosin heads to flex slightly at their hinge regions (Fig. 11.22). Since the myosin heads at the two ends of the myosin filament face in opposite directions, this would pull the actin filaments, to which the heads are attached, toward the center of the sarcomere. As the heads bend, they expose a region that has ATPase (enzyme) activity. ATP is broken down, and the released energy causes the myosin heads to detach from the actin sites, straighten, and recock. They then reattach to new adjacent actin sites. The numerous myosin heads attaching and reattaching act as a ratchet pulling the actin filaments deeper and deeper into the sarcomere. As this is going on, a calcium pump is driving free Ca^{2+} ions back into the terminal cisternae, and the enzyme **acetylcholinase** breaks down acetylcholine at the motor end plate. If muscle stimulation is not continued, the level of free calcium becomes so low that the troponin-tropomyosin complex again moves over the active sites on the actin filaments, blocks them, and contraction stops. In this model, ATP is needed to energetically charge the myosin heads and to detach them from the actin sites prior to the next cycle of attachment, pull, and detachment. ATP also drives the calcium pump. Energy is needed for relaxation, i.e., to detach the myosin heads and to remove free calcium ions. A muscle deprived of a source of ATP remains contracted. We know this as muscle cramps or, after death, as **rigor mortis**.

The attachment and detachment of each myosin head utilizes one molecule of ATP. When the small amount of ATP stored in muscle is used up, ATP is resynthesized from ADP by the transfer of a phosphate group from other stored energy-rich phosphates, called **phosphagens**: **creatine phosphate** in vertebrate and some invertebrate muscles and **arginine phosphate** in most invertebrates. These are only short-term energy stores. Energy sources for muscle contraction are discussed further in Chapter 13.

Smooth Muscle Contraction

The contraction of smooth muscle utilizes the same basic mechanism as that of striated muscle. Its fibers contain actin and myosin, but these myofilaments are scattered rather than aligned (Fig. 11.23a). Actin filaments are attached to dense bodies rather than Z lines. Since smooth muscle fibers are much smaller than striated fibers, calcium ions are not stored within the sarcoplasmic reticulum but are concentrated in the extracellular fluid. Activation involves the influx of calcium ions and then the formation of cross-bridges between actin and myosin, as in striated muscle.

Although smooth muscle fibers in different organs look very much alike, they have quite different properties. The smooth muscles of vertebrates can be grouped into two major types, multiunit and visceral. **Multiunit smooth muscle** fibers are found in the iris and ciliary body of the eye, form the arrector pili muscles of the hairs, and are present in the walls of many blood vessels. They are stimulated to contract by nerve impulses that reach each fiber at a motor end plate (Fig 11.23b). **Visceral smooth muscle** fibers that contribute to the wall of the digestive organs, uterus, and many other visceral organs are more tightly packed together (Fig. 11.23c). Only a few of the fibers have motor end plates, but the action potential generated in one fiber can flow to adjacent ones through **gap junctions** that tightly unite their plasma membranes at many points (p. 200). Visceral smooth muscle also has a degree of spontaneous activity and will contract automatically when stretched, as occurs when an organ fills with food or liquid.

Physiology of Muscle Contraction

Motor Units

Muscle contraction is an active state in which tension is developed within muscle. Contraction is initiated normally by a nerve impulse. In a typical vertebrate skeletal muscle, one nerve cell, or neuron, will branch in such a way as to have one or more motor end plates on each of a number of different muscle fibers. The neuron and the fibers it supplies constitute a **motor unit**. *A nerve impulse causes all of the fibers in the unit to contract at the same time.* The degree of fine control over the activity of a muscle is inversely proportional to the number of muscle fibers per motor unit. Ocular muscles that move the eyeball may have as few as two or three fibers to a unit, whereas some of the leg muscles that hold the body erect contain several hundred per unit.

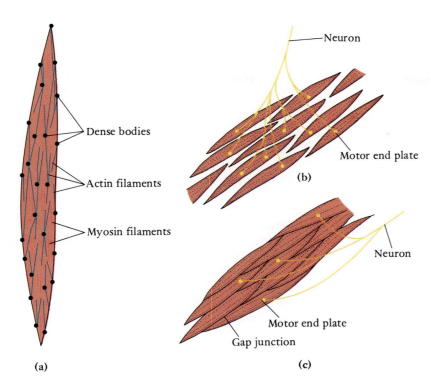

Figure 11.23
Smooth muscle structure and innervation. (a) A diagram showing the arrangement of actin and myosin filaments in a smooth muscle fiber. (b) The innervation of multiunit smooth muscle. (c) Innervation of visceral smooth muscle.

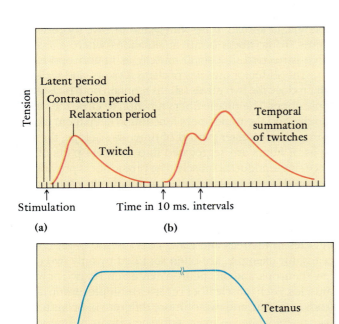

Figure 11.24
Isometric contraction in a motor unit. (a) A twitch; (b) temporal summation of two twitches; (c) tetanus. Vertical arrows indicate points of stimulation.

Stages in Contraction

Muscles or individual muscle fibers may be activated by a single nerve impulse, but two or more in rapid succession may be needed. The first impulse initiates changes that are built upon by subsequent ones, if they arrive before the effects of the first have worn off. The early impulse is said to facilitate activation.

Activation requires a few milliseconds, so there is a delay, or **latent period**, between the application of a stimulus and the onset of cross-bridge formation and the shortening of the contractile elements (Fig. 11.24a). The first effect is to stretch the elastic components in the muscle, including the hinge region of the myosin molecules, connective tissue, and tendon. Tension then develops (filament sliding) and the muscle may shorten during this **contraction period**. Tension decreases during the longer **relaxation period**.

If the muscle is working against a constant load, as when the biceps is flexing the arm in lifting a book, the muscle as a whole will shorten as tension develops, and the contraction is described as **isotonic** (Gr. *isos*, equal + *tonos*, strain) or **positive work**. If the biceps is pulling against an immovable bar gripped by the hand, internal shortening and tension develop as the elastic elements are stretched, but there is little or no overall shortening of the muscle. This is **isometric contraction** (Gr. *isos*, equal + *metron*, measure). In **negative work contraction** the length of the muscle increases as tension develops. As a human stands up from a sitting position, the hamstring muscles on the back of the thigh contract as they pull the thigh under the hips, yet at the same time, they increase in length as the lower leg, to which they also attach, straightens out.

Physiological Muscle Types

If muscles are to be efficient in movement, the tension they develop and their speed of contraction must match the conditions under which they operate. Not all muscle fibers are alike in their properties. Most skeletal muscles of the higher vertebrates are **twitch** or **phasic muscles**. Each fiber has a single motor end plate. When a motor unit is stimulated electrically by its neuron, a single stimulus of threshold intensity will elicit a simple **twitch** (Fig. 11.24a). This is an **all-or-none phenomenon** because the action potential spreads through the entire fiber, and the twitch is maximal; a stronger stimulus will not cause a stronger twitch. Gradation of contraction can, however, be accomplished in two ways. If a second stimulus reaches the motor unit before the effects of the first have worn off, a second twitch will be superimposed upon the first. This is possible because the first twitch does not last long enough to fully engage all the myosin heads or to stretch all of the elastic components. This phenomenon is called **temporal summation** (Fig. 11.24b). If impulses reach the motor unit so rapidly that no relaxation occurs between twitches, tension develops very rapidly to a plateau beyond which it will not go even when the rate of stimulation is increased. The muscle is then in **tetanus** and this state will be maintained until the impulses cease or the muscle tires (Fig. 11.24b). Graded contractile responses can also be elicited by recruiting more and more motor units to contract, a phenomenon called **spatial summation**.

Seldom are all motor units in a muscle active at the same time; some are relaxing as others are contracting. In this way a muscle can sustain a contraction for a long time without fatigue. Some fibers in a muscle are always in a state of contraction, a condition known as **tonus**. Tonus keeps muscles firm and in proper spatial relationship to each other.

Phasic muscle fibers differ in their speed of contraction and fatigue. **Slow phasic** fibers, used in sustained postural contractions and slow movements, are often red in color because they contain muscle hemoglobin, **myoglobin**. Since myoglobin has a higher affinity for oxygen than the hemoglobin in the blood (p. 287), it accelerates the transfer of oxygen from blood to muscle. Mitochondria are abundant in these slow fibers, since most of the energy of slow phasic fibers is derived from the complete oxidation of fuels, and the mitochondria contain the enzymes that catalyze most of these reactions. These fibers contract slowly; delivery of oxygen and food keeps up with the demand, hence they do not fatigue easily. **Fast phasic**, or **glycolytic**, **fibers** are used when quick, short bursts of activity are required. They lack myoglobin and are whitish in color. Mitochondria are few in number, since most energy is derived quickly from the anaerobic breakdown of stored glycogen (p. 263). These fibers contract rapidly but go into oxygen debt and soon fatigue (p. 265). Energy is replaced after the burst of activity. Slow and fast phasic fibers are sometimes segregated in different muscles (dark and light meat in a chicken or fish are a familiar example), but often there will be a mixture of both types in a single muscle. This two-geared system (slow and fast) is efficient where the muscles contain a relatively large number of fibers (cells), as in vertebrates, which are relatively large animals.

Tonic muscles resemble slow phasic fibers in contracting and fatiguing slowly, but differ in having multiple motor end plates on each fiber (Fig. 11.25c). They do not follow the all-or-none phenomenon. Activation by a single nerve impulse is limited to a few sarcomeres because the action potential does not spread far. Gradation in force is derived from more frequent impulses that cause the action potential to spread more widely and involve more sarcomeres. Tonic fibers occur in situations where a graded force is needed in a small muscle, as in arthropods.

The adductor muscles of clams, oysters, and scallops, which close the two valves (shells) contain both phasic and tonic fibers. The phasic fibers function to close the valves rapidly, and the tonic fibers function to keep them closed over a much longer period of time. In scallops, the two types of fibers are segregated into different parts of the large single adductor muscle. The phasic fibers provide for the rapid clapping of the valves, which produces the water jet for escape swimming.

Arthropod muscles differ greatly from vertebrate muscles in their organization (Fig. 11.25a and b). The muscles in the legs and claws of crabs and other crustaceans, for example, are often supplied by only two (or a few) neurons, and one neuron may extend to two or more muscles. Each muscle fiber receives multiple terminations from a neuron, and two or more different neurons terminate on each fiber. As many as five neurons may innervate a single fiber. At least one is inhibitory; the others induce varying degrees of fast or slow contraction (Fig 11.25d). The final response of a muscle depends on the rate of stimulation of the different neurons and the interaction between them. These complexities enable a few neurons and muscle fibers to act as a many-geared system and to generate as wide a range of speed, force, and duration of

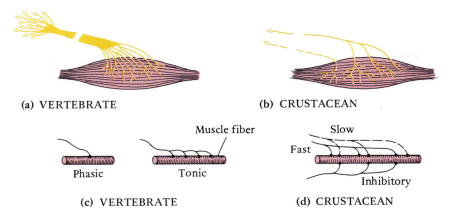

(a) VERTEBRATE

(b) CRUSTACEAN

Muscle fiber

Phasic   Tonic

Slow

Fast

Inhibitory

(c) VERTEBRATE

(d) CRUSTACEAN

Figure 11.25
Patterns of nerve supply in vertebrate and crustacean muscles. (a) Vertebrate muscles are supplied by large nerves composed of many neurons. (b) Crustacean muscles are supplied by small nerves that may contain as few as two neurons. The pattern of neuron endings on individual muscle fibers is shown in (c), vertebrate phasic and tonic muscle, and in (d) crustacean muscle.

muscle contraction as is brought about by the much larger nervous and muscular systems of vertebrates, in which speed and force of contraction are controlled by the type and number of muscle fibers activated.

Force, Work, and Power

The *force* that a muscle can generate is a function of the number of cross-bridges that can be formed at one time, and this, in turn, is determined by the number of fibers that are packed into the muscle. Since the number of fibers can be estimated by determining the cross-sectional area, force is proportional to the cross section of a muscle. The maximum force of contraction per unit of area is surprisingly similar for many muscles, ranging from about 4 to 6 kg/cm² for muscles as diverse as the adductor of a clam, the jumping muscles in a grasshopper, and the leg muscles of a human.

The *work* that muscles can do, which is the product of the force generated and the distance through which the force works, also is quite similar per unit of muscle (e.g., per gram), but obviously differs for entire muscles. Some muscles have their fibers arranged obliquely to the long axis of the muscle (Fig. 11.26b). Large numbers of fibers can be packed into a muscle with this arrangement, providing for great force, but the distance over which the muscle can contract is not as great as when the fibers are parallel to the long axis of the muscle (Fig. 11.26a). Large muscles, which have a large cross-sectional area and force and are long enough to contract a considerable distance (usually about one third of their resting length), do a great deal more work than do small, shorter muscles. Total work is proportional to the mass of a muscle.

One quality that is quite different on a unit basis is power output because *power* is the rate of doing work, i.e., work/time. The small muscles of a shrew can contract by one third of their length much faster than can the large muscles of an elephant, so their power, per gram of

muscle, is much higher. Because small muscles contract faster, the cycle of cross-bridge formation and release also is much faster in the shrew. Since a molecule of ATP is used for each cross-bridge, more ATP per gram of muscle is used.

The Vertebrate Muscular System

In describing the vertebrate muscular system and tracing its evolution, it is convenient to group the muscles into **somatic muscles**, associated with the body wall and appendages, and **visceral muscles**, associated with the pharynx and other parts of the gut tube. This grouping parallels the major divisions of the skeletal system. Somatic muscles

Figure 11.26
Muscle architecture. (a) In the fusiform pattern, the muscle fibers are oriented parallel to the long axis of the muscle. (b) In the pennate (feathered) pattern, the muscle fibers are oriented obliquely to the long axis of the muscle.

(a)   **(b)**

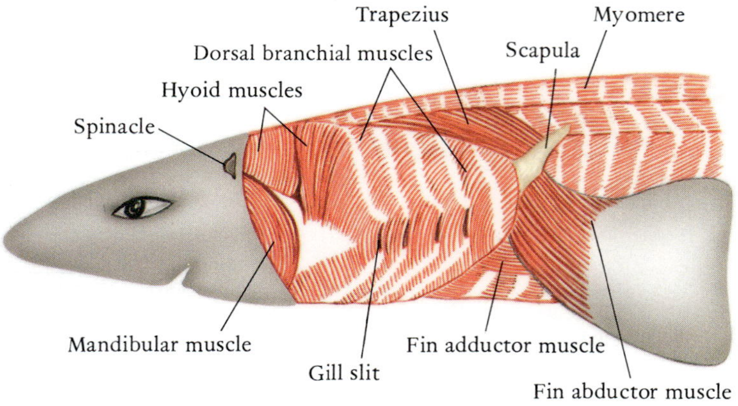

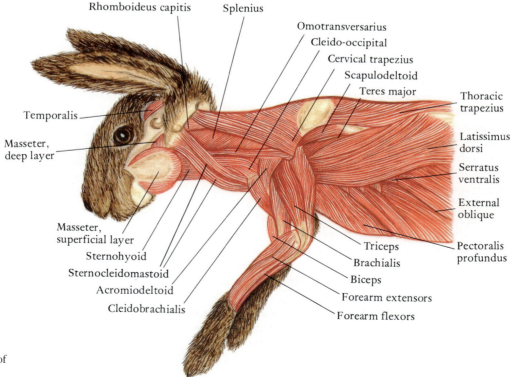

Figure 11.27
A lateral view of the anterior muscles of a dogfish.

are striated and most visceral muscles are smooth; however, the visceral muscles associated with the visceral arches, called **branchial muscles**, are striated.

Evolution of Somatic Muscles
Most of the somatic musculature of fishes consists of segmental **myomeres** (Fig. 11.27). This is an effective arrangement for bringing about the lateral undulations of the trunk and tail that are responsible for locomotion. The muscles of the paired fins are very simple and in many

fishes consist of little more than a single dorsal **abductor** that pulls the fin up and caudally and a ventral **adductor** that pulls the fin down and anteriorly.

The transition from water to land entailed major changes in the somatic muscles. The appendages became increasingly important in locomotion, and movements of the trunk and tail became less important. The segmental nature of the trunk muscles was largely lost (Fig. 11.28). Back muscles remain powerful, for they play an important role in supporting the vertebral column and body, but

Figure 11.28
Lateral view of the skeletal muscles in the anterior half of the body of a rabbit.

trunk muscles on the flanks form thin sheets, such as the **external oblique**, that help to support the abdominal viscera and assist in breathing movements. Some trunk muscles attach to the pectoral girdle and in a quadruped help transfer body weight to the girdle and forelimbs. The primitive single fin abductor and adductor became divided into many components, and these became larger and more powerful. The **latissimus dorsi** and **triceps** (Fig. 11.28), for example, are appendicular muscles that evolved from the fish abductor, whereas the **pectoralis** and **biceps** evolved from the adductor.

Evolution of Branchial Muscles

Branchial muscles are well developed in fishes and are grouped according to the visceral arches with which they are associated (Fig. 11.27): **mandibular muscles** and certain **hyoid muscles** of the first two arches are concerned with jaw movements; most of the rest, with respiratory movements of the gill apparatus. Branchial muscles obviously became less important in tetrapods, for the gills were lost and the visceral arches were reduced. Those of the mandibular arch remain as the **temporalis**, **masseter**, and other jaw muscles (Fig. 11.28). Most of the hyoid arch muscles moved to a superficial position and became the **facial muscles** that are responsible for movements of the scalp, nose, cheeks, and lips. Those of the remaining arches are associated with the pharynx and larynx, and some, e.g., the **sternomastoid**, **cleidomastoid**, and **trapezius**, are important muscles associated with the pectoral girdle (Fig. 11.28).

Nonmuscular Movement

Nearly all cells have some capacity to move and change shape because their cytoplasmic ground substance contains a cytoskeleton lattice (p. 80) of very delicate protein fibers, which, being composed partly of actin and myosin, are contractile. Additional myofilaments or microtubules often are associated with the cytoskeleton. Specialized contractile mechanisms emerge from this basic framework of the cell.

Amoeboid Movement

In many cells one or more processes, **pseudopodia**, develop on the cell surface, and the rest of the cell flows in their direction. This is known as amoeboid movement because it occurs in amoebas and related protozoa, but the phenomenon is widespread and is observed in the amoebocytes of sponges and in many of the coelomic and blood cells of other animals, including certain white blood cells of vertebrates. It also occurs in embryonic tissues, in wound healing, and in many cell types in tissue culture. The pseudopodia may be used to surround and engulf a food particle, or they may attach to the substratum and be used in locomotion.

Several pseudopodia may start to form on different parts of the cell, but usually one becomes dominant, and the cell advances in that direction. There is no permanent front end; the dominant pseudopod may appear on any surface. The cytoplasm of an amoeba can be divided into a semirigid **ectoplasm** beneath the cell membrane and a deeper, more fluid **endoplasm** (Fig. 11.29). According to the prevalent theory of pseudopod formation, the deepest, axial endoplasm flows forward, separated from the ectoplasm by a shear zone of endoplasm. It then spreads laterally in the fountain zone near the front of the pseudopod, becomes semirigid, and adds to the advancing sleeve of ectoplasm. Concurrently, at the posterior end of the cell, in the recruitment zone, ectoplasm is converted to the forward-moving stream of axial endoplasm.

The basis for amoeboid movement, as well as for other types of cell movement, is essentially the same as that in muscle contraction, for it involves contractile proteins. Actin molecules are in a diffuse unpolymerized state in the endoplasm, but in the fountain zone, where endoplasm is shifting to gelatinous ectoplasm, the actin microfilaments become polymerized and cross bonded. The process requires calcium, ATP, and probably an actin-binding protein. The gelatinous ectoplasm surrounding the cell thus forms a contractile jacket that squeezes the fluid

Figure 11.29
Amoeboid movement, with endoplasm flowing into an advancing pseudopod.

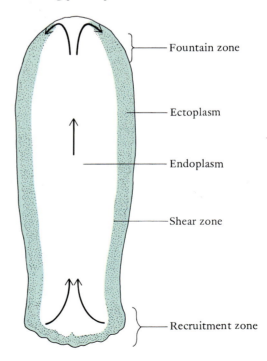

Fountain zone

Ectoplasm

Endoplasm

Shear zone

Recruitment zone

endoplasm forward. Actin becomes dissociated in the posterior recruitment zone and moves forward again in the endoplasm. Cell membrane is lost by endocytosis at the rear end of the amoeba and is gained at the front end by exocytosis.

Figure 11.30

Ciliary and flagellar movement. (a) Three successive stages in the undulatory movement of a flagellum. (b) Successive stages in the oarlike action of a cilium; effective stroke shown with solid arrows and recovery stroke with dashed arrows. In both (a) and (b), water is moved in the direction of the large arrow, and the cell moves in the opposite direction. (c) and (d) Diagrams of transverse and longitudinal sections of a cilium.

Ciliary and Flagellar Movement

Another type of motion involves the beating of slender, hairlike, cytoplasmic processes projecting from the cell surface. These are called **cilia** (L. *cilium*, eyelid) when each cell has one to many short processes and **flagella** (L. *flagellum*, whip) when each has one or a few longer processes, but the fundamental structure of cilia and flagella is the same. Cilia and flagella are nearly ubiquitous among protozoa and metazoa, where they are used to move a cell through liquid or to move liquid along the surface of cells. They occur in ciliates and flagellates among the protozoa; in the collar cells of sponges; on many epithelial surfaces, ranging from the gastrovascular

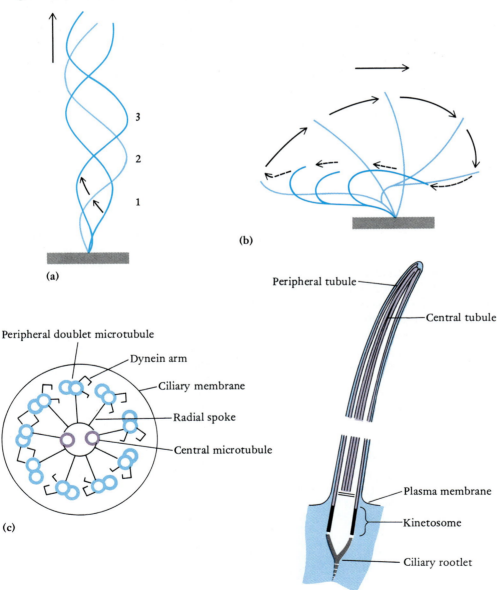

(a)

(b)

(c)

Peripheral doublet microtubule
Dynein arm
Ciliary membrane
Radial spoke
Central microtubule

(d)

Peripheral tubule
Central tubule
Plasma membrane
Kinetosome
Ciliary rootlet

cavity of cnidarians to the lining of vertebrate reproductive ducts; and in the tails of spermatozoa of most animals. Among the more familiar phyla, only roundworms and arthropods lack motile cilia, but ultrastructural comparisons reveal that processes of many of the sensory cells of even these animals are modified cilia.

Ciliary and flagellar movement, as analyzed by means of very-high-speed cinephotography, is very complex and also variable among different species. Undulatory waves move along the flagellum from base to tip (usually). In some species successive waves are in the same plane; in others they are oriented 90° to each other. The latter imparts a helical twist or rotary action to the flagellum, so that it acts like a boat propeller moving water parallel to the flagellum axis (Fig. 11.30a). Some flagellate protozoans have "hairy" flagella. The lateral branches have a hydrodynamic effect that causes the flagellum to pull rather than push (p. 562). The beat of a cilium is more planar; the action is oarlike, with a rather "stiff-armed" effective stroke and a "curling," or feathered, recovery stroke moving water parallel to the cell surface (Fig. 11.30b).

Each cilium contains a sheaf of nine double microtubules, situated just beneath the cell membrane, and a core of two single microtubules (Fig. 11.30c). Radial spokes extend between the outer doublets and the two central microtubules. The walls of the microtubules are made up of the protein **tubulin**. Arms composed of the protein **dynein** extend between the outer doublets and form temporary cross-bridges between them. The arrangement is similar in many ways to that in muscle filaments, although the contractile proteins are different. According to the sliding microtubule hypothesis, the release of energy from ATP and the successive forming and breaking of dynein bridges cause one or more microtubules to slide past others. In a sense, the dynein bridges of one double microtubule "walk along" an adjacent doublet. The resulting shearing forces are resisted by the radial spokes, so that some of the sliding action is converted to a bending action.

If there are many cilia upon a cell, their action must be integrated if movement is to be effective. Often traveling or **metachronal waves** (Gr. *metachronos*, done afterwards) pass along the surface (Fig. 11.31), but the beat appears to be integrated, or coupled, by hydrodynamic forces and not by any mechanism within the organism. Temporary beat reversal, which occurs in many ciliated protozoans, such as *Paramecium*, and even by the cilia of certain metazoans, is under internal control and involves a rise in intracellular calcium ions. For example, when swimming, *Paramecium* will reverse the ciliary beat, back away from an obstacle, turn, and then move forward.

Many small, multicellular animals, such as flatworms, and even larger forms, such as snails, use cilia to glide over the substratum. Mucus secreted by epidermal glands is essential for this type of gliding locomotion, although not

for ciliary swimming. The mucus functions as a blanket adhering to the substratum, over which the animal crawls. The cilia gain traction by gripping the mucus with the cilium tip during the effective stroke. Only the cilium tip makes contact with the mucus, the remainder of the cilium operating in a watery fluid believed to lie between the blanket and the animal's surface at the cilium base. The mucus is probably secreted as granules pass out into the blanket. A similar arrangement operates over many nonlocomotor ciliated surfaces, such as the gill of clams and the trachea of vertebrates. Here, however, the mucus blanket is moved by ciliary action, carrying with it trapped particles.

Cilia and flagella are associated with basal bodies (**kinetosomes**) (Fig. 11.30d) that are essentially similar to the centrioles associated with the mitotic spindles (p. 100). Basal bodies are related to the development and growth of the cilia and also appear to function as anchoring systems.

LOCOMOTION

The motion of an animal is derived from pushing against the surrounding medium or substratum, which, in accordance with Newton's third law of motion, generates an equal and opposite thrust against the animal. Part of this force moves the animal forward. For example, when you

Figure 11.31
Metachronal ciliary waves passing over the end of the ciliate *Spirostomum*.

jump up from a squatting position, the force of the contracting extensor muscles in the legs is directed against the ground. This produces an opposite thrust upward. Motion is resisted by inertia and adhesive forces. Inertia is the tendency of any object to remain in a motionless state if it is stationary or to keep moving if it is in motion. It is a function of the object's mass. Adhesive force results from the tendency of the molecules composing the surrounding medium to adhere to each other and to the moving object. Inertia is the primary factor resisting the motion of large animals and accounts for the fact that a large animal in motion retains momentum and glides or slides to a halt after propulsive forces cease. For animals less than a millimeter in length, adhesion is the predominant force resisting motion, for the surface area of the body on which adhesive forces operate is so great compared to its volume. A minute swimming animal does not glide to a halt when propulsive forces cease; it comes to an immediate halt because adhesive viscous forces so outweigh inertia, i.e., it loses momentum instantly.

The drag encountered by aquatic animals in swimming and by terrestrial animals, such as flying insects and birds, moving at high speeds reflects inertia and viscous force. An object placed in a current of water will cause the flow lines of the current to part in front of the object and converge behind it (Fig. 11.32a). The pressure in the region just behind the object and in front of the convergence will be reduced, and it will also be reduced if the flow lines are disrupted and there is turbulence behind the object. The tapered shape characteristic of large aquatic animals, such as fish, seals, and porpoises, represents a convergent streamlining adaptation to reduce turbulence and thus **pressure** or **inertial drag** during movement (Fig. 11.32b).

When air or water flows across a surface, the layer directly against the surface does not move because adhesive forces couple the fluid layer and the surface. Successive layers outward shear off with increasing speed until they equal the current velocity. This **boundary layer** of fluid exerts a **viscous drag**, also known as **skin friction**, on an object. Since viscous drag is a function of surface area, streamlining to reduce pressure drag increases viscous drag. However, for a large animal viscous drag is the lesser of two evils. As animals become smaller, the lower ratio of surface area to volume increases the importance of viscous drag relative to pressure drag. It is not surprising that small planktonic animals and larvae are not streamlined.

Among metazoan animals the initial thrust is usually generated in one of four ways: (1) by the beating of cilia, (2) by changes in body shape, (3) by body undulations, and (4) by the movement of appendages. In all but the first the force is produced by muscle contractions.

Ciliary Propulsion

Among metazoans, ciliary swimming is usually restricted to very small species, because above a certain body mass the sinking rate exceeds the propulsive force that can be generated by cilia. Large metazoans that swim with cilia, such as comb jellies, have gelatinous bodies with densities near that of sea water.

There are some metazoans, including certain flatworms and snails, that use cilia to creep over the bottom. This commonly involves the simultaneous secretion of a mucous blanket on which the cilia gain traction.

Propulsion by Changes in Body Shape

Thrust gained by changes in body shape has led to the most diverse modes of locomotion in animals. Both jellyfish and squids swim by water jets produced by body deformation: jellyfish by driving water from beneath the bell-shaped body (p. 602) and squids by expelling water from the mantle cavity and out a funnel (p. 685). Clams burrow into mud or sand with a muscular foot, which penetrates the substratum when it is blade shaped and anchors the animal when the tip is dilated with blood. Many snails, whelks, and slugs creep over rocks and other hard surfaces by means of body-deforming waves that sweep along the length of their muscular ventral surface (p. 666).

The metamerism of annelids (p. 694) is believed to have evolved as an adaptation for burrowing by means of shape change. The contraction of muscles in the body wall develops forces in the coelomic fluid of the hydroskeleton. Contraction of circular muscles in the body wall increases

Figure 11.32
(a) A spherical object is subjected to considerable pressure drag when placed in a water stream. (b) The flat, tapered body of a fish reduces pressure drag by permitting the flow lines of the current to converge behind the fish without turbulence.

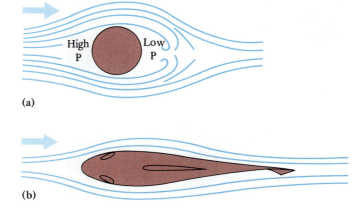

(a)

(b)

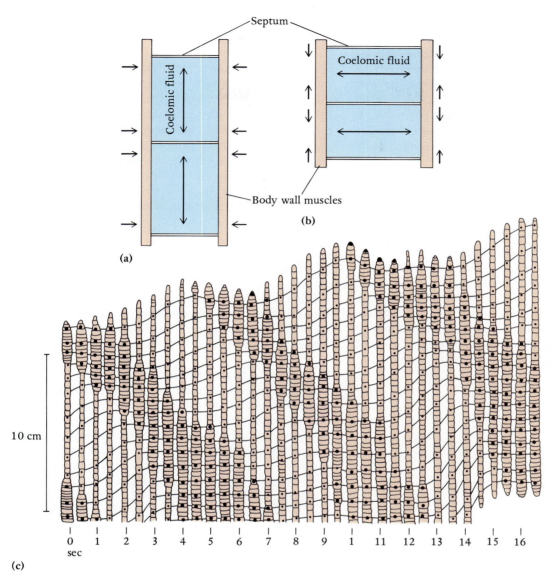

Figure 11.33

(a) and (b) Diagrammatic frontal section through two annelid segments. External arrows indicate direction of force exerted by body wall muscles. Internal arrows indicate direction of force exerted by coelomic fluid pressure during (a) circular muscle contraction and (b) longitudinal muscle contraction. (c) Diagram showing mode of locomotion of an earthworm. Segments undergoing longitudinal muscle contraction are marked with larger dots and drawn twice as wide as those undergoing circular muscle contraction. The forward progression of a segment during the course of several waves of circular muscle contraction is indicated by the horizontal lines connecting the same segments.

the pressure of the fluid, stiffens the body, and forces it to elongate, stretching longitudinal muscles in the body wall in the process (Fig. 11.33a). Contraction of longitudinal muscles also increases coelomic pressure but causes the body to shorten and widen as circular muscles are stretched (Fig. 11.33b). The partitioning of the coelom into discrete compartments in the earthworm and most other annelids enables body lengthening and shortening to be confined to a few segments at a time. Alternate waves of elongation and contraction move posteriorly along the length of the animal (Fig. 11.33c). First the front of the body is elongated and pushed forward. Then this anterior part shortens and widens, bristles (setae) in the body wall anchor it to the substratum, and more posterior parts of the body are pulled forward. The segmentation of annelids is basically a modification of the coelom, but blood vessels, nerves, and other internal organs are also segmentally arranged.

Propulsion by Body Undulations

A large number of animals—fish, whales, snakes, roundworms, rapidly swimming polychaetes, and leeches—are propelled by body undulations. The undulatory waves sweep down the body, although in most fish and whales they are largely confined to the tail region, and the backward-progressing loop accelerates, or displaces, the water parallel to (rearward) and at right angles to the long axis of the body. This produces an opposite motion or thrust. Since the lateral accelerations on each side of the body cancel each other out, the propulsive thrust is forward (Fig. 11.34).

Figure 11.34

Diagram of dogfish swimming to illustrate the thrust of the tail against the water and the reaction of the water on the tail.

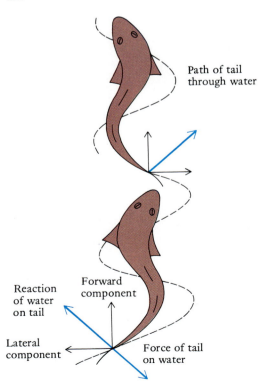

The bending of the body is brought about by waves of muscle contraction that alternate on the two sides of the body, and the counteracting force to straighten the body results from the cuticle and fluid pressure of the body cavity in roundworms, coelomic fluid pressure in polychaetes and leeches, and the vertebral column in fishes and whales. Although chordate metamerism probably evolved for shallow undulatory burrowing, it was soon adapted for undulatory swimming. In contrast to that of annelids, the primary segmental structures of chordates are the body wall muscles, and the coelom is not compartmented. The nerves and blood vessels supplying the muscle blocks are also segmentally arranged.

Appendicular Propulsion

A common mode of locomotion among animals is by the motion of appendages, which can function as legs in walking and running, as paddles in swimming, and as wings in flying. Typically, the skeleton and muscles of the appendage function together as a lever system to produce thrust against the ground or water (see Focus 11.1).

Walking and Running

During slow walking in terrestrial vertebrates and most arthropods the effective stroke of a leg on one side of the body alternates with the effective stroke of the leg on the opposite side of the body. The sequence of leg movement is such that in both groups there are at most points during the stride three limbs on the ground to form a triangle of support (Fig. 11.35). The center of gravity always remains within the triangle. At running speeds the gait changes, and the number and combination of feet placed on the ground at one time represents an interaction between the amount of support and thrust optimum for a particular body form and velocity.

Figure 11.35

Sequence of leg movements in an insect (a) and a tetrapod vertebrate (b). In both, the effective strokes alternate on the two sides of the body. The circled numbers are those appendages in contact with the ground and the dashed lines indicate the triangle of support formed by those legs.

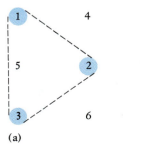
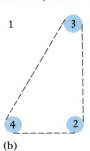

(a)   (b)

Flight

Four groups of animals have independently evolved the ability to fly: insects, birds, bats among mammals, and the extinct pterosaurs among reptiles. In all four, airfoils, or

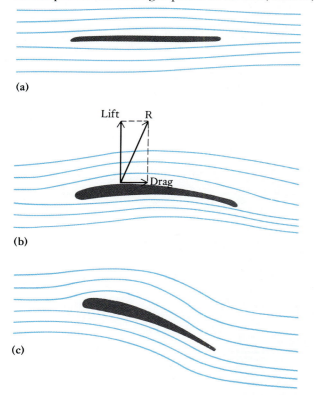

(a)

(b)

(c)

wings, generate lift as air moves across the surface, and the moving air is derived from the forward propulsion of the animal. The familiar sight of a piece of plywood or cardboard picked up from the ground by wind involves the same forces as those operating on animal wings or the wings of an airplane. If a flat plate is held horizontal to the air stream so that the speed of air moving across the upper surface is equal to that moving across the lower, the pressure on the two sides will be the same (Fig. 11.36a). However, if the upper surface is slightly arched, or cambered, as is true of bird and airplane wings, the distance from the front to the trailing edge of the wing will be slightly longer on the arched, or upper surface, as compared to the lower (Fig. 11.36b). When the air mass strikes

Figure 11.36

Flight aerodynamics. (a) A flat plate held parallel to air stream. Pressure is the same on both upper and lower surfaces. (b) Air stream over a cambered wing. Pressure is less on convex upper surface than on concave lower surface. This difference generates an aerodynamic force (R) on the upper surface consisting of a lift component perpendicular to direction of airflow and a drag component parallel to it. (c) Wing tilted slightly, increasing lift. (d) Wing profile path of a bat during upstroke and downstroke. Wing attack angles and relative lift (L) and drag (D) components are also shown.

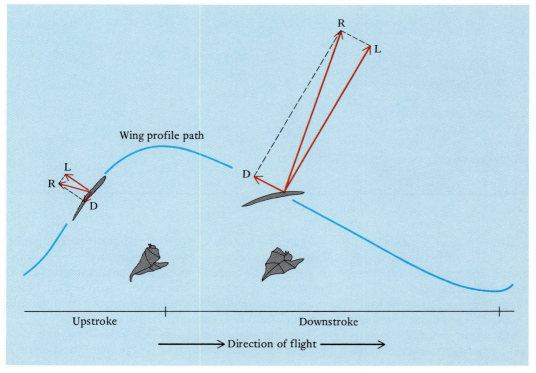

(d)

In a simple lever system like a seesaw, with the **fulcrum** in the center, the force applied (FA) at one end of the board results in an equal load force (FL) at the other end (Fig. a). The mechanical advantage (MA) of this lever system is one, because the two forces are equal and because the distances from the fulcrum to each end of the board (XA and XL) are equal:

$$MA = \frac{XA}{XL} = \frac{FL}{FA} = 1$$

Note that the distance (d) each end of the board travels when the seesaw moves is the same.

If we move the fulcrum so that it is twice as close to the load end of the board (Fig. b), the mechanical advantage goes up to 2:

$$MA = \frac{2\ (XA)}{1\ (XL)} = 2$$

Since the mechanical advantage in terms of force equals 2, the applied force need be only one-half that of the load force.

The distance the load end of the board moves is now less than that of the applied force end; the distance (d) is proportional to XL. This relationship is called the velocity ratio because one end will have to travel faster than the other in order to cover a greater distance in the same period of time.

$$Velocity\ ratio = \frac{d\ (load\ end)}{d\ (force\ end)}$$

In this case, the velocity ratio will be less than 1.

If we move the fulcrum in the opposite direction so that it is twice as close to the applied force end, the mechanical advantage changes to 1/2 and twice as much force as the load must be applied to the FA end of the board to move the opposite end. The velocity ratio will also be greater than 1. The fulcrum can be moved about with relationship to FA and

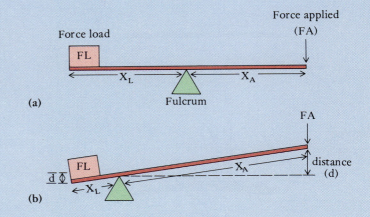

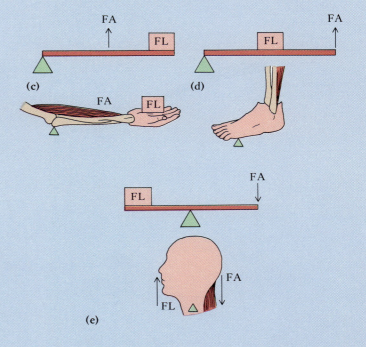

the front edge of the wing, part will move across the top surface and part below. The air moving across the upper surface will have to move faster to cover the slightly greater distance to the trailing edge. As a result the pressure across the upper surface will be lower than that across the lower surface, resulting in an upward-directed force called **lift**. Lift is also generated by elevating the front of the wing slightly above the horizontal plane (increasing the **attack angle**). This raises the pressure of the airstream beneath the wing and also, like camber, reduces the pressure on

the upper side. (Fig. 11.36c). Lift in birds is dependent on both wing camber and attack angle. Insects have rather flat wings, and the angle of attack is of primary importance in producing lift. However, many insects achieve a cambered wing surface in flight by raising or lowering the longitudinal veins and thus bending the wing.

Flying animals are propelled forward by wing flapping and in all the downstroke is of primary importance in producing thrust. This requires that the upstroke encounter less air resistance than the **downward power stroke**.

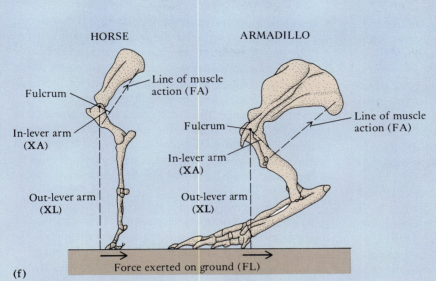

HORSE   ARMADILLO

Line of muscle
action (FA)

Fulcrum

In-lever arm
(XA)

Fulcrum

Line of muscle
action (FA)

In-lever arm
(XA)

Out-lever arm
(XL)

Out-lever arm
(XL)

Force exerted on ground (FL)

(f)

Lever systems. (*Left*) (a) Lever system in which the fulcrum is
equally centered between force load and force applied. (b)
Lever system in which fulcrum is closer to force load than
force applied. (c–e) The three types of lever systems as
determined by the relative positions of fulcrum, force load,
and force applied with an example of each from the human
body. (*Above*) (f) Forelimbs of a horse and an armadillo
drawn to the same size to show how changes in the in-lever
and out-lever arm relationships (lever ratio) adapt the limb
lever for velocity or force.

FL to produce other types of lever systems (Figs. c and d) but
the relationships determining mechanical advantage and
distance of movement are the same.

In the lever systems of arthropods and vertebrates, the
muscles provide the applied force at the points of insertion
on the bones or exoskeleton, and the load force either is
directed against the ground or water or is applied toward
lifting a weight. The fulcrum is a joint, which means that the
fulcrum is generally close to one end of the lever bar. All
three types of lever systems (Figs. c–e) are found in vertebrates and arthropods, but one in which the fulcrum is at one
end, the load at the other, and the applied force is in between
is very common (e.g., Fig. c, lifting your outstretched hand).
Shifts in position of the applied force in relation to the
fulcrum and the load have occurred as parts of adaptations
for various functions of arthropod and vertebrate limbs. For
example, in a horse limb the ratio of XL to XA (lever ratio) is
much greater than in an armadillo (Fig. f); therefore the limb
of the horse, which is adapted for running, travels through a
much greater arc than that of an armadillo. However, the
armadillo limb can deliver relatively more force, which is
important for burrowing.

This is achieved in various ways, principally by the position
in which the wing is held (Fig. 11.36d). Other aspects of
animal flight will be described in the chapters covering the
different groups of flying animals.

Energy Storage

The amount of energy required for locomotion tends to be
minimized by various devices. For example, as a part of the
body moves in one direction, some of the energy
expended may be stored as potential **strain energy** by the
stretching of a tendon or the deforming of elastic material
in a joint. This energy is then released and contributes to
the recovery movement. Wing movements of an insect
utilize the elasticity of the thoracic wall. When running
mammals land upon their toes, tendons in the foot and leg
are stretched because body weight, continuing to push
downward, flexes the leg and foot. The elastic recoil of this
strain energy helps raise the body and swing the leg
forward, and represents about 50% of the energy needed
for each bound.

Summary

1. Epithelial tissues are composed of closely placed cells, which may be flattened, cuboidal, or columnar in shape. They cover outer surfaces, line internal cavities, and form the bulk of multicellular glands. Connective tissues are composed of widely separated cells. The intervening matrix secreted by the cells varies and accounts for different types of connective tissue. Loose and dense connective tissues have a matrix composed largely of collagen and elastic fibers. The matrix of cartilage is dense but flexible and may contain a considerable number of fibers. The matrix of bone is rigid and contains calcium triphosphate in addition to organic constituents.

2. A body covering protects and separates the body from the outside world. The body covering of most invertebrates is an epidermis of simple columnar epithelium; that of vertebrates, a many-layered epithelium overlying a dense connective tissue. Pigment is present in the skin of many animals and frequently is contained in chromatophores.

3. Skeletons provide support, protection, mineral storage, transfer of muscle forces, and sometimes have other functions. Skeletons with hard parts may be composed of silicon dioxide, calcium carbonate, calcium triphosphate, or organic compounds, of which chitin is especially widespread. Many skeletons contain both mineral and organic constituents.

4. Many invertebrates have exoskeletons secreted on the body surface by the epidermis. They are very effective in small animals, but would become very cumbersome in a large animal. The arthropod exoskeleton does not grow with the animal and must be molted periodically. Endoskeletons occur in many invertebrates and in all vertebrates. The vertebrate endoskeleton is composed of two types of connective tissue—cartilage and bone. The matrix of cartilage contains varying numbers of fibers and considerable bound and unbound water. The matrix of bone contains calcium trophosphate in addition to organic components, such as fibers. A solid endoskeleton is more efficient for support in a large animal than is an exoskeleton. Fluid endoskeletons (hydroskeletons) depend upon the turgidity of fluids (commonly blood or coelomic fluid) confined within some body organ or space.

5. The vertebrate skeleton can be divided into a somatic skeleton in the body wall and appendages, and a deeper visceral skeleton associated with the gills and jaws. During evolution from fish to terrestrial vertebrates, the somatic skeleton becomes larger and stronger in connection with support and movement on land, and the visceral skeleton becomes reduced, although parts of it remain.

6. Muscle tissue is composed of elongate cells containing parallel myofilaments. In striated muscle tissue, characteristic of the vertebrate skeletal muscles, the alignment of the myofilaments is such that the cells have a striated appearance. In smooth muscle, characteristic of vertebrate internal organs, the myofilaments are not in alignment. Muscle cells usually are bound together by connective tissue and organized into muscles. Muscles only perform work by contracting and must act against an antagonistic force.

7. When a muscle cell is stimulated, calcium ions are released, and heads on the myosin myofilaments bind to the now unblocked active sites on actin myofilaments. The heads bend slightly, detach, and reattach to adjacent active sites, causing the actin and myosin myofilaments to slide relative to each other and the cell to shorten. One molecule of ATP is used for a cycle of attachment, bending, and detachment of a myosin head.

8. Most muscles of higher vertebrates are twitch or phasic muscles whose motor units respond in an all-or-none fashion. Gradation of tension (contraction) depends upon the number of motor units involved (spatial summation) and the frequency of impulses received by the motor units (temporal summation). Tonic muscle fibers do not respond in an all-or-none fashion. Gradation in force comes from the rate of stimulation and the distance the action potential spreads along the fiber.

9. In the evolution from fish to terrestrial vertebrates, the segmentation of trunk muscles is reduced, appendicular muscles increase in size and complexity, and branchial muscles become reduced.

10. Pseudopod formation during amoeboid motion depends on the polymerization of actin microfilaments. The bending in ciliary and flagellar movement results from the sliding of adjacent microtubules of tubulin, pulled by dynein bridges. Energy is supplied by ATP.

11. The propulsive force for metazoan locomotion is derived from either ciliary beat or most commonly from muscle contraction. Muscle contraction generates thrust by changing the shape of the body, producing undulatory waves, or by moving appendages. The storage of some expended energy as elastic strain energy in tendons and ligaments minimizes the total energy needs.

References and Selected Readings

Additional information on the skeleton and movement can be found in the general references cited at the end of Chapter 10.

Alexander, R.M. Walking and running. *American Scientist* 72 (1984): 348–354. The biomechanics of walking and running in mammals.

———. *Locomotion of Animals*. New York: Chapman and Hall, 1982. The biophysics of locomotion in animals.

Cameron, J.N. Molting in the blue crab. *Scientific American* 252(May 1985): 102–109. An account of the physiology of molting in the common commercial blue crab.

Chapman, A.J. Cartilage. *Scientific American* 251(Oct. 1984): 84–94. The chemistry of cartilage.

Curry, J. *Animal Skeletons*. Studies in Biology No. 22. New York: Crane, Russak and Co., 1970. An old but still useful account of the biophysics of skeletal materials.

Elder, H.Y., and E.R. Trueman. *Aspects of Animal Movements*. Cambridge, England: Cambridge University Press, 1980. A review by the Society for Experimental Biology of muscle contraction and skeletal factors in locomotion, swimming, flying, walking, and other types of movement.

Hildebrand, M., et al. *Functional Vertebrate Morphology*. Cambridge, Mass.: Harvard University Press, 1985. This multiauthored volume includes eight chapters on various aspects of vertebrate locomotion.

Lazarides, E., and J.P. Revel. The molecular basis of cell movement. *Scientific American* 240(May 1979): 100–113. A review of muscle contraction and the role of contractile proteins in other types of cell motility.

Pritchard, J.J. *Bones*. London: Oxford University Press, 1974. The microscopic structure, growth, remodeling, and architecture of bone are summarized in this Oxford Biology Reader.

Smith, K.K., and W.M. Kier. Trunks, tongues, and tentacles: Moving with skeletons of muscle. *American Scientist* 77(1989): 29–35. The organization of musculature and biomechanics of trunks, tongues, and tentacles.

Trueman, E.R. *Locomotion in Soft-Bodied Animals*. Bristol, England: Edward Arnold Press, 1975. An excellent review of crawling, swimming, and burrowing, and their evolution among soft-bodied invertebrates.

Webb, P.W. Form and function in fish swimming. *Scientific American* 251(July 1984): 72–82. Includes a good account of undulatory locomotion.

Yates, G.T. How microorganisms move through water. *American Scientist* 74(1986): 358–365. The physics of flagellar and ciliary locomotion.

12

Digestion and Nutrition

The double-crested cormorant
(Phalacrocorax auritus). *Cormorants are sea birds that dive for fish, which they can hold temporarily in their expandable throat.*

All organisms require energy and raw materials in order to synthesize new cells and cellular components, and to carry out the numerous metabolic activities associated with life. Green plants, algae, some protozoans, and some bacteria absorb light energy from the sun in pigments within their cells and use it to synthesize sugars and other high-energy compounds from carbon dioxide, water, and other simple inorganic materials. This process is called **photosynthesis**. Because they make their own food, these organisms are described as **autotrophs** (Gr. *autos*, self + *trophe*, nourish). Animals obtain their energy and most of the raw materials they need by eating other organisms or their products. They are described as **heterotrophs** (Gr. *heteros*, other).

KINDS OF DIETS

All animals must include in their diets water, many minerals, and a variety of organic compounds. The organic compounds, chiefly carbohydrates, fats, and proteins, provide energy and the building blocks needed to synthesize new protoplasm. Animals have exploited virtually all types of organic food sources (Fig. 12.1). **Herbivorous** animals, such as sea turtles and cows, utilize plants as food. Most herbivores feed on only a few, or in some cases on only one, plant species. Moreover, the diet may be restricted to certain parts of the plant, such as roots, leaves, sap, nectar, fruit, and seeds. There are also many species of animals that feed on algae or fungi. **Carnivorous** animals feed on the tissues of other animals. Lions and most other carnivores consume all or most of the bodies of their prey, but some eat only certain parts, such as blood, skin, and so on.

Omnivores (L. *omnis*, all + *vorare*, to eat) eat a wide variety of plants and animals. Human beings are a good example of omnivores.

Many animals utilize nonliving organic substances as food sources. Such specialized diets may be limited to hair, feathers, shed skin, feces, or decomposing animal bodies (carrion feeders). Nonliving plant products, such as cellulose, and other organic compounds of decomposing vegetation are important food sources.

In many cases the animals do not use plant or animal remains until they have been broken down into small fragments, or **detritus**, by bacteria, fungi, and other decomposers. For some animals the bacterial decomposers rather than the detritus itself may serve as food. In terrestrial habitats, such as a forest floor, detritus is called **humus** and consists mostly of plant remains. In the sea, detritus derived from both plant and animal sources is at first suspended in the sea water but gradually sinks to the bottom and becomes mixed with sand grains. The deposited material is an important food source for many species.

The diet of some animals is augmented by or dependent upon other organisms, or symbionts, that live within them. Algal symbionts in sponges, hydras, and corals provide their hosts with some organic nutrients that supplement the diets of their hosts (see p. 612).

STEPS IN FOOD PROCESSING

To be utilized, the food must first be taken into the digestive tract, a process called **ingestion**. Large particles may be broken down mechanically by grinding or churning actions that take place in the upper part of the digestive

Figure 12.1
Two of the many ways that animals obtain their food. (a) The larva or caterpillar of a polyphemus moth (*Antheraea polyphemus*) feeds upon the leaves of trees. (b) Young lions (*Panthera leo*) at a kill.

(a)   (b)

tract. Subsequently, the food molecules are broken down chemically into smaller molecules by a process called **digestion**. These small molecules can then be absorbed, that is, pass through the cells lining the digestive tract and enter the body. Within the body, the food is used as a source of energy or is assimilated and incorporated into the animal's own, unique protoplasm. The undigested residue is finally **egested** from the digestive tract as **feces**.

Digestion

You will recall that carbohydrates, lipids, and fats are synthesized by dehydration reactions in which the removal of molecules of water unites the single sugars, amino acids, and other small building units into larger ones (see Chap. 3). Digestion is the reverse of synthesis. The large organic compounds are digested by **hydrolysis**, a reaction in which water molecules are reintroduced between the building units, which then separate (Fig. 12.2). Although hydrolysis is an exergonic reaction, only the small amount of energy in the bond that held the building units together is released; the large store of energy in the units themselves is retained. Since these reactions are catalyzed by enzymes, they can take place at the body temperature of animals. Enzymes are more or less specific in cleaving certain types of bonds, and thus different digestive enzymes attack different types of food molecules or different parts of the molecule. However, many of these enzymes are not as substrate specific as are respiratory and many other types of intracellular enzymes. This is an advantage in digestion, for food generally includes a wide variety of similar but not identical compounds. Most animals, from sponges to humans, possess similar digestive enzymes, for their food is composed of the same types of organic molecules.

Carbohydrate Digestion

The enzymes that digest carbohydrates are collectively called **carbohydrases**. **Polysaccharidases** hydrolytically cleave high-molecular-weight carbohydrates (Table 12.1). **Amylase**, which is the principal polysaccharidase in animals, breaks plant starch and animal starch (glycogen) down into the disaccharide maltose (malt sugar). Maltose and other double sugars are then cleaved into individual single sugars or monosaccharides. Each double sugar requires a specific enzyme to cleave it. **Maltase** breaks maltose down into two glucose molecules; **lactase** cleaves lactose (milk sugar) into glucose and galactose, and **sucrase** cleaves sucrose (cane sugar) into glucose and fructose (Fig. 12.2a).

Cellulose is a structural carbohydrate that is a major component of the woody cell walls of plants. Some land snails and a few herbivorous arthropods can synthesize **cellulase**, a polysaccharidase that can hydrolyse the bonds between the beta-glucose molecules of cellulose to form disaccharide units known as **cellobioses**. Many animals that utilize cellulose in their diets as an energy source depend upon colonies of microorganisms in some part of their digestive tract for the synthesis of cellulase (p. 243).

Chitin, an important component of the exoskeleton of arthropods and some other invertebrates, is a nitrogen-containing polysaccharide. Only frogs and other animals having the enzyme **chitinase** can use chitin in their diet.

Lipid Digestion

Since lipids are not water soluble, they do not mix easily with the aqueous contents of the digestive tract. Large globules of fat are first emulsified by some agent, such as bile salts produced by the liver of vertebrates, that reduces their surface tension and breaks them up into many smaller droplets with a larger collective surface area. (Detergents that we use in dish water are a familiar example of an emulsifying agent.) The small droplets of fat are then digested hydrolytically by **lipase**, which cleaves the ester bonds that bind their building units together (Fig. 12.2b). If digestion is complete, each triacylglycerol yields three molecules of fatty acid and one of glycerol, but often digestion is incomplete and diglycerols and monoglycerols are formed (Table 12.1). The hydrophobic ends of the fatty acids attach to the bile salts to form very small, water-soluble micelles that are only 2.4 nm in diameter (see Fig. 5.5). The micelles float off of the fat droplets and expose new surfaces on which lipase can act.

Protein Digestion

Proteins are digested hydrolytically by a group of proteolytic enzymes collectively called **proteases** that cleave the peptide bonds (Table 12.1). The sequential action of an assortment of proteases is needed to break down a protein completely. **Endopeptidases** attack bonds within a polypeptide chain, splitting it into smaller fragments, or large peptides. The point of cleavage depends on the enzyme. **Pepsin** cleaves only those bonds adjacent to the amino acids phenylalanine and tyrosine. It is secreted in the stomach of vertebrates and is particularly effective in hydrolyzing collagen fibers in connective tissues and muscles. Hydrolysis of collagen breaks up chunks of ingested meat, allowing them to be acted on by other enzymes. **Trypsin** cleaves bonds beside the amino acids arginine and lysine. **Chymotrypsin** is the least selective of the endopeptidases and can act on bonds next to six other amino acids.

Exopeptidases split the large peptides into smaller ones and strip off some of the terminal amino acids. Exopeptidases known as **carboxypeptidases** act on the end of the chain containing a free carboxyl group (COOH); **aminopeptidases** act at the other end containing a free amino group (NH_2). Remaining small peptides are separated into their amino acids by **tripeptidases** and **dipeptidases** (Fig. 12.2c).

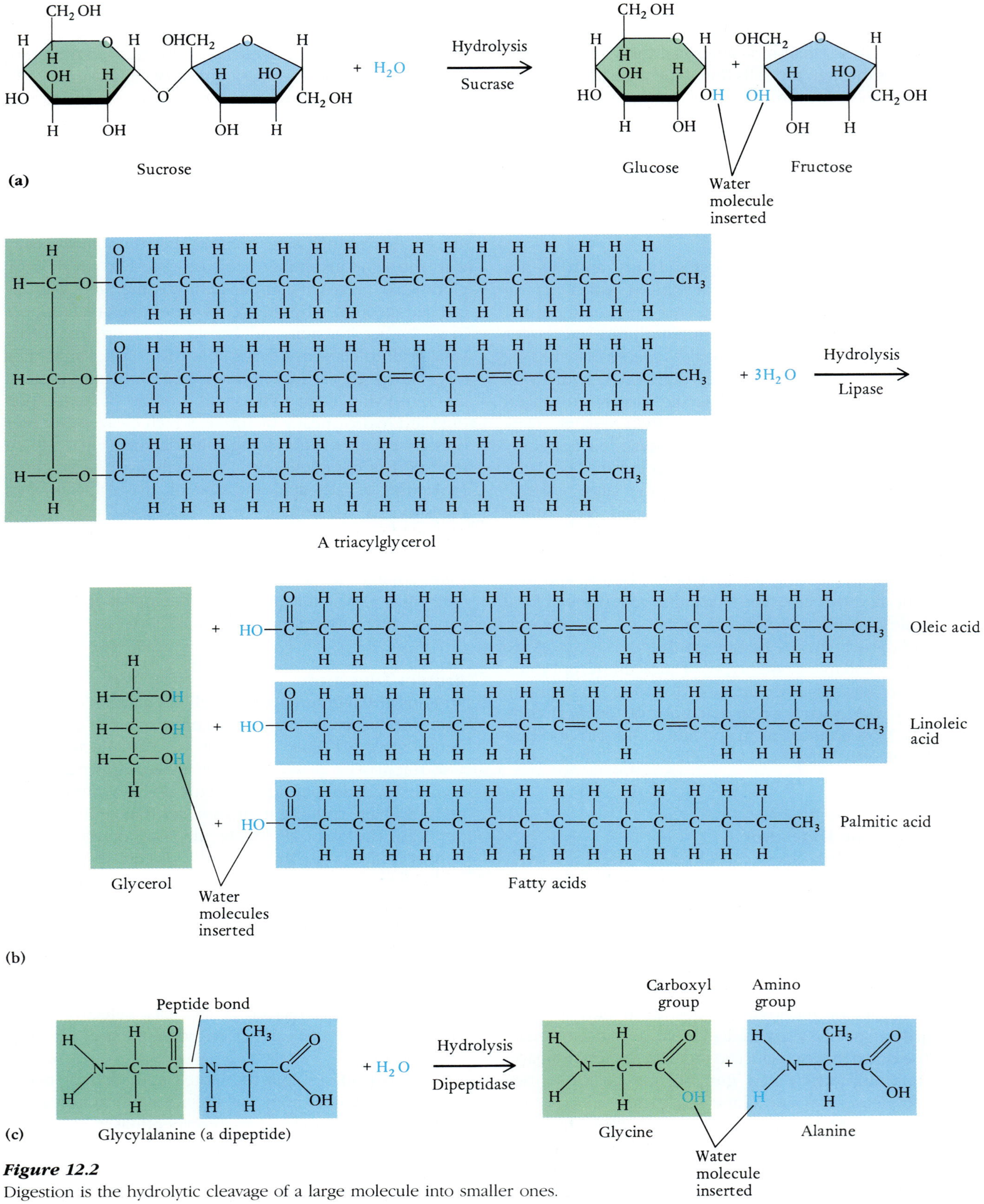

(a)

Sucrose + H_2O $\xrightarrow[\text{Sucrase}]{\text{Hydrolysis}}$ Glucose + Fructose

Water molecule inserted

A triacylglycerol + $3H_2O$ $\xrightarrow[\text{Lipase}]{\text{Hydrolysis}}$

Oleic acid

Linoleic acid

Palmitic acid

Glycerol

Water molecules inserted

Fatty acids

(b)

Peptide bond

Glycylalanine (a dipeptide) + H_2O $\xrightarrow[\text{Dipeptidase}]{\text{Hydrolysis}}$ Glycine + Alanine

Carboxyl group   Amino group

Water molecule inserted

(c)

Figure 12.2

Digestion is the hydrolytic cleavage of a large molecule into smaller ones.
(a) Digestion of sucrose; (b) digestion of a triacylglycerol; (c) digestion of a dipeptide.

Table 12.1
Major Digestive Enzymes

Enzyme	Substrate	Product	Source in Vertebrates
Carbohydrases			
Polysaccharidases			
Amylase	Starch, glycogen	Maltose	Salivary glands, pancreas
Cellulase	Cellulose	Cellobiose	Bacteria in part of gut
Disaccharidases			
Maltase	Maltose	Glucose	Small intestine*
Lactase	Lactose	Glucose, galactose	Small intestine*
Sucrase	Sucrose	Glucose, fructose	Small intestine*
Lipase	Triacylglycerol	Di- and monoglycerol, fatty acids, glycerol	Pancreas
Proteases			
Endopeptidases			
Pepsinogen → pepsin	Polypeptide bonds adjacent to phenylalanine or tyrosine	Large peptides	Stomach
Trypsinogen → trypsin	Polypeptide bonds adjacent to arginine or lysine	Large peptides	Pancreas
Chymotrypsinogen → chymotrypsin	Many other polypeptide bonds	Large peptides	Pancreas
Exopeptidases			
Carboxypeptidase	COOH end of large peptides	Small peptides, amino acids	Pancreas*
Aminopeptidases	NH_2 end of large peptides	Small peptides, amino acids	Pancreas*
Tripeptidases	Small peptides	Amino acids	Intestine*
Dipeptidases	Small peptides	Amino acids	Intestine*
Enterokinase	Trypsinogen	Trypsin	Intestine*
Nucleases	Nucleic acids	Nucleotides, nucleosides	Pancreas

* These enzymes are in the membranes of the microvilli on the surface epithelial cells lining the digestive tract.

Endopeptidases can act on all types of proteins including those in the cells that produce them and those in the cells that line the digestive tract. This is avoided by their synthesis in the secretory cells in an inactive form, their release only when food is present in the gut to be digested, and their chemical or enzymatic activation by conditions at the site of digestion. When food enters the vertebrate stomach, pepsin is secreted as an inactive **proenzyme** known as **pepsinogen**. Hydrochloric acid, also secreted in the stomach, provides the optimum pH for the cleavage of a fragment from the pepsinogen molecule, thereby converting the rest of the molecule into active pepsin. Trypsin and chymotrypsin are also secreted as inactive proenzymes: **trypsinogen** and **chymotrypsinogen**. The alkaline environment in the intestine and **enterokinase**, an intestinal enzyme, activates trypsinogen. Trypsin once formed continues to activate trypsinogen and also activates chymotrypsinogen.

Secretions of mucus, which are partly composed of carbohydrates and therefore are resistant to proteases, protect the gut lining to a great extent from autodigestion and other injuries. However, the cells lining the gut are constantly being injured to some extent. They are sloughed off and must be continuously replaced. The human stomach epithelium, for example, is gradually replaced about every three days.

Nucleic Acid Digestion

Nucleic acids are digested by **nucleases** into their constituent nucleotides and nucleosides. These, in turn, are hydrolyzed by other enzymes that release the phosphate groups, five-carbon sugars (pentoses), and nitrogenous bases.

Absorption

Digested organic materials, mineral ions, water, and other small molecules are taken into the body from the digestive tract by absorption. The absorbed materials must pass through cells to enter the body fluids and blood because tight junctions between epithelial cells lining the digestive tract prevent the passage of materials between them. Diffusion has a role in nutrient uptake so long as the materials to be absorbed have a higher concentration in the lumen of the digestive tract than in body fluids, but active transport of materials across cell membranes is also a very important mode of absorption. That this is the case is demonstrated by the fact that absorption of some materials continues after their concentrations in the lumen are less than in the body.

Different substances are absorbed by cells in different ways. Lipid-soluble materials, including fatty acids and the fat-soluble vitamins, dissolve in the lipid bilayer of the cell membrane and pass directly through it by simple diffusion following their concentration gradients (Fig. 12.3a). Water and water-soluble materials cannot pass directly through the cell membrane, but are absorbed through protein-bounded pores or channels in the membranes (Fig. 12.3b). As solutes are absorbed by the cells of the digestive tract, the osmotic pressure within the cells increases, and water diffuses through the small pores following its osmotic gradient. Alcohols such as glycerol and many ions also enter through minute channels (p. 69). Some miner-

Figure 12.3

A diagram of four important mechanisms of absorption. (a) Simple diffusion of a lipid-soluble material through the plasma membrane; (b) simple diffusion through pores in the membrane; (c) facilitated diffusion on a carrier molecule; (d) cotransport on a molecule also carrying sodium ions.

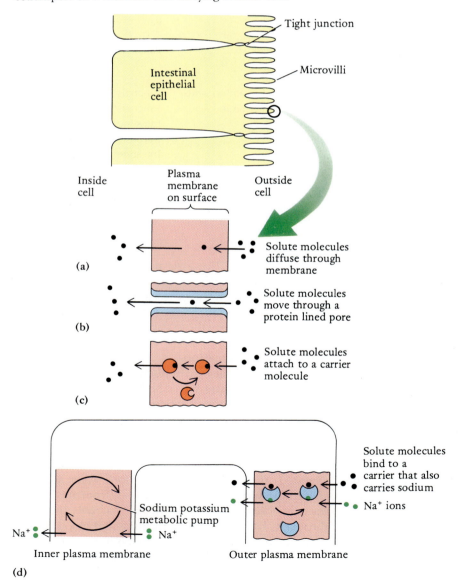

als, including calcium and iron, bind to carrier molecules that move across the membrane, and they are carried through by the carrier molecules. This method is called **facilitated diffusion** (Fig. 12.3c). Fructose also enters by facilitated diffusion.

Sodium ions also bind to carrier molecules and are transported into the cell by a gradient created by the active pumping out of sodium at the opposite end of the cell. As sodium is pumped out, potassium enters; hence this mechanism is called the **sodium-potassium pump**. Most of the monosaccharides and the amino acids also bind to the sodium carrier and are actively transported into the cell along with the sodium. This process is called **cotransport**

(Fig. 12.3d), and it depends on the continued input of energy needed to drive the sodium-potassium pump. Most single sugars and amino acids are not absorbed in the absence of sodium transport.

Most of the materials leave the cells lining the digestive tract and enter the body fluids and circulatory system by similar mechanisms. Single sugars and amino acids leave by facilitated transport. The fatty acids and glycerol reunite within the cells lining the digestive tract to form small globules of triacylglycerol. These globules leave the cells by exocytosis, and most are transported in part of the circulatory system (the lymphatic system) as small particles known as **chylomicrons**.

EVOLUTION OF THE ANIMAL GUT

With this background on how food is digested and absorbed, we can turn to the digestive tract, or **gut** (Anglo Saxon *gut*, channel), and examine where the different processes occur. We do not know what the first multicellular animals, or metazoans, were like, but many zoologists speculate that they were small, radially symmetrical, slightly elongate animals (Fig. 12.4a). A layer of monociliated cells covered the outside of the body, and a solid mass of cells filled the interior. It is quite probable that they had neither mouth nor gut. Small food particles, such as bacteria, algae, protozoans, and cell fragments, were engulfed by endocytosis (phagocytosis) by the exterior cells. The food particle became enclosed within a **food vacuole** located in the cell, and lysosomes within the cytoplasm secreted digestive enzymes into the vacuole, where digestion occurred **intracellularly**. The products of

Figure 12.4

The evolution of the digestive tract. (a) Ingestion of food particles by the surface cells of a hypothetical ancestral metazoan, and their digestion intracellularly. (b) The primitive digestive tract consisted of a simple gut, or enteron, that opened to the surface only through a mouth, as seen in this diagram of a freshwater flatworm, *Macrostomum*. (c) In planarian flatworms the pharynx can be protruded through the mouth during feeding. (d) Food can be processed continually in the digestive tract of earthworms and most other metazoans that have an anus in addition to the mouth.

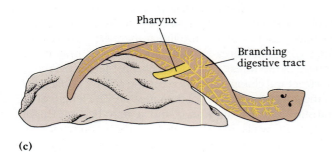

(a)

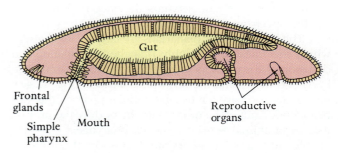

(b)

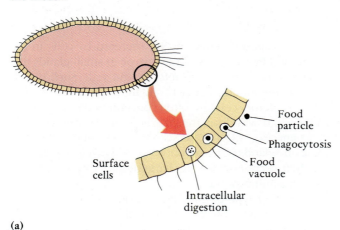

(c)

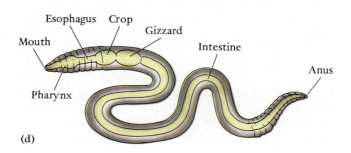

(d)

digestion were absorbed from the vacuole. Any undigested wastes were discharged to the exterior by exocytosis. Absorbed food material reached interior cells by diffusion. There are organisms living today that feed in this manner. Among multicellular animals, sponges have neither mouth nor gut, and the small particles on which they feed are digested intracellularly. Intracellular digestion is also characteristic of many protozoans, such as *Paramecium* and amoebas.

All other animals possess a mouth and gut, the lining epithelium of which is continuous with the epithelium on the body surface. Material within the gut cavity has not technically entered the body until it passes through this epithelial lining. If a gut is not present, as in some parasitic worms, the gutless condition is secondary, i.e., the gut has been lost. The ancestral gut was probably a simple sac with only one opening, the mouth. A digestive system of this type is found in many of the more primitive animals, such as the small flatworm *Macrostomum* (Fig. 12.4b). These little freshwater animals, less than 1 mm in length, swim or crawl about over debris in pools and ponds. The ventral mouth leads into a simple ciliated tube, the **pharynx**, that in turn opens into a large simple sac, the **intestine** or **enteron**. These flatworms feed on other small animals that they swallow whole. When the prey reaches the intestinal sac, certain cells produce digestive enzymes that pass into the cavity and digest proteins **extracellularly**. The prey soon fragments, and the resulting particles are engulfed by other intestinal cells, where digestion is completed intracellularly. Indigestible wastes are egested back out of the mouth. A gut and extracellular digestion have important selective advantages: larger food masses can be ingested, the animal need not feed so often, and those cells that secrete digestive enzymes can become specialized and synthesize only certain enzymes; they need not produce the full spectrum of enzymes.

The more familiar planarians are larger flatworms, in which the intestine is highly branched, facilitating the diffusion of absorbed food material to various parts of the body and increasing the absorptive surface. (Fig. 12.4c). Planarians also have a muscular, tubular pharynx that can be extended out of the mouth into the body of the prey or into dead animal remains. Special pharyngeal enzymes enable the pharynx to penetrate tissues, and the contents of the prey are sucked up into the intestine. Hydras, sea anemones, and corals also have a digestive cavity with only one opening. They feed on other small animals that, after being swallowed whole, are first partly digested extracellularly; digestion is completed intracellularly.

Most animals possess a digestive tract with separate intake and exit openings: the mouth and the anus (Fig. 12.4d). Such an arrangement permits continuous and sequential processing of food as it passes along the gut tube. The initial digestion of protein in an acid medium, for example, need not be halted to allow the continued digestion of protein in an alkaline environment farther along the gut. Parts of the gut tube become restricted to specific functions, leading to regional specializations, of which the most common ones are described briefly below.

Buccal or mouth cavity. The anterior region of the gut into which the mouth opens. The buccal cavity commonly contains teeth, jaws, salivary glands, and other structures concerned with ingestion, and sometimes with the mechanical breakdown of large food masses.

Pharynx. An anterior region of the gut that is usually muscular and, among invertebrates, is frequently specialized for sucking in food.

Esophagus. A tubular section that transports food to more posterior regions of the gut.

Crop. An area specialized for temporary food storage.

Gizzard. A region of the gut specialized for the mechanical breakdown, or **trituration**, of food into smaller particles prior to digestion. The walls are muscular, and the surface facing the lumen often bears hard plates, teeth, or ridges.

Stomach. A dilated, sometimes muscular, region of the gut, within which storage and digestion take place. In some animals the stomach may also be a site of absorption.

Intestine. A tubular region where most digestion and absorption occurs. Feces are usually formed in the posterior part of the intestine.

Rectum. A terminal part of the gut in which feces are formed and stored prior to egestion.

The digestive enzymes are secreted by cells in the lining of different parts of the gut, or in small glands that evaginate from the lining into the gut wall. Such glands occur in the mouth, stomach, and intestine of vertebrates. Other secretory cells are often concentrated in large glandular outgrowths from the gut, such as the vertebrate liver and pancreas. All of these secretory cells and glands are **exocrine glands** (Gr. *exo*, outwards + *krinein*, to secrete) for their secretions are discharged onto a surface, in this case the lining of the gut.

Another type of gut specialization is affected by the familiar surface-volume relationship. Since absorption rates are approximately proportional to the surface area through which the materials pass, animals encounter a problem as they increase in size. The mass to be nourished increases as the cube of the linear increases, whereas the internal surface area of their gut increases only as the square of the linear increases. This problem is overcome by structural changes that increase the relative surface area of the gut and especially of the intestine. The length of the intestine often is increased and it may become highly coiled. Its walls may be infolded, have outpocketings called ceca, or bear finger-like projections called villi (p.

253). Cells lining the intestine often have numerous, minute microvilli on their surface. Animals with high metabolic rates also need a relatively large absorptive surface and have evolved similar modifications.

FEEDING MECHANISMS AND GUT SPECIALIZATIONS

Few animals have all of the digestive tract specializations listed above. Whatever specializations are present depend largely upon the diet and the feeding mechanisms of the animals, which in turn are related to the animal's life style, i.e., whether it is free moving or attached, lives in burrows or in tubes, and so on. Although similar diets and feeding habits have evolved independently many times in different groups, the feeding processes are not identical because the evolutionary position of the group imposes constraints upon the type of features that can evolve. Each mechanism utilizes the special features of the body plan of the particular group of animals involved. We will consider a few examples of the many types of feeding mechanisms and their attendant gut specializations.

Sucking and Cropping Herbivores

The ingestive organs of some herbivores allow them to withdraw nectar from flowers, or pierce plant tissue and suck out just the sap or cell contents (Fig. 12.5). Most

Figure 12.5
Many insects obtain their food by sucking plant juices. (a) An aphid feeding on a plant; (b) The mouth parts of a sucking insect are adapted to penetrate plant tissues.

(a)

(b)

herbivores utilize more of the plant tissues, and their buccal cavities are equipped with a variety of structures, such as jaws, teeth, blades, and scrapers, that cut off, or crop, and ingest fragments of plant tissue.

Plant food is more difficult to digest than animal material, and herbivores have longer guts than closely related carnivores. This allows more digestive time and provides a larger absorptive surface. The difference in gut length associated with diet is especially striking in the development of frogs. Algae-feeding herbivorous tadpoles have a long, greatly coiled gut. During the metamorphosis of the tadpole to the insect-feeding adult, the gut becomes much shorter and is only about one third of its length in the tadpole.

Herbivores that utilize the cellulose in plant cell walls contain colonies of symbiotic microorganisms in some part of their guts that produce cellulase. Wood-eating termites and roaches harbor symbiotic protozoan flagellates in their guts that produce the cellulase, but bacterial colonies perform this function in most herbivores. Newly born or hatched individuals must eat some fecal material to "seed" their guts with these microorganisms.

In many species, the cellulose-digesting microorganisms are confined to special gut regions. In cattle and other ruminants, they are in the **rumen** and **reticulum**, two of the four chambers of the stomach (Fig. 12.6). These chambers have a capacity of up to 200 liters in a large cow. The partly digested hay or grass is regurgitated from time

Figure 12.6
Cellulose is digested by symbiotic microorganisms in the multichambered stomach of a cow. Food first enters the rumen and reticulum, where digestion begins. Periodically, food is regurgitated, chewed more, and reswallowed. Many products of digestion are absorbed from the rumen, but the rest of the food, together with some of the bacteria that multiply in the rumen and reticulum, pass through the omasum to the abomasum, where protein digestion begins.

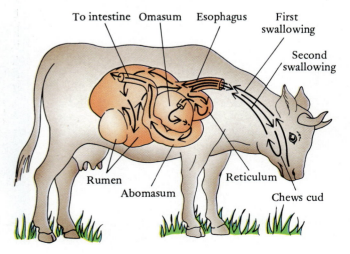

to time, and the animal ruminates or chews its cud. This reworking provides a further mechanical breakdown of the plant material. The microorganisms play a twofold role in the animal's nutrition. First, they produce digestive enzymes: cellulase that splits the cellulose into cellobiose, and other enzymes that convert the cellobiose to smaller units, mostly organic acids. The digesting cellulose and its products, some of which are acids, are buffered by the alkaline saliva that is secreted in great quantity by the salivary glands. An amount of saliva equal to about one third of the body weight of the cow is secreted daily!

Second, the microorganisms multiply and synthesize amino acids and proteins, some of which the cow harvests. Remarkably, the nitrogen source needed for this synthesis includes some of the cow's urea, which is normally a waste product. In ruminants, some urea diffuses into the rumen. Many of the products of digestion are absorbed in the rumen and reticulum. These chambers also slowly release liquified material through the valvelike **omasum** into the final chamber, the **abomasum**. The abomasum is the only chamber that contains gastric glands, and it is here that the digestion of the harvested microorganisms and proteins begins.

Horses, rabbits, rodents, and many other herbivorous mammals contain a bacterial colony in a long **cecum** at the beginning of the large intestine rather than in a part of the stomach. Such animals cannot regurgitate the food for further mechanical breakdown, so their feces contain a great deal more fiber than is found in ruminants feces. You may have noticed the difference in fiber content between the feces of a cow and those of a horse. Rabbits and some rodents capture the lost food content by eating feces and recycling material through the digestive tract.

Raptorial Feeding

Sharks, snakes, lions, and other raptorial feeders are carnivores that actively capture prey. Teeth and other buccal structures are often used to seize, kill, and hold prey, but many animals also utilize their limbs and other parts of the body. The prey either is swallowed whole or is torn apart and ingested in fragments. The digestive secretions con-

Figure 12.7
Parasites rob their hosts of nutrients.
(a) Medical leeches (*Hirudo medicinalis*) being used to withdraw blood that has accumulated beneath the skin near an incision. (b) The mouth lies within the anterior attachment sucker; another sucker occurs at the posterior end.

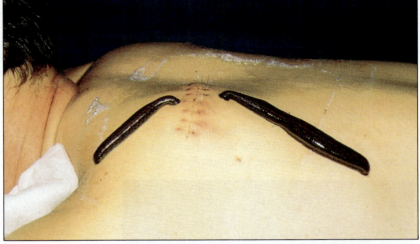

(a)

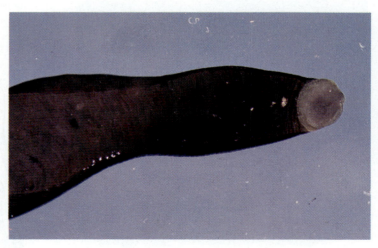

(b)

tain a high percentage of proteases, digestion is rapid, and the gut is shorter and simpler than in herbivores.

Parasitism

Animal parasites are small, specialized carnivores that do not normally kill their prey but rob it of nutrients. The prey, called the host, may become so weakened that it dies, but during the course of evolution most parasites and their hosts have adapted to each other and both survive. Leeches, ticks, and other species that attach to the outside of the host are called **ectoparasites**. They are adapted for clinging and generally have ingestive organs for sucking blood or tissue fluids that are obtained by piercing or biting the integument (Fig. 12.7). Species that live within the body of the host are called **endoparasites**. Many, including tapeworms and blood flukes, absorb simple food compounds directly through their body surface from the gut contents or body fluids of the host. The digestive tract of these parasitic species is reduced or completely lost.

Suspension Feeding

Many aquatic animals feed on minute plants, animals, and detritus suspended in water. The suspended plants and animals are collectively known as **plankton**. A common type of suspension feeding is **filter feeding**. Filter feeding typically involves three processes: (1) utilizing a water current to carry food particles; (2) separating, or filtering, the suspended food particles from the water; and (3) transporting the collected food from the filtering apparatus to the mouth. Fanworms, a group of marine tube-dwelling worms, are good examples of filter feeders (Fig. 12.8). In these worms, a funnel of feather-like structures, **radioles**, projects from the head. Cilia on the radioles produce a water current, and as water passes between the radioles, food particles are trapped in mucus on the radiole surface. Tracts of cilia, different from those producing the water current, transport food particles down the radioles to the mouth. Suspension feeding is a very effective way of gathering food and has evolved independently many times. Among vertebrates, many fishes, tadpoles, ducks, and the giant toothless whales are suspension feeders.

Figure 12.8
Filter feeding in the annelid fanworm, *Sabella*. (a) Water current passing through radioles projecting from the opening of the tube. (b) Water currents (*blue arrows*) and ciliary tracts (*black arrows*) over a section of one radiole.

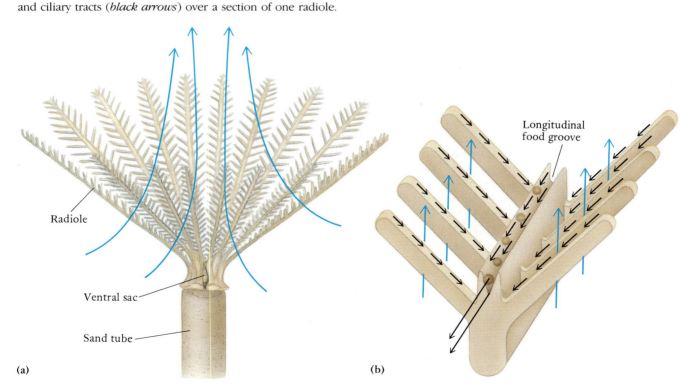

Radiole

Ventral sac

Sand tube

Longitudinal
food groove

(a)

(b)

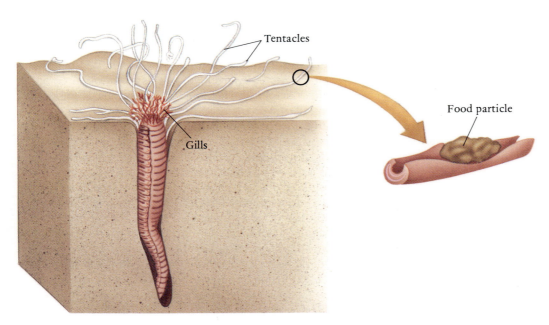

Figure 12.9
Selective deposit feeding in the annelid worm *Amphitrite*. Food particles are
gathered by the outstretched tentacles, as shown in the insert.

Deposit Feeding

The organic detritus deposited on the bottom of aquatic
habitats and mixed with sand grains is utilized as a food
source by many animals, called **deposit feeders**. Some,
such as the marine lugworms, are **nonselective deposit
feeders**. Lugworms live in burrows in the substratum and
ingest sand grains mixed with deposited detritus as they
extend their burrows. During passage through the gut, the
organic material is digested away from the sand grains.
Like many other nonselective deposit feeders, lugworms
periodically back up to the surface and defecate the mineral matter in conspicuous piles, called **castings**. Although
terrestrial, earthworms are in part nonselective deposit
feeders.

Some deposit-feeding animals ingest only organic
material. A group of marine annelids, the terebellid
worms, are such **selective deposit feeders** (Fig. 12.9). The
terebellids possess a large number of long, delicate head
tentacles that project out over the substratum from the
tubes or burrows in which the worms live. Small organic
particles adhere to mucus on the tentacles and are transported back to the mouth by cilia, as well as by contraction
of the tentacles.

THE VERTEBRATE PATTERN

The vertebrate digestive tract serves as a good example of
the integration of gut morphology with all of the processes
that take place during feeding, digestion, and absorption.
The basic pattern is similar in all vertebrate species.

Mouth

In very primitive species the mouth lacks jaws, but most
vertebrates have jaws supporting a set of teeth. Teeth vary
considerably among groups. A representative mammalian
tooth (Fig. 12.10a) consists of a **crown** projecting above the
gum, and often bearing two or more hillocks called **cusps**,
a **neck** surrounded by the gum, and one or more **roots**
embedded in sockets in the jaws. The crown is covered by
a layer of **enamel**, which is the hardest substance in the
body, consisting almost entirely of crystals of calcium salts.
Calcium, phosphate, and fluoride are the most important
mineral constituents of enamel. The rest of the tooth is
composed of **dentin**, a substance similar in composition to
bone. A **pulp cavity** in the center of the tooth is filled with
pulp, connective tissue containing blood vessels and
nerves. A layer of bonelike **cement** covers much of the root
and holds the tooth firmly in place in the jaw.

Mammals have different types of teeth for seizing and
mechanically breaking down food. There is a close correlation between the number and shape of tooth types and
the diet and feeding habit of the species (p. 912). Mammalian teeth, unlike those of other vertebrates, are not continuously replaced by new sets. Humans first develop a set of
deciduous, or **milk teeth**: two **incisors**, one **canine**, and
two **premolars** on each side of each jaw. These are later
replaced by **permanent teeth**. Three **molars** develop on
each side of each jaw behind the premolars. The molars
remain throughout life and are not replaced (Fig. 12.10b).

Once in the mouth, food must be transported through
it and swallowed. A fish can easily manipulate and swallow

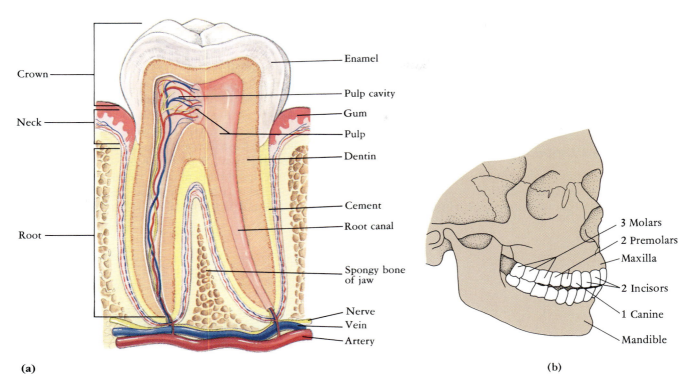

Crown ─

Neck ─

Root ─

─ Enamel
─ Pulp cavity
─ Gum
─ Pulp
─ Dentin
─ Cement
─ Root canal
─ Spongy bone of jaw
─ Nerve
─ Vein
─ Artery

(a)

3 Molars
2 Premolars
Maxilla
2 Incisors
1 Canine
Mandible

(b)

Figure 12.10
Human teeth. (a) A vertical section through a molar tooth. (b) The types of teeth in an adult.

food because the flow of water into the mouth and out the gill slits aids in carrying food back into the pharynx. Oral glands and a muscular tongue are poorly developed. The evolution of these structures accompanied the evolutionary transition of vertebrates from water to land, and they became most elaborate in the mammals. Many species, including frogs, chameleons, woodpeckers, and anteaters, also use a protrusible tongue to catch and ingest insects (Fig. 12.11). In most mammals, the tongue functions chiefly to manipulate food in the mouth and to transport it toward the pharynx, where it is swallowed. Food is pushed between the teeth so that it is thoroughly masticated and mixed with saliva. The food is then shaped into a ball, a **bolus**, and moved into the pharynx by raising the tongue. The tongue also has numerous microscopic taste buds, and the human tongue is of great importance in speech.

In addition to a liberal sprinkling of simple glands in the lining of the mouth cavity, mammals have evolved several pairs of large **salivary glands** that are connected to the mouth by ducts. The location of the human **parotid**, **mandibular**, and **sublingual glands** is shown in Figure 12.12. In primitive terrestrial vertebrates, oral glands simply secrete a mucous and watery fluid that lubricates the food, and this is still the major function of saliva. The saliva of most mammals and of a few other terrestrial vertebrates contains salivary amylase that splits starch and glycogen

Figure 12.11
Many amphibians such as this green frog (*Rana clamitans*) catch insects by rapidly extending their tongues.

into maltose (Table 12.1). Chloride ions present in the saliva are necessary to activate amylase. Its optimum pH is close to neutrality, so its action is eventually stopped by the acidic gastric juice of the stomach. Since it takes one-half hour or longer for the food and gastric juice to become thoroughly mixed, 40% or more of the starches are split before the amylase is inactivated.

Pharynx and Esophagus

The pharynx of vertebrates extends from the back of the mouth cavity to the beginning of the esophagus. In terrestrial vertebrates it is a rather short region in which the food and air passages cross, but in fishes it is a more extensive area associated with the gill slits. Part of the pharynx of mammals lies above the **soft palate** (Fig. 12.13) and receives the internal nostrils, or **choanae**, from the nasal cavities, and the openings of the pair of **auditory tubes** (eustachian tubes) from the middle ear cavities. Another part lies beneath the soft palate and receives materials from the mouth cavity. A third, in which food and air paths cross, lies just posterior to these parts and leads to the esophagus and larynx. The larynx is part of the passageway to the lungs, and the opening into it is called the **glottis**. Passage of food into the pharynx initiates a series of nerve reflexes: the muscular soft palate rises and prevents food

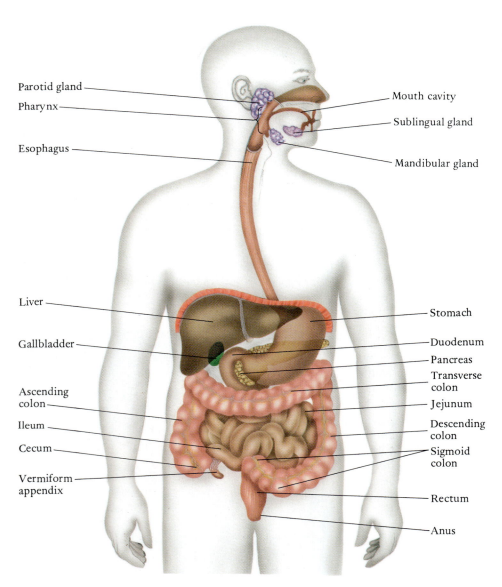

Figure 12.12
An overview of the human digestive tract.

Parotid gland
Pharynx
Esophagus
Mouth cavity
Sublingual gland
Mandibular gland

Liver
Gallbladder
Ascending colon
Ileum
Cecum
Vermiform appendix

Stomach
Duodenum
Pancreas
Transverse colon
Jejunum
Descending colon
Sigmoid colon
Rectum
Anus

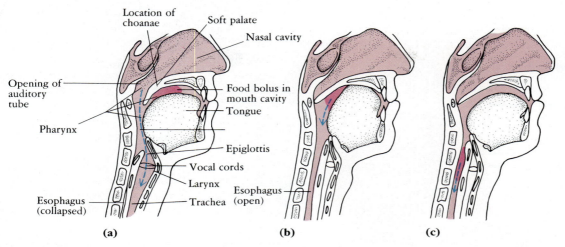

Figure 12.13
Drawings showing changes in the positions of the tongue, soft palate, and epiglottis as food passes through the pharynx.

from entering the nasal cavities; breathing momentarily stops; the larynx is elevated, and the flaplike epiglottis swings over the glottis, preventing food from entering the larynx; the tongue prevents food from returning to the mouth; and muscular contractions of the pharynx move the bolus into the esophagus. **Peristalsis**, which results from successive waves of alternate contraction and relaxation of muscles in the esophageal wall, propels the bolus down the esophagus to the stomach. The muscles relax in front of the food and contract behind it. When the food reaches the distal end of the esophagus, the cardiac sphincter that closes off the entrance to the stomach relaxes and allows it to enter.

Stomach

The stomach is usually a **J**-shaped pouch (Fig. 12.14) whose chief functions are the storage and mechanical churning of food and the initiation of the hydrolysis of proteins. Lampreys and some other primitive fishes do not have a stomach, and the absence of this organ is thought to have been a characteristic of ancestral vertebrates. The early vertebrates probably fed more or less continuously on minute food particles that they filtered from the water current passing through the pharynx and out the gill slits. These particles could be digested by the intestine alone. Presumably, the evolution of jaws and the habit of feeding less frequently and on larger pieces of food was accompanied by the development of an organ for storage and by the initial mechanical and chemical conversion of this food into a state in which it could be digested further.

After food enters the stomach, the **cardiac sphincter** at its anterior end and the **pyloric sphincter** at its posterior end close. Muscular contractions of the stomach churn the food, breaking it up mechanically and mixing it with the

gastric juice secreted by tube-shaped **gastric glands**. Pepsinogen secreted by the **chief cells** of the glands and hydrochloric acid produced by the **parietal cells** initiate protein digestion (p. 237). In addition, the enzyme **rennin** is particularly abundant in the stomach of young ruminants, causing the milk protein casein to coagulate and slowing down the passage of milk through the stomach. Rennin extracted from calves' stomachs is used to curdle milk in making cheese. Humans and many other mammals lack this enzyme.

Food in the stomach is reduced to a creamy material known as **chyme**. Peristaltic waves sweeping down the stomach cause the pyloric sphincter to open briefly, allowing acidic chyme to enter the intestine in small spurts. The acid food is quickly neutralized by the alkaline secretions flowing into the intestine from the liver and pancreas.

Liver and Pancreas

The **liver** and **pancreas** are large glands located anterior and dorsal to the stomach (Fig 12.15). Embryonically they develop as outgrowths from the anterior part of the intestine. Liver cells continually secrete **bile**, which passes through hepatic ducts into the **common bile duct** and then up the **cystic duct** into the **gallbladder**. Bile does not enter the intestine immediately, for a sphincter at the intestinal end of the bile duct is closed until food enters the intestine. Contraction of the wall of the gallbladder forces bile out. The bile that is finally poured into the intestine is concentrated, for a considerable amount of water and some salts are absorbed from it in the gallbladder.

Although bile contains no digestive enzymes, it nevertheless has a digestive role. Its alkalinity, along with that of the pancreatic secretions, neutralizes the acid food entering the intestine and creates a pH favorable for the

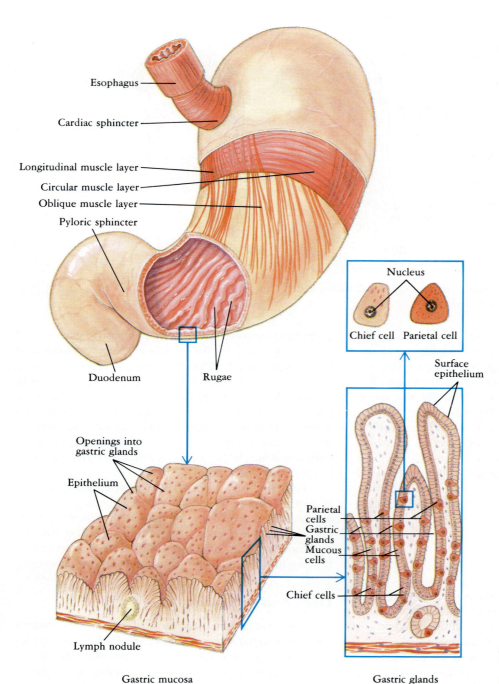

Esophagus

Cardiac sphincter

Longitudinal muscle layer

Circular muscle layer

Oblique muscle layer

Pyloric sphincter

Duodenum

Rugae

Nucleus

Chief cell   Parietal cell

Surface epithelium

Openings into gastric glands

Epithelium

Parietal cells

Gastric glands

Mucous cells

Chief cells

Lymph nodule

Gastric mucosa

Gastric glands

Figure 12.14
The human stomach and gastric glands.

action of pancreatic and intestinal enzymes. Its bile salts, as pointed out earlier, emulsify fats and facilitate the absorption of the products of fat digestion.

The color of bile results from the presence of **bile pigments** derived from the breakdown in the liver of hemoglobin from red blood cells. The bile pigments are converted by enzymes of the intestinal bacteria to the brown pigments responsible for the color of the feces. If the excretion of bile pigments is prevented by a gallstone

or some other obstruction of the bile duct, they are reabsorbed by the liver and gallbladder, the feces are pale, and the pigments accumulate in the skin, giving it the yellowish tinge characteristic of jaundice.

The pancreas is an important digestive gland, producing quantities of enzymes that act upon carbohydrates, proteins, and fats (Table 12.1). These enzymes enter the intestine by way of a pancreatic duct that joins the common bile duct. In some vertebrates an accessory pancreatic duct

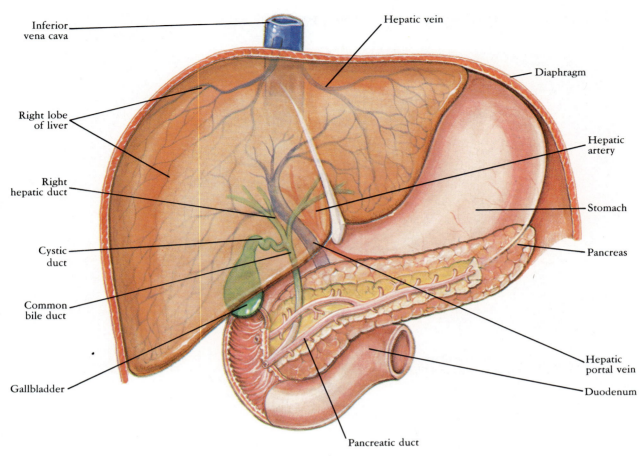

Inferior vena cava

Hepatic vein

Diaphragm

Right lobe of liver

Hepatic artery

Right hepatic duct

Stomach

Cystic duct

Pancreas

Common bile duct

Hepatic portal vein

Gallbladder

Duodenum

Pancreatic duct

Figure 12.15
A ventral view of the human liver and pancreas. Part of the stomach has been drawn as though it were transparent in order to show the pancreas, much of which lies behind the stomach.

empties directly into the intestine. The pancreas contains patches of hormone-producing tissue, the islets of Langerhans, which will be considered in Chapter 19.

Intestine

Most of the digestive processes, and virtually all of the absorption of the end products of digestion, occur in the intestine. The structural details of the intestine vary considerably among vertebrates. The intestine of terrestrial vertebrates has become differentiated into an anterior **small intestine** and a posterior **large intestine** (Fig. 12.12). The first part of the small intestine is the **duodenum** and, in mammals, the two succeeding parts are called the **jejunum** and **ileum**. The large intestine, or **colon**, of the frog and most vertebrates leads to a posterior chamber known as the **cloaca** (L. *cloaca*, sewer). The cloaca, which also receives the products of the urinary and reproductive sys-

tems, opens on the body surface by the **cloacal aperture**. In mammals, the cloaca has become divided into a ventral part that receives the urogenital products, and a dorsal **rectum** that opens on the body surface at the **anus**. A blind pouch called the **cecum** is present at the junction of small and large intestines of mammals. This is very long in many herbivores, such as the rabbit and horse, and contains the colonies of bacteria that digest cellulose. Humans have a small cecum with a **vermiform** (worm-shaped) **appendix** on its end. The wall of the appendix contains masses of white blood cells, so the organ appears to be a part of the immune system and not a simple vestige, as was once believed. An **ileocecal valve** located between the small and large intestines prevents bacteria in the colon from moving back up into the small intestine.

A transverse section of the small intestine of a mammal illustrates the microscopic structure of the digestive tract (Fig. 12.16). There is an outer covering of **visceral**

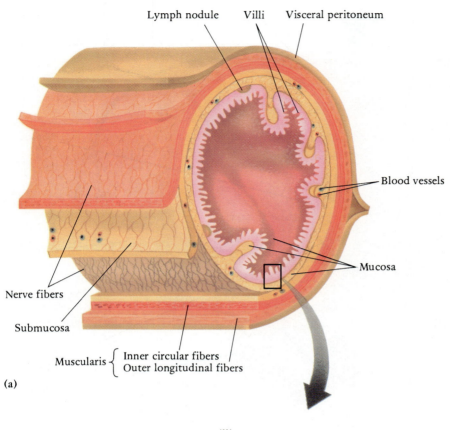

Lymph nodule   Villi   Visceral peritoneum

Blood vessels

Mucosa

Nerve fibers

Submucosa

Muscularis { Inner circular fibers / Outer longitudinal fibers

(a)

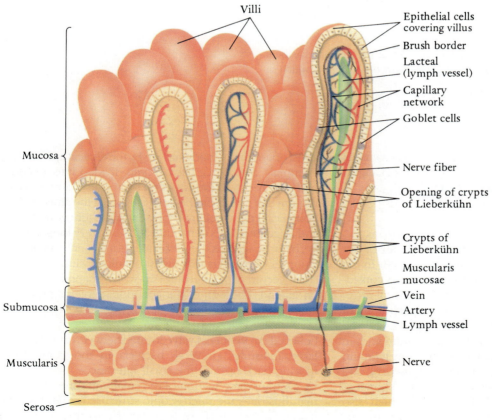

Villi

Mucosa

Submucosa

Muscularis

Serosa

(b)

Epithelial cells covering villus

Brush border

Lacteal (lymph vessel)

Capillary network

Goblet cells

Nerve fiber

Opening of crypts of Lieberkühn

Crypts of Lieberkühn

Muscularis mucosae

Vein

Artery

Lymph vessel

Nerve

Figure 12.16
(a) A section of the small intestine; (b) a detail of the mucous membrane.

peritoneum, the serous coat; a layer of smooth muscle, the **muscularis**; a layer of vascular connective tissue, the **submucosa**; and the innermost layer, the **mucosa** (or mucous membrane). The outer fibers of the muscular coat are usually described as longitudinal; the inner, as circular. Actually, both layers are spiral; the outer is an open spiral, and the inner, a tight one. When a section of the intestine is distended by its contents, the stretched muscle responds by contracting. Some of the contraction results in localized constrictions, called **segmentation**, that aid in mixing the contents. As the food is digested and absorbed, peristaltic contractions move it down the intestine. The mucosa consists of a layer of smooth muscle, the **muscularis mucosae**; connective tissue; and the simple columnar epithelium next to the lumen. Many **goblet cells** are present in the lining epithelium, and their secretion of mucus helps to lubricate the food and protect the lining of the intestine from abrasion and autodigestion. The mucosa of the small intestine of birds and mammals bears numerous minute, finger-shaped **villi** containing blood capillaries and small lymphatic vessels. Villi, **microvilli** on the surface of the epithelial cells (Fig. 12.17), circular and longitudinal folds in the intestinal mucosa, and the length of the intestine provide an enormous surface area for digestion and absorption. The gross cylindrical surface area of the human intestinal lining is approximately 0.4 m², but folds, villi, and microvilli increase it to nearly 300 m². Villi and microvilli have the effect of the pile on a good bath towel. The villi are moved about by the muscularis mucosae and by strands of smooth muscle that extend into them.

At the base of the villi are tubelike areas known as the **crypts of Lieberkühn** that extend into the mucosa. Mitotic division of epithelial cells in the bottom of the crypts continually produces new cells that migrate outward over the villi and are sloughed off. In the course of their outward migration, the cells differentiate into the mucus-producing goblet cells and other epithelial cells. There is a rapid turnover of epithelial cells, for an individual cell lasts only about 48 hours. The contents of the shed cells are the source of the watery intestinal juice and some of the enzymes.

Enzymes acting in the intestine are produced by the pancreas and by many of the epithelial cells lining the intestine (Table 12.1). Carbohydrate digestion is completed in the intestine. Starches that are not broken down in the stomach by salivary amylase are soon converted to maltose in the neutral environment of the small intestine by pancreatic amylase. Maltase, sucrase, and lactase and

Figure 12.17
An electron micrograph of part of two epithelial cells from a villus. The terminal bar is a tight junction between cells. Magnification ×30,000.

other enzymes produced by the intestinal cells are not completely free in the main part of the intestinal lumen, for cell-free extracts of intestinal juice contain very little enzyme. Rather, the enzymes are located in the cell membranes of microvilli that form the striated border of the absorptive cells of the villi (Fig. 12.17). The disaccharides are hydrolyzed into monosaccharides as they enter this border.

Protein digestion is completed in the intestine by endopeptidases and exopeptidases secreted by the pancreas, and by tripeptidases and dipeptidases in the membranes of microvilli.

Fats emulsified by bile salts are digested by lipases, most of which are secreted by the pancreas. The mixture of bile salts, fatty acids, and partly digested fats collectively emulsifies fats further.

The total volume of secretions by the digestive tract averages about 8 liters a day in an adult human. Although most of the water ingested and present in saliva and gastrointestinal secretions is absorbed in the small intestine, the material that enters the large intestine is still quite liquid. Most of the remaining water and many salts are absorbed as the residue passes through the colon. If the residue passes through very slowly and too much water is absorbed, the feces become very dry and hard, and constipation may result; if it goes through very rapidly and little water is absorbed, diarrhea results.

By the time food reaches the colon of humans most digestible compounds have been removed. Depending upon the diet, the residue includes a large amount of plant cellulose and other complex carbohydrates (pectins), large numbers of sloughed intestinal cells, and mucus, which is produced in great quantities along the entire length of the gut. A large colony of bacteria in the colon acts on this substrate, and, as in the rumen of cows, the products are organic acids and the gases methane and hydrogen. Although humans are not dependent upon these compounds, some appear to be absorbed because the organic acid content of the feces is less than expected. Most of the gases are reabsorbed, metabolized, and their breakdown products eventually eliminated through the lungs. The colon bacteria obtain some nitrogen from the host's urea, some of which diffuses into the large intestine, and multiply rapidly. As much as 25% of the volume of human feces consists of eliminated bacteria.

Regulation of Digestive Secretions

It should be noted that the processing of the food as it passes through the gut is carefully regulated. Each of the enzymes is secreted at the appropriate time and in the needed quantity when the food on which it acts is present in the lumen. We salivate when we begin to eat, gastric juice is released when food enters the stomach, and gastric secretion stops when acid food enters the duodenum and pancreatic secretion is initiated. This regulation is partly controlled by the nervous system and partly by gastrointestinal hormones, as we shall see in later chapters.

NUTRITION

We conclude this chapter by examining how the materials that are absorbed are used by the body. A study of these materials and their relation to the maintenance of good health constitutes the science of **nutrition**.

Water and Minerals

Water is a major constituent of protoplasm, and all of the biochemical reactions in cells occur in an aqueous environment. Animals can survive longer without organic foods than without water. Aquatic animals receive an ample supply from their environment, but terrestrial species drink water or obtain it from the moisture content of their food.

The most abundant minerals in the body are calcium, phosphorus, chlorine, sulfur, potassium, sodium, and magnesium (Table 12.2). These are known as **macrominerals**, for they are needed by humans in quantities ranging from a few tenths of a gram to a gram or two per day. Many are present as ions; others, as constituents of organic compounds. Calcium, for example, is a major component of shells, bone, and teeth. Its ion is essential for muscle contraction, blood clotting, and the secretion of many hormones. In addition to these macrominerals, over a dozen **microminerals** are needed in trace quantities. Copper, for example, is a cofactor, activating enzymes involved in the absorption of iron, which is incorporated into hemoglobin. Iodine is a component of the hormone produced by the thyroid gland. Zinc is a cofactor of carbonic anhydrase, an enzyme important in gas transport (p. 285).

Animals obtain most of the minerals they need in the water they drink and the foods they eat, but some go out of their way to ingest earth rich in mineral content (Fig. 12.18).

Organic Molecules

The organic molecules that are absorbed are used as sources of energy and as building blocks for the animal's own protoplasm. Most of an animal's energy needs are provided by the carbohydrates, especially glucose, which is metabolized in all cells. Excess carbohydrate usually is converted to glycogen for storage in liver and muscle cells or to fat. Cellulose in the food is not utilized by animals that are not herbivores; however, it is important as a fiber that gives bulk to the contents of the digestive tract, retains moisture, and facilitates passage of food through the tract.

Fatty acids, glycerol, and sometimes other lipids are recombined in the body to form the animal's own lipids. Phospholipids are structural components of all cells, especially of their membranes. Lipids are very important, but

Table 12-2 Important Minerals	
Mineral	**Function**
Macrominerals (in order of abundance)	
Calcium (Ca)	Component of shells, bones, and teeth; muscle contraction; blood clotting; hormone secretion
Phosphorus (P)	Component of bones and teeth; energy metabolism; component of nucleic acids
Chlorine (Cl)	Principal extracellular negative ion; water balance; acid-base balance; formation of gastric hydrochloric acid
Sulfur (S)	Component of many proteins
Potassium (K)	Principal intracellular positive ion; transmission of impulses in nerve and muscle; acid-base balance; protein synthesis
Sodium (Na)	Principal extracellular positive ion; transmission of impulses in nerve and muscle; acid-base balance; water balance
Magnesium (Mg)	Appropriate balance between magnesium and calcium needed for nerve and muscle function; activates enzymes
Microminerals (major trace elements, in alphabetical order)	
Cobalt (Co)	Component of cobalamin (vitamin B_{12}); synthesis of red blood cells
Copper (Cu)	Enzyme activation; synthesis of hemoglobin
Iodine (I)	Component of thyroid hormone
Iron (Fe)	Component of hemoglobin and respiratory enzymes (cytochromes)
Manganese (Mn)	Enzyme activation
Zinc (Zn)	Enzyme activation

Figure 12.18

A dik-dik (*Madoqua sp.*), a small African antelope, obtaining necessary minerals at a salt lick.

most fatty acids can be synthesized in the body from other compounds. However one, linoleic acid, which is abundant in vegetable oils and nuts, cannot be synthesized by humans. For this reason it is called an **essential fatty acid**. It is no more essential than others in the biochemistry of the body, but it is essential as a component of the diet. Its absence causes certain forms of dermatitis, increased water consumption and retention, and other impairments.

Triacylglycerols (fats and oils) are long-term energy stores. Many mammals and birds accumulate large fat reserves before they migrate or hibernate. Fat beneath the skin of birds and mammals also serves as important insulation needed to conserve body heat in these warm-blooded animals.

Absorbed amino acids may be used as a source of energy, as they are in carnivores, but most are recombined within the body to form the animal's own proteins. Not all 20 naturally occurring amino acids need be in the diet because animals can synthesize most of them from simpler materials. Humans require nine amino acids in their diet, because they lack the enzymes needed to synthesize them. These nine are the **essential amino acids**: histidine, isoleucine, leucine, lysine, methionine, phenylalanine, threonine, tryptophan, and valine. Plant foods often contain insufficient quantities, but diets that include dairy products, eggs and meat provide the needed mixture and quantities.

Many people living in less developed countries have diets high in carbohydrates and do not receive an adequate protein intake. Protein deficiency in young children reduces their ability to synthesize the proteins they must have, and both their physical and mental development are severely retarded. Often the liver cannot synthesize sufficient blood proteins to maintain a proper balance between the liquid in the blood and tissues (see p. 319). Some tissues accumulate water, which results in the swollen bellies characteristic of malnourished children (Fig. 12.19). Children with protein deficiency often die of measles, whooping cough, and other childhood diseases because they are not able to synthesize an adequate level of protective protein antibodies.

Energy Needs

The energy needs of an animal can be expressed in **calories (cal)**, one calorie being the amount of energy needed to raise the temperature of one gram of water 1° C.* Zoologists usually work with units of 1000 calories, which are called **kilocalories (kcal)**. An adult male of college age needs about 2500 to 3300 kcal per day, depending on his level of activity. A female of comparable age requires somewhat fewer, about 1700 to 2500 kcal per day.

*The scientific unit for energy is the joule (see Appendix B for its derivation and definition), but most people are more familiar with calories as units for energy. We will use calories in this text. One calorie = 4.184 joules.

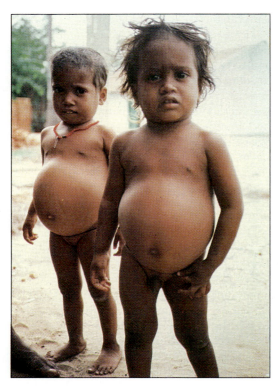

Figure 12.19
Children suffering from protein deficiency often have kwashiorkor, a disease characterized by swollen bellies and emaciated limbs.

Caloric needs decrease slightly as people grow older, and if their intake does not also decrease, they gain weight.

Although the dietary needs of people in all countries and cultures are the same, there is wide variation in how they are met. Citizens of the United States typically receive 45% of their calories from carbohydrates, 43% from fat, and 12% from protein. Fats and proteins are expensive to produce, and people in less developed parts of the world rely on carbohydrates in the form of rice, grains, and beans for most of their caloric needs because they are easier and cheaper to produce.

If a person's intake of calories falls below the minimum needed, then the body uses stored carbohydrate and fat. If these stores are not sufficient, protein is used. That is, the body begins to consume its own muscle and other proteins. Death by starvation usually results from a weakening of the heart muscles to the point where the muscles can no longer pump the needed volume of blood.

Vitamins

Vitamins are relatively simple organic compounds required in small amounts in the diet. They are called vitamins because they were originally believed to be derivatives of amino acids known as vital amines, but we now know that they differ widely in their chemical structure. All they have in common is their inability to be synthesized, or synthesized in adequate amounts, by an animal and hence must be present in the diet. Most organisms require these same substances to carry out specific metabolic functions. Plants can synthesize them, but animals differ in their synthetic abilities; thus, what is a "vitamin" for one animal is not necessarily one for another. Only humans, monkeys, and guinea pigs are known to require vitamin C (ascorbic acid) in their diets; other animals can synthesize it from glucose. All vitamins are molecules small enough to be absorbed directly.

Thirteen vitamins are now known to be needed by humans; others may yet be discovered (Table 12.3). Originally designated by letters, most are now known by their chemical names. The B vitamins are related only in being found in similar foods. The vitamins can be divided into two groups: those soluble in water and those soluble in fats. The water-soluble vitamins include ascorbic acid and the B complex. All function as coenzymes and are thus important in many metabolic reactions. The fat-soluble vitamins, A, D, E, and K, have diverse functions and appear to be vitamins only for vertebrates.

A diet deficient in a vitamin results in a deficiency disease with characteristic symptoms (Table 12.3). We now know that a deficiency in ascorbic acid causes the disease known as **scurvy**. Since ascorbic acid is needed to maintain intercellular materials, including collagen fibers, scurvy is characterized by such symptoms as slow wound healing, bleeding gums, and loosening teeth. The disease **beriberi** results from a deficiency of thiamine, one of the B vitamins. In the late 1800s Christian Eijkman discovered that chickens fed polished rice developed a disease rather like beriberi in humans. He found that both the chickens and beriberi patients could be cured by feeding them seed coats that had been removed from the rice grains in polishing.

Despite the American preoccupation with vitamins, an ordinary balanced diet will readily supply the daily minimum vitamin requirements. The water-soluble vitamins are not stored in body tissues so must be available in the diet on a regular basis. The fat-soluble vitamins are stored in fat deposits and in the liver. Deficiencies in these are more often associated with some problem in absorption rather than with diet. Although vitamin D can be obtained from a variety of foods, such as egg yolks, fish, and liver, it can be synthesized when the skin is exposed to ultraviolet light (p. 453). Its deficiency disease, **rickets**, is chiefly a consequence of inadequate diet during winter in northern climates, where there is little exposure of skin to sunlight. Because they are stored in the body, the fat-soluble vitamins A and D can in fact be toxic at high levels. Toxic concentrations can never be obtained through a normal diet, but may occur from overdoses of vitamin supplements.

Table 12.3
Common Vitamins

Vitamins	Common Sources	Function	Diseases and Symptoms if Deficient in Diet
Water soluble			
B Complex*			
B_1, Thiamine	Yeast, meat, whole grains, eggs, milk, green vegetables	Thiamine pyrophosphate coenzyme in decarboxylation reactions during carbohydrate metabolism	Beriberi: spasms or rigidity of legs, nerve and muscle degeneration
B_2, Riboflavin	Same as B_1: colon bacteria	Flavin mononucleotide and flavin adenine dinucleotide (FAD) are coenzymes for dehydrogenase reactions—electron transport in mitochondria and certain oxidations in the endoplasmic reticulum†	Similar to niacin deficiency, but only mild deficiency probably ever occurs
B_6, Pyridoxine	Same as B_1	Pyridoxal-phosphate is a coenzyme for many reactions involving amino acid metabolism: transamination, decarboxylation, and others	Dermatitis, gastrointestinal disturbances, but deficiency rare
Niacin, or nicotinic acid	Same as B_1	Nicotinamide adenine dinucleotide (NAD) and nicotinamide adenine dinucleotide phosphate (NADP) are coenzymes for many dehydrogenase reactions in cellular oxidation†	Pellagra: cracked, scaly skin, irritated mucous membranes, nervous disorders
B_{12}, Cobalamin	Synthesized by gut bacteria, found in animal foods only: meat, fish, eggs, milk	Cobalt-containing coenzymes involved in amino acid conversions and for DNA synthesis: cell division	Anemia (red cell precursors are most rapidly dividing cells)
Folic acid	Same as B_1	Coenzyme tetrahydrofolic acid involved in conversion of glycine to serine and in DNA synthesis; cell division	Anemia; probably most common vitamin deficiency worldwide
C, Ascorbic acid	Citrus fruits and fresh vegetables	Maintenance of intercellular substances: collagen fibers of connective tissue, capillary walls	Scurvy: bleeding gums, loosening teeth, slow wound healing
Fat soluble			
A, Retinol	Butter, eggs, fish liver oils; carotene in plants can be converted to vitamin A; stored in the liver	Component of the light-sensitive pigment, visual purple, in the retina; maintenance and growth of epithelial cells	Night blindness, inflammation of the eyes, scaly skin, easy infection
D, Cholecalciferol	Butter, eggs, fish oils, liver; produced in skin on exposure to ultraviolet light	Absorption and utilization of calcium and phosphorus	Rickets: weak bones and defective teeth
E, Tocopherol	Vegetable oils, vegetables, egg yolks, milk fat, liver; widely distributed in foods and stored in body	Not completely known, protects red blood cell membranes, maintains muscle	Red blood cells rupture, anemia, sterility (in rats)
K, Several quinone compounds	Green vegetables, colon bacteria	Synthesis of blood-clotting proteins in liver, hence normal blood clotting	Bleeding, especially in newborns, who lack colon bacteria

*Two other B-complex vitamins, pantothenic acid and biotin, are rarely deficient in human diets.
†See section on cellular respiration in Chap. 13.

Summary

1. Animals are heterotrophic, obtaining their organic compounds by the consumption of other organisms bodies or their products. Some species are herbivores, some are carnivores, and others are omnivores.

2. Ingested food often is broken up mechanically before it is digested, or broken down chemically, into molecules that are small enough to be absorbed. The hydrolytic reactions of digestion are catalyzed by enzymes; different enzymes attack different types and sizes of food molecules. Most animals possess the same general types of digestive enzymes.

3. Starches are digested by polysaccharidases, primarily amylase, into disaccharides. Each double sugar is acted on by a different enzyme (maltase, lactase, sucrase) that cleaves it into monosaccharides. Fats are first emulsified and then digested by lipase into fatty acids and glycerol. Proteins are acted on first by endopeptidases (pepsin, trypsin, chymotrypsin) that cleave the polypeptide chains into large peptides. Exopeptidases strip terminal amino acids off of large peptides. Nucleic acids are cleaved by nucleases into nucleotides and nucleosides.

4. Lipid-soluble materials can diffuse through the plasma membrane of the cells lining the digestive tract. Other molecules enter through minute pores, by binding with specific carrier molecules in the membrane (facilitated diffusion), and by binding to molecules that also transport sodium ions into the cells from the gut (cotransport). The last method is a process that requires metabolic energy that pumps sodium ions out of the opposite ends of the cells.

5. The earliest metazoans were probably gutless. Minute food particles were engulfed by surface cells and digested intracellularly. Digestion in most animals occurs within a gut cavity that permits extracellular digestion of large food masses. Primitively, as in hydras, sea anemones, and flatworms, the mouth is the only opening into the gut cavity. Digestion is initiated extracellularly and is completed intracellularly. Wastes are egested back out of the mouth. Most animals possess a gut with separate mouth and anus that permit a one-way passage and simultaneous processing of food in different parts of the gut tube. Digestion is primarily extracellular.

6. Symbiotic microorganisms contribute to the nutrition of many animals, either by sharing their products of photosynthesis (symbiotic algae in corals and some other animals) or by sharing their products of digestion (cellulose-digesting bacteria in ruminants and other herbivores).

7. Animals eat many different types of food and employ a great variety of feeding mechanisms. The structure of their digestive tracts correlates closely with these factors. Many herbivores crop plants or suck their juices. Raptors are carnivores that actively capture prey. Parasites do not usually kill the host they live on or in, but rob it of nutrients. Many animals feed on small food particles that may consist of living organisms (plankton, fine algae, bacteria) or the fragmented remains of organisms (detritus or deposit material). Such particulate food may be obtained by filter feeding or by deposit feeding.

8. Structures in the vertebrate mouth cavity ingest food, may provide some mechanical breakdown of food, and transport the food. Salivary secretions moisten food and may initiate digestion. Food is delivered by the esophagus to the stomach, where protein digestion is initiated. The acidic chyme leaving the stomach is neutralized in the duodenum by alkaline secretions from the pancreas and liver. A battery of pancreatic enzymes attacks all classes of large food molecules.

9. The products of gastric and pancreatic digestion are reduced in the small intestine to their basic units (simple sugars, amino acids, and so on) by enzymes located in the surface membrane of the absorptive cells. Most absorption occurs in the small intestine.

10. The large intestine, or colon, is a site of much water and mineral reabsorption, and feces formation. In mammals the large intestine supports a bacterial flora that ferments undigestible food residues, largely plant material, if this is part of the diet. The fatty acid fermentation products are an important food source for many herbivorous animals.

11. Animals must include in their diet water, many minerals, and a variety of organic compounds that provide the energy and raw materials they need to maintain themselves and synthesize new protoplasm. All animals require a certain amount of energy to sustain life. Adult humans need between 1700 and 3000 kcal/day. Vitamins are small organic molecules that must be in the diet because they cannot be synthesized. The water-soluble vitamins (C and members of the B complex) are constituents of coenzymes. The fat-soluble vitamins (A, D, E, and K) perform diverse functions.

References and Selected Readings

In addition to the titles listed below, many of the general references listed at the end of Chapter 10 cover digestion and nutrition.

Davenport, H.W. Why the stomach does not digest itself. *Scientific American* 226(Jan. 1972):86–93. An exploration of reasons why pepsin in an acid medium does not normally damage the stomach walls.

Eckert, R., D. Randall, and G. Augustine. *Animal Physiology*, 3rd ed. New York: W.H. Freeman and Co., 1988. Includes an excellent comparative chapter on feeding, digestion, and absorption.

Ensminger, A.H., et al. *Food for Health: A Nutritional Encyclopedia*. Clovis, Calif.: Pegus Press, 1986. An excellent reference for the general reader on all aspects of nutrition. Topics are listed alphabetically; many are short entries, but extensive discussions of many subjects are included.

Jennings, J.B. *Feeding, Digestion, and Assimilation in Animals,* 2nd ed. New York: Pergamon Press, 1973. A brief account of all aspects of animal nutrition.

Moog, F. The lining of the small intestine. *Scientific American* 245(November 1981):154–176. An account of digestion and absorption in the small intestine of mammals.

Sanderson, S.L., and R. Wassersug. Suspension-feeding vertebrates. *Scientific American* 262(March 1990): 96–101. Suspension-feeding vertebrates reap the abundance of the plankton and grow in large numbers or to enormous size.

Stevens, C.E. *Comparative Physiology of the Vertebrate Digestive Tract.* New York: Cambridge University Press, 1988. An authoritative review of the vertebrate digestive tract; includes structure, motor activity, digestion, absorption, and regulation of secretions.

13

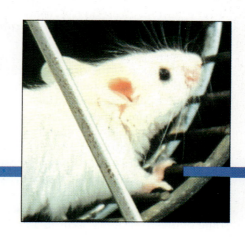

Cellular Respiration, Gas Exchange, and Metabolic Rates

Mouse running in a treadmill.

A major function of the organic nutrients consumed by animals is their use as fuels to power the many energy-demanding functions of cells. This, of course, is not their only fate. Organic nutrients are required for cellular synthesis essential for an animal's growth and maintenance, and the production of new individuals (Fig. 13.l). Indeed, an animal's nutritive budget must include enough food to supply both the demands of synthesis and the demands for energy.

Energy is required for many reactions within cells; some of these reactions occur in all cells and some, like muscle contraction, the beating of cilia, and the conduction of nerve impulses, are restricted to cells specialized for these activities. Collectively, the chemical reactions that take place within cells constitute **cellular metabolism**, and a particular series of reactions is **a metabolic pathway**. Metabolism can be divided into **anabolism**, the process by which compounds are synthesized from simple substances, and **catabolism**, the process by which compounds are degraded or broken down into simpler substances.

For all cells the principal catabolic reactions are those involved in the release of energy from organic fuels. These reactions are collectively called **cellular respiration**, since in most cells oxygen is one of the reactants and carbon dioxide is a product of fuel breakdown. The cells must therefore be supplied with oxygen and the carbon dioxide must be removed. For the organism this process involves **gas exchange** with the environment. In this chapter we will

first examine cellular respiration and then the exchange of gases that it demands. We will conclude with a discussion of metabolic rates, for which oxygen consumption is a common measure.

CELLULAR RESPIRATION

Cellular respiration is the transfer of energy in the bond structure of organic nutrients, i.e., sugars, glycerol, fatty acids, and amino acids, to adenosine triphosphate (ATP), the cell energy currency. The transferred energy within ATP is then available for energy-demanding reactions within the cell. The designation of organic nutrients as fuels is quite appropriate, for in the transfer of energy these compounds are oxidized or burned. Rapid oxidation of organic nutrients is readily apparent when a piece of bread is left too long in the toaster.

$$(C_6H_{12}O_6)_n + O_2 = CO_2 + H_2O + energy$$

This is an **oxidation-reduction** reaction because it involves a transfer of electrons (see Appendix A). In oxidation electrons are removed from the oxidized compound. In this case electrons as a part of the hydrogen atoms ($e + H^+$) are removed from the bread, which is degraded into small carbon fragments, the carbon dioxide. The electrons are added to molecular oxygen, which is thus reduced and appears to the right as water. The released energy is in the form of heat and light (if the toast ignites) and is dispersed. We say the energy is lost, because it is not available for

Figure 13.l

Fate of organic nutrients in animals.

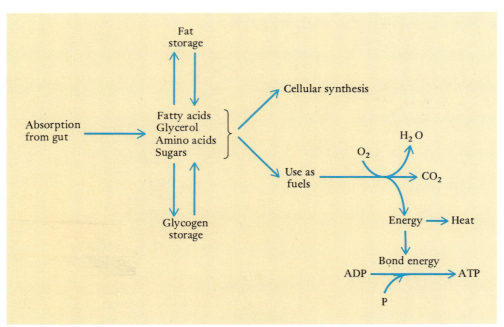

work. In a gasoline engine part of the energy is captured and drives the pistons; part is lost as heat.

The rapid oxidation described here would not be possible within a cell. The cell could not efficiently transfer a great burst of energy released at one time, and the inevitable energy lost as heat would raise the temperature too high for living systems. In **biological oxidation** the fuel molecule is slowly degraded through a long series of reactions. In the course of these reactions one molecule of the fuel glucose produces 36 molecules of ATP. Since a mole of ATP contains 7.3 kcal of energy, 262 kcal of useful energy are produced from the 686 kcal of each mole of glucose. The remaining energy is released as heat. In birds and mammals, this heat produces a constant, elevated body temperature. In other organisms, most of this heat is lost from the body surface. We will return to body heat production at the end of this chapter.

Cellular respiration can now be more precisely defined as the series of cellular reactions, i.e., the metabolic pathway, that transfers energy from organic fuels to ATP. It is termed respiration because in most cells molecular oxygen is the final acceptor of electrons in oxidation-reduction reactions.

Reactions of Cellular Respiration

It is possible to group the reactions of cellular respiration into several categories. Electrons are usually removed from the fuel along with hydrogen atoms, and reactions of this sort are called **oxidation** or **dehydrogenation**. Remember that in such reactions the compound losing the electrons, or hydrogen atoms, is oxidized and the compound receiving them is reduced. Therefore these reactions are typically **oxidation-reduction reactions**. As the fuel is degraded, carbon fragments are removed as carbon dioxide. These are **decarboxylation** reactions. Reactions involving the attachment of a phosphate group are **phosphorylations**. Finally, there are a number of reactions in cell respiration that could be called rearrangement reactions, because they involve some rearrangement of the fuel molecule preparatory to decarboxylation, dehydrogenation, or another reaction.

The starting point in cellular respiration is not the same for all fuels. Amino acids and fatty acids enter at later points along the pathway. We will therefore first examine the reactions that begin with glucose, and later consider the ways in which other fuels are utilized.

Glycolysis

The first part of cellular respiration is **glycolysis**, a series of reactions in which a molecule of glucose is initially phosphorylated, then split into two three-carbon units. Each three-carbon compound is decarboxylated to yield a two-carbon compound at the end of glycolysis. Only a very small part of the potential energy of glucose is released in glycolysis, and oxygen is not required. This phase of cell respiration occurs in most living organisms, even in those that live in habitats lacking oxygen.

The first two reactions of glycolysis are preparatory and require ATP (Fig. 13.2, steps 1 and 2). A molecule of glucose is phosphorylated to produce glucose-6-phosphate, the 6 referring to the particular carbon atom to which the phosphate group is attached. A molecule of ATP provides the phosphate group. This phosphorylation, moreover, traps the glucose inside of the cell: glucose can readily pass through the cell membrane into the cell, but its phosphorylated form cannot get out. The molecule of glucose is now converted to fructose (step 2) and then phosphorylated again by a second ATP to produce fructose-1, 6-diphosphate, a six-carbon sugar with a phosphate attached to each end (step 3). Note that these initial reactions of cellular respiration incur a cost of two ATPs.

Following these two preparatory reactions, the molecule is split into two interconvertible three-carbon compounds, one of which is called phosphoglyceraldehyde (PGAL) (step 4). Each has a phosphate group attached at the end. Each molecule of PGAL undergoes oxidative phosphorylation, i.e., hydrogen atoms (electrons) are removed and a high-energy, unstable, phosphate group is added (step 5). Inorganic phosphate is the phosphate source and not ATP. The receiver of the electrons is a coenzyme called nicotinamide adenine dinucleotide (NAD^+). When NAD^+ receives the electrons of hydrogen, its energy level increases, allowing it to act as a carrier and temporary store of energy. We will follow the fate of this stored energy shortly.

Following oxidation, each of the two rearranged molecules react with ADP, supplying a high-energy phosphate group to form ATP (step 6). The two molecules of ATP produced pay back the initial ATP debt incurred at the beginning of glycolysis. The other phosphate group is also given up to form two more molecules of ATP (steps 7–9). The remaining three-carbon compound is pyruvate.* The two residual molecules of pyruvate together contain 590 kcal/mol of the 686 kcal/mol of energy originally present in glucose. Much of the difference in the energy level of the two compounds resides in the two molecules of ATP and the two molecules of NADH.

Fermentation

Organisms that live in anaerobic environments rely entirely upon the glycolytic stage of cell respiration. This is also true of the skeletal muscle cells of vertebrates during

---

*The names pyruvate, acetate, citrate, ketoglutarate, lactate, and so on refer to the dissociated form of the organic acid. For example, $CH_3COCOOH$ is pyruvic acid and CH_3COCOO^- is pyruvate. We have used the name for the dissociated form throughout this discussion of cell respiration, because this is the form in which the compound largely occurs in living systems.

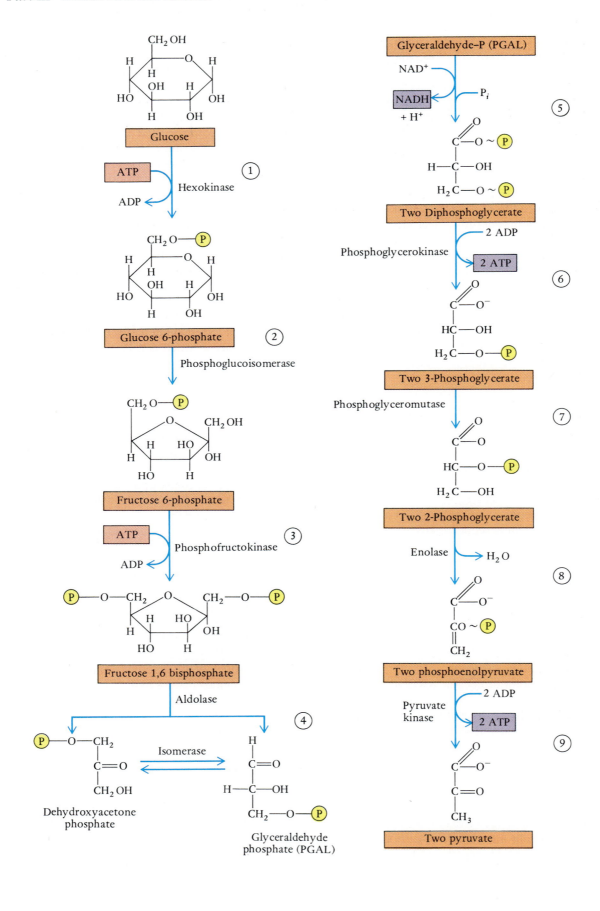

Figure 13.2

The reactions of glycolysis in which a single molecule of glucose is converted to two three-carbon molecules of pyruvate. The conversion involves phosphorylation and oxidation-reduction reactions, but no oxygen is required. Two molecules of ATP are used during the initial reactions to phosphporylate the glucose, but four molecules of ATP are produced in the last four reactions, to yield a net gain of 2 ATPs. Note that all steps require specific enzymes.

rapid muscle contraction, when the demand for oxygen outstrips the supply. In all of these cases pyruvate is reduced to some other terminal product using NADH as a source of electrons and energy for the reduction. The reduction is important because it frees the NAD$^+$ to participate again in glycolysis. Glycolysis together with the terminal reduction reactions is called **fermentation** or **anaerobic glycolysis**. In yeast, pyruvate is converted to alcohol. In many bacteria and animal cells, lactate is the terminal product (Fig. 13.3).

Although fermentation is relatively inefficient, in rapidly contracting muscle cells, lactic acid fermentation does have the advantage of producing some energy in the brief absence of oxygen. Lactate becomes a temporary store of potential fuel. Following exercise, when the demand for energy diminishes, the lactate is converted back to pyruvate. Some pyruvate is then used as a fuel for aerobic respiration, which provides sufficient ATP to convert most of the remainder back to glucose. This subsequent burning of pyruvate reflects what is often called the **oxygen debt**, i.e., the amount of oxygen needed to remove the lactate accumulated during muscle contraction.

Aerobic Respiration

For the great majority of organisms, cellular respiration does not end with glycolysis. The terminal products, i.e., the two pyruvate molecules, follow a biochemical pathway that releases the remainder and the greatest part of the potential energy of the fuel. In contrast to glycolysis, this pathway requires oxygen and is therefore referred to as **aerobic respiration**. For the sake of simplicity, we will show the reactions of only one of the pyruvate molecules,

Figure 13.3

Fermentation. In yeast, pyruvate (the end product of glycolysis) is converted to alcohol; in other organisms and in some animal muscle cells, pyruvate is converted to lactate. In both cases the conversion involves the oxidation of NADH, providing more NAD$^+$ for one of the reactions of glycolysis (see Fig. 13.2, step 5).

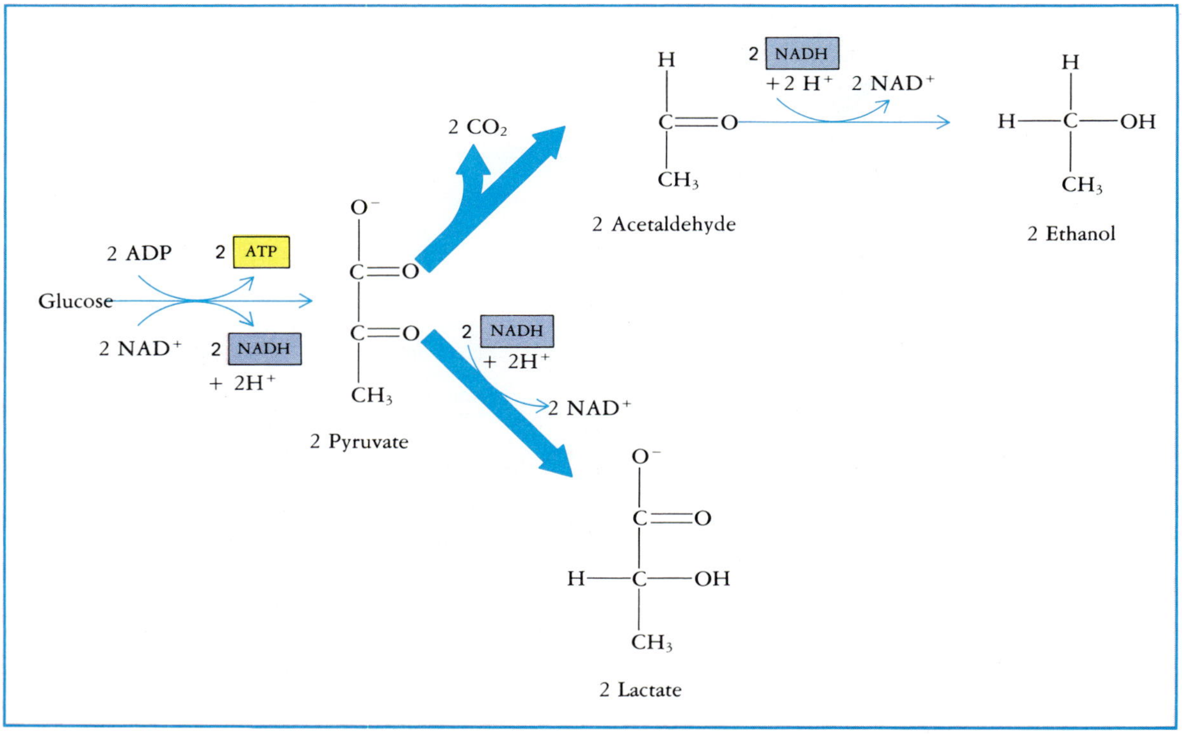

but keep in mind that two pyruvate molecules are derived from one molecule of glucose and therefore all products of aerobic respiration must be doubled.

The site of aerobic respiration is the mitochondrion (Fig. 13.4). Recall from Chapter 5 that the inner of the two membranes forming this organelle is arranged in a series of folds (cristae), which provide great surface area for the terminal reactions of respiration. The pyruvate molecule, which was produced outside, must enter the mitochondrion, crossing its membranes into the interior cavity, or matrix. It is here that the reactions resulting in the continued fuel breakdown occur.

The first step of aerobic respiration involves both oxidation and decarboxylation of the fuel. The removal of a carbon atom, which is released as a molecule of carbon dioxide, degrades the three-carbon pyruvate into a two-carbon product, acetate, which combines with a coenzyme to form acetylcoenzyme A (acetyl CoA). The electrons removed from pyruvate are passed to the carrier coenzyme NAD^+.

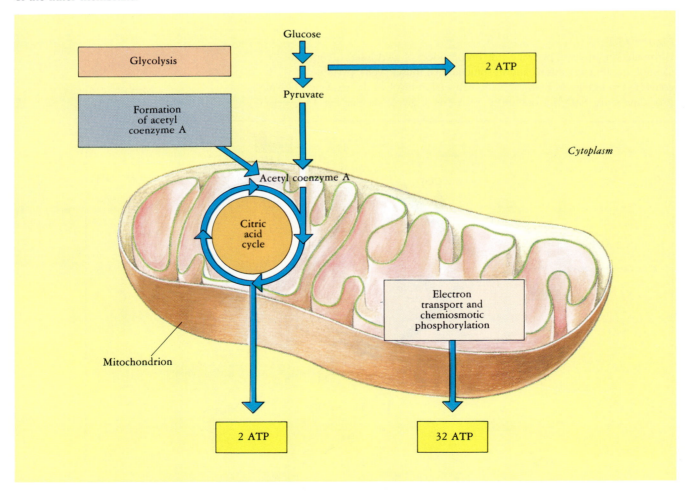

Figure 13.4

Sites of cell respiration. Glycolysis, and fermentation when it occurs, takes place out in the cytoplasmic solution. The remainder of cell respiration takes place in a mitochondrion. The pyruvate enters the mitochondrion and passes to the interior cavity, where it combines with acetyl coenzyme A to continue through the citric acid cycle. Electron transport and chemiosmotic phosphorylation occur in the folds of the inner membrane.

Citric Acid Cycle

The molecule of acetyl CoA now enters a cyclical series of reactions, called the **citric acid cycle** (also known as the tricarboxylic acid [TCA] cycle, or the Krebs cycle, after the Nobel laureate Hans Krebs, who discovered it). We will summarize it here in four steps, which will point up the most significant biochemical events (Fig. 13.5). The two-carbon acetyl CoA combines with a four-carbon ketoacid, oxaloacetate, to form the six-carbon citrate. Coenzyme A is released and thus available to pick up another molecule of acetate. An oxidation-reduction reaction and a decarboxylation reaction convert citrate to a five-carbon product, α-ketoglutarate. The electrons removed pass to the carrier NAD$^+$, and the cleaved carbon atom appears as a molecule of carbon dioxide.

α-Ketoglutarate is converted to the four-carbon succinate via further oxidation-reduction and decarboxylation reactions. Another molecule of carbon dioxide is formed and NAD$^+$ is the recipient of the electrons. There is also one molecule of ATP formed as a direct product of these reactions, i.e., it is a substrate-level production of ATP.

To complete the cycle, succinate is oxidized to oxaloacetate. The hydrogen atoms are passed to NAD$^+$ and

another carrier, flavin adenine dinucleotide (FAD). The four-carbon oxaloacetate can now react with another molecule of acetyl CoA to start the cycle over again.

With the completion of the citric acid cycle, the molecule of glucose, the fuel, has been completely degraded. One carbon atom was removed in the conversion of pyruvate to acetyl CoA, and two in the citric acid cycle, a total of three. But since two molecules of pyruvate are produced from each molecule of glucose, the total number of carbon dioxide molecules produced is six, which is the number of carbon atoms in glucose.

Where is the original energy of glucose now located? A small amount is located in the four molecules of ATP produced at the substrate level (a net production of two in glycolysis and one in the citric acid cycle, doubled). The remainder, and the greatest part, was transferred to the coenzyme carriers NADH and FADH$_2$.

Electron Transport and Chemiosmosis

The final stage of cell respiration is the harnessing of the energy in the electron carriers to produce ATP. This occurs separately from the reactions that degrade the fuel. It is here that oxygen finally comes onto the scene. Oxygen is

Figure 13.5

Summary of the citric acid cycle. Only four of the nine organic acids are shown.

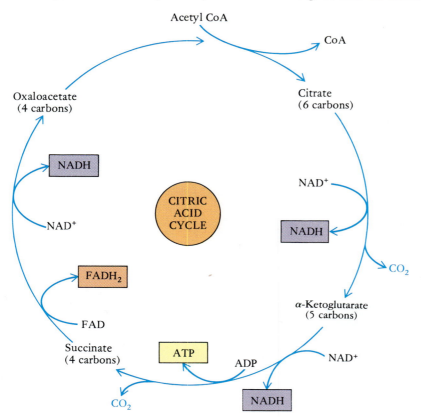

the final acceptor of the electrons held by the carriers NAD$^+$ and FAD. When oxygen is reduced (electrons added), water—one of the end products of cellular respiration—is formed.

If electrons were transferred directly from NADH to oxygen, far more energy would be released (59 kcal/mol) than could be efficiently utilized in ATP production. Instead, the electrons are passed along a series of carriers to oxygen, permitting the release of smaller increments of energy. All of the carriers, except for coenzyme Q, are proteins, although some, like the cytochromes, contain attached groups bearing a metallic atom. They are located in the inner membrane of the mitochondrion and are arranged in a series according to their affinity for electrons, i.e., the ease with which they accept electrons (Fig. 13.6). Each carrier is reduced when it receives electrons and then oxidized when it passes them on to the next carrier in line. The series is somewhat like a bucket brigade passing the

buckets downhill. The electron carriers are organized in three complexes, and there is one molecule of ATP generated for each complex. Thus for every pair of electrons that pass down the transfer series from NAD$^+$, sufficient energy is released to produce three molecules of ATP. The electrons from FAD join those from NAD$^+$ at coenzyme Q, which is partway down the series and below the site of generation of the first molecule of ATP. Therefore for each pair of electrons received from FAD, only two molecules of ATP are generated.

The mechanism by which electron flow down the carriers is coupled to, or brings about, ATP synthesis is becoming better understood. According to the **chemiosmotic theory**, the flow of electrons along the series of carriers in the inner membrane of the mitochondrion pumps protons (H$^+$) from the matrix of the mitochondrion across the membrane to the space between the inner and outer membranes. This sets up a proton gradient

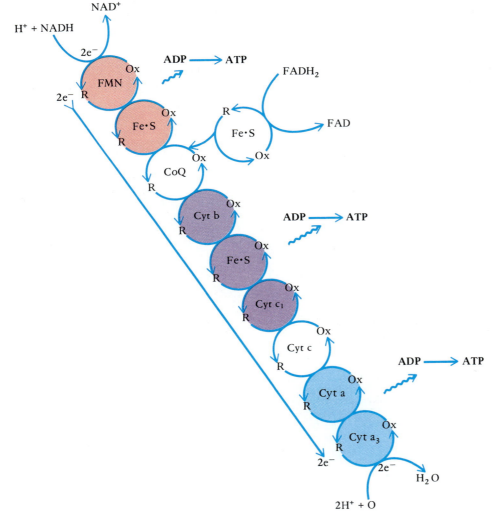

Figure 13.6

Carriers of the electron transport system. Each carrier complex is enclosed within a shaded circle. Sufficient energy is released by each complex to drive the formation of one molecule of ATP. The symbols for the carriers are: nicotinamide adenine dinucleotide (NAD$^+$); flavin mononucleotide (MN); flavin adenine dinucleotide (FAD); iron-sulfur protein (Fe·S); a quinone, coenzyme Q (CoQ); cytochromes (Cyt).

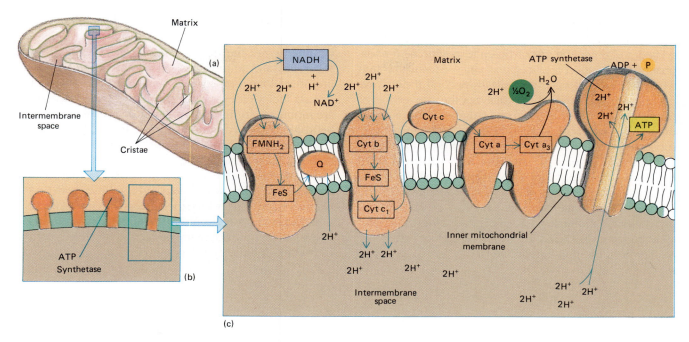

Figure 13.7

Chemiosmosis. The folded inner membrane, or cristae, of the mitochondrion, shown in (a), is the site of chemiosmosis. The formation of ATP occurs on the enzyme ATP synthetase depicted as flask-shaped molecules in (b) and (c). In the course of electron transport, the carrier complexes in the membrane (c) pump hydrogen ions into the space between the inner and outer membrane, creating a high concentration here relative to that in the matrix. ATP synthetase functions as a channel for the passage of hydrogen ions back to the matrix, and the energy in the gradient is used to drive the addition of a phosphate group to ADP, forming ATP.

that has sufficient energy to drive the phosphorylation of ADP to form ATP, a process known to occur in the inner membrane (Fig. 13.7).

ATP Yield

The complete oxidation of one molecule of glucose produces sufficient energy to synthesize 36 molecules of ATP. To account for them, we first tally the yield from one molecule of PGAL. This number can then be doubled, since two molecules of PGAL are formed from one molecule of glucose.

1 net substrate-level ATP produced in glycolysis (2 actually produced but 1 required to repay ATP used in initiating glycolysis)

1 substrate-level ATP produced in the citric acid cycle

12 ATPs derived from the 1 NADH produced in the formation of acetyl CoA from pyruvic acid and the 3 NADH produced in the citric acid cycle (3 ATPs from each)

2 ATPs from the single FADH produced in the citric acid cycle

2 ATPs from the NADH formed in glycolysis. (You would expect 3 here, but this molecule of NADH is formed outside of the mitochondrion, and it takes 1 ATP to provide the energy to move it into the site of electron transfer)

18 ATPs from each of the 2 molecules of PGAL = 36

Since each molecule of ATP contains 7.3 kcal in mole equivalents, the total yield of 36 ATPs represents about 262 kcal/mol of the original 686 kcal/mol of glucose. The difference was lost as heat. The energy transfer from glucose to ATP is therefore about 38% efficient in aerobic organisms.

Cellular Respiration Controls

Cellular respiration is regulated in a number of ways. We will describe three. First, the conversion of glycogen stores to glucose is under hormonal control and tied to cellular respiration through a series of reactions. In the course of glycogen breakdown, glucose components are phosphorylated and enter the glycolytic cycle as glucose-6-phosphate. The net effect is to mobilize glucose from glycogen when the glucose level in the cell falls. Second, the enzyme phosphofructokinase, which catalyzes the conversion of fructose 6-phosphate to fructose-l,6-diphosphate, is sensitive to ATP, citric acid, and fatty acids. A rise in any of these substances will inhibit the enzyme. On the other hand, elevated ADP levels will stimulate the enzyme. A third control is the availability of ADP in the mitochondrial membrane. Insufficient ADP to form ATP impedes the electron flow along the carriers. The backup extends into glycolysis and the citric acid cycle, because there is insufficient reduced NAD and FAD to accept electrons in the oxidation of the fuel compounds.

Utilization of Other Fuels

Glucose is not the only fuel molecule. Other forms of sugar, fats, and amino acids can also be burned. Fat is an important long-term storage compound and also an important fuel, yielding a large amount of energy. When fats are mobilized for fuel use, the triacylglycerol molecule is hydrolyzed to its component three fatty acids and a molecule of glycerol. Through a series of reactions, the three-carbon glycerol enters the glycolytic pathway at PGAL (Fig. 13.8). The fatty acids enter via acetyl CoA: two carbon units are removed from the fatty acid and combined with CoA to produce acetyl CoA, which then enters the citric acid cycle. Since a fatty acid is composed of a long chain of carbon atoms, it can produce many molecules of acetyl CoA and thus offers a high energy potential.

Figure 13.8

Diagram of metabolic pathways by which food compounds are interconverted, stored, or used as fuels. Keto acids are in yellow boxes; amino acids, in blue boxes.

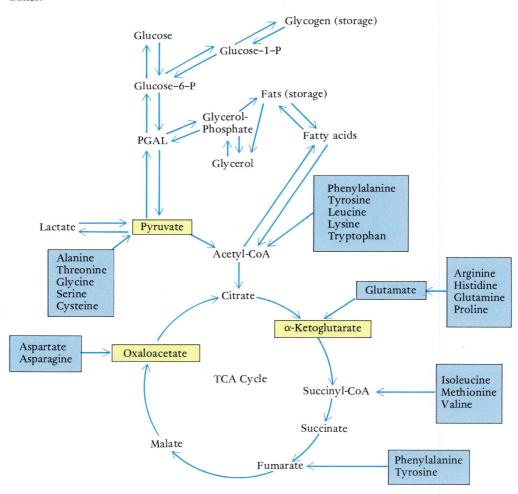

Amino acids enter the respiratory pathway at various points, depending upon the structure of the particular amino acid (Fig. 13.8). They are first deaminated, i.e., the amino group is removed, as shown below for alanine:

$$
\begin{array}{ccc}
\text{O} & & \text{O} \\
\parallel & & \parallel \\
\text{C—OH} & & \text{C—O}^- \\
| & \xrightarrow{\text{NAD}^+ \quad \text{NADH} + \text{H}^+} & \\
\text{H—C—NH}_2 + \text{H}_2\text{O} & & \text{C=O} + \text{NH}_4^+ \\
| & & | \quad \text{(waste)} \\
\text{CH}_3 & & \text{CH}_3 \\
\text{Alanine} & & \text{Pyruvate}
\end{array}
$$

The detached amino group (NH$_2$) is the principal source of nitrogenous wastes in animals (see p. 348). Deamination converts some amino acids to structurally similar keto acids, such as pyruvic, oxaloacetic, and ketoglutaric acids, and it is at these points that these particular amino acids enter the respiratory pathway (Fig. 13.8). Ordinarily, only excess amino acids in the diet are burned. However, during starvation, structural proteins, such as muscle proteins, will be metabolized.

BALANCING FOOD SUPPLY WITH CELLULAR DEMAND

Autotrophic organisms can regulate the acquisition of different food compounds by regulating their rates of synthesis. Regulation by controlling diet is much more difficult for heterotrophic organisms. Most animals feed discontinuously, which results in more food being acquired than can be utilized at one time, and the food ingested at a particular feeding may contain inadequate or just adequate amounts of some substances and excess amounts of others. These problems are overcome by storage and interconversion of the various types of food compounds.

Storage

Glycogen, the typical animal carbohydrate, commonly functions as a short-term storage product, one that is called up rapidly when available sugars in the blood or elsewhere are depleted. Glycogen is stored in most vertebrate cells but is especially abundant in liver and muscle. Significantly, all of the blood draining the intestine passes through the liver by way of the hepatic portal venous system. In the course of passage through the liver, some of the excess sugars absorbed from the intestine are taken up by the liver cells and converted to glycogen. The hormone **insulin**, produced in the pancreas, controls the cellular uptake of glucose and glycogen synthesis. A rise in blood sugar stimulates the pancreatic cells to produce insulin. Insulin is transported via the blood stream throughout the body, where it induces glycogen synthesis in muscle cells and in the liver (p. 448). The reverse reaction, the conversion of glycogen to glucose, is regulated by another pancreatic hormone, **glucagon**, and by the adrenal hormone

epinephrine. However, muscle cells lack the enzyme to convert glucose-6-phosphate to glucose, so that muscle glycogen can only be used as an energy store for muscle cells.

Fat serves as a long-term storage product in many animals. Within two or three hours after absorbing a fatty meal, the chylomicrons disappear from the blood of vertebrates. They are digested within the capillaries by lipoprotein lipase, which is produced in great quantities by the liver and fat cells of the body. It is believed that most of the hydrolyzed fat is quickly absorbed and resynthesized by these tissues. Fats stored in the liver or adipose tissue undergo constant breakdown and resynthesis, even though the total amount stored may change very little over long periods of time.

Proteins are not usually stored in animals, but there is continual turnover of tissue proteins; i.e., proteins are cleaved to amino acids, which may be transported elsewhere, and new proteins are constructed. In humans the protein turnover amounts to about 400 g/day, which is about four times the usual daily protein intake in the United States. During prolonged starvation many animals will utilize their tissue proteins to support maintenance metabolism, sacrificing muscles, gonads, and other organs.

Interconversion of Food Compounds

The metabolic pathways of cellular respiration should not be considered as only catabolic. They also have important anabolic functions. The glycolytic pathway and the fatty acid pathway have counterparts working in the opposite direction. For instance, in most animals the pathway from pyruvate to acetyl CoA is not reversible, but there is an alternate exergonic route that will convert acetyl CoA to pyruvate. These reverse pathways are of great importance for interconversion of food compounds; carbohydrates can be converted into glycerol and fatty acids and stored as fats, and fats can be converted to sugars. The organic substrates of the citric acid cycle can also be important points of synthesis. For example, amino acids can be converted to keto acids and keto acids to amino acids by moving the amino group from one to the other as shown in the **transamination reaction** below:

$$
\begin{array}{ccccc}
\text{O} & & & \text{O} & \\
\parallel & & & \parallel & \\
\text{C—OH} & & & \text{C—OH} & \\
| & & & | & \\
\text{H—C—H} & & \text{O} & \text{H—C—H} & \text{O} \\
| & & \parallel & | & \parallel \\
\text{H—C—H} + & \text{C—OH} & \rightleftharpoons & \text{H—C—H} + & \text{C—OH} \\
| & | & & | & | \\
\text{C=}\textcircled{\text{O}} & \text{H—C—}\textcircled{\text{NH}_2} & & \text{H—C—NH}_2 & \text{C=O} \\
| & | & & | & | \\
\text{C—OH} & \text{CH}_3 & & \text{C—OH} & \text{CH}_3 \\
\parallel & \text{Alanine} & & \parallel & \text{Pyruvic acid} \\
\text{O} & & & \text{O} & \\
\text{Ketoglutaric acid} & & & \text{Glutamic acid} &
\end{array}
$$

There are limits to interconversion of food compounds. As noted in Chapter 12, certain amino acids and fatty acids cannot be synthesized and must be obtained in the diet.

Note that in many of these metabolic pathways, pyruvate is at a sort of crossroads. It is the point of exit and entrance for the fermentation product lactate, and some amino acids enter the citric acid cycle here. Pyruvate connects glycolysis to aerobic respiration and is an important interconversion pathway between sugars and fats.

GAS EXCHANGE

In order to sustain cellular respiration, the cells of aerobic organisms must be continually supplied with oxygen, the final acceptor of electrons in electron transport. The carbon dioxide produced in the degrading of the fuel substrate (decarboxylation reactions) must be removed. The amount of oxygen available is quite different in aquatic and terrestrial environments, and animals also differ in their metabolic needs for oxygen. Many factors affect the type and location of body surfaces where gas exchange occurs and the ways water or air are moved across them. Gases move between the gas exchange surfaces and the cells and tissues of the body by some combination of diffusion and bulk flow (Fig. 13.9).

Availability of Oxygen

The availability of oxygen varies in different habitats. Dry air contains about 21% oxygen and 0.04% carbon dioxide; the remainder is nitrogen with traces of other gases. The availability of a gas is expressed as a **partial pressure** (P), that is, the pressure of that gas in a mixture of gases. This pressure is calculated by multiplying the total pressure of the mixture of gases by the percentage of the particular gas in the mixture. At sea level the weight, or pressure, of the dry air of the atmospheric envelope is sufficient to raise a column of mercury 760 mm and is 21% oxygen; hence the partial pressure of oxygen (P_{O_2}) at sea level is $760 \times 0.21 = 159.6$ mm Hg.* The partial pressure of oxygen is lower at high altitudes, where the total atmospheric pressure is reduced. At 6000 m (19,685 feet), the partial pressure of oxygen is reduced to 80 mm Hg, although the air at 6000 m is still 21% oxygen. The partial pressure of oxygen is also decreased when water vapor is a gaseous component of air, as it usually is. In human lungs, for example, where the air is saturated with water vapor, the partial pressure of oxygen is reduced to 104 mm Hg.

If a pan of water is exposed to air, oxygen from the air will dissolve into the water until the number of molecules of oxygen leaving equals the number entering, i.e., the air and water have reached equilibrium. The partial pressure of oxygen determines the availability of oxygen because it affects the amount that will go into solution. Oxygen must be in solution before it can be used by an organism or its cells. Aquatic organisms derive their oxygen from oxygen dissolved in water. In terrestrial animals, oxygen first dissolves in the moist surfaces of the body or respiratory organ and then enters the blood or body fluids. The higher the partial pressure of oxygen in air, the more oxygen will go into solution and be available to the cells, i.e., the amount of gas in solution is proportional to the partial pressure of the gas in the air with which it is in equilibrium. We can therefore speak of a gas in solution as having a partial pressure as well. However, it is important to realize that even when the rate of gas movement between air and water is in equilibrium, there is much more oxygen present in air than is dissolved in water. For example, at 20°C air at sea level contains about 210 ml of oxygen per liter, but sea water and fresh water contain only about 5 ml and 6.5 ml per liter, respectively. The quantity of a gas in solution is affected not only by partial pressure but also by the solubility of the particular gas and by factors that affect solubility, such as temperature and salinity (Table 13.1). Note that the concentration of oxygen in the environment can be expressed in terms of partial pressure or in ml per liter.

In some animals, known as **oxygen conformers**, the level of metabolism and oxygen consumption vary according to the availability of oxygen. In the annelid worm *Glycera*, for example, metabolism and oxygen consumption gradually fall as the water in its burrow becomes stagnant during low tide. As the tide comes in and the water is reventilated, oxygen consumption goes up. In the ab-

---

*We have stated pressure in the familiar unit of mm Hg, but scientific research now uses the standard international unit of kilopascal: 1 mm Hg = 0.1333 kPa (Appendix B).

Figure 13.9
A generalized aquatic gas exchange surface, showing the steps in the movement of gases.

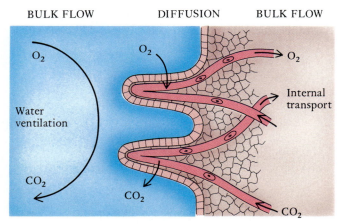

BULK FLOW   DIFFUSION   BULK FLOW

O_2

O_2

O_2

Water ventilation

Internal transport

CO_2

CO_2

CO_2

Table 13.1
Amount of Oxygen Dissolved in Water at Sea Level*

Temperature °C	Fresh Water	Sea Water
0	10.29	7.97
20	6.59	5.31

*Oxygen content is expressed in milliliters of oxygen per liter of water.

sence of oxygen some such animals depend solely on anaerobic glycolysis. Most animals are **oxygen regulators**. Their oxygen consumption, although varying with such factors as temperature, body size, and level of activity, is independent of the availability of environmental oxygen unless the oxygen level drops below a certain partial pressure known as the critical level. Below this, the animal's oxygen consumption necessarily declines rapidly, and death may result.

Steps in Gas Exchange

To obtain oxygen and eliminate carbon dioxide, an animal must have a **respiratory membrane**, a moist, thin, and permeable surface exposed to the environment, through which gases can move. The membrane may simply be the body surface, but often it is limited to parts of a **respiratory organ**, where the surface area has been increased and specialized for gas exchange (Fig. 13.9). There must also be a way of getting gases to and from the respiratory membrane, i.e., of ventilating its surface, and a way of transporting gases between the membrane and the cells of the body. The location of the respiratory membrane and modes of ventilation and transport vary greatly among animals, depending on their size, oxygen needs, and the environment in which they live.

Passage of gases across the surface of the respiratory membrane and in and out of the cells of the body is always by diffusion (Fig. 13.9). If the gases are not already dissolved in water, they will go into solution on the moist surface of the respiratory membrane and pass through it, following their concentration gradients. Since oxygen is being by the cells, there is always less in the cells and in the body than in the external medium—water or air—in which the animal is living. Conversely, since cells produce carbon dioxide, there is always a greater concentration of carbon dioxide inside the cells and in the body than there is in the surrounding medium.

Ventilation of the surface of the membrane and transport of gases within the body can also take place by diffusion if the animal is small enough. Since diffusion is a slow process (its rate is inversely proportional to dis-

tance), cells must be within 0.5 to 1.0 mm of an oxygen source, unless their metabolic rate is exceptionally low. Diffusion is inadequate in large and active animals, who need some system of **bulk flow**, whereby gases are carried passively in a medium (air or water) moving under pressure, both to ventilate the surface of the respiratory membrane and to transport respiratory gases in the body. We will first examine the location and nature of respiratory membranes and the ways they are ventilated, and then consider the transport of gases in the body.

Environmental Gas Exchange in Aquatic Animals

Most small aquatic organisms, such as protozoans, hydras, flatworms, and rotifers, have no special respiratory organs. Given their small size and very favorable ratio of surface area to volume, the only respiratory membrane that they need is their body surface, in some cases supplemented by the lining of the digestive tract or other body spaces. In general, animals several millimeters or less in length lack special gas exchange organs, but shape is an important factor. If the body were spherical, a 1 mm diameter might be the upper limit for adequate surface area for gas exchange. However, most animals are not spheres. Flatworms have no gas exchange organs and may reach several centimeters or more in length. This is possible because the body is very flat, greatly increasing the surface area to volume ratio and allowing all tissues to lie close to the surface epithelium.

As oxygen diffuses into the body and carbon dioxide diffuses out, the oxygen is rapidly depleted in the boundary layer of water next to the body surface, and carbon dioxide rapidly accumulates. If an animal is stationary and not living in a habitat where the boundary layer is replaced by moving currents, then ciliated cells (or appendages) move water across the organism's surface or through its cavities, ensuring a continual resupply of oxygen and removal of CO_2.

Gills

Larger and more active aquatic animals have respiratory organs, usually gills, to which the respiratory membranes are confined or that supplement other surfaces. **Gills** are organs composed of many delicate lamellae (plates), or filaments, that extend outward from some part of the body. They are usually modifications of the integument. Their delicate nature is possible in an aquatic environment because the density of water provides adequate support. Gills would tend to dry out, collapse, and clump together in air. Although primarily for gas exchange, gills may also be used for other purposes, including filter feeding, excretion, and ion exchange.

Gills must have a large surface area, and a large volume of water must be moved across them because oxygen diffuses so slowly in water. The weight of the water that must be moved is 100,000 times the weight of the available oxygen, whereas in air the weight of the inert medium, mostly nitrogen, is only 3.5 times the weight of the available oxygen. Thus ventilation in water is energetically much more costly than in air.

Usually water is moved in one direction across the gills by ambient water currents, the action of cilia, the beating of special appendages, the locomotion of the animal, or a muscular pump that carries water through a gill chamber. Gases diffuse rapidly between the water crossing the gills and body fluids circulating through them. The cuticle, if one is present, and the epithelial layers separating the water from the body fluids are very thin. The effectiveness of the gills is increased in some animals, such as molluscs and fishes, by the evolution of a **countercurrent exchange mechanism** in which water flows across the gills in a direction opposite to the flow of blood through them. This ensures that throughout the opposing streams the partial pressure of oxygen of the water will exceed that of the blood, thereby permitting the maximum extraction of oxygen by the gill (Fig. 13.10). Blood leaving a gill has become nearly fully saturated with oxygen. It is exposed to water just starting across it and flowing in the opposite direction. This water is fully saturated with oxygen, hence oxygen diffuses into the blood. At the other end of a gill, the water has given up most of its oxygen, but it is exposed to blood just entering the gill that contains even less oxygen. Countercurrent exchange mechanisms are extraordinarily efficient—as much as

90% of the oxygen dissolved in the water may be removed compared with the approximately 25% of oxygen that mammals extract from the air in their lungs.

Invertebrate Gills   Various sorts of gills and gill-like structures, some simple, others quite elaborate, are found among invertebrates, and they have evolved independently many times, even within the same class of animals. Polychaete annelid worms have paired lateral projections, the parapodia, that are used in locomotion, but parts may also be modified as gills (Fig. 30.3). Molluscan gills are composed of many flattened lamellae arranged along one or both sides of a central axis (Fig. 29.1). Most aquatic arthropods have gills that are modified appendages or outgrowths closely associated with appendages. In crayfishes and crabs, the gills arise from the base of the appendages and adjacent body wall and extend up into a **branchial chamber** (L. *branchia*, gills) that lies on each side of the animal between the body wall and a part of the exoskeleton (carapace) that has grown laterally and ventrally over the gills. Each gill consists of a central axis to which are attached either lamellae or filaments, depending on the species (Fig. 13.11). The ventilating current is produced by the rapid sculling action of the **gill bailer**, a semilunar-shaped process of one of the oral appendages (the second maxilla). The gill bailers drive two exhalant water streams out of the branchial chambers, one to each side of the mouth. Water enters the branchial chambers of the crayfish at the back of the carapace and between the legs. In crabs, the carapace is tightly sealed along its ventral

Figure 13.10
Theoretical diagrams to illustrate the effects of countercurrent and parallel flows on the exchange of oxygen between water and blood. The equilibrium attained in parallel flow would be somewhere above 50% saturation of the blood because of the presence of hemoglobin, yet the blood becomes less fully saturated than it would during countercurrent flow.

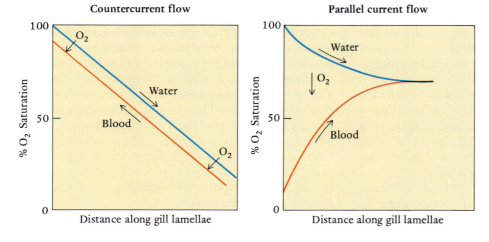

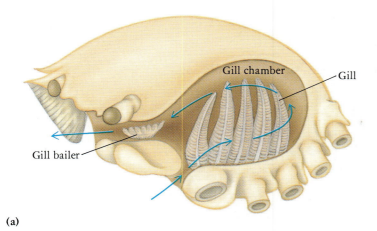

(a)

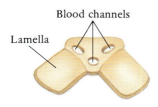

(b)

Figure 13.11
Gills of a crab. (a) Lateral view of gills within gill chamber.
Ventilation stream enters gill chamber at base of claw and
exits to either side of the mouth region, driven by the
beating of the two gill bailers. (b) Cross section of a gill.

margin, and the inhalant apertures are at the base of the
claws (Fig. 13.11). Channels in each gill axis carry blood to
and from spaces in the lamellae; gas exchange occurs
across the thin walls of the lamellae.

In many animals with gills, the general body surface
may also contribute to the exchange of gases. Blocking the
gill circulation of some clams, for example, results in a
reduction of oxygen intake by only 15% if the water is well
aerated. This is probably true of many marine annelid
worms as well.

Vertebrate Gills   Fishes and many amphibians exchange
gases with the environment by means of gills. Some fish
larvae and larval amphibians have **external gills**, filamentous processes that extend outward from the side of the
head near the openings of the gill slits. Adult fishes have
internal gills. In most species the gills are covered by a fold
of the body, the **operculum**, (L., lid) that forms an
opercular chamber lateral to them. The gills themselves
consist of lamellae attached to a branchial arch at the base
of each gill (Fig. 13.12a). Secondary folds perpendicular to
the lamellae contain capillary beds, and it is here that gas

exchange with the water occurs. Exchange is particularly
effective because of a countercurrent flow between blood
and water (Fig. 13.12b). As you would expect, active fish
have larger gill areas than slower-moving ones. The gill
area per gram of body weight of the fast-swimming mackerel is 50 times that of the goosefish, which spends most of
its time lying on the bottom waiting for food to come to it.

A pumping action of the mouth and pharynx causes a
ventilating current of water to flow across the gills (Fig.
13.12c and d). During inspiration, both the pharynx and
opercular chambers expand, causing a decrease in pressure within them relative to the surrounding water. These
chambers are acting as suction pumps. Water is drawn in
only through the mouth, for a thin membrane on the free
edge of the operculum acts as a valve, preventing entry in
this direction. Once in the pharynx, the water passes across
the gills into the opercular chamber as a result of the lower
pressure there. Expiration begins with the closure of the
mouth, or of oral valves, and a contraction of the pharynx
and opercular chamber, so that the pressure in them
exceeds that of the surrounding water. These chambers
are now acting as force pumps. The membrane at the edge

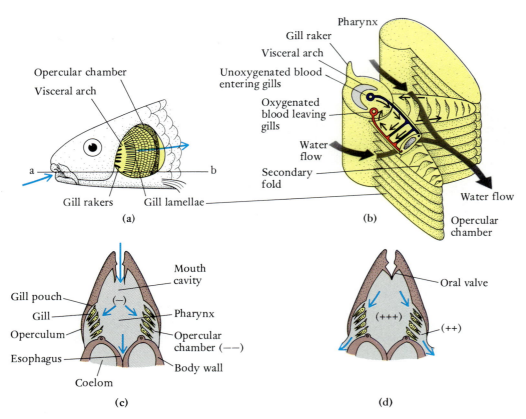

Figure 13.12
Gills of fishes. (a) The operculum has been cut away to show the gills in the opercular chamber. (b) An enlargement of a portion of one gill. (c) and (d) Frontal sections through the mouth and pharynx in the plane of line a–b in (a). Inspiration occurs in (c); expiration, in (d). Relative water pressures in the various parts of the system are shown by + and −.

of the operculum is pushed open, and water is discharged. During both inspiration and expiration there is a pressure gradient moving water across the gills. Water does not enter the esophagus, for this is collapsed except when swallowing occurs. Food and other particles are prevented from clogging the gills by gill rakers that act as strainers.

Some fishes that live in the open ocean keep their mouths open and use their forward motion to drive water across the gills, a process called **ram ventilation**. When a mackerel's speed exceeds 0.4 m per second, pumping movements of the operculum slow down. When speed exceeds 0.6 m per second, the operculum stops moving and the fish depends on ram ventilation.

Environmental Gas Exchange in Terrestrial Animals

A particular advantage of terrestrial life is that oxygen is proportionately far more abundant in air than in water. Terrestrial life, however, also presents certain problems: body water can be lost easily through any exposed surface that is moist, thin, permeable, and vascular enough to serve as a respiratory membrane. Most terrestrial animals have adapted to these conflicting conditions by evolving respiratory organs, either trachea or lungs, in which the respiratory membranes lie deep within the body in spaces with a relative humidity close to 100 % (saturation at the animal's body temperature). Moreover, because the oxygen content of air is so high, only a little of this air need be exchanged with each breath. All of these factors minimize water loss. No more air is moved in and out of these spaces than is needed to meet the animal's metabolic needs at a particular time and circumstance. The rate of ventilation is more carefully regulated than in animals utilizing gill ventilation. Air is much less dense than water and can be moved with little energy expenditure. In some animals diffusion is adequate. Also, oxygen diffuses 300,000 times faster in air than in water, so that some animals rely largely on diffusion to supply the exchange surface rather than bulk flow.

A few terrestrial animals whose metabolism is not high and who live in damp habitats, such as earthworms and certain salamanders, can exchange all or some of their gases through the general body surface and have no special respiratory organs. Gills can be used for gas exchange in air, provided that they lie in chambers in which a high humidity can be maintained or that the animals live in damp habitats. They are usually confined to animals, such as crabs and pillbugs, that live in damp places and whose exoskeletons can provide support for the gills, preventing their collapse.

Tracheal Systems

Tracheal systems are the most common gas exchange organs of terrestrial arthropods. Most body segments of insects have paired lateral apertures, the **spiracles** (L. *spiraculum*, air hole), that lead into a system of **tracheal tubules** (L. *trachea*, windpipe) (Fig. 13.13a,b). Spiracular filtering devices prevent small particles from clogging the system. Valves can control the degree to which the spiracles are open as the animal's metabolic needs vary. The tracheae have a more or less ladder-like pattern with interconnecting transverse and longitudinal trunks. They terminate in minute tubules, the **tracheoles**, that are generally less than 1 micrometer (μm) in diameter (Fig. 13.13c). The tracheoles are formed by special tracheole cells. They contain the respiratory membrane, and they permeate all of the tissues of the body. A small amount of liquid is present in the ends of the tracheoles, and gases dissolve in this liquid. As metabolites increase in active tissues, the osmotic pressure of the tissues increases, some of the fluid leaves the tracheoles, and more respiratory

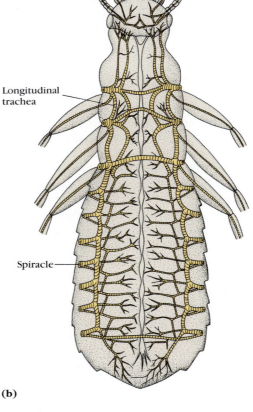

Longitudinal trachea

Spiracle

(b)

Figure 13.13

(a) A spiracle with filtering apparatus, atrium, and valve. (b) A tracheal system of an insect. (c) Diagram showing relationship of tracheoles with muscle. Fluid in tracheole tips is colored blue. (d) An air sac.

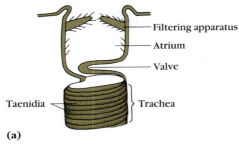

Filtering apparatus

Atrium

Valve

Taenidia

Trachea

(a)

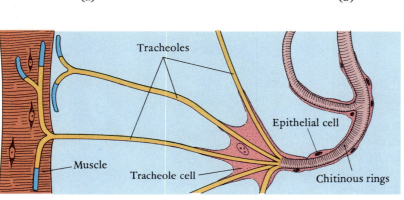

Tracheoles

Epithelial cell

Muscle

Tracheole cell

Chitinous rings

(c)

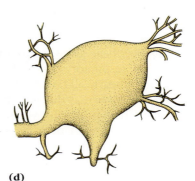

(d)

surface is exposed. In insect flight muscles, where oxygen needs are very high, tracheoles penetrate the muscle cells and frequently lie within 0.07 μm of the mitochondria, where aerobic respiration is taking place. A chitinous cuticle lines the entire system, but only the cuticle in the tracheae is shed during molting.

Cuticular rings, **taenidia** (L. *taenia*, ribbon), strengthen and hold the tracheae open (Fig. 13.13a). In small insects, diffusion is the only mechanism needed to exchange gases because it occurs so rapidly through the air-filled tubules. Larger and more active insects have a tracheal system that includes **air sacs** that can be compressed and expanded by the action of body muscles as the animal is moving (Fig. 13.13d). The opening and closing of spiracles is carefully regulated to permit adequate gas exchange while still preventing water loss. A synchronous pattern of the opening and closing of successive spiracles and coordinated compression-expansion cycles of the air sacs keeps a unidirectional flow of air moving through the larger tracheae.

A few insects have readapted to the aquatic environment and usually come to the surface to get air. Diving beetles have a dense mat of hairlike cuticular processes that are nonwettable and entrap a permanent bubble of air. The spiracles open only into this bubble, and gases diffuse between it and the water. The air bubble acts as a gill!

Note that the tracheal system delivers oxygen directly to the tissues, and blood is not important for gas transport. There are some exceptions, but this is true of insects, which comprise the greatest number of arthropods. Given the large role that diffusion plays, tracheal systems are suitable only for small animals: insects, centipedes, millipedes, and small spiders. However, they have the advantage of greatly reducing water loss by the very small openings into the system. Large spiders have lungs, or a combination of lungs and tracheae, and gases are transported by blood. Given the small size of some arthropods, such as many mites, which may be less than 0.5 mm in length, one might wonder why they need any special system of gas exchange. Insects, spiders, and mites all have a highly developed waxy layer on the outside of the exoskeleton that is very effective in reducing water loss. Since this layer also reduces the passage of gases, an internal gas exchange system like tracheae is necessary.

Diffusion Lungs

Lungs, which like tracheae are invaginations from some body surface, are the respiratory organs of many terrestrial invertebrates and of nearly all terrestrial vertebrates. Invertebrate lungs are known as **diffusion lungs**, for gases move in and out of them primarily by diffusion, although some body movements often help in ventilation. One group of molluscs, the pulmonate snails, have converted the mantle cavity, which contains the gill in their aquatic ancestors, into a lung, and scorpions and the larger spiders

have book lungs that have evolved from invaginations of the abdomen. Each book lung opens to the surface by a spiracle and contains many leaflike lamellae held apart by bars that enable air to circulate freely (Fig. 13.14).

Ventilation Lungs of Vertebrates

The lungs of vertebrates are called **ventilation lungs** because air is moved in and out of them by a pumping action of adjacent body parts. Secretions of phospholipids function as surfactants and reduce the surface tension of the film of liquid inside the lungs and hence the amount of energy needed to expand and fill them. Lungs usually develop as a bilobed outgrowth from the floor of the pharynx (Fig. 36.15), extend posteriorly, grow around the digestive tract, and come to lie in the dorsal part of the pleuroperitoneal cavity. This cavity is the part of the vertebrate coelom that contains the lungs and digestive organs; the pericardial cavity is the part that surrounds the heart.

Lungs probably evolved as an accessory air-breathing respiratory organ in swamp-dwelling fishes that lived in

Figure 13.14
The book lung of a spider is a type of diffusion lung.
(a) Position of one member of the pair of book lungs within the abdomen of a spider. (b) Diagrammatic section through a book lung showing diffusion of gases or air flow between lamellae.

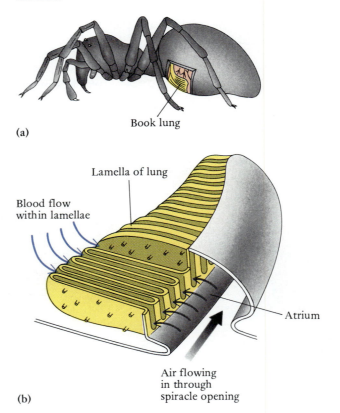

Book lung

(a)

Lamella of lung

Blood flow within lamellae

Atrium

Air flowing in through spiracle opening

(b)

ponds where the oxygen tension was low or that dried up occasionally. For example, African and South American lungfish (Fig. 36.25), although they have gills, obtain most of their oxygen through their lungs. In most fishes descended from ancestral lunged fish, the lungs have lost their connection to the esophagus and have been converted to a swim bladder, a primarily hydrostatic organ used to maintain buoyancy (see p. 207).

Amphibians evolved from primitive fishes with lungs. The aquatic larvae of most amphibians exchange gases through gills, but lungs develop and begin to function in late larvae that surface to gulp air. The lungs are sac-like, and their internal surface area is increased somewhat by pocket-like folds (Fig. 13.15a), but the surface area is far smaller than in terrestrial vertebrates such as mammals. The lungs are ventilated by up and down movements of the pharynx floor that first draw air through the nostrils into the pharynx and then force it into the lungs (Fig. 13.15b–e). Thus, the lungs are ventilated by a **force pump**, rather than by pressure changes that suck air into them, as in mammals. The quantity of air pumped into the lungs correlates with the size of the mouth cavity and pharynx,

which are large because of the width of the head relative to the rest of the body. After the lungs are filled, the glottis is closed, and a prolonged period without breathing (**apnea**) (Gr. *a*, without + *pnoe*, breathing) follows in which the lungs are not ventilated, and the oxygen in them is gradually consumed. The elastic recoil of the lungs is primarily responsible for expelling air, once the glottis reopens (Fig. 13.15c). Contemporary amphibians supplement pulmonary respiration by gas exchange through their thin, moist, and highly vascularized skin. Indeed, this is the primary method of gas exchange in many amphibians on land and in all, except gilled species, when they are under water. Most of the carbon dioxide is eliminated through the skin because the rate of lung ventilation is not sufficient to flush it out. Enough oxygen also enters through the skin to meet the animals' needs except during periods of increased activity in the spring and summer. Cutaneous respiration places a limit on the maximum size of amphibians; large organisms have less body surface relative to body volume mass than do small ones. Birds and mammals have evolved lungs with greater internal surface areas and more complex and efficient mechanisms for ventilating the lungs to

Figure 13.15

(a) A diagrammatic longitudinal section of the respiratory system of a frog. (b)–(e) The ventilation of a frog's lungs. In (b), the glottis remains closed and a lowering of the floor of the mouth draws air into a pocket in the ventral part of the mouth and pharyngeal cavity. In (c), the glottis opens, and the elastic recoil of the lungs drives stale air out over the fresh air being held in the ventral part of the mouth and pharynx. In (d), the raising of the floor of the mouth and pharynx pumps fresh air into the lungs. In (e), the pumping action continues until lungs are filled. The glottis closes and air is held in the lungs under a greater than atmospheric pressure.

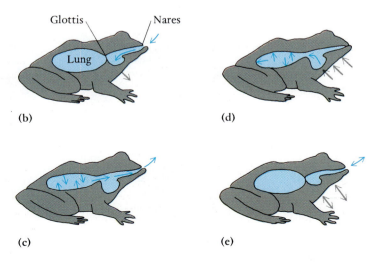

(a)

Pharynx
Eustachian tube
Esophagus
External nostril
Internal nostril
Tongue
Lung
Glottis
Bronchus
Stomach

Glottis
Nares
Lung

(b)

(c)

(d)

(e)

support their greater size and metabolic demands. They have thus been able to dispense with cutaneous respiration and the constraints that it imposes.

Mammalian Respiratory System

In mammals (Fig. 13.16), the air is drawn into the paired **nasal cavities** through the **external nostrils** or **nares**. These cavities are separated from the mouth cavity by a hard palate, allowing the animal to breathe while food is in its mouth. The surface area of the cavities is increased by a series of ridges known as **conchae** (Gr. *konche*, shell), and the nasal mucosa is vascular and ciliated and contains many mucous glands. In the nasal cavities the air is warmed and moistened, and minute foreign particles are entrapped in a sheet of mucus that is carried by ciliary action into the pharynx, where it is swallowed or expectorated. Inspired air is moistened in primitive terrestrial vertebrates, such as the frog, but cold-blooded tetrapods do not need as much conditioning of the inhalant air as do birds and mammals.

Air continues through the **internal nostrils**, or **choanae**, passes through the pharynx and enters the **larynx**, which is open except when food is swallowed. The

raising of the larynx during swallowing can be demonstrated by placing your hand on the Adam's apple, the external protrusion of the larynx. The epiglottis flips back over the entrance of the larynx when it is raised.

The larynx is composed of cartilages derived from certain of the visceral arches and serves both to guard the entrance to the windpipe, or **trachea**, and to house the **vocal cords** (Fig. 13.17). The vocal cords are a pair of folds in the lateral walls of the larynx. They can be brought closer together, or be moved apart, by the pivoting of laryngeal cartilages connected to their dorsal ends. The **glottis** of mammals is the opening between the vocal cords. When the cords are close together, air expelled from the lungs vibrates them, and they in turn vibrate the column of air in the upper respiratory passages, just as the reed in an organ pipe vibrates the air in the pipe. This phenomenon is called **phonation**, or the production of a vocal sound. Speech is the shaping of the vocal sounds into patterns that have meaning for us; it is accomplished by the pharynx, mouth, tongue, and lips. Muscle fibers in the vocal cords and larynx control the tension of the cords and the pitch of our voice. We normally speak when air is

Figure 13.16
The human respiratory system.

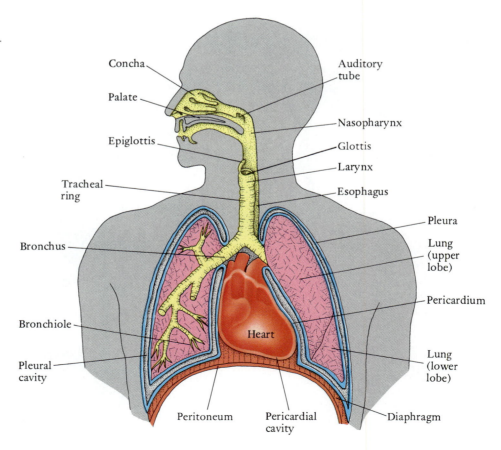

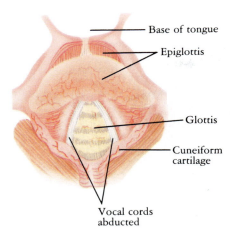

- Base of tongue
- Epiglottis
- Glottis
- Cuneiform cartilage

Vocal cords abducted

Figure 13.17
A view of the normal position of the vocal cords when looking into the larynx from above.

expired. Ventriloquists train themselves to speak during inspiration.

The trachea extends down the neck and finally divides into **bronchi** (Gr. *bronkhos*, windpipe) that lead to the pair of lungs. Unlike the esophagus, which is collapsed except when a ball of food is passing through, the trachea is held open by **C**-shaped cartilaginous rings, and air can move freely through it. Its mucosa continues to condition the air.

The lungs of amphibians lie in the anterodorsal part of the pleuroperitoneal cavity (Fig. 13.15a). This cavity has become divided in mammals. The pleural cavities lie within the chest, or thorax, and are separated from the peritoneal cavity by a muscular **diaphragm**. A coelomic epithelium, the **pleura**, lines the pleural cavities and covers the lungs.

The bronchi branch profusely within the lungs, and the walls of the respiratory passages become progressively thinner (Fig. 13.16). The tracheal and bronchial epithelium is ciliated (Fig. 13.18), and foreign material trapped in mucus is swept upward to the pharynx, where it is swallowed. The smaller passageways are called **bronchioles**. Each bronchiole eventually terminates in an alveolar sac

Focus 13.1   *Adaptations for Diving*

The ability to dive beneath the surface of fresh water or the sea has evolved a number of times in different birds and mammals. The duration and depth of the dives in many species are remarkable. The Weddell seal can dive to depths of 500 meters and can remain submerged for up to 70 minutes. However, most feeding dives last for only about 15 minutes. Sperm whales can remain submerged for 1½ hours. On the other hand, the little water shrew, with its high metabolic rate, dives for only about 30 seconds. Most birds also have short dives in shallow water, but the emperor penguin can dive to depths of over 200 meters.

Adaptations for diving involve a combination of respiratory and cardiovascular responses that are strikingly similar in different groups of vertebrates. In contrast to humans, seals and many other mammals *exhale* before diving, and at deep levels the lungs collapse, forcing air out of the alveoli and into the bronchi and trachea, where gas exchange does not occur. This has the advantage of reducing the danger of the bends, a condition that results from the dissolving of

nitrogen gas in the blood under the pressure of depth. If a human scuba diver returns to the surface too rapidly after a long, deep dive, the nitrogen gas in the diver's lungs comes out of solution as bubbles in the blood, causing pain, dizziness, paralysis, unconsciousness, and even death. The blood volume of diving mammals is commonly larger than that of nondiving species, enabling a large amount of oxygen to be stored in the hemoglobin. For example, the Weddell seal has a blood volume of 14% of total tissue volume compared to 7% in the human. There is also a large amount of oxygen stored in muscle myoglobin. A most important adaptation of diving mammals is oxygen conservation from cardiovascular responses. The heart beat decreases markedly and the distribution of blood is restricted largely to the heart and brain. The kidneys, for example, are virtually shut down. The skeletal muscles rely on stored oxygen and then anaerobic respiration (lactic acid fermentation). The metabolic rate decreases. A large oxygen debt is paid on resurfacing.

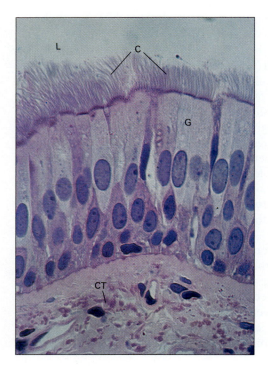

Figure 13.18
The ciliated epithelium lining a bronchus. L indicates lumen; C, cilia; G, gland; CT, connective tissue.

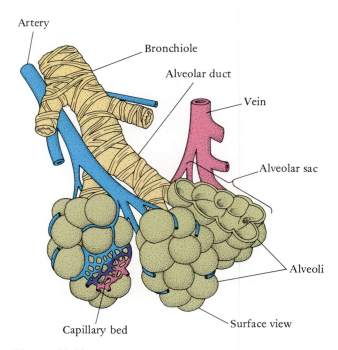

Figure 13.19
Termination of the respiratory passages in the mammalian lungs. Alveoli have a diameter of 0.2 to 0.3 mm.

whose walls are so puckered by pocket-shaped **alveoli** (L. *alveolus*, small hollow) that it resembles a cluster of miniature grapes (Fig. 13.19). The alveolar walls are extremely thin and relationships between them and the surrounding capillaries were not clear until they were studied by electron microscopy (Fig. 13.20). Alveoli are so densely packed that they abut against each other, with a dense capillary bed lying between them. Gases diffuse between the capillaries and the alveoli on each side of them. Some cells lining the alveoli have secretory properties and probably release the surfactant phospholipids. Large alveolar macrophages, called "dust cells" because they ingest and sometimes accumulate minute foreign particles, are abundant. Occasionally interalveolar pores extend through the septa separating alveoli (Fig. 13.20b) and may facilitate the movement of dust cells from one alveolus to another.

In order to sustain their higher levels of metabolism, warm-blooded animals, as one would expect, have larger respiratory exchange surfaces than cold-blooded animals. The alveolar surface of a normal adult human male, for example, is estimated to be 70 m². One might also expect that small mammals would have relatively more respiratory surface than large ones because their metabolism is much higher, but this is not the case. A shrew, a human being, and a whale all have lungs that are about 6% of body volume. The increased gas exchange needs of small mammals are met primarily by increased rates of ventilation and by changes in some characteristics of the blood, which will be described later in this chapter.

Mammalian Lung Ventilation

Mammalian lungs are ventilated by continuously changing the size of the thoracic cavity and, consequently, the pressure within the lungs. There are no prolonged periods of apnea, as in more primitive terrestrial vertebrates. The lungs follow the movements of the chest wall, for they are at once separated from it and united to it by the adhesive force of a thin layer of fluid that lies within the pleural cavities. During normal, quiet **inspiration**, the size of the thorax is increased slightly, intrapulmonary pressure falls to about 1 mm Hg below atmospheric pressure, and air passes into the lungs until intrapulmonary and atmospheric pressures are equal. Mammalian lungs, unlike those of a frog, are **suction lungs** because air is not pumped directly into them but enters passively as a consequence of the expansion of the thorax. During normal **expiration**, the size of the thorax is decreased, intrapulmonary pressure is raised to about 1 mm Hg above atmospheric pressure, and air is driven out of the lungs until equilibrium is again reached.

During inspiration, the thorax is enlarged by the contraction of the dome-shaped diaphragm and the

Chapter 13   Cellular Respiration, Gas Exchange, and Metabolic Rates   **283**

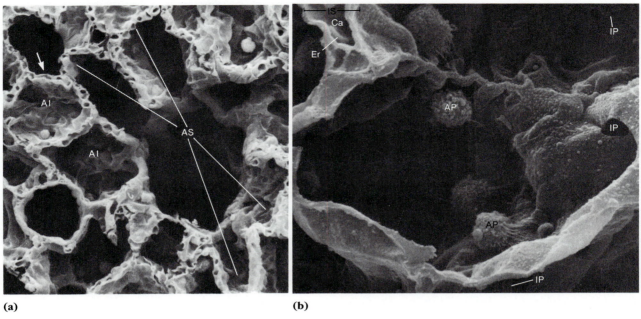

(a)   **(b)**

Figure 13.20
(a) Scanning electron micrograph of a group of alveoli (Al) branching from an
alveolar sac (AS). Numerous capillaries, some indicated by arrows, lie between
alveoli. (b) An enlarged view showing pores (IP) between alveoli, a capillary (Ca)
and an erythrocyte (Er) in the interalveolar septum (IS), and several alveolar
phagocytes (AP).

external intercostal muscles of the rib cage. The diaphragm pushes the abdominal viscera posteriorly and
increases the length of the chest cavity (Fig. 13.21); the
external intercostals raise the sternal ends of the ribs, push

the sternum forward, and expand the dorsoventral diameter of the chest. Expiration results from the relaxation of
the inspiratory muscles and the elastic recoil of the lungs
and chest wall, which are stretched during inspiration. But

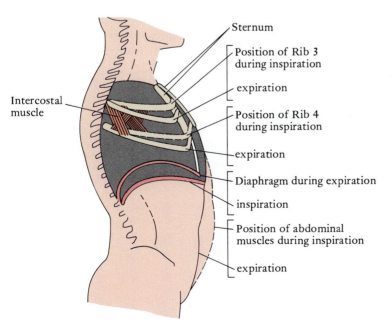

Figure 13.21
Mechanics and control of breathing. The elevation
of the ribs and depression of the diaphragm during
inspiration increase the size of the chest cavity,
indicated by the black area.

during heavy breathing, antagonistic expiratory muscles can decrease the size of the thoracic cavity. Contraction of abdominal muscles forces the abdominal viscera against the diaphragm and pushes it anteriorly; **internal intercostals** pull the sternal ends of the ribs posteriorly.

The lungs of an adult human male can hold about 6 L of air, but in quiet breathing the lungs contain only about half this amount, of which 0.5 L is exchanged in any one cycle of inspiration and expiration. This half-liter of **tidal air** is mixed with the 2.5 L of air already in the lungs. Vigorous respiratory movements can lower and raise the intrapulmonary pressure 80 to 100 mm Hg below and above atmospheric pressure, and under these conditions 4 to 5 L of air can be exchanged. This maximum tidal volume is known as the **vital capacity**. There is always, however, at least a liter of **residual air** left in the lungs to mix with the tidal air, for the strongest respiratory movements cannot collapse all the alveoli and respiratory passages. Alveolar air is always a mixture of tidal and residual air. If we exercise, more tidal air is exchanged, but more oxygen diffuses into the blood, and more carbon dioxide leaves the blood, so the composition of the alveolar air remains about the same as at rest. The composition of alveolar air is quite different from that of atmospheric air; it contains more water vapor, more carbon dioxide, and less oxygen (Table 13.2). The lungs of diving mammals such as seals and whales do not have a larger vital capacity than do other mammals. The ability of these animals to remain under water for prolonged periods is achieved through other adaptations (see Focus 13.1, p. 281).

The ventilation rate is regulated to keep the partial pressures of O_2 and CO_2 relatively constant despite differences in levels of activity and gas exchange rates. Respiratory movements are cyclic and are controlled by the respiratory center in the brain. Inspiratory neurons in this center have an inherent rhythm, becoming active every few seconds and sending nerve impulses out to the inspiratory muscles. After a few seconds of inactivity they again become active. The oscillatory activity of these neurons is all that is required for quiet breathing, but many factors can affect their activity. Of greatest importance is the level of carbon dioxide and hydrogen ions in the blood. Increase in blood levels of CO_2 when an animal becomes more active affects a chemosensitive area in the brain beside the respiratory center and this, in turn, affects the center. More impulses are sent to the inspiratory muscles and, if needed, to expiratory muscles, and the frequency and force of breathing movements increase. Sensory input from peripheral receptors also has an effect. If oxygen levels of the blood fall too low, chemoreceptors (the **aortic** and **carotid bodies**) associated with large arteries near the heart are affected. Impulses from them lead to an increase in the activity of the respiratory center. If an animal breathes very deeply, stretch receptors in the bronchi will be stimulated and they, in turn, will dampen the activity of the respiratory center.

Gas Transport

The tracheal system of insects brings oxygen from the body surface directly to the tissues and removes carbon dioxide. Other animals that are too large to exchange gases by diffusion alone have transport systems that carry gases between the respiratory membrane and the tissues. A small amount of oxygen and carbon dioxide can be carried in physical solution in blood. Most of these gases, however, are transported in other ways. Carbon dioxide reacts chemically with water and most ends up being carried as bicarbonate ion (HCO^-) and salts, as described below. Oxygen and some carbon dioxide bind loosely with colored compounds known as **respiratory pigments**. Hemoglobin, which gives the red color to our blood, is the most familiar one. If these gases were only in physical solution, human blood could carry only about 0.2 ml of oxygen and 0.3 ml of carbon dioxide in each 100 ml of blood. The reactions of carbon dioxide with water and the properties

---

Table 13.2
A Comparison of Atmospheric Air On a Cool, Dry Day at Sea Level with Alveolar Air*

	Atmospheric Air		Alveolar Air	
	P (mm Hg)	*Percent*	*P (mm Hg)*	*Percent*
Nitrogen	597.0	78.62	569.0	74.9
Oxygen	159.0	20.84	104.0	13.6
Carbon dioxide	0.3	0.04	40.0	5.3
Water vapor	3.7	0.50	47.0	6.2
Total	760.0	100.00	760.0	100.0

*From A. C. Guyton. *Physiology of the Human Body*, 6th ed. Philadelphia; Saunders College Publishing, 1984. *P* = partial pressure.

of hemoglobin enable blood to carry 100 or more times as much gas: 20 ml of oxygen and about 50 ml of carbon dioxide per 100 ml. *Thus, respiratory pigments enable the blood to carry far more oxygen than could be transported in simple solution.* Carriers are important for many animals. Transport in simple solution is adequate only for those animals, such as clams and some worms, that have low oxygen needs or large surface areas for gas exchange. Oxygen in most terrestrial arthropods (insects) is transported in the tracheal system and not by the blood. However, in spiders and mites, which have tracheae, the tracheae are bathed with blood, which then functions in transporting oxygen to the tissues.

Respiratory Pigments

The four types of respiratory pigments found in animals are all metal-containing proteins. Hemoglobin and hemocyanin are the most common. **Hemoglobin**, the most widespread of the respiratory pigments, is found in nearly all vertebrates and in many different invertebrates. The protein portion of a hemoglobin molecule is a globin consisting, in most vertebrates, of four peptide chains (Fig. 3.14). Each chain has attached to it a molecule of **heme**, a porphyrin composed of four carbon-nitrogen rings (pyrroles) with an iron atom at the center (Fig. 13.22). The heme is structurally the same in all hemoglobins. However, hemoglobins differ in the number of peptide chains and their amino acid composition, and these differences impart characteristic properties to the numerous varieties.

Hemoglobin is contained within the red blood cells of vertebrates but is in solution in the circulating fluids in most of the invertebrates. In those animals in which hemoglobin is dissolved in the plasma, it is always a giant molecule. For example, the hemoglobin molecule of *Arenicola*, a marine polychaete, is a polymer containing 96 heme units, whereas vertebrate hemoglobin, which is contained within cells, has 4 heme units.

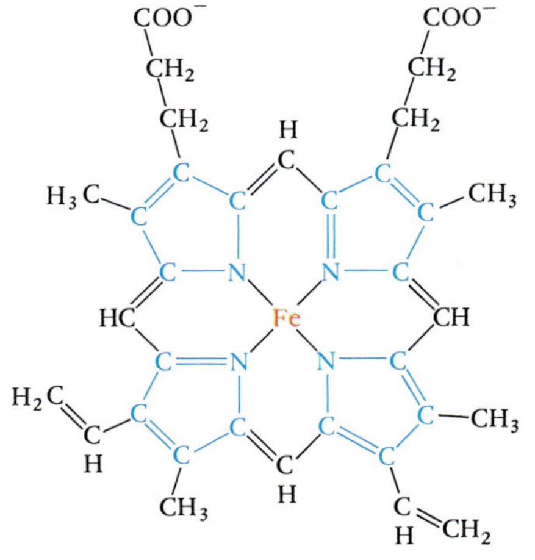

Hemocyanin is next in abundance to hemoglobin. It is found in molluscs (except for most bivalves), in the larger crustaceans, and in many arachnids and the related horseshoe crab, *Limulus*. It is a large molecule carried in solution and gives the blood a bluish tinge. Unlike hemoglobin, hemocyanin does not contain a porphyrin, and oxygen is carried between two copper atoms that are bound directly to the protein chain. Both hemoglobin and hemocyanin have evolved independently in a number of separate lineages.

Gas Reactions in the Blood

As oxygen diffuses from the respiratory organ into the blood, it combines with hemoglobin (Hb) to form **oxyhemoglobin** (HbO_2). During the formation of oxyhemoglobin in the lungs, where the partial pressure of oxygen (P_{O_2}) is high, one molecule of oxygen forms a loose bond with one of the iron atoms (part of the four heme groups) in the molecule of hemoglobin. Subsequently, additional molecules of oxygen bind to the iron atoms of the other three heme groups until the hemoglobin molecule is completely saturated with oxygen. Since a mammalian red blood cell contains as many as 265,000,000 molecules of hemoglobin, a great deal of oxygen can be carried. The reaction is reversible, and hemoglobin releases much of its oxygen when blood reaches the tissues where the P_{O_2} is low:

$$Hb + O_2 \underset{Low}{\overset{High}{\rightleftharpoons}} HbO_2$$

Oxygen partial pressure (P_{O_2})

As carbon dioxide diffuses into the blood from tissue cells, most of it combines with water to form carbonic acid which, in turn, dissociates into hydrogen and bicarbonate ions:

$$CO_2 + H_2O \rightleftharpoons H_2CO_3 \rightleftharpoons H^+ + HCO_3^-$$

In the lungs, the reaction above moves to the left and carbon dioxide is released. The ionizing reactions proceed very rapidly, but the reactions between carbon dioxide and water are much slower, taking several seconds. This is too long for much carbonic acid to be formed in the second or less that blood travels through a particular section of a capillary. However, the red blood cells contain the enzyme **carbonic anhydrase** that increases the rate of reaction between carbon dioxide and water by nearly 5000-fold. In addition to these reactions, 15 to 20% of the carbon dioxide entering the blood combines with hemoglobin. The binding site is not with the iron but with amino groups of the globulin. Most of the CO_2 is carried as a bicarbonate ion in the plasma. The bicarbonate ion leaves

Figure 13.22

Structure of a heme molecule. Metallic atom (Fe) is shown in red; the four pyrrole rings, in blue.

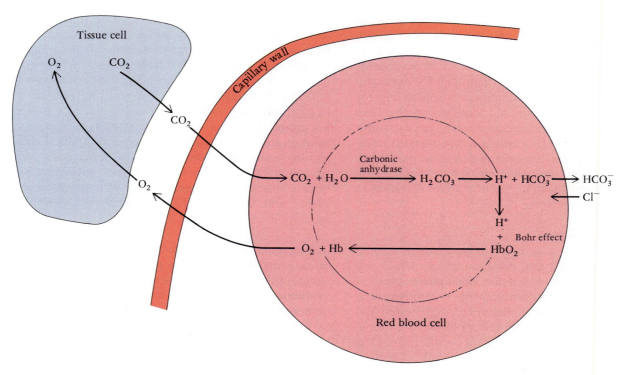

Figure 13.23
Reactions of carbon dioxide and their relationship to the release of oxygen by hemoglobin. In the lung the reactions are reversed.

the red blood cell after its formation in exchange for a chloride ion that enters the red blood cell from the plasma (Fig. 13.23). All of these reactions involving CO_2 are reversed in the lung.

The reaction of CO_2 with water in the tissue capillaries could greatly increase the acidity of the blood (decrease its pH) because carbonic acid is a fairly strong acid, i.e., it dissociates easily and releases its hydrogen ions. This acidification is prevented by the buffering action of blood proteins, of which hemoglobin is the most important because it is the most abundant. Oxyhemoglobin (HbO_2) combines with excess hydrogen ions to form reduced or **acid hemoglobin** (HHb). Since acid hemoglobin is a weaker acid than carbonic acid (it holds on to hydrogen ions to a greater extent), the pH of the blood does not fall very much (arterial blood has a pH of 7.6; venous blood, a pH of 7.2). As oxyhemoglobin takes up hydrogen ions, changes occur in its molecular configuration that cause it to reduce its affinity for oxygen, thus more oxygen molecules are released. More oxygen is driven off than would leave simply because of the low partial pressure of oxygen in the tissues. The release of oxygen caused by the entrance of carbon dioxide is called the **Bohr effect** (Figs. 13.23 and 13.24). The reverse reactions occur in the lungs.

As carbon dioxide leaves the lungs, the blood becomes less acid because hydrogen ions are used to form carbon dioxide (see equation in preceding paragraph). Acid hemoglobin gives up hydrogen ions, which permits more carbon dioxide to be formed and given off. Release of hydrogen ions in turn stabilizes the pH of the blood and increases hemoglobin's affinity for oxygen.

The reactions of gases in the blood enable the blood to carry a great deal more oxygen and carbon dioxide than could be carried in physical solution; minimize any change in the pH of blood as oxygen and carbon dioxide are transported; and facilitate the unloading of oxygen and the uptake of carbon dioxide in the tissue capillaries and the reverse movements of gas in lung capillaries.

Oxygen Dissociation Curves
Hemoglobin undergoes changes in shape, or configuration, when it binds with oxygen, and these changes affect the amount of oxygen it can carry. When one molecule of oxygen binds to hemoglobin, intramolecular changes occur that facilitate the uptake of the second molecule of oxygen by another heme, which in turn facilitates the binding of the remaining heme groups with oxygen. Reverse changes occur in the tissues, where the release of a

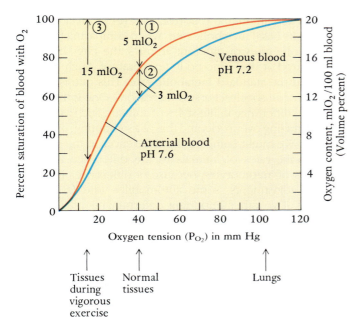

Figure 13.24

Oxygen dissociation curves for human arterial and venous blood. *Arrow 1* indicates the amount of oxygen delivered to normal tissues because of the reduction in oxygen tension between the lungs and tissues. *Arrow 2* indicates the extra amount of oxygen delivered because of the entry of carbon dioxide into the blood (the Bohr effect). *Arrow 3* indicates the amount of oxygen delivered by arterial blood to tissues during exercise. A small additional amount would be delivered because of the Bohr effect.

molecule of oxygen by a heme group facilitates the release of oxygen by other groups. Because of these properties, the amount of oxygen bound to hemoglobin is not directly proportional to the partial pressure of oxygen in the blood. By experimentally determining what percent of the hemoglobin is saturated with oxygen at different partial pressures, physiologists have determined oxygen dissociation curves for different hemoglobins. Because these curves are somewhat **S** shaped, or sigmoidal, more oxygen can be delivered to the tissues than if the relationship were a linear one. In humans, hemoglobin becomes nearly fully saturated in the lungs at a P_{O_2} of about 100 mm Hg (Fig. 13.24). Since the top of the curve is somewhat flat, nearly full saturation can be achieved over a broad range of pressures. In a sense, there is a wide margin of safety. Saturated human arterial blood holds 20 ml of oxygen per 100 ml of blood. No oxygen leaves the blood as it passes through the arteries to the tissues. But in the thin-walled tissue capillaries, the blood is in an environment where P_{O_2} is normally about 40 mm Hg. Because the dissociation curve is sigmoid, this is at a point on the curve where the blood can only be 75% saturated, so about 5 ml of oxygen is given off per 100 ml of blood. Oxygen dissociation curves are also affected by the pH of the blood. As carbon dioxide enters and the blood becomes more acid, the curve is shifted to the right, and an extra 3 ml of oxygen is given up at a pressure of 40 mm Hg (the Bohr effect). During vigorous exercise, the P_{O_2} falls to nearly 15 mm Hg. Under this condition, arterial blood can only be about 25% saturated, so it must give up nearly 15 ml of oxygen per 100 ml of blood.

Hemoglobins differ in the composition of the protein part of the molecule, and this difference imparts different oxygen carrying properties, i.e., different affinities for oxygen (Fig. 13.25). A hemoglobin with a high affinity for oxygen has an oxygen dissociation curve located to the left. This facilitates oxygen uptake because the blood can be-

Figure 13.25

Oxygen dissociation curves of representative animals.

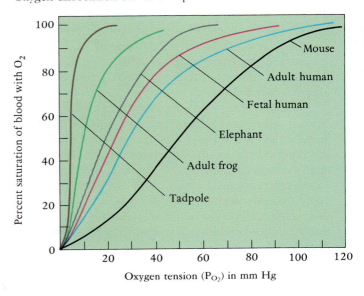

come fully saturated at lower oxygen pressures. Mammalian fetal hemoglobin has a curve of this type; its higher affinity for oxygen relative to maternal hemoglobin facilitates the transfer of oxygen from the mother to the fetus. Aquatic animals have curves located more to the left because oxygen partial pressure is usually less in water than in air. The curve for the hemoglobin of the aquatic tadpole is far to the left of the curve for the adult frog. Although a shift of the curve to the left makes it easier to take up oxygen, there is a trade off because less oxygen can be unloaded in the tissues at a given P_{O_2}. This is not a problem for a fetus or most aquatic animals because their metabolic rates are low relative to those of an adult mammal or many terrestrial animals. Saturating the blood with oxygen is not a problem for most terrestrial animals because air contains so much oxygen, but terrestrial animals with high metabolic rates need to unload a great deal of oxygen in their tissues. Small mammals, for example, with their relatively higher levels of metabolism, need more oxygen. They have evolved hemoglobins that place the oxygen dissociation curves to the right. At a tissue P_{O_2} of 40 mm Hg, an elephant's blood is about 85% saturated, whereas a mouse's blood is only 30% saturated; thus far more oxygen is delivered to the mouse's tissues.

FUEL UTILIZATION: METABOLIC RATES AND ENERGY

As we saw at the beginning of this chapter, foods that are not stored or utilized in the synthesis of new cell components are burned as fuel in the production of cellular energy. Fuels such as glucose and fatty acids are metabolized to carbon dioxide and water, oxygen is utilized as an electron acceptor, and some of the fuel energy is conserved in the energy-rich phosphates of ATP. These energy transfers are only about 40% efficient, and 60% of the energy produced is lost at the body surface as heat. Since the heat conversion factor is constant and since energy production is almost entirely an aerobic process, the amount of heat produced and the amount of oxygen utilized over a given time period are relatively precise reflections of the amount of fuel consumed, referred to as an animal's **metabolic rate.**

Metabolic rates can be measured in a number of ways. We can utilize gas exchange and can measure either the amount of CO_2 released or the oxygen consumed over a given period of time. Or we can measure the heat generated. Oxygen consumption and heat production are the most commonly used measures of metabolic rate, especially oxygen consumption, which is relatively easy to determine. Carbon dioxide is not generally used because the large bicarbonate pool into which the CO_2 molecule enters in the animal's body introduces variables that are not related to metabolic rate. For example, the bicarbonate ion may be incorporated into bone or shell.

When an animal is at rest, its level of metabolism is called its basal metabolic rate. When comparing the basal metabolic rates of different animals, the rate should be measured at the same body temperature. Such a rate is termed the **standard metabolic rate.** This rate is not the same for all species, for several reasons. One is the fundamental difference in life styles of animals, such as that of a snail grazing over rocks and that of a swimming, predatory squid. We might compare the differences in the standard metabolic rates of animals to the difference in idling speeds of automobiles; because of differences in carburetor adjustments, some automobiles idle faster than others. Two other important determinants of the metabolic rates of many animals are temperature and body size.

Temperature

Most animals are **ectothermic** (Gr. *ektos*, outside + *therme*, heat); their rate of heat production is so low, and their rate of heat loss is so high, that their body temperature and level of metabolism are determined primarily by the temperature of the external environment. They have high temperatures and metabolic rates when ambient temperatures are high, and become quite sluggish when temperatures are low. Because of these fluctuations in body temperature, these animals can also be described as **poikilothermic** (Gr. *poikilos*, varied). Ectothermic and poikilothermic describe different aspects of the same condition and are often used interchangeably. The term "cold blooded," which is often applied to ectothermic animals, is an inadequate description. The body temperature of ectotherms is not always cold.

The metabolic rate of most ectotherms varies directly with environmental temperatures because of the sensitivity of enzyme systems to temperatures. In general, the metabolic rate increases two or three times with each 10°C rise in temperature (Fig. 13.26a). The 10°C change, called the Q_{10} value, is an arbitrary interval used to compare metabolic rates. The upper limits of the metabolic rate are determined by the points at which proteins (enzymes) are destabilized, or denatured, and many vertebrates are handicapped well below this level because of the decreasing affinity of hemoglobin for oxygen at high temperatures. Many ectotherms die at freezing temperatures because ice crystals cause cells to rupture. Adaptations to survive low temperatures usually involve the presence of "antifreeze" substances that lower the freezing point and prevent the formation of intracellular ice crystals. One should not think of ectothermy as a "poor" metabolic strategy. Most animals are ectotherms, and their metabolic rate and activity are limited by environmental temperatures. On the other hand, they don't have to eat as much, and more of the food they consume can be directed toward reproduction and growth rather than toward maintenance activities.

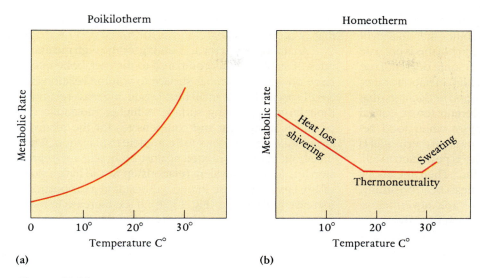

Figure 13.26

Metabolic rates and environmental temperature. (a) A generalized metabolic rate of a poikilotherm in which the rate doubles with each 10°C rise in temperature. The curves of various poikilotherms would only roughly approximate this curve. (b) Metabolic rate of a homeotherm. In the region of thermoneutrality the metabolic rate need not be increased to maintain a constant body temperature. Shivering below thermoneutrality and sweating or panting above it both require energy expenditure.

Birds and mammals are **endothermic** (Gr. *endon*, within). Their rate of internal heat production is greater than that of ectotherms, and they have greater control over heat loss from the body surface. Body temperature and rates of metabolism are maintained at high levels irrespective of environmental temperature changes and do not fluctuate greatly. Animals with a relatively constant body temperature can also be described as **homeothermic** (Gr. *homos*, the same). Birds and mammals are the only living animals that can be endothermic all of the time, but other animals, such as certain lizards, some active fish, and some insects, function as endotherms some of the time. These animals can also be called **heterotherms**.

Homeothermy is maintained in birds and mammals by regulating heat production and by controlling heat loss across the body surface. Fur and feathers act as insulating barriers because they entrap a still layer of air. Subcutaneous fat is also effective insulation. A particularly thick layer of subcutaneous fat (blubber) is found in aquatic mammals such as seals, porpoises, and whales. The evaporation of sweat secreted by sweat glands of some mammals serves to cool the body because the heat required for evaporation is removed from the skin. In birds and other mammals the evaporation of water from the buccal epithelium and tongue by panting functions in the same way. It is significant that sweat glands are particularly abundant in large mammals such as horses and camels, and in humans, all of which have little body hair. Very small and heavily furred species, such as rodents, moles, and shrews, lack sweat glands, but the small species possess a large surface area for heat loss relative to their heat-producing mass, and may be able to dissipate heat without resorting to evaporative mechanisms.

Various mechanisms controlling heat loss to maintain body temperature come into operation at temperature extremes (Fig. 13.26b). On a hot day, loss of excess internal heat must be facilitated. Blood flow into the skin is increased through the opening of arterioles, bringing warm blood close to surface tissues (the characteristic flushing in hot weather), and sweating and panting increase. Conversely, on a cold day, heat loss must be curtailed. Blood flow to the skin is reduced, and shivering—small involuntary muscle contractions that increase internal heat production—may occur. The hypothalamus, a part of the brain, is the control center of heat regulation. Changes in the temperature of blood passing through the hypothalamus elicit signals to sweat glands, blood vessels in the skin, muscles, and so on. Fever, an elevation of body temperature above normal, is caused by certain compounds produced directly or indirectly by pathogenic organisms that raise the "thermostat setting" in the hypothalamus.

Behavioral modifications are also important at extreme temperatures. Many animals increase or decrease their level of activity or avoid exposure by remaining in burrows or retreats. Bears and some other mammals will sleep for long periods during the winter, but this is not true hibernation for they do move about to some degree. True hibernators are small mammals with high metabolic rates, such as some rodents, hamsters, bats, and hedgehogs, that enter a marked state of torpor at certain lower critical temperatures, thus conserving energy when food resources are low. During torpor, their internal thermostat is set at a low level and body temperature is permitted to fall, but not below the reduced "set point." Their metabolic rate declines greatly, and their rate of heart beat, breathing, and other vital activities slow. There are, nevertheless, periodic, brief states of arousal, at which time the metabolic rate is elevated. The energy saving of torpor is very great, as the amount of heat generation required is greatly reduced, and the animal slowly utilizes a store of fat acquired before hibernation. Some birds also hibernate, and there are a few hummingbirds and bats that exhibit a daily state of torpor. Other adaptations for thermoregulation will be described in chapters dealing with specific groups of animals.

Body Size

Another critical variable determining standard metabolic rate is body size. The metabolic rates of mammals of different sizes can be compared by examining oxygen consumption per kilogram of body weight (Fig. 13.27). As you might expect, their metabolic rates vary inversely with size, with that of a shrew being many times greater than that of an elephant.

An important factor accounting for the high metabolic rates of small animals compared to the metabolic rate of large ones is the familiar surface to volume relationship. Small species have a small heat-producing mass relative to their large heat-losing surface. A higher metabolic rate produces more heat and compensates for the relatively larger loss. But this is not the complete explanation. When metabolic rates are plotted against body size, the regression line for an equivalent relationship would have a slope of 1.0, i.e., an increase in one will result in a proportional increase in the other (Fig. 13.28). If the increase in metabolic rate in small mammals were simply related to heat loss through increase in surface area, then the slope should parallel the slope for the relationship of surface area to volume, i.e., the slope should be 0.67 (volume increases by the cube, and surface area by the square). Instead, the actual slope is greater than expected, about 0.75. However, this much cited figure is based largely on domesticated animals. More recent studies on a wide range of mammals show great variation, the points of some species falling much below 0.67 and those of others falling well above 0.67. The variation seems to be related partly to life style and diet and partly to differences in relative amounts of tissues having different metabolic rates (e.g., bone versus muscle). For example, arboreal leaf-eating mammals, such as sloths, have lower metabolic rates than many other mammals. This reduces the volume of leaves, which tend to contain toxic compounds, that the animal must consume.

Figure 13.27
Rate of oxygen consumption of various mammals plotted against body weight. The ordinate has an arithmetic scale, but the abscissa is logarithmic.

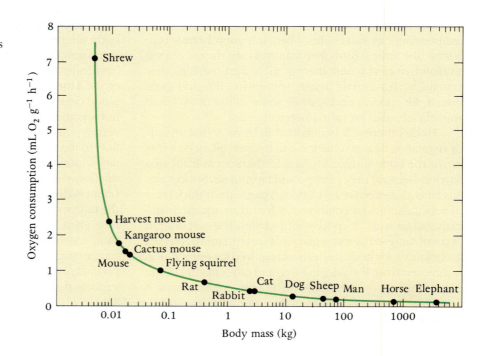

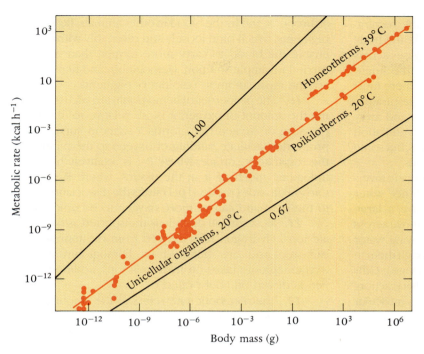

Figure 13.28
Metabolic rates of different organisms measured in the number of kilocalories consumed per hour, plotted against body weight, using logarithmic coordinates. The regression lines (slope) are about 0.75.

Summary

1. Cellular respiration is the transfer of energy from organic fuels to ATP, which is the direct source of energy for cellular reactions. The reaction occurs in small increments in the course of numerous reactions. Oxygen is required and the degraded fuel fragments are removed as carbon dioxide. The principal reaction pathways of cellular respiration are glycolysis, the citric acid cycle, electron transport, and chemiosmosis.

2. In glycolysis glucose is partially oxidized to yield a relatively small amount of energy and two molecules of pyruvate. No oxygen is required, and the pyruvate is converted to lactic acid or other end products, which oxidize the NADH produced in glycolysis. The conversion of pyruvate and the glycolytic reactions together are called fermentation.

3. The remaining pathways of cellular respiration are oxygen demanding and yield the greatest amount of energy from the fuel molecule. Pyruvate is oxidized and reduced to a two-carbon fragment, acetyl CoA, that then enters the citric acid cycle. Here, in the course of further oxidation and decarboxylation, the remaining two carbon atoms of the original fuel are removed as carbon dioxide. Electrons received by NAD⁺ and FAD in the course of the citric acid cycle are transferred along a series of electron carriers on the mitochondrial membrane. The energy released sets up a proton gradient that, by means of chemiosmosis, drives the formation of ATP.

4. When fatty acids are used as fuels, they are degraded to two-carbon units, which combine with CoA to form acetyl CoA and then enter the citric acid cycle. When amino acids are mobilized for fuel use, they are first deaminated and then enter the respiratory pathway via different organic acids. Food storage and the interconversion of food compounds, for which cellular respiration provides important pathways, are the principal means by which animals balance the irregularities of food intake with cellular demand.

5. The air from which animals obtain the oxygen they need to sustain cellular respiration is a mixture of gases containing about 21% oxygen and 0.04% carbon dioxide. The partial pressure of oxygen is the product of its percentage in the air and the pressure of the air. The amount of oxygen that will diffuse into water or the liquid environment of the body depends on its solubility and partial pressure. Some invertebrates are oxygen conformers and can adjust their level of metabolism to available oxygen supplies, but most animals are oxygen regulators and need a constant supply of oxygen.

6. Animals must have a moist and permeable respiratory membrane for gas exchange. This may be the body surface and/or a respiratory organ. Passage of gases across this membrane, and between body cells and the liquid bathing them, occurs by diffusion. In addition, there must be a way to bring gases to and from the respiratory membrane (ventilation) and to carry them to and away from the vicinity of the cells (gas transport). Ventilation and transport occur by diffusion in small animals, but some system of bulk flow is needed in most animals.

7. The respiratory organs of most large aquatic invertebrates and all aquatic vertebrates are gills. Since water contains less oxygen than air, gills must have a large surface area, and a large volume of water must move across them. Gas exchange is often facilitated by moving water across the gills in one direction and blood through them in the opposite direction (countercurrent exchange). Gills are ventilated by ambient water currents, ciliary action, body movements, movements of the gills, or a muscular pump that sends water through a branchial chamber.

8. Loss of body water is minimized in terrestrial animals by having the respiratory membrane located deep in a body chamber where the relative humidity is high and by moving no more air through the chamber than is needed to meet the animal's requirements. A few small terrestrial animals living in moist microhabitats utilize the body surface or gills for gas exchange. Most terrestrial arthropods have a system of tracheal tubes that lead from surface spiracles to all of the tissues of the body. Other terrestrial animals have lungs. Gas moves in and out of invertebrate lungs primarily by diffusion; vertebrate lungs are actively ventilated by a muscular pumping action of the floor of the pharynx (amphibians) or muscular movement of the ribs and diaphragm (mammals).

9. Transport of gases between the respiratory membrane and the cellular environment is by diffusion in small animals, by the tracheal system in most terrestrial

arthropods, and by the blood in other animals. Respiratory pigments, which bind loosely and reversibly with oxygen, enable the blood to carry more oxygen than could be carried in simple solution. Hemoglobin and hemocyanin are the most common respiratory pigments.

10. The properties of hemoglobin are such that plotting the amount of oxygen it binds relative to the partial pressure of oxygen (P_{O_2}) forms a sigmoid curve. These properties facilitate binding of oxygen to the pigment at the respiratory surface, where P_{O_2} is relatively high, and unloading much of it in the tissues, where P_{O_2} is low. Carbon dioxide (reduced pH) shifts the dissociation curve to the right and causes more oxygen to be unloaded.

11. Most carbon dioxide is transported as bicarbonate ions. Hydrogen ions derived from the ionization of carbonic acid are buffered by hemoglobin and, as oxyhemoglobin takes on hydrogen ions, additional oxygen is driven off (Bohr effect).

12. The metabolic rate refers to the amount of energy utilized by an organism over a given period of time. In animals the rate of energy utilization at a given temperature, excluding that resulting from muscle action, is called the standard metabolic rate. Metabolic rates of different animals can be determined and compared most easily by measuring oxygen consumption and heat production.

13. Animals that have relatively constant body temperatures are said to be homeothermic; they may also be called endothermic because the body temperature is largely determined by internal heat production. Animals with fluctuating body temperatures are said to be poikilothermic; they may also be called ectothermic because low heat production and high heat loss result in the body temperature being determined largely by the environment. There is a relationship between metabolic rate, size, and surface area of animals, with smaller animals, especially those that are endothermic, exhibiting higher metabolic rates.

References and Selected Readings

Additional information on gas exchange can be found in the general references on vertebrate organ systems cited at the end of Chapter 10.

Avery, M.E., N. Wang, and H.W. Taeusch: The lung of the newborn infant. *Scientific American* 228(Apr. 1973):75. An account of the discovery and importance of the surfactant that lowers surface tension in the lungs.

Becker, W.M.: *The World of the Cell*. Menlo Park, Calif.: Benjamin/Cummings Publishing Co., 1986. A textbook of cell biology that includes an extensive account of cellular respiration.

Feder, M.E.: Skin breathing in vertebrates. *Scientific American* 253 (Nov. 1985):126–143. A review of cutaneous gas exchange and some of the associated circulatory adaptations.

French, A.R. The patterns of mammalian hibernation. *American Scientist* 76(Nov.–Dec. 1988):569–575. The relationship of size to patterns of hibernation in mammals.

Hughes G.M. (ed.). Respiration in Amphibious Vertebrates. London, Academic Press, 1976. A series of papers presented at a symposium on the respiration of fishes and amphibians.

Perutz, M.F. Hemoglobin structure and respiratory transport. *Scientific American* 240(Dec. 1978):92–125. A discussion of the relationship between the structure of the hemoglobin molecule and oxygen transport.

Randall, D.J., et al. *The Evolution of Air Breathing in Vertebrates*. New York: Cambridge University Press, 1981. A thorough account of the physiological problems encountered in the evolution of air breathing in many vertebrate groups.

Schmidt-Nielsen, K. Countercurrent systems in animals. *Scientific American* 244(May 1981):118–128. The various ways countercurrent systems are utilized in the physiology of animals.

Schmidt-Nielsen, K.: *Scaling: Why Is Animal Size So Important?* Cambridge, Cambridge Univ. Press, 1984. The structural and physiological modifications of animals with changes in size.

West, J.B. *Respiratory Physiology: The Essentials*, 3rd ed. Baltimore: Williams & Wilkins, 1985. An analysis of the physiology of the mammalian lung and the transport of gases to the tissues.

Zapol, W.M. Diving adaptations of the Weddel seal. *Scientific American* 256(June 1987):100–105.

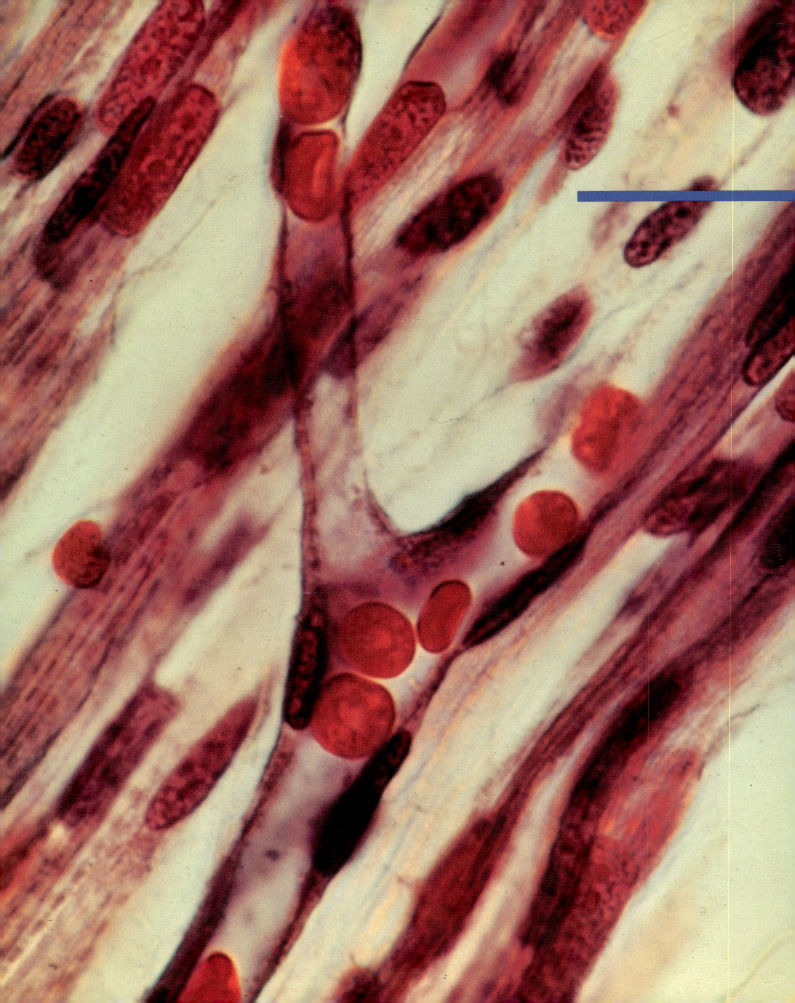

14

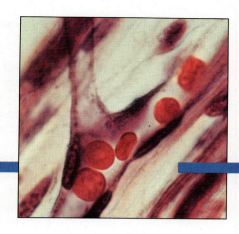

Internal Transport

A capillary bed where exchanges occur between the blood and the fluids bathing the cell.

As we have seen in Chapters 12 and 13, food, water, oxygen, and other essential materials enter and leave through various surfaces of the body by diffusion, supplemented in some cases by active transport. Similarly, carbon dioxide, nitrogenous wastes, and other products in excess of the body's needs are eliminated. Since most body cells of metazoans lie deep in the body, they do not exchange materials directly with these surfaces of the body, but with an **interstitial fluid** that lies in the minute spaces between the cells. Some mechanisms are necessary to move materials between the interstitial fluid and the exchange surfaces of the body. Diffusion is the primary means of transport through the body of very small metazoans, where internal distances are a millimeter or less, but in most species some system of active bulk flow in a transport system is required to move substances between the interstitial fluid and the sites where they enter and leave the body.

FUNCTIONS OF TRANSPORT SYSTEMS

The degree of development and complexity of the transport system is very much related to the size and metabolic rate of the animal. An animal cannot be large or sustain a high rate of metabolism without an efficient transport system that can both quickly supply the interstitial fluid with the multitude of materials the cells need and rapidly remove cellular waste products. By utilizing bulk flow, transport systems are essentially decreasing diffusion distance. Oxygen, for example, does not have to diffuse all of the way from the lungs to the tissues. It diffuses from the lungs into the blood, which carries it quickly by bulk flow to the tissues, where it then diffuses into the interstitial fluid and finally into the cells.

In addition to carrying gases, nutrients, and waste products, transport systems distribute metabolites from tissues that produce them to others that help to metabolize them (e.g., lactate from muscles to the liver). They carry chemical messengers, or hormones, from the glands that secrete them to their target organs (e.g., thyroid hormone from the thyroid gland to most of the cells of the body). Finally, the blood distributes heat through the body and to or away from sites where heat can be lost or gained. Transport systems not only serve to carry substances and heat around the body, but also do so in such a way that they help maintain a constant internal environment, or **homeo-**

stasis (Gr. *homoios*, alike + *stasis*, standing). Most cells cannot tolerate wide changes in their immediate environment. Components of the blood act as buffers, sopping up and holding excess hydrogen ions, hydroxyl ions, and other substances until they can be metabolized and removed from the body (p. 286).

TYPES OF TRANSPORT SYSTEMS

Many cavities within the body may function as transport systems. Currents created by flagella or cilia move fluid through the water canals of sponges and in and out of the gastrovascular cavity of sea anemones and other cnidarians (Fig. 14.1a, b) and within the coelom of sea stars. Con-

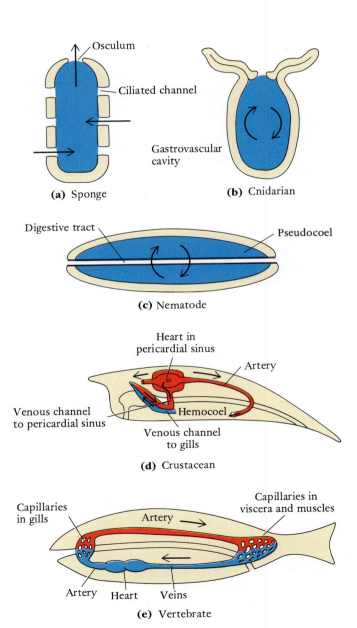

Figure 14.1
Types of transport systems. Water circulates through the cavities of a sponge (a), and the gastrovascular cavity of a cnidarian (b). (c) Movement of liquid in the body cavity of a primitive worm. (d) Open circulatory system of a crustacean. (e) Closed circulatory system of a vertebrate.

tractions of the body wall stir fluids in the body spaces of nematodes and many other worms (Fig. 14.1c). Most other animals have a distinct circulatory system through which blood is pumped to the tissues by the contractions of certain blood vessels, specialized hearts, or adjacent body muscles. Blood then returns to be pumped out again; in short, the blood circulates (see Focus 14.1). In the octopus and other cephalopod molluscs, and in annelid worms, vertebrates, and other animals with **closed circulatory systems**, blood is confined to vessels or vessel-like channels during its circulation. In the closed system of vertebrates, for example, blood moves from the heart through **arteries** to extensive networks of **capillaries** in the tissues and returns through **veins** (Fig. 14.1e). The capillaries are very numerous, small, thin-walled vessels through which small molecules diffuse easily between the blood and the interstitial fluid. Because most blood cells and proteins remain in the blood, blood and interstitial fluid differ in composition. Most molluscs and arthropods and some of the primitive chordates (tunicates) have **open circulatory systems** (Fig. 14.1d). Liquid moves from the heart through arteries directly into diffuse spaces among the cells. Collectively these spaces form a **hemocoel** (Gr. *haima*, blood + *koilos*, cavity), and the liquid moving through them is most appropriately called **hemolymph**, for the liquid in the vessels and interstitial spaces is indistinguishable.

Although the circulating liquids in animals with closed systems are typically confined to closed vessels, whereas those of species with open systems enter diffuse tissue spaces, the distinction is not always clear-cut. Blood enters tissue spaces in many worms, and blood in the spleen of vertebrates, where many blood cells are formed and stored, lies in a network of diffuse open spaces. Some arthropods and molluscs with open systems have organized channels through certain organs, including the gills and excretory structures.

A closed circulatory system is not necessarily more specialized than an open system, but a higher pressure can develop more easily because the blood is confined to vessels whose walls maintain much of the pressure developed by the heart. Materials are quickly distributed between organs, and blood returns more rapidly to the heart. The octopus and squid could not be very active molluscs without closed systems. Clams, snails, and other sluggish, slow-moving molluscs have open systems. Animals with closed systems are able to regulate the flow of materials to the tissues more precisely than animals with open systems. By dilating and constricting appropriate vessels, blood can be shunted to particularly active tissues—for example, to muscles during rapid locomotion—whereas less active tissues will have a reduced flow through them. The total volume of circulating blood can be less than would be the case if all tissues needed to receive a large supply at all times. Energy is saved. In contrast, circulating fluids and materials are by necessity distributed more evenly to all of the tissues in animals with open systems.

Whether a group of animals has an open or a closed system depends not only on factors affecting the distribution of materials, but also on the other functions of body fluids. Open systems are often favored in species in which the fluids perform important hydrostatic functions. The open system of many arthropods enables them to develop a high internal pressure that can rupture the exoskeleton during molting and provide the body with some support until the new skeleton hardens. The legs of jumping spiders are extended not by the action of muscles but by the sudden flow of hemolymph into them. High pressure can develop in the spider's leg because the exoskeleton prevents the leg from swelling as hemolymph is pumped in. Many molluscs, such as clams, use hydrostatic pressures to change the shape of their foot during locomotion and burrowing.

BLOOD AND INTERSTITIAL FLUID

Plasma

Adult human beings have about 5.7 liters of **blood**. About 55% of this is composed of a liquid **plasma**, and 45% of various formed elements that are carried in the plasma: red blood cells (erythrocytes), white blood cells (leukocytes), and platelets. The plasma itself has the following composition:

Water	90%
Soluble proteins	7–8%
Electrolytes	1%
Materials in transit:	1–2%

    Nutrients (e.g., glucose, amino acids, lipids)
    Metabolic intermediates (e.g., pyruvate, lactate)
    Dissolved gases
    Hormones

Many molecules in transit are free in solution; others, including some of the trace metals, are bound to specific transport proteins.

The primary plasma proteins are albumins, globulins, and fibrinogen. Albumins constitute 60% of all plasma proteins and function primarily to maintain the osmotic pressure of the blood. This is essential for normal fluid exchanges between the blood and interstitial fluid, as explained later (p. 319). The globulins are a large and diverse family of proteins, which include the immunoglobulins (antibodies important in the body's defense), transport globulins such as lipoprotein, blood clotting factors, and several other proteins of variable function. Fibrinogen is the source of the fibrin of blood clots. Most of the plasma proteins are synthesized in the liver, but immunoglobulins are made by certain white blood cells (lymphocytes) and tissue plasma cells derived from them (Chap. 15).

Focus 14.1   *The Circulation of Blood*

The notion that blood circulates may appear to us to be self-evident. But think a moment. How can this be demonstrated? Observers from Aristotle on were aware that the heart pumps blood into the arteries for this can easily be seen when the heart of a living animal is exposed and arteries leaving it are cut. But where does the blood come from and where does it finally go? Galen (ca 130–200), one of the great physicians of antiquity, proposed that blood was produced in the liver, moved to the heart, and was pumped to the tissues, where it was consumed. This was not an unreasonable idea for the period, and Galen's views prevailed for nearly 1500 years.

William Harvey, a British physician of the 17th century, began to make calculations and observations that led him to question this notion. He knew that a man's pulse rate was on average 72 beats per minute, and by dissecting cadavers found out that the left ventricle holds about two ounces of blood. Simple calculations showed that the heart would pump 8640 ounces of blood out to the tissues each hour ($2 \times 72 \times 60 = 8640$). This amount of blood weighs 580 lb, three times the weight of a heavy person! It seemed unreasonable that this amount of blood could be made and consumed in an hour. A continual circulation of the blood was a more logical hypothesis, even though connections between arteries and veins had not yet been discovered.

Observations that Harvey made on the veins in a human arm (see figure) supported this hypothesis because they showed that blood in the veins only moves toward the heart. If a tourniquet is applied to the arm, superficial veins distal to it fill with blood, and the location of valves in the veins appear as "knots." By placing a finger on a vein just distal to a knot, the blood can be moved back toward the hand. The segment of the vein thus emptied remains so because blood proximal to the knot does not back up. When the finger is released from the emptied vein, blood again flows proximally.

Harvey published his hypothesis in 1628 in a small book entitled *Anatomical Dissertation Concerning the Motion of the Heart and Blood*. Many consider this to be the beginning of modern biology, because for the first time a biological problem was analyzed quantitatively and experimentally. Investigators began to search for physical explanations of natural phenomena and were no longer content to assume mysterious "vital forces."

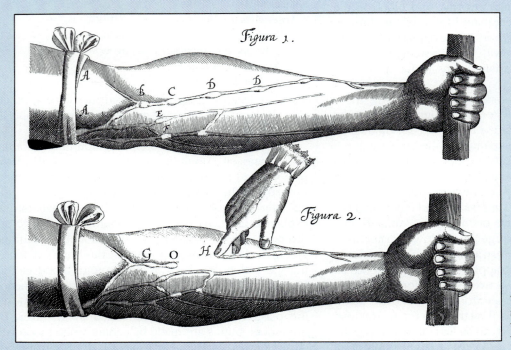

William Harvey's drawing illustrating the action of valves in the superficial veins of the arm.

The major electrolytes in the plasma and interstitial fluid are sodium, chloride, and bicarbonate ions. Only small amounts of other ions, including potassium, phosphate, and magnesium, are present. This is in sharp contrast to the intracellular fluids, in which the most abundant ions are potassium, phosphate, and magnesium.

Plasma is a complex liquid in dynamic equilibrium with other body fluids. Many substances move in and out of it, but its composition and properties remain remarkably constant. Plasma and interstitial fluid are separated by the endothelial wall of the capillaries, and they differ in composition only in the materials that do not readily pass through this wall. Red blood cells remain suspended in the plasma, but leukocytes, which are capable of amoeboid movement, squeeze between the cells of capillary walls and enter the interstitial fluid. Most of the large protein molecules remain in solution in the plasma, but some escape into the interstitial fluid. Water, gases, many ions, and other components of the plasma and interstitial fluid diffuse readily through the capillary wall, so their concentration in the two compartments tends toward equilibrium.

Red Blood Cells

The respiratory pigments of invertebrates are carried free in solution or within corpuscles of various kinds that circulate in the blood, hemolymph, or coelomic fluid (p. 284). The vertebrate respiratory pigment, hemoglobin, is always located within red blood cells, or **erythrocytes** (Gr. *erythros*, red + *kytos*, cell). Most vertebrates have relatively large oval-shaped and nucleated erythrocytes. Those of mammals are smaller and lose their nuclei, mitochondria, and many other organelles during their development (Fig. 14.2). They are more numerous and more densely packed with hemoglobin than those of ectothermic vertebrates. The biconcave shape they normally assume with the loss of the nucleus provides more surface area than a sphere of equal volume, thus increasing surface area for the passage of gases and other materials. As the erythrocytes traverse narrow capillaries, shear forces often distort their shape, but they are very pliant cells and resume their biconcave shape when not so confined. Red cells are the most numerous of the cellular elements of the blood, there being about 5,200,000/mm³ in the blood of an adult man and 4,700,000/mm³ in the blood of an adult woman.

By tagging mammalian erythrocytes with radioactive iron, it has been shown that they have an average life span of 120 days. Cells lining the blood spaces of the spleen and liver eventually engulf or **phagocytize** (Gr. *phagein*, to eat) the red cells and digest them. The iron of the heme is salvaged by the liver and reused, but the rest of the heme is metabolized into a bile pigment, **bilirubin**, and excreted by the liver and kidneys. Under normal circumstances, the

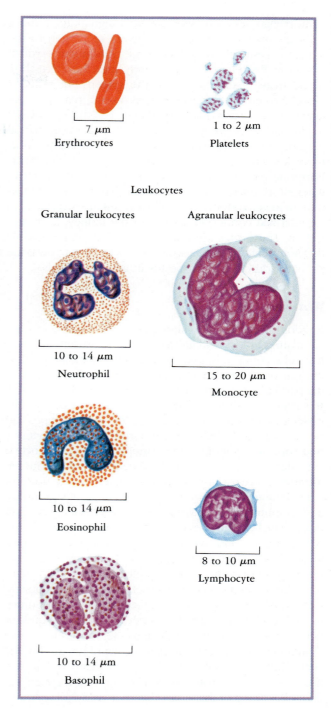

Figure 14.2
Human blood cells.

erythrocyte population is in a steady state, with new ones being synthesized as rapidly as the old are destroyed. One estimate of the turnover is 230,000 cells per second! If delivery of oxygen to the tissues is reduced, as during hemorrhage or when ascending to a high altitude, more cells are made available and delivery of oxygen to the tissues is restored to normal levels. Additional erythrocytes can be released immediately into the circulating blood from reserve stores in the spleen, but reduced blood oxygen tension also triggers a set of reactions that leads to an increased rate of synthesis of erythrocytes. Cells in the kidney in particular respond to a lower partial pressure of oxygen in the blood by synthesizing a hormone known as **erythropoietin** (Gr. *poiesis*, making). This hormone passes in the blood to the red bone marrow, where it stimulates the first stage in red blood cell production. In lower vertebrates, red blood cells are produced in vascularized connective tissues of the kidney, liver, and spleen. These sites are important during the embryonic development of mammals, but the red bone marrow is the primary source of erythrocytes in adult mammals.

White Blood Cells

In addition to cells containing respiratory pigments, the circulating fluids of all animals contain a variety of colorless cells, commonly amoeboid, that perform many different functions. The circulatory system transports them to the particular part of the body where they may be needed. In vertebrates, these cells are known as white blood cells or **leukocytes** (Gr. *leukos*, white). Five distinct types can be recognized (Fig. 14.2). Granular leukocytes—**neutrophils**, **basophils**, and **eosinophils**—have distinctive cytoplasmic granules and irregular-shaped nuclei. **Monocytes** have a large nucleus indented on one side and considerable cytoplasm that may contain a few granules, and **lymphocytes** have a nearly spherical nucleus and little cytoplasm. Lymphocytes originate in and are stored in the lymph nodes and other lymphoid tissues; the other leukocytes, in red bone marrow. Collectively, human leukocytes normally range in number from 5000 to 7000/mm³. Large numbers of leukocytes are synthesized during infections and released from storage. All have transit times in the blood of only a few hours because they can squeeze between cells in capillary walls and enter the tissues where their functions are performed. Some lymphocytes live for years, but most leukocytes are soon destroyed in the tissues or are lost from the body by passing through the epithelial lining of the lungs, digestive tract, kidneys, and reproductive passages.

Leukocytes play many important roles in the defense of the body. Lymphocytes are the key cells in the body's immune system, identifying bacteria and other foreign material and initiating responses that lead to their destruction (Chap. 15). Neutrophils are the most abundant leukocytes in the blood and phagocytose bacteria and dead and injured cells. They are the body's first line of defense, and large numbers collect at sites of injury and infection. Monocytes transform into large phagocytic macrophages that are very effective scavengers. Eosinophils respond primarily to parasitic invasions, collect around parasites, and release substances that destroy many of them. Eosinophils also contain enzymes capable of detoxifying toxins released by bacteria. Basophils release **histamine** in areas of tissue damage. Histamine dilates vessels and increases their permeability, and this increases blood flow and facilitates the movement of neutrophils into the damaged area. Basophils also produce heparin, an anticoagulant.

HEMOSTASIS

The loss of blood or hemolymph through an injury can severely impair the transport of materials to and from the body cells, and animals with circulatory systems have evolved mechanisms for **hemostasis** (Gr. *haima*, blood + *stasis*, standing); that is, for stopping the flow of blood after an injury. Hemostasis in all animals involves a short-term closure of the blood vessels, followed by the formation of a clot. Contraction of the clot holds the injured tissues together until fibroblasts penetrate the area and form fibrous material that permanently heals the injury. In animals with closed circulatory systems, nerve reflexes cause small broken vessels to constrict, thereby reducing blood flow. Many crustaceans with open systems can discard a limb that has been caught by a predator, a process known as **autotomy**. The limb is severed at an autotomy plane at the limb base, and a specialized membrane constricts around the perforation so there is very little loss of hemolymph.

Temporary closure of vessels in most animals is also achieved by a clumping of blood cells in the injured vessel or hemolymph space. In nonmammalian vertebrates **thrombocytes** (Gr. *thrombos*, clot), spindle-shaped nucleated cells, clump and initiate clot formations, but in mammals **platelets** play this role (Fig. 14.2). When they contact damaged vascular surfaces they swell and become sticky, adhering to each other and to the walls of the blood vessels (Fig. 14.3). Platelets are not complete cells but nonnucleated fragments of cytoplasm that bud off from giant **megakaryocytes** in the bone marrow. They number between 150,000 and 350,000/mm³ in human blood and survive for 8 to 10 days.

Whether a clot forms or not depends on a balance between 40 or more different substances in the blood and surrounding tissues. Normally blood does not clot within the vessels because the effects of **anticoagulating factors** outweigh those of **procoagulants**. When a tissue is injured and vessels rupture, the activity of the procoagulants increases and a clot forms. The injury leads to the formation of a complex called **prothrombin activator**, which is composed of phospholipids and several activated procoagulants. In the presence of calcium ions, prothrombin activator catalyzes the conversion of a plasma protein, **prothrombin**, to thrombin (Fig. 14.3). **Thrombin** is an enzyme that cleaves four small peptides from the soluble

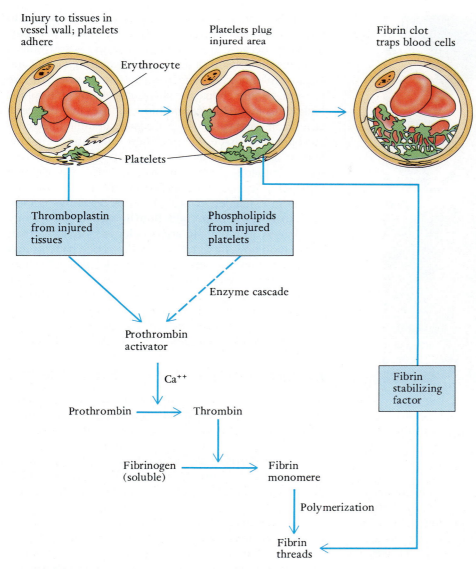

Figure 14.3
A diagram of clot formation in a small mammalian blood vessel and the factors
that cause it.

plasma protein, **fibrinogen**, thereby forming small threads of **fibrin monomer**. These monomers spontaneously polymerize, first forming loosely bound fibrin threads. Another enzyme, **fibrin stabilizing factor**, which is released by platelets entrapped in the developing clot, soon establishes stronger, covalent bonds between the monomers. The fibrin threads form a delicate network within which the formed elements of the blood are trapped (Fig. 14.4). A clot has formed. The clot subsequently contracts and squeezes out the liquid phase, called **serum**. Serum differs from plasma in that it lacks fibrinogen and most of the other coagulating factors.

The formation of prothrombin activator is obviously crucial to the initiation of a clot. Two pathways are involved

(Fig. 14.3). Firstly, the injured tissues themselves release a **tissue thromboplastin** that quickly stimulates the formation of a small amount of prothrombin activator. This reaction is rapid but does not form a large clot. Secondly, trauma to the platelets and their contact with connective tissue causes them to break up and release **phospholipids** that set off a chain reaction. One blood clotting factor is activated, which in turn activates a second, and so on through an "enzyme cascade." This chain reaction has a multiplying effect because an increased volume of factors is activated at each step. A large volume of prothrombin activator is produced, leading to the formation of a larger clot, but the process takes longer than the first pathway does.

Figure 14.4
Scanning electron micrograph of a portion of a blood clot showing an erythrocyte enmeshed in a network of fibrin.

After a few days, when healing is well under way, enzymes from the injured tissue and factors in the blood convert **profibrinolysin** (Gr. *lysis*, loosening), a blood factor that has been incorporated in the clot, into its active form. **Fibrinolysin** is a proteolytic enzyme that cleaves the fibrin molecules and breaks up the clot.

This same mechanism can destroy clots that start to form in vessels. Intravascular clots do not form easily, however, because the smoothness of the vessel linings prevents the release of the various activating factors and because inhibitors and anticoagulants, such as heparin, are present. Some heparin is produced by basophilic leukocytes in the blood, and more diffuses into the capillaries from mast cells, which are large, granule-filled cells in connective tissue. Despite these mechanisms, a clot, known as a **thrombus**, may develop and can be very serious if it plugs a vessel supplying a vital area, or if part of a thrombus in a larger vessel breaks off as an **embolus** and lodges in a smaller, critical vessel. A thrombus is most likely to form when the epithelium lining a vessel becomes roughened, as it does in arteriosclerosis. An abnormally slow blood flow, to which immobilized bed patients may be subject, may also lead to thrombus formation because procoagulants, which are always present in small quantities, may accumulate. In the hereditary disease **hemophilia** (p. 104) there is a deficiency of one of the blood clotting factors in the enzyme cascade, clots do not form well, and a slight scratch may lead to fatal bleeding.

INVERTEBRATE CIRCULATORY PATTERNS

Internal transport systems have evolved in those invertebrates, such as annelids, molluscs, arthropods, and echinoderms, that have a body size too large for diffusion alone to supply the tissues. Although the arthropod system may have evolved from that of annelids, those of molluscs and other invertebrates probably evolved independently. Flow patterns are closely correlated with the overall body design.

Most arthropods and molluscs have open systems, so blood-tissue exchange takes place in the tissue spaces, or hemocoel (Figs. 14.5 and 14.6). A central, pumping heart is present. In molluscs and gill-bearing arthropods, such as

Figure 14.5
A diagram of the circulatory system of a clam. One shell has been removed.

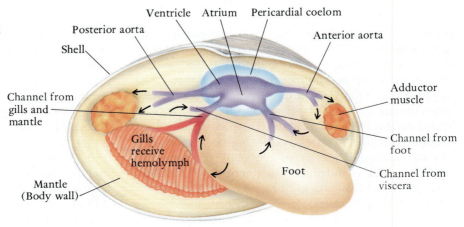

lobsters and crabs, at least part of the blood flows from the hemocoelic spaces to the gills before returning to the heart. The heart is therefore pumping at least partly oxygenated blood. In molluscs the space surrounding the heart is a true coelom, and usually a vessel from each pair of gills crosses the coelom to reach the heart atria, one of which is located on either side of the ventricle. Other channels enter the atria from the foot and viscera. The atria contract first, filling the ventricle; then the ventricle contracts. One-way valves between the chambers prevent backflow, and a similar valve is present at the junction of the ventricle and the aortas leaving the heart.

In arthropods the pericardial space is not a coelomic cavity but a blood-filled sinus. Blood from the gills collects here before entering the heart through paired, dorsolateral openings, or **ostia** (Fig. 14.6). Valves in the ostia prevent backflow when the heart contracts. In both molluscs and arthropods a system of arteries conducts blood from the heart to various parts of the body. However, in many arthropods such as insects, the arterial system is poorly developed. Only a short aorta at each end of the heart empties into the hemocoel. The reduction of the arterial system in insects is related to the transport of gases by the tracheal system and not by the blood.

The circulatory system has become closed in the octopus and squid, a condition correlated with their active, predatory swimming existence and higher metabolic rate, but the pattern of the vessels is basically the same as in other molluscs. Significantly, the blood on entering the gills passes through an accessory branchial heart, which boosts the blood pressure for the passage of blood through the gills and its return to the heart.

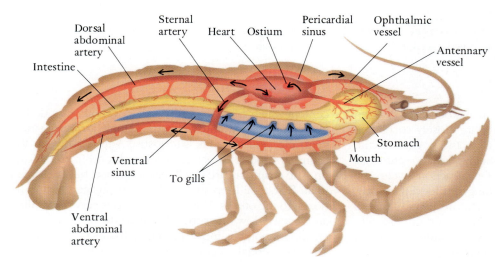

(a)

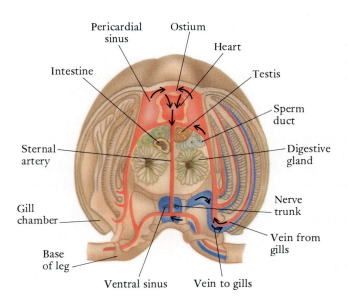

(b)

Figure 14.6

The circulatory system of a crayfish as seen in lateral view (a) and in a transverse section through the thorax (b).

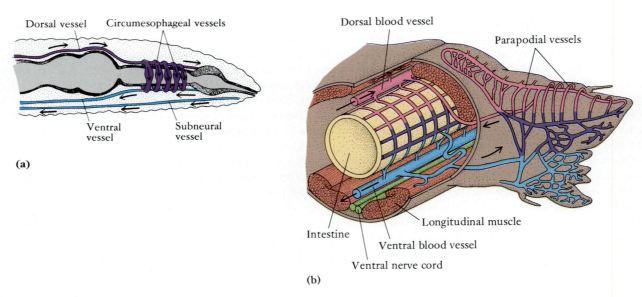

Figure 14.7

The circulatory system of annelid worms. (a) A lateral view of the anterior part of the system in an earthworm. (b) A transverse section through the trunk of a clam worm.

The annelid circulatory system may be closed or open to varying degrees. The pathways are closely correlated with the metameric body plan (Fig. 14.7). A contractile dorsal vessel, which extends through all of the segments, functions as the heart. Waves of contraction sweep the length of the dorsal vessel, propelling the blood anteriorly. Near the front of the body one or more pairs of lateral vessels carry blood around the gut to ventral and subneural vessels. In some annelids, such as the earthworm *Lumbricus*, the lateral vessels (circumesophageal vessels) are contractile and act as accessory hearts (Fig. 14.7a). Blood flows posteriorly in the ventral vessels and is distributed by branches in each segment to the gut and body wall, thence back to the dorsal vessel. Typically, paired branches also extend to the excretory organs (nephridia), body wall, and flaplike extensions of the body wall (parapodia) when present (Fig. 14.7b). Blood is usually aerated in the body wall or parapodia.

In some marine annelids and some leeches the coelom has taken over the function of internal transport, and the coelomic fluid may even contain blood corpuscles filled with hemoglobin. The coelomic fluid is also the principal means of internal transport in echinoderms.

VERTEBRATE CIRCULATORY PATTERNS

Primitive Fishes

All vertebrates have closed circulatory systems. In primitive fishes (Fig. 14.8), all of the blood entering the heart from the veins has a low oxygen and a high carbon dioxide content. Such blood is often called venous blood. The heart consists of a **sinus venosus**, a single **atrium**, a single **ventricle**, and a **conus arteriosus**.* Valves between the chambers and in the conus prevent blood from backing up. The chambers are in a linear sequence, but are folded in an **S**-shaped loop so the sinus venosus and atrium lie dorsal to the others. The sinus venosus is very thin walled and receives low-pressure blood from the veins. The other chambers contain more muscular tissue in their walls. Their sequential contraction increases the blood pressure and sends the oxygen-poor blood out through an artery, the **ventral aorta**, to five or six pairs of **aortic arches** that extend dorsally through capillaries in the gills to the **dorsal aorta**.† Carbon dioxide is removed and oxygen is added as the blood flows through the capillary beds in the gills. The dorsal aorta distributes oxygen-rich blood, often called arterial blood, through its various branches to all parts of the body.

Blood pressure decreases as blood flows along because of the viscosity of the blood and the friction within it, and between it and the lining of the vessels. Pressure is reduced considerably as the blood passes through the numerous gill capillaries. Blood pressure in the ventral

---

* Note that this is a four-chambered heart, but the atrium and ventricle are not divided as they are in the four-chambered heart of birds and mammals.

† Note that arteries are defined as vessels that carry blood away from the heart, and veins as ones that carry it toward the heart, not by the oxygen content of the blood within them.

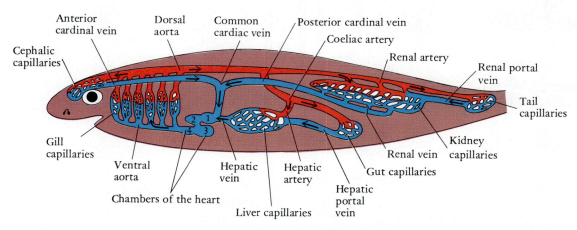

Figure 14.8
The major parts of the circulatory system of a primitive fish, such as a shark. The chambers of the heart are numbered: 1 = sinus venosus, 2 = atrium, 3 = ventricle, and 4 = conus arteriosus.

aorta of a dogfish during heart contraction, for example, is about 25 mm Hg; that in the dorsal aorta, about 18 mm Hg. This relatively low blood pressure in the aorta is reduced further as the blood passes through the capillaries in the tissues. Circulation in most fishes is rather sluggish.

Veins drain the capillaries and converge on larger ones that lead to the heart, but not all veins go directly to the heart. In fishes, blood returning from the tail first passes through capillaries in the kidneys before entering veins leading to the heart. Veins that drain one capillary bed and lead to another are called portal veins, and these particular veins are known as the **renal portal system**. Other veins, known as the **hepatic portal system**, drain the digestive tract and lead to capillary-sized sinusoids in the liver (p. 251). The liver is drained by **hepatic veins**. Since much of the blood returning to the heart has passed through one or the other of these portal systems, in addition to the capillaries in the gills and tissues, its pressure as it approaches the heart is near 0 mm Hg.

Evolution of the Double Circulation

As lungs replaced gills as the site for gas exchange during the transition of vertebrates from water to land, a **pulmonary circulation** through the lungs functionally replaced the **branchial circulation** through the gills. Birds and mammals have a double circulation, for the heart is completely divided into a right side that receives oxygen-poor blood from the body and sends it to the lungs, and a left side that receives oxygen-rich blood from the lungs and sends it to the body (Fig. 14.9d). The volume of blood flowing through the pulmonary circulation to the lungs and the **systemic circulation** to the body is always equal.

Unlike birds and mammals, lungfishes, amphibians, and reptiles do not ventilate their lungs continuously. There are long periods of apnea when the lungs are not ventilated. As oxygen in the lungs is used up, lung volume is reduced, the lungs collapse to some extent, and pulmonary resistance to blood flow increases. To continue to pump blood to the lungs under these circumstances would be futile and energetically expensive. An incomplete division of the atrium and/or ventricle enables different volumes of blood to go through the pulmonary and systemic circuits as circumstances warrant. The absence of a completely double circulation should not be viewed as an inefficiency, as was once believed, but as an adaptation of these animals to their environment and mode of life.

We know little directly about the circulatory system of the ancestors of terrestrial vertebrates, but like present day lungfishes, they lived in an aquatic environment in which the partial pressure of oxygen may have been reduced, at least during dry seasons. They probably resembled lungfishes in having a pulmonary circulation that could supplement branchial circulation, and a partly divided heart (Fig. 14.9a). The atrium and ventricle of a lungfish are partly divided into left and right chambers, and a complex spiral valve in the conus arteriosus shunts oxygen-poor blood from the right side of the heart to the posterior arches. These lead to gills, and the last pair also connects via **pulmonary arteries** to the lungs. Oxygen-rich blood, which returns from the lungs in **pulmonary veins** to the left atrium, is shunted to the more anterior arches, most of which go directly to the dorsal aorta without passing through gills. The dorsal aorta is paired in the pharyngeal region. The system is very flexible. When the water is well aerated and environmental conditions allow the use of gills, most of the blood is shunted through them, and the

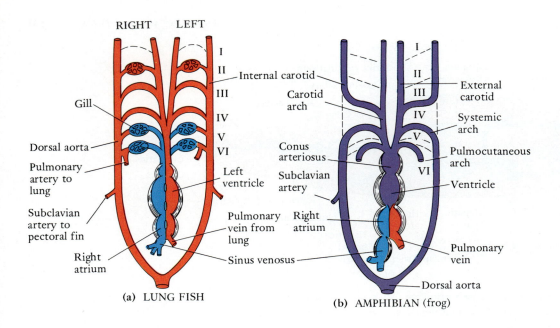

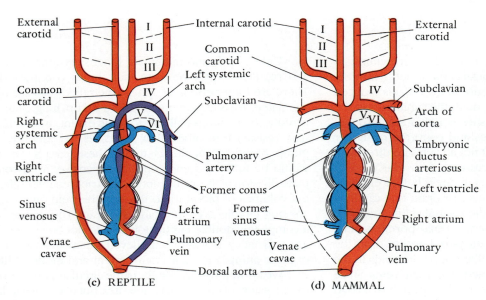

(a) LUNG FISH

(b) AMPHIBIAN (frog)

(c) REPTILE

(d) MAMMAL

Figure 14.9

Evolution of the heart and arteries in (a) a lungfish; (b) a frog; (c) a reptile; and (d) a mammal. These diagrams show the heart chambers drawn out in linear sequence, and the aortic arches as they would appear flattened out in a ventral view. The heart tube in reality is folded upon itself, and the aortic arches extend dorsally between pharyngeal pouches from the ventral to the dorsal aorta.

lungs are not ventilated very much. When environmental conditions require the use of lungs, they are ventilated more frequently and considerable blood passes through them.

Amphibians and Reptiles

Adult amphibians and all reptiles have lost the branchial circulation. Amphibians have a heart that is less completely divided than that of lungfishes for there is no division of the ventricle (Fig. 14.9b). Most reptiles have a completely divided atrium, but only a partial division of the ventricle (14.9c). Cineradiographic studies that trace the movement of opaque materials through the heart have confirmed that despite the absence of a complete division of the heart, oxygen-poor and oxygen-rich blood streams do not mix as much as might be expected. Reptiles have an unusual

tripartite division of the conus arteriosus and ventral aorta and a rather complex cardiac mechanism. When the ventricle begins to contract, oxygen-poor blood, which has been sequestered in a part of the ventricle, leaves first in the pulmonary artery to the lungs. Continued contraction of the ventricle narrows the opening between the left and right side of the ventricle, and oxygen-rich blood now starts to leave through both right and left systemic arches. The position of the entrances of the two systemic arches, and the pattern of their branches, allow for some mixing of blood streams, with more oxygen-poor blood going to the body than to the head.

The incomplete division of the heart of amphibians and reptiles allows for alterations in the volume of blood that flows through the lungs and body as conditions change. Sometimes these animals are submerged in water for long periods, but even on land the lungs are not

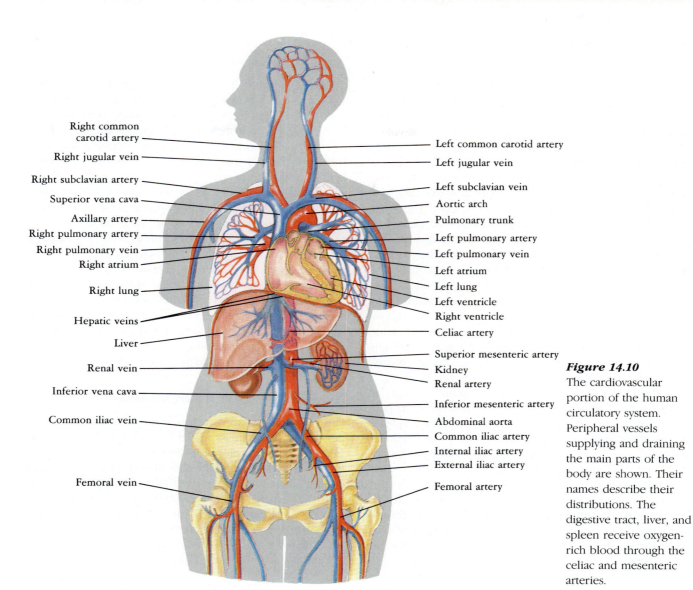

Right common carotid artery

Right jugular vein

Right subclavian artery

Superior vena cava

Axillary artery

Right pulmonary artery

Right pulmonary vein

Right atrium

Right lung

Hepatic veins

Liver

Renal vein

Inferior vena cava

Common iliac vein

Femoral vein

Left common carotid artery

Left jugular vein

Left subclavian vein

Aortic arch

Pulmonary trunk

Left pulmonary artery

Left pulmonary vein

Left atrium

Left lung

Left ventricle

Right ventricle

Celiac artery

Superior mesenteric artery

Kidney

Renal artery

Inferior mesenteric artery

Abdominal aorta

Common iliac artery

Internal iliac artery

External iliac artery

Femoral artery

Figure 14.10
The cardiovascular portion of the human circulatory system. Peripheral vessels supplying and draining the main parts of the body are shown. Their names describe their distributions. The digestive tract, liver, and spleen receive oxygen-rich blood through the celiac and mesenteric arteries.

ventilated continuously. The shunting of some venous blood to the body when the lungs are not in use does not necessarily mean that less oxygen is delivered to the tissues. Present day amphibians also supplement pulmonary respiration with cutaneous respiration. Reptiles do not exchange a significant amount of gas through their skin surface, but the increased carbon dioxide content of the blood will cause hemoglobin to unload a greater amount of its bound oxygen (the Bohr effect, p. 286), and this partly compensates for the mixture of bloods. When the lungs are being fully ventilated, some of the blood that has been through the lungs is recycled to the lungs. This fraction of the blood picks up additional oxygen, so the blood becomes more thoroughly aerated than otherwise would be the case.

Changes in the aortic arches correlate with the shift from a branchial to a pulmonary circulation. The first two and fifth aortic arches are lost by most adult amphibians and by all reptiles. Those that remain are no longer interrupted by gill capillaries. The third pair of aortic arches forms part of the **internal carotid arteries** supplying the head; the fourth, the **systemic arches** leading to the dorsal aorta; and the sixth, the pulmonary arteries leading to the lungs and, in amphibians, the skin. The loss of gills and gill capillaries, and the return of blood from the lungs to the heart where its pressure can be elevated again, make a slightly higher blood pressure possible in primitive terrestrial vertebrates than in primitive fishes.

Birds and Mammals

Adult birds and mammals ventilate their lungs continuously, making it unnecessary for them to alter the volume of blood flow through the pulmonary and systemic circuits. Their hearts are completely divided (Fig. 14.9d and 14.10). Oxygen-poor blood from the body enters the **right atrium**, into which the primitive sinus venosus has become incorporated. This blood enters the **right ventricle**, which pumps it through the pulmonary artery to the lungs. Oxygen-rich blood from the lungs enters the **left atrium** via the pulmonary vein. From there the blood enters the **left ventricle**, which pumps it through the aorta to the body. The primitive conus arteriosus has become completely divided, part contributing to the base of the pulmonary artery and the rest to the arch of the aorta. The complete separation of blood in the heart makes possible different degrees of muscularization of the ventricles. The right ventricle develops only enough pressure to drive the blood through the nearby lungs, and it is not much more muscularized than the reptile ventricle. Too high a pressure in blood going to the lungs would result in the loss of considerable water from the blood by hydrostatic flow from the capillaries, across the lung epithelium, and into the lung cavity (see p. 282). The left ventricle, which now pumps only to the systemic circuit, becomes heavily muscularized and develops a mean blood pressure of 100 mm Hg in the larger arteries of humans. This distributes materials rapidly throughout the body.

The sixth pair of aortic arches of mammals form the major part of the pulmonary arteries, and the third pair contribute to the internal carotid arteries. The left side of the fourth arch, known as the arch of the aorta, becomes enlarged and leads to the dorsal aorta. The right fourth arch contributes to the right subclavian artery to the shoulder and arm but does not connect posteriorly with the aorta. In birds, which evolved from reptiles independently of mammals, the situation is reversed, for the right fourth aortic arch persists and the left one is lost. Loss of one fourth arch and the enlargement of the other one reduces the resistance of blood flow to the body.

The major change in the veins of mammals is the complete loss of a renal portal system. Blood from the tail and posterior appendages enters an **inferior** or **posterior vena cava** (Fig. 14.10) that continues forward to the heart. It receives blood from the kidneys but does not carry blood to them. The advantages of the loss of the renal portal system are not entirely clear, but it does speed up the rate of venous return to the heart. A **superior** or **anterior vena cava** drains the head and arms. The hepatic portal system is still present, and the liver is drained by hepatic veins.

As circulatory systems have evolved greater complexity, blood pressure has increased; therefore, more liquids and plasma proteins escape from the capillaries into the interstitial fluid than are returned by the veins. A separate **lymphatic system** has evolved that returns excess fluid and plasma proteins from the tissues to the main part of the circulatory system. Lymphatic vessels develop as outgrowths from the veins and, in general, tend to parallel the veins and ultimately empty into them (Fig. 14.11). The lymphatic system reaches its greatest development in mammals. **Lymphatic capillaries** that have dead-ends and do not connect with arteries or veins occur in most of the tissues of the body. They are more permeable than other capillaries, and pressures within them are exceedingly low. Their high permeability also makes them the most likely route for the spread of microorganisms or cancer cells within the body. Lymph capillaries converge on small lymphatic vessels that convey lymph toward the heart. **Lymph nodes** lie at many points where small lymphatic vessels converge. They are important sites for the production of lymphocytes, and cells within them can phagocytize invading bacteria or respond to them by initiating antibody production (Chap. 15). A **thymus** on the ventral surface of the neck is an aggregation of lymphocytes. It is particularly large in young people, whose immune system is developing and regresses with age. The **spleen** is a vascular lymphoid organ located on the left side of the stomach. In addition to storing red blood cells, it is also an important site of lymphocyte production. Most of the small lymphatic vessels enter a larger **thoracic duct** or a **right lymphatic duct** before entering the large veins near the heart.

Fetal and Neonatal Circulations

Most mammalian species, when they become pregnant, develop a placenta in which maternal and fetal blood streams come very close together but do not mix (p. 525). It is here, rather than in the fetal digestive tract, lungs, and kidneys, that exchange of materials occurs. This, coupled with the fact that the blood vessels in the unexpanded lungs of the fetus are not developed enough to handle the total volume of blood that is circulating through the body, necessitates certain differences between the fetal and adult circulatory systems (Fig. 14.12a). Blood rich in oxygen returns from the placenta in an **umbilical vein**, passes rather directly through the fetal liver via the **ductus venosus**, and enters the posterior vena cava. The entrance of the posterior vena cava into the right atrium is directed toward an opening, the **foramen ovale**, in the partition separating the two atria, and most of the blood from the posterior vena cava passes through this foramen into the left atrium, thence to the left ventricle and out to the body through the arch of the aorta. The foramen ovale bypasses the lungs yet permits the left side of the heart, which otherwise would receive little blood from the collapsed lungs, to function and develop normally.

The rest of the blood from the posterior vena cava enters the right ventricle along with the blood from the anterior vena cava and starts out of the pulmonary artery toward the lungs. However, the high pulmonary resistance allows only a fraction of this blood to pass through the lungs to return to the left atrium; most goes through another bypass, the **ductus arteriosus**, to the dorsal aorta. This bypass of the lungs also allows the right ventricle to pump the volume of blood needed to develop normally even though little goes through the lungs. The ductus arteriosus represents the dorsal part of the left sixth aortic

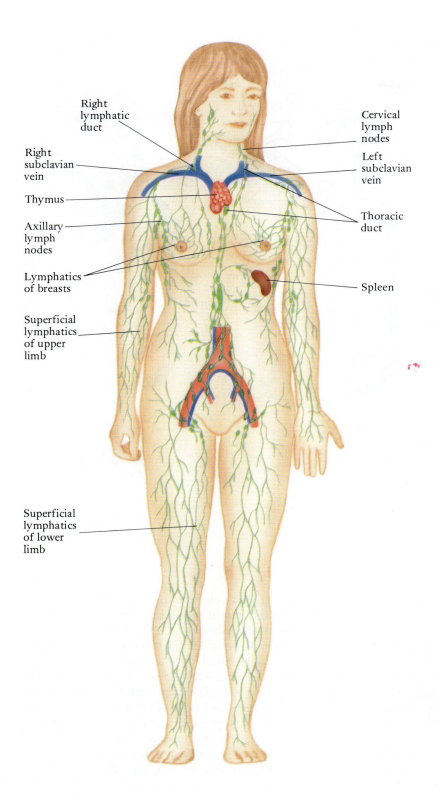

Right
lymphatic
duct

Right
subclavian
vein

Thymus

Axillary
lymph
nodes

Lymphatics
of breasts

Superficial
lymphatics
of upper
limb

Superficial
lymphatics
of lower
limb

Cervical
lymph
nodes

Left
subclavian
vein

Thoracic
duct

Spleen

Figure 14.11
The lymphatic portion of the circulatory
system.

arch (Fig. 14.9d). Since the ductus arteriosus enters the
aorta after the arteries to the head and arms have branched
off, these parts of the body receive the blood with the
highest oxygen content. After the entrance of the ductus
arteriosus, the blood in the aorta is highly mixed. This is

the blood that is distributed to the rest of the body and, by
way of **umbilical arteries**, to the placenta.

At birth, the placenta is expelled, and blood volume in
the systemic circuit is reduced (Fig. 14.12b). Carbon
dioxide accumulates in the fetal blood, activating the

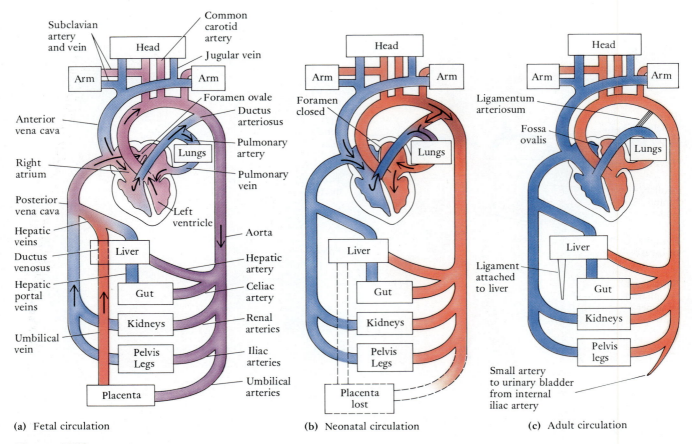

(a) Fetal circulation   **(b)** Neonatal circulation   **(c)** Adult circulation

Figure 14.12
Diagrams in ventral views of the mammalian circulation at different stages of life.
(a) Fetal circulation; (b) neonatal circulation; and (c) adult circulation.

respiratory system; the lungs fill with air, and the pulmonary vessels that were collapsed open up. Resistance to blood flow in the pulmonary circulation is now less than that in the systemic circulation. More blood flows through the lungs and returns to the left atrium than during fetal life. The valve in the foramen ovale is pushed against the interatrial septum and soon adheres to it by the growth of tissue. This bypass of the lungs is cut off. Its location continues to be represented in adults by a depression, the **fossa ovalis** (Fig. 14.12c).

The circulatory pattern is now very close to the adult condition except that the ductus arteriosus remains open. Since pulmonary resistance is less than systemic resistance, the direction of flow in the ductus arteriosus is reversed. Some of the blood that has already been through the lungs and is leaving the heart in the arch of the aorta flows back to the lungs through the ductus arteriosus. This pattern of circulation, which lasts from several hours to a day or two in the human infant, is known as the **neonatal**

circulation (Gr, *neo*, new + *natus*, born). Experiments on newborn lambs show that this reversal of flow, and the consequent double aeration of some of the blood, is of great significance because a changeover is taking place between the synthesis of fetal and adult types of hemoglobin. Although fetal hemoglobin has a higher affinity for oxygen than adult hemoglobin (p. 288), it also holds onto it to a greater extent than does adult hemoglobin. If the ductus arteriosus is experimentally tied off during this period, 10 to 20% less oxygen is delivered to the tissues.

Shunts in fetal and neonatal mammals perform the same functions as those in reptiles. They allow some blood to be diverted from the lungs when they are not in use and pulmonary resistance is high, as in fetal life, and they recirculate blood through the lungs when pulmonary resistance is low and a more thorough aeration of the blood is needed, as occurs immediately after birth.

Muscles in the wall of the ductus arteriosus eventually contract and stop all blood flow through it, and the adult circulatory pattern is established (Fig. 14.12c). Eventually, the duct becomes permanently occluded by the growth of fibrous tissue into its lumen and is converted into the **ligamentum arteriosum**.

THE PROPULSION OF BLOOD AND HEMOLYMPH

Hearts

Energy must be expended to move materials through transport systems. Cilia or flagella propel fluids in sponges, in gastrovascular cavities of cnidarians and flatworms, and in coelomic channels of echinoderms. They are effective in these situations because the cavities are generally small and a rapid flow is not required. If a simple liquid is to be moved rapidly through larger channels, muscle pumps, called **hearts**, are energetically more efficient. All animals with well-developed circulatory systems have hearts of some type. In its simplest form a heart is a pulsating vessel without valves in which blood is pushed along by waves of peristaltic contraction (Fig. 14.13a). **Peristaltic tubular hearts** of this type are found in many annelid worms.

Molluscs, arthropods, and vertebrates have **chambered hearts** composed of one or more enlarged chambers that receive and pump the blood or hemolymph (Fig. 14.13b), and valves that permit flow in one direction only. The heart frequently lies within a cavity, such as the pericardial sinus of arthropods or the pericardial coelom of molluscs and vertebrates, that facilitates its contraction and expansion.

Often the residual pressure in returning fluid is sufficient for the blood to enter a chambered heart. When two or more chambers are present, the first is thin-walled and easily expands as liquid enters. Sometimes, however, the pressure is so low in vessels or spaces just before the heart that the heart must actively expand. The pressure within it is reduced relative to that in the veins, and blood is "sucked" in. Arthropods have such a **suction heart** (Fig. 14.6). Thin muscle strands extend from the outside of the heart to the wall of the hemolymph-filled pericardial sinus in which it lies. Their contraction enlarges the single-chambered heart, and blood is sucked in through the ostia. Valves at the bases of the aortae leaving the heart close so that hemolymph is not drawn back into the heart from them. When the heart contracts, they open and valves in the ostia close.

The Mammalian Heart

The heart of mammals and other vertebrates lies within the pericardial cavity. It is covered with a smooth coelomic epithelium, the **visceral pericardium**, and is lined by the simple squamous epithelium, the **endothelium**, which lines all parts of the circulatory system. The rest of its wall is composed of **cardiac muscle** and dense connective tissue that forms a fibrous skeleton. Striated, cardiac muscle differs from striated skeletal muscle in important ways. Cardiac muscle fibers are not multinucleated, but consist of separate cells, each with a single, central nucleus (Fig. 14.14). The cells branch and unite with others exceptionally firmly, end to end, by transverse **intercalated discs**. Electron microscope studies have shown that the discs contain many gap junctions, i.e., regions of low electrical resistance that allow the rapid spread of action potentials from one cell to another. Atrial muscles are separated from ventricular muscles, but functionally each group acts as a unit. When one cell in the atrium or ventricle becomes active, the activity spreads rapidly to the others. Thus, the atria and ventricles follow the all-or-none phenomenon that applies to individual motor units of phasic skeletal muscles (see p. 218).

During a heart cycle, the atria and ventricles contract and relax in succession. Contraction of the ventricles is known as **systole** (Gr., drawing together); relaxation, as **diastole** (Gr., dilation). Since the muscle fibers of the ventricles are arranged in a spiral, the blood is not just pushed out when they contract but is virtually wrung out of them. When the ventricles relax, their elastic recoil reduces the pressure within them, and blood at first is drawn in from the atria. Atrial contraction completes the filling. The atria are primarily antechambers that accumulate blood during ventricular systole.

Figure 14.13
Major types of hearts. (a) A wave of contraction, 1, travels along a peristaltic tubular heart and is followed by a second wave of contraction, 2. (b) The chambered heart. The atrium is contracting and filling the ventricle.

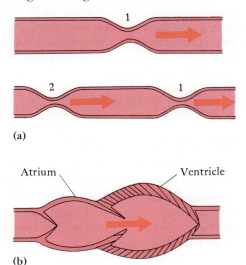

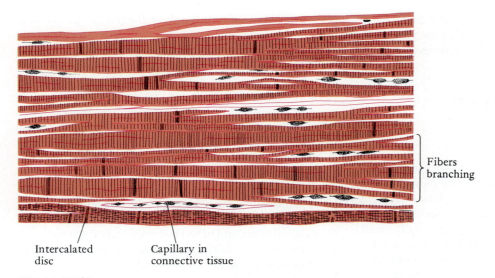

Fibers branching

Intercalated disc

Capillary in connective tissue

Figure 14.14
Cardiac muscle fibers as seen in a lateral view. Compare this with other types of muscle shown in Figure 11.19.

As we have seen, a complete division of the ventricle makes possible a heavier muscularization of the left ventricle than the right one, and it develops a higher pressure (Fig. 14.15a). Systemic blood pressure in a healthy adult is about 120 mm Hg during systole and drops to about 80 mm Hg during diastole. This is expressed as 120/80. Corresponding pulmonary pressures are 25/10. Systolic pressure, of course, measures the force with which the ventricles contract; diastolic pressure reflects the resistance of peripheral vessels to blood flow. An elevated diastolic pressure may indicate a disease in peripheral arteries such as a hardening of their walls.

Blood being pumped by the heart is prevented from moving backward by the closure of a system of valves (Fig. 14.15a). One with three cusps, known as the **tricuspid valve**, lies between the right atrium and ventricle; one with two cusps, the **bicuspid valve**, between the left chambers. These valves operate automatically as pressures change, opening when atrial pressure is greater than ventricular, closing when ventricular pressure is greater. **Tendinous cords** extend from the free margins of the cusps to the ventricular wall and prevent them from turning into the atria during the powerful ventricular contractions. When the ventricles relax, blood in the pulmonary artery and aorta, which is under pressure, tends to back up into them. This closes the **pulmonary** and **aortic semilunar valves** at the base of each of these vessels and prevents blood from returning to the ventricles. The pulmonary and aortic valves are each composed of three crescent-shaped pockets that fill with blood and press against each other. When blood first backs up against the closed valves, vibrations are set up that produce the characteristic "lub-dub"

sounds that can be heard with a stethoscope. "Lub" occurs at the beginning of systole when the atrioventricular valves close, and "dub" at the end of systole when the semilunar valves close. Abnormalities in the structure of the valves, occurring congenitally or produced by disease, may prevent their closing properly. Blood then leaks back during diastole; the leaking blood produces a "heart murmur."

Although a large volume of blood flows through the cavities of the heart, this blood does not provide for the metabolic needs of the heart musculature in many vertebrates. In mammals, a pair of **coronary arteries** arise from the base of the arch of the aorta and supply capillaries in the heart wall (Fig. 14.15b). This capillary bed is drained ultimately by a **coronary vein** that empties into the right atrium. Obviously, any damage to the coronary vessels, such as plugging of one of the larger arteries by a thrombus or embolus, could have serious consequences, for the heart muscles cannot function without a continuing supply of oxygen and food. In coronary bypass operations, blood is routed around the blockage.

Heartbeat and Its Integration

Heart musculature has an inherent capacity for beating. A heart, if properly handled, will continue to beat rhythmically when excised from the body. A group of specialized cardiac muscle fibers (nodal fibers) stimulates and integrates heart beat. Contraction is initiated in mammals by a **sinoatrial (SA) node** of these fibers located in that part of the wall of the right atrium into which the primitive sinus

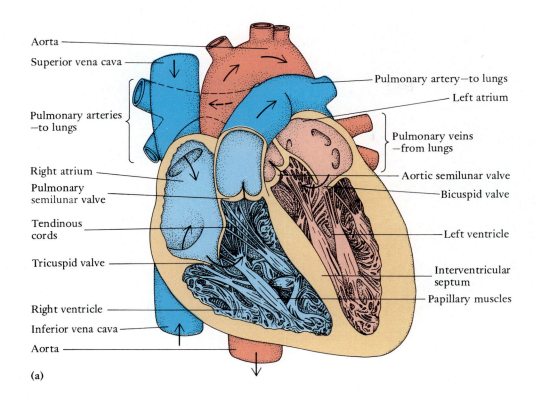

Aorta

Superior vena cava

Pulmonary arteries —to lungs

Right atrium

Pulmonary semilunar valve

Tendinous cords

Tricuspid valve

Right ventricle

Inferior vena cava

Aorta

Pulmonary artery—to lungs

Left atrium

Pulmonary veins —from lungs

Aortic semilunar valve

Bicuspid valve

Left ventricle

Interventricular septum

Papillary muscles

(a)

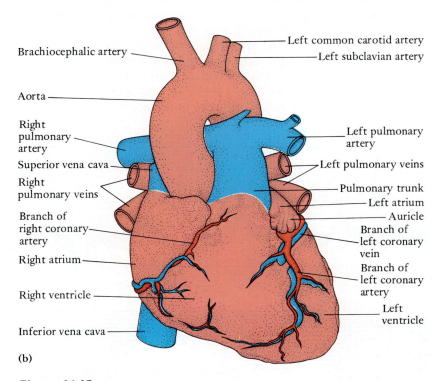

Brachiocephalic artery

Aorta

Right pulmonary artery

Superior vena cava

Right pulmonary veins

Branch of right coronary artery

Right atrium

Right ventricle

Inferior vena cava

Left common carotid artery

Left subclavian artery

Left pulmonary artery

Left pulmonary veins

Pulmonary trunk

Left atrium

Auricle

Branch of left coronary vein

Branch of left coronary artery

Left ventricle

(b)

Figure 14.15

The human heart. (a) The inside of the four chambers and their valves can be seen in this ventral dissection. (b) A surface view of the ventral surface showing the coronary arteries and veins.

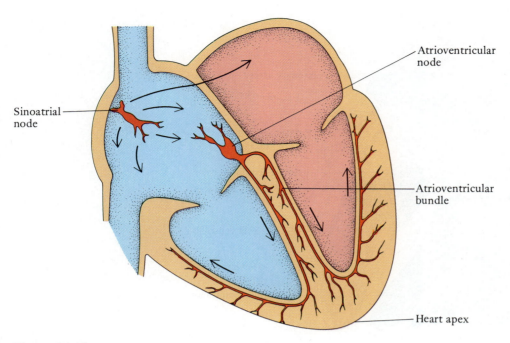

Figure 14.16
A diagram of the conducting system of the heart.

venosus is incorporated (Fig. 14.16). The impulse spreads through normal cardiac muscles to all parts of the atria and to another cluster of nodal fibers, the **atrioventricular (AV) node**, located near the ventricles. Since there is no muscular connection between atria and ventricles, ventricular contraction cannot begin until the atria have contracted and the impulse reaches the atrioventricular node. From the AV node, the impulse is picked up by the **atrioventricular bundle** and spreads very rapidly over the inside of the ventricles. Ramifications of the AV bundle are known as **Purkinje fibers**. Ventricular contraction begins at the apex of the heart and spreads toward the origin of the great arteries leaving it.

The electrical impulses generated by the contraction of cardiac muscle spread through the surrounding tissues and can be detected on the body surface with an electrocardiogram (Fig. 14–17). Departures from the normal pattern indicate diseases of the heart.

Figure 14.17
(a) A normal electrocardiogram. The P wave represents the contraction of the atria; the QRS wave, the contraction of the ventricles; and the T wave, the relaxation of the ventricles. (b) During atrial fibrillation, in which the muscles of the atria twitch rapidly, the ventricles also beat irregularly, and there is no T wave.

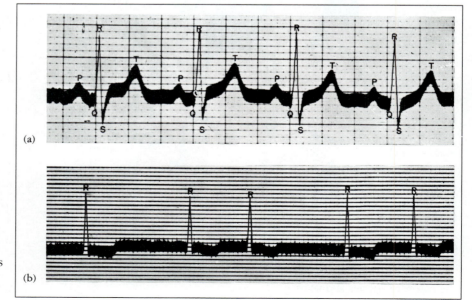

Cardiac Output

The heart of a normal adult human being who is not exercising sends about 70 ml of blood per beat out into the aorta. At the normal rate of 72 beats per minute, this is a total cardiac output of 5 L per minute, which is approximately equivalent to the total amount of blood in the body. As we have seen, small animals, with higher oxygen consumption per unit of volume, ventilate their lungs more frequently than large animals (p. 282). They also require more blood flow for each unit of tissue. During exercise all animals need an increased blood flow. Cardiac output must be adapted to these different needs. In theory, cardiac output could be increased by (1) having a large heart, (2) increasing the rate of heart beat (pulse rate), or (3) increasing the volume of output per beat or stroke (stroke volume).

One might expect that small mammals would have relatively larger hearts than larger mammals, but this is not the case. The heart of a shrew, a human, and an elephant each constitutes about 0.6 percent of body weight. Relative heart size is not an important factor in determining cardiac output in mammals. Changing the rate of heart beat is the major factor in adjusting cardiac output to differing metabolic needs of different species. The pulse rate of a shrew at rest is over 600 beats per minute, that of humans is about 70 beats per minute, and that of an elephant is only 25 beats per minute. Rate also increases in a given species during exercise, but the rate of increase is not directly proportional to the increased oxygen needs. The blood itself delivers more oxygen because the oxygen tension in active tissues drops, the pH rises, and a greater proportion of the oxygen carried by the blood is unloaded (p. 287).

The increased blood flow to the tissues needed during exercise is primarily met by increases in stroke volume. The more rapid return and increased pressure of oxygen-poor blood that occurs during exercise stretches the heart musculature. This causes the heart to contract with greater force and to send out the greater volume of blood received during each period of atrial diastole. Within physiological limits, the greater the tension on cardiac (or any other) muscle, the more powerful will be its contraction. The capacity of the heart to adjust its output per stroke to the volume of blood delivered to it is known as **Starling's law of the heart**. It is chiefly the increased stroke volume that enables the heart of a well-trained athlete to increase cardiac output many times over the normal resting output, e.g., from 5 L per minute to a range of 20 to 35 L per minute.

If cardiac output falls significantly, the tissues receive an insufficient supply of oxygen and nutrients, a condition known as **circulatory shock**. Many factors can lead to circulatory shock. The cardiac muscles may be damaged by an interruption of the coronary blood supply, preventing the heart from beating normally. A severe hemorrhage may result in such a great reduction of blood volume that there is not enough pressure in the peripheral vessels for an adequate return of blood to the heart.

Cardiac Control

Various control mechanisms adjust cardiac output to an animal's needs. The hearts of annelids and molluscs are primarily **myogenic**; heart beat initiates in the heart musculature itself, and rate of beat is influenced by extrinsic factors that affect the musculature directly. Increases in temperature or of the pressure of the fluid entering the heart cause an acceleration in heart rate. Inhibiting and accelerating neurons terminate on the molluscan heart but are of less importance in regulating beat.

The hearts of arthropods and vertebrates, although having an underlying myogenic rhythm, are **neurogenic**; nerve impulses greatly influence the activity of the heart. Sympathetic and parasympathetic neurons of mammals (p. 388), which respectively accelerate and decelerate the rate and force of heart beat, extend from a vasomotor control center in the medulla of the brain to the SA node, and often to other parts of the heart. The activity of the vasomotor center is affected by many factors. **Baroceptors** (Gr. *baros*, pressure + L. *capere*, to take) in the atrial wall, and in the walls of large arteries near the heart, detect changes in blood pressure and send signals on sensory neurons to the vasomotor center. By a combination of excitation and inhibition of appropriate motor neurons, heart rate will decrease if pressure is too high or increase if pressure is too low. Chemoreceptors in the carotid bodies at the junctions of the internal and external carotid arteries monitor oxygen and carbon dioxide content in the blood and its pH. In addition to affecting the rate of breathing (p. 284), they send impulses to the vasomotor center. If oxygen levels are too low, heart rate and blood pressure increase, thereby delivering more blood to the tissues. Many other factors, including hormones and emotional state, can also affect the vasomotor center.

THE PERIPHERAL FLOW OF BLOOD AND HEMOLYMPH

The Flow of Liquids in Pipes and Blood Vessels

Arteries, capillaries, and veins are in essence a set of pipes that supply and drain the tissues; the heart is the pump that drives the blood. The physics of flow through pipes determines to a large extent the design features of the circulatory system. The total flow through a system in a unit of time (F) is related to pressure and the resistance of the vessels by the formula

$$F = \Delta P/R$$

where ΔP is the pressure difference between the beginning and end of the set of pipes or blood vessels, and R is

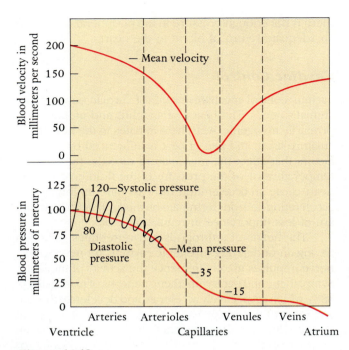

Figure 14.18
A graph of the variation in blood velocity and pressure in different regions of the circulatory system of a human.

the resistance to flow. (Those familiar with the physics of electric current flow will recognize this formula as Ohm's law, in which the current flow in amperes = the potential in volts/resistance in ohms.) Resistance to flow, which the heart must expend energy to overcome, is proportional to the viscosity of the blood, the length of the vessel, and the inverse of the fourth power of the radius of the vessel. Clearly vessel diameter is particularly important because it determines the surface area lining vessels. Blood encounters frictional resistance as it flows across this surface, just as water does as it flows across the surface of a swimming fish (p. 226). It costs much more energetically to move material through small vessels than large ones because of the increased wall surface relative to the volume of blood flowing through the small vessels. It is advantageous, therefore, to use large vessels (arteries and veins) to carry blood over long distances and confine the small vessels (capillaries) to the tissues where gas and material exchanges occur.

A repeated branching of the arteries as they approach the capillaries is helpful in ways other than simple distribution. If the total volume of circulating liquid in a pipe remains the same, liquid must travel faster when the pipe narrows. We have all seen this in the rapid movement of water in the narrow part of a river. Yet there would be

more time for the exchange of materials if the blood moved slowly through the capillaries. This is accomplished by the repeated branching of the arteries in such a way that there is an increase in their total cross-sectional area as the capillaries are approached. Velocity decreases as blood flows through the arteries and capillaries (Fig. 14.18 *top*). An analogy is a river broadening out into a pond, where the rate of flow decreases. The rate of flow again increases as blood passes from the capillaries to the venules, and as these small veins lead into fewer larger ones. The combined cross-sectional area of the veins decreases, so, like water flowing out of a lake into a narrowing river, the blood moves faster and faster. Pressure, unlike velocity, continues to decrease throughout the system because of the continued friction within the blood and between the blood and vessel walls (Fig. 14.18 *bottom*).

Blood vessels differ from pipes in that their walls are flexible, and those of the capillaries are permeable. All of the vessels of vertebrates are lined with endothelium. Capillaries have little more to their walls, but arteries and veins contain variable amounts of connective tissue and smooth muscle arranged in layers (Fig. 14.19). Since blood pressure is higher in them, arteries have much thicker walls than the accompanying veins. The composition of the walls has important functional consequences. The walls of the large arteries leaving the heart of vertebrates are richly supplied with elastic connective tissue. The force of each ventricular systole forces blood into the arteries and stretches them to accommodate it. During diastole, the elastic recoil of the first part of the artery to expand helps to push the blood into the adjacent part of the artery, which in turn expands. If the arteries were rigid pipes, they would deliver blood to the tissues in spurts that coincided with ventricular systole. The blood would pound like steam rushing into empty radiator pipes. The elasticity of the larger arteries conserves energy and transforms what would otherwise be an intermittent flow into a steady flow. The wave of alternate stretching and contracting of the arteries travels peripherally very rapidly (7.5 m per second) and can be detected as the **pulse**, but the blood itself does not move as fast.

Regulation of Peripheral Flow

An advantage of closed circulatory systems, as we have seen, is that the blood supply to active tissues can be increased at the expense of blood going to less active parts of the body. The volume of blood reaching capillary beds is regulated to a large extent by the degree of contraction of small arteries, known as **arterioles**, whose walls contain a great deal of smooth muscle (Fig. 14.19). Further control occurs within the capillary beds themselves. **Thoroughfare channels** carry some blood through the beds at all times, but tiny **precapillary sphincters**, which are located at the

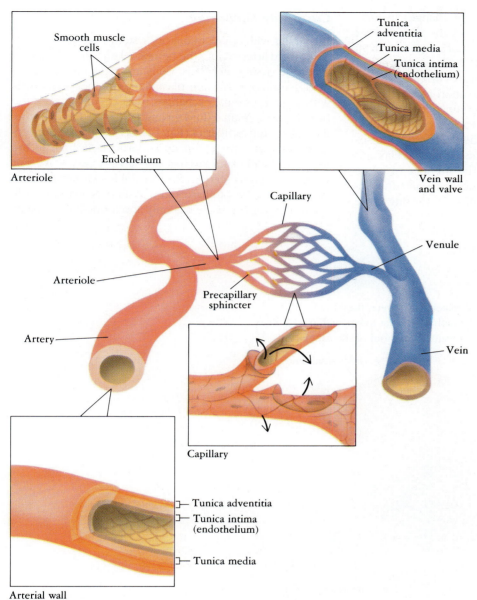

Figure 14.19
The structure of arteries, capillaries, and veins. The closure of precapillary sphincters cuts off the flow of blood to parts of a capillary bed.

beginning of capillaries that branch from the thoroughfare channels, control flow through particular parts of the beds. Small veins called **venules** collect the blood from capillary beds and lead to larger veins.

The total blood flow to the tissues can be altered by changes in the diameters of these vessels, which control peripheral resistance, and by the blood pressure developed by the heart ($F = \Delta P/R$). During exercise, for example, blood pressure rises due to the increased return of venous blood, which increases stroke volume and the force of heart contraction. Sympathetic nerve stimulation of the heart simultaneously increases heart rate and pressure. Peripheral resistance is lowered in the active tissues by the relaxation of muscles in the arterioles and precapillary sphincters. The degree of contraction of these muscles is controlled through the interaction of several factors. (1) An increase in blood pressure itself can dilate the small vessels. (2) The vessels have a high degree of autoregulation and respond directly to chemical changes in their environment. It is not certain whether the stimulus for vasodilation comes from a deficiency of oxygen or from the accumulation of metabolites. (3) The vessels are supplied by vasoconstrictor neurons that are controlled by the vasomotor center of the brain. These neurons have a base level of activity that maintains a certain degree of contraction and therefore peripheral resistance and blood pressure. During exercise the number of nerve impulses sent to these vessels decreases, so their muscles relax and peripheral resistance is lowered. The number of nerve impulses to the veins, however, increases and this increases the volume and pressure of blood returning to the heart.

Capillary Exchange

In animals with open circulatory systems, the hemolymph flows from arteries into the hemocoel and directly bathes all of the tissues and cells. Cells are separated from the hemolymph only by their plasma membranes. In animals with closed systems, exchanges must occur between the blood and interstitial fluid through the very thin endothelial walls of the capillaries (Fig. 14.20a). Capillaries are very numerous, and their density is particularly high in the tissues of small mammals that can become very active. For example, a cross section of mouse skleletal muscle contains about 2000 capillaries/mm², whereas a comparable section of horse muscle has only about one-half as many.

Figure 14.20
Capillary exchange. (a) The exchange of materials between the blood and interstitial fluid in a capillary bed. (b) A graph of the net forces within a capillary that are responsible for the movement of water in and out of the capillaries.

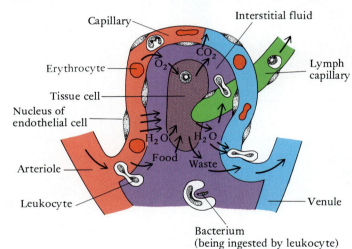

(a)

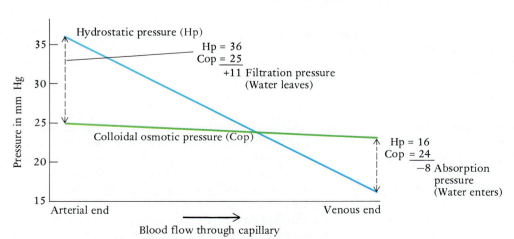

(b)

Oxygen, glucose, amino acids, various ions, and other needed substances diffuse easily through the endothelial walls of the capillaries into the interstitial fluid following their concentration gradients. Conversely, carbon dioxide, nitrogenous wastes, and other byproducts of metabolism diffuse easily into the blood. But capillary walls are more permeable to many of these materials than might be expected. Electron microscope studies have shown that the increased permeability results from more direct passages. Some of these passages may be present where endothelial cells come together and overlap, for the cells are not firmly united with one another over the full extent of these surfaces (Fig. 14.21a). In addition, small endocytotic or pinocytotic vesicles with diameters of about 50 nm frequently invaginate on either surface of endothelial cells, bud off, move through the cell, and release their contents on the opposite surface. In the parts of the kidney where materials leave the blood, in the intestinal villi, and in many glands, the capillaries are **fenestrated** with small pores up to 100 nm in diameter (Fig. 14.21b). These are closed only by the delicate protein fibers and glycoproteins that compose the basement lamina of an epithelial surface.

Water moves both ways across the capillary walls. Maintaining the proper amount of water in the interstitial fluid involves a combination of forces. As we have seen,

most of the large plasma proteins remain in the blood, where they exert an osmotic force called the **colloidal osmotic pressure** that draws water into the capillaries. An opposite force, the **hydrostatic pressure** of the blood, forces water out. At the arterial end of the capillary bed, hydrostatic pressure normally exceeds the colloidal osmotic pressure, so that there is a net force (called the **filtration pressure**) that moves water out of the capillaries (Fig. 14.20b). As blood moves through the capillary bed, friction causes a great reduction of the hydrostatic pressure. A smaller reduction in the colloidal osmotic pressure also occurs because of the loss of some plasma proteins. At the venous end of a capillary bed, colloidal osmotic pressure normally exceeds hydrostatic pressure so that there is a net force (the **absorption pressure**) that draws water into the capillaries. The absorption pressure is frequently a bit lower than the filtration pressure, but the tissues do not flood. Capillaries have somewhat more absorption surface at their venous end through which more water can enter. Terrestrial vertebrates, whose blood pressure is somewhat higher than that of fishes, also have a system of lymphatic capillaries that return excess interstitial fluid back to the venous system (Fig. 14.20a).

Factors that upset the balance between the hydrostatic and osmotic pressures of the blood can lead to an increase in the amount of interstitial fluid and a swelling, or **edema** (Gr. *oidema*, swelling), may result. Burns or other injuries may result in edema for they cause capillary permeability to increase, allowing more proteins to escape from the blood. Capillary permeability also increases when allergic reactions promote the release of histamine from mast and other cells (p. 300).

Venous and Lymphatic Return

In animals with open systems, the venous channels do not have an endothelial lining, although they may be lined with connective tissue. Their structure in vertebrates is fundamentally similar to that of arteries, though a vein is larger and has a much thinner and more flaccid wall than its companion artery (Fig. 14.19). Since they are larger, the veins hold more blood than the arteries and are an important reservoir for blood. Lymphatic vessels have even thinner walls. Valves present in both veins and lymphatics permit the blood and lymph to flow only toward the heart.

Though blood pressure is low in the veins (see Fig. 14.18) and lowest in the large veins near the heart, it is still the major factor in the return of blood to the heart in a mammal. Two other factors assist it. One is the fact that the elastic lungs are always stretched to some extent and tend to contract and pull away from the walls of the pleural cavities. This creates a slight subatmospheric or negative pressure within the thoracic cavity that is greatest when the thorax expands during inspiration. The larger veins pass through the thorax as they approach the heart, and the

Figure 14.21

Capillary structure as seen in cross section. (a) A typical capillary. Some material moves through the endothelial cells of capillaries in pinocytotic vesicles. (b) A fenestrated capillary is more permeable because of minute perforations in its wall.

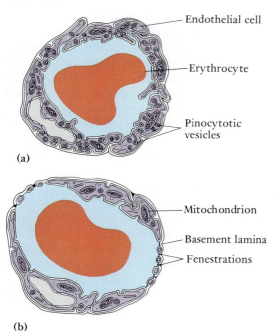

Endothelial cell

Erythrocyte

Pinocytotic vesicles

(a)

Mitochondrion

Basement lamina

Fenestrations

(b)

reduction of pressure around them decreases the pressure within them and increases the venous pressure gradient. The other factor is that the contraction and relaxation of body muscles exert a "milking" action on the veins, which, coupled with valves preventing a backflow, forces the blood toward the heart. These factors increase during exercise, which results in a more rapid return of blood and an increased cardiac output.

The return of lymph depends upon similar forces. The interstitial fluid itself has a certain pressure derived from the flow of liquid out of the capillaries. This establishes a pressure gradient in the lymphatics that is made steeper by the reduced intrathoracic pressure. The "milking" action of surrounding muscles and, for lymphatics returning from the intestine, the movement of the villi, help considerably. Some amphibians and other primitive terrestrial vertebrates have lymph "hearts," which are specialized pulsating segments of lymphatic vessels.

Summary

1. Transport systems provide a means for the bulk flow of materials between the interstitial fluid bathing the cells and the sites where materials enter and leave the body. They also help to maintain a constant internal environment (homeostasis). Although any body cavity can be used in transport, larger animals have well-developed circulatory systems. In open systems, hemolymph passes from the vessels into the interstitial spaces; in closed systems, the blood is confined to the vessels. In general, more active animals have closed systems.

2. The blood of all animals consists of some cells carried in a liquid plasma composed of water, plasma proteins, electrolytes, and materials in transit. Vertebrate blood carries platelets, leukocytes, and the red blood cells that transport gases.

3. Hemostatic mechanisms reduce blood loss after injury: the injured vessels constrict, a plug forms, and the blood clots. Platelets are necessary in mammals for the last two events.

4. Most arthropods and molluscs have open circulatory systems in which blood passes through the gills before returning to the heart. Active cephalopods have closed systems. The annelid system may be open or closed. Blood flows anteriorly in a pulsating dorsal vessel, ventrally around the front of the gut, and posteriorly in ventral vessels.

5. The heart of most fishes consists of a linear series of chambers that receive only oxygen-poor blood and pump it to the gills. Vessels leaving the branchial region distribute oxygen-rich blood to the body. With the loss of gills and the evolution of lungs during the evolutionary transition of vertebrates from water to land, a double circulation evolved in which the right atrium and ventricle receive oxygen-poor blood from the body and pump it to the lungs, and the left chambers of the heart receive oxygenated blood from the lungs and pump it to the body. The double circulation is incomplete in amphibians and reptiles, allowing the volume of blood passing through the lungs and body to vary with the degree to which the lungs are being used. The double circulation is complete in birds and mammals, for the lungs are ventilated continuously.

6. The fetal circulatory system of mammals enables the fetus to receive and eliminate materials through the placenta, bypass the functionless lungs, and supply both sides of the heart with enough blood to pump so that they can develop normally. Some blood is recirculated through the lungs during the brief neonatal period.

7. The most common type of heart is a chambered one with valves preventing the backflow of blood. Interconnection between cardiac muscle fibers allows those of the atrium and the ventricle to respond as units in an all-or-none fashion. The inherent rhythm of contraction of cardiac muscle originates in the sinoatrial node and spreads on cardiac muscle and specialized muscle fibers to other parts. Nerves terminating on the sinoatrial node and on cardiac muscle fibers increase or decrease their rate of contraction and the force of contraction. The heart beats faster in small mammals with high metabolic rates than in larger ones. Cardiac output increases during exercise because of increases in the rate of heart beat and stroke volume.

8. In conformity with the physical principles that govern the flow of liquids in pipes, arteries and veins are used to carry blood over long distances, and small capillaries are limited to sites of exchange. Blood pressure decreases throughout the circulatory system but its velocity varies with the total cross-sectional area of the vessels. It is slowest in the capillaries. Peripheral flow is regulated by the degree of contraction of the arterioles and precapillary sphincters.

9. Exchanges of solutes in the capillary beds occur by diffusion, by the passage of materials through pores between the cells of the capillary walls, and by the passage of pinocytotic vesicles across the cells. Water movements are regulated by the interaction between the hydrostatic pressure of the blood and its osmotic pressure.

10. Materials return from the capillaries and interstitial fluid through the veins and lymphatic vessels. The pressure in these vessels is low, and the contraction of surrounding body muscles helps in the return of blood and lymph. Valves in the vessels prevent backflow.

References and Selected Readings

The general references on vertebrate organ systems cited at the end of Chapter 10 contain considerable information on transport systems.

Durliant, M. Clotting processes in Crustacea Decapoda. *Biological Reviews of the Cambridge Philosophical Society* 60(1985):473–498. A detailed analysis of an example of clotting among invertebrates.

LaBarbera, M., and S. Vogel. The design of fluid transport systems in organisms. *American Scientist* 70(1982):54–60. A discussion of the physical principles that affect the design of gas and liquid transport systems in plants and animals.

Martin, A.W. Circulation in invertebrates. *Annual Reviews of Physiology* 36(1974):171–186. A scholarly review of the invertebrate circulatory system.

Robinson, T.F., S.M. Factor, and E.H. Sonnenblick. The heart as a suction pump. *Scientific American* 254 (June 1986):84–91. A review of the structure of the mammalian heart and an analysis of its suction mechanism.

Ruppert, E.E., and K.J. Carle. Morphology of metazoan circulatory systems. *Zoomorphology* 103 (1983):193–208. A review of the major types and distribution of transport systems throughout the animal kingdom.

Schmidt-Nielsen, K.: Countercurrent systems in animals. *Scientific American* 244(May 1981):118–128. A summary of the biological uses in the circulatory system and respiratory passages of the countercurrent principle.

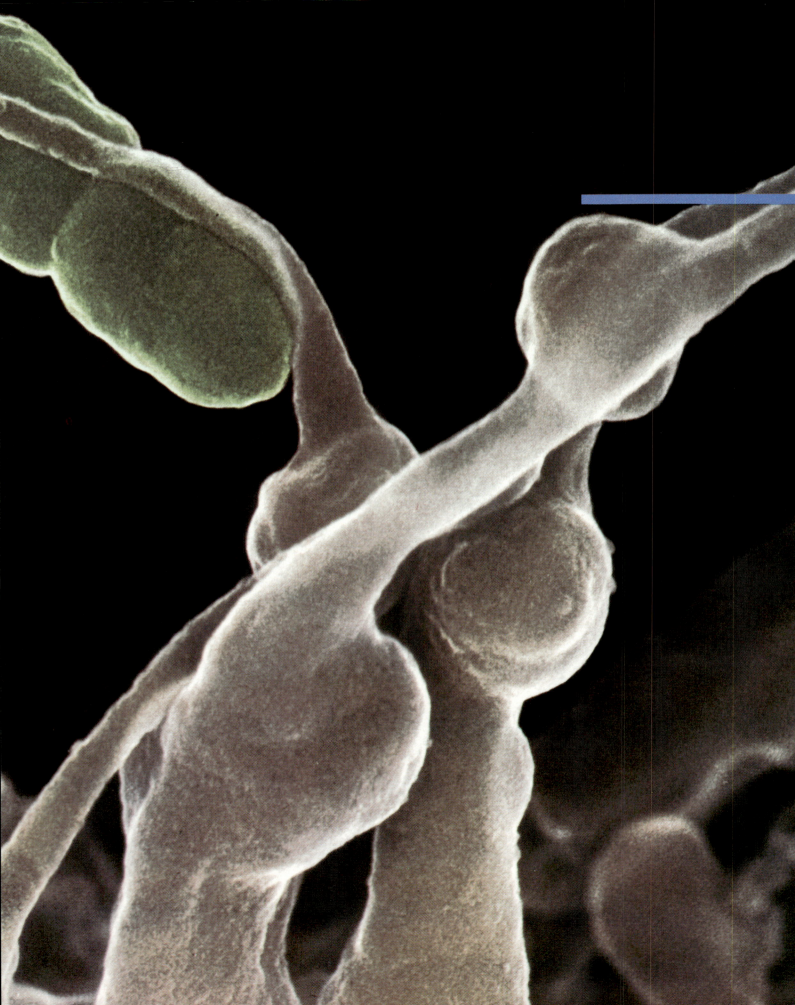

15

The Immune System

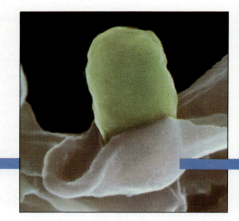

(Left) A macrophage, one of the cells of the immune system specialized for scavenging, mounts an attack against an invading bacterium shown in green. (Above) A small pseudopod extending from the macrophage envelops the invading bacteria.

Many of you will remember David, the boy in the bubble, forced to live in an environment of filtered air, masks, and gloves, unable to make actual physical contact with the outside world. For the boy in the bubble, the world was a hostile place. Bacteria, viruses and protozoans were not just other living organisms, part of the biosphere, but instead real, life-threatening adversaries. This boy suffered from severe combined immunodeficiency disease (SCID), a generalized failure of the immune system. His body could not defend itself against the outside world, and he died in 1984, at the age of 12, after emerging from his bubble for a short time (Fig. 15.1).

All of you have also heard of acquired immune deficiency syndrome, or AIDS. This syndrome is caused by a retrovirus, the human immunodeficiency virus (HIV), a virus whose genome encodes only seven proteins. This retrovirus acts insidiously, eventually destroying a subset of the host's T cells, a type of lymphocyte and a component of the immune response. Individuals with AIDS most frequently succumb to opportunistic infections, diseases that would have little chance of gaining a foothold in the presence of a full immune response. The AIDS virus, in fact, first came to be noticed in 1981 when physicians in Los Angeles, New York, and San Francisco reported several cases of a rare type of pneumonia, caused by the micro-organism *Pneumocystis carinii*, among young male homosexuals. Similarly, doctors noticed an unusually high incidence of a rare type of skin cancer, Kaposi's sarcoma. Somehow, the body's ability to patrol, identify, and eliminate infectious agents or cancerous cells seemed seriously impaired by the presence of HIV.

These two examples illustrate the catastrophic consequences of immune system malfunction. The immune response centers around the most fundamental aspect of biological identity: the notion of a biological "self." The task of the immune system seems inconceivably complex. The cells and molecules of the immune response must first make the distinction between "self" (the body's own cells and molecules) and "nonself" (foreign organisms, cells, or molecules). Upon recognizing an invader, the immune system must mount a rapid, effective response that destroys and disposes of the invader. The failure of either of these two aspects of the immune reaction—recognition or response—may have devastating consequences for the organism.

In the pages that follow, we introduce you to the main components of the immune system and detail their mode of action. We begin by examining **humoral immunity**, a response based on the production of antibodies that circulate in the blood stream. We then describe **cell-mediated immunity**, a response that depends on the concerted action of various cell types in the lymphatic and circulatory systems. We will then discuss how the immune system may have evolved, focusing on the phylogenetic distribution of its different components. Finally, we will look at the immune system from the target's viewpoint, describing how certain parasites and viruses have evolved mechanisms to trick or bypass the immune response of their hosts.

TYPES OF IMMUNE RESPONSE

Nonspecific Mechanisms

As we will see, vertebrates have evolved a complex and exquisitely specific set of mechanisms to deal with invaders. But these mechanisms do not come into play until the invading organism or substance has penetrated the tissues or fluids of the animal. In order to do so, the invader must first bypass a set of nonspecific defenses that protect the vertebrate body.

First among these defenses is the skin. In addition to its passive role as an elastic and largely impermeable body covering, the skin also protects the body by harboring a select bacterial fauna. This fauna is involved in a mutually beneficial partnership with its vertebrate host. The bacteria thrive on the skin cells sloughed off by the host, on the fatty secretions of the skin glands, and on the minerals present in sweat. In turn, the native bacteria prevent other

Figure 15.1

David, suffering from SCID (severe combined immunodeficiency), spent his whole life in the controlled, germ-free environment of this plastic bubble. In the absence of a functioning immune system, exposure to any bacterial or viral agent was potentially life-threatening.

disease-causing bacteria from becoming established, outcompeting them and inhibiting their growth. Such bacterial faunas are also important lines of defense in the esophagus, lungs, stomach, and intestinal tracts, all surfaces that are potentially in contact with the external world.

The body is also protected from infection and invasion by a variety of secretions: tears, saliva, mucus, and urine. These fluids act by flushing the invading bacteria, fungi, or viruses out of the body. In addition, these secretions often contain high levels of enzymes, such as lysozyme, that can break down some bacterial cell walls, and immunoglobulins, protein molecules that bind to and inactivate bacteria and viruses.

One final category of nonspecific immunity is conferred by the mother to the newborn baby. A newborn is said to be immunologically naive: its immature immune system has not been exposed to any foreign substance and thus cannot mount an effective response. A certain degree of **passive immunity** is provided to babies prior to birth by the transfer of maternal antibody molecules from the maternal circulation to the fetal blood stream. Breast milk is also very rich in certain antibodies as well as in generalized scavenger cells (macrophages and lymphocytes) that protect the newborn against infection.

The Humoral Immune Response

The humoral (L. *humor*, fluid) response involves a class of molecules, the **antibodies**, that circulate throughout the body in the blood stream. Antibodies function by attaching with great affinity to a foreign substance (the **antigen**), inactivating it, and creating a large aggregate (the **antigen-antibody complex**) that is recognized and disposed of by a subset of specialized cells in the immune system.

The Primary and Secondary Immune Responses

The pattern and timing of an antibody response almost always follow a predictable and stereotyped course. Let us examine the level of circulating antibody (the **antibody titer**) after a specific antigen has been experimentally injected. For the first three to five days, no antibody specific to the antigen can be detected. Beginning on day four or five, the titer climbs, reaches a peak 10 to 12 days after antigen injection, and subsequently declines to undetectable levels in two to six weeks. This orderly rise and fall of antibody titer is the **primary immune response** (Fig. 15.2). A second dose of the same antigen, however, unleashes a very different chain of events. In this **secondary immune response**, the antibody titer begins to rise 36 to 72 hours after antigen exposure, and reaches far higher levels than previously seen. The titers again begin to decline, but do so far more gradually than before.

The secondary immune response forms the basis of most vaccination therapies. In this process, a specific antigen is introduced deliberately to produce a primary immune response. On reexposure to the antigen, a rapid secondary immune response ensues, with its dramatic increase in the circulating titer of antibodies against the particular disease agent. Bear in mind that what is being injected as a vaccine is generally not a dose of live pathogens, the agents capable of producing disease, but avirulent strains, heat-killed bacteria, inactive protein extracts, or synthetic peptides capable of serving as antigens. These preparations induce the primary immune response and prepare the organism for a rapid response when the disease agent is actually encountered, but carry little or no risk of *causing* the disease.

Once exposed to an antigen, the immune system reacts quickly and dramatically upon reexposure. The speed and extent of the secondary immune response suggest some form of "memory" in the immune response. Experiments suggest that this **immunological memory** may last months, years, or the entire lifetime of an individual. Once seen, an antigen is seldom forgotten. The proof of this circulates in our blood streams: the blood serum of a human adult contains hundreds or thousands of different antibody molecules, each a record of a previously encountered bacterial, protozoan, or other infectious agent.

The principal feature of antibodies is their *specificity*. In effect, the action of antibody molecules depends on a tight, three-dimensional fit between particular features of the antigen (the **epitopes**) and the antibody. Every antigen will thus induce the production of a specific antibody, specifically fitted to some three-dimensional feature of the antigen (Fig. 15.3). The precise fit brings the antigen and antibody into close contact, allowing the formation of hydrogen and other weak bonds between them.

Figure 15.2

The increase of antibodies in the plasma in response to the administration of an antigen is shown in this graph.

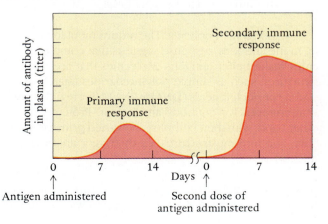

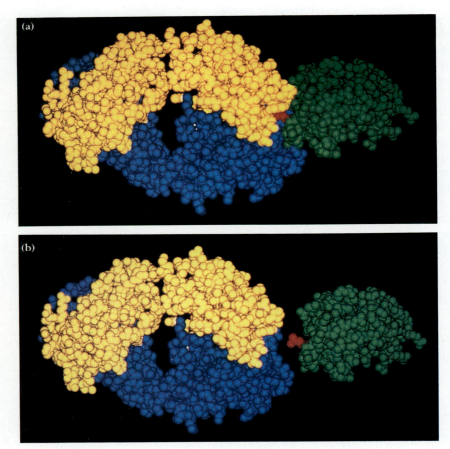

Figure 15.3
A model of an antigen binding with an antibody. (a) The antigen, shown in green, has an epitope shown in red that fits into a groove on the antibody. The heavy chain of the antibody is shown in blue, its light chain in yellow. (b) The antigen and antibody have been pulled apart to show the epitope and groove more clearly.

Antigens: The Triggers of the Immune Response
Practically any organic molecule can act as an antigen, although the degree to which a given molecule or structure can induce antibodies (its **antigenicity**) varies. The factors that affect antigenicity include:

1. Antigen complexity. The requirement for specificity in the antibody response suggests that complex molecules make better antigens. Molecules that are repeated structures, such as simple polysaccharides (starch or glycogen) or nucleic acids (DNA or RNA) offer few characteristic epitopes for antibody attachment. Proteins, on the other hand, with their complex and varied primary, secondary, and tertiary structures, easily trigger antibody production.

2. Antigen size. In general, the larger the antigen, the stronger the antibody response. This feature is somewhat puzzling, given that the actual epitope to which antibody binds is generally quite small. In the case of a protein antigen, for example, the antibody seldom contacts more than 6 to 8 amino acids directly. The requirement for large antigen size may simply reflect the larger number of distinct potential epitopes that exist on a larger molecule.

3. Antigen dose. A very large dose of antigen may overwhelm the immune response and prevent the occurrence of a secondary response. This mechanism may in part underlie the recognition of "self" by the immune response and explains the absence of immune response towards an animal's own tissues. The number of cells (and cell surface proteins) that make up a body is so immense that the immune system may be overwhelmed by the concentration of these "self" antigens, a process known as **high-dose tolerance**. The development of self-tolerance may be dose dependent. Conversely, exceedingly low concentrations of antigen may result in **low-dose tolerance**, a situation where the immune system is not triggered

to respond (Fig. 15.4). The concentration of antigen required to induce antibody formation depends on the nature of the antigen. Repeated exposure to low doses of an antigen (below the reaction threshold) eventually produces a long-term tolerance.

4. Antigen identity. Because the immune system must first recognize an antigen as "foreign," the identity of the antigen plays a crucial role. Thus, for example, certain proteins from other primate species (particularly chimps) differ little in amino acid sequence from the homologous protein in humans. These chimpanzee proteins will often induce little or no antibody production in a human. Our immune system, already self-tolerant to the human version of that protein, will not recognize the chimpanzee protein as a foreign invader.

The converse of this effect is the phenomenon of **cross-reactivity**. Despite the exquisite specificity of antibodies, structurally similar antigens may bind the same antibody (although with lower affinity). Cross-reactivity played a major role in the birth of the science of immunology. At the end of the 18th century, Benjamin Jesty, a dairy farmer, and Edward Jenner, a physician, independently discovered that children treated with cowpox scabs developed immunity against smallpox, a devastating and frequently lethal infectious disease. Jesty and Jenner prepared a puree made from the scabs of cows infected with cowpox and placed a small amount of this onto scratches on the skin or into the nostrils of children (the nostrils are heavily supplied with blood vessels) (Fig. 15.5). The

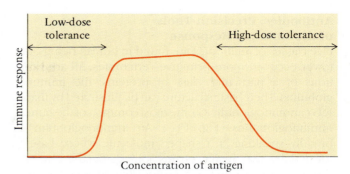

Figure 15.4

This graph illustrates how the concentration of the antigen affects the immune response. Both high-dose tolerance and low-dose tolerance inhibit the immune response.

cowpox virus entered the children's blood stream, triggering the formation of anticowpox antibodies. When these treated children were exposed to live smallpox virus, the virus would unleash a secondary immune response, protecting the children from infection. Even though cowpox and smallpox viruses are not identical, the cross-reactivity of the anticowpox antibody protected the children against smallpox. The eradication of smallpox from the human population in 1977, made possible by a massive immunization program, is a remarkable public health achievement.

Figure 15.5

A copper engraving of Edward Jenner performing the first vaccination against smallpox. Jenner used material he extracted from a pustule on the hand of Sarah Nelmes, who had contracted cowpox while milking. Vaccination with cowpox subsequently conferred immunity against smallpox, often a lethal disease.

Antibodies: Precision Tools of the Immune Response

Five classes of antibodies are found in human beings, but fewer occur in more primitive vertebrates. All are variations of a particular class of proteins, the **immunoglobulins**. Their basic structure can be illustrated by that of IgG (immunoglobulin G), the most common of the human immunoglobulins (Fig. 15.6). An immunoglobulin is, broadly speaking, a Y-shaped molecule whose base is formed by two **heavy chains** joined together by two disulfide bonds. The heavy chains diverge as they extend into the arms of the Y where each is bound by a single disulfide bond to a **light chain**. The connection of the two arms to the base is the **hinge region**, a particularly flexible string of amino acids that allows the arms to rotate like tentacles. The composition of the heavy chains determines the class to which immunoglobulins belong and the ways they perform their functions. IgG circulates in the blood and plays a major role in antibody defense mechanisms. Another immunoglobulin, IgA, can pass through epithelial surfaces with saliva, tears, sweat, and mucus and attack foreign materials at their point of entry into the body.

Both light and heavy chains consist of a **constant region** that shows little variation in its amino acid content from one organism to another, and a **variable region**, at the tips of the arms, that does differ from one organism to the next. One or more **hypervariable** segments are included in the variable region. Two different antibodies will almost always differ in their hypervariable regions. The tips of the arms, composed of the variable region of both the heavy and light chains, form the **antigen binding site**. Each immunoglobulin can thus bind two identical epitopes of antigens in the clefts or pockets formed by the paired variable region. An additional biologically important site, the **complement binding site**, lies on the constant region of the heavy chain.

The Generation of Antibody Diversity

Remember the challenge facing the immune system: it must be able to recognize a foreign substance, including compounds and molecular structures it has never previously encountered. It must then generate a molecule capable of combining specifically and tightly with this foreign substance, in order to inactivate and eventually eliminate it.

The structure of the immunoglobulin molecule provides important clues about its function. The antigen binding sites include the hypervariable regions and can therefore take on a vast array of three-dimensional shapes in order to bind with the epitope or the antigen. The immunoglobulin molecule as a whole is a sort of Tinker-Toy, assembled out of several kinds of constant regions and variable regions, with their hypervariable segments. The result of this combinatorial approach is an immune system capable, in mammals, of generating more than one million different antibody molecules.

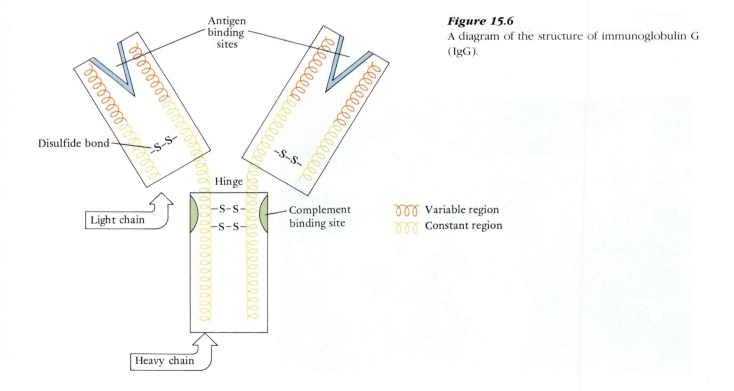

Figure 15.6
A diagram of the structure of immunoglobulin G (IgG).

But what is the mechanism responsible for generating antibody diversity? Immunoglobulins are proteins and are therefore encoded by DNA. There is nowhere near enough room in the human genome to house one million different immunoglobulin DNA sequences. How, then, is the genetic information required for immunoglobulin diversity encoded?

The answer comes from a mechanism you are already acquainted with: recombination. Unlike meiotic recombination, which occurs only during the formation of the germ cells, immunoglobulin variability is generated by **somatic recombination** occurring in the cells that will generate immunoglobulins: the **B lymphocytes** or **B cells**. These cells are known as B cells because they were found and isolated from the bursa of Fabricius, a small cloacal gland present only in birds. In mammals, B cells are produced principally in the bone marrow (Fig. 15.7).

The human genome (as well as all other mammalian genomes studied to date) contains several genes that code for the constant (C) region and a vast number (300 or

Figure 15.7

Both T cells and B cells of mammals develop from stem cells in the bone marrow. They undergo their maturation and processing in different tissues, lodge primarily in the lymph nodes and other lymphoid tissues, and have different roles in the immune responses of the body.

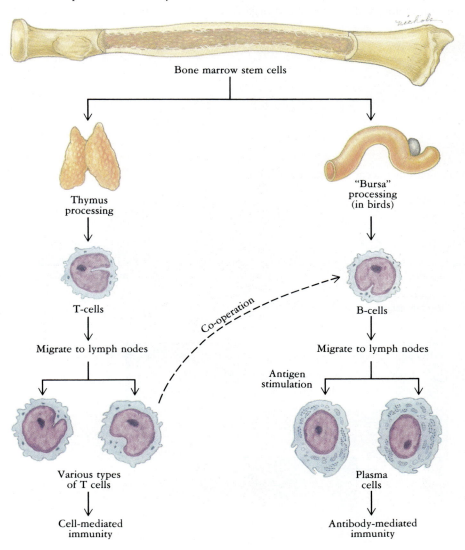

more) that code for the variable (V) region. In the formation of a light or heavy chain that takes place in B cells, a variable region gene segment is brought together with a constant region gene segment, *at the DNA level* (Fig. 15.8). This new composite DNA molecule is then transcribed, the corresponding mRNA spliced and processed, and the edited message translated to produce a particular immunoglobulin molecule. Additional mechanisms, such as imperfect recombination, deletion, insertion, aberrant splicing, and high levels of somatic mutation, all act to generate additional immunoglobulin variability.

These processes of somatic recombination take place in the immature B cell and irreversibly alter the DNA sequence of that lymphocyte. Interestingly, each lymphocyte generates only one kind of heavy chain and one kind of light chain. No additional or multiple rearrangements take place within a given cell. This phenomenon of **allelic exclusion** in effect means that every lymphocyte produces one and only one unique combination of immunoglobulin chains. The lymphocytes that arise by mitosis from that cell all carry that particular rearrangement; they are **clones** of the original cell and will only generate that one characteristic immunoglobulin.

The phenomenon of somatic recombination in B cells irreversibly alters the genetic instructions of the cell and results in the actual loss of certain DNA coding regions in mature B cells. In no other eukaryotic cell type is DNA so radically altered during differentiation. In contrast, liver cells differ from brain cells because the different genes are being expressed (or repressed) in the different cell types, not because they contain different DNA.

The Clonal Selection Theory

As a consequence of somatic recombination taking place during the course of embryonic development of a mammal, millions of different types of B cells are generated, each bearing a single distinctive type of antibody as a surface receptor. The body thus contains an immense dictionary of B cells capable of recognizing nearly any conceivable antigen. Most of these lymphocytes are located in the bone marrow, lymph nodes, spleen, and other lymphoid organs of the body, but many enter the blood

Figure 15.8
A diagram of the rearrangements of DNA segments that occur during the formation of antibody genes.

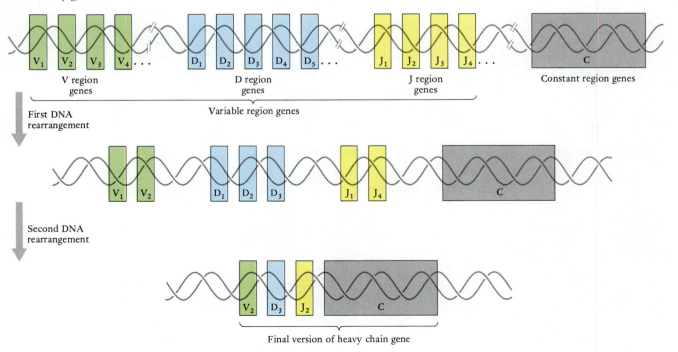

stream (Fig. 15.9a). In principle, we carry around an array of B cells so varied that every conceivable antigen will be met by an appropriately shaped antibody molecule.

Yet, there is still one aspect of the humoral response left to account for: how is the one specific immunoglobulin best able to bind with the invading antigen subsequently produced in large amounts? Is the central dogma, stating that information cannot flow from proteins (the immunoglobulins) to the DNA, being violated? The solution to that paradox was put forth in 1954 by Frank Macfarlane Burnet, an English immunologist. He proposed a version of the standard Darwinian paradigm, which he called the **clonal selection theory** (Fig. 15.10). In this model, a vast population of different B cells, each bearing a single, distinct immunoglobulin, is produced through somatic recombination. These cells are then exposed to the antigen. The B cell bearing on its surface the antibody best able to couple with the antigen becomes sensitized, or "competent" and is stimulated to divide far more rapidly than its competitors. The result is an explosion of clonal cells all bearing the appropriate, identical antigen on their surface. Bear in mind that the rearrangement is preserved only in the

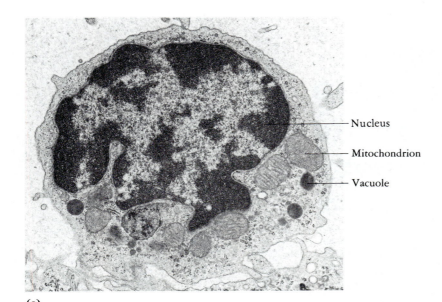

(a)

(b)

Nucleus

Mitochondrion

Vacuole

Mitochondrion

Nucleus

Rough endoplasmic reticulum

Figure 15.9
Electron micrographs of a B cell (a) and the plasma cell (b) into which it transforms on exposure to an antigen to which it is competent to respond. The plasma cell is at least twice the size of the B cell and the endoplasmic reticulum on which antibodies are synthesized is greatly enlarged. The large Golgi complex that secretes the antibodies is not in the plane of this section.

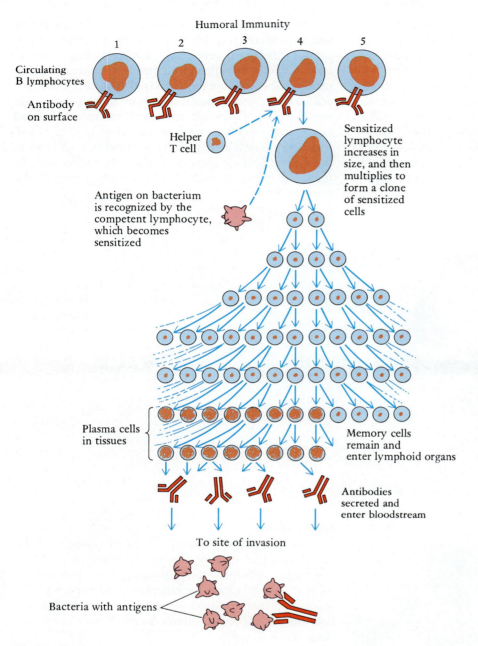

Humoral Immunity

Circulating
B lymphocytes

Antibody
on surface

Helper
T cell

Antigen on bacterium
is recognized by the
competent lymphocyte,
which becomes
sensitized

Sensitized
lymphocyte
increases in
size, and then
multiplies to
form a clone
of sensitized
cells

Plasma cells
in tissues

Memory cells
remain and
enter lymphoid organs

Antibodies
secreted and
enter bloodstream

To site of invasion

Bacteria with antigens

Figure 15.10

An overview of the humoral immune response and the clonal selection theory. An antigen on a bacterium "selects" the type of B cell bearing an antibody that matches the antigen. This cell rapidly multiplies to form a clone of genetically identical cells. Many clonal cells transform into plasma cells that synthesize the correct antibody and release it into the blood, where it binds with the antigens. Some of the clonal cells remain as memory cells, which can quickly mount a secondary response if the body is invaded again by the same antigen. Helper T cells greatly increase the efficiency of clone formation.

selected B cells; it is not fed back to the germ line. The DNA rearrangement of the competent B cell will not be inherited by the offspring of the organism.

Many of these clonal B cells transform into **plasma cells** and remain in the lymphoid tissues. Plasma cells have an extensive rough endoplasmic reticulum on which the appropriate antibodies are made and an enlarged Golgi apparatus for exporting them from the cell (Fig. 15.9b). The antibodies that the plasma cells produce enter the blood stream, eventually to bind with the antigen.

The stimulation of a specific B cell by antigen binding has one additional important consequence. A subpopulation of these cells differentiates into **memory B cells**, most of which also remain in the lymph nodes and other lymphoid organs or eventually lodge in the bone marrow. Extremely long-lived, these memory cells will spring into action when the same antigen is next encountered. The binding of a known antigen to the memory B cells will induce their differentiation and division into an active population of plasma cells, capable of producing antibodies in large quantities. These memory B cells form the basis of the secondary immune response.

A Summary of the Humoral Immune Response

Up to now we have been concentrating on the immunoglobulin molecule and its interaction with the antigen. But that interaction is only the initial step in the inactivation of a foreign invader. In this section, we follow the complete course of the humoral immune response to an invasion by bacterial agents.

For the sake of brevity, we assume that the organism has encountered this bacterial species previously. The initial event following the reinfection by the bacteria involves the activation of specific memory B cells. These B cells, you recall, have already undergone the appropriate DNA rearrangement and have survived the clonal selection process. As a result, when reexposed to the antigen, they transform into plasma cells capable of producing a single antibody molecule specifically directed to some external component of the bacterial cell such as the cell wall, flagella, or surrounding capsule. At the molecular level, the binding is taking place between the bacterial epitope and the two antigen binding sites of the immunoglobulin molecule. The antibody may bind to two epitopes on a single bacterial cell or to two epitopes on different cells (Fig. 15.11). In addition, more than one antibody molecule usually binds to a bacterial cell, resulting in the formation of large antigen-antibody complexes.

These antigen-antibody complexes are in effect covered with antibodies, making them readily identifiable to macrophages. The process by which foreign particles become more readily identified is called **opsonization**. Macrophages, which are derived from monocytes in the blood, are large phagocytic cells that are located in the lining of vascular sinusoids in the liver, spleen, and bone marrow and are also found in lymphoid organs and connective tissues (Fig. 15.12). The macrophages bind with the opsonized antigen-antibody complexes, engulf them, and digest them completely, thereby removing the antigen from the circulation.

More commonly, however, the binding of antibodies to an invading antigen triggers a second powerful defense system, the **complement cascade**. This system consists of series of proteins, most of which are enzymes that function, as the name suggests, in a cascade. The initial event is

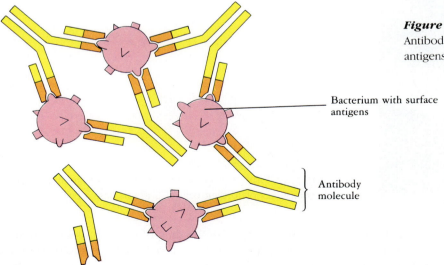

Figure 15.11
Antibodies typically bind with a group of antigens to form an antigen-antibody complex.

Bacterium with surface antigens

Antibody molecule

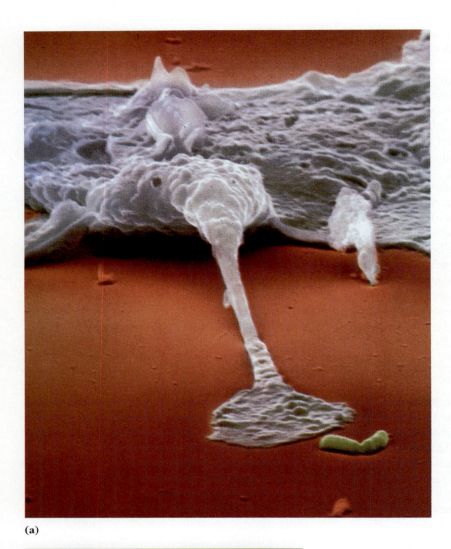

(a)

(b)

Figure 15.12
These false-color micrographs, taken in a culture, show that macrophages can form pseudopod processes (a) and engulf antigen-antibody complexes or foreign cells (b), in this case a bacterium.

the binding of one of the components of complement (C_1) to the constant region of several immunoglobulin molecules. This complement-binding site of the immunoglobulin molecule is only exposed when the antibody has bound to the antigen. The binding of C_1 to antibody activates a second protein, which in turn activates a third protein, and so on (Fig. 15.13). The complete cascade culminates with the assembly of a **lytic complex**. This lytic complex,

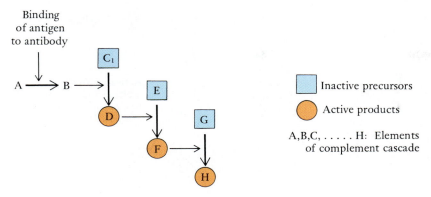

Binding
of antigen
to antibody

$\square$ Inactive precursors

● Active products

A,B,C, ... H: Elements
of complement cascade

Figure 15.13

Schematic diagram of the complement cascade. The binding of an antigen to an antibody exposes the complement binding site on the antibody (A → B). The first complement proenzyme (C_1) is transformed into an active enzyme (D), which in turn activates a second proenzyme (E) and so on through 11 steps in the case of the human complement cascade.

formed by several proteins of the complement cascade, acts essentially by punching and lining small holes in the cell wall of the bacteria. These breaches act as aqueous channels through the bacterial membrane, permitting the osmotic flow of water into the bacterial cell and eventually causing it to rupture. Certain activated components of the complement cascade also opsonize the antigen-antibody complex, attracting macrophages to the site where the bacteria are present.

The Cell-Mediated Immune Response

T Cells: The Main Component

An alternative type of immune response, one that does not involve the production and secretion of antibodies, is known as cell-mediated immunity. This type of immune response is directed primarily against certain viruses, fungi, and bacteria. It is also the basis of **graft rejection**, the reaction that occurs when tissue or organs taken from a different donor individual (or a different species) are transplanted onto a recipient.

The major player in this response is the **T lymphocyte** or **T cell** (Fig. 15.14). T cells, like B cells, are produced by stem blood cells in the bone marrow (Fig. 15.7), but they undergo a maturation process in the thymus, an organ that is best developed in the late fetus and young newborn (see Fig. 14.11). After maturation the T cells migrate into the blood stream and into lymphoid organs, and the thymus regresses.

T cells are involved in the identification, binding, and eventual destruction of foreign cells and antigens. Three major classes of T cells have been identified.

1. **Cytotoxic (killer) T cells**, whose function is to bind to foreign cells and destroy them.

2. **Helper T cells**, whose role is to bind foreign antigen and present it to the killer T cells. Helper T cells represent a bridge between the humoral and the cell-mediated response systems, as they, along with the macrophages, are also involved in presenting antigen to B cells, thus triggering the clonal selection process (see Fig. 15.10). These cells are the primary target of the HIV virus, accounting for the profound and devastating immune deficiency that characterizes individuals with AIDS.

Figure 15.14

A photograph of a T cell (*right*) and a B cell (*left*). These two cells differ primarily in their biochemical and functional properties; however, T cells have fewer surface microvilli.

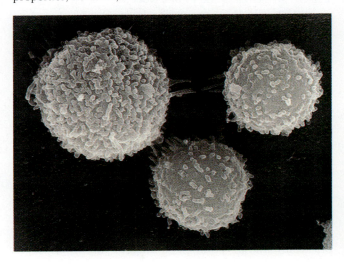

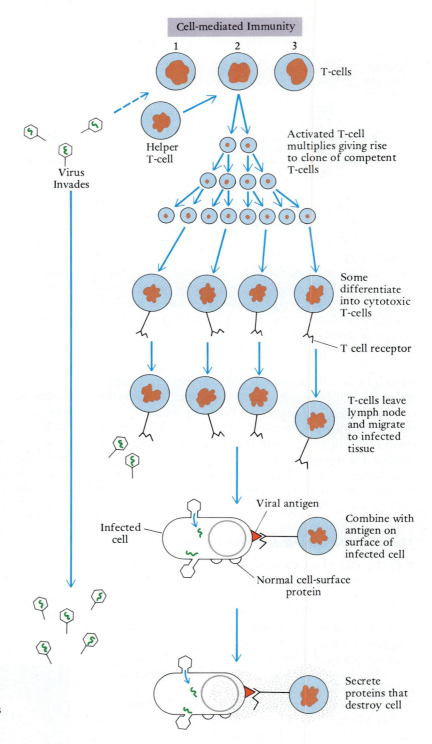

Cell-mediated Immunity

T-cells

Helper T-cell

Virus Invades

Activated T-cell multiplies giving rise to clone of competent T-cells

Some differentiate into cytotoxic T-cells

T cell receptor

T-cells leave lymph node and migrate to infected tissue

Viral antigen

Infected cell

Combine with antigen on surface of infected cell

Normal cell-surface protein

Secrete proteins that destroy cell

Figure 15.15

An overview of the cell-mediated immune response. T cells competent to respond to viral antigens are activated and form a clone of cytotoxic T cells. These recognize a body cell that has been infected with the virus, bind with it, and release proteins that destroy the cell. As in the humoral immune response, helper T cells increase the efficiency of clone formation.

3. **Suppressor T cells**, which assist in the regulation of the immune response by inhibiting the response of both cytotoxic T cells and B cells in the humoral immune response.

As with the humoral response, two key features characterize cell-mediated immunity: flexibility and specificity. The T cell must be able to recognize and respond to a practically infinite number of antigens that it has never previously encountered (flexibility), and it must tailor its attack to the particular antigen (specificity). The properties of a T cell arise from the presence, on the surface of the cell, of **T cell receptors**, molecules that bear a close evolutionary resemblance to antibody molecules (Fig. 15.15). Like antibodies, T cell receptors are composed of a variable and a constant region. The variable region is encoded by a series of genes that undergo frequent rearrangements and deletions (but little somatic mutation), as is the case with immunoglobulin genes. Finally, a process analogous to clonal selection takes place among T cells as they encounter particular antigens. The initial binding of antigen to a specific T cell surface receptor induces the differentiation and division of that cell into a large clonal population (with identical T cell receptors) as well as into a subset of memory T cells. When a given antigen is presented a second time, it triggers a faster secondary cell-mediated response.

Although different members of the T cell receptor family are found on the surface of different T cells, all these receptors operate by binding with great specificity and high affinity to the foreign antigen. Unlike the humoral response, however, no antibodies are being secreted in this system. Instead, the immune response of the organism depends on a variety of cell-cell interactions. In effect, these T cells swarm onto their target, cooperating in its identification and eventual destruction.

Recognizing Self and Nonself

The proper functioning of the cell-mediated immune response depends critically on the ability of T cells to distinguish foreign and infected cells from the normal cells of the body. Many of the factors involved in this identification emerged from research on the body's response to tissue grafts and organ transplants. Foreign grafts and organ transplants trigger a strong T cell response that results in the destruction of the transplanted tissue unless the donor and recipient are very closely related or unless high doses of immunosuppressant drugs are given to inhibit the T cell response.

Cells are identified as belonging to a particular individual and tissue by characteristic surface signature glycoproteins (see Fig. 5.6, p. 69). Those surface antigens that were found to be particularly significant in tissue transplants were called the **major histocompatibility complex (MHC)** antigens. Two main classes of MHC antigens have been identified: **class I**, found on the surface of most cells of the body, and **class II**, present only on the surface of certain B cells, T cells, and macrophages. The MHC antigens are structurally complex molecules whose synthesis in mammals is controlled by genes at three or more different loci, each of which may be represented in the population by several dozen multiple alleles. As a result, no two individuals, except for identical twins, are likely to be exactly alike in their MHC antigens.

The T cells are able to distinguish between the MHC antigens of normal body cells and those of foreign cells (a graft). They can also distinguish between a normal cell of the body and one infected by a virus because certain viral antigens appear on the surface of an infected cell (Fig. 15.15).

The T cells patrolling the body for signs of virally infected cells require a double password to spring into action: they must recognize a "self" antigen (now known to be a MHC class I molecule) *and* the viral antigen. One explanation suggested for the evolution of this double password mechanism is that it focuses the immune response on those cells that have been attacked by the virus, rather than on the large amount of free virus that may be circulating in the system. Once the double password is given, killer T cells secrete a class of proteins known as **perforins** that literally puncture the membrane of the target cell (Fig. 15.16). In addition, these T cells also secrete a class of chemicals, the **lymphokines**, that attract scavenger cells (macrophages) to the site of infection.

Self-tolerance and Autoimmune Disease

The immune system is extremely efficient in combating foreign invaders and diseased cells (Fig. 15.17), yet will not attack normal body cells. As noted earlier, this is partly because the number of potential self antigens is so large that the immune system is overwhelmed. Tolerance to the body's own cells and proteins develops during the late embryonic and postnatal periods, when B cells and T cells are being processed. This can be demonstrated by injecting cells from an embryo of one strain of mice (A) into a newborn individual of a different genetic strain (B). The strain B mouse will learn to accept these A strain cells as

Figure 15.16

A photograph of a portion of the surface of an erythrocyte showing the holes in the membrane that have been made by perforins secreted by cytotoxic T cells.

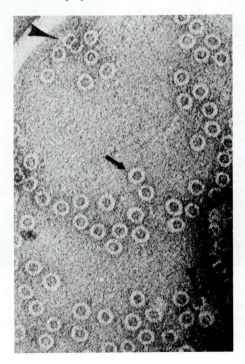

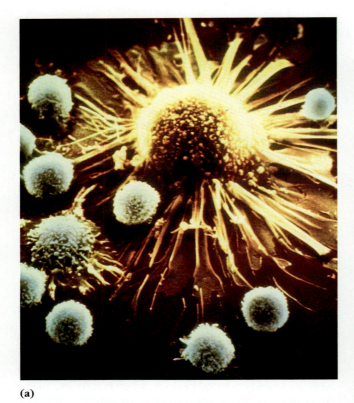

(a)

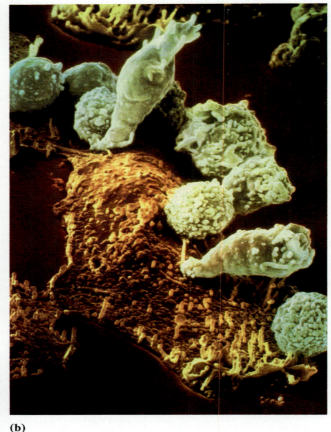

(b)

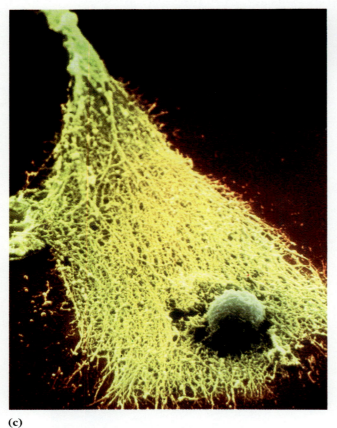

(c)

Figure 15.17
(a) Cytotoxic T cells recognize a cancer cell as nonself and attach to it. (b) Some T cells elongate and release perforins that break down the surface of the cancer cell. (c) Only the cytoskeleton remains of the destroyed cancer cell.

"self" along with its own cells. Later a transplant can be made from a mature A-strain mouse to a mature B-strain mouse and the transplant will not be rejected, as it would be if the B mouse had not developed a tolerance for cells from the A mouse.

Sometimes, however, tolerance to one's own tissues breaks down and the immune system attacks its own cells. It is not clear why, but in some cases it appears to be a byproduct of the defense of the body against a foreign antigen. If the foreign antigen is similar to some of the body's own proteins, antibodies or competent T cells that develop to combat the invader may at a later time go on to attack some of the body's own cells. A bout of rheumatic fever, for example, may lead to the formation of antibodies that later in life cause rheumatoid arthritis, an autoimmune disease in which tissues in the joints and heart are attacked.

Hypersensitivity: Allergies and Anaphylaxis

The immune system is remarkably well adapted to protect the body by producing appropriate antibodies and T cell lines. But the existence of immunological memory can result in **hypersensitivity**, an excessive immune response upon renewed exposure to certain antigens. Proteins on the surface of pollen grains, on worm parasites, and in certain insect venoms induce the formation in plasma cells of immunoglobulin E (IgE). If a person has become sensitized to the pollen grains that cause hay fever, for example, the plasma cells have become competent to produce IgE. A reexposure to the same pollen grains causes the release of large amounts of IgE. These antibodies attach by their constant region to **mast cells**, large granule-filled cells found in many connective tissues, including those underlying the nasal mucosa (Fig. 15.18a). When the variable ends of IgE bind to the pollen grain proteins, the mast cells rupture (Fig. 15.18b) and release the granules, which contain histamine and other chemicals that cause the capillaries to dilate and become more permeable. The tissues of the nose become red, swollen, and inflamed. Some degree of inflammation is protective, for it allows neutrophils and other phagocytic cells to escape from the blood along with immunoglobulins and comple-

Figure 15.18
(a) Mast cells in connective tissue are filled with large granules containing histamine. (b) When IgE antibodies attach to a mast cell, it releases its granules and the histamine induces an inflammatory reaction.

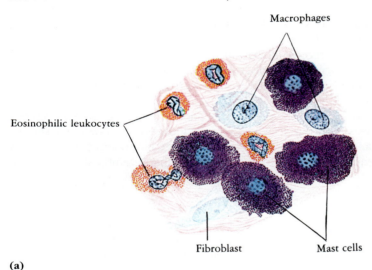

(a)

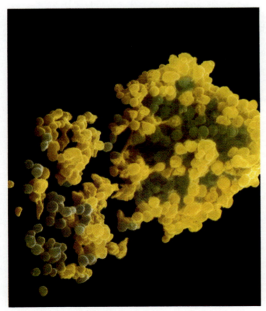

(b)

ment, and it attracts tissue macrophages to the area. In **allergic reactions**, the inflammation is intense and prolonged. The person may experience a "runny" nose, watery eyes, and congestion.

Anaphylaxis is a particularly dangerous allergic reaction that can occur in people sensitized to certain foreign substances such as the venom in bee stings. Another bee sting may unleash large numbers of IgE molecules and quantities of histamine from mast cells in many parts of the body. So much fluid may leave the blood that death from circulatory shock may occur. Antihistamines, drugs that block the action of histamine, can be used to treat allergic reactions.

Blood Groups

Erythrocytes, like other cells of the body, possess characteristic surface proteins (A, B, Rh, and Duffy factors) that act as antigens when blood from different individuals is mixed in transfusions. Mammals may also possess naturally occurring antibodies in the plasma directed against certain blood group antigens, although not against their own blood group antigens. The **ABO blood groups** were discovered when blood transfusions were widely used during World War I. Sometimes the transfusions were successful, but often they were not, and the donor's erythrocytes would clump (agglutinate) in the recipient's plasma. A careful analysis by Karl Landsteiner identified four pheno-

types with respect to blood types—O, A, B, and AB. An individual's group can be identified by taking blood and mixing it with sera containing either *a* antibodies (anti-A serum) or *b* antibodies (anti-B serum) and then observing which—if either—serum causes the erythrocytes to clump (Fig. 15.19). Landsteiner found that transfusions between members of the same group were safe and that transfusions between members of different groups also were safe provided that the donor's erythrocytes did not contain an antigen that would react with the recipient's antibodies. For example, members of group A, who have the antigen A on the erythrocytes and antibody *b* in the serum, cannot receive blood from members of groups B or AB. Although the donor's plasma may contain antibodies incompatible with the recipient's cells, the plasma becomes so diluted that its antibodies have no effect on the recipient unless a very large transfusion is given. Members of group O, who have neither of the antigens, can give blood to any group and are termed "universal donors." But since their plasma contains both of the antibodies, they can receive blood only from members of group O. In contrast, members of group AB, who have neither antibody, can receive blood from any group and are termed "universal recipients."

The presence of antibodies directed against a blood group to which an individual has not been exposed is a puzzling phenomenon. These blood group antigens contain structural components that resemble those found on

Figure 15.19
The ABO blood groups of humans. The two right-hand columns show the appearance of the erythrocytes when mixed with anti-A and anti-B serum. The blood group to which an individual belongs is identified by the antiserum in which the cells clump or agglutinate.

Blood group	Antigen (agglutinogen) on erythrocyte	Antibody (agglutinins) in plasma	Reaction with	
			Anti-A serum	Anti-B serum
O Universal donor	none	a and b		
A	A	b	agglutination	
B	B	a		agglutination
AB Universal recipient	A and B	None	agglutination	agglutination

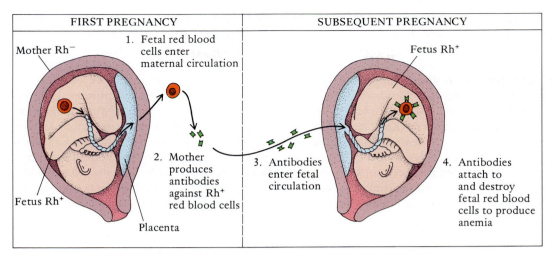

FIRST PREGNANCY

Mother Rh⁻

1. Fetal red blood cells enter maternal circulation

2. Mother produces antibodies against Rh⁺ red blood cells

Fetus Rh⁺

Placenta

SUBSEQUENT PREGNANCY

Fetus Rh⁺

3. Antibodies enter fetal circulation

4. Antibodies attach to and destroy fetal red blood cells to produce anemia

Figure 15.20
A fetomaternal blood incompatibility may develop if an Rh⁻ woman has an Rh⁺ child.

many bacteria and other foreign cells (to which an individual has very likely been exposed). One hypothesis proposes that blood group antigens develop in response to these foreign epitopes.

Many other blood groups have been discovered, but naturally occurring antibodies are not present so most are not of medical significance. One situation in which antibodies can develop and cause problems is the **Rh antigen**, first discovered on the red blood cells of rhesus monkeys. The synthesis of this antigen, which is present on the erythrocytes of many people, is controlled by a set of multiple alleles at one locus, but we can simplify the situation and assume that only two alleles are present: Rh and the recessive rh. An individual who is genetically RhRh or Rhrh has the antigen and is positive (rh⁺); rhrh individuals are negative (rh⁻). A fetus who has an rh⁻ mother (rhrh) and an rh⁺ father (RhRh or Rhrh) may have inherited the Rh allele from the father and be rh⁺ (Fig. 15.20). The placenta, where nutrient exchanges occur, does not normally permit the exchange of blood between mother and fetus, but small breaks in the placental membranes may arise and other breaks may occur during childbirth or abortion. If some rh⁺ blood from the fetus enters the mother, her immune system will respond by synthesizing antibodies directed against the rh⁺ antigen. The titer of antibodies is at first low because this is a primary response. If a subsequent fetus is also rh⁺ and some of its blood reaches the mother, a secondary response occurs and the titer of anti-rh⁺ antibodies increases rapidly. If some of these maternal antibodies cross the placental barrier and reach the fetus, many of the fetus's erythrocytes will be destroyed and a type of anemia known as **erythroblastosis fetalis** will develop. Fortunately, physicians have learned how to suppress the development of antibodies in the

mother by administering certain drugs or anti-(anti rh⁺) antibodies at childbirth or immediately following an abortion.

THE EVOLUTION OF THE IMMUNE RESPONSE

While the advantages of a fully functioning immune response are certainly obvious, how does a feature as complex as the immune system first evolve? We can approach this question in two complementary ways: (1) by examining the *phylogenetic distribution* of the various components of the immune system and (2) by looking at the *molecular homologies* between the different key participants in the immune response. Taken together, these two lines of information provide some clues about the pathways that have led to the complex and highly regulated immune system described in the previous sections.

Invertebrate Immune Responses

At least at the biochemical and cellular level, all multicellular animals appear to have some sense of self. Furthermore, in all cases studied thus far, all multicellular organisms can mount a response to the presence of foreign material within their body cavities or circulatory systems. In its simplest form, this response may only involve a few specialized phagocytic cells or the production of increased amounts of proteolytic enzyme. In one of the earliest experiments in immunology (1884), Eli Mechnikoff introduced a rose thorn fragment into the body cavity of sea star larvae. We now know that the presence of the thorn mobilized two types of specialized scavenger cells in the larva, the hemocytes (present in the hemolymph) and the coelomocytes (present in the body cavity, or coelom). These phagocytic cells targeted, engulfed, and eventually digested the thorn fragment. These cells, much like the macrophages described previously, act as generalized scavengers. Once foreign matter has been localized, these cells simply envelop the particle and attempt to digest it.

A certain degree of specificity is present in the humoral response mounted by cnidarians when they are exposed to foreign proteins. Sea anemones secrete a soluble factor that binds to the foreign protein and may remove it from the body, possibly by making it more accessible to phagocytizing cells. Similarly, both cnidarians and sponges are capable of rejecting tissue or cell grafts, implying both the ability to recognize foreign tissues and the capacity to mount a specific cell-mediated response. In fact, some sponges are capable of rejecting grafts from members of the same species. This suggests a capacity in sponges to recognize and react to the relatively slight genetic differences that exist from one individual to another.

The immune response has been well studied in certain invertebrate groups. In insects, a series of soluble factors is produced when foreign material (specifically bacterial cells) is introduced into the hemolymph. These factors are generally nonspecific proteases, capable of inhibiting bacterial growth and causing bacterial lysis. Many insects also produce a class of binding proteins that attach to glycoproteins commonly found on bacterial cell surfaces, causing the bacteria to clump. Such agglutinated bacteria are more easily targeted and phagocytized by scavenger cells. Molluscs, annelid worms, and arthropods are capable of rejecting tissues grafted from another species and may be capable of rejecting tissues from a genetically different individual of the same species as well. Graft rejection in these animals is a cell-mediated process that involves scavenger cells and lymphocyte-like cells.

The Vertebrate Immune System

The immune repertoire of vertebrates has undergone a thorough and systematic investigation that has yielded important clues about the evolutionary history of the immune response.

One of the key observations concerns the presence of circulating immunoglobulins in all vertebrate groups studied to date. While it is clear that antibody complexity and diversity have increased over evolutionary time, the most primitive vertebrates, the hagfish and lamprey (see p. 820), produce IgM when exposed to foreign antigens. Similarly, an early version of cell-mediated immunity is already in place in these vertebrates: lymphocytes, with their corresponding specific cell surface receptors, can be isolated from hagfish and lampreys. This observation is surprising, since these animals lack many of the specialized organs involved in the production and processing of these cells—bone marrow, thymus, spleen, and lymph nodes. Nevertheless, they are capable of graft rejection and exhibit a simplified form of immunologica memory.

Cartilaginous and bony fishes also produce IgM, and they are clearly capable of generating (by recombination and somatic mutation) the antibody diversity necessary for clonal selection to operate. In addition, a fully functional complement cascade is present in these vertebrates, albeit in a somewhat shortened form. At the cellular level, B-like cells are present in both cartilaginous and bony fishes, most likely produced by the thymus. Bony fishes produce both B cells and T cells.

Multiple kinds of immunoglobulins can already be detected in certain amphibians and reptiles, where bone marrow first appears. The MHC antigens, so important to cell-mediated immunity, are also present in these groups.

Birds, monotremes, marsupials, and placental mammals all display the full array of humoral and cell-based immune responses, although there are some slight variations in the diversity of immunoglobulins produced.

The Immunoglobulin Superfamily: Variations on a Theme

As we have already seen, selection can only act on available variation. As a result, evolution frequently makes novel use of available structures, modifying these structures to fulfill their new roles. One particularly striking example of this phenomenon involves the key molecules of the immune response. The immunoglobulins, the T cell receptors, and the MHC antigens all share remarkable similarity in certain aspects of their protein structure. In particular, all of them display the three-dimensional structure known as the **immunoglobulin fold**, which is part of the antigen binding site. This structure is the site of actual contact between antigen and antibody. Enormously flexible and specific, this fragment of protein appears in most of the molecules, where it is involved in some form of specific recognition. We can plausibly suggest that once this structure evolved, it conferred a function (recognition of three-dimensional structures) that could be put to advantageous use in a variety of different contexts. In fact, molecules that play no role in the immune response, but that are involved in cell-to-cell contact or recognition during development, also display this immunoglobulin fold. The current hypothesis proposes that this molecule first evolved as some form of cell surface receptor mediating cell-to-cell contact; only later, given its enormous versatility, did it become the basis of the immune response. The similarity between all of these receptors and cell-surface molecules is not coincidental. They are all members of a protein superfamily, related by common descent to an ancestral protein from which they have all diverged.

EVADING THE IMMUNE SYSTEM: A TARGET'S-EYE VIEW

In a Darwinian world, each organism must be understood on its own terms. From our standpoint, the immune system is an exquisite and versatile system, protecting us from the onslaught of pathogens. From the viewpoint of the infectious agent, our immune system is simply one more powerful selective force and a clear and present danger. As a result, heritable differences that confer upon a pathogen even a slight ability to evade or overwhelm the immune system will be strongly advantageous to their bearer. In evolutionary terms, pathogens and host are engaged in a permanent "arms race." In the section that follows, we briefly detail some of the mechanisms that viruses, bacteria, and protozoans have evolved to prevail against the immune response of their vertebrate hosts.

The Moving Target: Accelerated Antigenic Variation

One of the strategies that has evolved repeatedly and independently in viruses, bacteria, and protozoans involves a mechanism for constant changing in the three-dimensional structure of the pathogen's surface antigens. Given the precise fit between antibodies (or T cell receptors) and antigens, a pathogen capable of generating **antigenic variation** will likely be at a selective advantage. The best studied case involves the influenza viruses, responsible for the flu (Fig. 15.21). Two molecules on the

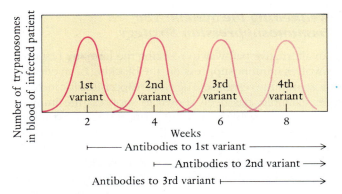

Figure 15.22

This graph shows the level of circulating trypanosomes in the bloodstream of affected patients. Note how the appearance of each new trypanosome surface antigen, the result of antigenic recombination, causes an increase in the levels of the parasite. Each time, the immune response of the host eventually catches up, resulting in a decrease in the circulating trypanosome levels.

surface of the influenza virus, hemagglutinin and neuraminidase, are the main targets of the immune response. In the type A influenza viruses of humans, 13 different types of hemagglutinin and 9 different types of neuraminidase are known. Each strain of influenza (e.g., Hong Kong Flu, Russian Flu, and Swine Flu) carries a particular combination of these surface antigens. More importantly, the genes encoding these antigens appear to mutate far more frequently than any other gene ever examined, and new surface antigens are constantly being generated. As a result of this process of **antigenic drift**, the immune system of the host is constantly facing a novel antigen and seldom can mount a secondary immune response. If this were not enough, human influenza viruses can also recombine with influenza viruses from other species (horses, pigs, and ducks), giving rise once again to completely novel surface antigens. The most widespread and devastating epidemics of influenza generally result from this process of interspecific recombination, or **antigenic shift**.

A variant of this strategy has evolved in a protozoan parasite, the trypanosome, responsible for sleeping sickness (Fig. 15.22). In this case, only a single antigen is present on the surface of the parasite. A single trypanosome, however, carries genetic information for over 100 different versions of this surface protein (known as variable surface glycoprotein or VSG), but expresses only one gene at a time. Every 10 to 14 days, the trypanosome switches on a new VSG gene. The immune system of its mammalian host cannot keep up with this antigenic variation. In this way, the trypanosome effectively avoids the surveillance of the immune system. Certain pathogenic bacteria have also evolved similar mechanisms for rapid antigenic variation.

Figure 15.21

The hemagglutinin and neuraminidase molecules on the surface of the influenza virus frequently change and elude the immune system of the host.

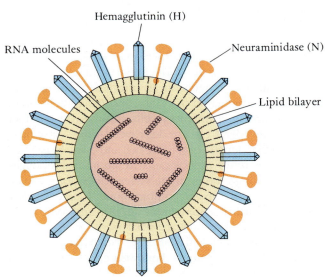

Weakening the System: The Immunosuppression Strategy

An alternative pathway to resisting the immune response involves **immunosuppression**, the blocking of an effective host immune response. The virus responsible for measles, for example, has been shown to suppress the immune response, possibly by attacking critical tissues involved in the processing of B and T cells. Other viruses, including Epstein-Barr virus (the cause of infectious mononucleosis), disrupt the immune response by attaching directly to B cell receptors, causing the B cells to divide in a haphazard manner as opposed to the normal clonal selection pathway. The HIV virus also acts by directly disrupting a critical component of the immune response. Finally, certain parasites may be capable of producing proteins that act as general immunosuppressants.

The Camouflage Strategy

This last method is most commonly associated with diseases caused by certain flatworms. Schistosomes, parasitic flukes that live in the blood stream and are responsible for the disease schistosomiasis (see p. 632), appear to incorporate the cell surface and MHC antigens of the host's red blood cells onto their own exterior surface, in effect concealing their own identity and making them invisible to the host's immune defenses. Certain viruses that infect cells may inhabit the host cell without exposing any viral surface antigen. Since the immune response is principally directed against surface features of cells, bacteria, or viruses, such concealment within the host cell is an extremely effective method of dodging the host's immune response.

Summary

1. In the humoral immune response, foreign antigens induce the production by the body of specific antibodies (immunoglobulins) that lead to the destruction of the antigen. The first exposure to an antigen causes a primary response in which the level (titer) of antibodies in the blood first rises and then falls off to low levels. A subsequent exposure to the same antigen causes a secondary response in which the antibody titer rises rapidly to a high level and drops off very slowly. This phenomenon is the basis for vaccinations.

2. Complex and large antigen molecules, such as proteins, are particularly effective in inducing antibody formation. Closely related foreign antigens lead to the production of similar antibodies. Vaccination was discovered when it was found that people who had been exposed to cowpox virus were immune to the similar but more virulent smallpox virus.

3. The ends of the two arms of the Y-shaped antibody molecules contain highly variable amino acid sequences and bear the antigen binding site. The stem contains a complement binding site. The great diversity of antibodies that B lymphocytes (B cells) can produce is attributed to somatic recombination of DNA sequences that occurs during the development and maturation of B cells. Each of the over one million types of B cells has a unique immunoglobulin-coding DNA sequence.

4. The clonal selection theory postulates that a close fit between foreign antigen and a specific cell-bound immunoglobulin present on the surface of a B cell renders that B cell competent. This competent cell then divides rapidly, producing a population of identical cells. Some of these then transform into plasma cells that synthesize the appropriate antibody and release it into the blood. Other B cells remain as memory cells that can quickly mobilize if the body encounters the same antigen again.

5. Many antibody molecules can bind with the foreign antigen, making the antigen-antibody complex susceptible to destruction by macrophages. In addition, the complement binding sites become exposed on the antibody molecules when antigen is bound, and this initiates an activation cascade whereby a group of complement proenzymes are now able to lyse foreign bacterial cells.

6. In the cell-mediated immune response, cytotoxic or killer T lymphocytes (T cells) are the ones that identify, bind with, and destroy foreign cells and those infected with viruses, bacteria, or other pathogens. Helper T cells assist in triggering the formation of clones of B cells and cytotoxic T cells, and suppressor T cells modulate the immune response. T cells have surface receptors that are as variable as the antibodies on B cells. Cells are identified as belonging to a particular individual and tissue by the major histocompatibility complex of antigens on their surface. Cytotoxic T cells recognize both a self-antigen and the viral antigen displayed on the surface of an infected cell. This focuses the immune response on infected cells rather than on free viruses. The initial binding of cytotoxic T cells with the antigen on an infected or foreign cell initiates the production of a clone of similar T cells, some of which remain as memory cells and the rest of which attack the invader. Cells are destroyed by the secretion of perforins that punch holes in the cell and cause lysis. The T cells also secrete lymphokines that attract macrophages.

7. Although the immune system does not normally attack the body's own cells, autoimmune diseases, such as rheumatoid arthritis, sometimes develop and some of the body's cells and proteins are destroyed.

8. Some individuals develop a hypersensitivity or allergy to foreign proteins. Renewed exposure can promote the synthesis of a particular class of antibodies (IgE) that attach to mast cells in the tissues and cause them to release histamine. Histamine may cause an uncomfortable local inflammation and constriction of respiratory passages, as in hay fever. Anaphylaxis, a more widespread allergic reaction, may be life threatening.

9. Erythrocytes also possess characteristic surface proteins that can act as antigens when blood from different individuals is mixed in transfusions. One set of such antigens, called A and B, is responsible for the A, B, AB, and O blood groups. Members of group O lack both of these surface antigens. A puzzling phenomenon in this case is the concurrent presence of antibodies (designated *a* and *b*) directed against the antigens an individual lacks. Members of group AB, who have both antigens, do not have these antibodies. A test of the compatibility of bloods must be performed before transfusions. The Rh factor is another blood cell antigen that assumes importance in pregnancies where the mother lacks the antigen (rh⁻) and the father has it (rh⁺). The fetus may inherit the Rh factor from the father, and antibodies formed in the mother's blood may reach the fetus and destroy many of its erythrocytes, causing severe fetal anemia.

10. At the biochemical and cellular level, all multicellular animals have a sense of self and can mount a response to the presence of foreign material within their bodies. The response ranges from the activation of phagocytic cells and the secretion of proteolytic enzymes to the elaborate immune responses of mammals. In the evolution of immune systems, one fragment of protein that is capable of recognizing surface proteins of other cells appears to have been modified and used in the many types of cells of the immune system, as well as in cell recognition and cell-to-cell interactions that occur during embryonic development.

11. Many viruses, bacteria, and parasites are able to evade the host's immune system by continually altering their surface antigens, by suppressing the immune response of the host, or by coating themselves with host antigens so they are not recognized as foreign.

References and Selected Readings

Buisseret, P.D. Allergy. *Scientific American* 246(Aug. 1982):86–95. A discussion of the cellular and biochemical changes that occur during an allergic reaction.

Darnell, J., et al. *Molecular Cell Biology.* New York: Scientific American Books, 1986. Includes an excellent chapter on the immune system.

Gallo, R.C. The AIDS virus. *Scientific American* 256(Jan. 1987):46–56. Describes the virus and its discovery.

Jaret, P. Our immune system: The wars within. *National Geographic* 169(1986):702–734. A superb pictorial essay of the action of the immune system.

Marrack, P., and J. Kappler. The T cell and its receptor. *Scientific American* 254(Feb. 1986):36–45. Describes the surface proteins that initiate the T cell response.

Scientific American 259(Oct. 1988). Many aspects of the AIDS epidemic are covered in the 10 articles of this single-topic issue.

Tizard, I.R. *Immunology: An Introduction,* 2nd ed. Philadelphia: Saunders College Publishing, 1988. A recent introductory textbook on immunology.

Tonegawa, S. The molecules of the immune system. *Scientific American* 253(Oct. 1985):122–131. A summary of the molecules and processes involved in the immune response.

16

Excretion and
Water Balance

(Left) Burchell's zebra (Equus burchelli) *of
Africa drinking at a watering hole. (Above)
Grevy's zebra* (Equus grevyi).

Living organisms are composed mostly of water. A 70-kg adult human contains about 49 L of water; about 35 L are within cells, and the rest is extracellular body fluids. Some of these extracellular fluids lie in the minute spaces between cells (interstitial fluid); others may be located in special cavities, such as the coelom. A large amount may be present in the blood vascular system. These extracellular fluids are a unique feature of multicellular animals and a dynamic part of their internal environment. There is nothing comparable, for example, in multicellular plants. The animal extracellular fluids contain salts and organic molecules; they are in osmotic balance with the intracellular environment (Fig. 16.1); and they are routes for the passage of food, gases, and wastes to and from cells. The composition of both intracellular and extracellular fluids—water and solutes—must be maintained within relatively narrow limits, regardless of the external environment.

Balancing water loss against water gain is a more complicated process than it might seem. Humans adrift in a lifeboat on a tropical sea, for example, are surrounded by water but cannot drink it. Their bodies are cooled by the evaporation of sweat, but in the process more water is lost. Without provisions, they might catch and eat fish, but the use of proteins as cellular fuel produces nitrogenous wastes that require water for removal. However, they could squeeze fluids out of the fish, which could be used as a source of water. The dilemma of these people in a lifeboat illustrates four important factors that may complicate the regulation of internal body fluids:

1. The salt content of the environment in which an animal lives.
2. The water demands of special physiological processes.
3. The need to maintain internal salt balance.
4. The elimination of nitrogenous wastes.

These complications are interrelated, but since excretory organs are involved with other aspects of regulation, in addition to the elimination of wastes, we will begin with excretion.

NITROGENOUS WASTES

Protein Metabolism

The continual breakdown of organic compounds within living systems (catabolism) results in **metabolic wastes**. A major waste is carbon dioxide, which is derived from the decarboxylation of organic fuels at various points in cell respiration. The degradation of nucleic acids and proteins produces nitrogenous wastes. When amino acids are catabolized (p. 271), the amino group (NH_2) is removed in a process called deamination and replaced by an oxygen atom. This process occurs primarily in the liver.

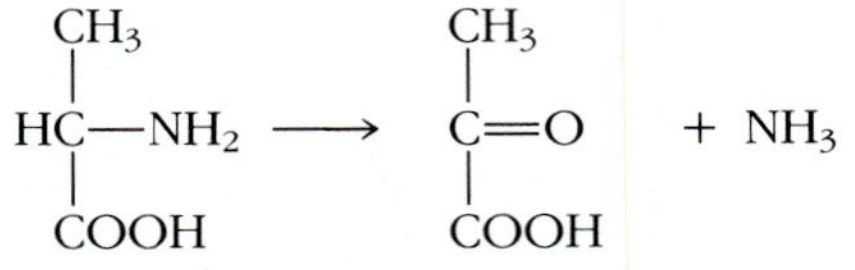

$$\underset{\text{Alanine}}{\begin{array}{c} CH_3 \\ | \\ HC-NH_2 \\ | \\ COOH \end{array}} \longrightarrow \underset{\text{Pyruvic acid}}{\begin{array}{c} CH_3 \\ | \\ C{=}O \\ | \\ COOH \end{array}} + \underset{\text{Ammonia}}{NH_3}$$

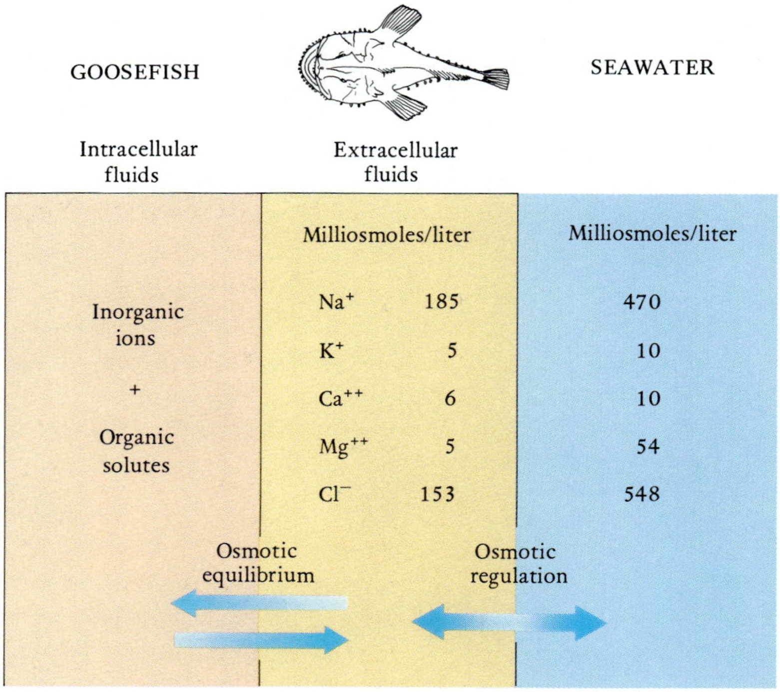

Figure 16.1

Ion content of a goosefish in relationship to that of seawater. The intracellular fluids are in osmotic equilibrium with the extracellular fluids, such as the blood plasma. However, the concentration of solutes in seawater is much greater than that in the fish. The fish will therefore tend to gain salts and lose water and must osmoregulate, i.e., expend energy to maintain its internal solute concentration at a constant level.

GOOSEFISH			SEAWATER
Intracellular fluids	Extracellular fluids		
	Milliosmoles/liter		Milliosmoles/liter
Inorganic ions	Na⁺	185	470
	K⁺	5	10
+	Ca⁺⁺	6	10
Organic solutes	Mg⁺⁺	5	54
	Cl⁻	153	548
	Osmotic equilibrium		Osmotic regulation

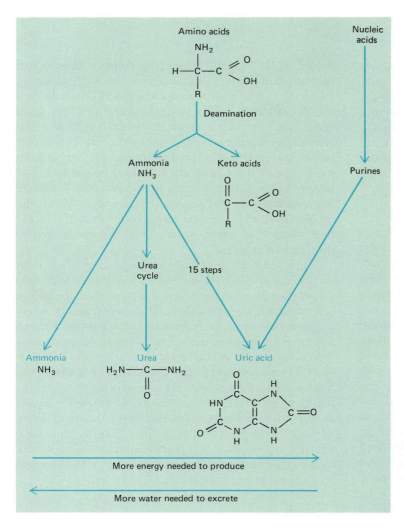

More energy needed to produce →

← More water needed to excrete

Figure 16.2
Sources of nitrogenous wastes and the forms in which they are excreted. Ammonia, which is very soluble and requires relatively little energy to produce, is shown to the left. The reverse is true for uric acid, which is shown to the right.

The carbon skeleton of the amino acid, now containing an oxygen instead of the amino group, forms a keto acid, i.e., an acid containing a central $C{=}O$ group with some radical (CH_3, COOH, etc.) on either side. The radicals are those of the particular amino acid, so each type of amino acid forms a different keto acid. Many of the keto acids can enter the citric acid cycle directly and be metabolized as fuels; others must pass through a series of intermediate steps before forming a compound that can be used. The removed amino group must either be eliminated or transferred to other compounds for the resynthesis of amino acids.

Forms of Nitrogenous Wastes

The removed amino group (NH_2) first picks up a hydrogen ion and appears as a molecule of **ammonia** (NH_3). Ammonia is a gas that is very soluble in water, and as it goes into solution some combines with hydrogen ions in the water to form ammonium ions (NH_4^+). Ammonia is the simplest form of nitrogenous waste but becomes a metabolic poison when its concentration rises. Since ammonia is very soluble, it can be quickly flushed out of the body if enough water is available. It takes about 500 ml of water to carry off 1 g of nitrogen in the form of ammonia. Ammonia and ammonium ion are the usual nitrogenous wastes of aquatic animals, including most fishes.

In terrestrial animals that must conserve water, ammonia is converted to less toxic urea or uric acid. **Urea** is a simple compound that combines two molecules of ammonia with a molecule of carbon dioxide:

$$2NH_3 + CO_2 \longrightarrow NH_2{-}\overset{\displaystyle O}{\underset{\|}{C}}{-}NH_2 + H_2O$$

Ammonia   Urea

Urea synthesis, which occurs primarily in the liver, involves a complex biosynthetic urea cycle. Only about 50 ml of water is required to carry off 1 g of nitrogen in this form, but its synthesis is energetically expensive, for three ATPs are used for the synthesis of each molecule. Urea is the nitrogenous waste of mammals, as well as some fishes and amphibians (Fig. 16.2).

Uric acid is the principal nitrogenous waste product of birds, terrestrial reptiles, insects, and land snails. It is a relatively insoluble purine and requires only 10 ml or less of water to remove 1 g of nitrogen in this form (Fig. 16.2). However, it is also energetically expensive to synthesize, about twice as expensive as urea. Uric acid is also produced in small amounts by humans and other mammals from the degradation of nucleic acids. Ammonia, urea, and uric acid are the most common forms of nitrogenous wastes eliminated by animals; spiders, however, excrete guanine (another purine).

EXCRETORY ORGANS

Most animals possess some sort of excretory organ for the removal of various sorts of excess substances, often including water. However, most aquatic animals do not use their excretory organs to get rid of ammonia. Aquatic mammals and reptiles, such as whales, porpoises, and sea turtles, are an exception to this rule, because they do not excrete ammonia. Very small aquatic animals have enough surface area in relation to volume to eliminate ammonia by simple diffusion to the exterior. Large aquatic animals with gills eliminate ammonia across the the gill surface.

Excretory organs are typically tubular or saccular structures adapted for eliminating wastes. Most belong to one of two general types. In one type, the inner end of the tubule opens to the coelom (Fig. 16.3a). Fluid filtered from the blood into the coelomic fluid passes in turn into the excretory tubule. As the coelomic fluid passes down the

tubule, it is subjected to varying degrees of **selective reabsorption**. Some organic substances, such as sugars and amino acids, may be reabsorbed, and, depending on the environment of the animal, there may be some reabsorption of salts and water. In addition to selective reabsorption, **secretion** of wastes by the tubule wall may occur. The secretion and reabsorption of materials from the lumen of the tubule are facilitated by the blood or blood vessels surrounding the tubule. As a result of selective reabsorption and secretion, the wastes within the tubular fluid may become more concentrated, and the final mixture of water, wastes, and salts is expelled to the exterior as **urine**.

In a second type of excretory organ, the tubule does not open into the coelom (Fig. 16.3b). Rather, the blind end of the tubule receives filtrate directly from the blood. The blood filtrate is modified and concentrated by selective reabsorption and secretion during its passage through the tubule. In neither type of excretory organ are wastes "screened out" from the blood, as is popularly believed. Remember that blood filtrate is not waste; it is simply the fluid portion of blood (minus large molecules and cells) that can pass through the capillary wall. *Excretory organs do not screen wastes; however, they may concentrate them.* Various modifications of these two general plans have evolved in a number of different animal groups. A common change has been the reduction or loss of filtration and an increase in the importance of secretion.

Nephridia

The excretory or osmoregulatory organs found in the largest number of different groups of animals are tubules called **nephridia** (Gr. *nephros*, kidneys), which open to the exterior of the body through a **nephridiopore**. There are two types of nephridia: protonephridia and metanephridia. Flatworms, rotifers, some marine annelids, and members of a number of other phyla possess blind tubules called **protonephridia** (Gr. *protos*, first). The inner blind end (or ends when branched) is composed of a terminal cell bearing one or more cilia, sometimes called a flame cell (Fig. 16.4). The beating cilia reminded early zoologists of a flickering flame, hence the name. The wall of the terminal cell consists of slender microvilli-like processes or interdigitating finger-like processes where the terminal cell joins the first cell forming the tubule. Either arrangement functions as a filter. The beating of the cilia or flagella drives water down the tubule and creates a negative pressure within the terminal cell. Surrounding water is then pulled inward, passing through the slits. The protonephridial tubule opens to the exterior by way of a nephridiopore. Although these tubules are excretory organs, they are usually not involved in the excretion of nitrogenous wastes; in many small animals they may function as pumps to remove excess water.

Figure 16.3

Models of the two principal types of excretory organs in animals, based on their function. In (a) the excretory tubule opens into the coelom, and wastes are received in coelomic fluid, which receives filtrate from the blood. In (b) the excretory tubule is closed, and wastes are received in blood filtrate

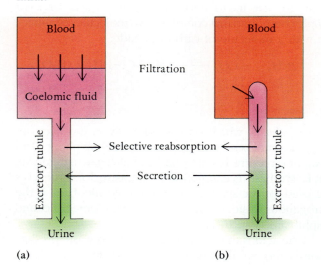

(a)   (b)

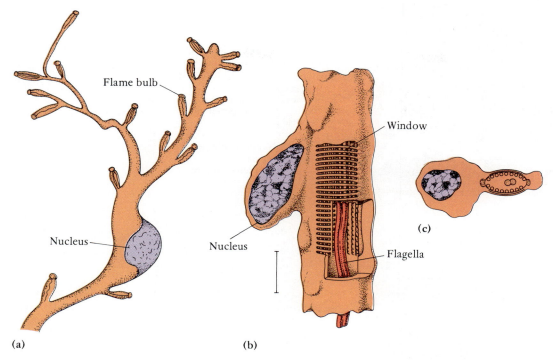

(a)   **(b)**   **(c)**

Figure 16.4

A protonephridium. (a) Protonephridium of the flatworm *Mesostoma*, in which a number of flame bulbs are a part of one cell. (b) Flame cell of the freshwater flatworm *Stenostomum*. (c) Cross section at level of nucleus.

Molluscs, most annelids, and the members of a number of small phyla possess a second type of nephridial tubule, called a **metanephridium** (Gr. *meta*, later). The inner end is not blind but bears a ciliated funnel, or **nephrostome** (Gr. *stoma*, mouth), that opens into the coelom. Filtrate leaves the blood and enters the coelomic fluid. The metanephridial tubule processes the coelomic fluid. Snails and clams possess one or two metanephridia, but annelids, such as earthworms, are metameric and have one pair of metanephridia per segment. Coelomic fluid enters the nephrostome in the coelomic compartment anterior to the one housing the tubule (Fig. 16.5).

Green Glands or Antennal Glands

The paired excretory organs of crayfish, shrimps, and crabs, called **green glands** or **antennal glands**, consist of a saccule located in the head and bathed in blood of the surrounding hemocoel (Fig. 16.6). The wall of the saccule contains specialized interdigitating cells, called podocytes, that facilitate the filtration of fluid from blood into the saccule. Similar cells are found in protonephridia and in the vertebrate kidney (see Fig. l6.11). The filtrate passes from the saccule into the tubule, the first part of which is

greatly expanded and folded (called the labyrinth). Selective reabsorption occurs in certain parts of the tubule, and the resulting urine is expelled to the outside through an excretory pore at the base of each antenna. Crayfish possess a bladder. Since the crustacean excretory saccule represents a remnant of the coelom, which is greatly reduced in adult arthropods, the tubule may be derived from a metanephridium inherited from the worm-like ancestors of crustaceans.

Malpighian Tubules

The **malpighian tubules** of insects and spiders are also blind-ending tubules bathed in blood (Fig. 16.7). Unlike crustacean green glands, the number of tubules varies from two to several hundred and they open into the intestine rather than to the exterior. Filtration does not occur in these organs. Uric acid, potassium, and sodium from the blood are secreted into the tubules and then pass from the tubules into the intestine and rectum. The rectum is an important part of the insect excretory system, because it is here that water is reabsorbed. By means of a complex ion pump, many insects are able to reabsorb sufficient water to produce a urine that is more concentrated than

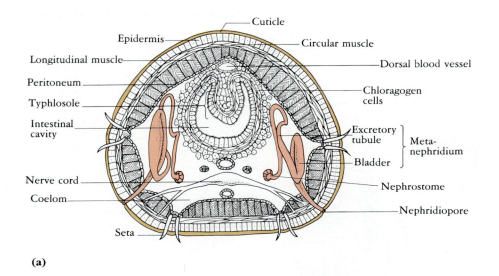

(a)

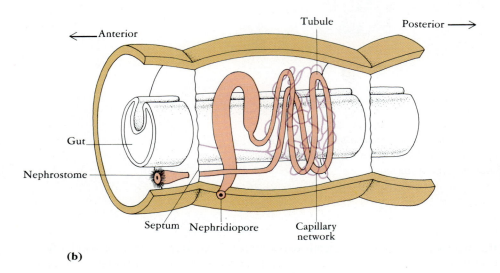

(b)

Figure 16.5
Metanephridia of an earthworm.
(a) Diagrammatic transverse view
of a segment, showing the pair of
metanephridia. (b) Longitudinal
view of three segments, showing a
metanephridium on one side.

Figure 16.6
Excretory organ (green gland) of a crayfish. The figure on the right shows the
organ unfolded and extended.

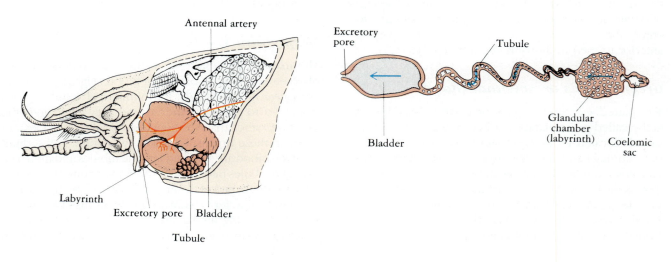

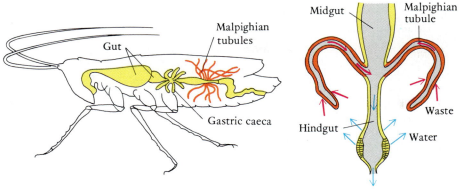

Figure 16.7
Malpighian tubules of an insect (grasshopper). Figure on the left shows their
position in relationship to the entire gut. Two tubules are shown on the right. The
red arrows indicate the passage of waste from the surrounding blood into the
tubule lumen as uric acid. The blue arrows out of the hindgut indicate the
reabsorption of water.

the blood (hyperosmotic to the blood), an ability shared only with birds and mammals. The pasty urine is eliminated through the anus with the fecal material.

THE VERTEBRATE KIDNEY

The kidneys of vertebrates are paired organs that lie dorsal to the coelom on each side of the dorsal aorta. All vertebrate kidneys are composed of units called renal tubules, or **nephrons**, that end blindly and receive a filtrate from the blood. The number and arrangement of the nephrons differ among the various groups of vertebrates. In ancestral vertebrates each kidney probably contained one nephron for each body segment that lay between the anterior and posterior ends of the coelom (Fig. 16.8a and b). These nephrons drained into an **archinephric duct** that continued posteriorly to the cloaca. Such a kidney is called a **holonephros** (Gr. *holos*, whole), for it extends the entire length of the coelom. A holonephros-like kidney is found today in the larvae of hagfishes but not in any adult vertebrate.

In the kidney of adult fish and amphibians (Fig. 16.8c), the anterior tubules have been lost, some of the middle tubules are associated with the testis, and there is a concentration and multiplication of tubules posteriorly. Such a posterior kidney is known as an **opisthonephros** (Gr. *opisthen*, behind). The original archinephric duct functions both as an excretory duct, and in males, as a sperm duct. The kidney of reptiles, birds, and mammals is known as a **metanephros** (Fig. 16.8d). All of the middle tubules not associated with the testis have been lost, and there is an even greater multiplication and posterior concentration of tubules. The number of nephrons is particularly large in birds and mammals; their high rate of protein metabolism

yields a large amount of wastes. It is estimated that humans have about 1,000,000 nephrons per kidney, whereas certain salamanders have less than 100. The tubules producing urine drain into a **ureter**, a tube that evolved as an outgrowth from the old archinephric duct. The archinephric duct has been taken over completely by the male genital system as the sperm duct.

The evolutionary sequence of kidneys is holonephros, opisthonephros, and metanephros. In the embryonic development of higher vertebrates—reptiles, birds, and mammals—there is a somewhat parallel sequence involving a posterior concentration of functional tubules. A transitory pronephros and mesonephros develop from the anterior and middle portions of the embryonic nephrogenic mesoderm. Nephrogenic mesoderm (kidney mesoderm) arises dorsally along the entire length of the embryonic coelom, but only the most posterior part develops into the adult metanephros (Fig. 16.8e and f).

Most adult tetrapods have a **urinary bladder** that develops as a ventral outgrowth from the cloaca. Generally, the excretory ducts from the kidneys lead to the dorsal part of the cloaca, and urine must flow across it to enter the bladder. However, in mammals (Figs. 16.8g and 16.9a) the ureters lead directly to the bladder, and the bladder opens to the body surface through a short tube, the **urethra**. In all but the most primitive mammals the cloaca becomes divided and disappears as such, with the dorsal part of the cloaca forming the rectum and the ventral part contributing to the urethra.

Urine is produced continually by the kidneys and is carried down the ureters by peristaltic contractions (Fig. 16.9). It accumulates in the bladder, for a smooth muscle sphincter at the entrance to the urethra and a striated muscle sphincter located more distally along the urethra

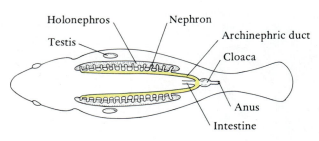

(a) Ancestral vertebrate
    (dorsal view)

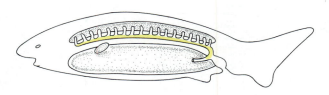

(b) Ancestral vertebrate
    (lateral view)

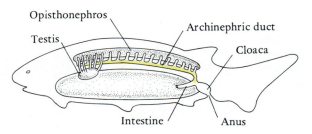

(c) Fish

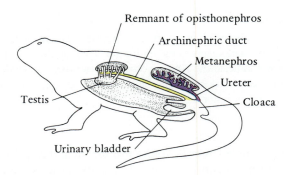

(d) Reptile

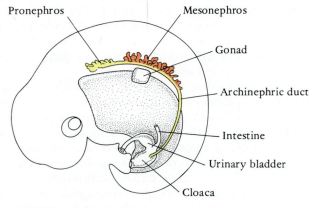

(e) Early reptilian embryo

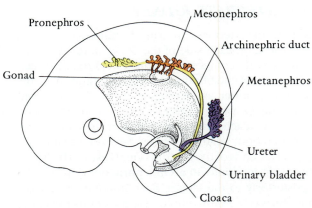

(f) Later reptilian embryo

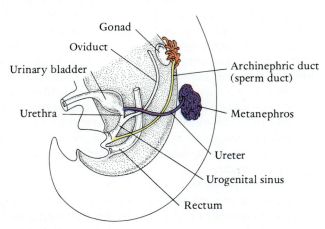

(g) Mammalian embryo

Figure 16.8

Comparison of the evolution (a–d) and embryonic
development (e–g) of the vertebrate kidney and its duct.

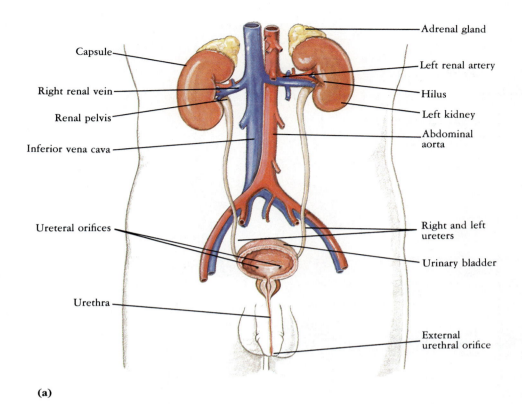

(a)

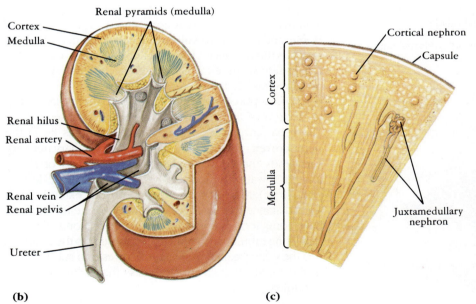

(b)   (c)

Figure 16.9
Human excretory system. (a) Ventral view of excretory system. (b) Three-dimensional sections of the kidney. (c) Section of kidney cortex and medulla, showing position of nephrons.

are closed. Urine is prevented from backing up into the ureters by valvelike folds of mucous membrane within the bladder. When the bladder becomes filled, stretch receptors are stimulated, and a reflex is initiated that leads to the contraction of the smooth muscles in the bladder wall and the relaxation of the smooth muscle sphincter. Relaxation of the striated muscle sphincter is a voluntary act that allows voiding of the urine externally through the urethra.

Nephron Structure

The proximal end of each nephron (Fig. 16.10), known as **Bowman's capsule**, is a hollow ball of squamous epithelial cells, one end of which has been pushed in by a knot of capillaries called a **glomerulus** (L. *glomus*, ball). Bowman's capsule and the glomerulus constitute a **renal corpuscle**. The rest of the nephron is a tubule, largely composed of cuboidal epithelial cells and subdivided in

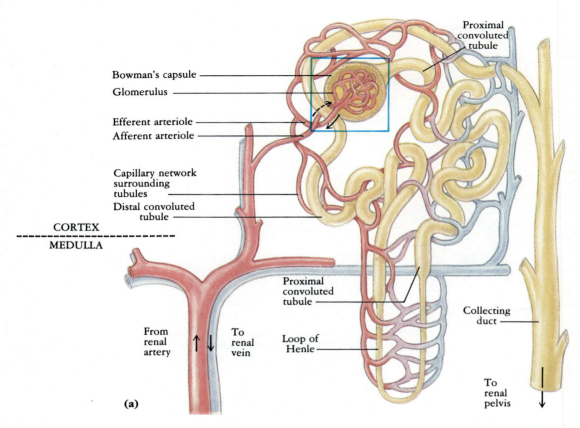

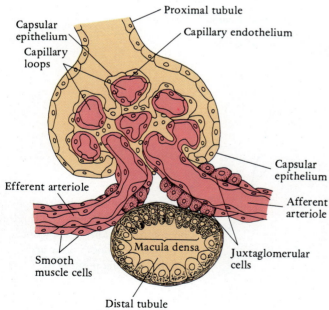

(b)

Figure 16.10

A nephron or renal tubule. (a) A complete nephron and its relationship to the surrounding blood vascular system. (b) An enlargement of the boxed area in (a) showing the juxtaglomerular apparatus.

mammals into a **proximal convoluted tubule**, a **loop of Henle**, and a **distal convoluted tubule**. A **collecting tubule** receives the drainage of several nephrons and leads to the **renal pelvis**, an expansion within the kidney of the proximal end of the ureter. The location of the different parts of a nephron within the kidney and their relationship to blood vessels have important functional consequences. The renal corpuscles and convoluted tubules lie in the outer **cortex** of the kidney, and a dense capillary network surrounds the convoluted tubules; the loops of Henle extend toward the center, or **medulla**, of the kidney (Fig. 16.9c). Most of the human nephrons extend only a short distance into the medulla, but about one-fifth of them have long loops of Henle that extend, along with the collecting tubules and capillary loops, far into the medulla. It is the convergence of these structures into subdivisions of the renal medulla (the **calyces**) that forms the **renal pyramids** (Fig. 16.9b).

Urine Formation

The kidney tubules produce urine, a watery solution containing waste products of metabolism removed from the blood. The most abundant nitrogenous waste in humans and other mammals is urea, but lesser amounts of ammonia, uric acid, and creatinine are present. Urine also contains excess potassium, sodium, hydrogen, and chloride ions. The yellowish color of urine is due to

urobilinogen, a pigment derived from the bile and reabsorbed in the intestine. In general, urine production involves first the production of a filtrate, followed by (1) the reabsorption of desirable substances from the filtrate back into the blood and (2) secretion of some unwanted substances from the tubule cells into the filtrate.

Glomerular Filtration

The first step in urine formation is **glomerular filtration**, which was demonstrated in the 1920s by Dr. A.N. Richards. He developed a micropipette technique for removing and analyzing minute samples of fluid from the lumen of Bowman's capsule. The wall, or **filtration barrier**, through which fluid must pass to reach the lumen of Bowman's capsule is composed of (1) the capillary endothelium, which is fenestrated and contains many large pores, (2) a middle, large meshed basal lamina, and (3) a layer of cells called **podocytes** (Gr. *podos*, foot + *kytos*, cell) that are derived from the infolded wall of Bowman's capsule. The podocytes are large cells with primary and secondary finger-like processes that interdigitate with each other (Fig. 16.11). The intervening slits are the passageways for the filtering fluid.

The endothelial pores and podocyte slits make the filtration barrier many times more permeable than other capillary beds but at the same time restrict passage to water, various ions, and small organic molecules, including simple sugars and amino acids as well as nitrogenous wastes. Blood cells and the large plasma proteins remain in the blood. The only small molecules to be held in the blood are those bound to plasma proteins: among these are hormones, fats, iron and other trace minerals, and certain vitamins. The resulting filtrate in Bowman's capsule is thus very much like plasma and is essentially the same as the interstitial fluid of other capillary beds.

Estimates of the filtration rate are obtained by the use of inulin, a polysaccharide derived from artichokes that is filtered easily and is not removed from or added to the filtrate as the fluid continues down the kidney tubules. For example, when inulin is injected intravenously until the plasma concentration reaches 1 mg per ml, 125 mg of inulin appear in the urine per minute. For this to occur, 125 ml of plasma must be filtered each minute. This amounts to 180 L of filtrate per day! The kidneys have a rich blood supply, usually receiving a little over 20% of the cardiac output in a mammal. This enables the kidneys to process the total blood volume every 5 minutes.

A glomerulus lies between an **afferent arteriole**, a branch of a renal artery, and an **efferent arteriole** that leads to capillaries around other parts of the tubule (**peritubular capillaries**) (Fig. 16.10). The efferent arteriole subjects blood leaving the glomerular capillaries to considerable peripheral resistance, because it has a smaller diameter than the afferent. This produces high pressure in the glomerular capillaries, favoring filtration, and low pres-

sure in the peritubular capillaries, favoring fluid return. The high filtration pressure in the glomerular capillaries drives water and small solute molecules from the blood. The filtration pressure exceeds the pressures within the Bowman's capsule and the osmotic pressure of the blood (both of which promote the return of fluid to the blood, p. 319). As a consequence, there is a substantial net production of glomerular filtrate, an amount equivalent to about 20% of the plasma entering the glomerular capillaries.

Since the volume of urine in humans is only about 1% of the glomerular filtrate, most of the water and nearly all of the glucose, amino acids, and ions present in the glomerular filtrate are taken back into the blood by **tubular reabsorption** as the filtrate passes down the tubules. Other substances may be added to the filtrate by **tubular secretion**. Each minute, human kidneys excrete an amount of urea equal to that present in 75 ml of plasma. It can be said that 75 ml of plasma has been cleared of urea, or that urea has a **clearance rate** of 75 ml per minute. Other substances have different clearance rates. From inulin studies we know that the filtration rate is 125 ml per minute. If the clearance rate of a freely filterable substance (such as urea) is less than this, some of this substance must

Figure 16.11
Scanning electron photomicrograph of part of a podocyte of a Bowman's capsule. These highly modified cells represent that part of the capsule wall that envelops the glomerulus and contributes to the filtration barrier through which fluid from the glomerulus must pass. From the central part of the cell (*CB*) containing the nucleus extend primary (*PB*), secondary (*SB*), and tertiary (*TB*) branches, which terminate in finger-like projections, called pedicels (*Pe*). The fine slits (*FS*) between adjacent interdigitating pedicels of the podocyte are the primary avenues of filtrate passage.

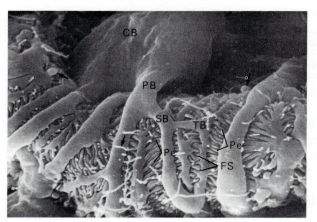

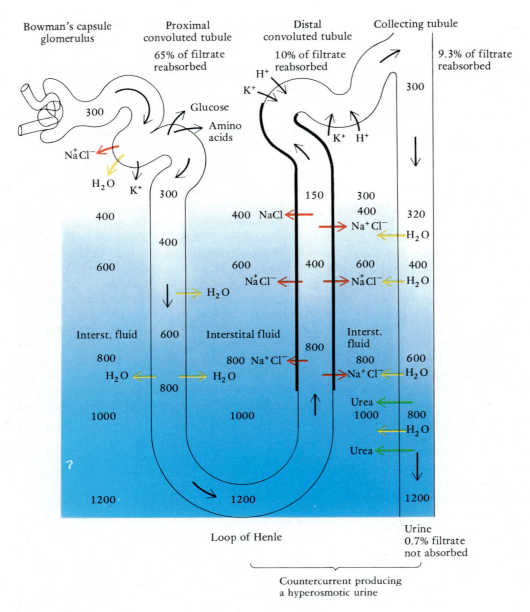

Figure 16.12

Diagram of the processes, including the countercurrent multiplying mechanism, that transform the glomerular filtrate into urine as it passes through a nephron and collecting tubule. The section of the ascending limb of the loop of Henle, shown with a heavy wall, is impermeable to water. The numbers indicate the relative concentrations of osmotically active solutes in milliosmols per liter.

be reabsorbed in the tubules; if the clearance rate exceeds 125 ml per minute, then some of the substance must be added by tubular secretion.

Tubular Reabsorption

Reabsorption involves both the passive diffusion of materials back into the capillaries surrounding the tubules and the active uptake of materials by the tubular cells and their secretion into the blood against a concentration gradient. Almost all of the glucose and amino acids and about two-thirds of the sodium ions are actively transported out of the proximal convoluted tubule (Fig. 16.12). Active transport is necessary, of course, because the concentration of these substances in the tubule is the same as in the blood, i.e., the fluid in the tubule is isosmotic to the blood. The active reabsorption of one ion can bring about the passive reabsorption of another with the opposite charge. For example, when sodium, which is positively charged, is removed from the filtrate, the remaining fluid will exhibit an excess of negative charges, i.e., a charge imbalance. As a result, negative chloride ions passively follow the reabsorbed positive sodium ions.

Those materials that can be actively reabsorbed are taken back in varying amounts, depending upon their concentration in the blood. If the concentration of one of these materials in the blood and glomerular filtrate rises above a certain level, known as the **renal threshold**, not all of it will be reabsorbed back into the blood from the tubule, and the amount present in excess of the renal threshold will be excreted with the urine. The quantitative value of the renal threshold differs for different substances. In **diabetes mellitus** the reduced ability to convert glucose to glycogen leads to a high concentration of glucose in the blood, exceeding the renal threshold for glucose (about 150 mg glucose per 100 ml blood). Thus high concentrations of sugar appear in the urine.

Water is reabsorbed passively. As solutes, of which sodium and chloride are the most common, are actively taken out of the filtrate in the proximal convoluted tubule, the water in the filtrate tends to become more concentrated than it is in the blood. However, water molecules passively diffuse out as fast as solutes are pumped out, hence the concentration of the filtrate remains the same as that of the blood. About 65% of all the water taken back is reabsorbed in this way in the proximal convoluted tubule, but the tubular filtrate cannot be made more concentrated (i.e., to contain less water) than the blood by this mechanism. Note that we are referring here only to the proximal convoluted tubule; the filtrate can be concentrated further along in the tubule.

Tubular Secretion

Potassium and hydrogen ions, ammonia, and organic acids and bases are the principal substances secreted by the renal tubule. Although most filtrated potassium ions are actively pumped back out of the proximal convoluted tubule, they can be either absorbed or secreted by the distal convoluted tubule and collecting tubule, depending upon the level of potassium in the blood (Fig. 16.12).

The distal convoluted tubule is also an important site for the secretion of H^+ and therefore contributes to the regulation of internal pH, i.e., acidity or alkalinity. Hydrogen ions enter the tubular fluid in exchange for Na^+ ions, which are reabsorbed. The acidity of the urine does not rise greatly because the H^+ ions combine with NH_3 to form NH_4^+ and with HPO_4^{2-} to form $H_2PO_4^-$. Neither NH_4^+ nor $H_2PO_4^-$ can diffuse out of the tubule.

Urine Concentration

Most of the lower vertebrates cannot produce a urine more concentrated than the blood, i.e., hyperosmotic to the blood. However, birds and mammals do produce a hyperosmotic urine. In mammals this concentrating ability is dependent upon the loop of Henle. We will examine the process in mammals. The descending and ascending limbs of the loop of Henle lie parallel to each other, so that the direction of flow of fluid in one is opposite to that in the

other. The descending limb, apex, and lower part of the ascending limb are thin walled; the upper part of the ascending limb is thick walled (Fig. 16.12). The active transport of sodium ions out of the thick-walled ascending limb into the interstitial fluid (with chloride ions following passively by charge attraction) raises the interstitial solute concentration. This causes the passive diffusion of water out of the adjacent descending limb, which is relatively impermeable to Na^+ and Cl^- and urea. Water does not leave the ascending limb (both thin- and thick-walled segments) because it is impermeable to the passage of water. This results in a concentration of Na^+ and Cl^- in both the descending and ascending limbs that increases toward the tip of the loop, as shown in Figure 16.12. This is also true of the surrounding interstitial fluid. The degree of accumulation depends upon the length of the loop; there is more opportunity to pump out Na^+ in a long loop. We have emphasized so far the role of Na^+ and Cl^- in the creation of this solute gradient. In fact, urea—which constitutes one-half of the solute—is a major contributor. Urea leaks out of the lower part of the collecting tubules, whose walls are somewhat permeable to this molecule and lie parallel to the loops of Henle. Realize that only some urea leaks out; much continues to the pelvis of the kidney as waste in urine.

The osmotic gradient established by these mechanisms is essential to the concentrating ability of the mammalian kidney. It makes possible the passive reabsorption of additional water from the collecting tubules, which lie parallel to the loops of Henle. Water simply follows the osmotic gradient, moving from an area of low osmotic pressure (low concentration of solutes) to one of high osmotic pressure (high concentration of solutes). The countercurrent flow of material through the nephron and the active pumping out of salts in the ascending limb of the loop of Henle combine to form a very effective countercurrent multiplier system.

Let us follow the changes in the glomerular filtrate as it passes through the renal tubule. On leaving the glomerulus the filtrate loses water in the proximal tubule, as water passively follows the Na^+ (and oppositely charged Cl^-) pumped out by this part of the tubule. Thus even though much water has been removed, the filtrate remains more or less isosmotic with the blood as it leaves the proximal tubule. The concentration of urea has risen to some extent because the proximal tubule is not very permeable to urea.

When the filtrate passes into the loop of Henle, it encounters the gradient of solute concentration in the interstitial fluid surrounding the loop. Water is continually lost from the filtrate in the descending limb, because the concentration outside the tubule is always greater as the filtrate moves toward the loop apex (Fig. 16.12). It does not regain this water as it ascends the loop of Henle, for the cells of the ascending limb of the loop have a low permeability to water. However, the filtrate becomes dilute be-

Focus 16.1   *Dialysis*

The large number of nephrons present in each kidney provides a great margin of safety in renal function, and an individual can get along quite well with only one kidney. More extensive kidney loss was formerly fatal. But our knowledge of kidney function has made possible the development of an artificial kidney machine utilizing the process of dialysis. Many persons with inadequate kidney function or even with complete kidney loss are able to survive with periodic dialysis treatments. The dialysis apparatus receives blood from an artery in an arm or leg and returns it to a vein in the same extremity. Within the machine the blood circulates through fine channels whose semipermeable membrane walls separate the blood from a dialysis fluid. All substances except large protein molecules and blood cells can diffuse across the membrane. The dialysis fluid contains most of the common plasma constituents in the same concentration but contains no wastes. As the blood circulates through the dialysis channels, wastes that have accumulated to high levels in the plasma—including urea, creatine, and potassium ions—rapidly diffuse across the membrane into the dialysis fluid following their concentration gradients. The dialysis fluid is continually being replaced. A 4- to 6-hour treatment three times a week is usually sufficient to maintain the patient in relatively good health.

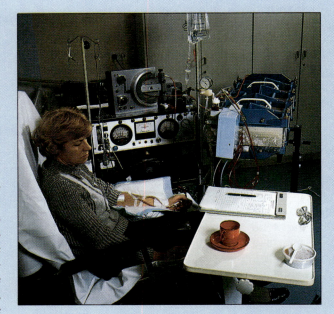

A patient undergoing dialysis.

cause of the large amount of sodium pumped out by the thick segment of the ascending limb. By the time the filtrate reaches the distal tubule, so much NaCl has been removed that the filtrate is hypoosmotic to the blood, i.e., it is less concentrated than the blood, and it loses water here.

The tubular fluid now descends through the medulla again, this time in the collecting tubule. It passes as a countercurrent to the adjacent gradient of NaCl in the ascending loop of Henle and the surrounding interstitial spaces and again loses water to the solution in those spaces. As it loses water it becomes more concentrated, but by then that parcel of fluid has traveled further along the collecting tube and is now adjacent to a more concentrated salt solution in the surrounding interstitial spaces; therefore it loses more water. Deep within the medulla the collecting tubule becomes permeable to urea, some of which leaks out into the surrounding interstitial space (Fig. 16.12). As we have already indicated, this urea is a major contributor to the solute concentration of the interstitial fluid; more water diffuses out of the collecting tubule. By

the time the fluid in the collecting tubule reaches the pelvis of the kidney, it can be very hyperosmotic and is now called urine. Note that water has been reabsorbed from the proximal tubule, the descending limb of the loop of Henle, the distal tubule, and the collecting tubule. All of this water is removed from the kidney by the peritubular capillaries. Remember that the filtration pressure is low in this capillary bed, and low pressure facilitates osmotic uptake of fluid into the capillaries. Why do these capillaries not take away the salts and urea as well? As blood flows in these capillaries down into the gradient, it does indeed pick up urea and salts, but then gives up these solutes on the outward passage through lower and lower concentrations of the solute gradient.

How concentrated is the urine? In mammals the concentration can vary, depending upon the need to reabsorb water. In humans the maximum concentration of salts is a little over twice that of the plasma; the concentration of urea may be 70 times greater; and the total osmotic concentration may be four times that of plasma.

Many mammals can produce a more concentrated urine than humans. The urine of the pocket mouse of American southwestern deserts can reach a concentration 25 times that of plasma (total osmolality). These animals have long loops of Henle, as you might expect, but only in comparison to other small mammals. The loops of Henle of a pocket mouse are not as long as those of a horse, which can produce a urine only five times as concentrated as its plasma. However, the mouse has a much more powerful sodium pump in the thick part of the ascending loop of Henle than does a horse. So in comparing the lengths of loops of Henle, we must first correct for the size and metabolic rate of the animals under comparison.

Control of Kidney Function

The amount of material reabsorbed or excreted by the kidney is regulated by adjusting three principal points of control: (1) the glomerular filtration rate through arteriole dilation and constriction; (2) the permeability of the collecting tubule to water; and (3) the tubular reabsorption of solutes. In the event of a severe drop in blood pressure, such as might occur in a hemorrhage, signals via sympathetic neurons will cause the afferent arterioles to constrict, reducing blood flow to the kidney.

Associated with the afferent and efferent arterioles and a part of the distal convoluted tubule that lies between them are special receptor cells (macula densa) and secretory cells that constitute the **juxtaglomerular apparatus** (Fig. 16.10b). By monitoring arteriole blood flow and the fluid passing through the distal tubule, the juxtaglomerular apparatus can detect changes in blood pressure, sodium concentration, and other physiological conditions. In response to a drop in blood pressure or sodium concentration, as would occur in blood loss or dehydration, the secretory cells release the enzyme **renin**. Renin secreted into the blood stream converts a plasma globulin into the hormone **angiotensin**, a powerful vasoconstrictor (p. 452). In the kidney, angiotensin causes vasoconstriction of the efferent arterioles, which decreases renal circulation while still maintaining glomerular filtration. The resulting low peritubular filtration pressure and high osmotic pressure would increase water reabsorption. In the brain, angiotensin increases thirst and thus water intake.

Angiotensin also promotes the release of another hormone, **aldosterone**, from the adrenal medulla (p. 452). Aldosterone promotes the reabsorption of sodium ions from the distal tubule and thus increases the reabsorption of water, which follows the sodium.

The amount of water excreted and hence the volume of body fluids are also very much affected by **vasopressin** (**antidiuretic hormone—ADH**), released by the posterior lobe of the pituitary (p. 445). The secretion of this hormone is also stimulated by angiotensin. Vasopressin increases the permeability of the cells of the distal tubule and collecting tubule to water so that more water is reabsorbed into the blood. On the other hand, if there is an excess of water present in the body fluids, the blood volume and pressure increase. This raises the glomerular filtration pressure, and more filtrate is produced. An increase in the amount of water in the tissue fluid inhibits the release of vasopressin, the permeability of the distal tubule and collecting tubule cells is lowered, and less water is reabsorbed. Thus dilute filtrate passes through these tubules, producing a large volume of hypoosmotic urine. Increased production of filtrate and decreased reabsorption of water rapidly bring the volume of body fluids down to normal. Both vasopressin and angiotensin feed back to the juxtaglomerular apparatus and inhibit secretion of renin.

Beer increases urination because, in addition to the intake of a large amount of fluid, alcohol inhibits production of ADH. Caffeine is a weaker diuretic; it probably increases filtration and reduces the tubular reabsorption of sodium.

OSMOTIC REGULATION IN MARINE ANIMALS

The amount of salt in the open ocean is about 35 parts per thousand, or about 3.5%. In estuaries and bays the salinity may be considerably lower, depending upon the inflow of fresh water from rivers and streams. The term **brackish** refers to waters that are more saline than fresh water but less saline than the open ocean.

Osmoconformers

The internal body fluids of most marine invertebrates are isosmotic with seawater. The total salt concentration of their blood, coelomic fluid, and intercellular fluids is about the same as that of their marine environment, although the concentration of specific ions is somewhat different. Their intracellular fluids are isosmotic with extracellular fluids, but their intracellular osmotic pressure is influenced by the large amount of organic compounds present.

If the salinity of the external environment changes slightly, the concentration of salts in the body fluids of the animal also changes. Such marine animals are said to be **osmoconformers**, i.e., the salt concentration of their internal environment conforms to that of the surrounding external medium. However, osmoconformers cannot tolerate a great drop in their internal salt concentration and therefore most must live in environments of relatively high salinity. They are **stenohaline** (Gr. *stenos*, narrow + *halinos*, saline), restricted to a narrow range of salinities, usually near that of the open ocean.

Osmoconformers, like aquatic animals in general, eliminate their nitrogenous wastes as ammonia. Most of this ammonia is eliminated across the body surface, especially gas exchange surfaces, and is swept away by ventilating and other currents. The excretory organs provide for fine adjustments of internal water volume and relative ion content. As might be expected, the urine of osmoconformers is isosmotic with both their internal body fluids and the surrounding seawater.

Osmoregulators

Not all marine animals are osmoconformers. If a spider crab and a blue crab are placed in an aquarium in which the salinity is slowly lowered, their blood salts will gradually fall, i.e., both act as osmoconformers (Fig. 16.13). With furthur decline in environmental salt concentrations, the spider crab dies, but the blue crab does not. If the blue crab's blood salts are analyzed, they are found to be at higher levels than the salts in the environment. It holds onto its blood salts and does not conform to the level of environmental salts. Such animals are said to be

Figure 16.13
The osmoregulatory ability of two crabs. The osmotic concentration of blood changes with the concentration of salts in the surrounding seawater. The spider crab cannot osmoregulate, and the osmotic concentration of its blood conforms to the salt concentration of the seawater. Below and above certain concentrations, the crab dies. The blood of the blue crab conforms to an upper range of external salt concentration, but at the lower concentrations the crab maintains higher osmotic concentrations in the blood by absorbing salts through the gills. The blue crab can live in brackish estuarine environments where the spider crab cannot.

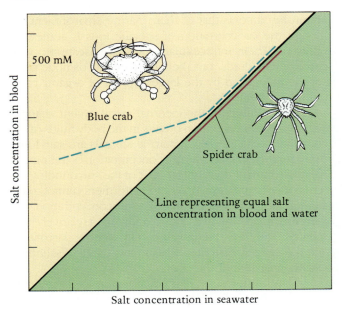

osmoregulators. Marine osmoregulators can invade or inhabit estuarine waters, where salinities are low. They can usually tolerate a wide range of salinities and are said to be euryhaline (Gr. *eurys*, wide).

When osmoregulating crabs are placed in dilute sea water, the urine flow may greatly increase as the animal eliminates the excess water diffusing inward. But in the process salts are lost, for most crabs can only produce a urine isosmotic or only slightly hypoosmotic to their blood. For example, in normal sea water the green crab *Carcinus* excretes a volume of urine equivalent to 3.6% of its body weight each day; in sea water diluted to 1.4% salinity it eliminates a urine equivalent to 33.0% of its body weight. Replacement of salt lost in the urine is provided by special chloride cells in the gills, which actively pick up salt from water ventilated over the gill surface. Thus, in osmoregulating crabs the green glands are the water pumps, and the gills are centers of salt replacement.

The Vertebrate Problem

The salt content of the internal body fluids of most vertebrates is about 1% and may reflect their invasion into fresh water early in their evolutionary history. In such a dilute environmental medium, vertebrates evolved a lower blood salt concentration. Later forms reinvaded the sea, but retained the low blood salt levels characteristic of their freshwater ancestors. In any event, the salt content of the body fluids of marine vertebrates (bony fish and aquatic mammals) is only about one-third that of sea water, and water tends to diffuse outward, especially across surfaces such as the gills. They are therefore living in a physiological desert, and water conservation is a major necessity.

Although marine bony fish have blood salts only one-third to one-half as concentrated as seawater, they cannot produce urine more concentrated than their blood, for the kidney tubules lack loops of Henle. Marine fish obtain their water with food and by drinking sea water. The excess salt (Na^+ and Cl^-) taken into the blood is excreted by active transport across the gill epithelium via chloride cells. The kidneys and gut remove calcium, magnesium, and sulfate ions, in addition to ammonia, which is the principal nitrogenous waste of most fish (Fig. 16.14a). However, most ammonia is eliminated through the gills rather than by the kidneys. The glomeruli of marine fish are usually reduced in size and in number. What would be the adaptive significance of such a reduction in glomeruli?

Other marine vertebrates solve the problem of living in a hyperosmotic medium differently. The body fluids of sharks contain urea and trimethylamine oxide in such high concentrations that the osmotic pull of water inward is in equilibrium with the outward diffusion resulting from the higher external salt concentration. Sharks produce urea as their nitrogenous waste and retain much of it in their internal body fluids. The gills of sharks are not permeable

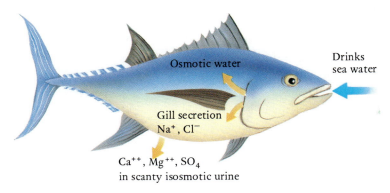

(a) SEA
(tuna)

Osmotic water

Drinks
sea water

Gill secretion
Na$^+$, Cl$^-$

Ca^{++}, Mg^{++}, SO$_4$
in scanty isosmotic urine

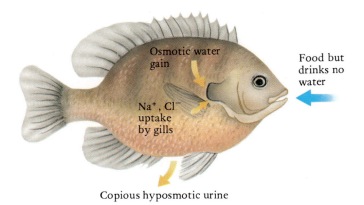

(b) FRESH WATER
(sunfish)

Osmotic water
gain

Food but
drinks no
water

Na$^+$, Cl$^-$
uptake
by gills

Copious hyposmotic urine

Figure 16.14
Osmoregulation in bony fish. The body fluids of fish have a salt content of about 1%. Marine fish (a) are thus hypoosmotic to seawater, and freshwater fish (b) are hyperosmotic to fresh water.

to urea, and the concentration of urea in the body fluids of sharks may reach 100 times that in mammals, a level far exceeding the tolerance of most other vertebrates. The sharks' tolerance is in part due to the counteracting effects of trimethylamine oxide. The salts that accumulate through food intake are removed by the gills, kidneys, and a special salt-excreting **rectal gland** attached to the caudal end of the shark intestine.

Sea birds and sea turtles excrete excess salt from a pair of **salt glands** in their heads (Fig. 16.15). In sea birds, such as gulls, these glands open into the nasal passageways; in sea turtles, they open into the eye orbits. Turtles literally weep salty tears. The glands excrete only in response to an internal salt load, and the concentration of salt in the excreted fluid can reach a level two to three times that of seawater. Only birds and mammals can excrete a urine saltier than the blood, but the urine of sea birds is only twice as concentrated as the blood, which is still less than the concentration of salt in seawater.

Seals, porpoises, and whales obtain most of their water from their food; some food (such as molluscs and

crustaceans) is isosmotic with seawater, and some (such as fish) is less concentrated than seawater. However, marine mammals can utilize the kidney for osmoregulation. They have very long loops of Henle that make it possible to produce a urine that has a salt concentration greater than that of seawater.

Humans cannot drink sea water because they cannot produce a urine with a salt content greater than approximately 2.2%. If a person were to drink 1 L of sea water, it would require $1\frac{1}{3}$ L of urine water to eliminate the excess salt. Moreover, the magnesium and sulfate ions in sea water cause diarrhea, and additional water is lost.

OSMOREGULATION IN FRESHWATER ANIMALS

The salt concentration in fresh water varies with the water source, but it is always extremely low. The external environment is thus very hypoosmotic to the internal body fluids of freshwater animals, and they must cope with the tendency of water to diffuse into the body, especially

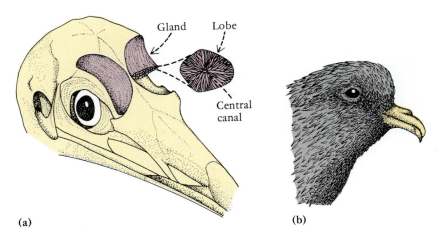

(a)   **(b)**

Figure 16.15
Salt-secreting glands of marine birds. (a) Skull of a gull showing the position of
the glands above the eyes. The glands drain into the nasal passageways. (b) A
petrel blowing salt droplets from the nasal openings.

through thinly covered areas such as the gills. Salts tend to
diffuse outward, and the internal body fluids lose salts
through excretion.

Given these conditions, the following generalizations
are not unexpected:

1. All freshwater animals are osmoregulators.
2. Excretory organs function as water pumps, and the
urine is usually hypoosmotic to the internal body fluids as a
result of selective reabsorption of salts.
3. The salt concentration of internal body fluids is
maintained at the lowest level compatible with the animal's metabolic needs. Thus, less energy is required to
maintain an internal salt balance.
4. Nitrogenous wastes are usually eliminated as ammonia.

Freshwater fish excrete a hypoosmotic urine but still
suffer some salt loss. Replacement of salt is provided by
food and by the gills that, in contrast to those of marine fish,
pick up salts from the ventilating current (Fig. 16.14b). (An
ammonium ion—waste—is exchanged for a sodium ion.)
Crayfish osmoregulate similarly. The green glands produce a hypoosmotic urine, and gills are also the site of salt
replacement.

The exoskeleton of crayfish and of some other crustaceans living in fresh or brackish water is less permeable
than that of marine forms. This reduces, but does not
eliminate, the inward diffusion of water.

The internal fluids of freshwater clams and snails have
extremely low salt concentrations, only about 0.16%, in
comparison with about 1% in freshwater fish. Their nephridia produce a copious hypoosmotic urine amounting

to approximately 50% of body weight per day, and the
concentration of salts in the urine is about one-half of that
of the blood. Salts are actively absorbed across the body
surface.

Many very small freshwater animals have no specialized excretory organs. The surface area is adequate to
permit elimination of ammonia by diffusion. Nevertheless,
they must osmoregulate. Freshwater sponges and protozoans, such as amoebas and *Paramecium*, possess water-
pumping organelles called **contractile vacuoles**. The
vacuole slowly accumulates water and, when full, collapses
to the outside through a temporary opening. *Paramecium*
possesses a contractile vacuole at each end of the body
(Fig. 24.22). The vacuoles fill through tiny radiating canals
and empty through a canal to the outside. The rate of
pulsation is governed by the rate at which water is taken in
by diffusion and feeding. Flatworms and rotifers use protonephridia to eliminate excess water, and the bladder of
rotifers, which are not much larger than most ciliated
protozoans but are multicellular animals, pulsates like a
contractile vacuole (Fig. 28.2).

OSMOREGULATION IN TERRESTRIAL ANIMALS

The danger of desiccation is a major threat to animals living
on land. Water is lost in urine and in feces, but evaporation
is the principal route of water loss. At any given temperature, a given volume of air is capable of holding a certain
amount of water vapor. The difference between the
amount of water vapor actually present and the amount
that would be held if the air were saturated is known as the

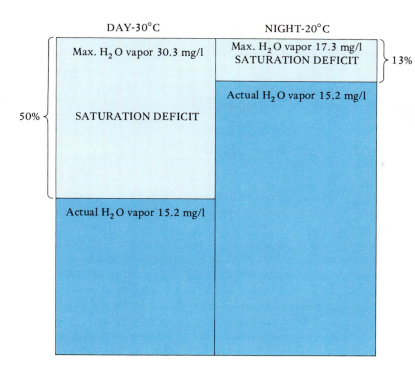

DAY-30°C | NIGHT-20°C

Max. H_2O vapor 30.3 mg/l

Max. H_2O vapor 17.3 mg/l
SATURATION DEFICIT

13%

Actual H_2O vapor 15.2 mg/l

SATURATION DEFICIT

50%

Actual H_2O vapor 15.2 mg/l

Figure 16.16
Comparison of the saturation deficit in the day and at night. At sea level a liter of air at 30° C could hold a maximum of 30.3 mg/L of water vapor. Assuming it contained only 15.2 mg/L, it would exhibit a saturation deficit of 50%. At night when the temperature drops to 20° C, the same volume of air will hold only 17.3 mg/L of water vapor. Assuming the same amount of water is present in the air at night as in the day, the saturation deficit will drop to 13%.

saturation deficit (Fig. 16.16). The saturation deficit determines the steepness of the evaporation gradient between the animal and the external environment and varies with season, climate, habitat, and time of day. However, of all the variables, movement of air across an evaporating surface, sweeping away the more saturated overlying layers of air, is the single most important factor increasing desiccation.

The general body surface and the gas exchange surface are the two principal sites of evaporation. In many snakes, for example, 64% of the water loss by evaporation may occur across the skin and 36% through gas exchange in the lungs. Water lost must be replaced with water gained by drinking and with water in food. A few terrestrial animals, such as certain insects, may take up water vapor from the surrounding air.

Adaptive Strategies

The many different groups of land animals, such as vertebrates, spiders, insects, and snails, are representatives of separate invasions of the terrestrial environment (Table 16.1). Within each of these groups various adaptations have evolved that increase the likelihood of maintaining a balance between water lost and gained. The principal adaptive strategies are:

1. Reduction of evaporation by:
   a. Internal gas exchange organs. The gas exchange organs of most land animals are located internally or at least within a protective covering where the thin, moist exchange surface is less likely to dry up by exposure to ambient air currents, and there is less water lost by evaporation.
   b. Modification of the integumental barrier. Loss of water by evaporation across the general body surface is reduced by various modifications of the skin that make the integumental barrier more impervious to the outward passage of water.
   c. Occupation of humid habitats. Subterranean burrows and spaces beneath stones and logs and in leaf mold, where saturation deficits are lower than in exposed locations and there is little air movement, are less stressful environments for maintaining water balance.
   d. Nocturnal activity. Many species move out of protective retreats only at night, when the danger of desiccation is greatly reduced.

Table 16.1
Major Invasions of Land Ranked by the Approximate Number of Living Species Generated

No. of Species	Independent Invasion
851,000	Centipedes, millipedes, insects
68,400	Scorpions, spiders, mites (arachnids)
22,000	Vertebrates
15,000	Pulmonate land snails
4,000	Operculate land snails (largely tropical)
3,500	Pill bugs (isopod crustaceans)
1,200	Earthworms

2. Reduction of water loss from excretion by:
   a. Production of insoluble nitrogenous wastes. Some land animals excrete their nitrogenous wastes in the form of uric acid (land snails, insects, birds, and most reptiles) or guanine (spiders), which have very low solubility and thus require little water for removal.
   b. Production of a hyperosmotic urine. Mammals, birds, and insects are able to excrete urine with an osmotic concentration greater than that of the blood; such a concentrated urine conserves water.
   c. Low urine output. Some terrestrial animals, such as many land crabs and pill bugs, conserve excretory water by greatly reducing urine flow and utilizing other avenues for the elimination of nitrogenous wastes.

3. Reduction of water loss from egestion. Some animals, especially those living in deserts, conserve water by defecating relatively dry feces.

4. Toleration of internal water loss. Most animals can tolerate relatively limited fluctuations in the levels of internal water. For example, if a human loses more than 12% of body water, death may result. But a few animals can tolerate the loss of more than 50% of body water. This can be replaced later when water is again available.

5. Utilizing the water of oxidation. Significant amounts of water are liberated in various reactions of cell respiration. The oxidation of 100 g of carbohydrate, for example, yields 56 g of water. A few desert animals survive entirely on this water source.

Adaptations of Animal Groups

Earthworms lack adaptations that permit exposure to the usual desiccating conditions of daylight hours. They are only active within their burrows in moist soil or at the surface at night. When the soil becomes dry, they move to deeper levels, where they may lose water and become dormant. Earthworms excrete ammonia and urea, and in the event of excess water the nephridia can produce a hypoosmotic urine.

Land snails and slugs are also subject to great evaporative water loss across the skin. Moreover, additional water is lost in the secretion of the mucous trail over which they crawl. However, these animals are able to tolerate considerable desiccation. The European land snail *Helix* can survive a water loss equivalent to 50% of its body weight, and the slug *Limax* has been claimed to survive an 80% loss. They excrete uric acid, which conserves some water. Nevertheless, snails and slugs are confined to humid habitats or to exposure only at night, when saturation deficits are low.

Many land snails living in deserts or in tropical regions subject to long dry seasons undergo a period of dormancy called **aestivation**. The body is withdrawn into the shell and sealed off with a secreted partition to reduce evaporation, although these snails can tolerate some inevitable water loss. Many other animals, including some vertebrates, living in similar arid environments are capable of aestivation (p. 848).

Spiders, insects, and pill bugs are all small terrestrial arthropods, each representative of a different past invasion of land. Spiders and insects are highly successful terrestrial animals, as their great numbers and density attest. Contributing to their success has been the evolution of a waxy layer on the surface of the exoskeleton. When exposed to the same saturation deficit, their evaporation per square centimeter is comparable to that of mammals and is only a fraction of that of snails and slugs. The internal tracheal tubules, which are the gas exchange organs of insects as well as of many spiders, reduce respiratory evaporation. Insects also conserve water by excreting uric acid and producing a hyperosmotic urine via the malpighian tubules and rectum. In fact, insects are the only animals other than birds and mammals that produce a hyperosmotic urine. Spiders excrete guanine, which, like uric acid, has relatively low solubility in water. Nevertheless, very small spiders, possessing a large surface area in relation to volume, are confined to leaf mold and other more humid habitats.

Pill bugs are not insects but crustaceans, relatives of crabs and shrimps. They are mostly nocturnal animals, living beneath stones and wood and in leaf mold. In contrast to spiders and insects, they lack a waxy outer covering and are subject to much greater water loss. They excrete ammonia, like most aquatic animals, but are remarkable in that they eliminate gaseous ammonia through the gills, and urine output from the excretory organs is very small.

Frogs and toads suffer the highest rate of evaporative water loss of all terrestrial vertebrates, a rate far higher than that of insects and spiders (Table 16.2). In spite of its warty skin, the toad loses water about as rapidly as the frog. Nocturnal habits, especially of toads, and protective environments are the principal defenses of these animals against desiccation. Both excrete urea, which is less toxic than ammonia but very soluble. Many amphibians can tolerate a much greater loss of body water than can most other vertebrates.

Frogs and toads do not drink water, and the skin is the principal avenue for the replacement of water lost through evaporation. Some species have areas of the skin on the hind legs that are especially adapted for water absorption when moisture is available. Some toads can store water in the bladder and reabsorb this water into the blood as needed.

Most reptiles are much better adapted for life on land than are amphibians. The skin of most snakes, for example, is more resistant to water loss from evaporation, and the

Table 16.2
Comparison of Water Loss by Evaporation from the Body Surface of Different Animals

	Evaporation (μg)*
Earthworm	400
Land snail, active	870
Land snail, inactive	39
Salamander	600
Frog	300
Human (not sweating)	48
Rat	46
Water snake	41
Pond turtle	24
Box turtle	11
Iguana lizard	10
Gopher snake	9
Desert tortoise	3
Cockroach	49
Flour mite	2
Tick	0.8

*Micrograms per cm² body surface per hour per mm Hg saturation deficit. (After Schmidt-Neilsen, K.: The neglected interface: the biology of water as a liquid gas system. *Q. Rev. Biophys.*, 2:283, 1969.)

Kangaroo rats, mammals abundant in southwestern North America, have been studied more extensively than any other desert animal (Fig. 16.17). These small rodents live in subterranean burrows and come to the surface only at night. They rarely drink, because standing water is almost never available. They eat relatively dry food, mostly seeds, and rely entirely on the water contained in food and that produced by the oxidation of food. Water is condensed from exhaled air as it travels through the nasal passageways, and this water is reevaporated into the cool dry air that is inhaled (such conservation is true of many other mammals). There are no sweat glands. The feces contain little moisture, and their kidneys can produce a highly concentrated urine, containing as much as 23% urea and 7% salt. Kangaroo rats would have no problem drinking seawater!

Camels are unable to store water (the hump contains stored fat) but can tolerate great water loss. They can drink an enormous amount of water at one time, as much as one-third the body weight in 10 minutes. A gradual loss then may follow that can take the animal through a prolonged period without water. Camels have sweat glands, but sweating commences at a higher internal body temperature than in other mammals. Moreover, camels begin the day with a lower early morning body temperature and do not regulate body temperatures as strictly as do other mammals.

snake's nitrogenous wastes are excreted as uric acid. Uric acid excretion probably first evolved in reptiles as an adaptation for reproduction on land, since it permits accumulation within the egg of nitrogenous wastes produced by the developing embryo.

Birds excrete uric acid, and as in reptiles, this insoluble waste produced by the embryo can be stored within the egg. Uric acid is probably of less significance in adult water conservation than in other ways. As an adaptation for flight, uric acid permits the elimination of the weight of the bladder and its contents. The integument of the bird, with its covering of feathers, reduces water loss, and the avian kidneys can produce a hyperosmotic urine.

Mammals have also adapted in various ways for life on land. The skin is an effective barrier against evaporation. Mammals excrete urea, which helps to conserve water, but most mammals live where sources of drinking water and water in food are adequate. Moreover, the mammalian kidneys can produce a hyperosmotic urine, providing considerable conservation of water. Although the kidneys can also produce a dilute (hypoosmotic) urine when it is beneficial, i.e., when overhydrated, under ordinary conditions the urine is hyperosmotic.

Figure 16.17
The kangaroo rat, a mammal highly adapted for living in the arid conditions of the southwestern North American deserts.

Summary

1. The extracellular body fluids (water plus solutes) of an animal are in osmotic equilibrium with intracellular fluids and are maintained, or regulated, at a relatively constant level, regardless of external environmental conditions.

2. The regulation of internal body fluids may be complicated by the excretion of nitrogenous wastes that require some water for their removal. Most nitrogenous waste products of animals result from the deamination of amino acids utilized as cellular fuels. Ammonia, the highly soluble and toxic waste product of deamination, is eliminated directly where sufficient water for dilution is available. Many animals convert ammonia to less toxic urea or uric acid, but this conversion requires energy.

3. Most animals possess tubular or saccular excretory organs. In aquatic species such organs regulate internal fluid and ion content, and nitrogenous waste (ammonia) is eliminated across the body surface, typically the gills. In terrestrial animals the excretory organs may concentrate wastes. The concentration process typically involves modification of coelomic fluid or blood filtrate through selective reabsorption and secretion. The principal excretory organs are protonephridia (flatworms, rotifers) metanephridia (molluscs and annelids), green glands (shrimps, crayfish, and crabs), malpighian tubules (insects and spiders), and kidneys (vertebrates).

4. In the mammalian kidney each tubular nephron, the structural and functional unit of the kidney, receives filtrate from a knot of capillaries, the glomerulus. The filtrate, essentially similar to plasma minus the large proteins, is received in a capsule (Bowman's capsule) and then passes through the tubule, where glucose and other organic nutrients, some salts, and most of the water are reabsorbed. The kidneys of mammals can produce a urine that is hypo- or hyperosmotic to the blood, depending upon internal fluid levels. A hyperosmotic urine is made possible by the loop of Henle, a hairpin loop of the tubule. A solute gradient is set up within and around the loop of Henle that permits the extraction of a large amount of water from the fluid in the counterflowing collecting tubule. The loops of Henle are found only in the nephrons of some birds and all mammals; other vertebrates (fish, amphibians, and reptiles) cannot produce a urine with a salt concentration greater than that of the blood.

5. The internal body fluids of most marine invertebrates are in osmotic equilibrium with the surrounding seawater, which in the open ocean contains about 3.5% salt. They are osmoconformers, for any decrease in environmental salts results in an equivalent decrease in the salts of their internal body fluids. Marine vertebrates have internal body fluids with a salt concentration of only about one-third that of seawater; they therefore live in a desiccating environment and must be osmoregulators. They drink seawater or eat salty food but eliminate excess salt by secretion through the gills (fish) or head glands (sea turtles and birds) or by excreting a urine saltier than seawater (whales, porpoises, and seals).

6. Freshwater animals and marine animals living in brackish estuaries are osmoregulators. They live in an environment with a salt content less than that of their internal body fluids and thus must hold on to salts and get rid of excess water. They excrete a hypoosmotic urine and may replace lost salts by absorption through certain body surfaces, such as the gills (fish), and by consuming food.

7. Desiccation, especially evaporation from the skin and gas exchange organs, is the principal problem in the regulation of internal body fluids in terrestrial animals. A variety of adaptive strategies have evolved that aid in balancing the limitations of water gained with water lost. Many species avoid desiccating environmental conditions by living in humid habitats or by nocturnal activity. Evaporative water loss is reduced in arachnids (spiders, scorpions, and mites), insects, reptiles, birds, and mammals by modifications of the integumental barrier and in almost all terrestrial animals by internal gas exchange organs.

8. A urine hyperosmotic to the blood can be produced only by insects, birds, and mammals. Terrestrial snails, insects, reptiles, and birds excrete uric acid, which is relatively insoluble and requires little water for elimination. Mammals excrete urea, which although very soluble, is less toxic than ammonia. A few animals, such as some land snails, slugs, and camels, have become adapted to survive great water loss for a short time.

References and Selected Readings

Many of the references listed at the end of Chapter 10 provide excellent coverage of the topic of regulation of internal body fluids and related subjects.

Crawford, C.C. *Biology of Desert Invertebrates.* Berlin: Springer-Verlag, 1981. Includes adaptations for water balance in various desert invertebrates.

Eckert, R., D. Randall, and G. Augustine. *Animal Physiology,* 3rd ed. New York: W.H. Freeman and Co., 1988. This comparative physiology text contains an especially good chapter on osmoregulation and excretion.

Fertig, D.S., and V.M. Edmounds. The physiology of the house mouse. *Scientific American* 221(Oct. 1969):l03–110. An account of adaptations of the common house mouse, many of which are shared with desert animals.

Hadley, N. Desert species and adaptation. *American Scientist* 60(1972):338–347. How desert animals cope with desert temperatures and aridity.

Little, C. *The Colonization of Land: Origins and Adaptations of Terrestrial Animals.* New York: Cambridge University Press, 1984. Includes adaptations for water balance among the various groups of animals that have invaded land.

Maloiy, C.M.O. (ed.). *Comparative Physiology of Osmoregulation in Animals.* 2 vols. New York: Academic Press, 1979. A comparison of osmoregulatory mechanisms in animals.

Ruppert, E.E., and P.R. Smith: The functional organization of filtration nephridia. *Biol. Rev.* 63(231–258), 1988. Common features in the functional design of animal excretory organs.

Schmidt-Nielsen, K. Countercurrent systems in animals. *Scientific American* 244(May 1981):118–128. This article covers the renal countercurrent system of mammals.

Schmidt-Nielsen, K. *Animal Physiology: Adaptation and Environment.* 4th Ed. Cambridge, England: Cambridge University Press, 1990. Osmoregulation and excretion are areas of particular interest to this author.

Taylor, C.R., K. Johansen, and L. Bolis (eds.). *A Companion to Animal Physiology.* Cambridge, England: Cambridge University Press, 1982. This volume contains four chapters on various aspects of osmoregulation in animals.

17

The Nervous System

(Left) Giraffes and the moon near Nanok, Kenya. (Above) A Masai giraffe (Giraffa camelopardalis tippelskirchi) browsing upon leaves. Feeding requires an intricate coordination of many body movements.

Most animals move about as they search for food and shelter, avoid predators, and look for mates. Their world is in constant flux, and they must adjust rapidly to the changes they encounter. The ability of animals to respond derives from a fundamental property of cells we call **irritability**. Irritability consists of three aspects: reception of a stimulus, the transmission of a signal, and the response of the target to the signal. The degree of development of these qualities correlates with the size and motility of the organism. Protozoans are small enough so that all of these tasks can be performed by a single cell. In larger and more complex metazoans, different groups of cells mediate these functions: **receptors** detect stimuli; nerve cells or chemical messengers known as hormones transmit signals; and target cells or **effectors** (muscles, glands, and chromatophores) respond to the signals. Reception, transmission, and response evolve together, and are most highly developed in the most motile animals.

The rapid transmission of signals to specific effectors is accomplished by a nervous system. Neural coordination is supplemented in most animals by endocrine integration in which hormones secreted by endocrine glands travel though the blood stream to certain effectors. Endocrine integration tends to be slower than nervous integration, and a specific hormone usually affects more effectors than a single nerve impulse. Nonetheless the two control systems have much in common and interact in an intricate communication network that allows the organism to adjust to both internal and external environmental changes in ways appropriate for its survival. In this and the next two chapters we will examine the nerve cells and the ways by which they integrate and coordinate the activities of the body, the receptors that detect the changes, and the means of chemical communication.

THE NEURON AND SURROUNDING CELLS

The nervous system of multicellular animals consists of many nerve cells, called **neurons** (Gr., nerve), and associated cells and tissues necessary for their functioning. In such radially symmetrical animals as hydras, jellyfish, and sea anemones, neurons are dispersed and form a nerve net. But in active, bilaterally symmetrical animals, many neurons become organized longitudinally in such a way that we can speak of a **central nervous system** (CNS) along the axis of the body (the spinal cord and brain of vertebrates), and a **peripheral nervous system** (PNS) consisting of neuron processes extending between the central nervous system and the receptors and effectors.

The neuron is the structural and functional unit of the nervous system (Fig. 17.1). It is a cell specialized for the rapid conduction of information, but it must also maintain itself, receive information from receptors or other neurons, and transmit information to other neurons or to

effectors. These diverse functions tend to be localized in different parts of the cell. A large **cell body** is the nutritive region, for it contains the nucleus and the metabolic machinery needed to sustain the cell. Usually the receptive region of the cell consists of many short, branching processes known as **dendrites** (Gr. *dendron*, tree), but information may be received directly by the cell body and occasionally by other parts of the cell. The **axon** (Gr., axis) is the conductive region. It is a long process specialized for the rapid transmission of nerve impulses, often over long distances, and without a weakening in the strength of the signal. A neuron has a single axon that does not branch as much as the dendrites do and often does not branch at all. A transmissive region at the distal end of an axon consists of minute, branching processes that transfer the nerve impulse to other neurons or to the effectors.

Zoologists define cell body, dendrites, and axons according to their functions. The location of these parts and their size and shape vary greatly according to the specific roles of the neurons (Fig. 17.2). **Unipolar neurons**, such as vertebrate sensory neurons, have the cell body set off on one side of a long axon. **Multipolar neurons** have many conspicuous dendrites that converge on the cell body and a single axon leaving it. Examples are vertebrate motor neurons (Fig. 17.1) and the Purkinje cells in the cerebellum of the brain. **Bipolar neurons** have a single process leading to the cell body and another away from it. Examples are the sensory neurons of invertebrates in which a short dendrite leads to a cell body in an epithelial layer and an axon leads from it, or the bipolar cells of the vertebrate retina in which the cell body is located near the middle of the axon.

The neurons of hydras, sea anemones, and other primitive invertebrates are naked, but most of the axons of the PNS of many invertebrates and all vertebrates become enveloped in a sheath formed by linearly arranged **neurilemma cells**, often called **Schwann cells** in vertebrates (Figs. 17.1 and 17.3a). Usually a Schwann cell is associated with only one axon, but some Schwann cells enfold a segment of several axons (Fig. 17.3b). Often the Schwann cells encircle the axon only once; these axons are said to be **unmyelinated**. Many Schwann cells continue to wind around the axon many times, forming a fatty **myelin sheath** (Gr. *myelos*, marrow) (Fig. 17.3c and d). Axons with such a sheath are said to be **myelinated**. The myelin sheath is not continuous, but is interrupted at 1- to 2- mm intervals by **nodes of Ranvier**, where successive Schwann cells come close together but leave a minute gap that exposes the plasma membrane of the axon. The myelin sheath and its nodes make possible a very rapid transmission of the nerve impulse, as we shall see.

Neurons within the CNS are interspersed with many nonconducting **neuroglial cells** (Gr. *glia*, glue) that perform many functions. Certain of them, the **oligodendrocytes**, form a myelin sheath around many

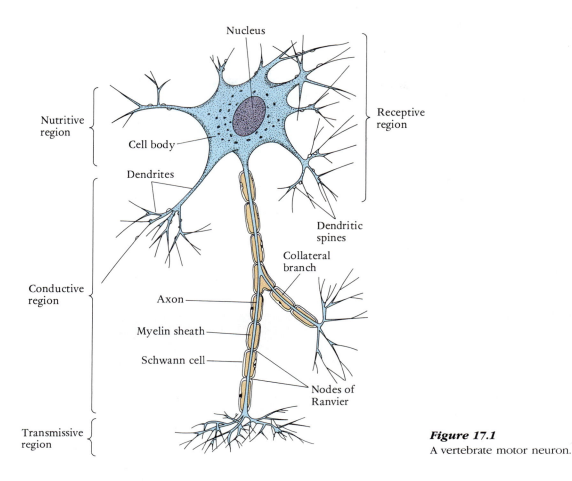

Nucleus

Nutritive
region

Receptive
region

Cell body

Dendrites

Dendritic
spines

Collateral
branch

Conductive
region

Axon

Myelin sheath

Schwann cell

Nodes of
Ranvier

Transmissive
region

Figure 17.1
A vertebrate motor neuron.

Figure 17.2
Representative nerve cells.

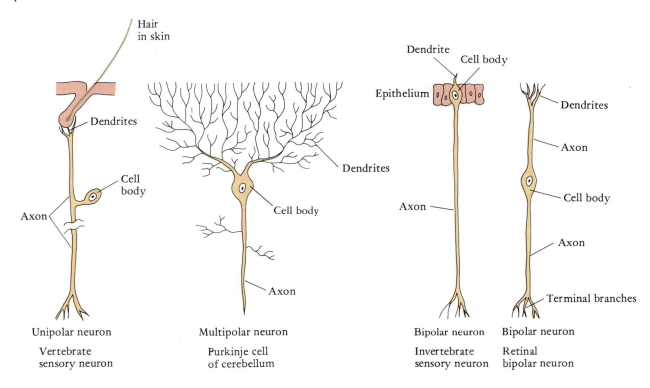

Hair
in skin

Dendrites

Dendrite   Cell body

Epithelium

Dendrites

Cell
body

Dendrites

Axon

Axon

Cell body

Cell body

Axon

Axon

Terminal branches

Unipolar neuron

Vertebrate
sensory neuron

Multipolar neuron

Purkinje cell
of cerebellum

Bipolar neuron

Invertebrate
sensory neuron

Bipolar neuron

Retinal
bipolar neuron

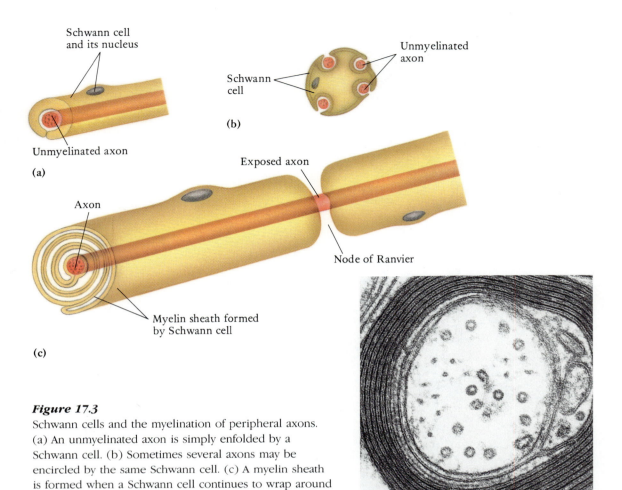

Figure 17.3

Schwann cells and the myelination of peripheral axons.
(a) An unmyelinated axon is simply enfolded by a
Schwann cell. (b) Sometimes several axons may be
encircled by the same Schwann cell. (c) A myelin sheath
is formed when a Schwann cell continues to wrap around
an axon. (d) An electron micrograph of a myelinated
axon.

axons of the CNS, performing the role that Schwann cells
play in the PNS. Other glial cells provide physical support
for the neurons, help regulate the ionic composition of the
surrounding interstitial fluid, synthesize some products
needed by the neurons, degrade metabolites released by
neurons, and play a phagocytic role.

THE NERVE IMPULSE

The plasma membrane of neurons, in common with that of
muscles and some other cells, carries a **resting potential**,
or electric charge. The potential can be measured by
inserting a microelectrode inside of an axon and another
in the interstitial fluid around it (Fig. 17.4a). The resting
potential varies with the axon. A value of -70 mV, with the
inside negative relative to the outside, is common in many
vertebrate neurons (Fig. 17.4b). When the neuron is stimulated, ions become redistributed across the membrane,
and the potential is momentarily reversed. As the inside
becomes less negative relative to the outside, an **action
potential** develops that raises the potential toward zero

and may reach $+35$ mV. The **nerve impulse** is a self-
propagating wave of reversed potential, or **depolarization**,
that spreads rapidly along the axon. This is followed by a
wave of **repolarization** that restores the resting potential.

The Resting Potential

The resting potential results from the unequal distribution
of ions inside and outside the cells and the unequal
movement of ions across the membrane. Nerve cells
contain many negatively charged proteins and phosphate
ions that are too large to diffuse out, and a very high
concentration of potassium (K^+) ions (Fig. 17.5). Sodium
ions (Na^+) are far more concentrated in the interstitial
fluid on the outside of the cells than within the cells.
Although potassium and sodium ions do not pass easily
through the lipid layers of the plasma membrane, they can
diffuse or "leak" through small passages called **sodium**
and **potassium channels** in the membrane following their
diffusion gradients. These channels are proteins that extend through the membrane. Because potassium ions
move through their leak channels nearly 100 times more

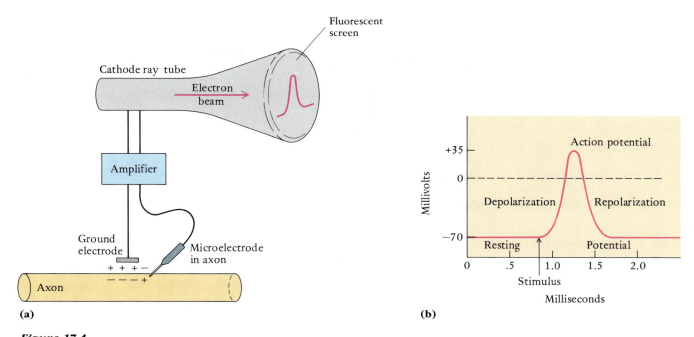

(a)   **(b)**

Figure 17.4

(a) The method of recording electrical potentials in a neuron. (b) A graph of an action potential.

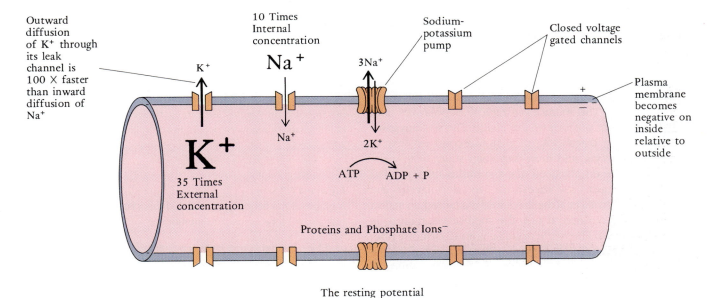

The resting potential

Figure 17.5

A diagram of the factors that lead to an unequal distribution of K⁺ and Na⁺ ions across the plasma membrane of an axon and the development of the resting potential.

easily than sodium ions move through theirs, and because of the very high concentration of potassium inside of the cell, more potassium ions move out than sodium ions move in. Because of this differential flow, the membrane becomes positive on the outside *relative* to the inside, even though large numbers of positive ions remain inside. The outflow of potassium ions slows down as the resting potential develops because the increasing positive charge on the outside of the membrane repels these positive ions, keeping them within the cell.

In addition to these leak channels, other protein complexes in the plasma membrane act as **sodium-potassium pumps**. The sodium-potassium pump is an active mechanism that uses energy derived from the breakdown of ATP to drive it (Figs. 5.9 and 17.5). When the pump is active, sodium ions are carried out and concurrently potassium ions are carried in. Again there is a differential movement, with three sodium ions being carried out for every two potassium ions that are carried in. The sodium-potassium pump helps to develop the resting potential and maintains it at a finely tuned equilibrium.

The Action Potential

In addition to the leak channels and sodium-potassium pump, the plasma membrane contains other proteins that form separate sodium- and potassium-gated channels. These are called **voltage-gated channels** because voltage changes across the membrane affect the molecular configuration of the channels and determine whether they are open or closed. They are closed when the cell is at rest (Fig. 17.5). When an axon is activated by a stimulus of sufficient strength from the receptive region of the cell (or experimentally activated by a direct stimulus) its sodium gates begin to open, and an action potential develops. Sodium ions start to move into the cell through their gates, and the membrane potential begins to rise from its negative value toward zero. A slight change in the membrane potential causes the sodium gates to open widely and sodium rushes in (Fig. 17.6a). The membrane rapidly depolarizes, and usually becomes positive on the inside. The sodium gates are open very briefly, $^2/_{10}$ of a millisecond or less in mammalian axons, and then they snap shut.

It must be emphasized that a stimulus must attain a certain strength, its **threshold**, to depolarize the membrane of an axon. If the threshold is attained, an action potential is generated; if not, the membrane retains its resting potential. The nerve impulse is an **all-or-none phenomenon**: it occurs or it does not. *The nerve impulse is the same in all neurons*. It does not vary with the magnitude of the stimulus or with the type of information being transmitted. Whether the information pertains to sight, touch, or the activation of a muscle depends not on the nature of the nerve impulse but on the specificity of the receptors and the neuronal connections within the nervous system.

The potassium gates also are closed during the resting period. As the membrane potential changes because of the inrush of sodium, the potassium gates open slowly. They are not fully open until the sodium gates close. The rapid exit of potassium through these channels quickly restores the resting potential (Fig. 17.6b). The action potential lasts for only $^1/_2$ of a millisecond or less. All of these changes are accomplished by rapid shifts in the normal sodium and potassium balances inside and outside the cell (Fig. 17.7).

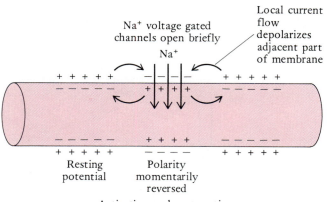

Activation and propagation

(a)

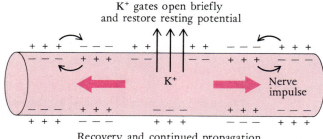

Recovery and continued propagation

(b)

Figure 17.6

A diagram of events occurring during the generation and propagation of a nerve impulse. (a) A stimulus given experimentally near the middle of an axon depolarizes this part of the membrane and initiates a nerve impulse. (b) As the impulse propagates itself in each direction from the point of stimulation, recovery occurs in the region initially stimulated.

Figure 17.7

A graph of the action potential and associated movements of Na⁺ and K⁺ ions across the membrane.

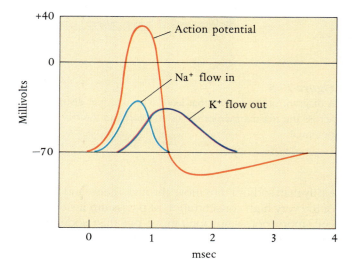

Impulse Propagation and Neuron Recovery

Since both the interstitial fluid outside of the cell and the cytoplasm within it contain electrolytes, depolarization of one part of the membrane sets up a local current flow between it and the adjacent polarized region. This activates the adjacent part of the membrane, the sodium gates open, and this part of the membrane becomes depolarized (Fig. 17.6). This in turn activates the next part of the membrane. The nerve impulse continues in this way along the axon membrane from the point of initial stimulation. This is usually from the dendrite and cell body, but if an axon is stimulated experimentally near its middle, a nerve impulse travels in both directions. Impulse propagation normally occurs in one direction within the nervous system, because impulses usually cross the junctions between neurons only in one direction, from axon to dendrite.

As a nerve impulse propagates itself, it depends on local current flows resulting from the distribution of ions at each level of the axon and so continues without weakening. The propagation of a nerve impulse is more similar to a burning fuse (except that the fuse does not reconstitute itself) than to the flow of electrons between the two ends of a wire. An electric current in the wire depends on potential differences between the two ends of the wire and is weakened as it continues by resistance in the wire.

The nerve impulse does not reverse upon itself because the sodium gates behind the propagating wave of depolarization cannot reopen until the resting potential is restored by the outward flow of potassium through its gates. There is a brief **refractory period** during which another impulse cannot be initiated. This lasts only a fraction of a millisecond in some mammalian neurons, so they can transmit several thousand impulses per second. Nerve impulse frequency is determined by the strength of the stimulus above its threshold level. A threshold stimulus may initiate only a single impulse, but a stronger one may initiate a train of impulses that follow each other in rapid succession.

Velocity of Nerve Impulses

Within different groups of animals, the speed with which a neuron can conduct a nerve impulse has been increased by increasing axon diameter and/or by the presence of a myelin sheath. Because velocity increase is approximately proportional to the cross-sectional area of an axon, large axons transmit impulses faster than small ones. Most vertebrate axons have a diameter of 10 μm, or less, but some fishes and amphibians have evolved large unmyelinated axons with a diameter of 50 μm. These extend from the brain down the length of the spinal cord and are used to activate many muscles during rapid escape movements. Giant axons also are common among many invertebrates, including earthworms, crayfishes, and squids. Some have evolved by the enlargement of a single cell. But the largest, those of the squid, develop by the fusion of the axons of many neurons, sometimes numbering in the hundreds. The largest have a diameter of 1 mm and a conduction velocity of 35 m/s. This compares with a velocity as low as 0.1 m/s in the axons of some sea anemones.

For any given diameter, the myelin sheath greatly increases velocity. Fatty myelin is an excellent insulator and effectively stops the movement of ions across it. Action potentials are generated only at the nodes of Ranvier, where the axon membrane is exposed to the interstitial fluid (Fig. 17.8). The depolarization of one node sets up, nearly instantaneously, a flow of electric current between it and the next node, which in turn becomes depolarized. Conduction is **saltatory** (L. *saltare*, to leap), proceeding by jumps from node to node.

Among vertebrates, myelination is the usual method of increasing conduction velocity. Some myelinated mammalian neurons with a diameter of 10 μm have a conduction velocity of up to 120 m/s. Myelination makes rapid conduction in small neurons possible and hence allows for the evolution of complex nervous systems that do not occupy an excessive amount of space. Myelination also conserves energy because the sodium-potassium pumps need to maintain ionic balances only at the nodes.

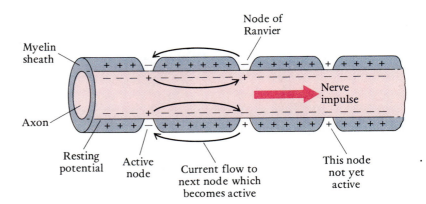

Figure 17.8

A diagram of saltatory conduction of a nerve impulse along a myelinated axon.

Node of Ranvier

Myelin sheath

Axon

Nerve impulse

Resting potential

Active node

Current flow to next node which becomes active

This node not yet active

SYNAPTIC TRANSMISSION

The axon terminals of one neuron come very close to the dendrites and cell body of another neuron at junctions known as **synapses** (Gr. *synapsis*, union). *Synapses are the control switches of the nervous system.* Information normally travels only from the axon knobs of the **presynaptic neuron** (the neuron bringing impulses to the synapse) to the membrane of the **postsynaptic neuron**. Synapses also enhance, retard, or stop information flow. Because of these properties, synapses play a key role in the integrative activity of the nervous system.

Electrical Synapses

In **electrical synapses**, the two cells are intimately united by gap junctions containing minute passages between them through which ions can pass. In these cases the action potential spreads directly from cell to cell with only a short delay. Electrical synapses are well adapted for rapid trans-

mission of the nerve impulse. They are found frequently between giant neurons. They also occur in the CNS of vertebrates, but their role here is not well understood.

Chemical Synapses

Chemical synapses are far more common. The terminations of the presynaptic neuron form enlarged **axon knobs**. Each knob abuts on a dendrite or cell body of the postsynaptic neuron, but a narrow **synaptic cleft**, about 20

Figure 17.9
Chemical synapses. (a) A scanning electron micrograph showing many presynaptic knobs synapsing with a neuron cell body and its dendrites. (b) A diagram of some common types of synapses: directly on the cell body, on the dendrite, and on the dendritic spine. A few axon knobs synapse on the other axon knobs. (c) A diagram of the events occurring during synaptic transmission and recovery. The numbers 1 to 5 refer to the sequence of events described in the text.

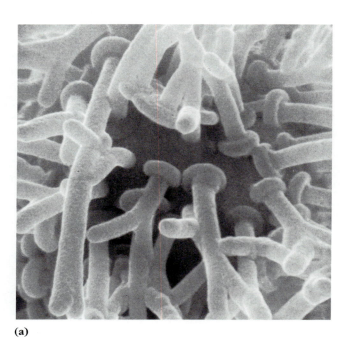

(a)

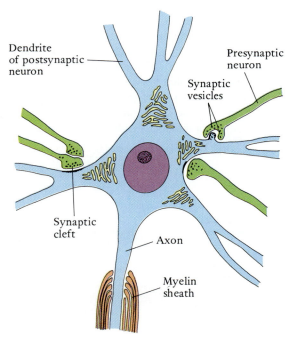

(b)

(c)

nm wide, separates them (Figs. 17.9a and b). An axon knob contains a great many membranous walled **synaptic vesicles** with a diameter of about 50 μm. These vesicles contain a chemical **neurotransmitter**, of which many types have been discovered in different neurons, especially in neurons of the CNS (p. 397). Figure 17.9c portrays the prevalent model of how **acetylcholine**, a neurotransmitter found in many neurons, is released, crosses the synaptic cleft and initiates an impulse in the postsynaptic neuron. On reaching an axon knob, an excitatory nerve impulse changes the voltage across the plasma membrane of the presynaptic neuron, causing voltage-gated calcium channels to open (1). The brief influx of calcium ions from the interstitial fluid causes some of the synaptic vesicles to unite with the plasma membrane of the presynaptic neuron, open, and discharge acetylcholine (2). Acetylcholine diffuses across the narrow cleft and unites with specific protein receptors in the postsynaptic membrane (3). These are chemically gated receptors, and acetylcholine causes a channel in them to open. Sodium ions rush into the postsynaptic neuron, and its membrane becomes depolarized. The enzyme **acetylcholinesterase**, which is present at the synapse, breaks down the acetylcholine into acetate and choline, potassium gates open, and the resting potential is restored in the postsynaptic neuron (4). Products of the breakdown diffuse back into the axon knob and are resynthesized into acetylcholine, using energy made available by the mitochondria (5). Many other types of neurotransmitters are not degraded but are instead pumped back into the presynaptic neuron to be used again. The neurotransmitters must somehow be removed from the synapse, otherwise the postsynaptic neuron would be continuously active.

Excitation and Inhibition

Most junctions between neurons involve numerous neurons, not just two. Many axon endings, sometimes as many as a thousand, cover the dendrites and cell body of a postsynaptic neuron (Fig. 17.9a). Each axon ending is one synapse, but collectively they can interact to convert a multineuronal junction into an elaborate switch.

If a nerve impulse in the presynaptic neuron causes some depolarization of the postsynaptic membrane, i.e., cause the postsynaptic resting potential to rise from say − 70 mV toward zero, it is said to be **excitatory**. The first

impulse to cross the synapse may not reach the threshold needed to generate a nerve impulse in the postsynaptic neuron, but it makes it easier for subsequent impulses to activate the neuron if they arrive before the excitatory effects of the first impulse have decayed. If new impulses cross the same synapse in a very short time period after the first impulse crossed, they may have an additive effect that leads to the generation of an impulse. This phenomenon is known as **temporal summation**. Since the effect of the first impulse also spreads a short distance along the postsynaptic membrane, an action potential may be generated if a second impulse crosses from a nearby synapse on the same postsynaptic membrane. This is **spatial summation**.

Other transmitter substances, which are released by different neurons, bind with their specific receptor sites, and some have an effect opposite to excitation. They cause channels to open that allow potassium ions to flow out of the postsynaptic neuron and hence bring about a **hyperpolarization** of its membrane. The voltage drops below the normal resting level, the excitability of the membrane is decreased, and it becomes more difficult for a nerve impulse to be initiated. This phenomenon is called **postsynaptic inhibition**.

Both the generation and frequency of nerve impulses in the postsynaptic neuron depend on the interaction of the excitatory and inhibitory impulses impinging on the postsynaptic neuron, and on their temporal and spatial distribution. The complexity of these interactions can be very great.

ORGANIZATION OF THE NERVOUS SYSTEM

While cooking, certain sounds and smells alert you to a pan boiling over. You grab the pot, burn your hand and jerk it away, and then turn down the heat. This scenario first involves an immediate response. Concurrent sensory signals from your hand, ears, and eyes reach the brain and are integrated into a total assessment of what is happening. This "picture" is compared with memories of what happened in similar circumstances in the past, all of this information is processed, and a set of motor command signals is initiated that will cause certain muscles to increase their rate and force of contraction and antagonistic muscles to slow down and relax. This can be summarized in the flow diagram below.

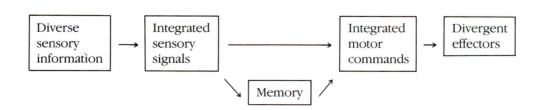

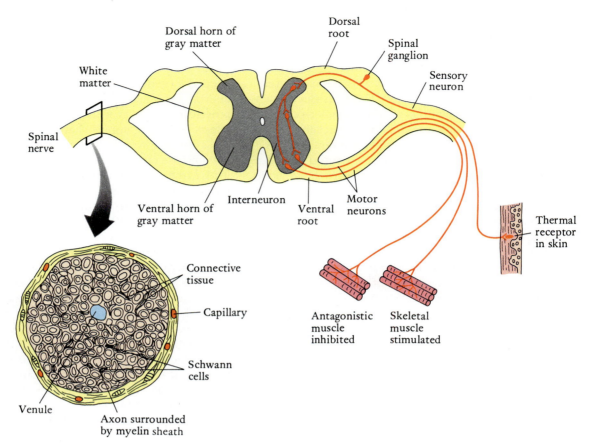

White matter

Dorsal horn of gray matter

Dorsal root

Spinal ganglion

Sensory neuron

Spinal nerve

Interneuron

Ventral horn of gray matter

Ventral root

Motor neurons

Thermal receptor in skin

Connective tissue

Capillary

Schwann cells

Venule

Axon surrounded by myelin sheath

Antagonistic muscle inhibited

Skeletal muscle stimulated

Figure 17.10
A drawing of the major types of neurons in the nervous system and their organization in the vertebrate spinal cord and nerve to form a reflex arc. The inset is a cross section of a nerve. Many more axons than shown would be present in this part of a spinal nerve.

The integration and coordination that characterize this set of responses depend on the interrelationships and organization of the neurons rather than the nature of the nerve impulses.

Types of Neurons and Their Arrangement

The basic organization of the nervous system can be appreciated by examining the neuronal pathways in a vertebrate spinal nerve and spinal cord that are involved in some of these actions (Fig. 17.10). **Sensory neurons** carry information from specific receptors into the CNS, and **motor neurons** carry signals from the CNS to specific effectors. The integration of the sensory signals and the coordination of the motor responses are performed by large numbers of **interneurons** that are confined to the central nervous system. These three categories of neurons form the entire nervous system and have precise relationships to each other.

Nerves are aggregations of the axons of many neurons; both sensory and motor neurons are combined in spinal nerves. Connective tissue surrounds groups of neurons, and small blood vessels in the nerve nourish them. Sensory and motor neurons tend to separate just lateral to

the spinal cord in the roots of a spinal nerve. The **dorsal root** of the spinal nerve of mammals contains only sensory neurons, and their cell bodies are clustered in a small swelling, the **spinal ganglion** (Gr., little tumor), on the dorsal root; the **ventral root** contains only motor neurons.

As seen in a transverse section, the spinal cord of vertebrates contains a roughly butterfly-shaped area of **gray matter** peripheral to a small central canal (Fig. 17.10). The gray matter is composed of the cell bodies of interneurons and motor neurons together with their unmyelinated processes. The absence of myelin gives the gray tone to this tissue. The **dorsal horn (column)** * of gray matter contains interneurons that receive signals from the sensory neurons and relay them to other parts of the cord or to the brain. The **ventral horn (column)** is related to motor functions, for it contains motor neurons whose axons leave through the ventral roots of spinal nerves. The tissue peripheral to the gray matter is **white matter** consisting of the myelinated axons of sensory and interneurons traveling to other levels of the cord or to the brain, and the

*Horn describes the two-dimensional appearance; column, the three-dimensional appearance.

myelinated axons of interneurons descending from the brain to end on the dendrites and cell bodies of motor neurons. Bundles of axons traveling together within the CNS are called **tracts** and not nerves.

In invertebrates such as crayfish and earthworms, the sensory, interneurons, and motor neurons have a relationship to each other that is similar to the vertebrate pattern. Cell bodies of invertebrate sensory neurons, however, lie in an epithelial layer near the receptors and the CNS is organized differently than in vertebrates.

Reflexes

The quick withdrawal of your hand after touching the hot pan is an example of a withdrawal reflex that depends on a specific pathway of neurons. A branch of the sensory neuron from your hand terminates on an interneuron in the dorsal horn (Fig. 17.10). One branch of this interneuron terminates on a motor neuron in the ventral horn, which stimulates certain muscles and leads to their contraction. Another branch of the interneuron goes to other interneurons that terminate on the motor neurons to antagonistic muscles, inhibits them, and causes the muscles to relax. **Reflexes** are involuntary acts that depend on preexisting neuronal pathways, the **reflex arcs**, and involve relatively few neurons. A particular stimulus always elicits the same response. Many of an animal's activities are controlled reflexively: respiratory movements, heart rate, salivation, and swallowing, to name just a few. Reflex behaviors are called **innate** or inborn, because the neuronal pathways responsible for them are rigidly established before birth.

Pathways between Cord and Brain

It is important to realize that while reflexes are important, complex motor activity—such as turning down the heat under the boiling pan—involves pathways to and from higher centers in the brain. Some branches of the sensory neurons returning from your hand terminate on interneurons that ascend in tracts of white matter in the spinal cord to the thalamus of the brain, from where they are relayed to specific parts of the cerebral cortex (Fig. 17.11). Other impulses reach the cortex from the ears and eyes. Many interconnections occur within the cortex. You become aware of the boiling pan and, having learned from past experience, decide to turn down the heat. Appropriate impulses leave the cortex on interneurons that descend the spinal cord through other tracts in the white matter, and stimulate and inhibit the appropriate combination of motor neurons to the muscles. Most ascending and descending impulses cross or **decussate** to the opposite side of the nervous system. Thus most sensory impulses originating on the right side of the body terminate in the left side of the brain, and motor impulses originating in the left side of the brain affect the right side of the body. The significance of decussation is not fully understood.

Neuron Pools

Interneurons in the cord and brain usually form more complex associations than the simple reflex pathways and tracts described. They are organized into complex **neuron pools** that receive sensory signals, process them in various ways, and initiate motor impulses to a great many muscles. Neuron pools are analogous in some ways to computer microchips that receive certain inputs, process them in characteristic ways, and initiate outputs. Often there is a spreading or **divergence** of an incoming impulse that is caused by a repeated branching of successive short neurons (Fig. 17.12a). A sensory impulse entering on a single neuron, for example, may affect many others and reach many parts of the brain. Conversely, impulses from several sources may **converge** on a single output neuron so its activity will be affected by a variety of excitatory and inhibitory influences from many sources (Fig. 17.12b). Other organizations of neurons in a pool may cause an **afterdischarge**, that is, a continuation of impulses from the output neuron that continue beyond the initial stimulus. Afterdischarges may result from neuron pathways of different length impinging on a single output neuron (Fig. 17.12c), or from feedbacks in which a branch of the output neuron feeds back upon itself and restimulates itself (Fig. 17.12d).

As one example, the rhythmic walking movements of terrestrial vertebrates are mediated by neuron pools in the spinal cord. When one leg is moved, sensory feedbacks from the contracting muscles enter the pool and initiate excitatory and inhibitory impulses to the muscles of the opposite leg, causing certain of its muscles to contract and others to relax. The movements of this leg, in turn, restimulate the first leg. Activity of the pool, in turn, can be modulated by signals coming from the brain that adjust locomotion to changes in the posture of the animal, changes in the visual field, decisions to move more rapidly or slowly, and so forth.

EVOLUTIONARY TRENDS IN THE NERVOUS SYSTEM

A nervous system could not evolve unless cells had a basic irritability and a capacity to respond directly to environmental stimuli. We see this irritability in protozoans, and some cells in more complex, multicellular animals, which are known as **independent effectors**, retain it. Among them are contractile cells (porocytes) of sponges, which regulate the amount of water entering a sponge, and the stinging cells of hydras, jellyfish, and other cnidarians.

Except for sponges and a very few specialized parasites, all animals have nervous systems. But how can neurons, which simply conduct or do not conduct impulses, evolve into systems with properties as complex as those we have been considering? An analogy from computers may help us understand how complex properties can emerge from simple units. Instructions to a computer are

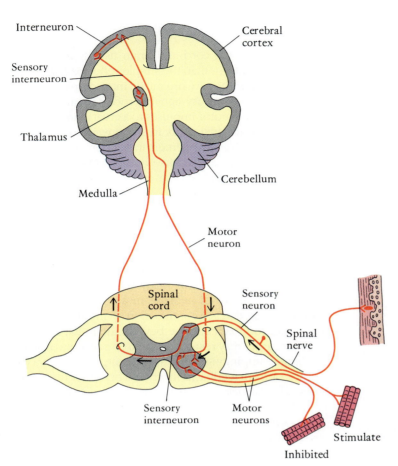

Figure 17.11

Interneurons also carry nerve impulses from the spinal cord to the brain and back to the cord. The reflex arc shown in Figure 17.10 has been omitted.

given by many electronic switches that, like neurons, can be either on or off. The switches represent the numbers 0 and 1 (on and off) in a binary numbering system. The zeros and ones are combined in groups of eight known as bytes. Eight on and off signals can be arranged into 256 combinations, or different bytes, representing the letters of the alphabet, the digits, and other symbols or instructions. Bytes, in turn, can be arranged in a vast number of ways that convey different meanings. Thus the complexity of a computer, like that of a nervous system, depends not on the complexity of its units, but on their number and the ways they are combined and interconnected. A personal computer with 640,000 bytes of memory has about 5,240,000 switches. Imagine the complexity that can be attained by the human brain, which has more than 100 billion neurons!

The nervous systems of simple animals have relatively few neurons. As the number of neurons and their interconnections increase, new abilities emerge. The simplest systems are found in hydras, sea anemones, and other cnidarians (Fig. 17.13). In some parts of the body one sensory neuron receives a stimulus and conducts an im-

pulse to an underlying contractile element, either a muscle cell or the elongated base of an epithelial cell. Early in this century George Parker proposed that nervous systems began as one-neuron systems of this type. A second level of complexity is seen in two-neuron systems in which the sensory neuron synapses with a motor neuron. The motor neuron, in turn, can extend to a group of contractile elements and a more complex response is attained. Most parts of the cnidarian body have three-neuron systems in which a network of branching interneurons is interposed between the sensory and motor neurons. Such a system is described as a **nerve net**. Sensory signals impinging on one part of the body can affect contractile elements over a wide area. Many cnidarians have two interconnected nets, one beneath the surface epithelium and another beneath the epithelium lining the gut. Neuron terminals on each side of most of the synapses in these nets can release neurotransmitters, so these synapses are not polarized. Impulses originating on one part of the body can spread in many directions.

The system is not perfectly symmetrical in cnidarians. Sensory cells are more abundant around the mouth and on

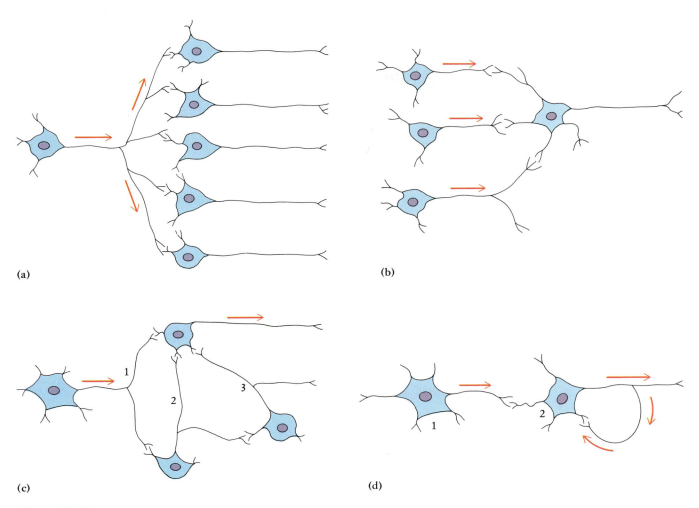

(a)

(b)

(c)

(d)

Figure 17.12

Some examples of the organization within neuron pools. (a) A divergent pathway.
(b) A convergent pathway. (c) A multiple chain circuit. (d) A feedback circuit. Both
multiple chain and feedback circuits can cause an afterdischarge. A given neuron
may be a part of more than one of these types of organization.

the tentacles than in other regions. Some neurons are
longer than others, some have a greater diameter and
conduct impulses more rapidly, and the processes of
several may be gathered into rudimentary tracts. As a
consequence, some regions of the body are more sensitive
than others, and impulses travel faster and more directly to
some effectors than to others. If a sea anemone is nipped
by a fish, the mouth closes and the tentacles withdraw
rapidly, but the body as a whole shortens more slowly. A
nervous system of this type is well adapted to the mode of
life of these radially symmetrical and largely sessile organisms. Despite the seeming simplicity of their nervous system, cnidarians have many behavioral responses, including withdrawal from dangerous stimuli, prey capture,
feeding, and, for jellyfish and a few other species, swimming.

As animals became more active and evolved a bilateral symmetry, neurons increased in number and the
nervous system became differently organized. Since the
front of the body encounters new environments first,
sensory neurons are concentrated here. Most of the interneurons aggregate into a central nervous system that
extends the length of the body (Fig. 17.14). Throughout the
animal kingdom, sensory neurons that both receive stimuli and transmit impulses lead into the CNS, but in many
vertebrate sense organs, such as the ear and eye, distinct
receptor cells initiate impulses that are then conveyed by
sensory neurons to the CNS. Interneurons in the CNS
coordinate the activities of the body and can transmit
impulses rapidly from one region to another. In most
animals motor neurons originate in the central nervous
system and extend to the effectors, but in a few animals,

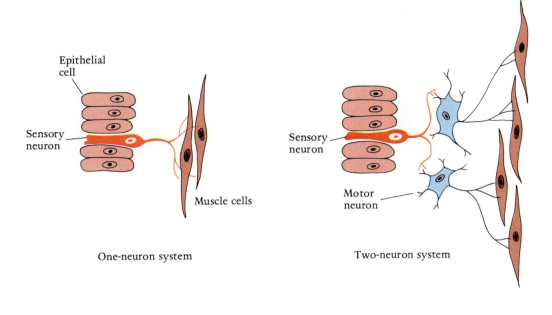

Epithelial
cell

Sensory
neuron

Muscle cells

One-neuron system

Sensory
neuron

Motor
neuron

Two-neuron system

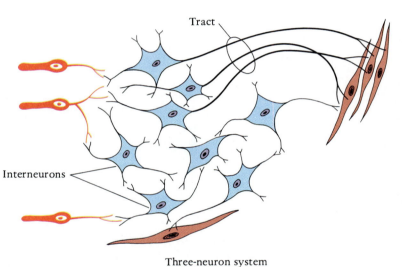

Tract

Interneurons

Three-neuron system

Figure 17.13
Drawings of primitive neuronal organizations found in cnidarians.

including roundworms and the protochordate amphioxus, slender, neuron-like processes of the muscle cells extend into the CNS. Although most of the nervous system becomes highly organized, some animals retain localized nerve nets beneath the surface epithelium or in other tissues. A nerve net coordinates many of the movements of the vertebrate gut.

The CNS of most bilaterally symmetrical animals consists of a dorsal aggregation of neuron cell bodies at the front of the body that forms one or more **cerebral ganglia** or a **brain** (Fig. 17.15a). Functionally the brain is a site of sensory integration and motor command that evolved in conjunction with the development of eyes and other sense

organs in the head region. Longitudinal strands, or **nerve cords**, extend caudally from the brain. Some flatworms have a number of cords located beneath the epidermis on the dorsal, lateral, and ventral sides. This may have been the primitive arrangement, but most animals have fewer nerve cords. Annelids and arthropods have a pair of ventral cords; vertebrates, a single dorsal cord. When the pair of cords are ventral they pass to either side of the anterior region of the gut to reach the brain, and they are connected along their length by transverse neurons that form **commissures**. However, in many species, such as the earthworm, the cords are united to a large extent. Primitively, the cell bodies of neurons are scattered along the

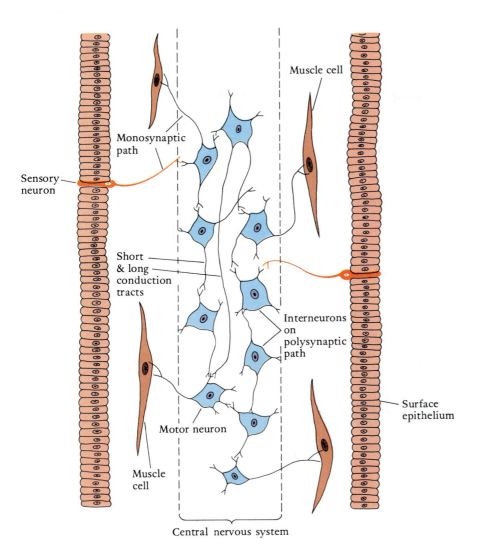

Sensory neuron

Monosynaptic path

Muscle cell

Short & long conduction tracts

Interneurons on polysynaptic path

Motor neuron

Muscle cell

Surface epithelium

Central nervous system

Figure 17.14

A drawing of the neuronal organization in a primitive worm with a central nervous system.

cords (Fig. 17.14), but in most animals they are aggregated into clusters, or **ganglia**. For example, in metameric annelids and arthropods, they are located one pair per segment along the length of the ventral cords (Fig. 17.15b). Ganglia fuse in many crabs, insects, and other animals in which body segments fuse or are reduced in number.

Rhythmic locomotor activity is mediated in the ganglia by neuronal pools and by the passage of impulses along the cords from segment to segment. Evidence suggests that there is a spontaneous rhythmic activity of control neurons in each ganglion such that stimulation of neurons to one group of muscles simultaneously inhibits the activity of neurons to the antagonistic muscles. Centers of this type are called **central neuronal oscillators**. Sensory feedback from contracting muscles is not as important in coordinating locomotion among invertebrates as it is in the vertebrates. Feeding, respiratory, and mating activities also are controlled in ganglia along the cords. The brain

does not mediate these functions directly, but does modulate them in relation to sensory information it receives and, in some species, memories of past experiences.

The octopus and other cephalopod molluscs have the largest and most complex brains among invertebrates, rivalling those of primitive vertebrates. This correlates with the development of large image-forming eyes, locomotor dexterity, carnivorous habits, and the high degree of cephalization of these creatures. Estimates of the number of neurons in an octopus brain approximate 200 million.

THE VERTEBRATE NERVOUS SYSTEM

The vertebrate nervous system is relatively larger, contains more neurons, and is more complexly organized than that of other animals. The central nervous system is always located dorsally, the spinal cord is a single and not a

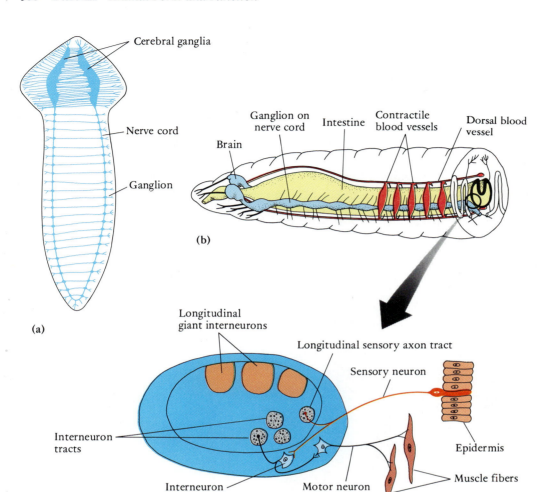

Figure 17.15

Some invertebrate nervous systems. (a) The nervous system of a planarian flatworm. (b) The anterior end of the nervous system of an annelid worm. (c) An enlarged insert of a cross section through a ganglion. Unlike the arrangement in the vertebrate nerve cord, the cell bodies of motor neurons and interneurons are located at the periphery of the ganglia, and axon tracts lie centrally.

double cord, and both the cord and brain contain a central cavity. These are among the distinctive features of vertebrates and other chordates that separate them from all other animals. Clearly the vertebrate nervous system is quite different from that of other advanced metazoans such as annelids and arthropods and evolved independently. The spinal cord lies in a vertebral canal of the vertebral column, and the brain is encased in a cranium.

The Peripheral Nervous System

Spinal and Cranial Nerves

Except for part of the head, the body of vertebrates is segmented. One pair of **spinal nerves** leave the cord through intervertebral foramina at each body segment, and if body regions are well defined, the nerves are named after them: cervical nerves, thoracic nerves, lumbar nerves, and so on. Each spinal nerve divides almost as soon as it is formed by the union of its roots into branches or **rami** (L. *ramus*, branch) that supply different body regions (Fig. 17.16).

The sensory neurons in the spinal nerves can be sorted out into **somatic sensory neurons** coming from the body wall and appendages, and **visceral sensory neurons** coming from deeper visceral organs. Similarly, the motor neurons can be grouped into **somatic motor neurons** that supply the striated, skeletal muscles of the body wall and appendages, and **visceral motor neurons** that supply visceral muscles of the gill region (most of which are also striated), cardiac muscles, and smooth muscles in the walls of visceral organs. These neurons have characteristic terminations or origins within the gray matter, as shown in

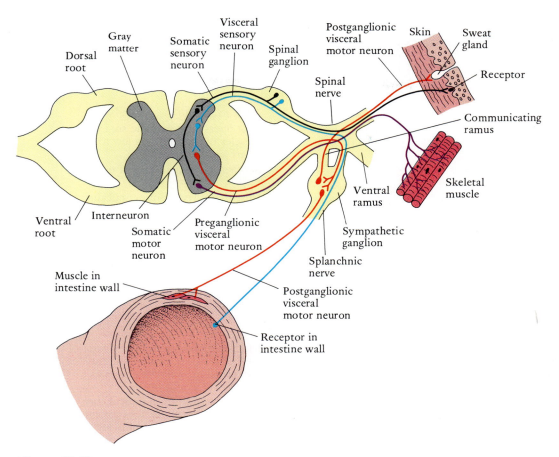

Figure 17.16

A diagram of a cross section through the spinal cord and a spinal nerve of a mammal, showing the types of neurons they contain and their locations and peripheral connections.

Figure 17.16. A few somatic sensory neurons (touch) project to the back of the brain before relaying, but most terminate in the spinal cord. Somatic sensory neurons terminate more dorsally in the dorsal horn than visceral sensory ones, and the cell bodies of somatic motor neurons are located more ventrally in the ventral horn than those of visceral motor neurons. Although the dorsal root of mammalian spinal nerves contains only sensory neurons and the ventral root only motor ones, this is not the case in primitive vertebrates. In some fishes, all of the visceral motor neurons travel with the sensory neurons in the dorsal roots, so their ventral roots contain exclusively somatic motor neurons.

The distribution and characteristic of the cranial nerves of reptiles, birds, and mammals are presented in Table 17.1. Their names reflect their distribution, and each is given a conventional number corresponding to their sequence in humans. Although the cranial nerves are not so clearly segmented as the spinal ones, numbers V, VII, IX, X, and XI appear to be serially homologous to the dorsal roots of primitive spinal nerves, for they contain visceral motor neurons to jaw, gill, and other visceral muscles, in addition to the expected complement of sensory neurons. Numbers III, IV, VI, and XII can be equated with primitive ventral roots, for they contain somatic motor neurons to the extrinsic ocular muscles that move the eyeball and somatic muscles in the tongue. The cranial nerves to the nose, eye, and ear have no relation to spinal nerves, for they evolved along with these special sense organs. Fishes and amphibians are usually described as lacking the accessory and hypoglossal nerves. They do lack the nerves as distinct entities, but the neurons in them are contained in other nerves. Accessory neurons travel with the vagus in fishes; hypoglossal ones, in a group of occipitospinal nerves at the back of the skull.

The Autonomic Nervous System

The subset of visceral motor neurons innervating glands, cardiac muscle, smooth muscles in the walls of blood vessels and visceral organs, and chromatophores forms

Table 17.1
The Distribution and Composition of the Cranial Nerves of Reptiles, Birds, and Mammals*

Nerve	Origin of Sensory Neurons	Distribution of Motor Neurons
I, Olfactory	Olfactory portion of nasal mucosa (smell) (S)	
II, Optic	Retina (sight) (S)	A few to retina (V)
III, Oculomotor	A few fibers from receptors in extrinsic muscles of eyeball (muscle sense) (S)	Most fibers to four of the six extrinsic muscles of eyeball (S), a few to muscles in ciliary body and pupil (V)
IV, Trochlear	Receptors in extrinsic muscles of eyeball (S)	Another extrinsic muscle of eyeball (S)
V, Trigeminal	Jaws, and skin receptors of the head (touch, pressure, temperature, pain); receptors in jaw muscles (S)	Muscles derived from musculature of first visceral arch, i.e., jaw muscles (V)
VI, Abducens	Receptors in an extrinsic muscle of eyeball (S)	One other extrinsic muscle of eyeball (S)
VII, Facial	Taste buds of anterior two-thirds of tongue (taste) (V)	Muscles derived from musculature of second visceral arch, i.e., facial muscles; salivary glands; tear glands (V)
VIII, Vestibulocochlear	Semicircular canals, utriculus, sacculus (sense of balance); cochlea (hearing) (S)	A few to cochlea (S)
IX, Glossopharyngeal	Taste buds of posterior third of tongue; lining of pharynx (V)	Muscles derived from musculature of third visceral arch, i.e., pharyngeal muscles concerned in swallowing; salivary glands (V)
X, Vagus	Receptors in many internal organs: larynx, lungs, heart, aorta, stomach (V)	Muscles derived from musculature of remaining visceral arches (excepting those of pectoral girdle), i.e., muscles of pharynx (swallowing) and larynx (vocalization); muscles of gut and heart; gastric glands (V)
XI, Accessory	Receptors in certain shoulder muscles (S)	Visceral arch muscles associated with pectoral girdle, i.e., sternocleidomastoid and trapezius (V)
XII, Hypoglossal	Other receptors in tongue (S)	Muscles of tongue (S)

*The types of sensory and motor neurons they contain are indicated: S = somatic, V = visceral.

the **autonomic nervous system**. These neurons leave the CNS through certain cranial and spinal nerves. Those in the spinal nerves leave the nerves through their communicating rami (Fig. 17.16) that connect (communicate) with the paired **sympathetic cords**. Many continue in nerves called **splanchnic nerves** (Gr. *splanchnon*, gut) toward the visceral organs. Visceral sensory fibers returning from the viscera in these nerves are usually not considered to be a part of the autonomic nervous system.

The autonomic nervous system is a motor system, but it differs in two important ways from the groups of visceral and somatic motor neurons supplying striated muscles. First, autonomic neurons do not extend all of the way from the central nervous system to their targets; they always end in a peripheral motor ganglion where they synapse with a second group of motor neurons that continue to the effectors. Zoologists call this a peripheral relay (Figs. 17.16 and 17.17). **Preganglionic fibers** extend from the central

nervous system to the ganglion, and **postganglionic fibers** continue to the targets.

Second, the autonomic nervous system is divided into a **sympathetic** and **parasympathetic nervous system**. Most, but not all, visceral organs are innervated by both. These two systems have antagonistic effects upon the organs because their postganglionic fibers release different neurotransmitters at their junctions with the effectors. Parasympathetic fibers release acetylcholine; sympathetic fibers release norepinephrine (also called noradrenalin). The effect of sympathetic stimulation is very similar to the effect of the secretion of the hormone epinephrine from the medulla of the adrenal gland (Chap. 19). Both enable a vertebrate to adjust to stress such as occurs in escaping a predator or chasing prey. Resources are mobilized to quickly expend a great deal of energy. The effects are often called the **fight-or-flight response**. Cardiac muscle contracts with greater force and speed, blood pressure rises,

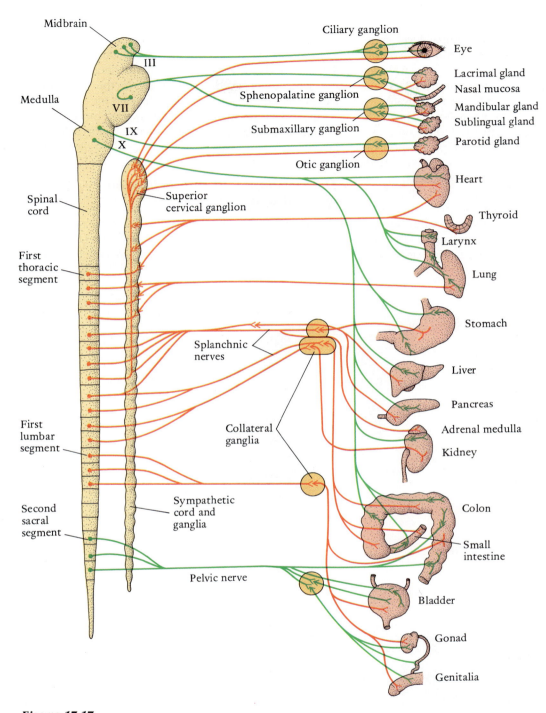

Figure 17.17

A diagram in lateral view of the human autonomic nervous system. Sympathetic fibers are shown in red; parasympathetic ones, in green. Sympathetic fibers to the skin are not shown.

blood flow to cardiac and skeletal muscles increases, bronchioles in the lungs dilate, blood sugar level rises, hairs become elevated, sweating increases, and the pupils dilate. Digestive functions are inhibited. Peristalsis decreases and gastrointestinal sphincters contract. Parasympathetic stimulation inhibits activities that are activated by sympathetic stimulation and promotes those that produce and store energy. More blood is directed to the digestive organs, gut motility and digestive enzyme secretion increase, bile flows, and sugars are stored. Under certain circumstances, sexual activity is also promoted.

The two parts of the autonomic system also differ morphologically (Fig. 17.17). Mammalian parasympathetic fibers to the head, thorax, and abdomen leave through the oculomotor (III), facial (VII), glossopharyngeal (IX), and vagus (X) nerves. Pelvic organs receive their parasympa-

thetic fibers through some of the sacral nerves. The parasympathetic fibers relay in small ganglia close to or in the wall of the organs being supplied. Sympathetic fibers leave through the thoracic and anterior lumbar nerves and enter the sympathetic cords. They relay in the **sympathetic ganglia** along the cords, or pass through the sympathetic cords, travel in the splanchnic nerves, and relay in **collateral ganglia** at the base of the major abdominal arteries.

The Development and Evolution of the Brain

The vertebrate brain develops as an expansion of the anterior end of the embryonic neural tube (Fig. 17.18). Three swellings can be seen early in development: a

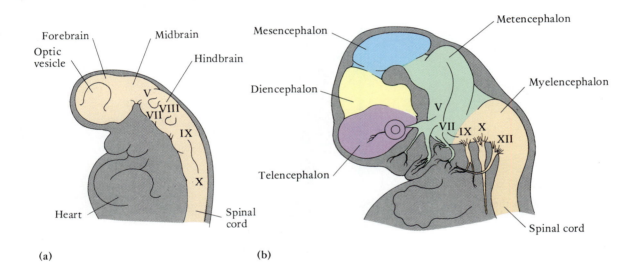

(a)

(b)

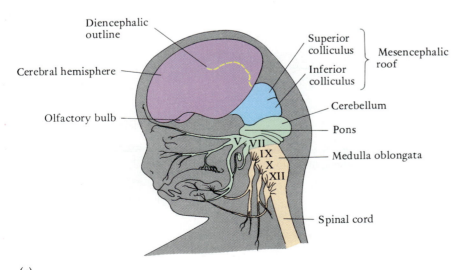

(c)

Figure 17.18
Three stages in the embryonic development of the human brain. (a) Three brain regions are present in the embryo 4 weeks old. (b) All five regions are evident in an embryo 7 weeks old. (c) The cerebral hemispheres and other features of the fully formed brain are beginning to differentiate in an embryo 11 weeks old.

forebrain, midbrain, and hindbrain. As development continues, the forebrain subdivides into a **telencephalon** and **diencephalon**, the midbrain (**mesencephalon**) remains undivided, and the hindbrain subdivides into a **metencephalon** and **myelencephalon**. Most of the telencephalon differentiates into a pair of **cerebral hemispheres**, together called the **cerebrum** (L., brain). The cerebrum is relatively small in primitive vertebrates, but enlarges greatly during evolution and becomes the dominant part of the brain in mammals. Other structures that differentiate from the five primary brain regions are shown in Figure 17.18 and are listed in Table 17.2. The cavity of the neural tube remains small in the spinal cord, but expands in the brain to form a series of interconnected chambers known as **ventricles** (p. 397).

The organization of neurons and of gray and white matter in the brain follows a modification of the basic pattern seen in the spinal cord. The dorsal and ventral columns of gray matter of the cord break up into small islands of gray matter called **nuclei** that are associated with the cranial nerves. As was the case with the columns in the spinal cord, dorsal nuclei are related to sensory functions; ventral ones, to motor functions. Additional nuclei have other roles, and, in mammals, extensive areas of gray matter migrate to the surface to form a cortex over the cerebrum and some other brain regions.

Myelencephalon

The myelencephalon differentiates into the **medulla oblongata** that connects with the spinal cord (Figs. 17.18 and 17.19). Most of the cranial nerves attach to the medulla and to the underside of the metencephalon and mesencephalon. These areas contain the sensory and motor nuclei of the cranial nerves and a diffuse column of gray matter known as the **reticular formation**. These nuclei and the reticular formation integrate feeding behavior, swallowing, and the rates of respiration and heart beat. Centers for respiration and heart beat have an inherent, rhythmic oscillation, but this is modulated by sensory impulses entering this region and by impulses descending from the cerebrum and other higher centers in the brain. The reticular formation also sends nonspecific impulses to the cerebral cortex of mammals. If the sensory input to the reticular formation is critical for the animal—for example, an unusual noise when the animal is sleeping—inputs from the reticular formation to the cerebrum will arouse the animal if asleep and keep it alert if already awake.

Metencephalon

A **cerebellum** (L., small brain) differentiates in the roof of the metencephalon (Figs. 17.18 and 17.19). It is an important center for motor coordination. Impulses from the inner ear related to body position and equilibrium are projected here, as are impulses from proprioceptive organs in muscles indicating their current degree of contraction (p. 672). Motor impulses from the cerebellum modulate the activity of motor neurons in relation to the current activity of the muscles and body position. For example, the cerebellum compares the motor commands from the cerebrum with actual muscle contractions and, if needed, initiates corrective impulses. As would be expected, the size and complexity of the cerebellum correlate with an animal's life style and its degree of motility and motor complexity.

The mammalian cerebellum is very large and involves a new area that interconnects directly with the motor parts of the cerebrum. Interaction between the cerebellum and cerebrum is essential for the precise timing and duration of muscle contractions involved in learned, voluntary actions. Most of the gray matter of the cerebellum has migrated to the surface and forms a gray cortex.

In most vertebrates, the ventral portion of the metencephalon is functionally a part of the medulla oblongata, but in mammals much of this area differentiates into the **pons** (L., bridge) (Fig. 17.18). The pons receives its name from a superficial, transverse band of neurons that interconnect the two sides of the cerebellum, but the pons also contains nuclei that relay impulses from the cerebrum to the cerebellum.

Mesencephalon

The roof of the mesencephalon forms a pair of large **optic lobes** in fishes, amphibians, and reptiles (Fig. 17.19). Neurons from the eye terminate in these lobes after decussating in an **optic chiasma** located on the underside

Table 17.2
Brain Regions and Major structures that Develop from Them

Forebrain	
Telencephalon	Olfactory bulbs
	Cerebral cortex
	Hippocampus
	Corpus striatum
Diencephalon	Pineal eye or gland
	Thalamus
	Hypothalamus
	Neural lobe of pituitary gland
	Retina
Midbrain	
Mesencephalon	Optic lobes or superior and inferior colliculi
Hindbrain	
Metencephalon	Cerebellum
	Pons
Myelencephalon	Medulla oblongata

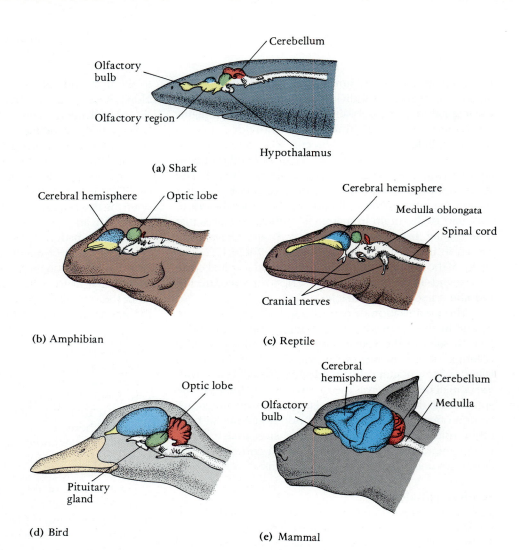

Olfactory
bulb

Cerebellum

Olfactory region

Hypothalamus

(a) Shark

Cerebral hemisphere

Optic lobe

(b) Amphibian

Cerebral hemisphere

Medulla oblongata

Spinal cord

Cranial nerves

(c) Reptile

Optic lobe

Pituitary
gland

(d) Bird

Cerebral
hemisphere

Cerebellum

Olfactory
bulb

Medulla

(e) Mammal

Figure 17.19
Lateral views of the brains of
representative vertebrates to
show major evolutionary trends.
Note in particular the great
enlargement of the cerebral
hemispheres and cerebellum.

of the diencephalon. The optic lobes are important centers for visual integration and for reflexes that control eye movements, blinking, pupil size, and focusing. Impulses from touch, hearing, and many other receptors also reach this area in primitive vertebrates, so it is the major integration center in these animals. Coordinated motor responses are initiated here, and impulses travel down the brain and cord to the appropriate motor neurons.

As the cerebrum becomes more important during vertebrate evolution, sensory pathways shift from the optic lobes and are projected to the cerebrum. The cerebrum gradually usurps many of the original functions of the optic lobes. The optic lobes of mammals, now called the **superior colliculi** (L. *colliculus*, little hill), are hillocks that mediate eye reflexes (Fig. 17.18). Another pair of **inferior colliculi** control auditory reflexes.

Diencephalon

The **thalamus** (Gr. *thalamos*, inner chamber) develops as a thickening of the lateral walls of the diencephalon. It is relatively small in primitive vertebrates, but enlarges

greatly in mammals because it is the major relay center to and from the cerebrum. All sensory information, apart from olfaction, first relays in dorsal thalamic nuclei (see Fig. 17.11). Sensory impulses are first processed here. Some are suppressed, others are enhanced, and then certain ones are relayed forward to the sensory areas of the cerebrum.

In all vertebrates, impulses for stereotyped motor activity related to locomotion and maintaining posture are relayed in ventral thalamic nuclei on their way from the motor centers in the cerebrum to the motor nuclei of the cranial nerves or motor columns of the cord. In mammals, impulses that mediate voluntary motor activity travel directly from the motor cortex to the motor neurons on long, uninterrupted neurons that do not stop in the thalamus (see Fig. 17.11).

The **hypothalamus** (Gr. *hypo*, under), which differentiates in the floor of the diencephalon, is the major center for visceral integration and homeostatic control in all vertebrates, and it changes little during the course of evolution (Fig. 17.19). It receives many sensory signals, and some of its cells respond directly to the concentration of glucose, water, salts, and many hormones in the blood. It regulates periods of rest and activity, the intake of food and water, gut movements and digestion, blood sugar levels, water and salt balances, and sexual activity. It interacts closely with centers deep in the cerebrum in performing many of these functions. Certain of its cells in warm-blooded birds and mammals monitor blood temperature and initiate responses needed to maintain body temperature. Its output is primarily through the autonomic nervous system, but it also controls much of the activity of the pituitary gland, part of which develops from the floor of the diencephalon (Chap. 19).

In many fishes, amphibians, and reptiles, either a small **parietal** or **pineal eye** attaches to the roof of the diencephalon. This parietal or pineal eye is not visual, but monitors ambient light that usually enters through a foramen in the roof of the skull. The median eye probably provides information that enables the body's activity and physiology to adjust to diurnal and seasonal changes. The pineal eye becomes the pineal gland in birds and mammals (Chap. 19).

Telencephalon

The **olfactory bulbs** (L. *olfactus*, smell) differentiate from the anteromost part of the telencephalon (Fig. 17.18). They receive the olfactory nerves from the nose and initiate the processing of olfactory signals. Impulses extend from them to the cerebral hemispheres and especially to areas of gray matter known as the **paleopallium** (Gr. *palaios*, ancient + L. *pallium*, mantle) and **archipallium** (Gr. *arche*, beginning). These areas are located deep in the cerebral hemispheres of primitive vertebrates above the lateral ventricles (Fig. 17.20a). The cerebral hemispheres are relatively small in fishes and amphibians, for they are chiefly olfactory integration centers (Fig. 17.19). Only a few other sensory impulses are projected to the cerebrum of primitive vertebrates, and they go primarily to a deep ventrolateral area of gray matter known as the **corpus striatum**.

The corpus striatum remains in mammals as a center for many involuntary activities, but the cerebral hemispheres become the dominant integration centers, for more sensory information is projected to them and more motor impulses are initiated here. As they enlarge, they extend caudally and cover much of the rest of the brain (Fig. 17.19). Most of the gray matter of the cerebrum moves to the surface in mammals to form the **cerebral cortex** (Fig. 17.20b). The larger area available on the surface accommodates the increased number of neuron cell bodies. Most of the cortex is a nonolfactory integration area known as the **neopallium** (Gr. *neos*, new) or **neocortex**. The neopallium evolved from an inconspicuous area of gray matter, the dorsal pallium, located between the paleopallium and archipallium of primitive species (Fig.

17.20a). As this area expanded, it pushed the ancient olfactory areas apart. The paleopallium migrated to the ventrolateral surface of the cerebrum and still receives the primary outflow from the olfactory bulbs. The archipallium migrated medially and ventrally and rolled into the lateral ventricle to form a band of gray matter known as the **hippocampus**. The hippocampus continues to receive olfactory signals, but it also receives impulses from taste, touch, visual, and other receptors. A large group of neurons extend from it to the hypothalamus. The hippocampus, part of the corpus striatum, and the hypothalamus form the major part of the **limbic system** (L. *limbus*, border) that influences motivational and emotional behavior relative to species survival: feeding, drinking, fighting, fleeing, and reproduction.

Figure 17.20

Cross sections of the left cerebral hemisphere of an amphibian and a mammal. The great expansion of the neopallium in mammals has pushed the primitive olfactory paleopallium and archipallium apart.

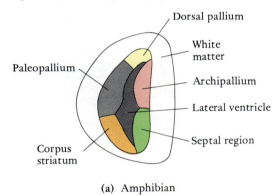

(a) Amphibian

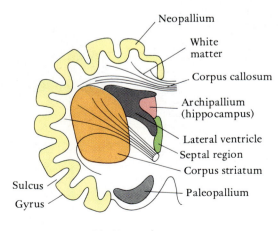

(b) Mammal

The Mammalian Cerebrum

The surface of the cerebrum of small and primitive mammals is relatively small and smooth, but it has increased greatly in size in large mammals and those with a wide range of motor activity and skills (Fig. 17.21). Many folds known as **gyri** (Gr. *gyro*, circle) further increase the surface area of the cortex in these mammals. Our universe centers in the cerebrum because our awareness of the world around us, consciousness, learning, memory, thought, and skilled voluntary actions are dependent on the complex interactions of the approximately 100 billion neurons that form the cortex.

Sensory, Motor, and Association Areas

Much remains to be discovered about the mechanisms of the cerebral cortex, although we have learned a great deal by tracing neuron pathways, administering drugs to experimental animals, searching for changes that occur during learning, observing the loss of function when certain areas are destroyed by lesions or trauma, probing and stimulating particular areas during neurosurgery, and studying the electrical and chemical events in the parts of the cortex that are active under different circumstances. An impressive method of studying sites of brain activity is by positron emission tomography. A patient is given a substance, such as deoxyglucose, containing an isotope that emits positron particles (the nuclei of hydrogen atoms). The parts of the brain that are metabolically very active will pick up the isotope and its positron emissions can be watched on a special screen. For example, the investigator can see what parts of the brain have increased rates of metabolism and rapidly take up the isotope when the subject is asked to read or sound a word or think about its meaning. Since the half-life of the isotope is very short, the effect does not last long.

By using some of these methods, the termination of sensory impulses has been localized in well-defined **somatic sensory**, **auditory**, **visual**, and **olfactory cortical areas** shown in Figure 17.21. The extent of different parts of the somatic sensory cortex correlates with the number of receptors in a body region that send impulses to it and not

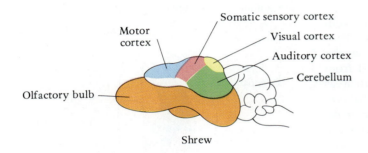

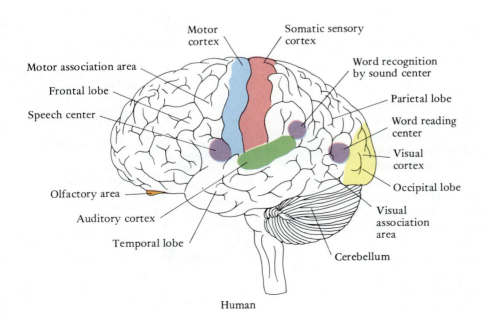

Figure 17.21

Lateral views of the cerebral hemisphere of a primitive mammal (a shrew) and a human. The hemisphere of an advanced mammal is characterized by the evolution of extensive association areas. Purple areas related to word recognition have been identified by positron emission tomography.

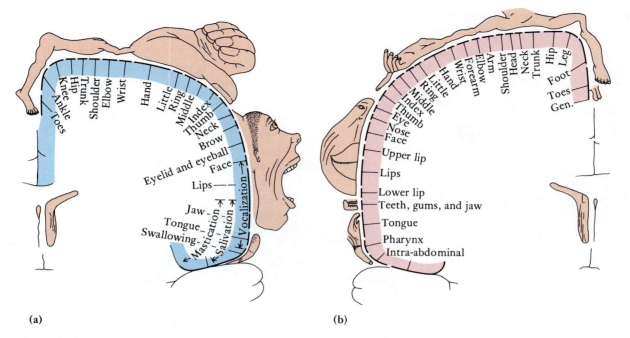

(a)

(b)

Figure 17.22
Vertical sections through (a) the motor cortex and (b) the sensory cortex. The area of the cortex associated with various body regions has been emphasized by drawing body parts in proportion to the area.

with the area of the region (Fig. 17.22). Similarly, motor impulses originate in a **motor cortex** the size of which correlates with the number of motor neurons going to the region and not with the size of the region. Thus our hand, which is particularly sensitive and has a complex motor control, occupies a disproportionate share of the somatic sensory and motor cortices.

The sensory and motor cortices occupy most of the cerebrum of primitive mammals, such as a shrew, but extensive **association areas** lie between them in more advanced mammals (Fig. 17.21). The sensory cortices receive sensory signals, but the interpretation of these signals requires the participation of adjacent association areas. Connections are made between the sensory and motor cortices and their association areas by association fibers that extend between gyri, and by impulses that travel to the hippocampus, parts of the corpus striatum, thalamus, and cerebral cortex. Thus deeper parts of the brain participate in cortical integration. The visual cortex, for example, registers a mosaic map of the visual field because each small area of the visual field projects by way of the thalamus to a small part of the primary visual cortex. Stimulating part of the visual cortex will cause a person to see only a flash of light, a color, or a line. As these impulses extend to the visual association area, they diverge widely to many cells and considerable convergence also occurs, so that a single neuron in the visual association area receives impulses from many parts of the visual cortex. "Images" of

words become synthesized in this manner in part of the visual association area. In a similar manner, stimulating part of the motor cortex may cause a finger to move, but the coordinated motor activity needed in skilled movements, such as playing the piano, requires the participation of the motor association area lying anterior to the motor cortex.

The two sides of the brain are interconnected by transverse bands of fibers known as **commissures** (L. *commissura*, seam), of which the largest is the **corpus callosum** (Fig. 17.23). This commissure makes possible a sharing of information between the neocortex on each side; it also allows for some specialization. The left hemisphere usually controls verbal symbolization, speech, logical thought, and manual dexterity. It is the dominant one in about 95% of human beings. Since motor fibers decussate, the left hemisphere controls the right hand, so most of us are right-handed. Many of these functions are lost if much of the cerebrum is injured on the left side, but musical and artistic skills, and a sense of spatial relationships remain, for they are controlled primarily by the right hemisphere.

Learning and Memory

The mechanisms of thought, learning, and memory are subjects of much current research, but we are just beginning to understand these complex processes. Two forms of memory, **short-term** and **long-term**, are recognized and appear to involve somewhat different mechanisms. If you

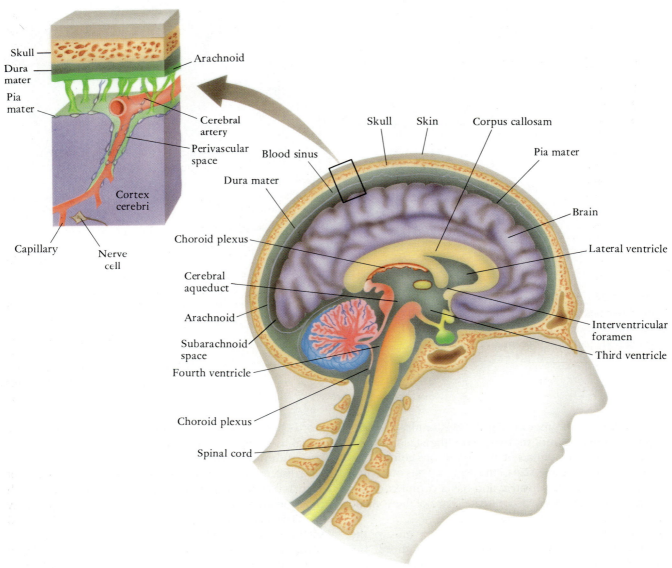

Figure 17.23
A sagittal section of the human brain showing the cerebrospinal fluid and its circulation.

wish to phone a recent acquaintance, you look up the number and store it in short-term memory long enough to place your call, but probably not much longer. If someone asks for your phone number, you have no problem recalling it from long-term memory even though you may not have had occasion to answer such a question for many months.

Short-term memory must utilize some neuronal mechanisms that can hold information for a few minutes. Probably feedback loops or a temporary facilitation of synaptic transmission occur. If someone distracts you as you are placing your phone call, the neuronal circuits involved may be interrupted and short-term memory is lost.

Continued practice and rehearsal are needed to shift memory from short-term to long-term. Long-term memory requires lasting changes that make it progressively easier for impulses to follow certain circuits. These must affect the morphology and biochemistry of the neuron and synapses, for memories are retained even after the electrical activity of the cortex has been stopped temporarily by anesthesia or cooling. Axon terminals may increase in number and size, more synapses may be established, and a semipermanent facilitation of synaptic transmission may occur. There is evidence for many changes of these sorts. Repeated passage of impulses has been shown also to increase the number of postsynaptic receptor sites. Certain patterns of neuronal activity affect genes such that the level of the mRNA that encodes for the synthesis of certain transmitter substances increases, and this change persists. The morphology and biochemistry of neurons are very plastic and they undergo changes that doubtless are the basis of long-term memory.

Memory appears to be stored in those neuronal systems that repeatedly participated in the processing and perception of the signals. These include parts of the cortex as well as deeper parts of the brain with which they connect directly or indirectly. Since most events we remember involve a combination of motor skills, and verbal, visual, and other sensations, extensive areas of the cortex become active in recalling past events. Although memories of past events can be recalled when the hippocampus is destroyed, the ability to learn new things is seriously impaired or lost. The hippocampus is essential for short-term memory and for the transfer of short-term to long-term memory.

Neurotransmitters in the Brain

Synaptic transmission occurs in the brain, as elsewhere, by the release of neurotransmitters. Acetylcholine was the first to be discovered and has been studied most thoroughly, but there are many others. Some neurotransmitters are excitatory, others inhibitory. Proper levels must be maintained or serious neurological disturbances occur. We will consider a few examples.

Acetylcholine is an excitatory neurotransmitter operating at the junctions between neurons and muscles and neurons and glands, and also in many autonomic synapses. It is a common neurotransmitter in the cerebral cortex, and in a group of neurons that travel to the cortex from nuclei deep in the cerebrum. Acetylcholine appears to be essential for our higher mental processes, including memory. The notion that acetylcholine has a role in memory is supported by the discovery that patients who have died with Alzheimer's disease, which causes memory loss, have far less acetylcholine in their cortex than other people of comparable age.

GABA (gamma-aminobutyric acid) is a neurotransmitter that has an inhibitory effect at synapses. It occurs in crustacean motor neurons that inhibit muscle contraction, and it is a common transmitter in the vertebrate brain. GABA has an effect that is very similar to that of benzodiazepine tranquilizers (e.g., Valium). Indeed, GABA and benzodiazepines reinforce each other's actions and lower the level of anxiety in patients. The more general significance of GABA is not well understood.

Norepinephrine (noradrenalin) and **serotonin** are derivatives of the amino acids tyrosine and tryptophan, respectively. The cell bodies of neurons that produce them are located in nuclei in the floor of the mesencephalon and metencephalon. Their axons terminate in many brain regions, but have a particularly high density in the limbic system, an area that affects emotional behavior. Patients suffering severe depression have inadequate levels of these neurotransmitters. Most antidepressant drugs block the metabolic breakdown of these neurotransmitters and thus help to maintain them at adequate levels.

Some patients with depression also suffer from mania, a mood disorder with symptoms opposite to those of depression, including hyperactivity, overconfidence, and sometimes delusions. The metal ion lithium has been found to control mania, so it must somehow act on the neurotransmitters associated with mood changes, but how it works has not yet been discovered.

Dopamine, another amine neurotransmitter, is particularly abundant in neurons in nuclei that are located deep in the cerebrum and have an important role in regulating movements. Low levels of dopamine cause Parkinson's disease, a disease characterized by muscle tremors and other motor disorders.

Some neurons in the brain release various peptides. The **endorphins** and **enkephalins** have actions similar to those of opium. These natural opiates suppress some pain, and they also occur in pathways that affect our emotions and moods. The presence of cell surface receptors for endorphins and enkephalins makes humans and some other vertebrates (e.g., rats) particularly sensitive to opium and its derivatives, including heroin and morphine, for these drugs attach readily to these receptors.

Brain Nutrition

Brain cells will not function normally in the presence of excess waste products of metabolism or when faced with fluctuating levels of sugars, amino acids, fatty acids, many hormones, and other substances that circulate in the blood and enter the interstitial fluid. The composition of the fluid around brain cells must be more carefully controlled than that bathing other cells of the body, because slight changes could upset the delicate balances of excitatory and inhibitory substances and bring about serious consequences. Control of the interstitial environment is effected by the nature of the capillaries in the brain and by the composition of the lymphlike **cerebrospinal fluid** that circulates around and through the CNS. The cells of brain capillaries, and those in the sites where cerebrospinal fluid is produced, are united by tight junctions that limit the passage of materials. This is often called the **blood-brain barrier**, but it is only a barrier to certain substances. Gases and many ions can diffuse through the lipid membranes of the cells, and other needed materials are actively transported.

Cerebrospinal fluid is produced in parts of the **ventricles** within the brain. Each of the two lateral ventricles of mammals lies in the cerebral hemispheres, the third in the diencephalon, and the fourth in the metencephalon and myelencephalon (Fig. 17.23). All are interconnected. One wall of each ventricle is very thin, for it consists of only the **ependymal cells** that line the CNS and a layer of vascular connective tissue that covers the brain's surface. Vascular tufts, known as **choroid plexuses**, protrude from these thin walls into the ventricles, where they secrete the cerebrospinal fluid.

Protective and nutritive layers of connective tissue cover the surface of the spinal cord and brain. A dense **dura mater** (L. *dura*, hard + *mater*, mother) forms the outermost layer (Fig. 17.24). In mammals, a deeper **arachnoid** (Gr. *arachnoeides*, cobweb-like) is connected by weblike strands of connective tissue to a very vascular inner **pia mater** (L. *pia*, soft), which intimately invests the surface of the brain and cord. Cerebrospinal fluid escapes through pores in the roof of the medulla oblongata and circulates slowly in a **subarachnoid space** between the pia mater and arachnoid. It enters the interstitial spaces of the brain through the perivascular spaces that follow the blood vessels entering the brain from the pia mater. The cerebrospinal fluid eventually reenters the circulatory system by way of vascular tufts called **arachnoid villi** that protrude into large veins in the dura mater. In addition to its nutritive role, the cerebrospinal fluid forms a protective liquid cushion around the CNS that buffers it from external blows and shocks.

Summary

1. Neurons, of which there are many types, are the structural and functional units of the nervous system. Schwann cells wrap around many neurons, forming a myelin sheath that allows for a rapid transmission of nerve impulses.

2. An unequal distribution of Na^+ and K^+ ions across the plasma membrane of a neuron results in a resting potential in which the outside of the membrane is positive relative to the inside. The resting potential is momentarily reversed when a neuron receives a threshold stimulus and a nerve impulse is initiated. The impulse is a self-propagating wave of membrane depolarization that spreads along the axon. The nerve impulse is an all-or-none phenomenon whose magnitude does not vary with the strength of the stimulus. All nerve impulses are also qualitatively alike. The ability to discriminate between different types of stimuli is a function of the receptors and neuron connections within the brain.

3. Impulses are usually transmitted across the synapses between neurons by the release of chemical neurotransmitters. Synapses are the control switches of the nervous system. They normally permit impulses to cross in only one direction. Some are excitatory and others are inhibitory.

4. Nervous systems are composed of only three functional types of neurons. Sensory neurons carry impulses from peripheral receptors to the central nervous system. They synapse within the CNS with interneurons that integrate many different sensory stimuli and initiate appropriate responses. Motor neurons carry these signals out of the central nervous system to the effectors. Some sensory neurons, interneurons, and motor neurons are organized into simple reflex arcs that allow the organism to make quick, stereotyped responses to certain stimuli. Other neurons have more complex interconnections that allow for the spreading of an incoming impulse to many regions, a continuation of the impulses after the initial stimulus has stopped, and the convergence of many signals upon a single center.

5. Sponges lack a nervous system, but other simple, radially symmetrical metazoans have a diffuse nerve net that mediates their limited array of responses. As animals became more active, neuron numbers and interconnections increased and new abilities emerged. In bilaterally symmetrical animals, the interneurons are aggregated into a central nervous system consisting of an anterior brain and one or more longitudinal nerve cords through which impulses can be transmitted rapidly from one region to another. Integration takes place within the central nervous system.

6. Vertebrate spinal nerves are segmented and carry both sensory and motor neurons. Sensory neurons enter the spinal cord through the dorsal roots of spinal nerves, have their cell bodies in a spinal ganglion, and terminate in the dorsal part of the gray matter of the cord. Motor neurons have their cell bodies in the ventral part of the gray matter and, in mammals, leave through the ventral roots of spinal nerves. Most cranial nerves are comparable to a dorsal or ventral root of a spinal nerve.

7. The autonomic nervous system consists of a special group of visceral motor neurons supplying cardiac muscles, smooth muscles, and glands. It is divisible into sympathetic and parasympathetic systems that have antagonistic effects upon the organs they supply. The sympathetic system promotes activities that expend energy and help an animal adapt to stress. Parasympathetic stimulation promotes activities that produce and store energy.

8. The vertebrate brain develops as an expansion of the cranial end of the embryonic neural tube. As it expands, five distinct regions differentiate and give rise to the parts of the adult brain: telencephalon, diencephalon, mesencephalon, metencephalon, and myelencephalon.

9. The cerebral hemispheres act primarily as olfactory centers in primitive vertebrates. During the course of evolution the cerebral hemispheres expand, increase in importance, and become the primary integration center of the brain. Many visceral functions, including swallowing, respiratory rate, and rate of heart beat, are controlled in the medulla oblongata. The cerebellum is a center for motor coordination in all vertebrates. It is very large in birds and mammals, for locomotor and other muscular activities have become very complex. The optic lobes are major integration centers for sight and many other senses in primitive vertebrates, but the enlargement of the cerebrum in mammals usurps many of these functions. However, the optic lobes remain centers for optic and auditory reflexes in mammals. The thalamus also expands greatly, for most sensory impulses going to the cerebrum relay here, as do many motor impulses leaving the cerebrum. The hypothalamus is the primary center for visceral and homeostatic control in all vertebrates.

10. Much of the human cortex forms association areas that interact with the primary sensory and motor areas in integrating different sensory stimuli, interpreting their significance, and initiating complex motor activities. Short-term memory appears to reside in temporary neuronal circuits. Long term memory requires lasting morphological and biochemical changes in the neurons and their synapses that make it progressively easier for nerve impulses to follow certain circuits.

11. Acetylcholine, GABA, norepinephrine, serotonin, dopamine, and certain peptides are examples of neurotransmitters in the brain. Some are excitatory, others are inhibitory. Disruption of the balances among them is thought to be responsible for certain types of memory loss and many mental disorders.

12. The composition of the liquids that bathe the brain cells is very carefully regulated because these cells do not function normally in the presence of certain metabolites and fluctuating levels of many substances in the blood. A blood-brain barrier limits the passage of certain materials from the blood. A lymphlike cerebrospinal fluid circulates in the ventricles of the brain and the subarachnoid space of the brain and spinal cord, protecting and providing nutrition to the cells of the CNS.

References and Selected Readings

Aoki, C., and P. Siekevitz. Plasticity in brain development. *Scientific American* 259(Dec. 1988):56–64. A review of the role of early experience in the development of neuronal connections in the optic system of the mammalian brain.

Bullock, T.H., R. Orkand, and A. Grinnell. *Introduction to Nervous Systems.* San Francisco: W.H. Freeman & Co., 1977. An excellent source book on both invertebrate and vertebrate nervous systems.

Goldstein, G.W., and A.L. Betz. The blood-brain barrier. *Scientific American* 255(Sept. 1986):74–83. A discussion of how brain capillaries act as gatekeepers.

Gould, J.L., and P. Marler. Learning by instinct. *Scientific American* 256(Jan. 1987):74–85. The distinction between instinct and learning is not as sharp as sometimes believed, and instinct often plays a role in learning.

Kuffler, S.W., J.G. Nicholls, and A.R. Martin. *From Neuron to Brain,* 2nd ed. Sunderland, Mass.: Sinauer Associates, Inc., 1984. An excellent advanced textbook on neurobiology. The discussions of cellular structure, biochemistry, and physiology are especially good.

Montgomery, G. The mind in motion. *Discover* (Mar. 1989):58–68. Positron emission tomography studies have shown the parts of the brain that are used in various aspects of language.

Nauta, W.J.H., and M. Feirtag. *Fundamental Neuroanatomy.* New York: W.H. Freeman & Co., 1986. A good introduction to the nervous system. Includes a discussion of the nervous system of primitive animals.

Sherrington, C.S. *Integrative Action of the Nervous System.* New Haven: Yale University Press, 1948. A true classic in neurobiology, written by a pioneer in the field.

Snyder, S.H. *Drugs and the Brain.* New York: Scientific American Library, Scientific American Books, Inc., 1986. A review of how drugs have been used as probes to help us understand brain functions and how psychoactive drugs are used in therapy to modify behavior.

———The molecular basis of communication between cells. *Scientific American* 253(Oct. 1985):132–141. A discussion of the molecular interrelationships between transmission of information in the nervous and endocrine systems.

Thompson, R.F. The neurobiology of learning and memory. *Science* 253(1986):941–947. The neuronal, biochemical, and biophysical substrates of memory are beginning to be understood in both invertebrates and vertebrates.

18

Receptors and Sense Organs

(Left) Close-up of a lion fish (Dendrochirus zebra). (Above) The squirrel fish (Holoceptrus rubrum) of the Red Sea. Most animals have well-developed sense organs enabling them to detect changes in their environment.

The ability of animals to respond in appropriate ways to important changes in their external and internal world is a prerequisite for survival. A grazing antelope can continue to graze with no particular behavioral or physiological adjustments, even with a pride of lions resting nearby. It may ignore a lion rolling over or stretching, but it must sense any change in a lion's behavior that suggests danger. If a lion attacks, the antelope must respond instantly. As it runs away, internal physiological adjustments must occur that provide the muscles with more energy, increasing the rates of breathing and circulation, and releasing sugar reserves from the liver.

Receptors detect significant changes in the external and internal environments. These receptors are biological transducers with the remarkable ability to detect minute changes in light, sound, or smell, transforming these changes into electrical nerve impulses. The integration of nerve impulses generated by sight, sound, and smell occurs within the central nervous system, and it is the central nervous system that determines whether or not a response to changing sensory information need be made, as well as the nature of that response.

Animals have evolved a wide assortment of receptors, each attuned to a particular type of energy. The basic physical characteristics of a given form of energy are the same, regardless of the organism receiving the stimulus. For example, the physical characteristics of light rays constrain the possible pathways evolution can follow in the generation of a functional eye. A variety of image-forming eyes have evolved that all, in some way, focus incoming light rays upon the photoreceptive cells. But beyond this, the detailed structure of eyes varies greatly because they have evolved independently in different lineages.

RECEPTOR MECHANISMS

Many receptors are composed of the dendrites of sensory neurons that end freely on the surface of the body or in its tissues. These sensory neurons combine reception and nerve impulse transmission. Their endings often are specialized and are located where they can be stimulated directly by specific types of stimuli. Many of those ending in vertebrate skin, for example, alert us to chemical irritants, burns, and cuts. We sense the stimuli as pain. Other receptors are specialized **receptor cells** that receive stimuli and initiate an impulse in a sensory neuron.

To be effective, receptors must be highly selective and discriminate between different **modalities** (L. *modus*, mode) or types of energy. Little useful information about environmental changes would be provided if a particular receptor could be activated by stimuli of equal intensity from a change in temperature and a change in light. The animal would be unable to distinguish between a fall in temperature or the approach of a predator. Each receptor

is thus particularly sensitive to one modality. **Chemoreceptors**, such as those for smell and taste, detect chemical changes. **Mechanoreceptors** detect changes in tension on a cell surface as occur in touch, hearing, and balance. **Photoreceptors** are sensitive to changes in light; **thermoreceptors**, to changes in temperature.

Receptors may also be classified by their location within the body. **Exteroceptors** occurring on or near the surface detect changes in the external environment. **Enteroceptors** deep in the body monitor changes in blood pressure, blood pH, the stretch of the lungs, and other qualities necessary to maintain homeostasis. **Proprioceptors** (L. *proprius*, one's own + *capere*, to take), located in muscles and tendons, monitor the extent and rates of change of muscle contraction.

Many receptor cells are scattered throughout the body, but others are aggregated with associated tissues to form complex **sense organs** such as the nose, ear, and eye. The associated tissues not only support and protect the delicate receptor cells, but also help to gather the energy being transduced, amplify it, and carry it to the receptors. Only the rods and cones in the retina of our eye are the actual receptors; other parts of the eye serve primarily to gather light and focus it on the retina.

Reception and Transduction of Stimuli

Receptors, whether they be neurons or specialized cells, have an electric potential across their plasma membrane that, as in a nerve cell, derives from the unequal distribution of sodium and potassium ions between the inside and the outside of the cell (see Figs. 17.5 and 17.6). Most receptors are **phasic** and maintain a resting potential when they are not stimulated. When such a receptor receives energy of the type to which it is attuned, ions are redistributed across the membrane, depolarizing it and allowing a **receptor potential** to develop. Just how a particular form of energy initiates the receptor potential is not entirely clear. A change in tension on the plasma membrane of mechanoreceptors apparently opens **stretch-activated ion channels**, and the exchange of ions through them initiates the opening of the voltage-gated sodium and potassium gates. (Voltage-gated channels also open in the generation of a nerve impulse, p. 376.) A receptor potential develops and spreads over the receptor membrane. On reaching the axon or synapse-like receptor-neuron junction, it may initiate a nerve impulse. Other phasic receptors presumably act in a similar way, although different forms of energy (certain chemicals, light, and temperature) initiate the opening of the channels. Proprioceptors are **tonic** receptors that generate receptor potentials continuously. Stimulation causes an increase or decrease in the magnitude of the receptor potential.

To be effective, receptors must exhibit a sensitivity and specificity appropriate to the type of energy to which they are attuned and not respond to other types of stimuli.

They must display a range of sensitivity adapted to the mode of life of each species. While our ears do not detect frequencies higher than about 20,000 cycles per second, bats can detect frequencies at least as high as 100,000 cycles per second. Receptors also must be sufficiently sensitive to respond to significant environmental changes, but they must not be so sensitive that they detect irrelevant information or generate background "noise" that interferes with the reception of a useful signal.

Sensory Coding and Sensation

The intensity or strength of a stimulus received by a sense organ is coded by the number of receptors activated and by the magnitude of the receptor potential in each one. The receptor potential, like the postsynaptic nerve potential, is graded. If it reaches the necessary threshold, a nerve impulse will be generated. If it goes above this, multiple nerve impulses will ensue (Fig. 18.1). Thus the frequency of impulses transmitted over a sensory neuron is directly proportional to the strength of the stimulus at the receptor.

Although receptors discriminate between different modalities and intensities, it should be emphasized that the receptor potential, like the nerve impulse, is an electrical phenomenon that does not vary qualitatively. Our perception of different qualities of sensation depends on the area of the brain to which the receptors project. Although normally stimulated by small light changes, the rods and cones in the eye may be activated by a sharp blow to the head, giving rise to the sensation of "seeing stars." Perception of subtle differences within a modality some-times depends on an assortment of qualitatively different receptor cells. Rods in the vertebrate retina are sensitive to weak light. Cones require stronger light, and different assemblages of vertebrate cones are attuned to the three primary colors of the spectrum (p. 429).

Perception of different smells and tastes is more complex because we can detect more qualities than there are types of receptors. There may be only a half-dozen different types of receptors in the nose, but hundreds of odors can be distinguished. Each receptor appears to be sensitive to a spectrum of odors, but they are not equally sensitive to each one. Moreover, sensory overlap occurs between receptors. A particular receptor may be very sensitive to a pungent odor, slightly sensitive to a musky one, and insensitive to a putrid one. Another receptor may be very sensitive to the musky odor, slightly sensitive to a putrid one, and insensitive to the pungent one. The sensation perceived depends not on a single receptor, but on the total pattern and degree of receptor activation and their projections to the brain. This phenomenon is called **cross-fiber patterning**.

Physiological Adaptation

The sensitivity to the continued presence of the same stimulus quickly decreases in most phasic receptors (Fig. 18.2). This phenomenon is known as **physiological adaptation**. We sense certain odors on first entering the kitchen, but lose our sensitivity to them after remaining there for a while. Touch receptors are also capable of physiological adaptation. We feel a pair of glasses as we put them on, but

Figure 18.1

Sensory coding. Stimuli below the threshold level may cause a weak generator potential but no nerve impulses. The magnitude of receptor potentials for stimuli at or above the threshold increases with the strength of the stimulus, and is translated into an increasing frequency of action potentials in the neuron.

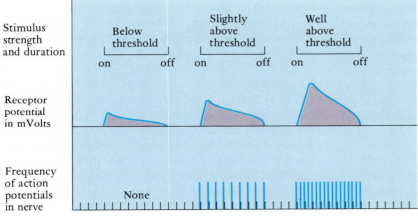

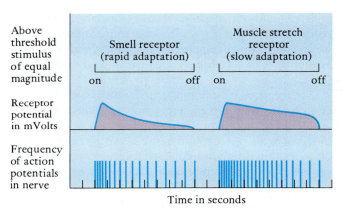

Above threshold stimulus of equal magnitude

Smell receptor (rapid adaptation)

Muscle stretch receptor (slow adaptation)

on   off   on   off

Receptor potential in mVolts

Frequency of action potentials in nerve

Time in seconds

Figure 18.2
Physiological adaptation in receptors. In the continued presence of a stimulus above the threshold, most phasic receptors (such as receptors for smell) adapt quickly. The initially high generator potential and frequency of nerve impulses fall off rapidly. Tonic receptors (such as proprioceptors) adapt very slowly and show little decay in their generator potential and frequency of nerve impulses. The responses of many other receptors fall between these examples.

do not continue to feel them after wearing them for a short time. No useful information would be provided by continuing to feel them. The majority of receptors are tuned to detect changes or contrasts in the environment. A new smell may mean the presence nearby of food or danger, but once detected, the continued sensing of this might obscure other important cues. Physiological adaptation facilitates detecting changes, for it "tunes out" a stimulus to which we have responded and increases our alertness to new stimuli. It also reduces the amount of unimportant information that reaches the central nervous system.

The degree of physiological adaptation varies among receptors. Tonic receptors adapt more slowly than most phasic ones. The proprioceptors in our muscles, for example, must constantly monitor changes in the tension and rate of contraction of skeletal muscles or else coordination of motor activity would not be possible. Proprioceptors are active all of the time, generating a base level of nerve impulses. A change in tension is detected quickly and translated into an increase or decrease in the number of impulses generated. Pain receptors, although phasic, adapt very slowly. It is important for an animal to continue to sense pain until its cause is identified and possibly eliminated.

CHEMORECEPTORS

Many chemicals affect the plasma membrane, triggering the opening and closing of ion channels, and influencing the transmembrane flow of ions. This basic responsiveness

of cells to chemical changes has made possible the evolution of a wide variety of chemoreceptive mechanisms. Small groups of chemoreceptive cells that monitor changes in the oxygen and carbon dioxide level of the blood of vertebrates, and the pH of their cerebrospinal fluid, initiate reflexes that control the rates of blood flow and breathing. Many animals find food and mates, sense prey, communicate with one another, and even navigate through their environment using their chemical senses of smell and taste. Smell is the dominant sense for many species. Humans sometimes find this hard to appreciate because our own sense of smell is so rudimentary.

We will examine more closely the chemoreceptors that Metazoa use to detect odors and tastes. Smell is usually regarded as the detection of particles from objects at a distance, and taste as the detection of molecules from objects in contact with some part of the body. But smell and taste have much in common, and the distinction between them is not always clear, especially in aquatic species.

Olfactory Receptors

Odors are detected by **olfactory receptors** (L. *olfactus*, smell). These are usually specialized bipolar sensory neurons whose cell bodies lie in some part of the surface epithelium and whose dendrites bear one or more modified cilia that greatly increase the receptive surface. In arthropods, in which the exoskeleton is not itself sensitive, nonmotile olfactory cilia lie in perforated pits or perforated extensions of the exoskeleton that have the form of bristles called setae (L. *seta*, bristle) or clubs (Fig. 18.3a). Structures of this sort that detect a variety of stimuli, other than light, are called **sensilla** (L. *sensus*, sense). The olfactory sensilla of insects are usually located on the antennae, which sometimes have so many branches that they resemble a feather (Fig. 18.3b). The males of moths and many other insects find the females by following odors the females produce. The system is so sensitive that males downwind can find females several kilometers away. Females often use their sense of smell to search out appropriate plants on which to lay their eggs.

Air moving through the nose of terrestrial vertebrates carries in odor-bearing molecules with the air going to the lungs. The olfactory receptors in a mammal's nose are located in an epithelium in the upper part of each nasal cavity and are interspersed with supporting and basal cells (Fig. 18.4). The latter differentiate into new olfactory cells. The delicate olfactory cilia are exposed to inhaled noxious agents, and a continuous turnover of the olfactory cells replaces those damaged by these materials. Multicellular glands secrete mucus that spreads over the olfactory cilia, keeping them moist, entrapping odor-bearing molecules, and possibly playing a role in the ionic exchange that occurs during the development of a receptor potential. The axons of the receptive cells aggregate into many

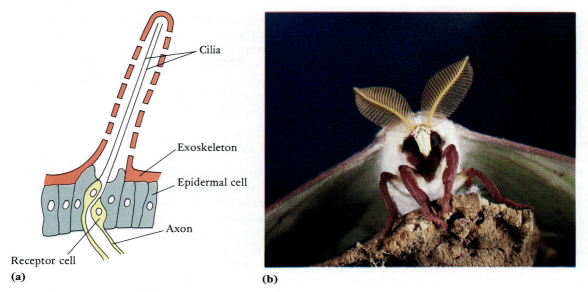

(a)

- Cilia
- Exoskeleton
- Epidermal cell
- Axon
- Receptor cell

(b)

Figure 18.3
Olfactory sensilla of an insect. (a) Diagram of an individual sensillum containing two cilia. (b) The antennae of a lunar moth. Minute sensilla are borne on the filaments.

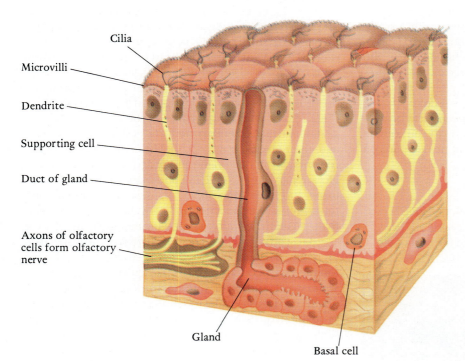

- Cilia
- Microvilli
- Dendrite
- Supporting cell
- Duct of gland
- Axons of olfactory cells form olfactory nerve
- Gland
- Basal cell

Figure 18.4
The mammalian olfactory epithelium.

bundles that enter the olfactory lobes of the brain. Collectively these bundles form the **olfactory nerve**, cranial nerve I (p. 388).

The plasma membranes of the olfactory cilia are densely packed with large receptor molecules that appear to be the sites to which odor-bearing molecules bind, initiating the flow of ions that leads to the receptor potential. Most odors are considered to be mixtures of primary olfactory stimulants, such as "camphoraceous," "musky," "floral," "pungent," and "putrid." Some investigators recognize a half-dozen primary stimulants; others, 50 or more. The number of different types of olfactory cells is unknown, but certainly there are far fewer than the number of odors that can be detected. As noted earlier, discrimination between odors appears to involve cross-fiber patterning (p. 403).

Most vertebrates have a very keen sense of smell that provides them with considerable information about their surroundings and neighbors. Young salmon, for example, become sensitized to the odor of the specific stream in which they hatched. Odors of streams vary because of differences in the surrounding vegetation and soil. After young salmon have spent many years in the ocean and return to a river system to spawn, they find the tributary where they were hatched by its distinctive odor. This has been demonstrated by capturing marked migrating salmon in their home stream and returning them near the river mouth. Those in which the olfactory sacs have been plugged redistribute themselves at random among the upper tributaries. Most of those with unplugged sacs return to their home stream.

In most terrestrial vertebrates, patches of olfactory epithelium, known as the **vomeronasal (Jacobson's) organ**, are set off from the rest of the nasal cavities. Often they connect with the mouth cavity through a pair of passages in the vomer bones located in the roof of the mouth. A snake darting its forked tongue is collecting odor-bearing molecules that are transferred to the entrances of the vomeronasal organs (Fig. 18.5). Many snakes, whose eyesight is poor, track prey in this way.

Many vertebrates and other animals secrete specific substances known as **pheromones** (Gr. *pherein*, to bear + *hormaein*, to excite). Some are carried by the air; others are deposited on the substratum. Pheromones are part of a

chemical communication system and serve as signals to other members of the same species. They are used to mark trails and territories, warn conspecifics of danger, indicate social status in a hierarchy, and signal sexual readiness. It is a pheromone secreted by female insects that attracts the males. Terrestrial vertebrates often detect pheromones through their vomeronasal organs.

Gustatory Receptors

Taste is an important sense that enables animals to find and sample food, and it often plays a role in courtship behavior. Tastes are detected by **gustatory receptors** (L. *gustare*, to taste). Gustatory sensilla of arthropods are setae and pegs located on their mouthparts, feet, and other regions likely to contact this type of stimulus. Some of these creatures literally taste with their feet. The sensilla have minute pores in their surface and usually contain several ciliary extensions of dendrites. Each sensillum on the mouthparts of the blow fly contains ciliary processes of five neurons: one is sensitive to sugars, one to anions, one to cations, one to water, and the last is tactile. If a thirsty fly senses water, it drinks; if it senses sugars, it feeds. Salty solutions are rejected.

Vertebrate gustatory receptors are specialized receptor cells that are aggregated with supporting and replacement cells in small barrel-shaped clusters called **taste buds** (Fig. 18.6a). A small pore leads to the body surface. Taste buds of terrestrial vertebrates are distributed through the oral cavity and pharynx, and are often especially abundant on the tongue, where they are located in pits or on papillae (Fig. 18.6b). The whisker-like barbels of a catfish are gustatory organs. Sensory neurons end on the taste buds, and their axons travel to the brain in cranial nerves returning from the tongue and pharynx: facial (nerve VII), glossopharyngeal (nerve IX), and vagus (nerve X). In many fishes and larval amphibians, the taste buds spread onto the body surface and are supplied by the facial nerve rather than by cutaneous nerves supplying the skin in these areas.

Gustatory receptors respond to a narrower spectrum of materials than olfactory ones. Areas of the human tongue have been shown to be particularly sensitive to salt, sour, sweet, and bitter substances (Fig, 18.7), but you do not have to be a gourmet to sense more tastes than this. It is probable that detecting many substances involves cross-fiber patterning. Our final perception of food depends on interactions among taste, olfaction, and tactile stimuli, and it is not easy to sort out the contributions of each. For example, we do not "taste" food well when we have a bad cold and cannot also smell it.

MECHANORECEPTORS

Although the plasma membrane can detect mechanical deformations directly, the detection of mechanical disturbances is enhanced by the evolution of cells with

Figure 18.5
A snake uses its forked tongue to collect odor-bearing molecules that are then transferred to the paired vomeronasal organs, whose entrances lie in the roof of the mouth.

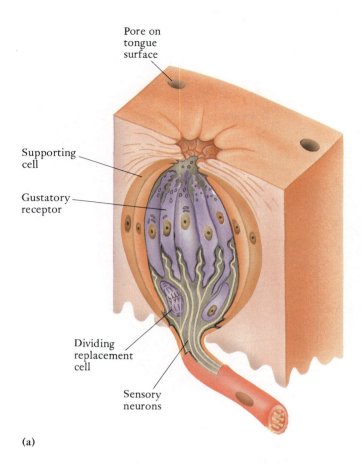

Pore on tongue surface

Supporting cell

Gustatory receptor

Dividing replacement cell

Sensory neurons

(a)

nonmotile cilia, microvilli, setae, encapsulated neuron endings, and other devices. These respond to touch, pressure, the pull of gravity, vibrations in the medium, and other mechanical stimuli.

Touch and Pressure

Like the chemical senses, the ability to detect and respond to tactile stimuli is universal. We gain a great deal of information from touching and handling objects. The **tactile sensilla** on the surface of the body and appendages of arthropods are good examples of touch receptors (Fig. 18.8a). Each sensillum is a thin, chitinous filament that is secreted by a specialized cell and articulates by a thin joint membrane with the exoskeleton in such a way that it can be moved easily. Movements at the base of the filament change tension on the dendrite of an adjacent neuron, and a receptor potential develops.

Vertebrate skin contains a variety of receptors that detect touch and pressure changes (Fig. 18.8b). Some dendrites wrap around the base of a hair follicle in mammal skin and are activated by the movements of the hair, which forms a small lever arm. Other dendrites in skin end in encapsulating layers of connective tissue and are particularly sensitive to a light touch. **Meissner's corpuscles,**

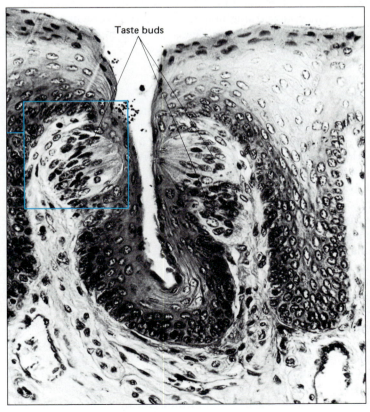

Taste buds

(b)

Figure 18.6

Mammalian taste buds. (a) An individual bud in the epithelium. (b) Many taste buds, one of which is outlined by the blue box, lie in grooves between tongue papillae, as seen in this photomicrograph.

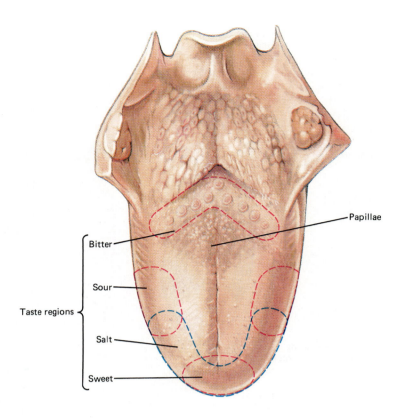

Figure 18.7
Sensory areas on the surface of the human tongue.

Labels on figure: Papillae, Bitter, Sour, Taste regions, Salt, Sweet

Merkel's discs, and **Ruffini's end organs** are examples. Deeper pressure on the skin is detected by **Pacinian corpuscles** that consist of many layers of fibroblasts and connective fibers arranged around a neuron terminal (Fig. 18.8c). Pacinian corpuscles also are found around joints and in many visceral organs.

Motion Detectors

An extension of the tactile sense allows some animals to detect movements in the water and air around themselves. This ability ranges from sensing gross movements of the medium to detecting vibrations and sound waves.

Hairlike sensilla of many arthropods are delicate enough to detect water and air currents. The long hairs, or **vibrissae**, on the snout of rats and many other mammals also can detect air movements. Fishes and larval amphibians have evolved a **lateral line system** that enables them to detect disturbances in the water (Fig. 18.9). Groups of receptor cells, called **neuromasts**, lie in water-filled canals in the skin that connect to the surface by pores. One canal extends along the flank (hence the term lateral line); others ramify over the head. Receptor cells in a neuromast are **hair cells**, each of which bears a single long, nonmotile cilium and a few microvilli that extend into an overlying, gelatinous cap or **cupula**. The hair cells are tonic receptors that generate a constant base rate of nerve impulses. Water movements that bend the cupula toward the longest process increase the receptor potential and the frequency of nerve impulses; movement in the opposite direction reduces the potential and the number of impulses. This

enables a fish to detect the direction of water movement. The system, which has been described as "distant touch," allows a fish to sense currents, pressure changes as it moves toward an obstacle, and low-frequency water vibrations. It is lost in terrestrial vertebrates, including metamorphosed amphibians. Specialized lateral line receptors in some fishes can detect changes in the electric field around the fish (Focus 18.1).

Gravitational Detectors

If parts of a motion detection system become isolated from the surrounding medium and modified slightly, they can detect changes in the direction of the pull of gravity. This enables motile animals to have a sense of their orientation in space and changes in their movements through the surrounding medium. Many groups of animals have independently evolved receptors that perform these functions.

Statocysts

Statocysts (Gr. *statos*, standing + *kystis*, bladder) occur in most active invertebrate groups including cnidarians, flatworms, molluscs, and arthropods. Although varying in number, location, and other ways, each is essentially a small, hollow sphere containing receptor cells bearing nonmotile cilia (Fig. 18.10a). One or more relatively heavy, calcareous **statoliths** (Gr. *lithos*, stone), which are secreted by certain of the cells, occupy the center of the sphere. The pull of gravity on a statolith stimulates the hair cells beneath it, thus indicating the body's current orientation, or state of **static equilibrium**. During body movements or

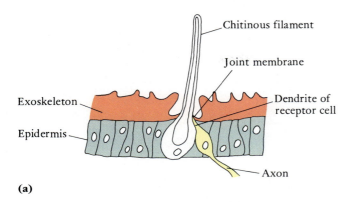

(a)

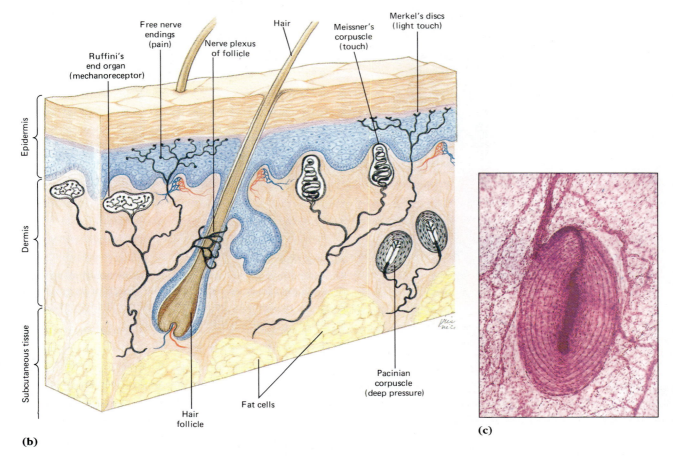

(b)

(c)

Figure 18.8

Touch and pressure receptors. (a) A tactile sensillum of an insect. (b) Receptors in human skin. (c) A Pacinian corpuscle.

changes in position, the weight of the statocyst decreases on certain cells and increases on others.

The statocysts of decapod crustaceans, such as crayfish and lobsters, are located at the base of each first antenna (Fig. 18.10b). They are lined with chitin and open to the surface. Sand grains gathered from the outside form the statolith. The chitinous lining, being part of the exo-

skeleton, is shed at each molt, so the statolith must be replaced. If a newly molted animal is provided with iron filings rather than sand, they form the statolith, and the animal will orient itself to a magnetic field!

Behavioral studies have shown that some vertebrates and other animals have a **magnetic sense** and may be able to detect the earth's magnetic field. Crystals of magnetite

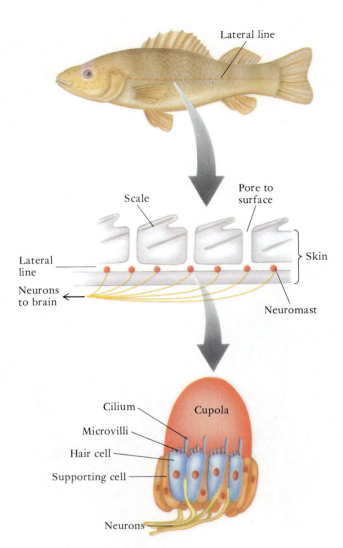

Lateral line

Scale

Pore to surface

Lateral line

Neurons to brain

Skin

Neuromast

Cilium

Microvilli

Hair cell

Supporting cell

Cupola

Neurons

Figure 18.9
The lateral line system of a fish. One canal extends along the flank; others ramify over the head. The groups of receptor cells, known as neuromasts, lie in water-filled canals in the skin that connect to the surface through pores.

Figure 18.10
Gravitation and equilibrium receptors. (a) A statocyst of a mollusc. (b) The statocyst in the antennal base of a crustacean.

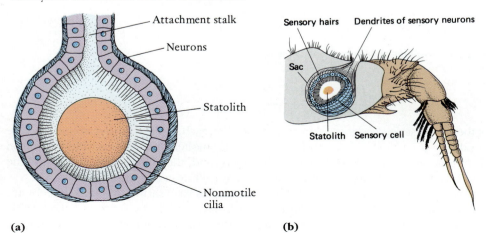

Attachment stalk

Neurons

Statolith

Nonmotile cilia

Sensory hairs

Dendrites of sensory neurons

Sac

Statolith   Sensory cell

(a)

(b)

Focus 18.1   *Electroreception by a Shark*

Sharks and many other primitive fishes have a passive electroreceptive system that enables them to detect and home in on the weak electric current produced by the activity of respiratory and other muscles in a prey organism. That electrical clues are more important to a shark than are olfactory ones was demonstrated convincingly in a series of experiments shown in this Focus figure. During normal feeding, a shark easily finds a flatfish (a plaice) buried in sand and attacks it (a). Either olfactory or electrical clues could have been used. When the plaice is buried within an agar chamber, which would stop olfactory clues, the shark attacks the chamber rather than homing in on odors leaving in a current flowing through the chamber (b). When the plaice is chopped up so electrical activity ceases and then buried in the same way, the shark homes in on the odors leaving the chamber (c). The shark ignores a chamber containing a plaice when it is electrically shielded and no water current flows through it (d). A shark will home in on electrodes that replicate the electrical activity of a plaice (e). Given a choice of a piece of plaice on the surface of the sand or an electrode replicating a plaice's electrical activity, the shark attacks the electrode (f).

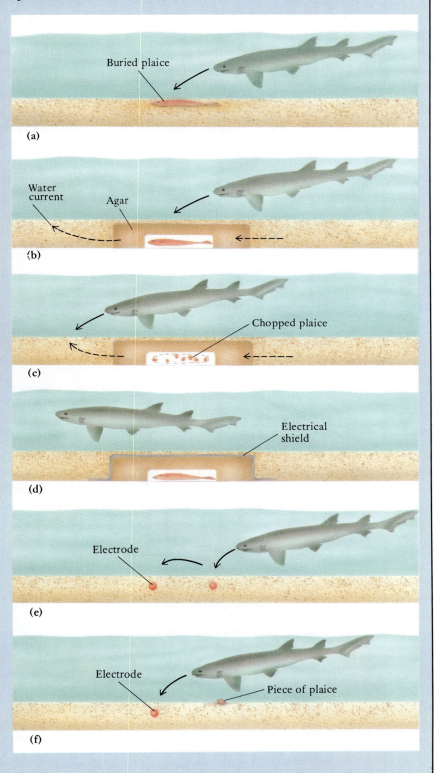

(a) Buried plaice

(b) Water current — Agar

(c) Chopped plaice

(d) Electrical shield

(e) Electrode

(f) Electrode — Piece of plaice

do orient in a magnetic field, and they have been found in some cells. Whether or not they act in a way analogous to iron filings that can occupy a lobster's statocyst is unknown. We will return to this puzzling phenomenon in Chapter 39 (p. 897).

Membranous Labyrinth

We usually think of the vertebrate ear as an organ of hearing, but much of it acts as a statocyst, for it receives stimuli that enable vertebrates to monitor their orientation and movements. The receptive cells of the ear resemble neuromasts of the lateral line, but they are located deep within the skull in an inner ear, or **membranous labyrinth**

(Fig. 18.11). Many zoologists believe that this part of the ear evolved when a section of the lateral line system folded inward and became isolated from external disturbances.

The parts of the membranous labyrinth related to equilibrium are remarkably constant among all vertebrates. Most species have three **semicircular ducts** attached to a chamber called the **utriculus** (L., small bag). A second chamber, the **sacculus** (L., small sac), lies beneath the utriculus. Patches of hair cells in the utriculus and sacculus form white spots called **maculae** (L., spot) that are overlaid by calcareous secretions similar to the statoliths of invertebrates, but they are called **otoliths** (Gr. *otikos*, pertaining to ear) in vertebrates. The one in the sacculus is

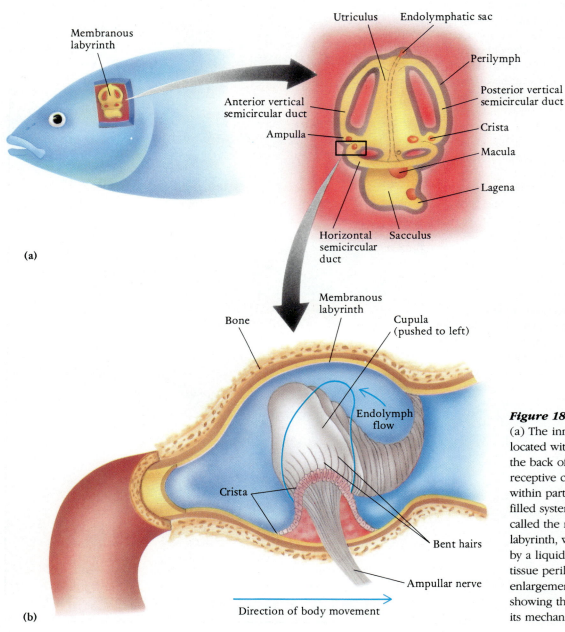

(a)

(b)

Figure 18.11

(a) The inner ear of a fish is located within the otic bones at the back of the skull. Its receptive cells are contained within parts of an endolymph-filled system of sacs and ducts called the membranous labyrinth, which is surrounded by a liquid and connective tissue perilymph. (b) An enlargement of one ampulla showing the receptive crista and its mechanism of action.

particularly large and important. Hair cells in each semicircular duct form a small cluster known as a **crista** in the enlarged **ampulla** at one end of the duct. These cells are capped by a gelatinous **cupula** that partly occludes the passage between the duct and utriculus.

Static equilibrium is detected, as it is in a statocyst, by the pull of gravity on the calcareous secretions. Movement, or **linear acceleration**, is also detected by the otoliths. If, for example, a vertebrate starts to move forward, the ear moves forward with the rest of the body. But the relatively heavy otoliths, which can move in the liquid endolymph, have more inertia and their movements lag. As they push back upon the hair cells, a change in acceleration is detected. The cristae in the semicircular ducts detect turns of the head, or **angular acceleration**, because the ducts lie at right angles to each other in the horizontal and two vertical planes of the body (Fig. 18.11). If a vertebrate turns in the horizontal plane to one side, the cristae in the vertical canals are not affected. The one in the horizontal canal moves at the same rate as the turn of the body, but the endolymph in the narrow duct lags and pushes upon the crista.

Sound and Its Detection

Sensing sound requires detecting vibrations, or waves of alternating pressure, with wave lengths ranging upward from about 20 cycles per second or Hertz (Hz) (Fig 18.12).

Lower frequencies are usually felt as vibrations and not perceived as sound. Frequency determines the sound's pitch, that is, whether the notes are low or high; loudness is determined by the pressure of the sound waves and is proportional to their height or amplitude. Only a few animals—certain arthropods and most vertebrates—can "hear."

Many ambient sounds provide important information to animals that can hear them: the roar of a lion or the snap of a twig may signal danger. Some animals also produce sounds that inform other members of the species about the location of food, predators, and mates. The calls of different species are quite distinct. Toothed whales, insect-eating bats, and some cave-dwelling birds emit very-high-frequency sounds and use the returning echoes to navigate and find food.

Sounds are produced in many ways. Crickets generate sounds by rubbing rough patches on their wings together, a method termed **stridulation**. Some fishes stridulate with parts of their fins. Cicadas produce sounds by the rapid muscular vibration of membranes on the surface of the abdomen. Many terrestrial vertebrates have vocal cords in their respiratory passages that vibrate as air crosses them.

Delicate, chitinous hairs on the antennae of many insects form sensilla that respond to sound waves. A male mosquito finds a female this way by detecting the high-frequency sounds generated by her beating wings, the same sound you hear before you swat. Cicadas, grass-

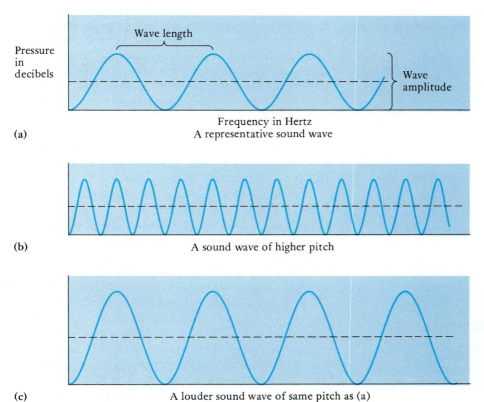

(a) A representative sound wave

(b) A sound wave of higher pitch

(c) A louder sound wave of same pitch as (a)

Figure 18.12
Properties of sound waves. (a) A sound wave has a characteristic wavelength and amplitude. (b) A sound wave of higher frequency (pitch) than that shown in (a), has shorter wave lengths but the amplitude (loudness) is the same. (c) A sound wave of the same pitch as that shown in (a) but with a higher amplitude, so it is louder.

hoppers, and some moths have evolved **tympanic ears** that consist of a delicate surface membrane stretched across an air-filled cavity. The cavity allows the membrane to respond easily to sound waves, and sensilla attached to it detect its vibrations. Moths preyed upon by bats have tympanic ears that can detect the very-high-frequency sound emitted by bats (100,000 Hz), allowing the moths to take evasive action when they detect sounds of their predators.

We usually think of sound waves as traveling through air, but they also travel easily and five times faster through water. A sound source in water generates two types of waves. Low-frequency **displacement waves** involve an orbital displacement of water particles. They are somewhat analogous to the ripples produced by dropping a pebble into a quiet pond. The lateral line system of fishes detects such waves, but only from a nearby source, because the waves weaken rapidly as they travel outward. Higher frequency **pressure waves** are formed when the water molecules in some bands move closer together and become compressed, whereas those in alternating bands move farther apart and become rarefied. Pressure waves do not attenuate rapidly and hence carry over much greater distances. Since the density of a fish is close to that of water, pressure waves travel through the fish as easily as through water. The problem for a fish is not in getting the waves to the inner ear, but in having a structure that can detect them as they pass through the head. In a sense a fish is transparent to sound. In many fishes a large, dense saccular otolith appears to respond to pressure waves differently from surrounding tissues, resulting in a movement of the otolith relative to the underlying hair cells. The swim bladder of carp, minnows, and related fishes, in addition to being a hydrostatic organ (p. 207), functions as a hydrophone. Pressure waves affect the gases within it, and its vibrations are transferred to the inner ear by a chain of small bones, the **Weberian ossicles** (Fig. 18.13). These fishes have the

keenest sense of hearing of all fishes and can detect frequencies as high as 5000 Hz. Although fishes hear, they cannot always detect the direction of a sound source because sound travels very rapidly in water and their ears are not far apart.

Terrestrial vertebrates have a different problem: how to get sound waves from air, which is not a dense medium, through the denser tissue of the head and amplify their force sufficiently to initiate pressure waves in the endolymph of the inner ear. Some salamanders, lizards, and snakes rely on conduction through bone. They are responsive only to ground vibrations and to low-frequency airborne sounds (below 1000 Hz). A more sensitive tympanic ear has evolved independently at least three times among terrestrial vertebrates: in frogs, in some reptiles and in birds that descended from them, and in mammals. As might be expected, its structure varies considerably among these groups. We will examine the mammalian tympanic ear as an example.

The **external ear** of mammals (Figs. 18.14 and 18.15) consists of a surface flap, the **auricle**, and an **external auditory meatus** (L. *meatus*, passage) that carries sound waves to the **tympanic membrane**, or ear drum, at the base of the meatus. In many mammals, not including humans, facial muscles can move the auricle and direct it toward the sound source. If the auricle is sufficiently large, as it is in dogs and many mammals, it concentrates sound waves on the tympanic membrane in the manner of an old-fashioned ear trumpet.

Vibrations of the tympanic membrane are carried across an air-filled **middle ear**, or **tympanic**, **cavity** by a chain of three small auditory ossicles: the **malleus** (L., hammer), **incus** (L., anvil), and **stapes** (L., stirrup). A foot plate of the stapes fits into a small **oval window** on the side of the bone containing the inner ear and transfers pressure waves to it. The reptilian homologues of the mammalian malleus and incus are the articular and quadrate, respectively, the bones that bear the jaw joint (p. 915). How these joint bones became additional auditory ossicles is one of the most fascinating stories in vertebrate evolution (Focus 18.2.)

Frogs, reptiles, and birds with tympanic ears have only the stapes, sometimes called the columella in these animals. In some salamanders and those reptiles without a tympanic ear, the stapes connects with superficial skull bones and often with those that form the jaw joint.

The tympanic cavity connects with the pharynx by an **auditory (eustachian) tube**, which evolved from the second pharyngeal pouch of fishes. The auditory tube allows air pressure on each side of the tympanic membrane to equalize. You may have noticed the change in pressure in your ears as you go up or down in an elevator and how it is equalized when the auditory tube opens and your ears "pop." The middle ear is a pressure-amplifying device. All of the energy that impinges on the tympanic membrane is

Figure 18.13
The Weberian ossicles that occur in carp, minnows, and related fishes connect the swim bladder with the inner ear, thus enabling the swim bladder to act as a hydrophone.

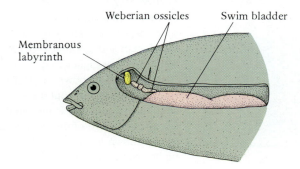

Membranous labyrinth

Weberian ossicles

Swim bladder

(a)

(b)

Figure 18.14
The auricle of many mammalian species helps to gather and concentrate sound waves to the tympanic membrane as did ear trumpets used as hearing aids in the 18th century (a). It is extraordinarily enlarged in bats, which have a very keen sense of hearing (b).

Figure 18.15
A dissection of the human ear shows the three parts characteristic of terrestrial vertebrates: an external ear that includes the tympanic membrane, a middle ear consisting of one or more small bones crossing an air-filled tympanic cavity, and an inner ear embedded within the temporal bone of the skull.

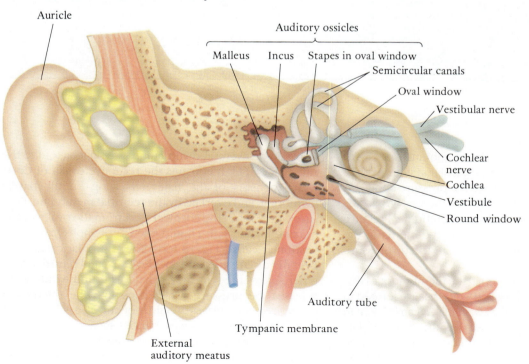

Biologists have long wondered how the articular and quadrate bones could be incorporated into an already functioning tympanic ear of the type found in many reptiles without disrupting ear function. The answer appears to be that the mammalian tympanic ear evolved independently and in a different way from that of modern reptiles. In the early reptiles on the line of evolution to mammals, the primary tooth-bearing bone of the lower jaw, the dentary, was a modest-sized element that did not come close to the squamosal of the skull (Fig. a). The jaw joint was between a large quadrate and articular. These reptiles did not have a tympanic ear, but they did have a deeply situated stapes that extended from the part of the skull containing the inner ear laterally to the quadrate. Presumably they received ground-borne and low-frequency sound waves of sufficient amplitude by bone conduction through their lower jaw, articular, quadrate, and stapes.

During the evolution of later reptiles on the mammalian lineage, feeding mechanisms changed, the bite force became stronger, and the tooth-bearing dentary enlarged considerably and approached the squamosal. Because these changes reduced the forces acting at the jaw joint, the quadrate and articular became smaller and presumably became more responsive in transmitting sound waves. In the very late reptiles that preceded the transition to mammals, the dentary had reached the squamosal and contributed, along with the articular and quadrate, to the jaw joint (Fig. b). The articular and quadrate were now very small, and a space that lay in front of them in the lower jaw presumably held a tympanic membrane.

In the transition to mammals, the dentary-squamosal component of the joint enlarged and the articular and quadrate became disassociated from the joint (Fig. c). The articular and quadrate were overspread by the tympanic membrane and became the malleus and incus. You will notice that throughout this sequence, the morphological relationships of the articular, quadrate, and stapes remained unchanged. At each stage these bones were involved with transmitting vibrations. Changes in feeding and jaw mechanisms caused them to become smaller, and this allowed them to become separated from the jaw and become specialized for sound transmission.

---

concentrated on the much smaller membrane that closes the oval window. In addition, the ossicular system acts as a lever system and further increases the pressure about $1\frac{1}{2}$ times. In humans, the total pressure on the oval window is about 22 times that on the tympanic membrane. Pressure amplification is essential to overcome the inertia in the relatively dense liquids in the inner ear and set up pressure waves in them.

Pressure waves are detected in the membranous labyrinth by specialized groups of hair cells that are located in the spiraled **cochlea** (Gr. *kokhlias*, snail). As can be seen in a cross section, the cochlea consists of three canals (Fig. 18.16a). The innermost **cochlear duct** is a part of the membranous labyrinth that evolved as an extension of the small posteroventral diverticulum, the lagena, of the fish ear (Fig. 18.11). It contains the receptive **organ of Corti**, which consists of the hair cells lying on the **basilar membrane** and overlaid by the **tectorial membrane**. Pressure waves reach the cochlear duct from the oval window by traveling through a liquid-filled perilymph channel, the **vestibular canal** (Fig. 18.16b). The vestibular canal loops over the distal end of the cochlear duct and returns on the other side as the **tympanic canal** to a **round window** that releases the pressure waves back into the middle ear cavity. Pressure waves would not travel easily through these liquid-filled passages without such a release point.

Reissner's membrane, separating the vestibular canal and cochlear duct, is so thin that pressure waves entering

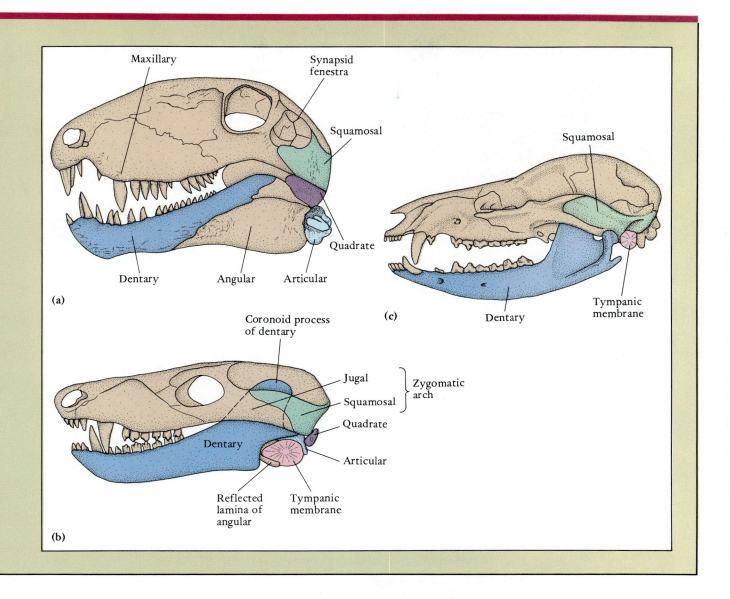

the vestibular canal immediately enter the cochlear duct (Fig. 18.16c). These two passages function as a unit. As pressure waves cross to the tympanic canal, they vibrate the basilar membrane. These movements set up shear forces between the processes of the hair cells and the tectorial membrane on which they impinge. Receptor potentials are generated and initiate nerve impulses in the neurons in the cochlear part of cranial nerve VIII.

The cochlea tapers in diameter from its base at the oval and round windows to its apex, so different amounts of liquid surround each level of the basilar membrane (Fig. 18.16b). Properties of the basilar membrane also change, it being stiffer near its base and more flexible near its apex. As a result of these features, pressure waves of different frequency do not affect parts of the basilar membrane equally. Notes of low pitch cause a maximum displacement of the membrane near its apex; notes of high pitch, near its base. Loudness is determined by the amplitude of basilar membrane displacement.

Proprioceptors

Animals with well-developed muscular systems have evolved tonic **proprioceptors** that constantly monitor the activity of the muscles, including their degree of contraction and rates of change in contraction, and the amount of tension created in tendons. This provides the nervous system with information necessary to integrate muscle

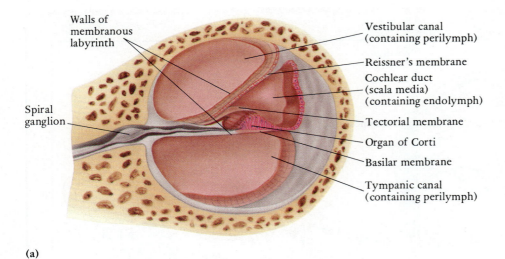

(a)

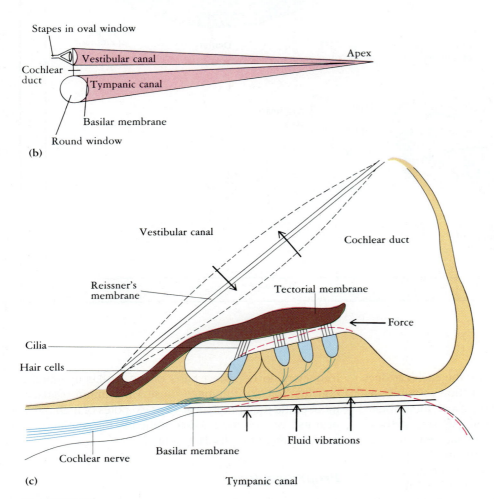

(b)

(c)

Figure 18.16

The mammalian cochlea. (a) A cross section showing the endolymphatic cochlear duct flanked by two perilymphatic canals, the vestibular and tympanic canals. The receptive organ of Corti lies on the basilar membrane on the floor of the cochlear duct. (b) A diagram of the three passages in an uncoiled cochlea. (c) A diagram of the action of the organ of Corti.

activity. Smooth, graceful movements require that the degree and rate of muscle contraction match the change in forces acting on the muscles. The simple act of picking up this book requires the sequential contraction of an assortment of muscles with just enough force to overcome the weight of the book and at an appropriate rate of speed. Stronger and more rapid contractions might cause the book to fly from your hands. Antagonistic muscles must decrease their force and rate of contraction by a corresponding amount. All of these factors change as you move your hand and arm because the angles between the limb segments change. Since terrestrial animals are subjected to greater gravitational forces than aquatic species, their proprioceptors are better developed.

Neuron endings on the tendons of vertebrates (**Golgi tendon organs**) detect the tension developed by muscles (Fig. 18.17). They provide information as to which muscles are active and the degree of their contraction. Many skeletal muscles also contain **muscle spindles**, receptors consisting of a group of encapsulated and specialized muscle fibers called **intrafusal fibers** (L. *fusus*, spindle). Both sensory and motor neurons end on them. Muscle spindles are stimulated during muscular activity and send information about changes in the length of muscles and their rates of contraction to the central nervous system. For example, if you are standing and begin slowly to sit in a chair, muscles on the back of your lower leg and thigh are stretched because the distance between their origins and insertions increases. Both the degree and rate of stretch are detected by different types of intrafusal fibers. Signals from them going to the spinal cord cause an increased output of motor impulses to the surrounding skeletal muscle fibers, which contract with a greater force and speed. If this did not happen, you would sit down rather abruptly.

PHOTORECEPTORS

The fact that complex organic molecules are altered by exposure to light has enabled many organisms to evolve a sensitivity to light based on these molecules. Most animals have specialized **photoreceptive cells** that contain variants of a carotenoid pigment, visual purple or **rhodopsin**. These pigments absorb radiant energy with wavelengths ranging from about 400 to 800 nm, and this induces photochemical reactions that lead to the development of a receptor potential. In most animals the pigments are located in microvilli, but those of some animals, including vertebrates and sea stars, are in the membranes of a modified cilium of a monociliated cell.

Some animals have dispersed photoreceptors in the skin. If a shadow, indicating a potential predator, passes over them, they withdraw. Earthworms withdraw into their burrows on exposure to light. However, in most animals the photoreceptors are clustered together as an organ,

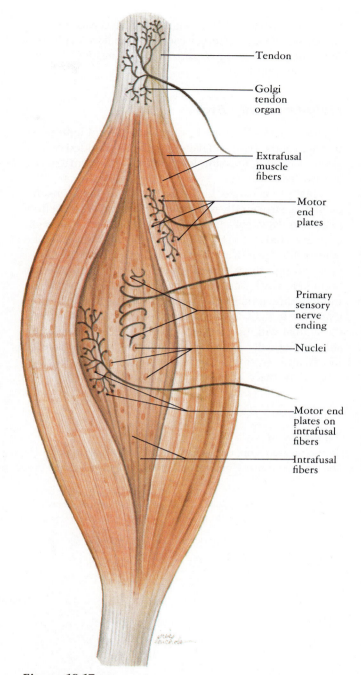

Figure 18.17

A diagram of a Golgi tendon organ and a muscle spindle embedded within a group of normal (extrafusal) skeletal muscle fibers.

which is called an eye or **ocellus** (L., little eye). An eye is capable of exploiting more of the information provided by the light source. Simple eyes only detect the presence or absence of light. More complex eyes detect its source and permit directional movements. The most complex ones form images that enable an animal to make a wide range of responses. An astonishing variety of eyes have evolved in different animal groups, and we can consider only the more common types.

Light-Orienting Eyes

Photosynthetic flagellated protozoans have a light-sensitive area at the base of a flagellum that is shielded on one side by pigment. This simple **eye spot** allows them to move toward light.

Many metazoans have one or more light-sensitive ocelli that structurally resemble a small cup. The ocelli of planarian flatworms are quite representative of **pigment cup (inverted) eyes** (Fig. 18.18) Sensory neurons enter the anterolateral part of a cup of pigmented cells, so light must pass through them to reach their receptive distal ends. Stimulation by light entering the front and sides of the cup induces directional movements that cause a reduction in exposure to light. If light comes in directly from the front, the animal will turn 180° so the photoreceptive cells are shielded equally from the light. In this way a planarian moves away from light and takes shelter under a rock or other dark place.

Many marine polychaete worms, relatives of the earthworm, have two to four pairs of ocelli that are **retinal cup (everted) eyes** (Fig. 18.19). The photoreceptive ends of sensory neurons point toward the light and form a light-sensitive layer called a **retina** (L. *rete*, net). These endings are interspersed with supporting cells and pigment cells that partially shield them from light. Secretions known as a "soft lens" fill the cavity of the cup, but unlike a true lens, the "soft lens" probably contributes little to concentrating light on the retina. The overlying epidermis is transparent. Eyes of this type also have evolved independently in many molluscs, annelids, and arthropods.

Image-Forming Eyes

Eyes that are capable of forming an image require optical mechanisms that can bend or **refract** divergent light rays coming from a point source so that they converge and form a single point upon the photoreceptive surface. In the eye of an aquatic animal that lacks a functional lens, divergent rays of light from points A and B in the visual field travel in straight lines and form many points of light on the retina; thus no image is formed (Fig. 18.20a). When light rays pass through a medium that is denser than the one they are in,

Figure 18.18

In the pigment cup, or inverted, eyes of a planarian flatworm, light must pass through the sensory neurons to stimulate their receptive distal ends.

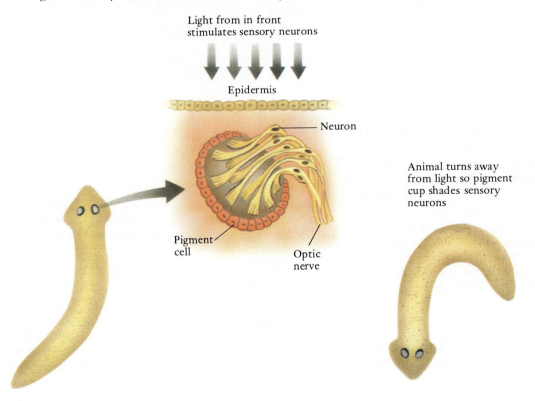

Light from in front stimulates sensory neurons

Epidermis

Neuron

Pigment cell

Optic nerve

Animal turns away from light so pigment cup shades sensory neurons

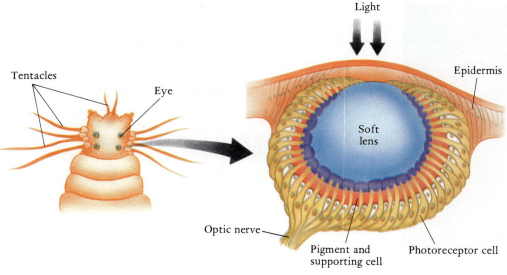

Figure 18.19

Light impinges directly upon the receptive ends of sensory neurons in the retinal cup, or everted, eyes of a polychaete worm.

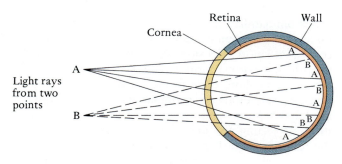

(a) Aquatic animal with no lens

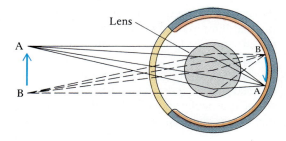

(b) Aquatic animal with spherical lens

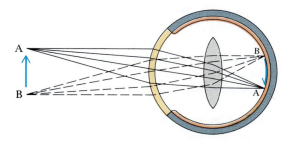

(c) Terrestrial animal; cornea acts as primary refractive element. Lens somewhat flattened

they are refracted because the light rays slow down in the denser medium. In an aquatic animal with a lens, divergent rays of light from point A are refracted and converge to form a single point upon the retina (Fig. 18.20b). The same applies to light from point B. Since each point in the visual field is represented by a single point on the retina, an image is formed.

The more curved the lens, the greater are its refractive powers. Aquatic animals usually have spherical lenses, for nearly all of the refraction must occur in them. Light is not bent as it passes through the transparent cornea at the eye surface, for its density is about the same as water. In terrestrial animals, the cornea is much denser than the air, so most refraction occurs here (Fig. 18.20c).

Rays of light from a distant point are sufficiently parallel relative to the eye that the cornea or lens provides adequate refraction. Light rays from objects close to the eye diverge greatly relative to the eye so they must be refracted more to bring them into a sharp focus. In fishes, the lens is moved farther away from the light-sensitive plane, as one does in focusing a camera; in mammals, the curvature of the lens is increased (p. 426).

Figure 18.20

The optics of eyes. (a) In the eye of an aquatic animal without a lens, divergent light rays from sources A and B do not converge, form multiple points of light upon the retina, and no image is formed. (b) The spherical lens of an aquatic animal causes light rays from A and B to converge, so single spots of each project to the retina and an image is formed. (c) The cornea acts as the primary lens in a terrestrial animal, and the less curved lens is used primarily for fine focusing.

(a)

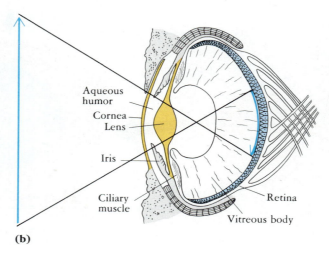

(b)

Figure 18.21

(a) The octopus has a pair of large image-forming eyes. (b) Each eye is a camera-like everted eye in which the receptive ends of the retinal cells are directed toward the light source.

In addition to optical mechanisms, image formation also requires a relatively large number of photoreceptors. There must be at least enough receptors to distinguish between spots of light of different intensity in different parts of the visual field. In the same way as small spots of light of different intensity form an image on a black and white television screen, so too spots of light from the retina are combined in the brain to form an image. Processing requires a complex neuronal circuitry. The evolution of image-forming eyes and that of the neuronal circuitry needed to take advantage of them are closely coupled.

Certain polychaete worms, molluscs, and wolf spiders that are very active and predatory have everted eyes capable of object discrimination. However, the image-forming capabilities of these eyes are poor because their eyes are very small and do not contain many photoreceptors. The large eye of the octopus and other cephalopods is an exception (Fig. 18.21). The retina of the octopus has as many as 70,000 photoreceptors/mm². A lens refracts light, a space between the lens and retina allows the convergence of light rays to form an image, and an iris controls the amount of light entering the eye. While superficially resembling the vertebrate eye, the octopod eye is an everted eye, for photoreceptive cells lie on the surface of the retina facing the lens. The vertebrate eye is inverted and photoreceptors lie on the back side of the retina (see Fig. 18.27). The image quality of the octopus eye is as good as that of fishes.

Compound Eyes

Insects and many crustaceans, such as shrimp and crayfish, have a different type of image-forming eye known as a compound eye, for it is composed of many distinct units called **ommatidia** (Gr. *ommatidion*, little eye). The exoskeleton of each ommatidium is modified as a transparent cornea, and most refraction occurs here (Fig. 18.22). Cells deep to the cornea form a long, cylindrical, and often tapered **crystalline cone** that concentrates the light on seven or eight elongated **retinular cells**. Each of these is a modified sensory neuron. Their adjacent surfaces bear many projecting and interdigitating microvilli that form the receptive, rod-shaped **rhabdome** (Gr. *rhabdos*, rod). **Pigment cells** flank the crystalline cone and sometimes the retinula cells. The pigment prevents light that enters an ommatidium from straying to others. In species that are diurnal, the pigment is migratory, screening more of each ommatidium in bright light than in dimmer light. In nocturnal species, pigment is not present around the rhabdome.

Since rays of light are directed separately down each ommatidium of a diurnal species, each detects the intensity of a particular spot of light in the visual field. The image that is perceived is a compilation of all the spots of light

Figure 18.22 ▶

The compound eye of an insect. (a) The many individual units, or ommatidia, can be seen in an enlargement of the eye surface. (b) An enlarged vertical section through the eye shows the numerous ommatidia. (c) Enlargements of the ommatidia of a diurnal and nocturnal species. The ommatidia of a nocturnal species are not screened as much by pigment cells. (d) The petals of *Nedelia trilobata* appear yellow to us, but blue to a bee or other insect sensitive to ultraviolet light. Note how the nectar-bearing center of the flower is "highlighted."

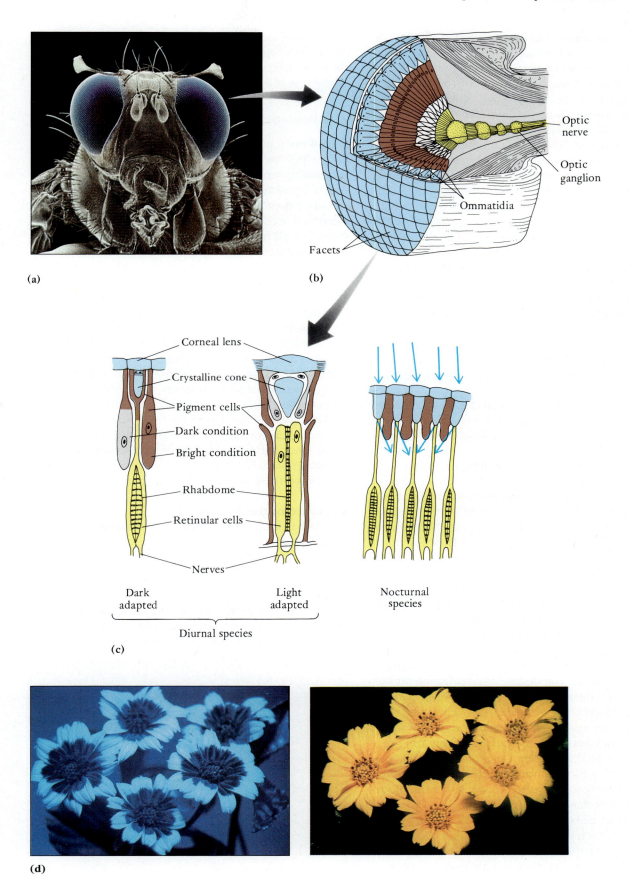

(a)

(b)

Optic nerve

Optic ganglion

Ommatidia

Facets

(c)

Corneal lens

Crystalline cone

Pigment cells

Dark condition

Bright condition

Rhabdome

Retinular cells

Nerves

Dark adapted

Light adapted

Diurnal species

Nocturnal species

(d)

from each ommatidium. This has been called a mosaic or aposition image, but it is no more mosaic than the image produced by the spots of light in a vertebrate eye. The difference is in the degree of resolution. The image of a compound eye is not as "fine grained" as that of a verte- brate eye because the compound eye is much smaller and has fewer receptive cells. Objects that are very close to- gether are not distinguished by a compound eye; nonethe- less this eye suits the needs of these small arthropods whose visual world extends only about 10 cm around them. In many ommatidia, the pigment separating om- matidia is retracted in weak light, thereby permitting light entering one ommatidium to cross over and stimulate the photoreceptive cells of adjacent ones, as is the case in nocturnal species (Fig. 18.22). In this case the image, a superposition image, is formed in much the same way as in an eye that is not compound.

While the compound eye cannot resolve images as well as the vertebrate eye, it is more sensitive in detecting motion, for a moving object stimulates many rhabdomes in succession. Images, or a flicker of light, that pass our eyes at a rate faster than about 24 per second fuse into a smooth picture. This is the basis of motion picture photography, in which separate pictures are taken of a object at 24 frames per second or higher. But flies can detect as distinct images flickers at a rate of nearly 300 per second. A fly can easily sense a slight movement in its visual field. It is not easy to sneak up on a fly.

Bees, and some butterflies and flies, can distinguish many colors because their retinula cells contain pigments that absorb light of different wavelength: ultraviolet, blue, and green in bees. Their view of the world is different from ours, for we have pigments that absorb in the blue, green, and red parts of the spectrum (Fig. 18.22). The compound eyes of some insects and other animals can also detect polarized light whose waves vibrate in only one plane. Mixed waves of light, which we sense as white light, vibrate in all planes. The blue appearance of the sky results from the reflection from molecules in the air of light of short wavelength (blue). Most of this is polarized, and it can be detected by some animals whose photosensitive pigments are so oriented that they absorb only light aligned with the plane of polarization. The degree of polarization and its plane depend on the animal's location with respect to the sun. Photographers know that the most striking effects of a polarized lens are reached at an angle of about 90° to the sun. If an animal that can sense polarized light can see a small patch of blue sky, even on a cloudy day, it can determine the sun's location and use this as a navigational aid. Bees returning to a hive are known to communicate the location of nectar- and pollen-bearing plants to other bees through an elaborate dance (p. 762). In one experi- ment Karl von Frisch placed a polarizing filter over a small opening in the hive through which the dancers could see a bit of sky. By rotating the filter and altering the plane of

polarization, he could change the orientation of the bees' dance. Some species of fish, reptiles, and birds have been shown to detect polarized light but it is not known to what extent this ability is used in navigation.

The Vertebrate Eye

Evolutionary processes have led to the development of many types of eyes, as we have seen. The vertebrate eye is yet another specialized light-sensing organ, and is in- verted. Light must pass through the photoreceptive cells to stimulate their distal ends. The eyes of different verte- brates are well adapted to their varied habitats and life styles, but all have the same fundamental structure. The surface of the eyeball of aquatic vertebrates is kept wet and clean by the surrounding water, and eyelids are not usually present. **Lacrimal glands** (L. *lacrima*, tear) around the eyeball of terrestrial vertebrates secrete a watery, isotonic fluid that bathes the eyeball surface, and moveable **eyelids** protect it. In addition to the familiar upper and lower lids, many terrestrial vertebrates have a third lid, the **nictitating membrane** (L. *nictare*, to wink), that can flick across the eyeball and clean it. This membrane is reduced to a vestigial **semilunar fold** in humans, who can rub the eyes to clean them (Fig. 18.23). Six small, strap-shaped **extrinsic ocular muscles** extend from the bony wall of the orbit to the periphery of the eyeball. Their actions move the eyeball as an animal looks in different directions or follows a moving object. For example, your eyes move in unison as you follow these words across the page.

Structure of the Eyeball

The eyeball itself is composed of three layers or tunics (L. *tunica*, coating). An outer **fibrous tunic** of dense connec- tive tissue forms the supporting wall (Fig 18.24). Most of

Figure 18.23
A surface view of the human eye. The semilunar fold, on the right, is a vestige of the nictitating membrane that can move across the eye surface in many terrestrial vertebrates.

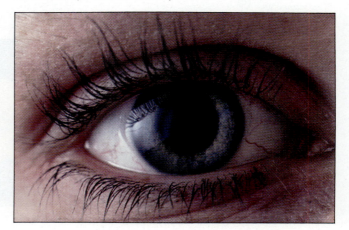

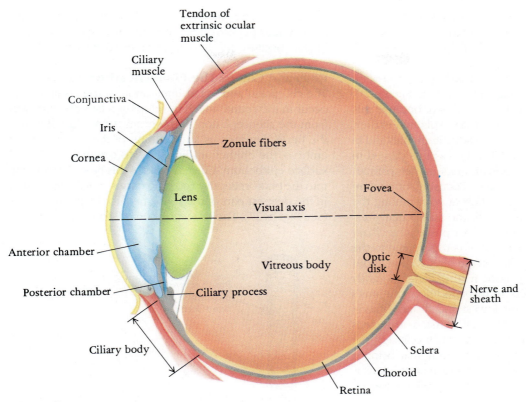

Figure 18.24
The structure of the mammalian eye as seen in a longitudinal section.

this is an opaque **sclera** (Gr. *skleros*, hard) and appears as the white of our eyes. The front of the fibrous tunic is modified as a transparent **cornea** (L. *corneus*, horny) through which light rays enter. A thin layer of modified epidermis, the **conjunctiva**, covers the surface of the cornea and turns onto the inside of the eyelids, when eyelids are present.

The innermost tunic is the **retina**. The retina develops from an embryonic optic cup with two distinct layers: pigment and nervous (p. 520). The nervous layer, from which the photoreceptive cells differentiate, is not present at the front of the eye.

A **vascular tunic** lies between the retina and fibrous tunic. The portion of the vascular tunic adjacent to the retina is the **choroid** (Gr. *chorion*, skin). Its rich supply of blood vessels helps to nourish the retina. Pigments within the choroid, which are derived from the pigment layer of the embryonic retina, absorb light that has passed through the nervous layer of the retina and prevents scattering of light rays and a blurring of the image. Many animals that can be active in dim light have a reflective layer in part of the choroid known as the **tapetum lucidum** (L. *tapete*, carpet + *lucidus*, shining), which reflects light back onto

the photoreceptive cells. The eye shine that you see when lights strike the eyes of a nocturnal animal such as a cat is a manifestation of the tapetum. The part of the vascular tunic adjacent to the margin of the **lens** forms the **ciliary body**. Delicate **zonule fibers** extend from it to the lens, help to hold it in place, and regulate its shape during focusing. The vascular tunic in front of the lens forms the **iris** (Gr., rainbow). Light passes through the **pupil** in its center. The iris of most vertebrates contains radial and circular smooth muscle fibers that are controlled reflexively by the autonomic nervous system. Their actions regulate the diameter of the pupil and the amount of light that enters the eye, and this adjusts the amount of light impinging on the retina to ambient conditions, preventing over- or underexposure. The iris is analogous to the diaphragm of a camera.

A gelatinous **vitreous body** (L. *vitreus*, glassy) occupies the large chamber between the lens and retina and helps to hold these structures in place. The **posterior chamber**, located between the lens and iris, and the **anterior chamber** between the iris and cornea, are filled with a watery **aqueous humor** that is secreted by the **ciliary processes** on the ciliary body. Aqueous humor is drained into a small canal, the **canal of Schlemm**, that extends around the eye at the junction of cornea and sclera. Over-production, or under-reabsorption, of this fluid leads to an increase in intraocular pressure and a group of eye diseases collectively known as **glaucoma**.

Light that enters the eye is refracted toward the optic axis and forms an inverted image on the retina (Fig. 18.20), but we do not see the world upside down, because this inversion is corrected in the brain. As explained earlier, most refraction occurs in the lens of aquatic animals and in the cornea of terrestrial ones. The lens is less curved in terrestrial vertebrates than in aquatic ones and acts primarily as a fine adjustment mechanism. Focusing, or **accommodation**, is accomplished in different ways among the groups of vertebrates. In mammals, the intraocular pressure generated by the aqueous humor pushes the ciliary body away from the lens. This force is transferred to the lens margin by the zonule fibers; the lens comes under tension and, because of its elastic nature, flattens slightly and its curvature is reduced. In this condition, the lens's refractive powers are at their lowest and distant objects come into focus. This is the state at rest. Tension on the lens is reduced during accommodation for a near object by the contraction of muscles in the ciliary body that bring it closer to the margin of the lens. As tension on the lens is reduced, the elastic lens bulges, its curvature increases, its

Figure 18.25

Photoreceptive cells of the retina. (a) A scanning electron micrograph of the rods and cones in the retina of the mudpuppy (*Necturus*), taken at 6000×. (b) The ultrastructure of a rod; its receptive outer segment is a modified, nonmotile cilium.

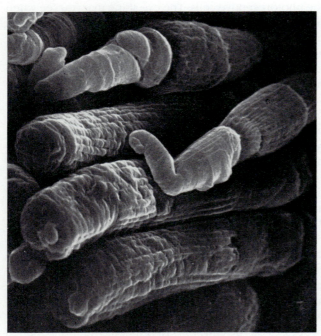

(a)

refractive powers increase, and close objects come into focus. Older people often have difficulty focusing on near objects because the lens loses its elasticity with age.

Rods and Cones: The Photochemistry of Vision

The reception and initial integration of the visual image occur in the retina, which develops as a lateral outgrowth from the diencephalon of the brain (Chap. 22). Because the brain develops as an infolding of the surface of the embryo, the epithelium lining it was originally a ciliated surface epithelium. Ciliated cells of this type differentiate into the photoreceptive **rods** and **cones**, which are located in the deepest part of the retina next to the pigment cells (see Fig. 18.27). Their receptive outer segments are modi-

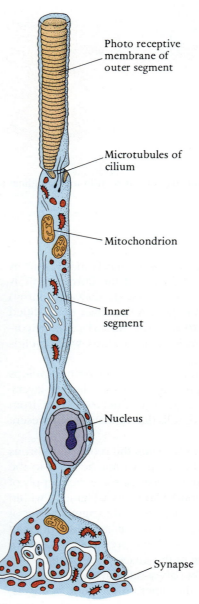

Photo receptive membrane of outer segment

Microtubules of cilium

Mitochondrion

Inner segment

Nucleus

Synapse

(b)

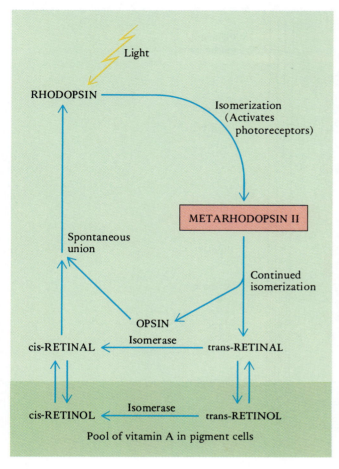

Figure 18.26
A diagram of the photochemistry of rod vision.

posing rhodopsin affects the permeability of the plasma membrane to sodium ions during the early stages of isomerization, and this activates the photoreceptive cell.

When light is not absorbed, the *trans*-retinal is reshaped into *cis*-retinal. This is an energy-consuming reaction catalyzed by the enzyme retinal isomerase. As the *cis*-retinal is reformed, it spontaneously recombines with opsin.

Trans-retinal and *cis*-retinal can also be synthesized from vitamin A (retinol), as shown in the Figure 18.26. Vitamin A is normally present in the adjacent pigment layer of the retina, and this supply forms a reservoir from which more retinal can be made. If an excess of retinal is present, it is stored as vitamin A.

Since the sensitivity of the rods to light depends on the amount of rhodopsin they contain, the interconversion of retinal and vitamin A provides a mechanism whereby rod sensitivity can adapt to changing ambient light. If the eye is exposed to a bright light for a while, much of the rhodopsin is reduced to opsin and *trans*-retinal, and excess *trans*-retinal is stored as vitamin A. The eye is said to be light adapted. If, on the other hand, the ambient light is weak, vitamin A reserves are utilized to form retinal and an increased amount of rhodopsin is present. The eye is dark adapted. You have experienced these shifts in adaptation when you first enter a dark theater on a bright afternoon and later emerge to the bright light again. When a shift is first made it is difficult to see for a few moments, but the eye soon becomes adapted. By shifts in the amount of rhodopsin present, the sensitivity of the eye to light can change nearly one millionfold!

Cones function in the same way. They contain a variant of rhodopsin known as **iodopsin**. The retinal component is the same, but it binds with a different opsin protein.

Retinal Structure and Integration

Several layers of neurons differentiate in the nervous layer of the retina (Fig. 18.27). **Bipolar cells** typically receive impulses from a group of rods and cones, and synapse with **ganglion cells** whose axons collect at the **optic disc** and leave the eyeball as the optic nerve (cranial nerve II), which continues to the brain. Since no receptive cells are found here, the optic disc is a blind spot and images falling on it are not detected. **Horizontal cells** form synapses horizontally among the receptor and bipolar cells, and **amacrine cells** do the same among bipolar and ganglion cells. The morphology of the retina permits the direct passage of some impulses from the rods and cones to the bipolar cells and on to the ganglion cells. At the same time divergence and convergence occur through the horizontal and amacrine cells.

Considerable integration occurs within the retina. Some impulses are amplified, others are suppressed, and yet others are combined to produce new messages. Al-

fied cilia whose membranes are infolded to form as many as 1000 disclike plates containing the photoreceptive pigments that receive light and initiate nerve impulses (Fig. 18.25).

The photochemical changes in vertebrate rods have been studied intensively (Fig. 18.26). Rhodopsin, which is bound to the membranes in the outer segments of the rods, consists of a pigment conjugated with an opsin protein. The pigment, known as **retinal**, is derived from vitamin A, also called **retinol**. The molecule of retinal in rhodopsin has a particular configuration known as ***cis*-retinal**. Absorption of light energy immediately initiates a change in shape, or **isomerization**, of the *cis*-retinal (Fig. 18.26). As its shape changes, it loses its ability to bind with opsin. Within less than 2 milliseconds, changes in rhodopsin have progressed through several intermediate compounds to **metarhodopsin II**. Within another few seconds, the isomerization is complete; *cis*-retinal is completely changed to ***trans*-retinal** and the opsin released. Decom-

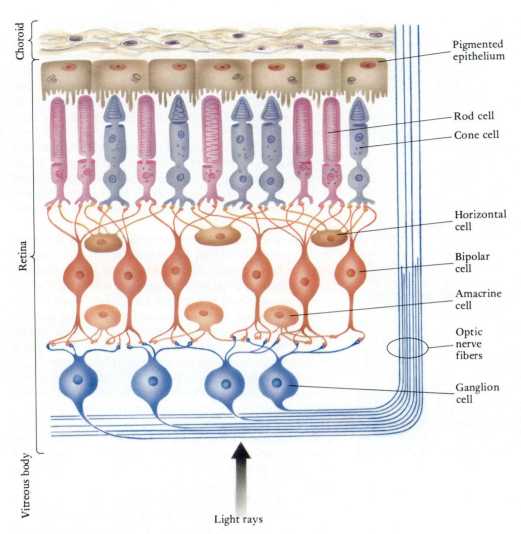

Choroid

Retina

Vitreous body

Pigmented epithelium

Rod cell

Cone cell

Horizontal cell

Bipolar cell

Amacrine cell

Optic nerve fibers

Ganglion cell

Light rays

Figure 18.27
A diagram of the mammalian retina showing the cell types and the synapses among them.

though rods directly adapt to the amount of light reaching the retina, additional light and dark adaptation of the eye occurs within the retina. Other interactions among retinal cells sharpen the boundaries of images and enhance motion detection.

Cones are particularly abundant in a slightly depressed area of the retina known as the **fovea** (L., depression) located at the posterior pole of the visual axis, a line perpendicular to the center of the lens. (Fig. 18.24). They decrease in number peripherally. Rods, on the other hand, increase in density toward the periphery of the retina, and none are present in the fovea. To see in dim light, one should look to the side so the image falls near the periphery of the retina. The human retina contains about 120

million rods and nearly 6 million cones, yet only 100 million ganglion cell axons leave through the optic nerve. A great many rods converge upon a single bipolar cell (Fig. 18.27), so that summation occurs. The stimulation of one rod may not activate the bipolar cell, but the simultaneous arrival of impulses from many will. Fewer cones synapse on a bipolar cell, and in the fovea most cones connect with only one bipolar and one ganglion cell. One consequence of this is that weak light can be sensed by a group of rods and their convergent pathway to the brain. Cones require brighter light, and they do not function in dim light. This circuitry also enables cones to produce a sharper image than rods because spots of light falling on them project more directly to the brain. Cones also distinguish colors, for different classes of cones contain pigments that absorb light of different wavelengths. Their absorption spectra overlap, but they absorb maximally at different points: 450 nm (blue), 525 nm (green), and 550 nm (red). To summarize, rod vision is analogous to a fast, coarse-grained black and white film; cone vision, to a slow, fine-grained color film.

Color vision occurs primarily in vertebrates that are active in brightly lighted habitats. Since light rays, especially the longer ones at the red end of the spectrum, are absorbed quickly by water, only a few reef fishes and others that live in brightly illuminated habitats have color vision. Secretive animals including salamanders and many mammals that either are nocturnal or evolved from nocturnal ancestors also lack color vision. In mammals, color vision is best developed among primates.

As impulses continue to the thalamus of the brain in the optic nerves, they decussate to some degree in a crossing of the nerves known as the **optic chiasma**. In most vertebrates the fields of vision of the two eyes overlap only slightly and the decussation is nearly complete. In those mammals, including primates, in which there is an extensive overlap of the two visual fields, only about one-half of the fibers decussate. The others remain on the side of the body where they originated. Thus half of each retina projects to the left side of the brain and half to the right side. As a consequence, images falling near the center of the retina project to each side of the brain. The processing of these images, in ways not fully understood, facilitates stereoscopic vision and depth perception.

THERMORECEPTORS

All animals have the ability to detect changes in the temperature of their surroundings and modify their behavior to avoid extreme cold or heat that could injure their tissues or kill them. The protozoan *Paramecium* collects in areas where the water temperature is moderate and avoids extremes. Sessile animals often contract and expose a minimum surface area when they encounter unfavorable temperatures.

The receptors appear to be free nerve endings that are strategically located to detect thermal changes, although some encapsulated endings in mammalian skin may also be receptors for heat or cold. Blood-sucking insects, such as mosquitos and lice, find their warm-blooded prey with sensilla that contain thermoreceptive neuron processes. These endings are often included in tactile or chemoreceptive sensilla. Boa constrictors, rattlesnakes, and other pit vipers have one or more pits on their head that are sensitive to temperature differences. The large facial pit of a pit viper (Fig. 18.28) contains a delicate membrane on which many neurons end. Nearly 1000 endings are present per square mm. This mechanism is so sensitive that a rattlesnake can find a rat 40 cm away that is only 10° C warmer than its surroundings. It can do so in 0.5 second, even when other senses have been blocked. Certain cells in the hypothalamus of warm-blooded birds and mammals also are thermoreceptors. They detect changes in blood temperature and initiate physiological changes that maintain body temperature.

Figure 18.28
The facial pit of a pit viper is a thermoreceptor. This is a yellow eyelash viper.

Summary

1. Receptors may be parts of neurons or specialized cells. Each type is particularly sensitive to one type of energy change. Sense organs are groups of receptors and associated cells that support and protect the receptors and frequently aid in gathering the type of energy to which the receptors are attuned.

2. When a receptor is stimulated, a change occurs in the flow of ions across its plasma membrane, and an electrical receptor potential develops or changes. A nerve impulse is generated if the receptor potential reaches the needed threshold. The magnitude of the receptor potential, and the frequency of nerve impulses it generates, are proportional to the strength of the stimulus. Perception of a stimulus depends on the part of the brain to which the receptor projects, and, sometimes, upon the pattern of activation of a group of receptors.

3. The ability to detect and respond to chemical stimuli in the environment is ubiquitous. Odors in advanced metazoans are detected by olfactory receptors; taste, by gustatory ones. The receptors are usually nonmotile cilia born on the dendrites of sensory neurons, but vertebrate taste buds are modified epithelial cells.

4. A wide variety of mechanisms detect mechanical displacement of one sort or another, including movement of the surrounding environment. The isolation of a motion-detecting system from the external environment enables an animal to detect the direction of the pull of gravity and changes in its own acceleration. Statocysts perform these functions in invertebrates; part of the inner ear performs them in vertebrates.

5. The sound receptors of arthropods and vertebrates are modified mechanoreceptors that detect waves of alternating high and low pressure with wavelengths ranging upwards from about 20 Hz. Sound waves reach the inner ear of a fish without difficulty and are detected primarily by the sacculus. The pressure of airborne sound waves is amplified in terrestrial vertebrates by the tympanic membrane and auditory ossicles, which set up pressure waves in the denser liquids of the cochlea, where they are detected.

6. Proprioceptors are mechanoreceptors located in tendons and muscles that monitor the degree and rate of contraction of the skeletal muscles. This information is essential for the integration of muscle activity.

7. Photoreceptive cells are frequently clustered to form eyes of various types. Some eyes detect the presence or absence of light; others also detect its source, which permits directional movements; still other eyes form images. Image-forming eyes require a lens that can refract divergent light rays from a point source and form a single point upon the the retina. They must also contain enough photoreceptors to distinguish between variations in light intensity in different parts of the visual field. Object discrimination also requires a complex neuronal circuitry in the brain.

8. Insects and many crustaceans have compound eyes composed of many individual ommatidia. Compound eyes do not have as much resolution as the eyes of vertebrates, but they are better adapted at detecting motion.

9. The vertebrate eye is a camera-like eye that forms a sharp image upon the retina. Considerable processing of the image occurs within the retina, which develops embryonically as an outgrowth of the brain.

10. The vertebrate photoreceptive cells are the rods and cones that contain the photosensitive pigment retinal conjugated with a protein. On exposure to light, the pigment isomerizes, activates the cell, and separates from the protein. These processes are reversed in the dark. Rods are sensitive to weak light; cones require brighter light and different populations of cones maximally absorb light in the blue, green, and red regions of the spectrum.

11. Temperature changes are usually detected by strategically placed free nerve endings or encapsulated endings. Boa constrictors and pit vipers have labial and facial pits that are particularly sensitive to thermal changes.

References and Selected Readings

Cronin, T.W. Photoreception in marine invertebrates. *American Zoologist* 26(1986):403–415. A review of the nature of photoreceptors and their functions among invertebrates.

Dethier, V.G. *The Hungry Fly*. Cambridge, Mass.: Harvard University Press, 1976. An account of the experiments dealing with the sense of taste in the fly.

Dowling, J.E. *The Retina: An Approachable Part of the Brain*. Cambridge, Mass.: Harvard University Press, 1987. A synthesis of major research on the structure and physiology of the vertebrate retina.

Hudsperth, A.J. The hair cells of the inner ear. *Scientific American* 248(Jan 1983):54–64. An analysis of the mode of action of these transducers.

Katsuki, Y. *Receptive Mechanisms of Sound in the Ear*. Cambridge, England: Cambridge University Press, 1982. A good sourcebook for information on the sense of hearing.

Koretz, J.F., and G.H. Handelman. How the human eye focuses. *Scientific American* 259(July 1988):92–99. A discussion of the mechanism of focusing and the geometric and biochemical changes that occur with aging.

Nilsson, D.E. Vision optics and evolution. *Bioscience* 93 (1989):298–307. A review of the diversity in eye design.

Northcutt, R.G. (ed.) *Neurobiology of Taste and Smell.* New York: John Wiley & Sons, 1987. A collection of papers written primarily for the specialist, but the first several review general aspects of chemoreception.

Sachs, F. The intimate sense. *The Sciences* (Jan./Feb. 1988):28–34. A review for the layperson of the mechanisms of touch.

Stryer, L. The molecules of visual excitation. *Scientific American* 257(July 1987):42–50. A review of the photochemistry of vision.

Waterman, T.H. *Animal Navigation.* New York, Scientific American Library, 1989. The use of various senses in navigation is discussed in this engaging little book.

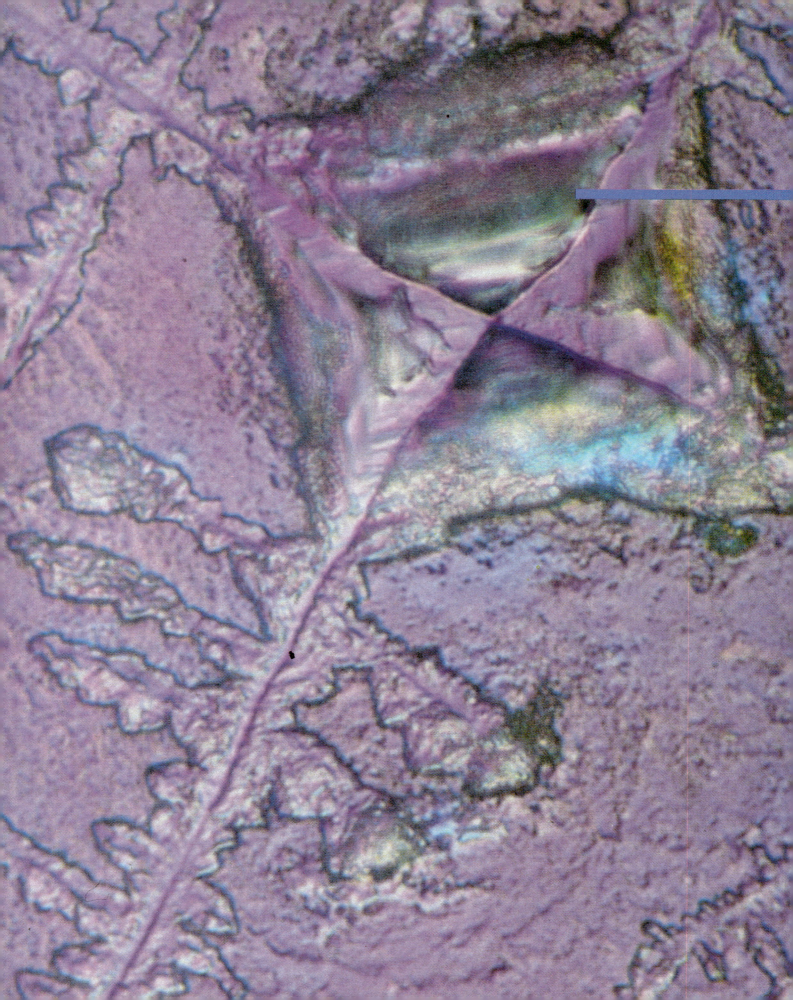

19

The Endocrine System

Whenever cells perform different functions, coordination among them is essential if their actions are to be combined into an integrated whole. Cells in close juxtaposition communicate directly by electrical and chemical interactions. As animals became larger, a communication network evolved that could convey messages over longer distances. The rapid response of specific effectors to environmental changes is modulated, as we have seen, by receptors and by the nervous system. Long-term processes that affect many parts of the body simultaneously are integrated by a subset of chemical messengers known as **hormones** (Gr. *hormaein*, to excite). These messengers are produced by a number of small **endocrine glands** that are scattered widely throughout the body (Fig. 19.1).

Endocrine glands differ from sweat glands, salivary glands, and other exocrine glands (p. 243) in lacking ducts, so their hormones are discharged directly into the blood or other body fluids, and are carried by these fluids to their targets, which may be a considerable distance away. Hormones are often defined as chemical messengers that are produced at one site and carried by the bloodstream to their site of action, or target. In general, endocrine integration controls long-term processes including growth and development, pigmentation, maintenance of the composition of body fluids within narrow limits, the conversion of food products into energy or storage, and reproduction. Most metazoans have both nervous and endocrine systems: the former are adapted for rapid, specific responses;

Figure 19.1

The location of the major endocrine glands in a human.

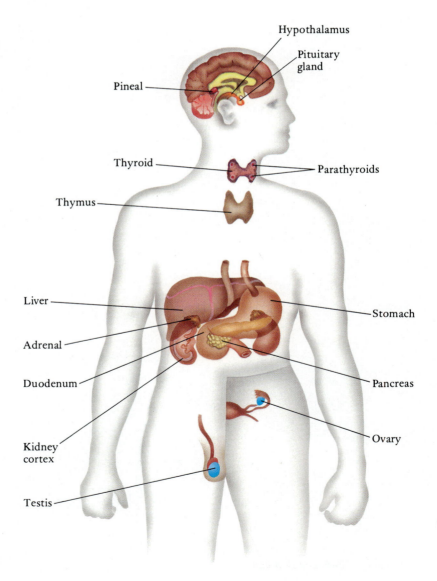

the latter, for slower, continuing processes that involve cells in many parts of the body. We will consider the major endocrine glands at this time, but postpone those that regulate reproduction until we consider the reproductive system (Chap. 21).

ENDOCRINE AND NERVOUS INTEGRATION

Correlated with differences in the nature of the processes that they control, nervous and endocrine integration also differ in how directly and rapidly their chemical messengers reach their targets (Fig. 19.2a and b). Autonomic neurons, which innervate endocrine glands, carry impulses to their targets at speeds ranging from 0.5 to 2.0 meters per second, and their neurotransmitters are stored in synaptic vesicles adjacent to the targets where they can be released quickly by the nerve impulse. Hormones are carried to their targets by the circulatory system. Their rate of transport is limited by the rate of circulation and the time it takes them to enter and leave the blood stream. Since the entire spectrum of hormones is carried by the

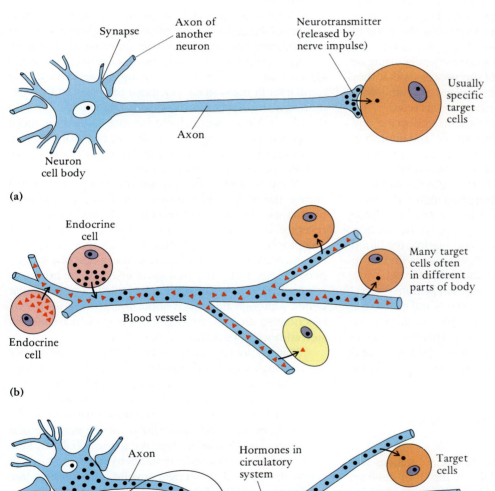

Figure 19.2
A comparison of neural and endocrine control systems. (a) The action of a neuron on its target cell. Neurotransmitters are released from the axon knob. (b) Transport of two different hormones through the circulatory system to their targets. (c) An intermediate condition in which hormones are released by neurosecretory cells into the circulatory system.

blood throughout the body, specificity comes not from specific connections but from the specificity of receptor molecules on the target cells that bind only with certain hormones. Many hormones, such as insulin (produced by islets of endocrine tissue in the pancreas), have receptors on most body cells and hence exert a global effect. Others, such as secretin (produced by certain intestinal cells), have a narrower range of targets. Secretin influences only those cells in the pancreas that secrete the aqueous and alkaline component of the pancreatic juice.

The effect of a nerve impulse is short-lived, a matter of milliseconds, because the neurotransmitter is rapidly degraded or reabsorbed by the neuron. Additional impulses are needed for a continued response. The effects of hormones usually last much longer: minutes, hours, or even weeks. Most hormones act by regulating the level of intracellular enzymes, increasing or decreasing the rates of ongoing biochemical processes. Many hormones are maintained in the blood at characteristic levels, rising or falling according to the needs of the processes that they mediate. As hormones are metabolized, or otherwise removed from the circulation, more are secreted.

Despite differences between them, endocrine and nervous integration have much in common and interact in many ways. Both exert their effects by chemical messengers that excite or inhibit target cells. Certain molecules act both as neurotransmitters and hormones. For example, norepinephrine is released by both postganglionic sympathetic neurons and certain cells of the adrenal gland. Some neurons, known as **neurosecretory cells**, synthesize hormones and release them from axon endings into the blood (Fig. 19.2c). A wide spectrum of hormones is known to affect the central nervous system and influence behavior. Some processes, such as color changes in primitive vertebrates, are controlled by both the nervous and endocrine systems. Finally, the two systems are intimately interconnected in vertebrates by the hypothalamus, whose neurosecretions control the action of the pituitary gland and, through it, much of the activity of the entire endocrine system.

THE NATURE OF HORMONES AND HORMONAL ACTION

Although the presence of hormone-like substances was suspected as early as the middle of the 19th century, William M. Bayliss and Ernest H. Starling first demonstrated unequivocally in 1902 that a vital process is hormonally regulated. During studies on factors responsible for the pancreatic secretion that occurred when acid food enters the intestine from the stomach, they found that the secretion was not a nervous reflex, as had previously been thought. Cutting the vagus nerve to the pancreas did not affect its secretion, as would have been the case if the secretion was controlled by a nerve reflex. Nor was secretion induced by the acid entering the duodenum, for

acid injected into the blood did not cause secretion. They reasoned that the acid food must cause cells of the intestinal mucosa to produce a product that, on reaching the pancreas via the circulatory system, induced secretion. Extracts of intestinal cells that had been exposed to the acid gastric contents did just this. They named the active factor **secretin** (p. 448), and Starling later introduced the term hormone for this type of material.

A diverse group of chemical compounds can act as hormones. They include (1) peptides and proteins, (2) steroids, and (3) amino acid derivatives. Nearly 50 hormones have been identified in vertebrates. Table 19.1 lists the major endocrine glands of mammals, the hormones they produce, their chemical characteristics, and their biological effects.

Hormones are synthesized by the metabolic machinery of the cell. Like neurotransmitters, they are stored in membrane-bound vesicles or granules and are released, on appropriate stimulation, by exocytosis. Endocrine glands have a very rich vascular supply, and fenestrated capillaries in many glands facilitate the entry of the hormones, many of which are relatively large molecules, into the blood. Most water-soluble hormones, such as those composed of peptides, are carried in physical solution. Most steroid hormones, which are hydrophobic and lipid-soluble, are bound to carrier plasma proteins.

Many steroid hormones produced by the adrenal cortex, the testis, and the ovary pass through the lipid plasma membrane of the receptor cells and bind with a receptor in the cytoplasm or nucleus (Fig. 19.3a). The receptor-hormone complex binds to acceptor sites on the chromosomes and increases the transcription of a specific gene (Focus 19.1). The biological effect of the hormone results from this transcriptional regulation of particular genes. Although this much about the process is known, biochemists are still studying the ways in which different steroid hormones activate different genes.

Water-soluble peptide and protein hormones cannot pass through the plasma membranes of cells. Rather, they bind with receptors on the cell surface, and the hormone-receptor complex activates an enzyme on the inside of the membrane (Fig. 19.3b). This enzyme, in turn, leads to the formation of a second intracellular messenger. The best understood process of this type is the hormonal activation of the membrane-bound enzyme **adenylate cyclase**, which forms the second messenger, **cyclic adenosine monophosphate (cAMP)**, from ATP. Cyclic adenosine monophosphate, in turn, activates a **kinase**. Kinases are a family of proteins that transfer a phosphate group from ATP to specific enzymes, thereby activating them. These enzymes, in their turn, catalyze cellular reactions responsible for the biological effects of the hormone. After cAMP has activated protein kinase, it is converted to inactive AMP by another enzyme. Again the basic mechanism is known, but much remains to be learned about the ways different hormones cause the activation of different enzymes. Evidence sug-

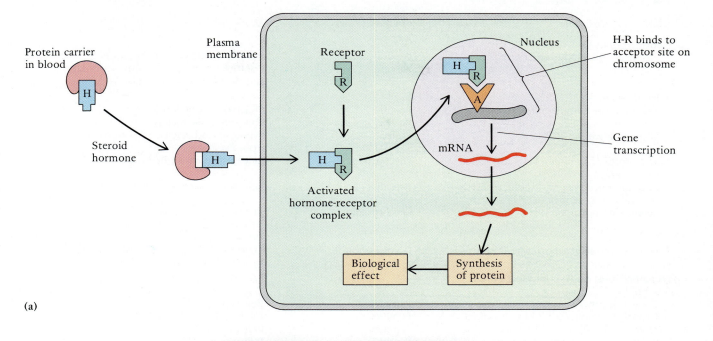

(a)

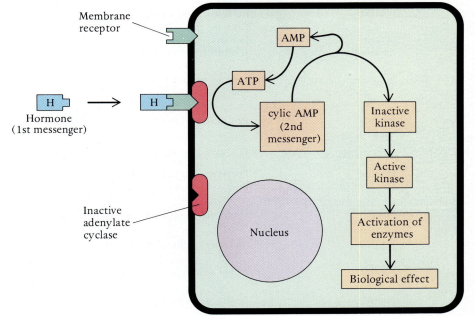

(b)

Figure 19.3
Methods of hormone action. (a) A steroid hormone enters the cell and binds with a receptor. The receptor may be in the cytoplasm or the nucleus. (b) A protein or peptide hormone binds with a membrane receptor, which in turn activates an enzyme (the second messenger) on the inside of the membrane.

gests that other intracellular messengers and a family of protein kinases are involved.

Hormone concentrations in the bloodstream are generally very low, ranging from 10^{-11} to 10^{-12} molar, yet they can exert profound effects because of the amplification cascade that occurs at the targets. Steroid hormones activate individual genes, bringing about the synthesis of many molecules of that gene's product, which is usually an enzyme. Similarly, with peptide and protein hormones, the second messenger may activate only a few protein kinase molecules, but these initiate a cascade of protein activations. Once proteins are formed and activated, the processes that they control are no longer dependent on the continued presence of the hormone. Hormone levels in the blood may fall, but their effects continue for some time.

Table 19.1
The Major Mammalian Hormones

Gland and Hormone	Chemical Class	Major Biological Effects
Neurohypophysis		
Vasopressin (Antidiuretic hormone)	Peptide	Water reabsorption from kidney
Oxytocin	Peptide	Contraction of smooth muscles in uterus and ducts of mammary glands
Adenohypophysis		
Melanophore-stimulating hormone (MSH)	Peptide	Melanin production
Prolactin	Protein	Milk synthesis, maternal behavior
Growth hormone (GH)	Protein	Synthesis and release of somatomedins from liver
Thyrotropic hormone (TH); also called thyrotropin	Glycoprotein	Release of thyroid hormones
Adrenocorticotropic hormone (ACTH); also called corticotropin	Peptide	Release of hormones from adrenal cortex
Luteinizing hormone (LH)	Glycoprotein	Production of testosterone, estradiol, ovulation, growth of corpus luteum which produces progesterone
Follicle-stimulating hormone (FSH)	Glycoprotein	Growth of seminiferous tubules and ovarian follicles, estradiol synthesis
Hypothalamus via Hypophyseal Portal System		
Growth hormone–releasing hormone	Peptide	Releases GH
Growth hormone–inhibiting hormone	Peptide	Inhibits release of GH
Thyrotropin-releasing hormone	Peptide	Releases TH
Corticotropin-releasing hormone	Peptide	Releases ACTH
Gonadotropin-releasing hormone	Peptide	Releases gonadotropic hormones (LH and FSH)
Prolactin-releasing hormone	Peptide	Releases prolactin
Prolactin-inhibiting hormone	Amino acid derivative	Inhibits release of prolactin
Pineal Gland		
Melatonin	Amino acid derivative	Inhibits gonadal development in some species
Liver		
Somatomedins	Peptide	Synthesis of DNA and protein, cell growth
Thyroid Gland		
Triiodothyronine / Thyroxine	Amino acid derivatives	Elevate metabolic level
Calcitonin	Peptide	Prevents excessive withdrawal of Ca^{2+} from bones
Kidney Cortex		
Erythropoietin	Protein	Synthesis of erythrocytes in bone marrow
Blood Platelets		
Platelet-derived growth factor	Peptide	Mitosis of cells near an injury
Thymus		
Thymosin	Peptide	Development of immune system
Mucosa of Stomach		
Gastrin	Peptide	Gastric secretion
Mucosa of Small Intestine		
Secretin	Peptide	Release of alkaline solution from pancreas
Cholecystokinin	Peptide	Release of pancreatic enzymes, contraction of gallbladder
Islets of Langerhans		
Insulin	Protein	Entry of glucose into cells, glucose utilization and storage
Glucagon	Peptide	Conversion of glycogen to glucose
Adrenal Medulla		
Norepinephrine / Epinephrine	Amino acid derivatives	Augment action of sympathetic nervous system
Adrenal Cortex		
Glucocorticoids (Cortisol)	Steroid	Protein and carbohydrate metabolism
Androgens	Steroids	Muscular development, pubic hair growth, libido
Mineralocorticoids (Aldosterone)	Steroid	Electrolyte and water balance, blood pressure

Gland and Hormone	Chemical Class	Major Biological Effects
Blood Plasma		
Angiotensin	Protein	Secretion of aldosterone
Parathyroid Glands		
Parathyroid hormone (PTH)	Peptide	Ca^{2+} reabsorption from kidneys and release from bones
Skin		
Vitamin D	Steroid	Promotes absorption of mineral ions from gut, release of Ca^{2+} from bone
Leydig Cells of Testis*		
Testosterone	Steroid	Growth of male reproductive organs, secondary sex characteristics, male behavior
Ovarian Follicles*		
Estradiol	Steroid	Growth of female characteristics, early development of uterine lining, growth of follicles
Corpus Luteum*		
Progesterone	Steroid	Maturation of uterine lining
Placenta*		
Chorionic gonadotropin	Glycoprotein	Maintains corpus luteum
Progesterone	Steroid	Maintenance of uterine lining, suppression of follicle development, development of glandular tissue in mammary glands
Estradiol	Steroid	Growth of uterine muscles, development of glandular tissue in mammary glands, labor contractions
Relaxin	Peptide	Relaxes cervix and pelvic ligaments
Placental lactogen	Peptide	Milk synthesis

*These hormones are discussed in Chapter 21.

Focus 19.1 *Hormonal Regulation of Genes*

In 1952, W. Beermann discovered that particular regions of the giant chromosomes of the midge, *Chironomus*, became enlarged or "puffed" at various stages of development (Focus fig.). These giant chromosomes are easy to see microscopically because they are composed of thousands of chromatid strands in permanent synapsis. The puffing results from the unwinding and slight separation of strands in particular regions.

Later studies showed that these puffs were correlated with the location of specific genes on the chromosomes. In one study a puff was associated with the ability to synthesize a particular salivary protein. Beermann discovered two different species of *Chironomus*, one of which could produce this protein and had a puff at a certain stage of development, and one that could not produce the protein and lacked the puff. By crossing these species he found that the ability to synthesize this protein followed the expected Mendelian pattern of inheritance: the puff appeared to represent the activity of the gene responsible for this protein.

Puffs appear in this midge and other flies when the genes become active and begin to synthesize mRNA. That a steroid hormone activates the transcription of specific genes has been demonstrated by labeling ecdysone (an insect molting hormone, p. 441) with fluorescent antibodies. Puffs appear when ecdysone binds to the chromosome.

REGULATION OF HORMONE LEVELS

The effects of hormones are proportional to their concentrations in the blood. The primary factor in determining their concentration is their rate of secretion, and this is carefully regulated by several feedback mechanisms. A familiar example of a feedback system is the control of room temperature by the thermostat and furnace. The thermostat is set for a certain temperature, called the set point, and a thermometer in it monitors temperature changes in the room. As the temperature falls below the set point of the thermostat, a signal turns on the furnace; as the temperature rises, another signal turns the furnace off. Room temperature oscillates slightly above or below the set point, but remains relatively constant. This is called a **negative feedback system** because room temperature and furnace output move in different directions: as room temperature falls, heat output increases, and as room temperature rises, heat output decreases.

The secretion of most endocrine glands is also controlled by negative feedback, but unlike the furnace, which is either on or off, most endocrine glands simply change their rates of secretion. In some cases the fall in the blood level of a hormone, or in the level of the substance it controls, is detected directly by the gland that produces the hormone and more is secreted; its increased concentration in the blood reduces the rate of secretion. Often two or more glands are involved in a negative feedback loop (Fig. 19.4). For example, the blood level of testosterone, a hormone produced by certain cells in the testis, is not controlled by the testis but by **gonadotropic hormones** secreted by the anterior lobe of the pituitary gland. As the blood level of testosterone falls, the level of gonadotropic hormone increases and more testosterone is secreted. A further complication comes about because testosterone does not feed back primarily to the pituitary gland, but to neurosecretory cells in the hypothalamus. These cells monitor the level of testosterone and vary their secretion of a **gonadotropin-releasing hormone** that then acts on the pituitary. This is an example of the close interaction between the nervous and endocrine systems.

In a few cases there is a nervous component to hormone regulation (Fig. 19.5). When an infant starts to suckle, sensory impulses are sent from the breast to the brain. On reaching the hypothalamus, the impulse causes the release of the hormone oxytocin from the posterior lobe of the pituitary gland. Oxytocin reaches the breasts through the circulatory system, causing the contraction of myoepithelial cells around the small sacs, or aveoli, where milk is stored, and milk is ejected. The release of milk is an example of a **neuroendocrine reflex**, in which part of the reflex arc is nervous and part is endocrine. Breast feeding continues the secretion of oxytocin, which also causes continued contractions of the uterus and the quick return of the uterus to normal size following delivery.

INVERTEBRATE HORMONES: ARTHROPOD MOLTING

When investigators searched for invertebrate hormones using the knowledge and techniques developed in vertebrate studies, they found that most invertebrates produce them. Neurosecretory cells may be a fundamental site of

Figure 19.4

Negative feedback control of endocrine secretion is illustrated by the regulation of testosterone secretion. A fall in testosterone production by the testis (−) promotes the secretion of a gonadotropin-releasing hormone by neurosecretory cells in the hypothalamus (+). This hormone, in turn, triggers the release of gonadotropic hormones by the anterior lobe of the pituitary gland, causing the production of testosterone to rise to normal levels. A fall in testosterone has a secondary effect (--→) directly on the pituitary gland.

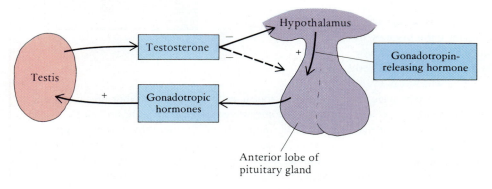

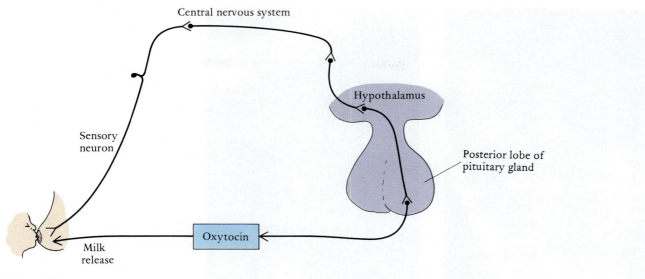

Figure 19.5
The release of milk to a suckling infant is a neuroendocrine reflex with both neural and endocrine components.

hormone production among invertebrates, for such cells are found in cnidarians, flatworms, nematodes, molluscs, annelids, arthropods, and echinoderms. Most of the hormones produced by neurosecretory cells affect reproduction and growth, but in some crustaceans they mediate color changes and several aspects of metabolism. Unlike the situation in vertebrates, only a few endocrine glands in invertebrates are derived from epithelial tissue, and they have been found only among arthropods. Endocrine integration among invertebrates has been studied most thoroughly in crustaceans and insects.

We will examine molting in arthropods as an example of hormonal action in the invertebrates. Arthropods are encased in a firm, chitinous exoskeleton that restricts an increase in body size until it is shed or molted, a process called ecdysis (p. 206). Periodic molting occurs throughout life in crayfish and some other crustaceans. Insects have a more complex life cycle that includes one or more feeding larval stages (called caterpillars in moths and butterflies), often a quiescent pupal stage or cocoon, and finally the reproducing adult (Fig. 19.6). Molting occurs between larval stages and between the larva and pupa, but the adult insect neither grows nor molts. Processes as complex as this require precise integration.

The tissues that respond during molting are stimulated in both crustaceans and insects by a steroid molting hormone known as **ecdysone** that is secreted by endocrine glands of epithelial origin: the **Y organ** in crustaceans, which is located near the base of the eyestalk (Fig.

19.7a), and the **prothoracic gland** in insects, located ventrally in the front of the first part of the thorax. These glands, in turn, are regulated by neurosecretory cells whose cell bodies are located in an **X organ** in one of the eyestalk ganglia of crustaceans, or in a part of the brain of insects. Neurosecretions produced by these cells are transported along axons and stored in the swollen ends of the axons in a **sinus gland** in the eyestalk of crustaceans, or in the **corpora allata** just posterior to the brain of insects. Upon appropriate stimulation, the neurosecretory hormones are released into the hemolymph.

The way the neurosecretions regulate molting differs between crustaceans and insects. In crustaceans the neurosecretion of the X organ–sinus gland complex is a **molting-inhibiting hormone** that stops the release of ecdysone from the Y organ (Fig. 19.7b, *left*). This was discovered by removing the eyestalk, which contained the sinus gland. Molting was induced. The activity of the neurosecretory cells must be inhibited by the nervous system for molting to occur. Crayfish molt seasonally in the spring, and an increase in day length is the stimulus that decreases the activity of the neurosecretory cells and allows the Y organ to produce ecdysone.

Two neurosecretions are produced in insects by different brain cells (Fig. 19.7b, *right*). **Ecdysiotropin** stimulates the activity of the prothoracic glands and the production of ecdysone. A **juvenile hormone** acts directly on the molting tissues and affects the way they respond to ecdysone. It tends to inhibit their transformation and so

(a)

(b)

(c)

Figure 19.6
Stages in the development of a monarch butterfly (*Danaus plexippus*). The larva or caterpillar actively feeds (a), molts several times, and transforms into a dormant pupa. Metamorphosis occurs in the pupal stage and the adult emerges (b). An adult monarch (c).

keeps them in the immature condition. The nature of the molt depends on the balance between these two hormones. When juvenile hormone is in excess of ecdysone, a larva molts into a larger larva; molting occurs, but the juvenile morphology is retained. Nearly equivalent concentrations of the two hormones cause a larva to metamorphose into a pupa. A large amount of ecdysone and little juvenile hormone induce a late larva or pupa to change into the adult. The stimuli for these changes vary among insects. Molting is controlled in many species by seasonal factors including day length and temperature. The larva of a blood-sucking bug, *Rhodnius*, feeds only when rare opportunities present themselves. It molts when a full meal activates stretch receptors in its gut.

Chemical analogues of juvenile hormone have been synthesized and some appear promising as insecticides that can be used to control insect populations without the use of toxic materials. Test sprays of these substances on caterpillars and the plants on which they feed prevented normal metamorphosis and caused the eventual death of the larvae.

VERTEBRATE HORMONES

The Hypothalamus and Pituitary Gland

Neurosecretion is also central to the control of much of the vertebrate endocrine system because neurosecretory cells in the hypothalamus affect the pituitary gland. The pituitary gland, technically known as the **hypophysis cerebri** (Gr. *hypo*, under + *physis*, growth), develops embryonically from two components (Fig. 19.8a). An **infundibulum** extends ventrally from the diencephalon and forms the

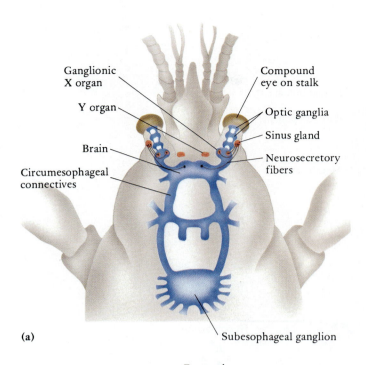

(a)

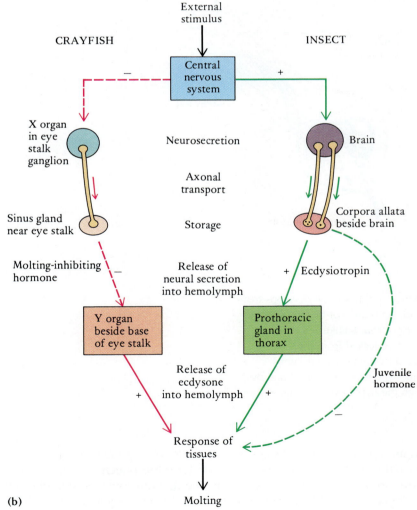

(b)

Figure 19.7

(a) The X organ, sinus gland, and Y organ of a crayfish are located in or near the eye stalk. (b) The control of molting in a crayfish (*left*) and an insect (*right*). In this and similar diagrams a + indicates stimulation; a − indicates inhibition.

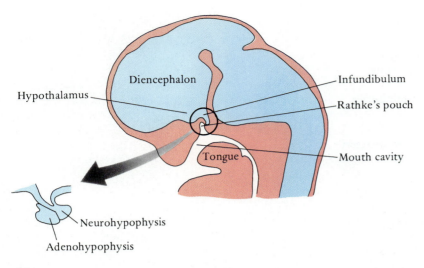

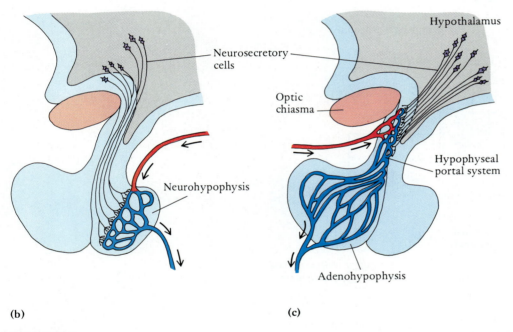

(b)   **(c)**

Figure 19.8

(a) The pituitary gland consists of two parts: a neurohypophysis that develops from an infundibulum that grows down from the part of the brain that becomes the hypothalamus, and an adenohypophysis that develops from Rathke's pouch that grows up from the roof of the mouth. (b) Secretions released from the neurohypophysis are produced by neurosecretory cells in the hypothalamus. (c) Secretions produced and released by the adenohypophysis are controlled by hypothalamic neurosecretions that reach it through the hypophyseal portal system.

posterior lobe of the gland, or the **neurohypophysis** (Fig. 19.8b). The neurohypophysis remains connected to the ventral surface of the diencephalon, which develops into the hypothalamus. The anterior lobe of the pituitary, or **adenohypophysis** (Gr. *aden*, gland), develops as a pouch-shaped outgrowth, called **Rathke's pouch**, from the roof of the embryonic mouth. Later this pouch loses its connection with the mouth. The adenohypophysis is supplied by blood vessels that descend from the adjacent hypothalamus. Capillaries in the hypothalamus converge on small veins that lead to another capillary bed in the adenohypophysis. This system of vessels is the **hypophyseal portal system** (Fig.19.8c).

Neurohypophysis

Since the neurohypophysis is an extension of the part of the embryonic brain that forms the hypothalamus, it should not be surprising that its hormones are neurosecretions produced by neuron cell bodies in the hypothalamus. They are transported along axons and stored in the terminals of the axons in the neurohypophysis (Fig. 19.8b).

Two neurohypophyseal hormones occur in all vertebrates. Each consists of only eight amino acids and they differ from each other by one or two amino acid substitutions. **Vasopressin**, or **antidiuretic hormone**, in mammals promotes water reabsorption from parts of the kidney tubules and hence prevents diuresis, the excretion of a copious and very dilute urine. It also raises blood pressure slightly by causing the smooth muscle in peripheral arterioles to contract. It is essential for the maintenance of water balance, and its release is controlled by the water content of the blood. A variant of vasopressin known as **vasotectin**, which differs by only one amino acid, is found in other vertebrates. It too promotes water reabsorption, but from the skin and urinary bladder of amphibians. The second hormone, **oxytocin**, differs from vasopressin in not affecting water balance and in having a stronger effect upon the contraction of the smooth muscles used in milk ejection (Fig. 19.5) and those of the pregnant uterus during childbirth. Oxytocin is sometimes used clinically to promote childbirth. Oxytocin variants present in other vertebrates stimulate contractions of the oviduct.

Adenohypophysis

Seven peptide, protein, and glycoprotein hormones are made by distinct cell types in the adenohypophysis (Table 19.1). **Melanophore-stimulating hormone (MSH)** causes the skin of ectothermic vertebrates to darken by dispersing organelles containing the pigment melanin throughout their chromatophores. The withdrawal of melanin-containing organelles toward the center of the chromatophore is caused by sympathetic stimulation. Although they lack chromatophores in which pigment migrates, birds and mammals have pigment-producing cells, and MSH promotes melanin production in them. **Prolactin**, so named because it stimulates the production and secretion (but not the release or ejection) of milk in mammals, is essential in many ways for reproduction in most vertebrates. It stimulates reproductive migrations, promotes nest building, and initiates maternal care. Roosters, if given prolactin, will brood the chicks! It also promotes fat deposition, which provides energy reserves for reproduction in many species. Finally, it has an influence on the salt- and water balances in the body, especially in those fishes that migrate between salt- and freshwater environments.

The five other adenohypophyseal hormones affect the production and release of hormones by other endocrine glands. The thyroid gland is the target of the **thyrotropic hormone (TH)** (Gr. *trophe*, nourishment). The adrenal cortex that forms the outer part of the adrenal gland is the target of **adrenocorticotropic hormone (ACTH)**. **Growth hormone (GH)** promotes body growth. Two gonadotropic hormones, **follicle-stimulating hormone (FSH)** and **luteinizing hormone (LH)**, promote the synthesis of reproductive hormones by certain cells in the ovary and testis.

The level of adenohypophyseal hormones that affects other glands is controlled by negative feedback systems. In some cases changes in the blood level of the hormones of the target glands affect the rate of secretion of certain adenohypophyseal cells directly, but most target hormones act on the adenohypophysis through the hypothalamus. Neurosecretory hypothalamic cells respond by changing their rates of production of hormones that are discharged into the origin of the hypophyseal portal system through which they are carried to their target cells in the adenohypophysis (Fig. 19.8c). Most of these hypothalamic hormones act as **releasing hormones**, but growth hormone and prolactin secretion appears to be regulated by an **inhibiting hormone** as well (Table 19.1). Various nerve impulses also affect the rate of production of releasing hormones, bringing much of endocrine activity under nervous control. For example, emotional states or changes in day length, which act through receptors and the nervous system, affect the endocrine balances of the body.

Pineal Gland

A **pineal gland** develops as an outgrowth from the roof of the diencephalon in most birds and mammals and remains attached to it (see Fig. 19.1). Cells within the pineal gland secrete **melatonin**, particularly at night or when the animal is kept in the dark. Light inhibits the enzymes needed for melatonin synthesis. In rats, light entering the eyes affects a hypothalamic nucleus that, in turn, sends inhibitory impulses to the gland via sympathetic neurons. Because melatonin production has a diurnal cycle, it may influence physiological processes such as color changes in some vertebrates, body temperature, and reproduction and adjust them to diurnal or seasonal changes in ambient light levels. It is not clear how widespread these effects are. Melatonin does have an inhibitory effect upon the development of the ovary and testis in rodents. An increase in light reduces this effect and allows the development of these organs. Pineal tumors in humans upset the onset of puberty.

In other vertebrates, the pineal gland is represented by a photoreceptive **pineal** or **parietal eye** on the top of the head that in some cases modifies the physiology and behavior of the animal in relation to light. Many lizards maintain an optimum body temperature during the

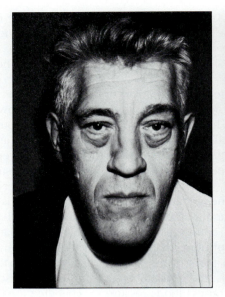
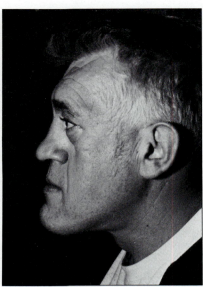

Figure 19.9
An individual with acromegaly. Note the disproportionately large size of the jaws.

daytime by moving between sunny and shaded areas. Changes in light that lead to this behavior are detected by the parietal eye.

Hormones Related to Growth and Maturation

Growth Hormone and Somatomedins
The hormones of the many other endocrine glands interact in complex ways, but it is convenient to simplify our discussion somewhat by grouping them according to the major activities that they regulate. Many hormones affect the growth of one or more tissues in the body. Of particular interest in controlling postnatal growth is **growth hormone** released from the adenohypophysis. Growth hormone is a protein with wide-ranging effects: it promotes skeletal growth by stimulating the proliferation of cartilage cells in the epiphyseal plates of bone (p. 209), it stimulates the growth of connective tissues, and it promotes the uptake of amino acids by cells of many types. The last characteristic stimulates protein synthesis in muscles and other tissues. Just how growth hormone acts is not entirely clear, but it is known that it causes the synthesis of **somatomedins**, a group of peptides produced by liver cells. The somatomedins promote cell division and differentiation, and mediate many of the effects of growth hormone.

The plasma level of growth hormone is regulated by hormones released by the hypothalamus. Low plasma levels of growth hormone and somatomedins induce the release of a **growth hormone–releasing hormone**. A high level of growth hormone and somatomedins induce the release of a **growth hormone–inhibiting hormone** (**somatostatin**) that dampens the activity of ad-

enohypophyseal cells secreting growth hormone. Appropriate levels of growth hormone and somatomedins are normally maintained by these feedback mechanisms, but imbalances sometimes occur. Oversecretion of growth hormone in childhood causes excessive growth and results in very tall but well-proportioned individuals. Undersecretion during childhood results in a stunting of growth and dwarfism. Oversecretion during maturity causes excessive bone growth in the face, hands, and feet, a condition known as acromegaly (Fig. 19.9).

Thyroid Hormones
The thyroid gland develops embryonically as an outgrowth from the floor of the front of the pharynx and then migrates posteriorly to lodge in the ventral surface of the neck (Fig. 19.1). It consists of many small epithelial follicles whose cells extract iodine and the amino acid tyrosine from the blood, and bind them to a protein to form **thyroglobin** (Fig. 19.10a). This protein is secreted into a colloid in the lumen of the follicles, where it is stored. Under the stimulus of thyrotropic hormone from the adenohypophysis, thyroglobin reenters the follicle cells. The active hormones, **triiodothyronine** and **thyroxine**, are cleaved from thyroglobin and enter the blood, where they bind loosely to plasma proteins. Although their level in the blood is controlled primarily by negative feedback mechanisms to the hypothalamus (Fig. 19.10b), nerve impulses also affect the release of hypothalamic **thyrotropin-releasing hormone**. Cold temperatures, for example, promote the production of the releasing hormone; stress reduces it.

Although they are not steroids, the thyroid hormones are small molecules that diffuse through the membranes of the cell and nucleus, bind with a nuclear receptor,

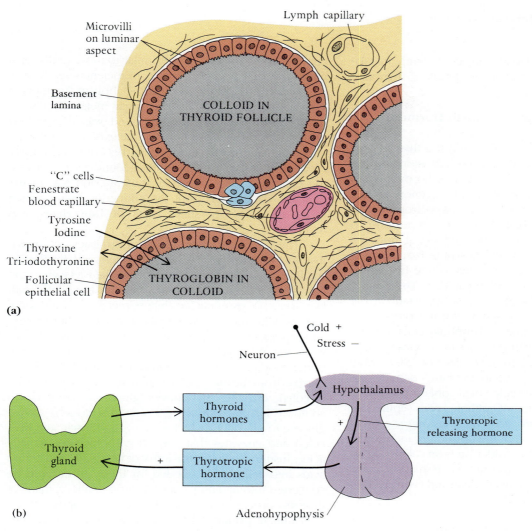

(a)

(b)

Figure 19.10
The thyroid gland. (a) Thyroid follicles are shown in a drawing of a microscopic section. (b) The control of thyroid secretion.

activate genes, and promote the synthesis of proteins. Many of the enzymes regulated by thyroid hormones are involved in cellular respiration. The high levels of metabolism and heat production of endothermic birds and mammals require adequate levels of the thyroid hormones. The thyroid hormones of mammals are needed for the normal absorption and utilization of glucose and fats. They interact with somatomedins and other hormones to promote protein synthesis and normal growth and differentiation. The skin and many of its derivatives, including scales, feathers, and hair, depend on thyroid regulation for normal growth.

Abnormally low levels of thyroid hormones in childhood cause retarded physical and mental development, a condition referred to as **cretinism**. Low levels in adults result in a poor utilization of food products, low metabolism, low body temperatures, and an increase in weight;

abnormally high levels cause an elevation of basal metabolic rate, excessive perspiration, weight loss, and other complications.

All vertebrates have a thyroid gland, but its function in ectothermic vertebrates is not fully understood. Thyroid hormones are needed for skin molting in amphibians and reptiles, and they play a very important role in amphibian metamorphosis. The transition from a tadpole to an adult is a period of elevated metabolism, and increasing levels of thyroid hormones are needed to sustain it. Early metamorphosis can be induced by providing extra doses of thyroid hormones, or, in some cases, iodine. Conversely, an insufficiency of thyroid hormones leads to an arrested development. Many salamanders are normally **neotenic**, that is, somatic development is slowed down relative to sexual development so they can reproduce while still retaining external gills and other larval features. Neoteny in salamanders is often the result of low levels of thyroxine (p. 852).

Other Protein and Peptide Growth Hormones and Growth Factors

Growth hormone, somatomedins, and the thyroid hormones have wide-ranging effects on cell division, growth, and differentiation, but a number of other hormones and growth factors have been discovered in recent years that affect the growth of particular tissues. **Erythropoietin**, a hormone synthesized in the kidney cortex in response to low levels of oxygen, promotes the synthesis of more red blood cells (p. 300). Blood platelets synthesize a **platelet-derived growth factor** that promotes wound healing by causing neighboring cells to undergo mitosis. **Thymosin**, a hormone secreted by the thymus, is important in the development of the immune system, for it promotes the proliferation and maturation of lymphoid cells. Certain **interleukins**, secreted by lymphocytes in response to antigens, promote the multiplication of a clone of cells capable of acting against the antigen (Chap. 15). Other peptides are known to stimulate the formation of fibroblasts (**fibroblast growth factor**), the division of epithelial cells (**epidermal growth factor**), and the growth of neurons (**nerve growth factor**). The primary action of the hormones and growth factors that we have been considering is on cell proliferation, growth, and differentiation, but, as we shall see, many of the hormones that affect metabolism and reproduction also influence growth.

Hormones Related to Metabolism and Homeostasis

Gastrointestinal Hormones

To be metabolized, food products must first be digested and absorbed. The sequential release of the digestive enzymes as food passes through the digestive tract (Chap. 12) is integrated partly by the autonomic nervous system and partly by hormones that are synthesized in the gastrointestinal mucosa in response to food in the gut lumen (Fig. 19.11). A number of polypeptide and peptide hormones have been identified, but the actions of only a few have been determined: **gastrin**, which stimulates secretion of gastric acid by the stomach; **secretin**, which promotes secretion of an alkaline solution from the pancreas; and **cholecystokinin**, which stimulates release of pancreatic enzymes and contraction of the gallbladder (and thus release of bile).

Glucose Metabolism and Regulation

In 1889, during studies on digestive functions, two German investigators, J. von Mering and O. Minkowski, surgically removed the pancreas from a dog. The caretaker of the animals noticed that flies were attracted to the dog's urine, and tests showed that it had a very high sugar content. The symptoms were similar to those of **diabetes mellitus**, a fatal disease known from ancient times. Clearly the pancreas had a role in sugar metabolism. Attempts to extract the factor that affected sugar metabolism were unsuccessful because the compound was digested by proteolytic enzymes produced by the exocrine part of the pancreas. Nearly 30 years later, in 1921, two Canadians, Frederick Banting and Charles Best, found that by tying off the pancreatic duct of a dog, the exocrine part of the pancreas degenerated, but the endocrine tissue, which forms thousands of small **islets of Langerhans**, was unaffected.

The islets produce several hormones, of which **insulin** and **glucagon** are the best understood. Insulin is a small protein secreted by beta cells in the islets; glucagon is a peptide secreted by alpha cells in the islets. Both hormones are essential for glucose metabolism and some aspects of fat and protein metabolism. Like other protein and peptide hormones, they activate intracellular enzymes via a second relay messenger, cAMP. For reasons not understood, most cells need insulin to be able to take up glucose from the blood, and insulin activates several of the glycolytic enzymes that convert glucose to pyruvate. In addition, insulin activates enzymes that convert excess glucose to glycogen (which is stored in muscle and liver cells) and enzymes that convert carbohydrates to fatty acids. Insulin facilitates the entry of fatty acids and amino acids into cells and is needed for these products to be converted into fat and protein. Glucagon affects many of the same processes as insulin, but its actions oppose those of insulin. Its primary target is liver cells, where it activates enzymes that convert glycogen to glucose, thus increasing blood sugar levels.

The primary regulator of insulin and glucagon is glucose itself (Fig. 19.12). Additional factors that have an effect include gastrointestinal hormones released during digestion, which stimulate insulin production, and sympathetic stimulation of the islets of Langerhans during stress or heavy exercise, which causes glucagon release and an increase in blood sugar. This is part of the "fight-or-flight response" that we discussed previously (p. 338). Glucose metabolism is also affected by many other hormones, including thyroid hormone, epinephrine, and glucocorticoids.

Low levels of insulin cause diabetes mellitus, which is the most common human endocrine disorder, affecting about 5% of the population of the United States. In diabetic patients, low insulin levels result in glucose not being adequately taken up by cells or effectively utilized. Blood sugar levels rise (**hyperglycemia**), excess sugar is excreted by the kidneys, and the osmotic effect of this causes a copious urine production accompanied by water and electrolyte loss. Fat and protein metabolism are also disrupted. The tissues, unable to use glucose as fuel, break down fat and protein, so a steady weight loss occurs. Fatty acids are

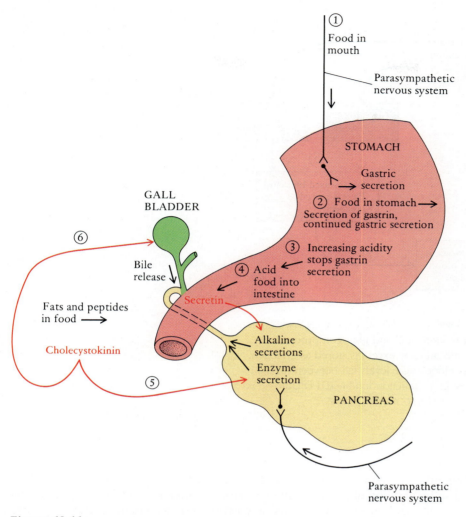

Figure 19.11

Digestive secretions of the stomach, pancreas, and gallbladder are controlled partly by nerve impulses (black arrows) and partly by the secretion of hormones (red arrows). (1) Food in the mouth initiates a reflex that sends impulses down the vagus nerve and starts gastric secretion. (2) When food enters the stomach, gastric secretion is continued by the production of the hormone gastrin. (3) As acidity levels in the stomach increase, gastrin production and gastric secretion decrease. (4) The entry of food into the small intestine stimulates intestinal cells to synthesize two hormones. Secretin production is induced by the acidity of the food, and promotes the secretion of a copious alkaline solution from the pancreas. (5) The release of pancreatic enzymes requires another hormone, cholecystokinin, whose production is initiated by the fats and peptides in the intestinal contents. Parasympathetic stimulation of the pancreas supplements this hormone in promoting enzyme release. (6) Cholecystokinin also stimulates contraction of the gallbladder and the release of bile, which is essential for fat emulsification and absorption.

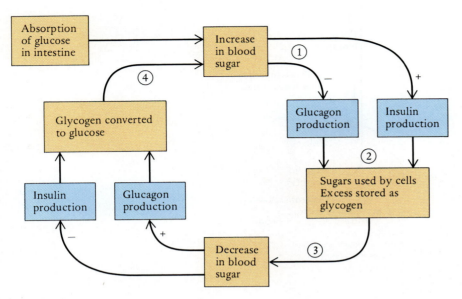

Figure 19.12
The hormonal regulation of blood sugar level. (1) As blood sugar level rises (after a meal, for example), insulin production is stimulated and glucagon production inhibited. (2) Glucose is utilized by the cells and any excess is stored as glycogen, primarily in liver and muscle cells. (3) As blood sugar levels fall between meals, glucagon production is stimulated and insulin production inhibited. (4) Glucose is released from storage and blood sugar levels rise to normal levels.

not completely metabolized and accumulate as partly oxidized ketone bodies. Although the insulins of different species vary slightly in their amino acid sequences, human patients can be treated by regular injections of insulin prepared from bovine and porcine pancreas because they are quite similar to human insulin. Recently it has become possible to synthesize human insulin using genetically engineered bacteria.

Adrenal Hormones

The adrenal glands of mammals are located adjacent to the superior end of each kidney (see Fig. 19.1) and consist of two quite distinct parts. The surface **cortex** is formed by islands and branching columns of epithelial cells; the deeper **medulla**, by **chromaffin cells** embedded in a connective tissue (Fig. 19.13).

Adrenal cortex and medulla are quite distinct in fishes, for the chromaffin cells form many small clusters that are separated from the cortical material. The cortical and medullary components of the gland come together to some extent in other vertebrates and are intimately associated in birds and mammals. The chromaffin cells of all vertebrates produce norepinephrine, a derivative of the amino acid tyrosine. Most of this is methylated in mammals to produce **epinephrine**, a similar but somewhat more

potent product. The methylation of norepinephrine requires the activation of a medullary methylating enzyme by the hormone **cortisol**, produced in the cortex. The close association of the two parts of the gland in mammals allows cortisol to reach the medulla directly. Norepinephrine and epinephrine are released when the chromaffin cells are stimulated by *preganglionic* sympathetic neurons. Embryological studies have confirmed that chromaffin cells are modified postganglionic sympathetic neurons, which explains why norepinephrine is produced by both the adrenal medulla and postganglionic neurons (p. 388).

Norepinephrine and epinephrine released by the chromaffin cells reinforce the effects of sympathetic stimulation in the fight-or-flight response. They act by binding with alpha and/or beta receptors on the membranes of the target cells (Fig. 19.14). Since these receptors activate different intracellular enzymes via different second messengers, the effects of norepinephrine and epinephrine on different target cells vary according to the distribution of the receptors on the targets. Some cells have either alpha or beta receptors, others a mixture of each. Binding with beta receptors activates the enzyme adenylate cyclase (p. 436), which affects many aspects of metabolism and causes an increase in the blood level of glucose and fatty acids. It also promotes many other reactions in the fight-or-flight

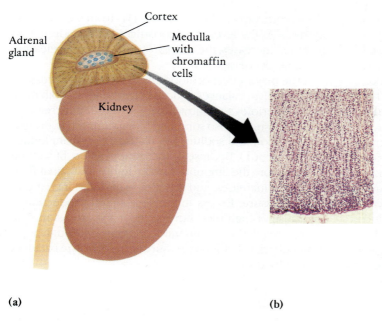

Figure 19.13

(a) The mammalian adrenal gland lies above the kidney and consists of two distinct parts: cortex and medulla. (b) A photomicrograph through the adrenal cortex of a rhesus monkey.

Figure 19.14

The effects of a particular hormone on a target can vary according to the receptors present on the target. The cell surface shown here has both alpha and beta receptors, but some target cells have only one or the other.

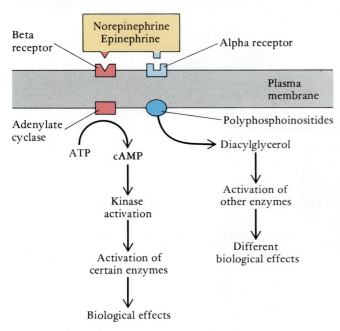

response including an increase in the force of heart contraction, a dilation of coronary arteries and those in skeletal muscles, a dilation of bronchioles, and a relaxation of smooth muscles in the gut with a concomitant decrease in peristalsis. Binding with alpha receptors hydrolyzes a membrane component (**polyphosphoinositide**), one of whose products is a different second messenger (**diacylglycerol**) that activates a different group of intracellular enzymes. This promotes constriction of blood vessels (other than those in the heart and skeletal muscles), induces sweating, and causes the contraction of gastrointestinal sphincters. The different distribution of receptors explains, for example, why epinephrine promotes the dilation of blood vessels in the heart and skeletal muscles, but the constriction of blood vessels in the gut. Some clinical drugs act by blocking one or the other of these hormone receptors.

Cells in the adrenal cortex produce many hormones that can be sorted into three groups: (1) **glucocorticoids**, of which **cortisol** is the most important, (2) **androgens**, which are similar to testosterone produced in the testis, and (3) **mineralocorticoids**, of which **aldosterone** is the most important. Cortisol has many metabolic effects. It stimulates the breakdown of protein in muscle into amino acids and aids their subsequent conversion into carbohydrates, which takes place in liver cells. This increases blood sugar levels and the formation of glycogen from excess glucose. It also enhances the effect of epinephrine in converting fats to fatty acids. These reactions help an animal adjust to long-term stress. Experimental animals subject to prolonged stress have enlarged adrenal glands. High levels of glucocorticoids suppress many of the body's normal inflammatory and immunological responses. They inhibit capillary dilation, decrease capillary permeability, and inhibit the mobilization of leukocytes. These reactions, too, help an animal adapt to certain aspects of stress.

Cortical androgens have little effect in males because they are less potent than the primary male hormone, testosterone, but in females they promote some muscle development and the growth of pubic hair, and lead to the development of sexual drive or libido. The level of glucocorticoids and androgens in the blood is regulated by a negative feedback mechanism. Hypothalamic **corticotropin-releasing hormone** promotes the release of adrenocorticotropic hormone (ACTH) from the adenohypophysis. As the level of glucocorticoids and androgen in the blood increases, the release of corticotropin-releasing hormone decreases.

The major effect of aldosterone is on blood pressure and electrolyte balances. Aldosterone causes smooth muscles in arterioles to contract, with a resulting increase in blood pressure. It also stimulates potassium and hydrogen ion excretion and sodium ion reabsorption from kidney tubules (p. 361). Because of the osmotic effect of sodium reabsorption, the amount of water in the interstitial fluid and blood increases, and this also leads to an increase in blood pressure. Excess mineralocorticoid production is one cause of high blood pressure, or **hypertension**. The primary regulation of mineralocorticoids is through negative feedback with blood pressure and electrolyte balances in the blood (Fig. 19.15).

Figure 19.15

The control of aldosterone secretion, mineral balances, and blood pressure involves interactions between blood pressure, a kidney enzyme, a circulating globulin, and the adrenal cortex. (1) An insufficient level of aldosterone causes a lowering of blood pressure, and less sodium is reabsorbed from the kidneys. Changes in blood pressure and sodium levels are detected by a group of kidney cells (the juxtaglomerular apparatus, p. 361). (2) They then produce renin, an enzyme that initiates the conversion of a globulin circulating in the plasma into angiotensin. (3) An increase in angiotensin stimulates the cells in the adrenal cortex that produce aldosterone. (4) Aldosterone, in turn, increases blood pressure and sodium reabsorption by the kidneys.

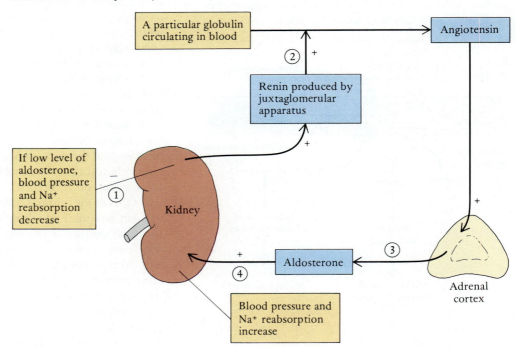

Calcium and Phosphate Homeostasis

As we have seen, calcium and phosphate ions are essential in a great many biochemical reactions. Three hormones interact to maintain the homeostasis of these ions. **Calcitonin**, a peptide, is secreted by **C cells** that develop as epithelial outgrowths from the most caudal pharyngeal pouch of the embryo but become embedded in the thyroid gland beside the follicles (see Fig. 19.10a). Calcitonin lowers plasma calcium and phosphate, but normally plays only a minor role in mammals. Its major effect appears to be to protect the skeleton from excessive calcium withdrawal when there is a large demand for calcium, as occurs during pregnancy and lactation.

Parathyroid hormone (PTH) is also a peptide that is secreted by the small **parathyroid glands**, which develop embryonically in mammals as epithelial outgrowths from the third and fourth pharyngeal pouches. They become embedded on the dorsal surface of the thyroid as it mi-

grates caudally (see Fig. 19.1). Parathyroid hormone acts primarily to maintain calcium homeostasis (Fig. 19.16).

Vitamin D (cholecalciferol) is synthesized in the skin when the skin is exposed to sunlight. It is called a vitamin because it was discovered as a dietary supplement, but technically it is a steroid hormone for it is produced at one site, travels in the blood, and is active at other sites. Vitamin D in the skin is inactive and must be hydroxylated to become the active form: 1,25-dihydroxycholecalciferol. One hydroxyl group is added in the liver; the second, in the kidneys. The presence of parathyroid hormone is necessary for the second hydroxylation of vitamin D (Fig. 19.16). The primary effect of active vitamin D is to increase calcium and phosphate absorption from the intestine, which insures that the intake of these ions balances their constant excretion.

Phosphate homeostasis is not as tightly controlled as that of calcium, and a response to changes to phosphate

Figure 19.16

Parathyroid hormone and vitamin D are the primary factors in controlling the level of calcium ions in the blood. (1) A drop in free plasma calcium stimulates the release of parathyroid hormone. This restores calcium homeostasis within minutes by (2) increasing calcium reabsorption from the kidneys and (3) activating bone-destroying osteoclasts that cause an efflux of calcium from reservoirs in bone. (4) Vitamin D assists this process by making the bones more responsive to parathyroid hormone. (5) Vitamin D also promotes the absorption of calcium from the intestine. (6) As calcium levels in the blood rise, parathyroid hormone release is inhibited. (7) Calcium is then stored in bone, and (8) its excretion increases.

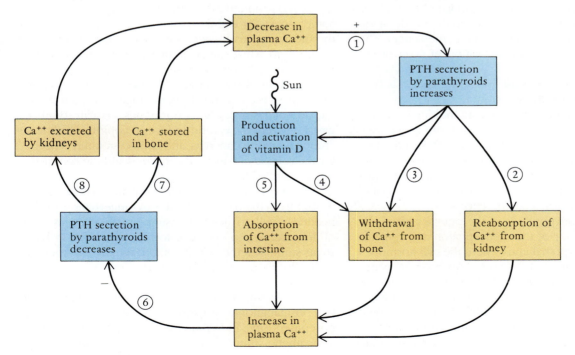

levels in the plasma may take days to correct. In general, phosphate levels fluctuate opposite to those of calcium. A decrease in the plasma level of phosphate ions decreases bone formation, and this leads to an increase in the plasma level of calcium. Calcium is excreted by the kidneys, but at the same time phosphate ions are reabsorbed. In addition, a decrease in plasma phosphate promotes the activation of vitamin D, and this increases phosphate absorption from the gut.

Other Hormone-like Substances

We have been discussing the major hormones of vertebrates, especially those of mammals, but many other materials with hormone-like properties exist and undoubtedly more remain to be discovered. Some of these substances may be hormones, others act in an ancillary way in conjunction with more conventional hormones. An example is the **prostaglandins**, so named because they were discovered in semen, which includes secretion from the prostate gland. We now know that the prostaglandins are synthesized by many tissues of the body in both males and females. Unlike true hormones they do not travel far, but they act as intracellular mediators of many other hormones. Those in the semen promote the contractions of the female reproductive passages that assist the movement of sperm toward the egg. Some of the prostaglandins in females, which are synthesized under the influence of increasing levels of female sex hormones during preganancy, appear to mediate the contraction of uterine muscles at childbirth. At the very least, these and other substances play important supportive roles in the activity of endocrine glands.

Summary

1. The nervous system enables animals to respond rapidly to specific changes in their environment. Slower, continuing processes such as growth, metabolism, homeostasis, and reproduction are integrated primarily by hormones secreted by endocrine glands. Despite differences between them, both neurons and endocrine glands exert their effects by the release of chemical messengers.

2. Hormones are steroids, peptides and small proteins, and amino acid derivatives. Steroid hormones pass through the plasma membrane, bind with receptors, enter the nucleus, and activate genes. Peptide and protein hormones bind with receptors on the surface of the plasma membrane. This activates an enzyme on the inside of the membrane that in turn activates a second messenger. The second messenger causes the activation of a set of enzymes.

3. The secretion of most endocrine glands is controlled by negative feedback loops.

4. Neurosecretions and a steroid molting hormone known as ecdysone control the periodic molting of crustaceans and insects, but the mechanisms differ in the two groups. Neurosecretions are also central to the vertebrate endocrine system because those of the hypothalamus control the numerous secretions of the pituitary gland, which, in turn, have far-reaching effects.

5. The neurohypophysis of the pituitary gland develops from the floor of the diencephalon and remains connected to the hypothalamus where the two hormones it releases are synthesized. Its hormones promote water reabsorption from part of the kidney tubules and the contraction of certain smooth muscles. The adenohypophysis develops from the roof of the embryonic mouth. Its seven hormones promote the production and dispersal of pigment, the production and secretion of milk, the release of thyroid hormones, the synthesis and release of adrenal cortical hormones, growth, and the synthesis by the ovary and testis of reproductive hormones. The level of these hormones in the blood is controlled to a large extent by releasing hormones that are synthesized in the hypothalamus and reach the adenohypophysis by way of the hypophyseal portal system.

6. Growth and maturation in vertebrates are affected by growth hormone released from the adenohypophysis, somatomedin produced by liver cells, and the secretions of the thyroid gland. Thyroid hormones activate certain genes and promote the synthesis of enzymes essential for some aspects of cellular respiration. Adequate levels of thyroid hormones are necessary to sustain the high level of metabolism of birds and mammals, and they play a key role in amphibian metamorphosis.

7. The release of digestive secretions at the appropriate times is controlled by parasympathetic neurons and a group of gastrointestinal hormones. Gastrin stimulates gastric secretion, secretin induces a copious alkaline secretion from the pancreas, and cholecystokinin is needed for the release of pancreatic enzymes and the contraction of the gallbladder.

8. A rise in blood sugar level promotes the release of insulin from the pancreatic islets of Langerhans. Insulin is needed by cells to metabolize sugar and for the conversion of any excess into glycogen for storage. A fall in blood sugar stimulates the release of glucagon from other islet cells. Glucagon is needed to release sugar from storage in liver cells.

9. Chromaffin cells of the adrenal medulla are modified postganglionic sympathetic neurons. Their products, norepinephrine and epinephrine, reinforce the effect of sympathetic stimulation in adapting the body to stress.

Cells of the adrenal cortex secrete three groups of hormones: glucocorticoids participate in carbohydrate and protein metabolism, androgens promote protein synthesis and have a masculinizing effect, and mineralocorticoids affect electrolyte balances and blood pressure.

10. Parathyroid hormone is secreted by the parathyroid glands in response to a drop in free plasma calcium. It increases plasma calcium levels by increasing its reabsorption by the kidneys and mobilizing calcium reserves in bone. Vitamin D affects calcium metabolism by promoting its absorption from the intestine.

References and Selected Readings

Bently, P. J. *Comparative Vertebrate Endocrinology.* Cambridge, England: Cambridge University Press, 1976. A discussion of the comparative morphology of the vertebrate endocrine glands is followed by detailed analyses of the hormonal regulation of many physiological processes.

Gern, W.A., D. Duvall, and J.M. Nervina. Melatonin: A discussion of its evolution and actions in vertebrates. *American Zoologist* 26(1986):985–996. A review of melatonin, from its function in primitive photoreceptive eyes to its role in the pineal gland.

Hedge, G.A., H.D. Colby, and R.L. Goodman *Clinical Endocrine Physiology.* Philadelphia: W.B. Saunders Co., 1987. An excellent short monograph on the physiology of human endocrine glands.

Lauffer, H., and R.G.H. Downer (eds.). *Endocrinology of Selected Invertebrate Types.* New York: Liss, 1988. An important source book in comparative endocrinology.

Orci, L., J.-D. Vassalli, and A. Perrelet. The insulin factory. *Scientific American* 259(Sept. 1988):85–94. An account of the synthesis and exocytosis of insulin.

Schally, A.V., A.J. Kastin, and A. Arimura. Hypothalamic hormones: The link between brain and body. *American Scientist* 65(1977):712–719. A review of hypothalamic control of the endocrine system.

Snyder, S.H. The molecular basis of communication between cells. *Scientific American* 253(Oct. 1985):132–144. A comparison of short- and long-range communication between cells.

Wilson, J.D., and D.W. Foster (eds.). *Williams: Textbook of Endocrinology,* 7th ed. Philadelphia: W.B. Saunders Co., 1985. A thoroughly documented sourcebook for human endocrinology with an extensive listing of references to the primary literature.

20

Behavior

Two king penguins preening each other.

Multicellular plants and animals are the conspicuous members of the world of living organisms, but their life styles are very different. The activity of a plant is most readily seen in daily or seasonal growth, such as the development of new shoots, flowering, and fruiting. The activity of animals is easily recognized by body movements that result from the response of effectors, especially muscle cells.

An animal's effector responses are not random but are highly organized and integrated patterns of activity that enable the species to interact with its environment and to meet the physiological demands of its existence. Such patterns of activity are called behavior, and the study of behavior is termed **ethology** (Gr. *ethos*, custom + *logos*, study). An animal's repertory of behavior patterns may be as simple as the feeding responses of a sea anemone or as complex as the homing of birds. Whatever the pattern, it fits the needs and life style of the animal.

The purpose of this chapter is not to describe animal behavior, although we will provide many examples. Rather, our task here is to examine the generalizations that have been established from comparative ethological studies and to explore those aspects of behavior that are most commonly encountered in animals.

THE NATURE AND DEVELOPMENT OF ETHOLOGY

Most animal behavior is a sequence of responses related to obtaining food, acquiring and maintaining living space, protection, reproduction, and other needs. For example, certain species of shrimp, called cleaning shrimp, pluck parasites and dead tissue from the bodies of certain fish (Fig. 20.1a). The shrimp may even extend its cheliped beneath the fish's operculum to remove parasites from the gills. The behavior benefits both parties of this symbiotic relationship; the shrimp gains food, and the fish is rid of harmful parasites. Some species of cleaning shrimp (*Periclimenes*) have cleaning stations to which "client" fish come. The shrimp signals an approaching fish by waving its long antennae and rocking its body. The fish comes closer to the shrimp, which extends its antennae toward the fish. The fish strikes a characteristic pose. The shrimp then

(a)

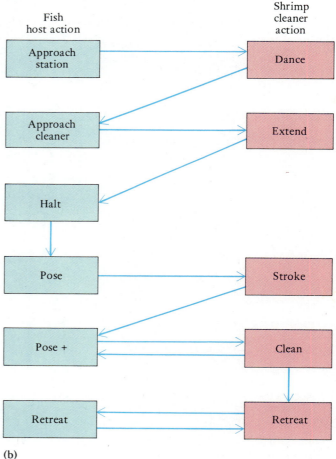

(b)

Figure 20.1

(a) The cleaning shrimp *Periclimenes* on a sea anemone. (b) An ethogram of the sequence of behavior in the interaction between a host fish and the cleaning shrimp *Periclimenes anthophilus*. Arrow from Approach station to Dance, for example, means that the approach of the fish to the cleaning station initiated dancing behavior in the shrimp. Extend refers to the shrimp directing antennae toward fish; pose + refers to fish holding characteristic pose for an additional period of time following stroking by shrimp.

strokes the fish with its antennae, climbs aboard, and proceeds to clean. A precise stereotyped sequence of behavior is characteristic of the species and is performed in the same manner by all members of the population. The steps can be depicted in a diagram called an **ethogram** (Fig. 20.1b).

The behavior patterns of either the fish or the shrimp could be studied in a number of ways. They might be analyzed to determine the components of the entire sequence, the specific stimuli that trigger responses, or whether the responses require certain physiological conditions in order to be triggered. An ethologist might ask how fixed or automatic the patterns are. Can they be altered or modified by environmental conditions? How are the patterns determined by the nervous and hormonal integrating systems of the animal? How do they arise in the course of development? How is one part of the pattern related to another? For example, why does the fish pose? All of these questions are sometimes called **proximate questions** because they deal with causes or are concerned with the way the pattern develops in the life of the animal.

But there are two equally important questions that might also be asked: What is the adaptive or ecological significance of the behavior, and how did it evolve? These are sometimes called **ultimate questions**. For the particular example we have given, the adaptive significance is not difficult to understand, but the steps that might have led to its evolution are more difficult to explain.

Thus, behavior can be examined from four perspectives: (1) **causation**, the causes of behavior, (2) **development**, the formation of the behavior in the development of the individual; (3) **evolution**, the origin of the behavior in the evolution of the species or group; and (4) **behavioral ecology** (**function**), the significance of the behavior in assuring the survival and reproduction of the animal.

In the history of ethology these four questions have not received equal attention. Most of the studies of animal behavior during the first half of the 20th century were experiments by physiologists and psychologists. Those studies treated laboratory animals, usually rats, largely as substitutes for humans. There was little interest in the relationship between behavior and the natural mode of life of that species. The Americans J.B. Watson and B.F. Skinner believed that most behavior was a consequence of stimulus, response, and reinforcement, and the school of

thought that developed around their ideas became known as behaviorism. It had great influence in the development of psychology. Contemporary comparative psychologists have a broader perspective and study more animal species, but they continue to focus largely on the causes and development of behavior.

In the 1930s, the German zoologist Konrad Lorenz laid the foundations for modern ethology and coined the term ethology. Lorenz was interested not only in causes and development of behavior but also in how they functioned and evolved in the species. For example, he would have wanted to know not just that rats can learn to run through a complex maze, but why this ability is important to the life style and evolution of rats. It is these two latter concerns that distinguish ethology from comparative psychology. The Dutch zoologist Nikko Tinbergen brought experimental design into ethological research.

CAUSATION

Elements of Behavior

The waving of the cleaning shrimp's antennae signals the fish to adopt a "ready to be cleaned" posture. Similarly, a male jumping spider responds to the visual presence of the female by "dancing" before her, elevating certain legs bearing colored hairs, and tilting his abdomen. A shiny band (a foreign object) around the leg of a baby bird will evoke the parents of some species to push the band out of the nest, even though the baby goes with it.

In each of these behavioral patterns, a **sign stimulus**—waving antennae, female frontal area, metal band—acts to release a specific behavioral response or **motor program**. The sign stimulus, or **releaser**, is the essential stimulus that sets off an **innate releasing mechanism**. The motor program in these examples are very rigid behavioral patterns, also known as **fixed action patterns**.

A classic example of a releasing mechanism is egg rolling in geese, first studied by Lorenz and Tinbergen. A nesting goose, on observing a nearby displaced egg, will use its beak to roll it back into the nest (Fig. 20.2). Lorenz and Tinbergen discovered that a goose will roll all sorts of inappropriate objects back into the nest, provided that the objects exhibit certain appropriate sizes and shapes. Thus, round objects of a particular size act as a visual stimulus (the sign signal) to trigger an innate releasing mechanism, initiating the egg-rolling motor pattern.

Figure 20.2

(a) Greylag goose retrieving an egg that is outside the nest. This movement is very stereotyped in form and used by many ground-nesting birds. (b) The goose attempts to retrieve a giant egg in precisely the same fashion.

(a)   **(b)**

Physiology of Behavior

The nervous and hormonal systems of an animal are the physiological basis of behavior. Although this has long been recognized, it has been difficult to make precise correlations between behavioral patterns, i.e., motor programs (not reflexes) and the firing of specific neurons. However, in recent years such correlations have been demonstrated in the nervous systems of leeches, crayfish, sea hares, and sea slugs. These animals have a few centrally located neurons that are large enough to be easily dissected and probed with electrodes.

The sea slug *Tritonia diomedia*, a shell-less marine snail found along the West Coast of the United States, can escape predators (such as certain sea stars) by a burst of swimming activity (Fig. 20.3). Arthur Willows at the University of Washington discovered the neuronal circuits that govern the swimming behavior of *Tritonia* by careful investigation of individual neurons. He first dissected the central ganglia of the nervous system and mapped the larger neurons. He then electrically stimulated each of these neurons with fine probes and recorded not only their action potential but also the specific motor responses each neuron evoked. He found that certain neurons governed specific turning movements or contractions on one side of the body; others could produce movement simultaneously on both sides of the body; and still others could bring about the swimming behavior in its entirety. Moreover, the swimming lasts some 30 seconds, long after the command neurons have ceased to fire. The command capability is made possible by junctions with other neurons along the length of the command neuron. Thus, an impulse generated in the command neuron is transmitted to all of the motor neurons involved in the swimming response and in the proper sequence. Apparently, reverberating (self-stimulating) circuitry will maintain the swimming behavior long after the initial command has ceased. Under natural conditions, contact with a predator

sea star is the stimulus that generates the escape swimming command.

Neuronal circuitry and the physiological regulation of neuronal junctions—threshold requirements, facilitation, spatial and temporal summation, and inhibition (Chap. 17)—are the underlying determinants of behavior, but each animal has a specific system adapted for its own life style. This applies not only to motor circuits, such as those described for *Tritonia*, but also to sensory neurons delivering data from outlying receptors. Many kinds of stimuli continually bombard an animal; only a few (those important for the species) are selected for internal reporting. To a large extent selection is determined simply by the specific receptors the animal possesses. The specific receptors act as a filter, recording some stimuli and ignoring others.

Different species often detect quite different sorts of sensory information. Crickets, which fly at night and are preyed upon by bats, have evolved auditory receptors that are particularly sensitive to the high-frequency sound waves emitted by bats. When the receptors detect these frequencies, nerve impulses are sent to interneurons in the central nervous system of the cricket. If the sensory nerve impulses from the auditory receptors have a sufficiently high frequency to produce a steady firing of one of the two interneurons, motor impulses will be initiated to the dorsal longitudinal muscle of that side. The muscle contracts; the abdomen, which acts like a rudder, bends; and the cricket is steered away from the sound source.

Web-building spiders detect a very different kind of signal. Slit sense organs located in the joint between the last two sections of the legs are very sensitive to vibrations transmitted through the web, and these vibrations are the spider's principal source of information about its surrounding world. The orb-weaver *Argiope argentata* sits in the hub of its web. It is alerted by vibrations of an ensnared insect and may pluck the web to determine the prey load

Figure 20.3

Escape response of the sea slug *Tritonia diomedia*. On contact with a predatory sea star, the sea slug withdraws, contracts its dorsal gills, and undergoes dorsoventral flexions. The flexions enable the animal to swim some distance away from the predator.

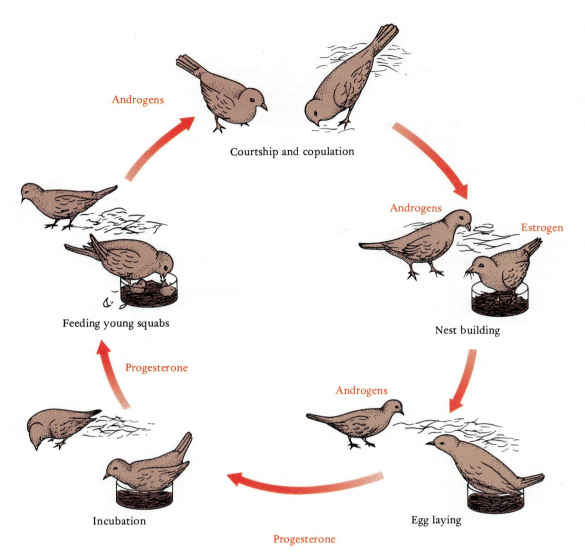

Androgens

Courtship and copulation

Androgens

Estrogen

Nest building

Androgens

Feeding young squabs

Progesterone

Incubation

Egg laying

Progesterone

Figure 20.4

Cycle of reproductive behavior in the ring dove, in which behavior results from an interaction of hormones and external environmental cues, including the behavior of the male. These doves are domesticated and they build their nests in containers placed in the dovecote. Under the influence of androgens, the male ring dove courts the female; the courting behavior stimulates the female pituitary (via visual stimulation) to secrete follicle-stimulating hormone (FSH). FSH causes the development of ovarian follicles that secrete estrogen. The pair soon begin nest building and copulate. Secretion of progesterone, either as a consequence or ovulation or perhaps as a consequence of nest-building stimuli, induces incubation behavior when the eggs are laid. Behavioral interactions continue in this manner through care of the young.

and position. The spider then runs out to the prey. Large prey are given a long bite and then wrapped; small prey are wrapped and given a short bite. If the prey is very small, it is simply seized and then wrapped at the hub. After wrapping, large prey is cut out and carried back to the hub and rewrapped prior to feeding. When a male orb-weaver enters the web, his vibrations can be distinguished by the female from those of prey.

From these two examples we can see that each group of animal species lives in a sensory world all its own. It responds to a complex of signals of which other animals, including humans, are largely unaware. Hormones are also part of the physiological basis of behavior. They control or set the physiological stage required for neuronal sensory, integrative, and motor processes to take place. Hormones may also have organizational effects, such as determining much of the structural and physiological course of sexual development in mammals and other animals. Hormonal and nervous processes, behavioral responses, and environmental stimuli all interact to control the sequence of behavior for a given species, as illustrated by the courting, nest-building, egg-laying, and brooding behavior of ring doves (Fig. 20.4). Hormones can evoke behavior; behavior can stimulate secretion of hormones; and other kinds of environmental cues, such as the nest or eggs, can evoke behavior or stimulate hormone secretion.

Timing of Behavior

Most complex behavior involves a sequence of actions. Neural mechanisms combined with environmental cues ensure that the actions follow each other in a coordinated sequence. Flight in blowflies is initiated by low blood sugar; the direction of flight is determined by the detection of airborne molecules from plant fluids or decomposing animal fluids upon which blowflies feed. When the stimulus from such odors becomes very strong, flight is inhibited, and the fly lands on the food source. Sugar receptors in the terminal tarsal segments of the legs are now stimulated, sending signals that cause the proboscis to extend into the fluid food (Fig. 20.5). Proboscis receptors bring about sucking, and fluid flows into the crop. When the crop becomes filled, stretch receptors in its walls are stimulated, which override stimuli from the sugar receptors, and feeding is halted. If the nerve between the stretch receptors and brain is cut, the fly will feed until it bursts!

The interaction of hormonal, neural, and environmental cues plays a primary role in controlling the timing of behavioral activity and the shifting of priorities in the life of the species. In their southern winter range, white-crowned sparrows are subjected to increasing day length after December (Fig. 20.6). The increasing photoperiod (detected by the eyes and affecting an underlying neural clock mechanism in the brain) causes the hypothalamus to secrete pituitary releasing hormones. These stimulate the production of pituitary gonadotropins that initiate the growth of the gonads from their small winter size. Feeding also increases, and the birds begin to deposit fat in preparation for the energy cost of the spring migration. The activity level rises, and when a threshold level is reached, the birds fly north to their summer range, where the reproductive phase of their life cycle begins.

Biological Rhythms and Clocks

Control mechanisms that regulate metabolic processes and behavior and synchronize them with cyclic changes in the external environment have evolved in a wide range of organisms. These control mechanisms, or clocks, follow cycles that may range in length from a day or less to a year

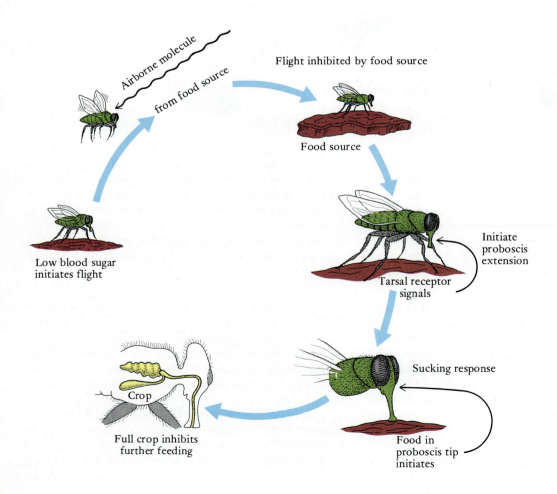

Airborne molecule from food source

Flight inhibited by food source

Food source

Low blood sugar initiates flight

Tarsal receptor signals

Initiate proboscis extension

Sucking response

Crop

Full crop inhibits further feeding

Food in proboscis tip initiates

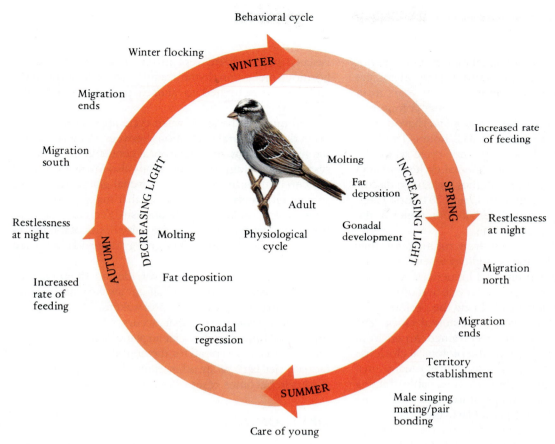

Figure 20.6

Seasonal changes in the physiology and behavior of the white-crowned sparrow.

or more. Frequently, the activity of a potential food source may regulate the activity of a predator: if prey is active in the early morning, the activity of the predator will begin shortly before dawn, even though dawn changes slightly from day to day. Periods of activity and sleep, feeding and drinking, changes in body temperature, and many other processes have cycles that are approximately 24 hours long, hence they are called **circadian rhythms** (L. *circa*, approximately + *dies*, day). Cycles may be shorter or longer. Many marine animals that live along the shore are tied to tidal (lunar) cycles. Fiddler crabs on the eastern coast of the United States emerge from their burrows to feed at each low tide (twice every 24 hours). Certain intertidal snails, on the other hand, release their eggs only at the very high bimonthly tides (spring tides). Swarming and egg release of some animals such as palolo worms (p. 702) and the California grunion, a fish, are so tied to the lunar cycle that the precise date of their reproductive events can be predicted a year in advance. Many animals migrate to and from breeding grounds twice a year, and many have annual cycles in reproduction, hibernation, and so forth. We will examine circadian rhythms as examples of cyclic phenomena because they have been found in virtu-

ally all organisms and many play an important role in behavior.

The existence of circadian rhythms has long been known, but the intrinsic nature of certain cycles was not recognized until 1729, when the French astronomer de Mairan reported that the sensitive plant *Mimosa* continued to open and fold up its leaves in a 24-hour cycle even when placed in constant darkness. Many investigators have argued that, although not controlled directly by light cycles, circadian rhythms must be **exogenous** (originating outside of the organism), controlled by some subtle environmental rhythm (e.g., light, temperature, or humidity) that the organism detects. Others have argued that circadian rhythms are **endogenous** (originating within the organism), regulated by an internal biological clock sensitive to environmental cues but capable of detecting the passage of time in their absence. The endogenous hypothesis received considerable support in the 1950s through the investigations of the behavior of honeybees. Researchers trained honeybees to collect sugar water in a Paris laboratory between 8:15 and 10:15 in the evening. They then transported the bees on a flight to New York and found that the bees continued to collect sugar water between 8:15 and

10:15 P.M. Paris time! Environmental influences could not have played a role in this case. Further support for the endogenous hypothesis has come in recent years from studies in many organisms that show a genetic component to biological rhythms. Fruit flies (*Drosophila*) have a clock that has a free-running period of 24.2 hours. The free-running period is the clock's repetitive cycle when the animals are isolated from environmental cycles and kept under constant conditions. Mutant flies have been discovered with free-running periods of 19 and 28 hours. These mutants turn out to possess time-different alleles at the same locus on the X chromosome.

Most biological clocks are free running for they continue to exhibit the same periodicity when the organism is isolated from environmental influences. An organism exposed to a 24-hour light-dark cycle and other environmental factors will eventually bring the period of the internal clock into phase with the external clock. The internal clock is said to be set, or **entrained**. An environmental cue capable of entraining the biological clock has been called a **zeitgeber** (German, time giver). Many environmental factors have been shown to be capable of acting as a zeitgeber, but light, temperature, and (for some marine organisms) tides are particularly important ones. Once the clock is set, it will continue to run on the set time for a while, even when environmental conditions are abruptly changed, as when the honeybees were transported from Paris to New York. As time goes on, the biological clock resets and comes into phase with the new environmental conditions. Normally the internal clock is adjusted slightly day by day as day length and other factors change. When the organism is isolated from external cycles and kept under constant conditions, the internal clock slowly drifts to its own free-running period.

The nature of the internal clock is not entirely clear, but it appears to depend on one or more self-sustained biochemical or physiological oscillators, similar to such oscillators as the respiratory center in the brain or the pacemaker in the heart. The complexity of the circadian timing system in mammals suggests that multiple oscillators are involved. Each of these oscillators probably has its own period; they must be coupled to each other and connected to certain sense organs, since they can be synchronized and entrained by external conditions. A group of nerve cell bodies located in the hypothalamus appears to be a part of the mammalian biological clock. Lesions in this region will destroy many circadian rhythms. A neuronal pathway from the eye that terminates in the part of the hypothalamus that lies above the optic chiasma has recently been discovered. There are probably other nerve regions involved in the biological clock, because lesions in this area in humans, although disruptive to the activity cycle, do not block circadian rhythms in body temperature. The pineal eye or gland also plays a role in the timing systems of rats, birds, and some other vertebrates (p. 445).

Isolated pineal tissue kept in the dark will secrete the hormone melatonin rhythmically following a diurnal cycle and can be entrained to altered periods of light and darkness.

In humans the sleep-wake cycle appears to depend upon diurnal neural oscillations between sleep centers in the hypothalamus and the arousal center in the reticular system of the brain. The oscillations also involve signals to the cerebral cortex and to the adrenal glands. The latter secrete glucocorticoid hormones that make glucose and amino acids available for metabolism. Secretion of these hormones reaches a peak in the early hours of the morning, preparing for the greater energy demands following waking.

The arousal center can override the sleep centers. Thus it is easier to make yourself stay awake when you are sleepy than to sleep when you feel wakeful. The disruption of the sleep-wake cycle that occurs when you cross a number of time zones in a jet airplane demands a shift in the cycle that is too rapid for the neural centers to accommodate (jet lag). As in other diurnal rhythms, the clock gradually becomes reset over a period of several days.

DEVELOPMENT

In a fundamental sense, behavior depends upon the existence and arrangement of neuronal and hormonal connections. As a result, the embryonic and postnatal development of nervous and hormonal systems, much of which are under genetic control, profoundly affect the range and variety of behavioral responses.

Psychologists and ethologists have long recognized two sorts of behavior, instinctive and learned. **Instinctive behavior** is innate and rigid, comprised of motor patterns that are carried out correctly the first time they are evoked, because they are preprogrammed, i.e., determined by sets of neuronal interconnections that develop under the influence of the genes. **Learned behavior** develops from an interaction between the neuronal base and repetitive inputs from the environment. The relative prevalence and importance of instinctive behavior versus learned behavior, "nature versus nurture," has been the basis of much past and present debate. The argument really centers on the way behavior develops. To what extent does behavior result from the interaction of the preprogrammed neural base with the environment? As knowledge of comparative behavior has grown, ethologists have come to recognize that most animals have some innate motor patterns (fixed action patterns) and some that result from environmental interaction.

Both instinctive behavior and learned behavior have advantages. Instincts permit immediate, accurate responses to appropriate environmental signals. Any delay (which learning might require) could reduce the chances of survival. Instinctive behavior is successful as long as the environmental conditions in which it occurs are relatively constant. In other words, the behavioral pattern must fit the circumstances most of the time. Learned behavior makes possible responses that will ensure a high rate of survival in the face of uncertain environmental conditions. A genetically based, predetermined component underlies both instinctive and learned behavior. However, the environmental component varies greatly, as the following examples show. The classification we have adopted is simply one possible way of grouping the variations.

Instincts

Closed instincts are preprogrammed, fixed motor patterns that are functional from the moment the neural circuitry is in place and are not modified by the environment. The first web of orb-weaving spiders, for example, is complete in all detail and repeatedly built in the same manner throughout the life of the spider (Fig. 20.7).

In many species, behavior that is functional when first performed is capable of modification as a result of interaction with the environment. Such **open instincts** are illustrated by herring gull chicks. They peck the beaks of the parents, which regurgitate partially digested food for the chick. The chick is attracted by a red spot on the beak and by the beak's shape and downward movement (sign stimulus). This "begging behavior" is sufficiently functional to get the chick its first meal, but there is much energy wasted in the pecking. Some pecks fail to reach the parent's beak and some are off target. However, the begging behavior becomes more efficient over time, and in experiments with models of the parent's beak, the chicks become increasingly selective about the shape necessary to evoke the begging response. Thus, the initial functional instinct is modified and refined by environmental interaction.

Open instincts are also suppressed under appropriate conditions. The escape response of many animals, for example, can be reduced as the animal learns through repetition that a particular stimulus is not dangerous. The animal is said to have become **habituated** to the stimulus, but the escape response still functions for other signals.

Restricted Learning

Learned behavior develops only after interaction with the environment. In **restricted learning**, also called **programmed learning**, a particular behavior pattern forms as a result of precise environmental stimuli. Two of the best-studied examples of restricted learning are imprinting and bird song learning.

For many ground birds and herd mammals—chickens, ducks, geese, horses, and sheep—the ability of the young to follow the parent is critical for survival. The development of this type of behavior was investigated by Lorenz in what is now considered a classic study of ethology. In chicks and goslings, any moving object of appropriate size will evoke an innate following response. This could be very dangerous if restrictions were not imposed on what the chicks should follow. For a brief critical period after hatching (12 to 14 hours, peak time), a chick is very sensitive to the characteristics of a moving object about it, which is usually the mother. After **imprinting** occurs, the goslings will only respond to moving objects with the particular characteristics of the imprint. In the absence of a mother, Lorenz was able to make goslings imprint upon him as a surrogate mother: the goslings continued to follow Lorenz for several weeks. The chick must actually follow for imprinting to occur, and the imprinting is irreversible. It is as if the characteristics of the parent (or moving object) were indelibly stamped on the nervous system of the chick during a brief receptive period of development. Depending upon the species, imprinting on sounds and odors may also take place, again usually those of the parent (Fig. 20.8).

In many species of birds, male recognition of the female of the same species is partly dependent upon imprinting during a critical period that follows parent/offspring imprinting. In some birds, such as those in which there is danger of receiving eggs or chicks of another bird in the nest because of crowded nesting conditions, the parents imprint on their own eggs shortly after laying them (or on their chicks immediately after hatching). They can thus differentiate their eggs from those of an intruder. However, there are parasitic birds (cuckoos) that have breached this defense and lay eggs that are indistinguishable in size and color from those of the host bird in whose nest they lay them.

The songs of male birds, which serve to advertise sexual maturity and territorial ownership, are innate in some species but learned in others. Sparrows, in which song learning has been most extensively studied, must hear a song in order to reproduce it and must have some contact with other members of its species. The juvenile bird "records" the adult song during a critical period of about two months (10 to 50 days in white-crowned sparrows). During this time, the nervous system of the young bird is primed to "memorize" the call. Certain qualities of the call trigger the memorization process, and out of the many surrounding calls the young bird encounters, only the call of its own species is selected for recording.

When the young males begin to sing some four months later, the song is at first just twitterings, called subsong; gradually the song becomes matched with the brain's recording, and the correct song of the species is reproduced (Fig. 20.9a). In white-crowned sparrows, the perfected song occurs by about seven months. Hand-reared sparrows kept in isolation cannot reproduce the species call (Fig. 20.9b), but those isolated after the early critical period can. However, birds deafened after the critical period cannot perfect the song (Fig. 20.9c), be-

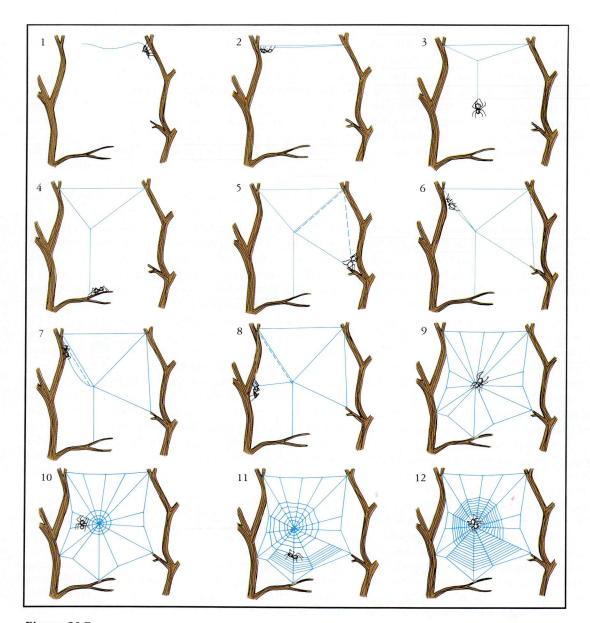

Figure 20.7

The familiar orb-webs are constructed by several families of spiders, although most of those seen in woods, meadows, and gardens are produced by the large, widespread orb-weaving family, the Araneidae. The spider begins the task of web construction by establishing a bridge line. A strand of silk released from the spinnerets is carried by air currents (1). When it strikes and attaches to some nearby object, the spider attaches the opposite end and may reinforce it with additional strands (2). Next the spider spins a vertical thread from one of the bridge lines, attaches it to a lower branch and then tightens it to form a Y (3–4). The fork of the Y will become the hub of the future web, and the arms and base of the Y, the first radii. When frame threads and additional radii have been added (5–8), the spider lays down a scaffolding spiral of dry threads from the inside out (9–10). Finally, starting at the outside, the spider removes the temporary spiral scaffold (eating it) and simultaneously replaces it with a permanent spiral of sticky thread (11–12).

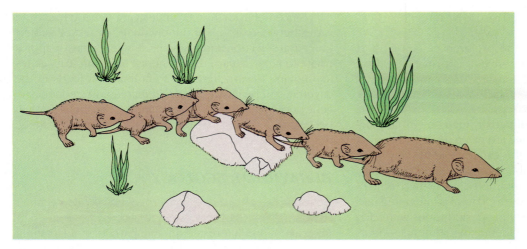

Figure 20.8
The young of the European shrew recognize their mother's odor and follow her as a result of early olfactory imprinting.

cause they cannot hear themselves sing, which seems to be necessary for matching the "memory" of the song.

Flexible Learning

The behavior of many animals undergoes continual modification as a result of interaction with the environment, and there may be a wide range of environmental data that can be learned. Flexible learning is perhaps best illustrated by the learning associated with exploration around a "home site" and with the development of food preferences.

Many animals that have dens, burrows, or nests learn the location of their home and important landmarks in the surrounding area. Deer mice (*Peromyscus*) have an extensive system of runways leading out from the nest. Under experimental conditions, they are capable of learning every detail of a maze containing the equivalent of one-quarter mile of runways in no more than three days. Some of the territorial information the mouse learns is of immediate importance, such as how to return from a foraging trip. Other information may be important at a later time, as when darting beneath a log to escape a predator. The application of unrelated stored information to a new experience is called **latent learning**.

For some animals, the time and conditions for exploratory learning can be very restrictive. A foraging bee learns the position of the hive only on the first trip out in the morning, learns the position of the floral nectar source only as it approaches the flower, and learns the flower color just as it lands. In fact, exploratory learning in bees

Figure 20.9
Songs of the white-crowned sparrow under varying experimental conditions. (a) A spectrogram of the song of a normal wild bird. (b) A spectrogram produced by a bird hatched and reared in isolation from other birds. (c) A spectrogram produced by a bird that was deafened after hearing other birds but before it began to sing itself.

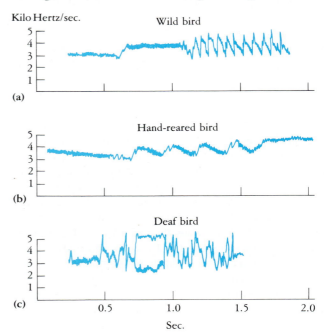

should probably be considered restrictive rather than flexible learning.

The development of food preferences is a type of flexible learning that involves a process called **conditioning**. As a result of "reward" or "punishment," a behavioral response becomes associated with a particular stimulus and thus becomes more frequently practiced (or avoided). Juvenile toads will initially feed on virtually any moving arthropod, but if they are given a bee or noxious species, they spit it out and thereafter avoid it. Many birds and mammals learn acceptable food by its systemic effects. A small amount of new food is sampled, and if there is no ill effect after a period of time, the food item is integrated into the diet. The length of the waiting period varies with each species, as well as the food-identifying stimulus—odor, color, shape, movement, and so on.

Much exploratory and food preference learning involves trial and error. A randomly selected pathway to shelter or choice of food that is correct becomes reinforced by reward.

GENETICS AND THE EVOLUTION OF BEHAVIOR

While strongly affected by environmental influences and learning, behavior also has a genetic component, for much of behavior is based on the way the nervous and endocrine systems develop. The genetic basis of behavior can be seen in classic experiments in which two separate species or subspecies with differences in mating calls, or other behavioral attributes, are crossed. Hybrids typically have an intermediate condition, and there is some segregation of the behavioral characters when hybrids are back crossed to the parents. Inbreeding experiments support this conclusion.

Although experiments show that complex behavioral patterns have a genetic basis, the patterns are based on the interaction of so many genes that it is difficult to determine the specific role of individual genes in behavior. Some idea of how changes in individual genes affect components of behavior can be gained by searching for individual mutant alleles that affect some aspect of behavior. The fruit fly, *Drosophila*, has proven to be a good experimental animal. Researchers have discovered mutant genes that change the flies' normal 24-hour circadian rhythm, their response to light and dark, various aspects of their courtship, and even their learning ability. Males that have a sex-linked mutant gene called "fruitless" are unable to distinguish between sexes, so they will court any fly, male or female, that they come upon. Sex recognition in flies is largely olfactory and depends upon a chemical substance or pheromone, secreted by the female. Analysis has shown that the "fruitless" males produce the female pheromone and so smell themselves when they approach another fly! In this case the allele for "fruitless" must be coding for some enzyme involved in the synthesis of the pheromone.

Since behavior has an underlying genetic basis and varies between groups and individuals, it must be subject to natural selection and adaptive change. Behavior appropriate to a particular environmental situation, e.g., genes for moving toward light for an animal that feeds in the daytime, will be favored and will increase in frequency in gene pools, whereas genes for inappropriate behavior will be selected against. In other words, behavior can be shaped by its effect on fitness.

BEHAVIORAL ECOLOGY

Behavioral ecologists are concerned with the correlations between particular environments and behavioral strategies. They are interested in the particular adaptations forged by evolution that enhance the fitness of individuals of a species, i.e., those adaptations that provide an advantage in the transmission of their genes. All animals must find a place to live, obtain food, avoid being eaten, and reproduce. They must perform these activities in ways that perpetuate themselves yet at the same time do not interfere with other species with which they interact and on which they may depend. In the following sections we present some examples of behaviorial ecology that are of fairly wide occurrence.

Foraging Strategies

Behavior involves not only interactions between members of the same species but also interactions between members of different species as they search for favorable habitats, living space, and food, and strive to avoid being eaten. Most animals are adapted structurally and physiologically to feed on a limited range of food and to gather it in specific ways. Herbivores select particular kinds of plants; carnivores select the type and size of prey. Even some detritus feeders sort out organic material from sand and other debris. Behavior is important in finding food of a particular type, gathering it in efficient ways, and deciding when and where to move on. A great deal of research is now being done to determine the extent to which **foraging strategies** have evolved to optimize net energy intake. The **net energy intake** represents the difference between the energy available in a food item and the energy expended in searching for the food, gathering or catching it, and eating it. We expect foraging behavior and food choices to vary with ecological conditions. Thus, for example, in the relatively unproductive marshes of northeastern United States, great blue herons must spend a great deal of time finding food, and they eat a fairly wide variety of small vertebrates. In contrast, in Florida, where marsh productivity is much higher and food is easier to find, the same herons select a more limited range of food types.

Food availability is not the only factor in foraging costs. For many animals there is the added cost of being preyed upon while they are foraging. Marmots spend considerable time looking for possible predators, time that is sacrificed from foraging and feeding, or they will avoid a richer feeding ground because of the distance it lies from the protection of their burrows.

Many strategies have evolved to optimize net energy intake. As we will see, natural selection promotes resource partitioning among related competing species (p. 986). Different species of Darwin's finches optimize net energy intake by feeding on seeds of different sizes. Some herbivores manipulate their plant resources so as to optimize production. Leaf-cutting ants of the American tropics carry bits of appropriate leaves into their subterranean galleries, chew them up, fertilize them with fecal droppings, and "seed" them with bits of mold. As the mold grows, they feed on it, in a form of invertebrate "agriculture." Some carnivorous animals construct traps, use sticks and other tools to gather food, wait in ambush, or engage in **aggressive mimicry**. The flower mantis (Fig. 20.10a) is a striking example of such mimicry as it lies in wait for insects that come to the orchid flower for nectar. In this

Figure 20.10
Concealment and deception. (a) A preying mantis from Costa Rica has the shape and color of a leaf on which it lies in wait for prey. (b) A katydid from the Peruvian Amazon has a body colored like that of a dead leaf. (c) The peanuthead bug *Fulgora laternaria* from the Peruvian Amazon has large eye-like spots on hind wings that may startle a predator when suddenly uncovered. (d) The large colored eye spots on this swallowtail caterpillar gives it the appearance of a larger animal, such as a snake.

(a)

(b)

(c)

(d)

case, the body shape, color, and behavior that lead the mantis to select flowers that it matches all interact to deceive the prey.

Defense

Many prey organisms have evolved interesting defensive measures and strategies that reduce the likelihood of their being eaten. Some, such as a mouse, simply retreat to a home refuge when danger threatens. A few, such as opossums, remain motionless, or freeze, when approached by a predator (hence the expression "playing possum"). This strategy is often enhanced by concealing or **cryptic coloration**. Some animals combine a cryptic coloration with a deceptive body shape. Many insects are shaped, as well as colored, to resemble twigs, leaves (Fig. 20.10b), or even fecal droppings. If concealment fails and predators become too inquisitive, some insects startle the predators by uncovering conspicuous colored "eye" spots or assuming the appearance and posture of a predator themselves (Fig. 20.10c and d).

A great many organisms are dangerous, toxic, or distasteful to predators. The caterpillar of the blue swallowtail butterfly discharges a toxic secretion from tentacle-like structures at its anterior end (Fig. 20.11a). The bombardier beetle has an unusually effective toxic mechanism. It synthesizes and stores hydrogen peroxide and quinones in separate compartments. When the beetle is threatened, these secretions are released and mixed with enzymes in

(a)

(b)

Figure 20.12

Batesian mimicry. (a) The red eft stage of the red-spotted newt, *Notophthalmus viridescens*, is avoided by predators who learn to associate its aposematic coloration and toxic skin secretions. (b) The palatable red salamander, *Pseudotriton ruber*, gains protection by its resemblance to the red eft.

(a)

(b)

an outer vestibule, where they heat quickly to the boiling point and are discharged as a caustic spray (Fig. 20.11b). Animals that have dangerous toxins, or whose tissues contain distasteful materials, frequently display a conspicuous appearance, or **aposematic coloration** (Gr. *apo*, away + *sema*, signal) that advertises their presence (see Fig. 20.11a). After several encounters, predators learn to associate this pattern with a disagreeable experience and avoid these animals. Aposematic coloration also provides an opportunity for certain tasteful animals to gain protection by mimicking the distasteful species. This strategy, termed **Batesian mimicry**, is quite common in insects and some vertebrates (Fig. 20.12). The terrestrial juvenile stage of a

Figure 20.11

Warning coloration. When threatened, the caterpillar of the blue swallowtail butterfly, *Battus philenor*, discharges toxic secretions from the pair of brownish-colored structures at its front end. Its conspicuous aposematic coloration advertises its danger. (b) A bombardier beetle aims a caustic spray between its legs at a predator.

newt of the eastern United States has a striking orange-red color and is called the red eft. The red eft is distasteful to bird predators, which soon learn to recognize the eft and avoid eating it. The similarly colored red salamander is perfectly palatable, but it too is avoided by birds that have had encounters with red efts.

In some cases several different distasteful and aposematic species resemble each other. **Mullerian mimicry** of this type acts to reinforce avoidance behavior by the predator and is found among several groups of Amazonian butterflies (Fig. 20.13). Since some individuals are sacrificed in educating the bird predators, sharing the burden of education among several species is advantageous to all.

Some animals decrease predator risks by living in colonies, herds, flocks, or schools. At least one member of a large group is more likely to spot a predator and sound an alarm than an individual feeding by itself. Once the group has been alerted, the predator may save its energy and try to sneak up on something else. But if not, any one individual in the group has a better chance of survival if it is not located on the periphery of the group, i.e., if there is another individual between it and the predator. Some studies on birds have demonstrated that the chance of predation goes up as the size of the flock goes down.

Agonistic Behavior and Dominance

The size of the population of many species, particularly those that live in habitats with relatively uniform conditions or conditions that change seasonally in a predictable way, is held at a nearly constant size by the resources of the environment (Chap. 42). Competition for those resources in shortest supply—food, shelter, mates, nesting sites, and

Figure 20.13
Mullerian mimicry. Several different, distasteful species of tropical American butterflies have a similar "tiger pattern" of aposematic coloration.

Figure 20.14
Agonistic behavior. African elephants fighting.

so forth—is keen. **Agonistic behavior** (Gr. *agonistes*, champion) is the pattern of aggressive behavior by which members of the same species adjust to conflicts that arise from competition for the same limited resource. Agonistic behavior may take the form of overt **aggression**, in which individuals fight and harm one another. But overt aggressive combat is far less common than one might expect. In part, this may reflect the high costs of actual combat. On the other hand, ritualized combat—encounters in which aggressive behavior is redirected towards displays or threats that seldom cause serious injury to the participants—is very common. Postures are taken, threats are made, and some combat may ensue, but before one individual is hurt, the one being bested exhibits a submissive signal and withdraws from the dispute (Fig. 20.14). The ritual conveys some information about the relative power of the opponents. For example, male red deer roar as a part of their rutting displays (displays of sexual excitement), and bigger, more powerful males can roar longer. Both combatants benefit from a ritualized battle. The combatant who submits lives to fight another day, perhaps with a weaker rival. The victor has established **dominance**, thus gaining access to mates, territory, or some other resource. Pursuing the conflict would have little benefit for either contestant and could result in the victor's injury.

In some social arthropods and vertebrates, ritualized combats lead to the establishment of a **dominance hierarchy**. This was originally observed in chickens by Schjelderup-Ebbe in 1922. One individual, *A*, is dominant over all others; a second individual, *B*, is dominant over all except *A*, and so on through a **pecking order**. Dominance hierarchies have also been studied in a number of mammals, such as baboons and elephant seals. Dominance hierarchies are not necessarily stable. They may also vary with the resource in question, with one individual having first access to water and another to mates. They may vary with age. The oldest male elephant seal is dominant and maintains a harem, but in the course of time he is challenged and eventually replaced by a younger male. Dominance hierarchies are a way to ritualize tensions that arise in a social group and to avoid the need for an agonistic encounter every time a limited resource is being contested.

Territoriality

In many cases agonistic behavior is directed toward the defense of a certain physical space. An area that is defended by an individual or pair for their *exclusive use* is known as their **territory**. Territorial behavior is not universal, but it has been observed in such diverse animals as limpets, crustaceans, insects, and most of the vertebrate classes. The size of the territory, which can vary from the size of a single leaf to large areas measured in acres, depends upon the size of the animal and the function of the territory. Territories may encompass a food source, mating site, or nesting site, or all of these. Some species in these groups may exhibit territoriality only during a critical part of the year, such as the breeding season. The defense of a territory requires constant vigilance, as well as considerable energy expenditure in advertisement, threats, and other agonistic displays. For territoriality to be effective, the resources must be worth fighting for (i.e., they must be in limited supply) and be concentrated in a defensible area. The importance of the cost of defending a territory versus the benefits it provides is illustrated by the golden-winged sunbird of Africa (Fig. 20.15), which can switch

Figure 20.16
Satin bowerbird in Queensland, Australia bringing tail feathers of another bird to decorate its bower.

from territorial to nonterritorial behavior. During the nonbreeding season, individuals of this species defend patches of mint whose nectar is a food source. Holding a territory can save its owner four hours of foraging time a day (2400 cal). However, one hour of accumulated defense flight burns 2000 calories. If the density of competing birds becomes too great and hence the caloric costs of defense too high, the defender will first reduce the size of its territory. If that is not sufficient, the defender will give up its territory and become nonterritorial.

The males of many songbirds use song and display to establish territories in the spring. These are small enough to be defensible yet large enough to provide good nesting sites, nesting materials, and the food needed to rear the young. Gulls and other sea birds can hardly defend their vast foraging area but do establish smaller, defensible territories around their nests. After the breeding season, birds abandon their territories.

A strange type of territoriality occurs in some birds and a few mammals in which a number of males share a courting area called a **lek**. Within the lek each male defends a small territory. The territory contains no food or nesting resources of value to either females or males; it is simply a small site where the male displays to the female. The leks of some species, such as the sage grouse, are cleared areas on the ground. In others, like the bowerbirds of New Guinea, the individual territory within the lek contains an elaborately decorated bower (Fig. 20.16). The females come to the lek to select mates and copulate. The male invests nothing in the care of the offspring, and some males are not selected by any females. The lek strategy appears to be correlated with the widely distributed and shifting food

Figure 20.15
An African golden-winged sunbird (*Nectarina reichenowi*). This species sometimes defends territories containing nectar producing flowers depending upon the energy costs of defense.

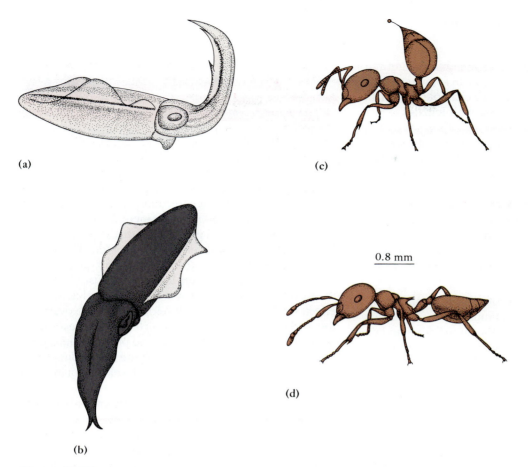

Figure 20.17
(a) and (b) Two chromatophoric color patterns and body and tentacle positions conveying alarm in the Caribbean reef squid, *Sepioteuthis sepioidea*. Squids and octapods have highly developed eyes that can distinguish between shapes.
(c) Alarm and defense posture of the ant *Crematogaster ashmeadi* compared with the resting posture (d).

resources of these species, such that it would be impossible for the male to maintain and defend a territory that would be of resource benefit to the female.

Mammals that forage widely have **home ranges**, areas in which they can be found 95% of the time. Home ranges of different individuals may overlap. In some cases a territory for nesting or some other special purpose may be established within the home range. Home ranges vary greatly in size according to the mobility and needs of the species. African hunting dogs forage over an area of 1500 square miles, whereas a mouse may have a home range of a fraction of an acre.

Communication

Behavioral interactions of all types require effective communication, for only by exchanging mutually recognizable signals can one animal influence the behavior of another.

Here we will only outline the five channels of sensory communication, the reception of which was discussed in Chapter 18.

1. *Visual signals*, such as body posture or color patterns, are widely used in animals possessing eyes and nervous systems capable of object discrimination (Fig. 20.17). Certain nocturnal animals and those that live in deep-water habitats communicate by light flashes and light patterns.

2. *Auditory signals* (via airborne sound waves) are common among many terrestrial animals but require both a means of producing the sounds and a sensory apparatus capable of receiving them. The frequencies employed may be very restricted in order to avoid confusion with other similar background sounds.

3. Some aquatic animals, such as species of fish and cetaceans (porpoises and whales), have become adapted for producing and receiving water-borne sounds. Communication by substrate vibrations occurs in some terrestrial animals. Some male fiddler crabs, for example, rap on the sand as a means of communicating with females.

4. *Touch* is an effective short-range form of communication. It is often employed in courtship but may have other functions. In primates, for example, grooming provides an opportunity for bonding that takes place between group members.

5. *Chemical signals* are likely the oldest and most widespread means of communication. Secreted compounds act as signals to other members of the same species. Any secretion that functions in this manner is called a pheromone. We will describe pheromones in more detail in the next section.

The behavioral repertoire of an animal may utilize several of these channels of communication, depending upon the environment in which the animal lives and the evolutionary constraints imposed upon it. For example, many spiders that respond to captured prey through web vibrations also utilize web vibrations in communicating with prospective mates. On the other hand, jumping spiders, which depend upon vision to stalk and capture prey, utilize visual signals in courtship. The costs of these avenues of communication are low, and they have evolved in conjunction with the particular life style of these species.

Pheromones

The secretion of pheromones, which act as signals to other members of the same species, is the only communication mechanism available to unicellular organisms and to many of the simpler invertebrates, since other sensory modes require rather complex sending and receiving mechanisms. Pheromone communication has been discovered in nearly all organisms. There is little energetic expense involved in synthesizing these simple but distinctive organic compounds. Conspecific individuals (members of the same species) have receptors attuned to the molecular configuration of the pheromone, which is most often inconspicuous to other species. Pheromones are effective in the dark, can pass around obstacles, can disperse over great distances, and may last for several hours or longer. The major disadvantages associated with pheromones include the slow speed of transmission and limited information content. The latter disadvantage is compensated for in some animals by secreting different pheromones with different meanings. Black-tailed deer release pheromones in their urine and feces, and from glands located on different parts of the feet and head. Pheromones may be deposited as scent markers on the ground or be carried by water or wind. Wind-blown pheromones are usually small, volatile molecules. Note that pheromones are defined by the way they are used and not by their chemical nature. There are many different kinds of substances that act as pheromones.

Many types of signals can be conveyed by pheromones. Most pheromones act as **releasers**, eliciting a very specific but transitory type of behavior in the recipient. Others act as **primers**, evoking slower and longer lasting responses. Some may act in both ways. In social insects some releaser pheromones supplement visual and tactile clues in leading members of the colony back to a food source, as in the trail pheromones of ants. Some enable members of a colony to recognize each other, and some warn the colony of foreign intruders. When termites from another nest enter a colony, some members of the resident colony release an alarm pheromone that attracts members of the soldier caste, who fend off the invaders. The composition of a beehive is controlled by an anti-queen pheromone secreted by the queen. It acts as a releaser, inhibiting the workers from raising a new queen. It also has a primer effect, since it inhibits the development of the ovaries in the workers. If the queen dies, or if the colony becomes so large that the inhibiting effect of the pheromone is dissipated, the workers begin to feed some developing larvae the special food that promotes their development into new queens (p. 762).

In many species, pheromones are important in attracting the opposite sex and in sex recognition. Many female insects produce pheromones that attract males of the appropriate species. Male butterflies, who mate in the daytime, are attracted to any flapping object that more or less resembles a female. After approaching the potential mate, the male engages in "hair penciling" behavior. Delicate pencil-like processes covered with pheromones are dangled in front of the prospective mate. If the flapping object is indeed a receptive female of the appropriate species, she will settle down and permit copulation. Another phenomone, bombykol, is secreted at night by female silkworm moths. Some sex attractant pheromones have been used as biological control agents. Male gypsy moths, for example, are lured to traps baited with synthetic analogues of the female pheromone.

Some aspects of the sexual cycle of vertebrates are affected by pheromones. In some species of mice the odor of a strange male will block the successful pregnancy of a recently impregnated female. Nerve impulses from the nose reach the female's hypothalamus and block the release of certain hormones necessary for implantation to occur. In some cases, the odor of a male mouse introduced among a group of females will cause their estrous cycles to become synchronized.

It is not known whether or not pheromones affect human behavior, but some curious phenomena suggestive of pheromone influences have been reported. Analysis of the menstrual cycles of the students in an American women's college showed a statistically significant tendency for the increasing synchronization of the menstrual cycles among roommates and close friends. The study rules out many possible explanations for the phenomenon and suggests that a pheromonal effect may be synchronizing the cycles of young women who are together much of the time.

Sexual Behavior and Reproduction

Natural selection has shaped the behavioral mechanisms that promote successful reproduction. For fertilization to occur, sperm and eggs must be released at about the same time. In many animals, environmental stimuli induce, often via hormones, the development of the gonads, ensuring that gametes are ready at the appropriate breeding season. Among many invertebrates and vertebrates, behavioral patterns then ensure the coming together of the gametes. In 1972, Robert Trivers proposed that males and females of most species are likely to evolve different courtship, mating, and rearing behavior if they are to optimize their reproductive success. He argued that the different strategies came about because of a fundamental difference in the parental investment of males and females. Females make a greater investment per gamete than males because the eggs are much larger than sperm and include enough stored energy to provide for at least the initial development of the embryos. The yolk stored in some individual bird eggs may be equivalent to 20% of the female's weight. Given a finite energy supply, females generally produce far fewer eggs than males do sperm. Trivers argued that a female would optimize her genetic contribution to succeeding generations by having her valuable eggs fertilized by the "fittest" males, a situation favoring the evolution of female choice of a breeding partner. Since sperm are relatively cheap to produce (energetically), a male would optimize his genetic contribution by mating with as many females as possible during his reproductive life.

In order to fertilize as many females as possible, many males compete intensely with one another for dominance and choice territories. In general, most females will be successful in becoming impregnated, but **sexual competition** among males of the same species means that some males will not be able to fertilize a female's eggs with their sperm. Thus sexual competition often has contributed to the evolution of larger body size, aggression, brilliant breeding colors, ornaments, antlers, and other features that give a male an advantage in establishing dominance and attracting females, a process referred to as **sexual selection**. It also has led to the evolution of strategies by which less successful males may occasionally be able to mate. In some fish species, a low-status male may mimic the behavior of a female, gain access to the dominant male's nesting territory, and spawn newly laid eggs of the resident female. Sexual selection has also led to strategies whereby a successful male will protect an inseminated female from copulation with other males. After copulation a male damselfly continues to grasp and fly with the female until she has deposited her eggs.

The more one-sided the parental investment, the more likely the investing parent will be the one to select a mate. Usually it is the female who chooses, and this **epigamic selection** (Gr. *epi*, upon + *gamos*, marriage) has affected those attributes that enhance a male's attractiveness. It has also affected those female characters that enable her to ascertain that the male is a member of her species and also that the quality of the male is worthy of her investment. Success of a male in dominance encounters with other males is an important indicator to the female of male fitness. Some females accept the first male that is able to reach her, but other females test the males by provoking encounters. Female baboons and chimpanzees in heat have brilliantly colored genital swellings that attract males and incite competition among them. A female frog grasped by a small young male during the breeding season may swim off in search of a larger male before shedding eggs.

The victorious male courts the female. A primary function of courtship is to ensure that the male is a member of the same species, but it also provides the female further opportunity to assess the quality of the male. As noted earlier, courtship in some cases is also needed as a releaser signal to trigger nest building or ovulation (see Fig. 20.4). Courtship rituals may be long and elaborate. The first display of the male releases a counter behavior of a conspecific female. This, in turn, releases additional male behavior, and so on until the pair are psychologically and physiologically ready for copulation.

Care of the young is an additional component of successful reproduction in many species, but it also requires a parental investment. The benefit of parental care is the increased likelihood of the survival of the offspring conceived, but the costs are a reduction in the number of offspring that can be produced and increased energy input by the parent or parents.

As in the case of mate selection, various behaviors have evolved to optimize the benefits of parental care versus the costs. Gulls, which nest on the ground close to each other, recognize their young and waste no energy in mistakenly feeding nearby chicks that are not their own. The related kittiwakes, whose nests on cliff ledges are fairly well separated from each other, cannot distinguish their young from others (but do not need to).

A female also has more to lose than a male does if the young conceived do not develop. Females are more likely to brood eggs than males, and usually the females invest more in parental care. A high investment in parental care usually is less advantageous to a male, for time spent in parenting is time lost in inseminating other females. Moreover, it may not be certain who fathered the offspring. Raising some other male's offspring is to the genetic disadvantage of a male. This is dramatically demonstrated in some species of animals, such as langur monkeys, in which a new male that takes over a band of females will kill all of the infants. The females then enter into a period of heat, and only the new harem master's genes will be transmitted to the next offspring.

Under some circumstances, it may be to the male's advantage to help rear the young. Receptive females may be few and far between, only a few young of certain paternity may be produced, food and other resources may be scarce and require more than one parent can provide, and the young may need protection against severe predation. In such cases, male investment in parental care increases the chances of offspring survival. When chances that the male will successfully mate again are low or that offspring from any mating are unlikely to survive without male care, the male optimizes his fitness by investing in the care of his offspring.

Reproductive Strategies

Male and female reproductive strategies enhance the perpetuation of their genes. The interaction of these strategies within the context of particular environments has determined the mating systems that have evolved. Male polygamy, known as **polygyny** (Gr. *polys*, many + *gyne*, female) is the condition in which a male mates with a number of females during the reproductive season and controls access to those females. Polygyny is especially favored in situations where dominant males can monopolize choice territories with ample resources to raise the young of several females, and the males need not make a large contribution to parenting. Males optimize their genetic contributions, and it is to the female's reproductive advantage to join a mated male in a good territory rather than an unmated one controlling an inferior territory. Polygyny is common among mammals, partly because only the females can nurse the young, and the males are released from some parenting. Sometimes a male may control a good territory, but the important factor may be simply his dominance over other males. Female elephants usually travel and feed together for protection. During the breeding season they are joined by a dominant male who adds little to their protection and can offer little more than his presumably superior genes.

In **monogamy** (Gr. *monos*, single + *gamos*, marriage), in which one male mates with one female, neither sex controls access to mates. Monogamy is favored in situations in which there are advantages to both parents in sharing in raising the young. The high survival rate of the young compensates for the males' inability to inseminate a large number of females without being involved in parenting. Most bird species are monogamous and have a close pair bonding that lasts for the breeding season, if not for the life of one of the parents. One parent alone could not brood the eggs, protect them against predation, or feed the voracious young, even when the male controls a favorable territory. The importance of both partners in parental care has been demonstrated by the reduced number of young reared by experimentally widowed snow buntings.

Female polygamy, or **polyandry** (Gr. *aner*, male), in which a female mates with a number of males during the reproductive season and controls access to those males, is not common, for it seldom is to the advantage of either sex. It is most likely to occur in situations in which the male assumes most, if not all, of the parenting duties. Some male fish, such as sticklebacks (Fig. 20.18), are better equipped to guard the nest than the females, and they need to do so because of high predation on the protein-rich eggs. The male cannot leave in search of other females or else he would lose his investment in the young, but after laying a clutch of eggs the female is free to go elsewhere and find another male to fertilize her next batch. Not surprisingly, in such polyandrous species, it is usually the female who courts and the male who selects the mate.

Social Behavior

Group living has evolved in many species because of the benefits conferred upon individuals of the group. Such associations vary greatly in complexity. The social interdependence of individuals composing fish schools and bird flocks, for example, is relatively simple; that of baboon troops and termite colonies, on the other hand, is complex. Common advantages of group living are a reduction in predation and an increase in foraging efficiency. By nesting together in colonies and synchronizing their egg laying, bank swallows reduce predation. A large, vulnerable population of nestlings is present for only a short time, and the adults band together to mob a predator that tries to gain access to the colony. In the season when zebras, wildebeest, and other large herbivores are abundant on the African plains, lions live together as a pride and cooperate in hunting (Fig. 20.19). A solitary lion can catch a zebra only about 15% of the time, but a group of five lions can fell one zebra 40% of the time, with the zebra providing food for the pride for several days. When the larger game migrate, the lions must prey on the much smaller gazelles, which only provide enough food for one lion for a day or two. During these periods, the prides break up and the lions hunt individually.

Social groups also have disadvantages. There often is increased competition within the group for food, mates, and other resources. Bank swallows sometimes steal nesting materials from each other, and neighboring males may copulate with a female whose partner is away. The risk of disease is higher. Investigators have shown that the level of nest infestation with bird lice increases with the size of bank swallow colonies. Before social groups evolve, the advantages must be greater than the disadvantages.

A comparison of the social blue-gill sunfish, a common freshwater fish, with the closely related, nonsocial

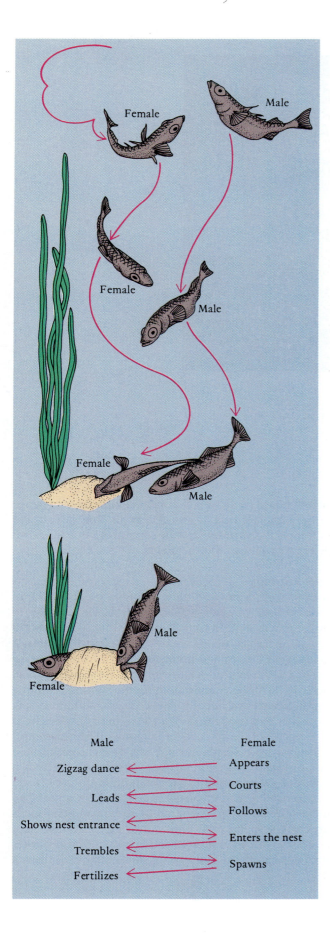

pumpkin-seed sunfish illustrates the relative costs and benefits of social versus solitary behavior. Blue-gills are social during the breeding season, when large numbers locate their nests close to each other and thereby gain mutual protection for their eggs from marauding catfish and other predators. The costs are interference in their courtship by other individuals, theft of eggs by nonbreeding males, and perhaps transmission of fish parasites and fungal pathogens of eggs. In contrast to blue-gills, pumpkin-seeds do not place their nests together. Their heavy jaws, adapted for a different diet than that of blue-gills, provide them easier defense against egg predators. Thus, for pumpkin-seeds, the cost of sociality would outweigh the benefits of cooperative egg protection.

Altruism

Social groups range in complexity from simple aggregations of unrelated individuals that may cooperate in chasing off a predator to the complex groups of ants, bees, and other social insects in which there is a close relationship within the group and a precise division of labor (p. 761). **Altruistic behavior, in which one individual appears to act in such a way as to benefit others, is frequently seen in the more complex social groups.** Altruistic behavior is easily understood where the helping behavior is obviously cooperative or reciprocal, and all the cooperating members share in the benefits, such as cooperative hunting by wild dogs, wolves, and lions. The selective advantages of altruistic behavior in which the helper does not seem to benefit from its actions have been harder to explain. A particularly clear case of altruistic behavior has been observed in the mating of wild turkeys. Several different groups of males, in each of which there is a dominance hierarchy, gather in a special mating territory and go through their displays of tail spreading, wing dragging, and gobbling in front of females who come to the area to copulate. Because of cooperation between the males within a group, one group attains a dominance over other groups, and its dominant member is the one to copulate most frequently with the females. Seemingly the males who helped establish the dominant group but who have low status within the group gain nothing. Close

Figure 20.18

Courtship in the three-spined stickleback. The appearance of the female in breeding condition elicits a zigzag dance by a territory-holding male (*top right*). The female follows the male, who leads her to the entrance of the nest he has built. When the female is in the nest (*bottom left*), the male's tactile prodding elicits egg laying.

Figure 20.19
Cooperative hunting. Lions killing a young wildebeest through cooperative hunting.

analysis has shown that members of a group are brothers from the same brood. Since they share many genes with the successful male, they are indirectly perpetuating many of their own genes. In this case altruism appears to have evolved by kin selection. Kin selection has also led to the evolution of the complex societies of social insects in which some individuals specialize for reproduction and other close relatives do the chores of the colony. Within the hives of honeybees, for example, all of the workers are daughters of the queen.

Altruism in the common vampire bat (*Desmodus rotundus*) of Central and South America is explained by both reciprocity and kin selection. These bats share blood meals with other members of the colony living together in hollow trees. About 30% of the immature bats and about 7% of the mature ones do not get a meal each night, either because they do not leave the roost or do not find a mammalian host, chiefly horses and cattle in ranching areas. These bats will starve to death if they do not feed within 60 hours. A bat that has fed will regurgitate parts of its blood meal to a hungry roostmate, giving it additional hours to find a meal on its own. Of what benefit is the altruistic behavior to the donor? First of all, there is a high probability that at some point a donor bat may need a shared blood meal itself. Second, a significant percentage of the recipient bats are related to the donor. Thus the donor bat is contributing to the survival of some of its own genes shared with the recipients.

Sociobiology

Sociobiology, a study of the evolutionary significance of social behavior by E.O. Wilson, was published in 1975. In his book Wilson dealt with all groups of social animals but drew much from his own extensive research on ants. A final chapter examined human behavior from an evolutionary and adaptive standpoint, i.e., that some human behavior evolved as adaptations for the hunting-gathering social groups that initially characterized the human species. This subjected the book to a storm of criticism from some parts of the scientific community. The old controversy of nature versus nurture was brought back to the boiling point. Critics argued that a genetic basis for human behavior (as opposed to an environmental one) is too indirect and remote to be uncovered and that any assignment of evolutionary significance to human behavior is unscientific speculation. The underlying concern of many critics was that if undesirable aspects of human behavior, such as aggression and warfare, were thought to have a genetic base or to be adaptive, they might be considered unchangeable and acceptable.

Much of the controversy over sociobiology has now declined. Most biologists and social scientists would agree that human behavior, like that of other animals, is a product of both nature *and* nurture. To recognize that there is a biological basis for parts of human behavior does not mean that we must condone all of the traits that are undesirable. Indeed, changing undesirable behavior is likely to be more successful through understanding and redirection than by claiming that there is no biological basis for such traits.

Our behavior is as much a part of our species character as that of any other animal, and it is difficult to under- stand why it would not be, in part, a product of evolution. A major difficulty in studying human sociobiology is that many basic adaptive features are now greatly obscured by the enormous technological and associated cultural over- lay the human species has accumulated in the last 10,000 years. We would be more successful if we could study the life style and social interactions of our hunting-gathering ancestors of 25,000 years ago. Remnants of that life style still exist among people in the very few parts of the globe relatively untouched by Western civilization. This is the task of anthropologists, but time is running out, since the influence of modern technology and culture is rapidly spreading to all of the world's people.

Summary

1. An animal's responses to its environment are highly organized and intricate patterns of activity that we call behavior, the study of which is termed ethology.

2. In behavioral patterns an animal first recognizes a sign stimulus given by another animal or some object. This releases a pattern of motor responses in the receiver. Frequently the responses are quite stereotyped. The signs an animal recognizes and the responses it makes depend upon the nature of its receptors and the activation of appropriate effectors by the nervous and endocrine sys- tems. Each animal has receptors attuned to signals of importance to it and a neuronal circuitry that generates responses adaptive to its particular needs. Hormones affect the development of sexual and other responses and set the physiological stage.

3. Many aspects of behavior as well as metabolism follow a 24-hour, or circadian, rhythm. Although these rhythms are based on an internal biological clock that has its own period, which is usually somewhat less or more than 24 hours, the internal clock becomes set so that it comes into phase and keeps in phase with daily changes in light and other environmental factors that follow a 24-hour cycle. The biological clock of mammals appears to be based on groups of neurons in the hypothalamus that have an inherent oscillating activity. The behavior of many marine animals also follows a tidal, or lunar, cycle.

4. Instinctive behavior depends on the activation of preprogrammed neuronal circuits. Appropriate fixed ac- tion responses are made the first time the animal sees the necessary release sign. Some instincts, such as a spider's web spinning, are closed and not modified by experience.

Other instincts are open and, although functional when first performed, are modifiable and improve with practice. Learning requires an interaction between an animal and its environment to become functional. The interaction occurs early in life in restricted learning and is not subsequently modified. Flexible learning continues to be modified as a result of continued interactions with the environment. Some flexible learning is accompanied by conditioning.

5. Evidence that behavior, like other attributes of organisms, is coded by the genes has been discovered by breeding experiments and by an analysis of the effects of individual mutations on behavior. Given an underlying genetic basis, it is to be expected that behavior is subject to evolutionary change and is adaptive. Behavioral ecologists study behavior in its ecological context to ascertain its evolution and adaptive nature.

6. Members of different species interact with each other as they search for habitats and food and try to avoid being eaten. Natural selection has favored the evolution of foraging strategies that optimize net energy intake. Defen- sive strategies include escape, cryptic coloration, deception of predators, poison mechanisms, the accumu- lation of distasteful materials in body tissues, aposematic coloration, Batesian mimicry, Mullerian mimicry, and group defense.

7. Members of the same species often use agonistic behavior to resolve conflicts for food, mates, and other limited resources. Combats tend to be ritualized and end with the dominance of one individual and the submission of another. Frequently a dominance hierarchy develops within social groups. In many cases agonistic behavior secures a certain territory for the exclusive use of an individual or pair, but territories are not established unless the value gained by controlling resources outweighs the costs of defending them.

8. Only by effective communication can one individual influence the activity of others. All sensory channels have been used by one species or another in communicating: visual signals, auditory signals, water-borne or substrate vibrations, touch, and chemical signals (pheromones).

9. Many types of behavior increase reproductive success and optimize an individual's genetic contribution to succeeding generations. Some ethologists believe that it is to the female's reproductive advantage to select superior males to fertilize her relatively few eggs, while it is to the male's advantage to fertilize as many eggs as possible. Sexual competition among males for access to females, and epigamic selection by females of superior males, have frequently led to the evolution of features such as larger body size, brilliant breeding colors, and ornaments in males and to elaborate courtship rituals.

10. Parental care of the young increases the probability of their survival but reduces the number of young produced. In general, females have a greater investment in the young than males and are more likely to assume the major care. Whether the mating system is polygyny, monogamy, or polyandry depends to a large extent on the amount of male involvement in parenting.

11. A social group is an aggregation of individuals of the same species that come together because of mutual advantages, such as a reduction in predation or an increase in foraging efficiency. Disadvantages are increased competition within the group and greater risk of disease. Advantages must outweigh disadvantages before social groups form. Social groups range in complexity from simple aggregations of unrelated individuals to complex societies of closely related individuals in which there is considerable division of labor. Altruistic behavior frequently is seen in the more complex groups. Such behavior may benefit the donor because at some time the altruism may be reciprocated or because the donor shares genes with the beneficiary.

References and Selected Readings

Aidley, D.J. (ed.) *Animal Migration*. Cambridge, England: Cambridge University Press, 1981. A collection of papers presented at a symposium of the Society of Experimental Biology dealing with the migration and navigation of insects, fish, birds, and whales.

Alcock, J. *Animal Behavior: An Evolutionary Approach*, 4th ed. Sunderland, Mass.: Sinauer Associates, Inc., 1989. A general text with an emphasis on the adaptive nature of behavior.

Altmann, S. The monkey and the fig. *American Scientist* 77(3):256–263, 1989. A delightful presentation of foraging ecology and its evolutionary interactions with other organisms.

Camhi, J.M. *Neuroethology: Nerve Cells and the Natural Behavior of Animals*. Sunderland, Mass.: Sinauer Associates, Inc., 1984. An exploration of the neurological basis of behavior.

Drickamer, L.C., and S.H. Vessey. *Animal Behavior: Concepts, Processes, and Methods*, 2nd ed. Boston: Willard Grant Press, 1986. A text dealing with the physiological basis of behavior and its ecological and evolutionary significance.

Heinrich, B. *Bumblebee Economics*. Cambridge, Mass.: Harvard University Press, 1979. A short, fascinating analysis of the cost and benefits of bumblebee foraging behavior.

Huber, F., and J. Thorson. Cricket auditory communication. *Scientific American* 253(Dec. 1985):50–68. An analysis of the auditory communication in the sexual behavior of crickets.

Lore, R., and K. Flannelly. Rat societies. *Scientific American* 236(May 1977):106. Interactions in a rat society that increase their chances of survival in a hostile environment are discussed.

Lorenz, K. *On Aggression*. New York: Harcourt, Brace and World, Inc., 1966. A classic volume discussing aggressive behavior.

McFarland, D. (ed.) *The Oxford Companion to Animal Behavior*. New York: Oxford University Press, 1982. A very useful reference to all aspects of animal behavior. Organized in the form of an encyclopedia.

Moore-Ede, M.C., F.C. Sulzman, and C.A. Fuller. *The Clocks that Time Us*. Cambridge, Mass.: Harvard University Press, 1982. A review of the circadian timing system, emphasizing the mammalian system and its importance to human beings.

Morris, D. *Dog Watching* and *Cat Watching*. New York: Crown Publishers, 1986. Two short volumes that discuss the significance of dog and cat behavior in terms of their wild ancestors.

Palmer, J.D. The rhythmic lives of crabs. *BioScience* 40(5):352–358, 1990. A review of the nature of tidal rhythms in crustaceans.

Tinbergen, N. *Social Behavior in Animals*. London: Science Paperbacks, Methuen and Co., Ltd., 1965. A reprint of Tinbergen's classic study.

Wilkinson, G.S. Food sharing in vampire bats. *Scientific American* 262(Feb. 1990):76–82. A study of the altruistic food sharing behavior in vampire bats.

Wilson, E.O. *Sociobiology*. Cambridge, Mass.: Harvard University Press, 1975. The biological basis of social behavior with a wealth of information about animal studies.

21

Reproduction

*(Left) Copulating checkerspot butterflies
(Melitaea sp.). (Above) The larva or
caterpillar of a checkerspot.*

If there is one feature of an organism that qualifies as the essence of life, it is the ability to reproduce. The survival of the species requires that its individual members multiply, that each generation produce new individuals to replace ones killed by predators and parasites or lost to aging. Reproduction is not necessary for the survival of the individual organism, but without reproduction the species becomes extinct. At the molecular level, reproduction involves the unique capacity of the nucleic acids for self-replication. Reproduction at the level of the organism ranges from the simple fission of unicellular organisms (a process that does not involve sex at all) to the incredibly complicated morphological, physiological, biochemical, and behavioral processes involved in the sexual reproduction of higher animals. The primary events of sexual reproduction in all metazoans are the formation of gametes, fertilization, and the transformation of the fertilized egg into a new individual. Many adaptations have evolved in different groups of animals that increase the likelihood that these primary events occur.

MODES OF REPRODUCTION

Asexual Reproduction

The origins of reproduction go back to the beginning of life, and in the billions of years since then many types of reproduction have evolved. **Asexual reproduction** is the simplest and undoubtedly the most primitive method. During asexual reproduction a single parent splits, buds, or fragments to give rise to two or more offspring. Since the process involves only the replication of cells by mitosis, the offspring have hereditary traits identical to those of the parent. Asexual reproduction is thus a kind of natural cloning. Although not all animals can reproduce asexually, the process is widespread among unicellular protoctistans and many different groups of animals.

Perhaps the simplest form of asexual reproduction is the splitting of the body of the parent into two and sometimes many more or less equal parts, each of which becomes a new, independent, whole organism. This form of reproduction, termed **fission**, occurs chiefly among the single-celled organisms (Fig. 21.1a), but some metazoans, such as planarian flatworms, divide transversely by fission, each portion regenerating the missing part (Fig. 21.1b).

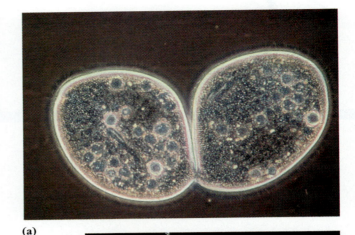
(a)

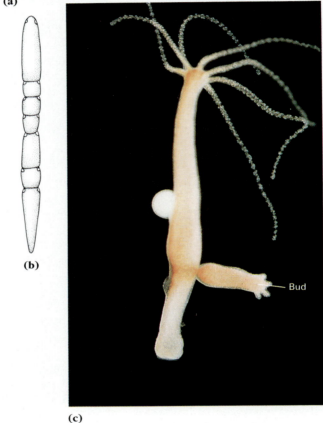
(b)

(c)

Bud

(d)

Figure 21.1

Asexual reproduction. (a) Late binary fission in the protozoan *Paramecium*. (b) Transverse fission in the flatworm *Stenostomum*. (c) Budding in *Hydra*. (d) A new individual of the tropical sea star (*Linckia multiflora*) can develop from a single arm and a portion of the central disc.

Hydras and certain other cnidarians reproduce by **budding** (Fig. 21.1c). A small part of the parent's body grows and becomes differentiated into a new individual that may take up independent existence, or the buds from a single parent may remain attached, forming a colony composed of many individuals.

Salamanders, lizards, sea stars, and crabs can grow a new tail, leg, or other appendage if the original one is lost. In some animals this regenerative ability makes it possible for a new individual to develop from a part of another. Certain sea stars, for example, have the ability to regenerate an entire new individual from a single arm and small part of the central disc (Fig. 21.1d), and many sea anemones can regenerate new individuals from fragments of tissue that sometimes are torn from the basal disc as the animal slowly moves across the bottom.

Asexual reproduction occurs during the embryonic period of some species. For example, the larval stages of many parasitic worms reproduce asexually, forming large numbers of other motile larvae, some of which may find a suitable host in which to develop (Chap. 27).

Sexual Reproduction

Sex is fundamentally the exchange of genes between two different individuals. It gives rise to new combinations, some of which may be superior in their fitness to the original combinations. Sex, which evolved first among the protoctistans, does not necessarily involve reproduction, that is, the formation of new individuals. In the protozoan *Paramecium*, for example, two individuals come together during conjugation, exchange nuclear material, and separate (see p. 580). Sex and its consequence, genetic recombination, has occurred, but no reproduction has taken place. The individuals may later reproduce by fission. In some other protoctistans and in metazoans, sex is combined with reproduction.

Sexual reproduction in animals involves the formation by meiosis of the **egg** and **sperm**, collectively called **gametes**, which contain the haploid chromosome number. The egg is typically large and nonmotile, with a store of nutrients to support the development of an embryo. The sperm is small and motile, adapted to swimming actively by beating its long, whiplike tail. The normal diploid chromosome number of the adult is restored when an egg and sperm unite at fertilization. In most species the sexes are separate, with distinct male and female individuals, so the genetic traits of two individuals also are combined at fertilization.

Parthenogenesis

The unfertilized eggs of some species can be stimulated to divide and develop, a process that is known as **parthenogenesis** (Gr. *parthenos*, virgin + *genesis*, origin).

Parthenogenesis is considered by many to be a special case of sexual reproduction because the offspring develop from gametes, the eggs. Changes in temperature, pH, the salt content of the surrounding water, or even mechanical stimulation may activate the eggs and initiate parthenogenetic development. A variety of marine invertebrates, frogs, salamanders, and even rabbits have been experimentally produced by parthenogenesis, but the resulting adult animals are generally weaker and smaller than normal and are infertile.

Parthenogenesis occurs naturally among many groups of metazoans, including a few vertebrates. In fact, there are some species of animals, such as certain microscopic freshwater animals called rotifers and certain lizard species, in which males are unknown. Parthenogenetic populations are often diploid because diploid eggs are produced by mitosis or because of a doubling of chromosome number occurs as haploid eggs begin to develop. But this is not always true. In some water fleas (crustaceans) and some aphids (insects), haploid females are produced parthenogenetically for several generations, and then some haploid males are produced that develop and mate with the haploid females.

In some species, sex of offspring is determined by whether the eggs are fertilized or develop parthenogenetically. The queen honeybee usually mates with half a dozen or more males during the few days she participates in nuptial flights. Approximately 5.5 million sperm are stored in a pouch connected with the genital tract and closed by a muscular valve. If sperm are released from the pouch as she lays eggs, fertilization occurs and the eggs develop into diploid females (queens and workers). If the eggs are not fertilized, they develop into haploid males (drones).

Advantages and Disadvantages of Asexual and Sexual Reproduction

Asexual reproduction is the simplest, quickest, and energetically least expensive way of forming offspring. Large numbers of individuals can be produced in a relatively short time. Only mitosis and growth are involved. True, all individuals usually are genetically alike, but this is not a disadvantage in a stable environment. Variability is introduced into the population by mutation.

Sexual reproduction is more complicated, and energetically it is very expensive. In most species large numbers of gametes must be formed to ensure that a reasonable number will come together in fertilization and that a few fertilized eggs will develop to maturity. The mortality of gametes and developing young in many species is extraordinarily high. A large codfish produces about 9,000,000 eggs in a season, but only a few hundred grow to adults. The eggs must also be provided with food reserves, and this requires a large expenditure of energy by the female. Some species, including mammals, produce far fewer gametes, but a great deal of energy is expended to increase the chance of fertilization and the probability that the young will develop to maturity. Females, in particular, invest considerable resources in nourishing the developing embryos and caring for the newborn.

Since the costs of sexual reproduction are so high, there must be offsetting advantages, otherwise it is unlikely that it would have evolved and become so widespread among both plants and animals. The increase in the genetic variability in a population resulting from the recombination of genetic material that occurs during meiosis and cross-fertilization would seem to be an obvious advantage. It is advantageous in a changing environment, but it can be a disadvantage in a stable one. Recombination, which occurs during meiosis, generates new combinations but may also break up favorable gene combinations.

Another advantage to sexual reproduction is the repair to damaged DNA that can occur during the first prophase of meiosis, when homologous chromatids line up in synapsis (p. 136). Often a defective or incorrect nucleotide in one strand can be detected and excised, and a new one inserted using the correct strand as a template. It has been argued by some investigators that DNA repair may have been the original advantage of sex because we see meiosis in many protoctistans that do not mate with other individuals. The increase in genetic variability may have been a byproduct of mechanisms for DNA repair. Variation would be enhanced by combining sex with reproduction. Sexual reproduction and the concomitant increase in variability may have been particularly advantageous in early stages of evolution when the diversity of organisms was increasing rapidly and the environment, partly because of the increase in diversity, also was changing rapidly. Some have argued that under these circumstances sexual reproduction became locked into various evolutionary lines and now there is no easy way to get rid of it. Others argue that, although sexual reproduction is expensive and sometimes disadvantageous, its capacity to generate new combinations is so powerful and beneficial under many circumstances that it has survived and continues to play a major role in evolution. The issues are not resolved, and biologists continue to disagree on the past and present advantages of sexual reproduction.

GAMETOGENESIS

Sexual reproduction begins with the formation of sperm (a process called **spermatogenesis**) and eggs (**oogenesis**). Collectively these processes are called **gametogenesis**. In most metazoans, spermatogenesis and oogenesis occur only in specialized gamete-forming organs, the **gonads**, that are restricted in their location. Sperm develop in testes; eggs, in ovaries.

Spermatogenesis

The **testes** usually protrude into part of the body cavity. They vary greatly in size and structure. In species with definite breeding seasons they enlarge as the breeding season nears, and spermatogenesis begins. At other times of the year they become small and inactive. In humans and most domestic animals the testes remain the same size, and spermatogenesis continues throughout the year once sexual maturity is attained.

Microscopically, the testes consist of small sacs, chambers, or tubules. The testes of mammals consist of numerous highly coiled **seminiferous tubules**, whose total length in the human male has been estimated at 250 meters (Fig. 21.2). This provides an area large enough for the production of billions of sperm. The wall of a seminiferous tubule is composed of clusters of sperm-forming cells interspersed with large **Sertoli cells**, which provide support, protection, and nourishment for the developing sperm (Fig. 21.3). Throughout embryonic development and during early postnatal development, the basal sperm-forming cells, the **spermatogonia**, divide mitotically, giving rise to additional ones. After sexual maturity, some spermatogonia begin to divide meiotically and form sperm, while others continue to divide mitotically and produce more spermatogonia for later spermatogenesis.

Spermatogenesis continues with the growth of the spermatogonia into larger cells known as **primary spermatocytes** (Fig. 21.4). These cells divide during the first meiotic division into two **secondary spermatocytes** of equal size that in turn undergo the second meiotic division to form four **spermatids** of equal size. The spermatid, a spherical cell with a generous amount of cytoplasm, contains the haploid number of chromosomes but is not motile.

A complicated process of growth and differentiation, though not cell division, converts each spermatid into a functional sperm or **spermatozoon** (Fig. 21.5). The nucleus of the spermatid becomes condensed, and most of its cytoplasm is phagocytosed by the Sertoli cells. Secretory granules from the Golgi bodies congregate at the front end of the sperm and form a modified lysosome, the **acrosome** (Gr. *akros*, tip + *soma*, body) that caps the nucleus. The acrosome contains enzymes and other materials that enable sperm to react to eggs of the same species and penetrate the outer coverings of the egg. The two centrioles of the spermatid move to a position just in back of the nucleus, and the more distal one gives rise to the **axial filament** of the sperm tail. This filament has the characteristic two central microtubules surrounded by nine double ones that occur in all flagella (see Fig. 11.30c). In many species, the axial filament is surrounded by a fibrous sheath that decreases in thickness toward the tail tip. Presumably, the sheath stiffens the tail somewhat, especially its proximal portion, and makes sperm movement less

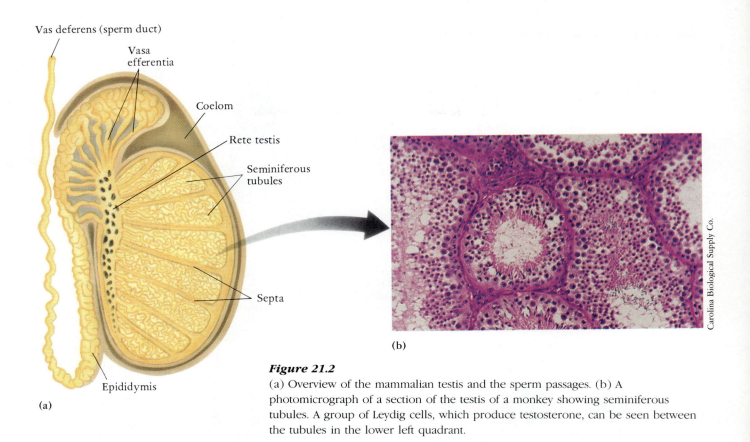

Figure 21.2

(a) Overview of the mammalian testis and the sperm passages. (b) A photomicrograph of a section of the testis of a monkey showing seminiferous tubules. A group of Leydig cells, which produce testosterone, can be seen between the tubules in the lower left quadrant.

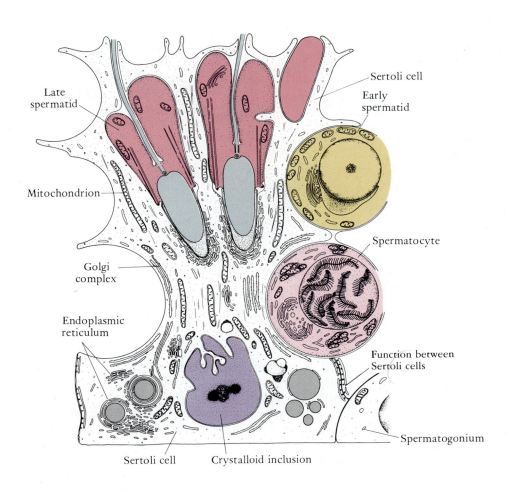

Figure 21.3

A drawing based on electron microscope studies of a cross section of a seminiferous tubule, showing a Sertoli cell and several stages of sperm-forming cells embedded in it. The Sertoli cell extends from the periphery of the tubule to its lumen.

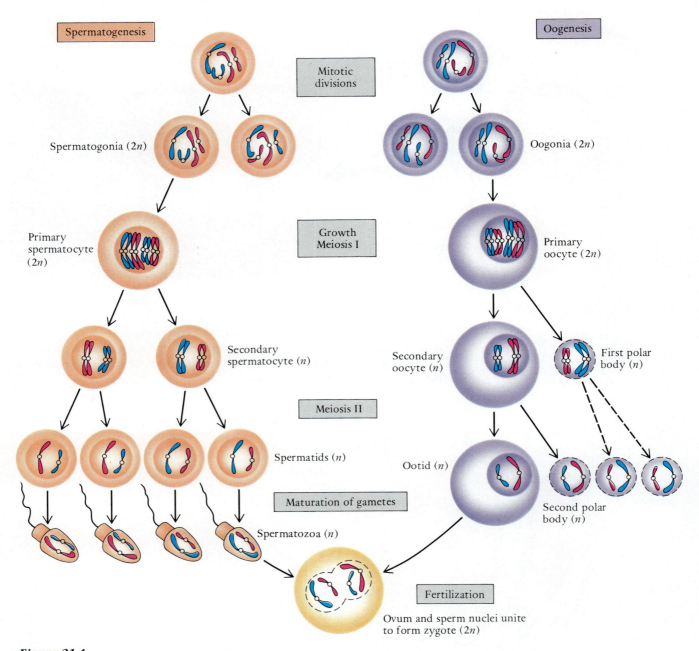

Figure 21.4

Diagrams comparing spermatogenesis and oogenesis in an animal with a diploid chromosome number of four. The two pairs of chromosomes are different in size, and those contributed by the male and female parents are colored differently.

erratic. The mitochondria move to the point at which head and tail meet and form a small **middle piece** that provides energy for the beating of the tail.

Spermatogenesis occurs in a cyclic fashion along all parts of a seminiferous tubule. In any one section of the tubule, the various stages will be passed through in succession. In adjacent sections of the tubule the cells will tend to be a stage behind or ahead. In human beings it takes 16 days for a mature sperm to develop from a spermatogonium.

There are great variations among animal species in the size and shape of the sperm tail and in the characteristics of the head and middle piece (Fig. 21.6). The sperm of a few animals, such as crabs, lobsters, and the parasitic

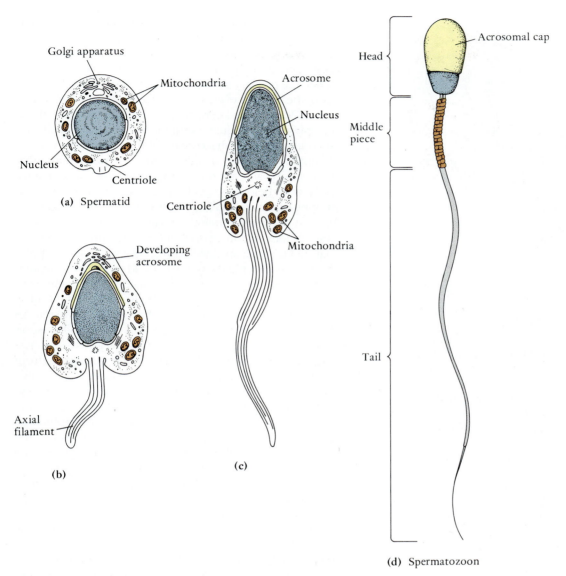

(a) Spermatid

Golgi apparatus
Mitochondria
Nucleus
Centriole

Developing acrosome

Axial filament

(b)

Acrosome
Nucleus
Centriole
Mitochondria

(c)

Head
Middle piece
Tail

Acrosomal cap

(d) Spermatozoon

Figure 21.5
(a–c) Three stages in the transformation of a mammalian spermatid into (d) a spermatozoon.

roundworm *Ascaris*, have no tail and move instead by amoeboid motion.

Oogenesis

The **ovaries** of most animals also protrude into the coelom. In seasonal breeders, they are very small and inconspicuous most of the year but enlarge as the breeding season approaches. The size that they attain varies greatly according to the number of eggs that are produced and the amount of food stored in them. The ovaries of fishes and amphibians, which produce hundreds or thousands of eggs, fill much of the body cavity. More advanced verte-brates produce fewer eggs, since fertilization is internal and the eggs are deposited in situations in which there is a greater chance for survival of the developing embryos. The ovaries of reptiles and birds are still large because the eggs contain much yolk. Mammals produce few eggs and they contain very little yolk, so their ovaries are quite small (2.5 cm in length in a human female).

The eggs develop in the ovary from immature sex cells called **oogonia**. In many animals, notably the verte-brates, each developing egg lies within a follicle that protects and nourishes the egg and secretes reproductive hormones (Fig. 21.7). Early in development the oogonia

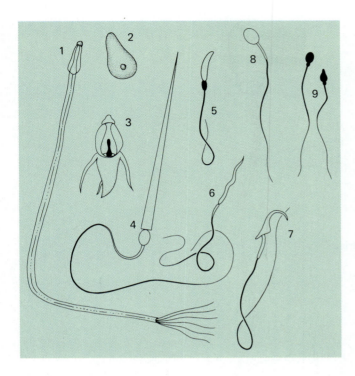

Figure 21.6
Sperm vary greatly in size and shape among different species. (1) Gastropod; (2) *Ascaris*; (3) hermit crab; (4) salamander; (5) frog; (6) chicken; (7) rat; (8) sheep; (9) man.

Figure 21.7
A photomicrograph through a mammalian ovary. The eggs develop in the walls of follicles, which gradually enlarge. Eggs are released, or ovulated, when a mature follicle ruptures through the ovary wall.

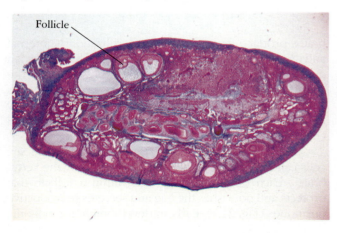

Follicle

undergo many successive mitotic divisions to form additional oogonia, all of which have the diploid number of chromosomes. Some or all of the oogonia begin to enlarge and develop into **primary oocytes** (see Fig. 21.4). When the primary oocyte undergoes the first meiotic division, the nuclear events are similar to those occurring in the male primary spermatocyte, but the division of the cytoplasm is unequal, resulting in one large cell, the **secondary oocyte**, that contains the yolk and nearly all the cytoplasm and one small cell, the first **polar body**, that consists of practically nothing but a nucleus. It was named a polar body before its significance was understood because it appeared as a small speck at one pole of the egg.

In the second meiotic division the secondary oocyte also divides unequally to yield a large **ootid** with essentially all the yolk and cytoplasm, and a small second polar body, both of which have the haploid number of chromosomes. The first polar body may divide at about the same time into two additional polar bodies. The ootid undergoes further changes, but no further cell division, to become a mature **ovum**. The small polar bodies soon disintegrate so that each primary oocyte gives rise to just one ovum in contrast to the four spermatozoa formed from each primary spermatocyte. The unequal cytoplasmic division ensures that the mature egg will have enough cytoplasm and stored yolk to survive and begin developing if it is fertilized. The egg has neatly solved the problem of reducing its chromosome number without losing the cytoplasm and yolk needed for development. When the eggs are mature, the follicles and ovaries rupture and the eggs are discharged into the coelom, a process termed **ovulation**.

The period in the life of a female when these events occur varies among animal species. In human females, the multiplication of oogonia occurs early in fetal development, and by the third month the oogonia begin to develop into primary oocytes. When a human female is born, her two ovaries contain about 400,000 primary oocytes that have attained the prophase of the first meiotic division. No more oocytes are produced during a woman's lifetime. The primary oocytes remain in prophase until the female reaches sexual maturity. During each monthly reproductive cycle thereafter one or more primary oocytes and follicles begin to enlarge. Not all reach maturity because many atrophy, but usually one follicle will. At the time of ovulation the egg will be in the secondary oocyte stage. The second meiotic division has begun, but sperm penetration is needed as a stimulus in most vertebrates for this division to go to completion and form an ootid.

FERTILIZATION

Mature gametes sometimes leave primitive animals through ruptures in the body wall, but in most species they are discharged from the body through specialized

gonoducts, called **sperm ducts** in males and **oviducts** in females. In many species, the eggs rupture the ovary wall and are ovulated into the coelom before being picked up by the oviducts. Eggs that are released from the ovary have reached a stage of maturation at which they are capable of uniting with a sperm in a series of reactions that we call **fertilization**. Fertilization must occur within a day or two, for the life span of released gametes is rather short. Various adaptations of animals, which we will describe presently, help to optimize the probability of fertilization by bringing sperm and eggs into proximity with each other. Fertilization begins with reactions enabling sperm and eggs of the same species to "recognize" and react to each other. Recognition reactions prevent sea star sperm from fertilizing sea urchin eggs or prevent successful fertilization if a male of one species copulates with the female of another. Initial recognition is accomplished in many species by the release of a **chemoattractant** from mature eggs into the environment. For example, sperm of one species of sea urchin swim at random over a small radius, but when a very small amount of a chemoattractant from mature eggs of the same species is introduced into the water, the sperm swim toward it and congregate there. This particular chemoattractant is a 14–amino acid peptide; another species of sea urchin has a different 10–amino acid chemoattractant.

A second recognition mechanism is the **acrosomal reaction** (Fig. 21.8). This was first studied in echinoderms, but the principles discovered appear to apply to many other groups, including mammals. When a sperm of a sea urchin contacts the jelly surrounding the egg, a polysaccharide in the jelly changes the permeability of the acrosomal membrane, and Ca^+ ions in the sea water enter (Fig. 21.8a and b). Acrosomal enzymes sequestered in small vesicles are released by exocytosis (Fig. 21.8a). These enzymes digest a path through the jelly. As the sperm penetrates the material surrounding the egg, actin molecules associated with the acrosome polymerize and form an **acrosomal process** that extends to the **vitelline envelope** that lies over the plasma membrane of the egg. This envelope is secreted around the egg while it is in the ovary. The acrosomal process contains species-specific proteins that unite with binding sites on an egg of the same species, but not with those on an egg of a different species.

If the acrosomal reaction is successful, the vitelline envelope breaks down at the point of contact, and the egg extends a cytoplasmic **fertilization cone** toward the sperm head. Sperm and egg plasma membranes unite, and the sperm nucleus and its proximal centriole, which was not used in forming the tail, are drawn into the egg. The sperm tail and middle piece are usually left behind.

Normally only one sperm nucleus with its haploid set of chromosomes enters the egg to unite with the haploid egg nucleus to form the diploid nucleus of the fertilized egg or **zygote** (Gr. *zygotos*, yolked together). Entrance of additional sperm nuclei could lead to the formation of triploid or other odd chromosome combinations and to abnormal development. Two reactions prevent multiple sperm entry. The entrance of one sperm immediately changes the permeability of the egg membrane to Na^+ ions. Na^+ rushes in (as in a nerve impulse), and the reversal of membrane polarity prevents the binding of additional sperm. A longer term block to polyspermy comes from a **cortical reaction** at the egg surface that spreads outward from the point of entrance of the first sperm. Sperm entry also triggers the release of Ca^+ ions, and their increase in the egg cortex causes mucopolysaccharides stored in thousands of minute cortical granules to be released into the space between the plasma membrane and vitelline envelope (Fig. 21.8a). The mucopolysaccharides have an osmotic effect and draw in water. As water enters this area, the vitelline envelope is raised from the egg surface as a **fertilization envelope**. This also prevents additional sperm from entering.

Once in the egg cytoplasm, the condensed sperm nucleus swells and forms a **male pronucleus** that moves toward the **female pronucleus** of the egg. Depending upon the species, these two pronuclei either fuse to form the zygote nucleus or each contributes its chromosomes to the mitotic spindle that leads to the first division of the zygote. One pole of this spindle is formed by the sperm centriole that entered with the sperm nucleus; the other, by the egg centriole.

Fertilization thus far has involved recognition, the entrance of one sperm, and nuclear union. The final event of fertilization is the **activation** of the egg cytoplasm, which until now has been very sluggish metabolically. The increase in free Ca^+ ions, which initiated the cortical reaction, also activates a series of reactions that throws the egg's metabolic machinery into high gear. The egg then prepares for the mitotic divisions that initiate embryonic development. Egg activation occurs in parthenogenesis in the absence of the other events of fertilization.

ADAPTATIONS FOR FERTILIZATION

Reproductive Synchrony

The synchronization of reproductive function in males and females inhabiting a given locality helps to ensure that eggs and sperm will meet. Reproductive synchrony requires first that gametes be *produced at the same time*, and second that they be *released at the same time*. In humans and some other species part of the population is always reproductively active, but in the majority of animals production of gametes is seasonal. Synchrony in gamete production is generally dependent upon environmental cues, such as changes in amount of daylight, air temperature, food supply, and lunar or tidal cycles, that work by activating internal neural and hormonal controls. In birds, increasing day length in spring stimulates via the nervous

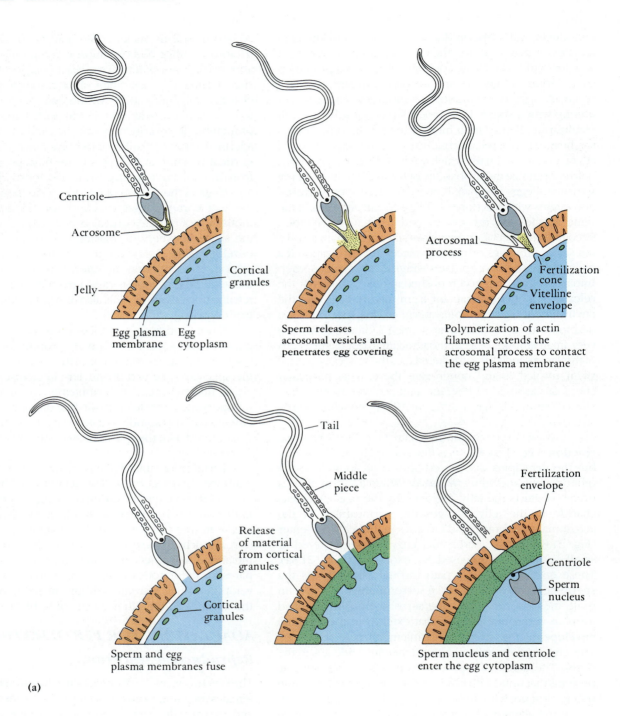

Centriole

Acrosome

Jelly

Cortical granules

Egg plasma membrane

Egg cytoplasm

Sperm releases acrosomal vesicles and penetrates egg covering

Acrosomal process

Fertilization cone

Vitelline envelope

Polymerization of actin filaments extends the acrosomal process to contact the egg plasma membrane

Cortical granules

Sperm and egg plasma membranes fuse

Tail

Middle piece

Release of material from cortical granules

Fertilization envelope

Centriole

Sperm nucleus

Sperm nucleus and centriole enter the egg cytoplasm

(a)

system the secretion of hormones that initiate migration, nesting, and egg production. Commercial poultry farms equip their houses with artificial lighting, for this has the same effect as increasing day length.

Gamete release can also be triggered by environmental cues. Some species of marine annelids, including palolo worms, emerge from their burrows and swarm near the surface at certain specific phases of the moon (p. 702). Similarly, the grunion, a small Pacific coast fish, is famous for its predictable nocturnal spawning during spring and summer high tides. Quite often the stimulus for gamete release is provided by the opposite sex. In many sessile or sedentary marine animals, the release of gametes into the surrounding water by one sex will stimulate the release of the gametes of nearby members of the opposite sex, sometimes leading to epidemic gamete release (Fig.

(b)

(c)

(d)

Figure 21.8
(a) A series of diagrams showing the entry of the sperm head into the egg. (b–d) Scanning electron micrographs, taken between 1700 and 37000 ×, of sperm entering a sea urchin egg. In (b), the acrosomal filament extends from the sperm head toward the egg surface. In (c), a fertilization cone elevates from the egg surface and enfolds the sperm. In (d), the sperm head is drawn into the egg. The small bumps seen on the egg surface are microvilli.

21.9a). Courtship between a male and a female is sometimes needed as a stimulus for gamete release (Fig. 21.9b and c). The courtship may be a very brief ceremony or may last for many days, as in some bird species. Courtship behavior serves two additional roles: it tends to decrease aggressive tendencies and it establishes species and sexual identification, i.e., it identifies a member of the same species but of the opposite sex.

Adaptations for External Fertilization

In addition to reproductive synchrony, which is found to some degree in all animals, additional adaptations increase the probability of fertilization. Primitive aquatic animals release their gametes into the surrounding water. Since sperm cannot swim far, the probability that eggs will be fertilized would seem to be low. You could probably predict a number of the following conditions that ensure that at least some sperm and eggs meet. More than one of the conditions may be operative for a particular species.

1. Large numbers of gametes are produced and released to compensate for the small percentage that are fertilized. For example, the American oyster releases from 15 to 115 million eggs at one spawning, of which only a small fraction will be fertilized. Of these an even smaller percentage will survive to settlement.

2. In many sessile aquatic animals, natural water currents carry gametes to neighboring animals of the same species.

3. The sperm of some sessile aquatic animals are brought to other individuals in their ventilating and feeding currents.

4. A favorable habitat will usually be occupied by more than one individual of the same species. There may be hundreds or thousands of individuals, and they may occur in dense aggregations, as on an oyster reef. Thus sperm and eggs do not need to be carried far to increase the chances of contact.

5. The individual members of some animal species

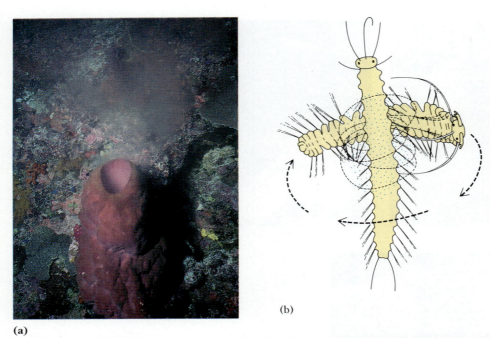

(a)

(b)

(c)

Figure 21.9
Spawning and gamete release. (a) A cloud of gametes being released from a sponge. (b) Male and female polychaete worms leave the bottom of the sea and come to the surface to court and spawn. (c) Red salmon ascending a stream in Washington State to spawn.

live relatively separated and solitary lives in holes or burrows but emerge and come together as a swarm during reproductive periods. The palolo worms mentioned earlier are a good example.

6. Many animals, including frogs, fish, and certain worms, come into close physical contact with each other at the time of gamete release, causing sperm and eggs to be discharged together. In fact, the contact, initiated by chemical signals (pheromones) or other factors, may trigger the gamete release.

Adaptations for Internal Fertilization

Fertilization within the body of the female has evolved in many marine and freshwater animals and in most terrestrial species. Desiccating conditions on land make external fertilization impossible. The likelihood of fertilization is increased by internal fertilization because sperm are placed in close proximity to the eggs. This allows a reduction in the number of gametes, especially eggs, that are produced, and energy need not be used in producing excess gametes. Internal fertilization offers still another advantage that has been exploited in many animals. Sperm can be stored in the female after being received from the male. Eggs can then be produced and fertilized continually or in batches without further need of the male.

Internal fertilization requires **copulation**, the deposition of sperm from the male system into the female system. Complex neural and hormonal mechanisms have usually evolved to bring about the necessary attraction and precopulatory behavior this requires. Sperm are usually transferred within a fluid medium, called **semen**, but there are many animals, such as leeches, octopods, lobsters, and some salamanders, in which great numbers of sperm are transferred together as a package, called a **spermatophore** (Fig. 21.10). The spermatophore disintegrates after being received by the female system. Spermatophores are usually formed in the terminal parts of the male system, and their shape varies from one group of animals to another.

Various modifications of the gonoducts have evolved in association with internal fertilization. In the male, a part of the sperm duct, or areas adjacent to it, may be modified for specific reproductive functions. A part of the duct may be given over for sperm storage; this part is often called a **seminal vesicle**. Glandular areas may be present for the production of seminal fluid, which serves as a vehicle for

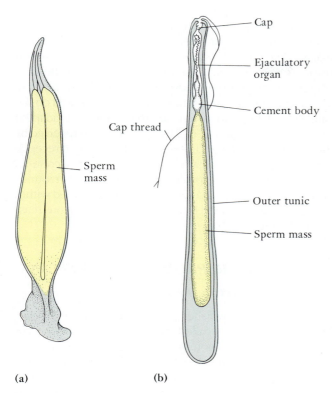

Cap

Ejaculatory organ

Cement body

Cap thread

Sperm mass

Outer tunic

Sperm mass

(a)   (b)

Figure 21.10
Sperm packages, or spermatophores, of a leech (a) and a squid (b).

the sperm and may also activate, nourish, and protect them. The terminal part of the sperm duct may open onto or into a copulatory organ, a penis, that provides for the transfer of sperm to the female.

In the female, the terminal portion of the oviduct sometimes is modified as a **vagina** for receiving the male copulatory organ, and a section of the oviduct, the **seminal receptacle**, may be modified for the storage of sperm after their transfer from the male.

Certain flatworms and leeches and a few other animals have an unusual mode of sperm transfer, called **hypodermic impregnation**, that bypasses much of the female reproductive tract. The sperm are injected by the penis into the body wall of the copulatory partner and make their way through the tissues to the eggs. Those flatworms with this mode of transfer have a dagger-like stylet associated with the penis, with which the partner is stabbed!

Hermaphroditism

The great majority of animals have separate sexes, i.e., an individual is either male or female, and this probably represents the primitive condition for animals. Animals

with separate sexes are sometimes described as **dioecious** (Gr. *di*, two + *oikos*, house).

In many animals both sperm and eggs can be produced in the same individual, a condition called **monoecious** or **hermaphroditic** (from Gr. mythology *Hermaphroditus*, the son of Hermes and Aphrodite, who became united in one body with the nymph Salmacis). Where both male and female systems are present at the same time, as in earthworms and flatworms, the animals are said to be **simultaneous hermaphrodites**, and there is typically a reciprocal transfer of sperm, i.e., each receives sperm from the opposite partner during copulation. However, some species of hermaphroditic animals are first one sex and then the other. These animals are **sequential hermaphrodites**. Most commonly testes develop first, and the individual functions as a male; later the male gonad atrophies, and a female gonad appears. This phenomenon is termed **protandry** (Gr. *protos*, first + *aner*, male). Among molluscs, some oysters and slipper shells are protandric. Individual slipper shells tend to live in clusters stacked one upon another, permitting the penis of the upper individual to reach the gonopore of the individual below (Fig. 21.11). The young specimens are always male, but after a period of transition, the male reproductive system degenerates, and the animal then develops into a female. The timing of the sex change appears to be influenced by the sex ratio of the entire cluster through substances secreted by its different members. The less common condition of individuals becoming first female and then male, which occurs in some species of fish, is called **protogyny** (Gr. *gyne*, woman).

Figure 21.11
Slipper shells (*Crepidula fornicata*) are protandric hermaphrodites. The young and small individuals in this cluster are males. As the males grow older and larger, they transform into females.

Some hermaphroditic animals, such as the parasitic tapeworms, are capable of self-fertilization. Since a particular host animal may be infected with but a few parasites, self-fertilization can be an important adaptation for the survival of certain parasitic species. But self-fertilization has the disadvantage of restricting the mixing of genetic material within the population, and most hermaphroditic animals utilize cross-fertilization, as do those with separate sexes.

The adaptive significance of hermaphroditism in groups of animals that are sessile is clear, for any individual that settles nearby is a potential mate. Thus, although most crustaceans are motile and have separate sexes, the sessile barnacles are hermaphroditic. However, there are many exceptions to this rule, and there are motile hermaphroditic forms and sessile forms with separate sexes. We have no good explanation as to why such animals as free-living flatworms and land snails are hermaphroditic.

EGG DEPOSITION

The eggs of some marine animals, such as certain worms, sea stars, and some fish, are simply dispersed into the water, and the developing eggs become part of the plankton. In many of these animals there is no shell or egg case, and the egg is covered only by a membrane. Mortality of eggs and young embryos of this type is very high.

Most animals reduce this mortality by enclosing the eggs singly or in groups within some type of envelope, usually secreted by the oviduct, and depositing them on the bottom or attaching them to some object. The envelope may be gelatinous, leathery, or horny. If the egg covering involves more than a mucous coat, fertilization must take place before the envelope is added. Thus many animals with egg cases have internal fertilization. In insects, the egg shell is produced before fertilization, but a hole in the shell, the **micropyle**, is present to permit entrance of the sperm. Because of the protection afforded to the developing embryo, species with egg envelopes produce fewer eggs.

PREDICTABILITY IN REPRODUCTIVE TRACT DESIGN

The reproductive tracts, or gonoducts, of animals reflect the reproductive habits characteristic of the species. If the reproductive habits are known, it is possible to predict a great deal about the design of the reproductive system. Let us examine two common animals, a sea star and a periwinkle, that have very different reproductive habits.

Sea stars utilize external fertilization, and the eggs, which lack shells or cases, are dispersed into the sea water. What would you predict about the reproductive systems of such an animal? Their gonoducts are adapted only for transport and are simple tubes extending from the gonads to the external gonopores (Fig. 21.12a).

Periwinkles, which are small intertidal snails, utilize internal fertilization and deposit their eggs within mucous masses. The coils of the long sperm duct can store sperm prior to discharge, and the more distal part of the duct contains glands for the production of semen (Fig. 21.12b). The duct terminates in a tentacle-like penis located on the right side of the head. In females, the gonopore is adapted for receiving the penis (Fig. 21.12c). Sperm are received by the female in a bursa and later stored in a seminal receptacle. Large glands associated with the oviduct secrete an albumen and an egg capsule that immediately surrounds each egg. Many egg capsules are in turn embedded within a mucous mass secreted by a jelly gland. The mucous mass is attached to a rock.

VERTEBRATE REPRODUCTIVE PATTERNS

Male Reproductive Tracts

In the males of many fishes and amphibians (Fig. 21.13a), microscopic tubules, called **vasa efferentia**, carry sperm from the seminiferous tubules of the testis through the mesentery supporting the testis to the anterior kidney tubules (the kidney is an opisthonephros, p. 353). Sperm pass through these tubules to the archinephric duct, which carries both sperm and urine to the cloaca, although not at the same time.

The testes remain in the coelom of reptiles and birds, but in most mammals, they undergo a posterior migration, or descent, and move out of the main part of the coelom into an external sac of skin and associated layers of the body wall known as the **scrotum** (Fig. 21.13b). Spermatogenesis cannot go to completion in most species of these endothermic vertebrates unless the testes descend because the temperature range at which it can occur is exceeded by the temperature in the abdominal cavity. Temperature in the scrotum is approximately 4° C lower than in the abdominal cavity, allowing spermatogenesis to proceed. The testes descend late in embryonic development and remain descended in the majority of mammals, but in rabbits and rodents they are migratory, descending into the scrotum during the breeding season, when temperature is critical, and withdrawing into the abdominal cavity at other times.

In the evolution of reptiles, birds, and mammals, the kidney changes from an opisthonephros drained by the archinephric duct to a metanephros drained by a ureter (p. 353). The pattern of sperm discharge remains the same as in amphibians, but the terminology is different because terms were given to the mammalian structures before their evolutionary relationship to primitive kidney structures was known. The anterior end of the primitive opisthonephros lies against the surface of the testis as a band known as the **epididymis** (Gr. *epi*, upon + *didymos*, testicle) (Figs. 21.2a and 21.13b). Since the epididymis is so close to the testis, the primitive vasa efferentia are very

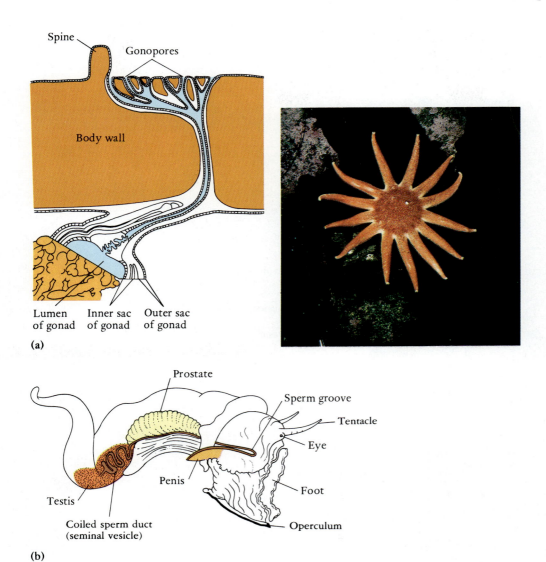

Spine

Gonopores

Body wall

Lumen
of gonad

Inner sac
of gonad

Outer sac
of gonad

(a)

Prostate

Sperm groove

Tentacle

Eye

Penis

Foot

Testis

Coiled sperm duct
(seminal vesicle)

Operculum

(b)

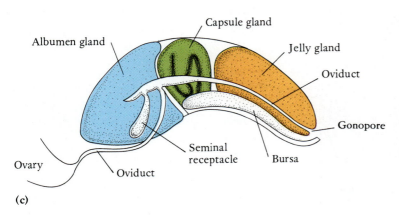

Capsule gland

Albumen gland

Jelly gland

Oviduct

Gonopore

Ovary

Oviduct

Seminal
receptacle

Bursa

(c)

Figure 21.12
Selected invertebrate reproductive systems. (a) The
gonads of sea stars, such as this tropical sunstar
(*Solaster dawsoni*) of the Northwest coast of the
United States, lie in the coelom of the arms. The
gonads open through multiple gonopores that lie
close to the central disc. (b) The male system of a
periwinkle. The shell has been removed. (c) The
female system of *Olivella*, a marine snail with a
reproductive tract similar to that of the periwinkle.
The bursa initially receives sperm from the male.
The albumen, jelly, and capsule glands provide
secretions around the egg as it is laid.

short and are incorporated into a network of microscopic
tubules, the **rete testis**, situated between testis and epididymis. Passages within the epididymis represent primitive
kidney tubules together with a highly coiled portion of the

archinephric duct. The rest of the primitive archinephric
duct passes as the sperm duct, the **vas deferens**, to join the
urethra, much of which evolved from the ventral part of the
cloaca.

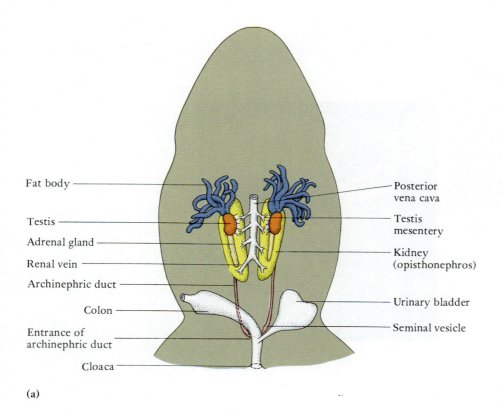

Fat body

Testis

Adrenal gland

Renal vein

Archinephric duct

Colon

Entrance of
archinephric duct

Cloaca

Posterior
vena cava

Testis
mesentery

Kidney
(opisthonephros)

Urinary bladder

Seminal vesicle

(a)

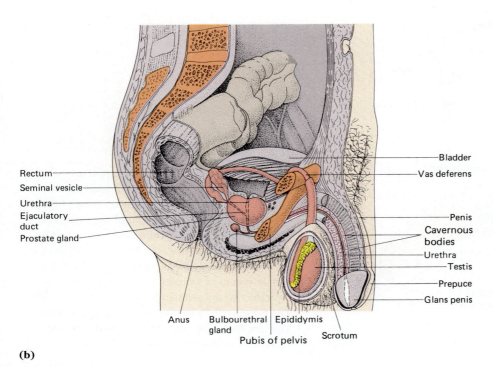

Rectum

Seminal vesicle

Urethra

Ejaculatory
duct

Prostate gland

Bladder

Vas deferens

Penis

Cavernous
bodies

Urethra

Testis

Prepuce

Glans penis

Anus   Bulbourethral Epididymis
              gland

Pubis of pelvis   Scrotum

(b)

Figure 21.13

The reproductive and associated urinary system of male vertebrates. (a) A frog;
(b) a man.

Sperm are stored in the epididymis and vas deferens. Secretions of cells in the epididymis create an environment in which the sperm mature physiologically and become motile. Sperm removed from the seminiferous tubules are nonmotile and incapable of fertilizing eggs.

Other differences between the male reproductive organs of vertebrates are correlated with whether fertilization is external or internal. In reptiles and mammals where fertilization is internal, the males have a penis. Fertilization is also internal in birds, but in most species the males lack a penis and transfer sperm by a brief cloacal apposition, sometimes called a "cloacal kiss." Accessory sex glands produce the seminal fluid, or **semen**, that transports the sperm. In mammals, the penis develops around the urethra and contains three **cavernous bodies** composed of spongy erectile tissue. Arterial dilatation coupled with a restriction of venous return causes the vascular spaces within the erectile tissue to become filled with blood during sexual excitement, making the penis turgid and effective as a copulatory organ. The accessory sex glands are a pair of **seminal vesicles** that connect with the distal end of the vasa deferentia, a **prostate gland** surrounding the urethra at the point of entrance of the vasa deferentia, and a pair of **bulbourethral glands** located more distally along the urethra. All secrete various components of the seminal fluid.

Female Reproductive Tracts

The oviducts of vertebrates are modified for various modes of reproduction. In fishes and nearly all amphibians, fertilization is external. They reproduce in the water, and the eggs develop into larvae that can care for themselves. Eggs are ovulated from the ovary into the coelom, from where they enter the anterior end of a pair of long, coiled oviducts through openings known as the **ostia** (Fig. 21.14a). The oviducts continue caudally to the cloaca. Their oviducts contain glandular cells that secrete layers of jelly about the eggs, and their posterior ends are expanded into **ovisacs** for temporary storage of the eggs, but they are not otherwise specialized.

Fertilization is internal in reptiles and birds. They reproduce on the land, and the free larval stage has been replaced by the evolution of a self-contained egg that provides food, water, and other needed materials until the embryo hatches as a miniature adult (p. 523). Most reptiles and all birds lay eggs that develop externally. Their oviductal glands, which secrete the albumin and a shell around the egg, are more complex than in those of amphibians and fishes, but in other respects the oviducts of reptiles have not changed greatly.

Most mammals and a few fishes and reptiles retain the fertilized eggs within the uterus until development is complete. In most mammals and a few nonmammalian species, a union of certain membranes surrounding the embryo with the uterine lining forms a **placenta** through which the embryo receives food and other needs from the mother and eliminates waste products of metabolism (p. 524). The oviducts are modified accordingly. In the human female (Figs. 21.14b), the fringed ostium lies adjacent to the ovary and may even partially surround it. When ovulation occurs, the discharged eggs are close enough to the ostium to be easily carried into it by ciliary currents. The anterior portion of each oviduct is a narrow tube known as the **uterine** or **fallopian tube**, and eggs are carried down it by ciliary action and muscular contractions. The remainder of the two primitive oviducts have fused with each other to form a thick-walled muscular **uterus** and part of the **vagina**. The terminal portions of the vagina and urethra develop from a further subdivision of the ventral part of the cloaca. The vagina is a tube specialized for the reception of the penis. It is separated from the body of the uterus, in which the embryo develops, by the sphincter-like neck of the uterus known as the **cervix**. The orifices of the vagina and urethra are flanked by paired folds of skin, the **labia minora** and the **labia majora**. A small bundle of sensitive erectile tissue, the **clitoris** (Gr. *kleitoris*, a small hill), lies just in front of the labia minora. Structures comparable to these are present in the embryo of a male and develop into more conspicuous organs. The labia majora are comparable to the scrotum; the labia minora and clitoris, to the penis. A pair of glands, homologous to the bulbourethral glands in the male, discharge a mucous secretion near the orifice of the vagina. A fold of mucous membrane, the **hymen**, partially occludes the opening of the vagina but is usually ruptured during the first copulation.

Mammalian Fertilization

During copulation, the sperm that have been stored primarily in the epididymis are ejaculated by the sudden contraction of muscles in and around the male ducts, and the accessory sex glands concurrently discharge their secretions. The seminal fluid that is deposited in the vagina may contain as many as 400,000,000 sperm in humans. Mucus in the seminal fluid serves as a conveyance for the sperm; proteolytic enzymes break it down into a more watery fluid after the semen has been deposited in the vagina, permitting the sperm to become highly motile. Fructose provides a source of energy, alkaline materials prevent the sperm from being killed by acids normally in the vagina, and certain fatty acids (prostaglandins) promote the contraction of the smooth muscle in the walls of the uterus and uterine tubes.

Sperm move from the vagina through the uterus and up the uterine tube in a little over one hour. How they do this is not entirely understood. Although sperm can swim by beating the tail, muscular contraction of the uterus and uterine tubes is primarily responsible for carrying them upward. Fertilization occurs in the upper part of the

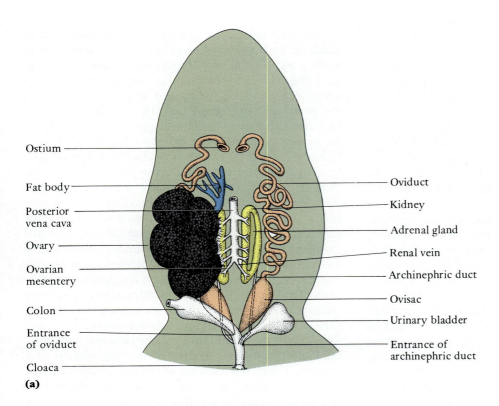

Ostium

Fat body

Posterior
vena cava

Ovary

Ovarian
mesentery

Colon

Entrance
of oviduct

Cloaca

Oviduct

Kidney

Adrenal gland

Renal vein

Archinephric duct

Ovisac

Urinary bladder

Entrance of
archinephric duct

(a)

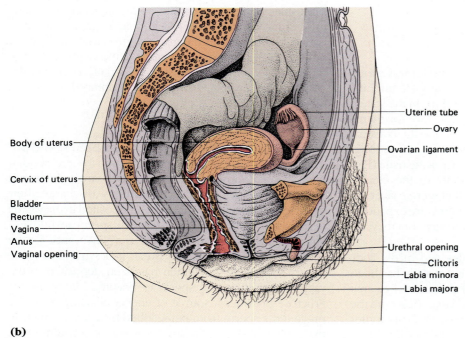

Body of uterus

Cervix of uterus

Bladder
Rectum
Vagina
Anus
Vaginal opening

Uterine tube

Ovary

Ovarian ligament

Urethral opening
Clitoris
Labia minora
Labia majora

(b)

Figure 21.14
The reproductive and associated urinary system of female vertebrates. (a) A frog;
(b) a woman.

uterine tube, but the arrival of an egg and the sperm in this region need not coincide exactly. Sperm retain their fertilizing powers for a day or two, and the egg moves slowly down the uterine tube, retaining its ability to be fertilized for about a day.

Only one sperm fertilizes each egg, yet unless millions are discharged fertilization may not occur. One reason for this is that only a fraction of the sperm deposited in the vagina reach the upper part of the uterine tube. The others are lost or destroyed along the way. When the egg enters the uterine tube, it is still surrounded by a few of the follicle cells that encased the egg within the ovary, and a sperm cannot penetrate the egg until these are dispersed. This requires the enzyme **hyaluronidase**, which can break down **hyaluronic acid**, a component of the ground substance between cells. Hyaluronidase is produced by the sperm themselves, and large numbers of sperm are apparently necessary to produce enough of it.

THE HORMONES OF REPRODUCTION AND PREGNANCY

Reproduction is a complex and highly integrated process. Sperm and eggs must be produced and released synchronously, and often their release must be integrated with seasonal changes or other external cues. Male and female morphology often are different, and some of these differ-

ences are correlated with reproductive behavior. In mammals, egg release must be synchronized with changes in the uterine lining that prepare for the implantation and nurture of the young embryo. Many changes in female physiology occur at the birth of the offspring and during lactation. All of these processes are controlled by a set of reproductive hormones, supplemented in some cases by nerve impulses. The endocrinology of reproduction has been studied most thoroughly in mammals, but the principles discovered in this group apply to most other vertebrates.

The production of reproductive hormones by the testes and ovaries can be viewed as a cascade of signals initiated by **gonadotropin-releasing hormone (GnRH)** secreted by the hypothalamus. Release of GnRH promotes the production of **follicle-stimulating hormone (FSH)** and **luteinizing hormone (LH)** by the adenohypophysis. These hormones are named for their effects in females, where they were discovered, but they are present in males as well. FSH and LH promote the synthesis of reproductive hormones by the testis and ovary. **Androgens** (primarily testosterone) are the male sex hormones; **progesterone** and **estrogens**, of which **estradiol** is the most important, are the female sex hormones. Hormones produced by the gonads are steroid derivatives of cholesterol, and the biochemical pathways for their synthesis overlap to a considerable extent (Fig. 21.15). Enzymes needed for their synthesis occur in both the testis and ovary, and in some

Figure 21.15
The reproductive hormones are steroid derivatives of cholesterol. Many intermediate steps in their synthesis have been omitted in this diagram.

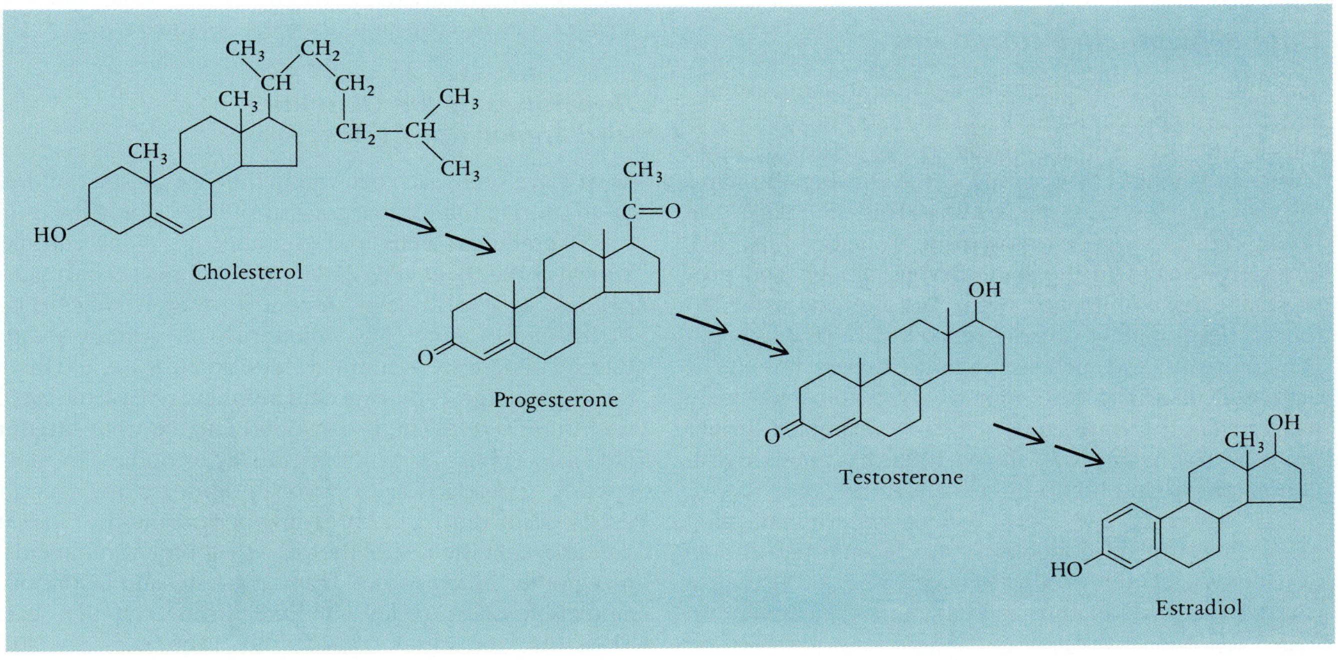

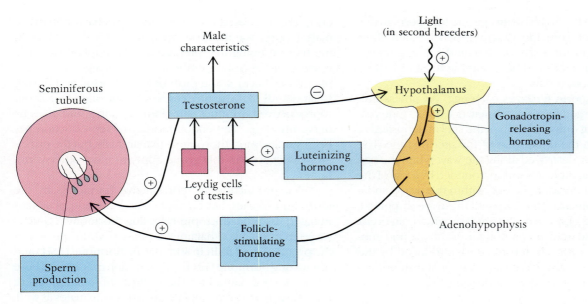

Figure 21.16

A diagram of the action of male reproductive hormones. Luteinizing hormone and follicle-stimulating hormone, which control the ovarian cycle, are also important in regulating testosterone and sperm production.

other tissues as well, so it should not be surprising that some androgens occur in females and that progesterone and estrogens occur in males. In fact, the immediate biochemical precursors for estradiol synthesis are androgens. Thus males and females differ not so much in the types of hormones produced, but in their quantity.

Male Reproductive Hormones

As a male matures, GnRH from the hypothalamus stimulates the synthesis and release of luteinizing hormone and follicle-stimulating hormone (Fig. 21.16). LH acts on small groups of **Leydig cells** that are located between the seminiferous tubules (see Fig. 21.2b). This stimulates the Leydig cells to produce testosterone. Together with FSH, testosterone acts on the seminiferous tubules and promotes the development of sperm. Testosterone is needed for the development of male reproductive passages and genital organs, and such secondary sex characteristics as increased muscular development, deep voice, the male pattern of hair growth, antlers in some mammals, bright plumage in some birds, and so on. It also has physiological effects, promoting libido or sexual desire.

Testosterone and sperm production are controlled primarily by negative feedback. As testosterone production increases, GnRH production decreases, and vice versa (Fig. 21.16). Many mature male mammals, including humans, reproduce throughout the year, and testosterone and sperm production continue. In many other species they are seasonal. Changes in day length affect the hypothalamus, the growth and descent of the testis, and the onset of the reproductive season. During the reproductive season, which may last for several weeks, testosterone and sperm production are controlled as they are in humans. After the mating season, GnRH is no longer synthesized, testosterone production decreases, and the testes become smaller and inactive until the next season.

Hormones of the Ovarian and Menstrual Cycles

As we have seen, each egg within the ovary is surrounded by an ovarian follicle that consists of inner epithelial cells called **granulosa cells** and a connective tissue capsule known as the **theca** (Fig. 21.17). Throughout sexual maturity, the ovaries undergo repetitive ovarian cycles (Fig. 21.18). One or more eggs, which are in the primary oocyte stage of oogenesis, accumulate food and mature, and their follicles enlarge, rupture at ovulation, and release the eggs. Each ruptured follicle is converted into a **corpus luteum** (see also Fig. 21.7), which eventually atrophies. In most species, reproduction is seasonal; one or a few ovarian cycles occur during the breeding period and the ovaries are quiescent the rest of the year. In contrast, in humans, many other primates, and many domestic and laboratory animals, ovarian cycles continue throughout the year. Their duration varies considerably. Laboratory rats and

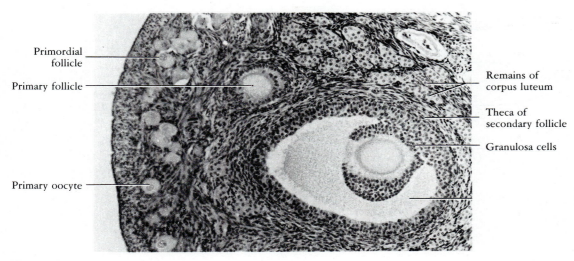

Primordial follicle

Primary follicle

Primary oocyte

Remains of corpus luteum

Theca of secondary follicle

Granulosa cells

Figure 21.17

A photomicrograph of a portion of the cortex of a cat's ovary showing stages in the growth of follicles. Note the granulosa and thecal cells that form the wall of the secondary follicles.

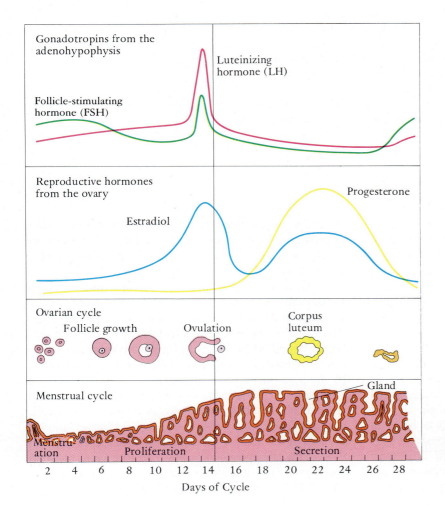

Gonadotropins from the adenohypophysis

Luteinizing hormone (LH)

Follicle-stimulating hormone (FSH)

Reproductive hormones from the ovary

Progesterone

Estradiol

Ovarian cycle

Follicle growth

Ovulation

Corpus luteum

Menstrual cycle

Gland

Menstruation

Proliferation

Secretion

2   4   6   8   10   12   14   16   18   20   22   24   26   28

Days of Cycle

Figure 21.18

Changes in the levels of reproductive hormones and the timing of the ovarian and menstrual cycles are intimately interconnected.

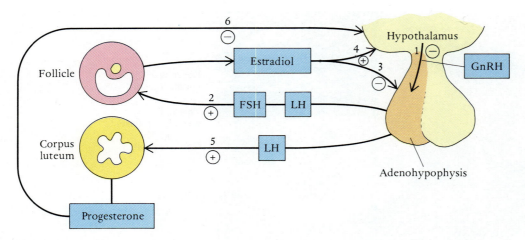

Figure 21.19

The hormonal control of the ovarian cycle. (1) GnRH from the hypothalamus stimulates (2) the release of FSH and LH from the adenohypophysis. These stimulate the initial growth of the follicle and (3) its secretion of a low-base level of estradiol. Low levels of estradiol inhibit the release of FSH, apparently by acting on the adenohypophysis and decreasing its response to GnRH. But once the follicle has enlarged and made the necessary enzymes, it continues to grow despite the reduction in FSH. Estradiol levels rise greatly shortly before ovulation. (4) The increased level of estradiol has a positive feedback effect upon the hypothalamus, more GnRH is produced, and the level of LH surges. (5) Continued secretion of LH after ovulation transforms the follicle into the corpus luteum. (6) The corpus luteum secretes progesterone, which inhibits the release of GnRH and hence the secretion of LH and FSH. Follicle growth is stopped as long as progesterone is produced. If fertilization does not occur, the corpus luteum regresses under the influence of changing levels of LH and FSH, and the inhibition by progesterone is removed. GnRH is again released and the ovarian cycle starts anew.

mice have a cycle of 4 to 5 days; humans have a cycle of about 28 days. Many mammals enter a period of heat or **estrus** (Gr. *oistros*, frenzy), about the time of ovulation and will receive males only at this time. Indeed, many females secrete pheromones at this time that attract males, as any person who has had a female dog in heat is aware. Cycles in which estrus is the most conspicuous feature are called **estrus cycles**.

Human females do not have these behavioral changes, but the ovarian cycle is accompanied by cyclic changes in the uterine lining or **endometrium** (Gr. *endon*, within + *metra*, uterus) that are known as the **menstrual cycle** (Fig. 21.18). The menstrual cycle begins with a period of **menstruation** during which superficial layers of the endometrium are shed and this is accompanied by bleeding. The endometrium then regenerates, thickens, and finally becomes very vascular and glandular in preparation for the implantation of a fertilized egg, should one

enter the uterus. Most other mammals do not have such conspicuous changes in the uterine lining, for either the lining does not increase in thickness to the same extent or it is reabsorbed rather than being shed.

The ovarian cycle and the accompanying estrus or menstrual cycles are controlled by the same cylical changes in the levels of reproductive hormones (Figs. 21.18 and 21.19). As a human female approaches puberty, GnRH begins to be secreted by the hypothalamus. The release of this hormone causes the secretion of FSH and LH by the adenohypophysis. These hormones promote the enlargement of one or more ovarian follicles. More food accumulates in the egg, and the granulosa and thecal cells multiply. The thecal cells have receptors for LH, and LH induces the synthesis of androgens within these cells, as it does in the Leydig cells of males. The androgens are then transferred to the granulosa cells, which have receptors for FSH. FSH is needed to activate the enzyme that converts androgens to estradiol. Thus a base level of estradiol secretion by the follicles requires the combined action of thecal and granulosa cells.

Ovarian cycles are irregular in frequency and length in young females for several years, but the buildup in estradiol promotes the development of most secondary sex characteristics including the enlargement of the reproductive passages, the widening of the pelvis, growth of the breasts, and the characteristic pattern of fat distribution. In human females as in males, androgens, rather than estradiol, are responsible for pattern of hair growth and for promoting libido. These androgens are secreted in females by the adrenal glands rather than by the gonads.

After puberty, estradiol is responsible for the restoration of the endometrium following menstruation, and for its initial increase in thickness, vascularity, and glandular development (Fig. 21.18). Estradiol acting together with FSH and LH promotes the further growth of the ovarian follicle, more estradiol secretion, and further development of the endometrium.

In the ovarian cycle, as the follicle matures it increases dramatically in size. Follicular liquid accumulates within it, and it bulges from the ovary surface. Estradiol levels increase greatly a few days before ovulation in humans and other **cyclic ovulators** (Fig. 21.18). This causes the granulosa cells, which previously have only had receptors for FSH, to develop ones for LH. Just prior to ovulation a great increase in LH secretion occurs. This is known as the **LH surge**. High levels of LH promote the maturation of the developing egg cell. Because all follicular cells can now respond to LH, the high levels of LH also cause a decrease in the adhesion between the follicular cells and their release of proteolytic enzymes that break down the follicular wall. Ovulation occurs. Ovulation typically takes place about the middle of the ovarian and menstrual cycles, on day 14 to 15 in humans.

The trigger for the LH surge in cyclic ovulators is positive feedback to the hypothalamus of the increased and sustained level of estradiol produced by one or more follicles nearly ready to ovulate (Fig. 21.19). In many species, the high estradiol level also causes estrus, which helps to synchronize copulation with ovulation, thereby maximizing the probability of the union of sperm and eggs.

Cats, minks, rabbits, and a few other mammals exhibit **induced ovulation**. In these species the LH surge and ovulation are initiated by nerve impulses originating from actual copulation. These nerve impulses affect the hypothalamus and immediately trigger the release of more GnRH and hence more LH. In some species repeated and frequent copulation is necessary to induce ovulation. A pair of lions copulates several times an hour for about three days.

Under the continued influence of high levels of LH, the ruptured follicle remaining in the ovary transforms into a small, yellowish mass known as the **corpus luteum** (L. *corpus*, body + *luteus*, yellow) (Figs. 21.18 and 21.19). LH production decreases after ovulation, but its continued presence at low base levels promotes the synthesis by luteal cells of the second female hormone, progesterone, and the continued synthesis of some estradiol. Progesterone is primarily a hormone of pregnancy, as we shall see, but at this time it has a twofold role. It inhibits the release of hypothalamic GnRH, which prevents the further growth of other follicles, the LH surge, and subsequent ovulation (Fig. 21.19). This is the basis for oral contraceptives that contain progesterone analogues. Progesterone also promotes the completion of the buildup of the uterine lining in preparation for the reception of a fertilized egg (Fig 21.18).

If pregnancy does not occur, the corpus luteum lasts for about two weeks in humans and then degenerates (Fig. 21.18). A low level of LH and an increase in estradiol appear to be the factors that cause this. As the corpus luteum regresses and progesterone levels decrease, the uterine lining cannot be maintained and it is sloughed off during menstruation in humans.

The Hormones of Fertilization, Pregnancy, and Lactation

Fertilization in mammals requires that sperm reach the eggs in the uppermost part of the reproductive tract within a few hours of ovulation. The high levels of estradiol that precede ovulation are also necessary for fertilization. Estradiol increases the water content of the cervical mucus, thus reducing its viscosity and making it possible for sperm to enter the uterus from the vagina. Estradiol also promotes a reversed peristalsis in the uterine tube that carries sperm upward against the downward beating of cilia in the tube.

The corpus luteum formed during the normal ovarian cycle produces enough progesterone to sustain the uterine lining until the embryo reaches the uterus 7 or 8 days after ovulation. In some species, the embryo simply lies on the uterine lining, but in most species the embryo penetrates, or implants in, the uterine lining. Certain of the cell layers that develop around the embryo secrete **chorionic gonadotropin**, a glycoprotein that is very similar to LH. Assays for chorionic gonadotropin are used as pregnancy tests. Chorionic gonadotropin prevents the degeneration of the corpus luteum and ensures the continued production of progesterone, at least for a while. Progesterone is essential for a normal pregnancy, for it maintains the uterine lining and placenta, and prevents uterine contractions that could abort the embryo. In humans and other species in which the placenta becomes well developed and produces progesterone, the corpus luteum eventually decreases in size and disappears. In pigs and many other species in which the placenta is less developed and does not produce sufficient progesterone, the corpus luteum enlarges and remains throughout pregnancy.

Prior to birth, or **parturition**, the increasing level of estrogens from the placenta promotes the growth of uterine muscles, and **relaxin**, a hormone also secreted by the placenta, dissociates and "softens" the tough collagen fibers in the cervix and in the ligaments uniting the pubic bones. It is probable that many factors acting in concert trigger the onset of birth. Among them appear to be (1) increased fetal size, which may stretch and activate the uterine muscles, (2) an increase in the glucocorticoids released by the fetal adrenal gland, which promotes placental estrogen synthesis, and (3) an increased level of estrogens relative to progesterone, for estrogens increase uterine contractions. Once labor contractions are under way, they are sustained and increased in intensity and frequency by the secretion of prostaglandins and by neuroendocrine reflexes. Sensory impulses from the dilating cervix feed back to the hypothalamus and promote the release of oxytocin from the neurohypophysis. One of the functions of oxytocin, as we have seen (p. 445), is to promote the contractions of the smooth muscles of the uterus, thus contributing to labor contractions.

Glandular tissue develops in the breasts during pregnancy under the influence of estrogens and progesterone, and the formation of the enzymes needed for milk production and secretion is stimulated by **placental lactogen**, a hormone secreted by the placenta, and by prolactin released by the adenohypophysis. Although the mother's breasts are ready to produce milk before she gives birth, milk secretion does not occur during pregnancy because of the inhibiting effect of high levels of estrogens and progesterone. This inhibition is removed with the loss of the placenta at birth. Two neuroendocrine reflexes cause milk secretion and release after birth. As an infant suckles, sensory impulses from the breasts are sent via neurons to the hypothalamus. This promotes the release of oxytocin from the neurohypophysis, which causes the ejection of milk that has accumulated in the alveoli of the mammary glands (p. 441, Fig. 19.5). The sensory impulses also remove the inhibition of a hypophyseal factor that has prevented the release of large quantities of prolactin, and presumably promote the production of an unidentified prolactin-releasing hormone. This stimulates the adenohypophysis to release prolactin, which in turn causes the secretion of more milk into the mammary alveoli.

Summary

1. Many animals can reproduce asexually by fission, budding, and fragmentation. Most reproduce sexually by producing haploid sperm and eggs that unite at fertilization to form a diploid zygote. Sexual reproduction increases genetic variability in populations and allows for DNA repair.

2. The gametes of most animals are produced in gonads—testes or ovaries. In most species, the gonads enlarge as the reproductive season nears and regress afterwards. The gamete-producing cells continue to multiply throughout reproductive life in most animals, but the egg-forming cells of humans and some other mammals cease to multiply by the time of birth. Four small, motile sperm develop from each spermatogonium. A single, relatively large egg, which contains the yolk, and three minute polar bodies develop from each oogonium. Mature gametes sometimes are discharged through ruptures in the body wall, but they usually leave the body through gonoducts, either sperm ducts or oviducts.

3. Fertilization is a complex process that includes recognition reactions between sperm and eggs of the same species, mechanisms that normally allow only a single sperm to enter the egg, the union of male and female genetic material, and the activation of the egg cytoplasm.

4. The synchronous production and release of gametes by males and females helps to ensure fertilization. Species in which the eggs are fertilized externally produce large numbers of gametes, often come together as a swarm at the time of gamete release, and have evolved other strategies that increase the probability of fertilization.

5. Internal fertilization within the body of the female is characteristic of many aquatic animals and of most terrestrial species. It requires copulation and various modifications of the reproductive tracts of both sexes, such as a copulatory organ (usually a penis), semen-producing glands, seminal vesicle, vagina, and seminal receptacle.

6. Primitively, animals have separate sexes, but many species are simultaneous or sequential hermaphrodites. However, cross-fertilization rather than self-fertilization is generally the rule. In simultaneous hermaphrodites, cross-fertilization is usually reciprocal.

7. The vertebrate reproductive tract varies greatly, reflecting different adaptations for fertilization and egg deposition. In mammals the male's penis deposits sperm in the vagina, and fertilization occurs in the upper end of the uterine tube. The large number of sperm released increases the likelihood that some will make the passage up the uterus and uterine tube and collectively contribute to the enzymatic dispersal of follicle cells surrounding the ovulated egg.

8. The reproductive hormones secreted by the gonads are steroid derivatives of cholesterol. Males and females differ not in the types of hormones that they produce, but in their quantity.

9. Luteinizing hormone (LH) secreted by the adenohypophysis of males stimulates the production of the androgen testosterone by Leydig cells in the testis. Both testosterone and follicle-stimulating hormone (FSH) from the adenohypophysis are necessary for sperm production.

10. LH also promotes the synthesis of androgens in certain cells of the ovarian follicle. FSH contributes to their conversion to the female hormone, estradiol. A follicle enlarges under the combined influence of estradiol, LH, and FSH. A surge in LH production causes ovulation. The ruptured follicle transforms into a corpus luteum, whose hormone, progesterone, stimulates the continued buildup of the uterine lining for the reception of a fertilized egg. If no pregnancy ensues, the corpus luteum regresses, the uterine lining sloughs off (in human females), and another follicle begins to enlarge. The ovarian cycle and cyclic changes in the uterine lining (menstrual cycle) are regulated by feedbacks between the reproductive hormones, the hypothalamus, and the gonadotropins of the adenohypophysis.

11. If a pregnancy occurs, chorionic gonadotropin is secreted by the developing placenta. This hormone maintains the corpus luteum until the placenta itself can synthesize the amount of progesterone needed for a successful pregnancy. Hormones of pregnancy cause the development of glandular tissue in the breasts and the synthesis of the enzymes needed for milk secretion, but secretion is inhibited until birth by the high level of estradiol and progesterone. The release of milk is a neuroendocrine reflex that requires the stimulus of suckling.

References and Selected Readings

Austin, J.R., and R.V. Short (eds). *Reproduction in Mammals*, 2nd ed. Cambridge: Cambridge University Press, 1982. Five short books covering all aspects of mammalian reproduction and development.

Clark, W.C. Hermaphroditism, a reproductive strategy for metazoans; some correlated benefits. *New Zealand Journal of Zoology* 5(1978):769–780. A review of hermaphroditism.

Gilbert, S.F. *Developmental Biology*. Sunderland, Mass.: Sinauer Associates, Inc., 1988. Contains an excellent chapter on the structure of gametes and on fertilization.

Halliday, T. *Sexual Strategy: Survival in the Wild*. Chicago: University of Chicago Press, 1982. The evolution of sexual strategies with an emphasis on vertebrate mating systems.

Hedge, G.A., H.D. Colby, and R.L. Goodman. *Clinical Endocrine Physiology*. Philadelphia: W.B. Saunders Co., 1987. An excellent short monograph on the physiology of human endocrine glands, including those that regulate reproduction.

Margulis, L., and D. Sagan. *Origins of Sex: Three Billion Years of Genetic Recombination*. New Haven: Yale University Press, 1986. Recombination became accidentally linked to reproduction early in the history of life.

Michod, R.E., and B.R. Levin (eds.): *The Evolution of Sex*. Sunderland, Mass.: Sinauer Associates, Inc., 1988. A collection of essays on the biologic significance and evolution of sex.

Wasserman, P. Fertilization in mammals. *Scientific American* (December 1988), 78–85. A consideration of the mechanisms controlling sperm penetration of the egg.

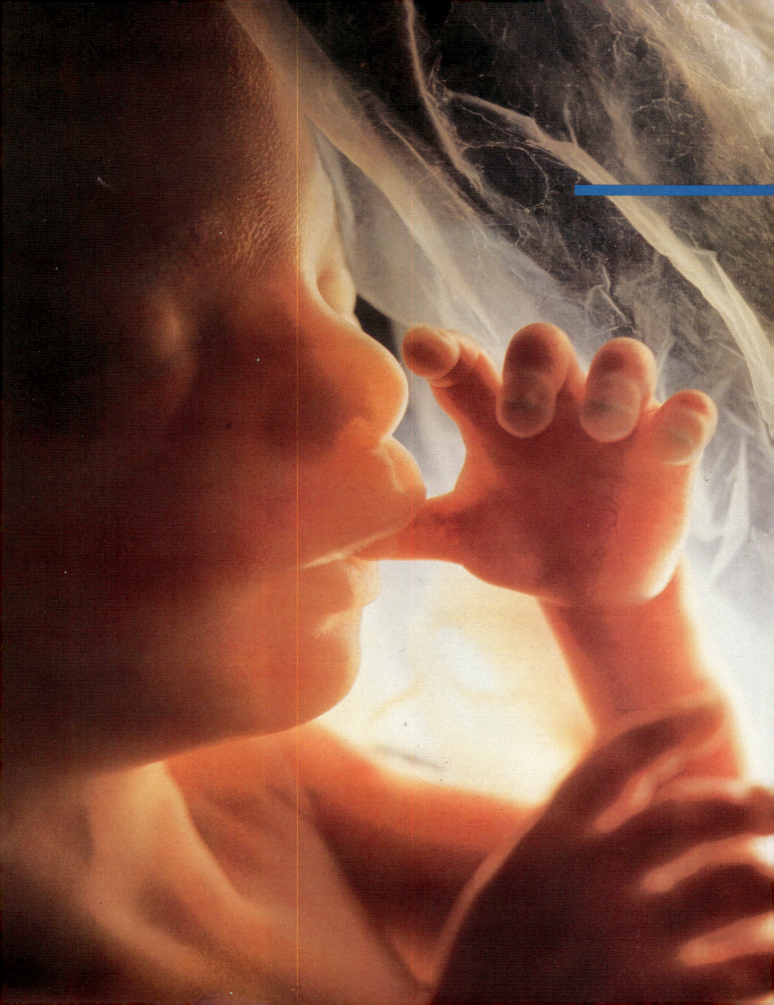

22

Embryonic Development

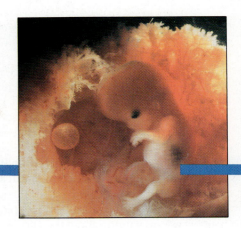

(Left) A human fetus at 18 weeks and (above) at 10 to 11 weeks.

Embryology, or the study of the development of a fertilized egg into the complex and interdependent system of organs that constitutes the adult animal, is certainly one of the most remarkable and fascinating subjects in biology. Cells multiply, migrate, aggregate, differentiate into many types, and become organized into levels of increasing morphological and physiological complexity. In this chapter, we will examine the morphological and physiological transformations that occur as a fertilized egg, or zygote, develops into an organism, and explore some of the insights we are gaining about the genetic control of development.

Throughout this chapter, keep in mind that inherited information—genes—does not directly encode the morphological features of the organism. There are no genes for legs, or fingers, or livers. Genes only direct the synthesis of proteins. The task of modern developmental biology consists of explaining how the controlled production of myriad proteins can lead to the formation of a complete individual with a clear species identity and bearing recognizable similarity to its parents. As with so many other key issues in biology, the answer comes from various levels in the biological hierarchy. To understand development, we need insight at the molecular level (the control of gene expression), at the cellular level (the ways in which cells and tissues interact), and at the physiological level (hormonal and environmental influences). Equally important, we need to consider the evolutionary context of development: like so many other aspects of the organism, development itself is the product of biological history.

EGG TYPES

Different amounts of **yolk**, which consists of proteins, fats, and other materials the developing embryo will need for energy and nutrition, are stored in the eggs while they are in the ovary. The amount of yolk in a particular species correlates with the duration of development required before the embryo can feed itself after hatching or before it establishes a placental attachment with a parent. Early development is very much affected by the amount and distribution of the yolk in the egg because yolk is an inert material that slows down cell division. The eggs of many echinoderms, amphioxus (a primitive fish-shaped chordate without a backbone), and mammals, which have a small amount of yolk evenly distributed within the egg cytoplasm, are called **isolecithal** (Gr. *isos*, equal + *lekithos*, yolk) (Table 22.1).

Many flatworms, most molluscs, marine annelids, and most vertebrates have eggs with more yolk that is concentrated at the lower or **vegetal pole**. Such eggs are termed **telolecithal** (Gr. *telos*, end). The upper or **animal pole** contains the nucleus and less yolky cytoplasm. Most molluscs and frogs have moderately telolecithal eggs, but in cephalopods, many fishes, reptiles, and birds, as much as 90% of the egg is composed of yolk. A cap of nonyolky cytoplasm, which contains the nucleus, occupies the animal pole.

The eggs of arthropods, especially insects, have a different pattern of yolk distribution and are termed **centrolecithal**. The yolk is concentrated in the center of the egg, and the cytoplasm is present as a thin layer on the entire surface of the egg. In addition, an island of cytoplasm in the center of the egg contains the nucleus.

CLEAVAGE

After fertilization, development of an activated zygote begins with a series of rapid mitotic divisions known as **cleavage** that convert a single cell into a multicellular embryo known as the **blastula** (Gr., a small bud) (Fig. 22.1). The blastula is a distinctive developmental feature of animals, and it is one of the defining features of the animal kingdom (p. 548). Mitosis is extremely rapid during early cleavage. A frog egg divides into 37,000 cells in less than two days. No growth occurs between divisions so the individual cells, known as **blastomeres**, become progressively smaller. Since the blastula is no larger than the

Table 22.1
Types of Eggs and Cleavage Patterns

Type of Egg	Basic Cleavage Pattern	Specific Cleavage Pattern	Examples
Isolecithal (Homolecithal)	Holoblastic	Radial	Many echinoderms, amphioxus, mammals
Telolecithal, moderate yolk	Holoblastic	Radial Spiral	Amphibians Some flatworms, most annelids, most molluscs
Telolecithal, much yolk	Meroblastic	Discoidal	Cephalopods, most fishes, reptiles, birds
Centrolecithal	Meroblastic	Superficial	Most arthropods

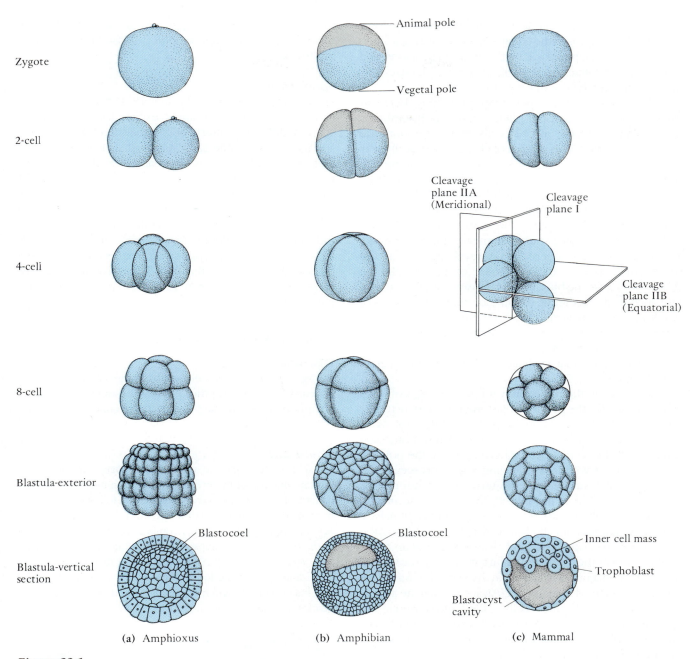

Figure 22.1
Radial cleavage and blastula formation in (a) amphioxus, (b) the frog, and (c) a mammal.

zygote, the ratio of cytoplasmic to nuclear volume decreases greatly.

In the lightly yolked isolecithal eggs and moderately yolked telolecithal eggs, the entire zygote cleaves. This pattern is known as **holoblastic** cleavage (Table 22.1). In heavily yolked telolecithal eggs and the centrolecithal eggs, cleavage is **meroblastic** or incomplete. Cells form only in the portion of the egg rich in cytoplasm, and most of the yolk does not cleave.

Major Types of Holoblastic Cleavage

Radial Cleavage
The embryo of amphioxus nicely illustrates the radial cleavage seen in many isolecithal eggs (Fig. 22.1a). The first division, which passes through the animal and vegetal poles forming two cells of equal size, is described as meridional, for like the meridian of a globe, it passes from pole to pole. The second cleavage is also meridional, but at

right angles to the first, and divides the two cells into four. The third cleavage is in the equatorial plane, at a right angle to the meridional planes of the first two cell divisions, and the embryo is split into eight cells of nearly equal size, four above and four below the equator.

Divisions continue, generally alternating between meridional and equatorial planes, resulting in embryos containing 16, 32, 64, and 128 cells, and so on, until the blastula is formed. This pattern of cleavage is described as **radial**, because the embryo can be bisected along any meridian and form two mirror-image halves. Early cleavages are synchronous, with each blastomere dividing simultaneously, but the timing of the divisions becomes more irregular later. Since the embryo uses materials stored in the yolk as a source of energy during cleavage, the embryo's mass decreases. A cavity, known as the **blastocoel**, appears in the center of the developing blastula because the blastomeres become oriented to, and are anchored together at, the surface where exchanges of gases and other materials between the embryo and environment occur. In embryos with little yolk, the wall of the blastula consists of a single layer of cells. The blastocoel provides a space that facilitates cell movements later in development.

Cleavage in the moderately telolecithal eggs of frogs is also radial for the first four blastomeres are structurally alike (Fig. 22.1b), although the polarity of the embryo is established at fertilization. At the time of fertilization, some pigment in the animal hemisphere opposite the point of sperm entry is withdrawn, leaving a **gray crescent**. The gray crescent lies at the future posterior end of the embryo. The first cleavage plane bisects the gray crescent and divides the embryo into prospective left and right halves. The larger amount of yolk slows down the formation of the cleavage furrow so the second cleavage has begun at the animal pole by the time the first furrow reaches the vegetal pole. The concentration of yolk in the vegetal hemisphere displaces the third, equatorial cleavage toward the animal pole, so the upper four cells, called **micromeres**, are smaller than the lower four **macromeres**. The blastocoel also is displaced toward the animal pole, and its wall is formed of several layers of cells.

Mammals evolved from reptiles with large yolked telolecithal eggs, but the reproductive pattern of most mammals is modified in association with intrauterine development and the early establishment of a placental relationship with the mother. The eggs have become secondarily isolecithal and cleave completely, but cleavage is slower and more irregular than in other animals with radial cleavage. The first plane is meridional. One of the resulting two blastomeres also divides meridionally in the second cleavage, but the other divides equatorially (Fig. 22.1c). As cleavage continues, the blastomeres become tightly compacted. The outer cells form a **trophoblast** (Gr. *trophe*, nourishment + *blastos*, bud) that initiates placenta

formation. The inner cells aggregate at one pole to form an **inner cell mass**. This modified blastula is called a **blastocyst**.

Spiral Cleavage

Many invertebrate animals, including flatworms, many annelids, and most molluscs except for cephalopods, share a different pattern of holoblastic cleavage called **spiral cleavage** (Table 22.1). As in radial cleavage, the first and second cell divisions are nearly meridional, forming four large macromeres of nearly equal size (Fig. 22.2). However, beginning with the third cleavage, the division of cells is unequal and the cleavage planes are oblique. The mitotic spindles are inclined to one side (one of them is indicated by the solid line in the eight-cell stage of Figure 22.2). As a result, four upper micromeres are formed and displaced circularly so that each upper cell lies over *two* lower cells. As viewed from the animal pole, the cleavage pattern appears spiral.

Types of Meroblastic Cleavage

Discoidal Cleavage

In the heavily yolked telolecithal eggs of cephalopods, fishes, birds, and reptiles, only the superficial disc of cytoplasm at the animal pole cleaves, hence this pattern is

Figure 22.2
Spiral cleavage as seen in a lateral view. Diagonal lines indicate the axis of the preceding mitotic spindle. The pattern of cleavage can be appreciated by following the progeny of the colored blastomere from the four-cell stage. Macromere A divides into macromere 1A and micromere 1a in the eight-cell stage. Subsequent divisions are also unequal and oblique, but in the opposite direction than that of the preceding ones. In the fourth division, macromere 1A divides into macromere 2A and micromere 2a, and micromere 1a divides into micromeres $1a^1$ and $1a^2$.

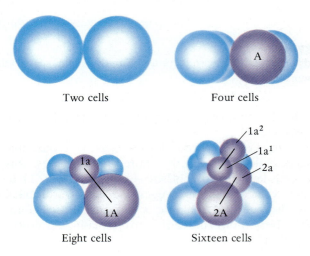

Two cells   Four cells

Eight cells   Sixteen cells

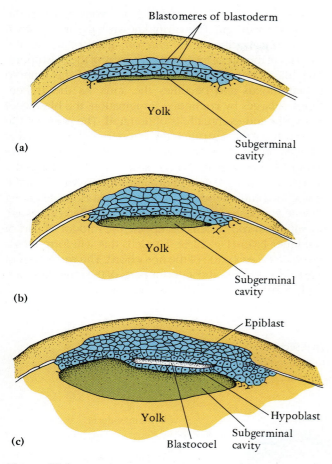

Figure 22.3
Successive stages in the discoidal cleavage of a hen's egg.

called **discoidal** (Table 22.1). Meridional cleavages divide the disc of cytoplasm into a single-layered **blastoderm**, and then equatorial cleavages divide this layer (Fig. 22.3). As cleavage continues in reptiles and birds, a **subgerminal cavity** appears beneath the blastoderm. Some of the blastoderm cells migrate into the subgerminal cavity and form a layer known as the **hypoblast**. The superficial cells of the blastoderm now constitute the **epiblast**, and the space between these layers is the blastocoel.

Superficial Cleavage

The nucleus of the centrolecithal eggs of insects and other arthropods at first lies in an island of cytoplasm in the center of the yolk (Fig. 22.4). Additional cytoplasm forms a superficial layer over the yolk. After several nuclear divisions without cytoplasmic division, the nuclei migrate out from the center of the egg, each surrounded by a bit of the original central cytoplasm. When the nuclei reach the surface of the egg, the cytoplasm surrounding them fuses with the superficial layer of cytoplasm. This layer is a syncytium, that is, a layer of cytoplasm containing many nuclei, for no cell membranes have formed around the nuclei. Cytoplasmic divisions subsequently convert the syncytium into a cellular blastoderm, which surrounds a mass of uncleaved yolk rather than a cavity. This is the blastula stage.

GASTRULATION AND COELOM FORMATION

Cell positions in the blastula are determined by the location of the cleavage planes that divide the zygote into a multicellular embryo. A period of **gastrulation** follows during which **morphogenetic movements** rearrange the cells of the blastula and bring those with different poten-

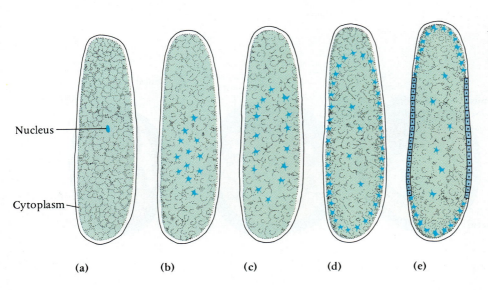

Figure 22.4
Successive stages in the superficial cleavage of an arthropod egg.

tials into parts of the embryo appropriate for their continued development (Fig. 22.5). Gastrulation changes the radial symmetry of the blastula into a bilateral symmetry, and converts the single layered blastula into a **gastrula** (L., little stomach) consisting of several epithelial layers. These layers are called the **germ layers**. **Endoderm** moves inward to form the lining of the gut and its derivatives. **Ectoderm** left on the outside gives rise to the epidermis and nervous system. A third germ layer, the **mesoderm**, forms in most metazoans between the ectoderm and endoderm during or after gastrulation, and gives rise to most of the skeleton, muscles, circulatory system, and other tissues.

Gastrulation in Isolecithal Eggs

In embryos that develop from isolecithal eggs, such as those of amphioxus, gastrulation begins with a flattening of the vegetal pole and an inward pushing of its cells, a process termed **invagination** (Fig. 22.5a–d). The inward-moving layer of cells bounds a new cavity, which is the precursor of the gut. This cavity, known as the **archenteron** (Gr. *arche*, beginning + *enteron*, gut), opens to the exterior at the site of invagination by way of the **blastopore**. The blastopore is at the posterior end of the embryo in amphioxus and becomes the anus. As the archenteron enlarges by continued invagination, the blastocoel regresses and eventually is obliterated. The outer of the two layers of the gastrula is the ectoderm. The inner layer of cells lining the archenteron is mainly future or presumptive endoderm, but a strip of cells along the archenteron roof, called **chordamesoderm**, will form the notochord and mesoderm. As the gastrula elongates along its anterior-posterior axis, the notochord differentiates as a longitudinal rod of cells occupying the middorsal part of the inner layer (Fig. 22.5e–h). Cells that will form mesoderm lie on each side of the notochord. The remainder of the lateral, ventral, and anterior parts of the inner layer are endoderm cells that will line the gut.

Figure 22.5

Gastrulation and mesoderm formation in amphioxus. *Top row*, Sagittal sections showing gastrulation. *Bottom row*, Transverse sections of mesoderm and coelom formation. In this and other embryonic figures ectoderm is shown in blue; endoderm, in yellow; mesoderm, in red; and the notochord, in green.

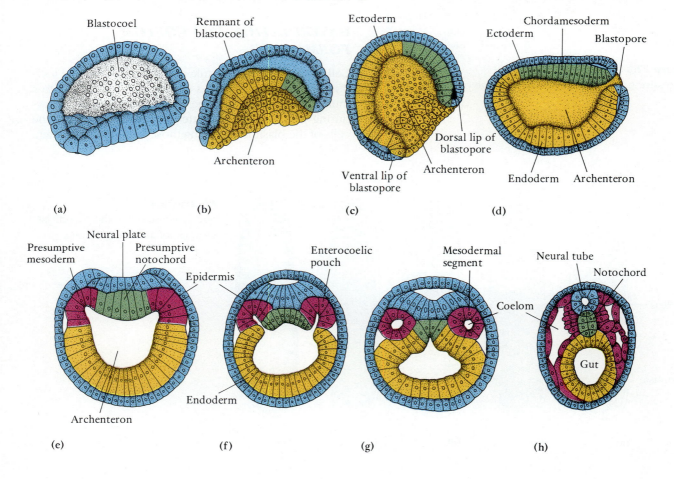

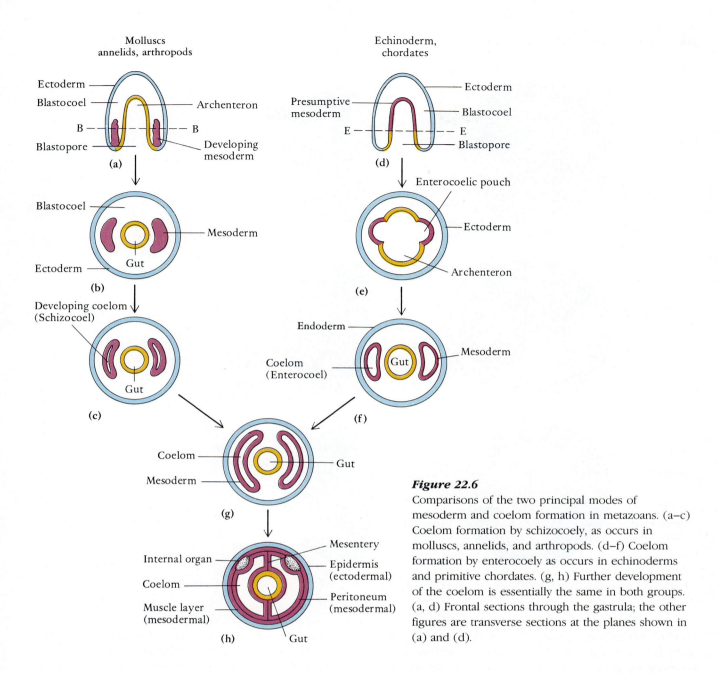

Figure 22.6

Comparisons of the two principal modes of mesoderm and coelom formation in metazoans. (a–c) Coelom formation by schizocoely, as occurs in molluscs, annelids, and arthropods. (d–f) Coelom formation by enterocoely as occurs in echinoderms and primitive chordates. (g, h) Further development of the coelom is essentially the same in both groups. (a, d) Frontal sections through the gastrula; the other figures are transverse sections at the planes shown in (a) and (d).

The definitive mesoderm separates in amphioxus as a series of bilateral pouches from the dorsolateral wall of the archenteron (Figs. 22.5e–h and 22.6d–f). These lose their connection with the gut and fuse one with another to form a connected layer. The fused cavities of the pouches form the coelom, called an **enterocoel** (Gr. *enteron,* gut + *koilia,* cavity) because it is derived indirectly from the archenteron.

Although their eggs are not isolecithal, the archenteron of annelids, molluscs, and many arthropods also is formed by invagination, but the blastopore typically gives rise to the mouth. The mesoderm does not come from cells that moved into the roof of the archenteron; rather, it develops from cells that differentiated early in cleavage. These migrate, or **ingress**, not as a sheet but singly into the blastocoel and come to lie on each side of the archenteron (Fig. 22.6a). They then multiply to form two longitudinal cords of cells that develop into sheets of mesoderm between the ectoderm and endoderm (Fig. 22.6b–c). The coelomic cavity of these animals originates by the splitting of the sheets and hence is called a **schizocoel** (Gr. *schizein,* to split).

Gastrulation in Amphibians

Gastrulation in the moderately telolecithal eggs of frogs and other amphibians begins with the invagination of a few cells at the margin of the gray crescent (Fig. 22.7a). The cleft formed is the beginning of the archenteron, and the cells above it are the dorsal lip of the blastopore. As

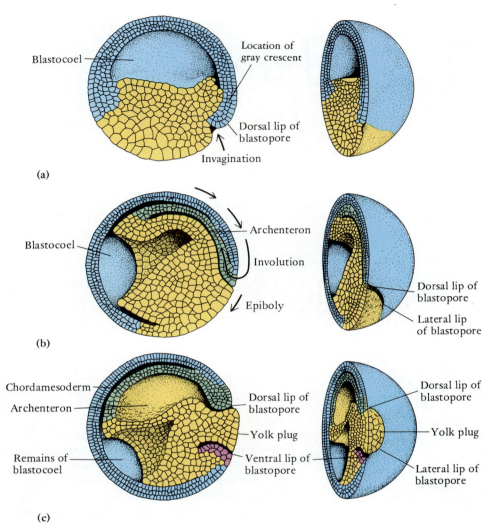

(a)

Blastocoel

Location of gray crescent

Dorsal lip of blastopore

Invagination

(b)

Blastocoel

Archenteron

Involution

Epiboly

Dorsal lip of blastopore

Lateral lip of blastopore

(c)

Chordamesoderm

Archenteron

Remains of blastocoel

Dorsal lip of blastopore

Yolk plug

Ventral lip of blastopore

Dorsal lip of blastopore

Yolk plug

Lateral lip of blastopore

Figure 22.7
Gastrulation and mesoderm formation in the frog. Sagittal sections are shown in the left column; posterior views, in the right column.

invagination proceeds, the sheet of cells covering the animal hemisphere expands and moves toward the dorsal lip of the blastopore, rolls over it, and turns inward, in a movement termed **involution** (L. *involutus*, rolled in) (Fig. 22.7b and c). The inward-turning cells deepen the archenteron, which gradually obliterates the blastocoel, and form the chordamesoderm. Endoderm develops from the yolk-laden cells of the vegetal hemisphere, which originally form the floor and sides of the archenteron. Since cells move toward the blastopore faster than they involute, the lip of the blastopore overgrows the mass of yolk-filled cells, a process called **epiboly** (Gr. *epibole*, a throwing on). All of these processes begin at the dorsal lip, but the lip spreads laterally and ventrally until a complete blastopore is formed that surrounds a mass of yolk-filled cells, the **yolk plug**. As processes of involution and epiboly continue after the completion of the blastopore, the yolk plug decreases in size. The blastopore eventually closes. A new anus later invaginates near the closed blastopore.

A notochord differentiates in the middorsal line. The prospective mesoderm of amphibians spreads anteriorly and laterally, and a cleft appears that separates it from the endoderm (see Fig. 22.9a). This method of separation is called **delamination**. The coelom develops as a cleft within the mesoderm. It is considered to be a modified enterocoel because the mesoderm came from the roof of the archenteron, as in amphioxus.

Gastrulation in Reptiles, Birds, and Mammals

In the heavily yolked eggs of reptiles and birds, and in the lightly yolked eggs of mammals, which evolved from reptiles, all of the germ layers develop from the epiblast of the blastoderm (called the **embryonic disc** in mammals). Morphogenetic movements carry epiblast cells toward the middorsal longitudinal axis of the blastoderm and form a thickened region known as the **primitive streak** (Fig. 22.8).

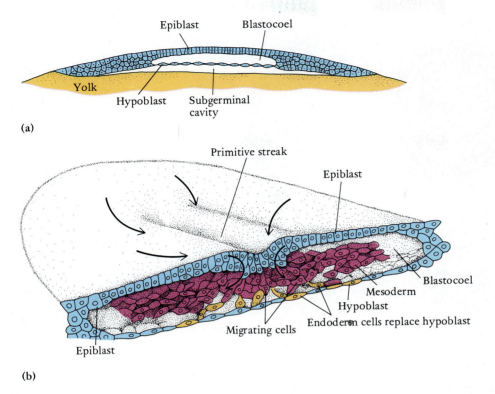

Figure 22.8
Gastrulation and mesoderm formation in a bird. (a) Longitudinal
section of a blastoderm in which the epiblast and hypoblast have
differentiated. (b) Three-dimensional diagram of a later blastoderm
in which endoderm and mesoderm cells are ingressing along the
primitive streak.

As more cells converge on the primitive streak, some begin
to move inward and spread out laterally into the
blastocoel. This is manifested on the surface of the
blastoderm by a longitudinal groove that appears in the
primitive streak. There is no opening through which
sheets of cells move inward, as in the blastopore of other
animals; rather, the cells move in individually, a process
termed **ingression**. The results, however, are similar, and
the primitive streak is considered to be the functional
equivalent of the blastopore. Some of the first cells to
ingress replace hypoblast cells and become endoderm.
The endoderm spreads out over the subgerminal cavity
and the large mass of uncleaved yolk. Other ingressing
cells form the mesoderm and notochord. The coelom
develops from a cleft that appears in the mesodermal layer,
as it does in amphibians. Cells that remain on the surface of
the epiblast constitute the ectoderm.

ORGANOGENESIS

By the end of gastrulation, the germ layers are formed and,
in vertebrates, a notochord is differentiating from the
middorsal strip of chordamesoderm. The stage is set for
the formation of the organ systems and the differentiation
of cell types. Because the nervous system is the first to
differentiate in vertebrates, early organogenesis is a period
of **neurulation**.

Ectoderm Differentiation

Interactions between the chordamesoderm and overlying
ectoderm cause the ectoderm cells to form a neural tube.
The process by which one group of cells affects the
differentiation of adjacent ones is known as **induction**. We
will return to the nature of induction at the end of this
chapter. Ectoderm cells overlaying the notochord first
thicken to form a **neural plate** (Figs. 22.9a and b). The
center of this becomes depressed and forms the **neural
groove**, and the outer edges of the plate rise in two
longitudinal **neural folds** that meet at the anterior end and
appear, when viewed from above, like a horseshoe (Fig.
22.9c). These folds gradually come together at the top,
beginning at the anterior end. The thicker, inner limbs of
the folds form a hollow **neural tube**. The cavity at the
anterior part of the neural tube becomes the ventricles of
the brain; that in the posterior part, the neural canal

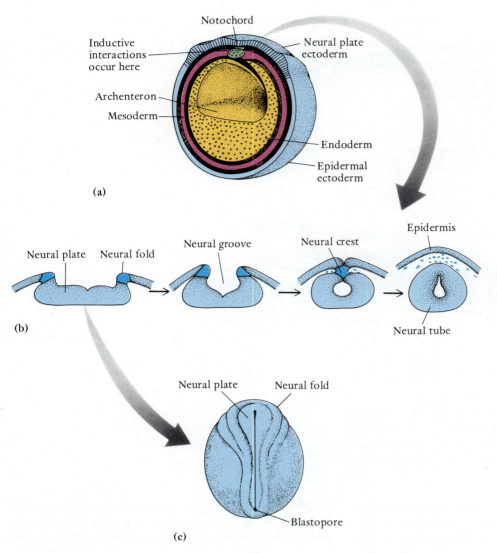

Figure 22.9
Formation of the neural tube in a frog embryo. (a) A cross section of the embryo
showing the formation of the neural plate over the notochord. (b) A series of
cross sections showing the development of the neural tube and neural crest. (c) A
dorsal view of the embryo.

extending the length of the spinal cord. As the folds come
together, tissue at the apex of each fold separates to form a
crest of cells overlying the neural tube. This is the **neural
crest**, a distinctive feature of vertebrates (p. 808). The
thinner outer limbs of the neural folds and the remaining
surface ectoderm become the epidermis of the skin.

The neural tube will form most of the nervous system,
including the motor neurons that extend from the neural
tube through spinal and cranial nerves to the effectors. As
the neural folds close anteriorly, three swellings begin to
appear. These are the primary parts of the brain: forebrain,
midbrain, and hindbrain (Fig. 22.10a and b).

Cells from the neural crest contribute, along with
certain mesoderm cells, to a population of star-shaped,
wandering cells known as **mesenchyme**, which differentiates into many cell types. Among the derivatives of neural
crest mesenchyme are the sensory neurons, postganglionic autonomic neurons, Schwann cells, cells of the pia
mater, chromaffin cells of the adrenal medulla, chromatophores, cells of much of the musculature and skeleton of the head, including the visceral arches (Fig. 22.10b),
and cells that induce the formation of scales and teeth.

The special sense organs of vertebrates (nose, eye,
and inner ear) form partly from thickened plates of surface

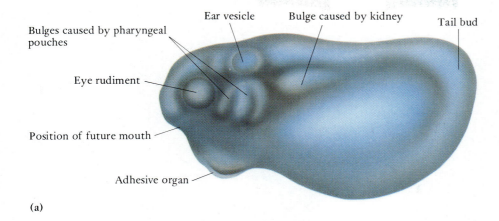

(a)

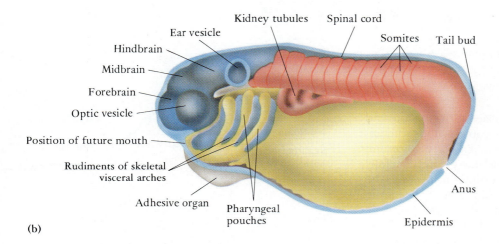

(b)

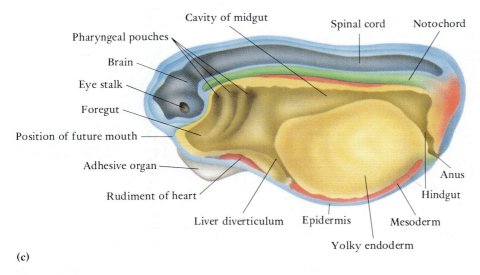

(c)

Figure 22.10

The tail bud stage of a frog embryo shortly before it hatches from the surrounding jelly layers and becomes a free swimming larva. (a) Lateral view; (b) the same with the skin removed; (c) a sagittal section.

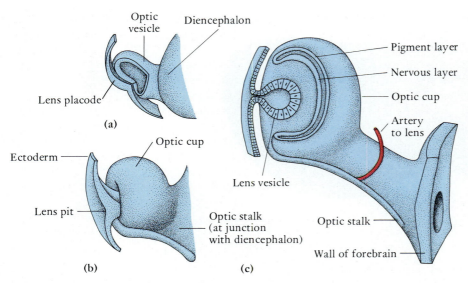

Figure 22.11

The development of the eye. (a) The optic vesicle approaches the epidermis and induces a lens placode. (b, c) The optic vesicle has become an optic cup that induces the formation of a lens vesicle.

ectoderm known as **placodes**. Placode formation and differentiation result from inductive interactions with underlying parts of the nervous system. Each eye, for example, is initiated by an **optic vesicle** that extends laterally from the part of the forebrain destined to become the diencephalon (Figs. 22.10b and 22.11). As it nears the surface, a **lens placode** is induced in the overlying ectoderm. The base of the optic vesicle narrows to become a stalk, and the distal part invaginates to form a two-layered **optic cup**, the outer layer of which forms the pigment layer of the retina; the inner one, the nervous layer. As the optic cup forms, the lens placode is induced to invaginate and form a **lens vesicle** that becomes the lens. The lens, in turn, induces surface ectoderm to differentiate as a cornea. That the optic vesicle induces the formation of the lens can be demonstrated by transplanting an optic vesicle beneath the abdominal ectoderm of a young gastrula. A lens and eye will appear on the abdomen!

Mesoderm Differentiation

As the neural tube forms, the paired sheets of mesoderm that have been expanding laterally from the notochord begin to differentiate. Segmental blocks of cells, called **somites**, develop in the mesoderm beside the notochord and neural tube (Figs. 22.10b and 22.12a). This process

Figure 22.12

Early organogenesis and cell movements in a generalized vertebrate embryo as seen in an early (a) and later (b) cross section.

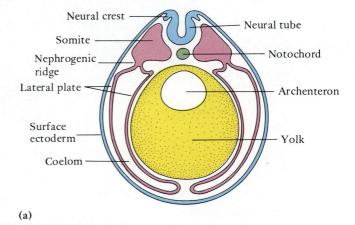

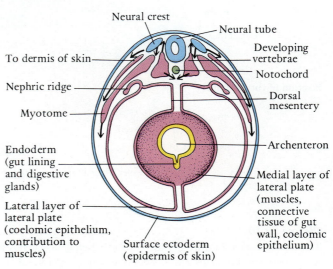

begins anteriorly and continues toward the posterior end. The most medial part of each somite forms mesenchymal cells that migrate around the neural tube and notochord to form much of the skull and all of the vertebral column (Fig. 22.12b). As the vertebral column develops, the notochord regresses. Cells on the dorsolateral surfaces of the somites also form mesenchyme that spreads out beneath the surface ectoderm and forms most of the dermis of the skin. The central and largest part of each somite differentiates as a muscle block, the **myotome**. Myotomes grow ventrally and laterally and give rise to most of the skeletal muscles of the body.

Mesoderm lying ventral to the somites and adjacent to the coelom forms a band of tissue, known as the **nephric ridge** (Fig. 22.12a and b), that differentiates into kidney tubules (Fig. 22.10b).

The rest of the mesoderm, called the **lateral plate**, lies on each side of the developing coelom (Fig. 22.12a and b). The lateral layer of the plate next to the ectoderm forms the epithelium that bounds the coelom laterally, and often contributes to the musculature and other tissues in the flank wall. The inner layer lies next to the endoderm. This layer and the endoderm gradually extend laterally around the yolk. The rest of the coelomic epithelium and visceral muscles of the gut wall are derived from this mesoderm.

Circulatory System

The growing embryo must obtain energy and materials needed for synthesis from stored yolk or from the placenta, and exchange gases with the environment or through the placenta. The early development of blood vessels to and from the yolk, and the formation of a functional heart, make this possible. Blood vessels first appear in the mesoderm over the yolk as clusters of epithelial cells surrounding other cells that will differentiate into blood cells (Fig. 22.13a). These **blood islands**, as they are called, coalesce to form a network of vessels over the yolk that penetrate the embryo (Fig. 22.13b).

A pair of **vitelline veins**, which return blood from the yolk or yolk sac, develop first (Fig. 22.13b and c). They fuse beneath the developing pharynx to form the heart, which begins to beat rhythmically right away. At first the heart is a

Figure 22.13

The formation of the circulatory system, (a) A blood island in the yolk sac area of a chick. (b) Blood vessels extending between the embryo and yolk sac as seen in the blastoderm of a two-day-old chick embryo that has been removed from the yolk. (c) A diagram of the circulation of an early mammalian embryo.

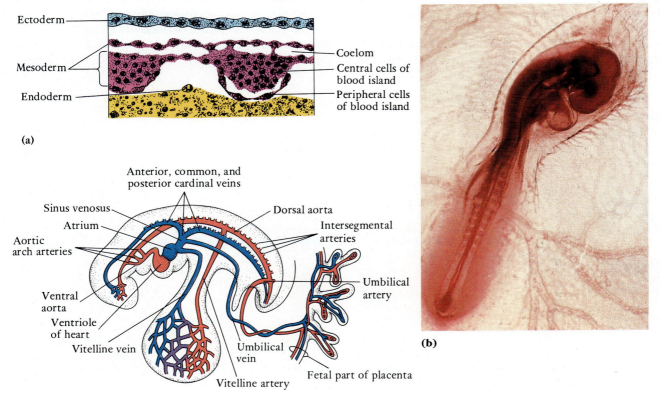

straight tube with the four linear chambers characteristic of a fish heart: sinus venosus, atrium, ventricle, and conus arteriosus (p. 304). Contractions are initiated in the sinus venosus and sweep to the other chambers. Because the heart grows in length faster than the cavity in which it lies, it folds upon itself, with the ventricle coming to lie ventral and posterior to the atrium, as in an adult. Divisions of the chambers occur later in the development of terrestrial vertebrates, but they are not complete until birth in mammals because the placenta remains the site for nutrition, gas exchange, and excretion.

A **ventral aorta** leads forward from the heart to **aortic arches** that extend dorsally between developing pharyngeal pouches to the **dorsal aorta**, which returns blood to the yolk by way of **vitelline arteries**. Other vessels develop later and carry blood to and from the body wall and, in mammals, the placenta.

Endoderm Differentiation

Most of the digestive tract is formed from the archenteron of the gastrula and elongates with the growth of the embryo (Figs. 22.10c and 22.12b). The lungs, liver, and pancreas originate as tubular outgrowths of the front part of the archenteron and hence are composed of endoderm, but these outgrowths are always associated with mesoderm from the medial part of the lateral plate, which forms the blood vessels, connective tissue, and muscles of these organs. The endoderm forms only the lining epithelium of the digestive tract and lungs, and the secretory cells of the pancreas and liver.

The most anterior part of the archenteron expands laterally and develops into the pharynx. A series of four or five paired **pharyngeal pouches** bud out laterally from the endoderm and meet a corresponding set of inpocketings from the overlying ectoderm (Fig. 22.10b). In fishes and larval amphibians, the two sets of pockets fuse to make a continuous passage from the pharynx to the outside, the **gill slits**, that function as respiratory organs. All of the gut lining and its derivatives arise from endoderm except for the mouth cavity and anal area of most animals, which arise from shallow pockets of ectoderm that grow in to meet the anterior and posterior ends of the archenteron (Fig. 22.10c).

DEVELOPMENT, ENVIRONMENT, AND LIFE STYLE

Planktonic Development and Larvae

The sea is the ancestral home of animals, and it is here that we find species with the most primitive life cycles. The eggs are fertilized externally in the sea water and begin development in the plankton (Fig. 22.14, *top*). Only enough yolk is present to supply the embryo until it develops into a **larva** that is motile and can feed for itself (Fig. 22.15). The larval stage also enables sessile species like barnacles and oysters to disperse and occupy new areas. Following planktonic life, the larva settles to the bottom and metamorphoses to the adult form.

A life cycle that includes a larval stage is called **indirect development** and is characteristic of more than half of the species of temperate and tropical marine bottom-dwelling invertebrates. The principal disadvantage of indirect development is the high mortality to which the planktonic eggs, embryos, and larvae are subjected. Many marine

Figure 22.14

A diagram showing variations that occur in the length of larval life during indirect development and comparing this with direct development in which the larval stage is omitted.

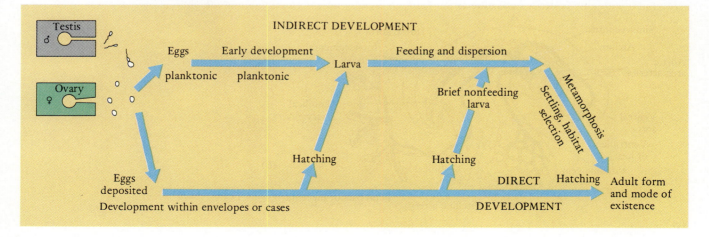

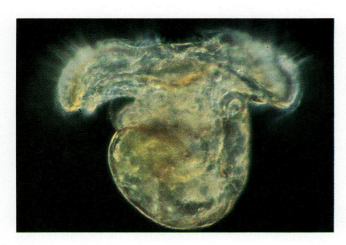

Figure 22.15
The veliger larva of a snail. The ciliated velum is used in locomotion and feeding.

animals have partly overcome this disadvantage by eliminating early planktonic development (Fig. 22.14, *bottom*). Eggs are deposited on the bottom within some sort of protective envelope, and hatching occurs at the larval stage. Some species have long-lived larvae that feed on diatoms and other minute organisms; others have evolved nonfeeding, yolk-laden larvae that live as planktonic larvae only long enough to permit adequate dispersion of the species.

Other species have dispensed with larvae altogether. The entire sequence of developmental events up to the adult body form occurs within a protective egg envelope

or case, and the young on hatching assume the adult mode of existence. Fewer eggs are deposited by such species and they contain large amounts of yolk. Such a developmental pattern is described as **direct development**, and is characteristic of many marine and most freshwater animals and of most terrestrial animals, except for insects (Fig. 22.14), most of which have a larval stage during development. Many groups of marine animals, such as snails and polychaete worms, exhibit a wide range of life styles. Some have motile feeding larva, some have nonfeeding larvae, and some have direct development.

Cleidoic Eggs and Extraembryonic Membranes

Many terrestrial animals that have direct development have eggs that are to a large degree self-contained systems. Such eggs are called **cleidoic eggs** (Gr. *kleidoun*, to lock up). The egg is provided with all of the necessary food material and with fats that yield water when metabolized (water is in short supply on land). The egg is enclosed within a covering or shell that affords protection and reduces the chance of desiccation yet permits gas exchange. Embryonic waste materials are stored in the egg. Well-formed cleidoic eggs are found in reptiles, birds, and insects, all of which are highly successful terrestrial groups.

Reptiles, birds, and primitive egg-laying mammals all have a similar type of cleidoic egg that first evolved in ancestral reptiles (Fig. 22.16). Their eggs contain a large amount of yolk, which is surrounded by albuminous materials and a protective shell that are secreted by the oviduct. In addition, four unique **extraembryonic membranes** evolved. These sheets of embryonic cells come to lie outside of the embryo proper and play an

Figure 22.16
A diagram of the cleidoic egg of a reptile or bird containing the embryo and its extraembryonic membranes. The albumin, shell membrane, and shell are secreted by the oviduct.

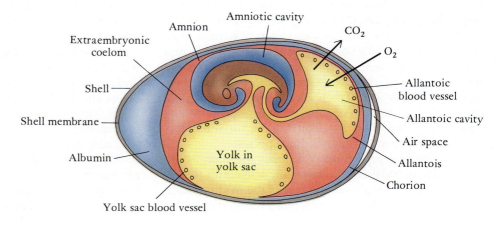

important role in protecting and maintaining the embryo.

One extraembryonic membrane, the **yolk sac**, surrounds the yolk mass. The **amnion** (Gr., fetal membrane) completely encloses the embryo, and its fluid-filled interior provides a protective aquatic environment in which the embryo develops. Reptiles, birds, and mammals are called **amniotes** because all have this membrane. The **chorion** (Gr., skinlike membrane) lies beneath the shell and surrounds both the yolk sac and the amnion. The fourth and last extraembryonic membrane to form, the **allantois** (Gr. *allas*, sausage + *eidos*, form), grows out from the posterior end of the gut and comes to lie against the inner side of the chorion like a great flattened balloon. Blood from the embryo is circulated through the wall of the allantois by allantoic vessels (called umbilical vessels in mammals, Fig. 22.13b). Gas exchange occurs here, oxygen diffusing inward through the shell and chorion, and carbon dioxide diffusing outward. The cavity of the allantois also is a repository for the deposition of nitrogenous wastes in the form of highly insoluble uric acid crystals. Storage of nitrogenous wastes as uric acid crystals conserves water.

The yolk sac, allantois, and amnion all connect to the center of the belly region of the embryo, and the connecting stalk is homologous with the umbilical cord of mammals. The yolk sac is largely reabsorbed during the course of development, and the connection with the other extraembryonic membranes is broken when the newborn hatches from the egg. The membranes are left behind with the shell.

Brooding

Those animals whose young develop from an egg outside of the body of the female are termed **oviparous** (L. *ovum*, egg + *parere*, to bring forth). Usually the female simply deposits the eggs in an environment favorable for their development. Many frogs, for example, attach their egg mass to sticks in the water. Many reptiles bury theirs in sand or place them under rocks, where they are protected from desiccation and where the radiant heat of the sun helps incubate them. Some parents care for their own eggs, a behavior termed **brooding**. Brooding reduces the mortality rate and occurs in a wide variety of animals. Most of those species that brood, for example, birds and octopods, do so externally. Birds lay their eggs in nests and incubate them by sitting on them; octopods attach their eggs to the rocky wall of their lair, cleaning and guarding them until they hatch. The number of eggs produced by brooding species is always small, but the survival rate is much greater than in nonbrooding species.

In almost every group of animals there are some species that brood their eggs internally. Usually the female broods eggs within a modified part of the oviduct called the uterus, although males sometimes brood the eggs in

their mouth (some marine catfishes), in an abdominal broodpouch (seahorses), or in other parts of the body. When development is completed, the young leave the parent. Animals of this type are described as **viviparous** (L. *vivus*, living). Gas exchange usually takes place between the embryo and a vascularized tissue within the parent. The source of energy and organic nutrients in viviparous development ranges along a continuum from complete dependence on material stored in the yolk (**ovoviviparity** or **aplacental viviparity**) to complete dependence on a transfer of materials from the mother through a placenta (**placental viviparity**). Ovoviviparity is the most common condition and is found in many invertebrates and in many sharks, lizards, and snakes. Placental viviparity is less common; scorpions and mammals are the best known but not the only examples.

Mammalian Placentation

In placental mammals, the embryos are completely dependent on the mother for nutrition, gas exchange, and excretion. Placental mammals evolved from primitive oviparous mammals. Presumably, there were intermediate stages in which a typical egg was simply brooded in the uterus rather than being laid. As the uterine lining began to provide for more and more of the embryo's needs, the shell was lost and the amount of yolk stored in the egg reduced. A small, empty yolk sac is retained. Its dorsal portion gives rise to the gut, and the first embryonic blood vessels develop in its wall.

Fertilization occurs in the upper part of the uterine tube, and the embryo cleaves as it passes down the tube (Fig. 22.17). In humans, the embryo is in the early blastocyst stage by the time it reaches the uterus on the fourth day after fertilization. The outer trophoblast cells multiply without cytoplasmic division to form a syncytial trophoblast (**syntrophoblast**); the inner part of the trophoblast remains cellular. The syntrophoblast performs many functions. In humans and many other species, it secretes enzymes that enable the embryo to digest its way into the uterine lining, a process called **implantation**. During this period, which lasts about three days, the embryo receives its nutrients from secretions produced by glands in the uterine lining.

As the human embryo implants, cavities form within the inner cell mass above and below a plate of cells, the embryonic disc, from which the embryo proper will develop. The upper cavity is the beginning of the amnion; the lower one, the yolk sac (Fig. 22.17, day 8). The trophoblast is homologous to the ectoderm of the chorion. During gastrulation, mesoderm cells migrate from the embryonic disc around the inside of the trophoblast, thereby completing the chorion, and also around the outside of the amnion and the endodermal yolk sac. The allantois, with its umbilical vessels, extends from the embryo and carries

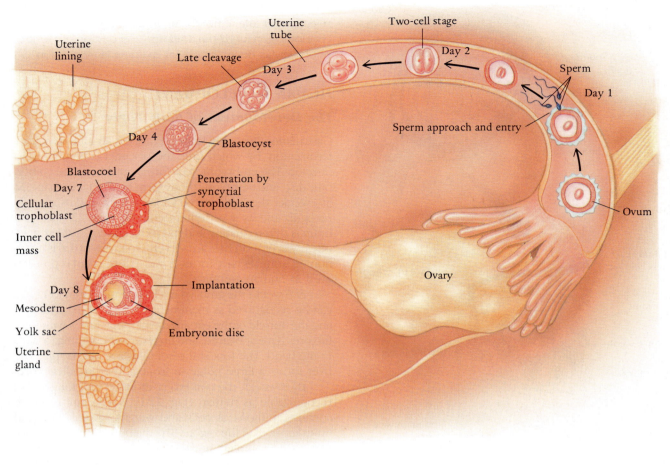

Figure 22.17

A diagram of fertilization and early development in a human embryo as it passes down the uterine tube to implant in the uterine lining. The zona pellucida, a membrane laid down around the egg while it is in the ovary, is the mammalian equivalent of the vitelline envelope.

fetal blood vessels to the chorion. Although the embryo itself has hardly begun to differentiate, all of the extraembryonic membranes characteristic of amniotes are present.

The intimacy of the union of the extraembryonic membranes with the uterine lining varies considerably among mammalian groups. Two modifications have evolved in most species that facilitate the fetal-maternal exchanges: (1) the exchange surface is increased by fingerlike projections of the chorion called chorionic villi, and (2) a reduction has occurred in the number and thickness of the tissue layers that separate maternal and fetal blood streams. In humans, the syntrophoblast continues to invade the uterine lining, especially on the side of the chorion closest to the uterine wall (Fig. 22.18a). It breaks open maternal blood vessels, and maternal blood enters

spaces, or **lacunae**, that develop within the syntrophoblast. As the lacunae enlarge, chorionic villi grow into them (Fig. 22.18b). The villi contain fetal capillaries brought to the chorion by the umbilical vessels. This union and region of exchange between the uterine lining and the fetal chorion and umbilical vessels is the structure called the **placenta**.

Maternal and fetal blood are separated by the endothelial walls of the fetal capillaries, some connective tissue, and the trophoblast. These tissues constitute a **placental barrier**. Unless there are breaks in some of the chorionic villi, which sometimes occur, no blood is exchanged, but materials, including oxygen, carbon dioxide, simple nutrients, and urea, easily diffuse between the fetal and maternal blood. Many other materials also cross the placental barrier, probaby by active transport by the syntrophoblast. Among them are vitamins, some proteins, antibodies,

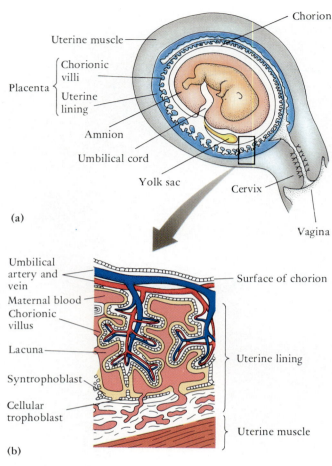

Figure 22.18

(a) A human fetus within the uterus. (b) An enlarged detail of the placenta.

some viruses (e.g., the AIDS virus), and many drugs. Because of the exchange of materials across the placenta, pregnant women who use nicotine, alcohol, and other drugs are endangering the health of the fetus.

Since one-half of the genes of the embryo are paternal, the invasion of the uterine wall by the trophoblast should bring about an immunological rejection from the mother. But this does not happen. You will recall that when a foreign graft is attempted, T cells of the host multiply rapidly and help other lymphocytes reject the graft (p. 335). In ways not fully understood, the syntrophoblast appears to inhibit the expected multiplication of maternal T cells. If T cells from one person are exposed in culture to lymphocytes from another person, they multiply rapidly, as expected, but their multiplication can be inhibited by adding extracts of syntrophoblast.

Thus the syntrophoblast has many complex functions. It enables the embryo to implant, plays an important role in forming the placenta, probably is involved in the active transport of materials across the placenta, prevents the rejection of the embryo, and produces the hormone chorionic gonadotropin. Clearly the evolution of a syntrophoblast has been critical for mammalian evolution.

HUMAN DEVELOPMENT

The Development of Body Form

Gastrulation occurs in human embryos about 8 days after fertilization, by which time the placenta is beginning to develop. The embryonic axis forms quickly on the embryonic disc, and embryonic body form rapidly takes shape. At $3\frac{1}{2}$ weeks the embryo is 3.4 mm long and is already recognizable as a vertebrate embryo (Fig. 22.19a). The brain is differentiating and has induced lens and ear placodes. Visceral arches are developing, the heart is functioning, somites have formed, and a distinct tail is present, even in human embryos.

At 6 weeks the embryo is a little over 13 mm long and can be recognized as mammalian (Fig. 22.19b). The face and the auricle of the ear are developing, limbs with joint flexures have appeared, and fingers are beginning to form in the hand. The foot still resembles a small paddle. The liver forms a large bulge in front of the umbilical cord.

By 8 weeks, the embryo is about 30 mm long and is recognizable as human (Fig. 22.19c). The embryo now contains millions of cells and weighs several thousand times more than the zygote. The major features of organogenesis are complete, and the embryo is now called a **fetus**.

Emphasis is on growth during the fetal period, although some differentiation continues, as indeed it does throughout life. Size increases greatly, and changes in external appearance and body proportions occur. External genitals differentiate, the limbs elongate, finger- and toenails appear, the neck elongates, eyelids develop, and hair appears. At first the fetus is quite wrinkled because the skin has grown faster than deeper tissues, but in the 7th month, subcutaneous fat is deposited and body contours smooth out. If birth occurs at this time, the fetus has a good chance for survival, although the total gestation period is usually about 266 days, just under 9 calendar months.

Birth

Hormones produced by the pituitary, ovary, and placenta have prepared the mother's body for birth (Chap. 19). Birth begins by a series of involuntary uterine contractions, termed labor, that gradually increase in intensity and push the fetus, generally head first, against the cervix of the uterus. The cervix gradually dilates, but in human beings as much as 18 hours or more may be required to open the cervical canal completely at the first birth. The sac of amniotic fluid that surrounds the fetus acts as a wedge and also helps to open the cervix. The amnion normally

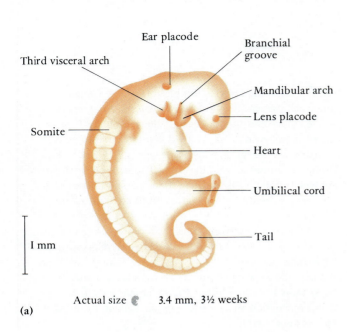

Third visceral arch
Ear placode
Branchial groove
Mandibular arch
Lens placode
Somite
Heart
Umbilical cord
Tail
I mm
Actual size   3.4 mm, 3½ weeks
(a)

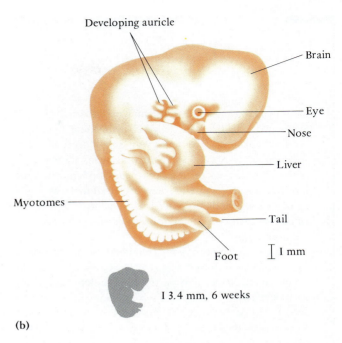

Developing auricle
Brain
Eye
Nose
Liver
Tail
Myotomes
Foot
I mm
I 3.4 mm, 6 weeks
(b)

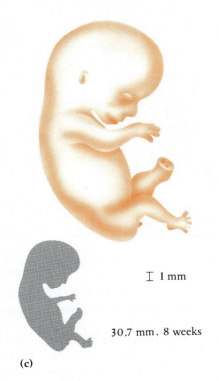

I mm

30.7 mm, 8 weeks

(c)

Figure 22.19

Three stages in the development of the human embryo from 3½ weeks to 8 weeks after fertilization. The shaded inserts depict the actual sizes of the embryos, and a 1-mm scale is shown for each.

ruptures during this process, and the amniotic fluid is discharged. When the head begins to move down the vagina, particularly strong uterine contractions set in, and the baby is usually born within a few minutes. A few more contractions of the uterus force most of the fetal blood from the placenta to the baby.

The birth process is very rapid in many mammals, including cats and dogs, and the young are passed to the outside surrounded by the placenta and all of the extraembryonic membranes, which the mother licks off and eats. When birth is slow, as in humans, the placenta remains in place until the infant is outside of the mother's body. Then uterine contractions continue for a while, and the placenta and remaining extraembryonic membranes are expelled as the afterbirth. The umbilical cord can be cut and tied, although tying is unnecessary, since contraction of the umbilical blood vessels would prevent excessive bleeding of the infant. Other mammals simply bite through the cord. Within a week the stump of the cord shrivels, drops off, and leaves a scar known as the **navel**.

Much of the uterine lining is lost at birth, for the human placenta is an intimate union of fetal membranes and maternal tissue. Uterine contractions prevent excessive bleeding at this time. Following the birth, the uterine lining is gradually reconstituted, and the uterus decreases in size, though it does not become as small as it was originally.

Most mammals have multiple births and some produce litters of offspring that may number a dozen or more. Humans and other primates, horses, whales, and a few other species normally have only one offspring, although twins sometimes occur (Focus 22.1).

Once in approximately every 88 human births in the United States, two individuals are delivered at the same time. More rarely, three, four, five, and even six children are born simultaneously. About 75% of the time, twins are the result of the simultaneous release of two eggs, both of which are fertilized and develop. Such **dizygotic (fraternal) twins** may be of the same or different sex and are genetically no more similar to each other than to brothers and sisters born at different times.

In contrast, **monozygotic (identical) twins** are formed from a single fertilized egg that at some early stage of development divided into two (or more) independent parts, each of which develops into a separate fetus. Such twins, of course, are of the same sex and are genetically identical. Monozygotic twinning may occur in several ways. The two blastomeres produced by the first cleavage may separate and each become an embryo, the blastocyst may contain two inner cell masses, or two embryonic axes may form on a single embryonic disc (Focus fig.). If two embryos form on the same embryonic disc, they will share an amnion and yolk sac. Although identical twins occur in about 25% of human births, a multiple subdivision of the early embryo (**polyembryony**) is the norm in some mammals. The armadillo, for example, always gives birth to four identical quadruplets.

When two embryonic axes form on the same embryonic disc, the identical twins sometimes develop without separating completely and are born joined together (conjoined or "Siamese" twins). All grades of union have been known to occur, from almost complete separation to fusion throughout most of the body, so that only the head or the legs are double.

Monozygotic twinning in mammals may occur by a subdivision of the inner cell mass (a) or by the development of two embryonic axes on a single embryonic disc (b). The latter type of monozygotic twins share a yolk sac and amnion (c).

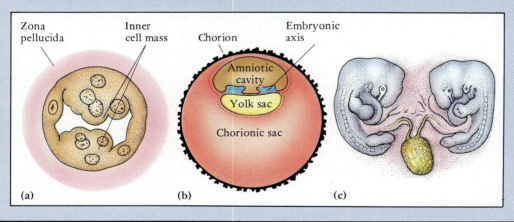

Zona pellucida · Inner cell mass · Chorion · Embryonic axis · Amniotic cavity · Yolk sac · Chorionic sac

(a)   (b)   (c)

Postnatal Development

The degree of maturity and self-sufficiency of the newly hatched bird or newly born mammal correlates with the life style of the species and varies widely from one group to another. The young of chickens and ducks, which nest on the ground, can run around and eat solid food just after hatching, but the young of robins, which nest in trees and shrubs, are blind, have very few feathers, and cannot stand. Newborn colts and antelope have fur and within an hour or so can run with the herd. Newborn rats and mice, which live in nests, and human infants, which are carried about, are quite helpless and require much parental care to survive.

Much development continues in humans after birth. At birth the teeth of the human infant and other placental mammals have not erupted from the gums. This obviously correlates with breast feeding. Body proportions of the newborn also are quite different from those of the adult. The head, shoulders, and arms of a newborn are disproportionately large, but their growth slows early in childhood. The arms attain their proportionate size shortly after birth, but the legs attain theirs only after some 10 years of growth. Neuron connections and tracts continue to form in the brain, and neuron myelination continues, as the infant learns to recognize objects, acquires motor skills, and begins to speak. The last organs to mature in humans are the gonads, reproductive passages, and external genitals, which do not begin to grow rapidly until puberty, 12 to 14 years after the infant is born.

CONTROL OF DEVELOPMENT

It should be clear from our description of development that the formation of a new organism is an intricate and carefully orchestrated affair. But how are the specialization of cells, the movement of cells, and the emergence of pattern controlled? Embryologists did not develop the needed experimental techniques to study questions of this sort until late in the 19th century, but this did not prevent speculation. Many embryologists in the 17th and 18th centuries, using the newly developed microscope and a good imagination, believed that they could see a miniature adult preformed in the egg or sperm (see Fig. 7.1). To **preformationists**, development was simply a question of growth and involved no differentiation. A logical corollary of this view was that the preformed embryos must contain within their germ cells another preformed generation, and so on to infinity, but, in the absence of a cell theory, there was no lower limit to size. Other embryologists, the **epigenesists** (Gr. *epi*, upon + *genesis*, development), believed that the organs gradually emerged from the formless fertilized egg. This view agreed with the growing knowledge of descriptive embryology, but was unable to account for the gradual differentiation of the embryo other than by postulating mysterious vital forces.

Mosaic and Regulative Development

Late in the 19th century and early in this century, embryologists began to perform experiments in a search for the factors that determine the fates of cells. It was soon recognized that the prospective fate of blastomeres, that is, the tissues to which they would normally give rise, could be identified very early. Indeed, prospective fates could be traced to regions of the zygote if the different parts of the zygote had a different pigmentation, as they do in some species of tunicates, a primitive chordate (Fig. 22.20), or if parts of the zygote were stained differentially with vital dyes.

The prospective fates are normally realized, and in some species they are rigidly determined during cleavage by the segregation of factors in the cytoplasm of the zygote. Edwin G. Conklin, working in the United States in 1905, found this to be the case in tunicates. If one blastomere was removed early in development, a specific defect, perhaps the absence of a notochord, would occur. Cleavage was described as **determinate**, for it parceled out factors in the zygote cytoplasm that controlled developmental fate to different blastomeres relatively early. Development was viewed as **mosaic**, with each blastomere forming a very specific part of the embryo. A mosaic pattern of development also occurs in snails and other species with spiral cleavage.

But a mosaic pattern of development does not apply to all animals, as Hans Driesch showed in Germany in 1892. He separated the blastomeres of early sea urchin embryos by vigorously shaking blastulae and discovered that each developed into a complete larva. The range of tissues to which a blastomere could give rise was greater than its normal prospective fate. Development is **regulative** because the fate of a blastomere can be regu-

Figure 22.20
Fate map of a tunicate zygote (a), and an eight-cell cleavage stage (b).

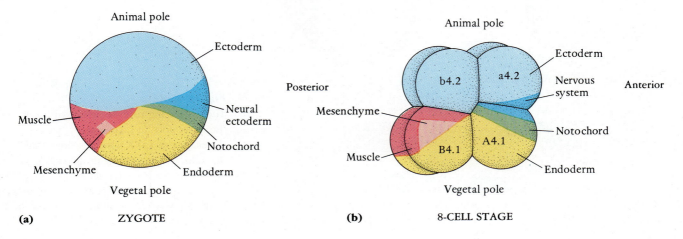

lated or changed according to others with which it is interacting. An early blastomere might normally give rise only to ectoderm, but form a whole embryo when it is separated from the other blastomeres. In regulative development the final determination of blastomere fate must occur later in development, for which reason cleavage is described as **indeterminate**. Indeterminate cleavage and regulative development is the pattern in all echinoderms and most of the chordates studied, the tunicates being an exception. Since humans too have regulative development, it has been possible in cases of *in vitro* fertilization to remove one blastomere from the 8-cell stage to test for certain genetic diseases. The rest of the embryo develops normally when implanted in the uterus.

Induction in Regulative Development

In mosaic development, factors that regulate gene expression are parceled out to different blastomeres during cleavage. The fates of cells are determined much later in regulative development by a series of inductive interactions. In the 1920s inductive interactions were discovered in amphibian embryos by Hans Spemann and his students in Germany. Spemann received a Nobel prize in 1935 for his pioneering work. By transplanting the dorsal lip of the blastopore from one embryo into the prospective belly of an early gastrula of another, Spemann induced host cells to form a second embryonic axis, complete with a notochord and a neural tube (Fig. 22.21). The dorsal lip of the blastopore was shown to be an **organizer** that could induce the formation of a neural tube and an embryonic axis in tissue that normally would have become belly epidermis.

The inductive influence of the organizer brought out certain latent capacities in these prospective belly epidermis cells. Apparently it activated genes that normally would have remained suppressed. Parts of the neural tube, in turn, can induce the formation of the eye and other sense organs, as we have seen.

The sequence of embryonic inductions requires tissues capable of producing the factors responsible for induction and also tissues that are *competent* to respond to these factors. As development proceeds, the competence of responding tissues changes. A transplanted dorsal lip of the blastopore can induce a second embryonic axis in an early host gastrula, but not in a late one. The surface ectoderm of a late neural tube stage embryo can no longer respond to chordamesoderm, but it is still competent to respond to the optic vesicle and form a lens placode. In regulative development, the responding tissues themselves gradually become committed to certain fates.

Differential Gene Expression

The work of Spemann and his followers began to provide insights into the control of development. Since this work was done before the chemical nature of genes and the ways genes control protein synthesis were known, early investigators could only guess at the nature of the factors that controlled development. With the recognition of the pervasive role of DNA, interest shifted to studying the genetic control of development. It was soon evident that gene expression in different tissues leads to the synthesis of some unique proteins. The protein crystallin occurs only

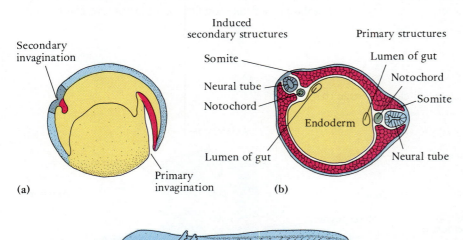

Secondary invagination

Induced secondary structures

Primary structures

Somite

Neural tube

Notochord

Lumen of gut

Primary invagination

(a)

Lumen of gut

Notochord

Somite

Endoderm

Neural tube

(b)

(c)

Figure 22.21
The organizing role of the dorsal lip of the blastopore in salamander development. (a) A transplanted dorsal lip induces a secondary invagination; (b) a transverse section showing that two embryonic axes develop on the same embryo; (c) resulting conjoined larvae.

in lens cells, hemoglobin occurs only in red blood cells and some muscle cells, keratin occurs only in epidermal cells, and so on. But many questions remained. Did differentiated cells contain the complete set of genes (genome) found in the zygote? If so, was only a small fragment of the genome expressed in a particular cell type? Did "unused" genes retain the potential of being expressed?

The techniques of molecular genetics have allowed investigators to address the question of gene expression. Their experiments have confirmed that nearly all cells (cells of the immune system being a notable exception) contain the full zygote genome. Normally only a small fraction of the genome is expressed and leads to characteristic protein synthesis in different cell types, but the other genes are there and under some circumstances can be expressed. This was demonstrated in 1952 in the classic experiments of King and Briggs in which they were able to show that the nucleus from a tadpole intestinal cell carried the genetic information needed for the development of a complete frog (see Fig. 7.3). In another set of experiments, a team of investigators was able to fuse a rat liver tumor cell with a mouse fibroblast (a connective tissue cell). The hybrid cell lived, multiplied, and synthesized protein. Not surprisingly, these cells could make rat liver protein, including albumin and several distinctive enzymes. However, the hybrid cells could also synthesize *mouse liver* proteins, something a mouse fibroblast would never do. The mouse fibroblast had retained the genes that direct the synthesis of liver proteins. Although they are not expressed in a fibroblast, the liver protein genes did become active under these experimental conditions.

These and other experiments have demonstrated differential gene expression in *space*: certain genes are expressed in some tissues; other genes, in different tissues. Other experiments have demonstrated a differential gene expression in *time*; some genes are active early in development, others are switched on later (Fig. 22.22).

Clearly genes are differentially expressed in different tissues and at different times of development. The different gene products, most of which are enzymes, have a profound effect on the ways cells differentiate and perform their functions. But what are the mechanisms that control gene expression in time and space? Not surprisingly it has been found that genes can be controlled at all of the numerous stages of protein synthesis: transcription of mRNA from DNA, processing of mRNA in the nucleus before export to the cytoplasm, and translation of the mRNA on the ribosomes to protein. Additional regulation occurs in the cytoplasm. The message (mRNA) may be present in the cytoplasm, but whether or not protein is formed, and the amount produced, depends on cellular conditions. Consider the synthesis of vertebrate hemoglobin, a complex molecule consisting of two alpha protein chains and two beta chains, each bearing a small iron-containing heme molecule (p. 44). These molecules must be synthesized in a developing red blood cell in the ratio of 2 alpha chains:2 beta chains:4 heme. Severe diseases result from any alteration in this ratio. Researchers have found

Figure 22.22

Activity of genes in a developing frog embryo (*Xenopus*). To test the hypothesis that different genes are active at different periods of development, mRNA (the product of gene transcription) was taken from frog embryos ranging in age from fertilization to 81 hours. The mRNA was spotted on strips of filter paper (*horizontal rows*). Each strip was then incubated with a radioactive, single-stranded DNA that was specific for a particular gene (DG10, DG17, and so on). If the DNA was complementary to the mRNA, it bound to it and formed a radioactive spot (*black spots*), if not, it washed off. For example, the gene represented by DG10 was most active from about 10 to 25 hours, and its activity then declined. The gene represented by DG17 began to become active at about 13 hours and remained active through 81 hours.

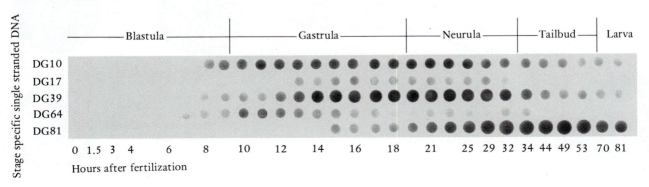

that if more heme is produced than globin with which it binds, the heme does two things: (1) it shuts down its own synthesis by inactivating an enzyme in the metabolic pathway that produces it and (2) it stimulates the synthesis of more globin.

Morphogenesis: Cell Movements and Interactions

Cell differentiation depends on differential gene expression, but the development of an organism also requires morphogenesis, that is, the movement and aggregation of cells to form tissues and organs. That embryonic cells somehow sense where they belong and move to their correct positions was clearly shown in 1955 by experiments of P.L. Townes and J. Holtfreter. In one experiment, they cut out a piece of presumptive epidermis from a heavily pigmented late gastrula of one amphibian species, and a piece of neural plate from a lightly pigmented species (Fig. 22.23). This enabled them to distinguish between the cells. Cells from the two tissues could be dissociated by placing them in an alkaline medium. Then they were mixed together and returned to a solution with a normal pH. The two types of cells first spontaneously reaggregated and then moved and segregated into positions close to the ones they would have assumed during the course of normal development. In this case, epidermal

cells came to lie on the outside of a sphere and neural tube cells on the inside.

During normal development cells also move about and rearrange themselves, as we have seen in the morphogenetic movements of gastrulation and early organogenesis. Movement is accomplished by the frequency and plane of cell division, by changes in cell size and shape, and by cell migration. But to assort themselves in an organized way requires some communication between cells. One communication system derives from interactions between "signature" proteins on the cell surfaces or interactions between the cells and the extracellular matrix. Neural crest cells do not migrate at random but appear to follow pathways laid out on the extracellular matrix. The inductive interactions between groups of cells that we saw in the formation of the neural tube, optic vesicle, and lens must involve a diffusible communicating substance that is produced by the inducing cells. Inducing cells have been placed in a filtering apparatus whose pores are so small that the cells cannot pass through, but the material that does go through is capable of causing induction.

Pattern Formation

Cells differentiate and move about during development, but they also become organized into patterns. Some cells in a developing vertebrate limb form muscle and others

Figure 22.23
Cells from different parts of the late gastrulae of two species of amphibians can be dissociated. They then reaggregate and move to positions they would have occupied during normal development.

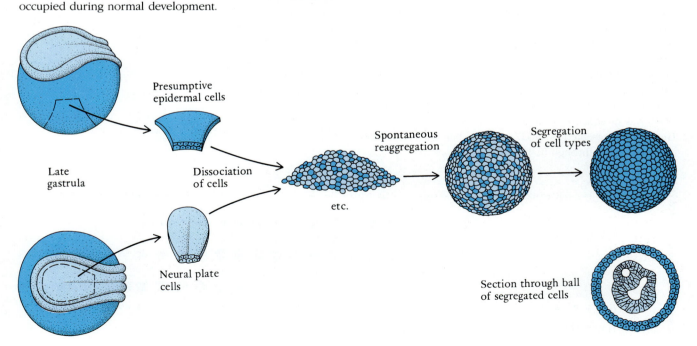

form bone, but according to their positions, they must contribute to specific muscles and bones, e.g., to a humerus or to a radius. Cells must somehow sense their position in a limb and differentiate with relation to this position. A widely held hypothesis suggests that diffusible substances (**morphogens**) are produced in one region and degraded in another, thus forming a gradient. Two or more overlapping gradients of different morphogens would give unique characteristics to each point in space.

A particularly interesting and well-studied example of pattern regulation is the control of body segmentation in developing fruit flies (*Drosophila*). The adult fly is composed of several fused head segments, three thoracic segments, and eight abdominal segments (Fig. 22.24). Some of these bear distinctive appendages: antennae on the head, a pair of legs on each thoracic segment, a pair of wings on the second thoracic segment, and halteres (small, club-shaped wings that act as balances) on the third thoracic segment.

Early development in *Drosophila* involves three distinct phases. First, the **fundamental geography** of the embryo must be established: the distinction between anterior and posterior ends, and between dorsal and ventral aspects, is crucial to successful development. Second, the **segmentation pattern** must be laid down: the developing embryo is divided up into many homologous segments. Finally, the fate of each segment, or **segment identity**, must be determined: the second thoracic segment (T2) will develop wings, three segments will develop legs (T1, T2, T3).

How are these three phases controlled at the molecular level? As in most genetic research, much of our understanding of the process comes from the study of **developmental mutants**, cases where normal development goes awry because one (or more) of the genes involved has undergone a mutation. Investigators have identified three broad classes of genes involved in early development: maternal effect genes, segmentation genes, and segment identity (or homeotic) genes.

Maternal Effect Genes. Immediately following fertilization, the fertilized zygote engages in a burst of protein synthesis. In all organisms, this burst involves the translation of stored maternal mRNAs that were already present in the cytoplasm; only later will the newly constituted diploid nucleus begin transcription of its own mRNA. Although this dependence on stored cytoplasmic messages varies in length from phylum to phylum (two cell stage in mammals, blastula stage in echinoderms), the basic geometry of the embryo is established very early in development. Current models suggest that the mRNA of a small number of **maternal effect genes** is distributed asymmetrically within the cytoplasm of the fertilized zygote. The products of these genes, in turn, influence the translation of other messages, resulting in asymmetrical protein gradients within the developing embryo. These asymmetries determine the polarity of the embryo. Mutations in these genes are always lethal, producing embryos with double body axes (e.g., two tails), no heads, or other gross developmental deformities.

Segmentation Genes. Once the basic coordinates of the embryo are established, normal development depends on the action of **segmentation genes** that control the development of the proper number of segments in the fly larva (several head segments, three thoracic segments, and eight abdominal segments). Approximately two dozen such genes have been identified, largely by studying mutants that affect segmentation. Once again, mutations at these genes are invariably lethal prior to the adult stage. Nonetheless, partial development does take place and can be studied. Segmentation gene mutations may result in larvae with precisely half the normal number of segments, with all even-numbered segments present (but lacking all odd-numbered segments), or with two or more fused segments.

Figure 22.24
A lateral view of *Drosophila* showing the segmentation of the body (T = thorax, A = abdomen) and which segments are controlled by the six homeotic genes.

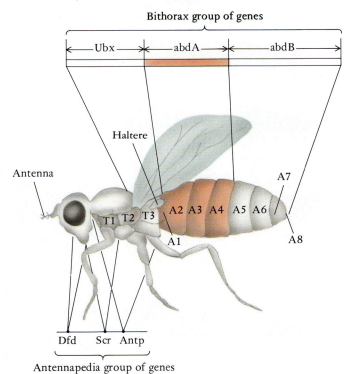

Segment Identity or Homeotic Genes. As their name suggests, these genes control the eventual fate and determine the types of structures that will develop from each segment. In effect, if we think of each segment as having the potential to develop along several different paths, these homeotic genes determine the path to be followed.

In contrast to the other classes of developmental genes, recessive mutations at homeotic genes are not necessarily lethal, even in homozygous conditions. Some of the most dramatic developmental mutations, the **homeotic mutants** (Gr. *homoios*, alike), affect segment identity, and result in the formation of complex structures at inappropriate locations. In effect, homeotic mutants "confuse" the segments, causing a leg to develop where antennae should be, or an additional set of wings on the third thoracic segment where normally halteres would develop (Fig. 22.25).

One particularly interesting experiment, performed by E.B. Lewis and his colleagues, sheds light on the possible evolution of these developmental genes. Lewis deleted entirely one of the segment identity gene complexes (the "bithorax" complex, Fig. 22.24) from the genome of a fly and then observed the developmental conse-quences of the deletion. The resulting larva developed apparently normal head segments and a normal first thoracic segment. All other segments, however, looked like the second thoracic segment, capable of producing both legs and wings! This experiment suggested that the second thoracic segment was the "default condition": in the absence of additional information from the **bithorax gene**, most segments developed into second thoracic segments. This further suggested that the second thoracic segment was most similar to the ancestral or primitive segment, the basis for the metameric body plan of the ancestor of all insects. The molecular organization of homeotic genes suggests that they too may have evolved by a process of "molecular metamerism," where an original gene, specifying the second thoracic (T2) segment identity, underwent several rounds of duplication. These new, additional copies of the "T2 gene" were now free to accumulate mutations without disrupting the function of the original gene. Eventually, these duplicated copies acquired new functions, and could now direct the segments to develop along new, unique, specified developmental pathways. The evolution of these gene complexes would eventually result in the differentiation of the various segments, and give rise to an immensely complex, successful, and morphologically diverse group: the insects (Fig. 22.26).

Figure 22.25
Dorsal views of *Drosophila*: (a) wild type with one pair of wings; (b) a mutant with two pairs of wings, and hence called bithorax.

(a)

(b)

Figure 22.26
A diagrammatic representation of insect evolution by the duplication and modification of genes in the homeobox. (a) and (b) ancestral arthropods; (c) a wingless insect; (d) a winged insect.

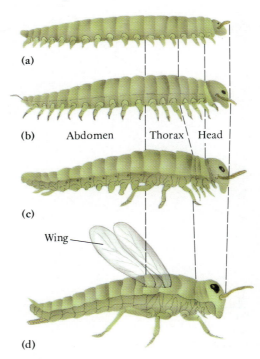

(a)

(b) Abdomen   Thorax Head

(c)

Wing

(d)

Once the relevance of homeotic genes to development became clear, a search began for such genes across the phylogenetic spectrum. Homeotic genes have now been found in a wide variety of organisms, from annelids to the vertebrates. The molecular analysis of these genes reveals a striking common feature: all of them contain a 180 base-pair segment of DNA, the **homeobox** or **homeodomain**. The amino acid sequence of this homeodomain is extraordinarily well conserved across vast evolutionary distances: the homeodomain in a given *Drosophila* homeotic gene is virtually identical (59 of 60 amino acids) with the homeodomain in the equivalent frog homeotic gene, even though 600 million years have elapsed since those two lineages last shared a common ancestor. Furthermore, a region homologous to the homeodomain has now been found in both maternal effect and in segmentation genes. We can conclude that the homeodomain is performing a crucial developmental function that depends critically on its amino acid sequence. Current models suggest that homeodomains may be acting as developmental switches, controlling the transcription of particular genes by binding directly to the DNA double helix. Although the repeated appearance and sequence conservation of homeodomains may be an example of evolutionary convergence, the evidence suggests that all homeotic genes (as well as other developmentally important genes) evolve from a common ancestral gene containing the homeodomain. Through repeated duplication, divergence, and coordination, these gene complexes have acquired the ability to regulate the increasingly intricate events of development.

SEX DETERMINATION AND SEX DIFFERENTIATION

Most animals have separate sexes, and during the course of development the embryo acquires either male or female anatomical features. The mechanism by which sex is determined varies greatly. In most animals, sex is determined by the chromosomal makeup of the zygote. As we saw in Chapter 6, mammals and certain insects, including fruitflies, possess two different sex chromosomes, X and Y. Females have two X chromosomes and males an X and a Y. But the way these chromosomes and the genes they contain affect sex is not the same in mammals and fruitflies. In mammals, a part of the Y chromosome carries a gene that directs the early embryonic gonad to differentiate as a testis. Thus an abnormal zygote containing a single X chromosome and lacking a Y chromosome would differentiate as a female, but would be sterile. In fruitflies a specific sex determining gene has not been identified on the Y chromosome; it is the ratio of sets of X chromosomes to sets of nonsex chromosomes (autosomes) that determines sex. Thus if the ratio is 1 (XX:AA) the fly is female; if it is 0.5 (XY:AA), the fly is male. An intermediate ratio in which the fly is XO:AA would give rise to a male.

In many groups of animals, including moths and butterflies, many reptiles, and birds, it is the female that is heterogametic, has two different sex chromosomes (ZW), and produces two different types of gametes. The male is homogametic, WW. In still other animals, such as honeybees, sex is determined by parthenogenesis. If the egg is fertilized it becomes a female; if it is not, the haploid condition determines the male sex.

Sex is affected by environmental factors in some animals. In certain marine worms (echiurans), the larva will develop into a female if it settles to the bottom without contacting another worm. But if the settling larva contacts a female, a substance liberated by the female induces the larva to develop into a male, which is minute and lives inside the reproductive tract of the female! In certain lizards and turtles, the sex ratio of broods is environmentally determined: female biased broods are produced by incubation temperatures up to 26°C and male biased broods above 28°.

The various mechanisms that determine sex are merely the triggers that bring about sex differentiation through a sequence of presumably genetically controlled events. In mammals, the early embryo possesses primordial gonads, reproductive tracts, and genitalia that are morphologically indifferent, i.e., the same precursor could become either a male or a female organ. A section of the Y chromosome contains a gene (ZFY, Zinc Finger Y) involved in the earliest events of sexual differentiation. That this is the case is indicated by the rare occurrence in human populations of XX phenotypic males and XY phenotypic females. But the XX males contain the small ZFY section of the Y chromosome that has been translocated to another chromosome, and the XY females completely lack this section of the Y chromosome. Individuals of this type are sterile. What might the ZFY gene be doing? At one time it was thought that ZFY controlled the synthesis of a product that caused testis differentiation. At the present time it is unclear as to just how ZFY affects testis differentiation.

After the formation of the gonads, subsequent differentiation of the male and female ducts and secondary sex characters are hormonally controlled. As the fetal testis develops, its Leydig cells secrete testosterone, and testosterone causes the embryonic archinephric duct (Wolffian duct) to differentiate into the vas deferens, and the genital swellings to form a penis and scrotum. A second hormone, **AMDF (Anti-Müllerian Duct Factor)**, produced by the fetal Sertoli cells causes the degeneration of the fetal oviduct (Müllerian duct). In the absence of a testis and these male hormones, maternal and placental estrogen cause the development in the fetus of the vagina, uterus, and uterine tube from the Müllerian duct and suppress the further development of the Wolffian duct. Estrogen is not produced by the developing female until puberty.

Still later in the course of sex differentiation, the sex hormones may affect the development of certain aspects of reproductive behavior in some species. The development of brain centers responsible for the ability to sing in some species of male song birds requires the presence of testosterone. The situation in mammals is not as clear because fewer behaviors occur only in one sex. Estrogen has been shown to affect the development of centers in the hypothalamus of rats that regulate some aspects of behavior. Finally, gender identification in humans following birth may be a factor leading to normal functioning of male and female roles.

Summary

1. Development begins with cleavage, a series of rapid mitotic divisions that converts the single-celled zygote into a multicellular blastula. A cavity, the blastocoel, develops within the blastula. No growth occurs between cell divisions, so the individual cells become smaller and smaller. The pattern of cleavage varies with the amount of yolk and the planes of the mitotic spindles. In eggs with little or a moderate amount of yolk, cleavage is complete (holoblastic), but in those eggs with a large amount of yolk, cleavage is incomplete (meroblastic). Holoblastic cleavage is radial or spiral; meroblastic cleavage is discoidal or superficial.

2. A period of gastrulation follows in which morphogenetic movements rearrange the cells and bring those with different potentials into parts of the embryo appropriate for their subsequent differentiation. The pattern of gastrulation is also affected by the amount and distribution of the yolk.

3. A new cavity, the primitive gut or archenteron, forms during gastrulation and opens to the surface by the blastopore. Three germ layers usually develop: endoderm, ectoderm, and mesoderm. The mesoderm of chordates and related animals develops from the roof of the archenteron, and the coelomic cavity within it is termed an enterocoel. The mesoderm of other coelomate animals develops from cells that migrate into the blastocoel, and the coelom, called a schizocoel, develops as a cleft within the mesoderm. The major morphogenetic movements of gastrulation are invagination, involution, ingression, and epiboly.

4. A notochord differentiates in chordates in the longitudinal axis of the archenteron roof, inducing the formation of a neural tube and neural crest from the overlying ectoderm cells. The neural tube differentiates into the central nervous system and motor neurons. Neural crest cells form other peripheral neurons, pigment cells, and many other structures. The developing brain induces the formation of sensory placodes in the surface ectoderm that differentiate into the nose, eye, and ear. Mesoderm forms most of the dermis of the skin, most of the skeleton, most of the muscles, the kidneys, and the circulatory system. Endoderm gives rise to the lining of the digestive tract and lungs, and to the secretory cells of the liver and pancreas.

5. In tunicates and animals with spiral cleavage, cleavage divides up the zygote in such a way that factors (morphological determinants) in the cytoplasm are confined to particular cells from the outset. Cleavage is described as determinate, and the resulting pattern of development as mosaic because a particular cell always forms a definite part of the individual. In most other animals, cleavage is indeterminate, and the pattern of development is regulative. The fate of cells is not determined until much later and can be changed. The morphological determinants appear to work through a succession of inductions in which one tissue affects the differentiation of an adjacent one.

6. Primitively, as in most marine animals, development includes a motile and feeding larval stage (indirect development) that provides for dispersal and an early source of nutrition outside of the egg. The high mortality of larvae can be partly overcome by shortening the length of their free-living period or by suppressing the larval stage completely (direct development).

7. Cleidoic eggs, which are more or less self-contained systems enclosed within a protective shell, have evolved in some animal groups, particularly in terrestrial groups. Extraembryonic membranes (yolk sac, amnion, chorion, and allantois) provide protection and maintenance for the developing embryo within the cleidoic eggs of reptiles and birds.

8. Parental care, or brooding of eggs, either outside or inside of the body of one parent, is a widespread adaptation enhancing the survival of the embryo. The embryo of most mammals is brooded within the uterus, where it arrives as a blastocyst following fertilization in the upper end of the uterine tube. The chorion and blood vessels of the allantois have become adapted for exchange of gases, nutrients, and waste between the embryonic and uterine blood streams. Those parts of the chorion, umbilical vessels, and uterine wall involved in the exchange constitute the placenta.

9. Most of organogenesis in human development is completed by the end of the 8th week. The embryo, now called a fetus, continues to grow rapidly and many changes occur in external appearance and body proportions. Postnatal development continues after birth with further changes in body proportions, the nervous system, and the reproductive system.

10. With the discovery of the nature of genes and their role in protein synthesis, investigators began studying the genetic control of development. The differential expression of genes in space and time leads to the differential synthesis of proteins, which has a profound effect on cell structure and function. Gene expression can be controlled at all levels of protein synthesis and by conditions within a cell.

11. Cells move about in an organized way during development through interactions between their surface signature proteins and through the production of diffusible substances. As cells shift positions, they become organized into patterns. It is hypothesized that diffusible morphogens establish gradients that enable cells to differentiate in relation to their position. The control of seg-

mentation and appendage differentiation during insect development is integrated by the products of maternal effect genes, segmentation genes, and homeotic genes. One feature of homeotic genes, the homeobox, is highly conserved in animal groups.

12. Sex is determined in different species of animals by special sex chromosomes, the balance between sex chromosomes and autosomes, or environmental factors. These mechanisms are triggers that bring about sex differentiation through a sequence of genetically controlled events and the production of sex hormones.

References and Selected Readings

Balinsky, B.I. *An Introduction to Embryology*, 5th ed. Philadelphia: Saunders College Publishing, 1981. A well-written and well-illustrated account of animal development.

Beaconsfield, P., G. Birdwood, and R. Beaconsfield. The placenta. *Scientific American* 243(Aug. 1980): 94–102. An account of the structure and physiology of the human placenta.

Corliss, C.E. *Patten's Human Embryology*. New York: McGraw-Hill Book Co., 1976. An updating of a classic morphological book on human embryology.

DeRobertis, E.M., G. Oliver, and C.V.E. Wright. Homeobox genes and the vertebrate body plan. *Scientific American* 263(July 1990):46–52. A review of the homeobox and how it controls the differentiation of segments in animals as remote as a fruit fly and mouse.

Edelman, G.M. Cell-adhesion molecules: A molecular basis for animal form. *Scientific American* 250(April 1984):118–129. An article discussing the importance of molecules on cell surfaces in the migration of cells during development.

Gilbert, S.F. *Developmental Biology*, 2nd ed. Sunderland, Mass.: Sinauer Associates, Inc., 1988. An excellent up-to-date and well-documented book. Experimental work and mechanisms of cellular differentiation are emphasized.

Laughon, A.S., and S.B. Carroll. Inside the homeobox. *The Sciences* (New York Academy of Sciences) (Mar./Apr. 1988):42–49.

Moore, J.A. Science as a way of knowing. IV: Developmental biology. *American Zoologist* 27(1986):1–159. An outstanding historical account of the growth of our knowledge of developmental biology.

Patashne, M. How gene activators work. *Scientific American* 260(Jan. 1989):40–47. Many of the mechanisms of gene regulation first discovered in bacteria are relevant to higher organisms.

Slack, J.M.W. *From Egg to Embryo*. New York: Cambridge University Press, 1983. The role of pattern formation in development.

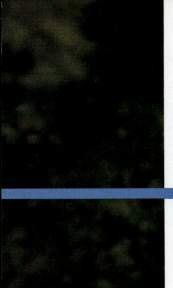

IV

Animal Diversity

23 Taxonomy, Phylogeny, and the Origin and Evolution of the Animal Kingdom

24 Protozoa

25 Sponges

26 Cnidarians

27 Flatworms

28 Aschelminths

29 Molluscs

30 Annelids

31 Arthropods I: Chelicerates and Crustaceans

32 Arthropods II: Uniramians

33 Bryozoans

34 Echinoderms

35 Hemichordates and Chordates

36 Fishes

37 Amphibians

38 Reptiles

39 Birds

40 Mammals

41 Primates and Human Evolution

The white tail deer of North America.

23

Taxonomy, Phylogeny, and the Origin and Evolution of the Animal Kingdom

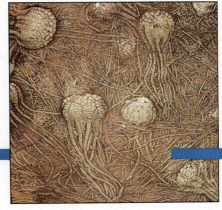

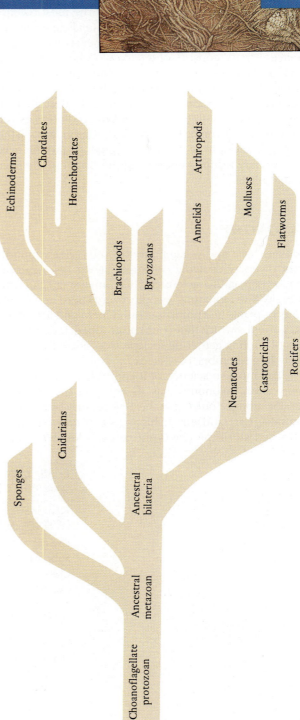

(Left) Fossil crinoids, Phanocrinus.
(Above) Fossil crinoids, Vintacrinus.

There are 32 animal phyla represented today by living species. Phyla are the largest divisions into which animals are classified or grouped. Some, such as the Mollusca, Arthropoda, and Chordata, have been known since humans began to give names to the animals that impacted on human life. Other phyla, like the Rotifera, became known soon after the first microscopes were invented. A few have been discovered only in this century. The most recent is the Loricifera, a phylum of minute animals first found in 1983 in gravel composed of shell fragments off the coast of France. Part IV consists of a survey of the major animal phyla and classes. This chapter will be a prologue to that survey. We will begin by examining the system used in the classification of species within phyla and by describing the ways in which zoologists recognize the evolutionary relationships that form the basis for classification. We will then explore the origin of the animal kingdom and the major evolutionary features underlying the origin of the major animal phyla.

TAXONOMY

Humans have always attempted to see order within the great diversity of living organisms. Long before the invention of magnifying lenses, when only a small part of the diversity was apparent, people everywhere recognized kinds of plants and animals around them and formulated categories by which they could be grouped. Indeed, members of hunting-gathering societies recognized an impressively large number of species.

The modern science of identification and classification, called **taxonomy**, has its roots in the system of classification formulated in the 18th century by a Swedish naturalist, Carl Linnaeus. In his *Systema Naturae*, Linnaeus described all of the plants and animals known to Euro-peans at the time. He utilized a binomial system in which the first name is the **generic name**, the name of the genus to which the species belongs, and the second is the **specific name**. The generic and specific names together designate the name of the species. For example *Terrapene carolina* is the common box turtle, and *Terrapene ornata* is the ornate box turtle. Although the system was not invented by Linnaeus, the *Systema Naturae* represented its first uniform application and is the starting point for the modern naming of species.

All taxonomic names must be unique, i.e., the same name cannot be applied to more than one species or higher taxonomic category. A specific name does not have to be unique, but it must be coupled to a different generic name, e.g., the box turtle *Terrapene carolina* and the rail *Porzana carolina*. When duplication occurs, the first species or category to which the name was applied takes priority in the claim to the name. This application of the **law of priority** for animals begins with the 10th edition of the *Systema Naturae* published in 1758.

Species are grouped together within a system of ascending ranks or **taxa**. Similar species are grouped together within a **genus** (pl. **genera**). Similar genera are grouped together within a **family**; similar families, within an **order**; orders, within a **class**; and classes, within a **phylum** (pl. **phyla**). (Fig. 23.1). As an illustration, the scientific names of the categories to which two familiar animals belong are listed in Table 23.1. Note that the family name ends in *-idae* and that the generic and species names are always italicized.

PHYLOGENY AND SYSTEMATICS

The species composing the animal kingdom are all related through some degree of common ancestry and thus their similarity in structure and in other features is a reflection of

Table 23.1
System of Classification of Animals

Taxonomic Category	Two Turtles	Two Skunks
Phylum	Chordata	Chordata
Subphylum	Vertebrata	Vertebrata
Class	Reptilia	Mammalia
Order	Chelonia (Turtles and tortoises)	Carnivora (Carnivores)
Family	Emydidae (Freshwater and marsh turtles)	Mustelida (Weasels, badgers, skunks, otters)
Genus	*Terrapene* (Box turtles)	*Mephitis* (Skunks)
Species	*Terrapene carolina* (Common box turtle) *Terrapene ornata* (Ornate box turtle)	*Mephitis mephitis* (Striped skunk) *Mephitis macroura* (Hooded skunk)

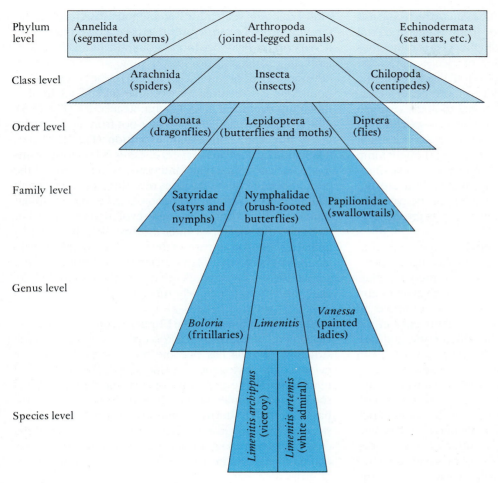

Figure 23.1
Part of the classification of arthropods showing the hierarchy of taxa and how lower ones are included within, or are nested in, the higher ones.

their evolutionary relationship or **phylogeny**. Early naturalists believed that each species was a special act of creation and that grouping the taxa was simply a way of recognizing the "divine order of nature." Evolution provides a scientific rationale for the grouping of species. Each taxon—genus, family, and so on—is **monophyletic**, i.e., it contains all species derived from the same common ancestor (Fig. 23.2). For example, all species of box turtles belonging to the genus *Terrapene* had a common ances-

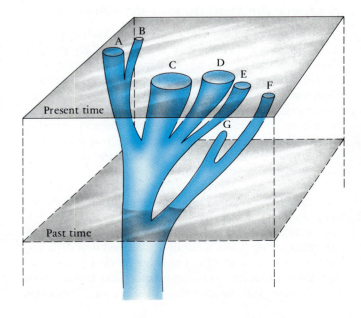

Figure 23.2
Diagrammatic representation of the evolutionary relationships of six (hypothetical) monophyletic species. Circular cross sections of the branches represent the species at the present time; junctions of branches represent points of common ancestry. Species G is extinct.

tor, and all members of the order Chelonia, which contains all of the turtles, should be derived from a single ancestral species of turtle. The grouping of higher taxa is sometimes more difficult than the recognition of species. For example, in Figure 23.2, should species A–F all be placed in one genus? Or should they be placed in three genera comprising A and B, C–E, and F and G? Some zoologists use gap size, i.e., the amount of difference between species, as a criterion, but a group of species that one zoologist might split up into several genera another zoologist might lump together within one. Also, as one looks across the classification of various groups of animals, higher taxa are not always comparable. The amount of difference between the orders of birds, for example, is not nearly as great as that between the orders of insects.

A group found to be **polyphyletic** (derived from more than one ancestor) is generally broken up. Its members have likely been erroneously grouped together because of convergent similarity rather than common ancestry. Cuckoo bees provide a good example of polyphyletic grouping. These bees lay their eggs in the nests of other species of bees, which then nurse and care for the cuckoo bee larvae along with their own. This form of parasitism evolved independently a number of times in the bees. The similar life styles involved in parasitism have led to certain similarities in structure. As a result, entomologists at first placed all cuckoo bees in the same family because they were so similar, not realizing that the grouping was polyphyletic. When the evolution of cuckoo bees was later understood, the assemblage was broken up into a number of different groups, each placed within or near the family of nonparasitic bees to which it was most closely related. Each group of cuckoo bees now represents a monophyletic unit.

Evidence for Evolutionary Relationships

Clearly most zoologists who are concerned with taxonomy must also be concerned with the evolutionary relationships of the species they are studying. This broader aspect of classification is called **systematic zoology**. Systematists determine relationships by searching for similarities between organisms. At all levels of biological organization, organisms resemble each other, sometimes to a startling degree. Zoologists use clues or evidence from many fields to determine evolutionary relationships and the phylogenetic history of a group.

Evidence from Comparative Anatomy   Similarities and differences in anatomical structures are the easiest to assess and have been widely used as evidence of evolutionary relationship. Early naturalists believed that each species was a special act of creation but since they too used structural similarities as a basis for grouping, much of their classification still stands today. In assessing similarities,

most biologists search for **homologous structures** in different species, i.e., structures that have a common evolutionary origin. Evidence for homology includes basic similarity in relationship to surrounding structures and in developmental origin (see Fig. 2.2). The presence of homologous structures in two different species implies **divergent evolution** from a common ancestor (Fig. 23.3). However, many similarities result not from a shared ancestry but from **convergent evolution** (Fig. 23.3), i.e., adaptation by unrelated or very distantly related organisms to similar environmental conditions, as in the case of the cuckoo bees. Similar structures resulting from convergent evolution are said to be **analogous**, i.e., they have a similar functional design but not the same evolutionary origin. For example, the marsupial mole of Australia, like the placental moles of other parts of the world, are blind, have no ear pinna, and have claws adapted for digging. These similarities evolved independently in the two mole lineages, which are not closely related.

Evidence from Comparative Ultrastructure   With the development of electron microscopy, comparative ultrastructure of cells has become an important source of evidence for phylogenetic relationships. The structure of mitochondria, the anchorage of cilia, and the structure of sperm are examples of the ultrastructural evidence that has been used to derive evolutionary relationships. Ultrastructure is really just another level of anatomy, and the concepts of homology and analogy are just as applicable to comparisons of ultrastructure as they are to comparisons of gross anatomy.

Figure 23.3
Diagram of divergent and convergent evolution. In divergence, an ancestral group breaks up into two or more lines of evolution that may lead far from the ancestral design. During this process distantly related groups (e.g., birds and bats) may converge and come to resemble one another in some ways as they adapt to similar modes of life (flying).

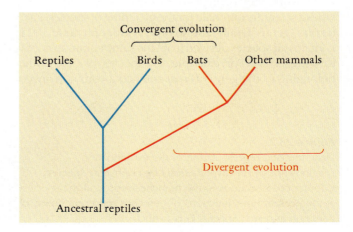

Evidence from Comparative Development Since closely related species share embryonic or larval stages, this can be an important source of evidence for common ancestry. Mammal embryos are similar to those of fishes in many ways. They have a tail, segmental muscle blocks, and pharyngeal pouches (see Fig. 22.19a). All of these structures persist in adult fishes but are lost later in the development of mammals. They are present early in the development of mammals because mammals inherited a basic pattern of embryonic development from ancestral fishes.

Evidence from Comparative Biochemistry Similarities in the sequence of amino acids in certain proteins or the base pairs of mitochondrial and nuclear DNA can be used to determine evolutionary relationships. Among vertebrates, for example, the more closely related the species, the more similar the sequence of animo acids composing the hemoglobin molecule and other proteins. This has been an important source of evidence in attempting to establish the phylogeny of the great apes (p. 944). Deriving evidence of evolutionary relationships from DNA involves a technique termed DNA **hybridization**. The two strands of the DNA molecule are separated and then the single strands of one species are placed with the complementary strands of another. Complementary bases on the strands of different species will bind, or hybridize, with each other, while those that are not complements remain separate. The degree or strength of binding between the two strands is measured by the amount of heat needed to separate them. The greater the degree of hybridization, the more closely related the two species.

Evidence from Paleontology Fossils can provide evidence of evolutionary relationships, but fossils that are ancestral to or intermediate between living forms are often lacking. As we shall see, fossil evidence is of more value for establishing or confirming relationships between taxa below the phylum level than between phyla.

Approaches to Systematics

Three main approaches to systematic analysis currently struggle for primacy: evolutionary systematics, phenetics, and phylogenetic systematics (cladistics). Much of the controversy between them centers on the question of **taxonomic characters**, the features of organisms that provide useful information for historical resconstruction. For example, should we use a character like body weight to assess relationships? Our intuition suggests that such a feature is probably not useful in classifying organisms, since similarity in body weight does not depend on common ancestry. On the other hand, the presence of a vertebral column is considered a reliable criterion on which to group species, because vertebral columns in different animal groups are probably homologous. The identity and variation of a particular character need to be explored before it can be confidently used in classification.

Another point of disagreement among systematists concerns the higher categories: phylum, class, and orders. Stated simply, the controversy centers on the reality of these higher categories and on their correct definition. All biologists agree that the living world is hierarchically organized. Cats, lions, and tigers form a coherent complex, as do dogs and wolves. These clusters in turn can be assembled into a larger inclusive unit that will also include bears and pandas, weasels and skunks, and so on. We can continue clustering the clusters until we encompass all living and fossil organisms. But there the consensus ends. As we will see, cladists consider many of the higher categories as they are now construed to be artificial and arbitrary. Evolutionary systematists, in turn, see the cladistic approach as unnecessarily rigid and potentially uninformative.

In order to understand the controversy, let us now turn to the basic features of these three approaches to classification.

Evolutionary Systematics

Evolutionary systematics, until recently the most commonly used systematic method, makes an explicit distinction between homology and analogy. Species are grouped only on the basis of shared homologies and evolutionary history. The recognition of homologous features depends to a large extent on the knowledge and experience of individual investigators, well acquainted with their particular group and therefore best able to determine the characters on which to base the classification. The branching diagrams prepared by evolutionary systematists are known as **phylogenetic trees** (Fig. 23.4a); the taxonomic categories above the species level (genus, family, and so on) that form the basis of evolutionary classification are also meant to reflect genealogical and genetic similarity. Sometimes this approach has included in one higher taxon a group that is more closely related to another taxon. For example, zoologists recognize that crocodiles share with birds a more recent common ancestry than crocodiles share with lizards and other living reptiles. Nevertheless, evolutionary systematists group crocodiles with reptiles because crocodiles share with reptiles important biological characters, such as being cold blooded, that they do not share with birds.

Phenetics

Phenetics is an explicitly quantitative approach to systematics that seeks to classify organisms on the basis of their overall similarity. Pheneticists use a method called **numerical taxonomy**, which measures and records similarities for large numbers of characters. This information is

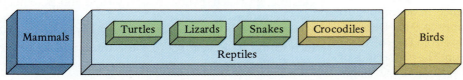

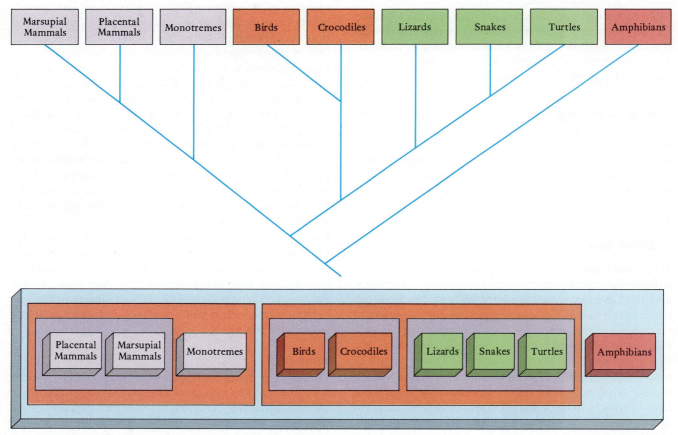

(a)

(b)

then fed into a computer and a single composite quantity, the distance measure, is derived. Species are then grouped, using this distance measure, on the basis of their overall similarity, assumed to be a reliable indication of common ancestry. By and large, pheneticists do not seek to make the distinction *a priori* between homologous and analogous similarities, in the belief that similarity due to common ancestry will always outweigh analogical similarity, provided that enough characters are examined. A **phenogram** of species is then constructed using these distance measures: more closely related species are likely to be more similar to one another. A phenogram is a grouping of organisms or taxa on the basis of overall similarity; it makes no explicit attempt to reflect the evolutionary history of the groups involved. In cases where extensive convergence between two taxa has occurred, a phenetic classification might group those two taxa together, despite their dissimilar ancestry.

Phylogenetic Systematics

Frustrated by the subjective nature of some aspects of evolutionary systematics in the selection of taxonomic characters and concerned by the pheneticists' merging of analogy and homology, a new approach to systematics arose in the mid-1960s. **Phylogenetic systematics** or **cladistics** (Gr. *klados*, branch) attempts to reconstruct the course of evolution by grouping organisms in terms of the relative recency of common ancestry. Cladists erect branching diagrams, or **cladograms**, that are hypotheses about the relationship among taxa based on certain shared characteristics.

In reconstructing the pattern of evolution, cladists use shared homologous structures between species, as do evolutionary systematists, but cladists do not treat all homolo-

◀ *Figure 23.4*

Two alternative arrangements showing the relationships among the major groups of terrestrial vertebrates. (a) depicts the evolutionary systematics approach, which uses both similarity and degree of difference to erect a classification. The designation "reptile" includes turtles, lizards, snakes, crocodiles, and dinosaurs, all of whom share a common set of highly distinctive features. A very different set of features delineate the two other major groups, the birds and the mammals. (b) depicts an alternative arrangement for the terrestrial vertebrates, arrived at by using cladistic methods. In this case the cladogram is constructed by identifying evolutionarily novel (derived) characteristics shared by two or more of the taxa. Since birds and crocodiles share certain derived features, crocodiles are grouped with birds rather than with the other reptiles. The mammals have also been subdivided into two groups.

gies equally. They make a distinction between a homologous feature, such as the use of DNA as the genetic material, that is shared with a distant common ancestor (a **shared primitive character**) and a homologous feature, such as hair in mammals, that is shared with a more recent common ancestor (a **shared derived character**). DNA is so widespread among organisms that it cannot be properly used to define a group such as mammals. Hair, on the other hand, is a derived character that all mammals and only mammals share as a consequence of a common ancestor. It allows all mammals to be grouped and separated from all other vertebrate groups. The presence of hair defines mammals as a monophyletic group. The distinction between a primitive and a derived character is not absolute but relative. Although hair is a derived character for mammals relative to other vertebrates, it is a primitive character for mammals, since all mammals have it. To sort out lines of mammalian evolution, one must look for other derived characters, such as early birth for marsupial mammals. The emphasis in this method is not on the degree of difference, but rather on the identification of shared derived characters that define particular groups. The result of a cladistic analysis is a set of fully nested monophyletic groups (Fig. 23.4b)

Cladists argue that classification can only be based on monophyletic groups. Like evolutionary systematists, cladists define a monophyletic group as one derived from a single common ancestor. However, they further restrict the definition to include all descendants of that ancestor. Under this rule, a single higher taxon cannot arise from more than one common ancestor, otherwise it is **polyphyletic** and not monophyletic. Nor can a monophyletic group arbitrarily exclude certain descendants of the ancestral group, otherwise it is **paraphyletic**. The group of vertebrates commonly called reptiles (turtles, lizards, snakes, and crocodiles) is paraphyletic because it excludes the birds and mammals, which arose from the same ancestral stock (see Chap. 38).

ORIGIN OF THE ANIMAL KINGDOM

There are over a million and a half described species of living organisms. A widely used scheme of classification divides them into five kingdoms. All of the prokaryotic organisms—bacteria and cyanobacteria—are placed in the kingdom **Monera** or **Prokaryota** (p. 66). The remaining four kingdoms contain eukaryotic organisms. The kingdom **Protoctista** contains the protozoa and algae. They are unicellular or multicellular, autotrophic or heterotrophic. However, they are united in having gametes produced in unicellular sex organs and the zygote does not pass through an embryonic stage. The kingdom **Fungi** contains heterotrophic organisms with bodies constructed of branching, multinucleate filaments. Spores germinate into filaments without passing through an embryonic

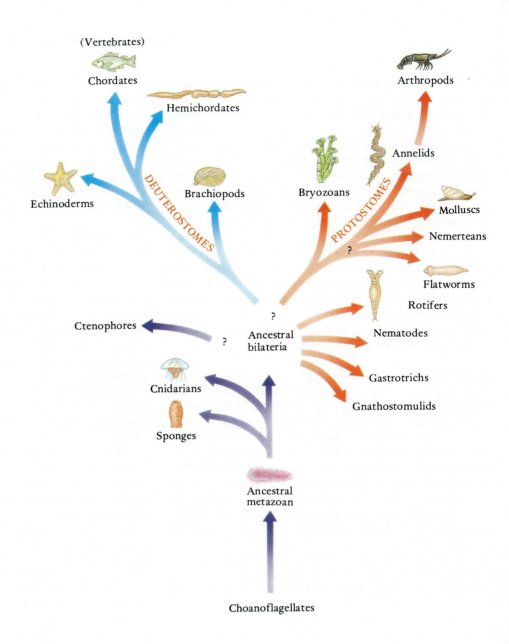

(Vertebrates)

Chordates

Hemichordates

Arthropods

Annelids

Bryozoans

DEUTEROSTOMES

PROTOSTOMES

Brachiopods

Molluscs

Echinoderms

Nemerteans

?

Flatworms

Rotifers

?

Ctenophores

Ancestral
bilateria

Nematodes

?

Gastrotrichs

Cnidarians

Gnathostomulids

Sponges

Ancestral
metazoan

Choanoflagellates

Figure 23.5
A possible phylogenetic tree of
the animal kingdom. A number
of minor phyla have not been
included.

stage. The kingdom **Plantae** contains multicellular, autotrophic organisms in which the gametes are produced in multicellular sex organs, and the zygote passes through an embryonic stage enclosed in maternal tissue. Plants as defined by this kingdom are primarily adapted for a terrestrial existence. The kingdom **Metazoa**, or **Animalia**, contains the multicellular animals. They are heterotrophic and motile, although many species have become secondarily sessile. Gametes are usually produced in multicellular sex organs, and the zygote passes through embryonic stages that include a blastula.

The Fossil Record

Most of the fossil record of past animal life extends from about 570 million years ago to the present, embracing what paleontologists designate as the Paleozoic, Mesozoic, and

Cenozoic eras (see Fig. 9.17). It would be logical to expect that different phyla would appear at different points over this great expanse of geologic time. But a striking and surprising feature of the fossil record is that most of the major phyla of animals, depicted in the phylogenetic tree of Figure 23.5, are represented virtually from the beginning of the record, i.e., all appear at some point in the Cambrian, the first period of the Paleozoic era. All classes of a phylum may not be represented, but at least one is present. For example, fossil sea stars are first found in the Ordovician period and brittle stars in the Mississippian period of the Paleozoic, but there are many other representatives of their phylum, the Echinodermata, in the Cambrian.

Much of what we know about Cambrian animals comes from fossils found in the Burgess Shale, a rich deposit located in southern British Columbia. All of these fossil species were marine animals living in shallow, coastal waters, and most were bottom dwellers. The Burgess Shale fauna includes sponges, segmented worms, molluscs, arthropods such as trilobites and crustaceans, echinoderms, and probably chordates (but no vertebrates) (Fig. 23.6c, f, g, and h). There are also fossil forms that do not belong within any existing class or even phylum. Indeed, some paleontologists believe that as many as 100 different animal phyla existed during the Cambrian, as compared to the 32 living today.

If all of the major phyla are present in the Cambrian, then much of the early evolutionary history of the animal kingdom must have taken place before this time. What is found in Precambrian rocks? Exposed, fossil-bearing Precambrian rocks are found in relatively few places in the world, and preservation of fossils tends to be poor. Nevertheless, some fossil Precambrian animals have been found, such as in the rocks of the Ediacara Hills of South Australia. This Ediacaran fauna lived from 630 million years ago (m.y.a.) to the beginning of the Cambrian (570 m.y.a.), and its record includes not only fossil remains of the animals themselves but also tracks and burrows. Some of these animals looked like jellyfish; others looked like segmented worms. A number appear to be different from any known animals (Fig. 23.6a and b). However, all tend to be soft bodied, and soft-bodied animals are less likely to be preserved than those with hard parts.

What conclusions can we draw from this early fossil record about the origins of the animal kingdom? Clearly, the first animals evolved in the ancient Precambrian oceans well before the scanty Ediacaran fauna makes its appearance, perhaps as early as 700 million to one billion years ago. Then there was an explosive evolution of diversity in the Cambrian, with a sudden appearance of all of the major phyla of animals that are known today, as well as many that later became extinct. Cambrian evolution is described as explosive and sudden, but remember this seems true only as we look back at the fossil record 500 m.y. later. Nevertheless, within the Cambrian time frame there was a great diversification of the animal kingdom at the level of phylum and class, which was reflected in the appearance of many different animal designs (Fig. 23.7). Such diversification through evolution of phyla never occurred again in later geological history. There have been periods of mass extinctions, as occurred at the end of the Paleozoic and Mesozoic eras (Fig. 23.7), which were followed by the evolution of new faunas, but these new faunas were simply variations on existing designs, i.e., they represented evolution within phyla rather than evolution of new phyla.

A number of explanations have been put forward to account for the great Cambrian diversification. The most plausible explanation is an ecological one. A world largely unpopulated by animals offered many niches or life styles, such as filter feeding, deposit feeding, burrowing, and so on, that could be filled with new designs. As faunas evolved, they in turn created more niches, such as various types of predation. Predation was certainly one factor in the appearance of protective skeletons, a striking innovation of Cambrian animals and one that accounts for the preservation of so many species. Some paleontologists believe that the Cambrian radiation (diversification) was correlated with a rise in the nutrient levels of the world's oceans (e.g., phosphate and nitrate levels). This led to increased primary production, i.e., photosynthesis, which in turn provided opportunities for increased heterotrophic nutrition.

Other paleonologists have argued that a major factor in the Cambrian radiation was an easily molded genome, or genetic base, on which selection was operating. The simple preexisting designs, and therefore simple developmental genomes, of the ancestral animals posed few restrictions to the evolution of new ground plans. Remember that the evolution of new form is a compromise between the restrictions of the ancestral design and the selection pressures of new opportunities. For example, the evolution of a flying snail is not possible because the snail ground plan will not permit it. This restriction may explain why the new faunas that evolved following the Permian and Cretaceous extinctions contained no new phyla.

The fossil record provides a partial account of the animals that have inhabited our planet and the sequences of their appearance and disappearance. It also provides clues about the life style of fossil species and the environments in which they lived. But note that the early fossil record tells us nothing about the origin of the animal kingdom, about the structure of the first animals, or about the common ancestors from which different groups of phyla evolved. Considering the entire fossil record of animals, there are many fossils that connect orders and families, but there are only a few fossil forms ("missing links") that connect different classes and none that connect phyla. For such connections we must turn to another source of information—comparative anatomy.

Evidence from Comparative Anatomy

Metazoan animals are motile, multicellular heterotrophs. The multicellular condition must have arisen from a unicellular one, so the ancestors of metazoans must have been some group of unicellular protozoan heterotrophs. The multicellular condition could have arisen in one of two ways: internal compartmentalization of the cell or differentiation of the cells of a simple colony of similar cells. A protozoan with many nuclei could have become internally compartmentalized, i.e., internal cell membranes could have formed around the nuclei converting the organism from a unicellular condition to a multicellular one. This scenario was proposed a number of years ago. Among living protozoans, the ciliates possess more than one nucleus, and it was suggested that a ciliated protozoan ances-

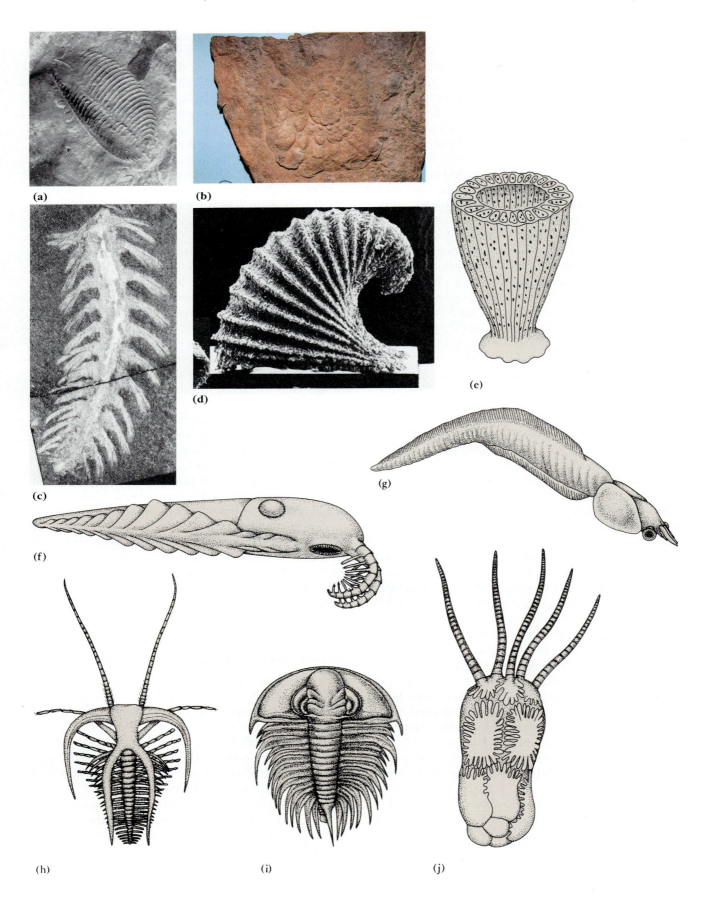

(a)

(b)

(c)

(d)

(e)

(f)

(g)

(h)

(i)

(j)

◀ **Figure 23.6**

Precambrian and Cambrian fossil animals. (a) *Pteridinium*, a Precambrian organism with triradially arranged fronds. It might have been a cnidarian or a member of a now extinct phylum. (b) *Mawsonites*, thought to be a Precambrian jellyfish. (c) *Burgessochaeta*, a polychaete annelid from the Burgess Shale. (d) *Latouchella*, an extinct class of molluscs from the Middle Cambrian. (e) A Cambrian archeocyathan, a member of an extinct phylum of sessile, calcareous pore-bearing organisms. (f) The Cambrian *Anomalocaris*, thought to be a predator not belonging to any living phylum. The animal was about 45 cm long. (g) *Nectocaris*, a member of an extinct phylum of animals from the Burgess Shale. (h) Reconstruction of *Marrella*, a member of an extinct class of arthropods abundant in the Burgess Shale. (i) *Olenellus thompsoni*, a Lower Cambrian trilobite arthropod. (j) *Lichenoides*, a stalkless Middle Cambrian eocrinoid echinoderm.

tor gave rise to the free-living flatworms, many of which are minute and ciliated. The flatworms would then have been the first metazoans. There are numerous problems with this idea, only two of which can be mentioned here. If flatworms are the most primitive animals, then bilaterality is the most primitive symmetry of animals. However, animals like jellyfish and sea anemones are radial, and there is no evidence that their radial symmetry is secondarily derived from a bilateral one. Ciliated protozoans have a peculiar mode of sexual reproduction in which a pair conjugate and exchange nuclei. If ciliated protozoans are the ancestors of flatworms, then the sperm and eggs of flatworms and other animals would have to have developed in the evolution of metazoans separately from their development in other organisms.

A second possible origin of the multicellular condition of animals could have been by differentiation of the cells of a colony, and this is the most widely held view at the present time. The ancestral protozoan would have been a colonial species, with the flagellated cells of the colony forming a sphere having distinct anterior and posterior

Figure 23.7

Numbers of described families of fossil metazoans over geologic time as a reflection of animal diversification. Arrows indicate periods of considerable extinction. The Cambrian and Ordovician diversification is shaded in red; the Mesozoic diversification following the great period of extinction at the end of the Permian is shaded in green. The letters and abbreviations stand for the geologic periods: Vendian (Precambrian), Cambrian, Ordovician, Silurian, Devonian, Carboniferous, Permian, Triassic, Jurassic, Cretaceous, Tertiary.

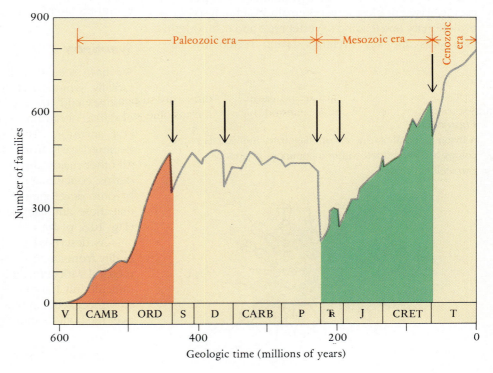

poles, i.e., it moved with the same pole always forward. Initially, every cell of the colony was similar and performed all functions—reproduction, locomotion, feeding, and so on—but by division of labor, the cells became specialized for different functions. Division of labor would be energetically advantageous, for each cell needs to expend only the energy necessary to fulfill its tasks. The large resources necessary to produce sperm and eggs, for example, could be allocated to reproductive cells .only. With differentiation, the cells became interdependent, and the aggregation shifted from being a colony to a multicellular organism.

The search for a group of protozoans containing colonial forms from which the multicellular condition of metazoans might have evolved has often focused on *Volvox*, but these spherical, colonial species exhibit too many plantlike characteristics to be good candidates. Indeed, *Volvox* and related forms are usually classified today among the algal protoctistans (p. 547). Studies on the ultrastructure of protozoans utilizing electron microscopy have revealed that the protozoans that are most similar to certain metazoan cells are members of a small group of flagellates called choanoflagellates. *Choano* means funnel, referring to a collar-like circle of microvilli around the single flagellum carried by these protozoa (Fig. 23.8). Sponges, sea anemones, and some other animals have cells with a ring of microvilli around a single cilium (Fig. 25.4). Remember that other than length there is no difference between a flagellum and cilium. In animals these organelles are usually designated cilia. It is now believed that a monociliated condition, i.e., one cilium per cell, is a primitive condition in animals. It is not only characteristic of most lower animal phyla but also of the evolutionary line leading to vertebrates and echinoderms (sea stars and sea urchins). Living vertebrates have

multiciliated cells, but there is evidence that, at least primitively, they possessed some monociliated cells. For example, the vertebrate photoreceptor cells (rods and cones) are derived from monociliated cells. The basal body and flagellary rootlet of choanoflagellates is very similar to that of metazoan cells, and the ultrastructure of choanoflagellate mitochondria is also like that of metazoans.

Thus, at the present time the best evidence suggests that multicellular animals evolved from a choanoflagellate colony through differentiation of cells. The first metazoan was probably a small marine planktonic organism that was shaped like a hollow ball. It was mouthless and covered by monociliated cells; the symmetry was probably radial, with similar parts arranged around an anterior and posterior axis. This hypothetical stage is called a **blastaea**. From some such ancestor the great diversity of animal phyla evolved. Most of that evolution must have occurred during the end of the Precambrian and the beginning of the Cambrian, because, as we have seen, by the end of the Cambrian all major animal phyla have appeared in the fossil record.

GROUPING OF PHYLA

Although most animals share such features as motility, cephalization, a gut cavity, and a similar basic histology, their ground plans differ in symmetry and internal organization and their development differs in cleavage pattern, in the mode of coelom formation, and in larval structure. These differences provide a means of grouping or organizing the animal phyla, and most zoologists believe that these groupings reflect evolutionary lines of descent within the animal kingdom (Fig. 23.9).

Symmetry

We have seen that the first metazoans were probably radially symmetrical. Two phyla, those containing the sponges and the cnidarians, are radial* and show no evidence of having been secondarily derived from bilateral ancestors. This suggests that these two groups departed early from the main line of animal evolution. All other major phyla are bilateral except for the echinoderms, containing sea stars and sea urchins, but the embryonic development of echinoderms clearly indicates that these animals are derived from bilateral ancestors (Fig. 10.3, p. 188).

As described in Chapter 10, the evolution of bilaterality from an ancestral metazoan that was radially symmetrical with anterior and posterior ends would not be a difficult shift. If some such ancestral form took up a benthic existence, keeping one surface down as it swam or crept over the bottom, there would be a tendency for the lower

*Many sponges have become asymmetrical.

Figure 23.8
A living colonial marine choanoflagellate. Individuals of the colony are embedded within a common, gelatinous, extracellular matrix.

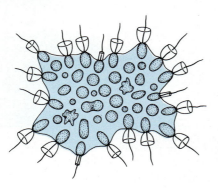

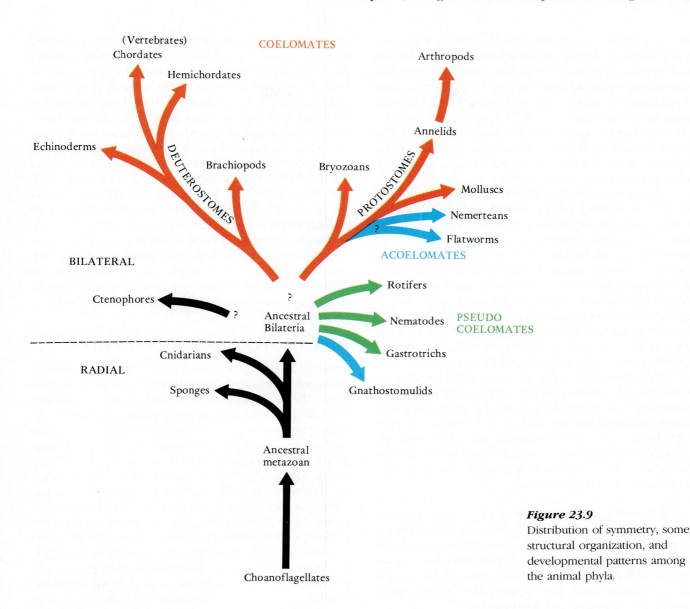

Figure 23.9
Distribution of symmetry, some structural organization, and developmental patterns among the animal phyla.

(ventral) surface to become different from the upper (dorsal) surface (see Fig. 10.2). The animal would then be bilateral. Free-living flatworms have long been considered the most primitive bilateral animals, but this idea is being challenged. Indeed bilaterality may have evolved more than once and the ancestral bilateral species may not have fit into any of the existing phyla. We can only speculate that this ancestor was soft bodied and covered with mono-ciliated cells.

Presence or Absence of Body Cavity

A fundamental feature of bilateral animals is the presence or absence of a body cavity between the gut and body wall. Flatworms and a few other phyla have a solid body construction and are said to be **acoelomate** (Gr. *a*, without + *koiloma*, cavity) (see Fig. 10.4a). Another group, princi-

pally including nematodes and rotifers, possesses a body cavity but of a type called a **pseudocoel** (Gr. *pseudes*, false), a misleading term. It is a body cavity that arises embryonically from the blastocoel. The embryonic blastocoel does not disappear following gastrulation but remains as the body cavity. A third group, containing all of the larger animals, such as segmented worms, molluscs, echinoderms, and vertebrates, have a coelom as the body cavity (see Fig. 10.4b and c). Some, such as arthropods, have lost the body cavity in the adult, but it is still present in the embryo.

Developmental Patterns

The coelom may have evolved more than one time, and the way the coelom forms in the embryo provides a means of grouping the many coelomate phyla (Fig. 23.9). In annelids, molluscs, and arthropods, the coelom forms as a cavity within a solid mass of mesodermal cells, the internal cells splitting apart from each other. This mode of coelom formation is called **schizocoely** (see Fig. 22.6a−c). At least some members of the schizocoelous phyla exhibit spiral, determinate cleavage (see Fig. 22.2), and in the course of development the mouth arises from the blastopore. This entire assemblage of phyla is believed to be related and to constitute what is called the **protostome** line of evolution (Gr. *protos*, first + *stoma*, mouth).

In echinoderms and chordates, the coelom forms from outpocketings of the archenteron, the cavity that results from gastrulation and gives rise to the gut lumen. The outpocketings separate from the archenteron; their cavities fuse and become the coelom and their walls, the mesoderm. Coelom formation by outpocketing of the archenteron is called **enterocoely** (see Fig. 22.6d−f). Radial, indeterminate cleavage characterizes this line of coelomate evolution (see Fig. 22.1), and the mouth arises as a new opening at the end of the embryo opposite from the old blastopore. These phyla constitute the **deuterostome** line of evolution (Gr. *deuteros*, second).

The developmental patterns described for protostomes and deuterostomes are the primitive patterns for these phyla. In every group there are species that have greatly modified patterns as a result of changes in their reproductive modes and consequent changes in the amount and distribution of yolk material in the egg. For example, typical enterocoely is found in primitive chordates, but not in vertebrates.

In recent years there has been increasing use of molecular evidence to determine evolutionary relationships of phyla. DNA hybridization and comparing amino acid sequences of certain proteins and nucleotide sequences of ribosomal RNA are sources of data. In general, the molecular evidence confirms the broad pattern inferred from comparative anatomy and development.

ADAPTIVE DIVERSITY

The animal kingdom displays a tremendous adaptive diversity. Every conceivable habitat within the ancestral oceanic environment has been exploited, and there have been numerous invasions of organisms into fresh water and onto land. The entry of animals into new habitats posed new problems for survival that were met with modifications of old structures and new systems for new functions. Since different animal groups invaded the same habitats, the evolutionary history of animals is filled with convergent adaptations.

Adaptation to new conditions never results in the loss of all of a species' ancestral characteristics. Some are modified to the demands of the new environment; some remain unchanged and are said to be primitive. Thus every species possesses some primitive and some specialized features. A primitive species is one that has retained many ancestral characteristics. But the term primitive is useful only in a comparative sense. One can speak of a primitive animal, a primitive vertebrate, or a primitive mammal. A primitive mammal is not, however, a primitive vertebrate.

Those groups of animals that diverged early in the history of the animal kingdom (i.e., from the lower part of the phylogenetic tree) are often described as being "lower animals." Others that arose later at a higher level of the tree are often said to be "higher animals." However, these terms are often carelessly used and seem to imply that all animals and animal structures fit into a hierarchy of simple to complex. Even the "lowest" animals may display certain specialized and unique features in addition to their primitive characteristics. For example, sponges are primitive in their lack of organs but specialized in the possession of a unique system of water canals penetrating the body.

HOW TO STUDY THE ANIMAL PHYLA

Chapters 24 to 41 will discuss each of the major phyla of animals in some depth. We will begin with the protozoan phyla. They are not metazoan animals, but they are motile heterotrophs and have traditionally been considered animals. The protozoans will be followed by the radiate metazoans. Among the bilateral animals, the acoelomate and pseudocoelomate phyla will be discussed first. They will be followed by the protostome coelomates and then the deuterostome coelomates. It is important to understand that this is not an evolutionary sequence but an arrangement of convenience. We are forced to present the phyla in a linear series, chapter by chapter. But the phyla are branches of a single tree, as indicated in Figure 23.5, and different branches evolve simultaneously, as indicated by fossil evidence. During the course of the following chapters, you will find it helpful to refer from time to time to Table 23.2 and Figure 23.9, noting how a particular phylum might be grouped with others and its position within a phylogenetic scheme.

The study of animal groups can be a tedious and painful task if it amounts to nothing more than committing to memory a list of characteristics of one group after another. A much more useful and interesting approach is one that emphasizes relationships and adaptations. This is the approach that will be utilized in the following chapters. Two questions should be frequently asked as these chapters are studied:

1. How is the structure and physiology of the group correlated with its mode of existence or the environmental demands of the habitat in which it lives?

2. How does the structure and physiology of the group reflect the group's evolutionary origin and its relationship to other groups?

These questions can provide a meaningful framework

into which facts about animal groups can be fit. Moreover, with a little knowledge of the structural ground plan of animal phyla and an understanding of the problems posed by different modes of existence and different environments, it is possible to make intelligent guesses about possible adaptations that might be encountered.

Table 23.2
Synopsis of the Phyla of Metazoa*

Subkingdom Parazoa.
Animals with poorly differentiated tissues and no organs.

Phylum Placozoa (1).
Microscopic, flattened marine animal (*Trichoplax adhaerens*) composed of ventral and dorsal epithelioid layers enclosing loose mesenchyme-like cells.

Phylum Porifera (5,000).
Sponges. Sessile; no anterior end; some primitively radially symmetrical, but most are irregular. Mouth and digestive cavity absent; body organized about a system of water canals and chambers. Marine, but a few found in fresh water.

Subkingdom Eumetazoa.
Animals with tissues and organs.

Radiata.
Radiate animals with tentacles and few organs. Digestive cavity, with mouth the principal opening to the exterior.

Phylum Cnidaria (9,000).
Cnidarians. Hydras, hydroids, jellyfish, sea anemones, and corals. Free swimming or sessile, with tentacles surrounding mouth. Specialized cells bearing stinging organoids called nematocysts. Solitary or colonial. Marine, with a few found in fresh water.

Phylum Ctenophora (90).
Comb jellies. Free swimming; biradiate, with two tentacles and eight longitudinal rows of ciliary combs (membranelles). Marine.

Bilateria.
Bilateral animals.

Protostomes.
Cleavage is determinate and commonly spiral; mouth arising from blastopore.

Acoelomates.
Area between body wall and internal organs filled with parenchyma. (This may be an artificial grouping. It is not at all certain that the Mesozoa and Gnathostomulida are closely related to the Platyhelminthes and Rhynchocoela.)

Phylum Platyhelminthes (18,500).
Flatworms. Body dorsoventrally flattened; digestive cavity (when not secondarily lost) with a single opening, the mouth. The turbellarians are free-living in the sea and fresh water, a few terrestrial. Flukes and tapeworms are parasitic.

Phylum Rhynchocoela or Nemertea (900).
Nemerteans. Long, dorsoventrally flattened body with a complex eversible proboscis apparatus. Digestive cavity with mouth and anus. Marine, with a few terrestrial and in fresh water.

Phylum Gnathostomulida (80).
Gnathostomulids. Minute wormlike animals. Body covered by a single layer of epithelial cells, each of which bears a single cilium. Anterior end with bristle-like sensory cilia. Mouth cavity with a pair of cuticular jaws. Marine.

Phylum Mesozoa (50).
Mesozoans. An enigmatic group of minute parasites of marine invertebrates. Organs absent, and body composed of few cells.

Pseudocoelomates.
Animals in which the blastocoel sometimes persists, forming a body cavity. Digestive tract with mouth and anus. Body usually covered with a cuticle.

(continued)

Table 23.2 (continued)
Synopsis of the Phyla of Metazoa*

Phylum Gastrotricha (460).
Gastrotrichs. Elongated body with flattened ciliated ventral surface. Few to many adhesive tubes present; cuticle commonly ornamented. Microscopic. Marine and freshwater species.

Phylum Nematoda (12,000).
Roundworms. Slender cylindrical worms with tapered anterior and posterior ends. Cuticle thick and complex. Free-living species usually only a few millimeters or less in length; many parasitic species. Marine, freshwater, and terrestrial species.

Phylum Nematomorpha (230).
Hairworms. Extremely long threadlike bodies. Adults free-living in damp soil, in fresh water, and a few marine. Juveniles parasitic.

Phylum Rotifera (1,500).
Rotifers. Anterior end bearing a ciliated crown; posterior end tapering to a foot. Pharynx containing movable cuticular pieces. Microscopic. Largely found in fresh water, some marine, some inhabitants of mosses.

Phylum Acanthocephala (1,150).
Acanthocephalans. Small, wormlike endoparasites of arthropods and vertebrates. Anterior retractile proboscis bearing recurved spines.

Phylum Kinorhyncha (100).
Kinorhynchs. Somewhat elongated body. Cuticle segmented and bearing posteriorly directed spines. Spiny, retractile anterior end. Less than 1 mm in length. Marine.

Phylum Loricifera (1).
Loriciferans. Body composed of a spiny, anterior introvert and a trunk encased within a cuticular armor (lorica). Microscopic in marine shelly gravel.

Phylum Priapulida (13).
Priapulids. Cucumber-shaped or wormlike marine animals, with a retractile anterior introvert. Body covered with spines and tubercles.

Schizocoelous Coelomates.
Body cavity a coelom, formed embryonically by a splitting of the mesoderm, or, if a body cavity is absent, the coelom has been lost. Digestive tract with mouth and anus.

Phylum Sipuncula (320).
Sipunculans. Cylindrical marine worms. Retractable anterior end, bearing lobes or tentacles around mouth.

Phylum Mollusca (60,000).
Molluscs. Snails, chitons, clams, squids, and octopods. Ventral surface modified in the form of a muscular foot, having various shapes; dorsal and lateral surfaces of body modified as a shell-secreting mantle, although shell may be reduced or absent. Marine, freshwater, and terrestrial species.

Phylum Echiura (140).
Echiurans. Cylindrical marine worms, with a flattened nonretractile proboscis. Trunk with a large pair of ventral setae.

Phylum Annelida (11,000).
Annelids. Segmented worms—polychaetes, earthworms, and leeches. Body wormlike and metameric. A large longitudinal ventral nerve cord. Marine, freshwater, and terrestrial species.

Phylum Pogonophora (80).
Pogonophorans. Deepwater marine animals, with a long body housed within a chitinous tube. Anterior end of body bearing from one to many long tentacles; posterior end segmented with setae. Digestive tract absent.

Phylum Tardigrada (400).
Water bears. Microscopic segmented animals. Short cylindrical body bearing four pairs of stubby legs terminating in claws. Freshwater and terrestrial in lichens and mosses; few marine species.

Phylum Onychophora (70).

Onychophorans. Terrestrial, segmented, wormlike animals, with an anterior pair of antennae and many pairs of short conical legs terminating in claws. Body covered by a thin cuticle.

Phylum Arthropoda (900,000 +).

Arthropods. Crabs, shrimp, mites, ticks, scorpions, spiders, and insects. Body metameric with jointed appendages and encased within a chitinous exoskeleton. Vestigial coelom. Marine, freshwater, terrestrial, or parasitic species.

Phylum Pentastomida (90).

Pentastomids. Wormlike endoparasites of vertebrates. Anterior end of body with two pairs of leglike projections terminating in claws and a median snoutlike projection bearing the mouth. Phylum status very questionable; preferably should be placed with arthropods.

Lophophorate Coelomates.

Mouth surrounded by a crown of hollow tentacles (a lophophore). An artificial but convenient grouping.

Phylum Phoronida (10).

Phoronids. Marine, wormlike animals with the body housed within a chitinous tube.

Phylum Bryozoa (4,000).

Bryozoans. Colonial, sessile; the body usually housed within a chitinous or chitinous-calcareous exoskeleton. Mostly marine, a few found in fresh water.

Phylum Brachiopoda (335).

Brachiopods or *lamp shells.* Body often attached by a stalk and enclosed within two unequal dorsoventrally oriented calcareous shells. Marine.

Phylum Entoprocta (150).

Entoprocts. Body attached by a stalk. Mouth and anus surrounded by a tentacular crown. Mostly marine.

Deuterostomes, or Enterocoelous Coelomates.

Cleavage radial and usually indeterminate; mouth arising some distance anteriorly from blastopore. Mesoderm and coelom develop primitively from outpocketings of the primitive gut.

Phylum Chaetognatha (70).

Arrow worms. Marine planktonic animals with dart-shaped bodies bearing fins. Anterior end with grasping spines flanking a ventral preoral chamber.

Phylum Echinodermata (6,000).

Echinoderms. See stars, sea urchins, sand dollars, and sea cucumbers. Secondarily pentamerous radial symmetry. Most existing forms free-moving. Body wall contains calcareous ossicles usually bearing projecting spines. A part of the coelom modified into a system of water canals with external tubular projections used in feeding and locomotion. Marine.

Phylum Hemichordata (85).

Hemichordates or *acorn worms.* Body divided into proboscis, collar, and trunk. Anterior part of trunk perforated with varying number of pairs of pharyngeal clefts. Marine.

Phylum Chordata (42,000).

Chordates. Pharyngeal pouches, notochord, and dorsal hollow nerve cord present at some time in life history. Marine, freshwater, and terrestrial species.

Subphylum Urochordata (2000).

Sea squirts or *tunicates.* Sessile or planktonic nonmetameric invertebrate chordates enclosed within a cellulose tunic. Notochord and nerve cord present only in larva. Solitary and colonial. Marine.

Subphylum Cephalochordata (45).

Amphioxus. Fishlike metameric invertebrate chordates.

Subphylum Vertebrata (40,000).

Vertebrates—fishes, amphibians, reptiles, birds, and mammals. Metameric. Trunk supported by a series of cartilaginous or bony skeletal pieces (vertebrae) surrounding or replacing notochord in the adult.

*The descriptions are limited to distinguishing characteristics. The approximate number of species described to date is indicated in parentheses.

Summary

1. Each species has a binomial name that combines the genus and specific name. Species are grouped into genera, families, orders, classes, and phyla. All generic and higher taxon names must be unique and the name first published has priority in usage.

2. Modern systems of classification strive to reflect the evolutionary relationships, or the phylogeny, of animals. In seeking to determine evolutionary relationships, zoologists utilize evidence from comparative anatomy, comparative development, comparative biochemistry, and paleontology.

3. There are currently three principal approaches utilized in the analysis of evolutionary relationships and its application to classification. Evolutionary systematics emphasizes the importance of shared homologies in determining relatedness. Phylogenetic systematics also emphasizes shared homologies and attempts to group species on the basis of recency of common ancestry. In contrast to evolutionary systematics, phylogenetic systematics demands that higher taxa embrace all descendants of the ancestral member of the taxon. Phenetics is a quantitative approach to the grouping of species, measuring similarity in a wide, nonselective array of characters.

4. The animal kingdom is thought to have evolved during the Precambrian period. It then underwent a great diversification during the Cambrian period, the first of the geological periods for which we have an extensive fossil record. All of the major animal phyla made their appearance by the end of the Cambrian.

5. Evidence from comparative anatomy (including comparative ultrastructure) and development supports the belief that metazoans may have evolved from some colonial choanoflagellate protozoan. The metazoan multicellular condition is thought to have been derived through a division of labor among the cells forming the colony of the flagellate ancestor.

6. The first metazoans were probably motile, radially symmetrical, and gutless. Sponges and cnidarians have retained the ancestral radial symmetry, and sponges are also gutless. The bilateral condition is believed to be derived from the ancestral radial symmetry. The two major evolutionary lines of bilateral phyla are reflected in differences in the patterns of development. Protostomes have spiral, determinate cleavage; the mouth is derived from the blastopore; and the coelom arises by schizocoely. Deuterostomes have radial, indeterminate cleavage; the mouth is formed at the end opposite the blastopore; and the coelom arises by enterocoely.

References and Selected Readings

Attenborough, D. *Life on Earth.* Boston: Little, Brown & Co., 1979. A fascinating and beautifully illustrated book that parallels the BBC television series on the emergence and diversification of life.

Boardman, R.S., A.H. Cheetham, and A.J. Rowell (eds.) *Fossil Invertebrates.* Palo Alto, Calif.: Blackwell Scientific Publications, 1987. A systematic account of the invertebrate fossil record.

Conway-Morris, S. Burgess Shale faunas and the Cambrian explosion. *Science* 246(1989):339–346. An excellent review of the Burgess shale faunas and the geological and evolutionary context of their interpretations.

Duellman, W.E. Systematic zoology: Cutting the Gordian knot with Ockham's razor. *American Zoology* 25(1985):751–762. A review of the revolution in systematic analysis.

Gould, S.J. *Wonderful Life: Burgess Shale and the Nature of History.* New York: W.W. Norton & Co., 1989. A rather lengthy semipopular account of the Burgess shale fauna and its significance in the history of animal evolution. The short review by Conway-Morris provides much of the same information.

———. The telltale wishbone. In *The Panda's Thumb.* New York: W.W. Norton & Co., 1980. An essay comparing traditional and cladistic systems of classification with reference to dinosaurs and birds.

Lake, J.A. Origin of the Metazoa. Proceedings of the National Academy of Science, USA, 87(1990): 763–766. A phylogeny of the animal kingdom based on similarities in ribosomal RNA.

Levin, R. A lopsided look at evolution. *Science* 241(July 1988):291–293. A brief review of current ideas about the way the fossil record reflects the evolution of animal diversity.

McMenamin, M.A.S. The emergence of animals. *Scientific American* 256(April 1987):90–102. The implications of the Precambrian and early Cambrian fossil record.

Margulis, L., and K.V. Schwartz. *Five Kingdoms.* 2nd ed. San Francisco, W.H. Freeman Co., 1987. A rationale for dividing living organisms into five kingdoms.

Wiley, E.O. *Phylogenetics: The Theory and Practice of Phylogenetic Systematics.* New York: John Wiley & Sons, 1981. A detailed consideration of the theory and procedures of the different schools of systematic analysis.

Willmer, P.G. *Invertebrate Relationships*. New York: Cambridge University Press, 1989. An overview of the ideas regarding the origin of the Metazoa and the phylogenetic relationships of the metazoan phyla.

The following invertebrate zoology texts and reference works contain general discussions of all of the invertebrate phyla, including the minor phyla only briefly covered in this text. Also included here are general field guides for identification. References that deal solely with particular groups will be listed at the end of each of the following chapters.

Barnes, R.D. *Invertebrate Zoology*. 5th ed. Philadelphia, Saunders College Publishing, 1987. A textbook of invertebrate zoology.

Barnes, R.S. K.P. Calow, and P.J.W. Olive. *The Invertebrates: A New Synthesis*. Blackwell Scientific Publications, 1988. A textbook covering invertebrates more by function than by a survey of groups.

Brusca, R.C., and G.J. Brusca. *Invertebrates*. Sunderland, Mass.: Sinauer Associates, 1990. A textbook of invertebrate zoology.

Geise, A.C., and J.S. Pearse. *Reproduction of Marine Invertebrates*, 5 vols. New York: Academic Press, 1974–1979. A work covering all aspects of reproduction in the various groups of marine invertebrates.

Gosner, K.L. *A Field Guide to the Atlantic Seashore*. The Peterson Field Guide Series. Boston: Houghton-Mifflin Co., 1979. This guide covers the northeastern Atlantic coast of the U.S. and Canada between the Bay of Fundy and Cape Hatteras.

Harrison, F.W. (ed.) *Microscopic Anatomy of Invertebrates*. New York: Alan Liss, 1990– . When completed, this multivolume series will cover all of the invertebrate groups.

Kozloff, E.N. *Marine Invertebrates of the Pacific Northwest*. Seattle: University of Washington Press, 1988. A guide for the identification of invertebrates along the Pacific coast of northwestern U.S. and Canada.

Kozloff, E.N. *Invertebrates*. Philadelphia, Saunders College Publishing, 1990.

Morris, R.H., D.P. Abbott, and E.C. Haderlie. *Intertidal Invertebrates of California*. Palo Alto, Calif.: Stanford University Press, 1980. A superb work that not only provides for identification but also summarizes information on the biology of the species included.

Parker, S.P. (Ed.): *Synopsis and Classification of Living Organisms*. Vol. 1 and 2. New York, McGraw-Hill Book Co., 1982. Descriptions of the families and higher taxa of all living organisms. Information is not restricted to morphology.

Pearse, V., J. Pearse, M. Buchsbaum, and R. Buchsbaum: *Living Invertebrates*. Cambridge, Mass.: Blackwell Scientific Publications, 1987. A textbook of invertebrate zoology.

Pennak, R.W. *Fresh-Water Invertebrates of the United States*, 3rd ed. New York: John Wiley & Sons, 1989. An excellent guide for the identification of freshwater invertebrates. There is a summary of biological information for each group and a description of methods for collecting and preserving them.

Ruppert, E.E., and R.S. Fox. *Seashore Animals of the Southeast*. Columbia, S.C.: University of South Carolina Press, 1988. A guide to the common shallow-water invertebrates of the southeastern Atlantic coast.

Smith, D.L. *A Guide to Marine Coastal Plankton and Marine Invertebrate Larvae*. Dubuque, Iowa: Kendall/Hunt Publishing Co., 1977.

Sterrer, W.E. (ed.) *Marine Fauna and Flora of Bermuda*. New York: Wiley-Interscience, 1986. This work is also a valuable reference for Florida and the Caribbean.

24

Protozoa

Characteristics of Protozoa

■ Protozoa are unicellular organisms belonging to a number of different phyla.

■ Most are motile and heterotrophic.

■ Food is digested within a food vacuole.

■ Excess water is eliminated by means of a contractile vacuole.

Phylum Sarcomastigophora

■ Flagella and/or pseudopodia are the locomotor or food-capturing organelles.

■ Skeletal structures are highly developed in many sarcodines.

Phylum Apicomplexa

■ Ringlike, tubular, and filamentous organelles at the apical end of the body are a distinguishing feature of these parasitic protozoa.

Phylum Microspora

■ These parasitic protozoa are characterized by a polar filament in the sporelike stage.

Phylum Ciliophora

■ Cilia are present at some stage in the life cycle.

■ The body is covered by a complex pellicle containing various types of organelles in addition to the ciliary basal granules.

■ Two types of nuclei are present: macronucleus and micronucleus.

(Left) The colonial ciliate Carchesium polyporum. (Above) The ciliate Euplotes patella.

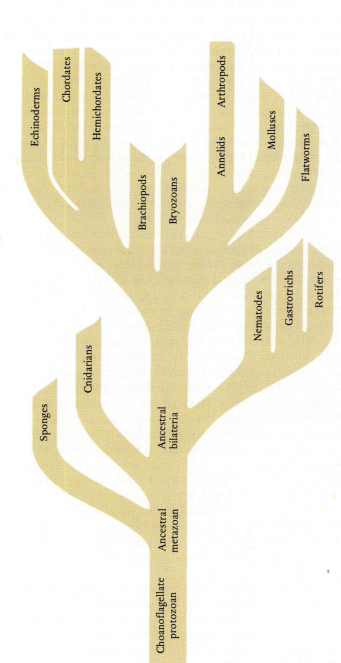

Protozoa are unicellular, eukaryotic organisms belonging to a number of different phyla. Although we have defined animals as multicellular, motile heterotrophs (**metazoans**), we include this chapter on protozoans because in the past the animal-like character of many protozoan groups (motility and heterotrophic nutrition) led to their being considered a part of the animal kingdom. Moreover, as we saw in Chapter 23, the metazoans must have arisen from some group of protozoans, most likely the choanoflagellates, and metazoan motility and heterotrophic nutrition are surely a protozoan heritage.

Most biologists today place the protozoan phyla along with the algae within the kingdom Protoctista. Even though some protoctistans are multicellular (certain algae), the gametes, when present, are never produced in multicellular gonads, and the eggs do not develop into embryos. The protozoan protoctistans are an enormously diverse assemblage of phyla, with over 60,000 species. The diversity is reflected in both their structural specializations and the great variety of habitats and life styles for which they have become adapted. The size range is very great. At one extreme some foraminiferan protozoans overlap in size small species of frogs and at the other extreme they probably reach the minimum size of eukaryotic cells, about 2 to 3 micrometers, and overlap the cell size of some prokaryotic cyanobacteria.

It is erroneous to think of protozoans as being primitive or simple just because they are single cells. The ciliates *Euplotes* and *Paramecium*, for example, are every bit as specialized as many metazoans, but the way they have become complex is very different from that of metazoans. The evolution of complexity in multicellular plants and animals has occurred through a division of labor among the cells, with certain cells becoming specialized for certain functions. Complexity in unicellular organisms has evolved through the specialization of different parts of the cell, for although protozoans are single cells, they are also complete organisms. The cell not only must perform specialized functions but also must retain the ability to perform all of the functions demanded of an organism.

The first protoctistans are believed to have been amoeboid and to have fed by engulfing other organisms. They were eukaryotes but lacked many of the common eukaryotic organelles. Subsequent evolution led to the acquisition of chloroplasts, flagella, Golgi apparatus, and mitochondria by either differentiation or endosymbiosis (p. 80). Living protoctistan groups probably stemmed from various points in this evolution, accounting for much of the great protoctistan diversity.

Not all biologists use the same system to classify the protozoa, and the current trend is to divide them into a larger number of phyla than we have utilized. Whatever the system, it is helpful to think of protozoa as being composed of four groups: those that are flagellated; those that are amoeboid; those that are ciliated; and those that possess sporelike stages.

Where do we look for evidence to sort out the complexities of protoctistan relationships? Comparative ultrastructure of mitochondria, flagellar roots, chloroplasts, and nuclear features is important anatomical evidence. Comparisons of ribosomal RNA and amino acid sequences of certain proteins are sources of molecular evidence. The evidence to date does not provide simple, clear answers to our questions about protoctistan evolution, but it is beginning to suggest some ancestries and groupings.

We will begin with the flagellated protozoa. They are the most diverse of all the protozoan assemblages and probably were the ancestors of most of the other protozoans and the multicellular plants and animals.

PHYLUM SARCOMASTIGOPHORA

The phylum Sarcomastigophora contains the largest number of protozoans, some 48,000 species. Although diverse, the species are united in having only one type of nucleus and in possessing flagella or pseudopodia or both as locomotor or feeding organelles.

Flagellates

The subphylum Mastigophora includes the flagellated members of the Sarcomastigophora: the phytoflagellates and the zooflagellates. In contrast to the lashing planar beat of a cilium, the beating of a flagellum typically involves undulations in one or two planes, which pass from the tip to the base or from the base to the flagellum tip (p. 224). If the flagellum is smooth, the undulations will drive the organism in the opposite direction of the waves. Since the undulations pass from base to tip in most flagellates, the smooth flagellum thus usually functions somewhat like a boat propeller or fan (Fig. 24.1c).

The flagella of many flagellates have fine lateral branches that affect the viscous drag in such a way that the base to tip undulations pull rather than push the flagellate (Fig. 24.1 a and b). Such a "hairy" flagellum may thus function somewhat like an aeroplane propeller.

Phytoflagellates

Most phytoflagellates possess chlorophyll and exhibit autotrophic nutrition; thus they are called plantlike flagellates. The ten orders are classified with different groups of algae by algologists. The body is commonly asymmetrical, and they possess one, or more commonly, two flagella, which are generally carried at the anterior end. A nonliving cell wall or envelope is often present. In most other respects phytoflagellates are very diverse. The nature of their food storage products is often an important characteristic separating the different orders.

The mostly freshwater **euglenids** are a good representative of the phytoflagellates. They have rather spindle-shaped bodies covered by a living pellicle (Fig. 24.2). One or two flagella arise from a deep recess of the anterior end.

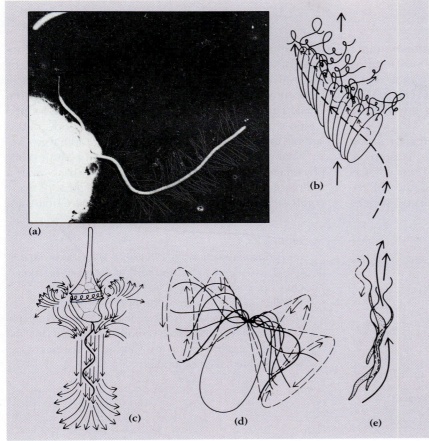

Figure 24.1

Flagellary locomotion. (a) A phytoflagellate with a long, hairy flagellum and a short, smooth one. (b) Movement in *Euglena viridis*. Actual path indicated by dashed arrows. (c) Locomotion in the dinoflagellate *Ceratium*. Arrows indicate the water currents generated by the transverse and posterior flagella. (d) Locomotion in the phytoflagellate *Polytomella*. Arrows indicate the spiral pattern of the flagellar beat. (e) Locomotion of the blood parasite *Trypanosoma*. Dotted arrow indicates movement of undulating membrane; solid arrow, the actual path of movement.

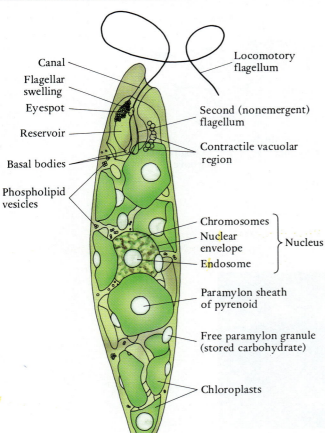

Canal
Flagellar swelling
Eyespot
Reservoir
Basal bodies
Phospholipid vesicles

Locomotory flagellum
Second (nonemergent) flagellum
Contractile vacuolar region

Chromosomes
Nuclear envelope } Nucleus
Endosome

Paramylon sheath of pyrenoid

Free paramylon granule (stored carbohydrate)

Chloroplasts

Figure 24.2

Structure of the phytoflagellate *Euglena gracilis*. Fine branches of principal flagellum not shown.

In *Euglena*, for example, there is a long principal flagellum and a short second one, which does not emerge from the recess and cannot be seen with the light microscope. In the same vicinity there is an eye spot (stigma) containing a red carotene pigment that apparently shields a light-sensitive area at the base of the flagellum. One or two contractile vacuoles are also present. Carbohydrate is stored in a form called paramylon.

Euglenids are typically green, but there are some colorless species, such as members of the genus *Peranema*. This common little flagellate swims against the bottom, and its large, conspicuous, forward-projecting flagellum undergoes lateral undulations in pulling the organism forward. Another smaller flagellum trails. *Peranema* is predaceous and feeds on other protozoans, including *Euglena*. The prey is ingested through an anterior cell mouth or **cytostome** (Gr. *kytos*, cell + *stoma*, mouth), which can be greatly distended in swallowing (Fig. 24.3a). The digestive process within food vacuoles will be described in the section on ciliates, where they have been most extensively studied.

The members of the **Volvocida** are small green phytoflagellates possessing two flagella and a glycoprotein wall. Although some species, such as *Chlamydomonas*, are solitary, the group as a whole tends to be colonial, forming platelike or spherical aggregations. The most highly developed colonies are the large hollow spheres of *Volvox* (Fig. 24.3d). The entire colony moves, rotating through the water.

The brownish or yellowish **dinoflagellates** (Gr. *dinos*, rotation) are common members of marine plankton, although there are also freshwater species. Their color is derived from accessory pigments that mask the chlorophyll. There are, however, some dinoflagellates, such as

the luminescent *Noctiluca*, that are heterotrophic. Dinoflagellates display a variety of shapes, but many are more or less oval or shaped like a top (Fig. 24.3b and c). The pellicle frequently contains deposits of cellulose, which is often divided into plates or valves. A "hairy" flagellum is located within a transverse groove that rings the body, and a smooth one is located posteriorly in a longitudinal groove.

Many species of dinoflagellates, as well as some other phytoflagellates, lose their flagella and pass into a nonmotile vegetative phase. The symbiotic algae of many corals, sea anemones, jellyfish, and other invertebrates are

Figure 24.3

(a) *Peranema* swallowing a *Euglena*. Associated with the cytopharynx are two hooked rods of uncertain function. Only one is shown here. Fine branches of leading flagellum not shown. (b) A freshwater dinoflagellate *Glenodinium cinctum*. (c) Scanning electron micrograph of the marine dinoflagellate *Gonyaulax digitale*. Only the equatorial girdle shows in this view. (d) A *Volvox* colony with daughter colonies inside.

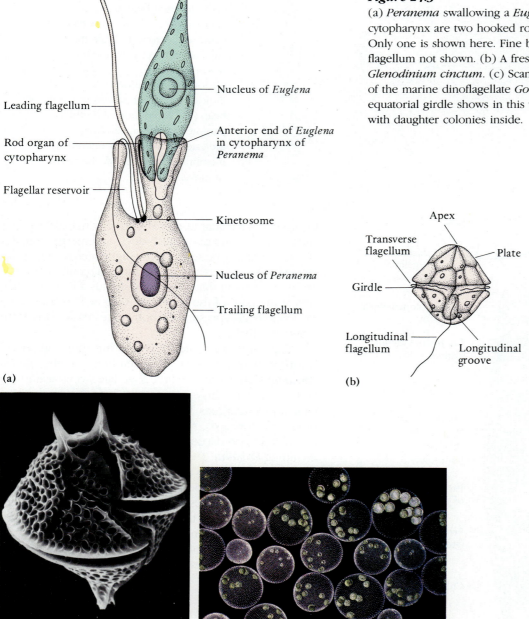

Leading flagellum

Rod organ of cytopharynx

Flagellar reservoir

Nucleus of *Euglena*

Anterior end of *Euglena* in cytopharynx of *Peranema*

Kinetosome

Nucleus of *Peranema*

Trailing flagellum

(a)

Apex

Transverse flagellum

Plate

Girdle

Longitudinal flagellum

Longitudinal groove

(b)

(c)

(d)

a species of dinoflagellate. The host is often colored yellow-brown by the large population of symbionts it harbors (p. 612).

The red tides that periodically plague coastal waters in various parts of the world are commonly caused by species of dinoflagellates. Optimum environmental conditions result in tremendous population growths, or blooms, of certain species. They occur in such great densities that certain of their metabolic wastes kill fish and other marine animals in the red tide or the toxins accumulate in filter-feeding oysters and clams, making them dangerous for human consumption. There are also certain dinoflagellates living on the surfaces of marine algae that produce ciguatoxin, a neurotoxin. This toxin can become concentrated in algae-feeding fish and then further concentrated in carnivorous fish. In certain areas the flesh of some fish is dangerous for humans to consume because of the levels of the ciguatoxin that has accumulated. The initial symptoms of poisoning are abdominal pain and nausea. More serious are abnormal skin sensations, such as numbness, tingling, creeping, and burning (in response to a cold object) that may last for several weeks. Although death is not common, severe poisoning can lead to hallucinations and coma.

Zooflagellates

The zooflagellates lack chlorophyll and are heterotrophic, and are thus called animal-like flagellates. Most species are either commensal or parasitic, and they constitute a much smaller part of the flagellate fauna than do the phytoflagellates. They are undoubtedly a polyphyletic assemblage; some evolved from different groups of phytoflagellates through loss of chlorophyll and au-

totrophic nutrition. The following examples will serve to illustrate a little of their diversity.

The **choanoflagellates** (Gr. *choane*, funnel) are a small group of free-living marine and freshwater forms having a collar of microvilli surrounding the base of the flagellum. The water current produced by the beating of the flagellum is filtered by the collar microvilli. At some point in the course of their life cycle they are either attached by a stalk or embedded within a gelatinous matrix. Colonial organization is common (Fig. 24.4a and 23.8). In the sea they can be an important part of the minute flagellates comprising the so-called nannoplankton (fine plankton). As described in Chapter 23, choanoflagellates are believed to have been the ancestors of the metazoans.

The **trypanosomid zooflagellates** are parasites responsible for a large number of diseases of humans and domesticated mammals, especially in the tropics (Table 24.1). They live in the blood stream and certain other tissues of the vertebrate host. Intracellular stages are aflagellate, but during the life cycle there are motile, extracellular flagellate stages. A single flagellum extends anteriorly and laterally along the side of the elongate body as an undulating membrane. These parasites are transmitted by blood-sucking insects, mostly various kinds of flies.

Leishmania is the agent of the widespread disease kala-azar and related diseases of Eurasia, Africa, and America, which cause skin and visceral lesions (Fig. 24.5a) and interference with immune responses, and can be fatal if not treated. Sandflies are the bloodsucking insect host.

Chagas' disease of tropical America is caused by *Trypanosoma cruzi* and is transmitted by blood-sucking

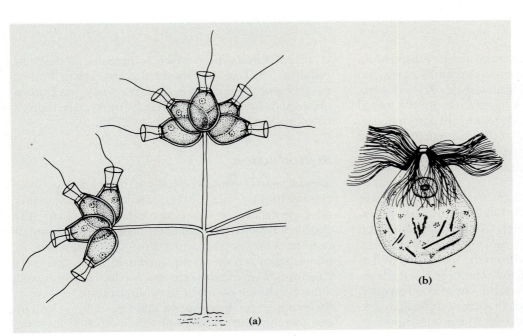

Figure 24.4
Zooflagellates
(a) *Codosiga botrytis*, a colonial freshwater choanoflagellate.
(b) *Barbulanympha ufalula*, a complex zooflagellate that lives in the gut of wood roaches and digests the cellulose of the wood fragments eaten by the roach.

Table 24.1
Pathogenic Protozoan Parasites of Humans

Parasite and Distribution	Adult Parasite's Location in Host	Pathology	Transmission
Flagellates			
Trypanosoma cruzi (Central and S. America)	Blood and endothelial tissues	Chagas' disease—damage to heart tissue, among other effects	Via bloodsucking bugs
Trypanosoma brucei gambiense and *T. b. rhodesiense* (Africa)	Blood, nervous, and lymphatic systems	African sleeping sickness—leads to lethargy, coma, and death	Via bite of tsetse fly
Leishmania spp. (Widely distributed outside of N. America and Europe)	Macrophages	Leishmaniasis (including kala-azar)—lesions of skin, mucous membranes, and viscera	Via bite of sandflies
Giardia intestinalis (Cosmopolitan)	Surface of intestinal epithelium	Gastrointestinal disorders, diarrhea; most infections are not pathogenic	Fecal cysts to mouth, usually via drinking water
Trichomonas vaginalis (Cosmopolitan)	Surface of vaginal mucosa (female) and urethral mucosa (male)	Vaginal discharge and mucosal erosion	By sexual intercourse
Amoebas			
Entamoeba histolytica (Widespread outside of Europe and N. America)	Intestinal mucosa	Amoebic dysentery—invasion and destruction of intestinal mucosa, but can attack other tissues as well; bloody, mucous stools; diarrhea	Fecal cyst to mouth
Sporozoans			
Plasmodium spp. (Tropical)	Liver and red blood cells	Malaria—chills, fever	Via bite of mosquito
Toxoplasma gondii (Cosmopolitan)	Any nucleated cell	Mild to severe infection of various organs; maternal infection can be transmitted to fetus	By eating infected raw meat
Ciliates			
Balantidium coli (Cosmopolitan)	Large intestine	Invasion of intestinal mucosa; diarrhea, nausea, vomiting	Fecal cysts to mouth

hemipteran bugs. Extensive damage may be caused in the human host if the parasite leaves the circulatory system and invades the liver, spleen, and heart muscles. *Trypanosoma brucei rhodesiense* and *T. b. gambiense* are the causal agents of African sleeping sickness and are transmitted by the tsetse fly (Fig. 24.5b and c). The parasite invades the cerebrospinal fluid and brain, producing the lethargy, drowsiness, and mental deterioration that mark the terminal phase of the disease.

The most complex flagellates, indeed among the most complex protozoans, are the hindgut symbionts of termites and wood roaches. The body is commonly saclike, and the anterior end bears a cap and rostrum complex. Numerous flagella may arise from the rostrum as well as from longitudinal grooves in the anterior half of the body (Fig. 24.4b). These flagellates engulf bits of wood ingested by the host and have enzymes that digest the cellulose to glucose within their food vacuoles. The product, glucose,

is shared with the host, which is incapable of digesting cellulose. The host loses its symbionts at each molt when the lining of the hindgut is shed, but a new gut fauna is obtained by licking other individuals, by rectal feeding, or by eating fecal cysts.

Reproduction

Asexual reproduction is typically by binary fission, but in contrast to ciliates, flagellates divide longitudinally to produce two more or less equal halves (Fig. 24.6). The flagellary basal bodies and other organelles duplicate or re-form before or after actual division. Sexual reproduction is unknown or poorly known in many groups of flagellates. Where sexual reproduction has been observed, it involves fusion of similar or dissimilar gametes. *Volvox*, for example, produces sperm and eggs. Life cycles vary considerably and in many groups meiosis follows the

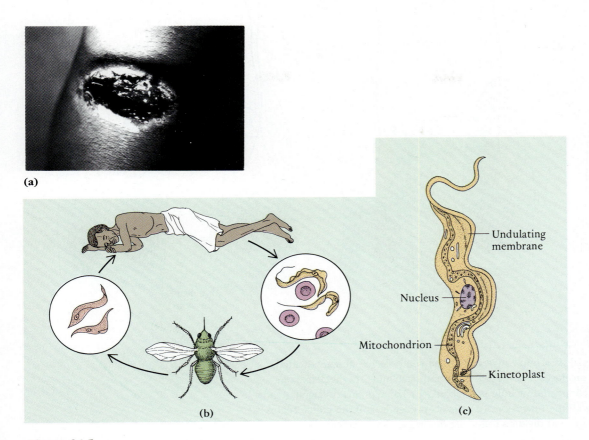

Figure 24.5
Trypanosomid flagellates. (a) Skin lesion on a boy's wrist (Chagas' disease) produced by a species of *Leishmania*. (b) Life cycle of *Trypanosoma brucei*, the causal agent of African sleeping sickness. The flagellate is transmitted by the bite of the tsetse fly. Parasite obtained by fly in human blood is shown to right of fly; parasite transmitted by fly is shown to left. (c) The structure of *Trypanosoma brucei*.

formation of the zygote rather than preceding gamete formation.

Sarcodines

The subphylum Sarcodina contains the amoebas and other protozoa that possess flowing extensions of the body known as **pseudopodia**. The amoeboid condition of some living sarcodines was probably derived from the ancestral eukaryote condition (p. 562). In other sarcodines the amoeboid condition appears to be a secondary development from flagellate ancestors, for many of these species have flagellate developmental stages. Clearly, the amoeboid protozoans cannot all be closely related.

Sarcodines possess fewer organelles than ciliates and flagellates and are therefore relatively simple in cytoplasmic structure. However, skeletal structures have reached a degree of development that is equaled by few other protozoans. The major groups of Sarcodina are distinguished by the form of their pseudopodia and their skeletons.

Amoebas

The most familiar sarcodines are the amoebas, which are found in the sea, fresh water, and soil. Some are naked and some are enclosed within a shell. Amoebas have straplike (filopods) or large blunt pseudopodia (lobopods) used in locomotion and in feeding (Fig. 24.7a and b), and sarcodines with such pseudopodia are often called **rhizopods**. In the shelled species, the shell is secreted (Fig. 24.7c) or is composed of mineral particles cemented together (Fig. 24.7b). A large opening in the shell permits extension of the body and the pseudopodia (Fig. 24.7c).

The pseudopodia and other parts of the body are bounded by a thick layer of gelatinous ectoplasm. Ectoplasm and the more fluid interior endoplasm are different molecular states of cytoplasm, and amoeboid flow involves a rapid change in the molecular organization at

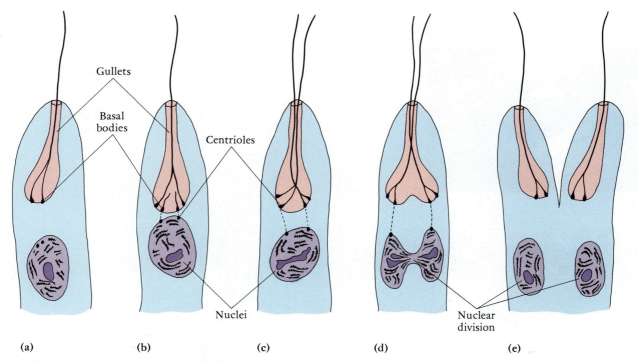

Figure 24.6
Details of longitudinal fission in *Euglena*. (a) The centriole has already divided.
(b) Each centriole produces a new basal body and flagellum. The nucleus is in
prophase, and the contractile vacuole is doubled. (c) The old pair of flagellar roots
separate and fuse with the new roots. (d) Mitosis proceeds, and the gullet begins
to divide. (e) Anterior end dividing following duplication of organelles.

the pseudopodial tip (p. 223). Amoebas feed on bacteria,
diatoms, algae, rotifers, and other protozoans. The food is
surrounded by pseudopodia and eventually enclosed
within a food vacuole (Fig. 24.7a), where digestion proceeds as will be described in ciliates. The vacuole containing undigestible residue ruptures at the posterior end.
Freshwater species possess a contractile vacuole for the
removal of excess water. Periodically, the vacuole fuses
with the cell membrane, and expels fluid to the exterior. A
new vacuole then forms from the fusion of small vesicles.

A number of commensal and parasitic amoebas inhabit the gut of different animals, including humans. Commensal species, such as the very common *Entamoeba coli*
of humans, feed on bacteria and intestinal debris. Parasitic
species, such as *E. histolytica*, the cause of amoebic dysentery, invade the intestinal tissues. They cause the death of
intestinal cells and engulf their contents. Some 480 million
people are believed to be infected, largely in tropical
regions. Both commensal and parasitic amoebas leave the
host as cysts in the feces, and reinfection occurs through
the mouth.

Foraminiferans

Foraminiferans (L. *foramen*, opening + *pherein*, to bear)
are marine rhizopods that secrete a chitin-like shell with
one chamber or, in most species, a multichambered cal-

careous shell (Fig. 24.8). They possess delicate branching
pseudopodia (reticulopods), which arise from cytoplasm
that flows out of the large shell opening and back over the
shell surface. In some species the shell contains tiny
perforations through which pseudopodia may protrude.
The pseudopodial cytoplasm forms an adhesive net and in
planktonic species, such as *Globigerina*, extends as a halo
around the shell (Fig. 24.8a). Most forams are benthic, and
the pseudopodial net projects over the bottom surface,
creating both a trap and a means by which the foram slowly
crawls. Any small organism contacting the net is restrained,
slowly surrounded by cytoplasm, and brought toward the
interior of the body within a food vacuole. Planktonic
forams and even some large shallow-water benthic species
of the tropics harbor symbiotic photosynthetic organisms,
most commonly dinoflagellates, within the cytoplasm.

In most forams the shell is composed at first of a single
chamber with an opening at one end. When the foram
outgrows the first chamber, it secretes another one. This
process continues, producing a multichambered shell that
is occupied entirely by one individual. Accumulations of
foram shells are an important constituent of fine ocean
bottom sediments. They have an extensive fossil record
that begins in the Cambrian, and there are great limestone
deposits, such as those forming the great chalk cliffs over-
looking the sea at Dover, England, composed largely of

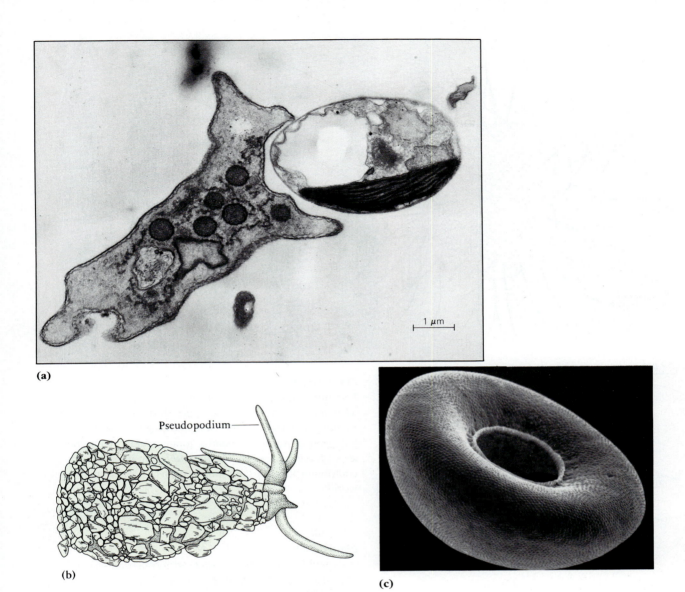

Figure 24.7
Amoebas. (a) A naked amoeba with large, blunt pseudopodia. Amoeba is flowing to the right. (b) *Difflugia oblonga*, an amoeba with a shell composed of cemented mineral particles. (c) The secreted helmet-shaped shell of *Arcella*. Scale bar equals 10 μm.

foram shells. One of the great pyramids of Egypt is composed of limestone derived largely from the accumulated shells of fossil forams (Fig. 24.8d).

Heliozoans

Heliozoans are freshwater and marine Sarcodina. Their spherical bodies are free or attached by a stalk. The body is composed of a central cytoplasmic core, or **medulla**, that contains one to many nuclei, and an outer **cortex** of highly vacuolated cytoplasm (Fig. 24.9). Radiating from the cortex are many needle-like pseudopodia, called **axopodia**, from which the name Heliozoa ("sun animals") is derived. Sarcodines with axopodia are commonly called **actinopods**. The axopods contain a central cytoplasmic rod of microtubules that extends from the medulla (Fig.

24.9b). Many heliozoans possess a skeleton of siliceous scales, tubes, spheres, or needles embedded in the cortex. Where the skeleton is composed of needles, they radiate out of the cortex like the axopods. Some species even have sand grains or living diatoms embedded within the cortex.

The axopodia function largely as food-trapping organelles. On contact, small organisms adhere to the axopods, which then withdraw or bend. The prey is covered by cytoplasm and is gradually withdrawn into the cortex to be digested within a food vacuole.

Radiolarians

The radiolarians (L. *radiolus*, little ray) are a group of planktonic marine actinopods in which skeletal structures are highly developed. The body is somewhat like that of

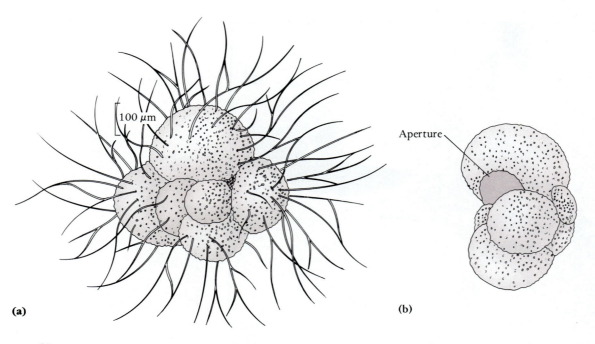

(a)

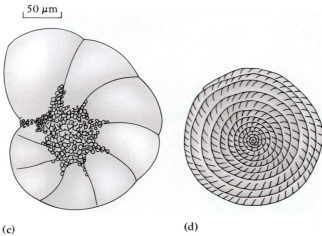

(b) Aperture

(c)

(d)

Figure 24.8
Foraminiferans. (a and b) *Globigerina*, a planktonic foraminiferan. (a) Shows the upper surfaces of the shell and pseudopodia; (b) is a side view, showing the aperture. (c) Side view of *Nonion*, a benthic foraminiferan. (d) Section through a fossil species of *Nummulites*, an important contributor to great limestone deposits in certain parts of the world.

Figure 24.9
(a) A multinucleate heliozoan, *Echinosphaerium eichorni*. (b) Electron photomicrograph of a section through an axopod. Note that the axial rod is composed of a double spiral of microtubules.

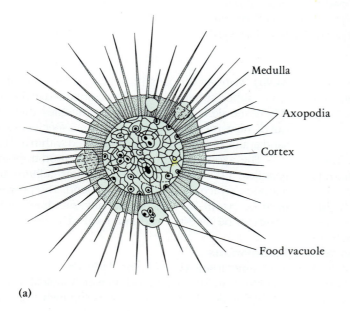

Medulla

Axopodia

Cortex

Food vacuole

(a)

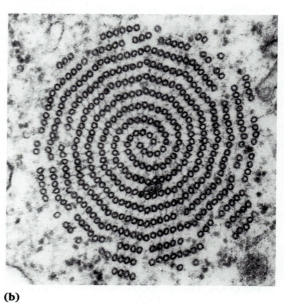

(b)

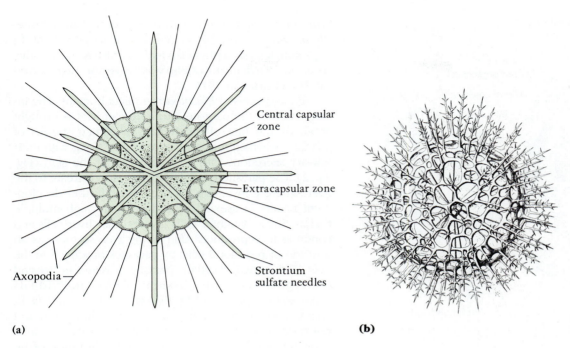

Central capsular
zone

Extracapsular zone

Axopodia

Strontium
sulfate needles

(a)

(b)

Figure 24.10
Radiolarians. (a) *Acanthometra*, a radiolarian with a skeleton of radiating strontium
sulfate rods. (b) A siliceous skeleton of a radiolarian.

heliozoans in being more or less spherical with a central
nucleated core of cytoplasm and a broad outer vacuolated
cortex (Fig. 24.10). The pseudopodia are axopods. Radiolarians differ from heliozoans in that the central
cytoplasm is encased within a membranous capsule that is
perforated to permit communication with the outer cortex
(the extracapsular zone).

Some radiolarians have a skeleton of radiating strontium sulfate rods (Fig. 24.10a), but most have a siliceous
skeleton arranged as concentric spherical lattices within
and outside of the cortical cytoplasm (Fig. 24.10b). Radiolarian skeletons accumulate on the ocean bottom and
may predominate in the fine, soft sediments (called ooze).
The fossil record of radiolarians extends back into the
Paleozoic, and they have contributed to sedimentary deposits.

The radiating pseudopodia project through the
skeletal openings and function as food-trapping structures
in the same way as in heliozoans.

Plankton samples reveal a distinct vertical stratification of radiolarians to depths of 5000 m, although the
greatest number occur in the upper 150 m of the water
column. The cortical cytoplasm of radiolarians living in
this lighted zone contains symbiotic photosynthetic organisms, mostly dinoflagellates. Radiolarians are capable of
some depth regulation through changes in the vacuolated
condition of the cortical cytoplasm.

Reproduction

Asexual reproduction is generally by binary fission. In
shelled amoebas, heliozoans, and radiolarians, the skeleton is divided or else one daughter cell gets the skeleton,
and the other secretes a new one. Sexual reproduction
usually involves the fusion of two morphologically similar
gametes (isogametes), which in radiolarians and foraminiferans are flagellated, but little is known about
sexual reproduction in radiolarians. Reproduction in forams involves an alternation of asexual and sexual stages.

SPOROZOANS

Sporozoans (Gr. *sporos*, seed + *zoon*, animal) are parasitic protozoa, living within or between cells of their
invertebrate or vertebrate hosts. They belong to the phylum Apicomplexa, which was formerly placed in an old
protozoan grouping, the Sporozoa, along with some other
spore-producing parasites. Sporozoan continues to be
used as a common name. The Apicomplexa contains a
number of parasites of great economic and medical importance. The phylum is so named because its members
possess a complex of ringlike, tubular, and filamentous
organelles at the apical end, visible only with the electron
microscope (Fig. 24.11). The apical complex is probably
involved in the entry of the sporozoan into the host cell.

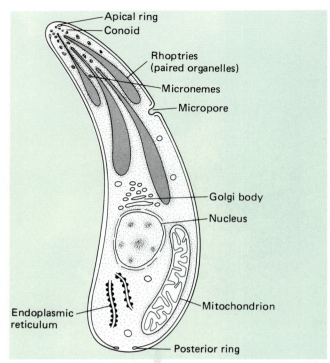

Figure 24.11

A lateral view of a generalized apicomplexan sporozoan. The apical complex components are shaded.

One or more feeding pores (micropores) are located on the side of the body.

The life cycle of apicomplexans typically involves an asexual and a sexual phase. An infective stage, called a **sporozoite**, invades the host and undergoes asexual multiplication by fission, producing individuals called **merozoites**. Merozoites can continue schizogamy (multiple fission) but eventually form gametes that fuse to form a zygote. The zygote undergoes meiosis to form sporozoites.

The nature and life cycle of apicomplexans can be illustrated by the coccidians, members of a subclass that include the parasites causing malaria in humans. Malaria is still widespread throughout the world and one of the worst scourges of humanity. About 300 million people are infected, and the annual death rate is about 1% of those infected. The death toll is especially high in children under six years of age. The untreated disease can be long-lasting and terribly debilitating. Malaria has played a major and often unrecognized role in directing the course of human history. The name means literally "bad air" because the disease was thought to be caused by the air of swamps and marshes. Although malaria had been recognized since ancient times, the causative agent was not recognized until 1880, when Louis Laveran, a physician with the French

army in North Africa, identified the coccidian parasite, *Plasmodium*, in the blood cells of a malarial patient. In 1897 Ronald Ross, a physician in the British Army in India, determined that the mosquito was the vector. He received the Nobel prize in 1911.

Four species of *Plasmodium* infect humans. The introduction of the parasite into a human host is brought about by the bite of the female of certain species of mosquitos, which inject the sporozoites along with their salivary secretions into the capillaries of the skin (Fig. 24.12). The parasite is carried by the blood stream to the liver, where it invades a liver cell. Here further development results in asexual reproduction through multiple fission. These daughter cells invade other liver cells and continue to reproduce. After a week or so there is an invasion of red blood cells by parasites produced in the liver. Within the red cell the parasite increases in size and undergoes multiple fission. These individuals (merozoites), produced by fission within the red cells, escape and invade other red cells. The liberation and reinvasion does not occur continually but occurs simultaneously from all infected red blood cells. The timing of the event depends upon the period of time required to complete the developmental cycle within the host's cells, and the timing differs in different types of malaria. The release causes chills and fever, the typical symptoms of malaria. Infected and less pliable red blood cells block many capillaries, causing a reduced blood flow to many tissues and hence more serious damage.

Eventually some of the parasites invading red cells do not undergo fission but become transformed into **gametocytes**. The gametocyte remains within the red blood cell. If such a cell is ingested by a mosquito, the gametocyte is liberated within the new host's gut. After some further development, the gametocytes form gametes, and then a male and female gamete fuse to form a zygote. The zygote enters the stomach wall and gives rise to a large number of spore stages (sporozoites). It is these stages, which migrate to the salivary glands, that are introduced into the human host by the bite of mosquitos.

The asexual stage of other coccidians occurs in blood cells or in gut cells. There are a number of diseases of domesticated animals caused by coccidians, such as *Eimeria* in chickens, turkeys, pigs, and sheep, and *Babesia* in cattle (red-water fever).

The phyla Microspora, Myxozoa, and Ascetospora contain a smaller number of spore-producing parasites formerly grouped with the Apicomplexa. Species of Microspora are found in most animal groups, especially arthropods. These intracellular parasites lack the apical complex of sporozoans, and the sporelike stage is characterized by a polar filament that is everted when this stage is taken into the host (Fig. 24.13). The everted filament provides a pathway by which the amoeboid sporoplasm escapes from the spore.

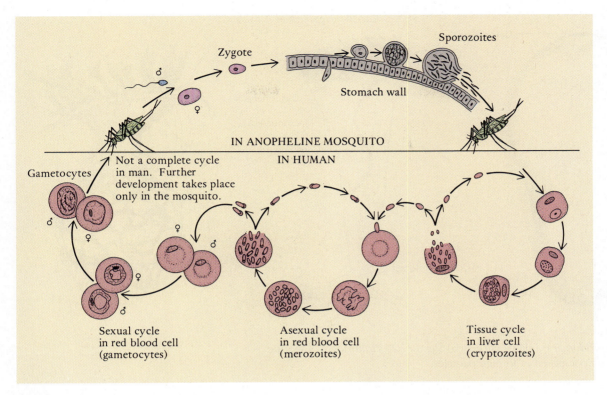

Figure 24.12
The life cycle of *Plasmodium*, the causal agent of malaria, in a mosquito and in a human.

Figure 24.13
Microspora. (a) Spore containing polar filament. (b) Sporoplasm escaping from everted polar filament.

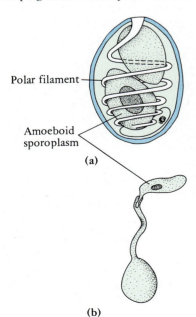

Polar filament

Amoeboid sporoplasm

(a)

(b)

PHYLUM CILIOPHORA

The phylum Ciliophora is a large homogeneous group of protozoans possessing at some time in their life history ciliary organelles for locomotion and feeding. They probably evolved from some group of multinucleated mastigophorans that possessed many flagella. The 7000 described ciliates are widespread in the sea, in fresh water, and in the water films around soil particles. There are a number of symbiotic ciliates, including species that live in the gut of vertebrates. Most are microscopic, but the largest (about 3 mm) can be seen with the naked eye.

Ciliates typically possess a distinct anterior end, and primitive species are radially symmetrical (Fig. 24.14a). However, most are asymmetrical. Their shape is maintained by a complex **pellicle**, a living outer layer of denser cytoplasm containing the peripheral and surface organelles.

The cilia arise from subsurface basal granules, or **kinetosomes** (Gr. *kinein*, to move + *soma*, body) (Fig. 24.15). The kinetosomes are connected together in longitudinal rows by fibrils, and all the fibrils and kinetosomes of a row make up a **kinety**. The kinetosomes and fibrils constitute the subsurface ciliature, or **infraciliature**, of the pellicle. The function of the infraciliature is still uncertain. It probably plays a role in the anchorage of the cilia, but there is no evidence that it is involved in coordinating ciliary beat.

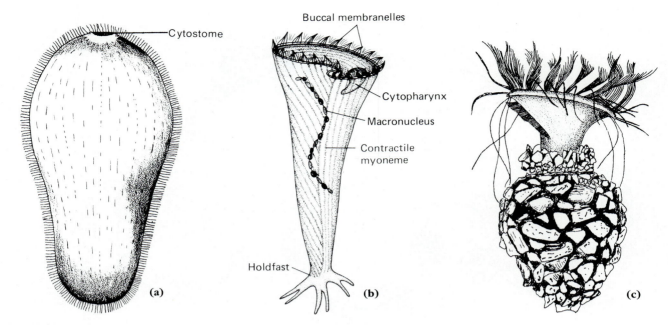

Figure 24.14
Ciliates. (a) *Prorodon*, a primitive ciliate; (b) *Stentor*; (c) *Tintinnopsis*, a marine ciliate with a test composed of foreign particles.

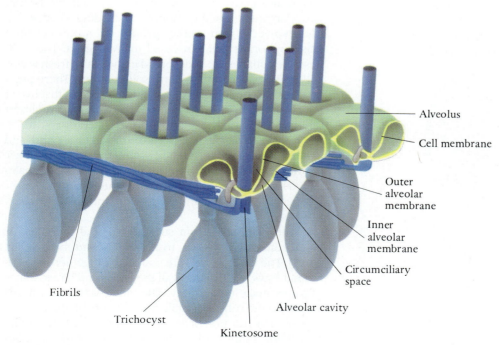

Figure 24.15
Pellicular system in *Paramecium*.

The cilia covering the general body surface are called the **somatic ciliature**. Primitively, longitudinal rows of somatic cilia cover the entire surface of the body, but in many species the somatic ciliature is reduced to girdles, tufts, or bristles or is lacking altogether (Fig. 24.16). However, even those species with no somatic cilia as adults possess an infraciliature persisting from cilia of earlier developmental stages.

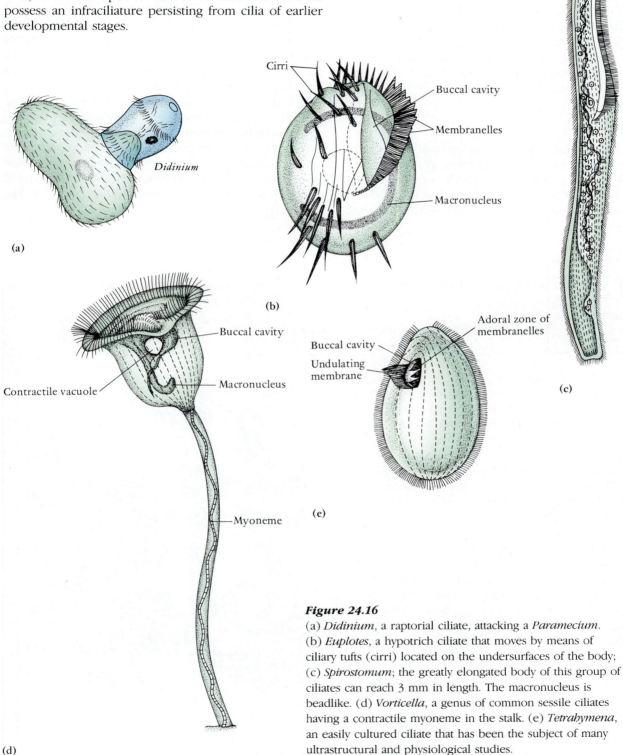

Cirri

Didinium

Buccal cavity

Membranelles

Macronucleus

(a)

(b)

(c)

Buccal cavity

Contractile vacuole

Macronucleus

Buccal cavity

Undulating membrane

Adoral zone of membranelles

(e)

Myoneme

(d)

Figure 24.16
(a) *Didinium*, a raptorial ciliate, attacking a *Paramecium*.
(b) *Euplotes*, a hypotrich ciliate that moves by means of ciliary tufts (cirri) located on the undersurfaces of the body;
(c) *Spirostomum*; the greatly elongated body of this group of ciliates can reach 3 mm in length. The macronucleus is beadlike. (d) *Vorticella*, a genus of common sessile ciliates having a contractile myoneme in the stalk. (e) *Tetrahymena*, an easily cultured ciliate that has been the subject of many ultrastructural and physiological studies.

The complex pellicle of ciliates is generally composed of three membranes, the outer one covering the body surface, including the cilia. The two inner membranes may be so folded, as in *Paramecium*, that they form large vesicles (alveoli) around the base of the cilia (Fig. 24.15). Other common pellicular organelles of ciliates are bottle- or rod-shaped bodies, termed **trichocysts** (Gr. *thrix*, hair + *kystis*, bladder). These can be discharged and transformed into fine threads that may serve in anchoring the organism during feeding, in defense, or in prey capture, although none of these functions have actually been demonstrated.

Locomotion

The majority of ciliates swim by ciliary propulsion (p. 225). In the forward swimming of such forms as *Paramecium*, the entire body of the organism spirals because the cilia beat somewhat obliquely to the long axis of the body. Beating occurs in synchronized waves down the length of the body (see Fig. 11.31). Some of the most specialized ciliates, members of the subclass Hypotrichia (Fig. 24.16b), have the somatic ciliature restricted to isolated tufts of closely placed cilia located in rows or groups on the side of the body that is kept against the substratum. All of the cilia of a tuft beat together.

Some ciliates are sessile. *Stentor*, a trumpet-shaped form, often becomes attached by the tapered end and shortens by the contraction of pellicular microfilaments (**myonemes**) similar to muscle myofibrils (Fig. 24.14b). *Stentor* can also release itself and swim about. *Vorticella* has a bell-shaped body connected to the substratum by a long stalk that contains a bundle of myonemes (Fig. 24.16d). The ciliate retracts, coiling the stalk like a spring, and extends by a sudden release and popping movement. Some related forms are colonial, and the individuals of the colony are connected together by a common stalk. Some other sessile ciliates live in tubes, which are either secreted or composed of foreign material cemented together (Fig. 24.14c).

Connections 24.1

The Spatial World of Microorganisms

The world of microorganisms occupies spaces and surfaces that are largely invisible to the naked eye. Yet these microhabitats have depth and distance relative to the size of their occupants that is comparable to the habitats of mammals. The accompanying drawing represents the surface of certain marine sands off the coast of Denmark. Compared to the size of the microorganisms, the sand grains are like boulders with much living space between them. Several species of filamentous cyanobacteria form a mat over and between the sand grains. Boat-shaped diatoms slowly move through the intervening spaces. A euglenoid can be seen in the upper right quadrant. All of these producer organisms serve as a direct food source for many other organisms, especially many ciliates. The long slender ciliate, *Tracheloraphis* (left), feeds on diatoms; the large bean-shaped ciliate, *Frontonia* (top middle), and the small hypotrich ciliate, *Diophrys* (upper right), feed on the cyanobacteria. A nematode worm moves through the spaces between sand grains in the lower right.

300 μ

Nutrition

Ciliates possess a mouth, or cytostome, that opens into a short canal, the **cytopharynx**. This in turn leads into the interior, more fluid cytoplasm where a **food vacuole** is formed. Most ciliates feed on other organisms in the surrounding water. The feeding mode of a particular species is related to the size of the food particles, which range from bacteria to other ciliates. Some ciliates, often described as raptorial, capture single organisms by direct interception. Others produce a water current from which food particles are filtered. Still others collect food particles that drift by on ambient currents. The feeding mode is also obviously correlated in part to the size of the ciliate. A small ciliate may capture by direct interception prey that in large ciliates is collected by filtering currents (Fig. 24.17).

Primitively, the cytostome is located at the anterior end (Fig. 24.14a), but in most ciliates it has been displaced posteriorly to varying degrees (Fig. 24.18). Species in which the oral apparatus is located at the anterior end usually capture prey by direct interception, and this may be the primitive mode of feeding in ciliates. The mouth can be opened to a great diameter to ingest prey (Fig. 24.16a). The prey or contents of the prey pass into a food vacuole, which forms within the fluid cytoplasm (endoplasm) at the end of the cytopharynx.

Those ciliates that are specialized filter feeders possess a more complex buccal apparatus. Typically, the cytostome lies at the bottom of a buccal cavity, which contains compound ciliary organelles (Figs. 24.14b and 24.16b and e). These organelles, which constitute the buccal as opposed to the somatic ciliature, consist of two types: **undulating membranes** and **membranelles**. An undulating membrane is a long row of closely placed cilia that beat together and thus form a functional membrane (Fig. 24.16e). A membranelle is a short row of closely placed cilia that form a plate. Membranelles are typically arranged in fairly large numbers, one behind another (Fig. 24.16b). The cilia of these organelles are not fused but rather are hydrodynamically coupled, i.e., they are so close together that they tend to adhere by the viscous forces of water. The function of the buccal ciliature is to produce a feeding current and drive suspended food particles into the cytopharynx, although the precise way in which this is accomplished varies in different groups of ciliates.

In the sessile *Vorticella* and the trumpet-shaped *Stentor*, the ciliary organelles wind around the distal end of the animal and spiral down into a pit on one side (Figs. 23.14b and 23.16d). In the familiar *Paramecium*, a buccal cavity, cytostome, and cytopharynx form a funnel located at the posterior end of a lateral oral groove. The somatic cilia within the oral groove produce a feeding current that sweeps from front to back, and the compound ciliary organelles of the funnel drive food particles down into the mouth (Fig. 24.18).

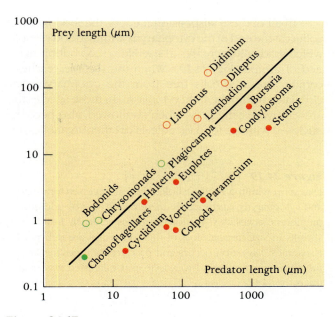

Figure 24.17

Relationship of prey size to protozoan predator size. Keep in mind that the lengths are plotted on a logarithmic scale. Red circles are ciliates; green circles are flagellates. Open circles are raptorial species; solid circles are filter feeders.

Figure 24.18

A photograph of a *Paramecium*, showing cilia, oral groove, and food vacuoles.

Carolina Biological Supply Co.

In all of these suspension feeders, the food particles collect at the end of the cytopharynx within a food vacuole, which gradually increases in size like a soap bubble at the end of a pipe (Fig. 24.19). When the vacuole reaches a certain size, it breaks free from the cytopharynx and circulates within the fluid cytoplasm. In those species that are carnivores, the prey, when swallowed, also becomes enclosed within a food vacuole. Like other organelles, the walls of food vacuoles are composed of lipid bilayer membranes, which can be synthesized or removed rapidly and recycled.

Studies on digestion in ciliates, especially species of *Paramecium*, have demonstrated that after formation, the vacuole first condenses and becomes acidic by the fusion of acidic vesicles to the vacuole membrane (Fig. 24.19). The pH drops from 7 to about 3. Then digestive enzymes

Figure 24.19

Digestion within a food vacuole. Vacuole receives food particles through cytostome and cytopharynx, and vacuole increases in size by addition of membrane from membranous vesicles. Vacuole becomes acidic by fusion of acidic vesicles. Vacuole shrinks in size through loss of fluid. Enzymes are delivered to vacuole by lysosomes. Digestion occurs and products of digestion are absorbed through membrane or removed by exocytosis. Undigested wastes are eliminated through the cytoproct.

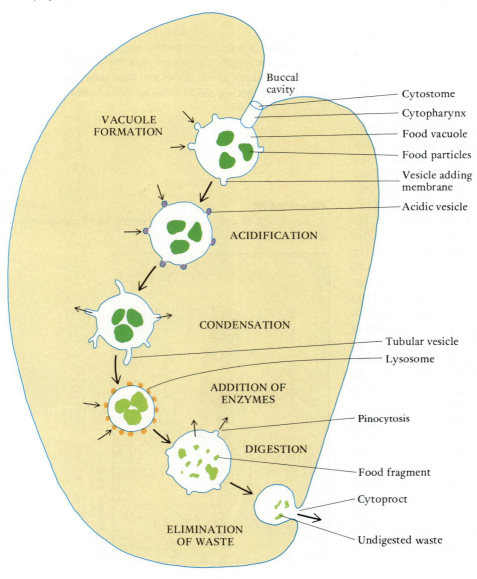

are delivered by lysosomes to the vacuole, which now expands. The pH of the vacuole contents shifts back toward neutral with the addition of enzymes. The products of digestion are absorbed from the vacuole into the surrounding cytoplasm by passage through the membrane or by exocytosis. The residual undigestible waste eventually connects with the **cytoproct**, a fixed cell anus in the pellicle. The contents of the vacuole are discharged, and the vacuole itself disappears. The life span of a food vacuole is generally about 20 minutes, and under optimum conditions a small ciliate can consume an amount of food equal to its own body volume in an hour.

About 15% of ciliates are parasitic, and many are mutualists or commensals. A number, including certain species of *Paramecium* and *Stentor*, harbor photosynthetic algae, as do some multicellular animals (see Connections 26.1, p. 612). Among the most interesting symbionts are the highly specialized species living within the stomach of some hoofed mammals, where they contribute to the digestion of cellulose (p. 244).

Within a microhabitat the assemblage of ciliate species, along with other protozoans, compose complex food webs just as do the larger multicellular plants and animals. There are microscopic producers that form the base of the food chain, and there are herbivores, carnivores, and decomposers (Fig. 24.20). Moreover, a microhabit may support a succession of species over a period of time, each occupying a particular niche and place in the food web at that particular point in the successional series (Fig. 24.21). Assemblages of certain protozoan species are often very good indicators of pollution. Their short generation time allows investigators to detect quickly population changes in response to changes in the environment.

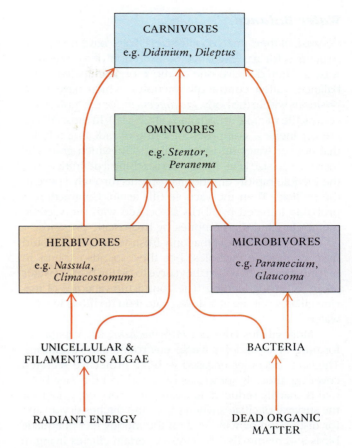

Figure 24.20

The relationship of ciliates to each other as food sources (trophic levels) in a protozoan community. All genera are ciliates except *Peranema*, which is a flagellate.

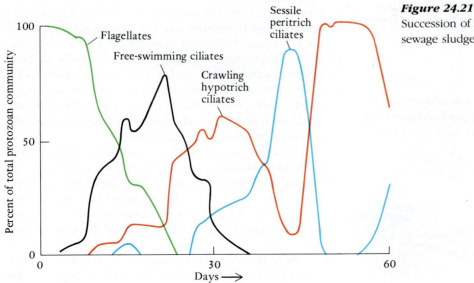

Figure 24.21

Succession of protozoans, chiefly ciliates, in sewage sludge over 60 days.

Water Balance

Because of their microscopic size, ciliates have no special structures for gas exchange or excretion of nitrogenous wastes. They do have one or more organelles for water balance, called **contractile vacuoles**, which have fixed positions within the body. *Paramecium*, for example, has a contractile vacuole at each end (Fig. 24.18). In most ciliates the organelle is composed of a ring of radiating tubules that deliver water into a central vesicle, which gradually increases in size (Fig. 24.22). On reaching a definitive size, the vacuole rapidly empties its contents through a pore in the pellicle. Then the vacuole fills again. Contraction is probably initiated by fibrils associated with the vacuole membrane and then empties by hydrostatic pressure.

One might expect that only freshwater ciliates would have contractile vacuoles, but they are also found in marine species, in which they serve to rid the body of water taken in during feeding. The contractile vacuoles of marine ciliates pulsate at a slower rate than do those in fresh water.

Most ciliates (but not *Paramecium*) are capable of forming cysts under adverse environmental conditions. The body becomes encased within a protective secreted covering; there is some loss of water; and the metabolic rate is sharply reduced. Encystment is very important for the survival of ciliates when pools and ditches dry up and for dispersion by wind and on the muddy feet of aquatic birds and mammals. The cysts of certain ciliates living in soil water films may survive for many years.

Reproduction

Ciliates are distinguished from other protozoa in possessing two types of nuclei. One type, the **macronucleus**, is large and governs the nonreproductive functions of the cell. It is highly polyploid and is the principal source of cytoplasmic RNA for protein synthesis. The shape and number of macronuclei vary greatly in different species (Figs. 24.16b and d).

Figure 24.22
Lateral view of a ciliate contractile vacuole adjacent to the unopened excretory pore. One lateral canal and associated tubules are shown on the right. Similar canals would encircle the vacuole.

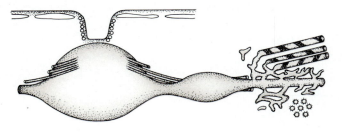

The small, round **micronuclei**, which range in number from one to 20, are typically located in the vicinity of the macronucleus. They are diploid and function in reproduction.

Asexual Reproduction

Ciliates reproduce asexually by means of transverse fission (Fig. 21.1a), which may take place every 3 to 20 hours, depending upon the species and the food supply. Each micronucleus and macronucleus undergoes division. Regeneration of organelles lost in fission is a complex process and depends largely upon replication of existing structures. In some highly specialized ciliates, such as the hypotrichs, all of the organelles are resorbed during division, and new organelles are formed from a small number of persisting "germinal" kinetosomes.

Sexual Reproduction

Sexual reproduction involves a process of **conjugation** and an exchange of nuclear material. Two individuals, called **conjugants**, meet, probably by random contact, and adhere, and their cytoplasm fuses in the region of adhesion (Fig. 24.23). The macronucleus is not involved in conjugation and is resorbed during the course of the process. All of the micronuclei undergo two meiotic divisions. Then all but one of these haploid micronuclei disappear. The remaining micronucleus divides mitotically to form two haploid nuclei: a stationary micronucleus and a wandering micronucleus. The wandering micronucleus of each conjugant migrates to the opposite conjugant and fuses with the stationary micronucleus to form a **zygote** micronucleus. The conjugants separate, and the zygote micronucleus undergoes a number of mitotic divisions to restore the number of micronuclei characteristic of the species. The macronucleus develops from a micronucleus.

Restoration of the nuclear number is commonly associated with cytosomal divisions. The process is highly variable and is best illustrated with an example. The adult *Paramecium caudatum* has one macronucleus and one micronucleus. Following conjugation the "zygote" nucleus divides three times to produce eight nuclei. Four become macronuclei and four become micronuclei. Three of the micronuclei degenerate. The remaining one divides twice in the course of two cytosomal divisions, which then provides each of the four daughter cells with one macronucleus and one micronucleus.

In some species of ciliates, including *Paramecium*, the individuals of the population belong to a number of different genetically determined mating types, and conjugation can only occur between individuals of different mating types. Adhesion of the cytoplasm apparently will not take place between individuals of the same mating type.

Note that conjugation itself does not result in any increase in the number of individuals but does provide for

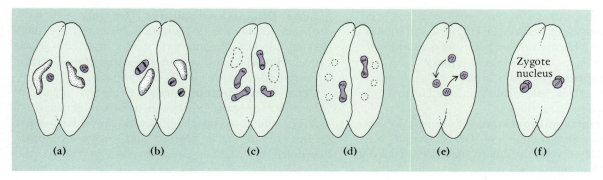

Figure 24.23
Sexual reproduction in *Paramecium*. (a–f) Conjugation. Micronuclei undergo three divisions, the first two of which are meiotic (b–c). One haploid nucleus divides; others degenerate (d). Wandering micronuclei are exchanged (e). They fuse with the stationary micronucleus of the opposite conjugant (f).

an exchange of genetic material within the population. All of the individuals that result from the asexual reproduction of an original parent without intervening conjugation are genetically identical and constitute a **clone**. Typically, a clone loses viability over time and conjugation has a rejuvenating effect on asexual reproduction. However, some species can undergo a fusion of their own haploid micronuclei (**autogamy**), which has a similar rejuvenating effect as outcrossing.

Classification of Protozoa

Phylum Sarcomastigophora (Gr. *sarx*, flesh + *mastix*, whip + *pherein*, to bear) Some 18,000 unicellular organisms having one type of nucleus and possessing flagella or pseudopodia as locomotor or feeding organelles.

  Subphylum Mastigophora Flagellates. Sarcomastigophorans possessing one or more flagella.

    Class Phytomastigophorea (Gr. *phyton*, plant) Phytoflagellates. Mostly autotrophic and usually with one or two flagella. (These organisms are also placed within the algal phyla.) *Euglena, Peranema, Volvox*, dinoflagellates.

    Class Zoomastigophorea (Gr. *zoon*, animal) Zooflagellates. Heterotrophic, with one to many flagella. Choanoflagellates, trypanosomes.

  Subphylum Opalinata (Gr. *opalus*, opal) Multiflagellated (ciliated) gut parasites of fish and amphibians possessing numerous similar nuclei. Mouthless. Produce flagellated gametes.

  Subphylum Sarcodina (Gr. *sarx*, flesh) Sarcodines. Sarcomastigophorans possessing pseudopodia.

    Superclass Rhizopoda (Gr. *rhiza*, root + *pous*, foot). Pseudopodia used for both locomotion and feeding.

      Class Lobosea (Gr. *lobos*, lobe) Shelled and naked amoebas. Pseudopodia large or straplike. *Amoeba, Difflugia, Arcella*.

      Class Granuloreticulosea (L. *granulum*, granule + *reticulum*, reticulum) Foraminiferans. Chiefly marine species with mostly multichambered shells and strandlike branching pseudopodia. *Globigerina, Nonion, Nummulites*.

    Superclass Actinopoda (Gr. *aktinos*, ray + *pous*, foot) Floating or sessile sarcodines with radiating rodlike pseudopodia (axopods) used in feeding only.

      Classes Polycystinea and Phaedarea Radiolarians with siliceous skeleton. *Thalassicola, Aulacantha*.

      Class Acantharea (Gr. *akanthos*, thorn) Radiolarians with a skeleton of strontium sulfate. *Acanthometra*.

      Class Heliozoea (Gr. *helios*, sun + *zoon*, animal) Heliozoans. Mostly marine and freshwater sarcodines lacking a central capsule. *Echinosphaerium*.

Phylum Apicomplexa (L. *apicalis*, apex + *complexus*, complex) Sporozoans. Some 3900 parasitic protozoans having an apical complex of organelles. *Plasmodium*.

Phylum Microspora (Gr. *mikros*, small + *sporos*, seed) About 1000 parasitic protozoans lacking an apical complex of organelles but having a coiled polar filament within the infective sporelike stage. *Nosema.*

Phylum Ciliophora (L. *cilium*, eyelash + Gr. *pherein*, to bear) Ciliates. About 7200 ciliated unicellular organisms possessing two types of nuclei.

> **Class Kinetofragminophorea** (Gr. *kinein*, to move + L. *fragmentum*, fragment + Gr. *pherein*, to bear) Ciliates lacking compound ciliary organelles in the oral region. *Didinium, Prorodon.*

Class Oligohymenophorea (Gr. *oligos*, few + *hymen*, membrane + *pherein*, to bear) Ciliates with a small number of compound ciliary organelles in the oral region. Such organelles are often hidden. *Paramecium, Vorticella, Tetrahymena.*

Class Polyhymenophorea (Gr. *polys*, many + *hymen*, membrane + *pherein*, to bear) Ciliates with a large number of conspicuous compound ciliary organelles in the oral region. *Spirostomum, Stentor,* hypotrichs.

Summary

1. Protozoans are unicellular or colonial organisms belonging to various protoctistan phyla. Most species are motile and heterotrophic, which accounts for their traditional placement in the animal kingdom. Protozoa are found in the sea, in fresh water, and in water films around soil particles, and there are many symbiotic species.

2. The phylum Sarcomastigophora contains protozoa that have only one type of nucleus and possess flagella or pseudopodia as locomotor or feeding organelles. The flagellate members are included in the subphylum Mastigophora and are divided into phytoflagellates and zooflagellates. The phytoflagellates are mostly biflagellated chlorophyll-bearing autotrophs and include euglenids and dinoflagellates. Zooflagellates bear one to many flagella and are heterotrophic. Although there are some free-living species, most are commensal or parasitic; the trypanosomids are of the greatest medical importance. Asexual reproduction is by longitudinal fission.

3. The subphylum Sarcodina includes sarcomastigophorans that possess pseudopodia. Although pellicular organelles are not highly developed, skeletons are a characteristic feature of most groups. Sarcodines reproduce asexually by binary fission; sexual reproduction is by fusion of isogametes but is poorly understood in some groups.

4. Amoebas and foraminiferans use their pseudopodia for both feeding and locomotion. The amoebas, which are mostly inhabitants of fresh water, are either shell-less or possess shells of secreted or foreign materials and have straplike or large blunt pseudopodia. Foraminiferans are largely marine, and most possess chambered, calcareous shells. The delicate, branching pseudopodia form a food-trapping net.

5. Radiolarians and heliozoans possess radiating, rodlike pseudopodia (axopods), which are used only in feeding. Radiolarians are planktonic marine sarcodines with well-developed skeletons of strontium sulfate or silicon dioxide. The mostly freshwater and marine heliozoans are free or attached and may or may not possess skeletons.

6. The phylum Apicomplexa consists of sporozoans, parasitic protozoans characterized by certain apical organelles and a spore-like stage. The life cycles of sporozoans are complex, with alternating asexual and sexual stages. Sporozoans parasitize both invertebrates and vertebrates, and some require two hosts. The coccidian apicomplexans, which live within blood or gut cells of their hosts, include most of the economically and medically important species, such as *Plasmodium*, the causative agent of malaria.

8. Ciliates, members of the phylum Ciliophora, possess complex organelles, especially as part of the pellicle (outer layer of the body). Cilia are used for swimming and in some organisms for feeding, and all ciliates possess an infraciliature of kinetosomes and connecting fibrils. Some ciliates are predatory, and others are suspension feeders. Food is ingested through a cytostome and cytopharynx; digestion occurs within a food vacuole. Contractile vacuoles provide for water balance. Ciliates possess two types of nuclei: the large polyploid macronucleus functions in cellular regulation; the one to many small micronuclei are involved in sexual reproduction. Following meiotic division the haploid micronuclei function as gametes and are exchanged during conjugation. The zygote micronucleus in each member of the pair then divides to form the number of macro- and micronuclei characteristic of the species.

References and Selected Readings

Accounts of the protozoa can be found in some of the references cited at the end of Chapter 23. The parasitology texts listed at the end of Chapter 27 also cover the parasitic protozoa. The references cited below are devoted exclusively to protozoa.

Anderson, O.R. *Comparative Protozoology*. New York: Springer-Verlag, 1987. A general biology of the Protozoa.

———. *Radiolaria*. New York: Springer-Verlag, 1983. A general account of the radiolarians.

Corliss, J.O. *The Ciliated Protozoa: Characterization, Classification, and Guide to the Literature*, 2nd ed. New York: Pergamon Press, 1979.

Farmer, J.N. *The Protozoa: Introduction to Protozoology*. St. Louis: C.V. Mosby Co., 1980. A textbook of protozoology with more emphasis on classification than that of Sleigh.

Fenchel, T. *Ecology of Protozoa*. Madison, Wisc.: Science Tech-Publishers, 1987. An excellent account of many aspects of protozoan physiology and ecology.

Hawking, F. The clock of the malaria parasite. *Scientific American* 222(Jun. 1970):123–131. The controlling mechanisms in the cyclical development of *Plasmodium*.

Jahn, T.L., E.G. Bovee, and F.F. Jahn. *How to Know the Protozoa*, 2nd ed. Dubuque, Iowa: W.C. Brown Co., 1979.

Lee, J.J., S.H. Hutner, and E.C. Bovee (eds.). *An Illustrated Guide to the Protozoa*. Lawrence, Kans.: Society of Protozoologists, 1985. An excellent introduction and guide to the protozoan groups.

Sleigh, M.A. *Protozoa and Other Protists*. New York: Routledge, Chapman and Hall, 1989. A good, well-balanced general account of the protozoa.

25

Sponges

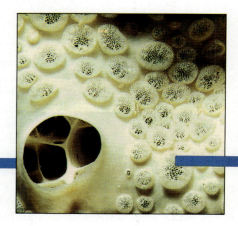

Characteristics of Sponges

■ Sponges are sessile marine or freshwater animals that are radially symmetrical or irregular in shape.

■ They have no organs, head, mouth, or gut cavity. Their body structure is organized around a system of canals and chambers through which water flows.

■ Flagellated collar cells that line the chambers not only create the water currents but also filter out fine food particles.

■ Support is provided by internal siliceous or calcareous spicules and/or spongin fibers.

■ Sponges are either hermaphroditic or the sexes are separate. Fertilization is internal and development leads to a free-swimming flagellated larva.

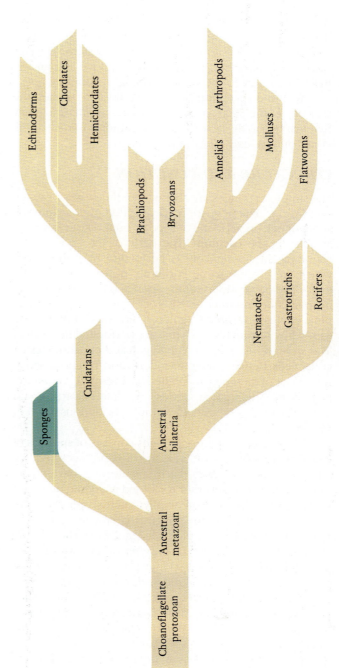

Echinoderms
Chordates
Hemichordates
Brachiopods
Bryozoans
Annelids
Arthropods
Molluscs
Flatworms
Nematodes
Gastrotrichs
Rotifers
Cnidarians
Sponges
Ancestral bilateria
Ancestral metazoan
Choanoflagellate protozoan

(Left) Purple tube sponge with orange encrusting sponge in the background.
(Above) Sponge surface showing two oscula with intervening incurrent pores.

Sponges, members of the phylum **Porifera**, are believed to be the most primitive of the large animal phyla. They lack organs and most of the typical animal features. There is no head or anterior end; there is no mouth or gut cavity; and the body is immobile. Moreover, they lack muscle cells and neurons. Indeed, not until 1765 were sponges generally recognized as being animals.

Although sponges possess primitive features, they are also specialized in many ways. Their sessile, or attached, condition is a specialization, and the body structure is built around a system of water canals and chambers. The physiology of sponges is dependent upon the flow of water through the body, and it is this water flow that accommodates the lack of organs. The evolution of sponges is shaped in part by flow dynamics. The system of water canals and chambers, an architecture unique among animals, is the key to the understanding of sponges.

With the exception of three freshwater families, the majority of the 5000 species of sponges are marine. They live attached to the bottom, most commonly to rock, shell, coral, pilings, and other hard surfaces.

STRUCTURE AND FUNCTION

The simplest sponges are shaped like little vases or tubes (Fig. 25.1). The interior cavity, called the **atrium** or **spongocoel**, opens to the outside through a large opening at the top, the **osculum**. The body of the sponge surrounding the spongocoel is perforated by pores, from which the phylum name Porifera ("pore-bearer") is derived. The outer surface of the sponge body is covered by the **pinacoderm**, which consists of a layer of flattened cells, **pinacocytes** (Gr. *pinax*, tablet + *kytos*, cell). Pores are formed by **porocytes**, cells that are perforated like a ring. The spongocoel is lined with flagellated **choanocytes** or collar cells. Between the pinacoderm and the choanocytes lies a layer of mesenchyme, called the **mesohyl**, containing amoeboid cells (amoebocytes) of different types and skeletal pieces embedded within a gelatinous protein matrix.

The skeleton of most sponges consists of **spicules** of calcium carbonate or silicon dioxide secreted by amoebocytes (Fig. 25.2). Each spicule may be a single needle-like ray or several rays joined at certain angles to each other. They are generally microscopic and unconnected, but in some sponges the spicules are fused together to form long strands (see Fig. 25.10). The size and shape of the spicule and the number of rays present vary greatly and each species of sponge possesses a characteristic combination of several different spicule types. Spicule structure is an important characteristic in the classification of sponges. Spicules have provided a long fossil record for sponges. The Burgess Shale, for example, contains spicules from two classes of sponges.

Many sponges possess an organic skeleton of coarse spongin fibers (Fig. 25.3), a substance similar to collagen.

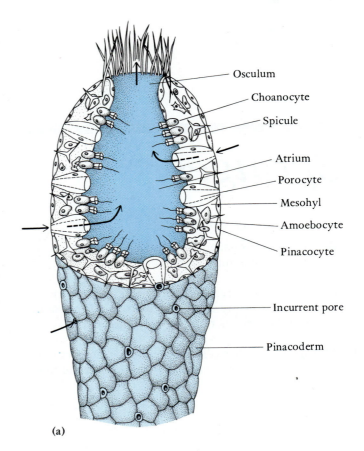

Osculum
Choanocyte
Spicule
Atrium
Porocyte
Mesohyl
Amoebocyte
Pinacocyte
Incurrent pore
Pinacoderm

(a)

(b)

Figure 25.1

(a) Diagram of a partially sectioned asconoid sponge. Arrows indicate direction of water flow. (b) *Tedania ignis*, fire sponge. This bright red sponge is a common shallow water species in the Caribbean. If handled, it causes in some people a skin reaction like that produced by poison ivy.

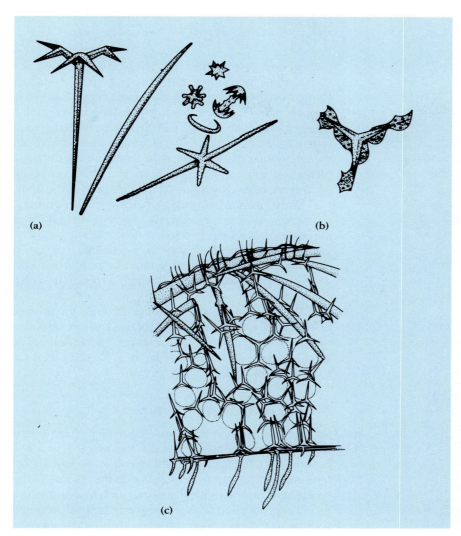

Figure 25.2
Sponge spicules. (a) A few different types, showing variations in shape and size. (b) Secretion of a calcareous triradiate spicule by amoebocytes. (c) Spicules in their natural position within a calcareous leuconoid sponge. The dotted circles are the location of flagellated chambers.

Some sponges have spongin fibers instead of spicules, and it is from such species that commercial sponges were once obtained (most commercial sponges now are synthetic). Many sponges have a skeleton of both spongin fibers and spicules. The skeleton, whether of different types of spicules or of spongin, is highly organized with regard to other features of sponge structure (Fig. 25.2c) and plays an important role in supporting the internal water canals and resisting deformation by surge and strong tidal currents.

Although some sponges are brown, many sponges are brightly colored red, yellow, green, blue, or purple (Fig. 25.1b). The color is derived from pigment located within the amoebocytes. Many of the freshwater species are green, due to symbiotic algae living within the cells. Such sponges often form large masses on stones or around submerged branches of fallen trees.

Water Flow and Feeding

The flagellum of the choanocytes drives water away from the cell (Fig. 25.1a). Since these cells line the atrium, they cause water to be sucked in through the pores and to be

Figure 25.3
Spongin fibers (they appear translucent in the photograph).

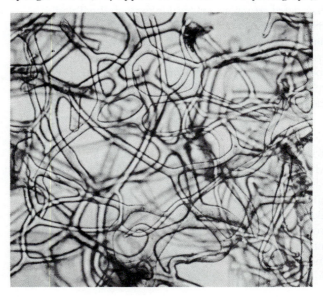

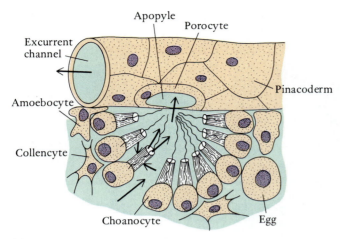

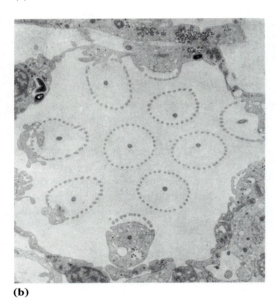

(b)

Figure 25.4
Flagellated chambers. (a) Section through flagellated chamber of freshwater sponge, *Ephydatia*. Arrows indicate direction of water currents. (b) Electron micrograph of a cross section through a flagellated chamber of the boring sponge *Cliona lampa*. The rings are cross sections of the collars of choanocytes, showing the circles of microvilli and the central flagellum.

driven up and out of the spongocoel through the osculum, which is large and offers less resistance to water flow.

Sponges are filter feeders, removing bacteria and other very fine suspended organic matter from the water. Eighty percent of the organic particles collected by some tropical sponges has been found to be invisible to a light microscope. An initial screening is provided by the pores and passageways, which permit only very small particles to enter. Some particles are engulfed by cells along the

passageways; very fine particles are collected by choanocytes, becoming trapped on the collar surface formed by the microvilli (0.1 μ apart) (Fig. 25.4b). The trapped particle passes down to the base of the collar, where it is engulfed by the cell. Strangely, evidence indicates that particles, even the vacuoles, are transferred from the engulfing cell to an amoebocyte at least for the completion of intracellular digestion. The products of digestion are passed by diffusion to other cells of the body.

Many marine sponges harbor photosynthetic symbionts, most commonly cyanobacteria. One-third of the body mass of some sponges consists of their symbiotic cyanobacteria. Many freshwater sponges contain the green alga *Chlorella*. All of these symbionts produce excess photosynthate, which is shared with the sponge host (see Connections 26.1, p. 612).

The current of water passing through the sponge body not only provides a source of food material but also functions as a system for gas exchange, for the removal of wastes, and for the transfer of gametes. No cell is very far from some part of the water stream.

Grades of Sponge Structure

Sponges with the simple vaselike structure are called **asconoid** sponges and are small, not more than a few centimeters tall (Figs. 25.1 and 25.5a). The asconoid structure imposes limitations in size, for as the volume of the spongocoel increases, the flagellated surface area does not increase proportionally. Consequently, a large asconoid sponge would contain more water than its choanocytes could efficiently move. In the evolution of sponges this problem was solved by repeated folding of the flagellated layer to increase its surface area. The first stage of folding is exhibited by **syconoid** sponges, in which the flagellated layer is evaginated outward into finger-like projections (Fig. 25.5b). The evaginations are called **flagellated canals**, and the corresponding invaginations of the external surface are called **incurrent canals**. Pores are located between the incurrent and flagellated canals. The flagellated canals of syconoid sponges open into a central atrium devoid of choanocytes.

In the great majority of sponges the surface area of the flagellated layer has been increased by the formation of many small chambers within which the choanocytes are located (Figs. 25.4, 25.5c, and 25.6). In these **leuconoid** sponges, water enters dermal pores on the body surface (formed by porocytes) and passes through a system of incurrent canals, eventually reaching the flagellated chambers. Many leuconoid sponges have no atrium; water leaves the body through converging excurrent canals opening to the exterior through an osculum. The development of flagellated chambers greatly increases the water-moving ability of a sponge. The number of flagellated chambers can range from 10,000 to 18,000 per mm³, and

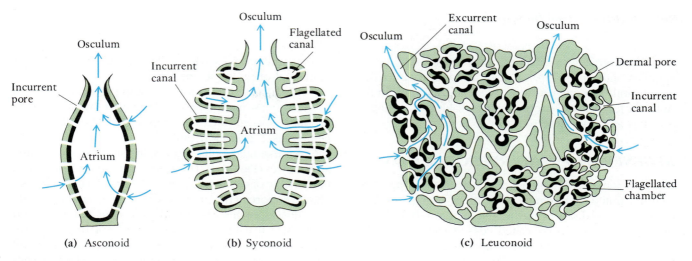

(a) Asconoid

(b) Syconoid

(c) Leuconoid

Figure 25.5

Types of sponge structure. (a–c) The three structural types of sponges. In each, the choanocytes are shown in black. Light arrows indicate the direction of water flow; heavy arrows indicate the exhalant flow from the osculum.

Figure 25.6

Scanning electron micrograph of a section taken from the freshwater sponge *Ephydatia fluviatilis*. In this section a number of flagellated chambers (FC) with choanocytes (Ch) surround a large current canal (eC) with a wall formed by pinacocytes (PC). The apopyles (aP), or openings, from the flagellated chambers pass through porocytes (P). Within the mesohyl can be seen archeocytes—a type of amoeboid cell—(A), spicules (S), and spongin (Sp).

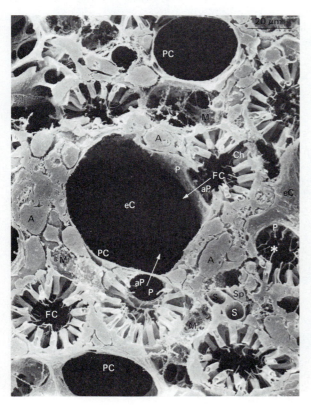

one leuconoid sponge 10 cm high × 1 cm in diameter was found to possess some 2,250,000 flagellated chambers, and could pass 22.5 L of water through its body in 24 hours.

Within the system of canals and chambers water always flows in the direction of the larger opening. Moreover, the flagella of the choanocytes are directed toward the larger opening of the flagellated chamber. The speed of the water current is very slow through the chambers because the total cross-sectional area of all of the chambers served by an incurrent canal is very large. The flow speed greatly increases as the water is carried in converging channels with decreasing cross-sectional area toward the osculum.

Most leuconoid sponges are irregular in shape, with many oscula located over the body surface (Figs. 25.5c). They attain far greater size than do asconoid and syconoid forms because any increase in bulk includes the addition of a large number of flagellated chambers necessary to drive the water current. The larger leuconoid sponges would more than fill a bushel basket, and some species reach over a meter in height. Most of the sponges commonly encountered in shallow water are leuconoid.

The shape of leuconoid sponges is related in part to the use of the substrate (Fig. 25.7). Erect sponges utilize relatively little substrate and their elevated oscula, often projecting like chimneys, enhance the exhalant water flow. Encrusting sponges, although requiring more substrate surface area, can utilize vertical and overhanging rock walls or confined spaces between and beneath stones.

The defenses of sponges appear to be largely chemical. Toxic substances produced by some species are known to reduce settling by the larvae of other animals or to inhibit grazing predators. Over 50% of Antarctic sponges and 64 to 75% of tropical sponges are toxic. Significantly, sponge-eating sea stars and sea slugs in the Antarctic feed mostly on nontoxic or weakly toxic sponges. Spicules are not a deterrent to predators.

REGENERATION AND REPRODUCTION

The truly remarkable regenerative powers of sponges are well illustrated by forcing a small bit of sponge through bolting silk and dissociating the cells. Within a short period of time, clusters of dissociated cells reaggregate in the proper relationship. Immunocompetence, i.e., the ability to distinguish between self and nonself, has been demonstrated in some sponges. In such species, a sponge will accept a graft from itself but will reject a graft from another individual of the same species.

Asexual reproduction by budding is not common in sponges, but some sponges, especially freshwater species, produce special asexual reproductive bodies, called **gemmules** (L. *gemmula*, little bud), consisting of an aggregate of essential cells, especially amoebocytes called archeocytes (Fig. 25.8). These amoebocytes are totipotent, i.e., capable of giving rise to any other type of cell, and they play an important role in sponge regeneration and growth.

Most sponges are hermaphroditic, but eggs and sperm are usually produced at different times. Sperm originate from choanocytes; the origin of eggs is still uncertain. Both develop within the mesohyl; they are not located within a gonad. Surprisingly, external fertilization

Figure 25.7

Relationship of leuconoid sponge form to utilization of substratum. The two massive sponges at the top right of the rock require an exposed surface, but their elevated form enables them to utilize water well above the substratum, and their attachment area is a relatively small part of the total body surface area. The encrusting sponges below the rock utilize much of their surface area for attachment, but their low encrusting form enables them to exploit the space of crevices and other confined areas. The sponge on the vertical surface at left utilizes space within the substratum. Small arrows indicate the movement of water in the sponge, large arrows indicate the exit of water from oscula.

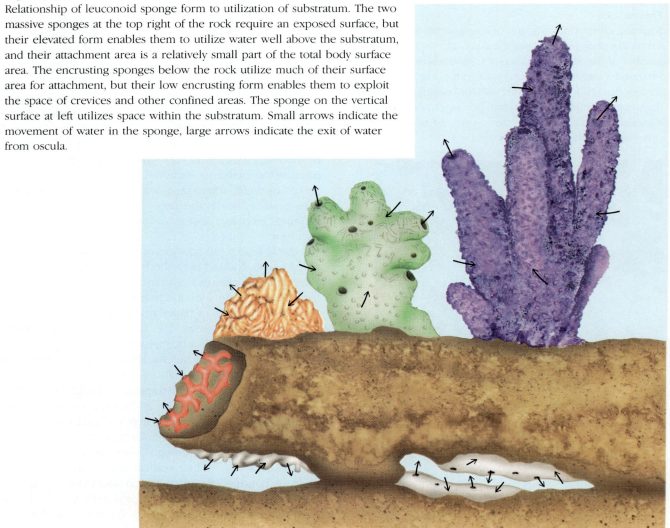

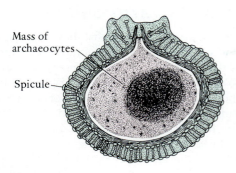

Figure 25.8
Section through a gemmule of a freshwater sponge.

Mass of archaeocytes

Spicule

does not occur in sponges; it always take place in situ, i.e., within the body of the sponge. Mature sperm are shed into the water canals and exit through the osculum. Some are carried into the water canals of neighboring sponges. There sperm are trapped and engulfed by choanocytes, which then move to an adjacent ripe egg. The carrier cell plasters itself against the egg and transfers the sperm. Fertilized eggs may be released into the water canals, carried out with the water stream, and undergo development in the sea water, but in most sponges the eggs are brooded and develop within the parental mesohyl.

Embryonic development leads to a free-swimming larva (Fig. 25.9), a stage that is important for species dispersal in sessile animals. In most species, the larval condition is reached at the blastula stage. The flagellated larva exits through the water canals, and after a brief free-swimming existence, settles to the bottom and develops into an adult sponge.

Connections 25.1   *Marine Borers*

Many different marine animals have evolved the ability to drill into hard substrates—shell, coral, limestone, sandstone, compact clay, and wood. In most cases the excavation provides a home for the borers, protecting them from most surface predators. However, some species, such as certain snails and octopods, are themselves predators and use their drilling ability to penetrate the calcareous shells of their prey. Some marine borers excavate mechanically, some chemically, and some by a combination of mechanical and chemical means.

Given the immobility of sponges, it may be surprising that the members of a common and widespread family (Clionidae) excavate tunnels in calcareous substrates—shell, coral, and coralline rock—which then become filled with the sponge body.

Dermal pores and oscula are located where the tunnels come to the surface (see Fig. 25.7, *left side*). The boring is a chemical process, in which chips of calcium carbonate are etched out by single sponge cells at the head of the boring (Fig. a). By secretion the margins of the cell cut down into the calcium carbonate and then toward each other until the cell completely embraces the chip, which is then eliminated through the excurrent canals and osculum. Boring sponges play an important role in breaking down coral and shell. Any shell-laden beach will contain many specimens perforated by boring sponges (Fig. b).

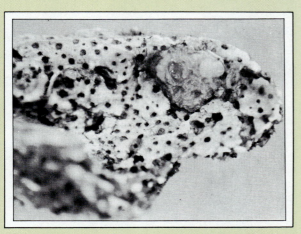

(a) Remains of a clam shell that has been riddled with boring sponge.

(b) Cacareous surface from which two chips have been removed; four more are partially etched.

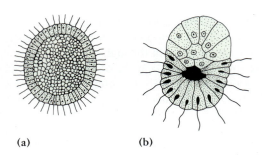

(a)   (b)

Figure 25.9
Sponge larvae. In the type shown in (a) (parenchymula larva), all of the external cells are flagellated. The larva shown in (b) (amphiblastula larva) has only one hemisphere of flagellated external cells.

Classification of Phylum Porifera

Class Calcarea (L. *calcarea*, limestone) Calcareous sponges. Spicules are usually separate and composed of calcium carbonate. Body form is asconoid, syconoid, or leuconoid. Calcareous sponges are largely inhabitants of shallow water and many species are small, not exceeding 10 cm in height. *Leucosolenia* and *Sycon*.

Class Hexactinellida (Gr. *hexe*, six + *aktinos*, ray) or Hyalospongia (Gr. *hyalos*, glass + *spongos*, sponge) Glass sponges. Spicules are siliceous and always include a six-pointed type. Some spicules are commonly fused together to form long strands and a highly organized skeleton. The flagellated layer has a somewhat synconoid organization. Both the mesh-like surface layer and the layer of choanocytes are syncytial, a distinctive hexactinellid feature. Many glass sponges are cup-, vase-, or urnlike in shape (Fig. 25.10), reaching a height of 10 to 100 cm. Most hexactinellid sponges occur in deeper water (200 to 2000 m) than do most other sponges and are thus less frequently encountered. *Euplectella* (Venus's flower basket).

Class Demospongiae (Gr. *demos*, bond + *spongos*, sponge) This class contains the great majority of sponge species (over 90%) and includes most of the commonly encountered sponges. The skeleton of Demospongiae is composed of separate siliceous spicules. But some species, e.g., the commercial sponges, possess a skeleton of spongin fibers and many members of the class possess both spongin fibers and siliceous spicules. The body structure is always leuconoid, and an irregular symmetry with many oscula is common. However, there are many large species shaped like baskets, vases, or tubes. Many Demospongiae are brightly colored. The small number of freshwater sponges belong to this class as well as the widespread marine boring sponges (see Connections 25.1, p. 591). *Haliclona, Microciona, Verongia, Cliona*; the freshwater sponges, *Ephydatia* and *Spongilla*.

Class Sclerospongiae (Gr. *skleros*, hard + *spongos*, sponge) A small group of tropical sponges with siliceous spicules and spongin fibers but encased within or resting upon a solid external skeleton of calcium carbonate. Body structure is leuconoid. Found within marine caves and tunnels associated with coral reefs.

EVOLUTIONARY RELATIONSHIPS

Although sponges are primitive in that they lack organs, including a gut, and have only a small number of different kinds of cells, they are highly specialized animals in other respects. The specializations are largely adaptations to a stationary mode of existence—the absence of an anterior end; the circulation of water through the body for filter feeding, gas exchange, and water removal; and the condition of hermaphroditism. Certainly sponges diverged early in the evolution of the animal kingdom, probably from some gutless, radially symmetrical, but motile metazoan ancestor. The many specializations of sponges make it highly unlikely that they gave rise to any other groups of animals.

Figure 25.10
Photograph of the siliceous skeleton of the glass sponge (hexactinellid) *Euplectella*. The spicules are fused to form intersecting girders.

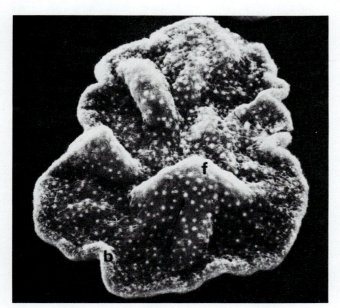

Figure 25.11
A scanning electron micrograph of the dorsal surface of
Trichoplax adhaerens. The labels b and f refer to a marginal
bulge and a fold, which characterize living specimens. The
arrow points to a lipid globule. Some cilia can also be seen.

ANOTHER PRIMITIVE PHYLUM

The phylum Placozoa contains a single species of a strange
marine animal called *Trichoplax adhaerens* that has been
found in many parts of the world. The flattened, irregular
body is 2–3 mm across and looks like a giant amoeba (Fig.
25.11). Monociliated epithelial-like cells cover the surface
and enclose an internal tissue rather like mesenchyme.
Trichoplax creeps over the surface with its cilia and feeds
on algae and bacteria, which are digested extracellularly. It
reproduces asexually by fission and also produces sperm
and eggs.

Except for its asymmetrical shape, *Trichoplax* certainly seems primitive. Like sponges, it may have evolved
from some very early metazoan stock.

Summary

1. Sponges, members of the phylum Porifera, are
sessile, aquatic animals, largely marine. The body structure
is built around a system of water canals, a specialization
correlated with their sessile nature. Sponges are primitive
in lacking a head or anterior end, a mouth and gut, and
other organs, and in having a low level of cell differentiation.

2. The skeleton of sponges is composed of coarse
spongin fibers or needle-like spicules of calcium carbonate or silicon dioxide, or of both spongin fibers and
siliceous spicules.

3. The most primitive sponges are vase-shaped
(asconoid form), with a central cavity or atrium. The body
wall is composed of an outer layer of pinacocytes; a middle
layer, or mesohyl, containing amoebocytes and the skeleton; and an inner layer of flagellated collar cells, or
choanocytes. The beating of the flagella of the choanocytes
sucks water into the atrium through pores in the body wall
and expels it from the atrium through a distal opening, the
osculum.

4. The surface area of the flagellated layer has been
increased in the evolution of sponges by folding of the
body wall (syconoid form) or by the formation of many
minute chambers to which the flagellated cells are
confined (leuconoid form). Most sponges are leuconoid,
and many lack atria and have an irregular shape.

5. Sponges are dependent upon the stream of water
flowing through the body for food, gas exchange, and
waste removal. Sponges are filter feeders on fine, suspended organic particles, the microvilli of choanocytes
being the final filter.

6. Most sponges are hermaphroditic, and eggs are
fertilized by sperm brought into the body by the water
stream. The eggs are usually brooded to the larval stage
within the parent body. The flagellated larval stage exits
through the water canals and, after a planktonic existence,
settles to form an adult sponge.

References and Selected Readings

Detailed accounts of sponges may also be found in the
references listed at the end of Chapter 23.

Bergquist, P.R. *Sponges*. London: Hutchinson & Co., 1978.
Excellent coverage of all aspects of the biology of
sponges.

Simpson, T.L. *The Cell Biology of Sponges*. New York:
Springer-Verlag, 1984. A detailed account of the structure, physiology, and reproduction of sponges.

Vogel, S. Organisms that capture currents. *Scientific American* 239(Aug. 1978):128–135. A description of the
way that sponges and some other animals utilize
ambient water currents.

26

Cnidarians

Characteristics of Cnidaria

◼ Members of the phylum Cnidaria are mostly marine animals that are free-swimming (medusa, or jellyfish) or live attached (polyp).

◼ The body is radially symmetrical, with the mouth and surrounding tentacles located at one end of the radial axis. The mouth is the only opening into the digestive cavity.

◼ There are few organs, and the body wall is composed of two principal layers: an outer epidermis and an inner gastrodermis, separated by the mesoglea, a membrane or a jelly-like layer.

◼ Explosive stinging or adhesive cells, called cnidocytes, that are used in prey capture are unique to the phylum.

◼ The nervous system is commonly in the form of a net with receptor cells dispersed over the body surface.

◼ The gonads are only aggregations of developing gametes, and there are no gonoducts. Fertilization is usually external, and development leads to a free-living planula larva.

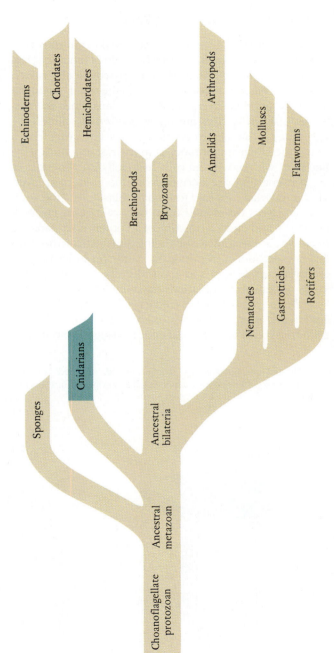

(Left) Strawberry anemone, Corynactis californica. *(Above) A jellyfish from the Antarctic.*

595

The phylum Cnidaria, formerly called Coelenterata, contains the familiar hydras, sea anemones, corals, and jellyfish.* They are radially symmetrical animals, with one end of the body bearing the mouth and tentacles. Some live attached, with the mouth directed upward; others are free-swimming, with the mouth directed downward. This chapter will first explain the general structure and physiology of cnidarians and will then examine the three cnidarian classes. We will see how the free-swimming and attached life styles have been exploited in each class and how the attached condition may have evolved.

STRUCTURE AND FUNCTION

Cnidarians are radially symmetrical, and the oral end of the axis terminates in the mouth and a circle of tentacles (Fig. 26.1). The mouth opens into a blind **gastrovascular cavity**. A layer of cells, the **epidermis**, covers the outer surface of the body; another layer, the **gastrodermis**, lines the gastrovascular cavity. The **mesoglea** (Gr. *mesos*, middle + *glia*, glue), located between these two layers, varies from a thin noncellular membrane, as in hydras, to a thick jelly-like layer with or without cells, as in jellyfish. Where cells are present, they are derived from the epidermal layer. Thus, cnidarians are said to be **diploblastic**, i.e., the adult cellular layers developed from only two germ layers—ectoderm and endoderm. Almost all other metazoans are triploblastic, the adult tissues being derived from the germ layers—ectoderm, endoderm, and mesoderm. The cnidarian epidermis and gastrodermis contain several kinds of cells, but cnidarians have few organs.

There are two types of cnidarian body form. **Polypoid** cnidarians, such as hydras and sea anemones, have a cylindrical body with the oral end (bearing the mouth and

tentacles) directed upward and the aboral end attached to the substratum (Fig. 26.1a).

Medusoid cnidarians, such as jellyfish (Fig. 26.1b), have bell- or saucer-shaped bodies with the aboral end convex and directed upward and the oral end concave and directed downward. Medusae are usually free-swimming.

Movement

Although many cnidarians such as hydras, corals, and sea anemones are sessile, the body and tentacles of cnidarians can be extended or contracted and bent to one side or the other. Movement is brought about by the contraction of longitudinal and circular muscle fibers. However, the fibers are located not in true muscle cells but in basal extensions of epidermal and gastrodermal cells called **epitheliomuscle cells** and **nutritive muscle cells**, respectively (Fig. 26.2). The contractile layers are variously developed in different cnidarians. The gastrodermal muscles of hydras, for example, are poorly developed, and movement is largely the result of contraction of the epidermal cells, operating against fluid within the gastrovascular cavity (a hydrostatic skeleton).

Nutrition and Nematocysts

Most cnidarians are carnivorous. Small forms, such as hydras and corals, feed upon planktonic organisms, but the larger jellyfish and sea anemones can consume small fish and clams. **Gland cells** lining the gut secrete proteolytic enzymes that rapidly digest the prey (Fig. 26.2). Mixing of the gut contents is aided by the beating of the cilia of the gastrodermal cells. Small fragments of tissue are then engulfed by nutritive muscle cells, and digestion is completed intracellularly. Undigestible waste materials are ejected through the mouth.

Cnidarians capture their prey with the aid of special stinging cells, called **cnidocytes**, located largely in the

*The old phylum Coelenterata, strictly defined, included both the Cnidaria and the Ctenophora, each of which is now placed in a separate phylum.

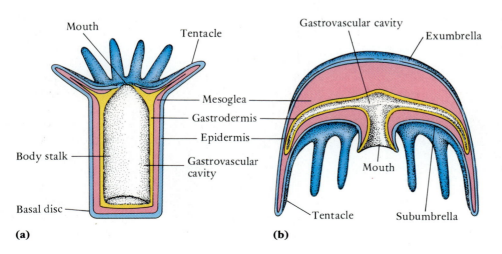

Figure 26.1

(a) Polypoid body form;
(b) medusoid body form.

(a) Mouth, Tentacle, Mesoglea, Gastrodermis, Epidermis, Body stalk, Gastrovascular cavity, Basal disc

(b) Gastrovascular cavity, Exumbrella, Mouth, Tentacle, Subumbrella

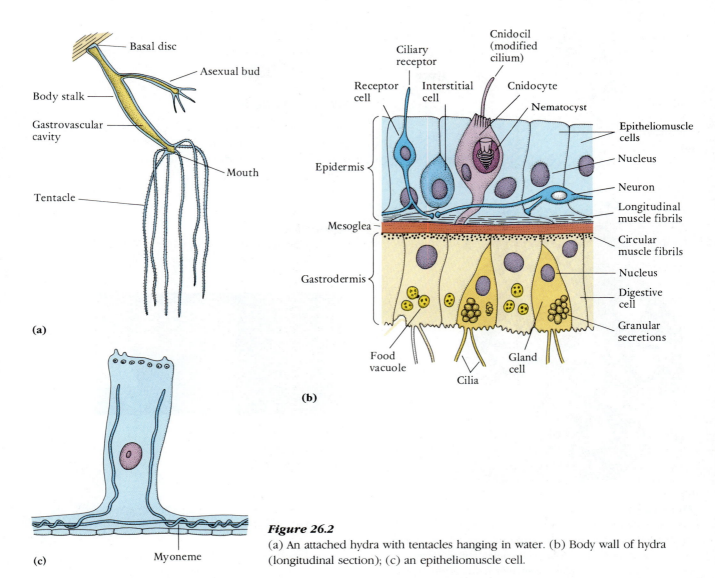

(a)

(b)

(c)

Figure 26.2

(a) An attached hydra with tentacles hanging in water. (b) Body wall of hydra (longitudinal section); (c) an epitheliomuscle cell.

epidermal layer (Fig. 26.2). It is from these cells that the name Cnidaria (L. *cnide*, nettle) is derived. A cnidocyte contains a surface cilium or modified cilium (the **cnidocil**) and a **nematocyst** (Gr. *nematos*, thread + *kystis*, bladder), the actual stinging element. The undischarged nematocyst is composed of a capsule and a long thread coiled within the capsule (Fig. 26.3). When discharged, the nematocyst is expelled from the cnidocyte, and the thread is everted out of the bulb in the process (the thread turns inside out). The mechanism of discharge is not completely understood, but it is believed that stimulation by the prey changes the osmotic pressure within the bulb so fluid rushes into the interior. The elevated fluid pressure both everts the thread and drives the nematocyst from the cnidocyte. Although many cnidocytes function as independent effectors, some may be innervated by motor neurons.

Some types of nematocysts function by entanglement; others are driven into the body of the prey and the open end of the thread may inject a protein toxin. It is these toxic

penetrants that produce the sting of jellyfish and other cnidarians. Meat tenderizer, which contains proteins, can be an effective remedy for cnidarian stings. Nematocysts are commonly ornamented with spines. Fine spines along the thread surface function to grip and entangle the prey (Fig. 26.4a). In hydras the heavy spines at the base of the nematocyst puncture the prey as the spines unfold in the course of nematocyst eversion (Fig. 26.4b). The puncture facilitates the penetration of the toxin-injecting thread. Each species of cnidarian typically possesses several types of nematocysts, which are closely correlated with the kind of prey upon which the species feeds. Hydras, for example, have four types of nematocysts. In addition to cnidocytes containing nematocysts, sea anemones have special adhesive spirocysts.

Cnidocytes are typically located between or embedded within the surface epidermal cells (Fig. 26.3c). Although they may be located throughout the epidermis, they are especially prevalent on the tentacles, where they

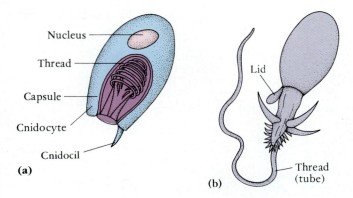

(a)

(b)

Lid

Thread
(tube)

Nucleus

Thread

Capsule

Cnidocyte

Cnidocil

There are no special systems for internal transport, gas exchange, or excretion. All of these processes take place by diffusion.

Nervous System

The cnidarian nervous system displays a number of primitive features. The neurons, located at the base of the epidermis and gastrodermis, are usually arranged as **nerve**

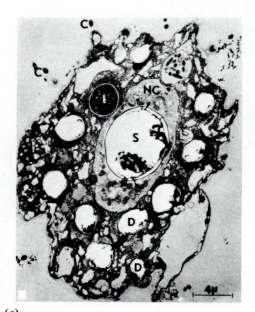

(c)

Figure 26.3

(a) Undischarged nematocyst within cnidocyte of hydra; (b) a discharged nematocyst; (c) electron micrograph of a cross section of an epitheliomuscle cell containing a central nematocyst (S) surrounded by smaller nematocysts (D and I). The letters C indicate cnidocils of adjacent cnidocytes, and NC is the central cnidocyte. The cnidocytes are enfolded by the epitheliomuscle cell rather than being intracellular.

Figure 26.4

(a) Spiny open-tubed nematocyst of the hydroid *Laomedea*. (b) Puncturing of prey integument by a barbed nematocyst of *Hydra*.

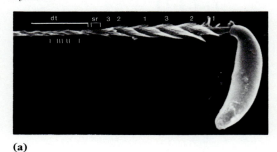

(a)

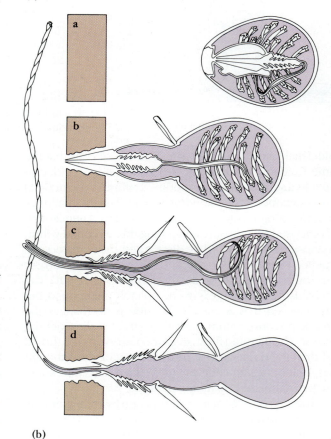

(b)

may be concentrated in ringlike or wartlike batteries. In hydras, contact with prey may result in the discharge of 25% of the tentacular nematocysts, which are replaced in about 48 hours. The old cnidocytes are resorbed, and new cnidocytes differentiate from **interstitial cells** (Fig. 26.2b), totipotent cells similar to certain amoebocytes of sponges.

After the prey has been quelled by the nematocysts, it is pulled by the tentacles toward the mouth, which often dilates enormously in swallowing (Fig. 26.5). Mouth opening is a response to certain amino acids and peptides in the prey tissues. Mucus secreted by gland cells in the mouth region facilitates the process of swallowing.

Figure 26.5
Hydra capturing a water flea, having already swallowed another.

nets rather than as nerve bundles (Fig. 26.6). The neurons may possess two or many processes, and the axons and dendrites of some interneurons are not differentiated. Conduction commonly occurs in a radiating manner. Thus transmission in some neurons and across some synaptic junctions can occur in more than one direction.

Regeneration and Reproduction

Polypoid cnidarians exhibit a high level of regenerative ability. For example, when a major part of the body, such as the oral end, is lost, the remaining part undergoes re-organization to form a new mouth region and tentacles. Hydras have been shown to have continual replacement of cells, so that none of the cells are ever very old.

Asexual reproduction is very common, especially in polypoid species. New individuals are usually formed by **budding**. A bud arises as an outpocketing of the body wall and thus contains an extension of the gastrovascular cavity

Figure 26.6
Diagram of the cnidarian epidermis, showing epitheliomuscle cells, sensory cell, and nerve net.

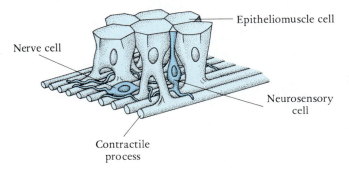

Nerve cell
Epitheliomuscle cell
Neurosensory cell
Contractile process

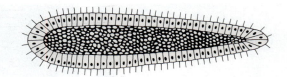

Figure 26.7
The planula larva of *Gonionemus*.

and all of the body wall layers (Fig. 26.2a). The bud separates from the parent, or in colonial species may remain attached as a new individual of the colony.

The sexes of most cnidarians are separate. The gametes develop from interstitial cells and form aggregations in specific locations in the epidermis or gastrodermis. There is no surrounding wall of somatic cells as in most other animals.

Fertilization is commonly external, with development occurring in the plankton. At the completion of gastrulation, a characteristic larval stage, termed a **planula**, is attained. The planula is slightly elongate and radially symmetrical (Fig. 26.7), composed of a solid interior mass of cells surrounded by an outer layer of ciliated cells. A posterior mouth (oral end) may be present or absent.

CLASS HYDROZOA

The class Hydrozoa includes the hydras and many colonial species called **hydroids**. While hydrozoan cnidarians are very abundant, they are usually small and not as conspicuous as the larger jellyfish, sea anemones, and corals. Most hydrozoans are marine, but the few species of freshwater cnidarians, such as hydras, are members of this class. Hydrozoans may exhibit a medusoid body form or a polypoid body form, or both, during the life history. The mesoglea contains few or no cells, cnidocytes are limited to the epidermis, and the gametes usually develop within the epidermis.

Four recurring features of hydrozoans provide a helpful basis for understanding the members of this class: (1) colony formation, (2) the presence of skeletons, (3) polymorphism, and (4) medusa reduction. The first three of these are interdependent.

Colony Formation

The hydras and the small hydrozoan jellyfish are solitary, but many species are colonial (Figs. 26.8a and 26.9). Imagine hydras budding but the bud remaining attached to the parent; this gives some notion of how a hydroid colony can be attained. The individuals, or polyps, of a hydroid colony are usually attached to a main stalk, which is in turn anchored to the substratum (algae, rock, shell, or

599

(a)

(b)

Figure 26.8
(a) The hydroid *Tubularia crocea*. Each
feeding polyp bears large numbers of
small attached reduced medusae.
(b) *Millepora alcicornis*, or fire coral, a
common West Indian reef-inhabiting
hydrozoan coral.

wharf piling) by a rootlike stolon (Fig. 26.9). The arrangement of polyps on the stalk and the branching of the stalk
vary with the species. In some hydroids, separate polyps
spring directly from the stolon. The tissue layers of the
stalks and stolons are continuous with the tissue layers of
the polyps. Thus all of the polyps are interconnected, and
there is a common gastrovascular cavity for the entire
colony.

Skeleton Formation

The solitary polypoid hydrozoans have no solid skeletons,
and some hydroid colonies in which the polyps arise
directly from the substratum have only anchoring skeletons (Fig. 26.10b). Most hydroid colonies, however, are 3
to 10 cm high, and support is provided by an external
chitinous skeleton secreted by the epidermis. In some

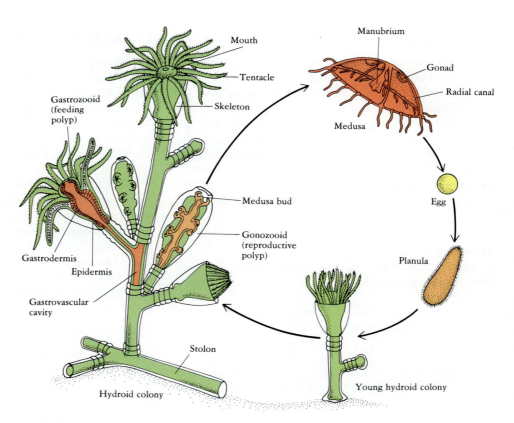

Figure 26.9
Life cycle of *Obelia*, showing structure of the hydroid colony.

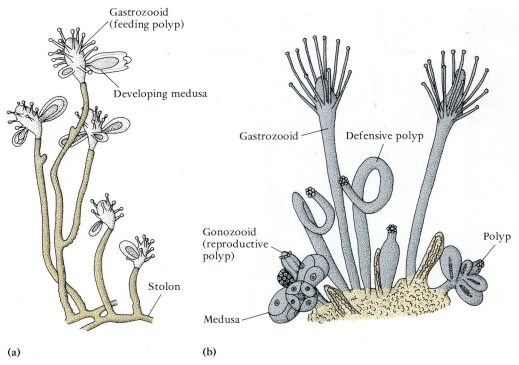

Figure 26.10
Hydroid colonies. (a) *Coryne tubulosa*, a dimorphic form in which medusoids are
formed directly on gastrozooids. (b) *Hydractinia*. A tetramorphic colony; polyps
arise separately from mat of stolons, and skeleton is limited to stolons.

species, the skeletal envelope is limited to the stalk (Fig.
26.10a); in others, the skeleton extends around the body of
the polyps but is open at the oral end for the emergence of
the tentacles and mouth (Fig. 26.9).

There is one group of colonial hydrozoans, called
hydrocorals, that has a solid skeleton of calcium carbonate.
Hydrocorals of the genus *Milleporina*, sometimes called
fire corals because of their sting, are common on coral
reefs throughout the world (Fig. 26.8b). They are much
larger than hydroids and some reach as much as a meter in
height.

Polymorphism

Many hydroid colonies have two or more structurally and
functionally different kinds of individuals within the same
species. This is termed **morphological polymorphism** and
is characteristic of many but not all colonial animals.
Among hydroids the commonest type of individual is the
feeding polyp called a **gastrozooid**, which resembles a
hydra (Fig. 26.9). These polyps capture food and carry out
the initial extracellular phase of digestion. Intracellular
digestion and absorption occur throughout the colony,
since the gastrovascular cavities are continuous.

In some hydroids, sexually reproducing individuals
known as medusae are budded from the body of feeding
polyps (Fig. 26.10a). In many hydroids, however, there has
been a further division of labor, and medusae are budded
only from special reproductive polyps called **gonozooids**
(Fig. 26.9). They lack mouth and tentacles and are no
longer involved in feeding. Some hydroid colonies include defensive individuals, highly modified polyps with
clublike bodies bearing great numbers of cnidocytes. Tentacles and mouth have disappeared (Fig. 26.10b).

Most colonial hydrozoans consist of groups of attached polyps. However, one group of colonial species,
the **Siphonophora**, includes large pelagic colonies composed of both medusoid and polypoid individuals. There
are individuals adapted as floats and pulsating swimming
bells, from which hang feeding polypoid individuals with
long tentacles. The most familiar siphonophoran, the
Portuguese man-of-war, has a single gas-filled float from
which the tentacles of the polyps may hang down several
meters (Fig. 26.11a,b). The nematocysts can produce painful stings, and entanglement with the tentacles can be a
dangerous encounter for a swimmer in deep water. However, most siphonophorans float below the surface and the
individuals are attached to a long stem (Fig. 26.11c).

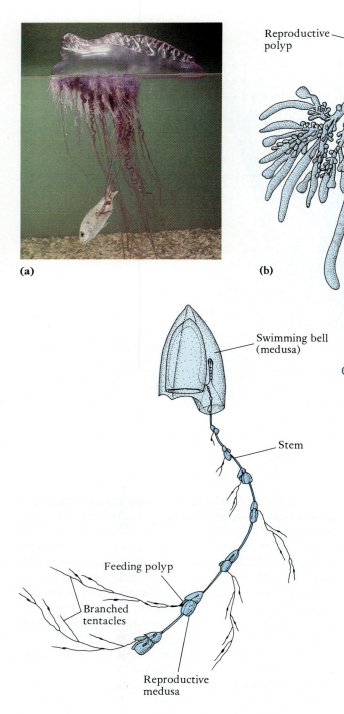

(a)

(b)

Reproductive polyp

Gastrozooid

Fishing tentacles

Swimming bell (medusa)

Stem

Feeding polyp

Branched tentacles

Reproductive medusa

(c)

Figure 26.11

(a) *Physalia*, the Portuguese man-of-war, with a captured fish.
(b) A cluster of polyps from *Physalia*, showing the various modifications of the individual polyps. (c) *Muggiaea*, a submergent, more typical siphonophore than *Physalia*. Clusters of feeding polyps and reproductive medusae are connected together by a long stem hanging beneath a swimming bell, which is about 1 cm long.

Medusae

Sexual individuals are usually medusae.* Hydrozoan jellyfish (**hydromedusae**) are small, usually not more than a few centimeters in diameter (Figs. 26.9 and 26.12). The large jellyfish often encountered at beaches belong to another class of cnidarians. A varying number of tentacles hang from the bell margin of the hydromedusa, and the **manubrium** (L., handle), a fold of body wall surrounding the mouth, hangs from the center of the concave undersurface (the subumbrella). The contraction of radial and circular sheets of epidermal muscle fibers of the subumbrella and bell margin reduces the diameter of the bell and drives water from beneath the animal. A shelflike inward projecting fold of the bell margin, the **velum** (L., veil), increases the force of the water jet. The gelatinous mesoglea resists deformation and restores the bell to its original shape between contractions. These pulsations of the bell propel the animal mostly in an upward direction and maintain the animal at a specific depth. Between the intermittent contractions, the little jellyfish drifts slowly downward. Even when swimming, they are too small to counter wind and tidal currents.

Located at the tentacle bases or between tentacles on the bell margin are **statocysts** (gravitational detectors, p. 408) and **ocelli** (clusters of photoreceptor cells), which

Medusa in Greek mythology was one of three Gorgons, or monsters, who had snakes for hair.

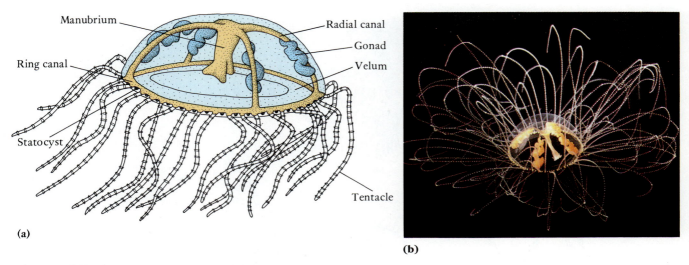

(a)

(b)

Figure 26.12
The hydromedusa *Gonionemus vertens*. (a) Lateral view, showing structure of
medusa. (b) Photograph showing rings of cnidocytes on tentacles and four bands
of gonads beneath radial canals.

orient the jellyfish to the pull of gravity from below and to
light penetrating the water column from above, respectively. In response to stimulation of statocysts, the bell will
contract asymmetrically, causing the body to return to an
upright position.

Hydromedusae feed on other small planktonic organisms. The gastrovascular cavity is not a single sac, as in
polyps. The mouth opens into a central stomach from
which extend four radial canals. The radial canals connect
in turn with a ring canal that extends around the bell
margin.

The gametes of hydromedusae develop in gonads
located within the epidermis beneath the radial canals.
The eggs are fertilized when shed into the sea water, and
development to the planula larva occurs in the plankton. In
some species, fertilization and early development occur
on the manubrium or even in the gonads.

Life Cycles

Hydrozoans exhibit great variation in their life cycles. In a
small number of pelagic species there is no polypoid stage.
The medusae produce planula larvae, which transform
into free-swimming, tentaculate **actinula** larvae (Fig.
26.13a). Then there are hydrozoans such as *Craspedacusta*
(Fig. 26.13b) and *Obelia* (Fig. 26.9), in which there is both a
free-swimming medusa and a sessile polyp. The eggs
produced by the pelagic medusa develop into a planula
larva, which settles and develops directly into a polyp. The
polyp (or polyps, in the case of hydroid colonies) gives rise
to medusae asexually by budding. In many hydrozoans, the
medusae are formed, but they remain attached to the

polyps and display varying degrees of reduction. Finally,
there are polypoid species, such as hydras (Fig. 26.13c), in
which there are no medusae at all. Eggs and sperm are
produced directly in the epidermis of the polyp.

How can we make sense out of this diversity of life
cycles? In the evolution of hydrozoans, which came first,
the medusoid or polypoid form? Many zoologists believe
that the medusa is the more primitive cnidarian body form.
As evidence they point to the fact that (1) motility, not
sessility, is the basic animal condition; (2) there are some
living hydrozoans with no polypoid stage; and (3) there are
hydroid species that display all degrees of reduction of the
medusa. According to this view, the hydrozoan life cycle
initially lacked a polypoid stage, and the planula developed first into an actinula larva and then into an adult
medusa:

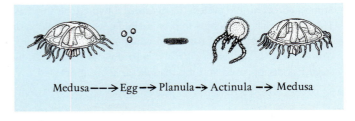

Medusa ---→ Egg → Planula → Actinula --→ Medusa

In some group of hydrozoans the actinula stage became
attached to the bottom by the aboral end. Here it fed,
perhaps with an advantage in the food supply or protection
in this benthic habitat. Note that in this attached position,
the actinula has the form of a polyp. And indeed, according
to the evolutionary scenario we are developing, this is
believed to have been the origin of the polypoid form. At

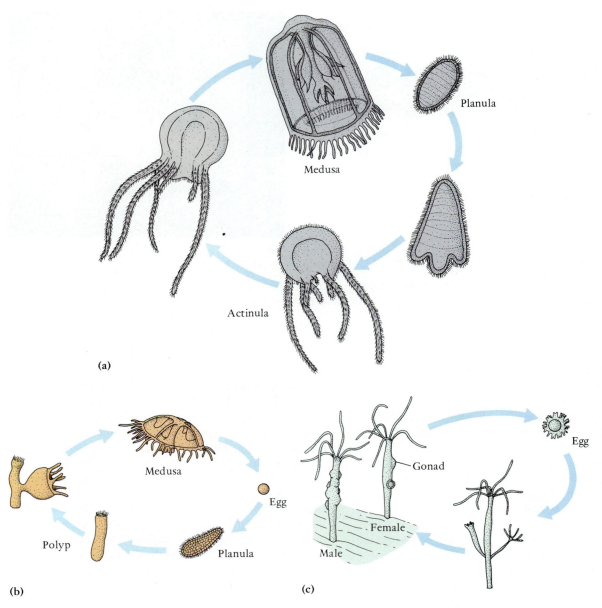

(a)

(b)

(c)

Figure 26.13

Some hydrozoan life cycles: (a) *Aglaura*, a hydrozoan that has no polypoid stage. The planula larva develops into an actinula, which develops directly into a medusa. (b) *Craspedacusta*, a hydrozoan in which the polyp is solitary. (Life cycle of *Gonionemus* is similar.) (c) *Hydra*, a freshwater hydrozoan in which the medusoid stage has disappeared and the planula larva is suppressed.

some point in the course of its life cycle, the attached actinula detached and transformed into a medusa. We will see when studying the class Cubozoa that there are cnidarians in which just such a transformation occurs.

In a later stage of evolution medusa were produced by budding from the attached actinula, which we will now call

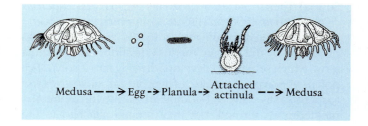

Medusa - - → Egg -> Planula -> Attached actinula - - → Medusa

a polyp. This is true of many living species and has the advantage that more than one medusae can be produced from one polyp.

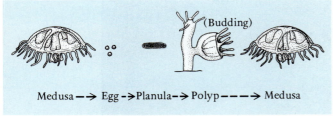

Medusa → Egg → Planula → Polyp ---→ Medusa

There now occurred varying degrees of reduction of the medusa in different groups of hydroid species. The medusa developed by budding but was not free or detached.

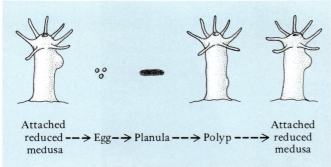

Attached reduced medusa ---→ Egg → Planula ---→ Polyp ---→ Attached reduced medusa

Finally, in some species medusoid tissues became so reduced that until the polyp epidermis itself produced gametes.

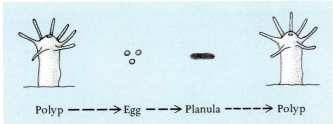

Polyp ----→ Egg ---→ Planula ----→ Polyp

Note that the evidence for this scheme of hydrozoan evolution is largely derived from comparative life cycles, for there are living cnidarians that have life cycles comparable to each of these hypothetical stages. Present evidence indicates that colonies evolved independently in different lines of hydrozoans, and this occurred independently of the evolutionary trend toward medusa reduction. Note that *Craspedacusta* (Fig. 26.13b) and *Obelia* (Fig. 26.9) have the same type of life cycle, but in one the polyp is solitary and in the other the polyp is colonial. A solitary polyp might produce more polyps by budding during certain times of the year and then produce medusa buds at another time. If the polypoid buds remained attached to each other, this could have led to the evolution of a hydroid colony. Colony formation did not evolve in polyps of all hydrozoans. The hydras, for example, probably evolved from some line of solitary polyps that underwent reduction of the medusoid form.

CLASSES SCYPHOZOA AND CUBOZOA

The medusae of the classes Scyphozoa and Cubozoa are the cnidarians to which the name jellyfish is usually applied; they are considerably larger and more conspicuous than hydrozoan medusae (Fig. 26.14). The medusoid body

Figure 26.14
Scyphomedusae. (a) The sea nettle, *Chrysaora quinquecirrha*, a common brackish water scyphozoan along the Atlantic coast of the U.S. (b) A sessile scyphomedusa. Members of this small order (Stauromedusae) live attached by the aboral end to algae. (c) *Stomolophus meleagris*, the cannon ball jellyfish of the southeastern coast of the U.S.

(a)

(b)

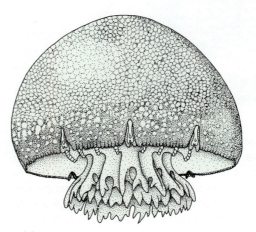

(c)

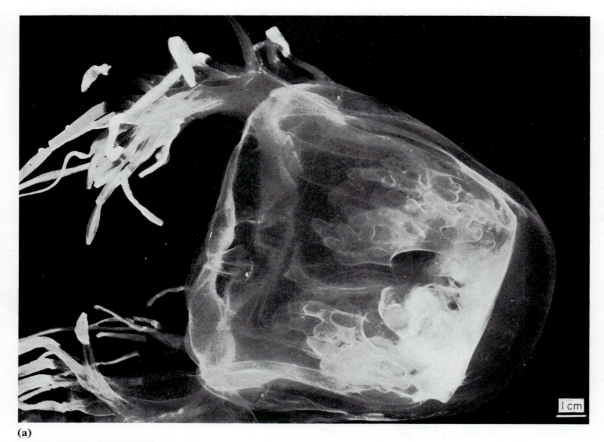

(a)

(b)

Figure 26.15
(a) The sea wasp *Chironex fleckeri*, a dangerous cubomedusan, or box jelly, of the Indo-Pacific. (b) Sea wasp warning and swimming enclosure at Townsville, Australia.

brief discussion of the larger jellyfish will focus on the Scyphozoa, which contains the bulk of the some 200 species. The relatively small number of cubozoans, or box jellies (the bell is somewhat cuboidal in shape), are all tropical and semitropical (Fig. 26.15).

Most scyphomedusae range in size from 2 to 40 cm, and many are about the size of a saucer or serving plate. However, a few reach a very large size; species of the genus *Cyanea* may have bell diameters of over 2 meters and very long oral appendages. The translucent body is often tinted with orange, pink, purple, and other colors. In most species, tentacles of varying numbers and length hang from the margin of the bell. Four long divisions of the manubrium, called oral arms, hang from the subumbrella and are often more conspicuous than the tentacles (Fig. 26.14a). The nematocysts, especially those in the tentacles and oral arms, can produce painful stings. The sea nettles,

form is dominant in the Scyphozoa and Cubozoa, the polypoid form being strictly a larval stage. In contrast to hydromedusae, the mesoglea of scyphomedusae may be cellular; some cnidocytes are present in the gastrodermis, and in both scyphomedusae and cubomedusae the gametes develop within the gastrodermis rather than within the epidermis. Unless otherwise indicated, this

such as the Chesapeake *Chrysaora*, can be a painful nuisance to swimmers at certain times of the year when they occur in large numbers (Fig. 26.14a). The sea wasps of the Indo-Pacific have such virulent nematocysts that their sting causes severe lesions and even death (Fig. 26.15).

As in hydromedusae, statocysts and ocelli are present, but in the scyphomedusae they are housed in projections called **rhopalia** (Gr. *rhopalon*, club) located between the scallops of the bell margin (Fig. 26.16b,c).

There is no velum in scyphomedusae, but swimming movements are accomplished by bell contractions like those of hydromedusae. However, some species can turn on their side and swim very rapidly in a horizontal direction. A velum-like structure is present in the cubomedusae, which are also rapid swimmers.

Scyphozoans feed on animals of various sizes, including fish, that come in contact with the tentacles and oral arms. Some species, such as *Aurelia* (Fig. 26.16a), feed on plankton that adheres to the subumbrella. The plankton is then swept by ciliated surface cells to the bell margin, which is wiped by the oral arms. There is a large central cavity, or **stomach**, which primitively is divided into four pouches by four vertical septa (Fig. 26.17). The free central margins of the septa bear filaments containing cnidocytes. In most of the common scyphomedusae of temperate

coastal waters, such as *Aurelia* and *Chrysaora*, the septa have disappeared and the filaments containing cnidocytes arise from the stomach floor (Fig. 26.16c). In these jellyfish numerous canals extend radially to the bell margin. A ring canal may or may not be present. The gastrodermal nematocysts perhaps function to quell any prey still alive when it enters the stomach.

Gametes develop in the gastrodermis of the stomach floor, and the "gonads" are often conspicuous within the transparent body (Fig. 26.16c). The eggs and sperm are shed through the mouth. Fertilization and early development may occur in the sea water, or the fertilized eggs may be brooded on the oral arms. In either case, a free-swimming planula larva is attained (Fig. 26.18). The planula develops into a polypoid larval stage, called a **scyphistoma**, that is about the size of a hydra and lives attached to hard bottom surfaces. The scyphistoma has a life span of one to several years, during which it feeds and produces more scyphistomae by budding. At certain seasons it ceases feeding and undergoes a special form of budding to produce young jellyfish. The buds are produced at the oral end of the body and in the course of formation in some species are stacked up like plates (**strobila**). When detached, the young jellyfish (**ephyra**) is tiny and displays only a rudimentary medusoid form. In the

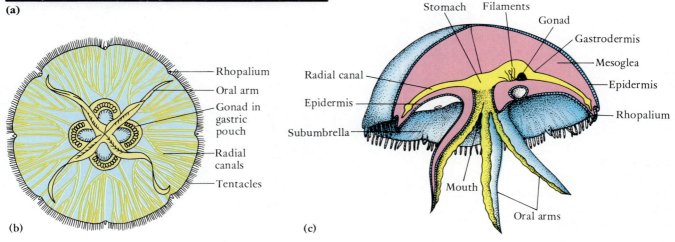

Figure 26.16

(a) *Aurelia*, a scyphozoan medusa. (b) Oral view. (c) Side view in section.

(a)

(b) Rhopalium — Oral arm — Gonad in gastric pouch — Radial canals — Tentacles

(c) Stomach — Filaments — Gonad — Gastrodermis — Mesoglea — Epidermis — Rhopalium — Radial canal — Epidermis — Subumbrella — Mouth — Oral arms

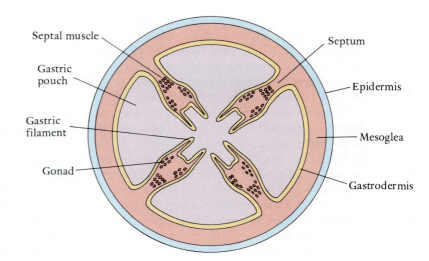

Figure 26.17
Highly diagrammatic transverse section through the bell of a scyphozoan, showing the primitive organization of the gastrovascular cavity (stomach) with septa, gastric filaments, and gastric pouches.

cubozoans the juvenile jellyfish are not produced by budding; rather the entire scyphistome larva transforms into a young jellyfish and swims away.

In describing the evolution of hydrozoans, we theorized that the polyps evolved from a larval stage. Note that the scyphozoan life cycle displays just such a polypoid larva, which by direct transformation or by budding produces medusae. Thus this theory of the origin of the polypoid stage is supported by evidence from the comparative development of scyphozoans and cubozoans as well as that of hydrozoans.

CLASS ANTHOZOA

The more than 6000 species of anthozoans constitute the largest class of cnidarians and include the familiar sea anemones and various types of corals. The class is entirely polypoid, with a cellular mesoglea, gastrodermal cnidocytes, and gastrodermal gametes. The distinctive feature of anthozoans is the presence of a **pharynx** and **mesenteries**. The pharynx is a tube that hangs from the mouth into the gastrovascular cavity like a sleeve (Fig. 26.19a). Since the pharynx is derived from an infolding of

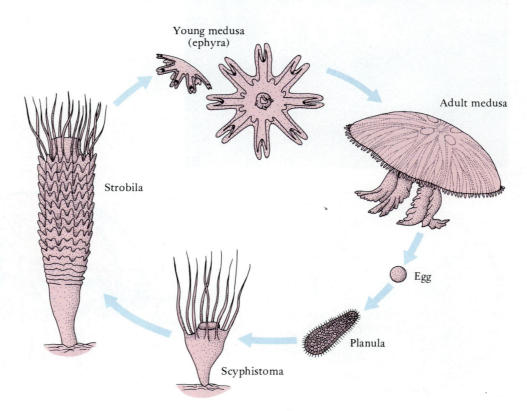

Figure 26.18
Life cycle of the scyphozoan *Aurelia*. The polypoid stage is a larva and produces medusae by transverse budding.

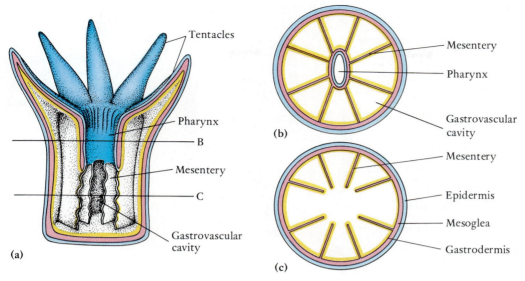

Figure 26.19

Structure of an anthozoan polyp: (a) Longitudinal section. (b) Cross section at the level of the pharynx. (c) Cross section below the pharynx.

the body wall around the mouth, it possesses the same layers as the outer body. The mesenteries are sheetlike partitions that extend from the outer body wall into the gastrovascular cavity. Each is composed of two layers of gastrodermis separated by a layer of mesoglea (Fig. 26.19b). At least eight of the mesenteries, called complete mesenteries, connect with the mesoglea and gastrodermis of the pharynx. In many anthozoans, there are additional incomplete mesenteries extending partway into the gastrovascular cavity. Below the pharynx all of the mesenteries are incomplete so that there is a common central region into which all of the sections of the gastrovascular cavity open (Fig. 26.19c).

The functional significance of the mesenteries is difficult to understand. One might expect that they would provide greater surface area for digestion and absorption, but studies have shown that only the free margin is involved in digestion and absorption. The mesenteries may serve in part as a form of support by limiting the diameter of the body when the gastrovascular cavity is filled with fluid.

Sea Anemones

The largest of the anthozoans are the solitary sea anemones (Fig. 26.20). The majority live attached to hard substrates. Most are one to several centimeters in diameter and are often brightly colored. However, there are some giants among the group. A species on the Great Barrier Reef of Australia and one on the north Pacific coast of the United States attain a diameter of 1 meter.

Figure 26.20

Sea anemones. (a) *Cribrinopsis fernaldi* from British Columbia. (b) *Urticina piscivora.*

(a)

(b)

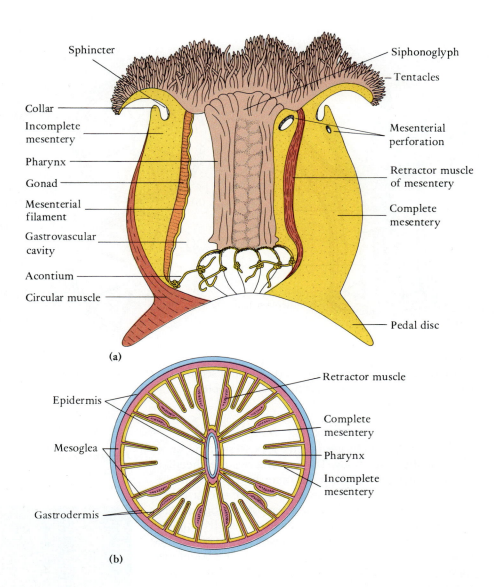

Figure 26.21
Structure of a sea anemone:
(a) Longitudinal section. (b) Cross
section at the level of the pharynx.

The body of a sea anemone is heavier than that of a hydrozoan polyp, and the oral end, bearing tentacles peripherally and a central slit-shaped mouth, is disclike (Figs. 26.20 and 26.21). At one or both ends of the mouth there is a ciliated groove, the **siphonoglyph**, which drives water into the gastrovascular cavity. The siphonoglyph grooves are open even when the mouth is closed. Under slight muscular tension and the shape limitations imposed by the mesenteries, the water in the gastrovascular cavity acts as a hydrostatic skeleton. In addition, the siphonoglyph current would provide some oxygen for gastrodermal tissue surfaces. A perforation in the upper part of each complete mesentery permits circulation of the gastrovascular fluid.

When sea anemones contract, the upper, collar-like rim of the body is pulled over the oral disc (Fig. 26.21a). The muscle fibers are entirely gastrodermal. Circular fibers are located in the gastrodermis of the body wall, and longitudinal retractor fibers are located in the mesenteries (Fig. 26.21b).

Sea anemones feed upon other invertebrates and small fish. The prey is passed down the pharynx and into the center of the gastrovascular cavity below the pharynx.

The numerous mesenteries are arranged in couples and the free edge of each mesentery bears a glandular ciliated band called the **mesenterial filament**. On reaching the aboral end of the mesentery, the mesenterial filament commonly extends back upward into the gastrovascular cavity as a free thread, called an **acontium**. In many sea anemones with rather transparent body walls you can often see the acontia within the hollow tentacles or even protruding out of the mouth. These mesenterial filaments are the location of the gastrodermal cnidocytes, of enzyme production for extracellular digestion, and of intracellular digestion and absorption. When prey is swallowed and the sea anemone contracts downward, the many free edges of the mesenteries with their gastric filaments would be pressed against the prey's body.

Sea anemones may seem like passive animal "flowers," but the appearance is deceptive. Many species can slowly creep over the bottom on their aboral end (pedal disc). Some show aggressive behavior against other sea anemones with which they may come in contact. Batteries of nematocysts on special sweeper tentacles or column tubercles can cause tissue lesions on the opponent

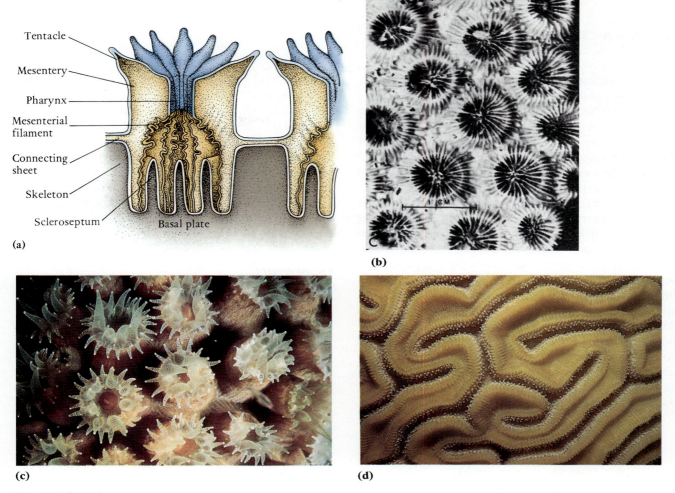

(a)

Tentacle
Mesentery
Pharynx
Mesenterial filament
Connecting sheet
Skeleton
Scleroseptum
Basal plate

(b)

(c)

(d)

Figure 26.22

Scleractinian corals. (a) A coral polyp within its skeletal cup (longitudinal section).
(b) Surface view of the skeleton of *Montastrea*. In life the polyps are in the large, distinct cups. Note the sclerosepta. (c) Surface view of a living coral colony (*Montastrea cavernosa*), showing expanded polyps. (d) The West Indian brain coral, *Diploria strigosa*, in its contracted state during the day. The underlying skeletal cups and troughs in which the polyps are located can be seen through the body tissues. The yellow-brown color of the living colonies is due to the presence of symbiotic algae (Zooxanthellae).

sea anemone. Many sea anemones produce a noticeable sting on contacting human skin, and a few are quite virulent. Sea anemones exhibit a considerable range of feeding adaptations. For example, some live just at the beginning of the subtidal zone on rocky shorelines and feed on mussels that have been knocked down by feeding sea stars higher in the intertidal region. Some live lodged within rock crevices from which long fishing tentacles emerge. There are species with very short tentacles that feed on fine particulate material trapped on the oral surface.

Despite the presence of nematocysts, some animals, such as certain shrimps and fish, live among the tentacles of sea anemones. The colorful clown fish of the Indo-Pacific live among the protective tentacles of certain large sea anemones (Fig. 26.20). Apparently, the mucus of the fish raises the firing threshold of the cnidocytes.

Some sea anemones reproduce asexually by splitting of the body or by the regeneration of fragments of tissues separated from the base of the body. Most sea anemones are hermaphroditic, but only one type of gamete is produced at any one time. The gametes develop in gastrodermal bands just behind the free edge of the mesenteries (Fig. 26.21a). Fertilization and early development may occur externally in the sea water or within the gastrovascular cavity. The planula larva develops into a ciliated planktonic polypoid larva in which mesenteries and pharynx make their appearance. The polyp soon settles and becomes attached as a young sea anemone.

Scleractinian Corals

Most other anthozoans are various types of corals and the great majority are colonial. The species with which the name coral is most generally associated are the **scleractinian** or **stony corals**. The polyps of scleractinian corals are similar to those of sea anemones but are usually smaller. Although there are some solitary scleractinian corals, most are connected together in colonies by means of a lateral fold of the body wall (Fig. 26.22a). A calcium carbonate skeleton is secreted by the epidermis of the undersurface of the connecting sheet and the lower part of the polyp. The living colony thus usually has the form of a sheet resting on top of or wrapped around a skeleton that is actually external to the coral tissues. The lower part of the polyp is situated within a skeletal cup, the bottom of which contains radiating septa that project up into folds in the base of the polyp (Fig. 26.22b).

Species of corals display various growth forms. Some are low and encrusting; others are upright and branching (Fig. 26.25c). The surface configuration of the skeleton depends upon the size of the polyps and how closely placed they are to each other. When the polyps are well separated the coral appears pockmarked (Fig. 26.22b). In brain coral, which has a surface configuration of troughs and ridges, the polyps are joined together in rows (Fig. 26.22d). No skeletal material is deposited between the polyps within a row, but skeleton is deposited between the rows. The rate of skeletal growth is about 0.5 to 2 cm (radial growth) per year in plate and domal corals and up to 10 cm or more (linear growth) per year in erect branching forms, such as staghorn coral.

Most reef-inhabiting scleractinian corals are yellow-brown in color because of the presence of symbiotic dinoflagellates known as zooxanthellae. The physiology of the coral is closely tied to its algal symbionts (Connections 26.1).

Reproduction in scleractinian corals is similar to that in sea anemones. The planula larva gives rise to the first polyp, which secretes the initial calcium carbonate platform. Additional polyps are produced by budding, and the colony slowly increases in size. Some domal corals reach ages of over 100 years.

Octocorals

Most other anthozoans, largely tropical, are octocorals. There are only eight mesenteries, all complete, within the body of the polyp and only eight tentacles, which have little side branches called pinnules (Fig. 26.23). Octocorallian polyps are usually very small but are organized into colonies that may attain sizes of over a meter in diameter or height. The interconnection of the polyps of a colony is quite different from that of scleractinian coral. A common

Connections 26.1   *Algal Symbiosis*

Symbiosis between animals and algae occurs in sponges, cnidarians, flatworms and molluscs, but cnidarians exceed all others in the number of species harboring algal symbionts. Some freshwater hydras and a few sea anemones contain the green alga *Chlorella* (zoochlorellae) (Fig. a), but for most marine cnidarians the algal symbiont is the dinoflagellate *Symbiodinium microadriaticum* (zooxanthellae) (Fig. b). The dinoflagellate is located in vacuoles within the gastrodermal cells, and a single host cell may contain more than 50 algal cells. As a symbiont, the dinoflagellate lacks flagella and has a reduced cell wall (p. 564). A remarkable feature of the symbiosis is that the algae are not digested by the host. Depending upon the species, the host obtains its initial symbionts via the egg or by larval or adult engulfment of free algal cells. The yellow-brown color of the dinoflagellate (produced by accessory pigments) is imparted to the host. Since most reef-inhabiting hydrocorals, scleractinian, gorgonian, and soft corals contain these algal symbionts, most tend to be shades of yellow-brown. Some biologists have speculated that there is more algal than cnidarian tissue in many corals, and certainly the algae account for the high

productivity of coral reefs, i.e., the production of large numbers of organisms. The coral bleaching that has occurred on some reefs in recent years results from the expulsion of its algal symbionts by the coral host, probably stimulated by elevated water temperatures.

Since the dinoflagellates require light for photosynthesis, their hosts must live in shallow clear water, a characteristic of coral reefs. Excess products of photosynthesis, as much as 50% of the algal production, may be transferred to the host. Most is received as glycerol, but glucose and alanine are also translocated. The dinoflagellates utilize the host's nitrogenous wastes and acetate also appears to be cycled between the two partners. In scleractinian corals, the secretion of calcium carbonate is enhanced by the symbiosis, for secretion is greatly reduced in the absence of light or the symbionts.

In addition to reef corals, a few hydroids and scyphozoan jellyfish and some sea anemones possess zooxanthellae. The tropical jellyfish *Cassiopeia* rests upside down on the bottom of quiet lagoons "sunning" the zooxanthellae in its frilly oral arms.

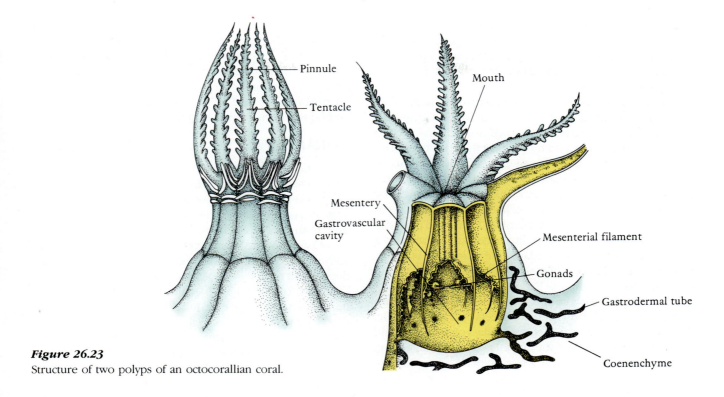

Figure 26.23
Structure of two polyps of an octocorallian coral.

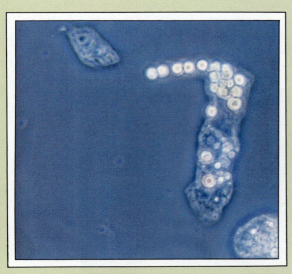

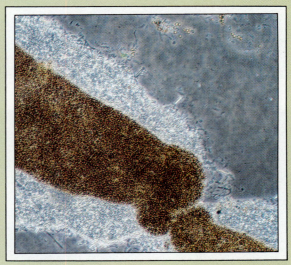

(a) Zoochorellae within an isolated gastrodermal cell of a hydra.

(b) Masses of zooxanthellae in the gastrodermal cells of a sea anemone tentacle.

mass of tissue composed of mesoglea (**coenenchyme**) and gastrodermal tubes unites the lower half of the colony. A skeleton of separate microscopic pieces and sometimes also an organic horny material may be present *within* the mesoglea.

The two most common groups of octocorallian corals on coral reefs are gorgonian and soft corals. Gorgonian corals include the sea whips, sea rods, and sea fans, and are especially conspicuous on the reefs of the Caribbean. The erect, rodlike design, which can exceed 2 meters in some species, provides for exploitation of the vertical water column with only a small area of substratum required for attachment (Fig. 26.24b). The skeleton composed of separate calcareous spicules and a central, horny rod provides support as well as the flexibility necessary to bend in a wave surge. Sea fans are similar to the sea whips, but their branches are all in one plane, often with lattice-like cross connections.

Soft corals are not common on Caribbean reefs but are very abundant and conspicuous on reefs of the Indo-Pacific. The polyps are located on a rubbery mass of coenenchyme containing spicules, and the colonies may be encrusting, erect, or cushion-like with lobes or peaks (Fig. 26.24a). Some look so much like scleractinian corals that you have to touch them to be certain they are soft corals.

Sea pansies and sea pens belong to a group of octocorallians adapted for living in soft bottoms. The shape of the colony, from which the common name is derived, is determined by the shape of the large, mouthless primary polyp, on which the smaller more typical polyps are budded (Fig. 26.24c). One end of this primary polyp is embedded in the sandy bottom.

CORAL REEFS

Coral reefs are tropical, shallow-water, calcareous formations that support a great diversity of marine plants and animals (Fig. 26.25). Moreover, certain of these plants and animals secrete the calcium carbonate that composes the underlying reef formation.

Contemporary coral reefs, located in the Caribbean and in the tropical parts of the Indo-Pacific oceans, have been built chiefly by scleractinian corals. Reefs are confined to warm, clear, shallow waters because of the light and temperature requirements of their symbiotic dinoflagellates (see Connections 26.1). Coral reefs can be classified into three major types, depending upon their location. (1) **Fringing reefs**, which are the most common type of reef, are located adjacent to the shores of islands or continental coasts (Fig. 26.25a). (2) **Barrier reefs** parallel the coast but are separated from the shore by a wider and deeper lagoon. The best known is the Great Barrier Reef, which parallels the northeast coast of Australia for over 1000 miles. (3) **Atolls** lie above old submerged volcanoes and are more or less circular reefs containing an interior lagoon (Fig. 26.25b). The reefs were first a fringe around an emergent volcanic cone. As the cone slowly subsided, the upward growth rate of the reef equaled the rate of cone

Figure 26.24
Octocorallians. (a) A soft coral (*Dendronephthya*) from Truk lagoon in Micronesia. (b) Gorgonian sea rods and boulder-like scleractinian corals from Bermuda. (c) Orange sea pen (*Ptilosarcus gurneyi*) from Vancouver Island, British Columbia.

(a)

(b)

(c)

(a)

(b)

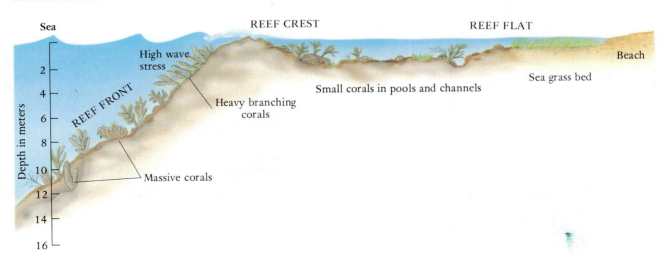

(c)

Figure 26.25
(a) Fringing reef on Aitutaki in the Cook Islands. Note the extensive reef flat that lies behind the reef front. (b) An atoll. (c) Underwater scene on a coral reef. Much of the reef surface is covered by scleractinian corals, especially species of *Acropora*.

subsidence. Darwin suggested such an origin for atolls from observations made during the voyage of the *Beagle* through the tropical Pacific, which contains the greatest number of atolls. Small patch reefs, ranging from 3 to 50 m across, commonly rise from the floor of lagoons of barrier reefs and atolls.

Most reefs have one side, the **reef front**, facing the open ocean (Fig. 26.26). Behind the reef front, which rises close to the surface, is a **reef flat** covered by only a meter or less of water (Figs. 26.25a and 26.26). The greatest growth of large massive corals is on the reef front, and in the West Indies elkhorn coral typically occupies the reef crest. The shallow reef flat may contain pools, sand beds, and rocky surfaces, and supports a very different fauna from that on the reef front.

The coralline rock that forms the underlying reef substratum is composed largely of chunks of dead coral,

Figure 26.26
Zonation of a coral reef.

riddled and broken by boring animals and encrusted with tubes and skeletons of other organisms. These larger pieces of calcium carbonate become filled in with fine calcareous sediment, like mortar between bricks.

As the reef increases in thickness, there is a gradual compaction of the lower deposits. Core drillings of coral reefs have disclosed coralline deposits of great depths. For example, on the Pacific atoll of Eniwetok, coral limestone extends downward for almost a mile before reaching basaltic rock. Since the deposition of coral by living colonies occurs only in the upper lighted zone, thick deposits of coral can only be explained by fluctuations in sea level and by subsidence of the bedrock upon which the reef is resting. Both have occurred at Eniwetok. The geologic history of Eniwetok began during the early Cenozoic Period, when a fringing reef developed around an emergent volcanic cone (Fig. 26.27a). This oceanic peak gradually subsided to below sea level. Coral deposition occurred at a rate equaling subsidence, and the original fringing reef was transformed into an atoll resting on vertical walls almost 1 mile in height (Fig. 26.27b). Fluctuating sea level periodically caused emergence and submergence of the atoll. The rise in sea level from the Pleistocene ice age low of 120 m below the present level gave Eniwetok its present emergent form.

Although existing coral reefs were formed by scleractinian coral, reefs in the past have been produced by other organisms. Two extinct groups of anthozoans, the tabulate and rugose corals, formed great reefs during the Paleozoic. Reefs have also been formed in the past by algae that secrete calcium carbonate. Coralline algae are still prevalent today and are common members of reef communities, but there are no existing reefs composed primarily of living coralline algae.

EVOLUTIONARY RELATIONSHIPS

Anthozoans appear to be much more closely related to scyphozoans than to polypoid hydrozoans. Both groups have a cellular mesoglea, some gastrodermal cnidocytes, gastrodermal gonads, and a septate gastrovascular cavity (only in primitive scyphozoans). But scyphozoans are medusoid and anthozoans are polypoid. How can we account for this apparent contradiction? The anthozoans may have evolved from some ancient group of scyphozoans in which the medusoid form became suppressed. Indeed, there are today a few scyphozoans that have very small medusae that are sexually mature as soon as they are budded from the scyphistome larva.

Structurally, hydrozoans are much simpler than either schyphozoans or anthozoans. For example, the hydrozoan mesoglea is acellular, and the gut of the hydrozoan polyp is not septate. This structural simplicity is important evidence for believing that the hydrozoans exhibit the most primitive features of the phylum. Our

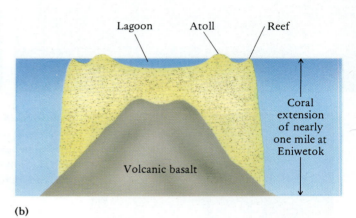

(a)

(b)

Figure 26.27
Formation of an atoll. (a) Fringing reef around an emergent volcano. (b) Continuous deposition of coral as the volcanic cone subsides leads to the formation of a great coralline cap; the emergent part of the cap is the atoll.

proposed phylogeny of the cnidarians could therefore be diagrammed cladistically in the following way:

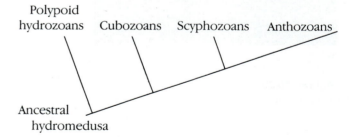

Compared to other metazoans, cnidarians are believed to be primitive because of their epitheliomuscle cells, lack of organs, lack of an anus, and lack of somatic gonadial walls. The radial symmetry is clearly primary, not secondary. Most zoologists, therefore, believe that cnidarians evolved early in metazoan evolution before the shift to a motile benthic existence and bilateral symmetry.

ANOTHER GELATINOUS PHYLUM

Members of the phylum Ctenophora, called **comb jellies**, **sea walnuts**, or more commonly **ctenophores**, are pelagic marine animals once thought to be closely related to cnidarians because of their similarity to medusae. The rather globose body of ctenophores is transparent and contains a jelly-like mesenchymal layer. Most of the some 90 described species range from the size of a pea to that of a golf ball, but some large melon-sized deep-water species have been observed. The digestive system is composed of canals arising from a central stomach. However, the similarity to cnidarian medusae is believed to be only superficial. The ctenophore body is markedly **biradial** rather than radial and in many species is flattened somewhat along one plane. An apple flattened a little from side to side would be biradial. You could not obtain equal pie-shaped sections from such an apple, but you could cut it along the long axis or short axis and in both cases obtain two equal halves. The name *ctenophore* ("comb-bearer") comes from the presence of eight longitudinal **ciliated bands**, which can be seen in Figure 26.28. These ciliated bands are a distinguishing feature of the phylum. The ciliated bands propel the animal through the water, oral end forward. Ctenophores are carnivores on other planktonic animals, especially crustaceans, and many including *Pleurobrachia* (Fig.26.28) possess long aboral tentacles with specialized adhesive cells used in prey catching. *Beroe*, which lacks tentacles, feeds on other ctenophores, which they can swallow in a gulp, when the large mouth contacts the prey surface. The cellular structure of ctenophores is more like that of higher metazoans than cnidarians. The outer surface is covered by a glandular epithelium, there are no cnidocytes, and the gelatinous mesenchyme contains smooth muscle cells. All ctenophores are bioluminescent (see Connections 29.3, p. 687). Ctenophores are hermaphroditic, and sperm ducts and oviducts transport sperm and eggs from the testes and ovaries to the exterior.

Figure 26.28
The ctenophore *Pleurobrachia bachei*. Species of this genus are less than 2 cm in diameter.

At certain times of the year, some species are very abundant in coastal waters. We do not know the evolutionary origin of ctenophores, but it appears to be remote from that of cnidarians.

Classification of Phylum Cnidaria

Class Hydrozoa (Gr. *hydra*, water serpent + *zoon*, animal) Common but mostly small cnidarians that are polypoid or medusoid, or exhibit both forms in their life cycle. Mesoglea is acellular, cnidocytes are entirely epidermal, and the gametes are usually formed within the epidermis. There are many colonial polypoid species, called hydroids, that are surrounded by a chitinous skeleton.
> **Order Trachylina** (Gr. *trachys*, rough + L. *linum*, flax) Planktonic hydromedusae; no polypoid stage present. This order contains perhaps the most primitive members of the class. *Aglaura*.
> **Order Hydroida** (Gr. *hydra*, water serpent + *eidos*, form) Hydrozoans with a well-developed polypoid generation. Medusoid stage is free-swimming (*Obelia, Craspedacusta, Gonionemus*), attached to polyps (*Hydractinia*), or absent (hydras). The majority of hydrozoans belong to this order.

> **Order Siphonophora** (Gr. *siphon*, tube + *pherein*, to bear) Pelagic hydrozoan colonies of polypoid and medusoid individuals. Colonies with float or swimming bells. *Physalia* (Portuguese man-of-war). *Abyla, Agalma*.
> **Order Hydrocorallina** (Gr. *hydra*, water serpent + *korallion*, coral) Colonial polypoid hydrozoans that secrete a calcium carbonate skeleton. Mostly tropical. *Millepora*, stinging coral, or fire coral, is a common member of coral reefs.

Class Scyphozoa (Gr. *skyphos*, cup + *zoon*, animal) Mostly large cnidarians in which the medusoid stage is conspicuous and the polypoid stage is small and larval. The mesoglea contains cells, some cnidocytes are located in the gastrodermis, and the gametes form in the gastrodermis and are shed through the mouth. Widespread marine forms, mostly free-swimming.

Order Stauromedusae (Gr. *stauros*, stake + *medousa*, one of the Gorgons of Gr. mythology having snakes for hair) Small sessile scyphozoans attached by a stalk on the aboral side of the trumpet-shaped body. Chiefly in cold coastal waters. *Haliclystus.*

Order Coronatae (L. *coronatus*, crown) Bell of medusa with a deep groove or constriction, the coronal groove, extending around the exumbrella. Many deep-sea species. *Periphylla.*

Order Semaeostomae (Gr. *semeia*, standard + *stoma*, mouth) Scyphomedusae with bowl-shaped or saucer-shaped bells having scalloped margins. Manubrium divided into four conspicuous oral arms. Gastrovascular cavity with radial canals or channels extending from central stomach to bell margin. Occur throughout the oceans of the world. *Aurelia, Chrysaora.*

Order Rhizostomae (Gr. *rhiza*, root + *stoma*, mouth) Bell of medusa lacking tentacles. Primary mouth lost and frilly projections of subumbrella bear numerous secondary mouths, which lead by canals to stomach. Tropical and subtropical littoral scyphozoans. *Cassiopeia.*

Class Cubozoa (Gr. *kybos*, cube + *zoon*, animal) Cnidarians that are similar to scyphozoans in having a large conspicuous medusoid stage and a small, larval polypoid stage. Gonads are gastrodermal. Like hydromedusae, the bell margin is not scalloped but turns inward as a shelflike projection and there are no rhopalia. Some species (sea wasps) are extremely virulent. Tropical and subtropical. *Chironex.*

Class Anthozoa (Gr. *anthos*, flower + *zoon*, animal) Solitary or colonial polypoid cnidarians; medusoid stage completely absent. Mouth opens into a pharynx and gastrovascular cavity partitioned by mesenteries.

Subclass Octocorallia (Gr. *okto*, eight + *korallion*, coral) Polyps with eight mesenteries and eight pinnate tentacles. All colonial; polyps connected by gastrodermal tubes and thick mesoglea (coenenchyme).

Order Telestacea (after the genus *Telestos*) Lateral polyps on simple or branched stems. Skeleton of calcareous spicules.

Order Alcyonacea (after the genus *Alcyonium*) Soft corals. Coenenchyme forming a rubbery mass and colony having a massive, mushroom, or variously lobate growth form. Skeleton of separate calcareous spicules. Largely tropical and very common on Indo-Pacific coral reefs.

Order Helioporacea (Gr. *helios*, sun + *poroa*, pore) Contains two species with massive calcareous skeletons. The Indo-Pacific *Heliopora* is blue.

Order Gorgonacea (Gr. mythology, the Gorgons, who had snakes for hair) Sea rods, sea whips, sea fans. Common tropical and subtropical octocorallians, having a largely upright growth form of branching rods. Skeleton consists of a central horny axial rod and calcareous spicules. They are often conspicuous members of coral reefs, especially in the Caribbean. *Gorgonia* (sea fan), *Leptogorgia* (sea whip).

Order Pennatulacea (L. *pennatulus*, winged) Sea pens and sea pansies. Colony having a fleshy, flattened, or elongate body, the shape determined by the primary polyp. Skeleton of calcareous spicules. Adapted for life on soft bottoms. *Renilla* (sea pansy).

Subclass Zoantharia (Gr. *zoon*, animal + *anthos*, flower) Polyps with more than eight tentacles and tentacles rarely pinnate. Six or more pairs of mesenteries present. Solitary or colonial.

Order Actiniaria (Gr. *aktinos*, ray) Sea anemones. Solitary anthozoans with no skeleton, with mesenteries in hexamerous cycles, and usually with two siphonoglyphs. *Metridium.*

Order Zoanthidea (Gr. *zoon*, animal + *anthos*, flower) Small, mostly colonial anemone-like anthozoans having one siphonoglyph and no skeleton. Some live on other invertebrates. *Palythoa, Zoanthus.*

Order Scleractinia (Gr. *skleros*, hard) or **Madreporaria** Stony corals. Mostly colonial anthozoans secreting a heavy external calcareous skeleton. Sclerosepta arranged in hexamerous cycles. Many fossil species. *Oculina, Montastrea, Fungia, Diploria, Acropora.*

Order Rugosa[†] (L. *ruga*, wrinkled) or **Tetracoralla** An extinct order of mostly solitary corals possessing a system of major and minor radiating sclerosepta. Cambrian to Permian.

Order Corallimorpharia (Gr. *korallion*, coral + *morphe*, form) Solitary or colonial. Tentacles radially arranged. Resemble true corals but lack skeletons. *Corynactis.*

Order Ceriantharia (Gr. *kerion*, honeycomb + *anthos*, flower) Sea anemone-like anthozoans with a greatly elongated body adapted for living in a secreted tube buried in sand. Mesenteries all complete. *Cerianthus.*

Order Antipatharia (Gr. *antipathes*, black coral) Black or thorny corals. Gorgonian-like species with upright, plantlike colonies. Polyps arranged around an axial skeleton composed of a black horny material and bearing thorns. Largely in deep tropical waters. *Antipathes.*

Subclass Tabulata[†] (L. *tabula*, tablet) Extinct colonial anthozoans with heavy calcareous skeletal tubes containing horizontal platforms, or tabulae, on which the polyps rested. Sclerosepta absent or poorly developed.

[†]Extinct group.

Summary

1. Members of the phylum Cnidaria are largely marine animals and include jellyfish, sea anemones, and corals. The familiar hydras are among the few freshwater species. The body is radially symmetrical, with the mouth and surrounding tentacles located at one end of the radial axis. Some cnidarians are columnar in shape (polypoid form) and live attached to the bottom, with the oral end directed upward. Others are bowl-shaped (medusoid or jellyfish form) and are free-swimming, with the oral end directed downward.

2. The body wall is composed of two principal layers: an outer epidermis and an inner gastrodermis, separated by the mesoglea. The epidermis and gastrodermis are composed of different cell types; the mesoglea may be an acellular membrane or a thick gelatinous layer with or without cells. Explosive stinging cells called cnidocytes are important in prey capture and are unique to the phylum. Most cnidarians are carnivorous. The mouth is the only opening into the digestive cavity, where digestion is both extracellular and intracellular.

3. The nervous system is commonly in the form of a net with receptor cells dispersed over the body surface. The statocysts and ocelli of jellyfish are the only cnidarian sense organs.

4. Many cnidarians have separate sexes; some are hermaphroditic. The gonads are only aggregations of developing gametes, and there are no gonoducts. Fertilization is usually external, and development leads to a free-swimming planula larva.

5. Members of the class Hydrozoa are polypoid or medusoid, or exhibit both forms in their life history. The mesoglea is acellular in all members of the class. Most polypoid hydrozoans are colonial and polymorphic (hydroids) and usually possess an external chitinous skeleton. Hydromedusae are small and are the sexual, or gamete-producing, individuals. Some hydrozoans, considered to be primitive, have no polypoid stages, and the planula develops directly into a medusa; others, such as the hydras, are entirely polypoid and produce sperm and eggs. Hydroid species possess both polyps and medusae. The medusae may be freed or remain attached to the polyps from which they are budded.

6. The classes Scyphozoa and Cubozoa contain the large medusoid cnidarians commonly called jellyfish. The mesoglea is cellular, and the manubrium is commonly divided into four long oral arms that hang down from the underside of the bell. The planula of scyphozoans develops into a small polypoid larva that buds off or transforms into juvenile medusae.

7. Members of the large class Anthozoa are entirely polypoid. A tubular pharynx leads from the mouth into the gastrovascular cavity, which is radially partitioned by gastrodermal mesenteries. The mesoglea is cellular. Sea anemones are large solitary anthozoans lacking a secreted skeleton. Scleractinian corals are mostly colonial and secrete an external skeleton of calcium carbonate. Octocorals, which include the sea rods, sea fans, sea pens, and sea pansies, are colonial and possess only eight tentacles and mesenteries. The skeleton usually consists of internal calcareous spicules.

8. Coral reefs are tropical, calcareous platforms supporting an array of marine plants and animals, some of which produce the calcium carbonate that composes the platform. Reefs are only found in clear shallow water because reef corals contain symbiotic zooxanthellae that require light.

9. Members of the phylum Ctenophora are marine, gelatinous planktonic animals superficially similar to medusae. They swim by means of eight ciliated bands arranged longitudinally around the biradial body.

References and Selected Readings

Detailed accounts of cnidarians may also be found in the references listed at the end of Chapter 23.

Ahmadjian, V., and S. Paracer. *Symbiosis: An Introduction to Biological Associations.* Hanover, N.H.: University Press of New England, 1986. A brief survey of all types of symbiotic relationships in living organisms, including invertebrate-algal symbiosis.

Colin, P.L. *Caribbean Reef Invertebrates and Plants.* Neptune City, N.J.: T.F.H. Publications, 1978. A field guide to the invertebrates and plants occurring on coral reefs of the Caribbean, the Bahamas, and Florida.

Goreau, T.F., N.I. Goreau, and T.J. Goreau. Corals and coral reefs. *Scientific American* 241(Feb. 1979):124. A good brief account of the biology and geology of coral reefs.

Kaplan, E.H.: *A Field Guide to Coral Reefs of the Caribbean and Florida.* The Peterson Field Guide Series. Boston: Houghton Mifflin Co., 1982. An excellent guide to the common invertebrates and fishes of this region, including considerable ecological information about reefs.

Muscatine, L., and H.M. Lenhoff. *Coelenterate Biology.* New York: Academic Press, 1974. Review papers on many aspects of the biology of cnidarians.

Wood, E.M. *Reef Corals of the World.* Neptune City, N.J.: T.F.H. Publications, 1983. A guide to scleractinian corals.

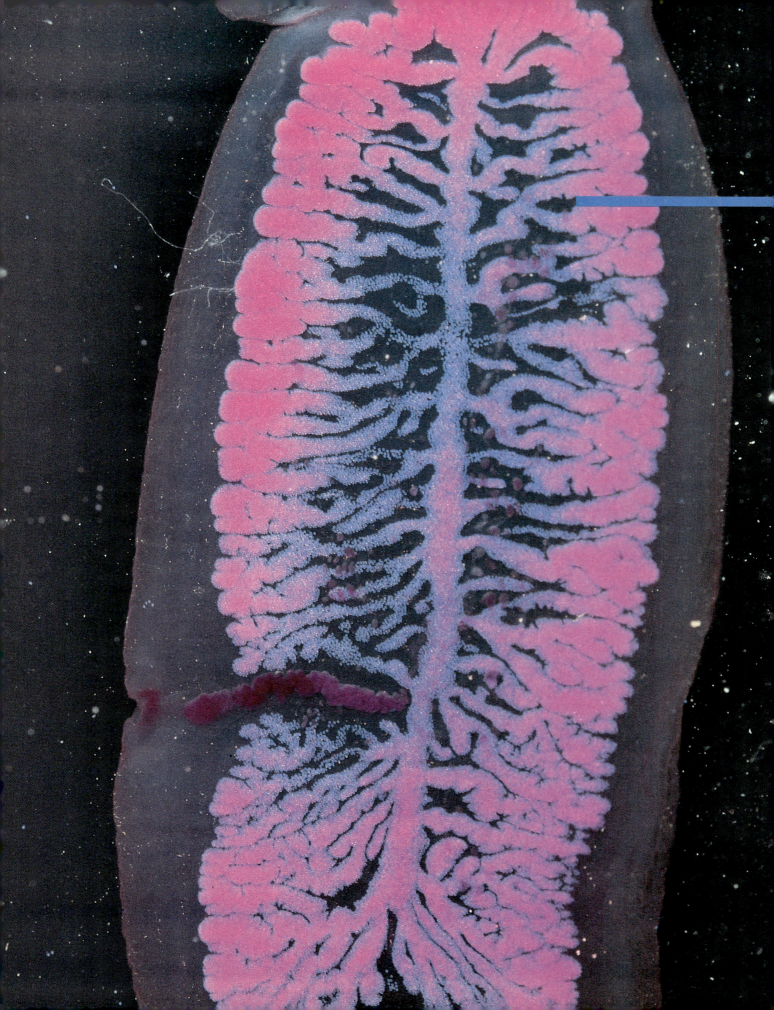

27

Flatworms

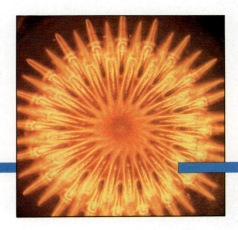

Characteristics of Flatworms

■ Members of the phylum Platyhelminthes include free-living marine and freshwater flatworms and parasitic flukes and tapeworms. Their bilateral bodies are greatly flattened dorsoventrally.

■ The body surface of free-living flatworms is covered by cilia, which are used in locomotion. The body of adult parasitic forms is covered by a nonciliated tegument.

■ The mouth is the only opening into the digestive tract. A digestive tract is absent in tapeworms.

■ There is no internal transport system, protonephridia are present, and the nervous system is composed of a varying number of longitudinal cords.

■ Most flatworms are hermaphroditic; development is usually direct.

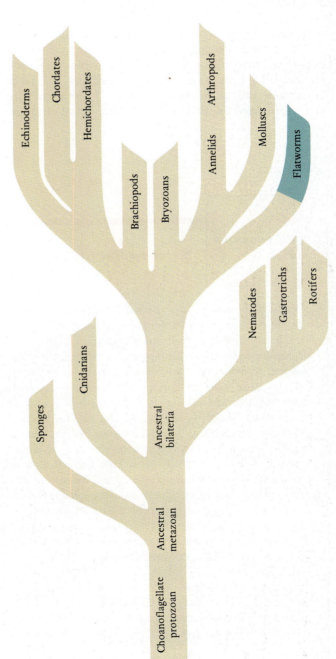

(Left) Segment of the beef tapeworm, Taeniarhynchus saginatus, *parasitic in humans. (Above) Array of hook at the attachment end of a dog tapeworm.*

The phylum Platyhelminthes (flatworms) and all the remaining members of the animal kingdom either are bilaterally symmetrical or, if radially symmetrical, have clearly evolved from bilateral ancestors. Bilaterality and cephalization (head development) are widespread animal features correlated with motility. As we have indicated earlier, bilaterality could have evolved from some radial ancestor that moved with the oral pole forward. If such forms began to swim or creep over the bottom, always keeping the same surface against the substratum, evolution would lead to the upper (dorsal) and lower (ventral) surfaces becoming different, and the animal would develop a bilateral form.

What were the first bilateral animals? Flatworms have long been thought to be the most primitive bilateral animals, but many zoologists are coming to doubt that position. The first bilateral animals probably possessed monociliated cells (one cilium per cell), as is true of cnidarians and a number of other bilateral groups. Flatworms have multiciliated cells, have spiral cleavage,

and exhibit some other features that make them unlikely to lie at the base of the assemblage that includes all bilateral animals. Among living groups of bilateral animals there are really none that neatly fill this position. Such a stem group is probably long extinct. However, in our survey of the bilateral animal phyla we must begin somewhere, so we will follow tradition and start with the flatworms, keeping in mind that this arrangement may not represent an evolutionary sequence.

The chapter will begin by examining the free-living flatworms found in the sea and freshwater, and we will see how the dorsoventrally flattened shape of the body, which is a distinguishing characteristic of the phylum, is of advantage to these animals. Indeed, many aspects of the biology of free-living flatworms are correlated with their size and flatness. We will then turn to the three classes of parasitic flatworms, asking the questions: how is their structure correlated with their parasitic habit and what are the adaptations of their life cycle that increase the chances that

---

Connections 27.1

The Interstitial Fauna

On the floor of the sea the interstitial spaces between sand grains are populated by an abundant, diverse fauna, a world of animals of which few humans bounding in the surf above are at all aware. A cup of sand scooped up from the bottom at wading depth usually contains 50 to several hundred marine animals. Most phyla are represented, but the most common groups are nematodes, copepod crustaceans, turbellarian flatworms, gastrotrichs, and annelids. All are small enough to live in interstitial spaces and, as would be expected, all tend to be somewhat elongate. Since wave and other currents in shallow water can throw sand into suspension, adaptations such as duogland systems and specialized cilia for temporary anchorage to sand grains are common. Other aspects of the biology of interstitial animals are diverse. For example, there are many carnivores but also many species that feed on particulate matter. Some interstitial animals graze on the diatoms, bacteria, or organic films covering sand grains.

The species composition of the interstitial fauna varies with the size of the sand particles and with location, such as intertidal versus subtidal sand and high-energy versus low-energy beaches (subject to little wave action). Most interstitial animals are restricted to the upper 2 cm of bottom sediments, because deeper sediments are poorly oxygenated. The finer the sediments and the greater the amount of organic deposits, the more limited the interstitial fauna, for the interstitial spaces become too small and the upper layer of oxygenated sediments is only a few mm deep.

Interstitial animals can be extracted from bottom sediment by swirling a sample in an isosmotic magnesium chloride solution, which narcotizes the animals, causing them to release their hold on the sand grains. The fluid containing the suspended animals is poured through a nylon sieve, and the collected animals are then reactivated with sea water in a Petri dish. If the animals are active, they are much easier to detect under a dissecting microscope than if they are narcotized and immobile.

The following reference discusses the biology of interstitial animals, describes the common groups that are encountered, and provides techniques for their extraction and study: R.P. Higgins and H. Thiel (eds.). *Introduction to the Study of Meiofauna*. Washington, D.C.: Smithsonian Institution Press, 1988.

some progeny of the parasite will eventually reach the host of the adult worm?

The phylum Platyhelminthes is composed of four classes containing about 18,500 species. The class **Turbellaria** contains the free-living flatworms. The classes **Monogenea** and **Trematoda** (**flukes**), and the class **Cestoda** (**tapeworms**), are entirely parasitic. The distinguishing characteristics of the phylum will be introduced with the following discussion of the free-living species.

CLASS TURBELLARIA

The approximate 3000 members of the class Turbellaria are mostly free-living flatworms and occur both in the sea and in fresh water; a few terrestrial species are found in humid forests. The aquatic turbellarians are bottom dwellers, living within algal masses and beneath stones and other objects. The common familiar freshwater

planarians inhabit the undersurface of stones and fallen leaves in springs, streams, and lakes. There are also marine turbellarians that live with other animals. *Bdelloura*, for example, lives on the gills of horseshoe crabs, and quite a number of other species are commensals or parasites in the guts of various invertebrates.

Many tiny marine turbellarians live in the spaces between sand grains. Animals that live in such spaces compose the **interstitial fauna** (Connections 27.1).

Turbellarians are largely colored shades of gray or brown, although some of the larger species are brightly colored (Figs. 27.1a, b). They range in size from microscopic species to ones more than 60 cm in length (land planarians) (Fig. 27.1b), but most are less than 10 mm long. The body tends to be dorsoventrally flattened, the condition to which the name Platyhelminthes (Gr. *platys*, flat + *helmins*, worm) refers. The larger the species, the more pronounced the flattened shape, and as you shall see, there are good reasons for this. The anterior end often bears

Figure 27.1
Turbellarians. (a) A marine polyclad flatworm, *Prostheceraeus bellostriatus*. (b) A land planarian from the Peruvian Amazon. (c) The catenulid *Stenostomum*, a common microscopic freshwater turbellarian. A simple ciliated pharynx connects the anterior mouth and the intestine, which is an elongated sac. (d) A marine polyclad with tentacles. (e) *Nematoplana*, a marine interstitial turbellarian. (f) *Polycelis*, a freshwater planarian.

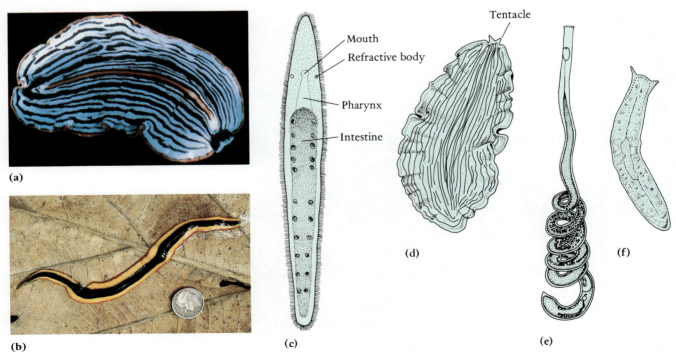

eyes and in some species tentacles (Fig. 27.1d). Planarians commonly possess lateral projections that give the anterior end a triangular shape (Figs. 27.1f and 27.2).

An epidermal layer of multiciliated cells covers the body, although in the planarians only the ventral surface is ciliated. Beneath the epidermis is a muscle layer of circular, diagonal, and longitudinal fibers (Fig. 27.2b). A network of loosely connected cells, called **parenchyma** and rather like connective tissue, surrounds the gut and other organs and fills the interior of the body. Flatworms are therefore said to possess a solid, or **acoelomate**, body structure (p. 190), there being no cavity between the body wall and the internal organs.

Locomotion

Very small flatworms swim or crawl about bottom debris by ciliary propulsion. Contractions of the muscle layer permit turning, twisting, and folding of the body. The movement of larger turbellarians also involves delicate

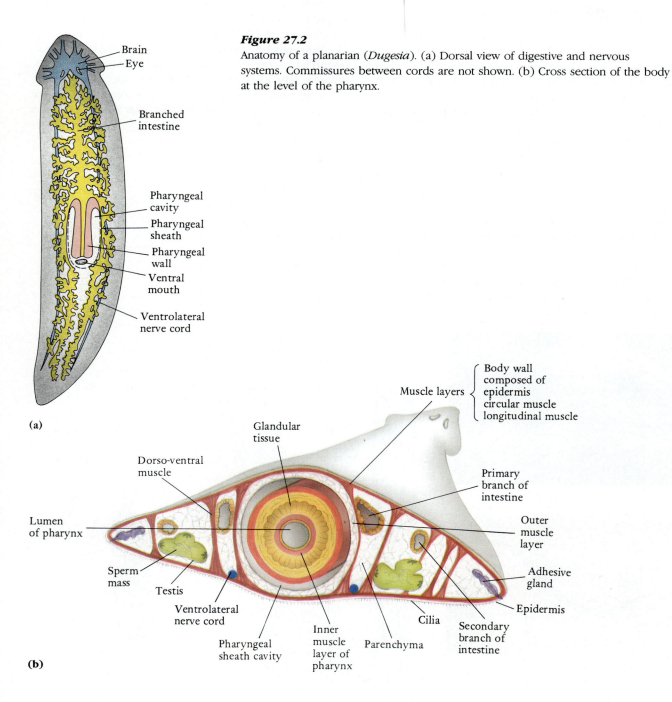

Figure 27.2
Anatomy of a planarian (*Dugesia*). (a) Dorsal view of digestive and nervous systems. Commissures between cords are not shown. (b) Cross section of the body at the level of the pharynx.

(a)

Brain
Eye
Branched intestine
Pharyngeal cavity
Pharyngeal sheath
Pharyngeal wall
Ventral mouth
Ventrolateral nerve cord

(b)

Muscle layers
Body wall composed of epidermis circular muscle longitudinal muscle
Glandular tissue
Dorso-ventral muscle
Primary branch of intestine
Lumen of pharynx
Outer muscle layer
Sperm mass
Testis
Adhesive gland
Ventrolateral nerve cord
Epidermis
Pharyngeal sheath cavity
Inner muscle layer of pharynx
Parenchyma
Cilia
Secondary branch of intestine

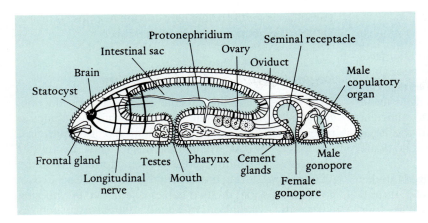

Figure 27.3
Lateral view of a generalized macrostomid turbellarian.

undulatory waves of muscle contraction. The dorsoventral flattening of the body is probably in part an adaptation for locomotion. With increased size, a flattened shape provides a large surface area upon which the body can be carried. Glands in the epidermis and underlying tissue secrete a mucous film over which the animal glides. In many species the mucus is derived from the disintegration of rod-shaped bodies, called **rhabdites**, produced by epidermal gland cells.

Many small turbellarians possess special **duogland systems** that provide for temporary anchorage. One gland of the pair secretes an adhesive substance; the other provides a secretion that breaks the adhesive bond, releasing the animal from anchorage. Glands of this type have now been found in a number of groups of small animals, and we shall encounter them elsewhere.

Nutrition

Most turbellarians are carnivorous, feeding upon other small invertebrates, and as in many other aquatic carnivores, chemical cues are important in finding prey. The mouth is located along the midventral line. The gut is differentiated into a **pharynx** adapted for ingestion and an **intestinal sac** that functions in digestion and absorption. No anus is present, and wastes are ejected through the mouth, as in cnidarians.

The structure of the intestinal sac is correlated with the size of the animal. In small turbellarians the intestine is a simple unbranched sac and the pharynx is a simple ciliated canal (Fig. 27.3; see also Fig. 12.4b) or a muscular bulb. Most of the larger flatworms have a highly branched intestinal sac and a tubular pharynx (Figs. 27.2 and 27.4). The tubular pharynx is muscular and, except for the end attached to the intestinal sac, lies free within a pharyngeal sheath cavity. The cavity opens to the exterior through the mouth. When the animal feeds, the pharyngeal tube is projected out of the mouth.

During feeding, most turbellarians crawl upon their prey, pinning them down and enveloping them with mucus. The prey is often swallowed whole. However, planarians, which have a tubular pharynx, extend the pharynx and insert it into the body of the prey or into dead animal matter with the aid of proteolytic enzymes secreted by pharyngeal glands (see Fig. 12.4c). Fragments of the body of the prey are then pumped into the intestine. The many divisions of the gut, which extend throughout much of the body, greatly increase the surface area available for digestion and absorption.

Digestion in turbellarians occurs in much the same way as in cnidarians: an initial extracellular digestion within the lumen of the intestinal sac is followed by intracellular digestion within the cells of the intestinal wall.

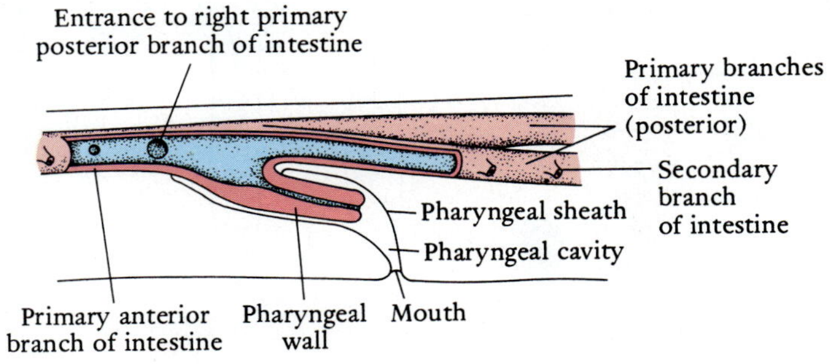

Figure 27.4
Longitudinal section of retracted pharynx and intestine of *Dugesia*.

Gas Exchange, Internal Transport, and Water Balance

Gas exchange takes place across the general body surface, which, because of its flattened shape, is large enough to meet oxygen demands. Moreover, the small vertical distances resulting from the flattened shape greatly facilitate internal transport by diffusion. Diffusion is also sufficient for the movement of food materials, and in the large flatworms the gut branches are so extensive that no tissue is very far from intestinal cells.

Most flatworms possess the system of paired tubules called protonephridia (p. 350). The number of pairs and associated nephridiopores varies. Planarians have four pairs but they are interconnected with many dorsal nephridiopores. The nitrogenous wastes of flatworms, which are chiefly ammonia, are removed by general diffusion across the body surface, and the protonephridia appear to function in the elimination of other kinds of metabolites and in water balance.

Nervous System and Sense Organs

In the majority of flatworms, neurons are organized in longitudinal bundles or nerve cords that lie just below the epidermis, but there are few ganglia. The most primitive arrangement of nerves in flatworms may be one in which there are four or five pairs of longitudinal cords radially arranged around the body (Fig. 27.3). The cords are connected laterally by cross connections; they merge at the anterior end, where the slight enlargement is sometimes called a brain.

In the majority of turbellarians, the radial arrangement of nerve cords has been lost through the predominance of certain pairs and the loss of others. Planarians have a very large ventral pair and a reduced lateral pair (Fig. 27.2). The other pairs are absent. Some large marine flatworms (polyclads) have the neurons organized as a nerve net instead of cords.

A statocyst is present in many turbellarians (Fig. 27.3), but not planarians. They may also have two or more cuplike simple eyes that function in detecting light intensity and direction (Figs. 27.1f and 27.2a; see also p. 420).

Turbellarians avoid bright light and the response is probably an adaptation to remain under cover.

Regeneration and Reproduction

Many turbellarians are capable of asexual reproduction, and this ability is closely correlated with the ability to regenerate. The most common method of asexual reproduction is by transverse fission. In many freshwater planarians, for example, a fission plane forms behind the pharynx, and during movement the animal breaks in two. Following fission, the two halves regenerate the missing parts. In a number of small flatworms, such as *Stenostomum* and *Microstomum*, fission planes and regeneration proceed more rapidly than the separation of the parts, and chains of individuals are formed as a result (Fig. 21.1b).

Parenchyma is the principal source of new cells for regeneration. As in cnidarians, the regenerative parts retain their original polarity (Fig. 27.5a). A piece taken from the middle of the body always regenerates a new anterior end at the severed anterior surface and a new posterior end at the posterior surface. There is also a lateral polarity; thus, a two-headed planarian can be produced by cutting the anterior end longitudinally along the midline (Fig. 27.5b). The rate of regeneration of pieces taken from different levels reflects a distinct metabolic gradient along the anterior-posterior axis. Anterior pieces regenerate more rapidly than do posterior pieces of equal size.

Most flatworms are simultaneous hermaphrodites but usually exhibit cross-fertilization. There is typically simultaneous and mutual copulation, the penis of each animal being received by the female system of the other animal (Fig. 27.6a,b), and the sperm are stored for a period of time. The eggs are fertilized internally and deposited on the bottom in jelly masses or in cocoons (Fig. 27.6c).

What we have just described—the reproductive pattern of the species—is the most important information to remember about the reproduction of any animal group, because from this you can predict much about the anatomy of the reproductive system. How many predictions can you make about the turbellarian system before reading the

Figure 27.5

Polarity and regeneration in *Dugesia*. (a) Each of the five pieces regenerates, but the rapidity with which the head develops depends upon the level of the piece. (b) A two-headed form produced by repeated splitting of the anterior end.

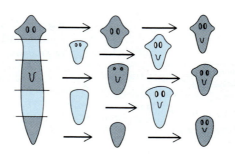

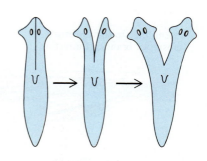

(a)   (b)

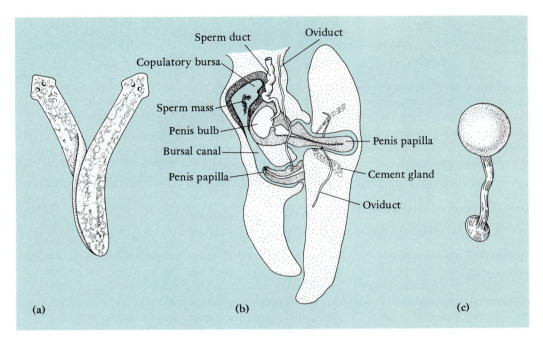

Figure 27.6
Reproduction in planarian flatworms. (a) Copulatory position. (b) Section through
a copulating pair. (c) An attached egg cocoon.

brief description that follows? (The correlations were
described in Chapter 21, p. 496.)

A representative male system may contain one testis
or many pairs of testes (Figs. 27.3 and 27.7a). A small duct
connects each testis with a main sperm duct that extends

along each side of the body. The sperm ducts join together
to form an ejaculation duct, which exits through a penis.
Commonly, the penis is located within a chamber, the
atrium, which may also contain the terminal part of the
female system. The atrium opens to the outside through a

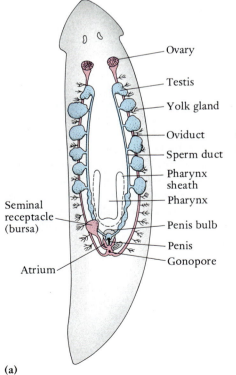

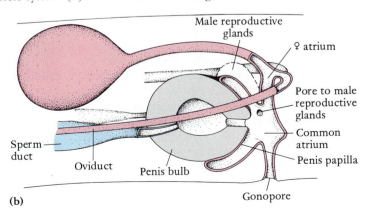

Figure 27.7
Reproductive system of the planarian *Dugesia*. (a) Dorsal view of
complete system. (b) Lateral view of atrial region.

The most common form of symbiosis between animals is **parasitism**, a relationship in which the host is harmed by the presence of the symbiont, or parasite. Parasites that live on the outside of their host are said to be **ectoparasites**; those that live on the inside, **endoparasites**. Most parasites can utilize several closely related or ecologically similar species as hosts. Some parasites require two or more hosts to complete their life cycles, in which case the host for the larval or developmental stage is termed the **intermediate host**; the host for the adult stage is termed the **definitive** or **primary host**.

Nutrition is the usual benefit derived by a parasite from its relationship with the host. The parasite feeds upon the host's tissues or body fluids or utilizes the food ingested by the host. When the parasite is enclosed by some part of the host's body, there may be other advantages, such as protection.

Barriers to Parasitism

A parasitic relationship, however, poses problems and is not without cost to the parasite. The host species represents a very small and discontinuous habitat, and a primary problem of parasites is that of reaching and penetrating new hosts. The host is not without defenses. It responds to invading parasites as it does to other invaders such as viruses and bacteria. Any tissue damage produces initial inflammation and phagocytic reactions of leukocytes or amoebocytes. The host's immune system is also stimulated by the antigen proteins on the parasite's surface. Parasites that get through these host defenses may still be walled off or isolated within the host's body. In vertebrates the HCl and proteases secreted by the stomach act as a formidable barrier that few potential parasites entering the gut can pass. The high body temperatures of birds and mammals and the low level of oxygen in the gut contents of many animals are also barriers to parasitism.

Adaptations for Parasitism

From these generalizations about barriers to parasitism, we can anticipate many of the adaptations encountered in parasites. A variety of structural adaptations may be present depending upon whether the parasite lives on or within the host. Ectoparasites that feed infrequently may have a part of the gut modified for storage. On the other hand, some endoparasites utilize digested food of the host and have lost the gut entirely. The problems of penetration and attachment to the host have resulted in the evolution of a variety of structures, such as suckers, hooks, and teeth. Enzymes may facilitate penetration in some species. The most common point of entrance for endoparasites is through the mouth of the host. The parasite may remain in the gut of the host or, in some species, break through the gut wall to reach other organs.

The problem of reaching new hosts has been met in most parasites through the production of enormous numbers of eggs or other developmental stages. The species is

---

gonopore. The sperm of most turbellarians and other flatworms are unusual in being biflagellate and in having a flagellar ultrastructure of 9:1 microtubules instead of the more typical 9:2 arrangement.

A number of turbellarians have the penis modified as a hollow stylet (Fig. 27.8a). Sperm transfer in such species is by **hypodermic impregnation**, whereby the copulating partners stab each other, injecting sperm through the body wall (Fig. 27.8b).

The female system contains either a single ovary or many pairs of ovaries, but only a single pair of oviducts is present (Figs. 27.3 and 27.7). A seminal receptacle or copulatory bursa that stores sperm, a vagina or atrium, and glands for the production of egg envelopes are typically present. There may be a separate female gonopore (Fig. 27.3), or there may be a common genital atrium and gonopore for both systems (Fig. 27.7b). In many turbellarians yolk glands are located along the length of the oviduct. Yolk cells are released as the eggs travel down the oviduct, and the eggs, when deposited, are surrounded by yolk. The deposition of yolk outside an egg cell (**ectolecithal**) is an unusual condition. In most animals, including some turbellarians, the yolk is contained within the egg cytoplasm (**endolecithal**).

Some freshwater turbellarians produce two kinds of eggs. Summer eggs have a thin shell, hatch within a short period of time, and make possible a rapid buildup of the population; winter or dormant eggs have a thick, resistant shell and are capable of withstanding cold and desiccation. Having two types of eggs is an adaptation shared with many other freshwater animals.

Spiral cleavage is present in turbellarians with entolecithal eggs, and some of these species have larvae.

PARASITIC FLATWORMS

The parasitic flatworms comprise three classes: the Trematoda, which contain the flukes, the Monogenea, and the Cestoda, which comprises the tapeworms. These three groups of flatworm parasites, along with the parasitic

perpetuated if only a few individuals reach the proper host and survive to adulthood. The reproductive system of parasites is highly developed for the production of great numbers of gametes. There are some parasites in which most of the body is concerned with reproduction.

The structure of endoparasites usually reflects the less rigorous demands of the limited and uniform environment in which they live. Locomotor processes may be reduced, and the sense organs found in free-living species are reduced or absent. The nervous system in turn is greatly reduced.

Parasites have evolved various means of dealing with the host's defenses. A common one is analogous to storming a castle: there is great loss of invading troops, but a few get through. To cope with the host immune system, some parasites simply avoid it by residing in sites, such as the gut, that are largely out of reach of the immune system. The developmental stages of some parasites are protected by cyst walls. Blood flukes have evolved the neat trick of concealing themselves immunologically by adsorbing the host's antigens on their surface so that the host's immune system does not recognize them as "foreign" (p. 343). Flagellate trypanosomes, which produce sleeping sickness and other diseases, keep changing their surface antigens so the host immune system cannot get a "fix" on them.

A "successful" parasite usually doesn't kill its host, for a dead host results in a dead parasite. Or if it does kill the host, the timing must be such that the parasite or its eggs get out before the host's death. A host may carry a small population of parasites without any serious consequences. Where the stress of a parasitic infection is manifested in the host, the condition is recognized as a disease. Parasitic disease may cause the destruction of host cells and tissues. Blood vessels, ducts, or the gut may be clogged. The host may be robbed of food. The parasite may produce wastes or substances that have a toxic or allergic effect on the host.

The Evolution of Parasitism

What have been the avenues by which parasitism evolved? Certainly many ectoparasites evolved from species that were initially commensals or occasional occupants of the host's surface. Some parasites evolved from ancestors that were preadapted for a parasitic existence. For example, certain parasitic flies that lay their eggs in skin wounds of cattle are closely related to species that lay their eggs in dead animal tissues (p. 760). The host's mouth has clearly been an important avenue for the evolution of many parasitic groups. The ancestors were accidentally ingested and *survived*. Survival would have led to rapid selection of characteristics that enhanced the parasitic relationship. In parasites that have intermediate hosts, the intermediate host, at least in some cases, was probably the original host. When the first host was eaten by the second host, the parasite survived in the new host, but the adaptations already evolved required the first host stage to be preserved in the life cycle of the parasite.

Figure 27.8
(a) Copulatory stylet of *Macrostomum*. (b) Hypodermic impregnation in *Stenostomum*. Worm at left is injecting worm on the right.

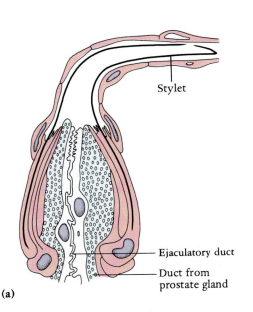

Stylet

Ejaculatory duct

Duct from prostate gland

(a)

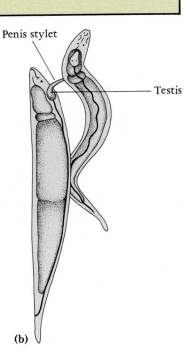

Penis stylet

Testis

(b)

Table 27.1
Principal Pathogenic Flatworm Parasites of Humans

Parasite and Distribution	Location in Host (Adult Parasite)	Pathology	Transmission
Trematodes			
Schistosoma japonicum	Veins of small intestine	Schistosomiasis—general debilitation, toxic reactions, enlargement of liver and spleen, destruction of intestinal tissue	Penetration of skin by cercariae (larvae) released from snails in fresh water
Schistosoma mansoni (Africa, parts of S. America)	Veins of large intestine	Similar to *S. japonicum* except that large intestine is more affected than small intestine	Penetration of skin by cercariae (larvae) released from snails in fresh water
Schistosoma haematobium (Africa, Turkey, Portugal)	Veins of urinary bladder	Bleeding and ulceration of bladder wall; painful urination	Penetration of skin by cercariae (larvae) released from snails in fresh water
Opisthorchis sinensis (Asia)	Bile ducts	Damage to bile ducts; gastrointestinal disorders	Eating encysted metacercariae in raw fish
Paragonimus westermani (Asia, parts of S. America)	Lungs	Some respiratory tissue damage with coughing, chest pain, high mucus production	Eating encysted metacercariae in freshwater crabs and crayfish
Cestodes			
Diphyllobothrium latum (North temperate parts of the world)	Small intestine	Anemia and various gastrointestinal disorders	Eating freshwater fish containing encysted intermediate stage
Hymenolepis nana (Cosmopolitan)	Small intestine	Toxic reactions usually absent or mild	Ingestion of embryonated eggs passed in rat, mouse, or human feces
Taenia solium (Cosmopolitan)	Small intestine	Cysticerci in brain, skeletal muscle, heart; reactions to adult worm absent or slight (diarrhea, weight loss); reactions to cysticerci infection in brain can be severe	Ingestion of cysticerci in rare pork (adult infection); ingestion of embryonated eggs passed in feces (cysticercus infection)
Taeniarhynchus saginatus (Cosmopolitan where beef is eaten)	Small intestine	Nausea, abdominal pain, diarrhea, weight loss	Ingestion of cysticerci in raw or rare beef
Echinococcus granulosus (Cosmopolitan)	Hydatid cysts in liver or other organs or tissues	Reactions to large, growing cyst and toxicity	Ingestion of embryonated eggs passed in feces

roundworms, include the majority of parasitic worms of economic and medical significance (Connections 27.2; Table 27.1). The flukes and tapeworms evolved independently from the free-living turbellarians.

Flukes (Class Trematoda)

Flukes are endoparasitic flatworms belonging to the class Trematoda. About 11,000 species have been described and many are of great economic and medical importance, because vertebrates are usually their primary hosts.

Flukes are flattened and commonly oval or somewhat elongate in shape (Figs. 27.9 and 27.10), the majority being not more than a few millimeters long. The body is covered by a nonciliated **tegument** that is syncytial (lacks cell membranes between nuclei), with part of the layer containing the nuclei sunken into the parenchyma (Fig. 27.9c).

On the outer surface is a layer of glycoproteins (the glycocalyx) that appears to be important in absorption and protection. The mouth is located at the anterior end. Adhesive suckers are usually present around the mouth

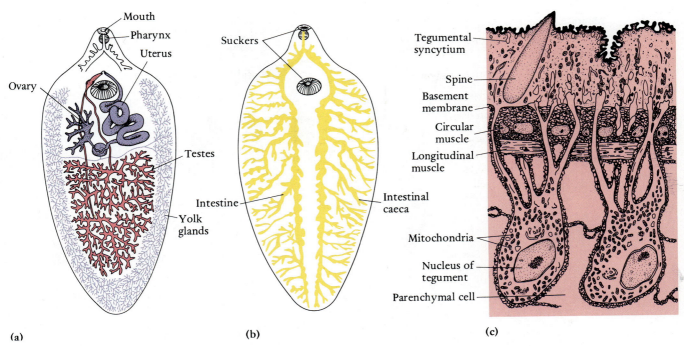

(a)   (b)   (c)

Figure 27.9

Trematode structure. (a) and (b) Structure of the sheep liver fluke, *Fasciola hepatica*. For clarity the digestive system is shown separately from the reproductive system. (c) Section through the body wall of the sheep liver fluke.

and may also be present midventrally. The digestive tract of flukes is composed of a muscular pumping pharynx behind the mouth, a short esophagus, and usually two long blind intestinal sacs (Fig. 27.9b). Flukes feed on cell fragments, mucus, tissue fluids, and blood of the host. Protonephridia are present, and the nervous system is similar to that of turbellarians, except that sensory structures are reduced.

Mutual copulation is the rule, and the reproductive system is adapted for the production of a tremendous number of ectolecithal eggs (Fig. 27.9a). Egg production has been estimated to be 10,000 to 100,000 times greater in trematodes than in the free-living turbellarians.

Life Cycles

Development requires at least one intermediate host. Eggs are passed out of the primary host and hatch as mobile aquatic ciliated **miracidium larvae**, which provide the means of reaching a new host, or the miracidium hatches from an egg ingested by the intermediate host (Fig. 27.10b). The free-swimming miracidium enters a molluscan host, usually a snail, and develops into a **sporocyst**. Within the host the sporocyst gives rise to a second generation of sporocysts or to a number of **rediae**, which migrate to the host's digestive gland or the gonad. A redia may produce more rediae, or develop into a tailed

cercaria larva (Figs. 27.10b and 27.11). The free-swimming cercaria (Gr. *kerkos*, tail) passes from the first intermediate to a second intermediate host or becomes attached to aquatic vegetation or sticks and stones, where it encysts and develops into a quiescent stage, called a **metacercaria**. If eaten by the primary host, the metacercaria is freed from the tissues of the intermediate host and develops into an adult. There are many variations of this generalized outline of the life cycle, but note that as a result of asexual reproduction during the course of development, one egg gives rise to numerous adults (several generations). Because of this multiplication trematodes are sometimes called digenetic flukes.

Opisthorchis sinensis   A great many trematodes infect the gut or gut derivatives of their primary host. Lungs, bile ducts, pancreatic ducts, and intestines are common sites. *Opisthorchis sinensis*, the Chinese liver fluke, lives in the bile ducts of humans, dogs, cats, and pigs (Fig. 27.10). The intermediate hosts for the miracidium and cercaria are a snail and a fish, respectively. Human infection has been common in the Orient because human feces are used to fertilize ponds, increasing the production of algae and water plants upon which the fish graze. The fish produced in these ponds are eaten raw, in part by preference and in part because of limited cooking fuel. A few worms cause no

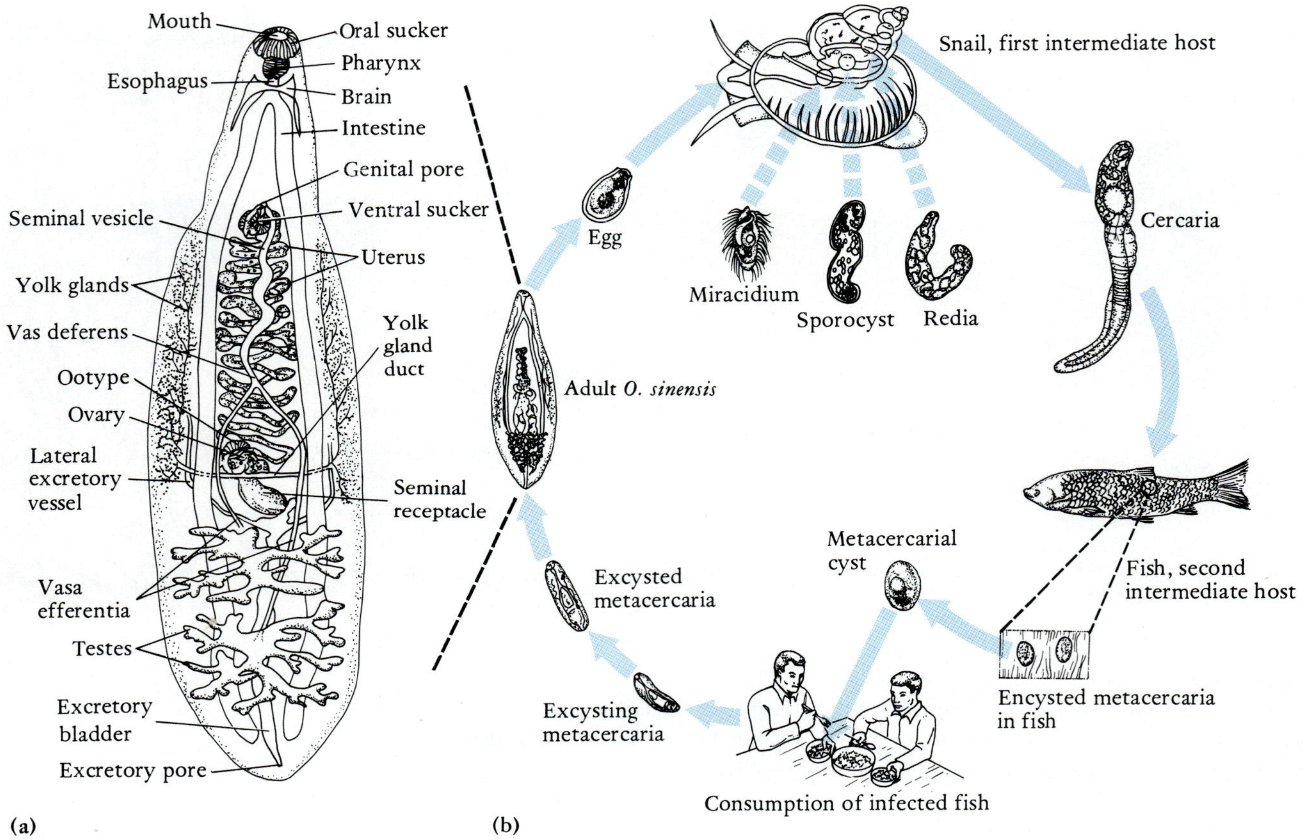

Figure 27.10
The Chinese liver fluke, *Opisthorchis sinensis*: (a) Dorsal view of adult worm.
(b) Life cycle.

disease symptoms, but several hundred can cause destruction of liver tissue, clogging of ducts, formation of bile stones, and hypertrophy of the liver. A single worm can live for as long as 8 years. Millions of people in Asia are believed to be infected.

Fasciola hepatica   Known as the sheep liver fluke, this species reaches 3 cm in length and is one of the largest trematodes (Fig. 27.9a). It is also an inhabitant of its host's bile ducts, and humans, cattle, pigs, rodents, and other mammals may be hosts in addition to sheep. Because of the large size and economic importance in domesticated animals, *Fasciola hepatica* has been one of the most extensively studied flukes. The life cycle is similar to that of the Chinese liver fluke, but a snail is the only intermediate host. The cercariae leave the snail and encyst as metacercariae on vegetation along the edges of ponds and streams, where they are eaten by grazing sheep or other mammalian hosts. Humans obtain the encysted metacercariae in drinking water, but the incidence of this parasite in human populations is low.

Schistosoma mansoni   Not all trematodes inhabit the gut and its derivatives. The members of three families live in the vertebrate circulatory system. Of these blood flukes several species of the genus *Schistosoma* parasitize humans and are responsible for the terribly debilitating disease **schistosomiasis**. Some 300 million people are believed to be infected in Africa and other parts of the tropics. Schistosomiasis ranks with malaria and hookworm infections as one of the three great scourges of humans. The adults of *Schistosoma* can be 2 cm long but not over 1 mm in width. The sexes are separate and the narrower female fits into a longitudinal groove on the male (Fig. 27.11). Depending upon the species, they live in the veins of the urinary bladder, small intestine, or large intestine. The female extends from or leaves the male to deposit eggs, which break out of the abdominal veins into the lumen of the intestine or bladder. Here the eggs are passed out of the host with feces or urine, and if deposited into water, hatch as a miracidium. The miracidium enters certain species of freshwater snails and via several generations of sporocysts gives rise to cercariae. Cercariae leave

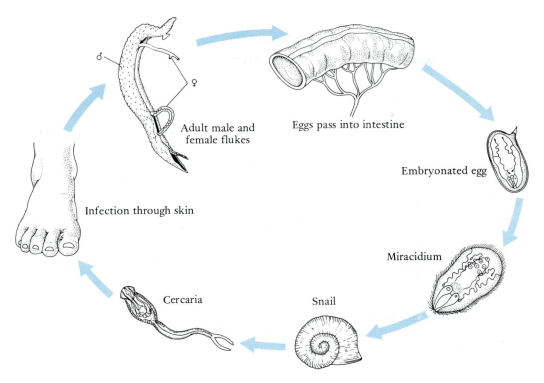

Figure 27.11

Life cycle of *Schistosoma mansoni*, a trematode causing schistosomiasis in humans.

the snail and on contacting a human wading or bathing in the stream or pond, penetrate the skin, using enzymes and muscular movements. They are then carried by the circulatory system to the lungs, liver, and eventually the intestinal veins. The passage of the eggs into the lumen of the intestine or bladder, the lodgement of eggs in aberrant sites, and larval development in the lungs and liver all cause damage to tissues and serious pathogenic responses.

Class Monogenea

The some 1100 species of monogeneans are mostly ectoparasites of aquatic vertebrates, especially fish. Although somewhat similar to trematodes in shape, they are distinguished from trematodes in having a large posterior attachment organ (**opisthaptor**) that bears suckers and hooks (Fig. 27.12). The monogenean life cycle is very different from that of trematodes. There is no intermediate host and one egg gives rise to only one adult worm, hence the name *monogenea*—one generation.

The various species of *Dactylogyrus* illustrate the life cycle of monogeneans. The members of this genus are common ectoparasites on the gills of various freshwater fish (Fig. 27.12).

Dactylogyrus can be a serious problem in fish hatcheries, causing high mortality of young fish from

Figure 27.12

Dactylogyrus vastator, a monogenean ectoparasitic on the gills of freshwater fish.

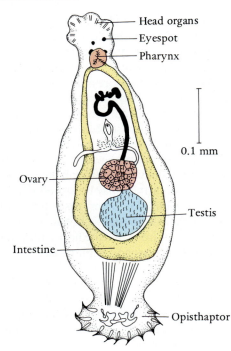

secondary infection, smothering by excess mucus production, or loss of blood. The eggs of the fluke drop to the bottom, where they hatch and liberate a free-swimming oncomiracidium larva. On contacting the host fish, the larva metamorphoses into the adult fluke. Egg production increases with rise of water temperature so that the population of flukes builds up over the summer. In the fall some eggs overwinter to begin new cycles of infestation in the spring.

Tapeworms (Class Cestoda)

About 3400 species of cestodes, or tapeworms, have been described. All are endoparasites, and the majority are adapted for living in the gut of vertebrates.

Although adult tapeworms are covered by a nonciliated tegument, the body form of adult tapeworms is quite different from that of turbellarians and flukes. The anterior end consists of a knoblike **scolex** containing suckers and, in some, hooks that anchor the worm to the host (Fig. 27.13a). A narrow neck region connects the scolex to the **strobila**, which makes up the greater part of the tapeworm body. The strobila is composed of flattened sections, called **proglottids**, arranged in a linear series. New proglottids are continually being formed in the neck region and old ones detach at the end of the strobila. Tapeworms are generally long, and some specimens, with thousands of proglottids, have been known to reach 25 meters!

The digestive system is completely absent in tapeworms, and digested food of the host is absorbed through the tegument, which is like that of trematodes but is covered with projections similar to microvilli. Longitudinal nerve cords and excretory ducts run the length of the strobila. Much of the body tissue is given over to the reproductive system, which is complete within each proglottid. A common gonopore is usually present on each proglottid. Copulation between the proglottids of two different worms often occurs, but copulation between proglottids on the same strobila and self-fertilization within one proglottid are more usual.

Eggs containing embryos, which are surrounded by a shell, may be continually shed through the gonopore into the host's intestine, or they may be stored in a blind sac, called the uterus. In the latter case, terminal gravid proglottids (packed with eggs) break away from the strobila and may rupture within the host's intestine, or they may leave with the feces and rupture later.

Life Cycles

One or more invertebrate or vertebrate intermediate hosts are required to complete the life cycle, which involves an **oncosphere larva** bearing three pairs of hooks, and one or more subsequent developmental stages.

Species of the family Taeniidae are among the best known tapeworms. *Taeniarhynchus saginatus*, the beef tapeworm, is one of the most common species in humans, where it lives in the intestine and frequently reaches a length of over 3 meters (Fig. 27.13b). Proglottids containing embryonated eggs (eggs with embryos) are eliminated through the anus, usually with feces. If an infected person defecates in a pasture, the eggs may be eaten by grazing cattle, sheep, or goats. On hatching within the intermediate host, an oncosphere larva bores into the intestinal wall, where it is picked up by the circulatory system and transported to striated muscle. In the muscle the larva develops into a **cysticercus** stage. The cysticercus, sometimes called a **bladder worm**, is an oval stage about 10 mm in length, with the scolex invaginated into the interior. If raw or insufficiently cooked beef is ingested by humans, the cysticercus is freed, the scolex evaginates, and the larva develops into an adult worm within the gut.

A severe infection of adult tapeworms may cause diarrhea, weight loss, and reactions to the toxic wastes of the worm. Despite folk legends to the contrary, persons harboring numerous tapeworms are not ravenously hungry; in fact, there is usually a loss of appetite. The worms can be eliminated with drugs. Much more serious is cysticercus infection. Fortunately, the cysticercus stage of the beef tapeworm will not develop within humans, but this is not the case with the pork tapeworm, *Taenia solium*, and one of the dog tapeworms, *Echinococcus granulosus*. The pork tapeworm has a life cycle rather like that of the beef tapeworm, except that pigs rather than cattle are the intermediate hosts. The adult *Echinococcus*, which lives in the intestine of dogs, is minute, with only a few proglottids present at any one time. Many different mammals, including humans, can act as intermediate hosts, although herbivores are the most important in completing the life cycle. The cysticerci of the pork tapeworm develop in subcutaneous connective tissue and in the eye, brain, heart, and other organs. The cysticercus, or hydatid, of *Echinococcus* usually develops in the liver or lung but can develop in many other sites as well. The bladder worms of both of these species can be very dangerous when growing in such places as the brain and can do much damage elsewhere. Hydatid cysts can reach a large size and contain a great volume of fluid (up to many liters!) that if released into the host could cause severe reactions. Bladder worm cysts can be removed only by surgery.

Copepods—tiny aquatic crustaceans—and aquatic annelids are the intermediate hosts for the tapeworms of many fish. Insects and mites are intermediate hosts for various tapeworms living in terrestrial vertebrates.

EVOLUTIONARY RELATIONSHIPS

As indicated earlier, flatworms have long been thought to be primitive bilateral animals, and there are many zoologists who still hold to this position. Others believe

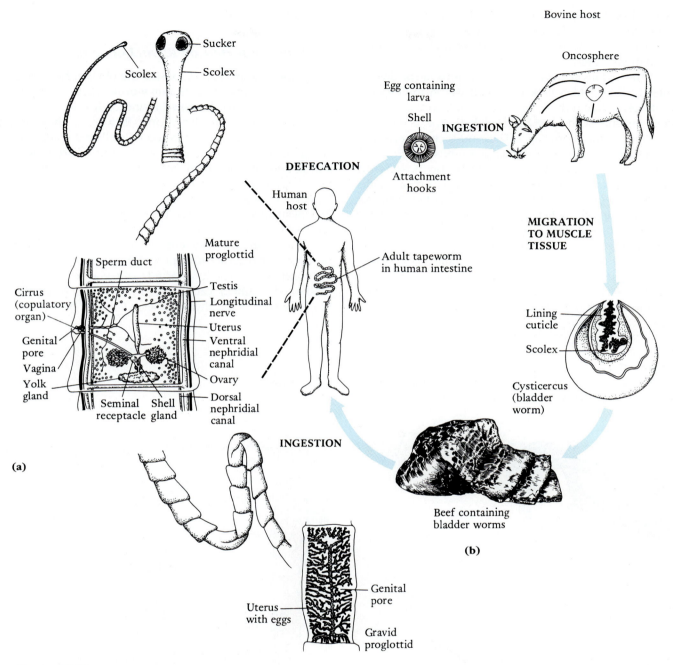

Figure 27.13
(a) The structure of the beef tapeworm, *Taeniarhynchus saginatus*, showing the
anterior end, a proglottid, and a terminal gravid proglottid. (b) Life cycle of
Taeniarhynchus saginatus.

flatworms are more closely related to coelomate protostomes. They are clearly protostomes, as evidenced by
their spiral cleavage and multiciliated cells. The debate
about their protostome position centers on the nature of
their acoelomate structure and their lack of an anus. Those
who believe that flatworms are primitive think that both
of these conditions are primary, i.e., the ancestors of
flatworms never possessed a body cavity or an anus. Those

who believe they are more closely related to higher protostomes think that the acoelomate condition and lack of an anus is secondary, i.e., that flatworms have lost the coelom and anus.

The various groups of living flatworms can be divided into those in which yolk is located within the egg cell (entolecithal) and those in which yolk is located in yolk cells outside of the egg (ectolecithal). The ectolecithal condition is the more specialized and is characteristic of triclad and rhabdocoel turbellarians (see Classification). The trematodes and cestodes, which also possess the ectolecithal condition, are believed to have evolved independently from the rhabdocoels. Because of their hooks, monogeneans are believed to be more closely related to cestodes than trematodes, despite their similarity to trematodes in general body form.

Of all the living flatworms, the little macrostomid turbellarians may be most primitive. They exhibit the entolecithal condition and have a simple unbranched intestinal sac and a simple ciliated pharynx.

RELATED PHYLA

Phylum Nemertea or Rhynchocoela

The some 650 species composing the phylum **Nemertea** are called **nemerteans** or **ribbon worms**. They are mostly marine animals that burrow in sand or live in algae or beneath stones. Like free-living flatworms, they are dorsoventrally flattened, covered by a ciliated epidermis and acoelomate. However, most nemerteans are longer and somewhat larger than flatworms. The burrowing ribbon-shaped species of *Cerebratulus* and *Lineus* reach lengths of several meters. The record is 30 m for one European species of *Lineus*. However, most nemerteans are less than 20 cm long.

The distinguishing feature of nemerteans is the long proboscis used in capturing prey (Fig. 27.14). The proboscis, which is usually unconnected to the gut, lies in a long, fluid-filled chamber called the rhynchocoel. When the fluid pressure of the rhynchocoel is elevated by the con-

Figure 27.14
(a) A nemertean, member of the phylum Rhynchocoela, with proboscis everted.
(b) *Prostoma rubrum*, a freshwater hoplonemertean.

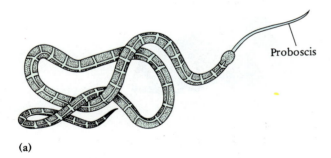

Proboscis

(a)

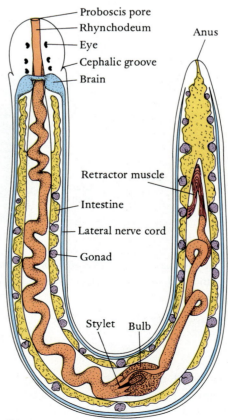

Proboscis pore
Rhynchodeum
Eye
Cephalic groove
Brain
Anus

Retractor muscle

Intestine

Lateral nerve cord

Gonad

Stylet   Bulb

(b)

tractions of surrounding muscles, the proboscis is shot out, turning wrong side out in the process. The everted proboscis of many nemerteans bears a poisonous, dagger-like stylet at the tip. The prey, which are mostly annelid worms and crustaceans, is stabbed and then brought to the mouth and swallowed whole or sucked from its external skeleton. A nemertean may leave its burrow to hunt for prey. *Cerebratulus lacteus* goes into the burrows of razor clams and swallows the owner. Prey mucus is important in locating prey, and the worm uses its own mucus trail to return to its burrow.

Nemerteans have long been thought to be closely related to flatworms, but some zoologists now believe that both groups evolved separately from some common coelomate worm and that the proboscis cavity is a remnant of the coelom (p. 190). Unlike flatworms, nemerteans possess an anus and a blood vascular system. They exhibit spiral cleavage and some species possess a larval stage that is somewhat like that of annelids.

Phylum Mesozoa

This phylum contains about 50 species of minute ciliated endoparasites of various marine invertebrates, especially cephalopods and bivalves (Fig. 27.15). Since they lack organs, their relationship to other animals is actually unknown. Their ciliation and parasitic nature have led some

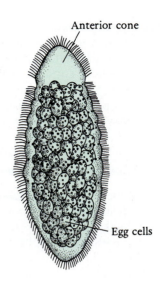

Anterior cone

Egg cells

Figure 27.15
A female mesozoan (*Rhopalura granulosa*) parasite in clams.

zoologists to speculate that they are related to flatworms but lost their organs as they became minute. Other zoologists think that the lack of organs is a primitive condition and that mesozoans departed early from the main line of metazoan evolution. We have included them here for purposes of comparison.

Classification of Phylum Platyhelminthes

Class Turbellaria (L. *turbellae*, disturbance) Mostly free-living ciliated flatworms. Of the nine orders of turbellarians, the following are commonly encountered and most easily described:

Order Acoela (Gr. *a*, without + *koiloma*, cavity) Small marine flatworms, usually measuring less than 2 mm in length. Mouth and sometimes a simple pharynx present, but no digestive cavity. Protonephridia absent. No distinct gonads. Oviducts and yolk glands absent. *Polychoerus, Convoluta.*

Orders Macrostomida (Gr. *makros*, large + *stoma*, mouth) and **Catenulida** (L. *catenula*, little chain) Small marine and freshwater species having a nonmuscular pharynx and a simple saclike intestine. *Macrostomum, Microstomum; Stenostomum* (Catenulida).

Order Polycladida (Gr. *polys*, many + *klados*, branch) Marine flatworms of moderate to large size, averaging from 2 to 20 mm in length, with a greatly flattened and more or less oval shape. Intestinal sac elongated and centrally located with many highly branched diverticula. Eyes numerous. Yolk glands absent. *Leptoplana, Prosthoceraeus.*

Order Rhabdocoela (Gr. *rhabdos*, rod + *koiloma*, cavity) A large group of small marine and freshwater turbellarians, having a bulbous pharynx, single intestine, and one pair of nerve cords. *Mesostoma.*

Order Tricladida (Gr. *treis*, three + *klados*, branch) Relatively large marine, freshwater, and terrestrial flatworms, ranging from 2 cm to more than 60 cm in length in the terrestrial forms. The freshwater and terrestrial species are called planarians. Digestive system has a tubular muscular pharynx and a three-branched intestinal sac. Both the mouth and the pharynx are located in the middle of the body. Eyes are present in most species. *Dugesia, Phagocata, Polycelis, Bdelloura.*

Class Monogenea (Gr. *mono*, single + *genea*, birth) Ectoparasites largely of fish. Body elongate or oval and covered by a nonciliated tegument. Posterior end of worm bears a large adhesive organ (opisthaptor) provided with suckers and hooks. Digestive system present, with mouth at anterior end. Life cycle does not require an intermediate host. *Dactylogyrus.*

Class Trematoda* (Gr. *trematodes*, pierced) Flukes, endoparasites, mostly of vertebrates. Body elongate or oval and covered by a nonciliated tegument. Anterior and more posteriorly located suckers present but no

*This class has commonly been treated as an order (Digenea), and under this system the class Trematoda also contains the monogeneans.

opisthaptor. Digestive system present, with mouth at anterior end. Life cycle requires one or more intermediate hosts. *Fasciola, Opisthorchis, Schistosoma.*

Class Cestoda (Gr. *kestos*, girdle) Tapeworms. Endoparasitic flatworms lacking a digestive system. Body covered by a nonciliated tegument and composed of a scolex and proglottids. *Taenia, Echinococcus.*

Summary

1. Members of the phylum Platyhelminthes, called flatworms, are free-living or parasitic animals that are dorsoventrally flattened and have an acoelomate body construction. Most possess protonephridial tubules and are simultaneous hermaphrodites.

2. Gas exchange and elimination of nitrogenous waste occur across the general body surface. Diffusion also provides for internal transport. Some flatworms possess a nerve net type of nervous system, but most have a varying number of longitudinal nerve cords. A statocyst and two or more simple eyes are the most comon sense organs.

3. The class Turbellaria contains the mainly free-living flatworms, found largely in the sea and in fresh water. The body is covered with cilia, which are used in locomotion. Turbellarians are carnivores or scavengers. The mouth is the only opening into the digestive tract, which consists of a pharynx and an intestine. The intestine, which is the site of extracellular and intracellular digestion, is a simple sac in small species and a highly branched sac in larger forms. Turbellarians exhibit copulation, reciprocal sperm transfer, and internal fertilization. Primitively, some species have spiral cleavage, and a few marine forms have a planktonic larva. Most have ectolecithal eggs and exhibit direct development.

4. The parasitic flukes belong to the large class Trematoda. The body of flukes is covered by a nonciliated syncytium. Suckers are typically present. Flukes are endoparasites of many different vertebrates, and the life cycle requires at least one intermediate host, commonly a mollusc. One egg gives rise to numerous offspring as a result of asexual reproduction at various points in development. Flukes are the cause of numerous parasitic diseases of humans and domesticated animals.

5. Members of the class Monogenea are largely ectoparasites of fish, and the life cycle usually involves a miracidium larva and only one host.

6. The class Cestoda contains the endoparasitic tapeworms, most of which live in the gut of vertebrates. The body is composed of an anterior section (scolex) modified for attachment and a long string of segment-like proglottids produced by budding from the scolex. Each proglottid contains complete male and female reproductive systems. Tapeworms are gutless, and food is absorbed from the host across a nonciliated tegument. To complete their life cycle tapeworms require one or more intermediate hosts, which may be invertebrates or vertebrates. The intermediate host usually obtains the parasite orally, and the primary host becomes infected by eating the intermediate host.

References and Selected Readings

Detailed accounts of flatworms may also be found in the references listed at the end of Chapter 23. The works listed below deal specifically with the biology of animal parasites and commensals and the general biology of symbiosis.

Ahmadjian, V., and S. Paracer. *Symbiosis: An Introduction to Biological Associations.* Hanover, N.H.: University Press of New England, 1986. A brief review of the types of symbiotic relationships found among living organisms.

Bogitsh, B.J., and T.C. Cheng. *Human Parasitology.* Philadelphia: Saunders College Publishing, 1990. A textbook of human parasitology.

Meyer, M.C. *Essentials of Parasitology*, 4th ed. Dubuque, Iowa: W.C. Brown Publishers, 1988. This and the following two references are textbooks of parasitology, surveying the parasitic groups of animals.

Noble, E.R., et al. *Parasitology*, 6th ed. Philadelphia: Lea & Febiger, 1989. A textbook of parasitology.

Schmidt, G.D., and L.S. Roberts. *Foundations of Parasitology*, 4th ed. St. Louis: C.V. Mosby Co., 1989. A textbook of parasitology.

Whitfield, P.J. *The Biology of Parasitism: An Introduction to the Study of Associating Organisms.* Baltimore: University Park Press, 1979. This volume deals with various aspects of the biology of parasitism rather than with a survey of parasitic groups.

28

Aschelminths

Characteristics of Aschelminth Phyla

■ The body is covered by a cuticle.

■ Adhesive glands for temporary anchorage are usually present.

■ The body cavity, when present, is a pseudocoel.

■ They are mostly minute animals (less than a few millimeters in length) and organs of gas exchange and internal transport are usually absent.

■ The sexes are usually separate.

Phylum Rotifera

■ Rotifers are mostly freshwater animals with somewhat elongate or saclike bodies.

■ An anterior ring of cilia (ciliated crown) functions in swimming and, in some species, in feeding.

■ A posterior foot with adhesive toes provides for temporary attachment in benthic species.

■ Food is chewed or ground with cuticular jaws located within a muscular pharynx (mastax).

Phylum Gastrotricha

■ Gastrotrichs are marine and freshwater animals with bottle- or strap-shaped bodies.

■ The ventral surface bears cilia with which the animal moves.

Phylum Nematoda

■ Nematodes are marine, freshwater, terrestrial, or parasitic worms with a long cylindrical body tapered at each end.

■ The body wall is composed of a thick outer elastic cuticle, an epidermis, and a layer of longitudinal muscle.

■ A large pseudocoel lies between the body wall muscles and the central tubular gut.

(Left) A colony of rotifers. (Above) The rotifer
Platyias quadricornis.

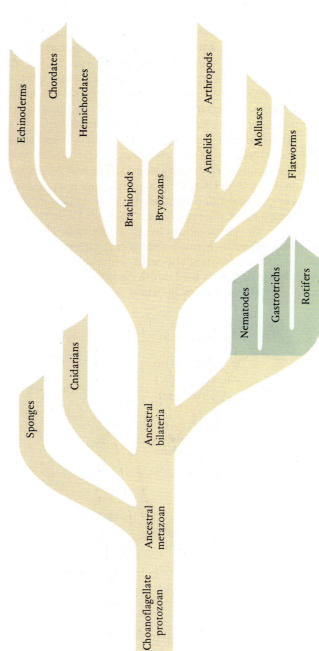

The aschelminths are an assemblage of eight phyla that appear to be related to each other to varying degrees. Although each of these phyla is distinctive in its own right, collectively there are only a few features that unite them. When Aschelminthes (Gr. *askos*, bag + *helminthos*, worm) was first proposed as a phylum to embrace all of these groups, it was jokingly called the "ash can" phylum, since it appeared to be a dumping ground for a heterogenous collection of animals. Most zoologists today consider each group a separate phylum and the aschelminths a convenient term to designate the entire assemblage.

Aschelminths vary greatly in shape but all are poorly cephalized, i.e., they lack well-developed heads with sense organs. The body is covered with a flexible organic **cuticle**, and many possess **adhesive glands** for temporary attachment. The gland openings commonly project from the cuticle as short adhesive tubes. The digestive tract of aschelminths is a tube with two openings; a mouth and an anus. A muscular pharynx and a stomach or intestine are the principal specializations of the gut tube.

Aschelminths are sometimes called pseudocoelomates because most species possess a type of body cavity called a **pseudocoel**. Unlike a true coelom, which develops within mesoderm (Fig. 22.6), a pseudocoel is a persistent blastocoel that is not obliterated by gastrulation or by the development of internal organs (Fig. 28.1). It is never bounded by peritoneum as is a coelom. However, *pseudocoelomates* is not a very good collective term for all of these animals, because although some, like roundworms, have a well-developed pseudocoel, others have only slitlike spaces between organs.

Most aschelminths are minute in size, and this is perhaps the most useful characteristic to associate with them. Many are microscopic or just barely visible to the naked eye. It is not surprising that large numbers compose the interstitial fauna (see Connections 27.1, p. 622), and a few of the aschelminth phyla are found only in this type of habitat.

Associated with their small size, aschelminths lack gas exchange organs and organs of internal transport. Their protonephridia are not involved in the excretion of nitrogenous wastes. Nervous and muscular systems are present, but members of several phyla utilize cilia in locomotion.

Some aschelminths exhibit the unusual condition, called **eutely**, of having organs with fixed numbers of cells. Mitotic divisions takes place only during embryogenesis, and growth following hatching occurs by increase in cell size alone. The rotifer *Epiphanes senta*, for example, possesses 958 nuclei, of which 35 form the stomach, 120 the musculature, and so on. The same numbers are found in all individuals of this species. There is a lower size limit for metazoan cells. Thus as animals become very small, their cells cannot become proportionately smaller. Rather, their bodies and organs become composed of fewer cells. Eutely may represent a means of fixing minimum cell numbers.

Reproduction in most pseudocoelomates is entirely sexual, and with few exceptions, the sexes are separate. Cleavage is determinate but not usually spiral. Larval stages are typically absent in free-living species.

Only three of the eight aschelminth phyla are discussed below. However, a brief characterization and illus-

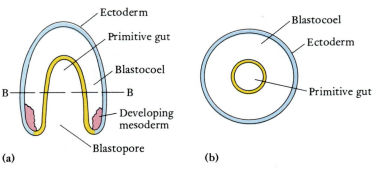

(a)

(b)

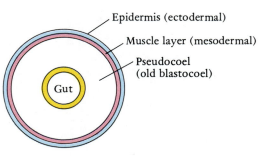

(c)

Figure 28.1

Embryonic origin of the pseudocoel. (a) Frontal section of the gastrula, showing the remains of the blastocoel and the developing mesoderm; (b) cross section of (a) taken at level B—B; (c) diagrammatic cross section of adult pseudocoelomate.

tration of each of the remaining five is provided in the classification at the end of the chapter.

PHYLUM ROTIFERA

The phylum Rotifera consists of some 1500 species, most of which live in fresh water, including the water films around soil particles and in mosses. In freshwater pools, ponds, and lakes, they are usually very common and abundant animals. Only a small number of species are found in sea water.

Most rotifers range in size from 0.1 to 1.0 mm, which overlaps the size range of ciliate protozoans. The elongate

Figure 28.2
Ventral view of a rotifer, *Philodina roseola*, showing many of the internal structures.

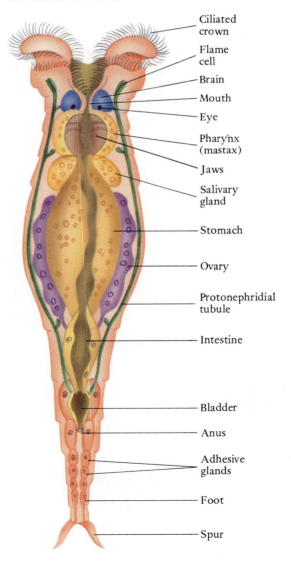

- Ciliated crown
- Flame cell
- Brain
- Mouth
- Eye
- Pharynx (mastax)
- Jaws
- Salivary gland
- Stomach
- Ovary
- Protonephridial tubule
- Intestine
- Bladder
- Anus
- Adhesive glands
- Foot
- Spur

or saclike body, which contains only about 1000 cells, is composed of a trunk and a posterior foot (Figs. 28.2 to 28.4). The anterior end of the trunk bears a **crown** of cilia (**corona**), a distinguishing characteristic of the phylum. The distribution of cilia varies in different rotifers. A common type of crown is one in which the cilia are arranged in the form of two discs (Fig. 28.2). The two discs of cilia beat in a circular manner, one clockwise and one counterclockwise, and look like two wheels spinning, from which the name *Rotifera* ("wheel bearer") is derived. In some rotifers the cuticle of the trunk is thickened to form an armor-like girdle called a **lorica** that commonly bears spines.

The foot is a narrow posterior extension of the trunk and in many species contains terminal adhesive glands, which open to the exterior through tubes, called toes or spurs, at the end of the foot (Fig. 28.2). The glands are not duogland systems. In some rotifers the foot is divided into ringlike sections, which can telescope into one another.

Rotifers swim by means of the ciliated crown and crawl in a leechlike fashion, using the adhesive glands as a means of attachment. Many species of rotifers live in algae and bottom debris. They swim and crawl intermittently. Some rotifers are planktonic and are never found on the bottom (Fig. 28.4b). In these species the foot is reduced or absent. A few rotifers are sessile (Fig. 28.4c).

A second distinguishing feature of rotifers is the pharyngeal apparatus, called the **mastax**. The mastax differs from the pharynx of the other aschelminths in possessing cuticular teethlike pieces projecting into the lumen (Fig. 28.2). The teeth vary in size and shape depending upon the feeding habit. Predatory rotifers have a mastax that can grasp and chew prey, mostly protozoans and other rotifers (Fig. 28.5b). In rotifers with disclike ciliated crowns, the

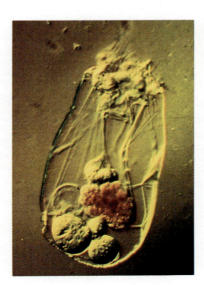

Figure 28.3
A planktonic rotifer; a species of *Asplanchna*.

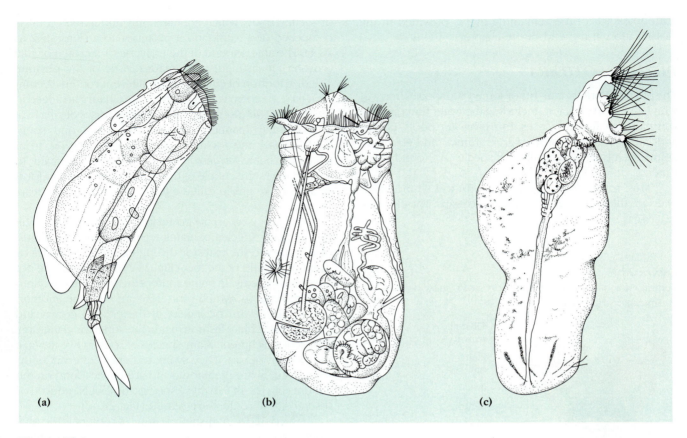

(a)   (b)   (c)

Figure 28.4
Three rotifers. (a) Lateral view of a species of *Euchlanis*. The body is
dorsoventrally flattened, and the foot possesses well-developed toes through which
the adhesive glands open. This rotifer swims and anchors intermittently. (b) A
species of *Asplanchna*. This predaceous planktonic rotifer has a transparent saclike
body with no foot. (c) A species of *Collotheca*, a sessile rotifer, in which the body
is enclosed within a gelatinous case. The ciliated crown is reduced, and the
anterior end, which has the form of a funnel, is adapted for trapping prey.

Figure 28.5
Two types of mastax teeth. (a) Mastax with teeth
adapted for grinding. Note the two large ridged
plates that provide grinding surfaces. (b) Mastax
with grasping, forceps-like teeth.

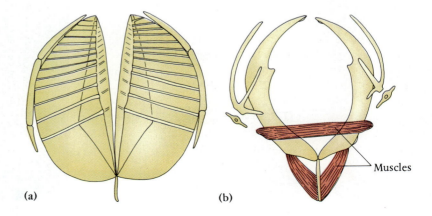

(a)   (b)

Muscles

Cryptobiosis: True "Suspended Animation"

Minute animals such as rotifers, gastrotrichs, and nematodes, which are found in soil, mosses, and lichens, are really aquatic and not terrestrial animals, since they live in the water films of their microhabitats. These habitats periodically dry up, but many of these animals have evolved remarkable abilities to survive extreme desiccation. With loss of water, the animals pass into a dormant state, called cryptobiosis.

One of the most notable examples of animals capable of cryptobiosis are the microscopic inhabitants of mosses called tardigrades (phylum Tardigrada) or water bears because of the presence of four pairs of stubby legs (Fig. a). They are not aschelminths but are related to arthropods (p. 765). When the moss becomes desiccated, the tardigrades pull in their legs, lose water, and become contracted and shriveled (Fig. b). Metabolism proceeds at a very low rate, and the animal can withstand extreme and unusual environmental conditions. For example, specimens have been recovered after submersion in liquid helium (−272 C), brine, ether, absolute alcohol, and other substances. When water is again present, the animal swells and becomes active in a few hours. There are records of tardigrades emerging from 7 years of cryptobiosis. The cryptobiosis of rotifers and nematodes is similar. Certainly these periodic cryptobiotic states must significantly increase the life span of the animal.

(a)

(b)

beating cilia drive small suspended particles into the mouth. The mastax of these rotifers contains broad flattened pieces adapted for grinding the particles (Fig. 28.2 and 28.5a).

A large stomach characterizes rotifers. A short intestine passes from the stomach to the anus, which is located on the ventral side at the end of the trunk. The terminal part of the intestine is called the cloaca because it receives the ducts of two protonephridia and in the female it receives the oviducts.

Reproduction is entirely sexual. In some species, the male, which is usually smaller than the female, stabs the female on any part of her body with his penis (hypodermic impregnation). In other rotifers, the male inserts his penis into the cloaca of the female. In either case, fertilization is internal. During the life of a female, 8 to 20 shelled eggs, one for each ovarian nucleus, are deposited singly on the bottom or are attached to the body of the female. On hatching, females have all of the adult features and attain sexual maturity after a growth period of a few days. The smaller males are sexually mature on hatching. Sessile rotifers are initially free-swimming.

Many freshwater species produce both rapidly hatching thin-shelled eggs and dormant thick-shelled eggs. Dormant eggs can withstand low temperatures and desiccation; they can be dispersed on the feet of birds or other animals or blown with dust. The production of rapidly hatching eggs is commonly coupled with parthenogenesis. In some species, both parthenogenesis and development from fertilized eggs occur, but in other species, there are no known males, and all individuals are females produced parthenogenetically. Parthenogenetic eggs are diploid and produce only females. When the population reaches a peak, haploid eggs are produced and males appear. The fertilized eggs are thick-shelled and undergo dormancy. Parthenogenesis and rapidly hatching eggs are perhaps adaptations for a rapid expansion in number of a population after it has been reduced in freshwater pools and streams by desiccation and other extreme environmental conditions.

As might be expected, the rotifers living on mosses or in soil are capable of withstanding extreme environmental conditions and can remain in an inactive state for as long as 3 to 4 years (Connections 28.1).

PHYLUM GASTROTRICHA

The phylum Gastrotricha constitutes one of the smaller phyla of pseudocoelomates, with about 460 species. Its members are found in the sea and in fresh water, and there are many more marine species than is true of rotifers. Many gastrotrichs live in the interstitial spaces between sand grains; others live among surface debris and algae and even the water films of soil particles. Like most rotifers, gastrotrichs are microscopic. The body is strap- or bottle-shaped (Figs. 28.6 and 28.7), and the ventral surface is flattened and ciliated, from which the name *Gastrotricha* ("stomach hairs") is derived. In some gastrotrichs this ciliation is formed by monociliated cells, and we will look at the possible evolutionary significance of this shortly. The anterior end may bear bristles or tufts of cilia. The posterior end is sometimes forked (Fig. 28.6). The cuticle of some gastrotrichs is modified in the form of scales, spines, or hooks. Adhesive tubes are present in rows along the sides of the body or in the forks at the posterior end (Fig. 28.7). The adhesive tubes contain a duogland system, as in turbellarians (p. 625).

Gastrotrichs glide over the bottom propelled by the ventral cilia and may temporarily attach by means of the adhesive tubes. They feed on bacteria, small protozoa, algae, and detritus. Food is swept into the anterior mouth by cilia or is pumped in by the muscular pharynx.

The pseudocoel consists of only slitlike spaces between organs. Thus gastrotrichs do not really possess a body cavity.

In contrast to most other pseudocoelomates, gastrotrichs are hermaphroditic. Fertilization is internal. Both dormant and rapidly hatching eggs are produced in freshwater gastrotrichs. In the few species that have been studied, sexual maturity is attained in about 3 days after hatching. In the laboratory, *Lepidodermella squamata* has a maximum life span of 40 days, which is certainly much longer than occurs under natural conditions.

PHYLUM NEMATODA

The approximately 12,000 known species of nematodes, or **roundworms**, constitute the largest phylum of aschelminths and one of the most widespread and abundant

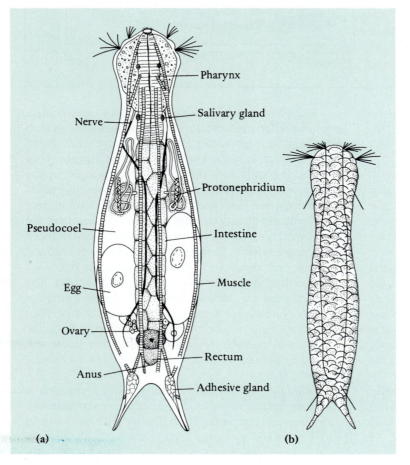

(a)   (b)

Figure 28.6

(a) Internal structure of a gastrotrich, ventral view.
(b) *Lepidodermella*, a freshwater gastrotrich in which the cuticle has a scalelike ornamentation.

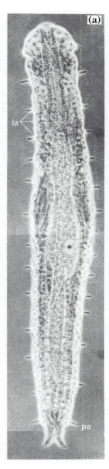

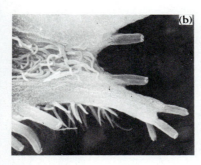

Figure 28.7
Adhesive tubes of *Turbanella*.
(a) Photograph of living animal,
showing lateral and posterior adhesive
tubes. Each lateral adhesive tube
contains a sensory cilium. (b) Enlarged
view of posterior adhesive tubes.

groups of animals. They occur in enormous numbers in the interstitial spaces of marine and freshwater bottom sediments and in the water films around soil particles. One square meter of bottom mud off the Dutch coast has been reported to contain as many as 4,420,000 nematodes. Several hundred billion nematodes may be present in an acre of good farmland. Decomposing plant and animal bodies often contain large populations of nematodes. For example, examination of one decomposing apple revealed some 90,000 nematodes belonging to several different species.

Food crops, domestic animals, and humans are parasitized by many species, and consequently roundworms are of tremendous economic and medical importance.

Most free-living nematodes are less than 2.5 mm in length, but species parasitizing the gut of vertebrates may attain much larger sizes; for example, the horse nematode *Parascaris equorum* reaches a length of 35 cm.

The bodies of nematodes are usually long, cylindrical, and tapered at both ends (Fig. 28.8). This threadlike shape and the minute size of most free-living nematodes reflect

Figure 28.8
(a) A free-living freshwater nematode.
(b) Female of the marine nematode
Pseudocella.

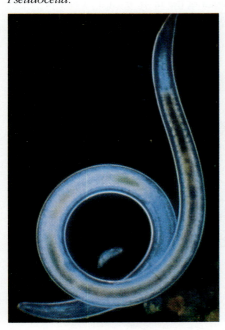

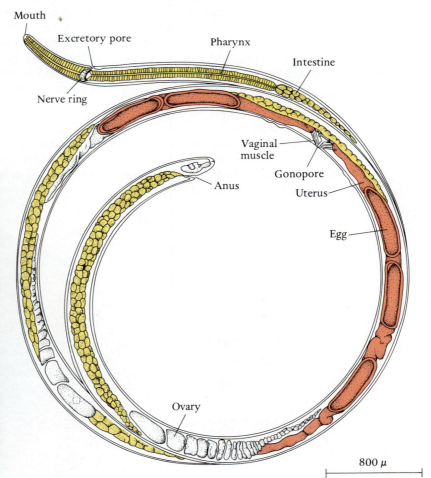

(a)

(b)

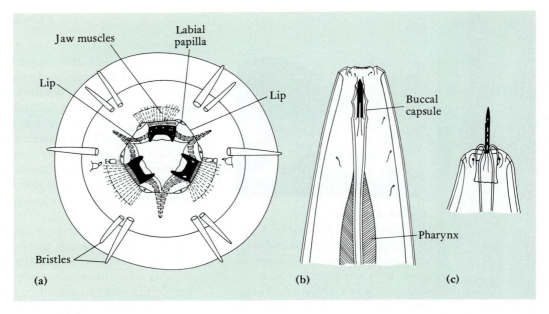

Figure 28.9

(a) Anterior end of a marine nematode with jaws and bristles; (b) anterior end of a predaceous nematode, with a retracted, spearlike stylet; (c) stylet projected.

adaptations for living in interstitial spaces. This is an important relationship to associate with nematodes. Lips, small sensory bristles, and papillae encircle the mouth at the anterior end (Fig. 28.9a). The anus is located a short distance in front of the posterior end, which in free-living species usually terminates in an adhesive gland (a duogland system).

The thick, complex, collagen cuticle is composed of several layers, the outer one often sculptured. The cuticle is shed four times in the life of a nematode, but unlike the exoskeleton of insects and crustaceans, the nematode cuticle grows between molts. Beneath the cuticle is an epidermis composed of cells whose nuclei, strangely, are restricted to dorsal, ventral, and lateral cords (Fig. 28.10).

Figure 28.10

Diagrammatic cross section through the body of a nematode at the level of the pharynx. Note that the muscles send extensions, or arms, to the nerve cord, instead of the more usual reverse arrangement. Only a few of the many muscle arms are shown. A cuticle lines the pharynx as well as covering the surface of the body.

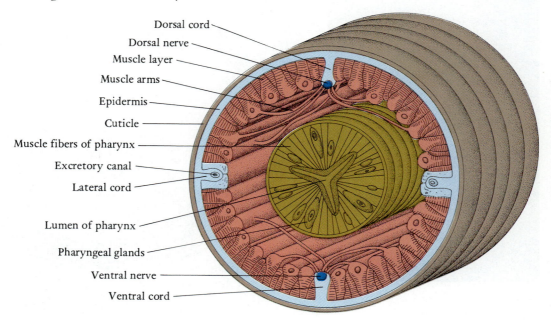

The subepidermal muscle layer is composed only of longitudinal fibers, which are also peculiar in each possessing a long process, or tail, that extends to a ventral or dorsal nerve cord in the body wall (in most animals, the motor neurons extend to the muscle cell). The pseudocoel is well developed and its fluid functions as a hydrostatic skeleton. Waves of contractions sweeping along the layer of longitudinal muscle fibers produce undulatory or thrashing locomotor movements that drive nematodes through the spaces between algae, sand, or soil particles. The contractions of the muscle cells are antagonistic to the elastic cuticle and the hydrostatic pressure of the pseudocoel. Note the structural features that are related to this undulatory locomotion, which in turn is correlated with an interstitial existence.

The mouth opens into a **buccal cavity**, which may be provided with teeth or a stylet (Figs. 28.9b,c). The buccal cavity is connected to a tubular muscular **pharynx**. The remainder of the gut, a long straight **intestine**, is the site of digestion and absorption (Fig. 28.8).

Many free-living nematodes are predaceous. Other worms, including other nematodes, are common prey, and are sucked out or swallowed whole. Plant-feeding nematodes suck out the contents of plant cells, chiefly cells of roots, and can be very damaging to the plant. For the many species that feed on decomposing organic matter, such as the dead bodies of plants or animals or dung (see p. 983), bacteria and fungi are probably the primary food source. A stylet is used to puncture prey or plant cells, although many carnivorous nematodes possess jaws. The pharynx pumps the food into the gut.

Nematodes lack protonephridia, but some possess a system of glands and tubules that open through a midventral pore. The system may have an osmoregulatory function. Many nematodes that live in the water films around soil particles, lichens, and mosses can withstand long periods of extreme desiccation (see Connections 28.1, p. 645).

The nervous system is composed of an esophageal nerve ring and a large ganglionated ventral cord located in the ventral epidermal cord. The anterior nerve ring also gives rise to nonganglionated longitudinal nerves located in the dorsal and lateral epidermal cords.

The female reproductive system is paired and tubular and includes two ovaries (Fig. 28.8b). Each oviduct empties into a uterus and then into a common vagina that opens to the exterior via a gonopore in the midregion of the body. Male nematodes are usually a little smaller than females. The male reproductive system is a long coiled tube composed of a testis, sperm duct, seminal vesicle, and muscular ejaculatory duct. The latter opens into the rectum so that the anus functions as a gonopore as well as for egestion of wastes. The rectum contains two short curved spicules, which can be projected from the anus. During copulation, the posterior end of the male is curled around the body of the female in the region of the gonopore. The spicules are used to hold open the gonopore of the female during sperm transmission. Although most nematodes have separate sexes, as do most other aschelminths, hermaphroditism and parthenogenesis are common in soil nematodes.

The sperm of nematodes are peculiar in lacking a flagellum (Fig. 21.6). The fertilized eggs possess a thick shell. Free-living nematodes deposit their eggs in the bottom debris and soil in which they live. The young have most of the adult features on hatching, including the adult number of cells for most organs, since eutely is common. As already indicated, nematodes undergo four molts before attaining maturity. Adults do not molt but continue to grow. The nematode *Caenorhabditis elegans* could be considered the best known animal from a developmental standpoint, since every cell has been traced throughout the course of its highly determinate development (see p. 529). This little nematode feeds on bacteria and is easily cultured in the laboratory. Hermaphroditic and self-fertilizing, it completes its life cycle in several days. *Caenorhabditis elegans* is the subject of extensive studies on biochemical and genetic controls of development, and its six pairs of chromosomes are being completely mapped.

Parasitic Nematodes

Parasitic nematodes attack both plants and animals, and exhibit all degrees of complexity in their relationship with the host and in their life cycle. Four groups containing species of significant medical importance are described here (see also Table 28.1).

Hookworms

The **hookworms** are small parasites of the digestive tract of vertebrates, usually being less than 12 mm in length. Most members of this group feed on the host's blood, and the mouth region is usually provided with cutting plates, hooks, teeth, or combinations of these structures for lacerating and attaching to the gut wall (Fig. 28.11).

Figure 28.11
Anterior end of a dog hookworm, showing buccal region and teeth.

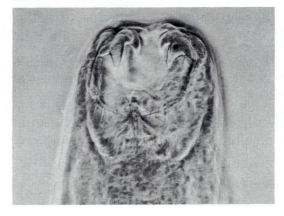

Table 28.1
Principal Pathogenic Nematode Parasites of Humans

Parasite and Distribution	Location in Host (Adult Parasite)	Pathology	Transmission
Strongyloides stercoralis (Largely tropical)	Mucosa of small intestine	Destruction of mucosa; diarrhea, abdominal pain	By skin penetration of larva
Ancylostoma duodenale—hookworm (Widespread in Old World outside of northern Europe)	Small intestine	Destruction of intestinal cells; anemia and general debilitation	By skin penetration (especially feet) by larvae
Necator americanus—hookworm (Tropics and semitropics)	Small intestine	Similar to *A. duodenale*	By skin penetration by larvae
Ascaris lumbricoides (Cosmopolitan)	Small intestine	Heavy infection—obstruction of pancreatic and bile ducts and intestine; peritonitis	Ingestion of eggs passed in feces
Enterobius vermicularis—pinworms (Cosmopolitan)	Small and large intestine; cecum	Itching of anal region; restlessness, abdominal pain	Ingestion of eggs deposited in anal region and passed in feces
Dracunculus medinensis—guinea worm (Africa, Mideast, and India)	Subcutaneous connective tissue	Skin ulcers produced by emerging worm, accompanied by nausea and diarrhea	Ingestion of infected crustacean intermediate host (copepod) in drinking water
Brugia malayi (Southeast Asia)	Lymphatic vessels and nodes	Lymphatic inflammation with fever and localized pain.	By bite of mosquito
Wuchereria bancrofti (Africa and tropical Asia and Pacific)		Blockage and enlargement of lymphatic system, especially in extremities (elephantiasis)	
Onchocerca volvulus (Africa, Central and S. America)	Subcutaneous connective tissue	Skin nodules, dermatitis; blindness	By bite of blackflies
Loa loa—eye worm (Equatorial Africa)	Subcutaneous connective tissues	Painful subcutaneous swelling; migration of worm across cornea	By bite of deer flies (*Chrysops*)
Trichinella spiralis (Cosmopolitan)	Mucosa of small intestine	Gastrointestinal disorders; muscle pain and fever from larval invasion	Eating encysted larvae in rare pork
Trichurus trichura—whipworm (Tropics and subtropics)	Intestinal mucosa	Damage to mucosa; abdominal pain and diarrhea; colitis	Ingestion of embryonated eggs passed in feces

An infection of more than about 25 worms may produce symptoms of hookworm disease, and a heavy infection of over 500 worms can produce serious danger to the host through loss of blood and tissue damage. An adult worm may live as long as 15 years in the intestine. Hookworms are one of the great parasitic scourges of humans. It is estimated that over 380 million people are infected with *Necator americanus*, the most important hookworm species, which is prevalent throughout the tropical regions of the world (despite the species name).

The life cycle of hookworms involves an indirect migratory pathway by the juveniles, as in ascaroids (see below). The fertilized eggs leave the host in its feces and hatch outside the host's body on the ground. The 3rd-stage larva (stage between 3rd and 4th molt) gains reentry by penetrating the host's skin (feet in humans). Waiting larvae will be erect and waving on the moist surface of soil and can survive for 3 weeks. On penetration of the skin the larva is carried in the blood to the lungs. From the lungs the larva migrates to the pharynx, where it is swallowed and passes to the intestine. Not all species of this group gain reentry by skin penetration. Some, such as dog hookworms, enter the gut directly through the mouth of the host.

Ascaroids

The ascaroid nematodes, which feed on the intestinal contents of humans, dogs, cats, pigs, cattle, horses, chickens, and other vertebrates, include the largest species of nematodes. They are entirely parasitic within a single host, and the life cycle typically involves transmission by the ingestion of eggs or larvae passed in the feces of another host. Thus, lack of sanitation systems is a major factor in the spread of human ascaroid parasites. The juvenile stages usually penetrate the intestinal wall to enter the circulatory system, where they are carried to the lungs. Here they break into the alveoli and migrate back to the intestine via the trachea and esophagus. This seems a strange, unneces-

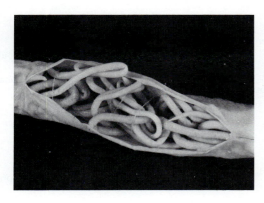

Figure 28.12
Adult specimens of *Ascaris suum* from the intestine of a pig. *Ascaris lumbricoides* of humans is very similar.

sary journey. It may indicate that the ancestral entrance into the host was not via the mouth but through the skin, as in hookworms.

The human ascaroid, *Ascaris lumbricoides*, reaches a length of 49 cm and is one of the best known parasitic nematodes. The species is widely distributed throughout the world, including the southeastern United States, particularly in children, who pick up eggs from soil on their hands and various objects placed in their mouths. Heavy infection causes malnutrition and sometimes intestinal blockage (Fig. 28.12). Adult worms that wander into the appendix or ducts of the pancreas and gallbladder can obstruct these structures, and large numbers of juveniles breaking out of the lungs can produce serious damage to lung tissue. The embryonated eggs are notoriously resistant to adverse environmental conditions and may remain viable in soil for up to 10 years. The very closely related species in pigs probably had a common evolutionary origin with the human species, the common ancestor being confined to one host or the other prior to human domestication of the pig.

Toxocara canis and *T. cati* are two small ascaroid species common in dogs and cats. It is these species for which puppies and kittens are usually wormed.

Filaroids

The **filaroid** nematodes have life cycles requiring an arthropod intermediate host. The filaroids are threadlike worms inhabiting the lymphatic vessels and some other tissue sites in the vertebrate host, especially birds and mammals. The female is viviparous, and the larvae are called **microfilariae**. Bloodsucking insects, such as fleas, certain flies, and especially mosquitoes, are the intermediate hosts. A number of species parasitize humans, producing **filariasis**.

The chiefly African and Asian *Wuchereria bancrofti* illustrates the life cycle. The male is 40 mm by 0.1 mm, and

the female is about 90 mm by 0.24 mm. Adults live in the lymphatic ducts adjacent to the lymph glands of humans, especially in the lower part of the body. The microfilariae are found in the peripheral blood stream. When certain species of mosquitoes bite the host, the microfilariae enter the mosquito with the host's blood. Development within the intermediate host involves a migration through the gut to the thoracic muscles and after a certain period into the proboscis. From the proboscis the microfilariae are introduced back into a primary host when the mosquito feeds. In severe filariasis, the blocking of the lymph vessels by large numbers of worms results in serious short-term lymphatic inflammation marked by pain and fever. Over a long period increase of connective tissue in affected areas may result in massive enlargement of the legs, arms, and scrotum. Such enlargement is called **elephantiasis** (Fig. 28.13) and can only be corrected by surgery. Fortunately, extreme cases of elephantiasis are no longer common.

Trichinellids

Trichinella spiralis, the nematode causing the disease **trichinosis** in humans, is a minute species, the females being only about 3 mm long. The adults live in the intestine of many carnivorous and omnivorous mammals. Following copulation, the female burrows into the submucosa of the intestinal wall of the host and gives birth to a large number of larvae. The larvae are carried by the blood stream to striated muscles throughout the body, where they form cysts (Fig. 28.14). The cysts must be eaten by another mammal to reestablish the adults in the intestine of a new host. Note that the life cycle is unusual in that the same animal is acting as primary and intermediate host for this parasite. This requires that there be numerous poten-

Figure 28.13
A victim of elephantiasis, which results from severe filariasis.

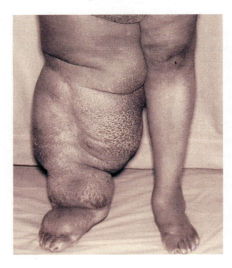

Figure 28.14
Larvae of *Trichinella spiralis* within calcareous cyst in striated muscle tissue of host.

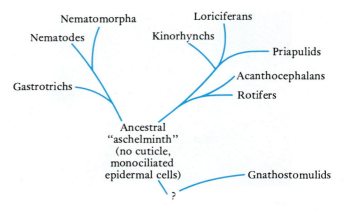

Figure 28.15
Possible phylogenetic relationships of the aschelminth phyla. The cuticle of the gastrotrich, nematode, and nematomorph line is extracellular; the cuticle of rotifers and acanthocephalans is intracellular but extracellular in the kinorhynchs and priapulids.

tial hosts. Humans usually obtain the nematode from eating insufficiently cooked pork. The cysts can be passed to pigs and rats in uncooked pork in garbage. Pigs can also obtain the parasite by eating rats. The human is obviously a dead end for those worms encysting in human muscle. The migration and invasion of tissues by the female and larvae can cause considerable tissue damage and can lead to death of the host. There is no cure for trichinosis at the present time.

Some other important nematode parasites are *Dirofilaria immitis*, the heartworm of dogs; *Enterobius vermicularis*, intestinal pinworms that often occur in children; and *Dracunculus medinensis*, the 120-cm-long guinea worm of the Mideast and Africa, which in the final part of its life cycle migrates to the subcutaneous tissue to release larvae and produces an ulcerated opening to the exterior.

EVOLUTIONARY RELATIONSHIPS

Having briefly surveyed this very diverse assemblage of aschelminths, you can now better understand why most zoologists are opposed to placing them within a single phylum. Aside from a cuticle and adhesive glands, there are few distinctive characteristics that unite them.

The ancestral aschelminths most likely possessed monociliated cells over most of the body surface. The ventral ciliation of gastrotrichs and the fact that some species have monociliated cells would suggest that gastrotrichs are the most primitive members of the assemblage.

There appear to be two major evolutionary lines of aschelminths (Fig. 28.15). Gastrotrichs, nematodes, and nematomorphs are characterized by an external cuticle and a pharynx composed of myoepithelial cells (cells combining both epithelial and contractile features, as in

cnidarians). Rotifers and acanthocephalans have an intracellular cuticle, and the pharynx of rotifers is composed of separate epithelial and muscle cells. In contrast to the monociliated cells of some gastrotrichs, the cells composing the crown of rotifers are multiciliated. Acanthocephalans, kinorhynchs, loriciferans, and priapulids all have a spiny anterior end.

How aschelminths are related to other phyla is very uncertain. We have placed them near the base of the protostome branch of the animal kingdom for they have determinate cleavage. But cleavage is not spiral. There is no coelom, and remember that a coelom is not derived from a pseudocoel.

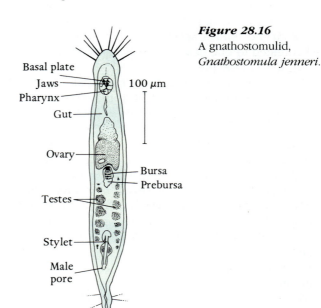

Figure 28.16
A gnathostomulid, *Gnathostomula jenneri*.

A RELATED PHYLUM?

Phylum Gnathostomulida (Gr. gnathos, jaw + stoma, mouth)

The Gnathostomulida is a phylum of some 80 species of minute animals living in the interstitial spaces of marine sediments. The elongate body is completely covered with an epidermis of monociliated cells, a primitive animal feature. The ventral lip of the mouth cavity bears a comblike cuticular plate used in scraping up bacteria and fungi, which are then ingested with the aid of a pair of pharyngeal jaws (Fig. 28.16). An anal pore has been reported. There is no pseudocoel but parenchyma is poorly developed. They are hermaphroditic and possess spiral cleavage and direct development. Gnathostomulids cannot be placed among the aschelminths nor are they flatworms. Their relationship to these groups is very uncertain. We have placed them here for purposes of comparison.

Classification of Aschelminths

The following synopsis summarizes all of the aschelminth phyla, with figures of those groups omitted from the text discussion. Characterization of the lower taxa (classes and orders) is technically difficult, and we have omitted them here.

Phylum Gastrotricha (Gr. *gastros*, belly + *trichos*, hair) Minute strap-shaped or bottle-shaped marine animals bearing ventral ciliated cells for locomotion; temporary attachment by lateral and/or terminal adhesive tubes. Marine and freshwater; all live on the bottom surface or are interstitial. *Lepidodermella, Turbanella.*

Phylum Nematoda (Gr. *nematos*, thread) Slender cylindrical animals with tapered anterior and posterior ends. Body adapted for moving through interstitial spaces. Free-living species in marine and freshwater sediments and in soils; many parasitic species. *Pseudocella, Caenorhabditis, Ascaris, Necator, Wuchereria, Trichinella, Loa.*

Phylum Nematomorpha (Gr. *nematos*, thread + *morphe*, form) Horsehair worms. Some 230 species of mostly freshwater animals having very long, hairlike bodies up to 36 cm in length (Fig. 28.17). They appear to be most closely related to nematodes. The nonfeeding adults live in freshwater or damp soil. The juveniles are parasitic in arthropods living around water. They inhabit the tissue spaces of the host, entering soon after hatching. The adult form emerges when the host is near water. *Gordius.*

Phylum Rotifera (L. *rota*, wheel + Gr. *pherein*, to bear) Minute animals with elongate or saclike bodies bearing an anterior ciliated crown and a posterior foot. Pharynx (mastax) bears cuticular pieces adapted for grasping or grinding. Epibenthic or planktonic, mostly in fresh water. *Philodina, Epiphanes, Euchlanis, Asplanchna, Collotheca.*

Phylum Acanthocephala (Gr. *akantha*, spine + *kephale*, head) A group of some 1000 species of endoparasites of the gut of marine, freshwater, and terrestrial vertebrates. The body, which is usually a few centimeters or less in length, bears a spiny anterior retractible proboscis, enabling the worm to hook to the gut wall of the host (Fig. 28.18). The gut of these parasites has been lost. Embryonated eggs are passed out within the host's feces. If the eggs are eaten by certain crustaceans or insects, the juvenile acanthocephalan develops within the tissue

Figure 28.17
A nematomorph or horsehair worm.

Figure 28.18
An adult acanthocephalan.

spaces of the intermediate host. The primary vertebrate host becomes infected by eating the intermediate host. *Andracantha.*

Phylum Kinorhyncha (Gr. *kinein*, to move + *rhynchos*, snout) A group of some 100 minute animals that burrow through marine sediments. The cuticle of the elongate body is segmented (Fig. 28.19). The anterior end of the body is spiny and can be withdrawn within the front segments. *Echinoderes.*

Phylum Loricifera (L. *lorica*, corset + Gr. *pherein*, to bear) The first representatives of the phylum Loricifera were collected in 1983 from shelly marine sediments off the coast of France, although they have now been found elsewhere. This is another group of interstitial animals, and the reason for the delay in their discovery is that they cling so tightly to the sediment particles that they were not collected by the usual extraction technique (see Connections 27.1, p. 622). Only about .25 mm long, the body of these little animals bears a spiny anterior end that can be retracted into the abdomen, which is encased within a cuticular girdle called a lorica (Fig. 28.20). The mouth is

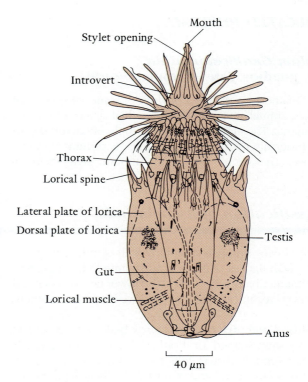

Figure 28.20
A loriciferan, *Nanaloricus mysticus.* Dorsal view of adult male.

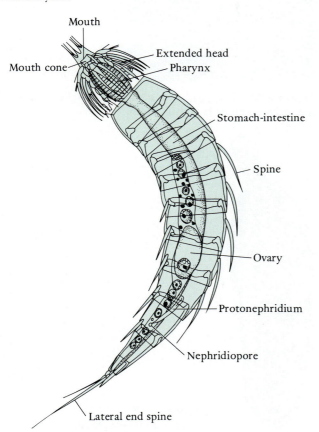

Figure 28.19
A kinorhynch.

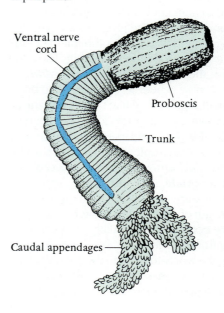

Figure 28.21
A priapulid.

surrounded by stylets, but little is known about how these animals feed or other aspects of their biology. *Nanaloricus* **Phylum Priapulida** (The Gr.-Rom. god of fertility, Priapus; or a phallus) A phylum of 13 species of wormlike or cucumber-shaped marine animals having a retractible proboscis (Fig. 28.21). Burrowers in sand and mud, they range in length from less than 1 mm to 20 cm. *Priapulus.*

Summary

1. Aschelminths are a heterogenous assemblage of eight phyla. The body is covered by a cuticle, and the body cavity when present is a pseudocoel. Most are minute animals and therefore lack organs of gas exchange, internal transport, or excretion. In most species the sexes are separate and development is direct.

2. Rotifers (phylum Rotifera) are largely freshwater aschelminths that possess an anterior ciliated crown used in swimming. Some rotifers are planktonic; others swim about bottom debris and algae, temporarily anchoring by means of adhesive glands opening at the posterior end of the body (foot). Rotifers are predaceous or suspension feeders, the latter using the ciliated crown to drive food into the mouth. The muscular pharynx, called the mastax, contains a number of cuticular pieces used in grasping prey or in grinding small particles.

3. Gastrotrichs (phylum Gastrotricha) are marine and freshwater animals with strap- or bottle-shaped bodies. They move by ventral cilia. Most species are interstitial but many live among surface debris and algae. They feed on fine particles pumped in by a muscular pharynx or swept into the anterior mouth by cilia. Gastrotrichs are hermaphroditic.

4. The phylum Nematoda, containing the roundworms, is the largest group of aschelminths. There are marine, freshwater, soil-inhabiting, and parasitic species. The long cylindrical and tapered body is adapted for living in interstitial spaces. The animal is propelled by whiplike movements produced by contractions of longitudinal muscles acting against the flexible cuticle and large fluid-filled pseudocoel. Nematodes have varied feeding habits, and jaws and stylet are characteristic of many predaceous and plant-feeding species. Food is ingested with a muscular pharynx and digested within a long straight intestine. The sexes are separate, and copulation and internal fertilization are the rule. The cuticle is molted four times in the life cycle of a nematode.

5. There are many parasitic nematodes that inhabit various parts of the bodies of a wide range of hosts. Some species utilize a single host (gut-inhabiting ascaroids, hookworms); others require two hosts (filaroids, trichinellids).

References and Selected Readings

Detailed accounts of the aschelminth phyla may also be found in the references listed at the end of Chapter 23. Parasitic forms are described in the parasitology texts listed at the end of Chapter 27. The following are a few works devoted to specific groups.

Croll, N.A., and B.E. Matthews. *Biology of Nematodes*. New York: John Wiley & Sons, 1977. A general account of the phylum.

Crowe, N.H., and K.A. Maden. Anhydrobiosis in tardigrades and nematodes. *Transactions of the American Microscopic Society* 93(1974):513–524. A study of the ability of nematodes to withstand extreme desiccating conditions.

Dropkin, V.H. *Introduction to Plant Nematology*. New York: John Wiley & Sons, 1980. The biology of plant-feeding nematodes.

Gilbert, J.J. Developmental polymorphosis in the rotifer *Asplanchna sieboldi. American Scientist* 68(Nov.-Dec. 1980):636–646. A study of the role of dietary vitamin E in adapting the body form to environmental changes.

Maggenti, A. *General Nematology*. New York: Springer-Verlag, 1981. A general biology of nematodes.

Nicholas, W.L. *The Biology of Free-living Nematodes*, 2nd ed. Oxford: Clarendon Press, 1984. A general biology of the nonparasitic nematodes.

Roberts, L. The worm project. *Science* 248(1990): 1310–1313. A summary of current research on *Caenorhabditis.*

Wharton, D.A. *A Functional Biology of Nematodes*. Baltimore: The Johns Hopkins University Press, 1986. A good introduction to the nematodes.

29

Molluscs

Characteristics of Molluscs

■ The body of molluscs is usually covered by a shell secreted by the underlying integument, called the mantle.

■ A ventral muscular foot is the locomotor organ of most molluscs.

■ The shell and mantle overhang the body, creating a mantle cavity that houses the gills of most aquatic species and forms a lung in land snails.

■ Except for bivalves, most species employ a radula as a feeding organ. The radula can be protruded to scrape or tear and pull food into the mouth.

■ The circulatory system is usually open, and the heart, which lies within a pericardial coelom, consists of a pumping ventricle and paired auricles that receive blood from the gills.

■ The excretory organs are metanephridia (usually one or two) that drain the coelom and empty into the mantle cavity.

■ The nervous system consists of a pair of pedal cords to the foot and visceral cords supplying the organs of the mantle and visceral mass. The pedal and visceral cords unite anteriorly in a cerebral ganglion. Sense organs include one or two statocysts, one or two pairs of head tentacles, a pair of eyes, and a pair of osphradia, the last being sense organs that monitor the ventilating current passing through the mantle cavity.

■ Molluscs may have separate sexes or may be hermaphroditic, and paired gonads are adjacent to the pericardial coelom. Primitively, the gametes exit through the nephridia, but separate and often complex gonoducts are usually present. The earliest larval stage is a trochophore, which is usually followed by a veliger, but in most marine snails with indirect development the hatching stage is a veliger.

(Left) Octopus cyanea, *from the North American Pacific coast.*
(Above) The market squid Loligo opalescens *laying eggs, from the California coast.*

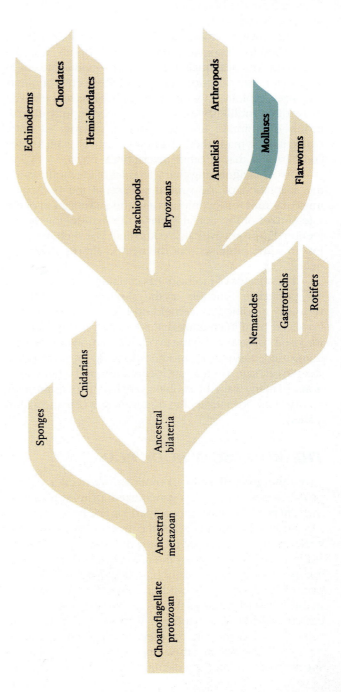

657

All of the animals we will be surveying in the remaining chapters of Part IV possess a coelom, the type of body cavity that arises within mesoderm (p. 515). In those cases where a coelom is absent in the adult, it has appeared in the course of development and has subsequently been lost. You may recall from Chapter 23 that there are two great evolutionary divisions of coelomate animals: the protostomes and deuterostomes. This chapter and the following three chapters cover the three main protostome phyla: the molluscs, the annelids, and the arthropods. Protostomes are characterized by spiral, determinate cleavage; a mouth derived from the blastopore; and a coelom arising from a splitting of mesoderm (schizocoely, p. 554).

The phylum **Mollusca** includes some 60,000 living and 35,000 fossil species. After the arthropods, it is the largest phylum of animals. Molluscs are found in the sea, in freshwater, and on land, and in all of these environments they have exploited a wide range of habitats and evolved many different life styles. They are a wonderfully diverse group of animals, and the long popularity of shell collecting has made their taxonomy relatively well known compared to that of other invertebrate phyla.

Two distinguishing characteristics of the molluscs are the adaptation of the ventral surface of the body as a muscular **foot** for locomotion and the modification of the integument of the dorsal surface, the **mantle**, for the secretion of a protective calcareous **shell**. The shell has been lost in some molluscs, but the mantle almost always remains.

The success of molluscs has certainly been related to the protective function of the shell, which has been variously modified as a shield or a retreat. The benefits, however, have been gained at some price. The shell has imposed constraints on the evolution of molluscs and has demanded numerous modifications to accommodate weight, carriage, ventilation, locomotion, and other conditions. In many ways the evolutionary history of molluscs is a story of the advantages and disadvantages of living within a shell.

THE MOLLUSCAN BODY PLAN

Given the great diversity of molluscs we need a starting point—a description of a generalized, primitive mollusc that will provide a basis for comparison and from which we can derive the members of the living classes. We can construct this generalized mollusc by looking for common features that appear across the many different groups that compose the phylum. Many of these features should be primitive, i.e., ones we would have expected to have found in the ancestral molluscs that crept over the bottom in the early Cambrian seas. Molluscs first appear in the fossil record of the Cambrian, but they are represented only by their shells. By examining the comparative anatomy of living species, we can infer the primitive structure of soft parts.

The ventral surface of a primitive mollusc forms a broad, flat, muscular foot on which the animal creeps (Fig. 29.1a). The head is poorly developed, although in many living species the head bears tentacles, eyes, or other specialized sense organs. The dorsal surface of the body is covered by a low shield-shaped shell secreted by the underlying mantle and composed of several layers of calcium carbonate covered on the outside by an organic layer, the **periostracum** (see Fig. 29.28b) (Gr. *peri*, around + *ostrakon*, shell). Much of the shell of living molluscs is secreted by the mantle margin, so that the shell grows in diameter as well as in thickness. The shell of this generalized mollusc provides protection as long as the animal is attached to the rocky substratum; pairs of retractor muscles extend from the foot to the shell permitting the shell to be pulled down over the body (Fig. 29.1a). The central hump of the body covered by the shell and mantle contains internal organs and is commonly referred to as the **visceral mass**.

The overhanging shell and mantle at the posterior end create a large mantle cavity containing a number of pairs of **gills** and nephridiopores, the excretory openings (Fig. 29.1). The anus is located at the dorsal side of the cavity opening. The gills are bipectinate, i.e., each gill consists of flattened filaments attached to either side of a longitudinal axis (Fig. 29.1b). Blood circulates through the filaments, flowing from a supply (afferent) to a drainage (efferent) channel in the gill axis (Fig. 29.1c). The filaments are ciliated and produce a ventilating current of water that enters the lower part of the mantle cavity, makes a U-turn across the gills, and then flows out of the cavity dorsally and posteriorly (Fig. 29.1a). Wastes discharged by the nephridia and anus are removed in the exhalant stream of water.

In most living molluscs, the mouth cavity contains a **radula** (L., a scraper), a unique rasping organ consisting of a beltlike membrane bearing a large number of teeth (Fig. 29.2). The radula rests upon a cartilaginous supporting skeleton, the **odontophore** (Gr. *odous*, tooth + *pherein*, to bear), which is supplied with protractor and retractor muscles. The end of the radula and odontophore can project to the mouth opening and lick the adjacent surface, like the tongue of a mammal. Imagine that your tongue has teeth on the upper surface. Now with your mouth open, press your lips against a surface and lick it with your tongue. This gives you a good idea of how the radula works. Notice that it would function not only as a scraper but also as a conveyor belt, bringing detached food particles back into the mouth (Fig. 29.2d). New teeth are formed at the posterior end of the radula as old teeth are worn away anteriorly.

The ancestral molluscs were probably **microphagous**—they fed upon small particles of algae scraped with the radula from rocks. However, among living molluscs the radula has become adapted for many diets.

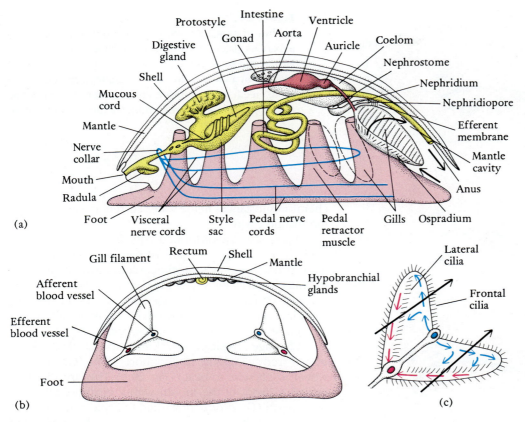

Figure 29.1
(a) Hypothetical ancestral mollusc (lateral view). Arrows indicate path of water current through mantle cavity. (b) Transverse section through body of ancestral mollusc at level of mantle cavity. (c) Cross section through one gill, showing two filaments. Large black arrows indicate direction of ventilating current produced by lateral gill cilia. Small red and blue arrows indicate flow of blood within gill. Frontal cilia on filament edge remove sediment.

Particles of food brought into the mouth are enveloped by mucus secreted by the salivary glands. In primitive, microphagous forms, a rope of mucus containing the algal particles then passes posteriorly through an esophagus to the stomach. The posterior end of the stomach, the **style sac**, is ciliated and rotates the mucous mass, winding in the string of mucus like a windlass winds in a rope (Fig. 29.1a). The mucus becomes less viscous in the more acid stomach medium. Particles dislodged from the mucus are sorted by a ciliated area. Large undigestible particles are conducted posteriorly to the intestine, and fine digestible particles are conveyed to the ducts of a pair of digestive glands located to either side of the stomach. Digestion occurs intracellularly within the digestive gland. The long intestine functions only in the formation of feces, which are usually compacted with mucus.

An open blood vascular system provides for internal transport in most molluscs. The heart, located within the coelomic cavity (also called the **pericardial cavity**), is composed of a **ventricle** and two lateral **auricles**, which receive blood from each gill (Fig. 29.1a). The contractile ventricle forces blood via an anterior aorta to the tissue spaces of the body—head, foot, visceral mass, and mantle. Blood then collects within larger sinuses from which it is returned to the heart by way of the nephridia and gills.

The pairs of excretory organs, often called kidneys, are **metanephridia**. In contrast to the blind protonephridia of flatworms and aschelminths, the inner ends of these nephridia open into the coelom by way of a ciliated **nephrostome** (Fig. 29.1a; see also p. 351). Fluid containing wastes collected by the blood from body tissues filters through special areas of the heart wall (in auricles) into the coelomic fluid. Coelomic fluid containing wastes passes into the inner end of the nephridium. The urine is expelled into the mantle cavity and flushed out by the exhalant ventilating current.

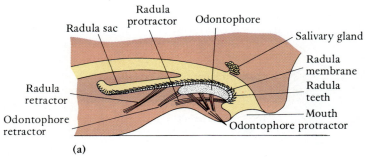

Radula sac • Radula protractor • Odontophore • Salivary gland • Radula membrane • Radula teeth • Radula retractor • Odontophore retractor • Mouth • Odontophore protractor

(a)

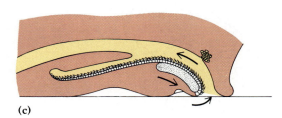

(b)

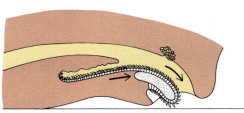

(c)

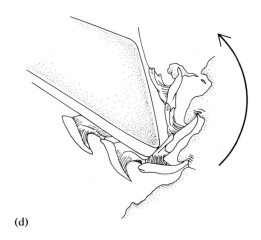

(d)

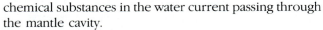

Figure 29.2

Molluscan radula. (a) Mouth cavity, showing radula apparatus (lateral view). (b) Protraction of the radula against the substratum. (c) Forward substratum is scraped by radula teeth. (d) The cutting action of the radula teeth when they are erected over the end of the odontophore. Large arrow indicates direction of cutting stroke.

chemical substances in the water current passing through the mantle cavity.

A pair of gonads is located to the front and sides of the coelom (Fig. 29.1a). Primitively, eggs and sperm are released from the gonads into the coelom and then carried to the mantle cavity by way of the nephridia. Fertilization is external either within the mantle cavity or in the surrounding sea water, but as we shall see, internal fertilization has evolved in many living species.

Cleavage is typically spiral, and the gastrula develops into a free-swimming **trochophore larva** (Gr. *trochus*, wheel + *pherein*, to bear) (Fig. 29.3). The larval body is ringed in front of the mouth with a girdle of cilia, the **prototroch** (Gr. *protos*, first), and the apical pole bears a tuft of cilia. A digestive tract is present. The beating cilia of the prototroch provide for locomotion and may also serve to collect fine plankton for food. Primitively, the trochophore develops directly into the adult body, but in many molluscs there is a later larval stage called a **veliger**.

Figure 29.3

The trochophore larva of *Patella*, a marine snail.

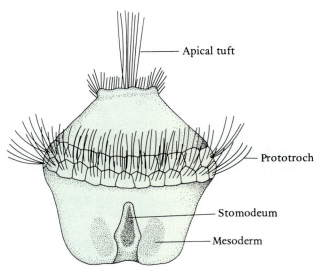

Apical tuft • Prototroch • Stomodeum • Mesoderm

Using this generalized form as a basis for comparison, we will now study five of the eight classes of molluscs.

CLASSES MONOPLACOPHORA AND POLYPLACOPHORA

Of all the living members of the phylum our generalized mollusc is most like certain species of the class known as the Monoplacophora. This is therefore a good group to begin with, even though they are less familiar to you than are snails and clams. Monoplacophorans had long been known from fossils that date back to the Cambrian, but they were thought to be extinct. Then in 1952, ten living specimens were dredged up from a great depth off the west coast of Central America. Since that time, additional specimens have been collected from other sites along the west coasts of the American continents and from the Atlantic and Indian Oceans. The 11 species belonging to three genera that have now been described all live at depths of 1800 meters or more, except for one species that has been found in several hundred meters of water off the California coast. They are all believed to be remnants of a class that once contained many more widely distributed species.

Monoplacophorans are only a few cm long. The dorsal surface is covered with a symmetrical shield-shaped shell, the apex of which is a little peaked and directed anteriorly (Fig 29.4). The ventral surface is broad and flat, with the mantle cavity in the form of two grooves located to either side of the foot. There are eight pairs of pedal retractor muscles and six or seven pairs of nephridia (Fig. 29.4d). The mantle groove contains five or six pairs of monopectinate gills (lamellae on one side only) drained by two pairs of auricles. A radula is present, and the digestive tract is similar to that described for the generalized mollusc. Because of the great depth at which monoplacophorans live, we know very little about their biology at the present time.

Compared to our generalized mollusc, notice that there is a greater repetition of parts in monoplacophorans. Also the gills of monoplacophorans are monopectinate, in contrast to the bipectinate type described as being primitive for molluscs. Zoologists believe this is a secondary specialization in these remnant monoplacophorans and that the gills of earlier members of the class were bipectinate. Why do we think that, since there is no fossil record of soft gill parts? Our speculation is based on comparative anatomy of living molluscs. Bipectinate gills are encountered in at least some species of all the other classes of molluscs that possess gills; therefore, the bipectinate structure is probably the original structure of the molluscan gill.

Most malacologists—specialists on molluscs— believe that the monoplacophorans were the ancestors of

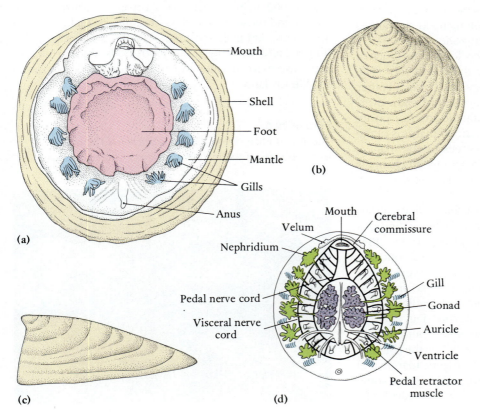

(a)

(b)

(c)

(d)

Mouth
Shell
Foot
Mantle
Gills
Anus

Mouth
Velum
Nephridium
Cerebral commissure
Pedal nerve cord
Visceral nerve cord
Gill
Gonad
Auricle
Ventricle
Pedal retractor muscle

Figure 29.4
The monoplacophoran *Neopilina*.
(a) Ventral view; (b) dorsal view of shell; (c) lateral view of shell; (d) internal anatomy, dorsal view.

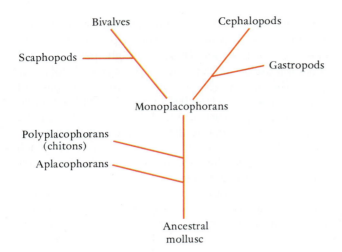

Figure 29.5
Phylogenetic tree reflecting the possible evolution of the molluscan classes.

four of the other molluscan classes (Fig. 29.5). We will return to this issue a little later. Let us now turn to a class of molluscs that most likely did *not* evolve from monoplacophorans and must have departed early from the main line of molluscan evolution: the polyplacophorans.

The class Polyplacophora includes about 500 species of molluscs called **chitons**. They range in size from a few millimeters to over 35 cm in length and are adapted for living on rocks and other hard substrata (Fig. 29.6). They are often common in shallow water along rocky shorelines and some extend into the intertidal zone.

The head of chitons is poorly developed, and much of the ventral surface is occupied by the broad foot (Fig. 29.7b). Chitons, however, are distinguished from all other molluscs in possessing a shell composed of eight overlapping plates arranged linearly along the anterior-posterior axis (Figs. 29.6 and 29.7a). The lateral margins of the plates are overgrown to varying degrees by the mantle, also called the girdle. The plates contain many channels that open to the surface and are occupied by sense organs, including eyes in some species.

It is these eight shell plates that tell us that chitons departed from the molluscan line prior to the evolution of the single shieldlike shell of monoplacophorans. It is difficult to derive the single shell of monoplacophorans from the eight shell pieces of chitons and vice versa. Thus chitons are believed to be the group derived from the opposite branch (sister group) from that leading to the monoplacophorans and their descendants (Fig. 29.5).

Chitons crawl about much like snails, but they are immobile when out of water at low tide. When the chiton is clinging to rocks and other objects, the mantle margin and foot adhere tightly to the substratum, so firmly that some can only be collected by prying them up with a heavy knife.

The mantle cavity of chitons is limited to two lateral troughs, or grooves, one on each side of the body between the foot and mantle edge (Fig. 29.7b). Within each groove lie many bipectinate gills. The ventilating current enters each groove through an opening created by the raised mantle margin. After passing the length of the groove and over the gills, the two currents exit as a single stream at the posterior end of the body, where the mantle edge is also raised (Fig. 29.7b).

Chitons feed largely on algae scraped from rock surfaces with a radula, whose teeth are capped with magnetite (iron). But as you would expect, the scraping is not very selective. The stomach contents of some chitons on the coast of Maine contained 14 different kinds of algae and other organisms but also a large amount of sludge (sediment).

Chitons have separate sexes, and each of the two gonads is provided with a gonoduct that opens near the nephridiopore in the mantle trough (Fig. 29.7b). The nephridia do not serve as gonoducts in chitons. The animals do not copulate, but sperm released by a male fertilize the eggs in the sea, or fertilization occurs in the mantle trough of the female. Eggs are shed singly or in strings, or they may be brooded within the mantle cavity. There is a trochophore larva, which develops directly into the adult.

CLASS GASTROPODA

The 35,000 living species of gastropods constitute the largest and most diverse class of molluscs. Included here are many familiar molluscs—snails, whelks, conchs, limpets, and sea slugs. Most gastropods are marine, but some live in fresh water and others have become adapted for life on land.

Figure 29.6
The West Indian chiton, *Chiton olivaceous*. The lateral exposed parts of the mantle is covered with calcareous scales.

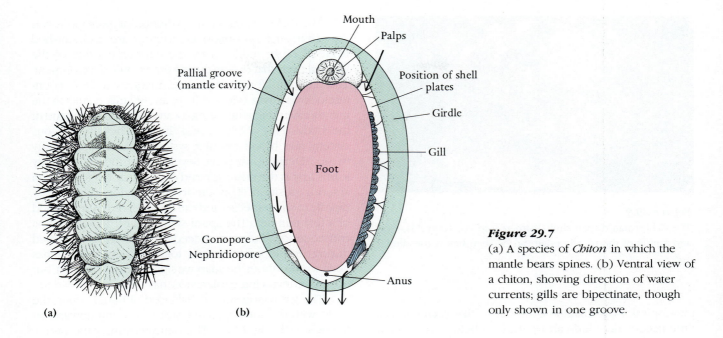

Figure 29.7
(a) A species of *Chiton* in which the mantle bears spines. (b) Ventral view of a chiton, showing direction of water currents; gills are bipectinate, though only shown in one groove.

Evolution of Gastropods

Most malacologists believe that gastropods evolved from monoplacophorans. If so, then three major changes occurred in that evolution. First, the shell became conical. The shieldlike shell of many monoplacophorans protected only their dorsal surface, and the animal would have been vulnerable when dislodged from the substratum. In the course of evolution the shell of the ancestors of gastropods became higher and more conelike, with a reduced aperture (Fig. 29.8a). Such a shell provided a protective retreat into which the entire body could be withdrawn. The spiral condition appears to have been an adaptation for making the shell more compact and less awkward to carry than if it were a long straight cone. Fossils of the earliest known gastropods and the group of monoplacophorans that may have been ancestral to gastropods had bilaterally coiled shells (planospiral shells) (Fig. 29.8b). The asymmetry characteristic of the shell of all modern gastropods provides additional compaction. Imagine coiling a large hose so that each coil lies on the ground and outside of the previous coil. The coiled hose would be flat, a symmetrical

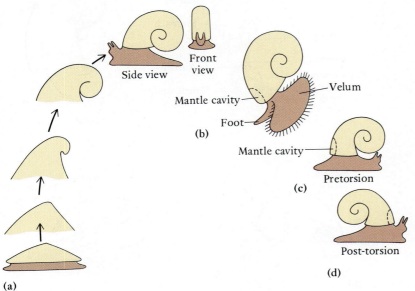

Figure 29.8
(a) Evolution of a planospiral shell. Height of the shieldlike shell of the hypothetical ancestral mollusc increases and a peak forms. The peak is pulled forward and coiled under. The aperture is reduced, and the animal can withdraw into the spiral shell, which is more compact and less awkward to carry than would be a straight conical shell. Note that the shell is bilaterally symmetrical.
(b) Diagram of a gastropod larva before torsion. The velum is the ciliated larval locomotor organ. (c) Hypothetical pretorsion gastropod with a planospiral shell. (d) Post-torsion gastropod. Torsion does not affect the planospiral shell except to place the coils of the shell posteriorly. The mantle cavity is now anterior.

Figure 29.9
The Californian turban shell, *Astraea undosa*. Tentacles, foot, and a little of the mantle edge can be seen below the shell.

planospiral. Now imagine coiling it so that each coil is a little inside and a little above the one below. The coiled hose would look like a pyramid. It would be an asymmetrical spiral like that of living gastropod shells. Such a shell is more compact because the diameter need not be as great as when each whorl lies outside of the previous whorl, as in a planospiral shell.

A second change in the evolution of gastropods was the increased development of the head, and this may be correlated with the more mobile, exposed life style of gastropods when compared to that of monoplacophorans. The head of most gastropods bears two tentacles with an eye at the base of each (Fig. 29.9).

The third change in the evolution of gastropods was the occurrence of **torsion**. Gastropods are distinguished from all other molluscs by the curious twisting of the visceral mass that occurs during embryonic development. This condition, called torsion, involves a 180° counterclockwise twist (viewed from above) that results in the mantle cavity and anus being located at the anterior end of the body (Figs. 29.8c, d and 29.10). The gut, nervous system, and blood vascular system are correspondingly twisted. During embryonic development, the mantle cavity forms first at the posterior end of the body but shifts to an anterior position. This embryonic torsion occurs very rapidly in some species and can be observed over a period of a few minutes. The spiral shell of gastropods is *not* a consequence of torsion; indeed the spiral shell is formed during development before torsion occurs. Much has been written about the adaptive significance of torsion, but no very convincing explanation has yet been provided.

The gastropod shell is balanced obliquely upon the body, with the spire directed posteriorly and upward on the right side (Fig. 29.11). The conical nature of the gastropod shell and its asymmetry result in an asymmetrical and partially occluded mantle cavity. Since the mantle cavity is larger on the left side than the right, most gastropods have retained only the left gill of an original single pair. The presence of only the left gill has in turn resulted in the reduction and loss of the right auricle of the heart. Some primitive gastropods still have two gills and two auricles, but most have only those on the left side.

Let us now review the gastropod ground plan. There is a broad, flat, creeping foot. A well-developed head bears eyes and tentacles. There is a shell of one piece that has the

Figure 29.10
Dorsal and lateral views of a hypothetical ancestral gastropod prior to torsion (a, b) and after torsion (c, d).

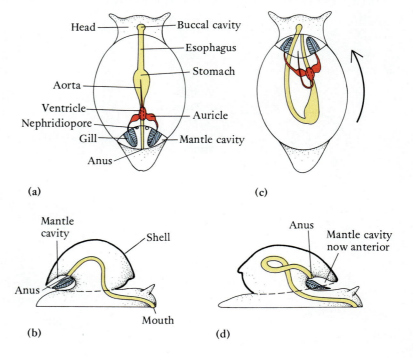

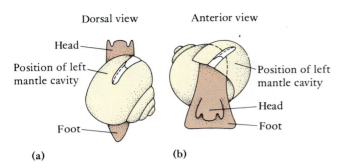

Figure 29.11
Position in which the shell is carried in gastropods: (a) dorsal view, (b) anterior view. The slot that is found in some primitive gastropods indicates the original mid-dorsal line of the mantle cavity. Note that the mantle cavity is restricted largely to the left side.

form of an asymmetrically spiraled cone. The body is twisted 180° between the head and foot and the visceral mass within the shell (torsion). The mantle cavity, which is larger on the left side, contains two gills or more commonly only the left gill.

Shell

The central axis of the asymmetrical spiral shell is the **columella** (Fig. 29.12a), to which the usually single large retractor muscle (columellar muscle) of the body is anchored. When the columellar muscle contracts, the gastropod withdraws into the shell by folding the middle of the foot (Fig. 29.12b−e). The anterior half of the foot and the head are withdrawn first, followed by the posterior half of the foot. Many gastropods have a round plate, the **operculum**, on the back of the foot, which plugs the aperture when the animal is withdrawn (Fig. 29.12e).

The shells of gastropods exhibit great variation in shape and coloring. Shell form and sculpture may contribute to protection, shell strengthening, or ease of carriage. We will mention only two common modifications of the shell. In some gastropods, commonly those living in surge or rapids, the shell has been secondarily converted into a dorsal shield. The spiral apex first develops in the typical manner during juvenile stages. Then the last whorl becomes extremely large and covers most of the body of the animal as a low convex plate. Abalones and limpets have shells of this type (Figs. 29.13 and 29.14b). The shell of limpets has even become secondarily bilateral.

Figure 29.12
(a) Longitudinal section through a shell. (b−e) Withdrawal into shell by folding of foot and closure with operculum.

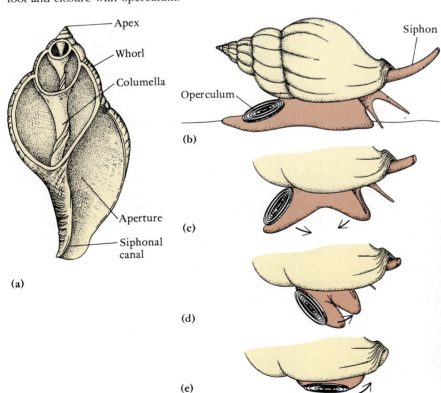

Figure 29.13
Barnacles and limpets on intertidal rock on the east coast of England.

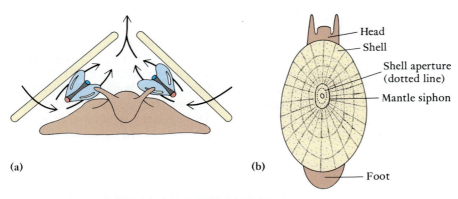

(a)

(b)

— Head
— Shell
— Shell aperture (dotted line)
— Mantle siphon
— Foot

(c)

Figure 29.14
(a) Diagrammatic section of a keyhole limpet, showing bipectinate gills and path of water flow. Opening at top of shell, from which the name keyhole is derived, provides for the exit of the water stream. Keyhole limpets are one of the small number of gastropods that have two gills. The two-dimensional nature of the diagram requires that the inhalant stream be shown entering laterally; it actually enters the mantle cavity over and to either side of the head. Afferent and efferent vessels of gills are shown in blue and red. (b) Dorsal view of a keyhole limpet. (c) Dorsal view of an abalone. Although the shell has become shieldlike, the spiral origin can still be seen. The ventilating current exits from the more posterior of the holes that are arranged in a row on the left side of the shell. The mantle margin bears many sensory tentacles; the two cephalic tentacles are much longer.

Another common modification is shell reduction and loss. This has occurred independently in a number of different groups, as we shall see. Gastropods without shells are commonly called slugs.

Locomotion

Gastropods creep upon their broad, flat, ventral foot by a process of body deformation. Small waves of contraction sweep along the length of the foot, one wave closely following another. The area of the foot in the contracted region is slid forward, performing, in a sense, a little step (Fig. 29.15c); the summation of these little steps gives the appearance of a gliding motion. The substratum is lubricated by large amounts of mucus produced by glands in the foot. In some gastropods, the waves sweep from back to front, in the same direction as the animal is moving (direct waves) (Fig. 29.15a). In many other species, the waves move from front to back, opposite the direction in which the animal is moving (retrograde waves) (Fig. 29.15b). Waves may extend all the way across the foot (monotaxic) (Fig. 29.15a), or those on one side of the foot may alternate with those on the other side (ditaxic) (Fig. 29.15b). In the latter case the movement of the snails looks like someone walking in snow shoes. The most common mode of muscular pedal creeping in gastropods is by ditaxic retrograde waves. If a gastropod, such as a periwinkle or a garden slug, is permitted to crawl on a glass plate,

these waves can easily be seen from the underside when the plate is flipped over.

Gastropods living on soft substrata, such as sand and mud, move entirely or in part by the cilia that cover the foot surface. Many soft-bottom forms use their foot to burrow into the substratum.

Many limpets that live in the intertidal zone return to the same spot, or home, at low tide (Connections 29.1).

Evolution of Water Circulation and Gas Exchange

Most of the gastropods with gills are members of the subclass Prosobranchia. This is the largest group of gastropods, about 18,000 species, and the greatest number are marine. The first gastropods were prosobranchs, and this class contains the most primitive living species. However, all prosobranchs are by no means primitive; indeed most are not.

Whatever the evolutionary significance of torsion may have been, the anterior position of the mantle cavity seems to have imposed fouling and ventilating problems on gastropods, because numerous changes occurred that better separate the inhalant and exhalant water streams. For example, in some primitive prosobranch gastropods, such as slit shells, abalones, and keyhole limpets, the shell contains a notch or one or more holes through which the exhalant water stream exits (Fig. 29.14a). In most

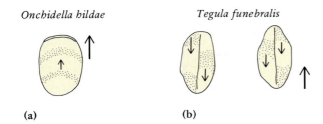

Onchidella hildae

Tegula funebralis

(a)

(b)

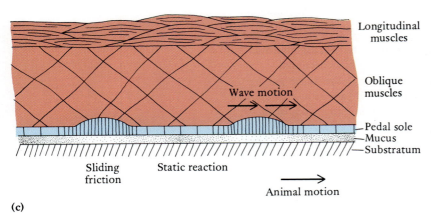

Longitudinal
muscles

Oblique
muscles

Wave motion

Pedal sole
Mucus
Substratum

Sliding
friction

Static reaction

Animal motion

(c)

Figure 29.15
Patterns of pedal waves in two gastropods:
(a) Direct monotaxic waves in the intertidal
pulmonate slug *Onchidella hildae*;
(b) retrograde ditaxic waves in *Tegula
funebralis*. Large black arrows indicate
direction of animal's movement; small black
arrows, the direction of the waves. (c) Direct
pedal muscular waves in the foot of a
gastropod. Waves of pedal contraction sweep
over foot in same direction as animal is
moving. In region of wave (close vertical
lines) sole slides forward over liquified
mucus.

The ability of some animals to consistently find their way
back to a home site is popularly associated with birds and
mammals, but other animals are capable of homing. Among
invertebrates, insects such as bees, ants, and wasps, and
molluscs such as limpets and terrestrial snails, are good
examples. Some intertidal chitons and limpets will leave
their home site to feed when splashed by the rising tide and
will return home before the tide falls again. The limpet
home scar is a slight depression in the rock onto which the
edges of the shell fit precisely. The home may provide
protection against predation and desiccation at low tide and
is the center of a feeding territory. Chemical cues in the
limpet's mucous trails appear to be of primary importance in
orienting to home. Many terrestrial snails and slugs will
return to shelters beneath stones and logs, again using cues
in mucous trails or airborne pheromones from fecal pellets.
In these terrestrial species, homing is enhanced by the
presence of other individuals sharing the home site.

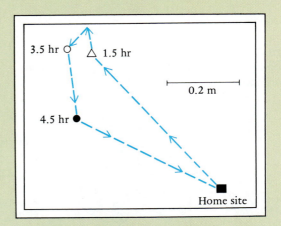

3.5 hr ○   △ 1.5 hr

0.2 m

4.5 hr ●

Home site

**Movements of a homing limpet *Patella vulgata*. Times refer to
times and positions after immersion by the rising tide.**

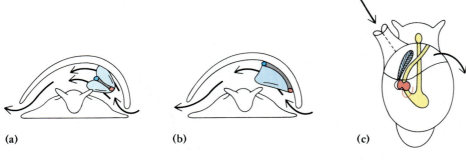

(a)   (b)   (c)

Figure 29.16
The left to right ventilating current of prosobranch gastropods. (a) Diagrammatic anterior view of a primitive prosobranch having a single but bipectinate gill. (b) Similar diagram of a prosobranch with the more usual monopectinate gill, in which the gill axis is anchored to the mantle wall. (c) Dorsal view of a prosobranch with a single monopectinate gill; mantle margin forms an inhalant siphon. Note that the siphon is formed by extension and rolling inward of the mantle margin. Afferent and efferent vessels in gill axis are shown in blue and red.

prosobranchs, however, the ventilating current enters the mantle cavity on the left side of the head, makes a turn across the single left gill, and exits to the right (Fig. 29.16). The anus is located near the right mantle margin so that feces are swept clear of the mantle cavity.

Primitive prosobranchs possess bipectinate gills (Figs. 29.14a and 29.16a). However, most prosobranchs have lost all of the filaments on one side of the gill axis, which is now attached directly to the mantle wall (Figs. 29.16b and 29.17). Commonly, the edge of the mantle on the left side is drawn out to form a siphon that receives the inhalant water stream (Figs. 29.16c and 29.17). The shell of such siphonate forms has a notch or spout in which the

Figure 29.17
Gastropod anatomy. Lateral view of the whelk *Busycon canaliculatum* with shell removed.

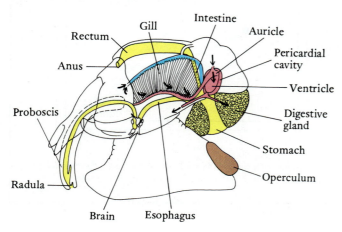

siphon is located (Fig. 29.12a). The siphon is related more to feeding than it is to ventilation, for it enables the animal to locate prey or food sources more easily from substances being monitored in the water stream.

The prosobranchs gave rise to two other subclasses of gastropods, the Opisthobranchia and the Pulmonata. The subclass Opisthobranchia, with some 2000 species of marine gastropods, is much smaller than the Prosobranchia, but it is a highly diverse group, containing bubble shells, sea butterflies, sea hares, and various sorts of sea slugs. Opisthobranchs are characterized by 90° of **detorsion**, or untwisting of the visceral mass. The mantle cavity and gill are on the right side (Fig. 29.18a). In many species detorsion has been accompanied by reduction or loss of the shell, mantle cavity, and gill. In such forms as sea hares and sea slugs, the body has become secondarily bilateral, at least externally (Fig. 29.18c−e). The mantle cavity has been lost in sea slugs (also called nudibranchs), and gas exchange occurs by means of a secondary gill located around the anus (Fig. 29.18c) or across the general body surface. In most of these sea slugs without the secondary anal gill, the dorsal surface bears a large number of clublike projections or **cerata** (Fig. 29.18d), which greatly increase the surface area. Despite their name, the flamboyantly colored sea slugs are among the most beautiful molluscs. At least in some species the color enables the animal to blend into an equally bright background, such as a sponge, or functions as a warning, for many sea slugs secrete noxious substances or contain functional nematocysts. The nematocysts, located in the cerata, are obtained from their cnidarian prey, but remarkably the nematocysts are not digested by the sea slug. They pass in the undischarged state from the stomach through the ducts of the digestive

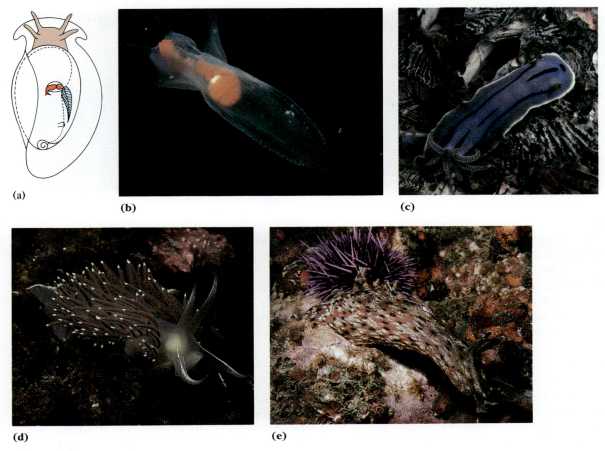

(a)   **(b)**   **(c)**

(d)   **(e)**

Figure 29.18

Opisthobranch gastropods. (a) Dorsal view of a bubble shell. Note that as a result of detorsion, the mantle cavity and gill are on the right side. (b) The Atlantic naked sea butterfly *Clione limacina*. It swims with finlike extensions of the foot. (c) An Indo-Pacific sea slug with a secondary gill around the anus. (d) A nudibranch sea slug from British Columbia in which the dorsal surface bears clublike projections (cerata). Note the two pairs of tentacles characteristic of most opisthobranchs. (e) A sea hare (*Aplysia californica*). This opisthobranch has a reduced shell buried within the mantle. The large foot folds, which are folded dorsally in the photograph, can be used for escape swimming. The anterior end with tentacles is to the right.

glands, which extend into each of the cerata. Here they become located in special sacs and can be used by the sea slug for its own defense.

Another highly specialized group of opisthobranchs are the sea butterflies (Fig. 29.18b). These are planktonic gastropods. Some have shells and some are shell-less. The gill is lost in most, and gas exchange occurs across the general body surface. They swim with a pair of finlike lateral extensions of the foot.

Members of the third subclass, the Pulmonata, containing the land snails and slugs, are adapted for gas exchange in air. The mantle cavity of the pulmonates evolved into a diffusion lung, and gills have been lost (Fig. 29.19a). Pulmonates probably evolved in estuarine habitats where they were often left out of water. Many pulmonates have remained aquatic, largely in fresh water.

Some of these freshwater species must come to the surface to obtain air; others have abandoned the lung altogether and acquired secondary gills outside of the mantle cavity. Since pulmonates lack an operculum, it is easy to distinguish freshwater pulmonates from the operculate freshwater prosobranchs in an aquarium.

Of the 16,000 species of pulmonates, the largest number are terrestrial. Desiccation is minimized by reducing the opening into the mantle cavity to a very small pore (Fig. 29.19b).

Circulation, Excretion, and Water Balance

Except for the asymmetry produced by torsion and the loss of the right gill, the circulatory system of gastropods is like that described for the generalized molluscs. The blood of

(a) (b) (c)

Figure 29.19

Pulmonate gastropods. (a) Dorsal view of a pulmonate, showing the vascularization of the wall of the mantle cavity, which forms the lung. Anus and excretory opening are on the right. (b,c) Two terrestrial pulmonates. (b) The yellow banana slug, *Agriolimax columbianus*, from the West Coast of the United States. The opening into the lung can be seen at the lower edge of the saddle-like mantle. (c) A tree snail. The eyes can be seen at the top of the longer second pair of tentacles. The first pair of tentacles are small and do not show in this photograph.

most gastropods contains the respiratory pigment **hemocyanin** (p. 285).

Ammonia is the excretory waste of aquatic gastropods, but terrestrial pulmonates excrete uric acid. By the addition of a **ureter** formed from the mantle wall, the excretory opening of land pulmonates is located at the opening to the mantle cavity, along with the anus. Thus, the lung is not fouled with wastes. Pulmonates are not especially adapted for avoiding water loss through desiccation, and large amounts of water are lost in the mucus secreted when crawling. Most pulmonates are therefore restricted to humid environments or must be nocturnal. During the winter or during very dry periods the animals hide in leaf mold or beneath wood or stones or attach the shell to vegetation by mucous cords (Fig. 29.20). The aperture of

Figure 29.20

Pulmonate snails estivating on vegetation in Israel.

the shell is then covered with a mucous film, which dries to form a protective **epiphragm**. In some species the epiphragm is even calcified. The animal is inactive during this time, and the metabolic rate drops to a very low level. Such a period of dormancy during dry periods (termed **estivation**) is a common adaptation of semitropical and tropical animals living in regions with long dry and wet seasons. Nocturnal activity and estivation permit some pulmonates to inhabit deserts and other arid regions.

Nutrition

Gastropods exhibit an enormous range of diets and feeding habits. There are herbivores, carnivores, scavengers, detritus feeders, plankton feeders, and even parasites. Almost all have retained the radula as a feeding organ, but it is modified for the particular diet of the gastropod. Some possess bladelike cutting jaws in addition to the radula. The radula of many carnivorous gastropods has large, heavy teeth and may be located at the end of an extensible proboscis (Fig. 29.21a). The prey—other invertebrates, especially other molluscs—may in some cases be smothered with the foot or wedged open with the edge of the shell (Fig. 29.22a). Some whelks can hold a clam in the foot and chip or erode the clam valves sufficiently to permit the proboscis to enter. Two families of carnivores use the radula as a drill to penetrate the shells of bivalve prey that have been previously softened with secretions from a gland on the foot or proboscis (Fig. 29.22b). The highly specialized tropical cone shells stab their prey—worms,

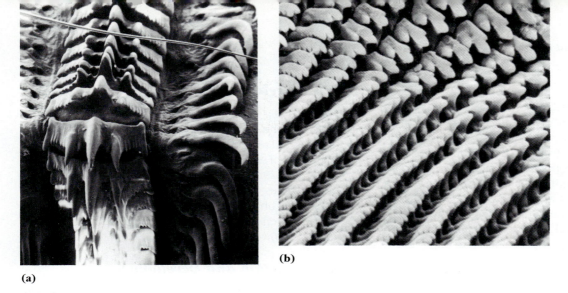

(a)

(b)

Figure 29.21
SEM of gastropod radulas. (a) Radula of *Urosalpinx*, a carnivorous drilling prosobranch. There are only three teeth to a transverse row, but the middle tooth bears several cusps. (b) Part of a radula of a herbivorous pulmonate. Pulmonate radulas have many teeth in each transverse row.

Figure 29.22
(a) The whelk *Buccinum* using the edge of its shell to pry open a cockle. (b) Photograph of a perforation through a bivalve shell produced by the radula of a drilling gastropod. Note the beveled edge. It takes a snail about 8 hours to penetrate 2 mm of shell. (c) SEM of the harpoon-like radula tooth of *Conus imperialis*. Note the folded structure and barbed end. (d) A cone shell swallowing a fish.

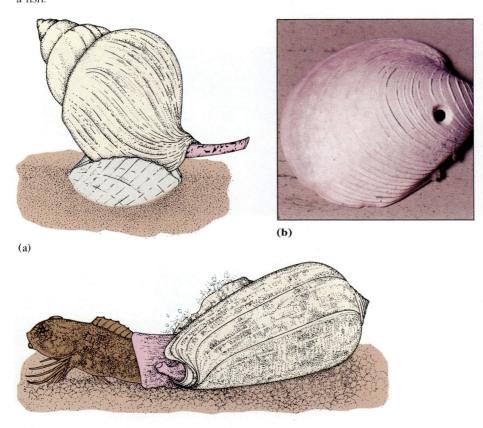

(a)

(b)

(d)

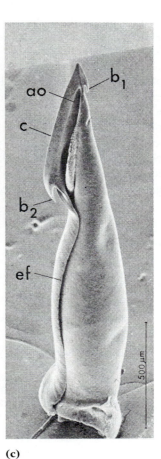

(c)

other snails, or fish—with a poisonous harpoon-like radular tooth (Fig. 29.22c)! Sea slugs, cowries, and some other groups are carnivores that graze on sessile animals, especially cnidarians.

Many aquatic gastropods are scavengers or detritus feeders. A few gastropods, like the semisessile slipper shells, use the gills to filter plankton from the ventilating current. Worm shells are common but largely tropical gastropods in which the shell has become unspiraled and plastered to rock, coral, or other shells (Fig. 29.23). These sessile gastropods use a mucous net spun out by glands in the foot to trap suspended food particles.

Marine herbivores feed on algae; freshwater and terrestrial herbivores feed on algae and vascular plants. Pulmonates are largely but not exclusively herbivores and have radulas with many teeth in each transverse row (Fig. 29.21b). The gut of many gastropod herbivores contains a gizzard. A remarkable group of herbivores are certain sea slugs that use their needle-like radula teeth to puncture algal cells, from which they suck the contents. The algal chloroplasts are not digested but are retained and continue to function in the production of photosynthate within the body of the sea slug!

Some gastropods are ectoparasites of bivalves, sitting at the edge of the shell and extending the little proboscis down into the edge of the clam's mantle from which they suck blood and tissue. A few gastropods are endoparasites in the body wall of echinoderms, such as sea stars, or in the coelom of sea cucumbers.

Primitive, microphagous gastropods have a stomach adapted for sorting fine particles as described for the generalized mollusc. However, the stomach of most gastropods, which are macrophagous, has the form of a simple sac in which extracellular digestion occurs. Enzymes are provided by the salivary glands, glands along the esophagus, and the digestive glands. Absorption occurs within the digestive glands, and the intestine functions primarily in the formation of feces (Fig. 29.17).

Sense Organs

The head of a gastropod bears one or two pairs of sensory tentacles: one pair in prosobranchs, two pairs in opisthobranchs and terrestrial pulmonates. The eyes, one located at the base of each tentacle (at the top of the tentacle in land snails, Fig. 29.19c), are not highly developed; indeed, in most species they function in orientation and are not capable of object discrimination. The osphradium is an important sense organ, and is especially well developed in aquatic carnivores.

Reproduction and Development

Most species of prosobranchs have separate sexes. Pulmonates and opisthobranchs are hermaphroditic. In primitive prosobranchs, the gametes are discharged through the nephridia, and fertilization is external. In most gastropods, however, a complex reproductive tract has developed, permitting copulation, internal fertilization, and the deposition of eggs within protective envelopes. The penis is located behind the right tentacle. The eggs of aquatic species are deposited in strings or masses or within special cases molded by the foot (Fig. 29.24). The large eggs of

Figure 29.24
(a) The whelk *Busycon carica* with egg case. The eggs are located in the small pill-box-like sections that are attached by a cord. (b) Collar-like egg case of a moon shell. The case is reinforced with embedded sand grains.

(a)

Figure 29.23
A cluster of sessile worm shells, prosobranchs in which the shell has become unspiraled and tubelike.

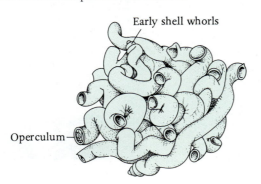

Early shell whorls

Operculum

(b)

pulmonates are laid singly and deposited in small clusters in the soil and leaf mold or beneath bark, logs, or stones.

Cleavage is typically spiral. In primitive marine prosobranchs the initial trochophore larva develops further into a veliger larva that displays many gastropod features, such as a spiral shell and foot (Fig. 29.25). Like the trochophore, the veliger is planktonic and swims by means of the **velum**, a large ciliated organ derived from the prototroch of the trochophore. The velum also collects fine suspended particles upon which the larva feeds. In many marine and in all freshwater and terrestrial species, development is direct, i.e., no larval stage is present. Little snails emerge from the eggs or cases.

CLASS BIVALVIA

The class **Bivalvia** contains some 20,000 species of marine and freshwater molluscs commonly called clams or bivalves; it includes the familiar mussels, cockles, oysters, and scallops. The smallest are less than 1 cm in length; the largest are giant clams (*Tridacna*) of the tropical Pacific; which attain lengths of 1.3 m. The distinguishing characteristics of the class represent adaptations for burrowing in a soft substratum. The body is greatly compressed laterally (Fig. 29.26). There is no head, the radula is lost, and the shell is composed of two lateral pieces, or valves, hinged together dorsally. Lateral compression has resulted in a great overhang of the mantle and shell, and the large mantle cavity extends to both sides of the body. The anteriorly directed foot is laterally compressed and somewhat bladelike, hence the older name of the class Pelecypoda—hatchet foot.

The bivalve design may seem remote from the ancestral monoplacophorans, but imagine folding a monoplacophoran into a right and left half and then hinging the crease along the middorsal line. You would have a rough approximation of a bivalve (Fig. 29.27). This is not quite as conjectural as it might seem, for there is an extinct fossil group of molluscs from the Cambrian, known as the Rostroconchia, in which the shell is folded into two lateral halves but the two halves are continuous, i.e., not hinged. The rostroconchs are believed to be either intermediate between monoplacophorans and bivalves or at least closely related to bivalves.

Mantle and Shell

The mantle margin possesses three folds (Fig. 29.28b). The muscular inner fold of the mantle edge of one side is pressed against that of the other when the valves are closed. Where these muscles are anchored to the shell, they produce a scar, which appears as a long line (pallial line) a short distance back from the shell margin (Fig. 29.26c). The middle fold is sensory. The outer fold secretes the organic periostracum and the outer part of the calcareous material of the shell. The remainder of the calcareous material is secreted by the general mantle surface. The calcareous portion of the shell consists of two to four layers laid down as thin sheets (nacre) or prisms or in more complex ways. Deposition is not continuous and is affected by environmental conditions (Connections 29.2). The periostracum functions to protect the underlying calcareous layers from dissolution and can be very thick in freshwater bivalves. Since it is the first part of the shell to be produced, secreted in the periostracal groove between the outer and middle folds, it is also important in creating a watertight space over the mantle into which the calcareous layers are deposited.

Pearls are formed by the deposition of concentric layers of calcareous material around a sand grain or some other foreign object that becomes lodged between the mantle and shell and then within a pocket of mantle tissue. Cultured pearls are produced by placing a shell bead in a mantle sac, which in turn is placed within the gonadal tissue of another pearl oyster (*Pinctada*). About 1 mm of nacre is laid down around the bead in several years.

The hinge ligament that connects the two valves dorsally is composed of elastic protein covered by a layer of periostracum (Fig. 29.28a). The valves are opened by tension resulting from compression and stretching of the elastic ligament and closed by anterior and posterior adductor muscles extending transversely between the valves. (If you are buying live clams at a seafood market, don't buy any that are gaping; they are probably dead.) The attachment points of the muscles are visible as scars on the inner surface of the valves (Fig. 29.26c). Commonly, this inner surface in the region of the hinge possesses teeth or

Figure 29.25
Lateral view of the veliger larva of a gastropod (slipper shell). The ciliated velum is seen on the right.

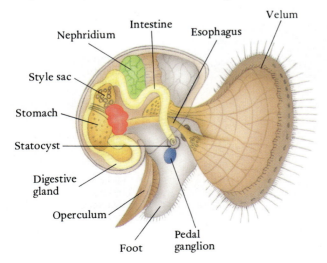

Nephridium
Intestine
Esophagus
Velum
Style sac
Stomach
Statocyst
Digestive gland
Operculum
Foot
Pedal ganglion

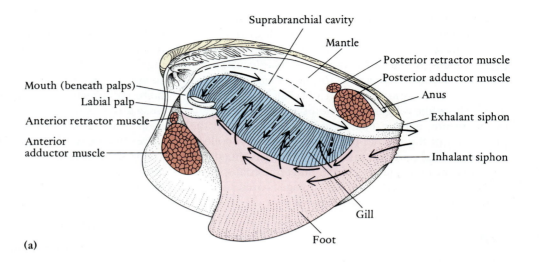

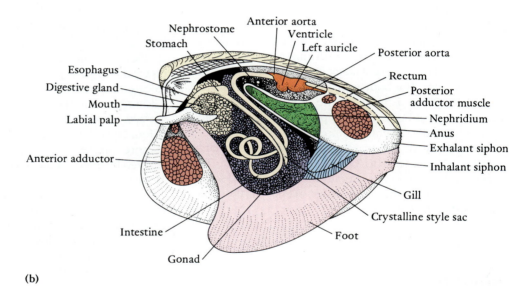

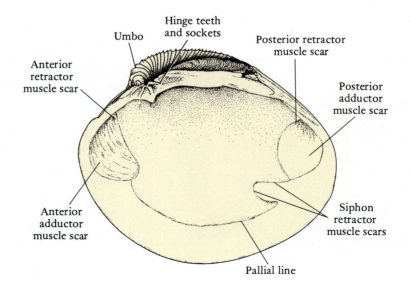

Figure 29.26
Anatomy of the quahog
Mercenaria mercenaria.
(a) Animal with left valve and
mantle removed. Heavy
arrows indicate path of water
current; dashed arrows, path
of filtered particles.
(b) Section through visceral
mass showing internal organs.
(c) Interior view of right valve
showing muscle scars.

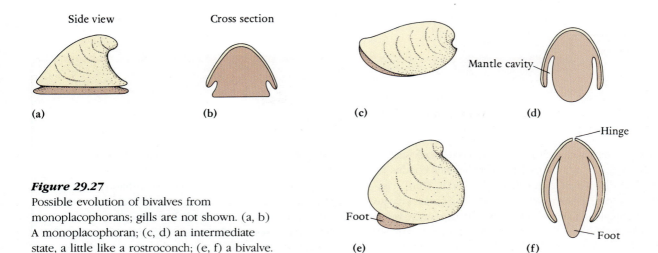

Figure 29.27
Possible evolution of bivalves from monoplacophorans; gills are not shown. (a, b) A monoplacophoran; (c, d) an intermediate state, a little like a rostroconch; (e, f) a bivalve.

ridges, which interlock with sockets and grooves on the opposite valve.

Nutrition

Protobranch Bivalves

Early bivalves belonged to the subclass Protobranchia, some species of which are still living today. The protobranchs possess a single pair of posterior-lateral bipectinate gills like those of primitive gastropods. The ventilating current enters the mantle cavity between the posterior and ventral gape of the valves, passes up through the gills, and exits posteriorly and dorsally (Fig. 29.29).

The early protobranch bivalves, like many living species, were probably selective deposit feeders. Such protobranchs utilize a pair of long tentacles for deposit feeding. Each tentacle is associated with two large flaplike folds, called **labial palps**, located to either side of the mouth (Fig. 29.29). During feeding, the tentacles are extended into the bottom sediments. Deposit material adheres to the mucous-covered surface of the tentacle and then is transported by cilia back to the palps. Each pair of palps functions as a sorting device. Light particles are carried by certain cilia to the mouth; heavy particles are carried by other cilia in grooves to the palp margins, where they are ejected to the mantle cavity.

The evolution of selective deposit feeding in protobranchs was correlated with the invasion of a soft-bottom environment by the ancestral bivalves. As the body became laterally compressed for burrowing, the mouth

Figure 29.28
(a) Diagrammatic cross section through a bivalve showing hinge ligament and antagonistic adductor muscle. (b) Cross section through margin of shell and mantle of a bivalve, showing mantle lobes.

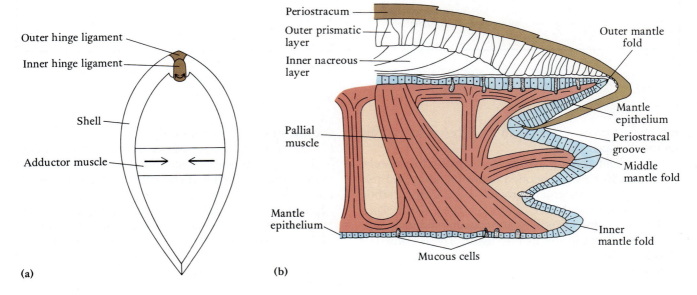

Records in Skeletons and Shells

In many animals with external calcareous skeletons, such as corals, molluscs, and barnacles, environmental conditions can affect the deposition of skeletal material. Thus, regularly recurring environmental events—seasonal climate changes, tidal fluctuations, and so on—will be recorded in the fine structure of the skeleton. For example, in many scleractinian coral and bivalve shells, the density of the calcium carbonate ($CaCO_3$) differs over the course of each year. The age of the animal can therefore be determined by counting the density bands in the coral skeleton or bivalve shell.

In some intertidal bivalves the interruption of $CaCO_3$ deposition when the animal is out of water at low tide produces a fine line in the shell ultrastructure. As a result, the shell contains a record of the tides, even the occurrence of spring and neap tides (bi-monthly strong and weak tides, respectively).

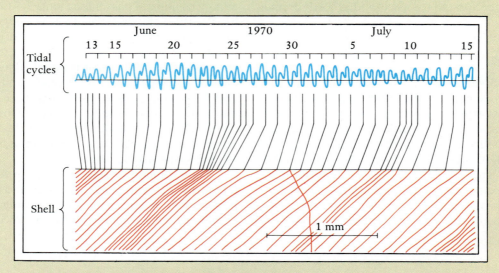

Microgrowth increments in the shell of the intertidal cockle (*Clinocardium nuttalli*) from Charleston, Oregon. The oblique growth lines of the shell shown at the bottom are correlated with the daily tidal cycles shown at the top. Spring tides are marked by one very low tide per day, neap tides by two equal tides of low amplitude. Cockles will be exposed when tides fall to level of horizontal line.

was shifted upward away from the substratum, and the radula disappeared.

In recent years it has been discovered that some protobranchs have abandoned deposit feeding and rely upon symbiotic bacteria to provide them organic carbon (see Connections 30.1, p. 612). The giant clams (*Tridacna*) of the tropical Pacific contain symbiotic algae (zooxanthellaer) in the mantle tissue. These clams live upside down in shallow water or within coralline rock, exposing the mantle margin to light.

Evolution of Lamellibranchs

In some group of early protobranch bivalves, filter feeding evolved. An explosive evolution followed, and the filter feeders, called **lamellibranchs**, came to dominate the bivalve fauna. The gills and ventilating current of protobranchs preadapted them for filter feeding. As the lamellibranchs evolved, plankton in the ventilating current came to be utilized as a source of food, the gills became the filters, and the frontal gill cilia (which originally served for the transport of sediment) became adapted for the trans-

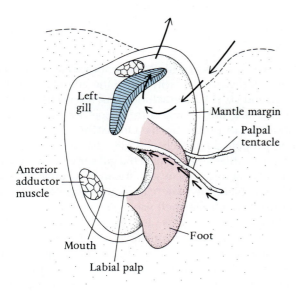

Figure 29.29
A generalized protobranch in feeding position. Large arrows indicate direction of water current; small arrows show food particles collected in sand moving along palpal tentacles to labial palps, where sorting occurs prior to ingestion.

Figure 29.30
Evolution of lamellibranch gills.
(a) Primitive protobranch gill (position relative to foot-visceral mass and mantle indicated in cross section). (b) Development of food groove to produce the lamellibranch condition. (c) Folding of filaments at food groove to produce the lamellibranch condition. (d) Small section of lamellibranch gill; arrows show direction of water flow.
(e) Tissue connections that provide support for the folded lamellibranch filaments.

port of the trapped plankton from the filter to the mouth.

The principal modification of the gills for filtering was the lengthening and folding of the gill filaments, which greatly increased their surface area (Fig. 29.30a−c). The long folded filaments are supported by the development of cross connections between the two halves, by connections between adjacent filaments, and by connection of the tips of the filaments to the foot or mantle wall (Fig. 29.30e). The development of these supporting connections varies greatly in different groups of lamellibranchs. The folding converted the original single gill into two gills, each new gill formed by one series of folded filaments. There is thus a pair of gills on each side of the body of lamellibranchs. The lengthened filaments and their attachment to one another give the gills a sheet- or platelike form, hence the name Lamellibranchia ("sheet gill"). In many lamellibranchs the surface area of the gills has been further increased by secondary lateral folding (plications) so that the gill looks pleated like a curtain (Fig. 29.31).

Where adjacent filaments are tightly connected, openings (**ostia**) remain for the passage of water between the filaments (Figs. 29.30d and 29.31). The interior space between the two folded halves of the filaments forms water

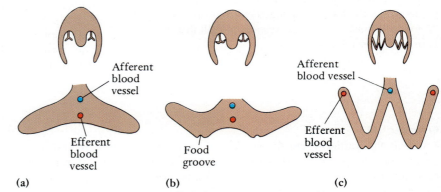

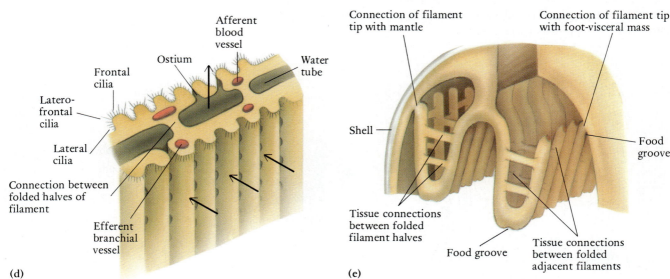

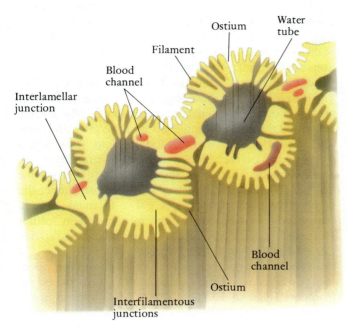

Figure 29.31

Cross section through a small part of a gill with secondary folding (plicated). Note that there are extensive junctions between filaments but interlamellar junctions are limited to points where the folds come together. Compare with Figure 29.30d, which shows a similar section of a gill that does not have secondary folding.

tubes, which connect with the **suprabranchial cavity**, the portion of the mantle cavity above the gills (Figs. 29.26a, 29.30d, and 29.31). In lamellibranchs, the ventilating current, now also the feeding current, enters posteriorly and ventrally as in protobranchs. On reaching the gills, the water, propelled by the lateral gill cilia, enters openings between the filaments on all of the gill surfaces. Within the interior water tubes, the water stream flows upward to the suprabranchial cavity, where it turns posteriorly and flows outward through the shell gape (Fig. 29.26a).

Suspended particles are filtered out by special laterofrontal cilia as the ventilating current passes between the filaments (Fig. 29.30d). The filtered particles are passed to short frontal cilia, become covered with mucus, and are transported downward or upward to food grooves, depending upon the species. The food grooves are located ventrally along the gill margins and dorsally adjacent to the points where the gill is attached (Figs. 29.30b and 29.32a and b).

The food grooves carry the collected particles to the labial palps, which retain their original sorting function (Fig. 29.32c). Fine particles, mostly phytoplankton, are conveyed to the mouth.

Digestive System

A cord of mucus filled with phytoplankton is carried down the esophagus and wound into the stomach (Fig. 29.33). The protobranch stomach has a style sac and protostyle like that of primitive gastropods. However, in lamelli-

Figure 29.32

Transport and sorting of filtered particles by lamellibranch gills. (a, b) Transverse sections of gills on one side (inner gill on right) showing direction of frontal cilia beat and position of anteriorly moving food tracts (black dots): (a) shows primitive condition with five tracts; (b) shows condition in many lamellibranchs. (c) One pair of palps (spread apart) and anterior section of gills of oyster. Arrows indicate ciliary tracts; x represents rejected material.

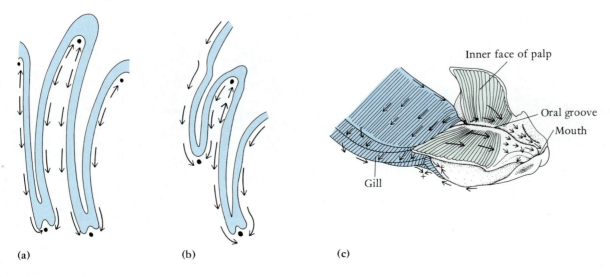

(a) (b) (c)

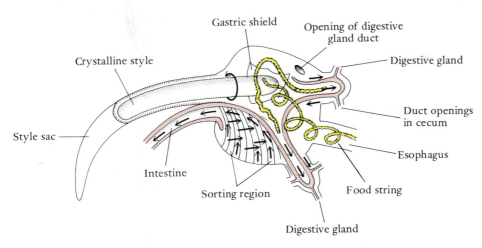

Gastric shield

Opening of digestive gland duct

Crystalline style

Digestive gland

Duct openings in cecum

Style sac

Esophagus

Intestine

Sorting region

Food string

Digestive gland

Figure 29.33
Diagram of the stomach and digestive gland duct of a lamellibranch, showing rotation of crystalline style and winding of mucous food string. Arrows indicate ciliary pathways.

branchs the mass of mucus within the style sac has become compacted into a stiff rod, the **crystalline style**. The rod is secreted by the style sac and rotated by the style sac cilia. In addition to the protein matrix, the style contains carbohydrate-splitting enzymes and lipase. The rotating end of the style is abraded against a chitinous piece in the stomach. The rotation of the crystalline style winds in the mucous rope from the esophagus and stirs the contents of the stomach; the abrasion of the end liberates enzymes and initiates the digestion of carbohydrates and fats within the stomach. The churning of the stomach mass throws particles against the sorting region, which separates fine particles from coarse particles. Fine particles are conveyed to the digestive glands, where digestion is completed intracellularly. The rejected coarse particles are carried along a groove to the intestine, where they are compacted into fecal pellets and eventually ejected.

Like primitive gastropods, bivalves possess a single pair of nephridia. The blood of most species lacks respiratory pigments, since gas exchange is accommodated by the large mantle and gill surface area and metabolic rates are low.

Adaptive Groups of Bivalves

The evolution of filter feeding enabled lamellibranchs to exploit habitats other than shallow depths in soft bottoms. Bivalves that occupy similar habitats and have similar life styles tend to exhibit similar adaptations. Collectively, we think of them as constituting an adaptive group, a very useful way of categorizing the diversity of animals.

Soft-Bottom Burrowers
Most lamellibranchs have continued to inhabit sand and mud bottoms, but the ability to utilize the ventilating

current in feeding has enabled many species to live at deeper levels in the substratum.

Burrowing is accomplished with the foot, which is extended anteriorly between the gape of the valves into the substratum (Figs. 29.26 and 29.34a). Initial extension of the foot is provided by a pair of protractor muscles, which extend from the foot to a point on the shell below the anterior adductor muscle. The substratum is simultaneously softened by water ejected from the mantle cavity as the valves are closed. The closing valves exert pressure on the water remaining within the mantle cavity, which in turn exerts pressure on the visceral mass, driving blood into the foot. The elevated blood pressure further extends the foot and dilates the distal end, anchoring it in the substratum. Two pairs of pedal retractors pull the valves down upon the anchored foot. The pedal retractors may contract somewhat alternately, causing the valves to rock and facilitating their movement through the substratum.

Burrowers in soft bottoms must cope with the tendency of the surrounding sediment to enter with the ventilating and feeding water stream. The problem becomes greater as the animal burrows deeper. Relatively permanent burrows with mucus-compacted walls are formed by many species, especially those that burrow deeply. There has also been a tendency for the opposing mantle edges to fuse together. The most common fusion points are around the apertures for inhalant and exhalant water currents, but fusion may also occur ventrally, leaving only an opening for the protrusion of the foot (Fig. 29.34b). Ventral mantle fusion undoubtedly aids in the elevation of hydrostatic pressure within the mantle cavity during burrowing. **Siphons**, fused tubelike extensions of the mantle around the inhalant and exhalant openings, are yet another adaptation to reduce the intake of sediment. The siphons are extended upward to the surface and permit the

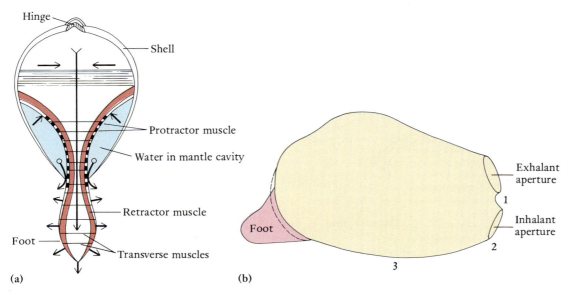

Figure 29.34

(a) Diagrammatic cross section of a bivalve, showing hydrostatic forces that produce dilation of the foot. Central vertical arrow indicates flow of blood into foot. (b) Areas of mantle fusion in bivalves: (1) between inhalant and exhalant apertures or siphons, the most common point of fusion, (2) below inhalant aperture or siphon, and (3) between inhalant aperture and foot aperture.

animal to obtain water relatively free of sediment (Fig. 29.35). You can tell whether or not a bivalve has siphons by examining the shell alone. The posterior part of the pallial line of a siphonate bivalve is recessed inward, the recess representing the scars of the siphon retractor muscles (Fig. 29.26c).

Attached Surface Dwellers

Many bivalves have abandoned burrowing and live on the surface of the sea bottom, especially on hard surfaces, such as rock and coral. The animal attaches itself in one of two ways. It may, like the oyster, lie on its side with one valve fused to the substratum (Fig. 29.35). The foot is absent, and the anterior adductor muscle is reduced or absent. Other bivalves, such as the common mussels, are attached by means of a horny **byssal secretion**, often in the form of threads, secreted by a gland in the reduced foot (Fig. 29.35).

Sediment is much less a problem for bivalves living attached to the surface of hard substrata. It is not surprising therefore that these species exhibit no mantle fusion and do not possess siphons. Some surface-dwelling bivalves, including both mussels and oysters, inhabit the intertidal zone and may be found attached to rocks, pilings, and sea walls that are exposed at low tide. During this period the animals are inactive and keep the valves closed to reduce desiccation.

Fouling organisms are algae and animals that attach to man-made structures. Bivalves can cause fouling problems when their larvae attach to the inner surfaces of circulating water pipes. The little European zebra clam, introduced into the Great Lakes through the ballast of freighters, has reached explosive population levels (750,000 per square meter) in some areas and is clogging the cooling systems of power plants and the water intake pipes of municipal water treatment plants.

Unattached Surface Dwellers

A few bivalves, such as scallops and file shells, rest unattached on the bottom or are attached only temporarily (Fig. 29.35). These bivalves can swim in a jerky manner for short distances by rapidly clapping their valves and driving a jet of water from the mantle cavity. Swimming is used primarily as a means of escape. Correlated with their surface habitat and their motility is the fact that the sensory lobe of the mantle margin is highly developed and may bear eyes and tentacles (Fig. 29.36).

Hard-Bottom Burrowers

Several groups of bivalves have evolved the ability to burrow into peat, clay, sandstone, coral, and even limestone rock. They use the anterior margins of the valves to drill. Cutting occurs when the two valves are opened, pulled down, or rocked against the head of the burrow. The

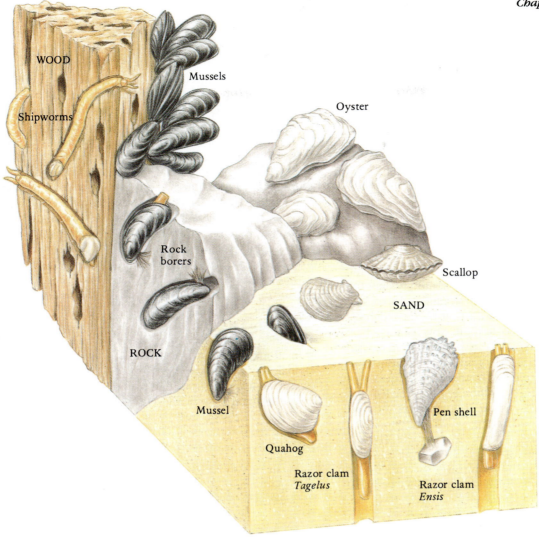

WOOD

Mussels

Shipworms

Oyster

Rock borers

Scallop

ROCK

SAND

Mussel

Quahog

Razor clam
Tagelus

Pen shell

Razor clam
Ensis

Figure 29.35
Bivalve life styles. All of these species would not found together in the same location as depicted here.

Figure 29.36
The North Atlantic scallop *Chalmys islandica*. The dark dots along the mantle margin are eyes.

drilling force is provided by pulling the body against the attachment of the suckerlike foot or against the attachment of byssal threads (Fig. 29.35). One group of drillers uses chemical secretions to soften the calcareous rock, coral, or shell they inhabit.

Shipworms are highly modified to drill into wood. The valves, which function as the drill, are very tiny and no longer cover the greatly elongated body and siphons (Fig. 29.35). The animal occupies the entire burrow, which is lined with calcium carbonate secreted by the mantle, and the ends of the siphons are located at the burrow opening. Shipworms ingest the sawdust that they excavate in drilling and digest it with cellulase. Shipworms can do extensive damage to marine timbers, and before the use of metal and fiber glass hulls and antifouling paints, shipworms greatly shortened the life of ships with wooden bottoms.

Reproduction and Development

Most bivalves have separate sexes. The two nephridia serve as gonoducts in the protobranchs, but separate gonoducts are present in most lamellibranchs. The gametes are shed in the exhalant water current, and fertilization occurs in the surrounding sea water. Some bivalves, including most freshwater species, brood their eggs within the water tubes of the gills.

Successive trochophore and veliger larvae are typical of the development of marine bivalves. The veliger is somewhat laterally compressed and covered by a bivalved shell (Fig. 29.37a). Many freshwater bivalves have a highly specialized developmental history. There is no trochophore or veliger, but a modified larval stage, a **glochidium**, is released from the brood chambers in the gill of the female (Fig. 29.37b). In some species the glochidia settle to the bottom of the stream or lake. When certain species of fish swim over the bottom, the glochidia become attached to the fins or gills by means of an attachment thread or a hook on the ventral margin of each valve. There are also species of freshwater bivalves in which the glochidia are released from the parent in masses that look like worms, which are then eaten by the fish and attach to the gills. In either case, the tissues of the host overgrow the attached glochidium, which now becomes a parasite for the remainder of larval development. The fish serves as a vehicle for the dispersal of these bivalves. When development is complete, the young bivalve breaks free of the host, drops to the bottom, and becomes a free-living adult.

CLASS CEPHALOPODA

The class **Cephalopoda** contains nautiluses, squids, cuttlefishes, and octopods. There are only some 600 living species, but there is a rich fossil record of more than 7500 species from past geological periods.

The cephalopod characteristics represent adaptations for a swimming, predatory mode of existence, although there are many species that have secondarily adopted other life styles. The dorsoventral axis has become greatly lengthened (Fig. 29.38). The foot has become divided into **tentacles** or arms, and has shifted somewhat anteriorly around the mouth, hence the name cephalopod ("head foot"). Cephalopods swim by means of jet propulsion. The force is generated by the contraction of the mantle, which becomes locked about the head, and water is expelled from the mantle cavity through a short funnel derived from a part of the foot.

The great increase in the dorsoventral axis and the assumption of a swimming habit have led to a shift in the orientation of the body. The tentacles, which represent the original ventral surface, are directed forward, and the visceral mass, which represents the original dorsal surface, is now directed posteriorly (Figs. 29.38c and 29.39).

Shell

Among living cephalopods only *Nautilus* possesses a well-developed external shell (Fig. 29.40), but a shell was characteristic of thousands of fossil species. The ceph-

Figure 29.37

(a) A fully developed veliger larva of an oyster. (b) Glochidium, the larva of a freshwater bivalve.

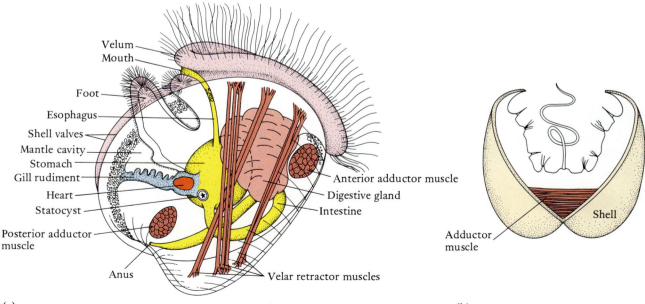

| Velum |
| Mouth |
| Foot |
| Esophagus |
| Shell valves |
| Mantle cavity |
| Stomach |
| Gill rudiment |
| Heart |
| Statocyst |
| Posterior adductor muscle |
| Anus |
| Anterior adductor muscle |
| Digestive gland |
| Intestine |
| Velar retractor muscles |
| Adductor muscle |
| Shell |

(a)   (b)

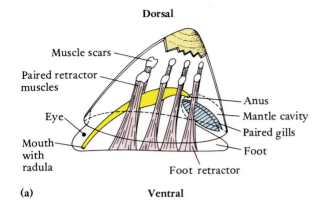

Dorsal

Muscle scars

Paired retractor
muscles

Eye

Mouth
with
radula

Anus
Mantle cavity
Paired gills
Foot
Foot retractor

(a) **Ventral**

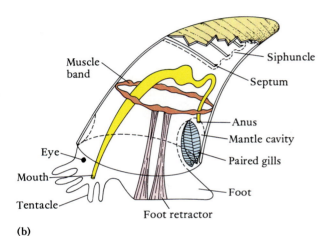

Muscle
band

Siphuncle
Septum

Eye
Mouth
Tentacle

Anus
Mantle cavity
Paired gills
Foot
Foot retractor

(b)

Figure 29.38

Evolution of the cephalopod form. (a) Reconstruction of a monoplacophoran with a cap-shaped shell. Note that the apex tilts slightly posteriorly. Some high cone monoplacophorans, from which the cephalopods may have evolved, had septate shells but no siphuncle.
(b) Reconstruction of an early cephalopod, such as the late Cambrian *Plectronoceras*. (c) An early straight cone cephalopod in swimming position.

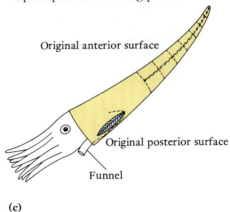

Original anterior surface

Original posterior surface

Funnel

(c)

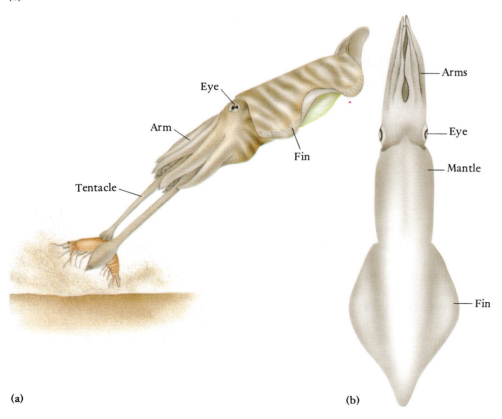

Eye
Arm
Tentacle
Fin

Arms
Eye
Mantle
Fin

(a) **(b)**

Figure 29.39

(a) The cuttlefish *Sepia* seizing a shrimp with its tentacles.
(b) Dorsal view of the squid *Loligo* in swimming position. Tentacles and arms are held together, acting as a rudder.

(a)

Figure 29.40
(a) *Nautilus*, the only living shelled cephalopod. (b) Sagittal section of *Nautilus*.

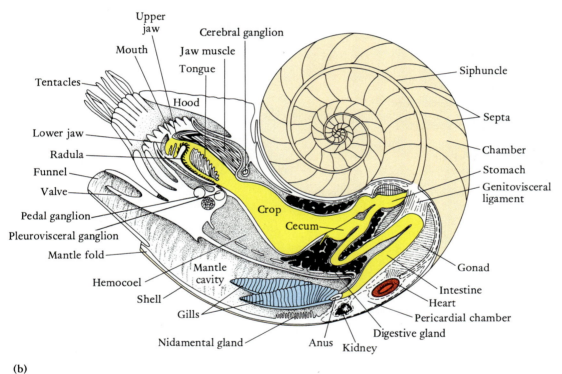

(b)

alopod shell, unlike that of other molluscs, is divided by transverse septa into interior chambers. The living animal occupies only the outer chamber, which opens to the exterior. The posterior chambers are filled with gas, which provides buoyancy for swimming. The gas is secreted by a cord of mantle tissue called the **siphuncle** that extends back through the septa.

In squids and cuttlefish, the shell has become greatly reduced and overgrown by the mantle. In octopods, the shell has disappeared completely.

Cephalopods reach the largest size of any invertebrate animal. The largest living species are giant squids of the genus *Architeuthis*, which may reach 16 m in length and are bottom dwellers at 300 to 600 m. Divers in the Sea of Japan have reported seeing octopods with arms as long as 10 to 15 m, but one of the largest described species, the Pacific *Octopus dofleini*, rarely has a mantle length greater than 36 cm, although the slender arms may be five times longer. The largest shelled fossil species have straight conical shells up to 5 m long or coiled shells up to 3 m in diameter.

Cephalopods are believed to have evolved from some group of Cambrian high-cone monoplacophorans (Fig. 29.38a). The apical end of the shell of some of these monoplacophorans was septate, which may have pre-adapted them for the cephalopod innovation of gas-filled chambers.

Locomotion

Squids and cuttlefish are the best cephalopod swimmers. The body of a squid is torpedo shaped (Fig. 29.39b); its posterior lateral fins are used as stabilizers in rapid swimming. The arms and tentacles are held together and function as a rudder. The siphon can be directed anteriorly or posteriorly to permit backward or forward swimming, but in rapid escape swimming, most cephalopods shoot backward. Some species of squids have been clocked at 40 km/hr, and the flying squids of the family Onycoteuthidae, which shoot out of the water in escape swimming, have been reported to land on the deck of a ship about 4 m above the water level!

The mantle of squids exhibits an array of adaptations for generating the water jet. There is one set of circular muscles used for powerful escape swimming and another set for slow cruising and ventilation. Contractions of the mantle muscles are synchronized by a system of giant neurons (p. 377), and the mantle cavity is expanded by the elastic recoil of collagen fibers in the mantle wall.

Cuttlefish have shorter bodies than squids and are very agile but not powerful swimmers (Fig. 29.39a). Their fins undulate and contribute to propulsion, which is also true of squids at low speeds. Many species live in shallow water, hovering and darting over the bottom. Some lie on the bottom covered with sand during the day or while waiting for prey. The shell of cuttlefish, the so-called cuttlebone, is buried within the mantle, but still contains gas-filled spaces. Buoyancy can be regulated by the replacement of fluid with gas, when the animal becomes active at night.

Nautilus swims backward with the shell held upright and the coils located over the head (Fig. 29.40). They have been collected down to depths of 600 m; the shell would likely collapse under pressure at any greater depth. The gas-filled chambers of the *Nautilus* shell function to counter the increased weight of the shell as it grows. There is no regulation of the gas with depth.

The body of octopods is rather globular (Fig. 29.41a). They are largely bottom dwellers and crawl about with the arms, swimming only to escape. They are most commonly found on rocky and coralline bottoms, where they establish lairs in holes and crevices. Some deep-sea octopods and squids have become pelagic. The arms are connected together by webbing, and the animals swim like jellyfish (Fig. 29.41b).

Feeding

Cephalopods are highly adapted for raptorial feeding and a carnivorous diet. Fish, shrimp, crabs, and other molluscs are principal foods. The prey is seized and held by the many arms and tentacles. Squids and cuttlefish have eight arms and two long prehensile tentacles, the latter providing for the initial prey capture (Fig. 29.39a); octopods possess only the eight arms. The arms and tentacles are provided with suckers, except in *Nautilus*, which has 38 ridged prehensile tentacles (about 90 tentacles altogether).

Figure 29.41
(a) Lateral view of the octopus *Eledone cirrhosa*. The large siphon projects to the right.
(b) Photograph of *Cirroteuthis*, an octopod with webbed arms. Scale mark equals 30 cm. Photograph taken with a deep-sea camera at 3000 meters in the Virgin Islands Basin.

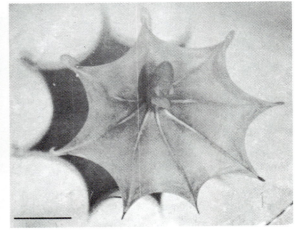

(a)

(b)

The principal ingestive organ is a pair of powerful, highly mobile, beaklike, horny **jaws**, which can cut and tear prey (Fig. 29.40). The radula functions as a tongue for pulling in pieces of flesh bitten off by the jaws. As a further aid to dispatching the victim, a pair of salivary glands have been modified as poison glands in some octopods and cuttlefish.

Internal Structure and Physiology

Many features of their internal structure and physiology reflect the active, raptorial life style of cephalopods. The stomach is quite different from that of other molluscs. Digestion is extracellular, and the digestive glands produce large amounts of powerful proteolytic enzymes. Cilia are no longer needed on the gills and mantle, since muscle contraction provides for the propulsion of the ventilating current, now also the locomotor current. A pair of **secondary hearts** elevates the blood pressure on entrance into the gills (Fig. 29.42), and blood is conveyed within closed vessels, increasing the rate of transport and reflecting a higher metabolic rate. Within the blood, oxygen is carried by hemocyanin. A single pair of nephridia provides for excretion.

Figure 29.42
Anatomy of the squid *Loligo*. Ventral view, mantle cut open.

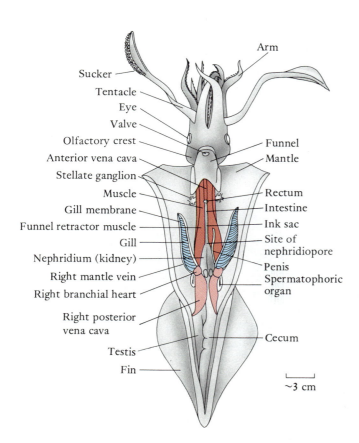

Sucker
Tentacle
Eye
Valve
Olfactory crest
Anterior vena cava
Stellate ganglion
Muscle
Gill membrane
Funnel retractor muscle
Gill
Nephridium (kidney)
Right mantle vein
Right branchial heart
Right posterior vena cava
Testis
Fin

Arm
Funnel
Mantle
Rectum
Intestine
Ink sac
Site of nephridiopore
Penis
Spermatophoric organ
Cecum

~3 cm

The nervous system and sense organs of cephalopods are among the most highly developed of invertebrates. The ganglia are concentrated at the anterior end to form a complex brain that is housed within a cartilaginous capsule. The behavior of the octopods and other cephalopods has been studied extensively.

The arms are provided with many tactile cells and chemoreceptors, and play an important role in the discrimination of surfaces. A statocyst is located on either side of the brain. The eyes are highly developed and parallel those of vertebrates to a striking degree (p. 422). Visual acuity is about the same as that of fish. Indeed, many malacologists believe that cephalopods with reduced or absent shells—squids, cuttlefish, and octopods—are the molluscan evolutionary response to competition from fish, explaining the many convergent features between the two groups.

An **ink sac** opens into the intestine near the rectum and is provided with an ejector mechanism that enables the animal to discharge a cloud of dark ink. It has been suggested that the cloud functions not as a smoke screen but as a dummy, confusing the predator and enabling the animal to escape. The alkaloids in the ink may narcotize the chemoreceptors of predators such as fish.

Yellow, orange, red, blue, and black chromatophores may be present, depending upon the species. The contraction of the muscle cells attached to them expands the chromatophores, and the color changes produced appear to function as background simulations, in the courting behavior of males, in aggressive responses between individuals, and in alarm reactions to intruders.

Deep-sea cephalopods do not have chromatophores but many are **bioluminescent**. The light is produced in special organs called **photophores**, which are located on different parts of the body, often in patterns and composed of different colors (Connections 29.3).

Reproduction and Development

The sexes are separate in cephalopods. The gonad in each sex is provided with a relatively complex gonoduct. Copulation involves the transfer of sperm in **spermatophores** (p. 494). Masses of sperm are cemented to one another or encased within special secretions in certain regions of the sperm duct.

During copulation, one especially modified arm of the male reaches into his mantle cavity and plucks a mass of spermatophores from the gonoduct. The copulatory arm of the male then deposits the spermatophores in the mantle cavity of the female near the oviduct opening or even within the oviduct itself (*Octopus*). Both the squid *Loligo* and the cuttlefish *Sepia* receive spermatophores outside of the mantle cavity in a fold beneath the mouth. In octopods only the copulating arm of the male contacts the female, but in squids and cuttlefish the male seizes the female head on and interlocks arms with her (Fig. 29.43).

The terrestrial fireflies are perhaps the most familiar example of animals capable of producing light, but bioluminescence is far more widespread among marine organisms. On a still night light may appear to trail behind a swimmer's arms or a slow-moving boat. In shallow coastal waters there are luminescent bacteria, dinoflagellates, jellyfish, ctenophores, crustaceans, brittle stars, and fishes but they are a small part of the total shallow-water fauna. On the other hand, in the mesopelagic zone of the open ocean (200 to 1000 m), where light is almost absent, 70% to 80% of the jellyfishes, ctenophores, shrimps, squids, and fishes are luminescent.

Bioluminescence has evolved independently in animals numerous times. The process involves the oxidation of a substrate substance, called luciferin, involving a molecule of oxygen and an enzyme luciferase. The luminescent product molecule is sufficiently excited to give off a photon. The light source of luminescent animals may be symbiotic bacteria (some fishes and squids), cellular secretions that are mixed extracellularly (ostracod crustaceans), or specialized cells within which light is produced. These specialized cells or the luminescent bacteria are commonly located within complex organs called photophores, which may be equipped with shutters, reflectors and lenses (shrimps, squids, and fishes). Light may be emitted as a flash or a sustained glow.

The most common and widespread function of bioluminescence appears to be protection from predators. The predator may be startled by a flash of light. Also, prey above a deeper swimming predator will stand out as a silhouette against light coming down from the surface. Glowing light organs located over the ventral surface can provide counterillumination to reduce or obliterate the silhouette. Another common function of bioluminescence is recognition between members of the same species.

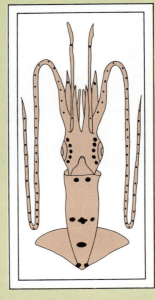

Distribution of light organs in *Nematolampas*. Eye and body light organs serve for counterillumination; those on arms and tentacles are of unknown function.

The spermatophores discharge their sperm shortly after deposition.

In some pelagic cephalopods, the eggs are planktonic. In most, however, the eggs are deposited in strings on the bottom or are attached to stones or other objects. Octopods remain with their eggs, guarding and cleaning them of sediment. Development is direct, and the young have the adult form on hatching.

EVOLUTIONARY RELATIONSHIPS

During the course of this chapter we have looked at the possible evolutionary relationships of the various molluscan classes to each other. Where does the entire phylum fit within the animal kingdom? The spiral determinate cleavage and schizocoelous mode of coelom formation (p. 554 and Fig. 22.6) clearly indicate that molluscs are protostomes. Their trochophore larva is strikingly similar to that of annelid worms, members of the next phylum we will survey. Molluscs and annelids are certainly closely related; perhaps they are sister groups, having been derived from a common ancestor. The annelid worms are segmented (metameric) and some zoologists believe that the repetition of parts found in some molluscs, such as the monoplacophorans, is additional evidence for a close evolutionary relationship between annelids and molluscs.

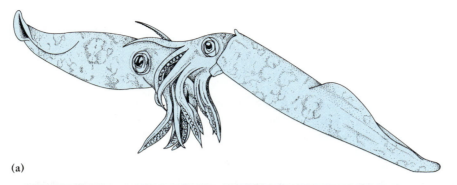

(a)

(b)

Figure 29.43
(a) Copulation in the squid *Loligo*.
(b) Copulating octopods. Male is on the right; only the male's copulatory arm contacts the female.

Classification of Phylum Mollusca

Class Caudofoveata (L. *cauda*, + *fovea*, small pit) A small class of shell-less molluscs that burrow in soft bottoms. The body is cylindrical and footless, and the posterior mantle cavity contains a pair of bipectinate gills. This and the following class are sometimes placed together in the class Aplacophora.

Class Solenogastres (Gr. *solen*, channel + *gaster*, belly) A small class of epibenthic shell-less molluscs. The body is cylindrical and the foot has the form of a ridge located within a longitudinal groove. The posterior mantle cavity lacks gills or contains secondary gills.

Class Polyplacophora (Gr. *polys*, many + *plax*, plate + *pherein*, to bear) Chitons. Body greatly flattened dorsoventrally. Head reduced. Shell composed of eight linearly arranged, overlapping plates. Marine. *Chiton*, *Chaetopleura*.

Class Monoplacophora (Gr. *monos*, single + *plax*, plate + *pherein*, to bear). Small group of marine deep-water species. Body flattened dorsoventrally. Shieldlike shell. Various organs replicated: eight pairs of retractor muscles, five or six pairs of gills, six or seven pairs of nephridia, and two pairs of auricles. *Neopilina*.

Class Gastropoda (Gr. *gaster*, belly + *pous*, foot) Snails, whelks, conchs, slugs. Mantle and visceral mass exhibit torsion. Shell typically spiraled. Foot flattened and head well developed.

Subclass Prosobranchia (Gr. *proso*, forward + *branchia*, gill) Gill-bearing species in which the mantle cavity and contained organs are located anteriorly. Shell is present and operculum is common. Most marine snails with well-developed shells belong to this subclass. There are some freshwater and terrestrial prosobranchs, the latter having lost the gill. Limpets (*Fissurella*, *Acmaea*), abalones (*Haliotis*), periwinkles (*Littorina*), worm shells (*Vermetus*), conchs (*Strombus*), cowries, moon shells, helmet shells, bonnets, whelks (*Busycon*), cones (*Conus*).

Subclass Opisthobranchia (Gr. *opisthen*, behind + *branchia*, gill) Visceral mass has undergone detorsion, and some degree of reduction of shell and mantle cavity is common. Hermaphroditic. Entirely marine. Bubble shells, sea hares (*Aplysia*), sea butterflies, and sea slugs (*Aeolidia*, *Doris*).

Subclass Pulmonata (L. *pulmonatis*, lung) Gills absent and mantle cavity converted into a lung. Shell usually present but an operculum is absent. Hermaphroditic. Terrestrial and freshwater species. Most land snails and slugs of temperate regions are pulmonates. *Physa, Helix, Limax.*

Class Bivalvia (L. *bi*, two + *valva*, valve) Clams, mussels, scallops, oysters. Body greatly flattened laterally. Shell composed of two lateral valves hinged dorsally. Head not developed. Radula absent. The following is an older, simpler but still useful classification of bivalves. A more contemporary system divides the class into six subclasses based on technical differences in shell structure.

Subclass Protobranchia (Gr. *protos*, first + *branchia*, gill) Primitive marine bivalves with one pair of unfolded, bipectinate gills. *Nucula, Solemya.*

Subclass Lamellibranchia (L. *lamella*, sheet or plate + Gr. *branchia*, gill) Marine and freshwater filter-feeding bivalves with folded gills. Most bivalves belong to this subclass.

Subclass Septibranchia (L. *septum*, wall + Gr. *branchia*, gill). A small group of carnivorous, marine species having the gills modified as a pumping septum. *Cuspidaria.*

Class Scaphopoda (Gr. *skaphe*, boat + *pous*, foot) Tusk or tooth shells. Burrowing marine molluscs having a tusklike shell open at each end (Fig. 29.44). *Dentalium.*

Class Cephalopoda (Gr. *kephale*, head + *pous*, foot) Nautiluses, cuttlefishes, squids, and octopods. Marine, mostly swimming molluscs, having the foot divided into tentacles or arms. Shell, when present, usually divided into chambers, of which only the most recent is occupied by the living animal.

Subclass Nautiloidea (Gr. *nautes*, sailer + *eidos*, form) Mostly fossil cephalopods with straight or coiled chambered shells. *Nautilus* is only living genus.

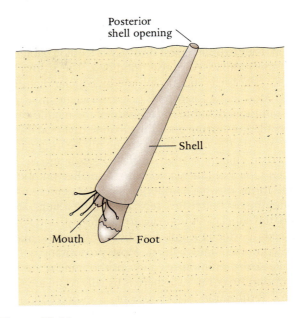

Figure 29.44
A scaphopod, or tooth shell, buried in the sand. Water enters and leaves mantle cavity through the posterior shell opening.

Subclass Ammonoidea[†] (like the horn of the Egyptian diety Ammon). Fossil cephalopods with coiled shells having the septum crinkled at its junction with the lateral wall of the shell.

Subclass Coleoidea (Gr. *koleon*, sheath + *eidos*, form) Shell internal or absent. Cuttlefish (*Sepia*), squids (*Loligo*), octopods (*Octopus*).

[†]Extinct group.

Summary

1. Members of the phylum Mollusca constitute one of the largest phyla of animals and are found in the sea, in fresh water, and on land. They are distinguished by the presence of a muscular foot, a calcareous shell secreted by the underlying integument, called the mantle, and the feeding organ, the radula. The blood vascular system is usually open, and the excretory organs are metanephridia. The usual sense organs are statocyst, eyes, tentacles, and osphradia. Primitively, the nephridia function as gonoducts, fertilization is external, cleavage is spiral, and a trochophore is the first larval stage.

2. The small number of living species of the class Monoplacophora are deep-sea remnants of a much larger and more widespread group of ancient molluscs, which probably gave rise to most of the other classes of this phylum. The repetition of both external and internal structures—gills, retractor muscles, auricles, and nephridia—is a distinctive feature of living monoplacophorans.

3. Chitons, members of the class Polyplacophora, are adapted for living on hard substrates, especially in the intertidal zone. The body is dorsoventrally flattened; the broad foot and mantle provide for gripping; and the shell is composed of eight linearly arranged overlapping plates. Food, especially algae, is scraped from rock surfaces by the radula.

4. The Gastropoda, containing the snails and slugs, is the largest and most diverse class of molluscs. The distinguishing feature is torsion, a twisting of the visceral mass during the development of the snail. The head is well developed, the foot is broad and flat, and the shell is asymmetrically coiled. Most gastropods creep by means of the foot, propelled by cilia or by waves of muscle contraction. They exhibit a great diversity of feeding habits, for which the radula is variously modified. The stomach is usually a simple sac and the site of extracellular digestion.

5. Most living gastropods possess a single monopectinate gill located within the left side of the mantle cavity. The ventilating current enters the mantle cavity on the left and leaves on the right. Most marine gill-bearing gastropods are members of the subclass Prosobranchia. The subclass Opisthobranchia contains a smaller number of marine forms that have undergone partial detorsion and includes the sea butterflies, sea hares, and sea slugs. Members of the subclass Pulmonata are freshwater and terrestrial snails in which the gills have been lost and the mantle cavity converted into a lung. Most prosobranchs have separate sexes; pulmonates and opisthobranchs are hermaphroditic. In either case there is usually copulation and internal fertilization. Development is indirect in many marine species, with a veliger the usual hatching stage.

6. Members of the class Bivalvia are adapted for burrowing in soft sediments, although many species have secondarily become adapted for other life styles. The body is laterally compressed and covered by two lateral shell valves. The valves are forced open by dorsal elastic hinge ligaments; the valves are closed by the contraction of two (one in some) adductor muscles that extend between the valves. The head is poorly developed. Fertilization is external, and development is usually indirect, with both a trochophore and veliger larva.

7. Some primitive bivalves (protobranchs) are selective deposit feeders, but most members of the class are filter feeders. In the adaptation of the gills for filter feeding, the filaments became long and folded, with all of the filaments on one side of the original gill forming two filtering surfaces. The gills are strengthened by tissue junctions between various parts of the gill. Trapped particles (primarily phytoplankton) are transported up or down the gill filament by frontal cilia and then anteriorly to the mouth in food grooves. Particles are sorted by the labial palps before ingestion. Bivalves have lost the radula. The stomach contains a crystalline style, which functions in mixing and in extracellular digestion. Much digestion occurs intracellularly within the digestive glands.

8. Bivalves burrow by extending the anteriorly directed and bladelike foot out into the sediment. The foot becomes anchored by engorgement with blood, and the body is then pulled down behind the foot. Most burrowing species have the mantle margins drawn out as siphons, permitting more direct access to surface waters. Bivalves have separate sexes or are hermaphroditic. Gonoducts transport gametes to the mantle cavity, and fertilization is external. Successive trochophore and veliger larvae are characteristic of marine species.

9. Members of the class Cephalopoda are designed for a pelagic raptorial existence. The foot has become modified as arms or tentacles arranged around the mouth. Among living species only *Nautilus* possesses an external shell; in all others the shell is reduced or lost. Cephalopods swim by ejecting water from the mantle cavity through a funnel. The gas content provides for buoyancy. Octopods are benthic and exhibit only escape swimming.

10. Prey is seized with the arms and tentacles, which are provided with suckers, and then bitten and torn with a pair of powerful beaklike jaws. Many features of cephalopods are correlated with their very active life and higher metabolic rate: absence of gill cilia, closed blood vascular system, branchial hearts, presence of hemocyanin, highly developed eyes, complex nervous system and behavior, chromatophores, and ink sac. During copulation, one of the arms of the male transfers spermatophores to the female. Development is direct.

References and Selected Readings

Accounts of molluscs may also be found in the general references listed at the end of Chapter 23. The references listed below are devoted solely to molluscs.

Burch, J.B. *How to Know the Eastern Land Snails.* Dubuque, Iowa: W.C. Brown Co., 1962. A field guide to the pulmonates.

Gosline, J.M., and M.E. DeMont. Jet-propelled swimming in squids. *Scientific American* 252(Jan. 1985):96–103. Structural and physiological adaptations for swimming in squids.

Hughes, R.N. *A Functional Biology of Marine Gastropods.* Baltimore: The John Hopkins University Press, 1986. A physiology of gastropods with special emphasis on energetics.

Jones, D.S. Sclerochronology: Reading the record of the molluscan shell. *American Scientist* 71(1983): 384–391. An account of the structural features in shell deposition that reflect age and environmental conditions.

Rehder, H.A. *The Audubon Society Field Guide to North American Seashells.* New York: Knopf, 1981.

Roper, C.F.E., and K.J. Boss. The giant squid. *Scientific American* 246(Apr. 1982):96–105. A review of our knowledge of giant squids from beached specimens and other sources.

Runham, N.W., and P.J. Hunter. *Terrestrial Slugs.* London: Hutchinson University Library, 1970. A general biology of pulmonate slugs.

Solem, A. *The Shell Makers: Introducing Mollusks.* New York: John Wiley & Sons, 1974. This little book is largely concerned with gastropods, especially land snails.

Ward, P.D. The extinction of the ammonites. *Scientific American* 249(Oct., 1983):136–147. An account of the evolution and demise of the cephalopod subclass Ammonoidea.

Ward, P.D. *The Natural History of Nautilus.* Winchester, Mass.: Allen and Unwin, 1987. A good summary of the biology of *Nautilus.*

Wells, M.J. *Octopus: Physiology and Behavior of an Advanced Invertebrate.* London: Chapman & Hall, 1978. An account of the physiology and behavior of *Octopus.*

Wilbur, K.M. (ed.). *The Mollusca.* New York: Academic Press, 1983–1985. Ten volumes covering the biochemistry, physiology, development, ecology and evolution of molluscs.

Yonge, C.M., and J.E. Thompson. *Living Marine Molluscs.* London: Wm. Collins & Sons, 1976. A general biology of the molluscs.

30

Annelids

Characteristics of the Annelids

■ The body is linear and metameric, and externally segments are usually demarcated by grooves and serial repetition of appendages. Internally, metamerism is reflected in the compartmentalization of the coelom by a septum between each segment.

■ The body wall is composed of an outer layer of circular muscles and an inner layer of longitudinal muscles.

■ A pharynx and intestine are usually the most conspicuous parts of the generally straight gut tube.

■ The excretory organs are paired segmental nephridia.

■ The circulatory system is at least partially closed and is composed of a dorsal vessel that pumps blood anteriorly and a ventral vessel that transports blood posteriorly. Within each segment vessels return blood to the dorsal vessel by way of the body wall and gut.

■ The nervous system is composed of a pair of dorsal anterior cerebral ganglia (brain), a pair of connectives around the gut, and a pair of longitudinal ventral nerve cords (commonly fused). Paired lateral nerves arise from a ganglionic swelling of the ventral nerve cords in each segment.

■ Marine annelids (polychaetes) have separate sexes, with gametes usually produced in many segments; freshwater and terrestrial annelids (oligochaetes and leeches) are hermaphroditic, with gonads and gonoducts in a few segments. Many marine annelids possess a trochophore larva; development is direct in freshwater and terrestrial species.

(Left) Sabellid fanworm. Spirally arranged radioles project from tubes. (Above) Projecting radioles of a fanworm.

Echinoderms
Chordates
Hemichordates
Arthropods
Molluscs
Annelids
Brachiopods
Bryozoans
Flatworms
Nematodes
Gastrotrichs
Rotifers
Cnidarians
Ancestral bilateria
Sponges
Ancestral metazoan
Choanoflagellate protozoan

Members of the phylum Annelida, the segmented worms, are common and widespread in the sea, fresh water, and soil. More than 11,000 annelid species have been described. In contrast to the layperson's general image of "worms," some marine species are very beautiful, brightly colored animals (see Fig. 30.9a).

The phylum is divided into three principal classes. The **Polychaeta**, the largest and most diverse of the three classes, contains the marine annelids; the **Oligochaeta** includes the freshwater annelids and earthworms; and the **Hirudinea** comprises the leeches. The latter clearly evolved from the oligochaetes, and the oligochaetes and polychaetes are believed to have evolved separately from some common marine burrowing ancestor.

METAMERISM AND LOCOMOTION

The structural design of annelids is closely related to the patterns of locomotion that have developed within the phylum. The most striking annelid characteristic is the division of the body into similar parts, or **segments**, arranged in a linear series along the anterior-posterior axis. Each segment is termed a **metamere** (Figs. 30.1 and 30.2), and this pattern of repeated segments is called **metamerism**. Only the trunk is segmented. Neither the head, or **prostomium**, anterior to the mouth, nor the terminal part of the body, the **pygidium**, which carries the anus, is a segment. New segments arise in front of the pygidium; the oldest segments lie just behind the head. Some longitudinal structures, such as the gut and the principal blood vessels and nerves, extend the length of the body, passing through successive segments; other structures are repeated in each segment, reflecting the metameric organization of the body.

Metamerism appears to have evolved twice in the evolutionary history of the animal kingdom; once in the evolution of annelids and arthropods, and once in the evolution of chordates. In both instances, the condition appears to have evolved as an adaptation for locomotion. The ancestors of the annelids were probably elongate coelomate animals that inhabited marine sand and mud; metamerism probably arose as an adaptation for peristaltic burrowing (p. 226).

The **coelom** is compartmentalized by transverse **septa**, and the longitudinal and circular muscles are organized as two continuous cylinders outside of the coelomic compartments. One coelomic compartment and its surrounding body wall constitute a segment, which can usually be recognized externally by transverse grooves in the integument encircling the body (Figs. 30.1 and 30.2).

It is important to note that annelidan metamerism is basically a modification of the coelom, making possible a localized hydrostatic skeleton (p. 227). The nervous, circulatory, and excretory systems are also metameric, i.e.,

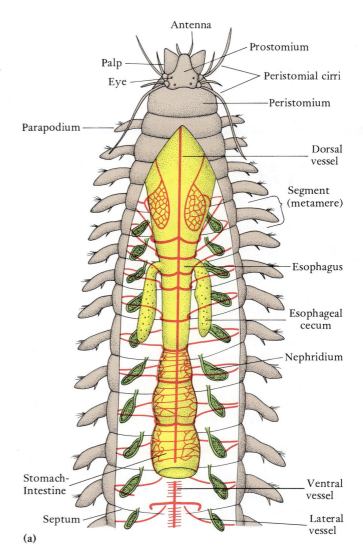

(a)

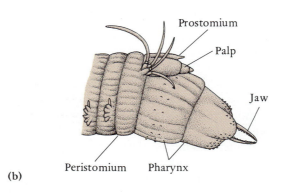

(b)

Figure 30.1

The polychaete *Nereis*. (a) Dorsal dissection of anterior end. (b) Lateral view of head with pharynx everted.

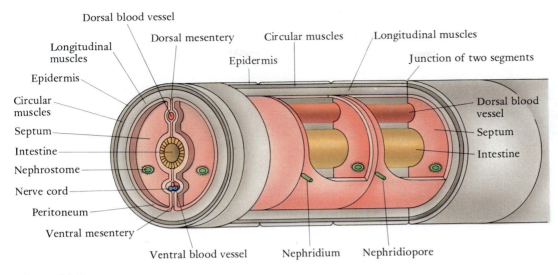

Figure 30.2
Diagrammatic lateral view of a series of annelid segments.

parts are repeated in each segment. But the metamerism of these structures probably reflects an adjustment of these supply systems to the primary metamerism of the coelom (Fig. 30.1).

The body surface of annelids is covered by a very thin cuticle overlying a layer of epidermal cells, which are ciliated in some species. The typical annelid has a nervous system with a dorsal brain and a pair of ventral longitudinal nerve cords, a straight tubular gut, and a more or less closed circulatory system (Figs. 30.1 and 30.2). The excretory organs are usually **metanephridia** (p. 350), typically one pair per segment.

CLASS POLYCHAETA

The members of the class Polychaeta are distinguished from other annelids by the presence of paired segmented appendages, **parapodia** (Gr. *para*, beside + *pous*, foot) (Fig. 30.1). The parapodia vary greatly in shape and size, but when well developed, each is composed of an upper **notopodium** and a lower **neuropodium**. Each division contains a large number of **setae**, chitinous bristles that provide traction against the substratum (Fig. 30.3). (This is the origin of the name Polychaeta—many setae). In those polychaetes with large movable parapodia, the parapodial muscles are anchored to an internal skeletal rod, called the **aciculum**, with one rod in each division.

The head, or **prostomium**, of polychaetes is often highly developed and may bear various sensory and feeding structures (Fig. 30.1). The mouth is located beneath the

prostomium, just in front of the first one or two modified trunk segments (the **peristomium**) (Gr. *peri*, around + *stoma*, mouth).

The digestive tract is usually composed of an eversible muscular pharynx and a long straight intestine. The pharynx is everted (turned inside out) through the mouth by special protractor muscles or by elevated coelomic fluid pressure (Fig. 30.1b).

The more than 8000 species of marine polychaetes exhibit great diversity both in form and in life style. An appreciation of their diversity can be gained by examining the major adaptive groups that compose the class.

Adaptive Groups of Polychaetes

Different families of polychaetes have frequently evolved similar life styles, i.e., similar modes of locomotion, habitation, and feeding. Such families constitute an adaptive group and provide a useful way to introduce the diversity of polychaete form and function.

Errant Polychaetes

Many polychaetes crawl about beneath stones and shells, in rock and coral crevices, and in algae and sessile animals. For want of a better name we will call these errant polychates, as opposed to those that live in burrows or tubes. These crawling, errant species most closely approximate the "typical" polychaete form described above. The head is well developed and carries several kinds of sensory structures: one to four pairs of eyes, up to five antennae, and a pair of palps (Fig. 30.1).

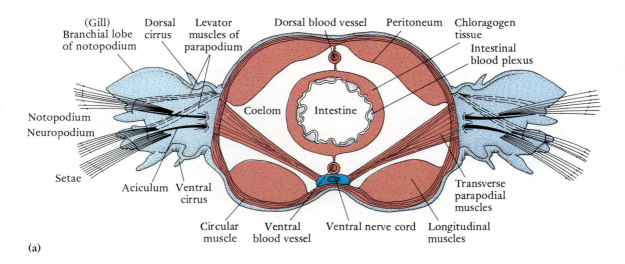

(a)

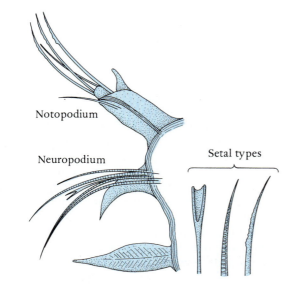

(b)

Figure 30.3

(a) Transverse section of *Nereis* at level of
intestine, showing a pair of parapodia.
(b) Parapodium and setae of *Scoloplos rubra*.
Note that the setae are not all of the same
type.

The parapodia are large and function like legs in crawling (Fig. 30.4b). Each makes a small step, and the movements of the parapodia occur in sequence as a result of waves of contraction passing along the length of the body. The waves of contraction on one side of the body alternate with those on the opposite side. In rapid movement, alternating contractions of the longitudinal muscles of the body wall throw the body into undulatory waves that generate additional thrust against the substratum.

The crawling mode of locomotion of errant polychaetes is reflected in the structure of the body wall and coelom. The longitudinal muscles are more highly developed than the circular muscles (Fig. 30.3a), and the septa are somewhat reduced since the coelom is less important as a localized hydrostatic skeleton than it is in peristaltic movement.

Many errant polychaetes are carnivores and feed on small invertebrates, including other polychaetes. Others are algae eaters, scavengers, or detritus feeders. The pharynx of these errant polychaetes is commonly provided with teeth or jaws, which vary in number from one family to another (Fig. 30.1b). The jaws typically swing open when the pharynx is everted and close when the pharynx is retracted.

Pelagic Polychaetes

Several families of polychaetes are adapted for living in the open ocean. They have well-developed heads, and the large parapodia are used as paddles in swimming. They are commonly carnivorous. Like most other small pelagic or planktonic animals, these polychaetes tend to be pale or transparent.

(a)

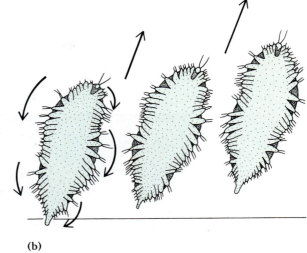

(b)

Figure 30.4

(a) A fireworm, an errant polychaete. The brittle setae release a toxin when broken off within an intruder. (b) Ventral view of parapodial movement in polychaete crawling. Waves of movement on one side of body alternate with those on the opposite side. Straight arrows indicate direction of movement of animal. Curved arrows indicate groups of parapodia undergoing the effective stroke.

Gallery Dwellers

Many polychaetes are adapted for burrowing in sand or mud. Some excavate extensive burrow systems, or galleries, that open to the surface at numerous points (Fig. 30.5b). The wall of the burrow is kept from caving in by its lining of mucus.

The prostomium is commonly conical or a simple lobe devoid of eyes and other sensory structures (Fig. 30.5a and c). However, there are numerous exceptions; for example, some species of *Nereis*, which have well-developed cephalic sense organs (Fig. 30.1), live in burrows but they emerge to feed.

Gallery dwellers may use the parapodia to crawl through the burrow system. However, many crawl by peristaltic contractions; the parapodia of these forms are somewhat reduced and function primarily to anchor the segments against the burrow wall. Septa and circular muscles are well developed. Some gallery dwellers are carnivores. Other gallery dwellers are nonselective deposit feeders and consume the substratum through which the galleries penetrate.

One of the best studied gallery dwellers is *Glycera*, the genus containing the bloodworms, polychaetes commonly used as fishing bait (Fig. 30.5a). These intertidal-to-subtidal carnivores lie in wait within their gallery systems, and when pressure waves created by a small crustacean or other prey moving across the surface are detected, the worm moves to a nearby opening. Bloodworms have a very long, armed proboscis (pharynx) that can be shot out with explosive force to seize the prey (Fig. 30.5c).

Sedentary Burrowers

Some burrowing polychaetes construct simple vertical burrows with only one or two openings to the surface (Fig. 30.6b). These sedentary burrowers, in contrast to gallery dwellers, move about relatively little. Peristaltic crawling is the rule, and part of the parapodium is reduced to a ridge that bears special hooklike setae that aid in gripping the burrow wall. The prostomium lacks most sensory structures, although special feeding appendages may be present.

Some sedentary burrowers are nonselective deposit feeders; others are selective deposit feeders. The lugworms (*Arenicola*) are examples of nonselective deposit feeders (Fig. 30.6). These worms live in L-shaped burrows and ingest the sand at the bottom with an unarmed eversible pharynx. Peristaltic contractions drive a ventilating current of water into the burrow. At rhythmically fixed intervals the worm backs out of its burrow and defecates mineral material at the surface as conspicuous castings. It then resumes feeding and ventilating. By means of a float attached to a stylus, the activity of worms in an aquarium can be recorded (Fig. 30.6c).

Selective deposit feeders have special head structures that pick up the organic matter from the surrounding sand grains. The head of *Amphitrite* is provided with a great mass of long tentacles. These spread over the surface from the opening of the burrow (see Fig. 12.9). Detritus material adheres to the mucus on the tentacles and is then conveyed to the mouth in a ciliated tentacular gutter and by tentacular contraction. There is no eversible pharynx.

697

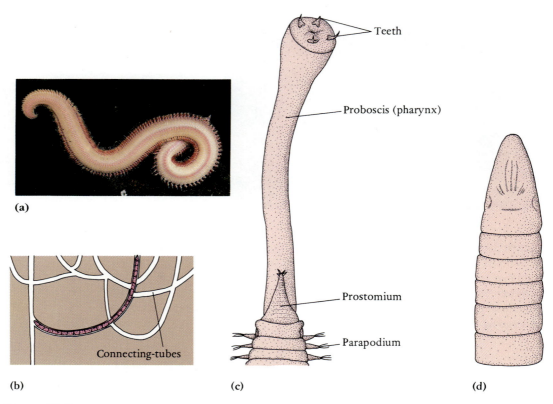

Figure 30.5
(a) Ventral view of a bloodworm, *Glycera*, a gallery-dwelling polychaete. The prostomium projects just beyond the looped part of the body. Gills associated with the parapodia appear red. (b) Burrow system of *Glycera alba*, showing worm lying in wait for prey. (c) Anterior end of *Glycera* with proboscis everted. (d) Anterior end of another gallery dweller, *Drilonereis*. Note that the conical prostomium lacks eyes and sensory appendages. Parapodia are poorly developed on the most anterior segments.

Tube-Dwelling Polychaetes

Many polychaetes live in protective tubes. In fact, tube dwelling is more widespread among polychaetes than in any other group of animals. In soft bottoms, the opening of the tube projects above the sand surface and thus provides the worm with access to water free of sediment (Fig. 30.7b). Tubes may also permit worms to live on firm exposed substrates, such as algae, rock, coral, or shell. The tube may be composed entirely of hardened material secreted by the worm, or it may be composed of foreign material cemented together. In contrast to a mucus-lined burrow, a tube will remain intact when it is dug out of the sand.

Some tube-dwelling (tubicolous) polychaetes are carnivores and extend from the tube opening to seize small invertebrate prey. As might be expected, these worms are very agile. Their adaptations are essentially like those of surface dwellers (Fig. 30.7a).

The majority of tube-dwelling polychaetes are sedentary and move about within the tube less actively. Their head usually lacks most special sensory structures, al-though feeding appendages may be present (Fig. 30.8). The animal moves within the tube by peristaltic contractions, and the parapodia tend to be reduced to ridges with hooked setae. Note that these adaptations are essentially the same as those for sedentary burrowers, for these two types of habitation are not that different, at least from the inside. Some families contain both burrow-dwelling and tubicolous species.

The structural diversity of tube-dwelling polychaetes is in large part correlated with their different modes of feeding. A few, such as the bamboo worms (*Clymenella, Axiothella*), are nonselective deposit feeders. The bamboo worm lives head down in sand grain tubes and ingests the substratum at the bottom of the tube with an unarmed eversible pharynx (Fig. 30.9b). Periodically, the worm backs to the surface and defecates the mineral material. Some sedentary tube dwellers are selective deposit feeders.

Filter feeding has evolved in several families of sedentary, tube-dwelling polychaetes. The beautiful fanworms, or feather duster worms, illustrate one type of filter feed-

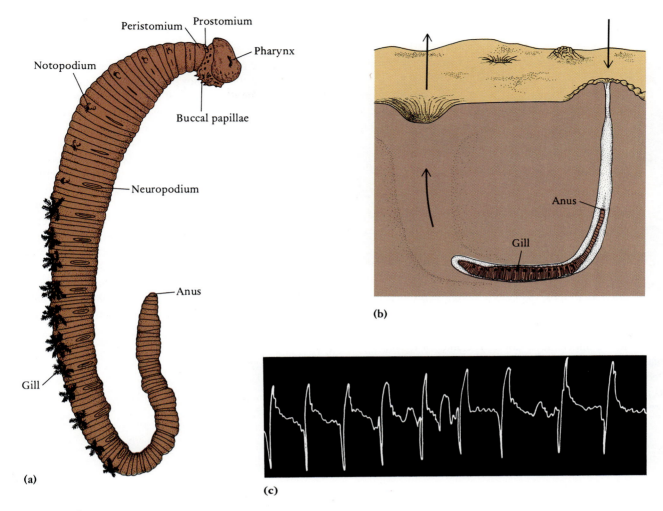

(a)

(b)

(c)

Figure 30.6

The lugworm, *Arenicola*. (a) Lateral view of worm. Pharynx is everted, obscuring the very small prostomium. (b) Worm in burrow. Arrows indicate direction of ventilating current produced by worm. (c) Inked stylus tracing of activity cycles of *Arenicola*. The downstroke reflects the worm backing up to the burrow opening to defecate; the sharp upstroke reflects the worm moving back down to the head of the burrow and vigorously resuming ventilation contractions and deposit feeding. Intervals between defecations are about 40 minutes.

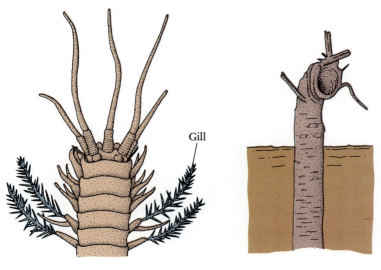

(a)

(b)

Figure 30.7

(a) Head and first two gill-bearing segments of *Diopatra cuprea*, a carnivorous tube dweller.
(b) Funnel-like parchment tube of *Diopatra*, to which pieces of shell and algae are attached.

Figure 30.8

A terebellid polychaete (*Eupolymnia crassiconnis*) from Bermuda. This large species, which reaches 30 cm in length, lives in tubes composed of shell fragments, commonly attached to the underside of stones. The anterior feeding tentacles stretch out in all directions for as much as a meter. It feeds in the same manner as *Amphitrite* (see Fig. 12.9). In this photograph the worm has been removed from its tube. Note the parapodial ridges used in gripping the tube wall.

ing. They have feather-like head structures called **radioles**. When feeding, the radioles project from the opening of the tube, in the form of a funnel or a spiral (Figs. 30.9a and p. 692). Collected particles are conveyed by cilia to the mouth (p. 245); there is no eversible pharynx.

Members of the Chaetopteridae feed by filtering water through a mucous bag. Species of *Chaetopterus* live in a large U-shaped tube composed of a secreted parchment-like material (Fig. 30.10). Three piston-like "fan" parapodia in the middle of the body drive water through the tube. A pair of long winglike anterior notopodia secrete a film of mucus, which is collected and rolled up by a ciliated cup. The water current being driven through the tube passes through the mucous film, which is held like a bag. Periodically, mucus secretion is halted, and the rolled-up ball of mucus containing trapped food particles is passed forward along a ciliated groove to the mouth.

Figure 30.9

(a) A caribbean Christmas tree worm. Each half set of radioles is spiraled; tube is buried in coral. (b) A bamboo worm, a polychaete that lives upside down in a sand grain tube and ingests the substratum at the bottom of the tube. Full extent of the tube is not shown.

(a)

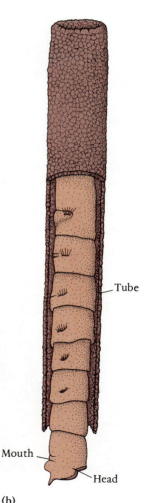

Tube

Mouth

Head

(b)

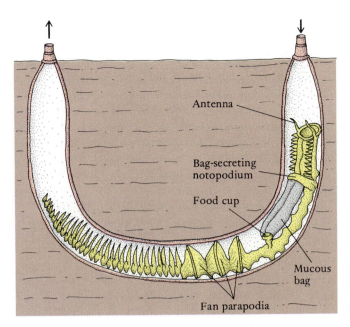

Figure 30.10
Lateral view of *Chaetopterus* within its tube.

Internal Transport, Gas Exchange, and Excretion

The closed blood vascular system of polychaetes has a relatively simple basic plan (p. 304). A contractile, longitudinal vessel located above the gut functions as the heart and conveys blood anteriorly. At the anterior end of the body, paired vessels carry the blood around the gut to a ventral longitudinal vessel in which blood flows posteriorly. In each segment the ventral vessel gives rise to a vessel supplying the gut and to paired vessels supplying the body wall, parapodia, and nephridia (Fig. 14.7b). These segmental vessels eventually return blood to the dorsal vessel.

The annelid system is commonly cited as an example of a closed blood vascular system. In actual fact, many annelids have parts of the system that are open or at least channels whose walls are formed by the tissue through which blood is flowing.

Coelomic fluid contributes to internal transport in all polychaetes, and in some, such as the bloodworm (*Glycera*), the blood vascular system is reduced and the coelomic fluid is the principal transporting medium.

Very small polychaetes and those with long, threadlike bodies have no gills. The ratio of surface area to volume makes simple diffusion an effective form of transport. The varying structure and the distribution of the gills

present in the larger species indicate that gills have evolved independently in different groups. Most commonly, the gills are modifications of the parapodia or outgrowths of some part of the parapodium (Figs. 30.3a, 30.6a, and 30.7a). However, gills are found in other parts of the body in some polychaetes. The radioles of fanworms serve as gills in addition to functioning as filters (see Fig. 12.8).

The ventilating current may be produced by cilia or by contractions of the gills. Gallery dwellers, sedentary burrowers, and tube dwellers drive a ventilating current through the burrow or tube, usually by undulations of the body or by peristaltic contractions.

Many polychaetes, especially those with gills, possess a respiratory pigment, usually hemoglobin, either dissolved in the blood plasma or contained within corpuscles in the coelomic fluid. Several families of polychaetes have a variant of hemoglobin (chlorocruorin) that is green rather than red in color. Polychaete excretory organs are protonephridia or metanephridia. A pair of nephridia is present in each segment, but the nephrostome of each nephridium opens through the septum into the segment anterior to the one containing the tubule (Fig. 30.1a).

Sense Organs and Nervous System

As we have seen, errant polychaetes have cephalic sense organs. The two to four pairs of eyes function primarily in light orientation, but there is one family of predaceous planktonic polychaetes (Alciopidae) with large, highly developed eyes capable of object discrimination. Antennae and palps are liberally supplied with receptor cells. The head of most polychaetes possesses a pair of ciliated slits or pouches, called nuchal organs. They are probably chemosensory organs, for they are most highly developed in predatory species. Many burrowers and tube dwellers have statocysts in the head region. The development of the polychaete brain is proportional to the development of cephalic sense organs. The ventral nerve cord is primitively double but is commonly single as a result of fusion. In many polychaetes the ventral nerve cord possesses giant axons, which facilitate rapid escape contraction of the body back into a burrow or tube.

Reproduction and Development

The sexes of polychaetes are separate. The gametes are produced by the coelomic peritoneum and mature within the coelomic fluid. Primitively, most segments produce gametes, but in many polychaetes gamete production is restricted to the segments in certain regions of the body. Gametes exit by special gonoducts, by the nephridia, or by rupture of the body wall.

Most polychaetes do not copulate. Instead, they shed their gametes into the sea water where fertilization occurs.

In some polychaetes, the likelihood of fertilization is increased by swarming behavior. Males and females with ripe gametes swim to the surface at the same time in large numbers, shedding their eggs and sperm simultaneously. Swarming usually occurs at specific times of the year, determined by annual, lunar, and tidal periodicities. The times at which swarming will occur in certain species can be predicted. The most notable example is the Samoan palolo worm. At the beginning of the last lunar quarter of October or November the worms release the posterior gamete-bearing regions of the body (Fig. 30.11), which come to the surface in enormous numbers and rupture.

The eggs, especially those of swarming polychaetes, are often planktonic. However, there are species that deposit their eggs within jelly masses attached to the tube or burrow.

Cleavage is typically spiral and leads to a trochophore larva like that of molluscs (Fig. 30.12; see also Fig. 29.3). During the course of larval life, the trochophore lengthens, and segments with parapodia and setae appear behind the ciliary girdle (prototroch) and mouth (Fig. 30.12c−f). However, there are many polychaetes in which hatching occurs following the trochophore or a comparable stage.

Figure 30.11
Eunice viridis, the Samoan palolo worm, with posterior region of gamete-bearing segments.

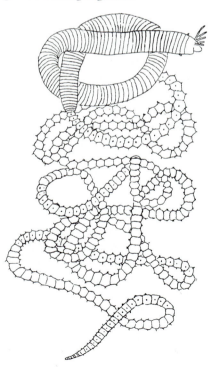

CLASS OLIGOCHAETA

In fresh water and on land, the annelids are represented by some 2900 species of oligochaetes. These are less diverse than the polychaetes, and the features of terrestrial oligochaetes in many respects parallel those of burrowing polychaete gallery dwellers. The simple conical or lobelike prostomium lacks sensory appendages (Figs. 30.13 and 30.14). Oligochaetes crawl by peristaltic contractions. Parapodia are completely absent, but setae serve to anchor segments in crawling. There are fewer setae per segment than in polychaetes, hence the name Oligochaeta—few setae. Correlated with their peristaltic locomotion, the coelom, septa, and circular and longitudinal muscles are well developed and reflect a high order of metamerism (Fig. 30.14).

Aquatic oligochaetes are usually less than 3 cm in length, and a few are almost microscopic. Many species have relatively long setae (Fig. 30.13b). They live in algal mats and in the bottom debris and mud of ponds, lakes, and streams, and there are even species that are found in marine sediments. Some oligochaetes are tube dwellers but this habit is not nearly as widespread as among polychaetes. Most aquatic species live in shallow water, but some Tubificidae, a large family of tube dwellers, occur in enormous numbers in the lower anaerobic regions of deep lakes and have been reported from deep ocean bottoms as well.

Many oligochaetes live in wet boggy soil and intergrade with the more terrestrial earthworms. The largest annelids are Australian earthworms of the genus *Megascolides*, which may attain a length of 3 m (Fig. 30.13a).

Earthworms are gallery dwellers with short heavy setae. They display few structural modifications for a terrestrial existence, for their microenvironment within the soil is usually quite moist. Like most other oligochaetes, they lack special organs for gas exchange. The high concentration of oxygen in air and the highly vascularized body wall permit adequate gas exchange across the general body surface despite the large size that earthworms may attain. The surface is kept moist for gas exchange by epidermal mucous glands and by coelomic fluid, which is released through a dorsal pore between adjacent segments.

Desiccation is minimized by behavioral adaptations. Earthworms come to the surface only at night or during rains. During dry periods or freezing winter weather, earthworms burrow deeper in the soil and become inactive. Water in moist soil can be absorbed across the body surface, and some tropical earthworms possess nephridia that open into the gut, apparently providing for recycling of urine water.

Well-drained soils with a large amount of organic matter provide the best environment for earthworms.

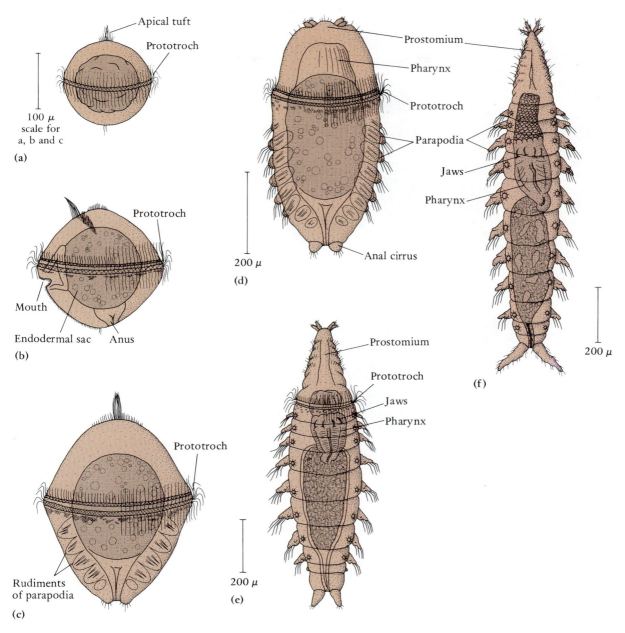

Figure 30.12
Larval stages of *Glycera convoluta*. (a) Early trochophore (15 hours). (b) Later trochophore (10 days). (c) Larva at 4 weeks. (d) Larva at 8 weeks. Although still a swimming stage, it frequently comes to rest on the bottom. (e) Postlarva at 8 weeks has become benthic and raptorial. (f) Young worm at two months.

Within such soils the earthworm population is vertically stratified. Small species and small individuals live in or near the upper humus layer; larger species, such as those of the genus *Lumbricus*, tunnel down to depths of 1 to 2 m.

Nutrition

Decomposing plant tissue is the principal food of most oligochaetes, although many freshwater species feed upon algae, and a few are raptorial on other small invertebrates.

(a)

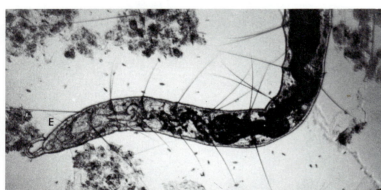

E

(b)

Figure 30.13
(a) A giant earthworm being removed from its burrow in Australia. (b) Anterior end of a freshwater oligochaete, *Stylaria*. In this genus the prostomium is tentacle-like. Note the long setae characteristic of many freshwater oligochaetes.

Earthworms are selective and nonselective deposit feeders. They consume bits of decaying vegetation at the surface and also cycle soil through the gut and digest the organic matter. The castings are commonly deposited at the surface, contributing to soil mixing. Five worms can thoroughly mix 500 cc of sand with 500 cc of soil in several months.

Food is ingested by means of a muscular pharynx, which can be everted as an adhesive pad or, in earthworms, functions as a pump (Fig. 30.14). From the pharynx food passes into the esophagus, which is often modified at different levels as a **crop** and one or more **gizzards**. The thin-walled crop functions in storage, and the muscular gizzard grinds the food into smaller particles. **Calciferous glands** over the dorsal esophageal wall excrete excess calcium taken in with food. The calcium is eliminated into the gut and out the anus as calcite crystals, which have a low solubility.

The remainder of the gut is a long straight intestine, the site of extracellular digestion and absorption. The surface area of the earthworm's intestine is nearly doubled by a dorsal longitudinal fold, the **typhlosole** (Fig. 30.14).

Surrounding the intestine is a layer of **chloragogen cells**. These, like the vertebrate liver, are an important site of intermediary metabolism. The synthesis and storage of glycogen and fat, the deamination of amino acids, and the synthesis of urea (in terrestrial species) occur in chloragogen cells.

Internal Transport, Excretion, and Nervous System

The circulatory system, nephridia, and nervous system of oligochaetes are essentially like those of polychaetes. The contractile dorsal blood vessel provides the principal force for blood propulsion. In some oligochaetes, certain anterior commissural vessels (five pairs around the esophagus in *Lumbricus*) are conspicuously contractile (Fig. 30.14). These "hearts" function as accessory pumps. Hemoglobin is present in solution in the plasma of the larger oligochaetes, including most of the earthworms.

The primary function of the metanephridia is probably water balance, for in freshwater species and in earthworms with a plentiful water supply the urine is

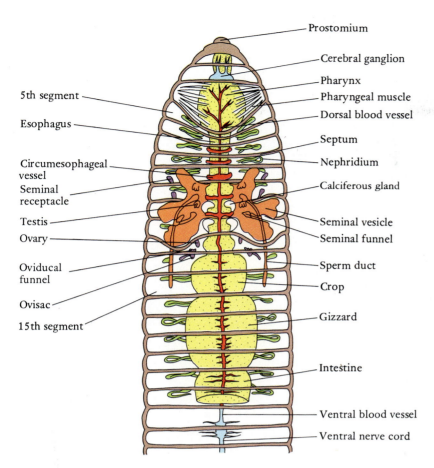

Prostomium
Cerebral ganglion
Pharynx
Pharyngeal muscle
Dorsal blood vessel
Septum
Nephridium
Calciferous gland
Seminal vesicle
Seminal funnel
Sperm duct
Crop
Gizzard
Intestine
Ventral blood vessel
Ventral nerve cord

5th segment
Esophagus
Circumesophageal vessel
Seminal receptacle
Testis
Ovary
Oviducal funnel
Ovisac
15th segment

Figure 30.14
Dorsal view of the anterior section of the earthworm *Lumbricus*.

hypoosmotic to the coelomic fluid. Like other freshwater animals, oligochaetes must eliminate the water entering by osmosis.

Giant fibers, present in the ventral nerve cord (fused pair), transmit impulses controlling rapid contraction of the body. Sense organs are lacking in most species, but the integument is richly supplied with sensory receptors, including photoreceptors.

Reproduction and Development

In contrast to polychaetes, oligochaetes are hermaphroditic. Moreover, there are distinct gonads, restricted to a few segments in the anterior third of the body. The paired ovaries are located within a single segment (Figs. 30.14 and 30.15). The eggs mature within the coelomic fluid of that segment and exit through a pair of simple oviducts. Semi-

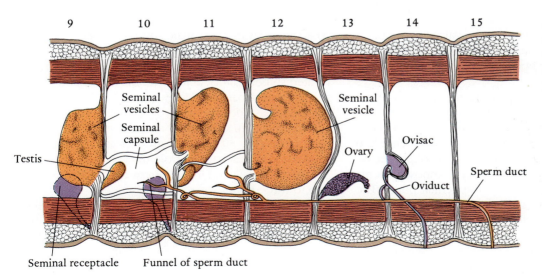

9 10 11 12 13 14 15

Seminal vesicles
Seminal capsule
Seminal vesicle
Ovary
Ovisac
Oviduct
Sperm duct
Testis
Seminal receptacle
Funnel of sperm duct

Figure 30.15
Reproductive segments of the earthworm *Lumbricus* (lateral view).

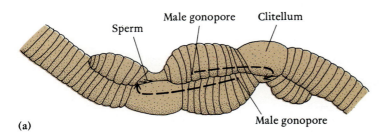

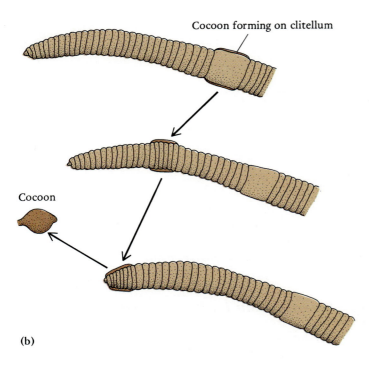

(a)

(b)

Figure 30.16
Reproduction in *Lumbricus*. (a) Movement of sperm during copulation. (b) Cocoon formation.

nal receptacles are present as inpocketings of the ventral body wall and are not connected to the other parts of the female system (Figs. 30.14a and 30.15).

There are one or two testicular segments, each containing a pair of testes (Figs. 30.14a and 30.15). The septum of the testicular segment is usually greatly evaginated to form a pouchlike seminal vesicle. The paired sperm ducts penetrate the posterior septum and may pass through several segments before opening onto the ventral surface (Figs. 30.14 and 30.15).

The epidermis of certain adjacent segments in the anterior third of the body contains many glands that produce secretions for various parts of the reproductive process. These segments, collectively called the **clitellum** (L. *clitellae*, packsaddle), are characteristic of oligochaetes. The clitellum of many aquatic species is composed of only two segments, those that contain the genital pores. The clitellum may be visible only during the reproductive period. The earthworms of the genus *Lumbricus*, in contrast, have a conspicuous, permanent, girdle-like clitellum of six to seven segments located behind the genital pores (Fig. 30.16a).

Copulation with reciprocal transfer of sperm is characteristic of oligochaetes. Two worms come together with ventral surfaces opposed and with their anterior ends facing in opposite directions. The worms are held together by mucus secreted by the clitellum and sometimes by special genital setae. In most oligochaetes, sperm are passed directly from the male gonopores into the seminal receptacles. However, in *Lumbricus* and other members of the same family, the seminal receptacles of one worm are located some distance anterior to the male gonopores of the copulating partner (Fig. 30.15). The sperm, on being released, pass down a pair of grooves on the ventral surface and then cross over at the level of the seminal receptacles (Fig. 30.16a). The grooves are arched over by mucus and thus separated from the grooves of the opposite worm. Copulation in *Lumbricus* is a process requiring two to three hours.

A few days after copulation the clitellum secretes a dense material that will form the **cocoon** (Fig. 30.16b). Albumin, serving as a food source for the embryos, is secreted inside the cocoon, especially in terrestrial species with little yolk in the egg. The secretions form an encircling band that slips forward over the body. Eggs from the female gonopores and sperm from the seminal receptacles are collected en route. The cocoon eventually slips off the anterior end of the worm, the two ends sealing in the process. Fertilization takes place within the cocoon, which is left in the bottom mud and debris or in the soil. Development is direct, and young worms emerge from one end of the cocoon. Only one egg completes development within the cocoon of *Lumbricus terrestris*.

CLASS HIRUDINEA

The members of the class Hirudinea constitute the smallest and the most aberrant of the three principal annelid classes. Only some 500 species have been described. Most live in fresh water, but there are some marine and terrestrial species. The leech design is correlated with a carnivorous diet and a looping, surface mode of locomotion. The body is dorsoventrally flattened, with the anterior segments modified as a small **sucker** surrounding the mouth and the posterior segments forming a larger sucker behind the anus (Fig. 30.17). The head is greatly reduced, setae are absent, and there is little external evidence of metamerism. The body is ringed with many grooves (annuli), but they do not correspond to the 34 segments that compose the leech body (Fig. 30.18a). There are no internal septa. Only the nephridia and nervous system indicate the original segmentation. The coelom is reduced by invasion of connective tissue to a system of interconnected spaces or sinuses.

Most leeches are 2 to 5 cm in length and none are as small as many polychaetes and other annelids. *Hirudo medicinalis*, the medicinal leech, may reach a length of 20 cm, but the giant is a tropical South American leech that can be as long as 30 cm.

Locomotion

Leeches crawl on the surface of objects in a looping manner, like an inchworm, a major change from the peristaltic burrowing of their oligochaete ancestors. The body is extended and the anterior sucker attached. Then the posterior sucker is released, pulled forward and reattached. Many leeches can swim by dorsoventral undulations of the flattened body. Correlated with these modes of movement, the longitudinal muscles of the body wall are powerfully developed.

Nutrition

Contrary to popular notion, not all leeches are bloodsuckers. Many are predaceous and feed upon other small invertebrates, especially oligochaetes, snails, and insect larvae. However, about 75% of the known species of leeches are bloodsucking ectoparasites on invertebrates, such as snails and crustaceans, and especially on vertebrates. All classes of vertebrates may be hosts, but fish and other aquatic forms, including waterfowl, are the most common victims. Mammals that come into the water to drink may be attacked. In humid areas of the tropics,

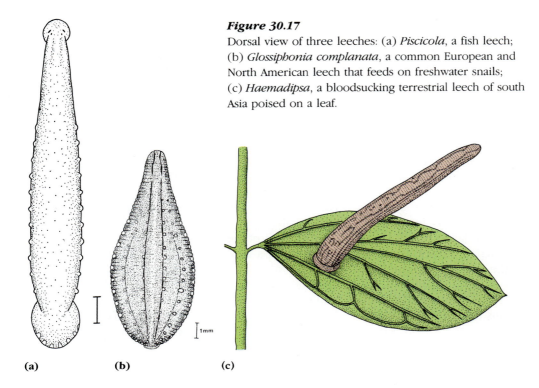

Figure 30.17
Dorsal view of three leeches: (a) *Piscicola*, a fish leech; (b) *Glossiphonia complanata*, a common European and North American leech that feeds on freshwater snails; (c) *Haemadipsa*, a bloodsucking terrestrial leech of south Asia poised on a leaf.

(a)   (b)   (c)

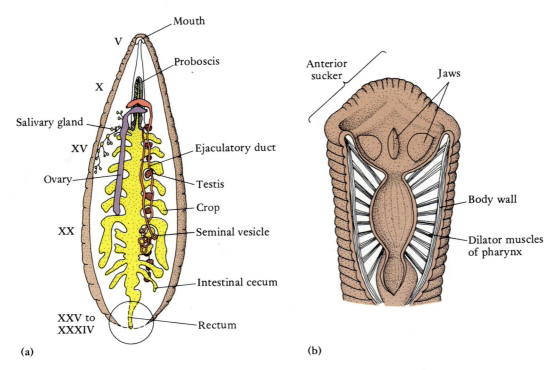

Figure 30.18
(a) Ventral view of internal structure of *Glossiphonia complanata*, a leech with a
proboscis. Roman numerals indicate segments. (b) Ventral dissection of anterior
end of *Hirudo*, showing three jaws.

leeches attack terrestrial birds and mammals, attaching to the moist thin membranes of the nose and mouth. They climb upon the host when it brushes through vegetation on which the leeches are located (Fig. 30.17c). Leeches detect the presence of a host by great sensitivity to changes in water pressure, host secretions, or body temperature.

Two principal types of ingestive organs occur in leeches: a proboscis and jaws. Some leeches ingest food through a tubular jawless proboscis that can be extended through the mouth from a proboscis cavity (Fig. 30.18a). Predatory species usually swallow the prey whole. In some bloodsucking species, the proboscis penetrates the skin, probably with the aid of enzymatic secretions, and also acts as an ingestive pump.

The pharynx of other leeches also serves as a pump but cannot be protruded. The mouth cavity in many of these forms contains three bladelike jaws (Fig. 30.18b). When the anterior sucker is attached in bloodsucking forms, the jaws slice through the skin of the host (Fig. 12.7a).

All bloodsucking leeches, both proboscidate and jawed, produce salivary secretions that contain an anesthetic, anticoagulants, and vasodilators. An esophagus connects the pharynx with a large crop (Fig. 30.18a). An intestine composes the posterior of the digestive tract and

opens through the anus just above the posterior sucker. Pouches (**ceca**) extend laterally from the crop and in some species from the intestine as well. The crop functions as a storage center and the intestine is the site of digestion and absorption. Many leeches may consume 2 to 11 times their weight in blood at a single feeding of up to 40 minutes. Water is removed from the ingested blood, and the remaining condensed material may require as long as 6 months to be digested. Leeches are unusual in possessing mostly exopeptidases as digestive enzymes (p. 237). Symbiotic bacteria provide the enzymes for a part of the digestive process and produce some substance that prevents decomposition of the blood cells. Thus a meal may be stored for a long time between infrequent feedings. Leeches have survived without feeding for as long as a year and a half.

The use of leeches in bloodletting, which was believed to be a cure for various disorders, goes back to ancient times but became especially popular during the 18th and 19th centuries. Between 1829 and 1836, five to six million leeches (*Hirudo*) were used each year in the hospitals of Paris. During the first third of the 19th century, as many as 30 million leeches a year were imported into the United States for medical use. Leeches still have some limited use in modern medicine, as in the removal of accumulated blood around sutures (see Fig. 12.7) or from certain hematomas (blood that has collected in tissue spaces).

Internal Transport, Excretion, and Sense Organs

Some leeches have a blood vascular system like that of their oligochaete ancestors. Other leeches have lost the original blood vessels, and the interconnecting coelomic spaces have been converted into a blood vascular system. Hemoglobin, dissolved in the plasma or coelomic fluid, provides for gas transport, but gills are present in only a few species. The nephridia of freshwater leeches produce a hyperosmotic urine, indicating a primary role in water balance; the nephridia of marine leeches are reduced. Although the head is reduced, some leeches possess eyes. Special sensory papillae are arranged in rings on the annuli.

Reproduction

The presence of a clitellum and a somewhat similar hermaphroditic reproductive system is taken as evidence that leeches evolved from oligochaetes. Copulation involves a reciprocal exchange of sperm, which in some leeches is similar to that in oligochaetes. In other species spermatophores are transferred by hypodermic impregnation. The spermatophores are injected into the ventral surface of the opposite leech, and the injected sperm make their way to the ovaries of the female system, where fertilization occurs.

The clitellum produces a cocoon that receives fertilized eggs from the female gonopores. Many fish leeches attach their cocoon to the host. Other leeches deposit the cocoon in water, beneath stones and other objects, and in soil. The members of one family of leeches brood their eggs within a modified cocoon attached to the ventral surface of the body.

EVOLUTIONARY RELATIONSHIPS

On the basis of superficial appearance leeches seem so different from polychaetes that you might wonder why they are in the same phylum. The carnivorous feeding habits and looping mode of locomotion of leeches have brought about marked structural changes. But as we have seen, the segmentation of their nervous system and nephridia reveal their metameric annelidan ancestry. And there is still another clue: one species of fish leech has setae on the anterior segments and still retains the coelom compartmented by septa. The reproductive system of leeches clearly relates them to oligochaetes. In fact, some systems of classification unite them within a common class, the Clitellata.

A zoologist who works with annelids once said that an ancestral annelid would have probably looked like an oligochaete but had a polychaete reproductive system. He reasoned that metamerism with coelomic compartmentation evolved in ancestral annelids as an adaptation for peristaltic burrowing in soft marine sediments. Since that mode of movement persists in most oligochaetes, the ancestral body form probably persists to a greater degree in oligochaetes as well, more so than in polychaetes, in which parapodia provide for a different mode of locomotion. However, the polychaete reproductive condition—separate sexes, with the peritoneum of most segments producing gametes—is certainly more primitive than that of the hermaphroditic oligochaetes.

Polychaetes probably represent a separate line of evolution from the burrowing ancestral annelids. The first simple parapodia may have evolved in some early annelids that moved from soft sediments to the superficial surface detritus of firmer bottoms. From here there was probably a radiating evolution of the various polychaete adaptive groups—errant forms, burrowers, and tube dwellers.

The spiral determinate cleavage and schizocoelous mode of coelom formation of annelids reflect their protostome position in the animal kingdom; indeed it is these features of annelidan and molluscan development that led to the definition of protostomes. As we indicated in the last chapter, the similarity between the annelid and molluscan trochophore larvae is believed to indicate a close relationship between these two phyla.

RELATED PHYLA

Phylum **Pogonophora** (Gr. *pogon*, beard + *pherein*, bearer) consists of deep-water tube-welling worms, of which some 80 species have been described to date. Most are 10–85 cm in length. A rear section of the body is segmented and bears setae. At the anterior end of a long middle trunklike region is a forepart bearing long ciliated tentacles. There is no mouth or gut (Connections 30.1).

Phylum **Echiura** (Gr. *echis*, viper + *oura*, tail) contains some 140 species of marine worms that possess a single pair of large setae on the underside of a cylindrical trunk. Anteriorly a large, nonretractable ciliated proboscis functions in deposit feeding (Fig. 30.19a). Echiurans burrow in sand and mud or live in spaces between shell and rock.

Sometimes called peanut worms, the approximately 320 species of the phylum **Sipuncula** (L. *sipunculus*, little pipe) are marine worms that burrow in soft bottoms, live in coral crevices or old mollusc shells, or bore into coralline rock. The unsegmented cylindrical trunk bears a retractable anterior end (the introvert) that contains the mouth (Fig. 30.19b). Sipunculans are deposit feeders, using tentacles, lobes, or fringes surrounding the mouth to collect sediment.

Symbiosis between Animals and Chemosynthetic Bacteria

Around the turn of this century some strange, gutless tube-dwelling worms were collected from the deep sea. With improved oceanographic techniques, some 80 species of these worms have now been collected from deep water throughout the world's oceans. The posterior part of the body is segmented and bears setae, but the rest of the body is quite different from that of annelids. They are placed in a separate phylum, the Pogonophora (beard bearer), so called because the anterior end bears one to many tentacles. The lack of a mouth and gut in these worms raised much speculation about how they fed. Then in 1977 the research submersible *Alvin* discovered a remarkable fauna living around deep hot-water vents on the Galapagos rift (2600 m)

off the northwest coast of South America. The fauna included clams and mussels and giant pogonophorans with spectacular red tentacular plumes. The pogonophorans reached 1.5 m in length and almost 4 cm in diameter. Subsequent studies revealed that these worms and the bivalves harbored chemosynthetic bacteria that are able to synthesize organic carbon. The bacteria can oxidize hydrogen sulfide and utilize the energy released from that oxidation to synthesize organic carbon from CO_2. Hydrogen sulfide is toxic to living systems, blocking one of the electron carriers in cell respiration and blocking the uptake of oxygen by hemoglobin. However, pogonophoran hemoglobin is remarkable in being able to carry both oxygen and hydrogen sulfide at the

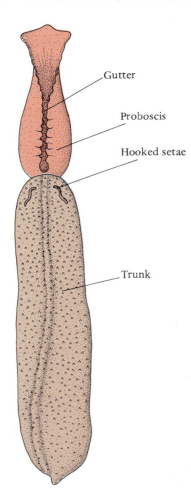

Gutter

Proboscis

Hooked setae

Trunk

(a)

Figure 30.19

(a) A species of *Echiurus* shown with its spoon-shaped proboscis out of its burrow. (b) A peanut worm (sipunculan) from the Australian Great Barrier Reef. This species lives in burrows excavated in coral.

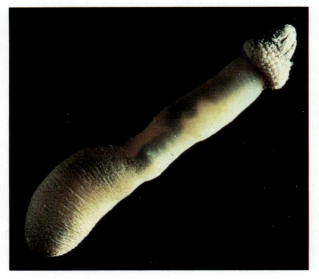

(b)

same time, delivering the hydrogen sulfide and some oxygen to the bacteria. Since the hydrogen sulfide is bound to the hemoglobin, it is also not free to poison the pogonophoran cells. The bivalves transport the hydrogen sulfide by a special blood protein, not by a respiratory pigment. The chemosynthate produced by the bacteria is shared with the pogonophoran and bivalve hosts.

A similar symbiotic relationship has now been found in other pogonophorans, which, of course, explains how they can be gutless. Certain protobranch bivalves found in other types of marine habitats have guts that are reduced or absent; studies have revealed that they too contain chemosynthetic bacteria.

Pogonophorans and mussels around the deep sea thermal vents along the Galapagos Rift.

Classification of Phylum Annelida

Class Polychaeta (Gr. *polys*, many + *chaite*, hair) Marine annelids in which setae are carried on lateral segmental parapodia. Metamerism usually well developed. Prostomium (head) variable, but commonly bears sensory or feeding structures. Sexes separate, and gametes produced by the peritoneum of numerous segments. *Nereis, Scoloplos, Glycera, Eunice, Diopatra, Arenicola, Amphitrite, Chaetopterus.*

Class Oligochaeta (Gr. *oligos*, few + *chaite*, hair) Freshwater annelids and the terrestrial species known as earthworms. Parapodia absent but setae present. Metamerism well developed. Prostomium is a simple lobe without sensory or feeding structures. Hermaphroditic, with gonads present in a few specific segments. Epidermis of certain segments modified as a clitellum for the secretion of a cocoon. *Stylaria, Lumbricus, Aelosoma.*

Class Hirudinea (L. *hirudo*, leech) Marine, freshwater, and terrestrial annelids known as leeches. No parapodia or setae. Body more or less dorsoventrally flattened with anterior and posterior segments modified as suckers. Metamerism reduced. Most ectoparasitic. Hermaphroditic, with a clitellum. *Piscicola, Glossiphonia, Hirudo.*

Class Branchiobdellida (Gr. *branchia*, gill + *bdella*, leech) A small group of little worms possessing only 14 or 15 segments, without setae. They show similarities to both oligochaetes and leeches. All live on freshwater crayfish, some existing as commensals on the exoskeleton and others being ectoparasitic on the gills. They attach by means of a posterior sucker (Fig. 30.20). *Branchiobdella.*

Figure 30.20
Stephanodrilus, a member of the leechlike class Branchiobdellida. These worms are parasitic or commensal on freshwater crayfish.

Summary

1. Members of the phylum Annelida, called segmented worms, are metameric animals in which the coelom is compartmentalized by transverse septa to form a localized hydrostatic skeleton. Metamerism is believed to have evolved in annelids as an adaptation for peristaltic burrowing. The segmental features of the excretory system (one pair of nephridia per segment), nervous system (segmental ganglia and lateral segmental nerves along the pair of ventral nerve cords), and internal transport system are probably accommodations to the primary segmentation of the coelom.

2. The class Polychaeta contains mostly marine species that possess lateral segmental appendages called parapodia. An eversible pharynx armed with jaws is commonly used in carnivorous and herbivorous feeding. Most errant polychaetes crawl with the parapodia and possess a well-developed prostomium bearing eyes, antennae, and other structures. Burrowing polychaetes live in galleries that have numerous openings to the surface, or they are more sedentary occupants of simple vertical burrows. The more active gallery dwellers usually have a simple prostomium with few sense organs and are carnivores or deposit feeders. The sedentary burrowers are selective and nonselective deposit feeders.

3. Tube-dwelling polychaetes construct tubes of secreted material or mineral particles cemented together or both. Active tube dwellers are carnivores and possess adaptations similar to those of errant polychaetes. Sedentary tube dwellers are selective or nonselective deposit feeders or are filter feeders. They move within the tube by peristaltic contractions and have ridgelike parapodia with hooked setae. ·

4. Large polychaetes possess gills, and the blood usually contains a respiratory pigment. The sexes of polychaetes are usually separate; gametes are produced by the peritoneum of many segments, mature in the coelom, and exit by way of gonoducts or nephridia. Fertilization is external, and development commonly leads to a trochophore larva.

5. The class Oligochaeta contains mostly freshwater species that live within bottom debris and terrestrial forms (earthworms) that inhabit galleries in soil. They move by peristaltic contractions and possess setae but no parapodia. The prostomium is usually a simple cone without sense organs. Most species are selective or nonselective deposit feeders. Oligochaetes are hermaphroditic, with only a few segments involved in reproduction. Two worms copulate and transfer sperm reciprocally. The eggs are fertilized within a cocoon secreted by certain glandular segments called the clitellum. Development is direct.

6. The class Hirudinea contains the leeches. Most are found in fresh water, but some are marine (certain fish leeches) and some live in tropical jungles. Correlated with their mode of locomotion, leeches have lost setae and developed anterior and posterior suckers. The coelom has been reduced to interconnected spaces, and septa have disappeared. The body of all leeches is composed of 34 segments. Leeches are predaceous on other invertebrates or are bloodsuckers, mostly on vertebrates. The feeding organs are jaws and a pumping pharynx or a protrusible proboscis. Reproduction in leeches is similar to that in oligochaetes.

References and Selected Readings

Detailed accounts of the annelids may also be found in many of the references listed at the end of Chapter 23. The following works are devoted solely to annelids.

Adams, S.L. The medicinal leech. *Annals of Internal Medicine*. 109(1988):399–405. The past and present medical use of the medicinal leech, *Hirudo medicinalis*.

Brinkhurst, R.O. Evolution in the Annelida. *Canadian Journal Zoology* 60(1982):1043–1059. The evolutionary relationships of the major annelidan groups.

Childress, J.J., H. Felbeck, and G.N. Somero. Symbiosis in the deep sea. *Scientific American* 256(May 1987): 115–120. The role of symbiotic bacteria in pogonophorans and deep sea bivalves.

Edwards, C.A., and J.R. Lofty. *Biology of Earthworms*, 2nd ed. London: Chapman & Hall, Ltd., 1977. This short volume covers the structure, physiology, and ecology of earthworms.

Klemm, D.J. *A Guide to the Freshwater Annelida of North America*. Dubuque, Iowa: Kendall/Hunt Publishing Co., 1985.

Mill, P.J. (ed.) *Physiology of Annelids*. London: Academic Press, 1978. A well-organized, multiauthored volume covering the major aspects of annelid physiology.

Sawyer, R.T. *Leech Biology and Behavior*, 3 vols. Oxford: Clarendon Press, 1986. A detailed account of the biology of leeches.

31

Arthropods I: Chelicerates and Crustaceans

Characteristics of Arthropods

■ The body is covered by an exoskeleton of chitin and protein, which is periodically molted, permitting growth, and which is divided into jointed plates and cylinders, permitting movement.

■ Arthropods are metameric, reflecting their annelidan ancestry, but most species exhibit some degree of reduction in metamerism.

■ The segments carry paired jointed appendages, which not only serve in locomotion but also have been adapted for many other functions.

■ All internal structures derived from invaginations of the body wall have a chitinous lining. The extensive foregut and hindgut regions possess such a lining. The midgut, derived from endoderm, is shorter than in most animals.

■ The blood vascular system is open, and the dorsal heart is primitively tubular.

■ The plan of the nervous system is very similar to that of annelids.

■ The sexes are usually separate. The eggs are generally centrolecithal, and cleavage is commonly superficial.

(Left) The acorn barnacle Balanus nubilus, *with cirri extended, is surrounded by sea anemones. (Above) Rock barnacles.*

The arthropods are a vast asemblage of animals. At least 750,000 species have been described, which is more than three times the number of all other animal species combined. They are found in virtually every type of habitat in the sea and in fresh water and are perhaps the most successful of all invaders of the terrestrial environment.

Most zoologists agree that there are three lines of arthropod evolution embracing living species, each constituting at least a different subphylum:

Subphylum **Chelicerata**—horseshoe crabs, scorpions, spiders, mites, and ticks
Subphylum **Crustacea**—shrimps, crabs, and allies.
Subphylum **Uniramia**—centipedes, millipedes, and insects.

In this and the following chapter, we will survey each of these subphyla, concluding with an exploration of the possible phylogenetic relationships between the three groups. We begin with a description of the general arthropod ground plan.

THE ARTHROPOD DESIGN

Annelidan Features

A number of features indicate that arthropods may have evolved from annelids. Arthropods are metameric and the segments bear lateral appendages (Fig. 31.1). However, we cannot be certain that arthropod appendages are homologous to the parapodia of polychaetes. The nervous system, with its pair of large ventral nerve cords, is essentially like that of annelids, and the dorsal tubular heart may be the homologue of the dorsal contractile blood vessel of annelids (see Figs. 30.2 and 14.7). There is evidence that certain of the excretory organs found in arthropods are derivatives of nephridia. Traces of spiral cleavages can be seen in some crustaceans but in most arthropods the egg is centrolecithal and cleavage is superficial.

The Arthropod Skeleton

The distinguishing feature of arthropods is the presence of a **chitinous exoskeleton**. The evolution of this nonliving protective and supporting covering is related in turn to numerous other structural and functional changes that make up the arthropod condition (Fig. 31.2). The arthropod exoskeleton is composed of glycoprotein (chitin and protein) organized in two layers, an inner **endocuticle** and an outer **exocuticle**. A thin protein **epicuticle**, commonly impregnated with waxy compounds, forms an external layer over the exocuticle (see p. 206 and Fig. 11.7). The exocuticle differs from the endocuticle in having the glycoprotein chains cross linked—a more rigid molecular arrangement called **tanning**. Certain parts of the exoskeleton of different arthropods are often especially hard and said to be highly **sclerotized**, which means the exocuticle in such an area is thick and greatly tanned.

Movement is possible because the exoskeleton is divided into plates over the body and numerous cylinders

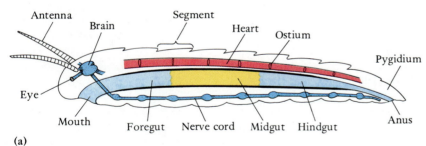

(a)

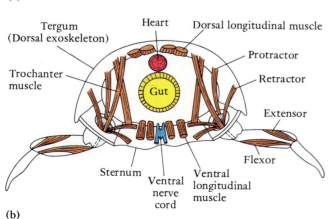

(b)

Figure 31.1
Structure of a generalized arthropod.
(a) Sagittal section; (b) cross section.

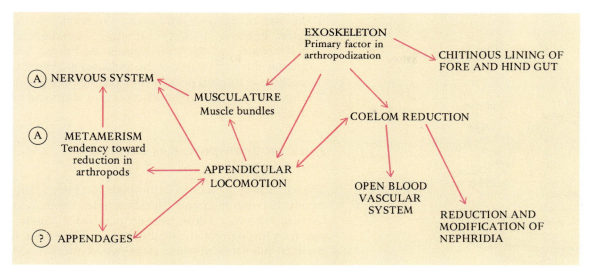

Figure 31.2
Interrelationship of changes resulting from the evolution of the arthropod condition (arthropodization). *A* refers to possible annelidan features retained by arthropods. Homology of appendages is questionable.

around the appendages like a suit of armor. At the junction point, or at joints between plates and cylinders, the exoskeleton is thin and folded (see Fig. 11.7b). The name Arthropoda (Gr. *arthron*, joint + *pous*, foot) refers to the jointed cylinders composing the appendages.

Despite the locomotor and supporting advantages of an external skeleton for small animals (p. 210), it poses problems for growth. The solution to this problem evolved by the arthropods has been the periodic shedding of the skeleton, a process called molting or **ecdysis** (p. 206).

The intervals between molts are known as **instars**, and the duration of the instars becomes longer as the animal becomes older. Some arthropods, such as lobsters, continue to molt throughout their life. Other arthropods, such as insects and spiders, have relatively fixed numbers of instars, the last being attained with sexual maturity.

Locomotion

The muscles of arthropods are distinct bundles of muscle fibers attached to the inner side of the exoskeleton (Fig. 31.1b). They function with the skeleton as a lever system, much as in vertebrates (p. 230). Only vestiges of the coelom remain in adult arthropods. The loss of the coelom is probably related to the shift from a fluid internal skeleton with peristaltic locomotion to a solid external skeleton with appendicular locomotion.

In arthropods the body tends to be suspended between the legs. Each leg performs an effective and a recovery stroke in which the leg is lifted, swung forward, and placed down upon the substratum. The legs on one side of the body carry out these movements in sequence, composing a locomotor wave; the locomotor waves on the two sides of the body alternate with each other (see Fig. 11.35a), as in crawling polychaetes.

Reduction of Metamerism

The primitive arthropod trunk was composed of a large number of segments, each bearing a pair of similar appendages. Such an arrangement was present in the fossil trilobites (Fig. 31.3), a group of extinct arthropods that were abundant in early Paleozoic seas. The evolution of different modes of existence with different types of locomotion and types of feeding behavior has led to a reduction in metamerism. Segments dropped out, fused together, or became specialized. Anteriorly, segments fused together to form a more complex head. Along both head and trunk, appendages became specialized for many functions other than locomotion—prey capture, filter feeding, food handling, gas exchange, ventilation, copulation, egg brooding, and so forth. In the crayfish, for example, the oral appendages are adapted for feeding; the large claws are specialized for grasping; the legs are used in crawling; the abdominal appendages are involved in egg brooding; and the last appendages form a flipper with the terminal section of the trunk (see Fig. 31.21a). Although these appendages are structurally quite different, they are all derived from originally similar segmental appendages. They are therefore said to be **serially homologous**. Certainly the great success of arthropods can be attributed in large part to the adaptation of segmental appendages to a wide range of functions.

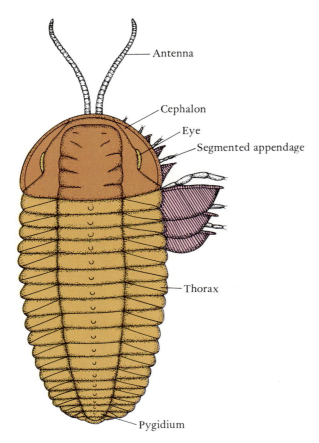

Figure 31.3
Dorsal view of the Ordovician trilobite, *Triarthrus eatoni*.
Each segment bears a pair of similar appendages, only a few
of which are shown in the figure. The name trilobite is
derived from the three conspicuous side to side divisions of
the dorsal body surface.

Digestive System

Arthropods have become adapted for a great range of diets,
and the appendages, especially those just beside and behind the mouth, are utilized in many different modes of
feeding. Commonly there is one heavy pair of appendages
flanking the mouth that bite or tear, followed by several
other pairs that are involved in food handling. The arthropod gut differs from that of most other animals in having
large anterior (stomodeum) and posterior (proctodeum)
ectodermal regions (Fig. 31.1a). The derivatives of these
portions are lined with chitin and constitute the **foregut**
and **hindgut**. The intervening region, derived from endoderm, forms the **midgut**. The foregut is chiefly concerned
with ingestion, mechanical breakdown, and storage of
food; its parts are variously modified for these functions
depending upon the diet and mode of feeding. The midgut

is the site of enzyme production, digestion, and absorption; however, in some arthropods enzymes are
passed forward, and digestion begins in the foregut. Very
commonly the surface area of the midgut is increased by
outpocketings forming pouches or large digestive glands.
The hindgut functions in the absorption of water and the
formation of feces.

Internal Transport, Gas Exchange, and Excretion

The circulatory system of arthropods is open; a large
pericardial sinus surrounds the dorsal heart (Fig. 31.1; see
also Fig. 14.6). Blood flows from the pericardial sinus into
the heart through small lateral openings called **ostia**. In
primitive tubular hearts there is one pair of ostia per body
segment. When the heart contracts, the ostia close and
blood is propelled anteriorly or posteriorly out of the
heart through arteries, which deliver the blood to tissue
spaces where various exchanges take place. From the
tissue spaces (collectively termed the hemocoel) blood
drains into a system of larger sinuses and eventually
returns to the pericardial sinus around the heart. The
arterial system varies greatly. Some arthropods, such as
insects, have little more than a short vessel leaving the
heart; others, such as crayfishes and crabs (Fig. 14.6), have
an extensive system of vessels delivering blood to tissue
spaces in various parts of the body.

Most arthropods have organs for gas exchange. Their
gills are usually modifications of appendages or outgrowths of the integument associated with an appendage
(p. 274). The gas exchange organs of terrestrial arthropods
are typically internal. However, they are derived from
invaginations of the integument and thus are lined with
chitin. The most common gas exchange organ in terrestrial
arthropods, and one that has evolved independently in
many different groups, is a system of air-conducting tubes
called **tracheae** (p. 277).

Those arthropods that have a respiratory pigment
most commonly have hemocyanin, a large molecule dissolved in the blood (p. 285). Hemoglobin occurs only
sporadically. Respiratory pigments are usually absent from
the blood of arthropods with tracheal systems, since the
blood in these animals plays only a small role in gas
transport.

Many arthropods possess one to several pairs of
excretory organs that are end sacs, i.e., the organs consist
of an internal blind sac connected to the outside through a
duct. Since the sacs represent a remnant of the coelom, the
duct may be derived from a nephridium (p. 351).

The excretory organs of insects and most arachnids,
such as spiders and mites, are not end sacs but consist of a
few to many blind **malpighian tubules** lying free in the
hemocoel and opening into the posterior section of the gut
(see p. 351 and Fig. 16.7).

Nervous System and Sense Organs

The nervous system of arthropods exhibits the same basic design as that of annelids—a dorsal anterior brain and a double ventral nerve cord (Fig. 31.1). However, the fusion and loss of segments is reflected in a corresponding forward migration and fusion of the ventral ganglia of many arthropods. For example, in a crab all of the ganglia of the ventral nerve cord have fused together anteriorly.

The external sensory receptors of arthropods are usually associated with some modification of the chitinous exoskeleton, which otherwise would act as a barrier to the detection of external stimuli (p. 407). These modifications, except for the eyes and their associated receptor cells, are called sensilla. Depending upon the kind of external signal being monitored, sensilla have various shapes, such as hairs or setae, pegs, pits, and slits.

Most arthropods have eyes, which can vary greatly in complexity. Some are simple and have only a few photoreceptors. Others are large, with thousands of retinula cells, and can form a crude image (p. 422).

Reproduction

Most arthropods have separate sexes with a single pair of gonads and gonoducts. Fertilization is internal in terrestrial species and is external or internal in aquatic forms. Copulation commonly involves the use of modified appendages. The eggs are typically rich in yolk, which is organized as a large sphere between the central nucleus and the periphery of the egg (centrolecithal egg) and cleavage is usually superficial (p. 513). Development may be indirect or direct, depending upon the group of arthropods.

SUBPHYLUM CHELICERATA

Members of the subphylum Chelicerata are distinguished from all other arthropods in lacking antennae. The first pair of appendages behind the mouth are **chelicerae** (Gr. *chele*, claw + *keras*, horn), which are modified in various ways for feeding (Fig. 31.4b). The body is composed of two regions, a cephalothorax and an abdomen (Figs. 31.4 and 31.8). The **cephalothorax** contains a number of trunk segments that have fused with the head, and the fused region is usually covered dorsally by a single skeletal piece, the **carapace**. Behind the chelicerae the cephalothorax carries a pair of appendages, called **pedipalps**, and four pairs of legs. There are no segmental abdominal appendages.

The chelicerates are an ancient group, first appearing in the fossil record in the Ordovician. The earliest chelicerates were marine, and one group, the Merostomata, still lives in the sea.

Class Merostomata

The members of the **class Merostomata**, called **horseshoe crabs**, are the only marine gill-bearing chelicerates (Fig. 31.4). The class has been known from the Ordovician, but there are only five species living today. *Limulus polyphemus*, the only species in the Western Hemisphere, is found along the Atlantic coast of the United States and in the Gulf of Mexico. They are among the largest living

Figure 31.4
A horseshoe crab: (a) Dorsolateral view; (b) ventral view, showing appendages.

(a)

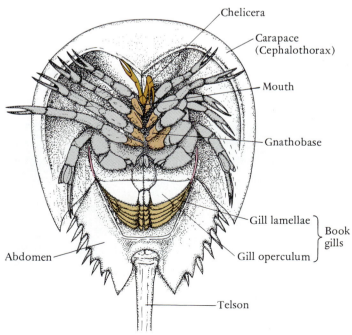

(b)

chelicerates, reaching a length of 60 cm. However, the biggest were certain fossil merostomes, the eurypterids, which were as long as 2 m (Fig. 31.5).

Dorsally, the carapace of horseshoe crabs bears a pair of compound eyes, and the underside of the cephalothorax bears the six pairs of appendages typical of most chelicerates: a pair of chelicerae, a pair of pedipalps, and four pairs of legs (Fig. 31.4b). However, the pedipalps are not markedly different from the posterior appendages and should properly be called legs. All except the last pair of appendages are **chelate**, that is, they possess pincers formed by two terminal segments of the appendage. The last pair of appendages is used for sweeping away sand and mud.

The abdominal segments are fused together and terminate in a long spikelike **telson**. The underside of the abdomen carries five pairs of **book gills**, each composed of a large number of thin plates (lamellae) protected by a flaplike cover.

Despite their appearance, horseshoe crabs are harmless animals. They crawl over or push through sand or mud in shallow water and can swim upside down using the gills as paddles. The telson is used to push and right the animal if flipped over. Grabbing the telson is an easy way to pick up a live horseshoe crab.

Horseshoe crabs are omnivores and scavengers, and their diet includes the soft-bodied invertebrates and algae they encounter as they plow through the bottom. The food is passed to the spiny medial region of the leg bases (gnathobases), where it is moved forward to the mouth.

During the reproductive period horseshoe crabs congregate in shallow water of sounds and bays to mate. In

Figure 31.6
Copulating horseshoe crabs. The male is in the rear.

Limulus polyphemus this occurs during the high tides of spring and summer full and new moons. The male clasps the dorsal side of the abdomen of the female with his modified second pair of appendages (the pedipalps of other chelicerates) (Fig. 31.6). Eggs are fertilized as they are deposited into shallow depressions in the sand, where they are covered and left. A **trilobite larva** (it looks like a trilobite) hatches from the egg and swims and burrows in the sand. It gradually acquires the adult form over a succession of molts. The little horseshoe crab is about 4 cm long after one year and does not reach sexual maturity before about 9 years. The life span may be around 19 years.

Class Arachnida

The great majority of chelicerates, some 60,000 species, are members of the class **Arachnida**. In contrast to merostomes, arachnids are terrestrial. They are widely distributed, most living in vegetation, in leaf mold, and beneath bark, logs, and stones. Contributing to the great success of arachnids as land animals has been the evolution of terrestrial gas exchange organs; a waxy epicuticle, which reduces evaporative water loss; and relatively insoluble nitrogenous waste products, which reduce excretory water loss.

The class is divided into 13 orders, 7 of which contain the most common and familiar species of temperate regions. **Scorpions** (order Scorpionida) are large arachnids with big chelate pedipalps and a long, segmented abdomen terminating in a sting (Fig. 31.7). Paired eyes are mounted on tubercles in the middle of the carapace, and two to five additional pairs of eyes may be present along the anterior lateral margins. Scorpions are secretive, largely nocturnal animals of the tropics and semitropics. In the United States they are common only in the Gulf and

Figure 31.5
Photograph of a fossil eurypterid from the Silurian. Note that these merostomes differ from horseshoe crabs in having a long segmented abdomen and the sixth pair of appendages in the form of large paddles.

Figure 31.7
A scorpion, *Vejovis spinigerus*, about to capture an insect.

small and leglike. Usually eight eyes are arranged in two rows of four each across the front of the carapace.

Pseudoscorpions (order Pseudoscorpionida) are only a few millimeters in length and are common inhabitants of leaf mold (Connections 31.1) in both tropical and temperate regions. These tiny arachnids have large pedipalps like scorpions, but the segmented abdomen is short and lacks a terminal sting (Fig. 31.9).

Harvestmen (order Opiliones), also known as daddy longlegs, are distinguished from other arachnids by their very long legs and segmented abdomen broadly joined to the cephalothorax (Fig. 31.10). A tubercle on the center of the carapace bears a single pair of eyes. The members of this order are common arachnids in both temperate and tropical regions.

The **mites** and **ticks** belong to three different orders and are sometimes collectively called the **acari**. The acari is the second largest and most diverse group of arachnids, and some acarologists believe that the 30,000 known species probably represent less than half of the total number in the orders. Most of the undescribed species will probably become extinct with the destruction of tropical rain forests and other habitats and will never be known.

Mites are found in all sorts of terrestrial microhabitats, and are especially abundant in leaf mold. They are the only arachnids that have reinvaded aquatic habitats, giving rise to species adapted for living in the sea and fresh water. Many species feed on human food products and on crop plants and are parasitic on humans and domesticated animals. As a consequence, mites are of great economic importance.

southwestern states. Scorpions are ancient arachnids, known from the Silurian, and probably were among the first terrestrial arthropods.

Spiders (order Araneae) compose the largest of the arachnid orders. About 32,000 species have been described, and they occur in far greater numbers than most people are aware of. An ungrazed meadow, for example, may support as many as 2,250,000 spiders per acre. The abdomen is unsegmented and connected to the cephalothorax by a narrow waist (Fig. 31.8). The pedipalps are

Figure 31.8
Internal anatomy of a spider.

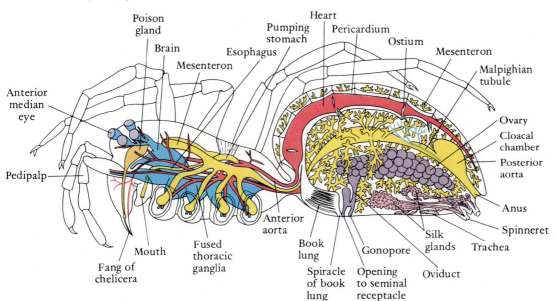

In soil and the accumulated leaves and decomposing leaf fragments (called humus or leaf mold) at the surface of the soil lives a terrestrial interstitial fauna. The interstitial spaces between leaves and leaf fragments are much larger than those between soil particles, and the animals that live there are also comparatively larger. An enormous number of animals can be found in leaf mold. A shopping bag full of decomposing leaves scraped from a forest floor contains hundreds of animals. Snails and small earthworms are commonly present, but the vast majority of leaf mold animals are arthropods, chiefly arachnids and insects, although crustaceans are represented by pill bugs. The most abundant group, both in numbers of species and numbers of individuals, is mites. One group of mites that is especially conspicuous in leaf mold is the oribatid or beetle mites. Their shiny, convex dorsal exoskeleton makes them look a little like beetles, but they are of course much smaller.

Leaf mold animals display a wide range of feeding habits. Many feed on decomposing vegetation, although precisely what elements they are utilizing is not always known. For example, many oribatid mites feed on fungus growing on decomposing leaves. There are also many carnivores, such as spiders, false scorpions, and some mites.

An important adaptation of many leaf mold arthropods is the presence of lipids in the epicuticle of the exoskeleton. A waxy epicuticle serves two functions. It reduces excessive water loss across the exoskeleton, a hazard to which small animals are especially subject. The lipids do not eliminate the danger, but the lower water vapor pressure deficit of the leaf mold also provides some protection. Many leaf mold animals are very sensitive to the amount of water vapor in the surrounding air and will position themselves at lower levels in the leaf mold when the vapor pressure drops.

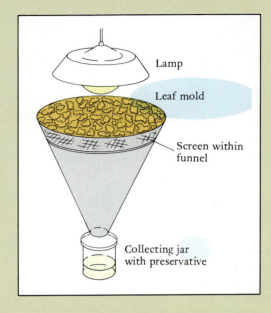

Lamp

Leaf mold

Screen within funnel

Collecting jar with preservative

A waxy epicuticle can also function as a water repellent. Following a rain, leaf mold will contain many water droplets and films. The hydrophobic surface of the epicuticle prevents the adhesion of water molecules and reduces the danger of the animals becoming completely submerged within water.

Leaf mold animals can be easily collected in a Tullgren funnel. A sample of leaf mold is placed in the funnel, which is in turn placed under a lamp. The heat and light of the lamp drive the animals down through the leaf mold, and they are collected in a jar containing preservative at the bottom of the funnel. They can be collected alive if precautions are taken against desiccation in the collecting jar.

Mites are usually less than a millimeter in length, and their adaptive diversity may in part be attributed to their small size, which has enabled them to exploit many types of microhabitats.

The abdomen of mites and ticks is unsegmented and broadly fused with the cephalothorax. The entire body is thus covered dorsally by a single skeletal piece (Fig. 31.11). The pedipalps are small and usually leglike.

Of the seven orders of arachnids we have introduced, scorpions—with their long, segmented abdomen—exhibit the greatest degree of external metamerism; mites and ticks, the least.

Silk

Silk is a fibroin protein in which a small number of certain repeating amino acids make up a major part of the protein polymer. Silk is produced by pseudoscorpions, spiders, certain mites, and some insects, such as the caterpillars of moths. However, of all these arthropods, spiders make the greatest use of silk.

Spider silk has about the same strength as nylon but can be stretched almost twice as much as nylon. The **silk glands** of spiders are located in the abdomen and open through conical **spinnerets** at the end of the abdomen, each spinneret bearing numerous spigots (Figs. 31.8 and

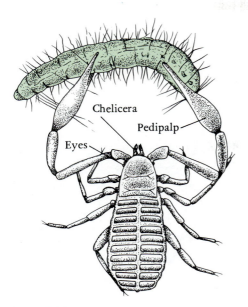

Figure 31.9
Dorsal view of a pseudoscorpion catching an insect larva with its pedipalps.

31.12a). A particular species may possess from two to six different kinds of silk glands (Fig. 31.8). The duct from each gland, of which there may be a large number belonging to any one type, opens at the end of one spigot. Silk hardens during the process of being drawn out, usually as a result of the spider moving away from the attached end of the line. The liquid silk changes from a water-soluble to a solid form as a result of changes in the molecular configuration and bonding.

Spiders utilize silk in many ways, but contrary to popular notion, only some spiders use silk to construct webs for trapping prey. Silk is used to build nests, which are used as retreats, for reproduction, or for overwin-

tering; in all spiders the eggs are encased within a silken **egg case**.

Most spiders lay down a **dragline** behind them, anchoring it at intervals to the substratum (Fig. 31.12b). The dragline is demonstrated when a spider appears suspended in midair after being brushed off clothing or some other object. The dragline not only functions as a safety line for the spider but also is an important means of communication between members of the species. A spider may determine chemically from another dragline whether its owner is male or female and whether it is immature or adult.

Small spiders and newly hatched spiders use the silk as a means of dispersal. They climb to favorable take-off points, tilt the abdomen upward, and release a strand of silk. When air currents produce sufficient pull, the spider lets go and sails out to whatever new habitat and fate the wind will take it. The wide distribution of many species of spiders is undoubtedly correlated with this ballooning phenomenon.

Aside from certain species of mites (spider mites), pseudoscorpions are the only other arachnids that make use of silk. The silk glands of pseudoscorpions open onto the chelicerae and are used in construction of nests or retreats.

Feeding

Most arachnids are predatory animals, and other arthropods are their usual prey. Arachnids detect prey by means of specialized sensory hairs, slit sense organs, and, in some species, eyes. Slit sense organs, many of which are located in the vicinity of joints, respond to stimuli, such as vibrations, which cause changes in the tension of the exoskeleton. As an aid to dispatching prey, certain arachnids have independently evolved **poison glands**. The poison glands

Figure 31.10
(a) Anterior view of a harvestman on a leaf. (b) Lateral view of a species of harvestman in which the first abdominal segment bears two large spines. Note the pair of eyes mounted on the dorsal tubercle of the cephalothorax. The chelicerae and pedipalps are shown, but the legs have been removed to make the body visible.

(a)

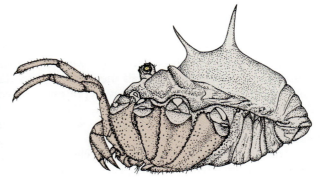

(b)

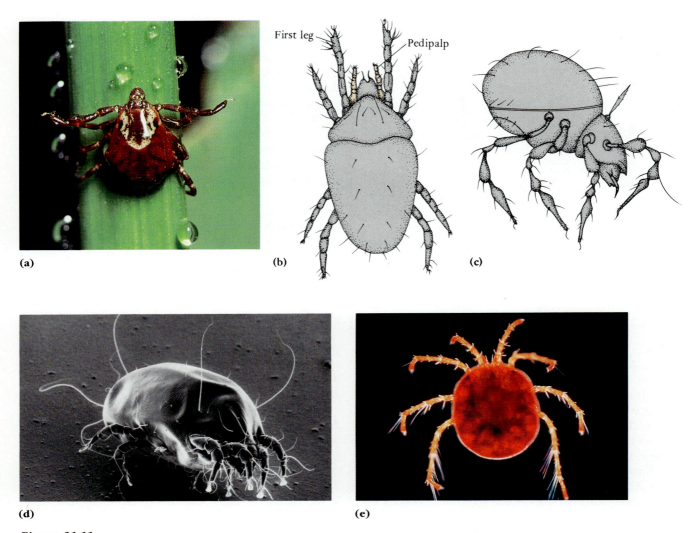

First leg   Pedipalp

(a)   (b)   (c)

(d)   (e)

Figure 31.11

Acari. (a) The dog tick, *Dermacentor variabilis*, on a plant stem. (b) *Tydeus starri*, a mite. (c) A beetle mite, a member of a large group of mites found in leaf mold. (d) A species of *Dermatophagoides*, a mite commonly found in house dust. When inhaled, the mite produces a house dust allergy in some people. The mite is probably normally parasitic on various animals found around human habitations. (e) Photograph of a water mite. Note the swimming hairs on the legs.

of scorpions are located in the terminal sting at the end of the abdomen (Fig. 31.7). Their prey is caught with the large pedipalps and then stabbed by the poison barb with a forward thrust of the abdomen. Although the sting of scorpions may be very painful, few species have a poison dangerous to human beings. The only dangerous North American scorpions are found in Mexico and the southwestern United States.

Pseudoscorpions catch prey with their large chelate pedipalps (Fig. 31.9), which have a poison gland opening at the end of one or both fingers.

Spiders have a poison gland associated with each chelicera, which consists of a terminal **fang** that folds down against a larger basal piece (Fig. 31.8). The gland opens by a duct through the end of the fang, and the poison is injected through a bite.

The poison of a very small number of spiders is dangerous to humans. Few tarantulas have a toxic bite, despite popular mythology. The members of the cosmopolitan genus *Latrodectus*, which includes the black widows, are perhaps the most notorious of the poisonous species. *Latrodectus mactans* is widely distributed in the

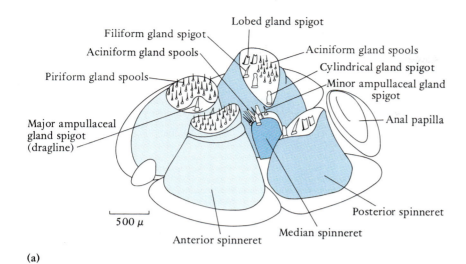

(a)

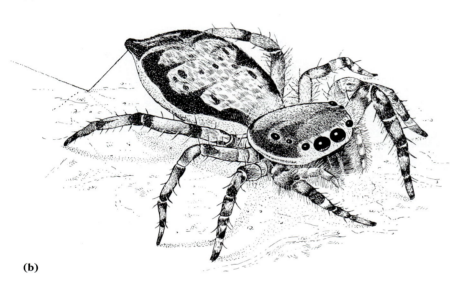

(b)

Figure 31.12
(a) Spinnerets of an orb-weaving spider showing distribution of spigots for different silk glands. (b) Jumping spider laying down a dragline. The arrangement of the eight eyes is a characteristic feature of this particular family of hunting spiders, which are capable of jumping by rapidly extending the legs.

United States (Fig. 31.13a). These spiders build webs in protected places, often in old lumber and trash around human habitations. Its venom, which contains seven different proteins, is neurotoxic, and symptoms include painful abdominal cramps. It would not ordinarily be fatal except to a small child, and if correctly diagnosed, it is easily treated. The brown recluse spider, *Loxosceles reclusa*, is another dangerous species most commonly reported in the midwestern and southeastern United States (Fig. 31.13b). Its bite produces an ulcerated wound that is difficult to heal.

The method of prey catching can be a basis for dividing spiders into two adaptive groups: **cursorial** and **web building spiders**. Cursorial spiders include the tarantulas, crab spiders (Fig. 31.13c), wolf spiders (Fig. 31.13d), jumping spiders (Fig. 31.12b), and others. They spin silk draglines, nests, and cocoons, but they do not use

silk to capture prey. Rather, the prey is stalked, pounced upon, and bitten. Cursorial spiders generally have heavier legs and more highly developed eyes than do web builders.

Web-building spiders construct various types of webs to trap prey, the web type—e.g., orb, horizontal sheet, funnel, irregular mesh—being characteristic of particular families. The first webs are believed to have been the irregular draglines laid down around the retreat of some primitive species. Among living spiders, the orb web is perhaps the most familiar type and the one most commonly associated with spiders.

Web-building spiders are aerialists and have rather slender legs for climbing about the silken lines (Fig. 31.13a). Eyesight is poor, but web builders are able to detect and interpret the various vibrations of the web with great facility. Web vibrations inform an orb weaver, for

(a)

(b)

(c)

(d)

Figure 31.13
Spiders. (a) Female black widow, *Latrodectus mactans*. The red hour glass marking on the underside of the abdomen is a distinguishing feature of the species. (b) Brown recluse spider. Note the violin-shaped marking on the dorsal side of the carapace. (c) A crab spider, a common cursorial spider, which catches insects visiting flowers. (d) A female wolf spider carrying egg case.

example, about the size of the struggling prey and whether it is securely caught. The spider approaches the prey and gives it a fatal bite, sometimes swathing it in silk before or after the bite.

Arachnids are unusual in that most begin digestion of their prey outside their bodies. While the tissues are torn by the chelicerae, enzymes are secreted by the midgut, passed forward through the foregut, and poured out of the mouth into the prey. The partly digested tissues are then sucked in by the pumping action of a part of the foregut. Digestion is completed and absorption occurs in the mid-gut, which may be greatly evaginated and ramify into various parts of the body (Fig. 31.8).

There are some exceptions to the predatory feeding habit of most arachnids. Harvestmen are omnivores and feed upon vegetable material and dead animal remains in addition to live invertebrates. The greatest diversity in feeding is displayed by mites. Some are predatory; others are herbivorous and have mouthparts adapted for piercing the cells of plants and sucking out the contents. Certain species of spider mites can be very destructive to plants. Some mites feed on plant products, decomposing plant material, and fungi. Members of several groups of mites are scavengers. Hair and feather mites spend their lives on the skin of mammals and birds, where they feed upon sloughed skin cells, gland secretions, and fragments of hair and feathers.

A number of groups of acari are parasitic for all or part of their life cycle. Ticks are bloodsucking ectoparasites of reptiles, birds, and mammals, and the chelicerae are adapted for penetrating and anchoring into the skin of the host. A tick feeds during each instar and usually prior to egg laying. Some ticks stay on the same host, but many, including the common dog tick (Fig. 31.11a), drop off following feeding engorgement. They molt and then seek another host. Although there can be direct pathogenic responses in humans to the bite of ticks, transmission of diseases, such as Rocky Mountain spotted fever, tularemia, and Lyme disease, are usually more serious concerns.

Chiggers, or redbugs, are ectoparasites on terrestrial vertebrates during the posthatching instar. In feeding, the minute mite secretes enzymes that produce a deep well from which it sucks out the digested contents. The mite secretions produce an irritating reaction resembling a mosquito bite but lasting much longer. Following feeding the chigger falls off the host and is predaceous as an adult.

There are numerous species of mites that attack the skin of birds and mammals and can cause severe dermatitis in their hosts. For example, the itch mite *Sarcoptes scabiei* produces scabies in various mammals, including humans. The mites spend their entire lives on the host, and the female tunnels through the host's skin, depositing eggs in the burrows.

Gas Exchange

Large arachnids have book lungs for gas exchange; small arachnids have tracheae (p. 278). Thus scorpions have book lungs, four pairs with slitlike openings on the ventral surface of the anterior abdominal segments. Pseudoscorpions, harvestmen, and mites possess tracheae. Primitive spiders, such as the large tarantulas, have two pairs of book lungs, but most spiders have one pair of book lungs and one pair of tracheae. Some very small spiders have only tracheae. The openings of both types of gas exchange organs are located on the ventral side of the abdomen (Fig. 31.8). Heart and blood vessels are best developed in those arachnids that possess book lungs.

Reproduction and Development

The paired gonads are located in the abdomen, and in both sexes a median gonopore opens onto the anterior ventral surface (Fig. 31.8).

Indirect transfer of spermatophores appears to be a primitive condition in arachnids and perhaps represents the early arthropod solution to the problem of sperm transmission on land. Spermatophore transfer in scorpions is preceded by a "courtship" dance, during which the large pedipalps of the male are locked with those of the female (Fig. 31.14). In the course of the dance, the male deposits a spermatophore on the ground and then maneuvers the female so that it is taken up into her gonopore.

Pseudoscorpions also utilize spermatophores and exhibit a wide range of behavioral modifications that increase the likelihood of the female finding and picking up the spermatophore. Our knowledge of reproductive behavior in mites is still relatively poor. Some species transfer sperm indirectly by spermatophores; most transfer sperm directly, utilizing a penis. Sperm transfer is also direct in harvestmen.

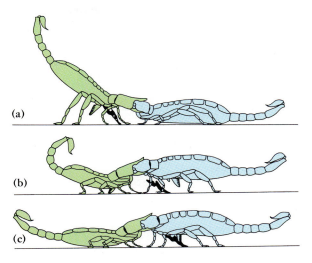

Figure 31.14
Sperm transfer in scorpions. (a) While holding the female's pedipalps with his own, male on left deposits spermatophore (black) on ground. (b) Female is pulled over spermatophore. (c) Spermatophore is taken up into female gonopore.

The process of sperm transfer in spiders is remarkable and is paralleled in few other animals. The copulatory organs of the male are the ends of the pedipalps, which resemble a pair of boxing gloves (Fig. 31.15). Prior to mating, the male spins a tiny web on which a droplet of semen is secreted. The two pedipalps are then dipped into the droplet until the semen is taken up within the reservoir of the palpal organ. The male now seeks a female.

The male is frequently smaller than the female, and the predatory nature of spiders makes it important for the male to ensure that the female does not mistake him for potential prey. Complex precopulatory behavior patterns have evolved utilizing visual, tactile, and chemical signals. Significantly, there is considerable difference in the precopulatory behavior of cursorial and web-building spiders, for the sensory cues that are important in prey catching are also the important ones in sex recognition and mating. Thus web vibrations are an important means by which the male signals a female in web-building spiders, and visual signals, such as posturing by the male, are important in many species of cursorial spiders.

Following various forms of precopulatory contact by the male, the palpal organ is locked onto the chitinous plate containing the female reproductive openings, and the ejaculatory process is inserted into the seminal receptacles. Sperm is transferred from one palp at a time (Fig. 31.15b). This unusual mode of sperm transfer in spiders probably had its origins in transfer by spermatophore. The male of some arachnid ancestral to the spiders may have used the palp to place a spermatophore into the female gonopore. In fact, there are certain living arachnids be-

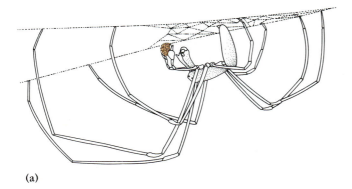

(a)

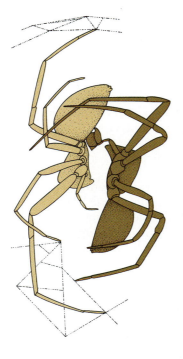

(b)

Figure 31.15
(a) Male tetragnath spider in sperm web, filling palps from globule of semen. (b) Mating position of *Chiracanthium* (male is the darker of the two).

Figure 31.16
A Costa Rican scorpion carrying her young.

their eggs within a membranous sac overlying the gonopore.

Most arachnids have direct development, and the young at hatching or at birth resemble the adult form.

A Related Class

Class **Pycnogonida** consists of a small group of long-legged marine arthropods called sea spiders (Fig. 31.17). They are like other chelicerates in lacking antennae and in having chelicerae and palps; but there may be four, five, or six pairs of legs, and there is an additional pair of append-ages (ovigerous legs), located behind the palps, that are

Figure 31.17
A sea spider, *Nymphon orcadense* (class Pycnogonida). The proboscis is paralleled by the small palps to the right.

longing to a small tropical and semitropical order in which the male uses his pedipalp to push the spermatophores into the gonopores of the female.

Harvestmen and free-living mites deposit their fertilized eggs in soil, in leaf mold, or beneath bark, but spiders place their eggs in silk cases, which are then usually left beneath stones, bark, or leaf mold or are attached to vegetation.

Brooding is common. Wolf spiders and fisher spiders carry their egg cases about with them (Fig. 31.13d). After hatching, wolf spiderlings are carried on the back of their mother. The eggs of scorpions develop within the body of the female. Following birth, the young are carried about on the female's back (Fig. 31.16). Pseudoscorpions brood

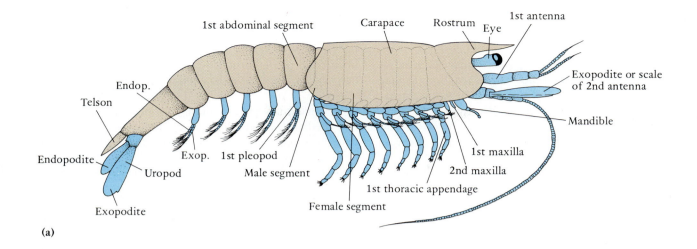

(a)

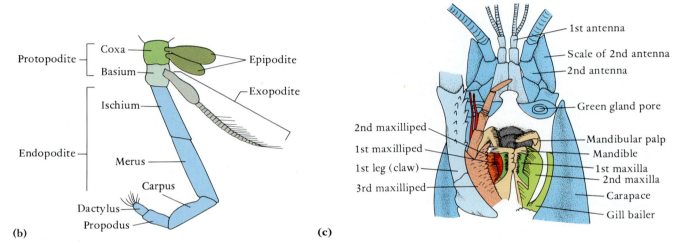

(b)   **(c)**

Figure 31.18
A generalized malacostracan crustacean. (a) Lateral view of body. (b) A thoracic
appendage. (c) Ventral view of the mouthparts of a crayfish. Maxillipeds (in shades
of red) have been removed from the left side to show the maxillae (shades of
green). The mandibles (beige) are large and heavy and bear overlapping teeth.
Each mandible also bears a short anterior finger-like palp. The median edge of the
flat gill bailer, which produces the ventilating current and is a part of the 2nd
maxilla, can be seen projecting from the front of the branchial chamber.

used in grooming and carrying eggs in the male.
Pycnogonids are commonly found slowly crawling over
sessile animals, such as hydroids and bryozoans. They feed
on the polyps or zooids of these animals or on the organic
detritus that accumulates on their surfaces.

SUBPHYLUM CRUSTACEA

The subphylum Crustacea contains the shrimp, lobsters,
crabs, barnacles, water fleas, and numerous other groups.
In contrast to the arachnids and insects, which are terrestrial, the crustaceans are primarily aquatic. Most are marine, but many species live in fresh water, and some,

principally the sow bugs, are adapted for a terrestrial
existence.

Crustaceans differ from all other arthropods in possessing two pairs of **antennae**. The first pair is probably
homologous to the antennae of insects, centipedes, and
millipedes, (p. 750), but the second pair is unique to
crustaceans. The feeding appendages, located just behind
the mouth, are a pair of heavy biting mandibles and two
pairs of small maxillae. Thus the head appendages, which
are constant for all members of the subphylum, are two
pairs of antennae, one pair of **mandibles**, and two pairs of
maxillae (Fig. 31.18a and c).

The trunk of crustaceans varies greatly from one class
to another. A carapace formed from a posterior fold of the

head is commonly present (Fig. 31.19). It may cover only a small part of the dorsal surface of the trunk, or it may greatly overhang the sides of the body. In some crustaceans, the entire body is enclosed within a carapace. The terminology of crustacean appendages is based on the assumption that, primitively, each appendage is a simple two-branched (**biramous**) structure. The outer branch (**exopodite**) and inner branch (**endopodite**) are attached to a basal piece (**protopodite**) (Fig. 31.18b). The trunk appendages have been adapted for a wide range of functions.

Crustaceans possess either compound eyes or a simple median eye, but rarely both. The median or **naupliar eye** is a little cluster of three or four pigment cups containing photoreceptors. It is characteristic of the naupliar larva (see Fig. 31.20), but in some groups the larval eye is retained in the adult. The naupliar eye functions primarily in orientation.

The crustacean epicuticle lacks the waxy waterproofing of arachnids and insects, and the procuticle usually contains calcium carbonate. The larger crustaceans are often brightly colored, the pigment being located in the endocuticle. The red color of boiled lobsters and crabs results from the denaturing of the protein portion of the pigment astaxanthin, which is bluish or green in the living state. The larger crustaceans also possess chromatophores. Pigment movement in the chromatophores enables the animal to adapt to the color of its background; a common change is simple darkening or lightening.

Although crustaceans have a wide range of diets and feeding habits, suspension feeding is very common, especially among small species. The collecting device is formed by closely spaced setae on certain head or trunk appendages.

Gills are typical of all of the larger crustaceans but they vary in structure and position from one group to another. The excretory organs are one or two pairs of end sacs, each with an excretory duct that opens onto the base of the second antenna or second maxilla or in a few crustaceans both (p. 351).

A number of generalizations can be made about crustacean reproduction. Copulation is the rule, and certain appendages are adapted for clasping the female and transferring sperm. Fertilization may be internal or it may

Figure 31.19
Decapods. (a) Shrimp. (b) A slipper lobster from Hawaii. There are no large chelipeds and the second antennae are large, short, and flat. (c) Hermit crab in a whelk shell.

(a)

(b)

(c)

Figure 31.20
Naupliar larva of a crustacean.

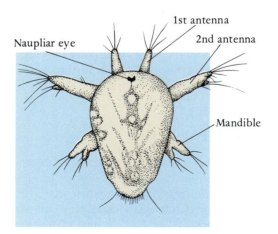

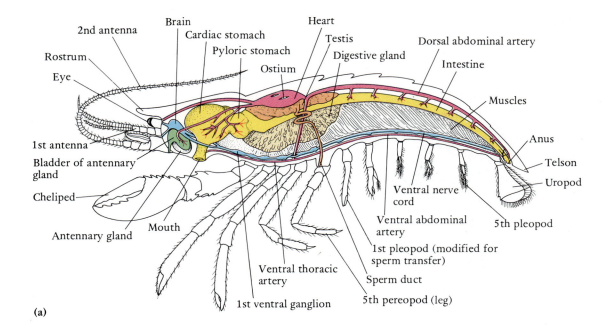

(a)

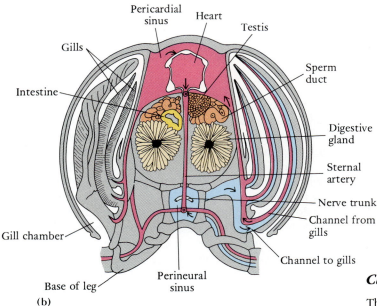

(b)

Figure 31.21

(a) Internal structure of a crayfish (lateral view). (b) Cross section of a crayfish just behind third pair of legs.

occur externally at the time of egg deposition, and the eggs are usually brooded on certain parts of the body. The earliest hatching stage, the **nauplius** (or **naupliar larva**), possesses only the first three pairs of appendages (Fig. 31.20). The gradual acquisition and differentiation of additional segments and appendages occur in subsequent developmental instars.

The structural diversity of crustaceans is very great, more so than in any other group of arthropods. While there are far more species of insects than crustaceans, the insect ground plan is relatively uniform. This is not true of crustaceans, and you will see when we have completed this brief survey that it is very difficult to describe a "typical" crustacean.

Class Malacostraca

The subphylum Crustacea is divided into classes, the largest of which is the **Malacostraca**. In these crustaceans, the trunk is composed of a thorax of eight segments, which bear the maxillipeds and legs, and an abdomen of six segments, which usually carry five pairs of biramous **pleopods** and a terminal pair of flattened **uropods** (see Fig. 31.18a).

Decapods

The **Decapoda**, the largest of the crustacean orders, contains over 10,000 species. It also contains the biggest and most familiar crustaceans: shrimp, lobsters, crayfish, and crabs (Fig. 31.19). The name Decapoda refers to the five pairs of legs, including the first pair, which are frequently modified as large claws (called chelipeds) (Fig. 31.21). In front of the legs are three pairs of smaller, forward directed appendages called **maxillipeds**, which, like the maxillae,

731

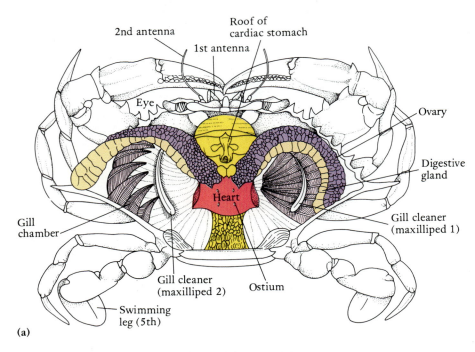

Anatomy of the blue crab, *Callinectes sapidus*, a portunid swimming crab. (a) Dorsal dissection.

Figure 31.22
Anatomy of the blue crab, *Callinectes sapidus*, a portunid swimming crab. (a) Dorsal dissection. (b) Frontal view, showing two forms of claws. The crushing claw is on the crab's right, and the cutting claw is on the left. Between the claw tips are the large third maxillipeds covering the other mouthparts.

function in food handling (Fig. 31.22b). The three pairs of maxillipeds and the five pairs of legs are appendages of the anterior trunk region, called the **thorax**. The thorax is covered by a carapace, and the head-thorax region together is usually called the **cephalothorax**. The front of the cephalothorax carries a pair of compound eyes that in decapods are located at the end of stalks (Fig. 31.21a).

Primitively, the abdomen is large, as in shrimp, lobsters, and crayfish, and carries six pairs of biramous appendages. The last pair, or **uropods**, are flattened and form a tail fan with the terminal telson (Fig. 31.21a). The first five pairs are called **swimmerets** or **pleopods** (Gr. *plein*, to swim + *pous*, foot). In a number of different decapod groups the abdomen is reduced and folded beneath the cephalothorax, resulting in a short-body form, called a **crab** (Fig. 31.23).

Most decapods are bottom dwellers, but many shrimp can swim, using their abdominal pleopods. The best swimmers, however, are the portunid crabs, which have the fifth pair of legs adapted as paddles (Fig. 31.22a).

Brachyuran crabs, the largest group of crablike forms, crawl sideways. In these crabs the evolution of the short-body form shifts the center of gravity forward beneath the cephalothorax and probably represents an adaptation for greater motility in their sideways gait.

The abdomen of hermit crabs is housed within an empty gastropod shell (Fig. 31.19c), which is gripped with the uropods and the fourth and fifth pairs of legs. The soft twisted abdomen bears reduced appendages. When the hermit crab becomes too large, it finds another shell. Only empty shells are used. Hermit crabs probably evolved from forms that backed the body into crevices and other retreats.

Decapods possess gills that project upward from near the base of the thoracic appendages and are enclosed within a protective gill chamber (Figs. 13.11, 31.21b and 31.22a). The ventilating current is produced by the rapid sculling motion of the **gill bailer**, a semilunar process of the second maxilla (see Fig. 13.12).

Several groups of crabs have invaded the land with varying degrees of success. Amphibious fiddler crabs (*Uca*) burrow in the intertidal zone. At high tide the crab remains within its flooded burrow, but at low tide it emerges from the burrow to feed. The related ghost crabs (*Ocypode*) are more terrestrial. These crabs live above the high tide mark and in many areas are common inhabitants

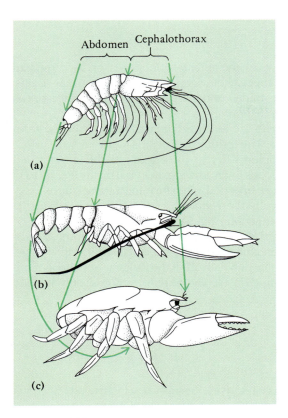

Figure 31.23

(a–c) Evolution of the short body form in decapod crustaceans. In forms with long bodies, such as shrimp (a) and lobsters (b), a large abdomen extends posterior to the cephalothorax. In forms with short bodies, such as crabs (c), the abdomen is reduced and folded beneath the cephalothorax.

of dunes. In the tropics, land crabs live in forests and thickets well back from the sea, as much as hundreds of miles inland. Almost all terrestrial species continue to use the gills as organs of gas exchange, which may explain the restriction of land invasions to crabs, for only in these decapods is the gill chamber sufficiently closed to make possible the retention of the moisture needed for gas exchange. Most land crabs are nocturnal and remain in burrows during the day.

Nutrition   Although the majority of decapods are predators or scavengers, there are many exceptions. Some marine species feed on algae, and recently fallen fruits and leaves are an important food source for many land crabs. The chelipeds are used in collecting and handling food in most decapods. Depending upon the species, these appendages may be adapted for scraping, tearing, cutting, or crushing, and some crabs that feed on snails and mussels have one claw adapted for crushing and one for cutting (Fig. 31.22b).

Some decapods are filter feeders. The mole crabs *Emerita talpoidea* found on surf beaches on the east coast of the United States and *Emerita analoga* on the west coast

are good examples. Mole crabs are washed up by the incoming wave. They then rapidly burrow backward in sand facing seaward and project their plumose antennae to filter the current of the receding wave with their second antennae (Fig. 31.24a).

The fiddler crabs feed on fine organic matter deposited onto the surface of the intertidal zone of protected beaches, estuaries, and marshes (Fig. 31.24b). A mass of

Figure 31.24

(a) A mole crab buried in sand, lateral view. The biramous first pair of antennae form a screening siphon for the ventilating current. The setose second antennae are extended into water current (*arrow*) and function in filter feeding. Eyes are on long stalks. (b) Fiddler crabs (*Uca minax*) from Delaware Bay. These amphibious crabs are deposit feeders in the large intertidal zone at low tide. The male has a large claw that is waved in a species-specific pattern in courtship.

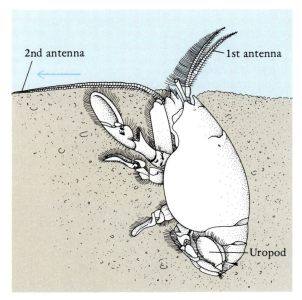

(a)

(b)

733

sand or mud is placed in the mouthparts, which are then flooded with water pumped forward from the gill chambers. Organic material is floated off and separated by fine setae on the feeding appendages. The remaining mineral material is ejected as a small spitball.

The decapod digestive tract is complex. The foregut includes a very large chitin-lined **cardiac stomach** containing one dorsal and two lateral teeth that form a **gastric mill**. The cardiac stomach opens posteriorly into the **pyloric stomach**, which ventrally contain the openings of the large, extensive pair of digestive glands, also called the **hepatopancreas** (midgut) (Figs. 31.21a and 31.22a). If you have opened steamed crabs, the conspicuous soft, cream-colored material you see, sometimes called "mustard," is the digestive gland.

The cardiac stomach functions as both a crop and a gizzard. The grinding action of the gastric mill plus the action of enzymes passed forward from the digestive glands reduces the food masses to fluid and fine particulate matter, which is then conducted along channels through the pyloric stomach to the digestive glands. Screens of setae prevent the passage of coarse particles. Absorption takes place in the digestive glands.

Excretion and Integration   The excretory organs of decapods are antennal glands, also called **green glands** (see p. 351 and Fig. 16.6). The organ consists of a blood-bathed sac in the head and a duct that opens to the outside by a pore at the base of the second antennae (Fig. 31.18c). Most decapods are marine and are osmoconformers, but some shrimp and crabs and the widely distributed crayfish inhabit estuaries and freshwater. The osmoregulatory ability of these species is described on page 362.

The location of the compound eyes at the end of stalks in decapods provides for a wide visual arc. Object discrimination by the eyes is well developed in many decapods, especially land crabs. In the base of each first antenna is a statocyst that opens by a pore to the exterior. These animals use sand grains as statoliths and must replace them following each molt (p. 408).

The hormones and neurosecretions of crustaceans are perhaps better known than those of any other invertebrates. There is experimental evidence of hormones that regulate chromatophores, molting, growth, sexual development, and reproduction. A number of hormones, such as those controlling chromatophores and molting, are released from a gland in the eye stalk (p. 441).

Reproduction   The single pair of testes or ovaries is located in the dorsal part of the thorax. The male sperm ducts open at the base of the fifth pair of legs (Fig. 31.21a). The oviducts open to the exterior at the base of the third pair of legs. During copulation the male assumes various positions astride the female and, using the greatly modified anterior two pairs of pleopods, transfers sper-

matophores to the female gonopores or to a median seminal receptacle between the fourth or fifth pair of legs. The tightly folded abdomen of brachyuran crabs opens during copulation.

The eggs are fertilized internally or on release from the oviduct. The fertilized eggs pass to the ventral surface of the abdomen, where in most decapods they are attached to the pleopods by an adhesive material on the egg surface. Here they are brooded, often forming a conspicuous mass. Even in crabs the reduced folded abdomen retains its brooding function (Fig. 31.25a).

Some shrimp hatch as a nauplius, but in most decapods the hatching stage is a later planktonic larval stage, called a **zoea**, in which all of the thoracic appendages are present (Fig. 31.25b). The adult form is gradually attained

Figure 31.25
(a) Female crab carrying eggs attached to pleopods. The right cheliped is missing. (b) A zoea larva of a crab.

(a)

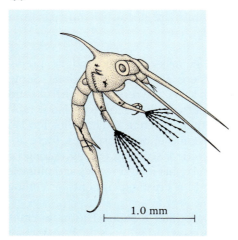

1.0 mm

(b)

through a series of instars, and the young shrimp, lobster, or crab settles to the bottom and takes up the adult mode of existence.

Copulation and brooding occur on land in terrestrial crabs, but all must go back to the sea (or to fresh water in some tropical families) to permit hatching of the eggs. Direct development takes place in most freshwater decapods, including the crayfish. In these species, the eggs are brooded on the abdomen throughout development.

Amphipods and Isopods

The two largest orders of malacostracans other than decapods are the **Amphipoda** and **Isopoda**. These malacostracans, most of which are only about a centimeter in length, share a number of features. The compound eyes are on the sides of the head and not on stalks, as in decapods. No carapace is present, and the thoracic and abdominal regions are not sharply demarcated on the dorsal side (Fig. 31.26). Of the eight pairs of thoracic appendages characteristic of all malacostracans, seven pairs are legs and one pair are food- handling maxillipeds. In both orders development is direct, and the young are brooded beneath the thorax in a chamber, the **marsupium**, formed by inward shelflike projections from the bases of the legs (Figs. 31.27a and 31.28b).

Isopods tend to be dorsoventrally flattened, and the pleopods are modified for gas exchange (Fig. 31.27a). Amphipods, in contrast, are laterally flattened and look somewhat like shrimp (Fig. 31.26). The gills of amphipods are simple processes from the bases of the legs.

The majority of isopods and amphipods are marine, and most are bottom dwellers, swimming only in-termittently with the pleopods. They live in bottom debris, in algae, and among sessile animals and often occur in large numbers. One common group of amphipods living on sessile animals are skeleton shrimp (Fig. 31.28a). Many amphipods burrow in sand or mud, and a number are tube dwellers. Some tube dwellers even carry the tube about with them. There are also some burrowing isopods, including a few that bore into wood. Most isopods and amphipods are omnivores and scavengers, but there are a considerable number of marine isopods that are parasitic on fish and other crustaceans.

Both orders include freshwater species, and both groups have also invaded land, but most terrestrial amphipods are limited to algae and the undersurfaces of boards washed up on the beach. Such amphipods are called beach fleas because they can jump.

The isopods known as pill bugs, also called sow bugs, wood lice, or roly-polies (Fig. 31.27c), represent the only truly successful crustacean invasion of land, which we believe they invaded directly from the sea, rather than by way of fresh water. Their success is in part related to their small size, which enables them to exploit protective terrestrial microhabitats, and to their brooding and direct development, which eliminates the need of an aquatic environment for reproduction, as is true of most land crabs. These isopods are widely distributed in both temperate and tropical regions, where they are commonly found beneath stones and wood and in leaf mold. They reduce desiccation by living in protected habitats, by avoiding light, and by the ability of some to roll up into a ball, covering the less highly sclerotized ventral surface. They have retained their gills (pleopods) but eliminate nitrogenous wastes as gaseous

Figure 31.26
Male of the amphipod *Gammarus*.

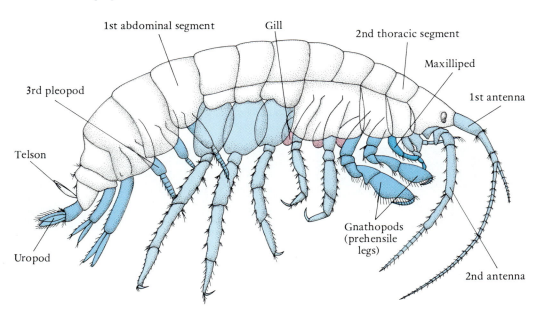

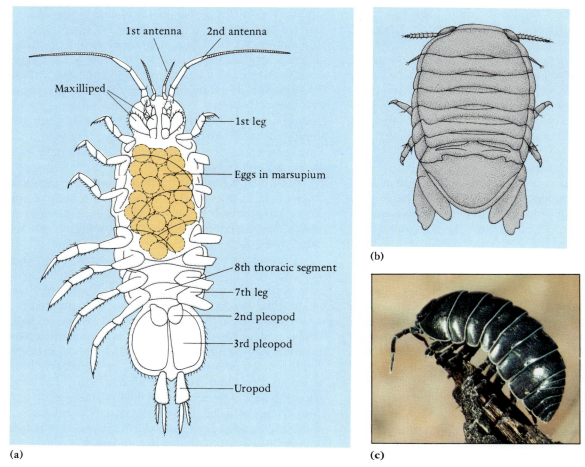

(a)

1st antenna   2nd antenna

Maxilliped

1st leg

Eggs in marsupium

8th thoracic segment

7th leg

2nd pleopod

3rd pleopod

Uropod

(b)

(c)

Figure 31.27

Isopods. (a) The freshwater isopod *Asellus* (ventral view). (b) Dorsal view of *Sphaeroma quadridentatum*, a common shallow-water marine isopod found on algae and pilings. All but one abdominal segment are fused with the telson. The members of this genus are capable of rolling into a ball, as are many terrestrial isopods. (c) A terrestrial isopod or pill bug on a twig. Head with conspicuous second antennae is to the left. The seven pairs of legs are visible.

Figure 31.28

(a) A skeleton shrimp, a caprellid amphipod on a piece of algae. The abdomen is greatly reduced. They are common on hydroids and bryozoans, which they slowly crawl over or hang on to with their long legs. (b) Lateral view of a female isopod (*Porcellio*) brooding her embryos.

(a)

(b)

ammonia. They also brood their eggs in the marsupium, like their aquatic relatives.

Some isopods that burrow at the high tide mark and certain species of amphipod beach fleas are able to return to their habitat when they move down to the intertidal zone to feed or are displaced for other reasons. Orientation is achieved by sensing the slope of the beach, sand moisture, horizon height, and the angle of the sun.

Mantis Shrimps and Krill

Two other groups of malacostracans deserve mention. Mantis shrimps (order **Stomatopoda**) are common tropical and subtropical crustaceans that live in burrows or within rock crevices. They are about the same size as decapod shrimps but have the second thoracic appendage enlarged and modified for prey catching. The terminal piece of the appendage folds back onto the second piece, like a jackknife. The size and shape of the appendages are somewhat like the raptorial appendages of a preying mantis, hence the name. Some mantis shrimps are spearers, rapidly unfolding the appendage to capture soft-bodied prey, such as worms and fish. Others are smashers. The appendage is heavy and is not unfolded in prey capture, but rather is used like a club to smash such prey as molluscs and crabs (Fig. 31.29a). Some mantis shrimps are powerful enough to shatter a flat piece of glass.

Mantis shrimps have the most highly developed eyes of all crustaceans. The movable eye stalks are so oriented at the front of the head that depth perception (stereoscopic vision) is possible.

Krill (order **Euphausiacea**) are shrimplike crustaceans closely related to decapods (Fig. 31.29b). They are planktonic and are found in all of the world's oceans, but certain filter-feeding species occur in enormous numbers in Antarctic waters and are an important link in the food chain of that region. Swarms of Antarctic krill may cover an area equivalent to several city blocks, and a blue whale may consume a ton at one feeding.

Class Branchiopoda

Most of the nonmalacostracan crustaceans are small, usually less than a centimeter in length. The majority belong to one of four classes. The class **Branchiopoda** contains the water fleas, fairy shrimp, and brine shrimp. All possess flattened leaflike trunk appendages bordered by fine setae (Fig. 31.30b), but the body form is quite variable. Water fleas (cladocerans) look like plump little birds (Fig. 31.31). The trunk, but not the head, is enclosed within a folded carapace that greatly overhangs the body. The head bears large biramous second antennae and a single median compound eye. Fairy shrimp and brine shrimp have a long trunk with numerous appendages, and there is no carapace (Fig. 31.30a).

The flattened foliaceous appendages of branchiopods certainly contribute to gas exchange, but given the small size of most branchipods, they probably should not be called gills, as the name branchiopod implies. Feeding is a primary function of these appendages in most branchiopods. It had long been thought that the fine setae

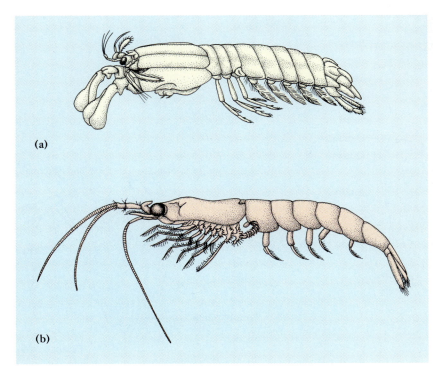

(a)

(b)

Figure 31.29

(a) A mantis shrimp (order Stomatopoda) with clublike maxillipeds adapted for smashing prey. (b) A krill (order Euphausiacea) is a planktonic shrimp-like malacostracan.

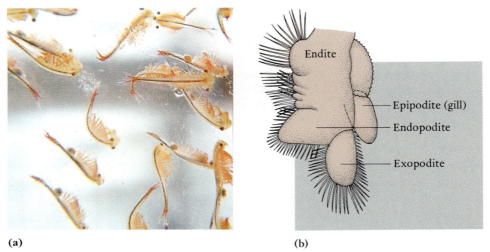

(a)   **(b)**

Figure 31.30
Fairy shrimp. These little crustaceans normally swim upside down. (a) Entire
animal. (b) A leaflike trunk appendage.

Figure 31.31
(a) Female of the cladoceran *Daphnia pulex* (lateral view). (b) Ventral view of
Daphnia.

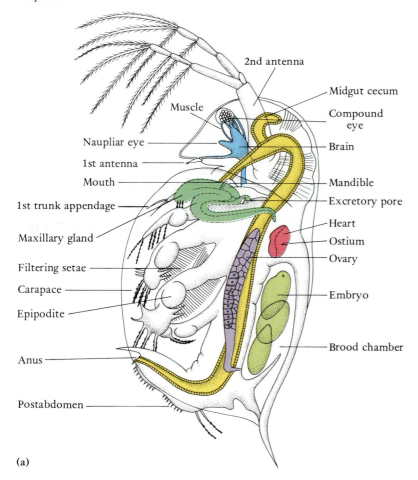

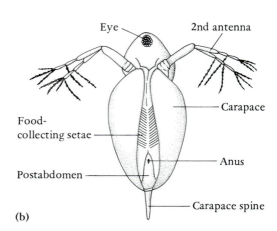

(a)   **(b)**

bordering these appendages functioned as filters, but recent studies have demonstrated that particles must be collected in some other way. Water does not actually pass between the setae, and if the animal is given polystyrene spheres of several sizes, some larger than the space between setae and some smaller, both sizes are collected (Fig. 31.32). Particles are probably collected by the setae because of the attraction of charges on their surfaces and not by sieving. Fairy shrimp also use the appendages for swimming, but water fleas swim with their antennae.

Figure 31.32
Particle collection in suspension-feeding water fleas. (a) Part of two setae, each bordered by setules (side branches). The space between setules is about 1 μm. (b) Polystyrene spheres collected by the setae. The diameter of the smallest is about one-half the distance between setules.

(a)

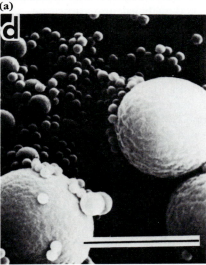

(b)

In contrast to most of the other classes of crustaceans, branchiopods are chiefly inhabitants of fresh water, especially ponds and temporary pools. Water fleas are an exception in including planktonic species that live in large lakes and also some that live in the sea. The brine shrimps (*Artemia*) are adapted for living in salt lakes and can tolerate salinities near the saturation point of salt.

Water fleas brood their eggs in the space beneath the carapace above the back of the trunk (Fig. 31.31a). Water fleas have direct development, but in other branchiopods the egg hatches as a naupliar larva. The retention of larval stages in freshwater branchiopods is an exception to the general rule of suppression of larval stages in fresh water (p. 522).

Branchiopods exhibit a number of reproductive and developmental adaptations to the environmental stresses common in freshwater lakes and ponds. Parthenogenesis is common. Many species produce thin-shelled eggs that hatch rapidly and thick-shelled eggs that remain dormant during periods of drought or freezing. These adaptations are the same as those exhibited by freshwater flatworms and rotifers.

Class Copepoda

The class **Copepoda** is the largest of the nonmalacostracan classes. The 7500 species are mostly marine, but there are also many freshwater forms. The copepod body, usually between 1 and 5 mm in length, is commonly cylindrical and tapered (Fig. 31.33). The head bears only a single median naupliar eye. The first pair of antennae are very large and are typically held at right angles to the body. The trunk is composed of an anterior thorax bearing biramous appendages and a narrower posterior abdomen, which lacks appendages but usually carries a pair of terminal processes. Thus the body of copepods looks a little like a bomb with a crossbar (the antennae) at the front end. As would be expected, given the small size of copepods, there are no gills.

Many copepods are planktonic and they are usually one of the most abundant and conspicuous animals in the plankton of the sea and freshwater lakes. Many species live near the bottom, where they swim or crawl about over debris, algae, and sea grasses. Pools and puddles often contain large numbers. Copepods may be interstitial (living in the spaces between sand particles) (see Connections 27.1, p. 622), and many are parasitic. The smaller second pair of antennae are the principal swimming appendages of many planktonic species; in others, the thoracic appendages are more important. The first antennae function largely to reduce sinking, acting somewhat like a parachute.

Although many copepods are grazers or are predatory, most planktonic forms are suspension feeders. Diatoms are the principal source of food, and copepods play

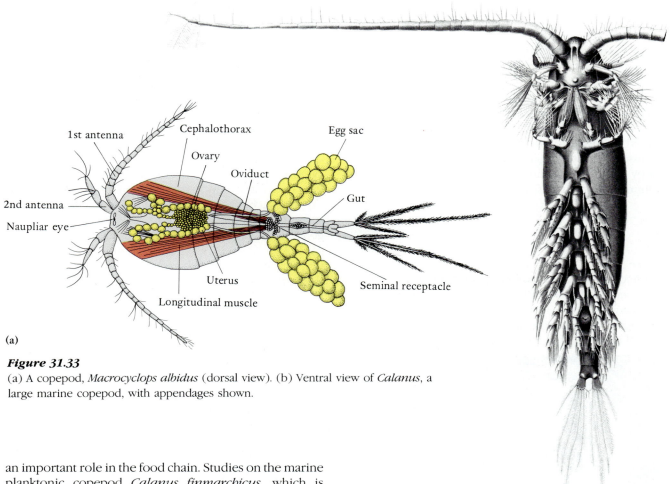

(a)

1st antenna
Cephalothorax
Egg sac
Ovary
Oviduct
2nd antenna
Gut
Naupliar eye
Seminal receptacle
Uterus
Longitudinal muscle

(b)

Figure 31.33
(a) A copepod, *Macrocyclops albidus* (dorsal view). (b) Ventral view of *Calanus*, a large marine copepod, with appendages shown.

an important role in the food chain. Studies on the marine planktonic copepod *Calanus finmarchicus*, which is 5 mm long, have shown that it may ingest as many as 373,000 diatoms every 24 hours. The diatoms are collected by the second maxillae, which clap together around the diatom.

About 25% of the copepod species are parasitic, attacking the skin, fins, and gills of marine and freshwater fish. All degrees of modification from the typical copepod body form exist, and some species are so aberrant that they no longer look like crustaceans. How do we know they are? They still have typical copepod larval stages.

Some copepods shed their eggs singly into the water, but many brood them within secreted ovisacs attached to the female gonopore (Fig. 31.33a). A naupliar larva is the hatching stage in both marine and freshwater forms. Dormant eggs and encysted developmental stages are produced by many freshwater species and may be transported from one body of water to another by the muddy feet and bodies of water birds and other animals.

Class Ostracoda

Members of the class **Ostracoda** are tiny marine and freshwater crustaceans with the body entirely enclosed within a bivalved carapace (Fig. 31.34a). Ostracods are sometimes called seed shrimp and parallel the bivalve molluscs in many ways. The two **valves**, usually 2 mm or

less in length, may have interlocking teeth and are held together dorsally by an elastic hinge. The valves are closed by a bundle of **adductor muscle fibers** extending transversely between the two valves (Fig. 31.34c). The chitinous skeleton is impregnated with calcium carbonate; as a consequence, ostracods have an extensive fossil record dating from the Cambrian. Over 7,000 living and 10,000 fossil species have been described (Fig. 31.34b).

Within the two valves, the ostracod body is mostly head, for the trunk and its appendages are greatly reduced. The two antennae are large, and the other head appendages are well developed.

There are some planktonic ostracods, but most species are benthic and scurry over the bottom, crawling and swimming with the large antennae. Along with copepods and water fleas, ostracods are very common crustaceans in freshwater ponds and small pools. Their feeding habits are diverse; there are carnivores, herbivores, scavengers, and filter feeders.

Eggs are released singly or brooded within the carapace. The juvenile has a bivalved carapace and molting is limited to the pre-adult instars.

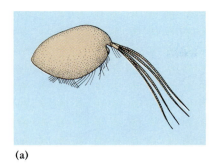

(a)

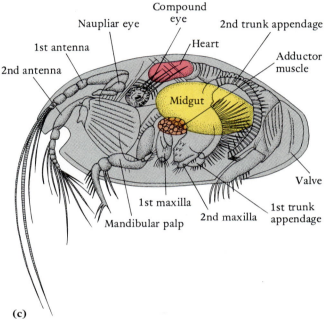

(c)

Figure 31.34

(a) A myodocopid ostracod with antennal notches in valves (lateral view). (b) A scanning electron micrograph of the valve of a fossil ostracod *Orionina*. (c) Female marine ostracod *Skogsbergia* with left valve and left appendages removed (lateral view). This is a marine scavenging carnivore that is able to swim over the bottom.

Class Cirripedia

The class **Cirripedia** contains the barnacles. These marine animals differ from other crustaceans in being attached to the substratum. Correlated with the sessile habit, the body is enclosed within a bivalved carapace covered with protective calcareous plates (Fig. 31.35).

If you can imagine an ostracod attached to the substratum by its antennae, and the valves covered by calcareous plates, you have some idea of the structure of barnacles. Indeed, the late larval stage of barnacles, called a **cypris larva**, looks very much like an ostracod. The cypris

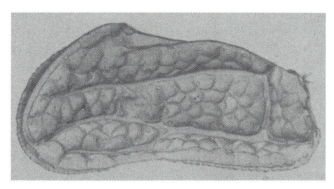

(b)

settles to the bottom and attaches head first by means of cement glands at the base of the first antennae.

All barnacles are attached, but the group may be divided into **pedunculate** (stalked) and **sessile** (stalkless) forms.* Stalked barnacles appear first in the fossil record and are believed to have given rise to the sessile types. The body of stalked barnacles is composed of a fleshy stalk, or **peduncle**, and a distal **capitulum** (Fig. 31.35). The peduncle represents the preoral part of the body and the capitulum the postoral part. It is the capitulum that is covered by the carapace, and the gape along one margin of the carapace can be opened and closed as in a clam.

Although some stalked barnacles are found on rocks, many species live attached to floating objects, such as timbers, or the bodies of swimming animals, such as whales, porpoises, sea turtles, sea snakes, crabs, and even jellyfish. Commensal species always exhibit some reduction in the size of their calcareous plates, since some protection is provided by the host.

The body of sessile barnacles, sometimes called **acorn barnacles**, consists mostly of the capitulum, for the peduncle is reduced to an attachment platform on which the capitulum rests (Fig. 31.36). The platform of sessile barnacles and the peduncle of stalked species both contain the antennal **cement**, or **adhesive glands**, indicating that the two structures are homologous. In sessile barnacles the basal plates of the capitulum form a rigid circular wall surrounding the movable lidlike terga and scuta.

Most sessile barnacles have become adapted for a life on rock or other hard substrata, and the heavy, somewhat fused ring of wall plates probably protects them against currents, pounding waves, and browsing fish. In the intertidal zone of temperate regions, barnacles commonly occur in enormous numbers, different species predominating at different levels between the high- and low-water mark (see Fig. 29.13). Like stalked barnacles, some sessile

---

*Unfortunately, the term sessile here means stalkless, not attached. All barnacles are attached.

Labels on figure (c): Naupliar eye, Compound eye, 2nd trunk appendage, 1st antenna, Heart, 2nd antenna, Adductor muscle, Midgut, Valve, 1st maxilla, 1st trunk appendage, Mandibular palp, 2nd maxilla, appendage

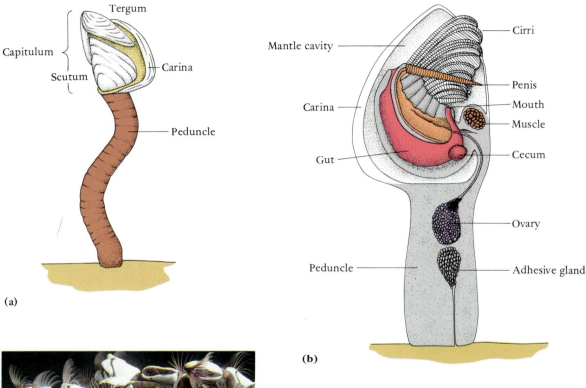

(a)

(b)

(c)

Figure 31.35

Lepas, a stalked barnacle: (a) External structure; The scutum, tergum, and carina are calcareous plates. (b) Internal structure. (c) A species of *Lepas* with extended cirri. Plates are white and stalk is greatly extended.

barnacles have become adapted for living on other animals, such as crabs and whales (Fig. 31.37).

The tendency of barnacles to utilize other animal bodies as substrata probably led to parasitism. Approximately one-third of them are parasitic and are so highly modified that they are recognizable as barnacles only by their larval stages. Other crustaceans are the principal hosts.

Barnacles are filter feeders, and the trunk bears six pairs of long, coiled, biramous appendages called **cirri**, from which the class name, Cirripedia, is derived. In feeding, the cirri unroll and project through the gape of the carapace as a large basket (Fig. 31.35c). The cirri perform a rhythmic scooping motion or the barnacle holds the cirri out like a fan, and the many fine setae of the appendages remove planktonic organisms.

Neither compound eyes, gills, heart, nor blood vessels are present. Correlated with their attached existence is the fact that barnacles, unlike most other arthropods, are hermaphroditic. At copulation, a long, extensible, tubular penis is projected from the gape of one individual into the carapace cavity of a neighboring barnacle. Fertilization and brooding occur within the carapace cavity, and a nauplius is the usual hatching stage. Following a number of naupliar instars (Fig. 31.38a), the cypris larva develops and is the stage at which settling occurs (Fig. 31.38b). Barnacles molt like other crustaceans but they do not shed the outer surface of the carapace bearing the calcareous plates. The plates increase in size by the secretion of additional calcium carbonate at their margins.

Barnacles are major fouling organisms on pilings, sea walls, buoys, and ship bottoms. Much research has been

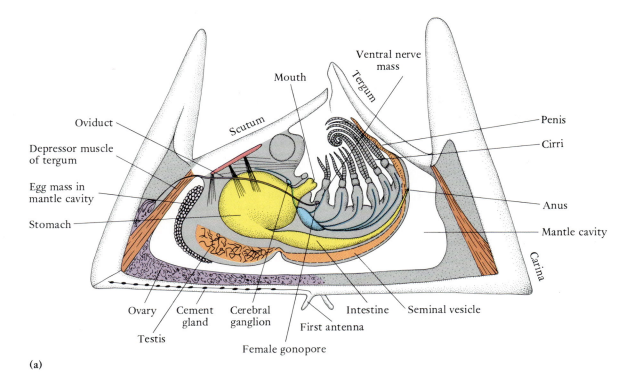

(a)

(b)

Figure 31.36
Sessile barnacles. (a) Diagram of the internal structure of a sessile barnacle, lateral view. (b) Cluster of sessile barnacles on a rock. The movable terga and scuta can be seen recessed within the opening of the ringlike wall of plates.

Figure 31.37
Commensal barnacles (*Coronula*) on a whale.

expended to develop antifouling measures, such as antifouling paint, for a badly fouled ship may have its speed reduced by as much as 35%.

Class Remipedia

Zoologists exploring marine caves in the Bahama, Turks and Caicos, and Canary Islands have collected what are probably the most primitive known crustaceans. First described in 1981 and placed in a new class, the Remipedia, these little crustaceans have long segmented bodies, each segment bearing a similar pair of appendages (Fig. 31.39). They resemble a polychaete annelid. Unlike most other crustaceans, their trunk appendages are all alike. Like the deep sea molluscan monoplacophorans, living remipedians may represent a remnant of a group of crustaceans that were more widely distributed in the past.

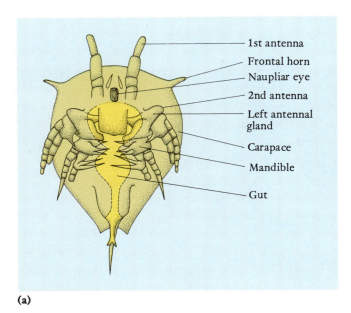

(a)

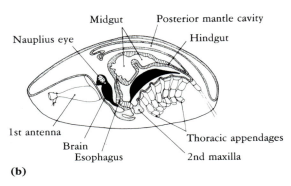

Labels: Nauplius eye, Midgut, Posterior mantle cavity, Hindgut, 1st antenna, Brain, Esophagus, Thoracic appendages, 2nd maxilla

(b)

Figure 31.38

(a) Naupliar larva of *Balanus*; setae omitted. (b) Cypris larva of *Balanus*; side view with left half of carapace removed.

Figure 31.39

Lasionectes entrichomas, a member of the newly discovered, very primitive class Remipedia. Note the long, wormlike body with numerous pairs of similar appendages.

Crustacean Radiation

At the outset of this section on crustaceans, we pointed out how difficult it is to describe a "typical" crustacean. Now you should have some appreciation of the great diversity of this subphylum. Crustaceans differ greatly in trunk design. Compare, for example, water fleas, copepods, and crayfish. They exhibit many different life styles: pelagic, benthic, burrowing, tube dwelling, and interstitial. They range in size from interstitial copepods less than 1 mm in length to spider crabs with a cheliped span of 4 m (i.e., from the tip of one cheliped to the tips of the other).

The great diversity of arthropods can be illustrated by noting the principal groups that are encountered in marine habitats. Most crustaceans are marine and benthic. Soft bottoms, sea grass beds, and algae are host to shrimps, crabs, isopods, and amphipods. Copepods and ostracods can be found swimming and scurrying over the sand surface, and some copepods live between the sand grains. Beaches are home to burrowers like mole crabs and to some burrowing amphipods, and to others (beach fleas) that live beneath drift at the high-tide mark. Barnacles are conspicuous crustaceans of rocky coastlines and tide pools, and subtidal rocks form protective retreats for many shrimps, crabs, and lobsters. Copepods, water fleas, and some shrimps can be found in plankton. Plankton also contains the larval stages of many benthic crustaceans.

The principal colonizers of freshwater habitats have been the small crustaceans: fairy shrimps, water fleas, copepods, amphipods, and some isopods. Crayfish are the most successful freshwater decapods, but there are some freshwater shrimps and crabs. Isopods known as pill bugs are the principal land-dwelling crustaceans.

First figure labels: 1st antenna, Frontal horn, Naupliar eye, 2nd antenna, Left antennal gland, Carapace, Mandible, Gut

Classification of Chelicerates and Crustaceans

Subphylum Chelicerata (Gr. *cheile*, claw + *keras*, horn)   Arthropods having the body divided into a cephalothorax and abdomen. Antennae lacking; first cephalothoracic appendages are chelicerae, followed usually by a pair of pedipalps and four pair of walking legs. Abdomen without appendages.

Class Merostomata (Gr. *meros*, part + *stoma*, mouth)   Horseshoe crabs. Marine chelicerates having book gills. *Limulus*.

Class Arachnida (Gr. *arachne*, spider)   Terrestrial chelicerates having book lungs and/or tracheae as gas exchange organs.

Order Scorpionida (Gr. *skorpios*, scorpion)   Scorpions. Large chelate pedipalps and long segmented abdomen with terminal sting. *Androctonus, Centruroides*.

Order Pseudoscorpionida (Gr. *pseudes*, false + *skorpios*, scorpion)   Tiny arachnids with large pedipalps and short segmented abdomens. *Chelifer*.

Order Araneae (L. *aranea*, spider)   Spiders. Abdomen unsegmented and attached to cephalothorax by a narrow waist. *Araneus, Latrodectus, Loxosceles, Lycosa, Phidippus*.

Order Opiliones (L. *opilion*, shepherd)   Harvestmen, or daddy-longlegs. Legs usually very long; the short, segmented abdomen is broadly joined to the cephalothorax.

Acari (collective common name for three orders) Mites and ticks. Small arachnids with unsegmented abdomen broadly fused with cephalothorax.

Order Opilioacariformes (L. *Opilio*, genus of harvestmen + *acarus*, mite + *forma*, form)   Large, brightly colored, primitive mites with segmented abdomens. Predatory and omnivorous species.

Order Parasitiformes (Gr. *parasitos*, parasite + L. *forma*, form)   Free-living and parasitic species, including the chiggers and ticks. Medium-size to large mites. Tracheal system with ventro-lateral spiracles. *Ioxdes, Dermacentor, Trombicula*.

Order Acariformes (L. *acarus*, mite + *forma*, form)   Free-living and parasitic species, including spider mites, follicle mites, water mites, mange mites, storage mites, beetle mites. Small mites. Spiracles near mouth parts or spiracles absent. *Sarcoptes, Dermatophagoides, Belba*.

Class Pycnogonida (Gr. *pyknos*, thick + *gony*, knee)   Sea spiders. Marine chelicerates with a narrow body and greatly reduced abdomen. Four, five, or six pairs of legs may be present, and there is a pair of ovigerous legs located behind the palps. *Nymphon*.

Subphylum Crustacea (L. *crusta*, hard surface)   Head with two pairs of antennae; first postoral appendages are a pair of mandibles, followed by two pairs of maxillae. Appendages primitively branched. Mostly aquatic arthropods.

Class Remipedia (L. *remus*, oar + Gr. *pous*, foot)   Primitive crustaceans having long segmented bodies, bearing numerous pairs of similar appendages. Inhabitants of marine caves. *Lasionectes*.

Class Branchiopoda (Gr. *branchia*, gill + *pous*, foot)   Fairy shrimp, tadpole shrimp, clam shrimp, and water fleas. Freshwater and marine crustaceans with leaflike setose appendages. *Daphnia*.

Class Ostracoda (Gr. *ostrakodes*, having a shell)   Mussel or seed shrimps. Tiny marine and freshwater crustaceans in which the entire body is enclosed within a bivalved carapace. *Cypris, Candona*.

Class Copepoda (Gr. *kope*, oar + *pous*, foot)   Copepods. Very small marine and freshwater crustaceans having a cylindrical tapered body with long first antennae held at right angles to long axis of body. *Cyclops, Calanus*.

Class Cirripedia (L. *cirrus*, curl + Gr. *pous*, foot)   Barnacles. Marine, sessile crustaceans in which the body is enclosed within a bivalved carapace that is typically covered with calcareous plates. *Lepas, Balanus*.

Class Malacostraca (Gr. *malakos*, soft + *ostrakon*, shell)   Trunk composed of an eight-segmented thorax, on which the legs are located, and a six-segmented abdomen.

Order Stomatopoda (Gr. *stoma*, mouth + *pous*, foot)   Mantis shrimps. Tropical and subtropical crustaceans having the second pair of thoracic appendages enlarged and adapted for prey capture. Compound eyes stalked; body somewhat dorsoventrally flattened. *Squilla, Gonodactylus*.

Order Amphipoda (L. *amphi*, on both sides + Gr. *pous*, foot)   Amphipods. Small marine and freshwater crustaceans in which the body is laterally compressed. Thorax, which is not covered by a carapace, carries one pair of maxillipeds and seven pairs of legs. Gills associated with legs. Eggs brooded within a marsupium. *Gammarus, Talitrus, Caprella*.

Order Isopoda (Gr. *isos*, equal + *pous*, foot)   Isopods. Marine, freshwater, and terrestrial (pill bugs) crustaceans in which the body is dorsoventrally flattened. Thorax, which is not covered by a carapace, carries one pair of maxillipeds and seven pairs of legs. Pleopods modified as gills. *Asellus, Sphaeroma, Lygia, Oniscus*.

Order Decapoda (Gr. *deka*, ten + *pous*, foot)   Shrimp, crayfish, lobsters, and crabs. Thorax covered by a carapace; thoracic appendages consist of three pairs of maxillipeds and five pairs of legs. *Pennaeus,*

Homarus, Cambarus, Callinectes, Cancer, Pagurus, Emerita.

Order Euphausiacea (Gr. *eu*, true + *phainein*, to show + *ousia*, substance)   Krill. Shrimplike crusta- ceans with thoracic appendages that are not differen- tiated into maxillipeds and legs. *Euphausia.*

Summary

1. The Arthropoda is the largest and most widely distributed of all animal phyla. An external chitinous skele- ton, divided into articulating plates and cylinders and periodically molted to permit growth, is the distinguishing feature of arthropods. Metamerism, a paired ventral nerve cord, and perhaps the appendages reflect an annelidan ancestry. The coelom has largely disappeared, nephridia are absent or modified with the coelom remnants as end sac organs, and distinct muscle bundles are attached to the inner side of the exoskeleton. The circulatory system is open. Although primitively the trunk appendages are nu- merous and similar, most arthropods exhibit varying de- grees of reduction of metamerism through differentiation of appendages and loss and fusion of segments.

2. The subphylum Chelicerata includes the only ar- thropods without antennae. The body is divided into a cephalothorax and abdomen, and the feeding appendages flanking the mouth are chelicerae. The small class Merostomata, containing the gill-bearing horseshoe crabs, reflects the marine origin of chelicerates, but the large terrestrial class Arachnida contains the scorpions, pseudo- scorpions, spiders, harvestmen, mites, and ticks.

3. Arachnids are largely predaceous on other arthro- pods, and scorpions, pseudoscorpions, and spiders utilize poison in prey catching. In most species digestion begins externally in the mouth region, with enzymes passed forward from the digestive gland. Spiders make the most extensive use of silk of any group of animals. A waxy epicuticle and gas exchange by book lungs or tracheae or both are important adaptations of arachnids for life on land. Indirect sperm transmission, often involving a sper- matophore, is a common feature of reproduction.

4. The subphylum Crustacea is a diverse assemblage of mostly aquatic arthropods, distinguished by having two pairs of antennae, one pair of mandibles, and two pair of maxillae. Larval stages are typical of many crustaceans, and the earliest stage is a naupliar larva, bearing only the first three pairs of appendages.

5. Members of the class Malacostraca include the largest crustaceans. Their trunk is composed of an eight- segmented thorax and a six-segmented abdomen, each segment bearing appendages. The order Decapoda, the largest in the Malacostraca, is distinguished by possessing three pairs of maxillipeds and five pairs of legs as thoracic appendages. The compound eyes are on stalks. Decapods include shrimp, lobsters, and crabs. Crabs have a short body form resulting from the folding of the abdomen beneath the thorax. Hermit crabs have unfolded abdo- mens housed within gastropod shells. Eggs are brooded on the pleopods, and the hatching stage is usually a zoea larva.

6. Members of the malacostracan orders Amphipoda and Isopoda brood their eggs in a marsupium beneath the thorax, a distinguishing feature. They possess one pair of maxillipeds and seven pairs of legs, but most amphipods are laterally flattened and isopods are dorsoventrally flat- tened. Both groups have stalkless compound eyes. They are largely marine but are also found in fresh water, and the widespread terrestrial isopods (pill bugs) represent the most successful crustacean invasion of land.

7. The class Branchiopoda contains the water fleas, fairy shrimp, and brine shrimp, most of which live in fresh water. They possess leaflike trunk appendages used in suspension feeding.

8. The class Copepoda includes many species of small planktonic, epibenthic, and interstitial crustaceans. They lack compound eyes but have a median nauplier eye; the large first antennae are held at right angles to the body. The trunk is composed of a thorax and abdomen, the latter lacking appendages. Many planktonic species are filter feeders on diatoms and are very important in marine food chains.

9. Members of the class Ostracoda are small, mostly benthic crustaceans. The body is completely enclosed within a hinged bivalve carapace. The trunk is greatly reduced and they swim or scurry over the bottom with the antennae.

10. Barnacles, members of the class Cirripedia, are sessile crustaceans with the body enclosed within a cara- pace covered with calcareous plates. Both stalked and stalkless barnacles are attached by cement glands that open at the base of the first antennae in the larva. Barnacles are suspension feeders, utilizing six pairs of long biramous setose trunk appendages as a scoop. Many barnacles are attached to the surface of other animals; about one-third of the described species are parasitic.

References and Selected Readings

Detailed accounts of arthropods may also be found in the references listed at the end of Chapter 23. The works listed below deal exclusively with specific arthropod groups.

Bliss, D.E. *Shrimps, Lobsters and Crabs*. Piscataway, N.J.: New Century Publishers, 1982. A semipopular account of the decapod crustaceans.

Bliss D.E. (ed.): *The Biology of Crustacea*. vols. 1–10. New York: Academic Press, 1982–1985. A multivolume work covering many aspects of crustacean biology.

Caldwell, R.L., and H. Dingle. Stomatopods. *Scientific American*, 234(Jan. 1976):80–89. An interesting account of stomatopods, emphasizing behavior.

Fitzpatrick, J.F. *How to Know the Freshwater Crustacea*. Dubuque, Iowa: W.C. Brown Co., 1983. A guide to the identification of freshwater crustaceans.

Foelix, R.F. *Biology of Spiders*. Cambridge, Mass.: Harvard University Press, 1982. A good general biology of spiders, with an emphasis on adaptive morphology and physiology.

Kaestner, A. *Invertebrate Zoology*, vols. 2 (Chelicerates and Myriapods) and 3 (Crustaceans). New York: John Wiley & Sons, Inc., 1968 and 1970. Good general accounts of the arthropods other than insects.

McDaniel, B. *How to Know the Mites and Ticks*. Dubuque, Iowa: W.C. Brown Co., 1979. A guide to the identification of mites and ticks.

McLaughlin, P. *Comparative Morphology of Recent Crustacea*. San Francisco: W.H. Freeman and Co., 1980. Detailed anatomical descriptions of representatives of the higher taxa of crustaceans.

Polis, G.A. (ed.) *Biology of Scorpions*. Stanford, Calif.: Stanford University Press, 1987. A multiauthored volume covering various aspects of scorpion biology.

Wicksten, M.K. Decorator crabs. *Scientific American* 242 (Feb. 1980):146–154. A study of the habit of certain spider crabs of growing algae and sponges on their backs.

Woolley, T.A. *Acarology: Mites and Human Welfare*. New York: Wiley-Interscience, 1987.

32

Arthropods II: Uniramians

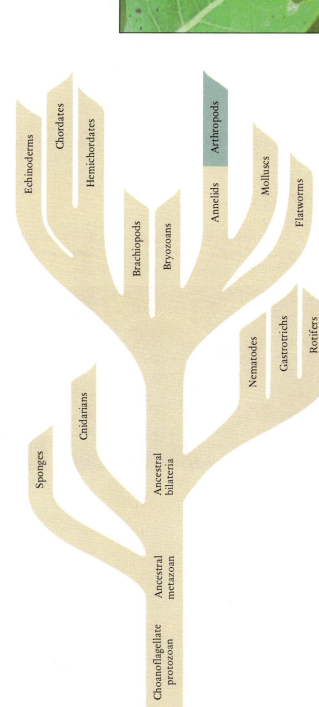

(Left) Dogbane beetles, Chrysochus auratus, *on a milkweed leaf.*

The subphylum Uniramia is the largest division of arthropods, containing the centipedes, millipedes, and insects. All uniramians possess a single pair of antennae, in contrast to the two pairs in crustaceans and the absence of antennae in chelicerates. The first pair of feeding appendages are mandibles, which are followed by one or two pairs of maxillae. The only other conspicuous segmental appendages are legs, which vary from three to many pairs. The name Uniramia ("one branch") refers to the unbranched nature of the appendages, which in uniramians is believed to be primitive. Many chelicerates and crustaceans possess some unbranched appendages, but they are believed to be derived from an original branched condition.

MYRIAPODS

The myriapodous arthropods consist of four classes of uniramians that were once placed within a single class, the Myriapoda. They are the Diplopoda, which contains the millipedes; the Chilopoda, which contains the centipedes; and two small classes, the Symphyla and the Pauropoda. Although the myriapodous arthropods belong to separate classes, they share a number of characteristics. All are terrestrial and secretive, living in soil and leaf mold and beneath stones, logs, and bark. The body is composed of a head and a long trunk with many segments and legs (Fig. 32.1). The eyes are not compound, except in a few species.

Gas exchange organs are **tracheae**, and a separate system of tubules and spiracles is present in each segment. Excretory organs are **malpighian tubules** (p. 351), and the heart is a long dorsal tube with ostia in each segment.

Reproduction parallels that in arachnids in that sperm transfer is indirect and, in centipedes, involves a spermatophore.

Adaptations for locomotion have been a primary theme in the evolutionary history of centipedes and millipedes. These animals have become adapted for running, climbing, pushing, wedging beneath objects, and

Figure 32.1
Millipedes. (a) A cylindrical millipede. (b) A flat-backed millipede. (c) Head and anterior segments of a cylindrical millipede. Maxilla are hidden behind mandibles. (d) Diplosegment of a flat-backed millipede.

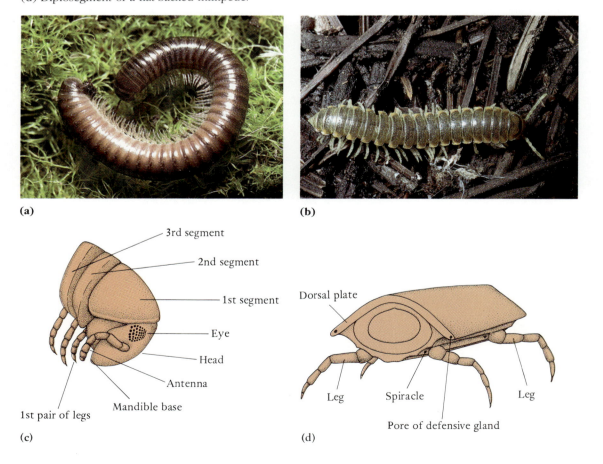

(a)

(b)

(c)

3rd segment
2nd segment
1st segment
Eye
Head
Antenna
Mandible base
1st pair of legs

(d)

Dorsal plate
Leg
Spiracle
Leg
Pore of defensive gland

burrowing in soil. Yet, in contrast to most other arthropods, in which increased motility is correlated with compaction of the body and a reduction in number of legs, myriapods have retained a long trunk with many appendages.

Millipedes

The 7500 species of millipedes, members of the class Diplopoda, have the body adapted for pushing, the force being generated by the large number of legs. As many as 52 legs may be involved in one cycle of movement sweeping down the length of the body (Fig. 32.1a). The large number of legs requires a large number of segments, but a large number of segments results in a weakened trunk column. This problem was solved by the formation of double segments, or **diplosegments**. Each trunk section of a millipede represents two fused segments, each with two pairs of legs, and other organs (Fig. 32.1d).

The common cylindrical millipedes are best adapted for pushing through leaf mold and other loose debris (Fig. 32.1a). The head is rounded, and the many legs are attached near the midventral line of the cylindrical smooth body. Some species are as big as pencils.

Flat-backed millipedes are adapted for wedging into confining places, such as beneath stones or bark. The body is dorsoventrally flattened, and the laterally projecting dorsal plates create a protective working space for the legs (Fig. 32.1b).

Both flatbacked and cylindrical millipedes possess glands on the diplosegments that secrete compounds containing iodine, quinone, or hydrocyanic acid, and are believed to serve a defensive function.

Most millipedes feed on decaying vegetation. They have only one pair of maxillae following the mandibles, in contrast to the two pairs found in centipedes and insects.

Centipedes

Centipedes, members of the class Chilopoda, are predaceous, and most are adapted for running. Behind the mandibles and the two pairs of maxillae are a pair of large poison claws used for seizing and killing prey (Fig. 32.2b). Some species can also pinch with the last pair of legs.

Figure 32.2
Centipedes. (a) A lithobiomorph centipede with unequal and overlapping dorsal plates. (b) Underside of head of a centipede, showing poison claws. (c) *Scutigera*, the common house centipede.

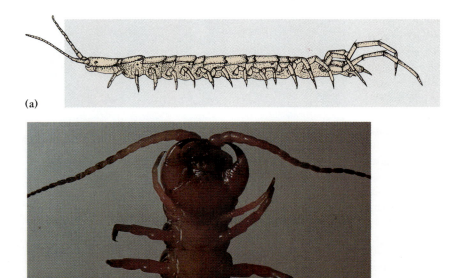

(a)

(b)

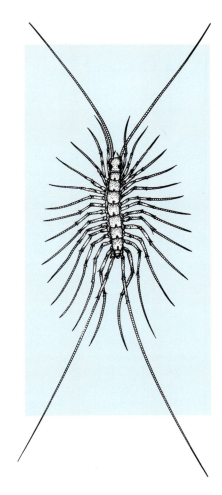

(c)

Centipedes feed mostly on other arthropods, and although large specimens can inflict a painful bite, only a few tropical species are actually dangerous to humans.

Only one pair of legs is present per trunk segment, for the segments are not doubled as in millipedes. To reduce the tendency to undulate or wobble when running, the trunk of many species is strengthened by having alternately long and short dorsal skeletal plates or overlapping plates of different lengths (Fig. 32.2a).

Scutigera, a genus of long-legged centipedes commonly found in bathtubs and sinks, contains the fastest running species. The leg length increases from front to back so that each leg moves to the outside of the preceding leg, reducing interference (Fig. 32.2c).

INSECTS

The Insecta is not only the largest class of arthropods but also the largest group of animals, including more than 750,000 recorded species. This may be just a small part of the actual number. Studies on the insects living in the tree canopy of tropical rain forests, a fauna that has only recently begun to be surveyed, reveal enormous numbers of new species. Terry Erwin, an entomologist studying tropical tree canopy insects, has postulated that there may be nearly 30 million undescribed species of insects, most in tropical rain forests (see Connections 32.1.) This is a mind boggling number.

The great adaptive radiation that the class has undergone has led to the occupation of virtually every type of terrestrial habitat, and some groups have invaded fresh water. A number of factors have contributed to the great success of insects relative to other terrestrial arthropods. Insects are small, permitting exploitation of a wide range of microhabitats. The evolution of flight provides advantages for dispersal, escape, and access to food or more favorable environmental conditions. The ability to fly evolved in reptiles, birds, and mammals, but the *first* flying animals were insects.

Like arachnids, insects evolved a waxy epicuticle, which not only reduces evaporative water loss but can act as a water repellent (see Connections 31.1, p. 722). This

---

Connections 32.1 — *Life in the Tree Tops*

The canopy formed by trees of tropical rain forests is a new frontier of animal biology. This unique aerial habitat harbors a world of plants and animals never encountered in the dim light and still air on the ground 150 feet below. Many species of birds, mammals, snakes, frogs, and insects and other invertebrates spend their entire lives in the tree tops. In addition to the microhabitats around the leaves and twigs of the trees themselves, the larger branches of many trees carry dense masses of orchids, ferns, and bromiliads. The large bromiliads usually contain a small amount of water trapped at the bases of the leaves, which are arranged in a rosette, forming a tank. Such bromiliad tanks are used as freshwater pools for frogs and other animals. In Jamaica there is even a freshwater crab that lives in bromeliad tanks!

To observe and collect canopy animals some zoologists have gone up into the canopy themselves, constructing special perches and even catwalks far above the ground. Others, like entomologist Terry Erwin of the Smithsonian, have sampled from below. Aerosol compounds that disturb, eject, and then narcotize insects are released in the canopy. The insects that fall down to the ground below are collected on an array of trays with funnels. Erwin's speculation of 30 million species of insects is based in part on such canopy samples, which contain large numbers of new species.

Moreover, he found that the insect fauna inhabiting canopies composed of certain types of trees is quite different from the fauna inhabiting others.

more protective outer layer of the exoskeleton enables them to live in a much wider range of terrestrial habitats than is possible for the terrestrial crustacean pill bugs, which lack wax in the epicuticle. Insects also evolved the ability to reabsorb considerable excretory water, an ability shared only with birds and mammals. They excrete uric acid, which is highly insoluble, as the nitrogenous waste.

Like birds and reptiles, insects possess a well-developed cleidoic egg (p. 523). A shell protects against desiccation and other adverse conditions and encloses all of the food necessary for embryonic development prior to hatching. As a consequence, many insects deposit their eggs on exposed surfaces.

Insects are of great ecological significance in the terrestrial environment. Two-thirds of all flowering plants are dependent upon insects for pollination. The principal pollinators are bees, wasps, butterflies, moths, and flies. All have an evolutionary history that is closely tied to that of the flowering plants.

Insects are also of enormous importance for humans. Mosquitoes, lice, fleas, bedbugs, and various flies can contribute directly to human misery. Some are vectors (carriers) of organisms causing human diseases or diseases of domesticated animals, including malaria, elephantiasis, and yellow fever (mosquitoes); African sleeping sickness (tsetse flies); typhus and relapsing fever (lice); bubonic plague (fleas); and, typhoid fever and dysentery (house flies). Our domesticated plants are dependent upon some insects for pollination but are destroyed by others. Vast sums are expended to control insect pests, which can greatly reduce the high agricultural yields necessary to support large human populations. But the overzealous use of pesticides can in turn be hazardous to the environment.

The Insect Ground Plan

Despite the great diversity of insects, their general structure is relatively uniform. Like arachnids, insects possess their complete complement of segments and segment groupings on hatching. The body is composed of a head, thorax, and abdomen (Fig. 32.3). A large pair of compound eyes occupies the lateral surface of the head. Between the

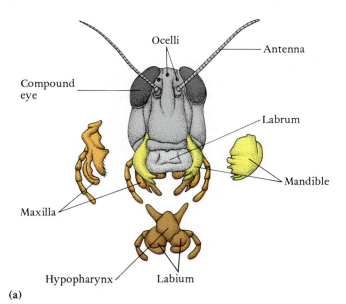

(a)

Figure 32.3
External structure of an insect. (a) Anterior surface of the head of a grasshopper showing mouthparts.
(b) Diagrammatic lateral view of a winged insect.

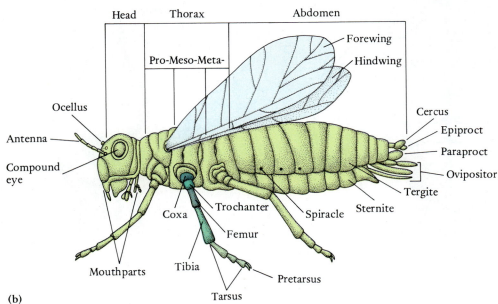

(b)

compound eyes are usually three small simple eyes, or **ocelli**, and a pair of antennae (Fig. 32.3a). The feeding appendages consist of a pair of mandibles and two pairs of maxillae. The second pair of maxillae are fused together and called the lower lip, or **labium**. An upper lip, or **labrum**, formed by a shelflike projection of the head, covers the mandibles anteriorly. Near the base of the labium a median process, the **hypopharynx**, projects from the floor of the oral cavity (Fig. 32.3a).

The thorax is composed of three segments (Fig. 32.3b). Each of the last two pairs of segments usually carry a pair of wings, but as we shall see, wings were not part of the original insect ground plan.

The abdomen is composed of 11 segments, which usually lack appendages; however, the terminal reproductive structures are believed by some entomologists to be derived from segmental appendages.

The excretory organs of insects are malpighian tubules. Two to several hundred arise from the hindgut-midgut junction (Fig. 32.4). They lie more or less freely in the hemocoel, and nitrogenous wastes picked up from the surrounding blood are secreted as uric acid into the tubule lumen (p. 351). Much of the water is removed in the rectum.

Tracheae are the gas exchange organs of insects, and the spiracles, or openings, to the tracheal system are located one on each side of the last two thoracic segments and the first seven or eight abdominal segments (Fig. 32.3b).

The heart is usually a long abdominal tube with nine pairs of ostia, and the only vessel is an anterior aorta leading into the thorax and head. Blood flows from posterior to anterior, although reversal of flow is known to occur in some forms.

The Radiation of Insects

The tremendous diversity of insects is related to three great events in their evolutionary history: the evolution of flight, wing folding, and complete metamorphosis (Fig. 32.5).

The Evolution of Flight

Ancestral insects were wingless, and a few primitive living species remain wingless. But of the 24 orders of insects existing today, 22 contain winged species or at least species, such as lice, fleas, and ants, derived from winged ancestors. These 22 orders constitute the subclass

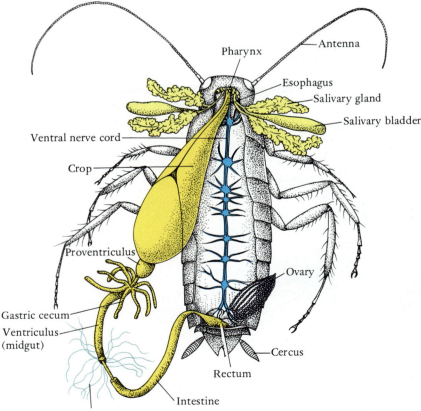

Figure 32.4
(a) Internal structure of a cockroach.
(b) Longitudinal section through the foregut and anterior part of the midgut of a cockroach.

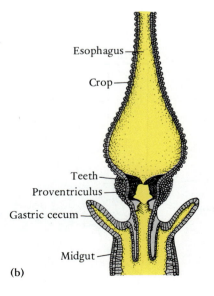

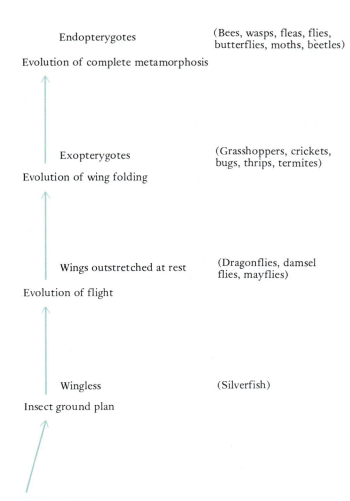

Endopterygotes   (Bees, wasps, fleas, flies, butterflies, moths, beetles)

Evolution of complete metamorphosis

Exopterygotes   (Grasshoppers, crickets, bugs, thrips, termites)

Evolution of wing folding

Wings outstretched at rest   (Dragonflies, damsel flies, mayflies)

Evolution of flight

Wingless   (Silverfish)

Insect ground plan

Figure 32.5
Major events in the evolution of insects.

Figure 32.6
A silverfish; a primitive, wingless apterygote insect.

Pterygota. The 2 wingless orders of the subclass Apterygota include the machilids (order Archeognatha) and the more familiar silverfish (order Thysanura) (Fig. 32.6). In the past, the six-legged and wingless collembolans, proturans, and diplurans, which live in leaf mold and soil, have been placed in the Apterygota, but their structure deviates in one or more ways from the insect ground plan just described, and most entomologists today exclude them from the Insecta (see p. 766). Insect wings are dorsolateral folds of the second and third thoracic segments, and hence are composed of an upper and lower layer of exoskeleton. They are supported by hollow strutlike thickenings called veins, which contain blood, nerves, and tracheae. The original function of the folds or flanges is uncertain. Gills, parachutes, and heat regulators have been proposed.

Each wing articulates with the edge of the thoracic roof (**tergum**), but its inner end rests on a dorsal pleural (lateral) process, which acts as a fulcrum (Fig. 32.7). The wing is thus somewhat analogous to an off-center seesaw. Vertical movements of the wings may be caused directly by muscles attaching onto their bases or indirectly by thoracic muscles, as shown below.

Wing Folding
The wings of the first fossil species, some of which reached a span of 77 cm, were held outstretched at rest, like those of dragonflies (Fig. 32.8a). Wings held in this position require considerable space even when the insect is not flying.

Figure 32.7
Diagrams of the indirect flight muscles of an insect as seen in cross sections of the thorax. (a) Contraction of vertical muscles lowers the tergum and raises the wings.
(b) Contraction of longitudinal muscles raises the tergum and lowers the wings.

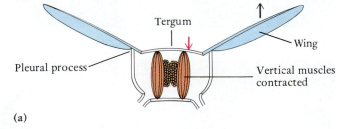

Tergum

Pleural process

Wing

Vertical muscles contracted

(a)

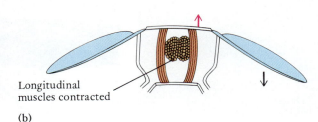

Longitudinal muscles contracted

(b)

Damselflies and mayflies reduce this problem by holding the wings upright when at rest (Fig. 32.8b). A better solution was provided with the second great event in the evolution of insects—wing folding. The development of small articulating sclerites (plates) at the wing base, as well as modifications of the wing venation, enabled the wings to be folded and placed over the abdomen (Fig. 32.8c). Folding the wings when at rest enabled insects to exploit many confining microhabitats, such as the narrow spaces within vegetation and cracks and crevices beneath bark, wood, and stones. Twenty orders of insects can fold their wings.

The basic aspects of insect flight were described earlier (p. 230), and only a few additional details need be mentioned here. In most insects there is only one functional pair of wings, because two independently beating wings would cause too much interference with air flow. Indeed, given the indirect way that most insect wings are moved up and down, it would be difficult for them to function out of phase. Bees, wasps, butterflies, and moths lock the two pairs of wings together in flight with special setae located on one wing or another. The hindwings of flies are reduced to knobs, called **halteres** (Fig. 32.9a), and the forewings of beetles and grasshoppers have become modified as protective wing covers (Fig. 32.9b). Flying ability varies greatly. Many butterflies and damselflies have a relatively slow wing beat and limited maneuverability. At the other extreme, some flies, bees, and moths can hover and dart. The fastest flying insects are hummingbird moths and horseflies, which may attain speeds of 25 mph. Gliding, an important form of flight in birds, occurs in only a few large insects, such as dragonflies and butterflies.

There is no flight control center in the insect nervous system, but the eyes and the sensory receptors on the antennae, on the wings, and in the wing muscles themselves provide continual feedback information for flight control. The modified second pair of wings (halteres) of flies, gnats, and mosquitoes beat with the same frequency as the forewings and function as gyroscopes for the control of flight instability (pitching, rolling, and yawing).

Complete Metamorphosis

The newly hatched young of most primitive wingless insects are similar to the adults, except in size and sexual maturity, i.e., there is no metamorphosis. In contrast, the young of the members of some winged orders—grasshoppers, crickets, dragonflies, leaf hoppers, bugs,

(a)

Figure 32.8
(a) A dragonfly. Wings are held horizontally and outstretched at rest. (b) A mayfly. Wings are held dorsally, but outstretched at rest. (c) A stonefly. Wings folded back over abdomen at rest.

(b)   **(c)**

(a)

(b)

Figure 32.9

(a) A robberfly. The modified club-like second pair of wings (halteres) show as yellow clubs just below the anterior wing bases. (b) A barkbeetle. The heavy forewings are modified as covers for the hindwings.

and many others—are similar to the adults but lack the wings and reproductive system, which gradually develop during the course of subsequent instars (Fig. 32.10). This type of development is called **gradual** or **incomplete meta-morphosis (hemimetabolous development)**, and the immature stages are called **nymphs** or, when aquatic, **naids**. Aquatic development with incomplete metamorphosis is characteristic of dragonflies, stoneflies, and mayflies.

The higher orders of insects, such as beetles, butterflies, moths, bees, wasps, and flies, undergo **complete metamorphosis (holometabolous development)**, involving more radical changes between the immature and adult forms (Fig. 32.11). The hatching stage is an active, feeding, wormlike larva called a **caterpillar** (butterflies and moths), **grub** (beetles), or a legless **maggot** (flies). Wing development is suppressed. The larva increases in size through successive instars, and at the end of the larval period, passes into a quiescent pupal stage, in which a radical transformation takes place. The adult structures are rapidly developed from special groups of cells, called **imaginal discs**, which had been inactive in the larva. Certain of the imaginal discs, such as those that give rise to the wings, are actually invaginations of the body wall. Thus the wings develop internally. Only 9 of the 24 orders of insects possess holometabolous development, but these 9 orders—sometimes grouped together as the Endopterygota, as opposed to the Exopterygota, where the

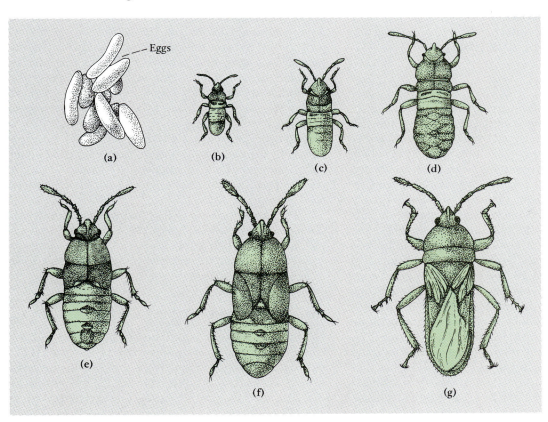

(a)   (b)   (c)   (d)

(e)   (f)   (g)

Eggs

Figure 32.10

Stages in the gradual metamorphosis (hemimetabolous development) of a chinch bug.

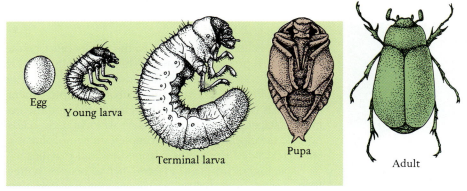

Figure 32.11
Stages in the complete metamorphosis (holometabolous development) of a beetle.

Egg
Young larva
Terminal larva
Pupa
Adult

wings gradually develop externally—contain the greatest number of insect species. The transformation of immature insects into reproducing adults is under endocrine control (p. 442).

Larvae are a specialized developmental condition and, as we have seen, are absent in primitive insects. Larvae are able to utilize food sources and habitats that would be unavailable to the adults and vice versa. For example, the caterpillars of butterflies are usually chewing herbivores, in contrast to the nectar-feeding habit of the adults. In certain moths, the short-lived adults have completely lost the ability to feed and only serve to reproduce. In some insects the pupa functions as a resistant dormant stage, taking the life cycle through adverse environmental conditions during certain seasons of the year. Such a period of arrested development in insects, which may occur at any stage (not just the pupal), is called **diapause**.

Nutrition

Primitive insects were herbivores with heavy jawlike mandibles for chewing plant material (Fig. 32.3a), and this feeding habit has been retained by members of a number of major orders, such as grasshoppers, crickets, and beetles. But in the evolution of various groups, the mouthparts have become adapted for many other modes of feeding, and a wide range of diets is utilized. Sucking, piercing-sucking, and chewing-sucking insects have the mouthparts elongated to form a tube, but since this feeding habit evolved independently a number of times, the beak is formed in different ways. For example, in moths and butterflies, which commonly feed on flower nectar, parts of the two maxillae have become modified to form a long sucking tube or proboscis, which is rolled up when not in use (Fig. 32.12a). In mosquitoes, all of the mouthparts are elongated and encased by the labium (Fig. 32.12b and c). The piercing beak of hemipterans (true bugs) has food and salivary channels enclosed within the stylet-like maxillae and mandibles (Fig. 32.12d−f). In the housefly, the labium is modified as a sponge. Food is first dissolved with salivary

secretions and then sucked back up with the labial sponge (Fig. 32.12g).

Salivary secretions play an important role in feeding, serving in different groups to lubricate food, digest sugar and pectins, act as an anticoagulant, and produce venom.

The foregut of insects is variously modified to suit the diet and mode of feeding, but usually includes a **pharynx**, **crop**, and **proventriculus** (Fig. 32.4). The proventriculus may function as a gizzard or as a valve into the midgut.

The insect midgut (the **ventriculus** or stomach) is the site of enzyme production, digestion, and absorption, as in other arthropods (Fig. 32.4). In those species that ingest solid foods, the foregut-midgut junction or the midgut lining secretes a thin chitin-protein membrane, the **peritrophic membrane**, which surrounds the food mass as it passes through the midgut. The peritrophic membrane probably protects the delicate midgut walls from abrasion by the food mass and perhaps conserves enzymes by dividing the gut lumen into two compartments. The membrane is permeable to some enzymes and the products of digestion. The initial products of digestion pass through the membrane, where they are attacked by a second order of enzymes that are restricted to the space between the gut wall and peritrophic membrane. Outpocketings of the midgut, called **gastric ceca**, are characteristic of many insects, and are an important site of food absorption.

The hindgut, composed of an intestine and rectum, opens to the exterior at the end of the abdomen. In many insects, the hindgut is an important site of water absorption, and some species have special rectal structures to facilitate the process.

Insect-Plant Interactions

Plants are a food source for thousands of insects and, as a consequence, plants and insects have been engaged in a virtual arms race during the course of their evolutionary history. Plants have evolved various chemical substances that are either toxic or unpalatable to insect herbivores or

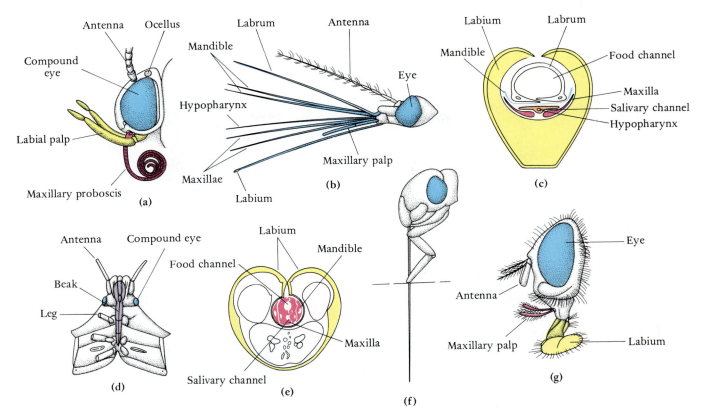

Figure 32.12

Sucking mouthparts of insects. (a) Lateral view of the head of a moth. Two maxillae have become modified to form a proboscis, which is rolled up when not in use. (b) Lateral view of the head of a mosquito, showing separated mouthparts. (c) Cross section of the mouthparts of a mosquito in their normal functional position. (d) Ventral view of anterior half of a hemipteran (a bug), showing piercing beak. (e) Cross section through a hemipteran beak, showing the food and salivary channels enclosed within the stylet-like maxillae and mandibles. (f) A hemipteran penetrating plant tissue with its stylets. (g) Lateral view of the head and mouthparts of a housefly. Labium is modified as a sponge.

that make the plant tissues difficult for the insect to chew or penetrate. Insects have, in turn, evolved enzymes to detoxify plant tissues or have come to utilize young tissues or only certain plant parts. This coevolution, whereby each group has contributed to the selection pressures acting on the other, has led to the development of a wide range of feeding and defense strategies.

Insect-plant coevolution has not been limited to plant chemical defenses. As the primary pollinating agents of plants, insects have been a major selection force in the evolution of various types of flowers, which have in turn played a major role in the evolution of the principal insect pollinating groups—bees, wasps, butterflies, moths, and flies.

Parasitism

There are many parasitic insects, and the condition has undoubtedly evolved a number of times within the class. Insect parasitism represents an adaptation to meet the habitat-nutrition needs of different stages in the life cycle. For some insects, parasitism provides a new food source and habitat for the adults; for others, a new food source for the larvae.

Most adult fleas and lice are bloodsucking ectoparasites on the skin of birds and mammals. The eggs and immature stages of fleas usually develop off the host in the host's nest or habitation. The wingless, laterally compressed body of fleas facilitates living between hairs and feathers, and the ability to jump is an adaptation to get on, off, or between hosts (Fig. 32.13a). Lice spend their entire life cycle on the host and are beautifully adapted for clinging to hairs or feathers (Fig. 32.13b).

Many species of wasps and flies illustrate larval parasitism. Certain small wasps insert an egg into the leaves and stems of plants. The surrounding plant tissue reacts to form a large mass or gall (Fig. 32.13c). The egg develops within the gall, and the larval stage feeds upon the surrounding plant tissue. Other wasps deposit their eggs in the bodies of insects, especially insect larvae (Fig. 32.13d). Two

Carolina Biological Supply Co.

Carolina Biological Supply Co.

(a)

(b)

(c)

(d)

Figure 32.13
Parasitic insects. (a) A flea (order Siphonaptera). (b) A human body louse (order
Phthiraptera). (c) Insect galls on a wild rose. The injection of the egg into the
plant stimulates the surrounding tissues to form the gall, on which the larva will
feed. (d) A caterpillar parasitized with the pupae of a wasp.

groups of wasps, spider wasps and dirt daubers, put their
eggs on spiders, which they paralyze and bring back to
their nests.

The screw-worm fly—a species of blowfly and a pest
of domestic animals—lays its eggs in the wounds and
nostrils of mammals, and the larvae feed on living tissue.
The parasitic condition was probably preceded by the
deposition of eggs in decaying flesh of dead animals, for
this is the habit of many nonparasitic species of blowflies.

Communication

Both social and nonsocial insects utilize chemical, visual,
and auditory signals as methods of communication, but
chemical signals are the most common and widespread.
Many examples of chemical communication by phero-

mones are now known. For example, the males of some
moths can locate females from a considerable distance by
means of airborne substances, and such pheromones are
now used for monitor trapping or in the biological control
of some insect pests. Ants returning from foraging trips
deposit substances on the ground that serve as trail
markers for other individuals in the colony.

Among the more unusual visual signals are the luminescent flashings of fireflies, which play a role in sexual
attraction. In species of *Photinus*, for example, flying males
flash at definite intervals. Females located on vegetation
will flash in response if the male is sufficiently close. The
male will then redirect his flight toward her and further
flashing will occur.

Sound production is especially notable in grasshoppers, crickets, and cicadas. The chirping sounds of the

first two are produced by rasping. The front margin of the forewing or the hind legs acts as a scraper and is rubbed over a file formed by veins of the forewing. In crickets, where both scraper and file are on the forewings, these wings cross over during sound production. Each species of cricket produces a number of songs that differ from the songs of other species. Cricket songs function in sexual attraction and aggression. The static-like sounds of cicadas, which serve to aggregate individuals, are produced by vibrations of chitinous abdominal membranes oscillated by special tymbal muscles.

Social Insects

Colonial, or social, organization has evolved to a remarkable degree among two orders of the social insects: the Dictyoptera, which contain the termites, and the Hymenoptera, which includes the ants, bees, and wasps. In all social insects, an individual cannot exist outside of the colony in which it developed. All social insects exhibit some degree of **morphological polymorphism** (p. 192), and the different types of individuals of a colony are termed **castes**. The castes of termites are shown in Figure 32.14.

Termites have been called social cockroaches, for the two groups are related and in both the gut harbors symbiotic flagellates that are important in the digestion of cellulose. The contact necessary between individuals for the transfer of the symbionts may have been a factor in the evolution of termite social behavior.

Termites live in galleries constructed in wood or soil, and in some species, the colonies may be huge and structurally very complex. The colony is built and maintained by workers and soldiers. The soldiers possess large heads and mandibles and defend the colony. Workers and soldiers, which include both males and females, are sterile and wingless. Wings are present in the fertile males and queens only during a brief nuptial flight. The male, or king, remains with the queen, copulating intermittently and aiding her in the construction of the first nest.

The colonies of ants, bees, and wasps differ from those of termites in being essentially female societies, for all the workers are sterile females. Caste determination of the female is regulated by the presence or absence of certain substances provided in the immature stages by other members of the colony. Males have a brief existence, functioning only in the copulatory nuptial flight. Unlike termite males, they neither contribute to the construction of the first nest nor remain with the queen.

Ant colonies resemble those of termites and are housed within a gallery system in soil, in wood, or beneath stones. There may be a soldier caste in addition to workers. Wings are present only during the nuptial period.

Different species of bees and wasps exhibit a gradation between solitary and social organization. The honeybee, *Apis mellifera*, is the best known social insect. This species is believed to have originated in Africa and to be a recent invader of temperate regions. Unlike other social bees and wasps of temperate regions, the honeybee colony survives the winter, and multiplication occurs by the division of the colony, a process called **swarming**. Stimulated at least in part by the crowding of workers (20,000 to 80,000 in a single colony), the mother queen leaves the hive along with part of the workers (a swarm) to found a new colony. The old colony is left with developing queens. On hatching, a new queen takes several nuptial flights during which copulation with males (**drones**) occurs, and she accumulates enough sperm to last her lifetime. The male dies following copulation, when his reproductive organs are literally exploded into the female. A

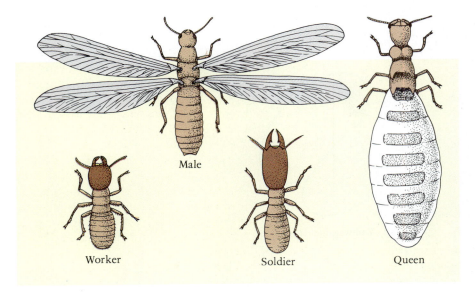

Figure 32.14

Castes of the common North American termite, *Reticulitermes flavipes*.

Male

Worker

Soldier

Queen

new queen may also depart with some of the workers as an after-swarm, leaving the remaining workers to yet another developing queen. Eventually the old colony will consist of about one-third of the original number of workers and their new queen.

Honeybee colonies are large. The workers' life span is not long, and a queen may lay 1000 eggs per day. Nursing workers provide these larvae with a diet that results in their developing into sterile females, i.e., additional workers. The nursing behavior of the workers is a response to a pheromone ("queen substance") produced by the queen's mandibular glands. At the advent of swarming or when the vitality of the queen diminishes, the production of this pheromone declines. In the absence of the inhibiting effect of the pheromone, the nursing workers construct royal cells into which eggs and royal jelly are placed. The exact composition of royal jelly is still unknown, but those larvae fed upon it develop into queens in about 16 days. At the same time that queens are being produced, unfertilized eggs are deposited in cells similar to those for workers. These haploid eggs develop into drones.

A remarkable feature of honeybee social organization is the temporal division of certain tasks of the workers. The first activities of the worker after hatching are maintenance tasks within the hive, such as secreting wax for comb construction, storing food, and caring for larvae. After about three weeks, the bee begins a period of foraging outside of the hive. A large amount of time is spent by the older worker bees in resting and patrolling.

Communication between members of a honeybee colony is highly evolved; some aspects, such as the tail-wagging dance, set the honeybees apart from all other social insects. A successful foraging scout returns to the hive and communicates to other workers the nature, direction, and distance of a food source. The nectar and pollen and the scout's body provide the information about the kind of food that has been found. The scout bee also executes an excited dance that is a ritualization of the flight path. The dancing bee circles to the right and to the left, with a straight-line run between the two semicircles (Fig. 32.15). During the straight-line run, the bee wags her abdomen and emits audible pulsations. In the middle of this century Karl von Frisch, a pioneer in the study of communication in bees, discovered that the orientation of the circular movements shows the direction of the food and that the frequency of the tail-wagging runs indicates the distance. The closer the food source is to the hive, the greater the frequency of tail-wagging runs. Bees use the angle of the sun and light polarization as a means of orientation, and the dance of the scout bee indicates the

Figure 32.15

Diagram illustrating the inclination of the straight wagging run by a scout bee to indicate the location of a food source by reference to the sun. The food source is located at an angle 40° to the left of the sun. The tail-wagging run of the scout bee is therefore upward (indicating that food is toward the sun) and inclined 40° to the left (indicating the angle of the food source to the sun).

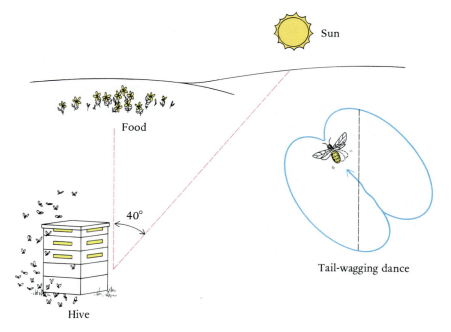

Sun

Food

40°

Hive

Tail-wagging dance

location of the food in reference to the sun's position. If the tail-wagging run is directed upward, the food is located toward the sun; if the tail-wagging run is directed downward, the food is located away from the sun. The inclination of the run to the right or to the left of vertical indicates the angle of the sun to the right or to the left of the food source. An internal "clock" compensates for the passage of time between discovery of the food and the start of the dance, so that the information is correct even though the sun has moved during the interval. On cloudy days, the polarization of the light rays and ultraviolet light act as indirect references. If the food source is closer than 80 m, the clues provided by chemoreception are sufficient for finding the food, and the tail-wagging dance is not performed by the scout bee.

Although the tail-wagging dance has been decoded, the sensory modality by which it is transmitted to other bees cannot be vision. The hive is too dark for this to be possible. The surrounding bees must receive the dancer's vibrations through their antennae or legs. The sound pulsations of the dancer apparently also indicate the distance of the food source from the hive.

Reproduction

In insects, the single pair of gonads is located in the abdomen. Gonoducts from each side unite posteriorly before opening at the end of the abdomen through a short median vagina in the female or an ejaculatory duct in the male. In addition, the female system usually includes a seminal receptacle (spermatheca) and accessory glands associated with the vagina, and the male system includes paired seminal vesicles, accessory glands, and a penis (**aedeagus**) (Fig. 32.16).

The sperm of many insects are transferred within spermatophores. In the primitive apterygotes the spermatophores are deposited on the ground and then picked up by the female. In most insects, however, transfer is direct, and the tubular aedeagus is inserted by the male into the vagina of the female. The posterior abdominal segments of the males of many moths, butterflies, true flies, and other insects bear clasping structures that are used to hold the abdomen of the female during copulation.

Sperm are stored in the spermatheca, and fertilization occurs internally as the eggs pass through the oviduct. The egg leaves the ovary encased within a hard shell, but a tiny opening, or **micropyle**, at one end of the egg permits entrance of the sperm.

Certain parts of the terminal segments of the female form an **ovipositor**, by means of which eggs are buried in soil, excrement, or carrion; injected into plant tissue; or applied to twig, leaf, soil, water, or other surfaces. The eggs are generally deposited in batches, cemented to each other and to the substratum by secretions from the accessory glands.

Parthenogenesis occurs in a number of insect groups. The condition in aphids closely parallels that of the crusta-

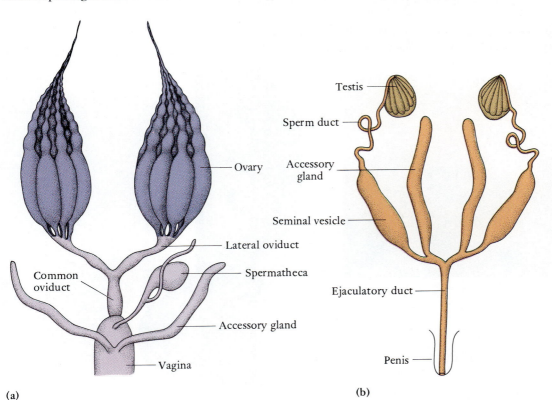

Figure 32.16
Dorsal views of the reproductive systems of insects. (a) Female; (b) male.

Testis

Sperm duct

Accessory gland

Seminal vesicle

Ejaculatory duct

Penis

Ovary

Lateral oviduct

Spermatheca

Common oviduct

Accessory gland

Vagina

(a)

(b)

cean water fleas, where there are successive generations of parthenogenetic females followed by the appearance of males. In bees, unfertilized eggs produce males; fertilized eggs produce females.

EVOLUTIONARY RELATIONSHIPS

Arthropods are clearly related to annelids. Their metamerism, nervous system, dorsal tubular heart, and determinate cleavage all point to an annelidan ancestor or at least a common ancestry with annelids.

Zoologists are in less agreement as to how the three living evolutionary lines of living arthropods—chelicerates, crustaceans, and uniramians—are related to each other. The presence of mandibles, antennae, and compound eyes in both crustaceans and uniramians was in the past the reason for placing these two groups within a single subphylum, the Mandibulata. The Chelicerata, having chelicerae and no antennae, constituted a second subphylum. In recent years, however, many zoologists have come to believe that the uniramians evolved from terrestrial ancestors independently of the chelicerates and crustaceans, both of which had marine ancestors. In this hypothesis, the ancestors of uniramians are thought to have been members of the phylum Onychophora, a group of terrestrial caterpillar-like animals (Fig. 32.17a). Fossil onychophorans are known from the beginning of the

Paleozoic, and living species are found in the tropics and the southern hemisphere. Onychophorans show many features that are intermediate between annelids and arthropods. If uniramians indeed evolved from onychophorans, then the phylum is polyphyletic, i.e., all arthropods do not have a common arthropod ancestor and the arthropod condition evolved more than one time.

Not all zoologists accept a uniramian origin separate from that of chelicerates and crustaceans. Some entomologists question the primitive "unirame," or unbranched, nature of insect appendages and claim that certain fossil insects have branched appendages similar to the primitive appendages of chelicerates and arthropods.

The relationship of chelicerates to crustaceans is also obscure. They may have had a common marine ancestor or even separate origins from annelid-like forms. The fossil record is not helpful. Indeed, the early fossil record contains a number of arthropods that do not belong to any of the living groups.

RELATED PHYLA

The phylum **Onychophora** (Gr. *onychos*, claw + *pherein*, to bear) consists of ancient wormlike animals that live beneath stones, logs, and leaves in the tropics and south temperate parts of the world (Fig. 32.17a). They have a body wall and nephridia like annelids but have a reduced

Figure 32.17
Related phyla. (a) A species of *Peripatus,* from Costa Rica (phylum Onychophora).
(b) Two water bears (phylum Tardigrada) on a filament of algae.

(a)

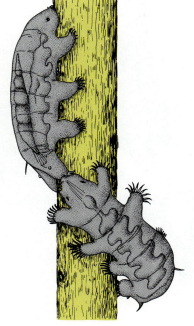

(b)

coelom, open internal transport system, and tracheae like arthropods. The cuticle contains chitin. A pair of antennae and numerous pairs of short peglike legs make them look a little like caterpillars. Most onychophorans are predaceous on other small invertebrates and catch their prey and defend themselves with adhesive secretions squirted from glands in the mouth region.

The phylum **Tardigrada** (L. *tardus*, slow + *gradus*, step) is composed of minute animals called water bears (Fig. 32.17b). They are found in soil and in fresh water, but most live in the water films of mosses. The name water bear comes from their short body and four pairs of stubby legs with claws. The body is covered by a thin chitinous cuticle that is periodically molted. Tardigrades feed on the contents of algal and moss cells pierced by buccal stylets. Many moss-inhabiting species are able to survive long periods of extreme desiccation (see Connections 28.1, p. 645).

Classification of Uniramians

Subphylum Uniramia (L. *unus*, one + *ramus*, branch) Head with one pair of antennae; first postoral appendages are a pair of mandibles. Appendages unbranched. Mostly terrestrial arthropods.

Class Diplopoda (Gr. *diploos*, double + *pous*, foot) Millipedes. Elongate trunk composed of many similar doubled segments, which bear two pairs of legs. *Polydesmus, Julus.*

Class Pauropoda (Gr. *pauros*, small + *pous,* foot) Minute animals that inhabit leaf mold. Elongate trunk of 11 segments, 9 or 10 of which bear legs. *Pauropus.*

Class Chilopoda (Gr. *cheilos*, claw + *pous*, foot) Centipedes. Elongate trunk of many leg-bearing segments. First trunk segment carries a pair of large poison claws. *Scolopendra, Lithobius, Scutigera.*

Class Symphyla (Gr. *syn*, with + *phyle*, species) Small centipede-like mandibulates that inhabit leaf mold. Trunk contains 12 leg-bearing segments. *Scutigerella.*

Class Insecta (L. *insectus*, segmented) Insects. Body composed of head, thorax, and abdomen. Thorax bears three pairs of legs and usually two pairs of wings. The abdomen is composed of eleven segments.

The following classification follows that presented by Evans (see references), and some of the less common orders have been omitted.

Subclass Apterygota (Gr. *a*, without + *pteron*, wing) Primary wingless insects.

Order Archeognatha (Gr. *archaios*, primitive + *gnathos*, jaw) Machilids. Small rock-inhabiting insects with cylindrical or laterally compressed bodies and well developed compound eyes and ocelli.

Order Thysanura (Gr. *thysanos*, bristle + *oura*, tail) Bristletails, silverfish. Body dorsoventrally flattened. Compound eyes small or absent; ocelli commonly absent.

Subclass Pterygota (Gr. *pteron*, wing) Winged insects, or if lacking wings, the wingless condition is secondary.

Infraclass Paleoptera (Gr. *palaios*, ancient + *pteron*, wing) Wings with a network of veins and not folded at rest.

Order Ephemeroptera (Gr. *ephemeros*, short-lived + *pteron*, wing) Mayflies. Triangular membranous forewings with many cross veins. Wings unfolded at rest. Hemimetabolous aquatic development.

Order Odonata (Gr. *odon*, tooth) Dragonflies and damselflies. Membranous fore- and hind-wings of equal size and held outstretched at rest. Hemimetabolous aquatic development.

Infraclass Neoptera (Gr. *neos*, new + *pteron*, wing) Wings folded over abdomen at rest.

Series Exopterygota (Gr. *exo*, outside + *pteron*, wing) Insects with hemimetabolous development.

Order Plecoptera (Gr. *plecos*, plaited + *pteron*, wing) Stoneflies. Membranous wings with numerous veins, but hindwings capable of being folded beneath forewings.

Order Orthoptera (Gr. *orthos*, straight + *pteron*, wing) Grasshoppers, crickets, katydids, walking sticks. Somewhat leathery forewings covering wide, folded, membranous hindwings.

Order Dictyoptera (Gr. *dictyon*, net + *pteron*, wing) Cockroaches, preying mantids, and termites. Characteristics somewhat similar to those of orthopterans but coxae large and placed close together. Termite fore- and hindwings are membranous. (Mantids and cockroaches are often placed within the order Orthoptera, and the termites in a separate order, the Isoptera.)

Order Dermaptera (Gr. *derma*, skin + *pteron*, wing) Earwigs. Short padlike forewings; forceps-like cerci at end of abdomen.

Order Hemiptera (Gr. *hemi*, half + *pteron*, wing) True bugs, aphids, cicadas, and leaf hoppers. Forewings membranous or partially thickened. Mouthparts modified as a piercing beak folded back under the thorax.

Order Thysanoptera (Gr. *thysanos*, fringe + *pteron*, wing) Thrips. Minute insects. Wings fringed with hairs.

Order Phthiraptera (Gr. *phtheir*, louse + *a*, without + *pteron*, wing) Lice. Wingless biting and sucking ectoparasites of birds and mammals. Legs are adapted for gripping both hairs and feathers.

> **Series Endopterygota** (Gr. *endo*, inside + *pteran*, wing) Insects with holometabolous development.

Order Megaloptera (Gr. *megalo*, large + *pteron*, wing) Dobsonflies and snakeflies. Large membranous wings with many veins and capable of being folded back over abdomen. Holometabolous development commonly aquatic.

Order Neuroptera (Gr. *neuron*, nerve + *pteron*, wing) Lacewings and antlions. Membranous wings similar to those of megalopterans. Holometabolous development typically terrestrial.

Order Coleoptera (Gr. *koleos*, cover + *pteron*, wing) Beetles. Sclerotized forewings modified as protective covers for hindwings.

Order Trichoptera (Gr. *trichos*, hair + *pteron*, wing) Caddis flies. Membranous wings with few cross veins. Aquatic larvae build cases.

Order Lepidoptera (Gr. *lepidos*, scale + *pteron*, wing) Butterflies and moths. Wing surfaces covered with scales.

Order Diptera (Gr. *di*, two + *pteron*, wing) Flies, midges, and gnats. Forewings membranous; hindwings modified as clublike halteres.

Order Siphonaptera (Gr. *siphon*, tube + *a*, without + *pteron*, wing) Fleas. Wingless, laterally compressed ectoparasites of birds and mammals.

Order Hymenoptera (Gr. *hymen*, membrane + *pteron*, wing) Bees, wasps, and ants. Membranous wings with few veins; forewings larger than hindwings.

The following three classes have in the past been considered orders of apterygote insects. All possess three pairs of legs.

Class Collembola (Gr. *kolla*, glue + *embolos*, peg) Springtails. These small arthropods are usually very common in leaf mold, where they feed on plant or decomposing material. Abdomen contains only six segments and terminates in a lever-like device.

Class Protura (Gr. *protos*, first + *oura*, tail) Proturans. These small, blind, soil-inhabiting arthropods lack antennae and on hatching lack the full complement of body segments.

Class Diplura (Gr. *diploos*, double + *oura*, tail) Diplurans. Small, blind, soil-inhabiting arthropods with antennae.

Summary

1. The subphylum Uniramia, which contains the centipedes, millipedes, and insects, is the largest group of arthropods. Most living uniramians are terrestrial and have one pair of antennae and, as feeding appendages, a pair of mandibles and one or two pairs of maxillae. All of the appendages are primitively unbranched. Gas exchange organs are tracheae, and excretory organs are malpighian tubules. Primitively, sperm transmission is indirect with spermatophores.

2. The uniramian classes Chilopoda (centipedes) and Diplopoda (millipedes) have long trunks of many leg-bearing segments. Centipedes, which live beneath stones and logs, are predaceous, using a pair of large poison claws located behind the second maxillae. Many species run rapidly, and trunk stability is increased by alternating large and small or overlapping tergal plates. Trunk stability in millipedes has been increased by the evolution of diplosegments, each of which bears two pairs of legs.

3. The class Insecta contains over 750,000 species. The body usually consists of a head with three ocelli and a pair of compound eyes, a thorax with two pairs of wings and three pairs of legs, and an abdomen without conspicuous segmental appendages. Primitive species are wingless, but most insects are capable of flight. The evolution of flight, wing folding, and complete metamorphosis have been important factors in the great success of the class.

4. In insects the mouthparts—one pair of mandibles, one pair of maxillae, and a labium (fused second maxillae)—are primitively adapted for chewing vegetation but have become modified for a wide range of other diets and feeding modes.

5. In primitive insects sperm transmission is indirect, i.e., transferred by a spermatophore, but most species copulate and have direct sperm transmission. The newly hatched young of many species are more or less similar to the adults but lack wings, which are gradually acquired during the course of subsequent instars (incomplete metamorphosis); other insects hatch as a wormlike larva that undergoes a radical metamorphosis (complete metamorphosis) to attain the adult form.

6. Intraspecific communication by pheromones, sound, and light is utilized by many species. Varying degrees of social organization have evolved in termites, ants, wasps, and bees. Such colonies commonly contain reproductive and worker castes.

References and Selected Readings

Detailed accounts of insects may also be found in the references listed at the end of Chapter 23.

Borror, D.J., D.M. De Long, and C.A. Triplehorn. *An Introduction to the Study of Insects*, 5th ed. Philadelphia: Saunders College Publishing, 1981. A general entomology text emphasizing insect taxonomy.

Borror, D.J., and R.E. White. *A Field Guide to the Insects*. Boston: Houghton-Mifflin Co., 1970. A good guide for the identification of common North American insects.

Chapman, R.F. *The Insects: Structure and Function*. 3rd ed. Cambridge, Mass.: Harvard University Press, 1982.

Evans, H.E. *Insect Biology: A Textbook of Entomology*. Reading, Mass.: Addison-Wesley Publishing Co., 1984. An excellent introduction to the biology of insects, emphasizing the relationship of insects to the world around them.

Holldobler, B.H., and E.O. Wilson. *The Ants*. Cambridge, Mass.: Harvard University Press, 1990. A superb general biology of ants and their social organization.

Prestwich, G.D. The chemical defense of termites. *Scientific American,* 249 (Aug. 1983):78−87. The role of chemical weapons of termite soldiers.

Romoser, W.S. *The Science of Entomology*, 2nd ed. New York: Macmillan, 1981. A well-balanced general biology of insects.

Wilson, E.O. *Insect Societies*. Cambridge, Mass.: Harvard University Press, 1971. A detailed account of social insects.

33

Bryozoans

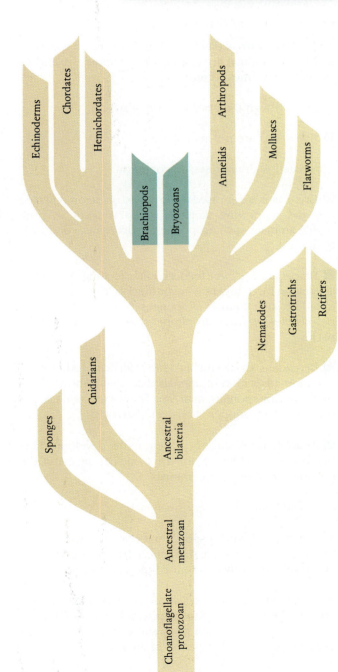

(Left) Leaf-like colony of the marine bryozoan, Hippodisplosia insculpta. (Above) Protruded horseshoe-shaped lophophores of the freshwater bryzoan, Cristatella mucado.

This short chapter will explore a phylum of minute colonial animals, called the Bryozoa. Let's begin by using the characteristics of bryozoans to illustrate the predictive value of some basic zoological principles you have already studied.

The members of this phylum are *sessile*, *colonial*, and *coelomate*, and individuals are *small, usually less than 0.5 mm in length*. On the basis of these four facts, what other features—structural or functional—would you expect to find in bryozoans? Think a few minutes and make some educated guesses before you continue reading.

Characteristics of Bryozoans

The following bryozoan features are correlated with their sessile life, colonial organization, and minute size and are therefore not unexpected:

Correlated with their colonial organization:

■ Bryozoans are polymorphic.

Correlated with their sessile condition:

■ Bryozoans are encased within an exoskeleton. A protective and supporting skeleton is characteristic of many sessile animals, such as sponges, hydrozoans, corals, and barnacles.

■ Bryozoans are filter feeders. Filter feeding, a common adaptation for a slow-moving or immobile animal, is illustrated by such animals as sponges, sedentary tubicolous polychaetes, worm shells, slipper shells, bivalves, and barnacles.

■ Bryozoans are hermaphroditic. Hermaphroditism is a common adaptation of animals that cannot move about, such as many sponges, many sea anemones, some corals, some bivalves, slipper shells, and barnacles.

■ A motile larval stage is present in the development of bryozoans. Such larvae are the principal means of dispersion for most sessile animals.

Correlated with their very small size:

■ Bryozoans have no special internal transport system, for internal distances are short enough to permit transport by movement of the coelomic fluid and by diffusion.

■ Bryozoans have no special gas exchange and excretory organs. The ratio of surface area to volume is sufficiently favorable for the general body surface to suffice for gas exchange and the elimination of ammonia.

While the members of the phylum Bryozoa are very common marine organisms, by virtue of their small size they are unknown to most people. Bryozoans live attached to rocks, shells, pilings, jetties, and ship bottoms, and some species occur in fresh water. The name Bryozoa (moss animal) refers to the texture they give to the surfaces they colonize. Approximately 4000 living species have been described and there is a very rich fossil record that dates back to the Ordovician.

STRUCTURE OF A BRYOZOAN INDIVIDUAL

Individuals (**zooids**) of a bryozoan colony are shaped something like a little box or cyclinder, in which each of the sides represents one of the usual morphological surfaces—anterior, posterior, ventral, and so on (Figs. 33.1 and 33.2). The body of most bryozoans is covered with an exoskeleton composed of chitin. A layer of calcium carbonate is commonly found just beneath the chitin; both are secreted by a single layer of epidermal cells. In most marine bryozoans a layer of peritoneum lies immediately beneath the epidermis; the rigidity of the exoskeleton would make ordinary layers of body wall muscles useless. The interior of the body is occupied by a spacious coelom.

The principal organ projecting into the coelom is the gut, which is U-shaped to avoid the obstruction resulting from the attachment to other members of the colony. At the anterior end, the skeletal housing is perforated by a circular orifice through which the two ends of the gut project (Fig. 33.1).

The feeding organ of bryozoans is a crown of ciliated tentacles, the **lophophore** (Gr. *lophos*, tuft + *pherein*, to bear). When the lophophore is protruded, the outstretched tentacles usually form a funnel with the mouth in the center of the base and the anus projecting outside of the base (Figs. 33.1b and 33.2). Because of the position of the anus outside of the lophophore, the name *Ectoprocta* (Gr. *ektos*, outside + *proktos*, anus) is sometimes used for the phylum. The hollow tentacles contain an extension of the coelom. When the lophophore is protruded, a sheath of the body wall extends from the rim of the orifice up to the base of the lophophore. On retraction of the lophophore into the orifice, this tentacle sheath is reversed and surrounds the bunched tentacles within the skeletal housing (Fig. 33.1a).

The lophophore is protruded by increased pressure in the coelomic fluid. The pressure is elevated by the inbowing of the body wall on contraction of certain transverse muscle bands (Fig. 33.1). In many bryozoans the muscle bands attach to the inner side of one surface, called the frontal surface. This surface is thin, membranous, and easily depressed, rather like a flexible windowpane (Figs. 33.1 and 33.2). Withdrawal of the lophophore back into the orifice is brought about by special retractor muscles. In many species a lid (operculum) closes over the orifice when the lophophore retracts.

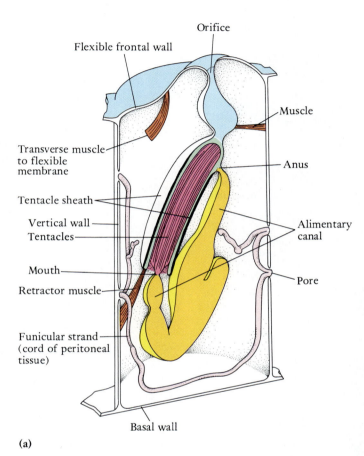

(a)

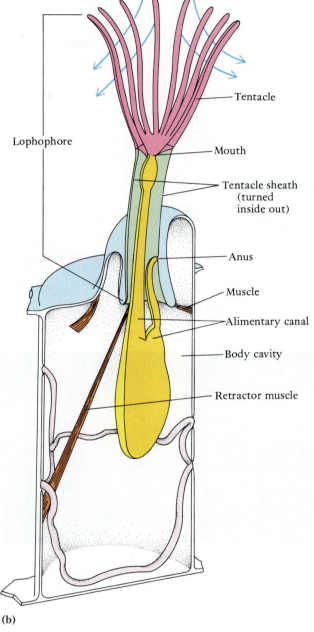

(b)

Figure 33.1

Two individuals in a colony of a generalized bryozoan. The lower surface is attached to the substratum; only upper (frontal) surface is exposed. (a) Individual with lophophore retracted. (b) Individual with lophophore protuded. Arrows indicate direction of water flow.

During feeding, the lateral cilia on the outstretched tentacles drive water down into and out of the sides of the lophophore funnel (Fig. 33.1b). When the lateral cilia are struck by suspended food particles in the incoming water stream, there is a rapid reversal of ciliary beat. The particle is tossed back upward and toward the center of the funnel. In this manner, the particle is bounced down the tentacle toward the mouth. The tentacle tips are always held 110 μm apart, creating a uniform filter regardless of the number of tentacles or their length.

The nervous system consists of a **ganglion** and **nerve ring** around the anterior end of the gut. Fibers extend from the nerve ring to the tentacles and other parts of the body.

No special systems for internal transport, gas exchange, or excretion are present, but certainly the large surface area provided by the extended lophophore is an important site of gas exchange and the elimination of ammonia.

ORGANIZATION OF COLONIES

The many zooids composing bryozoan colonies are derived by asexual budding. Primitively, the members are connected together by a stolon, but in most bryozoans the individuals are attached directly to each other (Figs. 33.1 and 33.2). In either case, the resulting colonies form encrusting sheets or erect plantlike growths (Figs. 33.2b

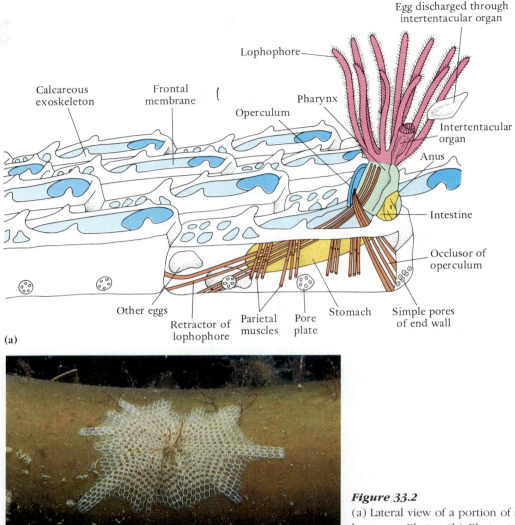

Egg discharged through
intertentacular organ

Lophophore

Calcareous
exoskeleton

Frontal
membrane

Operculum

Pharynx

Intertentacular
organ

Anus

Intestine

Occlusor of
operculum

Simple pores
of end wall

Other eggs

Retractor of
lophophore

Parietal
muscles

Pore
plate

Stomach

(a)

(b)

Figure 33.2
(a) Lateral view of a portion of a colony of an encrusting bryozoan, *Electra*. (b) *Electra pilosa* growing on a marine alga from the coast of Maine.

and 33.3a). To visualize an encrusting colony, imagine a large number of rectangular boxlike individuals lying on their backs on a rock. The dorsal surface is attached to the substratum; the lateral, anterior, and posterior surfaces are attached to surrounding individuals; and the ventral surface is exposed (Fig. 33.2). The orifice of such encrusting species has shifted over to the exposed ventral surface, since the anterior end is connected to adjacent individuals. The colony follows the contours of the rock, shell, or large alga (kelp) on which it is growing and may reach many centimeters in width.

Erect colonies may be leaflike, shrublike, or straplike, but the ventral surface of each zooid remains the one that is exposed. Individuals are attached together by the dorsal or lateral and posterior surfaces. The anterior end may be free. *Bugula*, one of the most widely distributed bryozoans and a common marine fouling organism, illustrates the erect growth form (Fig. 33.3).

Members of a colony are connected to each other more intimately than by the fusion of their exoskeletons. Pores in the walls permit diffusion of substances from one individual to another. Some species possess a cord of peritoneal tissue (funiculus) along which food compounds are moved (Fig. 33.1).

Bryozoans are polymorphic, but the most common members of the colony are the feeding individuals already described. The colonies of some species also include highly modified, bristle-like individuals (vibracula) or pincer-like individuals (avicularia) (Fig. 33.3c). These protect the colony by sweeping away or grasping settling larvae or small animals crawling over the colony. Some species have vegetative stoloniferous individuals that grow over and anchor to the substratum. Reproductive individuals with special brooding chambers called **ovicells** are still another type of zooid found in many bryozoan colonies (Fig. 33.3c).

(a)   **(b)**   **(c)**

Frontal
membrane

Ovicell

Avicularium

Figure 33.3
Bugula, a common widespread genus of erect branching bryozoans. (a) A species of *Bugula* from
Vancouver Island, British Columbia, having erect spirally arranged colonies. (b) Part of a colony of
Bugula neritina, showing the protruded funnel-shaped lophophores. This species is a common
bryozoan along the East Coast of the United States. (c) Section of colony showing defensive individuals
(avicularia) and brood chambers (ovicells).

REPRODUCTION

Most bryozoans are hermaphroditic, and the gonads shed
their gametes into the coelom. Sperm exit through a pore
at the end of two or more tentacles, and the eggs exit
through a special pore between the tentacles, which is
sometimes elevated as an intertentacular organ (Fig. 33.2).
Sperm swept in with the feeding current usually fer-

tilize the eggs as they are released. Sperm probably come
from other individuals in the same colony and other
colonies.

Most bryozoans brood their few eggs during the early
stages of development, commonly in ovicells. The developing eggs are released from the ovicells as larvae that are
a little like the trochophore of annelids and molluscs (Fig.
33.4a). Following a planktonic existence of varying length,

Figure 33.4
Development of the colony of an encrusting bryozoan. (a) Motile larva. (b) First individual of colony
following settling, attachment, and metamorphosis of larva. (c–d) Formation of first four individuals of a
colony by budding. (e–f) More individuals added to colony by further budding.

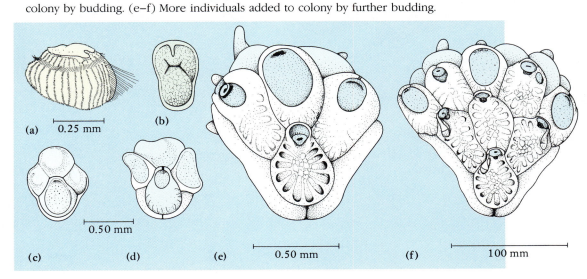

(a)   0.25 mm   **(b)**

0.50 mm

(c)   **(d)**   **(e)**   0.50 mm   **(f)**   100 mm

Figure 33.5
Lophophores projecting from the tubes of the phoronid
Phoronis vancouverensis from the Pacific Coast of the United
States.

the larva settles to the bottom to become the parent
member of a new colony. The body wall evaginates and
separates as a new individual, with the particular way in
which subsequent budding occurs accounting for the
growth form of the colony (Fig. 33.4).

OTHER LOPHOPHORATE PHYLA

Bryozoans are not the only animals having a lophophore.
Two other phyla, the **Phoronida** and the **Brachiopoda**, are
lophophorates. The phoronids (from the name of a
priestess in Greek mythology) are tube-dwelling marine
animals with wormlike bodies. The lophophore tentacles
are arranged like a horseshoe, with the mouth in the center
of the part connecting the two arms (Fig. 33.5). The anus
opens at the anterior end, as in bryozoans. Only about 15
species are known.

The brachiopods, (L. *brachium*, arm + Gr. *pous*,
foot), although represented today by only about 335 species, have a rich fossil record that dates back to the
Cambrian. Brachiopods superficially resemble bivalve
molluscs in having the body enclosed within two calcareous valves and in being about the same size. However, in
brachiopods the valves are dorsoventrally oriented, and
the ventral valve is commonly larger than the dorsal one. A

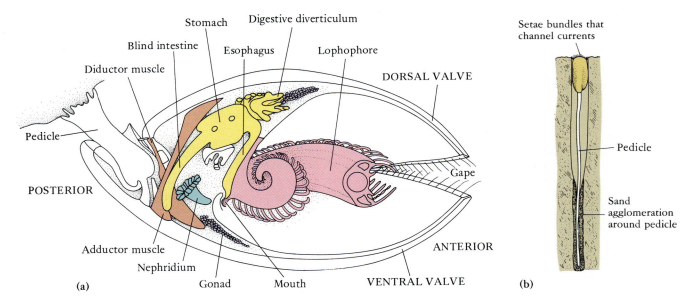

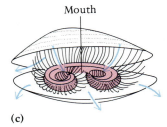

Figure 33.6
Brachiopods. (a) Section through a brachiopod. (b) *Lingula* in feeding position
within burrow. (c) A brachiopod with gaping valves. Arrows show direction of
water flow over spiral lophophore.

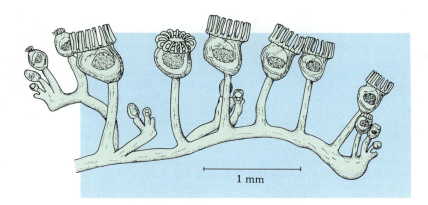

Figure 33.7
Part of a colony of a marine
entoproct.

fleshy stalk (pedicel), which emerges from a hole in the back of the ventral shell, attaches these animals upside down to the bottom (Fig. 33.6a). Because this ventral valve looks somewhat like an ancient oil lamp, the brachiopods are also known as lamp shells. A small number of living species (*Lingula*) live vertically in burrows attached to a long pedicel that emerges from between two equal-size valves (Fig. 33.6b).

The lophophore is horseshoe shaped, as in phoronids, but the arms are typically spiraled, thereby greatly increasing the filtering surface. Water enters and leaves the gape between the two valves following a flow pattern determined by the position of the lophophore (Fig. 33.6c). The sexes of most brachiopods are separate. There is a larval stage, but it is not at all like a trochophore.

The phylum **Entoprocta** is a small phylum of sessile, mostly marine animals. Some are solitary and some are colonial (Fig. 33.7). Entoprocts are rather similar to bryozoans and have often been included within that phylum. However, along with other differences, entoprocts, as the name implies, have the anus opening within the crown of tentacles, not outside, as in the bryozoans.

EVOLUTIONARY RELATIONSHIPS

The three lophophorate phyla may not be closely related, despite the similarity of their feeding organ. The presence of multiciliated cells and a trochophore larva in bryozoans indicates that they are protostomes and they may have had a common origin with the entoprocts. Phoronids and brachiopods have monociliated cells, and their embryology is more like that of deuterostomes. Some zoologists have suggested that phoronids may have been the ancestors of brachiopods.

Summary

1. Members of the phyla Bryozoa, Phoronida, and Brachiopoda all possess a tentacular filter feeding organ called a lophophore. The tentacles contain an extension of the coelom.

2. Bryozoans are colonial, mostly marine lophophorates in which the individuals are very small. The colonies can be quite large. Most bryozoans are covered by an exoskeleton. The lophophore is typically funnel shaped and is protruded from the skeletal encasement by elevated coelomic fluid pressure.

3. Brachiopods are enclosed within two calcareous valves oriented dorsoventrally to the body. They are about the size of bivalve molluscs. Phoronids are tube-dwelling lophophorates.

References and Selected Readings

Detailed accounts of bryozoans, phoronids, and brachiopods can also be found in many of the references listed at the end of Chapter 23. The works listed below deal exclusively with the biology of two of these lophophorate phyla.

Rudwick, M.J.S. *Living and Fossil Brachiopods*. London: Hutchinson University Library, 1970.
Ryland, J.S. *Bryozoans*. London: Hutchinson University Library, 1970.

34

Echinoderms

Characteristics of Echinoderms

■ Echinoderm symmetry is usually radial, typically pentamerous, or five-parted, but secondarily derived. The larva is bilaterally symmetrical.

■ The body wall contains an endoskeleton of calcareous ossicles, which are usually movable against one another but may be fused together (sea urchins and sand dollars) or reduced to microscopic size (sea cucumbers). External spines are commonly present and are a part of the skeleton.

■ Echinoderms possess a unique water vascular system of internal canals and external appendages called tube feet or podia. The system functions in locomotion, gas exchange, feeding, and sensory reception.

■ Movement is by means of tube feet, spines, or arm movement. Sea lilies are attached, and many fossil echinoderms also lived attached.

■ The mouth is in the center of one side of the radial body, and in most echinoderms this oral surface is directed toward the substratum. The anus is usually aboral.

■ Coelomic fluid is the principal means of internal transport, and exchange of gases and wastes between sea water and coelomic fluid takes place across various surface structures, including the tube feet.

■ The nervous system is pentamerous and closely associated with the integument on the oral side of the body.

■ The sexes are separate and the gonoducts are simple. Development typically involves a larval stage having ciliated bands.

(Left) The seastar Dermasterias imbricata *from the coast of Oregon. (Above) A sun star* Pycnopodia helianthoides *from the coast of British Columbia.*

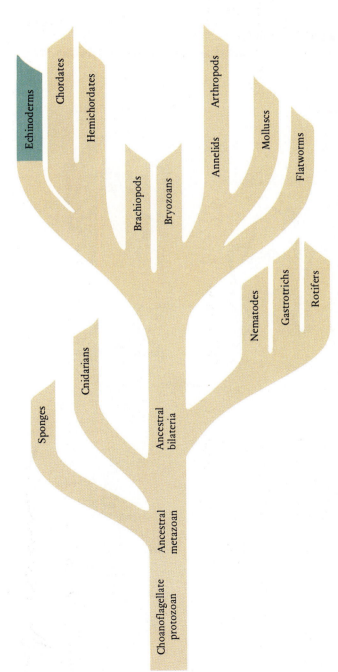

12

Echinoderms, hemichordates, and chordates are deuterostomes, the second major evolutionary line of the animal kingdom (p. 554 and Fig. 23.5). In contrast to the protostomes, the mouth arises as a new opening and the anus forms from or near the blastopore. Cleavage is radial, not spiral (Fig. 23.9), and the fate of the blastomeres is generally fixed much later in development than in protostomes. Primitively, the mesoderm and coelom arise as paired outpocketings of the embryonic gut; the coelom is an enterocoel (p. 515 and Fig. 22.6), in contrast to the schizocoelous mode of coelom formation characteristic of most protostome groups.

13

The 6000 species of the phylum Echinodermata are entirely marine and include the familiar sea stars (starfishes), brittle stars, sea urchins, sand dollars, sea cucumbers, and sea lilies. They are distinguished by possessing a **pentamerous** (five-part) radial symmmetry. The body wall contains an endoskeleton of small calcareous pieces, or **ossicles**, which commonly include surface spines, hence the phylum name Echinodermata—spiny skin. There is a unique **water vascular system** of canals and appendages that function in locomotion, feeding, sensory reception, and gas exchange. The larvae of echinoderms are bilateral and swim and feed by means of ciliated bands that wind over the body (Fig. 34.1). At the end of planktonic life the larval symmetry changes from bilateral to radial. Note then that the echinoderm radial symmetry is secondary, not primary as is true of jellyfish and sea anemones. After surveying the living echinoderm groups, we will examine how this radial symmetry may have evolved.

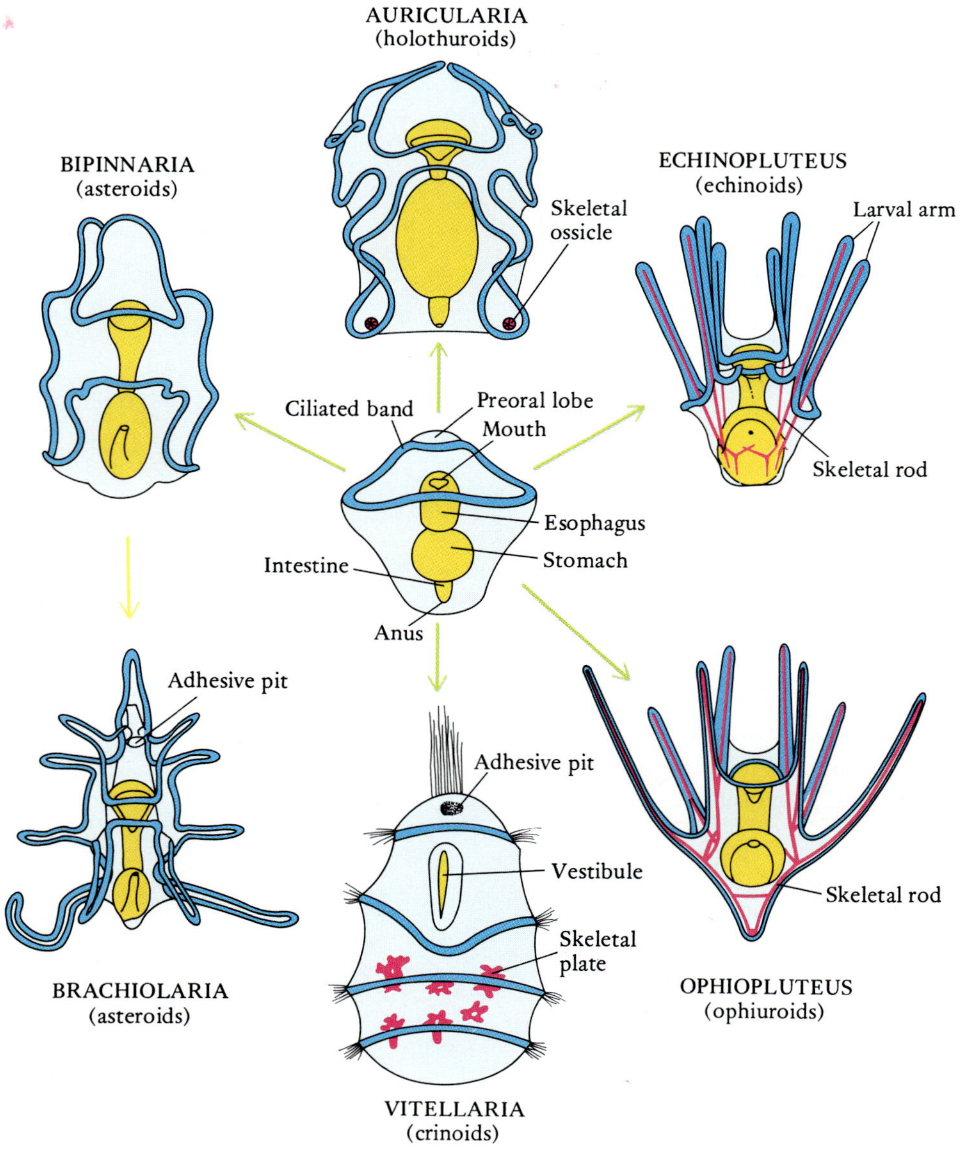

Figure 34.1
Comparison of different larval types of echinoderms. Arrows indicate possible evolutionary relationship to a hypothetical ancestral form. Note that all are bilateral; all bear one or more ciliated bands; and the ciliated bands of some run out onto larval arms.

AURICULARIA
(holothuroids)

Skeletal ossicle

BIPINNARIA
(asteroids)

ECHINOPLUTEUS
(echinoids)

Larval arm

Ciliated band

Preoral lobe
Mouth

Skeletal rod

Esophagus
Stomach

Intestine

Anus

BRACHIOLARIA
(asteroids)

Adhesive pit

Adhesive pit

Vestibule

Skeletal plate

VITELLARIA
(crinoids)

OPHIOPLUTEUS
(ophiuroids)

Skeletal rod

Echinoderms are almost entirely bottom dwellers, and hard substrates such as rock, shell, and coral were, and are, the habitats of many extinct and contemporary forms. But within each class of echinoderms are species that live on soft bottoms and are adapted for life in sand.

CLASS STELLEROIDEA

Asteroids

The class Stelleroidea contains those echinoderms in which the body is drawn out into **arms**. The most familiar stelleroids are members of the subclass **Asteroidea**, which includes the sea stars. In this group the arms are not sharply set off from the **central disc**. There typically are five arms, but the sun stars have many (Fig., top of p. 777). Most sea stars are 12 to 24 cm in diameter, and many are brightly colored in hues of red, orange, or blue.

The mouth is located in the center of the oral surface, which is directed downward. An **ambulacral groove** radiates from the mouth to the tip of each arm and contains the **tube feet** or **podia** (Fig. 34.3d). The aboral surface may appear smooth or granular or may bear prominent spines (Figs. 34.2 and 34.3a). In most sea stars a conspicuous button-like body, the **madreporite**, is located to one side of the central disc (Fig. 34.2).

Body Wall

The surface of asteroids is covered by a monociliated epithelium. A thick layer of connective tissue (dermis) beneath it secretes the endoskeleton of small calcareous ossicles, which are perforated by irregular canals filled with cells. The asteroid ossicles are arranged as a flexible lattice (Fig. 34.4a), and a muscle layer below the dermis enables the arms to bend. A ciliated peritoneum lines the very large coelom (Fig. 34.5b). Circulation of the coelomic fluid is the principal means of internal transport.

All asteroids bear calcareous spines that may be projections of the dermal endoskeleton or special ossicles resting on the others. The spines vary greatly in size and prominence but are usually covered by the outer epithelium. The spines of most sea stars are blunt, but those of the coral-eating crown-of-thorns sea star, *Acanthaster planci* (Fig. 34.2), which has devastated many coral reefs in the Indo-Pacific, have long sharp spines. The tissues covering the spines contain a toxin, and an accidental puncture is very painful to humans.

Located between the spines are small finger-like projections of the body wall called **papulae**, which function in gas exchange and excretion (Figs. 34.3 and 34.5b). Oxygen, carbon dioxide, and ammonia easily cross the papula wall between the inner circulating coelemic fluid and the sea water flowing over the body surface. Waste-laden coelomocytes also accumulate in the papula and are liberated by the pinching off of the papula tip. In those sea stars that live on soft bottoms, the papulae are protected from the surrounding sediment by special table-like spines that create a cover under which the papulae are located and a ventilating current flows (Fig. 34.4b). The table-like spines account for the smooth appearance of some sea stars (Fig. 34.2b).

Figure 34.2

Asteroids. (a) Crown-of-thorns sea star (*Acanthaster*), a common Indo-Pacific sea star that preys on the polyps of stony corals. High populations on some reefs have caused extensive damage to the coral cover. (b) A species of the sea star *Linckia* widely distributed in the Indo-Pacific.

(a)

(b)

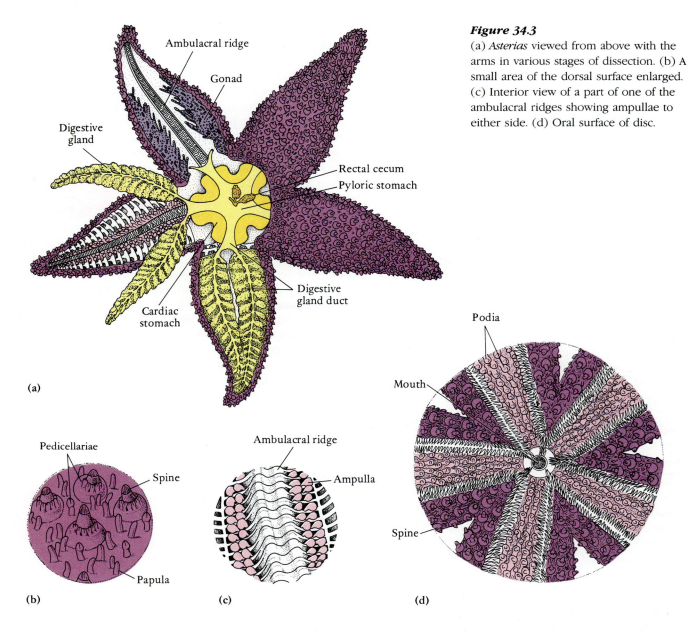

Figure 34.3

(a) *Asterias* viewed from above with the arms in various stages of dissection. (b) A small area of the dorsal surface enlarged. (c) Interior view of a part of one of the ambulacral ridges showing ampullae to either side. (d) Oral surface of disc.

Other structures frequently found on the body wall of some sea stars are minute jawlike appendages called **pedicellariae** (L. *pediculus*, small foot). A pedicellaria contains two ossicles that form the jaws, along with opening and closing muscles (Figs. 34.3b and 34.4c). The pedicellariae are believed to function in the killing of small organisms that might settle on the body surfaces. Such organisms, as well as sediment, are swept clear by the ciliated surface epithelium.

Water Vascular System

The water vascular system, unique to echinoderms, is composed of tubular outpocketings of the body wall—the podia—and an internal system of canals derived from the coelom. The system opens to the exterior through the button-like aboral madreporite, which is perforated by tiny canals. The canals from the madreporite converge into a vertical **stone canal** that extends orally to a **ring canal** embedded in the ossicles around the mouth (Fig. 34.5a). The inner side of the ring canal bears four to five pairs of projections called **Tiedemann's bodies**, which may function as a pathway for fluid between the water vascular system and the body coelom. From the ring canal a **radial canal** extends into each arm, passing between the ossicles at the top of the ambulacral groove. At frequent intervals, **lateral canals** leave the radial canals. Each lateral canal terminates in an **ampulla**, located in the body coelom, and a podium, which projects into the ambulacral groove.

Muscular contraction of the ampulla forces fluid into the podium and at the same time closes a valve in the lateral

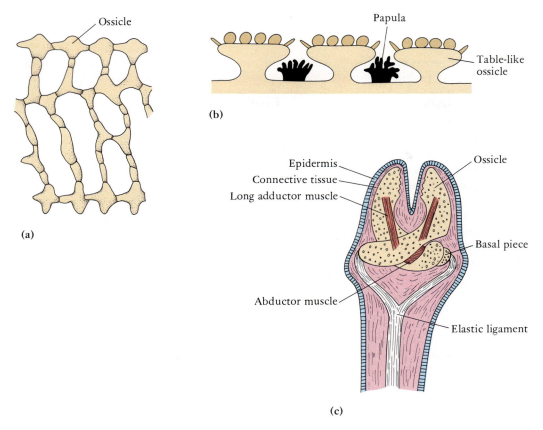

(a)

(b)

Papula

Table-like ossicle

Ossicle

Epidermis
Connective tissue
Long adductor muscle

Ossicle

Basal piece

Abductor muscle

Elastic ligament

(c)

Figure 34.4

(a) Lattice-like arrangement of skeletal ossicles in the arm of a sea star. (b) Diagrammatic cross section through several paxillae of *Luidia*. The raised, table-shaped ossicles bear small rounded spines on the surface and flat movable spines along the edge. Dendritic papulae (black) are located in the spaces between the projecting edges of the paxillae and associated spines. (c) Distal ends of a scissors-type pedicellaria from *Asterias*.

Figure 34.5

(a) Diagram of the asteroid water-vascular system. (b) Diagrammatic cross section through the arm of a sea star.

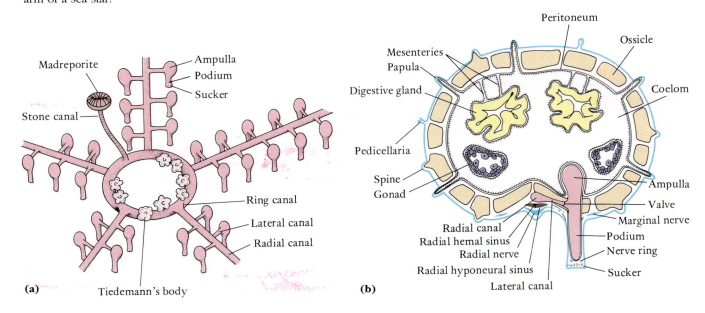

Madreporite

Stone canal

Ampulla
Podium
Sucker

Ring canal

Lateral canal

Radial canal

Tiedemann's body

(a)

Peritoneum

Ossicle

Mesenteries
Papula

Digestive gland

Coelom

Pedicellaria

Spine
Gonad

Ampulla

Valve

Marginal nerve

Radial canal
Radial hemal sinus
Radial nerve
Radial hyponeural sinus
Lateral canal

Podium

Nerve ring

Sucker

(b)

canal, preventing backflow. Hydraulic pressure and asymmetrical muscle contraction extend the podium, and swing it forward. The sucker at its tip is then brought into contact with the substratum. Following adhesion, longitudinal muscles of the podium contract, shortening the podium, forcing water back into the ampulla, and pulling the body forward. Each podium performs a little step, and the combined activity of all the podia enables a sea star to grip objects tenaciously and to crawl about. Recent studies have demonstrated that the attachment and release of the sucker result from the secretions of adhesive and releasing glands similar to those found in flatworms, gastrotrichs, and nematodes. The podia do not move synchronously, but they are coordinated to the extent that they all step in the same direction. One arm acts as the leading arm and exerts a temporary dominance over the other arms. The water vascular canal system appears to function in pressure regulation of the podia, although blocking the madreporite does not cause any immediate impediment to locomotion. Since the podia are thin walled, they also contribute to gas exchange along with the papulae.

Podia with suckers represent an adaptation for life on hard substrates such as rock, shell, and coral. Soft-bottom sea stars have doubled ampullae that provide greater pressure for thrusting the suckerless pointed podia into the sand.

Nervous System

The nervous system of asteroids, like that of most other echinoderms, is primitive in being poorly ganglionated and intimately associated with the epidermal layer. A nerve ring encircles the mouth, and a radial nerve extends into each arm (Fig. 34.5b). Fibers in the nerve ring and the radial nerve function as a rapid conducting system and make connection with neurons of a general epidermal nerve plexus.

An intact nervous system is necessary for the coordination of the podia. If a radial nerve is cut, the podia distal to the cut will continue to move but may not step in the correct direction with those in other arms.

An eye spot at the tip of each arm, composed of a cluster of photoreceptor and pigment cells, constitutes the only sense organ, but individual receptor cells are present in the general body epidermis and are especially concentrated in the epidermis of the podia and the margins of the ambulacral groove.

Nutrition

The mouth of sea stars opens into a large, thick-walled cardiac stomach that fills most of the central disc (Fig. 34.3a). The cardiac stomach opens in turn into a smaller aboral pyloric stomach. A pair of digestive glands located in each arm discharge enzymes into the pyloric stomach. A short intestine extends from the top of the pyloric stomach to an inconspicuous anus opening on the aboral surface.

Associated with the intestine are rectal ceca, outpocketings that function as pumps to expel fecal wastes through the anus (Fig. 34.3a).

Most sea stars are carnivores and scavengers, but there are also many detritus feeders. Crustaceans, molluscs, and other echinoderms are common prey. Some species, especially those with short arms, swallow their prey whole; others evert the cardiac stomach over the prey. *Asterias* and related species feed largely on bivalve molluscs, penetrating their prey in a remarkable way. The sea star humps over the mollusc, the mouth directed over some part of the edges of the valves. The continual pull exerted by the arms produces a very slight opening between the valves, and through this opening the stomach of the sea star is slithered. These animals can pass the cardiac stomach through a cleft of no more than 0.1 mm. It is this ability, rather than the ability to pull, that enables sea stars to prey upon bivalves. Many bivalves cannot close the valves tightly enough to prevent entrance of the sea star's stomach. Sea stars can be a serious pest of commercial oyster beds. They are commonly removed with a special mop to which the sea stars attach by their pedicellariae when the mop is dragged across the oyster bed.

Enzymes produced by the digestive glands flow into the cardiac stomach, which also secretes enzymes. The pyloric stomach is simply a center for the convergence of the digestive gland ducts. In those species that evert the stomach, the enzymes flow out onto the everted surface and initiate digestion outside the body. Digestion of the bivalve's adductor muscle causes the valves to gape widely. The stomach is later retracted, bringing with it the partially digested prey. The digestive glands appear to be the principal site of nutrient absorption.

Although the circulating coelomic fluid is the principal means of internal transport, sea stars also possess an inconspicuous, radially arranged system of sinus channels, called the hemal system, that parallels the water vascular system. The hemal system has been demonstrated to play a role in the distribution of food material.

Reproduction and Development

Like most echinoderms, sea stars have separate sexes; there are usually two gonads in each arm (Figs. 34.3a and 34.5b), with a simple gonoduct from each leading to an inconspicuous gonopore at the base of the arm. The eggs and sperm are shed freely into the sea water, where fertilization takes place. Development usually occurs in the plankton, but some species brood their eggs.

In development, the appearance of ciliated bands on the body surface indicates that development has reached a larval stage called a bipinnaria (Figs. 34.1 and 34.6a and b). The bipinnaria larvae, like larvae of all other echinoderms, are distinctly bilateral and are believed to reflect the original symmetry of ancestral echinoderms. Internally there is a gut and three pairs of coelomic vesicles; the first

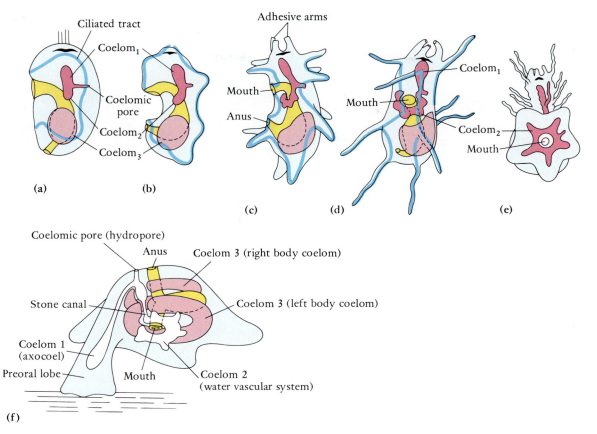

Coelomic pore (hydropore)

Anus

Coelom 3 (right body coelom)

Stone canal

Coelom 3 (left body coelom)

Coelom 1 (axocoel)

Preoral lobe

Mouth

Coelom 2 (water vascular system)

(f)

Figure 34.6

Diagrammatic lateral views of larval development and metamorphosis of a sea star, showing development of coelom and water vascular system. (a,b) Early bipinnaria larva; (c) brachiolaria larva; (d) attached metamorphosing larva; (e,f) young sea star developing from posterior part of old larva.

two pairs are connected (coelomic vesicles 1 and 2 in Fig. 34.6a). The bipinnaria gradually develops body projections, called **larval arms**, over which the ciliated bands extend. The larval arms, which are not the same as the adult arms, are probably an adaptation for increasing the ciliated locomotor and feeding surface. Toward the end of the bipinnaria stage anterior adhesive structures form (three short arms and a sucker), and the larva, now called a **brachiolaria** (Figs. 34.1 and 34.6c), settles to the bottom and attaches. A radical and complex metamorphosis ensues. The larval arms and much of the gut degenerate, and the radially symmetrical adult body develops (Figs. 34.6d–f). The original left side of the larva becomes the oral surface and the right side, the aboral surface. The water vascular system develops from the left anterior two larval coelomic vesicles (coeloms 1 and 2 in Fig. 34.6), and the body coelom develops from the posterior pair of larval vesicles (coelom 3). After metamorphosis the little sea star detaches and crawls away.

The regenerative ability of sea stars is well known. Any fragment of the body that contains a portion of the central disc is capable of complete regeneration of an entire individual (see Fig. 21.1d). But the process is slow and may take as long as a year. Only in some sea stars is regeneration coupled with asexual reproduction, when the central disc is cleaved along certain predetermined lines and each part grows into a new sea star.

Ophiuroids

Closely related to the Asteroidea is the largest group of echinoderms, the subclass **Ophiuroidea**, which contains the basket stars and brittle stars (serpent stars). The great success of ophiuroids is certainly related to the versatility of their feeding habits, their mobility, and their small size, which enable them to exploit habitats unavailable to other echinoderms.

The ophiuroid body, like that of the asteroids, is composed of arms and a central disc, but the arms are long, slender, and sharply set off from the disc (Fig. 34.7). The

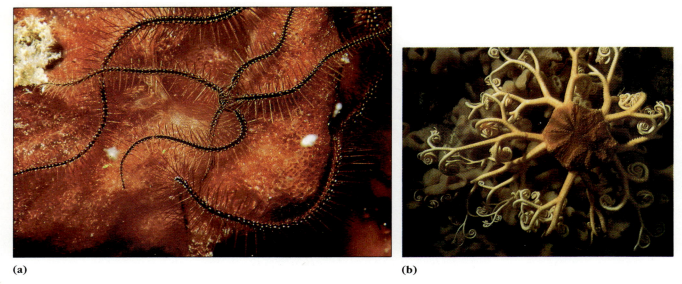

(a)   **(b)**

Figure 34.7

Ophiuroids. (a) A brittle star on the surface of a sponge. (b) The basket star
Gorgonocephalus arcticus from the coast of Maine.

arms are highly mobile and easily broken, characteristics
that led to the names serpent star and brittle star. Most
brittle stars are small and have a central disc no larger than
several centimeters; basket stars are larger, with discs
4.5 cm across, and the arms are branched (Fig. 34.7b).

The arms of ophiuroids are nearly filled by a series of
large **vertebral ossicles**, so called because of their superficial similarity to the bones in the vertebrate backbone (Fig.
34.8). Each vertebra is covered by flattened ossicles called
shields, one on each side of the oral, aboral, and lateral
surfaces (Fig. 34.9). The shields are visible externally and
account for the jointed appearance of the arms. There is no
ambulacral groove, but podia are located along the length
of the oral side of the arm. The articulation of the arm
ossicles and their musculature give the arms great mobility; brittle stars move by flexing and pushing with their

arms rather than by means of the podia. Spines along the
sides of the arms increase traction. An arm can be broken
off at any point if seized by a predator. This ability to
undergo self-amputation, or **autotomy**, is an escape adaptation also found in some arthropods, such as crabs.

Ophiuroids are especially common in rock and coral
habitats, where they live beneath stones and in crevices
and holes, but some live in burrows on sandy bottoms.
Basket stars and some brittle stars can coil their arms and
can climb, and are especially common on branched corals
and sea rods (gorgonian corals).

Brittle stars may be scavengers, deposit feeders, or
suspension feeders, and many species utilize more than
one mode of feeding. In scavenging, the looping motion of
the arms and the five jawlike ossicles that frame the mouth
are used to ingest the bodies of dead animals (Fig. 34.9). In

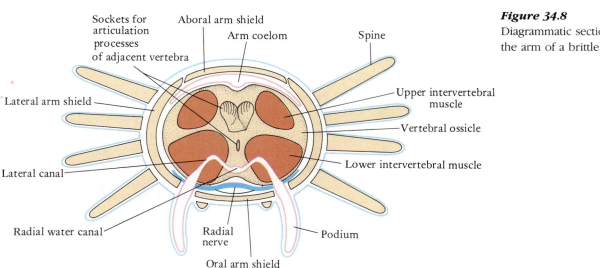

Figure 34.8

Diagrammatic section through
the arm of a brittle star.

Sockets for articulation processes of adjacent vertebra
Aboral arm shield
Arm coelom
Spine
Lateral arm shield
Upper intervertebral muscle
Vertebral ossicle
Lower intervertebral muscle
Lateral canal
Radial water canal
Radial nerve
Podium
Oral arm shield

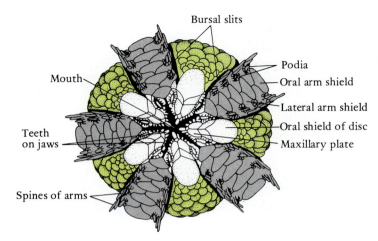

Bursal slits
Mouth
Podia
Oral arm shield
Lateral arm shield
Oral shield of disc
Maxillary plate
Teeth on jaws
Spines of arms

Figure 34.9
Oral view of disc of a brittle star.

deposit feeding, the podia are used to pick up organic particles from bottom detritus and pass them down the oral side of the arms to the mouth. In suspension feeding, the arms are lifted up into the water current, and suspended plankton and detritus are trapped on podia or on strands of mucus draped between spines. The material is collected as mucus-bound balls and pushed by the podia along the length of the arms to the mouth.

The podia of brittle stars are suckerless and lack ampullae. This may be correlated with their mode of locomotion and the great reduction in the arm coelom. Fluid pressure is derived from contractions of the lateral canals. An ossicle on the oral surface of the disc serves as the madreporite.

Ophiuroids do not have papulae; their gas exchange organs are pouches, called **bursae**, that are infoldings of the body wall. The bursae open to the surface of the oral disc by slits at the base of each arm (Fig. 34.9). The ventilating current is produced by cilia or pressure changes within the disc. Coelomic fluid is the medium for internal transport.

Most brittle stars have separate sexes, and the gonads are attached to the coelomic side of the bursal sacs. When mature, the gametes rupture into the bursae and exit out the bursal slits. Development is similar to that of asteroids, but the larva, called an **ophiopluteus** (Fig. 34.1), has a different arrangement of larval arms, and metamorphosis occurs before settling. There is no attachment phase as in asteroids. Many species brood their embryos in the bursae, and a little brittle star crawls out of one of the slits.

CLASS ECHINOIDEA

The class Echinoidea contains the sea urchins, most of which are adapted for life on hard bottoms, and the sand dollars and heart urchins, which are adapted for burrowing in sand. Both groups lack arms and the body is spherical or discoidal (Fig. 34.10). The skeletal ossicles are flattened plates fused together to form a rigid internal shell, or **test**, and the body surface is covered with movable **spines** mounted upon tubercles.

Sea Urchins

The body of a sea urchin is spherical, with the oral pole bearing the mouth directed downward (Fig. 34.10a, b). Although there are no arms, five ambulacral plates bearing the tube feet radiate out and upward over the test to the aboral pole (Fig. 34.11b). Thus the body is divided around the equator into alternating ambulacral and interambulacral meridians. The anus opens at the aboral pole within a circular area filled with small plates, called the **periproct**. Around the periproct are five large **genital plates**, each containing a conspicuous gonopore. One of the genital plates also functions as the madreporite.

The long movable spines are usually covered by the surface epithelium and articulate on a tubercle on the test surface (Fig. 34.12). Members of the tropical genus *Diadema* possess extremely long needle-like spines that break off in the flesh of a punctured intruder. Toxic substances in the epidermis around the spine may produce an initially painful puncture in humans; the fragment of calcium carbonate is gradually resorbed.

Many echinoderms possess an unusual type of connective tissue capable of rapidly shifting from a soft to a rigid state. In sea urchins, connective tissue of this type forms a ring around the base of the spines and when the tissue becomes rigid, the spine is erected (a protective state). A ring of muscle fibers enables the spine to be moved about. Stalked pedicellariae are located between the spines, and some species of sea urchins possess pedicellariae with poison glands (Fig. 34.12). Such pedicellariae are too small to be dangerous to humans.

Many sea urchins use the podia, which can extend beyond the spines, to move about in the same manner as asteroids, and the water vascular system is basically similar.

(a)　　　　　　　　　　　　(b)

(c)　　　　　　　　　　　　(d)

Figure 34.10
Echinoids. (a) The purple sea urchin (*Strongylocentrotus purpuratus*) from the
west coast of the United States. (b) The Hawaiian slate pencil urchin,
Heterocentrotus mammillatus. (c) Sand dollar, *Dendraster excentricus*, from the
Pacific Coast of the United States. (d) Heart urchin, *Lovenia cordiformis*, removed
from its burrow; from the Pacific Coast of the United States.

However, each ampulla on the inside of the test and the podia on the outside are connected by double perforations through the ossicle (Figs. 34.11 and 34.12). Locomotion may also be aided by the pushing movement of the spines, and some sea urchins move entirely with the spines, using the podia only for gripping the surface.

Sea urchins feed on algae and encrusting animals, which are scraped up or chewed with a complex movable apparatus known as **Aristotle's lantern** that contains five projecting teeth (Fig. 34.13). The gut is tubular and loops about inside the test (Fig. 34.14).

Five pairs of gills, which are highly branched outpocketings of the body wall, are located to either side of the ambulacral areas at the oral pole and provide for gas exchange (Fig. 34.11a). Coelomic fluid is pumped into and out of the gills. As in all echinoderms, the podia are thin extensions of the body wall and contribute in some degree to gas exchange.

Sand Dollars and Heart Urchins

Sand dollars and heart urchins burrow in sand. These animals, when moving, always keep the same meridian forward and thus have a definitive anterior end. Shifts in the position of the oral center or anus, or both, have led to a degree of bilateral symmetry. In sand dollars, the oral-aboral axis is so depressed that the body is flattened (Fig. 34.10c). The mouth is still in the center of the oral surface, but the anus has shifted out of the aboral center and is located orally and eccentrically toward the posterior end. Heart urchins are somewhat egg shaped, and not only has the anus shifted out of the aboral center, but the entire oral center has also shifted forward, making these animals even more strikingly bilateral (Fig. 34.10d).

These animals crawl slowly through the sand by the movement of the tiny spines that cover their surface. Sand dollars burrow into the surface layer of sand; heart urchins construct burrows well below the surface.

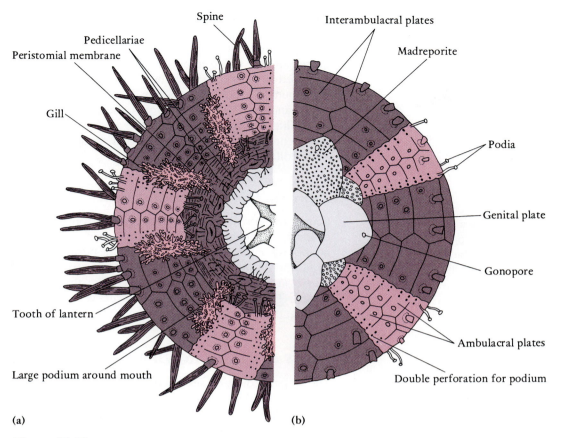

Figure 34.11
The common Atlantic coast sea urchin, *Arbacia punctulata*. (a) Oral view.
(b) aboral view.

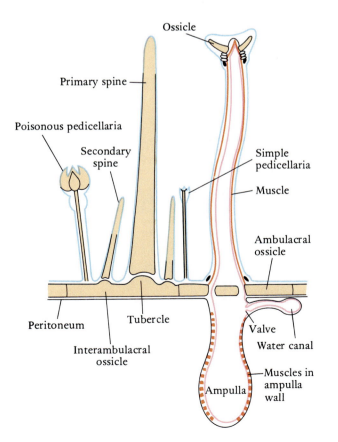

Figure 34.12
Diagrammatic section through the body wall of a sea urchin,
showing one ambulacral and one interambulacral ossicle and
associated structures.

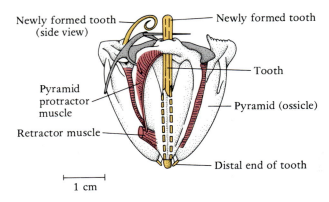

Newly formed tooth (side view)　　Newly formed tooth

Pyramid protractor muscle

Retractor muscle

Tooth

Pyramid (ossicle)

Distal end of tooth

1 cm

Figure 34.13
Lateral view of Aristotle's lantern of a sea urchin. The five large ossicles shaped like arrow heads (pyramids) form the principal framework of the apparatus. The five beltlike teeth beneath them converge at the mouth. As the teeth are eroded away at the oral end, new teeth material is added at the aboral end. Muscles enable the apparatus to be protracted or retracted slightly from the mouth and also rocked from side to side.

On the aboral surface, the podia are very wide and flattened as a modification that increases the surface area for gas exchange. The dense covering of spines extends just beyond the podial gills and prevents the gills from being smothered by sand. Each of these five areas of specialized podia is called a **petaloid**, since they are shaped somewhat like the petal of a flower (Fig. 34.15a).

Soft-bottom echinoids are selective deposit feeders and use the podia to collect food particles. In sand dollars, the oral podia pick up organic particles among the sand grains and then pass them down a branching system of food grooves that converge into five principal grooves corresponding to the ambulacral areas. The food grooves are lined by podia that push the collected food masses to the mouth. The conspicuous slots (**lunules**) in the test of some sand dollars (Fig. 34.15a) are believed to function as an antilift mechanism, reducing the tendency of a rapid water current flowing over the aboral surface to lift the sand dollar off the substratum.

Reproduction and Development

Four or five gonads are suspended radially from the inside of the test, and the short gonoducts open through the gonopores in the genital plates around the periproct (Figs. 34.11 and 34.14). Fertilization is external, and the larva, called an **echinopluteus**, looks much like that of ophiuroids (Fig. 34.1). Metamorphosis occurs very rapidly on settling; there is no attachment phase.

CLASS HOLOTHUROIDEA

The holothuroids, or **sea cucumbers**, are similar to sea urchins in lacking arms, but in holothuroids the oral-aboral axis is greatly lengthened, so that the animal has a worm-like or cucumber-like shape and lies on its side (Fig. 34.16). Unlike other echinoderms, the skeleton is reduced to microscopic ossicles and the body wall has a leathery texture. Most species range from 6 to 30 cm in length and are commonly colored black, brown, or white.

Figure 34.14
Internal structure of the sea urchin *Arbacia* (lateral view).

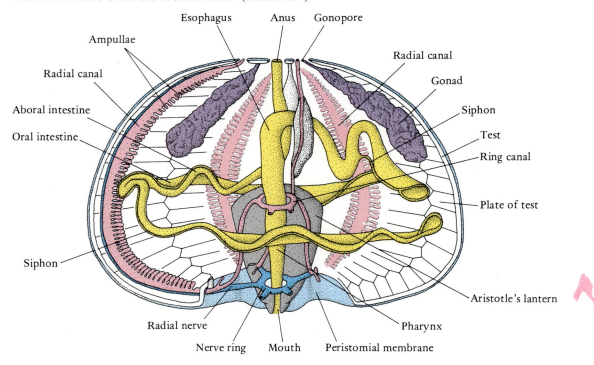

Esophagus　　Anus　　Gonopore

Ampullae

Radial canal

Aboral intestine

Oral intestine

Radial canal

Gonad

Siphon

Test

Ring canal

Plate of test

Siphon

Aristotle's lantern

Radial nerve

Nerve ring　　Mouth　　Peristomial membrane

Pharynx

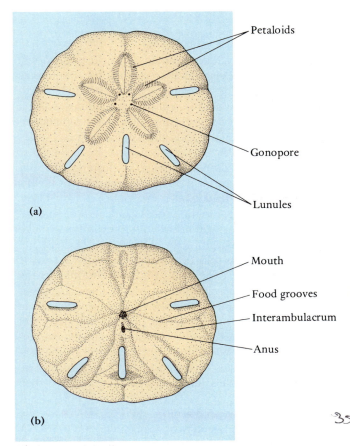

Petaloids

Gonopore

Lunules

(a)

Mouth

Food grooves

Interambulacrum

Anus

(b)

Figure 34.15
The Atlantic coast five-slotted sand dollar, *Mellita quinquiesperforata*. (a) Aboral view; (b) oral view.

Many sea cucumbers live on hard substrata. A few live exposed on the surface; others lodge themselves beneath and between stones or within coral crevices. The body wall connective tissue can shift rapidly from a flaccid to a rigid state, as described in connection with sea urchin spines, making it difficult for a predator to dislodge the animal from its crevice. Hard-bottom sea cucumbers move by means of podia; in some species, three ventral ambulacra are kept against the substratum as a sole and the two dorsal ambulacra have reduced podia (Fig. 34.17). Note that this differentiation of ambulacra makes the body bilateral in symmetry, but the derivation is different from that of the sand dollars and heart urchins.

Sea cucumbers that live in sand or mud generally construct burrows with one or two openings to the surface. The very wormlike synaptid sea cucumbers burrow by peristaltic contractions; their podia have completely disappeared.

At the oral end of the body, a circle of **tentacles** representing modified podia surrounds the mouth (Figs. 34.16b and 34.17). The tentacles are outstretched and collect plankton or detritus from the surrounding sea water or sea bottom. They are then retracted and stuffed one at a time into the mouth. The deposit feeders may be selective or nonselective, and some nonselective deposit feeders leave conspicuous castings on the bottom surface.

The gut of sea cucumbers is tubular and leads, before the anus, into a muscular cloaca that is involved in gas exchange (Fig. 34.18). The actual gas exchange organs are unusual tubular branching structures called **respiratory trees** that arise as evaginations of the cloacal wall and extend into the coelom. The pumping action of the cloaca moves a ventilating current of sea water into and out of the respiratory trees. Some sea cucumbers possess the most highly developed hemal system (p. 782) of any echinoderm, with conspicuous vessels and regions of peristaltic contractions propelling the blood. The hemal system functions in both food and gas transport in these sea cucumbers and certain of the blood cells contain hemoglobin.

Figure 34.16
Holothuroids. (a) Sea cucumbers on a reef flat at Guam. Both are species of *Holothuria*, the white color of one species comes from foram shells (p. 568) that stick to its surface. (b) A synaptid sea cucumber from Guam with tentacles extended. The body of these cucumbers is long and wormlike and lacks podia.

(a)

(b)

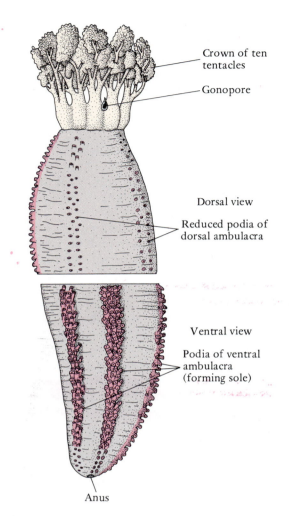

Crown of ten tentacles

Gonopore

Dorsal view

Reduced podia of dorsal ambulacra

Ventral view

Podia of ventral ambulacra (forming sole)

Anus

Figure 34.17
The North Atlantic sea cucumber *Cucumaria frondosa*.

In the tropics, a little commensal pearlfish (*Carapus*) uses the base of the respiratory tree as a home. To enter its home, the fish nudges the anus of the sea cucumber with its snout and then backs in, tail first, sometimes twisting against the pressure of the closing anus. As an adaptation to entering and leaving the anus, the fish has reduced scales and no pelvic fins.

Some sea cucumbers possess a cluster of tubular evaginations from the base of the respiratory trees. Called **tubules of Cuvier**, they can be shot out of the anus (Fig. 34.19). They elongate in the process and are adhesive. An intruder or predator that disturbs a sea cucumber enough to evoke the discharge of the tubules can become enmeshed in a death trap of adhesive threads. Not to be confused with the expulsion of adhesive tubules is the phenomenon of evisceration, in which a sea cucumber will rupture at the anterior or posterior end and discharge the gut and some other internal organs. Although most commonly seen in stressed laboratory animals, eviscerated animals have been found under natural conditions, and for some species evisceration may be a seasonal removal of waste-laden internal organs, which are then regenerated.

There is only one gonad in sea cucumbers, and the gonopore opens between two of the tentacles (Fig. 34.17). The bilateral larva, an **auricularia**, possesses ciliated locomotor bands but never develops the long larval arms characteristic of other echinoderm larvae (Fig. 34.1). Transformation to a young sea cucumber occurs before settlement.

CLASS CRINOIDEA

The echinoderms we have examined thus far have the mouth directed downward and the oral surface placed against the substratum or, in the case of holothuroids, lie on their sides. However, in one living class of echino-

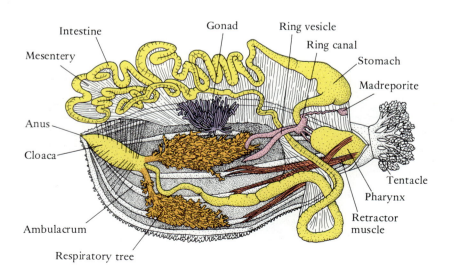

Intestine

Mesentery

Anus

Cloaca

Ambulacrum

Respiratory tree

Gonad

Ring vesicle

Ring canal

Stomach

Madreporite

Tentacle

Pharynx

Retractor muscle

Figure 34.18
The sea cucumber, *Thyone briareus*, cut open along one side. The digestive tract has been moved to one side to show the respiratory trees, retractor muscles of the anterior end, and the internal surface of the body wall with its five ambulacra. In holothurians, the madreporite lies in the body cavity, so that the water vascular system is not filled with sea water but with coelomic fluid.

Figure 34.19
A species of *Holothuria* from Hawaii releasing defensive adhesive tubules of Cuvier.

derms, the **Crinoidea**, the oral surface is directed upward. Moreover, many members of this class are attached to the substratum by a **stalk**. The attached crinoids are believed to be the most primitive of living echinoderms and probably illustrate the phylum's ancestral mode of existence.

Crinoids consist of two groups: the sessile **sea lilies**, in which the aboral surface is connected to the substratum by a stalk (Fig. 34.20a), and the free-moving **feather stars**, in which a stalk is absent (Fig. 34.21). Sea lilies are usually less than 70 cm in length, and most occur in relatively deep water. The stalk, which is composed of ossicles and can bend, is attached to the aboral surface of the body proper, or **crown**. The pentamerous crown bears arms that fork repeatedly in many species. All along the length of the arms are side branches called **pinnules**. A ciliated ambulacral groove with flanking suckerless podia runs the length of the arms and up onto the pinnules (Fig. 34.20b). The grooves from all the arms converge at the mouth in the center of the oral surface. The arms are composed mostly of ossicles, but the articulation and musculature permit considerable movement.

Feather stars are similar to sea lilies, from which they clearly have evolved. They are stalked and attached in the last stages of larval development. Then the crown breaks free and swims away as a tiny feather star. Although the animal is unattached, its oral surface is still directed upward (Fig. 34.21). They perch for long periods of time by means of an aboral ring of clawlike projections, called **cirri**, on rocks, on coral, or even on soft bottoms. Feather stars swim intermittently by rapidly raising and lowering the arms.

Feather stars are often abundant on coral reefs, especially in the Indo-Pacific Oceans. None occur in very shallow water in the Western Atlantic.

Crinoids are suspension feeders. Indeed this mode of feeding may have been the original function of the water vascular system. On contact, rapid movements of the podia

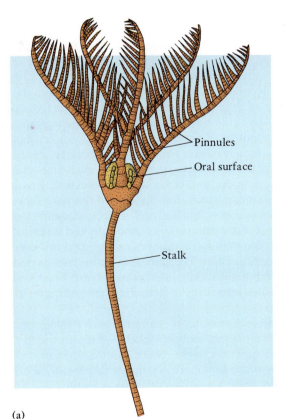

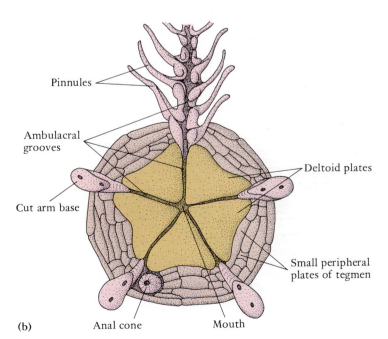

Figure 34.20
(a) *Ptilocrinus pinnatus*, a stalked crinoid (or sea lily) with five arms.
(b) Oral surface of a crinoid crown, including a section of one arm. Podia not shown.

Pinnules
Oral surface
Stalk

(a)

Pinnules
Ambulacral grooves
Cut arm base
Deltoid plates
Small peripheral plates of tegmen
Anal cone
Mouth

(b)

Figure 34.21
A feather star from the Red Sea perched on a coralline rock.

whip suspended food particles into the ambulacral grooves, where they are entrapped in mucus and conveyed within the ciliated groove down to the mouth. The branching arms and pinnules greatly increase the surface area available for collecting food (Fig. 34.22). When feeding, crinoids unroll their arms and hold them in various positions, e.g., like a fan or an umbrella, depending upon the direction of the water currents.

The anus of crinoids is located on a short elevation to one side of the central oral surface. Feces can thus be more readily swept away without fouling the ambulacral grooves. Podia are the gas exchange organs.

Gametes develop from coelomic epithelium in the arms and pinnules, and spawning takes place by rupture of the body wall.

Some species retain their embryos in brood chambers on the arms. The larva, a **vitellaria**, is barrel shaped with encircling ciliated bands (Fig. 34.1).

Figure 34.22
Extended position of the podia along the pinnules of the feather star *Florometra serratissima*.

CLASS CONCENTRICYCLOIDEA

In 1983 and 1984 representatives of a new class of echinoderms were discovered living on sunken wood in about a thousand meters of water off the coast of New Zealand. Measuring only 2–9 mm across, these little echinoderms look like medusae, hence their name *Xyloplax medusiformis*. The body is somewhat disc-shaped with no arms and is covered by small plate-like ossicles and a ring of short marginal spines (Fig. 34.23a). Unlike that in other echinoderms, the water vascular system is composed of two ring canals located near the margin, the outer one connected by short lateral canals to podia (Fig. 34.23b). There is no gut, but the ventral surface is thought to represent an everted stomach. There are five bursae, each associated with a pair of gonads, To date little is known about the biology of these echinoderms.

EVOLUTIONARY RELATIONSHIPS

From our brief survey of echinoderms, you can see that crinoids, with their oral surface directed upward, are very different from sea stars, brittle stars, and sea urchins, in which the oral surface is directed downward. Clearly, there is no way that the podia of crinoids could have the locomotor function of the podia of sea stars. Which is the more primitive function? What were the first echinoderms like? And why among all of the bilateral animals have the members of this phylum secondarily shifted to a radial symmetry?

There is a rich fossil record of echinoderms that provides some answers to these questions. The phylum made its appearance in the Cambrian, and a number of classes, now extinct, flourished during the Paleozoic. Most of these Paleozoic echinoderms were attached with the oral surface directed upward (Fig. 34.24a), as is true of crinoids today. This tells us that such an orientation is probably primitive. Echinoderm fossils consist only of ossicles; we have no direct record of the podia. However, podia almost certainly were present, since all living echinoderms have podia, and pores through which they pass are found in many fossils. But they could not have functioned in locomotion, for the oral surface was directed upward. Most of the Paleozoic echinoderms had spherical bodies with the ambulacral grooves extending down over the body surface, rather like spineless sea urchins turned upside down. Significantly, the grooves were bordered with **brachioles**, structures very similar to crinoid pinnules (Fig. 34.24b).

All of this suggests that these early fossil echinoderms fed in a similar manner to that of modern crinoids and that the original function of the water vascular system was feeding. The brachioles increased the surface area for food collection as do the pinnules of living crinoids.

The larvae of all living echinoderms are bilateral and coelomate, and in the course of development the symme-

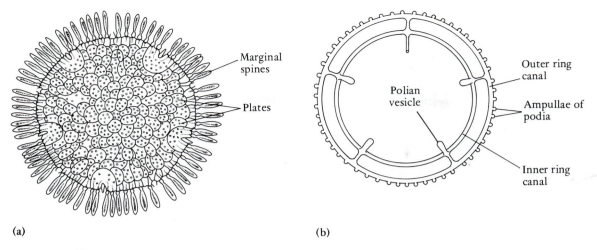

(a)   (b)

Figure 34.23

Xyloplax medusiformus (class Concentricycloidea). (a) Dorsal view; (b) water vascular system.

try shifts to radial. In both feather stars and sea stars, the larva becomes attached during the transformation. From this we postulate that the ancestors of echinoderms were some group of motile bilateral coelomates. The ancestral stock took up an attached existence, which resulted in a shift from a bilateral to a more adaptive radial symmetry (see Fig. 10.3). Also correlated with an attached mode of existence was the evolution of the calcareous skeleton and suspension feeding.

Modern free-moving echinoderms appeared late in the fossil record, and there are no fossil groups bridging the gap between those species that have the oral surface directed upward and the modern forms that have the oral surface directed downward. The functional shift is so great, however, that some intermediate conditions must have occurred. In any event, locomotion appears to be a basic function of the podia in modern forms (sea stars, sea urchins, and sea cucumbers). The exceptions to this pattern are the brittle stars and sand dollars, although they probably evolved from forms that used the podia for locomotion.

From a locomotor function, at least some of the podia in different groups of modern echinoderms have become adapted for gas exchange, sensory reception, selective deposit feeding, nonselective deposit feeding, suspension feeding, and food transport.

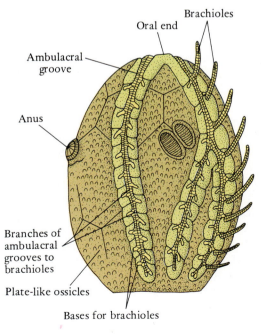

Figure 34.24

(a) Reconstruction of a Paleozoic cystoid, member of an extinct class of stalked echinoderms, attached to an erect bryozoan. (b) Lateral view of a cystoid.

(a)   (b)

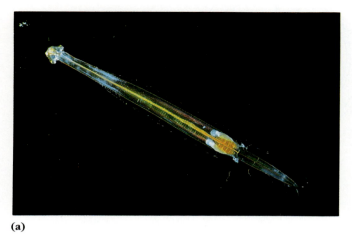

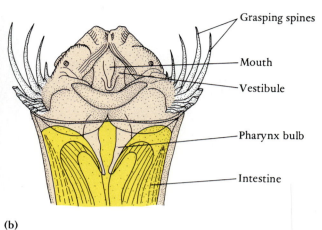

(a)

(b)

Figure 34.25
Phylum Chaetognatha. (a) A marine planktonic arrowworm.
(b) Anterior end of a chaetognath, showing grasping spines.

ANOTHER DEUTEROSTOME PHYLUM

The phylum **Chaetognatha** (Gr. *chaite*, hair + *gnathos*, jaw) contains a small group of common planktonic animals called arrowworms (Fig. 34.25). They appear to be deuterostomes, but their relationship to echinoderms is obscure. Arrowworms have transparent, torpedo-shaped bodies, about 1–3 cm long, bearing paired fins. They alternately float and swim with a rapid dartlike motion. Arrowworms feed on other planktonic animals, which they capture with anterior chitinous grasping spines.

Classification of Phylum Echinodermata

Subphylum Echinozoa (Gr. *echinos*, spiny + *zoon*, animal) Radially symmetrical, commonly globoid echinoderms without arms or armlike appendages. Includes three extinct and two living classes.

 Class Holothuroidea (Gr. *holothourion*, or sea polyp) Sea cucumbers. Mouth and anus at opposite ends of an orally-aborally elongated body. Oral end bearing tentacles derived from podia. Ossicles microscopic. *Holothuria, Cucumaria, Thyone.*

 Class Echinoidea (Gr. *echinos*, spiny + *eidos*, form) Sea urchins and sand dollars. Body globose or greatly flattened dorsoventrally. Ossicles fused to form a rigid test that is covered with movable spines. *Arbacia, Lytechinus, Strongylocentrotus* (sea urchins); *Mellita, Dendraster* (sand dollars); *Echinocardium, Spatangus* (heart urchins).

Subphylum Crinozoa (Gr. *krinon*, lily + *zoon*, animal) Radially symmetrical globoid echinoderms with arms or brachioles. Oral surface directed upward. Includes the extinct cystoids and blastoids, several other extinct classes, and one living class.

 Class Crinoidea (Gr. *krinon*, lily + *eidos*, form) Sea lilies and feather stars. Living and fossil echinoderms dating from the early Paleozoic. Body attached by a stalk (sea lilies) or free (feather stars). Ambulacra located on arms, which are typically branched. *Cenocrinus* (sea lily); *Antedon, Florometra* (feather stars).

Subphylum Asterozoa (Gr. *aster*, star + *zoon*, animal) Radially symmetrical, unattached star-shaped echinoderms. Oral surface directed downward.

 Class Stelleroidea (L. *stella*, star + Gr. *eidos*, form) Body composed of a central disc and radiating arms.

  Subclass Asteroidea (Gr. *aster*, star + *eidos*, form) Sea stars. Arms grade into central disc. Tube feet are located within a groove on the underside of the arms. *Asterias, Astropecten, Pisaster, Patiria; Heliaster, Pycnopodia* (sun stars).

  Subclass Ophiuroidea (Gr. *ophis*, serpent + *eidos*, form) Brittle stars and basket stars. Arms sharply set off from a central disc. Arms with central vertebral ossicles and lacking an ambulacral groove on the oral side. *Ophiothrix, Ophioderma* (brittle stars); *Astrophyton* (basket star).

 Class Concentricycloidea (L. *concentricus*, concentric + *cyclus*, ring) Minute disc shaped echinoderms having a water vascular system composed of two peripheral ring canals and no radial canals. Despite the absence of arms, these echinoderms have been placed in the Asterozoa because of certain features of growth and the fact that the podia pass between the plates.

Summary

1. The phylum Echinodermata contains marine animals having a pentamerous radial symmetry, an internal skeleton of calcareous ossicles, which often includes surface spines, and a water vascular system composed of internal canals and surface appendages (podia) used in feeding, locomotion, gas exchange, and sensory reception. The radial symmetry arises secondarily following the metamorphosis of a bilateral larva, bearing one or more ciliated bands used in locomotion and feeding.

2. The class Stelleroidea contains those echinoderms in which the body is composed of a central disc and arms. In the subclass Asteroidea, which includes the sea stars and sun stars, the arms and disc are not sharply set off from each other, and the podia are located within an ambulacral groove on the oral surface of the arms. The body surface bears short spines, many finger-like evaginations of the body wall (papulae) that function in gas exchange and excretion, and in some species minute defensive grasping structures (pedicellariae). The podia of asteroids serve in locomotion and gas exchange. Most asteroids are carnivores, and the prey, commonly molluscs and other echinoderms, is swallowed whole or partly digested externally by an everted stomach. A pair of digestive glands in each arm is the principal source of digestive enzymes and also the site of absorption. An oral nerve ring in the disc and a radial nerve on the oral side of each arm constitute the major part of the nervous system. Usually there are two gonads in each arm, fertilization is external, and development is planktonic.

3. Brittle stars and basket stars, members of the subclass Ophiuroidea, are stelleroids in which the arms are sharply set off from the disc. They move rapidly by pushing with the arms, and their small size enables many species to live beneath stones or in crevices and holes, and even to burrow in sand. Brittle stars are scavengers, deposit feeders, and suspension feeders, and the podia are used to collect and transport food. The bursae, a pair of pocket-like invaginations of the disc wall at the base of each arm, are sites of gas exchange, excretion, gamete release, and brooding.

4. The class Echinoidea includes sea urchins, sand dollars, and heart urchins. The body is not drawn out into arms, and the ossicles are fused to form a test on which are mounted movable spines. The spherical sea urchins move by means of their podia and in some species by their long spines. Protection is provided by spines and pedicellariae. Sea urchins scrape rocks or chew algae with a toothed feeding apparatus (Aristotle's lantern). Sand dollars and heart urchins inhabit soft bottoms and display some degree of secondary bilateral symmetry. The many short spines are used in burrowing. They are selective deposit feeders.

5. The class Holothuroidea contains the sea cucumbers. There are no arms, and the body is greatly elongated along the oral-aboral axis. The ossicles are microscopic, and the anterior podia are modified as tentacles around the mouth. Many sea cucumbers creep on the podia of three ambulacra (the sole). The tentacles are used in suspension feeding and selective and nonselective deposit feeding. Gas exchange is provided by water pumped into respiratory trees.

6. Members of the class Crinoidea are the only living echinoderms in which the oral surface is directed upward. Sea lilies are attached to the bottom by a stalk; feather stars are unattached but perch on the substratum with aboral cirri. All crinoids have five to many arms bearing lateral pinnules. Podia on the oral surface of the arms and pinnules collect plankton and other suspended particles, which are transported to the mouth by the cilia of the ambulacral groove. At settling, the larva of crinoids attaches to the substratum and metamorphoses into the adult form. During development, young feather stars are attached to a little stalk like that of sea lilies from which they break free to take up the free-living habit of the adult.

References and Selected Readings

Detailed accounts of the echinoderms can also be found in many of the references listed at the end of Chapter 23.

Binyon, J. *Physiology of Echinoderms*. Oxford: Pergamon Press, 1972. A general physiology of echinoderms.

Birkeland, C. The Faustian traits of the crown-of-thorns starfish. *American Scientist* 77(1983):154–163. The biology of the sea stars that have been very destructive to the corals on many reefs.

Jangoux, M., and J.M. Lawrence (eds.). *Echinoderm Nutrition*. Rotterdam: A.A. Balkema, 1982. Chapters on feeding and digestion for each of the echinoderm groups.

Lawrence, J. *A Functional Biology of Echinoderms*. Baltimore: The Johns Hopkins University Press, 1987. A short general biology of the echinoderms.

Macurda, D.B., and D.L. Meyer. Sea lilies and feather stars. *American Scientist* 71(1983):354–364. A beautifully illustrated introduction to the crinoids.

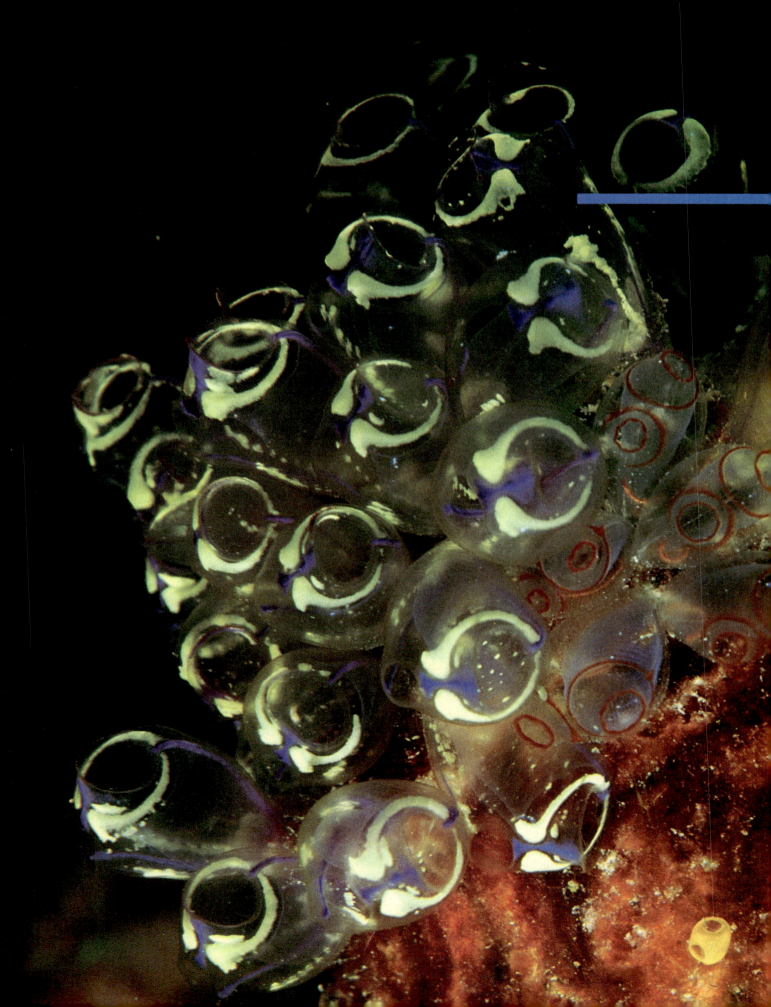

35

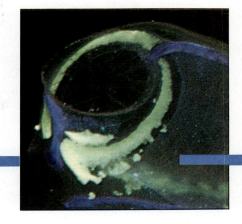

Hemichordates and Chordates

Characteristics of the Hemichordates

■ Hemichordates are marine animals, most of which (acorn worms) have wormlike bodies and burrow or live beneath stones or algae.

■ The body is divided into an anterior proboscis, a short middle collar region, and a long posterior trunk.

■ In most species, pharyngeal pouches evaginate from each side of the pharynx and open as pores on the body surface.

■ Most hemichordates are deposit feeders.

■ The sexes are separate, fertilization is external, and development includes a tornaria larva that resembles the sea star bipinnaria.

Characteristics of the Chordates

■ The phylum Chordata is a large and diverse assemblage of marine, freshwater, and terrestrial animals, and includes sea squirts, lancelets, fishes, amphibians, reptiles, birds, and mammals.

■ At some stage of their life cycle all possess pharyngeal pouches that primitively opened as slits to the body surface.

■ Primitive species are filter feeders and have a mucus-secreting endostyle as part of their feeding mechanism. Vertebrates lack the endostyle but their thyroid gland may have evolved from an endostyle-like organ.

■ A dorsal rodlike notochord is present in the developmental stage of all species and in the adults of the more primitive species.

■ A single, dorsal, hollow nerve cord occurs in the embryonic stage of all and in the adults of most species.

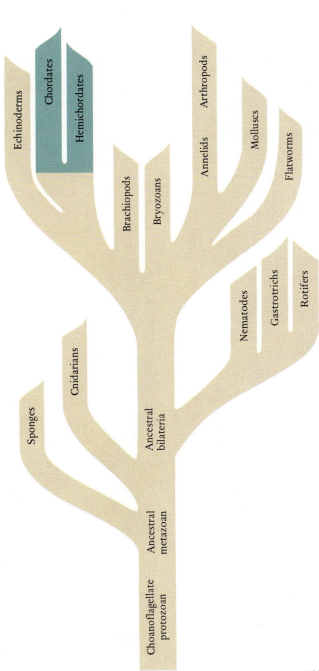

(Left) Although sessile adult ascidians (Clavelina picta) *do not resemble chordates, their motile larvae display all of the chordate characteristics. (Above) A close-up of* Clavelina picta.

The phylum Chordata comprises the backboned animals, or vertebrates, such as fishes, frogs, birds, and humans, and includes two soft-bodied, marine groups: the tunicates, or urochordates, and the lancelets, or cephalochordates. The hemichordates are a seldom seen group of marine deuterostomes, but they are important in showing some affinities to both the echinoderms and the chordates. After examining the hemichordates, we will then consider the urochordates and the cephalochordates and justify their placement in the same phylum (chordates) as the vertebrates.

PHYLUM HEMICHORDATA

The phylum **Hemichordata** is a little known group of about 85 species of deuterostomes. Despite their obscurity, the hemichordates are of considerable phylogenetic interest. Most members of the phylum are wormlike marine animals that range in length from 2 cm to 2 m. They live beneath stones and algae, in burrows in sand and mud, and a few species occupy tubes. Other hemichordates are small colonial species.

Class Enteropneusta

The worm-like species belonging to the class **Enteropneusta** are commonly called acorn worms because the front of the body of some species resembles an acorn in its cup. Their body is composed of three regions: an anterior **proboscis**, a middle, bandlike **collar**, and a long posterior **trunk** (Fig. 35.1a.) The mouth is located ventrally just in front of the collar, which can constrict around it (Fig. 35.1b). The trunk makes up the greater part of the body, and ends in a terminal anus.

Acorn worms move primarily by peristaltic contractions and by the action of ciliated cells over their entire surface. Some species also use part of the coelom as a hydroskeleton during burrowing. Their coelom is divided into three compartments, as it is in developing echinoderms, with a portion being located in each body region. The portions of the coelom in the proboscis and collar contain sea water that can enter and leave through minute pores (Fig. 35.1b). When the worm burrows, the collar coelom is first inflated with sea water, anchoring the worm. The contraction of circular muscles in the proboscis then pushes it forward. The proboscis, in turn, is inflated with

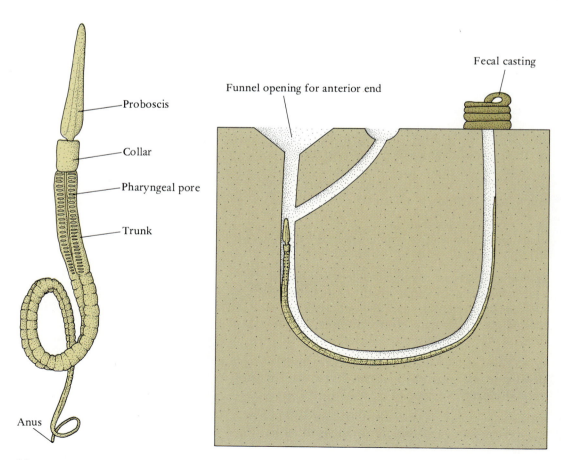

Proboscis

Collar

Pharyngeal pore

Trunk

Anus

(a)

Funnel opening for anterior end

Fecal casting

water and becomes anchored. Finally, the contraction of longitudinal muscles in the proboscis pulls the rest of the worm forward. A **buccal diverticulum** that extends into the base of the proboscis from the buccal cavity contains a rod of vacuolated cells that helps to stiffen the proboscis. At one time zoologists considered this buccal diverticulum to represent a notochord, a stiffening rod characteristic of chordates (p. 802), but this is no longer the prevailing view. The term hemichordates (Gr. *hemi*, half + *chorda*, cord) derives from this belief.

Nutrition and Gas Exchange

Acorn worms collect suspended and deposited material that adheres to mucus on the surface of their proboscis and is carried by cilia toward the mouth (Fig. 35.1c). Coarse particles are rejected, but a considerable volume of sand enters the mouth along with food particles and water. Most of the water enters the dorsal part of the pharynx and escapes through **pharyngeal pouches** or **slits** that develop as lateral evaginations from the pharynx. Some species have only a few such slits; others have a 100 or more. The pouches become partly subdivided by a **tongue bar** that

grows down from above so the pharyngeal openings are U-shaped, whereas the external openings form a pair of rows of pores on the dorsal part of the trunk. The pharyngeal pouches do not contain gills, but their lining is vascular and gas exchange probably occurs as water passes through them. Additional gas exchange occurs across the body surface. The pouches eliminate excess water entering with the food. Food and sand continue along the ventral part of the pharynx and enter the intestine. The organic material is digested as the material continues down the intestine. The fecal castings of these animals are often a conspicuous feature on sand flats exposed at low tide (Fig. 35.1a). Studies made off the coast of Japan have found that acorn worm castings amount to 10,000 kg per square kilometer per day. Acorn worms play an important ecological role by helping to turn over sea floor sediments, as earthworms do on land.

Circulation and Excretion

Hemichordates have an open circulatory system with contractile dorsal and ventral vessels connected by open spaces. As in many invertebrate phyla, hemolymph flows

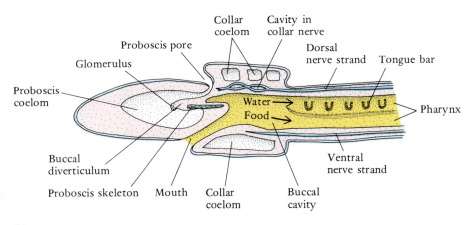

(b)

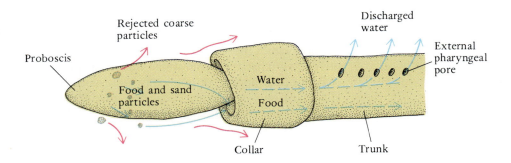

(c)

Figure 35.1
Acorn worms. (a) A sketch of a worm in its burrow. (b) The structure of a worm as seen in a sagittal section of the front end. (c) Ciliary feeding currents.

forward in the dorsal vessel and posteriorly in the ventral one. A group of vascular sinuses, called a **glomerulus**, protrudes from the buccal diverticulum into the proboscis coelom (Fig. 35.1b). While the glomerulus may be an excretory organ, most nitrogenous wastes (ammonia) appear to be lost by diffusion through the body surface.

Nervous System

As in echinoderms, the nervous system of hemichordates consists of a subepidermal nerve net that connects with solid nerve cords, one dorsal and one ventral (Fig. 35.1b). The dorsal cord of some species expands within the collar and contains one or two cavities. This feature is suggestive of the dorsal, hollow nerve cord of chordates, but the nervous system as a whole is simpler than in most chordates.

Reproduction and Development

Acorn worms have separate sexes, and the numerous paired lateral gonads, each with a separate gonopore, are located within either side of the trunk. Fertilization is external. Development follows the typical deuterostome pattern and leads in many species to a **tornaria larva** (Fig. 35.2), which is strikingly similar to the bipinnaria larva of some echinoderms (see Fig. 34.1). After a planktonic existence, the tornaria larva lengthens and settles to the bottom as a young worm. Asexual reproduction by fragmentation also occurs.

Class Pterobranchia

A small number of species constitute another hemichordate class, the **Pterobranchia**. Most are deep-water hemichordates that live in attached colonies within tubes that they secrete (Fig. 35.3). Individuals in the colonies are only

1 to 2 mm long. The proboscis is shield shaped and helps to secrete the tube. Pterobranchs are suspension feeders, and their collar bears two or more branched feeding tentacles. Food caught on the tentacles is carried by cilia to the mouth. The trunk is short, and the intestine loops upon itself, bringing the anus close to the mouth. A stalk interconnects individuals in the colony, and its contraction draws them into their tubes when danger threatens. Species of one genus, *Cephalodiscus*, have one pair of pharyngeal slits that open to the surface; others lack them. Pterobranch species are capable of both sexual reproduction and asexual reproduction by budding.

PHYLUM CHORDATA

The phylum Chordata is the largest of the deuterostome phyla, for it includes the vertebrates and two groups of invertebrate chordates that lack a vertebral column: the urochordates (tunicates), and the cephalochordates (lancelets). Collectively we call these invertebrate chordates **protochordates**.

Chordate Characteristics

The characteristics that define the chordates can be seen quite clearly in the larval stage of an ascidian, one of the urochordates (Fig. 35.4). Although the larva does not feed, it is beginning to develop the feeding mechanism that characterizes protochordates and possibly ancestral vertebrates. In adult protochordates, cilia draw a current of water bearing minute food particles through the pharynx, and a longitudinal ciliated groove on the pharynx floor, known as the **endostyle**, secretes a film of mucus that traps the food particles. Some of the cells of the endostyle also bind and secrete iodinated proteins, but the significance of this to these animals is not known. The endostyle is a unique derived feature of chordates that occurs in all protochordates. Although vertebrates lack an endostyle as such, the vertebrate thyroid gland is homologous to the iodine binding cells of the endostyle.

The feeding current leaves through pharyngeal slits (Fig. 35.4). The slits of protochordates open into a chamber, the atrium, that flanks the pharynx rather than opening directly to the outside. Some gas exchange may occur as water passes through the slits, but their primitive function in chordates is related to feeding. Gills, however, do develop within the slits in fishes. Adult terrestrial vertebrates do not have pharyngeal slits, but transitory pharyngeal pouches appear during embryonic development. Pharyngeal slits or pouches occur only among the hemichordates and chordates.

Other distinctive chordate features are associated with the method of locomotion of primitive species. They can be seen in the locomotive organ of the ascidian larva, the tail (Fig. 35.4). A dorsal, longitudinal rod of turgid cells

Figure 35.2
The tornaria larva of an acorn worm.

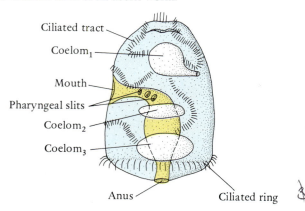

Ciliated tract

Coelom₁

Mouth

Pharyngeal slits

Coelom₂

Coelom₃

Anus   Ciliated ring

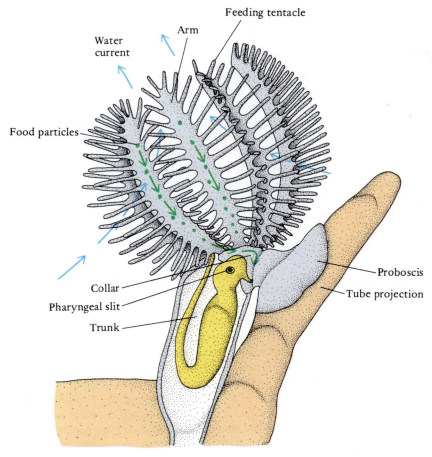

Figure 35.3
Pterobranchs. A sagittal section of an individual of *Cephalodiscus*.

Figure 35.4
The derived features that characterize chordates show clearly in the motile larva of
an ascidian.

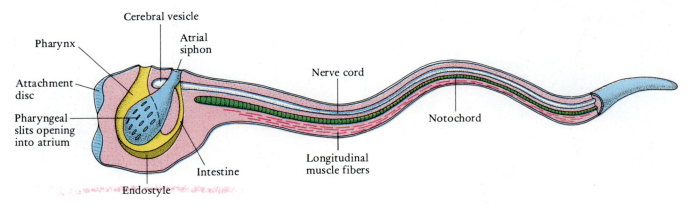

that is encased in a fibrous sheath forms a **notochord** (Gr. *notos*, back + *chorde*, cord). The notochord acts as a hydrostatic skeleton. It is flexible but resists compression so that contraction of longitudinal muscle fibers in the tail cannot shorten it, as contraction of similar fibers shortens the body in many worms; rather contraction causes lateral bendings or undulations of the tail that propel the animal. The integration of muscle activity occurs in a **single**, **dorsal**, **hollow nerve cord** that lies above the notochord. This cord is unique in being single, dorsal, and hollow rather than double, ventral, and solid, as in many invertebrates. Although patterns of locomotion have changed dramatically during chordate evolution, all chordates have a notochord and a single, dorsal, hollow nerve cord at some stage of their development.

Subphylum Urochordata (Tunicata)

About 2000 species, commonly called tunicates, constitute the subphylum **Urochordata**. All but a few are sea squirts, or ascidians, belonging to the class **Ascidiacea**. Ascidians are sessile as adults, living attached to rocks, shells, pilings, and ship bottoms (Fig. 35.5). Solitary species have more-or-less spherical or tubular-shaped bodies, which range in size from a pinhead to a small potato. The body is covered by an envelope, or **tunic**, which is composed of fibers of **tunicin** embedded in a mucopolysaccharide matrix. The presence of tunicin, a form of cellulose, makes this a unique group in the animal kingdom. Many other species are colonial. Some of these ascidians resemble a vine creeping over the substratum, others have bodies arranged in clusters, and still others are so intimately connected that numerous individuals are embedded within a common tunic.

Nutrition and Gas Exchange

One end of the body of a solitary ascidian is attached to the substratum; the opposite end contains two openings, or siphons, through which water enters and leaves (Fig. 35.6). Adult ascidians are filter feeders. A stream of water, created by cilia in the pharynx, carries water and plankton through the **buccal siphon** and into a huge **pharyngeal basket**, which occupies most of the body. The plankton is trapped in a sheet of mucus secreted by the endostyle and is carried by cilia along the interior of the two sides of the pharynx. The film of mucus collects in a **gutter** on the opposite side of the pharynx and is carried by cilia into the stomach. Water escapes through numerous minute pharyngeal slits into the **atrium** from whence it is discharged through the **atrial siphon**. The atrium surrounds and protects the very delicate, perforated pharyngeal region from damage by drifting debris and other hazards. Some colonial tunicates, such as *Botryllus*, are composed of symmetrically arranged individuals whose atria open into a common atrial siphon (Fig. 35.7). The body below or to the side of the pharynx contains the stomach, intestine, heart, and other internal organs (Fig. 35.6a). Food collected in the pharynx enters the **U**-shaped gut and is carried along by ciliary action. As with most other invertebrates, muscles are not present in the intestinal wall. The intestine opens into the atrium, and fecal matter is swept away by the water current in the atrial siphon.

Figure 35.5
Representative ascidians. (a) The solitary blue and gold ascidian of the Philippines. (b) Members of the colonial species, *Botryllus schlosseri*, from the Eastern Atlantic share a common atrial siphon.

(a)   (b)

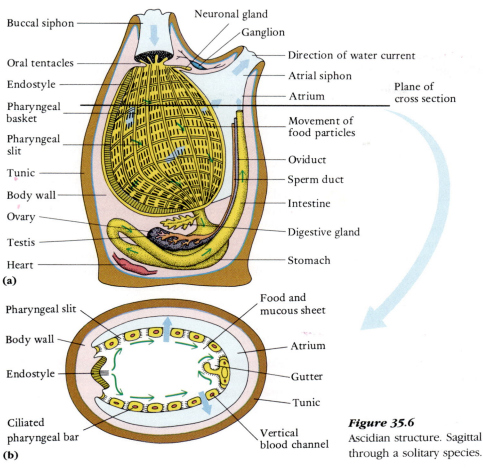

(a)

Buccal siphon
Oral tentacles
Endostyle
Pharyngeal basket
Pharyngeal slit
Tunic
Body wall
Ovary
Testis
Heart

Neuronal gland
Ganglion
Direction of water current
Atrial siphon
Atrium
Plane of cross section
Movement of food particles
Oviduct
Sperm duct
Intestine
Digestive gland
Stomach

(b)

Pharyngeal slit
Body wall
Endostyle
Ciliated pharyngeal bar

Food and mucous sheet
Atrium
Gutter
Tunic
Vertical blood channel

Figure 35.6
Ascidian structure. Sagittal (a) and transverse (b) sections through a solitary species.

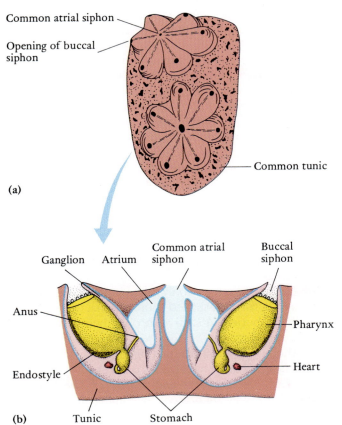

Common atrial siphon
Opening of buccal siphon

Common tunic

(a)

Ganglion
Atrium
Common atrial siphon
Buccal siphon

Anus
Pharynx
Endostyle
Heart

(b)

Tunic
Stomach

Figure 35.7
Surface view (a) and vertical section (b) of the colonial tunicate *Botryllus*.

An enormous amount of water passes through the pharynx, which serves not only for feeding but also for gas exchange. The current can be regulated or halted by the degree of opening or closing of the siphons. Exposed intertidal species may suddenly contract and eject a spurt of water from the siphons, hence their common name—sea squirts.

Excretion and Internal Transport

There are no special organs for excretion of nitrogenous wastes. Most waste escapes by diffusion; some accumulates as inert pigments and crystals, including uric acid, in various parts of the body. Ascidians have an open internal transport system, but the blood follows distinct channels, which course through the pharyngeal basket, abdominal organs, and, in some species, even the tunic. The heart, located near the stomach, periodically reverses its beat and the direction of blood flow, a most unusual phenomenon.

Reproduction and Development

Most ascidians are capable of asexual reproduction by budding. All species are hermaphroditic with a single abdominal ovary and testis (Fig. 35.6a); the oviduct and sperm duct open into the atrium. Fertilization may be external with planktonic development, especially in solitary species. Alternatively, fertilization may take place within the atrium and the eggs may be brooded there through early embryonic stages.

Fertilized eggs of ascidians develop quickly into small, motile larvae (Fig. 35.4). A larval stage, which does not feed, allows for the dispersal of species where the adults are sessile. The larval tail, as we have seen, contains the characteristic locomotor features of chordates: muscles, notochord, and a distinctive nerve cord. An expanded **cerebral vesicle** at the rostral end of the nerve cord is not a brain; rather, it contains a statocyst and a light-sensitive eye spot. At first the larva is attracted to light and swims toward the ocean surface. After one or two days of pelagic life, its reactions to light and gravity are reversed, and it moves to the bottom to attach to a suitable substratum by means of secretions from its attachment discs. It then metamorphoses into the sessile adult. During metamorphosis, the tail is reabsorbed, as are the notochord and most of the nerve cord. Part of the cerebral vesicle transforms into a ganglion located between the two siphons (Fig. 35.6a).

Planktonic Tunicates

Ascidians are the most abundant group of tunicates, but there are two other pelagic classes, some species of which are common animals in the plankton.

The class **Thaliacea**, often called salps, are tunicates that occur primarily in subtropical and tropical oceans. In common with many other planktonic animals, they are transparent. They differ from ascidians in having the buccal and atrial siphons at opposite ends of the body so the feeding current can also be used for locomotion (Fig. 35.8a). Bands of muscle in the body wall assist cilia in moving a current of water through the siphons. *Doliolum* has a solitary sexual phase in its life cycle that alternates with a colonial asexual phase.

The class **Larvacea**, or appendicularians, occurs in surface ocean waters throughout the world. Their anterior end somewhat resembles an adult ascidian, but they retain the larval tail in a sexually mature individual (Fig. 35.8b). Individuals lack a tunic, but they secrete elaborate mucous houses in which they live. The beating of the tail draws a current of water through the house, and several mucous filters trap only the smallest of the planktonic organisms on which the animal feeds. The houses soon become clogged with larger planktonic organisms and fecal pellets, and are replaced every few hours.

Chordate Metamerism

Although members of the subphylum **Cephalochordata** lack a backbone, they do share body segmentation, or **metamerism**, with vertebrates. Metamerism appears to have evolved in chordate history in the lines leading to the cephalochordates and vertebrates, after the divergence of the urochordates. Cephalochordates and vertebrates propel themselves by the contraction of muscle fibers that are arranged in segmented blocks, the **myomeres** (Fig. 35.9a and c). Segmentation developed as an adaptation for undulatory swimming or perhaps for rapid burrowing into sand.

In contrast to the segmentation of annelids, it is the musculature of chordates rather than the coelom that exhibits the primary segmentation. The segmentation of the nervous system and blood vascular system represents an adjustment of these systems to supply the muscle blocks. Segmentation in chordates does not involve the coelom. The coelom in chordates is not utilized as a localized hydrostatic skeleton as it is in many annelids; rather it is the notochord against which the muscle blocks are indirectly pulling.

Subphylum Cephalochordata

The subphylum **Cephalochordata** includes only one type of animal, the lancelets. The most common genus, *Branchiostoma*, is commonly called amphioxus. Lancelets are superficially fish-shaped chordates that can swim, but usually lie buried in sand with only their anterior end protruding.

The body of amphioxus (Fig. 35.9a and b) is elongated, tapering at each end, and compressed from side to side. Small median fins and a pair of lateral finlike metapleural folds are present. Swimming and burrowing are accomplished by the contraction of the myomeres.

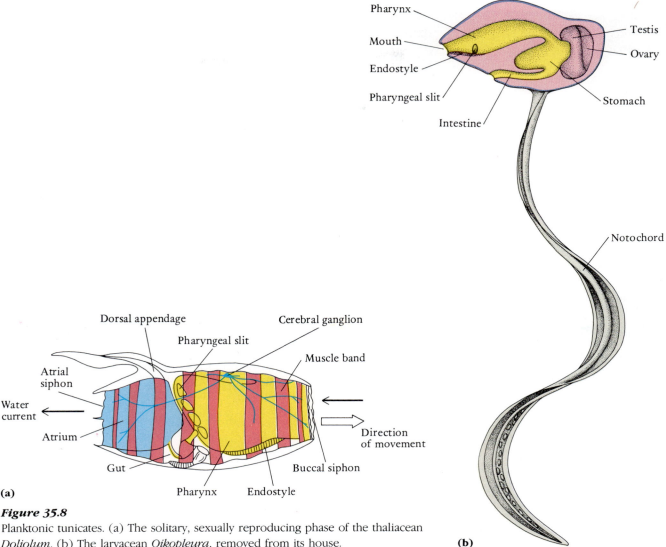

(a)

Figure 35.8
Planktonic tunicates. (a) The solitary, sexually reproducing phase of the thaliacean *Doliolum*. (b) The larvacean *Oikopleura*, removed from its house.

(b)

Shortening of the body is prevented by an unusually long notochord that extends farther anteriorly than in any other chordate, an attribute after which the subphylum is named (Gr. *kephale*, head + *chorde*, cord). A special type of striated muscle fiber (**paramyosin muscle**), which has the ability to remain partially contracted for long periods of time, occurs in the notochord, and by controlling the turgidity of the notochord allows it to function as an adjustable hydrostatic skeleton. Amphioxus can swim and burrow in either direction, and the turgidity of the notochord in the propelling end is adjusted to allow a greater oscillation of the body in this region.

Nutrition
Amphioxus, like other protochordates, is a filter feeder. Water and minute food particles are taken into the mouth through the **oral hood**, whose edges bear a series of delicate projections, the **cirri**, that act as a strainer and

exclude larger particles (Fig. 35.9b). The inside of the oral hood is lined with bands of cilia, the **wheel organ**, which, together with cilia in the pharynx, produce a current of water that enters the mouth. The mouth is located in the center of a transverse partition known as the **velum**.

Food is entrapped within the pharynx in a film of mucus secreted by an endostyle on the floor of the pharynx, just as it is in urochordates. Water in the pharynx escapes into an atrium through 100 or more pharyngeal slits. The slits become divided during development, as they do in hemichordates, by the downward growth of tongue bars. Some gas exchange occurs in the pharynx, but the integument is the major respiratory surface.

The pharynx is primarily a food-gathering device. Food particles trapped in the mucus are carried back into the intestinal region where they are sorted out by complex ciliary currents. Large particles continue posteriorly, but small ones are deflected into the **midgut cecum**, a ventral

(a)

Figure 35.9

The lancelet, *Branchiostoma* (amphioxus). (a) The cephalochordate, amphioxus, normally lies partly buried in the sand; notice the V-shaped myomeres used in swimming to new sites and burrowing. (b) Its structure as seen in lateral view. (c) Its structure in transverse section.

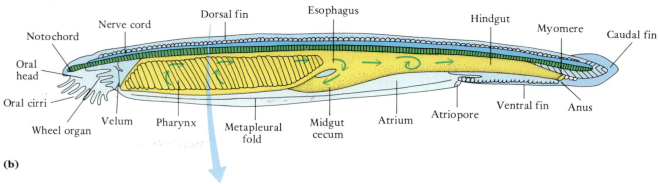

Notochord · Oral head · Oral cirri · Wheel organ · Velum · Nerve cord · Pharynx · Dorsal fin · Metapleural fold · Esophagus · Midgut cecum · Atrium · Atriopore · Hindgut · Myomere · Caudal fin · Ventral fin · Anus

(b)

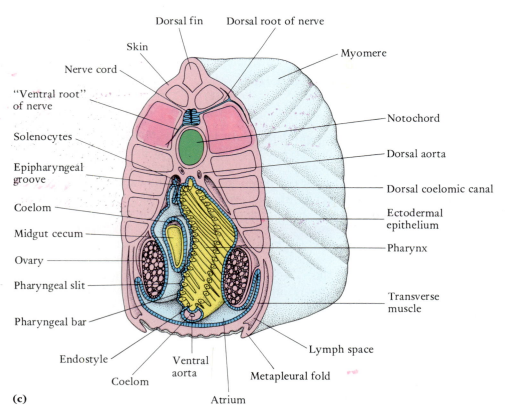

Dorsal fin · Skin · Nerve cord · "Ventral root" of nerve · Solenocytes · Epipharyngeal groove · Coelom · Midgut cecum · Ovary · Pharyngeal slit · Pharyngeal bar · Endostyle · Coelom · Ventral aorta · Atrium · Metapleural fold · Lymph space · Transverse muscle · Pharynx · Ectodermal epithelium · Dorsal coelomic canal · Dorsal aorta · Notochord · Myomere · Dorsal root of nerve

(c)

outgrowth from the floor of the gut. Many of these particles are ingested by cecal cells and digested intracellularly. Undigested material, together with enzymes secreted by certain cecal cells, is carried posteriorly to join the mass of larger food particles. As this mass is passed down the gut and rotated by ciliary action, it is degraded further by enzymatic action. Material not broken down in the midgut is carried posteriorly. Fecal material is discharged through an anus, which, as in vertebrates, lies slightly anterior to the posterior tip of the body. There is a short **postanal tail**.

Internal Transport and Excretion

Absorbed nutrients are distributed by a circulatory system. A series of veins returns blood from the various parts of the body to a sinus that is located ventral to the posterior part of the pharynx. No muscular heart is present; blood is propelled by the contraction of the arteries. A ventral aorta extends from the sinus forward beneath the pharynx (Fig. 35.9c) and connects with branchial arteries that extend dorsally through the pharyngeal bars to a pair of dorsal aortae. The dorsal aortae, in turn, carry the blood posteriorly to spaces within the tissues. True capillaries are absent, but the general direction of blood flow, i.e., anteriorly in the ventral part of the body and posteriorly in the dorsal part, is similar to that of vertebrates and different from that of other animals.

The excretory organs are segmentally arranged groups of **solenocytes** that lie in the **dorsal coelomic canals** (Fig. 35.9c). Solenocytes are similar to the terminal cells of protonephridia and transfer excretory products from the blood and from the dorsal coelomic canals into the atrium. Additional excretion occurs by diffusion.

Nervous System

The nervous system of amphioxus consists of a tubular nerve cord located dorsal to the notochord (Fig. 35.9c). Its anterior end is differentiated slightly but is not expanded to form a brain. Paired dorsal and ventral nerve roots extend into the tissues. The "ventral roots," which supply the myomeres, are not composed of neurons, but of axon-like processes of the muscle cells. A similar condition occurs in nematodes (p. 649). Amphioxus is sensitive to light and to chemical and tactile stimuli, but no elaborate sense organs are present.

Reproduction and Development

The sexes are separate in amphioxus, and numerous testes or ovaries bulge into the atrial cavity (Fig. 35.9c). The gametes are discharged into the atrium upon the rupture of the gonad walls. Fertilization and development are external.

Subphylum Vertebrata

The subphylum **Vertebrata** is by far the largest of the chordate subphyla, for it contains approximately 40,000 living species. Vertebrates share with the protochordates

the diagnostic characteristics of the phylum. All embryonic vertebrates have a series of pharyngeal pouches, but they break through to the body surface as gill slits only in fishes and larval amphibians (Fig. 35.10). All have a thyroid gland that develops from the floor of the pharynx from cells homologous to the iodine-binding cells of the endostyle (p. 800). A single, dorsal, hollow nerve cord is present in the embryos of all species, and differentiates into the brain and spinal cord in adults. A notochord, which lies ventral to the neural tube and extends from the middle of the brain nearly to the posterior end of the body, develops in the embryos of all vertebrates and persists in the adults of some fishes.

Vertebrate Characteristics and Their Origin

Investigators have suggested for a long time that vertebrates evolved as a group of chordates that became more active and predaceous than other chordates. J.A. Ruben and A.F. Bennett in 1980 presented evidence that primitive living vertebrates differ from protochordates and most invertebrates in having the capacity to support bursts of activity using anaerobic metabolism with the concomitant temporary increase in the level of lactic acid in the blood (p. 265). The lactic acid is later metabolized aerobically. They suggest the ability temporarily to accommodate an increased level of lactic acid is a fundamental derived feature of vertebrates that evolved in their earliest ancestors.

Carl Gans and R. Glen Northcutt proposed in 1983 that the other derived characteristics of vertebrates are correlated with a more active life style. A series of well-developed myomeres provided the locomotive force (Fig. 35.10). A segmented vertebral column replaced the notochord as a stronger compression strut and allowed ancestral vertebrates to propel themselves more rapidly through the water. As in other active groups, sense organs and nervous tissue became concentrated at the front of the body. Vertebrates evolved nasal cavities, eyes, and ears, and primitive species featured a lateral line system (p. 408). A brain integrated the activities of these sense organs with those of the rest of the body. A skeletal cranium encased the brain and protected many of the sense organs. Some parts of the internal skeleton undoubtedly were cartilaginous in ancestral species, but enamel, dentine, and bone, all unique calcium phosphate compounds, were present in the dermal scales and plates that encased the body surface of early vertebrates. A recent hypothesis correlates the presence of a skeleton of calcium phosphate with the ability to sustain periods of anaerobic metabolism with the concomitant accumulation of lactic acid. An increased acidity of the blood leads to a slight dissolution of bone. A skeleton of calcium phosphate is more resistant to this dissolution than one of calcium carbonate, which is the type typically found in invertebrates. Superficial bones may have evolved as a way of

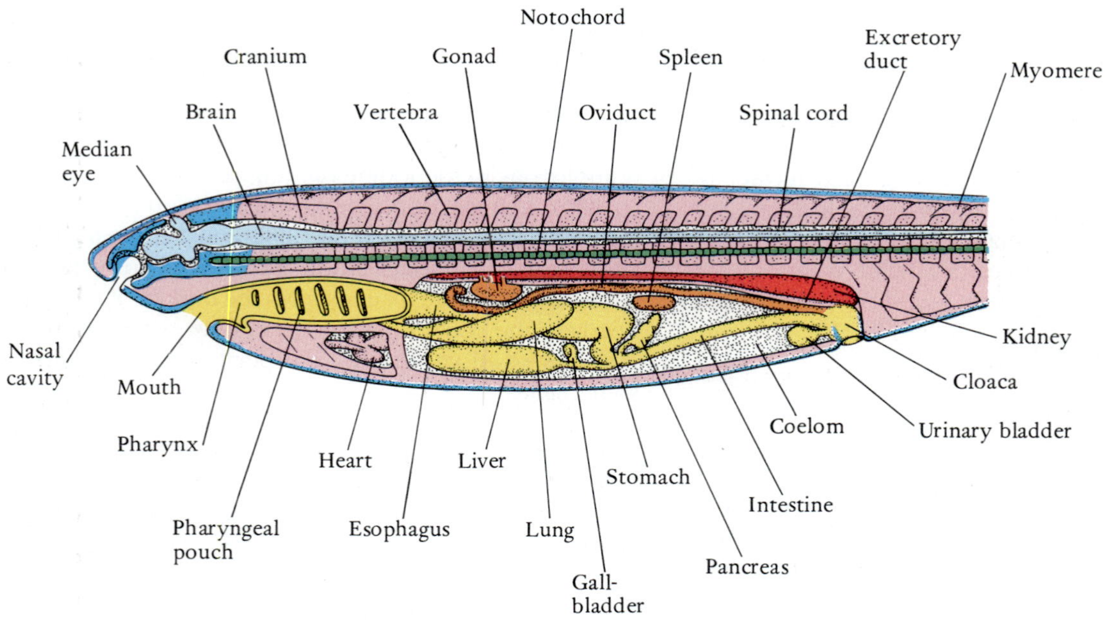

Figure 35.10
A diagrammatic sagittal section of a vertebrate showing most of the distinctive features of the subphylum.

housing the lateral line system, especially those parts of the system adapted for electroreception.

Increased activity is not possible without a corresponding increase in metabolism. The gut became muscularized, and a series of elastic visceral arches evolved between the pharyngeal pouches. These features enable vertebrates to contract and expand their pharynx by means of muscles and thus draw a larger volume of water and food through it than could be accomplished by the ciliary currents used by protochordates. The larvae of ancestral vertebrates probably continued to be filter feeders, as are the larvae of contemporary lampreys. However, the combination of a greater degree of motility, an anterior array of sense organs acting as distance receptors, and a muscular pharynx probably enabled adults to ingest small prey organisms and tear off soft organic material from dead plants and animals. Later in their history, vertebrates evolved jaws and became efficient predators. Gills evolved in the pharyngeal pouches and provided for increased and more efficient gas exchange. The gut became muscularized, and peristaltic contractions rather than ciliary action carried food caudally. A large digestive gland, the pancreas, secreted a copious supply of enzymes, and a liver, among its many functions, stored excess food as glycogen or lipid. A muscular heart developed that pumped blood effectively through a closed circulatory system. Hemoglobin carried in erythrocytes helped trans-

port gases. Many wastes were excreted by a pair of kidneys composed of numerous tubules quite unlike any invertebrate excretory organ. All of these features enabled vertebrates to become the most successful and dominant group of chordates.

The evolution of three embryonic features made possible the development of most of the derived characters of vertebrates.

1. *Sensory placodes*, which are thickened discs of neuroectoderm, give rise by invagination to the nose, lens of the eye, ears, and lateral line system.
2. Cells that migrate from the *neural crest*, which is a pair of columns of neuroectoderm that flank the developing neural tube (Fig. 22.9, p. 518), give rise to or induce the formation of the front part of the head, visceral arches, bony scales and plates, teeth, pigment cells, sensory neurons, Schwann cells, postganglionic autonomic neurons, endocrine cells of the adrenal medulla and gut epithelium, and other vertebrate features.
3. *Muscle fibers* develop from the inner surface of the lateral plate mesoderm and muscularize the gut.

These features are unique to vertebrates and are their most important derived characters. It is probable that the sensory placodes and neural crest initially evolved as concentrations of nerve cells from the primitive subepidermal nerve net present in other deuterostomes. Vertebrates no longer have such a nerve net.

Vertebrate Evolution

The features vertebrates acquired early in their evolution enabled them to diversify rapidly, occupy many habitats, an adopt many modes of life. Zoologists now recognize five classes of fishes (two of which are extinct) and four classes of terrestrial vertebrates. A cladogram showing the classes, their presumed interrelationships, and some of the major events in vertebrate evolution is presented in Figure 35.11. We will consider each of the vertebrate classes in detail in Chapters 36 to 41.

EVOLUTIONARY RELATIONSHIPS OF DEUTEROSTOMES AND CHORDATES

Basic similarities in the early embryonic development of echinoderms, hemichordates, and chordates constitute an important reason for recognizing the deuterostome line of evolution, but there is other evidence of a close evolutionary relationship between these three principal deuterostome phyla. A tripartite division of the coelom, a subepidermal nerve net, and, especially, strikingly similar larval stages indicate a connection between hemichordates and echinoderms. On the other hand, the presence of pharyngeal slits, and perhaps similarities between the

cavities in the collar nerve cord of hemichordates and the hollow cord of chordates, relate hemichordates and chordates. Hemichordates and cephalochordates also share a developmental process in which pharyngeal slits multiply by the growth of tongue bars.

The nature of the common ancestral form, as well as the origins of each phylum, are obscure. Vertebrates and cephalochordates are the most motile chordates and quite likely evolved from some motile filter-feeding ancestor (Fig. 35.12). Cephalochordates and vertebrates doubtless evolved independently because some of their adaptations for motility are quite different. As one example, the myomeres on the two sides of amphioxus alternate, whereas those of vertebrates are paired. Cephalochordate-like creatures may have evolved very early, for an amphioxus-like creature, *Pikaia*, has been found in the Burgess shale of the mid-Cambrian (Fig. 35.13). Vertebrates appear later in the fossil record and became the more active group.

The ancestors of amphioxus and vertebrates are unknown, but both groups may have evolved from motile ascidian larvae. The ascidian larva is quite different from any other primitive deuterostome larva, for it is larger and propelled not by cilia but by its tail, which contains a notochord, a distinctive nerve cord, and muscle fibers. These features of the ascidian larva are retained in cephalochordates and vertebrates, but the tunic and other features that adapt adult acidians to a sessile life do not

Figure 35.11
A cladogram of vertebrate classes and major events in vertebrate evolution.

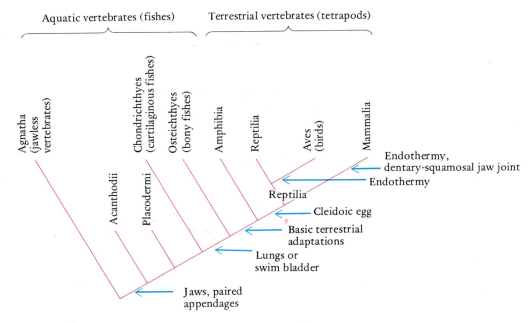

develop. ==The evolutionary retention in sexually mature adults of ancestral larval features is called **paedomorphosis**==. In evolution through paedomorphosis, many of the adult features that normally appear after the larval period do not develop, and new features can evolve. The larvacean group of urochordates may also have evolved through paedomorphosis.

Urochordates appear to be the most primitive chordate group, but their origin is even more obscure. Some researchers have suggested that they evolved from sessile pterobranch hemichordates. These species gather food with ciliated feeding tentacles, but one species has a pair of pharyngeal slits. A hypothetical stage has been postulated in which an increase in pharyngeal slits made possible a shift from the tentacles to pharyngeal slits as the site for filter feeding and the loss of the feeding tentacles. A straightening of the gut and an elongation of the body of such a creature would lead to acorn worms. Retention of the U-shaped gut, evolution of an endostyle, a further increase in the number of pharyngeal slits, and the devel-

Figure 35.12
Possible evolutionary relationships of the major deuterostome phyla.

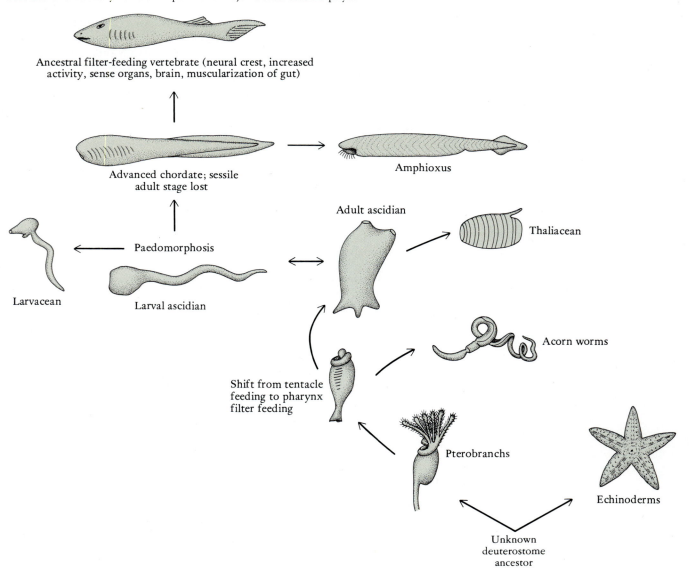

opment of a protective atrium and tunic would lead to adult ascidians. Such a scenario implies that the first chordates were sessile animals, but this is by no means certain. The first chordates may have been motile and the attached condition secondary. While little is known about more remote ancestors, these three deuterostome phyla (chordates, hemichordates, and echinoderms) share so many developmental features that they must at some point have shared a common ancestor.

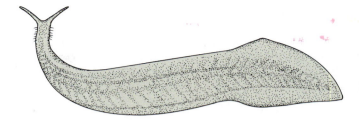

Figure 35.13
Pikaia, the oldest known chordate, found in the Burgess shale.

Classification of the Phyla Hemichordata and Chordata

Phylum Hemichordata (Gr. *hemi,* half + *chorde,* cord) Acorn worms and their allies.

Class Enteropneusta (Gr. *enteron,* intestine + *pneustikos,* breathing) Acorn worms. Wormlike hemichordates having a cylindrical proboscis and collar and a long trunk. Pharyngeal slits are present. Many species burrow in sand or mud; others live beneath stones or among algae. *Balanoglossus.*

Class Pterobranchia (Gr. *pteron,* wing + *branchia,* gills) Minute hemichordates having a shield-shaped proboscis and a collar bearing tentaculate arms. Most are colonial and live in secreted tubes in deep water. *Cephalodiscus.*

Phylum Chordata (Gr. *chorde,* cord) Protochordates and vertebrates.

Subphylum Urochordata (Tunicata) (Gr. *oura,* tail + *chorde,* cord) Mostly sessile, nonmetameric invertebrate chordates enclosed within a tunic containing a cellulose-like material. Pharynx highly developed and used in filter feeding; notochord and nerve cord present only in the larva.

Class Ascidiacea (Gr. *askidion,* little bag) Sea squirts or ascidians. Sessile; solitary or colonial. This class contains most species of urochordates. *Molgula.*

Class Thaliacea (*Thalius,* a German naturalist) Salps. Free-swimming planktonic urochordates; solitary or colonial. *Doliolum, Salpa.*

Class Larvacea (L. *larva,* ghost) Small, paedomorphic, planktonic urochordates living within or attached to a delicate gelatinous house. *Oikopleura.*

Subphylum Cephalochordata (Gr. *kephale,* head + *chorde,* cord) Lancelets. Small fish-shaped invertebrate chordates. Metameric; body supported by well-developed notochord; no vertebrae, no brain, no anterior array of sense organs. Jawless filter feeders; mouth surrounded by an oral hood. *Branchiostoma* (amphioxus).

Subphylum Vertebrata (L. *vertebratus,* having a backbone) Vertebrates. More active metameric chordates with the trunk supported by a linear series of cartilaginous or bony skeletal pieces (vertebrae) surrounding or replacing the notochord in the adult. Neural tube differentiated into a brain and spinal cord. Head well developed with the brain encased in a cranium and major sense organs located on the head. Most vertebrates have a mouth supported by jaws. The gut is muscularized, and a pancreas and liver are present. (A detailed classification of vertebrate classes is given in Chapters 36 to 41.)

Class Agnatha Extinct ostracoderms, present day lampreys and hagfishes.

Class Placodermi Extinct primitive jawed fishes.

Class Chondrichthyes Sharks and rays.

Class Osteichthyes Bony fishes.

Class Amphibia Frogs, toads, salamanders.

Class Reptilia Turtles, lizards, snakes, alligators.

Class Aves Birds.

Class Mammalia Mammals.

Summary

1. Most members of the phylum Hemichordata are burrowing acorn worms, but pterobranchs are sessile filter feeders.

2. The phylum Chordata contains a diverse assemblage of animals united in having, at some time in their life history, pharyngeal slits, which evolved as part of a filter-feeding mechanism, and a notochord and single, dorsal, hollow nerve cord, both of which evolved in connection with undulatory locomotion. Although most chordates belong to the subphylum Vertebrata, in which a vertebral backbone surrounds or replaces the notochord, there are two subphyla of invertebrate chordates without backbones, collectively called protochordates.

3. The subphylum Urochordata contains mostly sessile marine animals, called sea squirts or ascidians. The notochord and dorsal nerve cord are present only in the motile larval stage. Much of the interior of the body is occupied by a large pharynx perforated by many slits. The body of urochordates is covered by a protective tunic containing a cellulose-like material (tunicin) and secreted by the epidermis. The many colonial species of urochordates have the individuals connected together by a stolon or have their tunics fused together.

4. The subphylum Cephalochordata contains a small group of fish-shaped chordates, the lancelets, that swim and burrow in marine sand by rapid body undulations produced by the contraction of metameric muscle blocks. A long notochord and nerve cord are present in adults, but the anterior end of the nerve cord does not form a brain. Cephalochordates are also filter feeders and collect food in a way similar to the urochordates.

5. The subphylum Vertebrata contains the majority of chordates. Like cephalochordates, they are metameric, with segmented muscles. The notochord is surrounded or replaced by vertebrae. A well-developed head containing sense organs and a brain is present. Muscles, rather than cilia, are used to collect food and move it along the gut. Vertebrates are more active animals than other chordates, and early in their evolution shifted from filter feeding to exploiting larger types of food.

6. The first chordates are believed to have been ascidians. They may have evolved from creatures that resembled pterobranchs but had lost the feeding tentacles and acquired an increased number of pharyngeal slits. The free-swimming larva developed among the ascidians as a dispersal mechanism. The cephalochordates and vertebrates evolved through retention of the motile larval body form into adulthood and acquired additional adaptations for increased motility.

References and Selected Readings

Hemichordates and Chordates

Accounts of the hemichordates and chordates can also be found in many of the references listed at the end of Chapter 23.

Alldredge, A. Appendicularians. *Scientific American* 235(Jan. 1976):94−102. An interesting account of the planktonic urochordates (larvaceans) that secrete gelatinous houses for filter feeding.

Barrington, E.J.W., and R.P.S. Jefferies (eds). *Protochordates*. London: Academic Press, 1975. A series of papers presented at a symposium on many aspects of the biology of these animals.

Goodbody, I. The physiology of ascidians. *Advances in Marine Biology* 12(1974):1−149. A good coverage of all aspects of tunicate physiology.

Lester, S.M. *Cephalodiscus* sp.: Observations of functional morphology, behavior and occurrence in shallow waters around Bermuda. *Marine Biology* 85(1985): 262−268. A report on the biology of a pterobranch.

Northcutt, R.G., and C. Gans. The genesis of neural crest and epidermal placodes: A reinterpretation of vertebrate origins. *Quarterly Review of Biology* 58(1983):1−28. A presentation of the authors' hypothesis of the origin of vertebrates as particularly active chordates.

Ruben, J.A., and A.A. Bennett. The evolution of bone. *Evolution* 41(1987):1187−1197. The advantage of a calcium carbonate skeleton is its ability to resist dissolution in the presence of lactic acid resulting from glycolysis.

————. Antiquity of vertebrate pattern of activity metabolism and its possible relation to vertebrate origins. *Nature* 286(1980):886−888. Evidence that ancestral vertebrates evolved the ability to support a burst of activity by glycolyis.

Young, J.Z. *The Life of Vertebrates*, 3rd ed. Oxford: Clarendon Press, 1981. This textbook of zoology contains two excellent chapters on the anatomy and physiology of protochordates.

Vertebrates

Alexander, R. McN. *The Chordates*, 2nd ed. Cambridge, England: Cambridge University Press, 1981. An outstanding discussion of the major vertebrate groups in which biomechanical, physiological, and ecological factors are skilfully integrated with structure.

Carroll, R.T. *Vertebrate Paleontology and Evolution*. New York: W.H. Freeman and Co., 1988. A valuable source book on vertebrate evolution.

Fishbein, S.L. (ed.) *Our Continent, A Natural History of North America* Washington, D.C.: National Geographic Society, 1976. Five chapters in this superbly illustrated and authoritative book deal with the evolution of vertebrates.

Orr, R.T. *Vertebrate Biology*, 5th ed. Philadelphia: Saunders College Publishing, 1982. A valuable text and reference on many aspects of vertebrates; a chapter is devoted to each major group of vertebrates and to such general topics as territory, dormancy, and population dynamics.

Radinsky, L.B. *The Evolution of Vertebrate Design*. Chicago: University of Chicago Press, 1987. A discussion of vertebrate evolution with an emphasis on functional anatomy.

Stahl, B.J. *Vertebrate History: Problems in Evolution*. New York: McGraw-Hill Book Co., 1974. Reprint ed., New York: Dover Publications, 1985. A fascinating account of vertebrate history with some emphasis upon alternative interpretations of the data and unresolved problems.

Young, J.Z. *The Life of Vertebrates*, 3rd ed. Oxford: Clarendon Press, 1981. The anatomy, physiology, and evolution of all groups of vertebrates are explored in detail.

36

Fishes

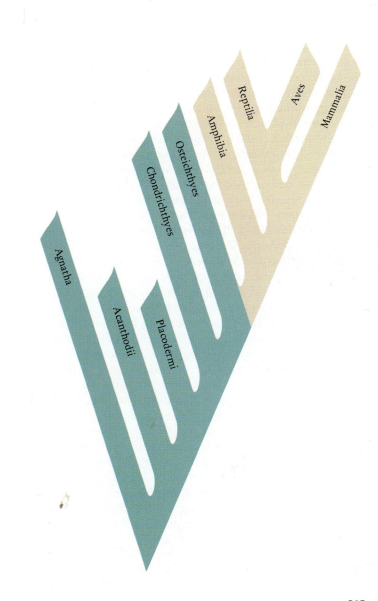

Agnatha

Acanthodii

Placodermi

Chondrichthyes

Osteichthyes

Amphibia

Reptilia

Aves

Mammalia

(Left) A school of scalefin anthias, Anthias squamipinnis, *a sea bass that lives on coral reefs. (Above) An individual from the Red Sea.*

There are nearly 25,000 living species of fishes, more species than all terrestrial vertebrates combined. Fishes exhibit tremendous adaptive diversity. They inhabit most bodies of fresh water and all oceans including shallow coastal areas, the open ocean, and the deep sea. A few even make brief excursions upon the land. Fishes can be found even in extreme environments, from the polar oceans to hot springs, and from brightly lit coral reefs to the total darkness of caves. Fishes are not a natural phyletic group, but rather an assemblage of five distinct classes (Fig. 36.1), yet all share common features suiting them for life in the water.

ADAPTATIONS OF FISHES

Locomotion and Buoyancy

Most fishes have a streamlined body shape that enables them to move easily through the water. Integumentary mucous glands are abundant. Discharged mucus spreads over the body surface, where it helps to protect the body from ectoparasites, prevents undue exchanges of body water with the environment, and reduces friction between the body surface and water. Their skeleton is not as strong as that of terrestrial vertebrates, for the buoyancy of the water provides considerable support. Trunk muscles are segmented myomeres whose contractions cause the lateral undulations of the trunk and tail by which the animal swims (p. 228). Undulations pass posteriorly along most of the trunk and tail in eels, sharks, and other long-bodied species, but are confined to the posterior part of the trunk and tail in shorter bodied and faster swimming species, including tuna and swordfish (Fig. 36.2a and b). The anterior myomeres of the faster swimming species have tendinous extensions that reach the tail, so the tail undulates with considerable force. The myomeres of most fishes are composed partly of slow phasic (red) muscle fibers that are used in cruising and partly of fast phasic (white) muscle fibers that are used when a sudden burst of speed is needed. Fishes that move slowly back and forth in dense aquatic vegetation or other confined spaces often have short bodies and swim by undulating their fins.

A fish must also be able to remain at a particular depth and maintain its stability in the water as it swims. The forces responsible can be seen by examining a shark (Fig. 36.2c). The water that a fish displaces exerts a lift force that acts upward through its center of buoyancy, but since flesh is denser than water, the lift is less than the weight of the fish, which acts downward through its center of gravity. When a shark swims, an additional lift force is generated by a hydroplaning effect of its ventrally flattened head and broad pectoral fins attached low on the body. The tendency to pitch upward is counteracted by the lateral movements of its tail, which has a large upper lobe into which the end of the vertebral column extends. An asymmetrical tail of this type, which gives an upward lift to the posterior end of the body, is called a **heterocercal tail** (Gr. *heteros*, different + *kerkos*, tail) (Fig. 36.3a). You will notice from Figure 36.2c that all of the lift forces equal the weight of the fish, so the fish remains stable in the water column.

The livers of sharks and some other primitive fishes contain considerable oil, which being lighter than water reduces the density of the fish somewhat and provides buoyancy. However, this alone does not overcome the high density of most primitive fishes. Most of these fishes also require muscular effort to remain stable in the water column. When not moving, many sharks lie on the bottom. More advanced fishes have evolved a swim bladder (p. 279), a sac of gas located dorsal to the body cavity, that gives them a density close to that of water, so they can remain stable in the water column with little muscular effort. These fishes do not have large pectoral fins, a flat head and other lift-generating features found in more primitive species. Their tails have become superficially symmetrical, or **homocercal**, and so do not generate an upward thrust. Although the tail is superficially symmetrical, the tip of the vertebral column still turns upward, reflecting its heterocercal ancestry (Fig. 36.3c).

Dorsal fins and other median fins, as well as the paired pectoral and pelvic fins, provide stability against rolling from side to side and yawing, that is, a tendency for the front of the body to move from left to right (Fig. 36.2d). The paired fins are also used in turning, braking, and changing depth.

Sense Organs

The sense organs of fishes are adapted for receiving stimuli in water. Since the eyes are bathed by the surrounding water, fishes have no need for movable eyelids and tear glands. Light does not travel far in water, so fishes can only see clearly those objects that are relatively close to them. Their eyes accommodate only to a slight extent, and this is accomplished by the lens moving rather than by changing its shape. The lens must remain spherical because a thick lens is needed to refract light rays entering the eye from the water. Little refraction occurs as light rays pass from water through the cornea (p. 421).

Except in lungfishes, the nose is only an olfactory organ and is not used in ventilation. Water does not pass through it on the way to the pharynx. The sense of smell and other chemical senses are very important to fishes in finding food, mates, and identifying their surroundings.

An inner ear is present, and pressure waves in the water pass easily through the tissues of the fish to reach it (see Fig. 18.11). Fishes lack the external and middle ear structures that in terrestrial vertebrates capture and transmit airborne sound waves.

Low-frequency vibrations and certain water movements are detected by the lateral line system, a unique aquatic sensory system that responds to changes in water

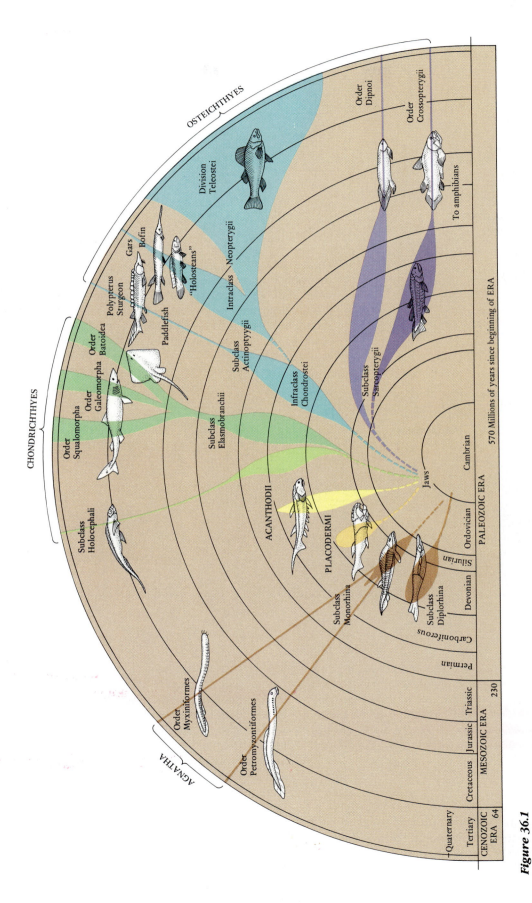

Figure 36.1

The distribution of fish groups through geological time and their probable evolutionary interrelationships are shown in this phylogenetic tree.

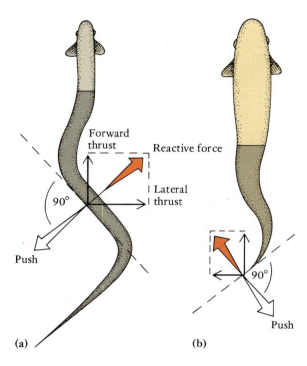

(a)   (b)

Figure 36.2
Locomotion, buoyancy, and stability. Undulations sweep down most of the trunk and tail of long-bodied fishes (a), but are confined to the posterior part of the trunk and tail in shorter-bodied ones (b). (c) A diagram of forces acting on a shark that enable it to float. (d) The median and paired fins provide stability.

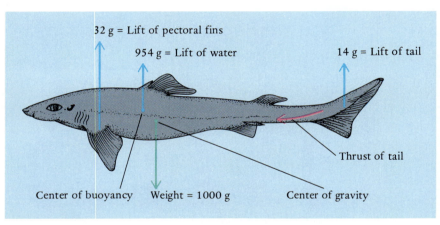

32 g = Lift of pectoral fins
954 g = Lift of water
14 g = Lift of tail
Thrust of tail
Center of buoyancy   Weight = 1000 g   Center of gravity

(c)

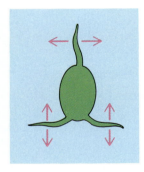

(d)

movements and pressures (see Fig. 18.9). With it a fish detects objects it is approaching or objects moving toward it. Some lateral line receptors become modified in many fish groups as **ampullary organs**, which allow the fish to detect and subsequently home in on the weak electric activity generated by muscle contraction in prey organisms (see Focus 18.1).

Metabolism

Because of the thermal stability of water, fishes are not subjected to the extreme temperature fluctuations encountered by many terrestrial animals. Although tuna and some other rapidly swimming species have circulatory countercurrent mechanisms that retain heat in muscle tissue (p. 835), most fishes are ectothermic vertebrates with rather low levels of metabolism. They thus require less food and lower levels of gas exchange than animals with a higher metabolic level. Most fishes have jaws, and a few use them to seize food, but most fishes ingest food by suddenly opening their jaws, expanding their pharynx, and sucking food in. A muscular tongue is not needed to manipulate and transport food in the mouth. Food is held in the mouth and is sometimes cut or crushed by teeth that may be located on the jaws, on the roof of the mouth and pharynx, and often on the gill arches.

Gases are exchanged between water and blood by diffusion across the gills. Water typically enters the phar-

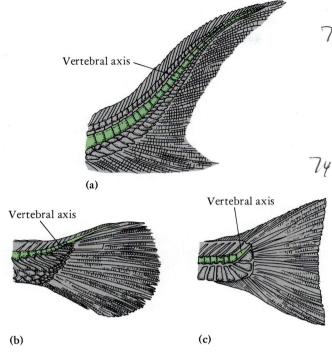

Vertebral axis

(a)

Vertebral axis

Vertebral axis

(b) **(c)**

Figure 36.3
The evolution of fish tails. (a) The heterocercal tail of primitive species. (b) An intermediate abbreviated heterocercal tail. (c) The homocercal tail of teleosts.

ynx through the mouth and is discharged through the gill pouches (see Fig. 13.12). A countercurrent flow between the blood passing through the gills and the water across them makes for a very efficient exchange of gases.

The heart receives blood low in oxygen content from the body. Cardiac contraction drives the blood forward and through the gill capillaries, where considerable pressure is lost. Thus oxygenated blood is distributed from the gills to the body under relatively low pressure.

Excretion and Water Balance

Nitrogenous metabolic wastes are easily eliminated in water by diffusion through the gills, supplemented by an opisthonephros, the kidney of adult fishes (p. 353). Maintaining salt and water balance is more of a problem. Depending on whether fishes live in sea or fresh water, water must be conserved or eliminated. Most marine fishes drink sea water to compensate for lost body water, and eliminate excess sodium, potassium, and calcium ions by salt-excreting cells on the gills or by special glands. Calcium, magnesium, and sulfate ions are eliminated by the kidneys. Freshwater fishes must eliminate excess water as a dilute urine and absorb lost salts from the medium through cells on the gills.

Reproduction and Development

Most fishes are oviparous and fertilization is usually external. As the female lays eggs, the male discharges sperm over them. Development typically includes a free-swimming larval stage. Some species have evolved internal fertilization and some degree of viviparity. Their young are born as fully formed juveniles.

CLASS AGNATHA

Characteristics of Agnathans

■ Jaws are absent.

■ Although pectoral spines or fins were present in some extinct species, paired fins are generally absent.

■ Early species had heavy bony scales and plates in their skin, but these have been lost in living species.

■ The cranium was ossified in some extinct species, but in most cases the deeper parts of the skeleton are cartilaginous. The embryonic notochord persists in the adult.

■ Seven or more paired gill pouches are present.

■ The branchial arches supporting the gill pouches lie close to the body surface.

■ A median, light-sensitive pineal eye is present.

■ The inner ear lacks a horizontal semicircular duct.

■ Living and many extinct species have a single median nostril located in front of the pineal eye, but some extinct groups had paired nostrils.

The Ostracoderms

The class **Agnatha** is represented today by about 60 species of lampreys and hagfishes, but these are just a remnant of a large and diverse group of Paleozoic fishes. Extinct agnathans are commonly called **ostracoderms** because the skin of most species contained a shell-like skeleton of bony plates and scales (Gr. *ostrakon*, shell). Fragments of these scales have been found in marine deposits of the Cambrian period, which demonstrate that vertebrates are very old, as old as most invertebrate groups. They also indicate that vertebrates arose in the ocean.

Ostracoderms had radiated widely by the Silurian and Devonian periods, and by this time were primarily freshwater fishes. Paleontologists recognize nearly 40 families grouped into two subclasses: the **Diplorhina**, in which the nasal sacs are paired, and the **Monorhina** with a single median nasal sac. By looking at two examples we can gain some understanding of the structure and probable mode of life of these early vertebrates.

Both *Pterapsis*, a diplorhine, and *Cephalaspis*, a monorhine, were somewhat flattened, small fishes that reached a maximum length of about 15 cm (Fig. 36.4). They were heavily armored with thick bony plates and scales in the skin. These were covered by dentine-like and enamel-like materials. Such heavy scales certainly offered some mechanical protection, possibly against aquatic scorpion-like eurypterids (p. 720). In addition, the layers of dentine and enamel may have offered some protection against an excessive inflow of water from the freshwater environment in which most of these fishes lived. As in all vertebrates, bone was an important reservoir for calcium and phosphate ions. Cavities and canals within the scales indicate the presence of lateral line and electroreceptive systems.

Jaws were absent, and the fins were poorly developed. *Cephalaspis* had paired pectoral fins. The tail was heterocercal in *Cephalaspis* but reversed heterocercal in *Pterapsis*, whose ventral lobe was enlarged and stiff. Ostracoderms had from seven to ten gill pouches that were supported by a branchial skeleton that lay close to the body surface rather than next to the pharynx (see Fig. 36.8). Pouches on each side opened through a common duct to the surface in *Pterapsis*, but independently on the underside of the head in *Cephalaspis*. In specimens in which the braincase is well preserved, we can tell that the inner ear lacked the horizontal semicircular duct characteristic of other vertebrates. Paired eyes and a median pineal eye were present. Three peculiar groups of small plates on the top of the head of *Cephalaspis* were underlain by large nerves and may have been electric organs or specialized electroreceptors.

The absence of well-developed fins, the flattened body shape of most species, and the heavy bony scales and plates suggest that ostracoderms were not agile swimmers but slow-moving, bottom-dwelling species. The absence of jaws would have limited these fishes to filter feeding, scavenging, or preying upon small, soft-bodied organisms that they may have located by means of an electroreceptive system.

Living Jawless Vertebrates

Ostracoderms became extinct by the end of the Devonian period, but the lampreys and hagfishes, descendants of monorhine ostracoderms (Fig. 36.5), have survived because of highly specialized feeding mechanisms. Unlike

Figure 36.4
A Devonian freshwater scene. Two ostracoderms and a eurypterid are shown.

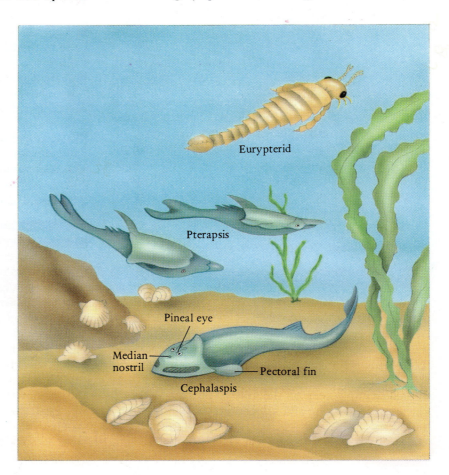

(a)

(b)

Figure 36.5
Living agnathans. (a) Sea lampreys, *Petromyzon marinus*, attached to a carp from which they suck blood and bits of tissue. (b) The Pacific hagfish, *Bdelostoma stouti*. Hagfishes feed upon soft-bodied invertebrates and dead or injured fishes.

ostracoderms, lampreys and hagfishes have an eel-like shape and slimy, scaleless skin. Like the extinct agnathans, they are jawless and have more gill pouches than other living fishes, their branchial skeleton is close to the body surface, they lack paired appendages, and they retain a pineal eye. They resemble the monorhines in having a single median nostril. Besides leading to an olfactory sac, this nostril in lampreys opens into a **hypophyseal sac** that passes between the brain and pharynx (Fig. 36.6). Respiratory movements of the pharynx alternately compress and expand the hypophyseal sac, helping to circulate a current of water across the olfactory epithelium in the sac. The anterior lobe of the pituitary gland is derived from an embryonic hypophysis. Lampreys and hagfishes also lack a horizontal duct in the inner ear. Despite these similarities, lampreys and hagfishes differ from each other in so many ways that they are now placed in separate orders. A few

fossil lampreys have been discovered; less is known about the evolution of hagfishes.

Most lampreys (order **Petromyzontiformes**) live in fresh water, but some spend their adult life in the ocean and return to fresh water only to reproduce. A familiar example of the group is the sea lamprey, *Petromyzon marinus*, of the Atlantic Ocean and eastern United States. The chief axial support for the body is a notochord that persists throughout life and is never replaced by vertebrae (Fig. 36.6). The mouth lies deep within a **buccal funnel**, a suction-cup mechanism by which the lamprey attaches to other fishes. The mobile tongue, armed with horny "teeth," rasps away at the prey's flesh, allowing the lamprey to suck blood and bits of tissue into its mouth. The lamprey secretes an anticoagulant that keeps the blood of the prey flowing freely. From the mouth cavity, the food enters a specialized esophagus that bypasses the pharynx to lead

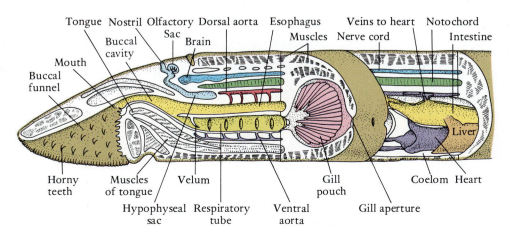

Tongue Nostril Olfactory Dorsal aorta Esophagus Veins to heart Notochord
Buccal Sac Brain Muscles Nerve cord Intestine
cavity
Mouth
Buccal
funnel

Liver

Horny Muscles Velum Gill Coelom Heart
teeth of tongue pouch
Hypophyseal Respiratory Ventral Gill aperture
sac tube aorta

Figure 36.6
The structure of the anterior part of a lamprey.

into a straight intestine. There is no stomach. The respiratory system consists of seven pairs of gill pouches that connect to a modified pharynx known as the **respiratory tube**. The separation of the pharynx from the food passage enables a feeding lamprey to pump water in and out of the gill pouches through the external gill slits. The gill pouches are supported by a branchial skeleton located lateral to them.

The eggs are laid on the bottom of streams in a shallow nest, which the lampreys make by removing the larger stones (Fig. 36.7). Fertilization is external, and the adults die after spawning. Developing sea lampreys pass through a larval stage that lasts five to six years. The larva is so different in appearance from adult lampreys that originally it was believed to be a different kind of animal and was named *Ammocoetes*. This name has been retained for the larva. The **ammocoetes larva** is eel shaped but lacks the specialized feeding mechanism of the adult. It lies within burrows in the mud at the bottom of streams and sifts minute food particles from water being pumped through the muscular pharynx. As in protochordates, a mucus-producing endostyle helps to trap food.

Sea lampreys reached the Great Lakes in the 1920s, presumably having circumnavigated Niagara Falls through the Welland Canal. They did considerable damage to commercial fisheries in the upper Great Lakes during the late 1940s and 1950s, but the depletion of fish on which they fed, together with control programs using poisons that selectively kill the larvae, have greatly reduced the lamprey problem in recent years.

Brook lampreys probably evolved from the predaceous sea or lake species. They too spend most of their lives as filter-feeding larvae. During metamorphosis their digestive tract atrophies so they cannot feed. After a few weeks they reproduce and die.

The hagfishes (order **Myxiniformes**) are exclusively marine. The best known species, *Myxine glutinosa* of the Atlantic Ocean and *Bdelostoma stouti* of the Pacific Ocean, are primarily scavengers feeding at the bottom on dead or injured fishes and soft-bodied invertebrates. They often enter the body through the anus or gill pouches, and then feed upon the soft internal organs. Horny, toothlike plates on either side of a protrusible tongue can act like pincers to tear off bits of flesh. Their median nostril connects by a duct to the pharynx. This allows water to enter the pharynx and gill pouches when food is in the mouth. Rather than opening independently on the surface, the gill pouches lead to one or more ducts that continue to the body surface. Hagfishes are the only vertebrates with a blood that is isosmotic to sea water, the usual condition in marine invertebrates. This suggests that hagfishes have always been a marine group. All other living vertebrates are osmotically independent of their environment, a characteristic that probably evolved during a freshwater stage of their ancestry.

EARLY JAWED VERTEBRATES

In contrast to the jawless agnathans, all other vertebrates have jaws, for which reason they are collectively called **gnathostomes** (Gr. *gnathos*, jaw + *stoma*, mouth). Jaws enabled fishes to feed upon a wider variety of food sizes and types than could the jawless ostracoderms. Jaws evolved from an anterior gill, or visceral arch. The visceral

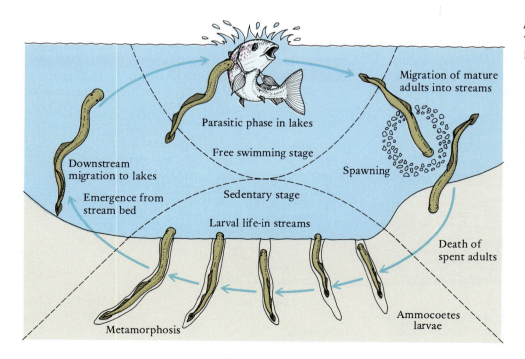

Figure 36.7
The life cycle of the sea lamprey in the Great Lakes.

Migration of mature adults into streams

Parasitic phase in lakes

Free swimming stage

Spawning

Downstream migration to lakes

Emergence from stream bed

Sedentary stage

Larval life-in streams

Death of spent adults

Ammocoetes larvae

Metamorphosis

arches of gnathostomes are located deeply next to the mouth and pharynx and median to the gills (Fig. 36.8). This is in sharp contrast to the visceral arches of agnathans, which lie close to the body surface and lateral to the gills. The visceral arches appear to have evolved independently and somewhat differently in agnathans and gnathostomes. It is possible that both groups diverged shortly after the origin of vertebrates, although gnathostomes do not appear in the fossil record until the early Silurian, well after the first ostracoderm fossils (Fig. 36.1).

Paired pectoral spines or fins appear sporadically among the agnathans, but the earliest known jawed fishes had both pectoral and pelvic fins. These paired fins and a streamlined body shape indicate that early jawed fishes could be active swimmers, not just sluggish bottom dwellers. Jaws and mobility made possible an efficient, predatory mode of life.

Class Acanthodii

There is a bewildering array of primitive jawed fishes from the Devonian, and paleontologists are not in agreement as to how they should be classified. Some are obviously related to contemporary sharks and rays, others to bony fishes, and still others appear to be unique, extinct groups without living descendants. Jaws and paired fins first appear in the fossil record in the acanthodians, often called "spiny sharks" because of their sharklike shape and numerous median and paired spines (Fig. 36.9a). We hasten to add that acanthodians, which had ossified skeletons, were not sharks. They may be a subclass of bony fishes, but the absence of a regular pattern of tooth replacement seen in bony fishes and other technical differences lead many paleontologists to place them in their own class, the **Acanthodii**. The visceral arches of acanthodians are hidden by superficial dermal bones, but can be seen when these are removed (Fig. 36.9b). The **mandibular arch**, which formed the core of the upper and lower jaws, is part of a series of visceral arches and can be thought of as the enlarged first one. The second visceral arch, known as the **hyoid arch**, lies close behind the mandibular arch and extends as a prop from the braincase to the posterior end of the upper jaw, as it does in many contemporary fishes (see Fig. 11.14). The remaining visceral arches are **branchial arches** that supported the gills.

Acanthodians had a series of paired spines protruding from the side of their trunk (Fig. 36.9a). Internal skeletal rods behind the pectoral and pelvic spines indicate that fins were present at these sites, but apparently the intermediate paired spines were not associated with fins. The pectoral and pelvic fins were attached to the trunk by broad bases. Such fins would have been effective stabiliz-

Figure 36.8

Frontal sections through the oral cavity and pharynx of an agnathan (a) and a gnathostome (b). Note the difference in the location of the gill chambers and gills relative to the gill arches.

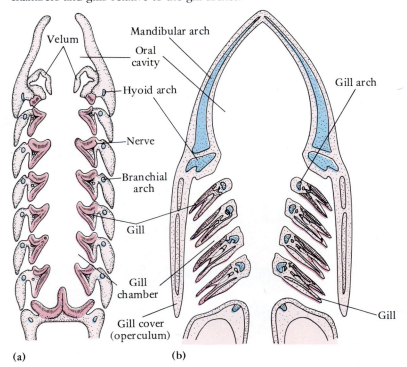

(a)   (b)

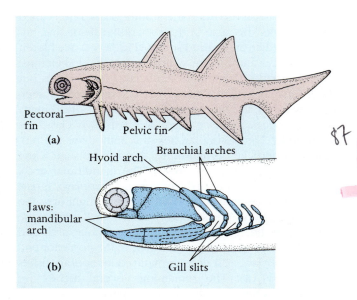

Figure 36.9

(a) Acanthodians, represented by the spiny shark *Climatius*, were among the first jawed vertebrates. (b) Removal of dermal bones on the head reveals that the jaws were the enlarged first visceral arch.

ing keels but could not have changed planes to help the fish maneuver.

Early acanthodian fossils occur in coastal marine deposits, but most of these fishes later adapted to fresh water. Acanthodians were an abundant group during the Devonian, but became extinct during the Permian (see Fig. 39.1).

Class Placodermi

The Devonian seas also contained a large and diverse assortment of jawed fishes belonging to the class **Placodermi**. Most were heavily armored, had a heterocercal tail, and were slightly flattened dorsoventrally—features indicative of a benthic life. Paired fins were present. The most spectacular genus, *Dunkleosteus* (Fig. 36.10), was a monster of the late Devonian sea living where Cleveland, Ohio, is now located. It attained a length of 2 m or more. Its head and anterior trunk were covered by bony plates, but the rest of the body was naked. It lacked teeth, but had formidable jaws composed of large cleaver-like bony plates that were attached to the underlying mandibular arch. *Dunkleosteus* was the dominant marine predator of its time, but it probably laid on the bottom and waited for prey to come to it. When resting on the sea bottom, it could not open its mouth by dropping its lower jaw far. Instead, a unique joint between the thoracic armor and skull permitted it to raise the top of its head.

CLASS CHONDRICHTHYES

Characteristics of Cartilaginous Fishes

■ They resemble bony fishes and terrestrial vertebrates in having jaws, paired appendages, visceral

Figure 36.10

The placoderm, *Dunkleosteus*, was the dominant predator of the Devonian seas.

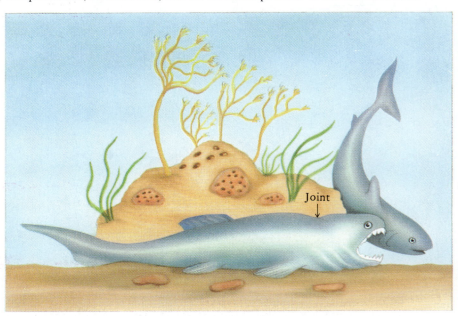

arches located deeply next to the pharynx, an inner ear with three semicircular ducts, and paired nasal cavities.

■ An electroreceptive system is well developed.

■ Bony scales are either tiny placoid scales or are lost completely.

■ The internal skeleton is entirely cartilaginous.

■ They are heavy-bodied fishes without lungs or a swim bladder. The tail is usually heterocercal.

■ The five pairs of gill pouches open independently on the body surface in most species.

■ Their intestine is short, but its surface area is increased by a spiral valve.

■ Males have a clasper on the pelvic fins with which sperm are transferred to the female. Fertilization is internal.

All remaining fishes belong to two classes: the **Chondrichthyes**, or cartilaginous fishes, and the **Osteichthyes**, or bony fishes. As we shall see, these classes differ in many ways, but they share a very similar type of mandibular arch and a similar pattern of continuous tooth replacement. This suggests a common ancestry separate from that of other jawed fishes. Both groups can be traced back to the Silurian (Fig. 36.1).

Basic Features of Cartilaginous Fishes

The class Chondrichthyes includes about 750 species of contemporary sharks, rays, and similar fishes. The earliest known species were marine, and the group has remained primarily marine ever since, although one extinct group and several contemporary species have secondarily adapted to fresh water.

Cartilaginous fishes retain the cartilaginous embryonic endoskeleton throughout their life (see Fig. 11.14). It is not replaced by bone, but calcium salts are sometimes deposited within it and strengthen it. No evidence indicates that the ancestors of cartilaginous fishes ever had the extensive dermal skeleton seen in ostracoderms and placoderms. Minute dermal denticles, or **placoid scales**, are imbedded in the skin (Fig. 36.11). Each is a toothlike structure containing a pulp cavity surrounded by dentine and capped by an enamel-like material. The triangular teeth of sharks closely resemble enlarged placoid scales and have undoubtedly evolved from them.

Sense organs are well developed in the cartilaginous fishes. Olfaction and especially electroreception play

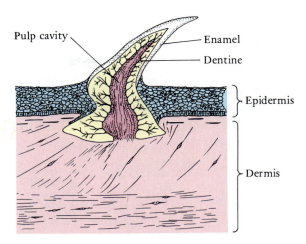

Figure 36.11
A placoid scale of a shark.

prominent roles in feeding behavior. The electroreceptors are ampullary organs. They lie at the ends of long tubes filled with a gelatinous material. These tubes are distributed over the head in such a way that the fish can detect the source of weak electric currents generated by the muscular activity of prey organisms (see Focus 18.1).

The mouth cavity is continuous posteriorly with a long pharynx (Fig. 36.12a). A **spiracle**, containing a vestigial gill, and the gill pouches, containing functional gills, open from the pharynx to the body surface. An esophagus leads from the back of the pharynx to the stomach. A short, straight valvular intestine receives secretions from the liver and pancreas. It contains an elaborate helical fold known as the **spiral valve** that serves both to slow the passage of food and to increase the digestive and absorptive surface of the intestine (Fig. 36.12b).

The circulatory system is of the primitive type in which the heart is undivided and pumps only oxygen-poor blood forward to the gills, where it is aerated (see Fig. 14.8).

The kidneys are of the primitive opisthonephric type (Fig. 36.12a; see also p. 353). The problem of conserving body water in the marine environment is met by retaining considerable urea, so the body fluids are slightly hyperosmotic to sea water (p. 362). Some ions are eliminated by the kidneys, but most excess salt taken in with the food is eliminated by a salt-excreting **rectal gland** that empties into the end of the intestine and by salt-excreting cells in the gills. Intestine and urogenital ducts discharge into a common cloaca that opens on the underside of the body.

Specialized parts of the male pelvic fins form **claspers** that are used to transfer sperm to the female. The eggs are fertilized in the upper part of the oviduct, and a horny protective capsule is secreted around them by certain oviductal cells. Skates are oviparous, but there is no free

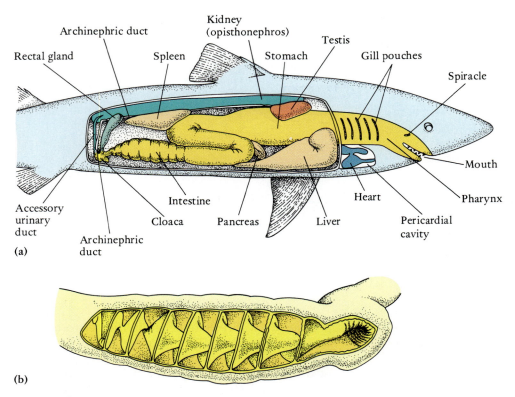

Figure 36.12
(a) The visceral organs of the dogfish, *Squalus*; (b) the intestine has been cut open to show the spiral valve.

larval stage as there is in many fishes. The eggs are heavily laden with yolk, and the embryos develop within the protective capsule. Some sharks are also oviparous, but most sharks brood their young internally. The fertilized eggs develop in a uterus, a modified portion of the oviduct. A few species are viviparous and develop an intimate association between each embryo's yolk sac and the uterine lining, forming a **yolk sac placenta**. Embryos of this type have a greater dependence for their nutrient requirements upon the mother than upon food stored in the yolk. However, most shark embryos are ovoviviparous (p. 524). Although developing within a uterus, they have a greater dependence upon food stored in the yolk. In some cases, a portion of the nutritional requirements is derived from materials secreted by the mother into the uterine fluid or by certain embryos consuming other embryos. In these cases a placental relationship is not established.

Adaptive Radiation of Cartilaginous Fishes

Cartilaginous fishes are grouped into two subclasses (Fig. 36.1). The **Holocephali** are an evolutionary side branch that includes the present day ratfish, *Chimaera* (Fig. 36.13a). Their gill pouches do not open directly to the outside but into a common chamber covered by a flap of skin. This is the **opercular chamber**. Their upper jaw is fused to the cranium. The solid jaw construction and crushing tooth plates are adaptations that enable holocephalans to feed upon molluscs, crustaceans, and similar hardbodied animals.

The **Elasmobranchii** include the sharks and skates and are characterized by having separate gill slits for each gill pouch. Ancestral elasmobranchs had broad-based fins and could not protrude their jaws, hence they were not as efficient in swimming and feeding as contemporary species with narrow based fins and protrusible jaws. Contemporary elasmobranchs are grouped into two orders of sharks (**Squalomorpha** and **Galeomorpha**) and one that includes the skates and rays (**Batoidea**).

Sharks

Squalomorph sharks, which include the dogfish (*Squalus*) used in comparative anatomy laboratories, tend to be cold- or deep-water species (Fig. 36.13b). The more familiar active, predaceous sharks of subtropical and tropical oceans are galeomorphs. A few—including the great white shark (*Carcharodon*), which attains a length of 11 m—will attack humans as they would a seal, large fish, or other prey organism (Fig. 36.13c) . Whale sharks (*Rhiniodon*), which may reach a length of 15 m, have minute teeth and feed

(a)

(b)

(c)

(d)

(e)

(f)

Figure 36.13

Representative cartilaginous fishes. (a) Unlike other cartilaginous fishes, the gill slits of *Hydrolagua colliei* and other ratfishes have a common opening just in front of the pectoral fin. (b) The chain dogfish is a small species of shark. (c) The great white shark, *Carcharodon carcharias*, of tropical oceans feeds primarily on other large fishes but this species also attacks humans. (d) A tropical spotted eagle ray, *Aetobatis narinari*, swims with its enlarged pectoral fins, and its tail is reduced to a slender whip. (e) The sawfish, *Pristis pectinata*, lives in tropical and subtropical seas but occasionally enters the lower reaches of rivers. (f) A manta ray, *Manta*, of the Atlantic and Indo-Pacific Oceans. Notice the huge pectoral fins with which it swims and the forward protruding, hornlike head lobes that help gather plankton.

entirely upon small crustaceans and other planktonic organisms. They gulp mouthfuls of water, and as the water passes out through the gill pouches, the food is kept in their pharynx by a sieve at the pharyngeal entrance to the gill pouches that is composed of many slender filaments. Whale sharks are the largest living fishes.

Skates and Rays

Most skates and rays are bottom-dwelling fishes that are dorsoventrally flattened. Trunk and tail muscles are usually reduced, and these fishes swim along the bottom by undulations of their enormous pectoral fins (Fig. 36.13d). When resting on the bottom, their mouth is often buried in the sand or mud, and water for respiration enters the pharynx via the pair of enlarged spiracles. Most species have crushing teeth and feed upon molluscs and crustaceans. The electric ray (*Torpedo*) can stun prey and discourage predators by releasing pulses of an electric current at an output of up to 2500 watts. The electric organ of *Torpedo* consists of modified gill muscles in which the action potential of each unit is magnified and the output of many units is summed. The sawfish (*Pristis*) has an elongated, blade-shaped snout armed with toothlike scales (Fig. 36.13e). By thrashing about in a shoal of small fishes, it can disable many and eat them at leisure. Bathers are sometimes injured if they startle a sawfish. The largest member of the group is the devilfish, *Manta*, which sometimes attains a width or "wing spread" of 6 m (Fig. 36.13f). Devilfish have abandoned a bottom-dwelling mode of life and feed upon plankton. As with many other plankton-feeding vertebrates, their teeth have been reduced.

CLASS OSTEICHTHYES

Characteristics of Bony Fishes

■ Well-developed bony scales usually are present.

■ Primitive species retain electroreceptive organs, but these are lost in most contemporary species.

■ The deeper skeleton always contains some bone; in most species it is nearly completely ossified.

■ Lungs or a swim bladder are present, except in a few bottom-dwelling species in which they have been secondarily lost. The body is more buoyant than in cartilaginous fishes, and the tail has become homocercal in most living species.

■ The gill pouches open into a common chamber covered by an operculum.

■ The intestinal spiral valve is absent in all but the most primitive species. Surface area is increased by a longer, coiled intestine and by pyloric ceca.

■ The sperm ducts usually bypass the kidneys and lead directly to the cloaca; eggs are discharged directly into oviducts.

■ Most species are oviparous and fertilization is external. The copulatory organ of the few viviparous species with internal fertilization is a modified part of the anal fin.

Over 24,000 living species of bony fishes are known, and they comprise over 95% of all fishes and over half of all vertebrate species. They are grouped into two subclasses: the **Actinopterygii**, or ray-finned fishes, and the **Sarcopterygii**, or lobe-finned fishes. Sturgeons, salmon, minnows, perch, and most other species of bony fishes are actinopterygians. Although once abundant, sarcopterygians are represented today only by three genera of lungfishes and one crossopterygian species.

Basic Features of Bony Fishes

Much or all of the embryonic cartilaginous skeleton is replaced by bone during development. Relatively heavy bony scales were present in the skin of primitive species. They were similar in some ways to the bony plates of ostracoderms and placoderms, but were not as large. Heavy scales were certainly protective in many ways, but they must have been cumbersome and may have impeded lateral undulations and swimming.

The spiracle is lost except in the most primitive bony fishes. Remaining gill pouches open into a common opercular chamber. A sac of gas, either **lungs** or a **swim bladder**, is an important feature of bony fishes (Fig. 36.14). A swim bladder is primarily a hydrostatic organ that allows the fish to adjust its density to that of the surrounding water and maintain a neutral buoyancy at different depths (p. 279). Since the gas within it is chiefly oxygen, a swim bladder can also be a reservoir of this gas. Lungs and a swim bladder are homologous organs, and they appear to have evolved very early in the history of bony fishes. The ancestral organ was probably more like a lung than a swim bladder. Early bony fishes probably had lungs similar to those of the living African lungfish (*Protopterus*), in which a pair of saclike lungs develops as a ventral outgrowth from the posterior part of the pharynx (Fig. 36.15a). Lungs enable *Protopterus* to survive conditions of stagnant water and drought. The rivers and lakes in which the African lungfish live may completely dry up, but the fish can survive curled up within a mucous cocoon that it secretes around itself in the dried mud (Fig. 36.16). A small opening from the cocoon to the surface of the mud enables the fish to breathe air during this period.

Air breathing evolved in fishes as a supplement to gill respiration, but just when this occurred is uncertain. Scales recovered from late Silurian marine deposits indicate that

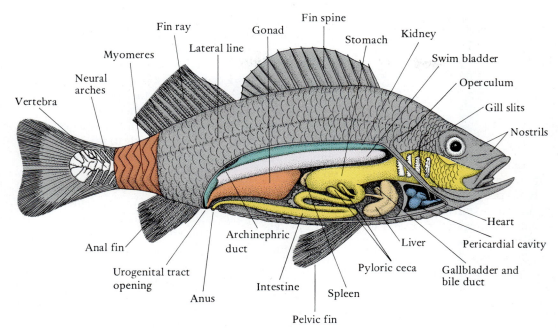

Figure 36.14
A representative actinopterygian, the perch, showing internal anatomy.

Figure 36.15
A diagram illustrating the evolution of lungs and the swim bladder. Both a cross section through the lungs and gut and a lateral view are shown. (a) Ancestral bony fishes probably had lungs similar to those of the African lungfish. (b) Terrestrial vertebrates retain lungs. (c) An intermediate condition seen in one contemporary fish in which the lungs are shifted dorsally. (d) The swim bladder of a primitive teleost still connects with the gut, but this connection is lost in more advanced teleosts.

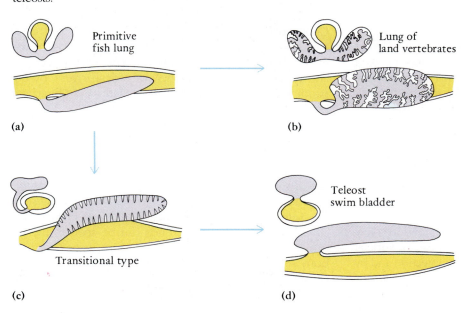

Figure 36.16
An estivating African lungfish, *Protopterus aethiopicus*. Notice
the cocoon that it secretes around itself (partly cut away),
and the air shaft in the dried mud to the surface.

some early bony fishes lived in the ocean, and we would
not expect them to have had lungs. But lungs would
certainly have been a useful adaptation for the majority of
species that lived in the unreliable freshwater habitats of
the Devonian period. During certain seasons, many bodies
of fresh water undoubtedly either became stagnant
swamps with a low oxygen content or dried up completely.
Fishes with lungs could survive these conditions, but
others became extinct or adapted to life in the sea, as did
many placoderms. Groups of bony fishes that have remained in fresh water throughout their history tended to
retain lunglike organs. Terrestrial vertebrates evolved
from one such group (Fig. 36.15b). Fish that migrated back
to the ocean did not retain lungs. Most marine habitats
contain a constant and adequate oxygen supply. The lungs
of most marine fishes were transformed into hydrostatic
swim bladders (Fig. 36.15d). What are presumed to be
intermediate stages in this shift can still be seen in certain
contemporary species (Fig. 36.15c). Later, when conditions in fresh water were more favorable, many salt-water
bony fishes reentered fresh water but retained their swim
bladders.

The spiral valve in the intestine is retained in a few
primitive species but is lost in others. Intestinal surface
area is increased by having a long, coiled intestine and by
short **pyloric ceca** that extend outward from the anterior
end of the intestine just behind the stomach (Fig. 36.14).

The body fluid of freshwater bony fishes is hyperosmotic to their environment. Excess water entering the
body is excreted by the kidneys as a urine hypoosmotic to
the blood. Lost salts are replaced in part by salts in the food
and in part by special salt-absorbing cells in the gills.
(Similar cells excrete salt in marine species.)

Unlike in other fishes, the sperm of most bony fishes
do not pass from the testes to the kidneys, or to structures
derived from the kidneys, but directly to sperm ducts that
join the posterior ends of the urinary ducts. The eggs of
most bony fishes are discharged from the ovary directly
into an oviduct rather than into the coelom. Nearly all
species are oviparous and fertilization is usually external.

Both actinopterygians and sarcopterygians can be
traced back to the late Silurian (Fig. 36.1). Since both share
such features as a partly ossified internal skeleton, well-
developed bony scales, an operculum, and continuous
tooth replacement (a combination of characters not present in other classes of fishes), it is assumed that they had a
common evolutionary origin.

Nonetheless, actinopterygians and sarcopterygians
differ from each other in a number of ways (Fig. 36.17).
The paired fins of actinopterygians are described as **ray-
fins** because they are fan shaped and supported by delicate
skeletal rays that evolved from rows of bony scales. Sarcopterygians, in contrast, have **lobe fins** supported by a
central axis of bone and muscle. Actinopterygians also
have an incurrent and an excurrent opening to each nasal
sac that allows a current of water to flow through the sac.
Most sarcopterygians have a single external nostril for each
nasal sac, and the sac discharges through an internal nostril
in the roof of the mouth.

Actinopterygians

Actinopterygians underwent two major adaptive radiations. The earlier radiation, giving rise to the infraclass
Chondrostei, occurred during the Paleozoic and early
Mesozoic eras (Fig. 36.1). Chondrosteans are now nearly
extinct, being represented only by the bichirs (*Polypterus*)
and reedfish (*Calamonichthys*) of African rivers, by the
paddlefish (*Polyodon*) of the United States and China, and
by the widespread sturgeons (*Acipenser*) (Fig. 36.18). A
second radiation, infraclass **Neopterygii**, has continued to
expand since the early Mesozoic. Primitive neopterygians
are sometimes called holosteans, but this is not a natural
group. Primitive neopterygians are represented today by
such fishes as the gars (*Lepisosteus*) and bowfin (*Amia*)
(Fig. 36.19). Most neopterygians belong to a large division,
the **Teleostei**, that includes nearly all living bony fishes.

Several evolutionary trends can be seen in the actinopterygian groups. Ancestral chondrosteans had thick
bony scales overlain by layers of a dentine-like material
(**cosmine**) and an enamel-like material (**ganoine**). They
are called **ganoid scales** because the ganoid layer was

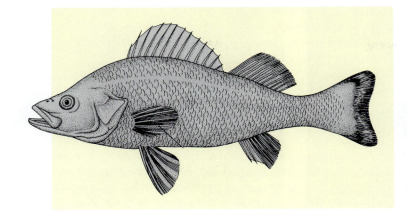

Figure 36.17
(a) The perch, *Perca*, an actinopterygian.
(b) *Osteolepis*, an early sarcopterygian. The nature of their paired fins and their external nostrils are among the important differences between sarcopterygians and actinopterygians.

(a)

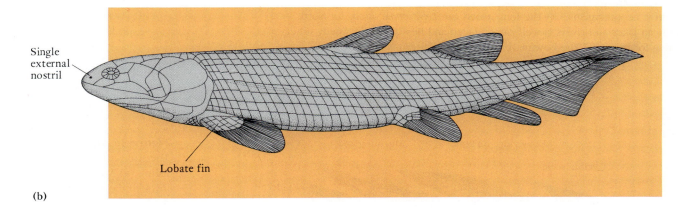

Single external nostril

Lobate fin

(b)

particularly thick. Ganoid scales are retained to some extent in surviving chondrosteans and in gars, but the scales have become thinner and lighter in most neo-pterygians. The ganoine and cosmine layers have been lost, and the underlying bone reduced to a thin disc that develops in the dermis of the skin (Fig. 36.20). Such a scale

Figure 36.18
Representative chondrosteans. (a) The bichir, *Polypterus ornatipinnis*, of tropical African rivers retains primitive ganoid scales. (b) The Volga sturgeon, *Huso huso*, from the Caspian Sea retains the primitive heterocercal tail. The barbels hanging from its snout are used in searching for food.

(a)   (b)

(a)   **(b)**

Figure 36.19
Primitive neopterygians. (a) The long-nosed gar, *Lepisosteus osseus*, of North American lakes and rivers is well camouflaged and lies quietly in weedy waters waiting for prey. (b) The bowfin, *Amia calva*, of Eastern North America.

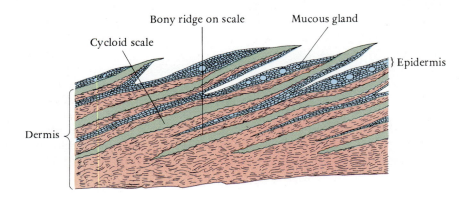

(a)

Figure 36.20
The scales of modern bony fishes. (a) A vertical section through the skin showing the location of the scales. Surface views of a cycloid scale (b) and a ctenoid scale (c).

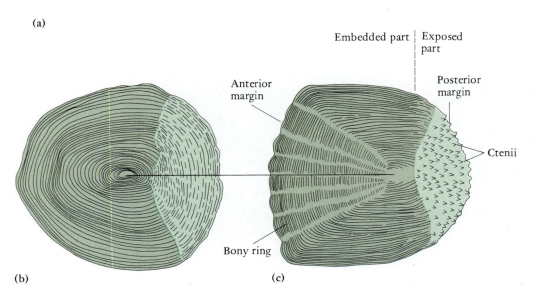

(b)   **(c)**

is termed **cycloid** when its surface is smooth, and **ctenoid** when the exposed portion bears minute spiny processes (ctenii). As scales became thinner, the trunk and tail could undulate more freely. Swimming efficiency improved, but the vertebral axis had to resist increased stresses. The notochord, which persists in adult chondrosteans, is largely replaced in neopterygians by a well-ossified vertebral column.

The functional lungs of early actinopterygians give way to swim bladders. As control over buoyancy improved and the body became more streamlined, the primitive heterocercal tail of most chondrosteans became superficially symmetrical or homocercal in teleosts, but as we have seen, the caudal skeleton still shows indications of the upward tilt of the vertebral column (Fig. 36.3c). Some primitive neopterygians (gars and bowfins) have an intermediate condition, the **abbreviated heterocercal** tail (Fig. 36.3b). Increased buoyancy relieved the paired fins of their primitive hydroplaning function. The point of attachment of the pectoral fins shifted dorsally so that they were closer to a horizontal plane passing through the center of gravity, and the pelvic fins shifted anteriorly. These changes enhanced the braking and turning functions of the fins.

The great success of advanced actinopterygians has been attributed in large measure to an improvement in their method of feeding. In chondrosteans, the premaxilla and maxilla, bones on the margin of the upper jaw, are firmly united with adjacent skull bones (Fig. 36.21a). Their mouth has a wide gape when the jaws open to seize food or suck it in. A squirming fish could escape relatively easily. In teleosts, the premaxilla has enlarged and excluded most of the maxilla from the jaw margin (Fig. 36.21b). The maxilla is free of adjacent skull bones and is articulated in such a way that it and the premaxilla swing forward as the mouth

is opened. The mouth opening advances toward the prey, and the prey cannot escape through a wide gap between upper and lower jaws.

The approximately 24,000 living species of teleosts form a large and diverse group. Many investigators follow a cladistic analysis of Lauder and Liem (1983) in sorting them into nine groups, here called superorders. Each is characterized by one or more unique, derived features. These superorders are listed and characterized in the Classification (p. 838).

Adaptive Radiation of Teleosts

Few groups of vertebrates have undergone an adaptive radiation as extensive as that in teleost fishes. They are found in every conceivable aquatic environment and have become specialized to feed in nearly every possible way. Some are filter feeders; many are carnivores and herbivores. Some have jaw and tooth modifications enabling them to crush coral, and some are specialized to feed on the scales or nibble the eyes of other fishes! Here we examine only a few of the fascinating adaptations of this interesting group.

Eels have long, snakelike bodies and have lost their pelvic fins and usually their scales (Fig. 36.22). This body form is particularly well adapted to living in crevices in coral reefs, as many marine species do, or in waters with considerable mud and vegetation, as freshwater species do. Freshwater eels have a particularly interesting life cycle, which was determined in the early part of this century by Johann Schmidt, a Danish investigator. With the cooperation of sea captains, he was able to catch thousands of larval eels (**leptocephali**) in plankton nets from various regions of the Atlantic Ocean. By correlating larval size with site of capture, he determined where they bred and their migration routes. When ready to breed, the European

Figure 36.21
Mouth action in a primitive actinopterygian (a) and in a teleost (b).

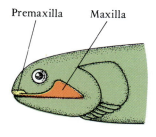

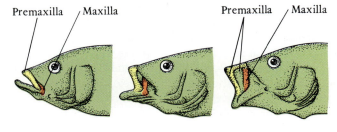

(a)

(b)

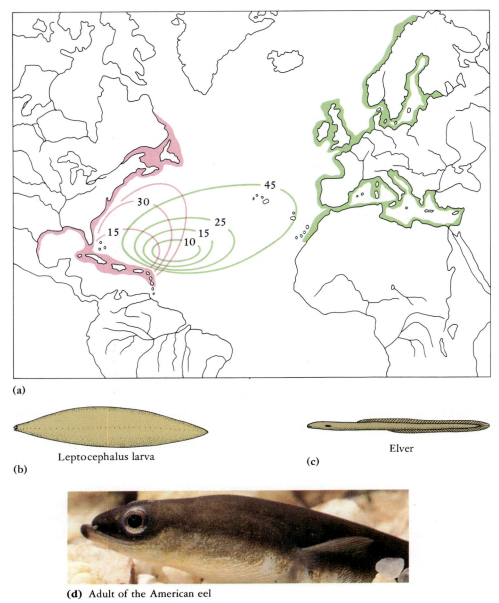

(a)

Leptocephalus larva

(b)

Elver

(c)

(d) Adult of the American eel

Figure 36.22
The life cycle of two species of eel: the North American species (*Anguilla rostrata*) and the European-North African species (*Anguilla vulgaris*). Curved lines and numbers in the ocean indicate larval distribution and size in millimeters. Colored coastal areas show where the elvers enter rivers.

and North African species, *Anguilla vulgaris*, and the American species, *Anguilla rostrata*, leave the rivers in which they have been living and migrate at considerable depth to an area of floating seaweed southeast of Bermuda, the Sargasso Sea, where they spawn and die. The young larvae move to the surface and are carried by the Gulf Stream back towards the coasts of Europe and North America. Larvae of the two species can be distinguished by having slightly different numbers of vertebrae and by

differences in their mitochondrial DNA and other molecules. The European species takes three years to develop into a young eel (**elver**), by which time it has drifted to the coastal rivers and is ready to ascend them. The American species has a shorter distance to travel and reaches the elver stage in one year. The two species breed in different parts of the Sargasso Sea and drift in somewhat different parts of the Gulf Stream. Should a larva of the European species be carried to North America, it would reach the coast long before it had developed into an elver and would die; if one of the American species were carried toward Europe, it would become an elver long before reaching the coast and also would die. Both species mature and live in fresh water for 10 to 15 years before migrating back to the ocean.

The area of the Sargasso Sea was probably the ancestral breeding ground for these two closely related species. The difference in the duration of their larval stages probably evolved as plate tectonics carried Europe and North America farther apart. North America remained closer to the ancestral breeding ground than Europe.

Mackerel and tuna are fishes of the open ocean. They are highly streamlined and are very fast and powerful swimmers (Fig. 36.23a). They seldom rest; indeed, they depend upon their forward motion to ventilate their gills (ram ventilation, p. 276). If they are caged so they cannot swim, the oxygen tension in their blood drops to low levels. They have a larger proportion of dark or red muscle than other fishes, as you may have noticed at a fish market. When temperature probes are inserted into this musculature in freshly caught specimens, it is found that much of this musculature is maintained at a temperature nearly 10° C above that of the water and body surface. Fish blood is cooled to water temperature as it flows through the gills, so how is a high temperature maintained in these muscles? Blood is distributed to the red muscles not by branches of the dorsal aorta, as in other fishes, but by cutaneous arteries traveling close to the skin (Fig. 36.23b). This blood, which at first is the same temperature as the ocean, enters

Figure 36.23

(a) The albacore tuna, *Thunnus alaluga*, and related species are fast-swimming species of the open oceans. (b) Trunk muscles are supplied by a cutaneous artery and drained by a cutaneous vein rather than being supplied and drained by the dorsal aorta and posterior cardinal veins, as is usually the case in fishes. (c) A cross section through a cutaneous artery and vein and adjacent muscles. Arteries and veins in the muscles travel side by side. The vascular arrangement maintains muscles at an elevated temperature because warm blood leaving the active muscles in the veins transfers heat back to the colder arterial blood entering the muscles.

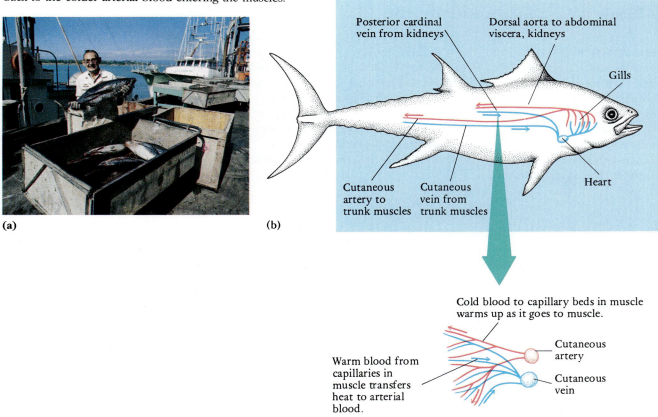

(a)　　　　　　　　　　　　　(b)

(c)

(a)

(b)

(c)

Figure 36.24

Some unusual adaptations of teleosts. (a) The cryptic coloration of the Caribbean striped anglerfish, *Antennarius striatus*, allows it to lie concealed and entice prey by dangling a flesh-like lure, which can be seen extending from its forehead. (b) Flounders are flattened fish that swim along the bottom on one side of the body. The eye that would have been on the bottom migrates during development to the upside. (c) The mudskipper, *Periophthalamus*, of East Africa and Southeastern Asia lives in shallow water from which it frequently emerges. Notice the strong, muscular pectoral fins.

the muscle through many small arteries that parallel, but flow in an opposite direction to, many small veins leaving the muscles to drain into a cutaneous vein (Fig. 36.23c). Heat generated by muscular activity is absorbed by the venous blood and transferred back to the cooler arterial blood. Because of this countercurrent heat exchange system, high temperatures can be maintained in the red muscles, increasing their power output.

Fishes that live on the bottom tend to reduce or lose their swim bladder, and they are frequently flattened. They can hug the bottom as they feed and hide. Teleosts have evolved a flattened shape in two different ways. Angler fish are flattened dorsoventrally, have a concealing color pattern, and lie on the bottom with their large mouths directed upward (Fig. 36.24a). In some species, the anterior dorsal fin spine bears a lure that overhangs the mouth. Some other species use a red, wormlike fold in the floor of the mouth as a lure. Flounders, sole, and halibut are flattened from side to side and lie on or swim slowly over the bottom on one side as they feed on invertebrates (Fig. 36.24b). The downward side, which may be either the left or the right side, according to the species, has lost its pigment, but the upward side has a concealing color pattern. The larvae are perfectly symmetrical, but during the course of embryonic development, much of the head is remodelled and the eye that would be on the downward side migrates to the upward side. Mouth position, however, does not change.

(a)

(b)

(c)

Figure 36.25

A group of sarcopterygians. (a) Ancestors of this coelacanth are thought to have given rise to the amphibians. The paired fins show the basic plan of a jointed series of bones that could evolve into the limbs of a terrestrial vertebrate. This coelacanth, *Latimeria chalumnae*, was photographed near the Comoro Islands, in deep waters off the coast of East Africa. (b) The Australian lungfish, *Neoceratodus fosteri*, retains lobate paired fins and scales. (c) Scales are lost and the paired fins reduced to tendrils in the South American lungfish, *Lepidosiren paradoxa*.

A few teleosts, such as the mud-skipper of Southwest Asia, even leave the water for brief periods and climb up on stones or tree roots with muscular pectoral fins as they chase a crustacean or escape a predator (Fig. 36.24c). By closing their operculum they can maintain a high humidity in the opercular chamber and obtain oxygen from the air with their gills.

Sarcopterygians

Terrestrial vertebrates did not evolve from teleosts but from sarcopterygians of ancient Devonian swamps. Primitive sarcopterygians had lungs, internal nostrils, strong lobate fins, and other structures of the type that could give rise to the features that characterize terrestrial vertebrates (Fig. 36.17b). Primitive species also had cosmoid scales, two dorsal fins, and a heterocercal tail. Contemporary species have evolved cycloid scales independently of actinopterygians and a symmetrical tail, though of a type that tapers to a point (**diphycercal**) (Fig. 36.25).

Sarcopterygian evolution diverged early into two lines: crossopterygians (order **Crossopterygii**), and the lungfishes (order **Dipnoi**). Crossopterygians have a well-ossified internal skeleton, a unique jointed braincase, and small conical teeth suited for grasping prey. Those belonging to the suborder **Rhipidistia** were the largest and most predatory freshwater fishes of the Paleozoic, and it is from this group that amphibians probably arose. Unfortunately none have survived. Members of the related suborder **Coelacanthiformes** specialized early for marine life, and one genus, *Latimeria*, survives in the ocean near the Comoro Islands between Africa and Madagascar (Fig. 36.25a).

Throughout their evolutionary history, lungfishes have had a weak skeleton with little ossification of the vertebral column, and they have shown tendencies toward reduction of the paired appendages. They developed specialized crushing tooth plates, which enabled them to feed effectively upon crustaceans and molluscs. Three genera of lungfishes survive in tropical rivers and lakes, one each in South America, Africa, and Australia (Fig. 36.25b and c). As we have seen (p. 828), the retention of lungs allows these fishes to live in unstable waters that are subject to seasonal drought and stagnation. Living lungfishes provide clues as to how vertebrates met the respiratory and circulatory problems associated with the transition from water to land, but they have too weak a skeleton and too specialized a dentition to have been the ancestors of terrestrial vertebrates.

Classification of Fishes

Class Agnatha (Gr. *a*, without + *gnathos*, jaw)   Jawless fishes.

> **Subclass Diplorhina† (Gr.** *diploos*, double + *rhis*, nose). Two orders of ostracoderms with paired nasal sacs. *Pterapsis.*

> **Subclass Monorhina** (Gr. *monos*, single + *rhis*, nose) Agnathans with a single median nasal sac.

>> Three orders of ostracoderms†. *Cephalaspis.*

>> **Order Petromyzontiformes** (Gr. *Petromyzon*, a genus of lamprey + L. *forma*, shape)   Lampreys (*Petromyzon*).

>> **Order Myxiniformes** (Gr. *Myxine*, a genus of hagfishes + L. *forma*, shape)   Hagfishes (*Myxine, Bdelostoma*).

Class Acanthodii† (Gr. *akanthodes*, thorny)   The "spiny sharks." (*Climatius*).

Class Placodermi† (Gr. *plax*, plate + *derma*, skin) *Dunkleosteus.*

Class Chondrichthyes (Gr. *chondros*, cartilage + *ichthys*, fish)   Cartilaginous fishes.

> **Subclass Elasmobranchii** (Gr. *elasmos*, a thin plate + *branchia*, gills)   Gill slits open independently on the body surface; upper jaw not fused to braincase; placoid scales present.

>> **Order Squalomorpha** (L. *Squalus*, the dogfish + *morphe*, shape)   Sharks with a well-developed spiracle; no anal fin. The dogfish *Squalus.*

>> **Order Galeomorpha** (L. *galea*, helmet + *morpha*, shape)   Sharks with a reduced or absent spiracle; anal fin present. Subtropical and tropical sharks. The great white shark (*Carcharodon*), the whale shark (*Rhiniodon*)

>> **Order Batoidea** (Gr. *batis*, a skate + *eidos*, form) Sawfishes, skates, and rays. Body dorsoventrally flattened, trunk and tail muscles reduced, pectoral fins enlarged. Sawfish (*Pristis*), skate (*Raja*), electric ray (*Torpedo*).

> **Subclass Holocephali** (Gr. *holos*, whole + *kephale*, head)   Gill slits covered by an operculum; upper jaw fused to cranium; scales absent. The ratfish, *Chimaera.*

Class Osteichthyes (Gr. *osteon*, bone + *ichthys*, fish) Bony fishes.

> **Subclass Actinopterygii** (Gr. *aktin*, ray + *pterygion*, fin)   Ray-finned fishes. Paired fins fan shaped, supported by radiating bony rays or spines.

>> **Infraclass Chondrostei** (Gr. *chondros*, cartilage + *osteon*, bone)   Thirteen extinct and two living orders of primitive ray-finned fishes. Scales usually ganoid; mouth opening large; tail usually heterocercal. The bichir (*Polypterus*), paddlefish (*Polydon*), sturgeon (*Acipenser*).

>> **Infraclass Neopterygii** (Gr. *neos*, new + *pterygion*, fin)   Advanced ray-finned fishes. Scales usually cycloid or ctenoid; mouth opening usually advances when mouth is opened; tail usually homocercal.

>>> "Holosteans" (Gr. *holos*, whole + *osteon*, bone) Five extinct and two living orders of primitive neopterygians with an abbreviated heterocercal tail. Gar (*Lepisosteus*), bowfins (*Amia*).

>>> **Division Teleostei** (Gr. *teleos*, end + *osteon*, bone)   More specialized neopterygians.

>>>> **Superorder Osteoglossomorpha** (Gr. *osteon*, bone + *glossa*, tongue + *morphe*, shape)   One order of freshwater fishes, most of which live in the southern hemisphere. Bite force between roof of mouth and bony tongue. The North American mooneye (*Hiodon*), the African elephant-snouted fish (*Mormyrus*).

>>>> **Superorder Elopomorpha** (Gr. *elopos*, a larval fish + *morphe*, shape)   Three orders of

fishes with a leptocephalus larval stage. Tarpons (*Megalops*), eels (*Anguilla*).

Superorder Clupeomorpha (L. *Clupea*, a herring-like fish + Gr. *morphe*, shape) One extinct and one living order of primitive teleosts with extensions of swim bladder reaching inner ear. Common herring (*Clupea*).

Superorder Protoacanthopterygii (Gr. *protos*, first + *akantha*, thorn + *pterygion*, fin) One order of primitive teleosts, some members of which are beginning to acquire certain of the advanced features of acanthopterygians. Pike (*Esox*), salmon (*Salmo*), trout (*Salmo*)

Superorder Ostariophysi (Gr. *ostarion*, small bone + *physa*, bladder) Four orders of primarily freshwater teleosts; primitive in most respects, but have Weberian ossicles. Carp (*Cyprinus*), suckers (*Catostomus*), catfish (*Ictalurus*), minnows (*Notropis*).

Superorder Stomatiiformes (Gr. *stoma*, mouth + L. *forma*, shape) One order of specialized teleosts; live at moderate depths, photophores present. Light fishes (*Gonostoma*).

Superorder Scopelomorpha (Gr. *Scopelus*, a genus + *morphe*, shape) Two orders of specialized, mostly moderate to deep-sea teleosts. The lantern fish (*Myctophus*).

Superorder Paracanthopterygii (Gr. *para*, beside + *akantha*, thorn + *pterygion*, fin) Six living orders of specialized teleost fishes with advanced characteristics; scales usually ctenoid; spines usually present in fins; pelvic fins located far forward. Pirate perch (*Aphredoderus*), clingfish (*Gobiesox*), anglers (*Ceratioidei*), cod (*Gadus*).

Superorder Acanthopterygii (Gr. *akantha*, thorn + *pterygion*, fin) Fifteen orders of advanced, spiny teleosts; scales ctenoid; head and operculum usually scaled; spines present in fins; trunk often short; often two dorsal fins; pelvic fin thoracic in position; swim bladder without a duct. Sea horses (*Hippocampus*), sculpins (*Cottus*), perch (*Perca*), bass (*Centrachidae*), halibut (*Hippoglossus*), and many others.

Subclass Sarcopterygii (Gr. *sarkodes*, fleshy + *pterygion*, fin) Lobe-finned fishes. Paired fins lobe shaped with a central axis of flesh and bone; primitive members had cosmoid scales.

Order Crossopterygii (Gr. *krossoi*, tassels + *pterygion*, fin) Crossopterygians. Primitive fleshy-finned fishes, conical teeth present; median eye usually present.

Suborder Rhipidistia† (Gr. *rhipis*, fan) Primitive freshwater crossopterygians; includes the ancestor of amphibians. *Osteolepis*.

Suborder Coelacanthiformes (Gr. *koilos*, hollow + *akantha*, thorn + L. *forma*, shape) More specialized marine crossopterygians; one living genus (*Latimeria*).

Order Dipnoi (Gr. *di*, two + *pnoia*, breath) Lungfishes. Specialized lobe-finned fishes; teeth forming crushing tooth plates; median eye usually absent; three living genera confined to Australia (*Neoceratodus*), Africa (*Protopterus*), and South America (*Lepidosiren*).

† = Extinct groups.

Summary

1. Fishes are well suited for an aquatic life: they have a streamlined shape, their skeleton is not as strong as that of terrestrial vertebrates, the segmented muscles provide the thrust for locomotion, and fins provide stability and maneuverability. A combination of forces developed by the tail, fins, and swim bladder enables fishes to maintain their position in the water column. The structure of their sense organs enables them to detect changes beneath the water. Fishes are for the most part ectothermic vertebrates. Their heart pumps only unoxygenated blood, and blood pressure is greatly reduced as blood first flows through the gills before reaching the body.

2. The earliest fishes, which go back to the Cambrian period, were heavily armored ostracoderms of the class Agnatha. Most lived in fresh water and fed on the bottom with jawless mouths. They lacked well-developed paired fins and were not very active fishes. The only living agnathous vertebrates—the lampreys and hagfishes—also lack jaws and paired appendages.

3. Jaws, which are first found in the extinct class Acanthodii, evolved from an enlarged visceral arch, the mandibular arch. Acanthodians had both pectoral and pelvic fins, which were supported by spines, and additional spines lay between them. Many placoderms, another extinct class of primitive jawed fishes, had cleaver-like jaws.

4. Cartilaginous fishes of the class Chondrichthyes are characterized by the absence of bone in the skeleton, the absence of lungs or a swim bladder, small placoid scales usually present, a heterocercal tail, a spiral valve in the intestine, and a pelvic clasper in the male. Fertilization is internal. They may be oviparous or brood their young internally, with varying dependence upon yolk or maternal nutrition. In sharks and rays (subclass Elasmobranchii), each gill pouch opens independently on the body surface. In ratfish (subclass Holocephali), an opercular fold covers the gill pouches. Most sharks are predaceous; skates and rays are flattened, bottom-dwelling species that feed on molluscs and crustaceans.

5. Most living fishes are bony fishes belonging to the class Osteichthyes. Bony scales usually are present. The deeper skeleton is partly or nearly completely ossified. Lungs or a swim bladder are usually present. The tail is usually homocercal. The spiral valve has been lost in most species, and pyloric ceca are present. The gills are covered by an operculum. Fertilization is normally external, and development is usually oviparous.

6. Most early bony fishes lived in fresh water subject to seasonal stagnation and drought. Lungs probably evolved as an accessory respiratory organ. Lungfishes and other species that have remained in fresh water have retained lungs. Others became marine, and the lungs were transformed into a hydrostatic swim bladder. Many of these fishes reentered fresh water and retained the swim bladder.

7. The class Osteichthyes is divided into two subclasses. The Actinopterygii (minnows, perch, and similar species) have fan-shaped paired fins supported primarily by soft rays. The Sarcopterygii (lungfish and crossopterygians) have lobe-shaped paired fins supported by a central axis of flesh and bone.

8. The subclass Actinopterygii can be divided into two infraclasses: the Chondrostei, which are represented by a few relict species (bichir, paddlefish, and sturgeon); and the Neopterygii. Neopterygians include a few primitive forms (gars and bowfin), but most belong to the teleost division, which includes most living species of fishes. During evolution from the more primitive chondrosteans to the teleosts, the lungs became a swim bladder, the tail shifted from heterocercal to homocercal, the scales changed from ganoid to cycloid or ctenoid, and the mouth became highly protrusible. Teleosts have undergone an extensive adaptive radiation.

9. The sarcopterygians are grouped into two orders. The Dipnoi (lungfishes) have poorly ossified skeletons and crushing tooth plates that enable them to feed on crustaceans and molluscs. Three genera survive, one each in tropical South America, Africa, and Australia. The crossopterygians have a stronger skeleton and many conical teeth. Most are extinct, but a marine coelacanth has survived. Terrestrial vertebrates evolved from early freshwater crossopterygians.

References and Selected Readings

Bemis, W.E., W.W. Burggren, and N.E. Kemp (ed.). *The Biology and Evolution of Lungfishes.* New York: Alan R. Liss, Inc., 1987. Papers presented at a symposium that bring together what is known about these fascinating animals.

Bond, C.E. *Biology of Fishes.* Philadelphia: Saunders College Publishing, 1979. A general text covering the relationships, structure, physiology, and behavior of fishes.

Carey, F.G. Fish with warm bodies. *Scientific American* 288 (Feb. 1973):36. An account of the countercurrent heat exchange system of tuna and mackerel.

Eastman, J.T., and A.L. DeVries. Antarctic fishes. *Scientific American* 255(Nov. 1986):106–114, 1986. Some antarctic fishes have evolved an "antifreeze" that enables them to live at temperatures at which others would perish.

Hall, M. The survivor. *Harvard Magazine* (Jan.-Feb. 1989):36–41. A brief history of the discovery of the surviving coelacanth, *Latimeria*, and the status of research on this "living fossil."

Hardisty, M.W., and I.C. Potter (eds.). *The Biology of Lampreys.* New York: Academic Press, 1971. All aspects of lamprey biology are thoroughly explored in chapters written by experts in the field.

Lauder, G.V., and K.F. Liem. Patterns of diversity and evolution of ray-finned fishes. *Bulletin of the Museum of Comparative Zoology* 150(1983):95–197. A cladistic analysis of actinopterygian evolution.

McCosker, J.E. Great white shark. *Science* 81(1981):42–52. Great white sharks are attracted to the electric fields generated by their prey.

Marshall, N.B. *Explorations in the Life of Fishes.* Cambridge, Mass.: Harvard University Press, 1971. Deep-sea fishes and convergent evolution among fishes are emphasized.

Norman, J.R. *A History of Fishes*. 3rd ed. Revised by P.H. Greenwood. London: Benn, 1975. A revision of a classic treatise on all aspects of fish biology.

Partridge, B.L. The structure and function of fish schools. *Scientific American* 246(June 1982):116. Both the lateral line system and eyes play important roles in schooling behavior, which reduces the risk of being eaten.

Policansky, D. The asymmetry of flounders. *Scientific American* 246(May 1982):116. A study of migration of the eye from one side of the head to the other.

Webb, P.W. Form and function in fish swimming. *Scientific American* 251(July 1984):72−82. Examines the relationship between patterns of swimming and fish structure.

37

Amphibians

Characteristics of Amphibians

■ Amphibians are ectothermic vertebrates

■ Contemporary species lack scales except for traces in caecilians. The skin contains only a small amount of keratin, mucous glands are abundant, and poison glands are sometimes present.

■ Vertebral centra are ossified in modern species, articular processes are present, and ribs are short and frequently fused to the vertebrae. The skull tends to be short, broad, and incompletely ossified.

■ Amphibians have movable eyelids and tear glands, and usually an ear sensitive to airborne or ground-borne vibrations.

■ Amphibians have a muscular tongue that is protrusible in many species. Their intestine is divided into small and large segments.

■ Larval gills are lost at metamorphosis, and gas exchange with the environment is through moist membranes in the lungs, skin, and buccopharyngeal cavity.

■ The heart atrium is divided into right and left chambers that receive primarily venous and arterial blood, respectively. These bloodstreams remain separated to a large extent in their passage through the single ventricle.

■ Nitrogen is eliminated primarily as urea. A urinary bladder is present.

■ Most amphibians return to the water to reproduce. Fertilization usually is external. The eggs develop into aquatic larvae that later metamorphose into terrestrial adults.

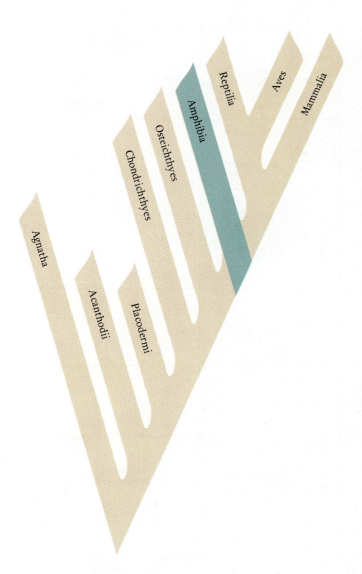

(Left) The American bullfrog, Rana catesbeiana. *(Above) A bullfrog can still watch for danger even when most of its body is submerged.*

Most of us have seen a toad hopping about, a frog jumping from a stream bank into the shelter of the water as we approach, and salamanders in a pond or under a log in the woods. These animals, together with the less familiar, worm-shaped caecilians of the tropics, are representatives of the three contemporary orders of the class **Amphibia** (Fig. 37.1). There are over 4000 living species of frogs, salamanders, and caecilians, nearly all of which must live in damp terrestrial habitats or in water. Most return to the water to reproduce, whence their class name (Gr. *amphi*, double + *bios*, life). Amphibians, descendants of crossopterygian fishes, were the first terrestrial vertebrates. In this chapter, we will consider the extent to which amphib-ians are adapted to terrestrial life and examine their evolution and diverse life styles.

THE TRANSITION FROM WATER TO LAND

The transition from fresh water to land was a momentous step in vertebrate evolution. It opened up vast new ecological opportunities for vertebrates to exploit, but successful adaptation to the terrestrial environment necessitated important structural and functional changes throughout the body. The land is a stressful and inhospitable environment for an aquatic animal; physical conditions on the land are very different from those in water.

Figure 37.1

A phylogenetic tree showing the relative abundance, geological distribution, and presumed interrelationships of amphibian groups.

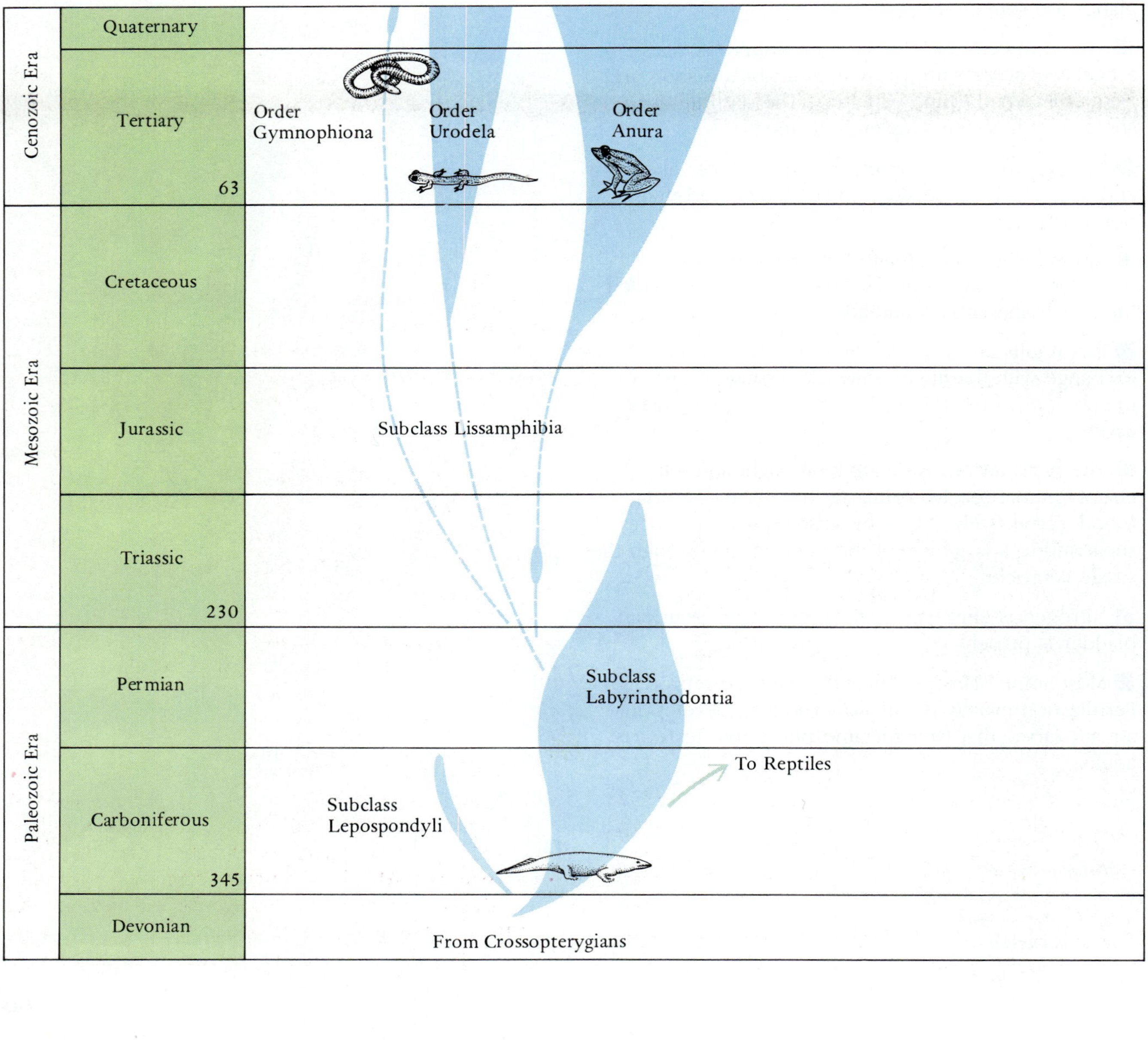

1. First, water absorbs many wavelengths of light that air does not, so more solar radiation impinges on terrestrial animals than on aquatic ones. Ultraviolet rays are harmful to living tissues, and they are not absorbed by air as they are by water. Terrestrial vertebrates need more surface protection than fishes.

2. Air is less dense and viscous than water and thus affords less support and offers less resistance to movement. Support on land requires a strong skeleton and strong muscles that brace the bones. Terrestrial locomotion requires the thrust of legs upon the ground rather than the thrust of a tail against the water. Ingesting food and manipulating it in the mouth is not as easy as in the aquatic environment where food is often sucked into the mouth and pharynx in a stream of water.

3. Receiving sensory cues in air is quite different from receiving them in water. Sense organs on or near the body surface must be protected and kept moist. Light travels farther in air than in water, but with a different refractive index. Sound waves in the thin air could not reach the inner ear if they were not amplified and transported through the denser tissues of the head. All sense organs have been modified during the transition of vertebrates to terrestrial environments. Changes in the nervous system were a natural corollary of the more complex movements, changes in the muscular system, and altered sense organs.

4. Air has a lower thermal stability than does water. Temperatures in terrestrial habitats, therefore, vary over a wider range and changes occur more rapidly. Methods of regulating body temperature in the face of changing ambient temperatures are needed for successful adaptation to the terrestrial environment.

5. Oxygen is more abundant in air than in water, but gases must be exchanged with the air without undue loss of body water. Changes in the site and methods of gas exchange accompany terrestrial life, with profound effects upon the circulatory system.

6. Water is a limited resource on land and needs to be conserved. Reducing water loss has required modifications in the skin, nitrogen metabolism, and the excretory system.

7. Delicate sperm and eggs cannot be released on land, and larval stages are very vulnerable. Truly terrestrial vertebrates have evolved methods for internal fertilization and for suppressing a free larval stage.

In view of the magnitude of these changes, it is not surprising that the transition from water to land took millions of years. Many groups of vertebrates have adapted in different degrees to life on land. The crossopterygian fishes unwittingly made the first steps in the transition. In adapting to their own aquatic environment, they evolved features that would coincidentally prove useful in a new and different environment: they became **preadapted** to certain terrestrial conditions. Most probably they lived, as

do modern lungfishes, in shallow bodies of water that were often choked with aquatic plants and subjected to seasonal stagnation and drought. The development of lungs in crossopterygians enabled them to breathe air when needed, and their sturdy paired fins would have helped in pushing them along the bottom through the dense vegetation. These features may also have allowed crossopterygians to make brief forays onto the land in pursuit of prey or to move from one drying swamp to an adjacent one with more water. Crossopterygians should not be seen as rudimentary amphibians. Their features made crossopterygians a successful and long lived group. The first amphibian fossils are found in strata that were formed nearly 50 million years later than those containing the first crossopterygians.

ADAPTATIONS OF AMPHIBIANS

Amphibians retained the lungs and sturdy paired appendages of their crossopterygian ancestors, and they evolved many additional features that adapted them to life on land. But since they also have retained many fishlike characteristics, most amphibians cannot withstand the full rigors of the terrestrial environment. They confine many of their activities to the water and to shady, damp habitats on land.

Integument

Traces of fishlike bony scales remain embedded in part of the skin of caecilians, but scales are completely lost in other modern groups. The epidermis remains thin, but a small amount of keratin, a water-insoluble protein, is produced and accumulates in the outer layers of the epidermis. This keratinized layer of cells affords some protection against water loss and abrasion. The outer layers of cells die and are shed periodically, often in large sheets. Pigment cells, or chromatophores, present in the outer part of the dermis, absorb light rays and protect against the more intense solar radiation (Fig. 37.2). Combinations of different types of chromatophores are responsible for the brilliant colors of some species. One type of chromatophore contains a pigment that reflects yellow light rays, another contains refractive granules that scatter light rays resulting in a bluish color. Acting together, the yellow and blue rays produce the green color of many frogs.

The dermis of amphibian skin contains many large mucous glands (Fig. 37.2) whose secretions help protect the body surface and keep it moist. The dermis is also very vascularized. The combination of a thin and moist epidermis and a vascular dermis allows for considerable gas exchange across the skin. This combination also allows the loss of considerable water through the skin, a factor that limits most amphibians to habitats with a high humidity. The dermis of many amphibians also contains poison

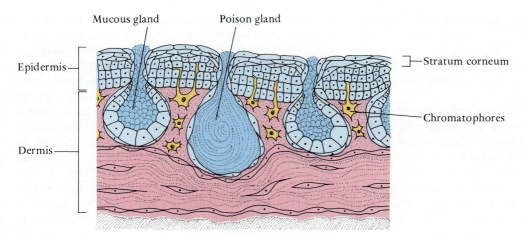

Figure 37.2
A vertical section through amphibian skin.

glands that secrete a noxious or poisonous material protecting them to some extent from predation.

Skeletomuscular System

Amphibians have evolved a strong skeleton and associated musculature needed to support the weight of the body on land. A strong skeletomuscular system is also needed to raise the head during feeding and to raise the body upon the limbs during locomotion. Vertebral centra are thoroughly ossified, and successive vertebrae interlock securely by means of overlapping articular processes on the vertebral arches (Fig. 11.16, p. 213). The lobe-shaped paired fins of crossopterygians are transformed into limbs supported by strong girdles (see Fig. 37.7a). The hip or pelvic girdle attaches to the vertebral column by a pair of sacral ribs, but the shoulder or pectoral girdle is suspended from the trunk by muscles. The body is not raised as high upon the limbs during locomotion as it is in mammals, for the proximal segments of the legs (humerus and femur) move back and forth primarily in the horizontal plane rather than vertically. The muscles of the trunk are powerful, and lateral undulations of the trunk and tail help to advance and retract the feet of tailed amphibians such as salamanders. Their pattern of foot placement on the ground follows naturally from the positions assumed by the paired fins of a fish as a result of the undulations of the trunk (Fig. 37.3). During most stages of a stride, just one foot is advanced, leaving three feet to support the body.

Sense Organs and Nervous System

Many changes occurred in the sensory apparatus during the evolution of the amphibians. Larval amphibians retain the primitive lateral line system, but this aquatic sensory system is of no use on land and it is lost in terrestrial adults.

Electroreceptors are present in some caecilians, but otherwise they too are lost.

The fishlike eye of aquatic larvae is transformed at metamorphosis. The lens flattens as the cornea assumes more importance in light refraction, and the eye is protected and cleansed by eyelids and tear glands.

The evolution of a tympanic membrane and a single auditory ossicle (the **stapes** or **columella**) that transmits

Figure 37.3
A comparison of locomotion in a sarcopterygian fish (a) and a salamander (b). Fishlike undulations of the trunk and tail in salamanders assist limb muscles in advancing and retracting the limbs.

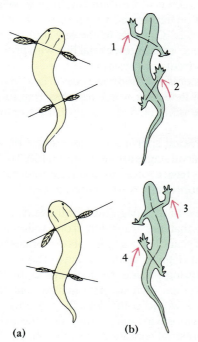

(a)   (b)

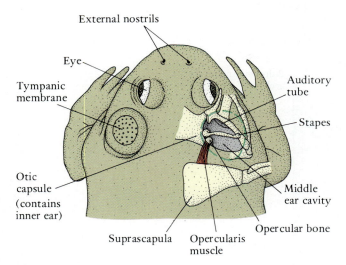

External nostrils

Eye

Tympanic membrane

Otic capsule (contains inner ear)

Suprascapula

Opercularis muscle

Opercular bone

Middle ear cavity

Stapes

Auditory tube

Figure 37.4
The frog ear detects and transmits high-frequency airborne vibrations by the tympanic membrane and stapes, and ground-borne and other low-frequency environmental vibrations through the opercular apparatus.

vibrations across the middle ear cavity to the oval window on the side of the inner ear enable frogs to detect airborne sound waves (Fig. 37.4). This type of tympanic ear is quite different from that of other terrestrial vertebrates for it is located rather high on the skull and not behind the jaw joint. The tympanic ears of the different groups of terrestrial vertebrates evolved independently. The frog tympanic membrane detects primarily the high-frequency vibrations of mating calls. Low-frequency environmental sounds are detected by an **opercular apparatus**. An **opercular bone** fits into the oval window beside the stapes, and connects by a slender **opercularis muscle** to the **suprascapula** of the pectoral girdle. Experiments have demonstrated that the opercular apparatus is necessary for the animal to detect the sound of rushing water, the rustle of leaves, or other low-frequency environmental vibrations, but just how it works is unclear. Salamanders, who have no mating calls, lack a tympanic membrane but have the opercular apparatus.

Lateral line centers are lost in the brain of adult amphibians along with the lateral line system, and there are other subtle changes related to changes in the sense organs and pattern of locomotion, but for the most part the brain of amphibians is very similar in size, morphology, and function to that of fishes.

Metabolism

Amphibians are ectotherms, and thus their body temperature fluctuates with the environmental temperature. They do, however, avoid high temperatures and dry conditions

by retreating to cool, moist habitats. Their metabolic rate is relatively low, so food and oxygen needs are not great. Movements of the head, jaws, hyoid apparatus, and a newly evolved muscular tongue are used to seize small invertebrates and transport them through the mouth and pharynx. Frequently the tongue is protruded rapidly, catching an insect on its sticky surface, and then pulled back into the mouth (Fig. 12.11). Amphibians also practice **inertial feeding**. After prey is seized and held by teeth located on the jaw margins and palate, the mouth is opened slightly and the head advanced rapidly relative to the prey, whose inertia keeps it stationary. A succession of such movements carries the food back into the mouth cavity. Mucus secreted by buccal and pharyngeal glands lubricates the food. The gastric glands and pancreas of amphibians produce **chitinase**, an enzyme that hydrolyzes the chitinous cuticle of insects and other small arthropods that compose much of the diet. The intestine has lost the spiral valve present in primitive fishes and is longer than in most fishes. Usually it is divided into small and large intestines. A cloaca is still present.

Larval amphibians have pharyngeal gill slits and gills, but these structures are lost in adults. The lungs, which are the primary site of gas exchange in the adults of most species, are not suction lungs but force pump lungs similar to those of lungfishes (p. 279). Air is pumped into them by movements of the floor of the mouth and pharynx (Fig. 13.15). The quantity of air pumped into the lungs correlates with the size of the mouth cavity and pharynx. These cavities are large in most contemporary species because the head is wide relative to the size of the body. Contemporary amphibians, most of which are rather small animals with a large surface relative to their volume, supplement pulmonary respiration by **cutaneous respiration** in which gases diffuse across their thin, moist, and vascular skin. Some species also practice **buccopharyngeal respiration**, in which gases also diffuse across the vascular lining of the mouth cavity and pharynx. Cutaneous respiration is particularly important for eliminating carbon dioxide because the lungs are not ventilated frequently enough to flush this gas out as fast as it accumulates.

The substitution of lungs and skin for gills as the major sites of gas exchange had important consequences in the circulatory system (p. 305). When the lungs are in use and pulmonary resistance is low, the division of the heart atrium into right and left sides keeps oxygen-depleted blood from the body and oxygen-rich blood from the lungs separate (Fig. 37.5). Although the ventricle is not divided, its spongy wall and differences in the timing of blood entering from the two atria keeps the two blood streams separate to a large extent. When the lungs are not in use and therefore pulmonary resistance to blood flow is high, the absence of a division of the ventricle probably allows the lungs to be bypassed. Since blood leaving the heart on its way to the body does not first go through gills,

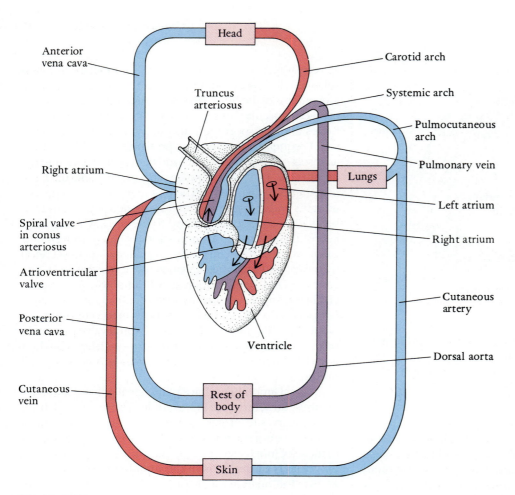

Figure 37.5

A diagram in ventral view of blood circulation in the frog when both lungs and skin are used in gas exchange. Oxygen-depleted blood from the head and body and oxygen-rich blood from the skin enter the right atrium via a sinus venosus on the dorsal surface of the heart. Oxygen-rich blood from the lungs enters the left atrium. Various mechanisms keep the two blood streams separate to a large extent in the single ventricle. The blood streams are sorted out by the spiral valve in the conus arteriosus and the division of the truncus arteriosus. Blood richest in oxygen is sent to the head, oxygen-depleted blood is sent to the lungs and skin, and a somewhat mixed blood goes to the rest of the body.

blood pressure in the dorsal aorta of a frog is nearly double that in the dorsal aorta of a dogfish, which receives its blood from the gills, where resistance has lowered blood pressure. In this respect, amphibians have a more efficient circulatory system than that of fishes. The presence of a lymphatic system returning some fluid from the tissues to the blood is a correlate of increased blood pressure (p. 319).

Since they are ectotherms, amphibians living in temperate regions of the world cannot be active on land at the low temperatures of winter, for their metabolic rate would be too low to sustain activity. In the fall of the year, they burrow into the mud at the bottom of ponds or into soft ground below the frost line and enter a dormant state known as **hibernation**. Metabolic activity is at a minimum during hibernation. Only food stored in the body is utilized, and rates of respiration and circulation are very low. Most amphibians cannot survive freezing, but one investigator has found that several North American frog species, whose ranges extend far north, accumulate glycerol in their body fluids and can survive temperatures as low as −6° C for 5 days. Some tropical amphibians go into a comparable dormant state, known as **aestivation**, during the hottest and driest parts of the year.

Excretion and Water Balance

The features we have been considering adapt amphibians reasonably well to terrestrial life, but amphibians also retain many features characteristic of freshwater fishes. These features restrict amphibians to moist habitats and prevent them from fully exploiting the terrestrial environment. Their ability to conserve body water is rudimentary. Considerable water is lost by evaporation through their skin. Most species excrete a copious and dilute urine because a large volume of water is needed to flush out toxic excretory products through their kidney tubules. Species that live in water as adults and the larvae of most species resemble freshwater fishes in that they excrete a large part of their nitrogen as ammonia. Terrestrial adults eliminate most nitrogen as urea, which, because it is less toxic than ammonia, does not need to be flushed out of the body by water as rapidly (Table 37.1). Several tree frogs conserve water by eliminating up to 80% of their nitrogen as uric acid, the least toxic of nitrogenous excretions.

Reproduction and Development

A few amphibians can reproduce in damp terrestrial habitats, but most species must return to the water. Frog reproduction follows this pattern (Fig. 37.6). Large numbers congregate in favorable ponds during the spring. Males arrive first, and their distinctive mating calls attract females of the same species. Males grasp the females in an embrace termed **amplexus** and discharge their sperm as the eggs are laid. The eggs develop into aquatic larvae known as **tadpoles**. Tadpoles have fat, roundish bodies and powerful muscular tails, and lead an active and independent life. External gills are present at first, but they are soon covered by a fold of skin. Internal gills develop later. Most tadpoles are herbivorous, scraping up small algae from the substratum with specialized mouthparts. Hind legs develop early; front ones develop somewhat later beneath the skin covering the gills. Larval features are lost rapidly during metamorphosis as terrestrial features are acquired. Among the many changes, the mouth is reorganized and the intestine shortens as the animal shifts from a herbivo-rous to a carnivorous mode of life. Frogs do not feed during this period; they are nourished by material being reabsorbed from the tail as it regresses.

EVOLUTION OF AMPHIBIANS

The earliest amphibians, belonging to the subclass **Labyrinthodontia**, can be traced back to the late Devonian period, about 350 million years ago (Fig. 37.1). Labyrinthodonts were large, ponderous creatures; some attained lengths of over 1 meter. They were abundant throughout the late Paleozoic and early Mesozoic eras, but none have survived. The earliest labyrinthodont, *Ichthyostega* (Fig. 37.7), had well-formed limbs and girdles that supported and allowed it to move about on land. However, the retention of a fishlike tail, together with lateral line canals in the skull bones, suggest that this animal spent a great deal of time in the water. *Ichthyostega* had long, strong ribs, unlike modern amphibians, in which the ribs are very short and often fused to the vertebrae. Ribs of this type may have been necessary to prevent the weight of the body from collapsing the lungs when the animal was resting on land. The vertebral column, although having articular processes, was not as strong as in later species of labyrinthodonts. Vertebral centra consisted of small blocks of bone flanking a persistent notochord. Many early labyrinthodonts also retained small, fishlike bony scales. A scaly skin and large size suggest that cutaneous respiration, if used at all, was less efficient than in contemporary species.

Most investigators believe that labyrinthodonts evolved from the freshwater rhipidistian group of crossopterygian fishes with whom they shared many skeletal details. Among other features, both share a peculiar infolding of the enamel of the teeth into the dentine in a labyrinthine pattern. The name labyrinthodont derives from this character. Labyrinthodonts probably gave rise to another extinct subclass, the **Lepospondyli** (Fig. 37.1). Lepospondyls differed from labyrinthodonts in having spool-shaped vertebral centra surrounding the notochord. They became extinct at the end of the Carboniferous period without leaving descendants.

Contemporary amphibians include salamanders (order **Urodela**), frogs and toads (order **Anura**), and caecilians (order **Gymnophiona**). Herpetologists—zoologists who specialize in the study of amphibians and reptiles—group them into a common subclass, the **Lissamphibia** (Fig. 37.1), because they share several unique, derived characters, including teeth in which the crown and base are separated by fibrous tissue, an opercular bone in the middle ear, and short ribs that are sometimes fused onto the side of the vertebrae. It is probable that lissamphibians evolved from labyrinthodonts, but the fossil record of modern groups is very sparse.

Table 37.1
Types of Nitrogen Excretion in Representative Vertebrates*

Animal	Ammonia	Urea	Uric Acid	Other
Freshwater minnow, *Cyprinus*	60.0	6.2	0.2	33.6
Bullfrog, *Rana*	3.2	84.0	0.4	12.4
Europian tortoise, *Testudo*	4.1	22.0	51.9	22.0
Hen	3.0	10.0	87.0	0.0
Adult human	3.5	85.0	2.2	9.3

*Data from Prosser and Brown. Figures in percent of total nitrogen excretion.

Figure 37.6
The life cycle of *Rana temporaria*, the common European frog.

Embryos

Hatching from jelly

Egg laying

Early tadpole

Metamorphosis

Late tadpole

SALAMANDERS

In many ways salamanders (order Urodela) are closer in their structure to ancestral amphibians than the more specialized frogs and caecilians. Although much smaller than labyrinthodonts, they retain the primitive body form with a long trunk and tail and usually well-developed limbs (Fig. 37.8). In other major anatomical and physiological respects, salamanders resemble other amphibians. They are distributed primarily in the temperate regions of the world, and most species occur in North America. They are secretive creatures living beneath stones and logs in damp woods; some species are entirely aquatic. Most are small animals ranging in length from 5 to 15 cm. Nearly 370 species have been described.

Adaptations of Salamanders

The most abundant of the North American salamanders are woodland species belonging to the family Plethodontidae.

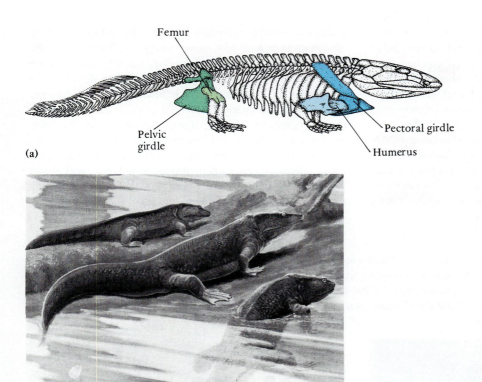

Femur

Pelvic
girdle

Pectoral girdle

Humerus

(a)

(b)

Figure 37.7

The transition from water to land. (a) A reconstruction of the skeleton of
Ichthyostega, the earliest known amphibian. (b) An artist's conception of
labyrinthodont life in a Carboniferous swamp about 345 million years ago.

Familiar examples are the red-backed salamander, *Pleth-
odon cinereus*, that lives under rocks and logs, and the
stream-dwelling two-lined salamander, *Eurycea bislineata*
(Fig. 37.8). A particularly interesting feature of

Figure 37.8

The two-lined salamander, *Eurycea bislineata*, a lungless
plethodontid.

plethodontids is their complete loss of lungs; gas exchange
occurs entirely across the moist membranes lining the
mouth, pharynx, and skin. The skin is a more effective
respiratory organ than in other salamanders because the
epidermis is exceptionally thin and capillaries are close to
the body surface. Loss of lungs may seem to be a curious
adaptation for a terrestrial vertebrate. A widely held view is
that early in their evolution plethodontids became adapted
for life in rapidly flowing mountain streams in the Appala-
chian mountains. Since the oxygen content of cold water is
rather high, and low temperature reduces the rate of
metabolism, sufficient oxygen could be obtained through
the skin. Air in the lungs would be disadvantageous under
these conditions, for the animals would float and be
washed away. Many plethodontids still live in streams;
others have moved onto the land but never regained the
lost lungs.

This scenario has been challenged recently by J.A.
Ruben and A.J. Boucot on the basis of geological data
indicating that Appalachia was a relatively flat area with a
subtropical to tropical climate in the late Mesozoic, the
probable period of plethodontid evolution. They propose
instead that changes in feeding mechanisms led to a
narrowing of the head relative to body size. Since a wide

head is necessary in amphibians for an effective force pump mechanism of lung ventilation, a narrowing head must have been accompanied by an increasing reliance on cutaneous respiration. A continuation of this trend could lead to the loss of lungs in small terrestrial salamanders with low levels of metabolism. Their invasion of mountain streams occurred later.

Several families of salamanders have become paedomorphic (p. 810), that is, they retain many larval features as adults and do not complete their metamorphosis. Paedomorphosis among salamanders has evolved by **neoteny**, a process in which somatic development slows down relative to sexual maturation so the animals become sexually mature before they have completed their metamorphosis. Neoteny is an evolutionary strategy that enables these animals to remain aquatic and escape the rigors of terrestrial life. As one example, most mole salamanders (family Ambystomatidae) have a normal amphibian life cycle. Populations of the tiger salamander (*Ambystoma tigrinum*) that live in cool mountain ponds of the western United States, however, retain external gills and many other larval features, reproduce as a larva, and never metamorphose completely (Fig. 37.9a). This reproducing larva is called an **axolotl**. Populations of the tiger salaman-

102

Figure 37.9

A group of neotenic salamanders. (a) An axolotl, or sexually mature larva of a salamander, such as the tiger salamander, *Ambystoma tigrinum*, that in some localities does not metamorphose. (b) The terrestrial adult of the tiger salamander; (c) the mud puppy, *Necturus maculosus*; (d) the dwarf siren, *Pseudobranchus striatus*.

(a)

(b)

(c)

(d)

der living at lower elevations metamorphose normally (Fig.37.9b). The hormone thyroxine, secreted by the thyroid gland, is necessary for amphibian metamorphosis (p. 447). Its secretion appears to be inhibited at the low temperatures in the mountains, but not at the higher temperatures found at lower elevations. The tiger salamander is an example of a **facultative neotenic species**, one in which metamorphosis can take place if conditions change.

The mud puppy, *Necturus* (family Proteidae), is an **obligatory neotenic species**, one in which metamorphosis does not occur under any circumstance (Fig. 37.9c). Thyroxine is produced in *Necturus*, as evidenced by the fact that when its thyroid gland is transplanted to a frog tadpole it hastens the tadpole's metamorphosis. The failure of *Necturus* to metamorphose appears to result not from the absence of thyroxine, but from the inability of its tissue to respond to thyroxine, because the species possibly lacks the receptors for this hormone.

The sirens (family Sirenidae) are also obligatory neotenic species found only in ponds and rivers of the southern United States (Fig. 37.9d). They have an eel-like body shape, external gills, no hind legs, and only vestiges of front legs. They have many unusual features, including a heart with a nearly completely divided ventricle that is more like a reptile's heart than an amphibian's. If the pond in which they live dries up, they can secrete a slimy mucous cocoon around themselves and aestivate in the manner of an African lungfish.

Salamanders are preyed upon by many animals including snakes, birds, and shrews. Many species have evolved adaptations that discourage predation. Skin glands often produce noxious secretions, and this ability is sometimes coupled with warning or aposematic coloration (p. 470). The terrestrial stage (eft) of the red-spotted newt, *Notophthalmus viridescens*, is a brilliant red and has a particularly toxic skin secretion (Fig. 37.10a). Birds soon learn to recognize this color and avoid eating the eft. Some salamanders can discard their tails in a process called **autotomy**. The tail thrashes about, attracting the attention of the predator, while the rest of the animal crawls off, eventually growing a new tail.

Reproduction and Life Cycles

Neotenic aquatic species, of course, spend their entire life cycle in the water. Most terrestrial species return to the water to reproduce. Some travel considerable distances to reach their breeding sites. The red-bellied salamander, *Taricha rivularis*, has been experimentally displaced 15 km from its breeding site and successfully returns to it even if it must cross small mountains and bypass other favorable breeding sites. Its method of navigation is not clear, but may involve magnetic navigation and the use of olfactory clues.

Males of most salamanders package their sperm in small capsules called **spermatophores** and often engage in an elaborate courtship ritual as they deposit them. The females take the spermatophores into their cloacas. Fertilization is internal. Eggs hatch into aquatic larvae that resemble the adults more than tadpoles do frogs. The larvae are carnivorous and retain external gills until metamorphosis.

The red-spotted newt has a similar cycle, but only the postmetamorphic juvenile efts are terrestrial. The adults return to the water when sexually mature and remain there (Fig. 37.10b). The European fire salamander, so called because it was once believed to be able to walk through fire, retains fertilized eggs in the oviduct until they reach the larval stage, at which time the female enters the water to give birth to the larvae.

Many of the woodland salamanders reproduce on land, laying small clutches of eggs in damp sites under

Figure 37.10

(a) The juvenile red-spotted newt, *Notophthalmus viridescens*, is terrestrial and has a conspicuous aposematic coloration. (b) The adult has a cryptic coloration and lives permanently in the water, where it reproduces.

(a)   (b)

Figure 37.11
The Black Mountain dusky salamander, *Desmognathus walteri*, brooding her eggs in a damp spot under a log.

stones or logs. A parent often broods the eggs, protecting them and keeping them moist (Fig. 37.11). The eggs are provided with a large store of yolk and hatch as miniature adults.

FROGS AND TOADS

Frogs and toads (order Anura) occur on all continents except Antarctica, but they are absent from many oceanic islands. Most species cannot tolerate salt water, so they could not survive a long passage to oceanic islands on natural rafts of drifting vegetation. They are particularly diverse in the subtropics and tropics. About 3500 species are known. Most range in size from 2 to 12 cm, but one Cuban species, *Phyllobates libatus*, is only 1 cm long and the West African *Gigantorana goliath* attains a length of nearly 1 meter.

Adaptations of Frogs and Toads

Anurans are amphibians specialized for terrestrial jumping. They have lost their tail, their trunk is short and compact, embryonic ribs fuse to the vertebrae, and their hind legs are long and strong. Adaptations for jumping are very evident in the skeleton (Fig. 37.12). The vertebral column is stiff and never contains more than nine vertebrae. The leg bones—femur and tibio-fibula—are quite long. Elongation of two tarsal bones increases foot length and makes, in effect, another limb segment for jumping. Fusion of the tibia and fibula strengthens the leg. Elongation of the ilium provides an additional lever that can be

Figure 37.12
Many adaptations for jumping can be seen in the frog's skeleton.

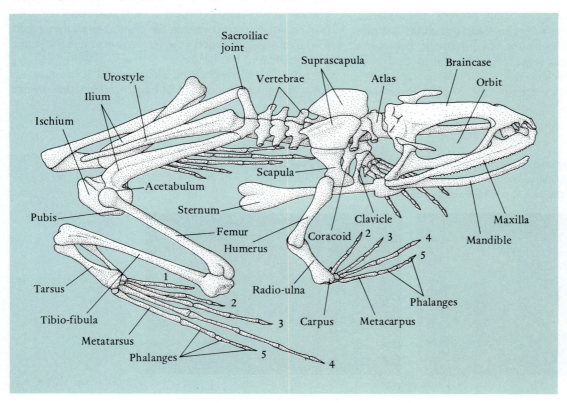

extended because the pelvic girdle can swivel at the **sacroiliac joint**. When the frog is at rest, its hind legs are flexed, and the ilium and the **urostyle**, which represents several fused caudal vertebrae, are also flexed, thus giving the frog a "humped-back" appearance (Fig. 37.13a). The absence of a tail allows the hind legs and feet to be synchronously extended during a leap, and additional thrust is given by the extension of the urostyle and ilium (Fig. 37.13b). The front legs act as shock absorbers when the animal lands (Fig. 37.13c), and their strength has been increased by the fusion of the radius and ulna (Fig. 37.12). Muscles are also modified, and those used in the extension of the hind legs, feet, and pelvis are particularly large.

Anurans can see and breathe even when the rest of the body is submerged because their eyes and nostrils are elevated on top of the head. As we have seen (p. 427), interconnections of cells within the vertebrate retina are very complex and allow for considerable integration of images. The visual integration that occurs in the frog's retina is particularly suited for its mode of life. For example, some ganglion cells in the frog's retina are activated by sharp edges that outline major objects in the visual field (**edge detectors**); others are activated by a sudden decrease in light intensity, such as may be caused by the shadow of a predator (**dimming detectors**); and yet others are stimulated by images of small moving insects and other food-sized objects (**bug detectors**).

Frogs and toads are not as secretive as salamanders, and their greater exposure causes a high evaporative loss of body water. They make up for this when in the water by absorbing water and salts through the skin. Some can even absorb water through their belly skin when sitting on damp ground. Water can be stored in the urinary bladder.

Frogs and toads are preyed upon by snakes, raccoons, and many other animals. Many have cutaneous poison glands, which sometimes are aggregated to form conspicuous warts. A predator that has tried eating a toad will not soon repeat this mistake. As with salamanders, some frog species combine poison glands with a conspicuous aposematic coloration. South American frogs of the genus *Dendrobates* are brilliantly colored (Fig. 37.15b) and produce a highly toxic skin secretion traditionally used by Indian tribes to poison their arrow tips.

The terms frog, toad, and tree frog have no taxonomic meaning but refer loosely to adaptations for aquatic, terrestrial, and arboreal habitats. Several different families of anurans have adapted independently to these modes of life. The most common frogs in North America are semiaquatic species belonging to the family Ranidae. Examples are the leopard frog (*Rana pipiens*) often used in biology classes, the green frog (*Rana clamitans*), and the bullfrog (*Rana catesbeiana*) (Fig. 37.14a). Ranids live in water, on the shores of ponds and streams, and in damp meadows and woods. Their skin is thin and moist, and

Figure 37.13
Three stages in the leap of a bullfrog. Notice that the pelvic girdle is flexed when the frog is resting, and this gives it the characteristic humped back (a). The hind legs, feet, and pelvis extend at take off (b), and the frog lands on its front legs (c).

(a)   (b)

(c)   (d)

Figure 37.14

Representative frogs and toads. (a) The bullfrog, *Rana catesbeiana*, sitting in a pond. (b) The African clawed frog, *Xenopus laevis*, is entirely aquatic. (c) The American toad, *Bufo americanus*, is primarily terrestrial but returns to the water to sing and reproduce. (d) The grey tree frog, *Hyla versicolor*, lives in the trees.

their long hind legs, with webbed feet, are very effective for swimming and jumping.

Clawed frogs and Surinam toads of the family Pipidae are the most aquatic of all frogs and seldom leave the water. The African clawed frog, *Xenopus laevis*, has adapted to an aquatic life partly by becoming neotenic and retaining many larval characteristics (Fig. 37.14b). Among them is the presence of a lateral line system and the absence of many adult features, including a muscular tongue, eyelids, and a tympanic membrane. Its webbed feet are exceptionally large. *Xenopus* is widely used in experimental studies in this country because it is so easy to breed and raise in the laboratory.

Familiar North American toads belong to the family Bufonidae (Fig. 37.14c). They are typically found in woods,

fields, and gardens. Compared to frogs, their hind legs are relatively shorter and so they cannot leap as far. A web may be present on their hind feet, but it frequently is reduced in size. Their skin is drier and hornier than in frogs, but toads also lose a lot of water through it. There is less cutaneous gas exchange in toads than in frogs, and the internal surface area of their lungs is increased slightly. Toads are most active early in the morning and in the evening when the humidity is high. They take shelter during the day by using horny spurs on their hind feet to burrow backwards into the soil.

The Australian water-holding toad, *Cyclorana platycephalus* of the family Myobatrachidae, lives in very arid deserts. It soaks up water during the brief rainy spells and becomes bloated with water stored in its urinary bladder and large, subcutaneous lymph sacs. It then burrows deeply into the ground, where it can survive years of drought if need be.

Most North American tree frogs belong to the family Hylidae. They are small anurans with a low body weight. These features make it easier for them to cling to small branches and leaves with the **digital pads** on the tips of their toes (Fig. 37.14d). The adhesive properties of the pads come not from the secretion of sticky substances but from the friction of numerous minute folds and ridges that catch in irregularities in the substratum. Frequently the skin of tree frogs is dry and warty, as in toads. Some tropical tree frogs retreat to small holes in the trees during dry periods. They retain moisture in their holes by plugging the holes with the horny tops of their heads.

Reproduction and Development

Most anurans return to the water to reproduce, but developmental patterns have changed in many tropical frogs, probably as a protection against seasonal drought and numerous aquatic enemies, such as predaceous insect larvae. South African grey tree frogs (*Chiromantis*) deposit their eggs shortly before the rains in foam nests attached to tree branches overhanging pools and ponds (Fig. 37.15a and b). Females secrete a fluid that they then whip into a

(a)

(b)

(c)

(d)

(e)

Figure 37.15
Some modifications of amphibian reproduction that protect the larvae from drought or predators. (a) South African tree frogs, *Chiromantis xerampelina*, cooperate in beating a secretion into an arboreal foam nest in which the eggs are deposited and develop into small tadpoles. (b) When the seasonal rains come, the tadpoles drip from the nest into temporary pools on the ground where development is completed. (c) A male *Dendrobates* of the Neotropics carrying tadpoles on his back. (d) The eggs and larvae of the aquatic Surinam toad, *Pipa pipa*, develop in pits on the female's back. (e) The eggs of the South American marsupial tree frog, *Gastrotheca*, develop within a brood chamber on the female's back.

froth by vigorously kicking with their hind legs. Sometimes one or more males assist in this process. The surface of the froth hardens, much like a meringue, and protects the developing eggs. When the rains come, the nest dissolves and the tadpoles drop into pools below. A more striking means of protecting the young is seen in the South American poison arrow frog, *Dendrobates*, in which the male carries the eggs and tadpoles upon his back until they are young froglets able to fend for themselves (Fig. 37.15c). In the aquatic Surinam toad, *Pipa pipa*, the male pushes the fertilized eggs into the puffy skin on the back of the female (Fig. 37.15d). The eggs gradually sink deeper into the skin, and the outer membrane of each egg forms a protective cap over its temporary skin pocket. The young emerge as little toadlets. The eggs and larvae of all these frogs are equipped with a large supply of yolk, but in other respects the larvae resemble typical frog tadpoles and simply develop in a sheltered environment.

In certain species, the vulnerable larval stage is omitted, and the embryo develops directly into a miniature adult. Among these are the marsupial frog, *Gastrotheca*, and *Eleutherodactylus*, both of the New World tropics. The former carries her eggs in a dorsal brood pouch (Fig. 37.15e); the latter lays eggs in protected damp places, such as beneath stones or on leaf stalks in the area where the leaf meets the stem. The jelly layers about the egg of *Eleutherodactylus* help to prevent desiccation; sufficient yolk is stored within the egg for the nutritive requirements of the embryo; such larval features as horny teeth, gills, and opercular folds are vestigial or absent in the larvae; the fins of the larval tail are expanded, become highly vascular, and form an organ for gas exchange; and the period of development is shortened.

CAECILIANS

Caecilians (order Gymnophiona) are the least familiar group of living amphibians. They occur only in the tropics and lead a subterranean life, pushing through damp soil with their trowel-shaped heads as they search for the small invertebrates on which they feed. About 160 species are known. All have an elongated body form without legs or girdles (Fig. 37.16a). On a quick glance a caecilian could be confused with a large earthworm. Many species are about the same size, and annular folds in their skin give the

(a)

(b)

Figure 37.16
Tropical caecilians. (a) *Gymnophis m. mexicana*, removed from its burrow. (b) The reduced eye, sensory pit, and a nostril are evident on the head *Dermophis mexicanus*.

appearance of segments. However, caecilians have a head with small jaws, nostrils, and a pair of vestigial eyes partly hidden beneath the skin (Fig. 37.16b). A pair of protrusible tentacles, lying in a pit between each nostril and eye, are important sense organs. Their cloacal aperture is not quite at the end of the body, so there is a short tail. Small bony scales are embedded in the skin. Fertilization is internal; the male's cloaca is everted to form a penis-like organ. Most species are oviparous and lay their eggs in damp soil. The eggs hatch and the larvae complete their development in nearby bodies of water. Other species brood their eggs externally and the embryos develop directly into miniature adults. A few species are ovoviviparous, nourishing the embryos with uterine secretions; there is no placenta.

Classification of Amphibians

Class Amphibia (Gr. *amphi*, double + *bios*, life)
  Subclass Labyrinthodontia† (Gr. *labyrinthos*, labyrinth + *odontos*, tooth) Several orders of extinct and primitive amphibians collectively called labyrinthodonts; vertebral centra consisting of two or three arches of bone encasing the notochord. *Ichthyostega*.
  Subclass Lepospondyli† (Gr. *lepos*, husk + *spondylos*, vertebra) Several orders of extinct amphibia with spool-shaped vertebral centra

pierced by a longitudinal canal housing a persistent notochord.

Subclass Lissamphibia (Gr. *lisso*, smooth + amphibia) Modern amphibians, probably evolved from labyrinthodonts, having teeth with a fibrous segment between crown and root, an opercular bone, short ribs, never more than four toes on front foot.

>**Order Urodela** (Gr. *oura*, tail + *delos*, visible) Salamanders. Tail well developed; legs usually present. The tiger salamander (*Ambystoma*); red-backed salamander (*Plethodon*), mud puppy (*Necturus*).

Order Anura (Gr. *a*, without + *oura*, tail) Modern frogs and toads. Short trunk; tail absent; caudal vertebrae form a urostyle; legs specialized for jumping. The leopard frog (*Rana*), American toad (*Bufo*), tree frog (*Hyla*).

Order Gymnophiona (Gr. *gymnos*, naked + *ophioneos*, snakelike) Caecilians. Wormlike trunk; limbs absent; tail very short; vestiges of dermal scales in the skin. Confined to the tropics.

† Extinct.

Summary

1. In adapting to life on land, vertebrates evolved a more protective integument, strong body support, different methods of locomotion, ways of receiving sensory cues from the air, and methods of obtaining oxygen without an undue loss of body water. They evolved features enabling them to regulate their body temperature in the face of wide fluctuations of ambient temperature and to reproduce upon land.

2. Amphibians are well adapted to terrestrial life with respect to support, locomotion, and their sensory-nervous systems. They ventilate their lungs with a buccopharyngeal pump. Cutaneous gas exchange supplements pulmonary exchange. The heart atrium is divided into left and right chambers, and there is little mixing of the blood streams from the body and lungs in the single ventricle. Amphibians are limited to moist habitats by their thin, moist, and scaleless skin; by their production of a copious and dilute urine; by their inability to regulate their body temperature; and by the necessity of laying their eggs in water or in very damp terrestrial locations.

3. Salamanders (order Urodela) retain a long tail, and most species have short legs. Some terrestrial salamanders are lungless and depend upon cutaneous gas exchange. Many species are neotenic.

4. Frogs and toads (order Anura) are highly specialized for jumping. Although most species are aquatic, toads are quite terrestrial, and tree frogs are arboreal. Some tropical frogs have evolved interesting reproductive modifications that protect or bypass the delicate aquatic larval stage.

5. Caecilians (order Gymnophiona) are tropical wormlike amphibians specialized for a burrowing mode of life.

References and Selected Readings

Many of the general references on vertebrates cited at the end of Chapter 35 also contain considerable information on the biology of amphibians.

Conant, R. *A Field Guide to Reptiles and Amphibians of Eastern and Central North America,* 2nd ed. Boston: Houghton Mifflin Co., 1975. A very useful guide for the field identification of amphibians and reptiles; similar to the Peterson bird guides

del Pino, E.M. Marsupial frogs. *Scientific American* 260 (May 1989):110–118. A fascinating account of the adaptations of different kinds of marsupial frogs.

Duellman, W.E., and L. Trueb. *Biology of Amphibians.* New York: McGraw-Hill Book Co., 1986. An up-to-date source book on the anatomy, physiology, ecology, and other aspects of anuran biology.

Halliday, T., and K. Adler (eds). *The Encyclopedia of Reptiles and Amphibians.* New York: Facts on File Inc., 1986. A collection of authoritative and beautifully illustrated essays on the biology of amphibians and reptiles. Summaries are included of all of the living families.

Panchen, A.L., and T.R. Smithson. Character diagnosis, fossils, and the origin of tetrapods. *Biological Reviews of the Cambridge Philosophical Society* 62 (1987):341–438. A cladistic study of sarcopterygian fishes and amphibians that concludes that the amphibians arose from the rhipidistian group of crossopterygians.

Ruben, J.A., and Boucot, A.J. The origin of the lungless salamanders. *American Naturalist* 134(1989): 161–169.

Stebbins, R.C. *A Field Guide to Western Reptiles and Amphibians.* Boston: Houghton Mifflin Co., 1966. A companion to Conant's *Field Guide.*

38

Reptiles

Characteristics of Reptiles

■ Reptiles are amniotes that have evolved a cleidoic egg.

■ Although ectothermic, reptiles maintain a relatively high body temperature during their periods of activity by controlling their exposure to the sun. They are more active than amphibians.

■ Their body surface is covered by horny scales or plates, and few cutaneous glands are present.

■ Their skeleton is well ossified and strong, and two sacral vertebrae support the pelvic girdle.

■ A tympanic ear has evolved in most reptiles independently from amphibians. It has been secondarily lost in some.

■ Cutaneous respiration is negligible. Lung surface area is increased, and the lungs are ventilated by changing the size of the body cavity.

■ A more complete division of the heart ventricle and conus arteriosus prevents a mixing of oxygen-rich and oxygen-depleted blood when the lungs are in use, but the lungs can be bypassed during prolonged periods of apnea.

■ Considerable nitrogen is eliminated as uric acid, thus conserving water.

■ Fertilization is internal. Most reptiles are oviparous, but some species brood their young internally.

(Left) A den of red-sided garter snakes,
Thamnophis sirtalis pariatalis, *of the Western plains of the United States. (Above) Red-sided garter snakes.*

Agmatha

Acanthodii

Placodermi

Chondrichthyes

Osteichthyes

Amphibia

Reptilia

Aves

Mammalia

As we have seen, several amphibian groups evolved features that allow some degree of terrestrial reproduction. Terrestrial reproduction prevents exposure of delicate eggs and larvae to aquatic predators and to the unpredictable water supply experienced by many species that reproduce in water. One lineage of labyrinthodonts gave rise to vertebrates that evolved a reproductive pattern that freed them completely from dependence on external water sources. This was made possible by the evolution of a cleidoic egg containing enough yolk to provide for the nutrition of the embryo throughout its development, and extraembryonic membranes that protect the embryo and allow for gas exchange and excretion. The larval stage is suppressed, and the embryo develops within the safety of a shelled egg until it hatches as a miniature adult. Reptiles, birds, and mammals are the descendants of this group of labyrinthodonts. All are called **amniotes** because of the shared presence of one of the extraembryonic membranes, the amnion, that forms a protective liquid cushion around the embryo.

Amniotes are a monophyletic group that arose near the middle of the Carboniferous. By the late Carboniferous, about 285 million years ago, amniotes had diverged into three well-defined lineages: (1) a line leading to mammals, (2) one to turtles, and (3) one to all of the other reptiles (snakes, lizards, crocodiles, dinosaurs, and other extinct groups) and to birds. Reptiles, as the term is commonly used, includes parts of all three lineages (Fig. 38.1). From the standpoint of cladistics, reptiles, therefore, are not a natural coherent group, but a paraphyletic one

Figure 38.1
The phylogeny and geological distribution of amniotes. Groups commonly assigned to the class Reptilia are colored a bluish green. Note that reptiles are not a monophyletic group, for they encompass parts of three distinct evolutionary lines.

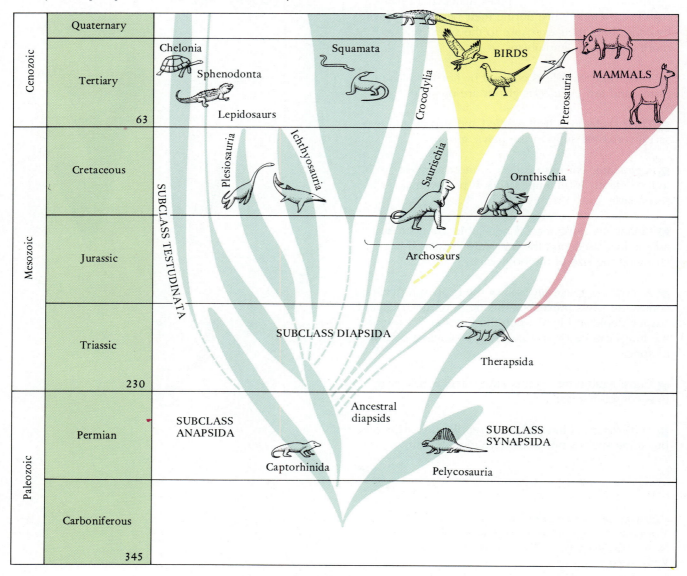

(p. 547). Reptiles can be defined as those amniotes that are not birds or mammals. In this chapter, we will examine contemporary reptiles, of which there are about 6500 species, and important related extinct groups. We will defer considering those reptiles that gave rise to mammals until we examine the mammals in Chapter 40. Living reptiles are sorted into four orders: turtles (**Chelonia**), a single species of tuatara (**Sphenodonta**), lizards and snakes (**Squamata**), and alligators and crocodiles (**Crocodylia**).

Reptiles continued the colonization of land begun by the crossopterygian fishes and amphibians, and are well adapted to terrestrial life. They are more active than amphibians and can live in drier habitats. Many are found in deserts; indeed, reptiles are the dominant group of vertebrates in deserts, where they outnumber birds and mammals. Some reptiles have also readapted to an aquatic life, but all must lay their eggs upon the land unless, like the sea snakes, they can brood their young internally. Even sea turtles leave the water to lay their eggs on beaches. Reptiles are particularly abundant in the tropics and subtropics, since they are still ectothermic animals and gain their body heat from their surroundings. They are less numerous in temperate regions because they cannot maintain an elevated body temperature when the sun sets or on cold days. They are not found in arctic regions for their metabolic rate could not sustain activity at such low temperatures.

ADAPTATIONS OF REPTILES

Integument

Much of the success of reptiles can be attributed to modifications of the integument. The epidermis is dry and horny, with large amounts of keratin, a water-insoluble protein, deposited in epidermal cells. The surface cells form **horny scales** or plates that protect the body from harmful radiation, abrasion, and reduce the loss of body water (p. 203, Fig. 11.5b). As new cells develop deep in the epidermis, the heavily keratinized layers wear away in turtles or are shed as a unit in snakes and some lizards. Unlike fishes and amphibians, reptiles have very few glands in their skin. Some reptiles have scent glands that produce odiferous secretions used in species and sexual recognition. In certain snakes and lizards other glands secrete irritating or sticky materials that afford some protection from predators. Bone may develop in the dermis beneath the horny scales, e.g., the bony plates in a turtle shell and small nodules of bone in some lizards and crocodiles.

Skeletomuscular System, Locomotion, and Nervous System

Reptiles have a well-ossified and stronger skeleton than amphibians. During locomotion, the proximal segment of their front and hind leg moves back and forth close to the horizontal plane, as it does in labyrinthodonts and salamanders. Many of the dinosaurs became bipeds. The thrust of the hind legs is transferred to the vertebral column through two sacral vertebrae rather than through a single one, as in amphibians. Claws on the toes enable the feet to get a good grip upon the substratum. The muscular system is more complex than in amphibians, and the central nervous system is better developed. The cerebral hemispheres have enlarged somewhat by the growth of deep masses of gray matter, called the corpus striatum (Fig. 38.2). Reptiles do not have a cerebral cortex.

Metabolism

Reptiles are ectotherms, gaining body heat from their surroundings and not, as endothermic birds and mammals do, from metabolic heat production. The body temperature of species living in temperate climates falls greatly at night and they become very sluggish. When the sun rises,

Figure 38.2
The major visceral organs of a lizard.

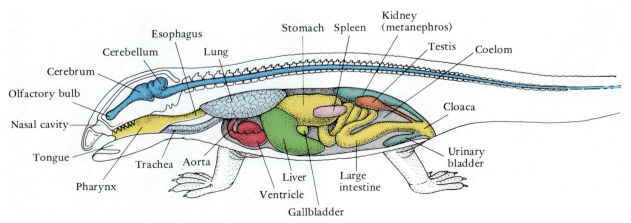

they seek out sunny spots and warm their bodies up by solar radiation and by conduction of heat from the solarly warmed rocks and soil on which they are basking (Fig. 38.3). Absorption of solar energy is augmented early in the morning by a dispersal of pigment in the chromatophores so the skin becomes dark and absorbs more light energy, and by an increased blood flow through the skin. When they become too warm, the skin lightens, cutaneous blood vessels constrict, and they retreat to the shade. A few species, such as the chuckwalla (*Sauromalus*), can also eliminate body heat by panting. Reptiles maintain a high and relatively constant body temperature and remain active during the daytime by behavioral thermoregulation, adjusting their exposure to the sun. Some of the lizards in the southwestern United States maintain a temperature close to 34° C (93° F) even when air temperature is substantially lower. This is close to the body temperature of mammals. Reptiles living in temperate climates cannot maintain their temperature during prolonged cold periods, and they, like the amphibians, hibernate. Tropical reptiles live in a more thermally favorable and more uni-

form environment, thus they have little difficulty in regulating their temperature.

Body temperature is regulated in reptiles, as it is in mammals, by a center in the hypothalamus of the brain. The amount of solar radiation being received is monitored in many lizards by a well-developed **parietal eye** on the top of the head (Fig. 38.4). Since exposure to sunlight by behavioral changes plays such a large role in reptile thermoregulation, they are sometimes called **heliotherms** (Gr. *helios*, sun).

Digestion

Unlike endotherms, who rely on metabolic heat obtained from the oxidation of food, reptiles rely on ambient temperature to maintain body temperature. One mouse or frog will provide the energy a small snake needs for a week or more; in comparison, an endothermic shrew consumes food equivalent to its body weight *each day*. Low food requirements enable reptiles to flourish in deserts and other habitats where there is not enough food to sustain the elevated energy demands of many mammals. Most reptiles are carnivores. Some have specialized ways of catching and swallowing food, but in most respects the structure of their digestive tract is very similar to that of amphibians. Digestion is more rapid than in amphibians, however, because reptiles maintain a higher body temperature much of the time. Captive reptiles that are not kept warm or given access to the sun often cannot digest their food.

Tortoises and some lizards are herbivores. Their large intestine usually is longer than the small intestine and

Figure 38.3
Reptiles maintain a body temperature close to 34° C during their periods of activity by behaviorly controlling the amount of heat they gain or lose to their environment.

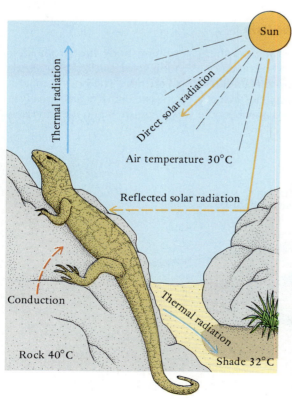

Figure 38.4
The parietal eye by which many lizards monitor the amount of sunlight they receive can be seen in the depression on the top of the head between the two lateral eyes of this South American iguana, *Iguana*.

is divided by internal folds into compartments. Food passes slowly through the large intestine, and a bacterial colony within it digests cellulose into fatty acids that are absorbed there.

Respiration

The dry scaly skin of reptiles reduces cutaneous respiration to a negligible level. This is compensated for by an increase in the respiratory surface of the lungs (Fig. 38.2). A large lung surface area also provides for the increased volume of gas exchange necessitated by the general increase in metabolic activity. Reptiles, like mammals, have aspiratory lungs that are ventilated by changing the size and pressure within the body cavity that contains them. Air is retained in the lungs during a period of apnea. The contraction of trunk muscles and the elastic recoil of the lungs force the air out.

Circulation

The heart is more completely divided in reptiles than in amphibians by a partial or, in crocodiles, a complete interventricular septum. There is an unusual tripartite division of the conus arteriosus leaving the heart (see Fig. 14.9c). As a consequence of these features, venous and arterial blood streams are nearly completely separated, reducing the mixing of oxygen-depleted blood and oxygen-rich blood. Certain persistent interconnections, either between the two ventricles or between the left and right sides of the conus arteriosus, permit the lungs to be bypassed when they are not being used.

Excretion

Reptiles, in common with other amniotes, have evolved metanephric kidneys (Fig. 38.2). Less water is needed to remove nitrogenous wastes from the blood than in amphibians because a large portion of the wastes is excreted as nontoxic uric acid in the form of solid crystals that do not require water for their removal (see Table 37.1, p. 849). Much of the water that is removed from the blood is later reabsorbed by other parts of the kidney tubules, the urinary bladder, and cloaca.

The development of the chemical pathway that produces uric acid appears to be correlated with the evolution of the cleidoic egg. In the terrestrial environment in which the eggs are laid, little water is available to carry off ammonia and urea, and nitrogenous wastes cannot be eliminated as gaseous ammonia. Conversion of nitrogenous wastes to insoluble uric acid crystals enables ammonia to be detoxified and nitrogenous wastes to be temporarily stored as uric acid by the embryo in the allantois, one of the extraembryonic membranes. The embryonic capacity to produce uric acid was then retained in the adult.

Reproduction and Development

Fertilization is internal and most males have a copulatory organ. Reptiles lay fewer eggs than most fishes and amphibians, but the cleidoic eggs protect the embryos and provide for food and other metabolic needs, so the mortality of embryos is low. A collared lizard lays only 4 to 24 eggs; a leopard frog, 2000 or more.

Most reptiles are oviparous and bury their eggs in soil, sand, or leaf mold, where heat from solar radiation or plant decomposition will incubate them. Some lizards and snakes retain their eggs in a uterus, and the embryos develop there, primarily using food stored in the yolk. Only a few species have developed a placental relationship with the mother.

EVOLUTION AND ADAPTATIVE RADIATION OF REPTILES

Early Reptiles

The remains of early reptiles have been found in the fossilized stumps of primitive trees dating back to the middle Carboniferous, about 320 million years ago. These creatures, which are placed in the order **Captorhinida**, resembled lizards superficially (Fig. 38.5). Their skeleton was more thoroughly ossified than that of labyrinthodonts living at that time, and their limb proportions suggest that they were more agile. They probably fed upon small terrestrial arthropods that were becoming abundant at about the same time.

Reptiles were the first vertebrates that could penetrate the terrestrial environment far beyond the shores of bodies of fresh water; they therefore encountered few competitors upon the land. They multiplied rapidly, spread into the terrestrial niches available to them, and specialized accordingly. Much of their adaptive radiation involved different methods of locomotion and feeding.

Figure 38.5

A captorhinid from the Permian, about 285 million years ago. This specimen was about 25 cm long.

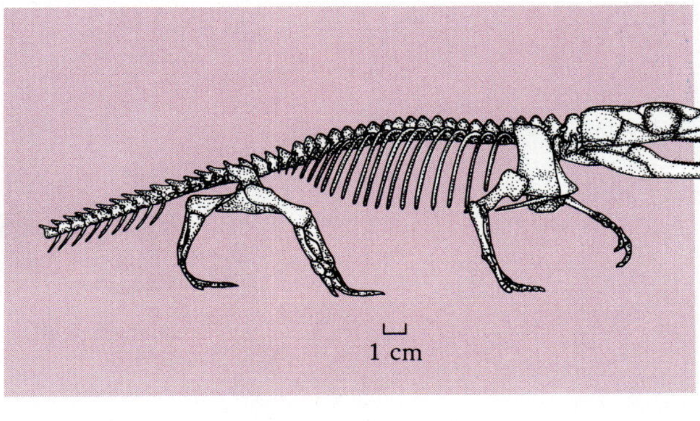

1 cm

Different feeding patterns entailed, among other things, modification of the jaw muscles, and this in turn affected the structure of the temporal region of the skull on which the muscles originate. Skull morphology, therefore, provides a convenient way to sort out the various lines of reptile evolution. The captorhinids had a solid roof of dermal bone covering the jaw muscles dorsally and laterally (Fig. 38.6a). This is referred to as the **anapsid** (Gr. *a*, without + *apsis*, bar) condition, for there are no openings in the temporal roof bounded by bony arches. Captorhinids and their close relatives are placed in the subclass **Anapsida** (Fig. 38.1). As feeding patterns and jaw muscles changed, various types of openings or fenestrae evolved in the temporal roof overlying the jaw muscles. Some parts of the dermal roof were apparently not stressed, and bone does not develop in such areas. Muscles can attach more firmly to the edges of openings and to bony ridges than they can to flat surfaces, and they can bulge through the openings when they contract.

Three lines of evolution diverged from captorhinids. An anapsid skull was retained in the line leading to turtles, but turtles differ from other reptiles in so many ways that they are now placed in their own subclass, the **Testudinata**. A second line, the mammal-like reptiles (subclass **Synapsida**), led to mammals. Synapsids are characterized, as are mammals, by a synapsid skull with a single, laterally placed temporal opening (Fig. 38.6b), which originally lay ventral to the postorbital and squamosal bones. The third line (subclass **Diapsida**) includes the remaining reptiles and led on to birds. Diapsids originally had two temporal openings, one above and one below the postorbital-squamosal bar (Fig. 38.6c). Some diapsids secondarily lost one or both of the bars of bone bounding the temporal openings as they evolved jaw mechanisms that enabled them to open their mouths very widely (Fig. 38.6d).

Turtles

Although turtles (order **Chelonia**) do not appear in the fossil record until the Triassic, they probably diverged earlier from captorhinids. Teeth have been lost and the jaws are covered by sharp, horny plates. They have evolved a tympanic ear. Their short, broad trunk is encased in a protective shell composed of bony plates overlain by horny ones. The bony plates have ossified in the dermis of the skin and some have fused with the vertebrae and ribs. The portion of the shell covering the back is known as the **carapace**; the ventral portion is called the **plastron** (Fig. 38.7).

Turtles compensate for an inflexible trunk by having a long flexible neck that facilitates feeding. Ancestral turtles were unable to retract their heads, but modern species can withdraw theirs into the protection of the shell. This is accomplished in most species by bending the neck in an **S**-shaped loop in the vertical plane (Fig. 38.7), but one group of species in South America, Africa, and Australia retract their neck by bending it in the horizontal plane, and therefore are known as the side-necked turtles (Fig. 38.8a). Limbs too are withdrawn into the shell when danger threatens. Box turtles have a plastron hinged both anteri-

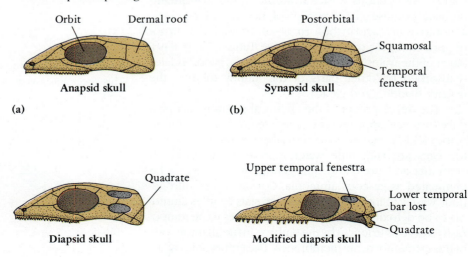

Figure 38.6
Temporal openings in reptiles: (a) the anapsid skull of a captorhinid; (b) the synapsid skull of a mammal-like reptile; (c) the diapsid skull of a sphenodontid; (d) the modified diapsid skull of a lizard in which the lower bar bounding the lower temporal opening has been lost.

Orbit   Dermal roof

Anapsid skull

(a)

Postorbital

Squamosal

Temporal fenestra

Synapsid skull

(b)

Quadrate

Diapsid skull

(c)

Upper temporal fenestra

Lower temporal bar lost

Quadrate

Modified diapsid skull

(d)

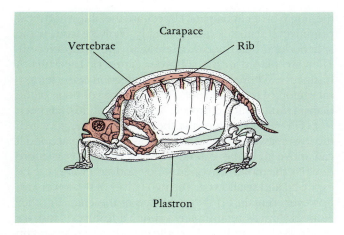

Figure 38.7
The skeleton of a turtle. Most species withdraw their head into the shell by bending their neck in the vertical plane into an **S**-shaped loop.

orly and posteriorly that can be pulled up to the carapace, enabling them to close their shell completely. The plastron is smaller in snapping turtles and the head and limbs are not withdrawn as far into the shell. The animals' strong bite and ferocious temperament provide adequate protection.

A rigid shell precludes lung ventilation by a change in the diameter of their rib cage; instead, muscles that line the skin pockets into which the limbs can be withdrawn act as diaphragms. Aquatic species can also exchange gases underwater through the vascular lining of their mouth and pharynx and through sacs that bud off the cloaca.

The shell affords protection, but it also restricted the evolutionary options open to turtles; one would not expect to find jumping and climbing species. But turtles are a successful group, and have undergone their own modest adaptive radiation. Nearly 250 contemporary species are known (Fig. 38.8b−d). Most are semiaquatic, and these

Figure 38.8
Adaptive radiation among turtles. (a) Geoffroy's sideneck turtle, *Phrynops geoffroyanus* of Argentina, folds its neck under its shell in the horizontal plane. (b) The semiaquatic Eastern painted turtle, *Chrysemys picta picta*, has a flattened shell. (c) The pectoral appendages of the green sea turtle, *Chelonia mydas*, are modified to form long flippers. (d) The terrestrial Galapagos tortoise, *Geochelone elephantopus*, and other tortoises have dome-shaped shells.

(a)

(b)

(c)

(d)

species tend to have flattened shells that offer less resistance to swimming. Several groups are entirely aquatic. The front legs of sea turtles are modified as flippers, and these turtles spend most of their lives in the water. Females come on shore only to lay their eggs. Tortoises are completely terrestrial and have strong, high-vaulted, dome-shaped shells that predators have difficulty biting, crushing, or swallowing. This shape also makes it easy for the turtle to right itself if it accidentally turns over during its activities.

Primitive Diapsids: Lepidosaurs

Primitive diapsids appeared in the late Carboniferous and adapted early to different life styles (Fig. 38.1). Of several evolutionary lines, only the tuatara (*Sphenodon*) and the lizards and snakes have survived. Collectively these reptiles are placed in the superorder **Lepidosauria**.

Sphenodontids

Sphenodontids (order **Sphenodonta** or **Rhynchocephalia**) were abundant in the early Mezosoic, but a single species, *Sphenodon*, survives today on several islands off the coast of New Zealand. *Sphenodon* is the most primitive of surviving lepidosaurs (Fig. 38.9). It remains a quadruped, and lateral undulations of the trunk and tail continue to be an important part of its locomotor pattern. The skull remains diapsid, and a well-developed parietal eye is present on the top of the head. The males lack distinct copulatory organs and transfer sperm to the females by everting their cloacas. *Sphenodon* is sometimes described as a surviving "fossil," for it has not changed greatly from species that were living 150 million years ago.

Lizards

The familiar lizards and snakes, though superficially different, are similar enough in basic structure to be placed in

Figure 38.9
The New Zealand tuatara, *Sphenodon punctatus*, is the sole surviving species of the order Sphenodonta.

the single order **Squamata**. Lizards are the oldest and most primitive squamates and probably evolved from some sphenodontid-like ancestor early in the Mesozoic era. They are placed in the suborder **Sauria** or **Lacertilia**. A distinctive feature of lizards is the reduction of the temporal region of the skull roof. The lower bar of bone of the diapsid skull, which is present in sphenodontids, has been lost (Fig. 38.6d). As a result, the quadrate bone, to which the lower jaw attaches, is not held as firmly in place and it can move to some extent. The jaw apparatus is more flexible, the mouth can open wider, and larger prey can be captured and swallowed. The two halves of the lower jaw, however, remain firmly articulated at the chin. In other respects lizards retain many primitive features. Most are quadrupeds, most have movable eyelids, and many retain the parietal eye. Lizards evolved a tympanic ear independently of turtles, although it has been secondarily lost in some species.

Lizards are a very successful group and have undergone an extensive adaptive radiation. One group of lizards, the mososaurs of the late Cretaceous, even became specialized for a marine life. They probably swam in an eel-like manner, for their bodies were long and slender and their paired appendages were reduced to small fins. Some reached a length of 10 m. About 3850 species of contemporary lizards occur throughout the tropical and temperate regions of the world. Most species feed upon insects and other small animals during the daylight hours, but tropical geckos forage at night. They often scamper about buildings, clinging to walls and ceilings with expanded digital pads that resemble those of tree frogs (Fig. 38.10a). The arboreal African chameleons have a prehensile tail that can wrap around a branch, and an odd foot structure in which the toes of each foot are fused together into two groups that oppose each other like the jaws of a pair of pliers (Fig. 38.10b). They catch insects by a rapid flick of their sticky tongue that can be protruded nearly a full body's length. Some lizards are adapted to very arid conditions. The moloch of the deserts of central Australia has a grotesque shape and coloration that provides a very effective camouflage (Fig. 38.10c). Microscopic striations on its head scales trap and carry dew by capillary action to its mouth. The Komodo dragons, now restricted to certain Indonesian islands, are the largest living lizards, reaching lengths of 3 m or more (Fig. 38.10d). They feed upon small deer, wild pigs, and goats. Some other lizard groups have become herbivores. The marine iguanas of the Galapagos islands, for example, feed upon seaweed (Fig 38.10e). Some lizards, especially burrowing ones, have reduced limbs or have lost them entirely. The "glass snake" of the Southeastern United States is one example.

The only poisonous lizards are the two species of beaded lizards, including the Gila monster of the southwestern United States (Fig. 39.10f). Modified glands in the floor of the mouth discharge a neurotoxic poison, which is

(a)

(b)

(c)

(d)

(e)

(f)

(g)

Figure 38.10

Adaptive radiation among lizards. (a) The nocturnal tokay gecko of Sri Lanka clings to surfaces by the friction of expanded digital pads bearing hundreds of thousands of minute setae. (b) An African chameleon, *Chamaeleo chamaeleon*, grasps a branch with its prehensile tail and plier-like feet as it flicks its tongue at an insect. (c) The moloch, *Moloch horridus*, is a desert species of central Australia. (d) The Komodo dragon, *Varanus komododensis*, of Indonesia attains a length of 3 m. (e) A male marine iguana of the Galapagos islands, *Amblyrhynchus cristatus*. (f) The gila monster, *Heloderma horridum*, of the deserts of the Southwestern United States is a poisonous species. (g) The American blue-tailed skink, *Eumeces fasciatus*, has a conspicuous tail that can be autotomized when the animal is attacked.

injected into the victim by means of grooved teeth. This is a relatively inefficient method, so their bite is not as dangerous as the bite of most poisonous snakes.

Many lizards, like some salamanders, can discard (autotomize) their tail when threatened. The tail often is conspicuously colored (Fig. 39.10g), and anaerobic metabolism of its muscles sustains tail thrashing for a considerable period, thus distracting the predator and increasing the time it needs to subdue and swallow the tail. In the meantime the lizard moves safely away. Tails break at a special autotomy plane. New ones regenerate from this point, but they are supported by a cartilaginous rod and not by regenerated vertebrae. Tail loss may save a lizard's life, but the lizard also loses food reserves that are often stored in the tail.

Amphisbaenids

The worm lizards (suborder **Amphisbaenia**) appear to have been an offshoot from very early squamates. Their body form is wormlike, and they live an exclusively subterranean life burrowing after worms and small arthropods (Fig. 38.11). The name of the group (Gr. *amphi*, both + *baino*, to go) refers to their ability to move easily both forward and backward through the soil. Their skull and tail are used to push through the soil and are spade shaped and exceptionally strong. About 140 species are known from the tropics and subtropics. One genus has short front legs, but the others lack legs and the girdles are vestigial. Although they lack a tympanic membrane, they can detect ground vibrations by an extension of the stapes to the lower jaw. Their rudimentary eyes are partly concealed beneath the skin, and they track prey primarily by hearing and smell.

Figure 38.11
A tropical amphisbaenid, *Amphisbaena alba*, that has been removed from its burrow.

Snakes

Snakes are strange creatures, feared and loathed by many people, but revered in many cultures as symbols of rejuvenation, power, and healing. The medical profession is symbolized by the caduceus, two snakes entwined around a staff. Snakes (suborder **Serpentes** or **Ophidia**) branched from the lizards during the Cretaceous, and differ from them primarily in feeding mechanisms, sense organs, and patterns of locomotion. Lizards retain one bar of bone bounding the temporal opening (Fig. 38.6d) but snakes have lost this bar, so their jaws are exceptionally flexible. Snakes have five jaw joints where motion occurs during feeding (Fig. 38.12). Snakes feed only upon animals, and this specialization of jaw structure enables them to swallow whole prey that is several times their own diameter. Each side of the lower jaw can be moved independently as prey is gripped by sharp, recurved teeth and pulled into the mouth.

Snakes lack movable eyelids, and the eye is covered instead by a transparent **spectacle** (Fig. 38.13). Arboreal species have good vision, but sight is limited in many others. The tympanic membrane has been lost, but snakes can detect low-frequency ground vibrations through their skull bones. They compensate for reductions in the senses of sight and hearing with a very sensitive forked tongue, which is often seen darting from the mouth. The tongue is an organ of touch and smell. Particles adhere to it, the tongue is withdrawn into the mouth, and its tip is projected toward a pair of openings in the roof of the mouth that lead into specialized parts of the nasal cavity (Jacobson's organ) where the odor of particles can be detected (p. 406). Although this organ is present in many other terrestrial vertebrates, it is particularly well developed in snakes, and it is needed by snakes to follow prey trails and for sex recognition. Some species, including boas, pythons, and pit vipers, that feed on warm-blooded prey have sensory pits that detect infrared rays (p. 429, Fig. 18.28). This enables the snakes to seek out and strike accurately at objects warmer than their surroundings.

Pythons and boa constrictors retain spurlike rudiments of hind legs that are used in courtship, but otherwise limbs and girdles are absent. Snakes move primarily by lateral undulations of their exceptionally long trunk and tail that together can contain 200 or more vertebrae (Focus 38.1).

Their elongated body form and degenerative changes in the eye and ear suggest that snakes evolved from a burrowing group of lizards. A few primitive species, the worm snakes, still burrow, but during their subsequent evolution most have readapted to life above ground. They underwent an extensive adaptive radiation. Nearly 2400 species now live in the temperate and tropical regions of the world.

Those that prey upon very active animals have evolved methods of prey immobilization. Boa constrictors,

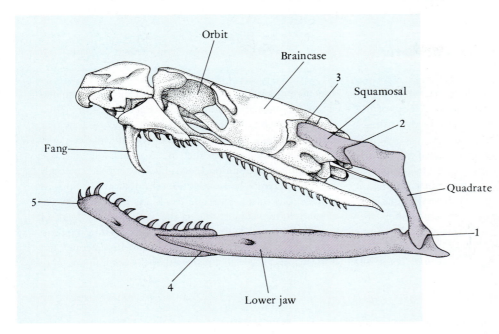

Figure 38.12

The skull of a cobra. The bones that move during feeding have been colored, and sites of movement have been numbered: (1) the usual jaw joint between quadrate and lower jaw, (2) a joint between quadrate and squamosal, (3) one between squamosal and braincase, (4) one about halfway along the lower jaw, and (5) one at the chin, for the two halves of the lower jaws are not firmly united at this point.

pythons, and many rodent-eating snakes, such as king snakes, quickly entwine their prey in loops of the trunk and suffocate the prey by stopping their respiratory movements (Fig. 38.14). Several other groups have independently evolved poison glands, which are modified salivary glands. The poison is injected with grooved or hollow hypodermic-like teeth, the **fangs**. Old World vipers and New World pit vipers (rattlesnakes, copperheads, and cottonmouths) have a pair of large hollow fangs at the front of the mouth that are articulated to bones of the upper jaw and palate in such a way that they are folded against the roof of the mouth when the mouth is closed and automatically brought forward when the mouth is opened (Fig. 38.15). Coral snakes, cobras, and sea snakes have short, immovable fangs at the front of the upper jaw (Fig. 38.12). Rear-fanged species have grooved teeth at the back of their jaws. Snake venom contains many polypeptides and proteins, including neurotoxins that block the transmission of nerve impulses to muscles and hemolytic enzymes that destroy blood-clotting factors, produce internal bleeding, and destroy some tissues. The exact composition of the venom varies with the species, but hemolytic enzymes are

Figure 38.13

Sense organs of snakes; the forked tongue, the sensory pit in front of the eye, the spectacle covering the eye, and the absence of a tympanic membrane are evident on the head of Pope's pit viper.

Figure 38.14

A white-lipped python feeding on a mouse that it has subdued by constriction.

Snakes have four patterns of locomotion. During **undulatory locomotion**, the most typical mode, loops of the trunk form behind the head and move posteriorly. As these loops meet protuberances from the ground, they push upon them, and the resultants of these forces move the animal forward. Snakes can also move slowly in a straight line as they sneak up on prey by **rectilinear locomotion**. Some of their large, transverse ventral scales catch on irregularities on the ground and are pulled posteriorly while others are pulled forward to gain a new contact. Many snakes, especially those that climb among rocks and trees, use **concertina locomotion** (Focus fig. a). The body is folded together much like an accordion, and the posterior end wraps around a small branch or is wedged in a crevice. Then the body is extended and gains a new contact, and the rear part of the body is pulled forward. The sidewinder rattlesnake and other species that move across soft sand use **sidewinding locomotion** (Focus fig. b). Loops of the body are pressed into the sand at no more than two points. As undulations travel down the trunk and tail, the points of contact move laterally across the sand. As one point of contact leaves the tail, a new one is established by the head and neck. The overall direction of travel is laterally and forward, and a series of **J**-shaped tracks is left in the sand.

(a) Concertina locomotion is used by many climbing species. (b) Sidewinding locomotion is used by some rattlesnakes crossing sand.

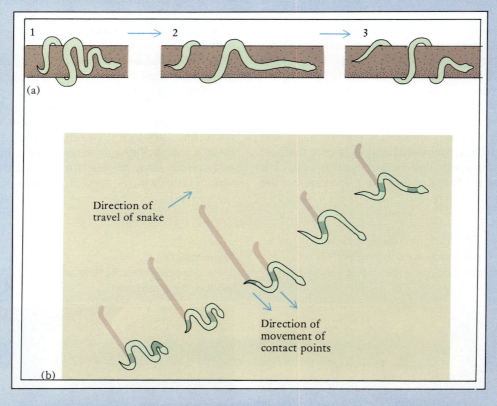

particularly abundant in viper and pit viper venoms, and neurotoxins abundant in the others.

Many snakes vibrate their tails when they are in danger. Rattlesnakes have increased the efficiency of this habit, turning it into a warning, by developing rattles on the ends of their tails (Fig. 38.15). The rattles are simply remnants of dry, horny skin that have been prevented from being completely shed by being caught on an enlargement

Figure 38.15
The Western diamond-back rattlesnake, *Crotalus atrox*,
striking. As it strikes, it vibrates its rattle and opens its mouth,
which automatically rotates the fangs forward.

on the end of the tail. The center of rattlesnake evolution
probably was the western plains of the United States,
where the rattles may have developed as a warning mecha-

nism that prevented the snakes from being trampled by
bison.

Blind Alleys: Extinct Marine Reptiles

The extinct plesiosaurs (order **Plesiosauria**) were highly
specialized marine reptiles that diverged from primitive
diapsids (Fig. 38.1). They had a short broad trunk and a
long flexible neck, and propelled themselves by means of
large paddle-shaped appendages (Fig. 38.16a). Some species reached a length of 12 m. Plesiosaurs, like lizards, lost
the lower bar of the diapsid skull, but they differ from
lepidosaurs in so many other ways that they are placed in
their own superorder, the **Sauropterygia**.

Ichthyosaurs (order **Ichthyosauria**) are another extinct group of marine reptiles. They were porpoise-like in
size and shape, and probably in habits (Fig. 38.16b). They
moved with fishlike undulations of the trunk and tail, and
their paired appendages were reduced to finlike flippers.
They are characterized by a single, dorsal temporal opening and may also have evolved from primitive diapsids.
However, the first known fossils are already so highly
specialized that we cannot be certain of their ancestry.
They are placed in their own infraclass, the **Ichthyopterygia**.

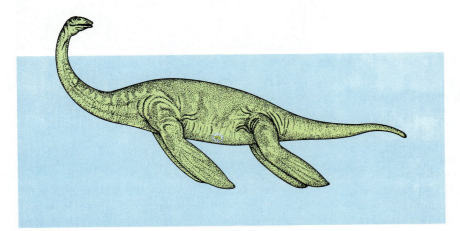

(a)

Figure 38.16
Marine reptiles of the Mesozoic. (a) The
plesiosaurs swam with enlarged pectoral
and pelvic flippers. (b) Ichthyosaurs were
more fishlike and swam by lateral
undulations of their trunks and tails.

(b)

Plesiosaurs probably came onto the beaches to lay their eggs, but the extreme aquatic adaptations of the ichthyosaurs would have precluded their doing so. We believe that they were viviparous because a fossil has been found with several small ichthyosaur skeletons lodged in the uterine region of the mother's abdominal cavity, and one is part way out the cloaca.

Plesiosaurs and ichthyosaurs flourished during the Mesozoic, competing with the more primitive kinds of fishes. Just why they became extinct near the close of this era is unclear, but their extinction coincides with the evolution and diversification of the teleosts. Possibly they could not compete successfully with these fishes.

Advanced Diapsids: Archosaurs

Archosaurs include the most spectacular and fearsome of reptiles: the dinosaurs, flying pterosaurs, and crocodiles (Fig. 38.1). Birds evolved from one group of dinosaurs and are technically archosaurs, although most zoologists continue to place them in their own class. Archosaurs diverged from primitive diapsids in the Permian. Most retain both temporal bars of the diapsid skull, but they display a suite of skeletal specializations, many of which, like relatively long hind legs, are associated with moving their hind legs fore and aft in the vertical plane, rather than horizontally. Many archosaurs became bipeds. Archosaurs were the dominant terrestrial vertebrates throughout the Mesozoic era.

Crocodiles

Apart from birds, the alligators and crocodiles (order **Crocodylia**) are the only surviving archosaurs. They are the largest living reptiles, some attaining lengths of 5.5 m. Ancestral species were terrestrial, but contemporary ones have become specialized for an amphibious mode of life. Their tail is laterally compressed and is a very effective swimming organ. Eye, ear, and nose openings are situated on elevations on the top of the head, so they protrude above water. Most of the animal can be submerged as it stalks prey. Valves close the nose and ear when beneath water.

Twenty-two species of crocodilians occur in the tropics and subtropics. All are predatory, armed with strong jaws with sharp teeth. Snout length and shape vary considerably with their diet. Some crocodiles and the gharials of India and parts of Asia have very long, narrow snouts that can slice through the water to seize fish (Fig. 38.17a). Alligators, the caimans, and other crocodiles have shorter, broader snouts that enable them to feed on larger prey including turtles, aquatic birds, and mammals that may venture too far into the water. A secondary palate and a flap of flesh at the back of the mouth enable them to tear prey apart beneath water without shipping water into the respiratory passages. Several African, Australian, and Asian species attack humans.

(a)

(b)

Figure 38.17
Crocodilians. (a) The long narrow snout of the Indian gharial, *Gavialis*, is adapted for seizing fish. (b) The American crocodile, *Crocodylus acutus*.

Only two species occur in the United States: the American alligator and the American crocodile. Many people living in the South are familiar with the loud bellowing mating call of the male alligator in the spring. The crocodile has the more pointed snout, and its fourth mandibular tooth is visible when the jaws are closed (Fig. 38.17b). The American crocodile is an estuarine species that tolerates salt water. Its range extends from the southernmost tip of Florida through the Caribbean islands to the coasts of South and Central America.

Dinosaurs

Dinosaurs were active animals that probably maintained an elevated level of metabolism (Focus 38.2). Most became very large. They are grouped into two orders: the Saurischia and Ornithischia. Saurischian dinosaurs (order **Saurischia**) retained a primitive lizard-like pelvis (Fig. 38.18a). They evolved from ancestors that were less than 1

(a) Saurischian pelvis / Ornithischian pelvis

Anterior ←

Ilium

Pubis

Socket for hind leg

Ischium

(b)

(c)

(d)

Figure 38.18
Representative dinosaurs. (a) Saurischian dinosaurs had a lizard-like pelvis with the pubis lying in front of the socket (acetabulum) for the hind leg; in ornithischian dinosaurs, part of the pubis extended posteriorly beside the ischium. Unlike other reptiles, the acetabulum of dinosaurs and other archosaurs contained a large hole. (b) *Tyrannosaurus* confronts two *Triceratops*. (c) The giant saurischian herbivore, *Diplodocus*, may have browsed upon high vegetation. (d) A large stegosaur, *Diracodon*, protects itself from the attack of a large saurischian carnivore, *Ceratosaurus*.

Many contemporary reptiles can travel very rapidly for a short distance, but structurally and physiologically they are incapable of sustaining rapid movements: stride length is relatively short, for the proximal segment of their limbs moves back and forth in the horizontal plane; they are ectotherms and derive their energy for muscle contraction primarily from glycolysis (p. 263). Lactic acid accumulates rapidly, so bursts of activity must be followed by long periods of recovery. Mammals, in contrast, carry their limbs beneath the body, they are endothermic, and most of their energy is derived from oxidative metabolism.

Dinosaurs also carried their legs beneath the body and had long strides. Structurally they were capable of sustained rapid locomotion. Were they endothermic and did they maintain a high and constant body temperature because of a greatly increased rate of metabolism? Paleontologists are not in agreement. Some argue that they were truly endothermic. Dinosaurs had a highly vascularized bone structure that resembles that of mammals. An endothermic carnivore also needs a larger food supply than an ectothermic one, and an analysis of predator-prey ratios in carnivorous dinosaur communities is close to that seen in mammalian communities.

Other paleontologists see many exceptions to these correlations, and argue that dinosaurs could have maintained a high and uniform body temperature while retaining the reptilian level of metabolism. If the climate was warm and uniform, as it appears to have been during most of the Mesozoic, they would retain heat gained from the environment because they had such a large mass relative to their heat-losing surface. Calculations have shown that a body 1 meter in diameter would maintain a temperature close to 34° C in a subtropical climate. Dinosaurs may have been **inertial homeotherms**. Once warmed up to the environmental temperature, they could have maintained a high and constant body temperature and had an oxidative metabolism without the energetic cost of endothermy. They would, however, have been very vulnerable to the general climatic cooling that occurred toward the end of the Cretaceous because heat, once lost, could not be regained from the environment.

m long, but later ones became giants. *Tyrannosaurus* was the largest terrestrial carnivore that the world has ever seen (Fig 37.18b). It stood about 6 m high and had large jaws armed with dagger-like teeth 15 cm long—a truly formidable creature! It had long hind legs, reduced front ones, and a heavy tail that acted as a counterbalance to hold the body upright. Other saurischian dinosaurs were herbivores that reverted to a quadruped gait, but their bipedal ancestry is reflected in their retention of a heavy tail and in having hind legs larger than the front ones. Some, such as *Diplodocus*, were enormous, attaining a length of nearly 30 m and a weight of 45 metric tons (Fig. 37.18c). Their huge size led to the hypothesis that they lived in swamps where they would be supported partly by the buoyancy of the water, but they did not have the specializations, such as a compressed tail, that we associate with an aquatic life. On the contrary, their elephant-like legs suggest a terrestrial habitat. They may have fed in a giraffe-like fashion upon high vegetation.

Dinosaurs belonging to the order **Ornithischia** were more specialized and characterized by a suite of derived characters including a more birdlike pelvis (Fig. 38.18a). All were herbivores. Teeth at the front of the jaws were replaced by horny beaks, and the teeth in the back of the jaws were modified for grinding. Some lived in swamps; others, in the uplands. Many reverted to a quadruped gait and increased in size. These animals undoubtedly formed much of the diet of saurischian carnivores, and many, such as *Stegosaurus* and *Triceratops*, evolved protective devices, such as spiked tails, bony plates on the body, and horned skulls (Fig. 38.18b and d).

Dinosaurs and many other groups of terrestrial and marine organisms became extinct toward the end of the Cretaceous period (Fig. 38.1). One compelling explanation for this mass extinction was proposed in 1980 by L.A. Alvarez and his coworkers, who discovered a thin layer of iridium at the Cretaceous-Tertiary boundary in clays in many parts of the world. Since iridium is normally rare in the earth's crust but more abundant in meteorites, Alvarez concluded that a very large meteorite or asteroid struck the earth. Such an impact would have thrown up large dust clouds that circled the earth, shutting out sunlight for months and perhaps years, reducing photosynthesis and the mean annual temperature. The synchrony between the iridium layer and the mass extinction of marine organisms supports a causal relationship. But most dinosaurs had become extinct by the time the iridium layer was formed, and turtles and squamates showed no significant extinction at the end of the Cretaceous. Dinosaurs may have been driven to extinction by a gradual climatic cooling that occurred in the late Cretaceous. Large inland seas that would have buffered temperature fluctuations disappeared and extensive mountain ranges began to form. Seasonal and diurnal temperature changes would have been more pronounced than earlier in the Mesozoic era. Large dinosaurs would have been particularly vulnerable, especially if they were inertial homeotherms (see Focus 38.2). Small terrestrial vertebrates that could take shelter would not have been as severely affected.

Pterosaurs

Wings have evolved independently three times among vertebrates: in the flying reptiles (order **Pterosauria**), in birds, and in bats. In each case the wings evolved from pectoral appendages that had been freed from use in terrestrial locomotion, but each type of wing is different. Pterosaurs evolved from early bipedal archosaurs and their wings consisted of a membrane of skin supported by a greatly elongated fourth finger (Fig. 38.19). Impressions of the membrane show that additional support was provided by stiff fibers that radiated from the wrists to the edge of the wing. The fifth finger was lost, and the others probably were used for clinging to perches. A broad keeled sternum for the attachment of flight muscles, and other features of the skeleton, indicate that pterosaurs were capable of active flapping flight, as are birds and bats.

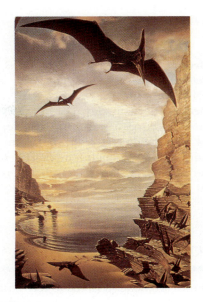

Figure 38.19
Pteranodon, a large pterosaur of the late Cretaceous. Recent studies indicate that the wings were narrower than shown in this reconstruction and did not include the hind legs.

The large Cretaceous species, one of which had a wing span of 12 m, must have glided and soared a great deal of the time, as do large birds today.

Classification of Reptiles

Class Reptilia (L. *reptare*, to crawl) The reptiles.
  Subclass Anapsida (Gr. *a*, without + *apsis*, bar) Reptiles with a solid temporal roof.
    Order Captorhinida† (L. *caput*, head + Gr. *rhis*, nose) Ancestral reptiles. The captorhinids.
  Subclass Testudinata (L. *testudo*, tortoise) Anapsid reptiles encased in a plastron and carapace.
    Order Chelonia (Gr. *chelone*, turtle) Turtles and tortoises.
  Subclass Diapsida (Gr. *di-*, two + *apsis*, bar) Reptiles with a diapsid or modified diapsid skull.
    Infraclass Lepidosauriomorpha (Gr. *lepis*, scale + *sauros*, lizard + *morphe*, form) Primitive diapsids; usually quadruped; skull often loses one or both temporal bars.
      Superorder Lepidosauria
      Order Sphenodonta (Gr. *sphen*, wedge + *odontos*, tooth) Retains a diapsid skull. *Sphenodon*.
      Order Squamata (L. *squama*, scale) One or both temporal bars lost. Lizards, amphisbaenids, snakes.

Superorder Sauropterygia† (Gr. *sauros*, lizard + *pterygion*, little fin or wing)
    Order Plesiosauria† (Gr. *plesios*, near + *sauros*, lizard) Marine reptiles with paddle-like appendages. The plesiosaurs.
  Infraclass Ichthyopterygia† (Gr. *ichthys*, fish + *pterygion*, little fin or wing)
    Order Ichthyosauria† (Gr. *ichthys*, fish + *saurus*, lizard) Mesozoic marine reptiles with fishlike paired appendages. The ichthyosaurs.
  Infraclass Archosauromorpha (Gr. *archon*, ruler + *sauros*, lizard, + *morphe*, form) Advanced diapsid reptiles; diapsid skull retained; often show tendencies toward bipedalism. The ruling reptiles.
    Superorder Archosauria (Gr. *archon*, ruler + *sauros*, lizard)
      Order Crocodylia (L. *crocodilus*, crocodile) The alligators and crocodiles.
      Order Saurischia† (Gr. *sauros*, lizard + *ischion*, pelvis) Dinosaurs with a reptile-like pelvis. *Tyrannosaurus, Diplodocus.*

Order Ornithischia† (Gr. *ornis*, bird + *ischion*, pelvis) Dinosaurs with a birdlike pelvis. *Stegosaurus*, *Triceratops*.
Order Pterosauria† (Gr. *pteron*, wing + *sauros*, lizard) Mesozoic flying reptiles.
Subclass Synapsida† (Gr. *synapsis*, union) The mammal-like reptiles; a single temporal opening on the lateral surface of the skull. This subclass will be discussed in Chapter 40.

Order Pelycosauria† (Gr. *pelyx*, wooden bowl + *sauros*, lizard) Primitive mammal-like reptiles retaining many primitive reptilian features.
Order Therapsida† (Gr. *ther*, wild beast) Advanced mammal-like reptiles beginning to show many mammalian characteristics.

† Extinct.

Summary

1. Reptiles are better adapted to terrestrial life than are amphibians. Their nearly glandless skin is covered with horny scales, and during periods of activity they can maintain a high and nearly constant body temperature by regulating their exposure to the sun. Gas exchange occurs only through the lungs, which are ventilated primarily by rib movements. The heart is nearly completely divided into left and right sides. Nitrogen is excreted primarily as uric acid rather than as ammonia or urea so little water need be excreted. Reptiles have evolved a cleidoic egg that must be laid on the land or retained within the uterus. A few species of lizards and snakes are viviparous.

2. As might be expected of a successful group of terrestrial vertebrates, reptiles underwent an extensive adaptive radiation, and they were the predominant terrestrial vertebrates during the Mesozoic era. Three evolutionary lines are recognized: one line led to mammals, one to turtles, and one to all of the other reptiles and to birds.

3. Turtles (order Chelonia) retain the primitive anapsid skull and are encased in a bony shell. Living species can withdraw their neck and limbs.

4. The tuatara of New Zealand is a primitive diapsid reptile with an unmodified diapsid skull. It is the only surviving species of the order Sphenodonta.

5. Lizards, snakes, and the burrowing amphisbaenids of the order Squamata have a modified diapsid skull that increases jaw flexibility. This flexibility is most developed among snakes, enabling them to swallow prey several times their own diameter.

6. The two groups of dinosaurs and the flying pterosaurs, all now extinct, were advanced diapsid reptiles. They were very active animals that walked with their legs beneath their body and probably maintained a high level of metabolism.

7. Crocodiles and alligators (order Crocodylia) are closely related to the extinct dinosaurs and pterosaurs. They are adapted to an amphibious mode of life.

References and Selected Readings

Many of the general references on vertebrates cited at the end of Chapter 35 also contain considerable information on the biology of reptiles.

Alexander, R. McN. *Dynamics of Dinosaurs and other Extinct Giants*. New York: Columbia University Press, 1989. A fascinating account of the lives of dinosaurs as deduced from physical principles.

Alvarez, L.W., et al. Extraterrestrial cause for the Cretaceous-Tertiary extinction: Experimental results and theoretical interpretations. *Science* 208 (1980): 1095.

Bakker, R.T. *The Dinosaur Heresies*. New York: William Morrow and Co., Inc., 1986. Bakker brings the dinosaurs to life in this fascinating book. He is an advocate of the view that they were warm-blooded.

Conant, R. *A Field Guide to Reptiles and Amphibians of Eastern and Central North America*. 2nd ed. Boston: Houghton Mifflin Co., 1975. A very useful guide for the field identification of amphibians and reptiles; similar to the Peterson bird guides.

Dunkle, T. A perfect serpent. *Scientific American* 81 (Oct. 1981):30−35. The adaptations of the rattlesnake as a four-speed, self-propelled, spring-loaded, heat-seeking hypodermic.

Gore, R., and J. Blair. Extinctions. *National Geographic* 175 (June 1989): 662−695. A beautifully illustrated review of the many extinctions that have occurred throughout the history of life.

Halliday, T., and K. Adler (eds.) *The Encyclopedia of Amphibians and Reptiles*. New York: Facts on File, 1987. Contains excellent and beautifully illustrated essays on many aspects of reptilian biology.

Newman, E.A., and P.H. Hartline. The infrared "vision" of snakes. *Scientific American* 246:(Mar. 1982) 166. A fascinating analysis of the sensory pits of rattlesnakes and pythons and of how information from the pits and eyes is integrated in the brain to provide a unique picture of the world.

Padian, K. The flight of pterosaurs. *Natural History* 97 (Dec. 1988): 58–65. An analysis of pterosaur flight.

Shipmann, P. Bringing up baby. *Discover* (Aug. 1988). Evidence for maternal behavior among the dinosaurs.

Stebbins, R.C. *A Field Guide to Western Reptiles and Amphibians*. Boston: Houghton Mifflin Co., 1966. Similar to Conant's book listed above, but deals with western species.

Thomas, D.K., and E.C. Olson (eds.). *A Cold Look at the Warm-Blooded Dinosaurs*. Washington, D.C.: AAAS Symposium, 1980. A review by authorities on the subject of metabolic levels among the dinosaurs.

39

Birds

Characteristics of Birds

■ Birds are endothermic vertebrates.

■ Horny scales are retained on the feet, but feathers cover most of the body. Cutaneous glands are absent except for a uropygeal (oil) gland.

■ Skeletal bones are exceptionally light and usually pneumatic. The pectoral appendages are wings, the sternum is broad and usually keeled, and the reduced number of caudal vertebrae fuse to form the pygostyle.

■ The eyes and visual centers in the brain are large and very important. The inner ear contains a cochlea, but it is not long and coiled.

■ The narrow jaws form a horn-covered beak in contemporary species. Teeth are absent. Villi are present in the small intestine.

■ The lungs are relatively small, but an unusual pattern of air passages and air sacs produces an exceptionally efficient one-way passage of air across the respiratory surfaces.

■ The atrium and ventricle of the heart are completely divided so oxygen-depleted blood and oxygen-rich blood are separated.

■ Nitrogenous wastes are eliminated primarily as uric acid. The urinary bladder has been lost.

■ Birds are oviparous. One ovary and oviduct are lost during development. The eggs are cleidoic with a large amount of yolk. Albuminous materials and a calcareous shell are secreted around the eggs as they pass down the oviduct.

(Left) The bald eagle, Haliaeetus leucocephalus, *attains a wingspread of nearly 2 meters. (Above) Close-up of the bald eagle.*

Agnatha

Acanthodii

Placodermi

Chondrichthyes

Osteichthyes

Amphibia

Reptilia

Aves

Mammalia

Birds, class **Aves**, contain more species (about 8800) than any other terrestrial vertebrate class. Birds adapted to flight early in their evolution, and most species are excellent fliers. Even the few species that have reverted to a completely terrestrial life show anatomical and physiological features that reflect their evolution from flying ancestors. Adaptation for flight has imposed a certain uniformity in bird structure and physiology; thus birds show less anatomical diversity than do species in other vertebrate classes. In addition to feathers and wings, or vestiges of wings in certain terrestrial species, flight requires a high expenditure of energy. All birds are endothermic and have developed ways of achieving high metabolic rates in a body of light weight. Endothermy and the powers of flight have enabled birds to have a wide distribution from the polar regions to the equator; they live in mountains, deserts, forests, and jungles. Some species spend most of their lives on the ocean, returning to land only to nest.

BIRD FLIGHT

Since flight has played such a key role in the evolution and adaptations of birds, we begin by examining the ways birds fly. The movement of air across the surface of a bird's wing generates lift and drag forces, as it does in all wings (p. 229 and Fig. 11.36). Wing shape allows a smooth flow of air across the surface and minimizes lift-reducing eddies, but some of the air flow does roll up as a vortex, which is shed from the trailing margin and tips of the wings as a pair of vortex lines. Vortex lines often can be seen in high-flying aircrafts appearing as a pair of vapor trails as the air within them condenses.

Many features of the wing affect lift. Lift increases in direct proportion to the surface area of the wing. Wing area differs among different species of birds. Individual birds can also control their wing area by altering the degree to which the wing is stretched out or unfolded. Lift also increases greatly as the speed of airflow across the wing increases, for lift is proportional to the square of the speed. Fast flying birds, such as the swift, have relatively smaller wings than slower flying species. Additional lift is generated during take off and landing, when air speed is low by increasing the wing's angle of attack (p. 230). Compensation for the lift-reducing turbulence that is also produced is accomplished by separating certain feathers to produce slots through which the air moves very rapidly. A small group of feathers, the **alula** (L. diminutive of *ala*, wing), can produce a slot at the front of the wing (Figs. 39.1 and 39.2). Additional slots are often formed along the trailing margin of the wing and at the wing tip. The latter slots reduce the turbulence known as **tip vortex**. Some birds obtain additional lift on landing by fanning out the tail feathers and bending them down. The tail, then, acts both as a high-lift, low-speed airfoil and as a brake.

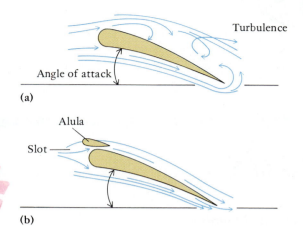

(a)

(b)

Figure 39.1
Air flow across a wing becomes turbulent when the angle of attack is increased (a), but slots such as that formed by the alula increase the air speed, reduce turbulence, and maintain lift (b).

The simplest kind of flight is **gliding**, in which the wings provide lift and the forward motion comes from falling through the air. Altitude is lost in a glide of this type, but altitude can be maintained or even increased if the bird

Figure 39.2
Air speed is low as a red-shouldered hawk, *Buteo lineatus*, lands, but lift is increased by several slots in the wing: (1) the alulae, (2) separation of feathers near the wing tip, (3) separation of feathers on the leading edge of the wing, and (4) separation of some feathers on the upper surface of the wing.

also **soars**. Land birds, such as the turkey vulture or the osprey, circle and fall within a rising column of warm air or in an air current that is deflected upward by a bluff (Fig. 39.3 a and b). Birds that engage in **static soaring** of this type have relatively short, broad wings that enable them to maneuver easily in the capricious air currents. Such wings have a low **aspect ratio** (wing length/wing width). Flight is slow so the wings need a large surface area to provide a lift adequate to support the bird. This is called a low **wing-loading ratio** (weight of bird/surface area of wing). Additional lift is generated by considerable slotting of the wing, particularly near the tip. Oceanic birds engage in **dynamic**

Figure 39.3

Soaring. (a) Static soarers maneuver in rising warm air masses or in air that is deflected upward by a cliff or other large obstacles. (b) The Galapagos hawk, *Buteo galapagoensis*, and other static soaring birds have short, broad wings with prominent slots between the feathers at the wing tip. (c) Oceanic birds, such as the great frigate bird, *Fregata minor*, are dynamic soarers with long, narrow and pointed wings. (d) Dynamic soarers utilize the increasing velocity of the air above the ocean surface.

Wind velocity

(a)

(b)

(c)

(d)

soaring, which makes use of the increase in air speed with increasing elevation above the ocean surface. Friction with the ocean causes air speed to be slowest at the ocean surface (Fig. 39.3c and d). Starting at a high elevation, these birds glide rapidly downward with the wind. Just above the ocean, they wheel into the wind and use the momentum gained in the glide to regain altitude. As they gain altitude, they encounter increasingly fast air speeds that in turn generate additional lift. In this way, the birds regain their original altitude. Since they are making use of air speed for much of their lift, dynamic soarers have higher wing-loading ratios than static soarers. They also have long, narrow wings (and a high aspect ratio) with little slotting. This reduces the tip vortices and keeps the vortices far apart.

Flapping flight is more complex and energetically expensive than gliding or soaring because the wings are not stationary. One way of viewing flapping flight is illustrated in Figure 39.4. The wings are extended and move downward and forward during the downstroke. They are also inclined from the horizontal plane, with their leading edge lower than the trailing edge. This changes the direction of the local lift and drag forces acting on each part of the wing, giving the lift force a forward component and also reducing the retarding component of the drag force. In this way the wing (or at least the part distal to the wrist, which inclines from the horizontal to a greater extent than the rest of the wing) has a propeller effect. Many of the individual flight feathers on the distal part of the wing are also separated, and to some extent they act as individual propellers. On the upstroke the wing is flexed and moves upward and backward. This is simply a recovery stroke and generates no useful aerodynamic forces. Other lifting and propelling forces appear to be generated by the reaction of the air movements caused by the wings (Focus 39.1). In all types of flight, the tail helps to support and balance the body and is used as a rudder.

The **hovering flight** of hummingbirds is a type of flapping flight. A hummingbird holds its body nearly vertically when hovering over a flower to gather nectar (Fig.

Figure 39.4

Flapping flight. (a) Stages in a wing cycle showing how the forward and downward tilt of the wing on the downstroke generates a forward thrust. (b) The little owl, *Athene noctua*, in flight. The stages shown from left to right are (1) end of downstroke, (2) beginning of upstroke, (3) end of upstroke, (4) beginning of downstroke, (5) middle of downstroke.

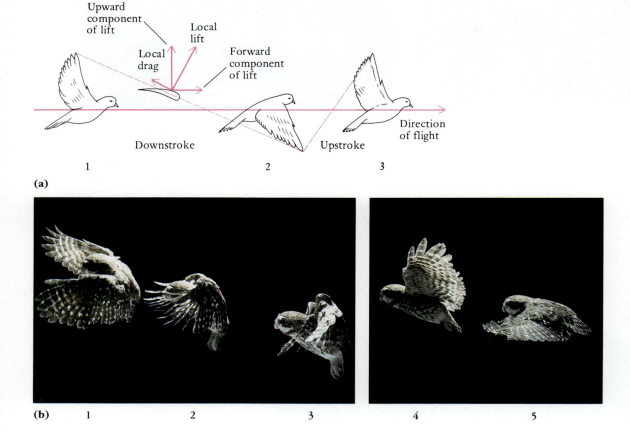

J.M.V. Rayner of the University of Bristol has proposed that the primary lift and propelling forces in flapping flight are generated not so much from the local forces acting on the wings as from the reactions of air motions generated by the wings against the bird (see Focus fig.). As the air stream flows across the wings, some of the air leaving from beneath their undersurface is drawn into the reduced pressure area above them. This air rolls up as a vortex that is shed from the wing tips. Since the wing tips come close together at the top of the upstroke and at the bottom of the downstroke, the vortices at

the wing tips come close together and fuse at these points. The vortices leave the bird at the bottom of the downstroke as a vortex ring. A smoke ring is a familiar example of a vortex ring. Circulation of air in the vortex ring itself induces a downward flow of air through the center of the ring. The ring is drawn downward and backward with a certain momentum (mass times velocity). In accordance with Newton's third law of motion, there is an opposite and equal reaction to the shedding vortex rings that pushes upward and forward against the bird, thereby generating lift and thrust.

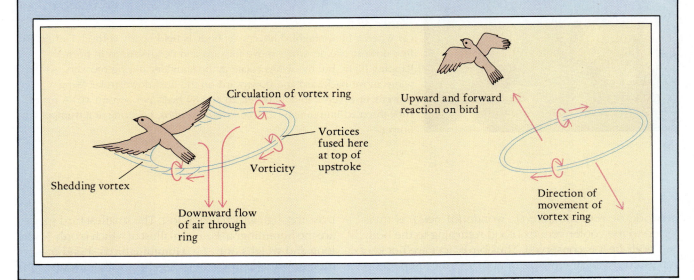

Circulation of vortex ring

Upward and forward reaction on bird

Vortices fused here at top of upstroke

Vorticity

Shedding vortex

Downward flow of air through ring

Direction of movement of vortex ring

39.5). Lift is generated in one downstroke in the usual way, by moving the wings downward and forward. After recovery the wings rotate at the shoulder joint so their dorsal surface pushes backward and downward upon the air in the next downstroke.

ADAPTATIONS OF BIRDS

Most features of birds can be related directly or indirectly to flight. They are adapted structurally and functionally to provide a high energy output in a body of low weight.

Endothermy

The flight of birds requires a very high energetic output, and they have evolved true endothermy. Their level of metabolism is many times that of reptiles. Heat is produced

internally, and its loss is controlled at the body surface. Insulation is provided by subcutaneous fat and by feathers. Feathers evolved from the horny scales of reptiles (p. 203), but scales are retained on parts of their legs, on their feet and, in modified form, as a covering for their beaks. Water, a very good conductor of heat, is prevented from penetrating the feathers by an oily secretion produced by the uropygeal gland (Gr. *oura*, tail + *puge*, rump) located on the back near the tail base (see Fig. 39.11). When a bird preens, or draws the feathers through its bill, it spreads an oily secretion from this gland over the feathers. Water fowl have very large uropygeal glands.

Heat loss at the feet is reduced in many birds by a vascular countercurrent mechanism. Arteries carrying blood down the legs break up into a network of small vessels that are entwined with veins returning from the feet. Heat flowing peripherally in the arterial blood of this

Start of
first downstroke

Downstroke
1

Recovery

Downstroke
2
(Wing has
rotated at
shoulder)

(a)

(b)

Figure 39.5

(a) When a hummingbird hovers, its body is held nearly vertical. During the first downstroke, its wings move forward and downward as in normal flapping flight, but the wings rotate at the shoulder during recovery, so the wings move backward and downward on the next downstroke. Wing movement is very rapid, between 20 and 80 beats per second. (b) A male ruby-throated hummingbird, *Archilochus colubris*, hovering at a trumpet vine flower.

network, or **rete mirabile** (L. wonderful nets), is transferred to the cooler venous blood returning to the body, and body heat is conserved. Thus birds' feet are not kept very warm. At an air temperature of 18° C, the temperature of a pheasant's feet is 27° C, but body temperature remains close to 40° C.

Control of body temperature involves several interacting mechanisms. When temperatures fall, the feathers are fluffed out; this increases the thickness of the insulating layer of air trapped between them. At very low temperatures, the bird must produce more heat by raising its metabolic level, as mammals do. When excess heat is to be lost, the feathers are held closer to the body, more blood is directed through the skin (and especially to uninsulated areas, such as the legs), and panting starts. Birds have no sweat glands that would allow evaporative loss of heat. Sweat glands would be of little use in a body densely covered with feathers. These mechanisms enable birds to maintain their body temperature constant and at relatively high levels, 40° to 43° C.

The need to maintain a high level of metabolism and body temperature imposes a lower limit on bird size, for small animals have a large surface area relative to their mass through which heat is lost. The smallest bird species are the hummingbirds, the smallest of which is only 5.8 cm long and weighs only 2 g. Hummingbirds spend most of their time in the tropics, visiting the north only during the summers. To meet their energetic needs and maintain body temperature, hummingbirds must take in food (primarily nectar) equivalent to over one-half their body weight each day. Many species conserve energy on colder nights by permitting their body temperature to fall from its daytime level of about 41° C to that of the air temperature, and they enter a torpid state not unlike that experienced by hibernating animals.

Feathers

More than any other single feature, feathers characterize birds. Feathers, like horny scales, are epidermal outgrowths of the integument whose cells have accumulated large amounts of keratin and are no longer living. Pigments deposited in these cells during the development of the feather, together with surface modifications that reflect certain light rays, are responsible for the brilliant coloration of some birds. Although feathers cover a bird, they fan

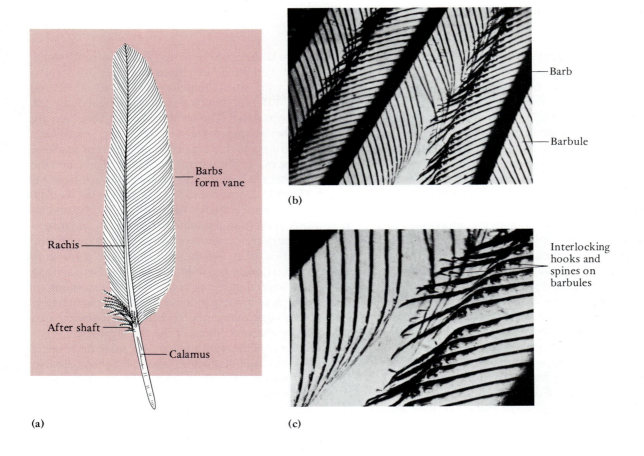

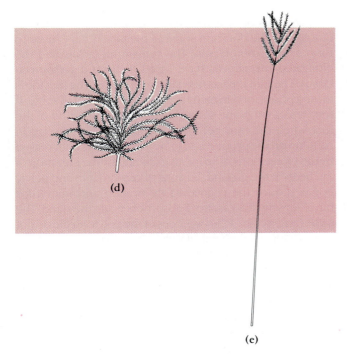

Figure 39.6
(a) A contour feather of a bird. (b and c) Photomicrographs of barbs and interlocking barbules in a contour feather. (d) A down feather. (e) A filoplume.

out in most species from localized bands, called **feather tracts**, rather than growing out uniformly from all of the body surface.

The **contour feathers** that cover the body and provide the flying surface consist of a stiff central shaft, whose base, the **calamus** (Gr. *kalamos*, quill), is embedded in a follicle in the skin (Fig. 39.6a–c). The distal part of the shaft, the **rachis** (Gr. *rhakhis*, spine), bears a **vane** composed of numerous side branches, the **barbs** (L. *barba*, beard). Each barb bears minute hooked branches, **barbules**, along its side that interlock with the barbules of adjacent barbs. If the barbs separate, the bird can preen the feather with its bill until the barbules hook together again; thus, the vane is a strong, light, and easily repaired surface ideal for both insulation and flight. In birds that have lost the power of flight, such as the ostrich, hooks are not present upon the barbules and the feather is very fluffy, functioning only in insulation. A small **aftershaft**, which is often reduced to a tuft, may arise from the distal end of the calamus.

The contour feathers on the posterior border of the arm, hand, and tail form the flying surfaces and are called the **flight feathers**. In birds that are good fliers, the rachis of a wing feather is close to the leading edge, thus thickening this part and giving the feather properties of an airfoil.

Down is a type of feather that covers young birds and is found under the contour feathers in the adults of certain species, particularly aquatic ones. It is unusually good insulation, for it has a reduced shaft and long fluffy barbs arising directly from the distal end of the calamus (Fig. 39.6d).

A **filoplume** consists of a slender, whiplike rachis bearing a few barbs at its tip (Fig. 39.6e). Filoplumes are interspersed among the flight feathers, and their follicles are richly supplied with nerve endings, which suggests that they may serve as sense organs that help regulate the movements of the flight feathers.

Bristles are stiff, vaneless feathers often found around the eyes and nose where they help keep out dirt. Some insect-catching birds, such as nighthawks, have long, slender bristles around their mouths that act as insect nets.

Since feathers are nonliving structures that fray and break, most birds shed them at least once a year during a **molt** that usually occurs after the breeding season, when birds are not under stress. Species with special breeding plumages also molt before the breeding season. Feathers are lost and replaced in a sequence characteristic for each species. The process is gradual in most species, and the birds can move about normally during molting. Many water birds, including ducks, geese, and loons, have difficulty flying when only a few flight feathers are missing. Their adaptive strategy is to complete the molt as quickly as possible. They shed the large flight feathers on their wings so rapidly that they are unable to fly for a while. Prior to the molt, they retreat to sheltered bodies of water where food is plentiful and where they can escape enemies by swimming.

Skeleton

Many adaptations for flight are apparent in the skeleton of birds (Fig. 39.7a). Among the most important is reduction of weight in the skeleton, thus reducing the effort needed to sustain flight. Bird bones are thin, hollow, and very light. Extensions from the lungs enter many bones, and these bones are described as **pneumatic** (Gr. *pneuma*, air). The skeleton of a frigate bird having a wingspan of over 2 m weighs only 115 g, which is less than the weight of its feathers! The skeletons of all birds weigh less in relation to their body weight than do the skeletons of mammals. Many bird bones are strengthened by internal struts of bone arranged in a manner similar to the trusses inside the wing of an airplane (Fig. 39.7b).

The skull is notable for the large size of the cranial region, the large orbits, and the toothless beak. A toothless beak reduces weight in comparison with toothed and heavy jaws. The neck region is very long, and the cervical vertebrae are articulated in such a way that the head and neck are very mobile. Since the bird's bill is used for feeding, preening, nest building, and defense, freedom of movement of the head is very important. The trunk region, in contrast, is shortened, and the trunk vertebrae are firmly united to form a strong fulcrum for the action of the wings and a strong point of attachment for the pelvic girdle and

hind legs. In the pigeon, 13 of the posterior trunk, sacral, and caudal vertebrae are fused together to form a **synsacrum**, with which the pelvic girdle is fused. Several free caudal vertebrae, which permit movement of the tail, follow the synsacrum. The terminal caudal vertebrae are fused together as a **pygostyle** (Gr. *puge*, rump + L. *stylus*, a writing instrument) that supports the large tail feathers.

The last two cervical vertebrae of the pigeon and the thoracic vertebrae bear distinct ribs. The thoracic basket is firm yet flexible. Extra firmness is provided by the ossification of the ventral portions of the thoracic ribs and by posteriorly projecting **uncinate processes** (L. *uncinus*, a hook) on the dorsal portions of the ribs, which overlap the next posterior ribs. Flexibility, needed in respiratory movements, is made possible by the joints between the dorsal and ventral portions of the ribs. The **sternum**, or breastbone, is greatly expanded and, in all but the flightless birds, has a large midventral **keel** that increases the area available for the attachment of the flight muscles.

The bones of the wing are homologous to those of the pectoral appendage of other tetrapods, but those distal to the wrist have been greatly modified. Three short fingers arise from a fused **carpometacarpus**. The most anterior of these (digit A in Fig. 39.7a) supports the alula. The pectoral girdle, which supports the wing, consists of a narrow dorsal scapula, a stout coracoid extending as a prop from the shoulder joint to the sternum, and a delicate clavicle, which unites distally with its mate of the opposite side to form what is commonly called the wishbone.

The legs of birds resemble the hind legs of bipedal dinosaurs. The tibia and some of the tarsals have fused to form a large **tibiotarsus**. The remaining tarsals and the elongated metatarsals have fused to form a **tarsometatarsus**. The fifth toe has been lost in all birds. The first toe is turned posteriorly in many species and is absent in some. It serves as a prop and increases the grasping action of the foot when the bird perches. The efficiency of the leg in running on the ground and jumping at takeoff is increased by the elongation of the metatarsals and by the elevation of the heel off the ground. Fusions of certain limb and pelvic bones reduce the chance of dislocation and injury, for birds' legs must act as shock absorbers when they land. The pubes and ischia of the two sides of the pelvic girdle do not unite to form a midventral pelvic symphysis as they do in other terrestrial vertebrates. This permits a more posterior displacement of the viscera, which, together with the shortened trunk, shifts the center of gravity of the body over the hind legs.

Muscles

The strong and intricate movements of the wings and the support of the body by a single pair of legs entail numerous modifications of the avian muscular system. The flight muscles include the **pectoralis**, which originates on the

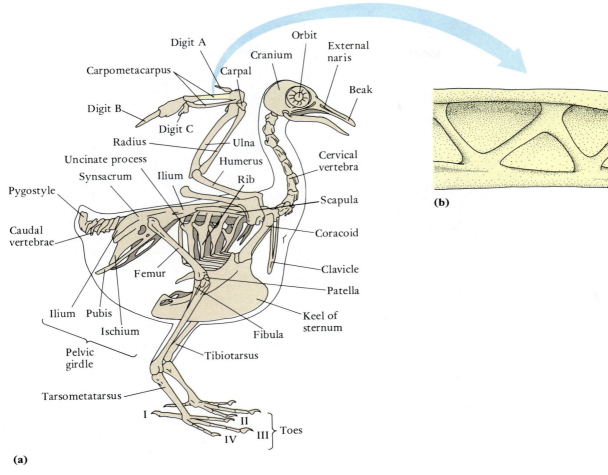

Figure 39.7
(a) The skeleton of a pigeon. The distal part of the right wing has been omitted to
show the trunk skeleton. (b) A longitudinal section of the metacarpal bone
showing internal trussing that is similar to that in an airplane's wing.

sternum and inserts on the ventral surface of the humerus.
It is responsible for the powerful downstroke of the wings
(Fig. 39.8). One might expect that dorsally placed muscles
would be responsible for the recovery stroke, but instead
another ventral muscle, the **supracoracoideus**, is responsible for the upstroke. The origin of the supracoracoideus is
on the sternum dorsal to the pectoralis. Its tendon passes
through a pulley-like canal in the pectoral girdle near the
shoulder joint and inserts on the dorsal surface of the
humerus. These two muscles are exceptionally large and
together make up as much as 25 to 35% of body weight in
birds that are powerful fliers. In ducks and other birds that
fly a great deal, the flight muscles consist mostly of aerobic
slow phasic fibers, which are rich in muscle hemoglobin
(myoglobin) and dark in color. In chickens and other birds
that beat their wings rapidly but intermittently, the flight
muscles are primarily fast phasic or glycolytic fibers and
are whitish in color.

Muscle strength is roughly proportional to muscle
mass (p. 221). This imposes an upper limit on the weight
of flying species because, as birds become larger, muscle
mass increases at a faster rate than wing surface area. Large
birds have compensated for this by evolving relatively
larger wings than have small species, but a point is
reached, at a weight of about 15 kg (33 lb), when the mass
of the muscles needed to move increasingly large wings is
simply more than can be sustained. The great bustards,
swans, and condors all approach this upper limit, as did the
large extinct pterosaurs.

Sense Organs and Nervous System

In animals that spend much of their life off the ground, the
sense of smell is less important than many other senses,
and the olfactory organ and olfactory parts of the brain are
reduced in most species of birds. Nevertheless, carrion

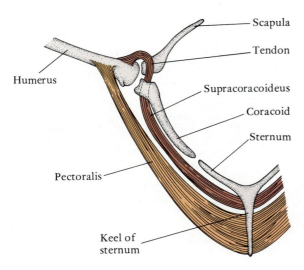

Figure 39.8

A diagrammatic cross section through the shoulder region and sternum of a bird showing how ventrally placed muscles can both lower and raise the wing.

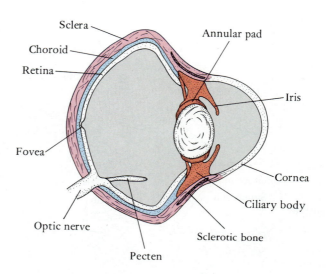

Figure 39.9

The eye of an owl.

feeders and ground-dwelling species have a well-developed sense of smell. The nocturnal kiwi (see Fig. 39.19b) of New Zealand has nostrils at the tip of a long bill and finds earthworms by smelling them as it probes the ground. Sight is very important to flying animals, thus the eyes of birds are relatively large, constituting 15% or more of the weight of the head. (They are only 1% of head weight in humans.) Color vision is well developed. Rods and cones are packed more closely in the bird retina than in the mammalian retina, making the visual acuity of birds—the ability to distinguish objects as they become smaller and closer together—several times greater than that of humans. The fovea (p. 428), where cones are particularly abundant, is the area of greatest acuity. Hawks and some other predaceous species have an extra laterally placed fovea in addition to the central one. The avian eye resembles that of reptiles in containing a peculiar vascular fold, the **pecten**, that probably has a nutritive function (Fig. 39.9).

Birds can also accommodate or change focus very rapidly, since they must change quickly from distant to near vision as they maneuver among the branches of a tree or swoop to the ground from a considerable height. To focus on a near object, the ciliary body contracts and squeezes upon an annular pad around the periphery of the lens. Lens thickness increases. Owls, hawks, and some other species also focus with small muscles that extend from a ring of peripheral **sclerotic bones** to the lens and pull it forward, and with other muscles that can change the curvature of the cornea.

The placement of the eyes on the head correlates with the birds' mode of life. Ducks, for example, have laterally placed eyes and can see behind themselves as well as forward. The eyes of hawks and owls are directed forward,

and those of bitterns, a species that searches for food in marshes, are directed downward.

The sense of hearing, too, is highly developed in most birds, as one would expect from the importance of songs in the behavior of many species. Although the cochlea of their ear is not as long as it is in mammals, experiments have shown that birds detect as wide a range of frequencies as humans do, and their ability to detect rapid changes in frequency is greater.

Birds have large brains in which the cerebrum, optic lobes, and cerebellum are particularly well developed (see Fig. 39.11). The large cerebrum results from the enlargement of a deeply situated mass of gray matter rather than from an enlarged cortex, which makes up much of the brain of mammals. The cerebral cortex of birds is thin, and removal of it has little effect on behavior. On the other hand, removal of parts of the gray matter within the cerebrum seriously affects eating, locomotion, vocalization, and reproductive behavior.

Digestion

Birds eat a variety of high-caloric foods, primarily insects and other arthropods, small vertebrates, fruits, and seeds rich in digestible organic foods. They seldom eat bulky, low-caloric foods such as leaves and grass, which contain a high proportion of cellulose and other materials that are difficult to digest. The structure of the bills of birds is highly specialized for the type of food they eat (Fig. 39.10).

The compact and efficient avian digestive system reduces weight and allows for the processing of the large volume of food needed to sustain a high metabolic rate. Pigeons, finches, game birds, and similar seed- and grain-eating species have a **crop** that has developed from the

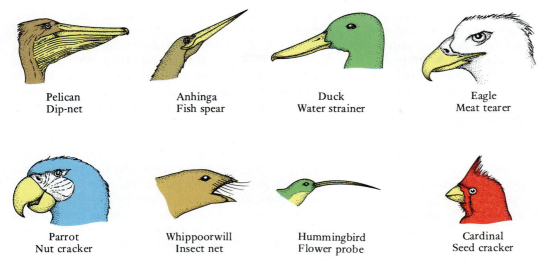

Figure 39.10
Examples of the specializations of bills for different types of feeding.

lower end of the esophagus (Fig. 39.11). Seeds are temporarily stored here and softened by the uptake of water. Food is mixed with peptic enzymes in the **proventriculus** (Gr. *pro*, in front of + L. *ventriculus* stomach), which is homologous to the proximal part of the stomach into which the esophagus leads in other vertebrates, and then passes into the **gizzard** (Old Fr. *gezier*, cooked entrails), the highly modified distal part of the stomach charac-

Figure 39.11
The major internal organs of a pigeon as shown in a lateral dissection.

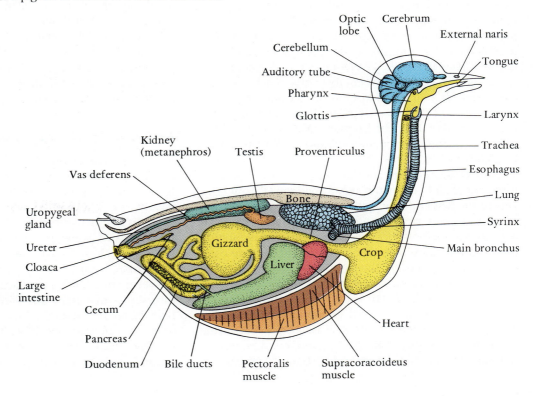

terized by thick muscular walls and modified glands that secrete a horny lining. Many species swallow small stones or grit that lodge in the gizzard and aid in grinding the food to a pulp and mixing it with the gastric juices. The gizzard is not as large in owls and other carnivorous species as it is in seed-eating species. It acts as a trap in many carnivorous species, preventing sharp bones, hair, and similar indigestible material from entering the intestine. This material is then regurgitated as pellets. Digestion continues in the intestine, which is relatively shorter and not as clearly differentiated into small and large intestines as in mammals. Villi are present on the lining of the small intestine. A pair of small ceca are present in some species at the junction of small and large intestines. The intestine enters the cloaca.

Gas Exchange

The lungs of birds are comparatively much smaller than those of mammals, but they are more efficient because air flows through them in one direction rather than back and forth. One-way air flow through the lungs functions to maintain a much greater concentration of oxygen at the epithelial exchange surfaces than in other terrestrial vertebrates that ventilate their lungs bidirectionally. A unidirectional flow is made possible by a unique pattern of airways through the lungs that connects with a system of anterior and posterior **air sacs** (Fig. 39.12a). The air sacs act as bellows and extend into many parts of the body, some entering the bones through small openings. Each **main bronchus** extends through the lungs to the posterior air sacs (Fig. 39.12b). A branch of the main bronchus also leads to groups of small, parallel passages, the **parabronchi**, that make up much of the lung. Minute branching and anastomosing **air capillaries** extend from the parabronchi (Fig. 39.12c). Distances are very short, and gases diffuse easily between the parabronchi and air capillaries. The air capillaries are surrounded by blood capillaries, and it is here that gas exchange with the blood occurs. Another bronchial branch leads from the parabronchi to the anterior air sacs and back to the main bronchus (Fig. 39.12b).

Two cycles of inspiration and expiration are required for a unit of air to move through the system (Fig. 39.12b). During the first inspiration the sternum is lowered and the air sacs expand, drawing air directly through the main bronchus to the posterior sacs. During the first expiration the sternum is raised and the posterior air sacs are compressed, forcing air into the parabronchi from whence it diffuses into the air capillaries. On the second inspiration the air in the parabronchi is pulled into the expanding anterior sacs. On the second expiration the air in the anterior sacs is expelled to the outside. Notice that during inspiration both sets of air sacs expand, but they receive different types of air: the posterior sacs receive oxygen-rich air from the outside; the anterior sacs, oxygen-depleted air from the lungs. During expiration both sets of sacs are compressed but send air to different places: the posterior sacs, to the parabronchi; the anterior sacs, to the outside. Thus with each expiration and inspiration air is drawn anteriorly through the parabronchi, maintaining a high oxygen concentration at the respiratory surface between the air capillaries and blood.

Because air flows through the lungs in only one direction, the air at the gas exchange surfaces has a higher oxygen content than is the case in the blind alveoli of a mammal's lung. Birds can obtain adequate oxygen even when flying at high altitudes, where the partial pressure of oxygen is low. In an experiment simulating an altitude of 20,000 feet, a sparrow was able to fly but a similar-sized mouse was barely able to crawl owing to the great reduction in available oxygen.

The air sacs also have other advantages. To the extent that they enter the bones and replace marrow, they lighten the bird. They also provide a large surface area through which water evaporates, making them important in the evaporative dissipation of body heat.

A mechanism for the production of sounds is associated with the air passages. Membranes are set vibrating by the movement of air in a **syrinx** (Gr., shepherd's pipe) at the posterior end of the trachea (Fig. 39.11). Muscles associated with the syrinx vary the pitch of the notes.

Circulation

The circulatory system of birds is also very efficient. As in mammals, the heart is relatively large and completely divided into left and right sides, keeping oxygen-depleted and oxygen-rich blood from mixing. Most birds are small and their heart beats very rapidly, up to 400 to 500 times per minute in a sparrow. This is comparable to the heart beat of a mammal of similar size. Blood vessels supplying the flight muscles are very large.

Excretion and Water Balance

Birds, in common with other amniotes, have metanephric kidneys, but the urinary bladder found in reptiles has been lost in birds, possibly as one adaptation that reduces body weight (Fig. 39.11). The high rate of metabolism of birds requires a large number of kidney tubules, yet water must be conserved. Some, but not all, of the kidney tubules of birds have loops of Henle that enable water to be reabsorbed from the tubules, as it is in mammals (p. 359). However, birds cannot produce as concentrated a urine in this way as mammals can. Birds conserve most of their body water by excreting 75 to 90% of their nitrogenous wastes as uric acid (see Table 37.1, p. 849). Since most of the uric acid is secreted into the tubules rather than entering by filtration, a great deal of water need not be

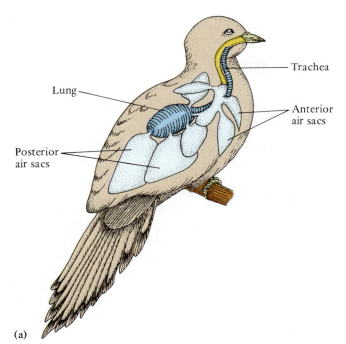

(a)

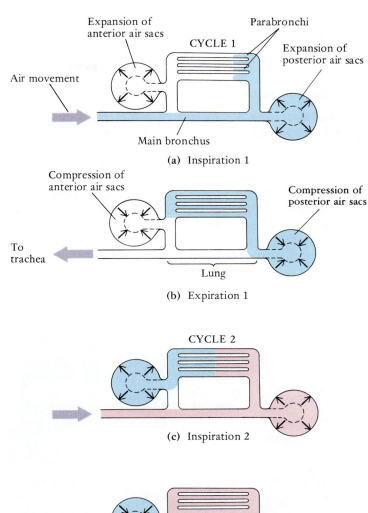

(a) Inspiration 1

(b) Expiration 1

CYCLE 2

(c) Inspiration 2

(d) Expiration 2

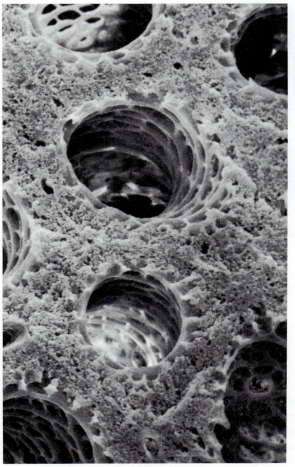

(b)

(c)

Figure 39.12

Respiration in birds. (a) The lungs and major air sacs of a bird. (b) Diagrams of the movement of a volume of air (blue) through the lungs and air sacs in two cycles of inspiration and expiration, and the entry of a new volume of air (red) during the second cycle. (c) Scanning electron micrograph of a section of a chicken's lung showing parabronchi in cross section and associated air capillaries.

filtered by the renal corpuscles. The renal corpuscles of birds are small. Water needed to carry off uric acid and other waste products is reabsorbed in the cloaca, and the uric acid is discharged as a white, crystalline material mixed with the feces. The excess salts that sea birds gain are eliminated by salt-excreting glands that are located above the eyes and discharge into the nasal cavities (p. 364).

Reproduction and Development

All birds are oviparous, have cleidoic eggs, and they are the largest group of externally brooding animals. Fertilization is internal. A few male birds, including the ostriches, ducks, and geese, have a copulatory organ for the transfer of sperm, but sperm is transferred in most species by the two sexes briefly bringing their cloacas together. Possibly as an adaptation for weight reduction in a flying animal, birds lose the right ovary and oviduct during development. The remaining ovary is small, except during the reproductive season, when it enlarges greatly as the eggs accumulate their store of yolk. The absence of a pelvic symphysis facilitates laying large, fragile eggs.

As is the case in many other animals, a great deal of the activity of birds is focused on reproduction and rearing young. Prior to the breeding season, the males of most species establish a mating and nesting territory, an area that the male will defend aggressively against the incursions of other males of the same species. The brilliant plumage and colorful songs of many males serve both to warn other males to stay away from their territories and to attract potential mates. Territorial behavior spaces the birds and prevents the disadvantages of overcrowding.

After a territory has been established and a female has taken up residence, the birds engage in species-specific courtship behavior (Fig. 39.13a). Some of these courtship displays are very elaborate. Courtship establishes that the two birds are in fact members of the same species, leads to the establishment of a strong pair bond between them, and prepares the partners physiologically for effective copulation. Ancestral birds may have buried their eggs as

(a)

(b)

(c)

Carolina Biological Supply Co.

Figure 39.13
Reproduction in birds. (a) The mating ritual of a pair of gannets in a New Zealand nesting colony. (b) A group of woven nests of the African weaver, *Pseudonibrita cabanisi*. (c) A young chinstrap penguin, *Pygoscelis antarctica*, being fed by a parent.

crocodiles do and depended upon environmental temperature to incubate them. The Egyptian plover still does this today, and many species cover their eggs with vegetation when they leave them to forage for food. However, most species build nests that may be as simple as a depression on the ground used by whippoorwills or as elaborate as the woven nests of African weaver birds (Fig. 39.13b). After eggs have been laid, adults brood and incubate them. Many bird species lose some feathers on the underside of their abdomen at this time and form **brood patches** that facilitate heat transfer from the parents' bare skin to the eggs. In gulls and many other sea and aquatic birds, the two parents share equally in brooding, but in most songbirds the female generally is the chief or sole brooder. The male brings her food.

Near the end of the embryonic period, the chick develops powerful dorsal neck muscles and a horny thickening, or **egg tooth**, on the end of its bill that it uses to break through the shell at hatching. Ducks, shore birds, chickens, and quail are examples of species in which the young are well developed at hatching, that is, they are **precocial** (L. *praecox*, ripening before its time). Their eyes are open and the birds are covered with down at hatching. Such chicks are able to locomote and follow their parents right away, and the parents help them to find food and may feed them for a short time. In contrast, other birds are **altricial** (L. *altrix*, nourisher). Their eyes are closed, and they are featherless at hatching. The parents must continue to brood and feed them until they are ready to leave the nest (Fig. 39.13c). The nestlings' first attempts at flight are often rudimentary, and they alight on the ground. The parents remain close to the young and try to scare off or distract potential predators.

Raising a brood of young is a very time- and energy-consuming activity for most birds. Some cuckoos, cowbirds, and a few other species avoid the energy-consuming activity of brood raising by practicing **brood parasitism**. These species lay their eggs in the nests of other birds. The foster parents usually are tricked into accepting the eggs and treat them as their own, allowing the parasitic species to avoid the major metabolic costs and dangers associated with nesting and raising young. The European cuckoo helps to ensure this by laying eggs that resemble those of the host species. In a further adaptation for parasitism, the young cuckoo hatches before the host's own eggs and then rolls the unhatched eggs out of the nest, thereby insuring that it will receive the exclusive care of the host parent that does not recognize the cuckoo as an intruder.

MIGRATION AND NAVIGATION

The energy costs of reproduction are very high. Birds need a large territory in which to find enough food to rear their young, and they must have the ability to find their way back to their nesting area. Many species take advantage of their powers of flight to migrate from winter quarters in temperate or tropical regions to breeding areas in the north, where the days are long in the summer and where a large food supply, for which there are few other competitors, develops for a few months. Birds can establish territories with a minimum of effort, and they have long hours of daylight to obtain food. As conditions become inclement, the birds return to winter quarters in warmer southern ranges. Migration also has the advantage of reducing predation; predator populations do not have time to build up to large numbers when the birds move out at intervals. But there are hazards as well as advantages to migration, for many migrants are caught in storms and perish.

Many factors interact in disposing certain birds to migrate. Increasing day length in the spring of the year in the northern hemisphere is a major factor, for it initiates a series of reactions that lead to an increased secretion of gonadotropic hormones and an increasing size of the gonads (p. 445). There is also a rapid increase in fat deposits and an increased activity, or restlessness, of the birds. Hereditary factors, including an innate rhythm that approximates a year in length, also appear to be involved. In northern California, certain populations of the Oregon junco accumulate fat and become restless prior to their spring migration, whereas other populations, living under the same environmental conditions, do not show signs of migratory behavior and remain sedentary. Once birds are in a migratory condition, favorable weather and other external factors trigger the onset of migration.

Most songbirds migrate at night, stopping to feed and rest during the day. Some may fly several hundred kilometers during a single night, but then may rest for several days. Many larger birds, including hawks and herons, migrate by day. Ducks and geese may migrate either by night or by day. Flight groundspeeds of migrants, as determined by radar tracking, range from about 30 to 70 km per hour. Many species tend to follow the advance of certain temperature lines, or **isotherms** (Fig. 39.14) so their arrival at certain locations is affected by the weather. Other species have a nearly calendar-like regularity. Swallows reach Mission San Juan Capistrano in California on nearly the same date each year.

The length of migration and the route taken are consistent for each type of bird but vary from species to species. The Canada goose winters in the United States from the Great Lakes south, breeds in Canada as far north as the Arctic coast, and migrates along a broad front between the two areas. Other species follow narrower paths, and in some cases these are different for the spring and fall migrations. The golden plover breeds in the Arctic (Fig. 39.15). In the fall it flies south through the Canadian Maritime Provinces and northern New England before crossing the Atlantic to its winter quarters in southeastern South America. It flies north in the spring through Central America and the prairies of the United States and Canada.

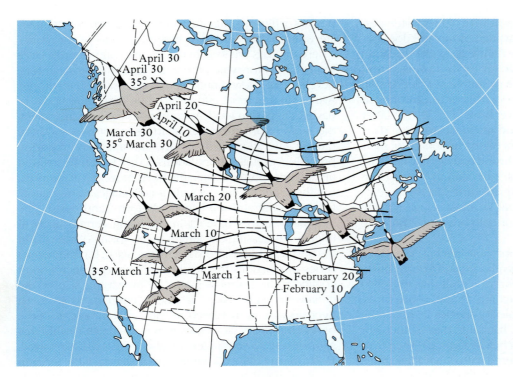

Figure 39.14
The northward migration of the Canada goose keeps pace with the advance of spring, following approximately the northern displacement of the 2° C (35° F) isotherm.

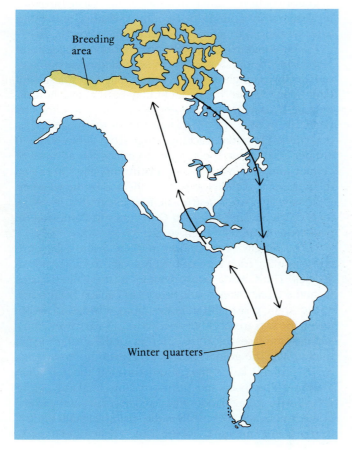

Figure 39.15
The golden plover has different routes on its spring and fall migrations, following available food supplies.

Availability of food is undoubtedly a factor in determining this migration route. The birds fatten on late summer berries in the Maritime Provinces and New England prior to their flight across the Atlantic, but this route is inclement and devoid of appropriate food in the spring. A route through the grasslands of mid-America is better at this time of year. The longest migration route is that of the arctic tern. This species breeds in the Arctic, then migrates to its winter quarters in the South Atlantic, about 20,000 km away.

The ways in which birds navigate and find their way during migration remain an intriguing, incompletely solved problem of animal behavior. Obviously, the birds must have some notion of their destination. There must be some feature of the environment that is related to the destination, and the bird must have some way of perceiving this feature. It is likely that some combination of methods is used in navigating and that the specific choices vary among bird species or even within a species according to the bird's experience and environmental conditions. Certain species have an innate sense of direction and migrate along a predetermined compass course, but a correction can be made in the course if it is needed. This was demonstrated by banding thousands of European starlings in Holland just before their fall migration and then transporting them to Switzerland, about 600 km southeast, where they were released. Young birds continued in their "innate compass" migratory direction toward the south and ended up far to the east of their normal winter quarters. Experienced older birds made an appropriate correction for their displacement and ended in their normal winter range.

Many species use visual landmarks to some extent when migrating or in finding their way home if they have been displaced. The visual clues used by birds include topographical features, such as coastlines and mountain ranges, and ecological features, such as deserts, prairies, and forests.

Some birds use the position of the sun as a compass to find their way. Nesting lesser black-backed gulls have been displaced many miles from their nests, then released and their direction noted as they vanished from view. Most birds took off in the direction of home on sunny days but became disoriented on cloudy days.

Other studies show that night migrants use the star pattern to navigate. Spring migrants heading north were caught and placed in a cage in a planetarium where they oriented themselves to the north of the artificial sky regardless of true north. The use of celestial features, sun or stars, in navigation requires that birds have a keen sense of time and an internal clock of some sort because celestial clues for direction vary with the time of day. This was demonstrated by subjecting birds to artificial light-dark cycles out of phase with natural day and night. Their clocks become reset. In one experiment, starlings with their clocks advanced 6 hours assumed it was dawn when in reality it was midnight. Their flight direction was shifted about 90° counterclockwise from what it should have been.

Investigators have shown that some species can perceive the earth's or other magnetic fields and that magnetic fields might be used in navigation. Caged European robins caught during migration oriented appropriately to their migratory direction in the absence of any celestial clues, but their choice of direction was altered by subjecting them to artificial magnetic fields. We do not know how birds sense magnetic fields.

EVOLUTION AND ADAPTIVE RADIATION OF BIRDS

The Origin of Birds

Birds most certainly evolved from early archosaurian reptiles. The earliest known species, *Archaeopteryx lithographica*, comes from Jurassic deposits in Germany, and shows a mosaic of reptilian and avian features. Three

nearly complete specimens and several partial ones have been found. Some are preserved with remarkable clarity in fine-grained limestones that resemble lithographs (hence the species name). *Archaeopteryx* was about the size of a pigeon (see Fig. 2.16). Its skeleton is reptilian in having relatively thick-walled bones, toothed jaws, no fusion of trunk or sacral vertebrae, and a long tail. Birdlike tendencies are evident in the enlarged orbits, in some expansion of the braincase, and particularly in the winglike structure of the hand. As in modern birds, the hand is elongated, and only three fingers are present; however, there is no fusion of carpal and metacarpal bones and each finger bears a claw. All but one of the specimens show clear impressions of feathers. The unfeathered specimen was at first misidentified as a small dinosaur. *Archaeopteryx* probably could not fly well, since it had a large body relative to its wing area and since the sternum, to which flight muscles attach in modern species, was absent or too poorly developed to be fossilized. However, it must have engaged in powered flight to some extent, because the location of the rachis close to the leading edge of the wing feathers is characteristic of the feathers of flying birds rather than of those of flightless birds, in which the rachis is close to the center of the feather. These most primitive birds are placed in the subclass **Archaeornithes**.

Archaeopteryx probably evolved from a group of early saurischian dinosaurs with which it shared many derived features (Fig. 39.16). These small saurischians were bipedal creatures with a long tail, a long flexible neck, and arms bearing only three clawed fingers. It is likely that the ancestors of birds were becoming more active and possibly endothermic, and feathers may have first been of selective value in helping to conserve body heat. Some paleontologists argue that bird ancestors were originally arboreal climbers. Later, the feathers along the posterior edge of the forelimb elongated as an adaptation for gliding. Others believe that bird ancestors ran rapidly along the ground, and feather enlargement enabled them to use their forelimbs as nets to seize insects (Fig. 39.17). Further enlargement of wing and tail feathers may have conferred stability in running rapidly, jumping, and rudimentary gliding from low branches. Aerodynamic analysis suggests that these feathers would be very important in providing stability in a small running and jumping biped. Feathers and wings may have evolved in this manner, acquiring new functions along the way: first as insulation, then for insect catching and stability, and finally for flight. They would have been adaptive at all stages of their evolution.

Cretaceous Toothed Birds

The next fossil birds, found in Cretaceous deposits, had lost the long reptilian tail and evolved a well-developed sternum. A true pygostyle had not yet evolved, and teeth

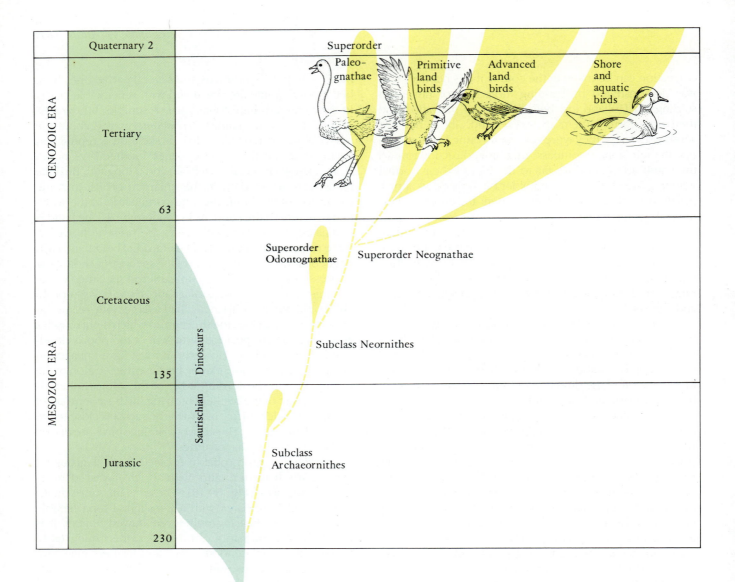

Figure 39.16
Phylogeny and geological distribution of the major groups of birds.

were still present. *Hesperornis* (Fig. 39.18) was a large diving species with powerful hind legs and vestigial wings. *Ichthyornis* was a flying species about the size of a small sea gull. Although they are placed in the subclass **Neornithes** along with modern birds, the more primitive nature of these Cretaceous species is recognized by placing them in a distinct superorder, the **Odontognathae**.

Ratites

Bird bones are very fragile and do not preserve well, so the fossil record of birds is sparse. Nonetheless, it is clear from fossils that have been found that all later birds lacked teeth and shared other features that indicate a common ancestry. Birds radiated widely in the late Cretaceous and early

Tertiary periods (Fig. 39.16). By mid-Tertiary, 15 to 20 million years ago, most contemporary groups had evolved. We cannot be certain of the interrelationships of the 28 living and extinct orders, but many ornithologists sort them into two superorders, the **Paleognathae** and **Neognathae**. The size and configuration of bones in the roof of the mouth of paleognathous birds remain archosaurian, and little motion occurs between the bones; the palate of neognathous birds contains several joints and is more flexible.

Most paleognathous birds are flightless and adapted for a terrestrial life. These flightless species are often called **ratites** (L. *ratis*, raft) because of their keel-less flat sternum. The group includes the extinct moas of New Zealand, the kiwis of New Zealand, the cassowaries and emus of Austra-

(a)

(b)

Figure 39.17
Some paleontologists hypothesize that the ancestors of birds were running bipedal predators that caught insects with their scaled hands (a). The enlargement of these scales into feathers on the hands and tail would improve stability and the hand could more easily grasp insects (b).

Figure 39.18
Reconstruction of *Hesperornis*, a toothed and flightless diving bird of the Cretaceous.

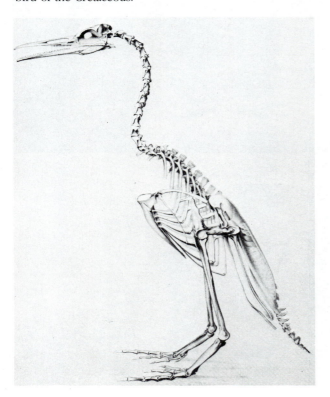

lia, the ostriches of Africa, and the rheas of South America (Fig. 39.19). Their height ranges from 25 cm for the kiwi to 3 m for an extinct moa (*Dinornis*). The legs of ratites are large and powerful, their feet are adapted for running on the ground, and their wings are vestigial. Although their sternum lacks a keel, its breadth indicates that they evolved from flying ancestors. Members of one paleognathous group, the tinamous of tropical America, resemble quail. A tinamou is a ground-dwelling species, but does have a keeled sternum and can fly short distances. A tinamou-like species may have been ancestral to the flightless ratites. We cannot be certain, however, that the shared features of ratites reflect a common origin; they could represent convergent adaptations for terrestrial life among unrelated primitive birds.

Land Birds

The flexible palate of neognathous birds is a shared, derived feature that indicates a common ancestry. Cladistic analysis suggests that neognathous birds diverged early into two large assemblages: one group adapted for life on the land; the other, for life near the shore and in the water (Fig. 39.16).

(a)   **(b)**

Figure 39.19
Flightless ratites. (a) Ostriches, *Struthio camelus*, attain a height of 2.5 meters.
(b) The great spotted kiwi, *Apteryx haastii*, of New Zealand is a smaller bird that
reaches a height of only 35 cm.

Figure 39.20
Primitive terrestrial birds. (a) A roadrunner, *Geococcyx californianus*. (b) A male
ruffed grouse, *Bonasa umbellus*, on a drumming log. (c) A black kite, *Milvus
migrans*, carrying a fish. (d) A group of white-backed vultures, *Gyps africanus*, at a
kill on the Serengeti plain.

(a)   **(b)**

(c)   **(d)**

The cuckoos (Fig. 39.20a) and roadrunners appear to be among the most primitive of neognathous land birds, but other primitive groups include pigeons, grouse, parrots, and the hawks and eagles. Most of these birds fly very well. The grouse (Fig. 39.20b), chickens, and similar birds (order **Galliformes**) are primarily ground-dwelling species, although they can fly short distances. Hawks, eagles, falcons, and vultures (order **Falconiformes**) are diurnal birds of prey (Fig. 39.20c). Their toes bear strong claws, or talons, with which they seize prey that they then tear apart with strong, hooked beaks. Vultures are scavengers. They often gather in great numbers near a lion kill awaiting their turn at the carcass (Fig. 39.20d).

An assortment of anatomical features, including the loss of one thigh muscle (the ambiens) that is present in reptiles and more primitive land birds, characterizes the advanced group of neognathous land birds. This group includes the kingfishers, swifts, hummingbirds, owls, woodpeckers, and the numerous songbirds. Owls (order **Strigiformes**) are a particularly interesting group. They are predaceous birds that evolved talons and hooked beaks independently of the falcons (Fig. 39.21a). They differ from falcons in being adapted for night hunting. They have large, forward-turned eyes located in facial discs of radiating feathers. The sounds of mice and other animals direct owls to their prey. The ear openings of owls are located in the facial discs and also face forward. Owls locate the prey by turning their head until the sound intensity is the same in each ear. Then they glide silently down upon it. Some of the smaller species of owls are insect eaters.

The birds most closely related to owls are another group of nocturnal insect eaters, the nighthawks and whippoorwills (order **Caprimulgiformes**). They catch insects in flight with large, gaping mouths that are surrounded by a net of bristles (Fig. 39.21b)

Among the most advanced of land birds are the perching songbirds (order **Passeriformes**). Passerines are the

Figure 39.21
Advanced land birds. (a) A long-eared owl, *Asio otus*, feeding on a mouse. (b) The whippoorwill, *Caprimulgus vociferus*; (c) the Eastern meadowlark, *Sturnella magna*. (d) A diagram of the perching mechanism of a songbird.

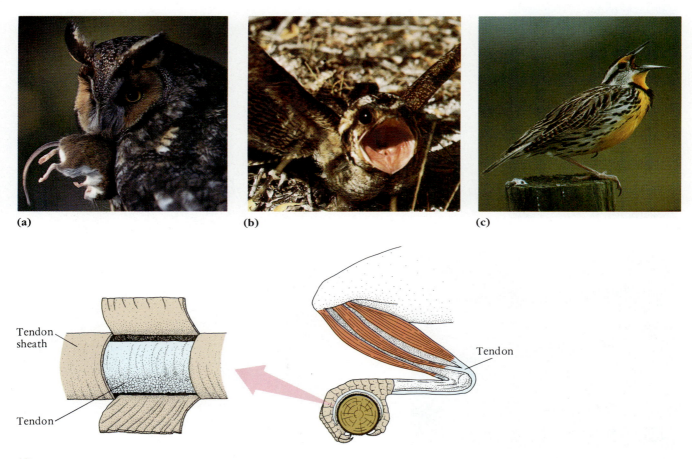

(a)

(b)

(c)

Tendon sheath

Tendon

Tendon

(d)

largest and most diverse group of birds, including over 5000 species. Swallows, wrens, thrushes, robins, warblers, sparrows, blackbirds, and bluejays are examples (Fig. 39.21c). They have diverged widely in feeding habits, eating seeds, worms, insects, and many other small animals. All have a well-formed syrinx and a foot adapted for perching. When a passerine alights and lowers itself on a branch, the tendons in its feet are automatically pulled by the weight of the body in such a way that the toes flex and grip the branch (Fig. 39.21d). The bird must exert muscular effort to raise its body and extend its foot in order to release the perch.

Shore and Water Birds

Although aquatic birds appear to have had a common origin, they adapted for shore and water life in many different ways. They probably diverged from early land birds, and the cranes, coots, and rails (order **Gruiformes**) appear to be the primitive group (Fig. 39.22a). Many cranes are terrestrial, but most live in swampy environments and have legs and feet that enable them to wade or swim. One, the jacana of the tropics, has exceedingly long toes and claws that distribute its weight over a large surface, enabling it to walk on aquatic plants (Fig. 39.22b). During the early Cenozoic, a South American rail, *Phorusrhacus*, became a large, flightless predator that stood 1.5 m in height (Fig. 39.22c). More advanced groups of shore birds include the flamingos, herons, storks, sandpipers, gulls, and terns.

Ducks and geese (order **Anseriformes**) are more aquatic than the shore birds (Fig. 39.23). They are good swimmers and divers, and, like most swimming species, have a webbed foot that includes the three front toes. The flattened bill characteristic of most species has strainers along its margin and is adapted for filtering food.

Figure 39.22
Primitive aquatic birds. (a) White-naped cranes, *Grus vipio*, in a marsh. (b) A jacana, *Actophilornis*, walking on lily pads. (c) *Phorusrhacus*, a large flightless rail from the Tertiary period, was 1.5 m tall.

(a)

(b)

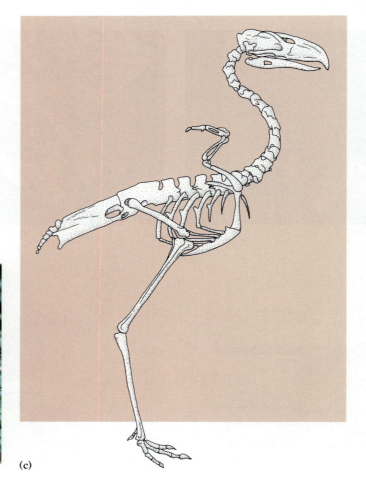

(c)

Figure 39.23
The webbed foot and flattened bill characteristic of ducks show in a male mallard, *Anas platyrhynchos.*

Figure 39.25
A Laysan albatross, *Diomedea chrysostoma*, from the Hawaiian islands shows the tubular nostril characteristic of the Procellariiformes.

Other aquatic groups are more highly adapted for life in the water. Boobies, cormorants, and pelicans (order **Pelecaniformes**) are relatively large, fish-eating birds with a characteristic throat pouch (Fig. 39.24). Their hind toe is turned forward and included in the web.

Petrels and albatrosses (order **Procellariiformes**) are characterized by a tubular nostril through which a salty secretion is discharged from their salt-excreting glands (Fig. 39.25). They are strong fliers and spend most of their lives at sea, coming ashore only to breed. The wandering albatross has a wingspan of 3.5 m, the largest of any living species. It is a soaring bird and needs strong winds to stay aloft.

Loons (order **Gaviiformes**) are chiefly foot-propelled divers. Their tail is short, and powerful hind legs with large feet are placed near the posterior end of the body. Grebes (order **Podicipediformes**) are smaller foot-propelled divers. Unlike loons, which have webbed feet, each toe of grebes bears enlarged, paddle-like lobes (Fig. 39.26).

Penguins (order **Sphenisciformes**) are a unique group (Fig. 39.27). Their wings are modified as flippers, and they figuratively fly beneath water. They probably evolved from a group similar to diving petrels, which can use their wings both in the air and under water. Antarctica

Figure 39.24
The large throat pouch of the white pelican, *Pelicanus erythrorhynchos*, is used as a dip net to catch fish.

Figure 39.26
The eared grebe, *Podiceps nigricollis.*

was their center of evolution, and most species still live there. One species has followed the cold Humboldt current along the west coast of South America as far as the Galapagos Islands, located on the equator.

Figure 39.27
Crested penguins, *Eudyptes chrysocome*, cavorting in the Antarctic sea; note their flipper-like wings.

Classification of Birds

Class Aves (L. *avis*, bird) The birds.
  Subclass Archaeornithes† (Gr. *archaios*, primitive + *ornis*, bird) Ancestral birds of the late Jurassic period. They retain many reptilian features including jaws with teeth, a long tail, unfused carpals and metacarpals, and three fingers, each bearing a claw. *Archaeopteryx*.
  Subclass Neornithes (Gr. *neos*, new + *ornis*, bird) Birds with a reduced number of caudal vertebrae; metacarpals and some carpals are fused together; the three fingers are reduced and clawless.
    Superorder Odontognathae† (Gr. *odous*, tooth + *gnathos*, jaw) Cretaceous birds retaining teeth. *Hesperornis, Ichthyornis*.
    Superorder Paleognathae (Gr. *palaios*, ancient + *gnathos*, jaw) Modern birds without teeth, but retaining a primitive archosaurian palate. The ratites.
      Order Tinamiformes (*tinamus*, a native name + L. *forma*, shape) Tinamou. Largely a ground-dwelling group, but with weak powers of flight; sternum retains a keel. Mexico to South America.
      Order Struthioniformes (L. *struthio*, ostrich + *forma*, shape) Ostriches. Large flightless birds with small wings; unkeeled sternum; head and neck largely devoid of feathers; large powerful legs with only two toes. Africa and western Asia.
      Order Rheiformes (Gr. *Rhea*, the mother of Zeus + L. *forma*, shape) Rheas. Large flightless birds with unkeeled sternum; head and neck feathered; heavy legs with three toes. South America.
      Order Casuariiformes (L. *casuarius*, cassowary + *forma*, shape) Cassowaries and emus. Large flightless birds with small wings and unkeeled sternum; long hairlike feathers with long aftershaft; heavy legs with three toes. New Guinea and Australia.
      Order Dinornithiformes† (L. *dinornis*, generic name + *forma*, shape) Moas. Largest of the flightless birds, attained a height of 3 m. Became extinct about 300 years ago. New Zealand, *Dinornis*.
      Order Apterygiformes (Gr. *apterygos*, without wings + L. *forma*, shape) Kiwis. Hen-sized flightless birds with unkeeled sternum and vestigial wings; four toes on feet; long bill with nostrils near the tip used in probing soft ground for food; nocturnal in habits. New Zealand, *Apteryx*.
    Superorder Neognathae (Gr. *neos*, new + *gnathos*, jaw) Modern toothless birds with a flexible palate. (We are following Olson [1985] in listing neognathous bird orders by phylogenetic position and major adaptive radiations rather than in the usual linear sequence from aquatic through terrestrial groups.)
The Terrestrial Radiation
      Order Cuculiformes (L. *cuculus*, cuckoo + *forma*, shape) Primitive members of the terrestrial radiation: cuckoos and roadrunners. Foot with fourth toe capable of being turned back beside the first toe. Tail long. Worldwide.
      Order Falconiformes (L. *falco*, falcon + *forma*, shape) Vultures, kites, hawks, falcons, and eagles. Diurnal birds of prey with strong hooked bill; sharp curved talons. Wordwide.
      Order Galliformes (L. *gallus*, rooster + *forma*, shape) Grouse, quails, partridges, pheasants, turkeys, and chickens. Seed- and plant-eating, largely ground-dwelling birds; short stout bill; heavy feet with short strong claws, adapted for running and scratching in the ground; wings relatively short; often sexually dimorphic. Worldwide.

Order Columbiformes (L. *columba*, dove + *forma*, shape) Pigeons and doves. Short slender bill with fleshy pad at its base overhanging slitlike nostrils; short legs. Worldwide.

Order Psittaciformes (Gr. *psittakos*, parrot + *forma*, shape) Parrots. Feet adapted for grasping, with the fourth toe capable of being turned back beside the first toe; bill heavy and hooked; often brilliantly colored plumage. Worldwide.

Order Coliiformes (Gr. *kolios*, a type of woodpecker + L. *forma*) The colies. Small birds with long tails; first and fourth toes can be turned posteriorly. Africa.

Order Coraciiformes (Gr. *korax*, raven + L. *forma*, shape) Kingfishers, trogons, mousebirds. Strong sharp bill; grasping foot; feathers often forming a crest on the head; plumage often brilliant. Worldwide.

Order Strigiformes (Gr. *strix*, owl + L. *forma*, shape) Owls. Nocturnal birds of prey with strong hooked bills; sharp curved talons; feathers arranged as a facial disc around the large, forward-turned eyes. Worldwide.

Order Caprimulgiformes (L. *caper*, goat + *malgeo*, to suck + *forma*, shape) Nighthawks, goat suckers, and whippoorwills. Twilight flying birds with small bills, but large mouths surrounded by bristle-like feathers that help net insects; legs and feet small. Worldwide.

Order Apodiformes (Gr. *apous*, without foot + L. *forma*) Swifts and hummingbirds. Fast flying birds with long narrow wings; legs and feet very small. Worldwide.

Order Bucerotiformes (Buceros, a generic name + L. *forma*, shape) Hoopoes and hornbills. A diverse group of African and South Asian birds that typically nest in holes in trees.

Order Piciformes (L. *picus*, woodpecker + *forma*, shape) Woodpeckers and toucans. Bill chisel-like (woodpeckers) or very large (toucans); the fourth toe is permanently turned posteriorly (zygodactylous foot). Worldwide.

Order Passeriformes (L. *passer*, sparrow + *forma*, shape) The perching birds and songbirds. The largest order of birds, it comprises about 60 percent of all birds and includes the flycatchers, larks, swallows, crows, jays, chickadees, nuthatches, creepers, wrens, dippers, thrashers, thrushes, robins, bluebirds, kinglets, pipets, waxwings, shrikes, starlings, vireos, wood warblers, weaver finches, blackbirds, orioles, tanagers, finches, and sparrows. Syrinx well-developed; foot adapted for perching; three toes in front opposed by one well-developed toe behind. Worldwide.

The Aquatic Radiation

Order Gruiformes (L. *grus*, crane + *forma*, shape) Primitive members of the aquatic radiation: cranes, rails, gallinules, and coots. Marsh birds; feet not webbed, but toes sometimes lobed; legs elongate in some groups; the portion of the head between the eye and nostril (the lores) feathered. Worldwide.

Order Podicipediformes (L. *podex*, rump + *pes*, foot + *forma*, shape). Grebes. Legs located far back on the body; lobate toes; reduced tail; very good divers. Worldwide.

Order Charadriiformes (Gr. *charadrios*, a cleft-dwelling bird + L. *forma*, shape) Plovers, woodcock, snipe, sandpipers, stilts, ibises, phalaropes, gulls, terns, skimmers, auks, and puffins. A diverse group of shore birds. Worldwide.

Order Phoenicopteriformes (Gr. *Phoenicopterus*, a generic name + L. *forma*, shape) The flamingos. Worldwide in tropics.

Order Anseriformes (L. *anser*, goose + *forma*, shape) Ducks, geese, and swans. Short-legged, web-footed swimming and diving birds; bill usually broad and flat with transverse horny ridges adapted for filtering. Worldwide.

Order Ciconiiformes (L. *ciconia*, stork + *forma*, shape). Storks. Long-legged and long-necked wading birds; feet broad, but usually not webbed; the area between the eye and base of the bill (the lores) usually devoid of feathers. Worldwide.

Order Pelecaniformes (Gr. *pelekan*, pelican + *forma*, shape). Pelicans, gannets, cormorants, water-turkey, and man-o-war bird. Totipalmate swimmers with four toes included in the webbed foot; tendency for the development of a throat pouch. Worldwide in tropics and subtropics.

Order Procellariiformes (L. *procella*, tempest + *forma*, shape) Albatrosses, shearwaters, fulmars, petrels, and tropic birds. Webbed feet; fourth toe vestigial; long narrow wings; hooked beak; tubular nostrils. Oceanic, worldwide.

Order Gaviiformes (L. *gavia*, a sea mew + *forma*, shape) Loons. Legs located far back on the body; webbed feet; reduced tail; long, compressed and sharply pointed bill; very good divers. North America, Eurasia.

Order Sphenisciformes (*Spheniscus*, generic name of a penguin + L. *forma*, shape) Penguins. Flightless oceanic birds with four anteriorly directed toes with a web between three of them; wings modified as paddles; excellent divers. Confined to the southern hemisphere, chiefly Antarctica, one species occurs on the Galapagos Islands.

† Extinct.

Summary

1. Adaptation for flight is the key theme in bird evolution. Birds use their wings both as airfoils to provide lift and as propellers. Some species also glide and soar.

2. Flight requires low body weight and a high energy output. The bones of birds are light in weight yet strong. The arm is modified as a wing, and the hind legs function as levers for takeoff and as shock absorbers on landing. A large keeled sternum increases the area for the origin of the powerful flight muscles used for the downstroke and upstroke of the wings.

3. The senses of sight and hearing are important in birds, but the sense of smell is reduced. The brain is large and has a well-developed cerebrum and cerebellum.

4. Birds are endothermic and maintain a high level of metabolism. Feathers insulate the body surface as well as forming the flying surfaces. Teeth are absent. Food is broken down mechanically and mixed with digestive enzymes in the gizzard. Bird lungs and air sacs are so arranged that air moves only in one direction across the respiratory surfaces, thus maintaining a high level of oxygen at the gas exchange surfaces. Oxygen-depleted blood and oxygen-rich blood are completely separated by the double circulation through the heart. Water is conserved by eliminating most of the nitrogenous wastes of metabolism as uric acid. The urinary bladder has been lost.

5. Most male birds transfer sperm to the female by means of cloacal apposition. Females have only one ovary and oviduct. Males establish a territory. When a female takes up residence, courtship and nest building begin. Usually the female broods the eggs and young.

6. Not all birds migrate, but there are certain advantages to migration, including ease of establishing territories, obtaining food, and rearing young. Migration also prevents a great increase in predator populations. Many factors interact to dispose certain birds to migrate: hereditary factors, day length, hormonal changes, fat deposition, and increased restlessness. Migratory paths are specific for each kind of bird. Birds use some combination of clues in navigation: an innate sense of direction, visual landmarks, a sun compass, star patterns, and magnetic fields. Those using celestial clues also use an internal clock.

7. *Archaeopteryx*, a Jurassic bird, had teeth, a long tail, clawed fingers, and many other reptilian features. It probably evolved from a species of small, running saurischian dinosaur. *Hesperornis* and *Ichthyornis* from the Cretaceous retained teeth but resembled modern birds in most other respects. Later birds have lost their teeth. Ostriches and other paleognathous birds retain a primitive archosaurian palate, but an increased capacity for intrapalatal movement between bones has evolved in neognathous species. Neognathous birds underwent extensive terrestrial and aquatic radiations and now occupy most habitats on land, shore, and water.

References and Selected Readings

Many of the general references cited at the end of Chapter 35 also contain considerable information on the biology of birds.

Burton, R. *Bird Behavior*. New York: Alfred A. Knopf, Inc., 1985. A summary of major aspects of bird behavior.

Editors of Scientific American. *Birds. Readings from Scientific American*. San Francisco: W.H. Freeman and Co., 1980. An assortment of articles published in *Scientific American*, including ones on flight and migration.

Farner, D.S., J.R. King, and K. Parkes (eds.). *Avian Biology*. New York: Academic Press, 1971 to present. A multivolume treatise covering most aspects of the biology of birds.

Feduccia, A. *The Age of Birds*. Cambridge, Mass.: Harvard University Press, 1980. An excellent account of the evolution and adaptive radiation of birds.

Knudsen, E.I. The hearing of the barn owl. *Scientific American* 245(Dec. 1981):112, 1981. An analysis of the uncanny ability of the barn owl to find prey in the dark.

Olson, S.L.: The fossil record of birds. In D.S. Farner, J.R. King, and K. Parkes (eds.), *Avian Biology*, vol. 87. New York: Academic Press, 1985.

Perrins, C.M., and A.L.A. Middleton (eds.). *The Encyclopedia of Birds*. New York: Facts on File Publications, 1985. A beautifully illustrated account of the biology of all living bird families.

Peterson, R.T. *A Field Guide to Birds*, 4th ed. Boston: Houghton Mifflin Co., 1980. The standard and widely used guide for the field identification of birds from the Great Plains to the East Coast.

Peterson, R.T. *Field Guide to Western Birds*, rev. ed. Boston: Houghton Mifflin Co., 1961. A companion to the preceding volume, it covers the birds from the Pacific Coast to the western parts of the Great Plains.

Rayner, J.M.V. Vorticity and animal flight. In H.Y. Elder and E.R. Trueman: *Aspects of Animal Movement*. Cambridge, England: Cambridge University Press, 1980. A summary of the vortex theory of flight.

Terres, J.K. *The Audubon Society Encyclopedia of North American Birds*. New York: Alfred A. Knopf, 1980. This is probably the most comprehensive, single-volume treatment of birds in print. It includes complete descriptions of the natural history of all North American birds along with outstanding color photographs of most of them. Excellent accounts of the anatomy, physiology, flight, migration, and many other important topics are also included.

Waterman, T.H. *Animal Navigation*. New York: Scientific American Library, 1989. An outstanding analysis of how some crabs, butterflies, fishes, birds, whales, and other animals find their way about.

Wellhofer, P. Archaeopteryx. *Scientific American*. 262(May 1990):70−77. An analysis of the known specimens of *Archaeopteryx* and the degree to which this species could fly.

Welty, J.C. and L.B. *The Life of Birds*, 4th ed. Philadelphia: Saunders College Publishing, 1988. A comprehensive one-volume work on all aspects of the biology of birds.

40

Mammals

Characteristics of Mammals

■ Mammals are endothermic vertebrates.

■ Hair and subcutaneous fat form an insulating layer. Cutaneous glands are abundant, secreting sweat, oil, and pheromones.

■ The limbs of most mammals are carried in a position to a lesser or larger extent beneath the body. The skull is of the synapsid type and has a relatively large braincase. The jaw joint lies between the dentary bone of the lower jaw and the squamosal (temporal) bone of the skull.

■ There are three auditory ossicles in the middle ear and a spiral cochlea in the inner ear.

■ The cerebrum is large and has a gray cortex. Large cerebellar hemispheres are present.

■ Teeth are heterodont and have a precise occlusion, and their replacement is limited. The small intestine has numerous multicellular intestinal glands and microscopic villi. Most species lack a cloaca.

■ Respiratory and digestive passages are nearly completely separated in the oral and pharyngeal regions by a secondary palate. Numerous lung alveoli greatly increase respiratory surface area. A muscular rib cage and diaphragm play major roles in lung ventilation.

■ Oxygen-depleted and oxygen-rich blood are completely separated as they move through the heart.

■ Nitrogenous wastes are eliminated primarily as urea by metanephric kidneys. Long loops of Henle in the renal tubules make possible the production of a urine hyperosmotic to the blood.

(Left) The American buffalo, Bison bison, *on its winter range. (Above) An American buffalo.*

Agnatha
Acanthodii
Placodermi
Chondrichthyes
Osteichthyes
Amphibia
Reptilia
Aves
Mammalia

■ The testes of most mammals either lie permanently within a scrotum or descend into the scrotum during the reproductive season. Males have a penis, and fertilization is internal.

■ Except for primitive egg-laying mammals, the ovaries are small and produce few eggs; little yolk is deposited in the eggs. The oviducts have differentiated into vaginal, uterine, and uterine tube regions.

■ Monotremes are oviparous; other mammals are viviparous. The uterine lining and certain extraembryonic membranes unite to form a placenta. Mammary glands are always present in females.

Of all the vertebrate groups, mammals (class **Mammalia**), are of particular interest to us because we are mammals, as are most of our domestic animals, which help us in our labors and provide us with wool, leather, and much of our food. Although mammals are not a large class (there are only about 4500 species), their evolution has not been constrained, as has that of birds, by adaptation to a particular mode of life. Mammals are a very diverse group. Most are terrestrial, but some, such as the whales, are highly adapted for an aquatic life, and others, the bats, have evolved flapping flight. Mammals range in size from a small species of bat that weighs only 1.5 g to the giant whales that exceed 100 tons. The class includes the egg-laying duck-billed platypus and spiny anteater (monotremes) of the Australian region; the opossum, kangaroo, and other pouched marsupials; and a wide variety of true placental mammals. They all produce few young but invest considerable time and energy in caring for them. All are endothermic, and most can maintain their body temperature at nearly constant levels irrespective of changes in the external environment. This ability has allowed mammals to occupy a wide range of habitats including the polar seas, high mountain ranges, deserts, and tropical jungles. Endothermy and greater care of the young have been the touchstones of mammalian evolution, and most of their characteristics are related to these attributes.

ADAPTATIONS OF MAMMALS

Endothermy and Temperature Regulation

Endothermy in mammals probably evolved gradually, and ancestral mammals likely had a lower metabolic rate and a simpler set of thermoregulatory controls than contemporary species. We can make some inferences as to how endothermy evolved by reconstructing the mode of life of ancestral mammals. The earliest mammals of the late Triassic and early Jurassic periods were small, mouse-sized creatures whose dentition was adapted for feeding on insects (see Fig. 40.6). Their skull shows a great elaboration of the cochlear region of the ear and of the olfactory organ. These conditions suggest that early mammals were nocturnal insect eaters because well-developed senses of smell and hearing would be essential for this mode of life.

It would not have been difficult for small mammals to be active at night at ambient temperatures of 25° to 30° C, provided they had insulating layers of subcutaneous fat and fur that would reduce the loss of body heat produced by a modest metabolic level. Ancestral mammals probably did not maintain a body temperature much higher than that of their nocturnal surroundings, and they probably retreated to cool shelters during the daytime. The European hedgehog and the tenrecs of Africa and Madagascar (Fig. 40.1) are contemporary nocturnal insectivores of this type that probably have occupied this niche throughout their evolutionary history. They maintain their body temperature only a few degrees above the nocturnal ambient temperature and can do so with no more oxygen consumption than a reptile of similar size and activity at the same temperature.

When diurnal ecological niches became available to mammals as a result of the late Cretaceous extinction of the dinosaurs, some early mammals evolved the ability to become active during the daytime, when ambient temperatures would be much higher. To maintain a body temperature as low as that of their nocturnal ancestors, i.e., 25° to 30° C, would require considerable evaporative cooling and loss of precious body water. They evolved a higher body temperature of approximately 35° to 40° C. But for a small animal to maintain this temperature at cooler times of the day required a greater energy expenditure than that of reptiles. Their energetics presumably changed in this context, and their metabolic rate became three to five times higher than that of reptiles of similar size and under similar conditions.

Contemporary mammals maintain their body temperature by regulating the rate of heat loss at the body

Figure 40.1
The mouse tenrec of Madagascar.

surface and the rate of heat production. Each mammalian species has a characteristic **thermal neutral zone**, a range of ambient temperature at which body temperature can be maintained with little change in metabolic rate (p. 289, Fig. 13.26b). Substantially more energy needs to be ex-expended to maintain body temperatures when ambient temperatures fall below or go above the thermal neutral zone. Most mammals do maintain their body temperature, even when the thermal neutral zone is exceeded, but some species permit body temperature to rise or fall over a limited range.

Adaptations for Cold Stress

Mammals that live in areas where the environment may become very cold have evolved adaptations that supplement thermal regulation. Many mammals with a thick coat of hair (the **pelage**) molt twice a year. In early fall, they gradually lose their thin summer pelage and develop a much thicker **undercoat** for the winter. A second molt to a summer pelage occurs in the spring. In addition to an undercoat, the pelage of many mammals includes long, coarse **guard hairs** that protect the undercoat. A thicker fur for winter extends the thermal neutral zone to a lower temperature and decreases the steepness of the slope of the ambient temperature–metabolic rate curve. Less energy need be expended to maintain body temperature than would be the case if a thicker pelage did not develop.

As in birds, the appendages cannot be insulated as well as the rest of the body. Their temperature is permitted to fall below that of the body core. Arteries carrying blood to the limbs are sometimes closely intermeshed with the veins returning blood so that a countercurrent exchange mechanism is established in which much body heat moves from the arteries to the veins and is not lost. Enough heat must be permitted to enter the appendages, however, to keep them from freezing. Nerves and other organs in the distal part of the limbs are adapted to function at lower temperatures.

Some arctic and temperate mammals, notably many insectivores, bats, and rodents, adjust to winter weather by going into a period of dormancy known as hibernation (p. 290). During this period, the thermostat in the hypothalamus is turned down, in some species as much as 20° C below the normal body temperature. Unlike amphibian and reptilian hibernators, in which body temperature falls to ambient levels, mammalian hibernators maintain body temperature, but at a low thermostat set point. Energy is conserved because metabolism is very low during hibernation yet high enough to sustain life and to keep the body from freezing. There are certain advantages to hibernation for a small endotherm. In many regions, insects and certain types of plant food are not available in sufficient quantity during the winter, making it difficult for the animal to get enough food to sustain the necessary metabolic rate. By permitting its body temperature and metabolic rate to

drop, the animal can get by on food reserves stored within its body.

Some other mammals avoid the problem of cold temperatures by retreating to more sheltered or warmer climates. Many small rodents remain active all winter beneath the snow cover, where the temperature seldom falls far below 0° C. Occasionally they venture forth on the snow surface. A few of the larger mammals undergo extensive seasonal migrations. For example, the caribou of Alaska and Canada spend the summer on the arctic tundra, but retreat south to the more sheltered forests in the winter.

Adaptations for Heat Stress

Mammals living in very hot climates also have special adaptations that help keep them cool without an excessive loss of body water due to evaporative cooling. Water often is in short supply. Small desert rodents avoid overheating by being nocturnal. During the day they burrow so they are sheltered in a cool and moist microhabitat.

Camels and some gazelles have thick fur coats that reduce heat loss in cold periods and heat gain in warm periods. Elephants, on the other hand, have very little body hair. Their large body size provides thermal stability, and their large ears function as efficient radiators.

Some mammals can allow their body temperature to rise and avoid the necessity of evaporative cooling. Camels can tolerate a body temperature as high as 41° C in the daytime. Their body cools down at night, when ambient temperatures fall. Permitting body temperature to rise is possible in many large mammals living in hot, open habitats because they can keep the critical brain temperature lower by a countercurrent exchange mechanism (Fig. 40.2). Arteries supplying the brain first break up into

Figure 40.2
The countercurrent circulatory mechanism that keeps the brain of gazelles and some other ungulates cool.

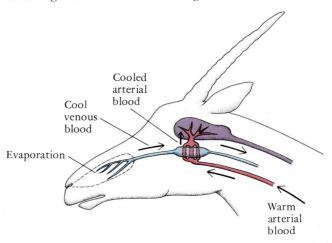

Cooled arterial blood

Cool venous blood

Evaporation

Warm arterial blood

minute passages that are entwined with veins returning cool venous blood from the nasal passages. Considerable heat passes from the warm arterial blood to the cooler venous blood before the arterial blood reaches the brain.

Locomotion and Coordination

Changes in all of the mammalian organ systems are closely correlated with the increased activity made possible by endothermy. Greater activity and agility are reflected in the skeletal system of even the earliest mammals of the Triassic period (see Fig. 40.6). The arched back, the posterior inclination of the vertebral spines of the thoracic verte-brae, and the anterior inclination of the spines of the lumbar vertebrae are typical of quadruped mammals and are correlated with the abandoning of lateral trunk undula-tions in locomotion. The elbow and knee have moved in close to the trunk, with the elbow pointing posteriorly and the knee anteriorly, so that the legs extend down to the ground more or less under the body. This position pro-vides better mechanical support and the potential for a longer swing of the appendage, increased stride length, and greater speed. Primitively, the feet were placed flat upon the ground, a posture called **plantigrade**. Most mam-mals have a sacrum formed by the fusion of three sacral vertebrae that forms a strong articulation between the pelvic girdle and vertebral column. Many species use the tail for balancing, as do cats, and it plays a major role in the propulsion of whales, but in most mammals the tail has lost its primitive locomotor function and is frequently reduced in size or is absent.

More complex patterns of locomotion, and probably the increased exploratory behavior and agility of mam-mals, require more complex muscular, sensory, and ner-vous systems. The senses of smell and hearing are very acute in primitive mammals. The eyes are well developed in most species, but are reduced in burrowing species and some nocturnal species. The brain is extraordinarily well developed. The cerebrum is greatly enlarged, with a gray cortex containing centers associated with sensory input and important motor centers (p. 394). The cerebellum also enlarges as motor coordination becomes more in-tricate.

Digestion

To sustain their high level of metabolism, mammals must obtain large supplies of food. The lower temporal opening of the skull (synapsid fenestra) is enlarged and houses powerful jaw muscles (compare Figs. 38.6b and 40.6). Mammals cut and crush their food within their mouths, and the teeth are firmly rooted in sockets in the jaws. The teeth of mammals also have differentiated into types that perform different functions during food processing (see Fig. 12.10). Such teeth are described as **heterodont** (Gr.

heteros, other + *odous*, tooth) in contrast to the uniform **homodont** (Gr. *homos*, same) dentition of most other vertebrates in which all of the teeth are very similar, usually simple cones of different sizes, that are used primarily to seize and hold the food. Chisel-shaped incisors at the front of each jaw are used for nipping and cropping (Fig. 40.3a). Next is a single canine tooth, which is primitively a long, sharp tooth with a piercing and tearing action. A series of premolars and molars follows the canine. These cheek teeth may have a sheering action (carnivores) or a grinding action (herbivores). In primitive living mammals, each molar tooth bears three small cones, called **cusps**, that are arranged in a triangle. The triangle of cusps is called a **trigon** in an upper molar and a **trigonid** in a lower molar (Fig. 40.3b and c). Trigon and trigonid are mirror images of each other, so good shearing action occurs as parts of the trigon and trigonid slide past each other. Crushing of food occurs when the primary cusp of the trigon falls upon a low heel, or **talonid**, located on the posterior surface of the trigonid. Additional shearing action occurs as the pre-molars slide past each other.

In order to perform these functions, upper and lower teeth must come together, or occlude, in a very precise manner. Young mammals are suckled and are born with-out teeth. A first set of incisors, canines, and premolars develops as the young begin to feed for themselves. As the jaw grows in size, good occlusion is maintained as these so-called milk teeth are gradually replaced by larger perma-nent teeth. The molar teeth appear sequentially as a mam-mal matures and the jaw enlarges further. They are not replaced. Reptiles, in contrast, frequently lose teeth, and gaps are present in the tooth row where replacement teeth are growing in. Tooth replacement in reptiles is continu-ous throughout life.

As mammals chew their food, they mix it with saliva that lubricates the food and often contains amylase, an enzyme that begins the digestion of starches. Digestion continues in the stomach and intestine. Numerous micro-scopic villi line the small intestine and increase the surface area available for digestion and absorption.

Respiration

The greater exchange of oxygen and carbon dioxide required to maintain a high level of metabolism is made possible by the evolution of pulmonary alveoli that greatly increase the respiratory surface of the lungs and by the evolution of a diaphragm that increases the efficiency of ventilation. The palate of mammals is a horizontal partition of bone and flesh that separates the air and food passages in the mouth cavity and pharynx (Fig. 40.4). The palate permits nearly continuous breathing. Mammals can ma-nipulate food in their mouths while still breathing, for food and air passages cross only in the laryngeal part of the pharynx. The mammalian palate is quite different from that

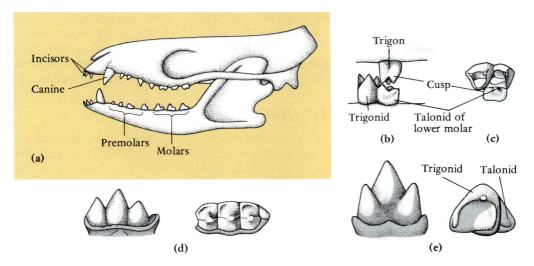

Figure 40.3
The teeth of primitive mammals. (a) The teeth of an insectivore. (b and c) Lateral and top (crown) views of the left upper and lower molars of an insectivore to show their occlusion. The upper molar in (c) is drawn as though it were transparent so as to show how its apex comes down upon the talonid of the lower molar. (d) Lateral and crown views of a triconodont molar showing the linear arrangement of the three cusps. (e) Lateral and crown views of the lower molar of a trituberculate in which the three cusps have assumed a triangular configuration.

of other terrestrial vertebrates in which the nasal cavities open into the front of the mouth cavity. The mammalian palate is technically a **secondary palate**, a new shelf of bone and flesh that divided the primitive mouth cavity and anterior part of the pharynx into passages for gas transport and food processing.

Circulation

Mammals, like birds, have evolved a heart that is completely divided internally so there is no mixing of oxygen-depleted blood with oxygen-rich blood. The complete separation of blood streams going to the lungs and body

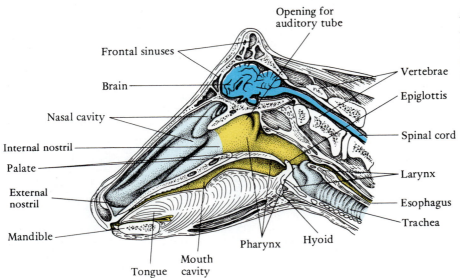

Figure 40.4
A sagittal section of the head of a cow. Notice that the respiratory and digestive systems are separated by the palate and cross only in the laryngeal region.

allows the musculature of the left ventricle to become more massive and powerful than that in the right ventricle (p. 312). Systemic blood pressure is much higher than it is in reptiles, and blood is circulated rapidly throughout the body.

Excretion and Water Balance

The high metabolic rate of mammals results in the formation of a larger amount of nitrogenous wastes than in ectothermic vertebrates. As in amphibians (and unlike reptiles and birds), most of the nitrogenous waste is urea, which, being toxic, requires a considerable volume of water to carry it off (see Table 37.1). Mammals have many kidney tubules and a high filtration rate because the glomeruli are large and blood pressure is high. Large amounts of water leave the blood along with urea, but little water is lost because the loops of Henle (p. 359) make it possible for approximately 99% of the water that starts down the kidney tubules to be reabsorbed. Mammals produce a very concentrated urine that is hyperosmotic to the blood.

Reproduction and Care of the Young

Upon hatching from its egg, or upon birth, a young reptile must feed and fend for itself. Some are successful, many are not. Mammals have evolved a different reproductive strategy. Fewer young are produced, but considerable maternal energy and care are invested in the few that are conceived and raised. The embryos of endothermic mammals must themselves develop in a warm and closely controlled environment. The platypus and spiny anteater resemble most reptiles in being oviparous, but the eggs are carefully brooded, and the newborn are fed milk secreted by the **mammary glands** or breasts. Mammary glands are unique to mammals, and the class name, Mammalia, is derived from their presence (L. *mamma*, breast). They probably evolved from either sweat or sebaceous glands that occur in mammalian skin (p. 204).

All other mammals are viviparous and retain their embryos in a uterus. All of the extraembryonic membranes characteristic of amniotes are present (p. 523). The yolk sac contains little or no yolk, and a placental relationship develops between the mother and embryo(s) allowing the transfer of maternal nutrients to the embryo and embryonic wastes to the mother. The placenta is transitory in the opossum and other marsupials; the young are born at a

very small size and complete their development attached to the mother's nipples, which are usually located in an external pouch. The placental relationship with the mother lasts for a longer time in other mammals and the young are born at a more advanced stage of development.

THE ORIGIN OF MAMMALS

The features that characterize mammals began to appear in the **synapsid**, or mammal-like, reptiles. This line of evolution diverged from other amniotes very early in the late Carboniferous (see Fig. 38.1). Indeed, the earliest captorhinids and synapsids appear in geological deposits of the same age. The transition from these early ectothermic and scaly reptiles to small endothermic and furry mammals spanned a period of about 150 million years. It is the best documented of all of the major transitions in vertebrate evolution. Many of the changes that occurred in behavior, soft anatomy, and physiology can be inferred from the structure of the skeleton.

The earliest mammal-like reptiles, which are placed in the order **Pelycosauria**, were medium-sized, somewhat clumsy, terrestrial quadrupeds. As in other primitive terrestrial vertebrates, their limbs were sprawled out at right angles to the body and not carried under the body. Some species, such as *Dimetrodon* (Fig. 40.5a), had bizarre nonmoveable sails upon their backs supported by elon-

(a)

(b)

Figure 40.5
Mammal-like reptiles. (a) *Dimetrodon*, a pelycosaur with a thermoregulatory sail; (b) *Lycaenops*, a late therapsid that carried its limbs in a mammal-like fashion.

gated vertebral spines. Sail size increased at a faster rate than body size as members of this evolutionary line became larger, which helped maintain a constant ratio of body surface to mass. The rapid increase in sail size suggests that the sail may have been a primitive thermoregulator that permitted these rather large animals to absorb heat, warm up, and become active early in the day. *Dimetrodon* attained lengths of 3 m.

Pelycosaurs were replaced by more advanced synapsids of the order **Therapsida** late in the Permian (see Fig. 38.1). Therapsids came to resemble mammals more closely (Fig. 40.5b). Their limbs were more nearly beneath the body, where they could provide better support and move more rapidly back and forth. Limb structure suggests that therapsids were capable of sustained locomotion, as are mammals. The synapsid fenestra in the skull (Fig. 38.6b) enlarged greatly and accommodated powerful jaw muscles. The jaws were strong, the bite powerful, and their teeth specialized into ones suited for cropping, stabbing, and cutting. Therapsids were capable of processing more food than other reptiles. A secondary palate separated the mouth and nasal passages, so that breathing could continue as the animal chewed its food. Long ribs were limited to the thorax, as in mammals, and lumbar ribs were very short, indicative of a diaphragm. Therapsids could have exchanged a large volume of oxygen and carbon dioxide in the lungs. The existence of numerous pits for blood vessels and nerves in the snout bones suggests a bare and moist nasal area (rhinarium), seen in mammals, and possibly the presence of modified facial hairs, the whiskers or vibrissae. Therapsids obviously were becoming very active animals, and their muscle metabolism may have shifted from a reliance on glycolysis to oxidative metabolism. Body temperatures must have been elevated, but therapsids may not have yet evolved true endothermy.

The major osteological characters that distinguish late therapsids from mammals are the bones that bear the jaw joint and the number of auditory ossicles (see Focus 18.2, p. 416). The final incorporation of the articular and quadrate (bones bearing the jaw joint in reptiles) into the middle ear as additional auditory ossicles (the malleus and incus) and the restriction of the jaw joint to the dentary and squamosal bones occurred at the synapsid-mammal transition.

PRIMITIVE MAMMALS: PROTOTHERIANS

Early Mammalian Evolution

Mammals separated from therapsids late in the Triassic period, about 240 million years ago (Fig. 38.1). Early species were contemporaries of the dinosaurs, who dominated the terrestrial fauna during most of the Mesozoic era, but these early mammals remained small, secretive creatures until the late Cretaceous extinction opened up many ecological opportunities for them. Considerable diversity is seen among Mesozoic mammals, but the first ones were mouse-sized creatures whose endothermy probably permitted them to exploit a nocturnal insect-eating niche that was not available to the ectothermic reptiles (Fig. 40.6). They had the mammalian pattern of delayed tooth eruption, which indicates that their young fed upon milk. Their molar teeth differed from those of later mammals in having

Figure 40.6
The skeleton of a primitive Triassic triconodont.

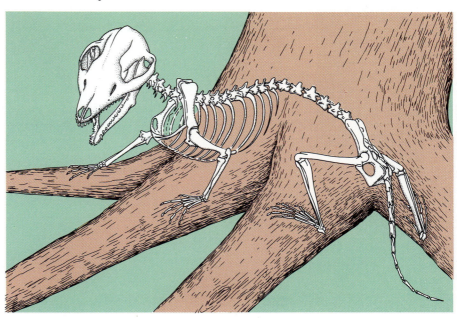

three conical cusps arranged in a linear series (Fig. 40.3d). Because of this pattern of cusps, they are placed in the order **Triconodonta** (Fig. 40.7).

The fossil remains of triconodonts have been found in many parts of the world. It would not have been difficult for primitive mammals to move from one continent to another early in the Mesozoic era for the present continents were clustered into one gigantic land mass known as Pangaea (see Focus 9.3, p. 174).

Monotremes

The many lines of mammalian evolution that developed during the Mesozoic era may all have diverged from triconodonts, but the fossil record is not complete enough to be certain. Most of these lines became extinct. One line probably led to contemporary **monotremes**, the duck-billed platypus and the echidnas, or spiny anteaters, of the Australian region. These are certainly the most primitive mammals living today, but their relationships are somewhat uncertain. Few fossil monotremes are known. Teeth have been used extensively in sorting out lines of mammalian evolution, but the echidnas lack teeth and the ones that begin to develop in the platypus are lost before maturity.

They lay and brood their eggs, and retain many reptilian characteristics, including a cloaca. The ordinal name for the group, **Monotremata** (Gr. *monos*, single + *trema*, hole) refers to the presence of a single opening for the discharge of feces, urinary, and genital products. In other mammals, the cloaca has become divided, and the opening of the intestine, the anus, is separate from that of the urogenital ducts. Because of their primitive nature, monotremes and triconodonts are grouped in the subclass **Prototheria** (Fig. 40.7).

Fossil monotremes and contemporary species are confined to New Guinea, Australia, and Tasmania. When Pangaea began to break up by continental drift later in the Mesozoic era (see Focus 9.3), the ancestors of monotremes must have been restricted to the part of the southern continent (Gondwana) that was to become Australia. The living species have survived here, partly because of their isolation and a lack of competition with most other mammalian groups.

The platypus (*Ornithorhynchus*) is confined to Australia and Tasmania. It is a semiaquatic species with webbed feet, short hairs, and a bill like a duck's used in grubbing in the mud for soft-bodied invertebrates (Fig. 40.8a). Claws are retained, and they dig long burrows in

Figure 40.7
The geological distribution and phylogeny of the major groups of mammals. Most of the extinct Mesozoic groups have not been shown.

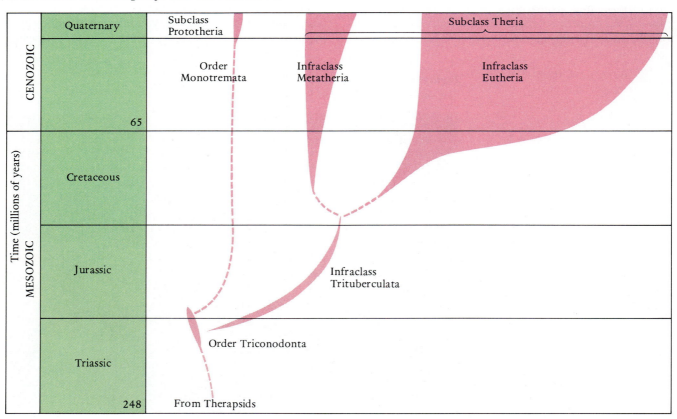

(a)   **(b)**

Figure 40.8
Monotremes. (a) The platypus, *Ornithorhyncus anatinus*, of Australia grubs for
aquatic invertebrates with a duck-like bill. (b) The short-nosed echidna,
Tachyglossus aculeatus, of Australia, Tasmania, and New Guinea probes for insects
and other invertebrates with its elongated snout.

muddy banks. One genus of echidna (*Tachyglossus*) occurs in New Guinea, Australia, and Tasmania; the other
(*Zaglossus*) is confined to New Guinea (Fig. 40.8b). Echidnas have large claws and a long beak adapted for digging
and probing for ants and termites. These insects are
crushed between sets of spines on the back of the tongue
and the roof of the mouth. Echidnas can burrow very
effectively. Many of their hairs are modified as quills. Facial
muscles are poorly developed in monotremes, and they
do not have fleshy lips. Infants cannot suckle in the usual
mammalian way; rather, they lap up milk discharged from
nippleless mammary glands onto tufts of hair.

THERIANS

In another line of evolution from triconodonts, the three
linearly arranged molar cusps shifted to a triangular pattern, and a low heel on the posterior edge of each lower
molar later evolved (Fig. 40.3e). These changes, first seen
in the extinct infraclass **Trituberculata**, improved the efficiency of the molar teeth in cutting and crushing insects
and other small invertebrates. Primitive marsupials and
placental mammals living today have similar molar teeth.
Because of similarities in their tooth structure, trituberculates, marsupials, and placental mammals are grouped
together in the subclass **Theria** (Fig. 40.7).

Reproductive Strategies

Marsupials (infraclass **Metatheria**) and true placental
mammals (infraclass **Eutheria**) separated from a common
trituberculate ancestor early in the Cretaceous period and

have evolved quite different reproductive strategies. In
eutherians, an intimate placental relationship develops between the embryonic chorioallantoic membrane and the
uterine lining. Trophoblast cells that form the surface of
the fetal portion of the placenta play an important role in
establishing this union and in preventing the rejection of
the embryo by the mother's immune system (p. 526). The
gestation period of eutherians is relatively long, ranging
from 45 days for some shrews to 650 for the elephant. The
young are born at a relatively advanced stage of development. Because of their well-formed placenta, eutherians
are commonly called placental mammals, but this term
does an injustice to the marsupials, who also have a
placenta, albeit a more transitory one.

The marsupial egg contains more yolk than that of
eutherians and is surrounded by a **shell membrane** of
maternal origin. The embryo is nourished for more than
half of its brief gestation period by the stored yolk and by
uterine secretions. The shell membrane acts as an immunological barrier. It is absorbed near the end of gestation
and a placenta is established, usually between the yolk sac
and uterine lining. The placenta is not an intimate union,
and it lasts only a few days before birth occurs, probably
because of an immunological rejection. The total gestation
period of marsupials is relatively short, ranging from 13
days in the North American opossum to 35 days in an
Australian wallaby (Fig. 40.9). Newborn marsupials are
very small and developmentally immature but have special
adaptations, including well-developed forelimbs, that enable them to crawl up the belly of the mother and attach to
mammary gland nipples, which usually are located within
a pouch or **marsupium** (L., a small purse). Pouch life is

(a)

(b)

Figure 40.9
(a) The Virginia opossum, *Didelphis virginiana*, is the only North American marsupial. (b) A group of recently born young attached to nipples in their mother's pouch.

several times longer than the gestation period, and development is completed there.

There are advantages and disadvantages to each reproductive strategy; one is not necessarily inherently superior to the other. Placental mammals are developmentally more mature at birth, but their maturity requires the investment of considerable maternal energy. A placental mother is committed to carrying her embryos until their relatively late birth, even at the expense of her own fitness if food and other environmental resources fail. During the lactation period the young can be abandoned if need be, but then the mother's reproductive effort will have been lost. Marsupial young are less developed and more vulnerable at the time of their early birth, but few maternal resources have been invested in them. Indeed, the life of the mother has hardly been affected in any way. If environmental resources fail, the pouch young can be aborted easily and the mother will have lost little reproductive effort. She will have a better chance to survive and reproduce again.

Kangaroos and wallabies have an additional way of adjusting their reproductive effort to environmental conditions. Unlike other mammals, they can become pregnant again shortly after giving birth and during their lactation period, but the development of the new embryo is arrested in the blastocyst stage. This phenomenon is called **embryonic diapause**. As lactation of the young in the pouch diminishes, development of the blastocyst resumes. Environmental conditions in the "outback" of Australia are often harsh, but when the rains come and the grass grows, these marsupials can reproduce rapidly. They may simultaneously have one offspring (joey) at heel, who occasionally crawls back into the pouch and sometimes drinks some milk; a recently born young in the pouch, who is attached to a different nipple; and an arrested blastocyst in the uterus (Fig. 40.10).

Adaptive Radiation of Marsupials

Fossil evidence indicates that marsupials originated in the early Cretaceous period in North America and Western Europe when Pangaea was intact (see Focus 9.3, Fig. a). They spread from there through South America, Africa, and across Antarctica to Australia, for these southern lands were still broadly interconnected, forming the supercontinent of Gondwana (see Focus 9.3, Fig. b). The discovery of marsupial fossils in Antarctica supports the notion of

Figure 40.10
The Australian red kangaroo, *Macropus rufus*. A joey frequently takes refuge in its mother's pouch a few months after its nursing period.

their spread into Australia by way of Antarctica. Although a few primitive eutherian groups reached South America, none spread as far as Antarctica. As Gondwana broke up, Antarctica moved south and became glaciated, Africa became connected to Europe and Asia, and South America separated from North America and Africa (see Focus 9.3, Fig. c).

South America and Australia were effectively isolated as havens for marsupial evolution. Marsupials radiated widely and occupied most of the ecological niches exploited elsewhere by eutherian mammals. When a connection was reestablished between North and South America late in the Tertiary period (see Focus 9.3, Fig. d), many eutherian groups spread into South America, and many marsupials and other primitive species living there became extinct. Over 20 species of opossums survive in South America, and the Virginia opossum has reentered Central and North America.

The Australian region remained as a haven for marsupial evolution, since the only placental mammals to reach there before human beings were bats and a few rodents. Australia, Tasmania, and New Guinea still harbor a tremendous diversity of marsupials, such as the carnivorous Tasmanian devil, anteating species, arboreal gliders, koala bears (the original "teddy bear"), plains-dwelling kangaroos, and rabbit-like bandicoots (Fig. 40.11). There is such a wide variety, and different groups have been evolving independently for so long, that some authorities divide the marsupials into several distinct orders.

Adaptive Radiation of Eutherians

Eutherians probably diverged from trituberculates early in the Cretaceous, but the group did not undergo an extensive adaptive radiation until late Cretaceous (Fig. 40.7). Their radiation occurred rapidly and most lines of evolution were established by the early Tertiary. Eutherians are currently the dominant group of terrestrial vertebrates; several groups have exploited lake, river, and ocean habitats; and bats have evolved flapping flight.

Ant- and Termite-Eating Mammals

Specialists do not agree on how to group the numerous eutherian orders. We are following a cladogram based on an analysis of 104 cranial characters that was developed in 1986 by Michael J. Novacek (Fig. 40.12).

The earliest Cretaceous eutherians appear to have been small insect-eating species not unlike contemporary shrews in general appearance. The South American anteater (Fig. 40.13a) and other species of the order **Xenarthra** diverged from primitive eutherians of this type in the late Cretaceous period and underwent an extensive radiation in South America while it was isolated from North America. Many unusual and now extinct types have been discovered. A xenarthran has large claws that enable it to tear open ant and termite nests and a long snout with which to probe the nest tunnels for its prey. An anteater lacks teeth and laps up insects with its long tongue, which is covered by a very sticky saliva secreted by enlarged salivary glands. The insects are swallowed whole and are crushed and ground up in the gizzard-like pyloric region of the stomach. The armadillos and leaf-eating tree sloths (Fig. 40.13b) retain reduced teeth without a coating of enamel. Armadillos entered North America after its reconnection with South America late in the Tertiary period.

The pangolins, or scaly anteaters (order **Pholidota**), of Africa and southeastern Asia appear to have diverged from xenarthrans while Africa was still connected to South America. They independently evolved similar adaptations for feeding on ants and termites. The protective horny scales of the pangolin are composed of hairs cemented together (Fig. 40.13c). When disturbed, the animal rolls into a ball.

(a)

(b)

Figure 40.11
Some Australian marsupials. (a) A squirrel glider, *Petaurus norfolkensis*, feeding on beech nuts. The folded gliding membrane can be seen on the left side of the photograph between the animal's legs. (b) Koala bears, *Phascolarctos cinereus*, are marsupials specialized to live in trees. Their diet is restricted to leaves of a very few species of eucalyptus. This young koala has outgrown the pouch and will ride on its mother's back until it is old enough to go out on its own.

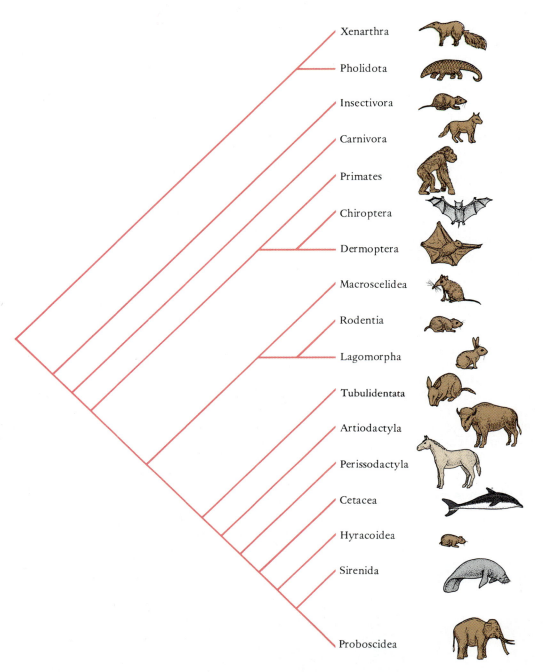

Figure 40.12
A cladogram showing the relationship between the living eutherian orders.

The aardvark of Africa (order **Tubulidentata**) is another species specialized for feeding on ants and termites (Fig. 40.13d). It has large claws, a pointed snout, and peg-shaped cheek teeth. Enamel has been lost, but the teeth do not wear out because they grow continuously. These adaptive features represent a convergence with the xenarthrans and pholidotans and do not reflect a common ancestry. Cladistic analysis points to the origin of the aardvark from early hoofed mammals or ungulates.

Insectivores
The tenrecs of Africa and Madagascar (Fig. 40.1), the European hedgehog, the shrews and moles (Fig. 40.14a and b), and other species of the order **Insectivora** have

(a)

(b)

(c)

Figure 40.13

Several groups of mammals have specialized independently for feeding upon ants and termites. (a) The South American giant anteater, *Myrmecophaga tridactyla*. (b) The related neotropical brown-throated three-toed sloth, *Bradypus varigatus*, feeds primarily upon leaves. (c) The Chinese pangolin, *Manis pentadactylus*. (d) The South African aardvark, *Orycteropus afer*.

(d)

Figure 40.14

Insectivores. (a) A short-tailed shrew, *Blarina brevicauda*, feeding upon a frog. (b) A star-nosed mole, *Condylura cristata*, burrows with its enlarged forelimbs and its nose bears many sensitive fleshy tentacles.

(a)

(b)

retained many of the features of ancestral eutherians. Shrews are nocturnal creatures with small eyes, but other senses are well developed. Their brain, too, is relatively small and the cerebrum has a smooth surface. Limb structure is quite primitive. Five clawed toes are retained, and the foot posture is plantigrade. Insectivores have a dentition consisting of three incisors, one canine, four premolars, and three molars in each side of the upper and lower jaw (Fig 40.3a). This can be expressed as a **dental formula** : I 3/3, C 1/1, Pm 4/4, M 3/3. Except for the toothed whales, no living eutherian mammal has more teeth than this, and the number of teeth is reduced in many groups. Humans, for example, have a dental formula of I 2/2, C 1/1, Pm 2/2, M 3/3. Insectivores feed voraciously on worms and insects and other small arthropods. Since they are small endotherms with a relatively large surface area, they must consume a great deal of food to sustain the high metabolic level required to maintain their body temperature. Shrews eat the equivalent of their body weight each day. Some species have become specialized for unique modes of life. Moles are burrowing species with enlarged and very strong front feet and legs.

Carnivores

Only a small adaptive shift is needed for a primitive insect-eating eutherian to adapt to eating other forms of animal prey. Weasels, dogs, raccoons, bears, and cats (Fig. 40.15a–b) are familiar members of the order **Carnivora**. Seals, sea lions, and walruses are more specialized carnivores that have adapted to a marine life. Most terrestrial carnivores have teeth specialized for killing and cutting prey. Canines are large, and in contemporary species the cusps on the last upper premolar tooth and first lower molar have enlarged and moved into the same plane to form a set of **carnassial** teeth well adapted for slicing through meat (Fig. 40.16). The jaws are hinged in such a way that they close like a pair of scissors. The posterior teeth come together before the front teeth.

Terrestrial carnivores have well-developed claws and a limb structure that enables them to run rapidly and catch prey. The limbs are long, which increases stride length, and most carnivores have shifted from the primitive plantigrade to a **digitigrade** foot posture (Fig. 40.17a and b). They stand upon their toes (though not on the tips), with the rest of the foot raised off of the ground in the

(a)

(b)

(c)

(d)

Figure 40.15

A group of carnivores. (a) The short-tailed weasel or ermine, *Mustela erminea*, a native of the northern forests and tundra, has a white coat in winter and a brown one in summer. (b) The cub of the common black bear, *Urus americanus*. (c) A group of walruses, *Odobenus rosmarus*, on Arctic ice. (d) A young California sea lion, *Zalophus californianus*. Notice that its tail is very short and its hind legs can be turned back to form propulsive flippers during swimming.

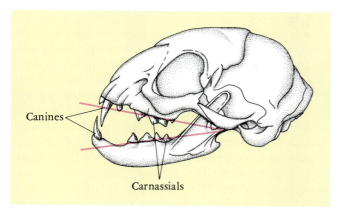

Figure 40.16
Carnivore teeth and jaw specializations as seen in a cat. Lines through the upper and lower tooth rows intersect at the joint and illustrate the way the teeth come together when the jaws close.

manner of a sprinter. The foot lengthens to a greater extent than the proximal limb segments. This pattern keeps most of the muscle mass, which is concentrated in the proximal limb segments, close to the body and on a part of the limb that moves through a shorter arc during a limb cycle. The proximal parts of a limb, therefore, have a lower velocity than the distal part. Since the force needed to move a limb segment is equal to its mass times its acceleration, energy is conserved by keeping most of the mass proximal.

Sea lions, walruses, and seals are carnivores that became specialized for feeding on fishes and other resources of the sea, but they return to land or to ice flows to rest and reproduce (Fig. 40.15c and d). Their premolar and molar teeth often are simple cones used to seize fish that are swallowed whole or in large chunks. Their limbs have evolved into flippers. Sea lions and walruses have very large pectoral flippers with which they propel themselves, making simultaneous sweeps through the water. Seals turn their enlarged pelvic flippers posteriorly and move them from side to side in the manner of a fish's tail. Seals and sea

Figure 40.17
The hind leg of representative mammals drawn with a constant femur length to show that the distal part of the leg and especially the foot lengthen as the limb posture changes from plantigrade (a) to digitigrade (b) to unguligrade (c). (a) An armadillo (a xenarthran) has a plantigrade foot but one that is specialized for digging; (b) a coyote (a carnivore); (c) a pronghorn (an ungulate).

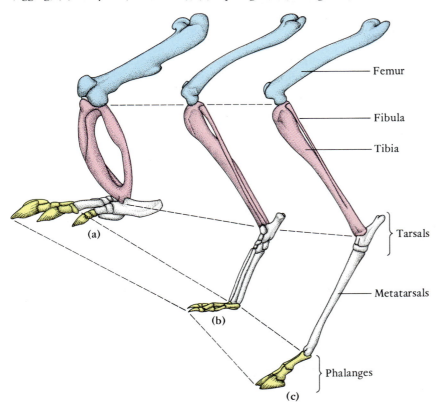

lions feed primarily upon fish. Walruses gather bivalved molluscs and other invertebrates by rooting in the bottom, in piglike fashion, with their tough, cornified snouts. They also can squirt jets of water to uncover clams. Their upper canine teeth are enlarged to form **tusks**. Tusks are not used by walruses for digging, as was formerly believed, but are "social organs" that convey information about the sex and age of the individual. Males with the largest tusks establish dominance in a social hierarchy.

Primates

Lemurs, monkeys, apes, and other species in the order **Primates**, together with bats (order Chiroptera) and the gliding colugos (order Dermoptera) are closely related groups that evolved from primitive insect-eating species (Fig. 40.12). They adapted early for an arboreal life, first searching for insects in the trees and then utilizing flowers, fruits, nuts, and other resources in this habitat. Because primates are of particular interest to us (human beings are primates), we will examine their radiation and evolution in more detail in Chapter 41.

Flying Mammals

Bats (order **Chiroptera**) have evolved true flight. Their wings (Fig. 40.18a) are structurally closer to those of pterosaurs than to those of birds because the flying surface is a leathery membrane. However, a bat wing is supported by four elongated fingers (the second to fifth) rather than by a single one, as in pterosaurs. The wing membrane is very large; its attachment extends along the trunk and incorporates the hind legs and, in many species, the tail as well. Bat wings are more arched or cambered than bird wings, which increases their lift at low flight speeds. The wings are well adapted for slow flight and for providing a

high degree of maneuverability. Muscle fibers within the wing membrane control its tension during flight and help fold it when the bat rests. Some species net insects with their wings. The first finger is free of the wing, bears a small claw, and is used for grasping and clinging. The hind legs are small but they are effective grasping organs and are used for clinging to a perch from which the bats hang upside down when at rest. Some bats, including the vampire bat, have strong legs and can run rapidly and leap quickly, but most species are rather clumsy on the ground. While they are active, bats maintain their body temperature between 30° and 35° C, but permit it to drop to 20° C or lower when they rest. A lowering of body temperature economizes on the amount of food they must gather. Temperate species either hibernate or migrate to warm climates during the winter.

The most common North American bats are insect eaters that fly about at dusk in search of their prey. Some bats are specialized to feed on fruit, pollen and nectar, blood (vampire bats), and many kinds of small vertebrates, including fishes. Fish-eating bats catch their prey near the surface of the water by means of hooked claws on powerful feet.

Some other mammals stretch a loose skin fold between their front and hind legs and glide from tree to tree. The colugos (order **Dermoptera**) of the East Indies and Philippines can execute long, controlled glides (Fig. 40.18b). Gliding enables these animals to forage efficiently on leaves and flowers in the tropical tree canopy without the high energetic costs of climbing up all of the way from the ground when they move from tree to tree.

Echolocation   Insectivorous bats have evolved a system of **echolocation** that enables them to search for their prey

Figure 40.18
Flying and gliding mammals. (a) A mouse-eared bat, *Myotus myotus*, using its wings and tail to help catch an insect. (b) A colugo (flying lemur) from Malaysia.

(a)   (b)

at dusk, avoid obstacles, and find their way in dark caves. As early as 1793 it was observed that a blinded bat could find its way about, but that one in which the ears had been plugged was helpless. However, it was not until the availability of sophisticated electronic apparatus about the time of World War II that investigators demonstrated that bats emit ultrasonic sounds that bounce off objects and return as echoes. By analyzing the echoes, many bats can determine the distance, direction, size, and possibly the texture of the object. As bats fly about at dusk searching for insects, they emit ultrasonic pulses that range from 25 to 150 kHz, well above our upper level of hearing. Sounds at these frequencies have short wavelengths and hence can produce sharp echoes from small objects. Different species of bats have evolved somewhat different sonar systems. Most species emit sounds through their open mouths. Species with elaborate noses, such as the leaf-nosed bat, emit sounds through their noses. Nose structure focuses the sound and produces a concentrated beam. In many species the sound pulses are frequency-modulated and drop about an octave (i.e., from 60 to 30 or from 80 to 40 kHz) during their 1- to 4-millisecond duration. While the bat is searching, pulses are emitted at the rate of about 10 per second, but the frequency of emission increases tenfold and the duration of the pulses shortens when a bat detects an insect and homes in on it. The emitted sounds are at very high energy levels, over twice that in a boiler factory. Tiny muscles in the middle ear contract during sound emission, thereby damping the movement of the auditory ossicles and protecting the inner ear. These muscles relax as the echo returns. Distance is perceived in many species by the time interval between the emitted pulse and the echo, but less well understood mechanisms are involved in other species. Directionality appears to be determined by a comparison of the differences in intensity of the echo between the two ears. Bats can perceive meaningful signals in the presence of considerable extraneous noise.

Adaptations for Herbivory

Many mammals have become highly specialized for a plant diet. Eating plant food has entailed a considerable change in their dentition, for plant food must be thoroughly ground by the teeth before it can be acted upon by the digestive enzymes. The molars of primitive herbivorous mammals (and those of omnivorous species, such as human beings) have become square or rectangular, as seen in a crown view. An extra cusp is added to the primitive trigon and trigonid, so that four primary, rounded cusps are present on each molar (compare Figs. 40.3b and c with 40.19a). A molar of this type is called a **bunodont** tooth (Gr. *bunos*, mound + *odous*, tooth). In more advanced species of herbivores the cusps form a pattern of ridges (**lophodont** teeth; Gr. *lophos*, crest) or crescents (**selenodont** teeth; Gr. *selene*, crescent). Herbivores that feed upon leaves and other soft vegetation retain the primitive low-crowned molar teeth, but those that graze upon coarser grasses have evolved high-crowned molar teeth (Fig. 40.19b). A high-crowned tooth extends a considerable distance above the gum line, and cement has grown from the roots up over the surface of the tooth and into the "valleys" between the elongated cusps. More tooth is provided to wear away, and the tooth is more resistant to wear. In some groups, the premolar teeth assume the form of the molars so all of the cheek teeth (premolars and molars) are effective crushing or grinding teeth.

Figure 40.19

Modifications of the molar teeth and jaws in herbivorous mammals. (a) Crown views of common cusp patterns; (b) vertical sections of a low-crowned and a high-crowned tooth; (c) the jaws of a rabbit and other herbivores are hinged in such a way that the molar teeth come together simultaneously.

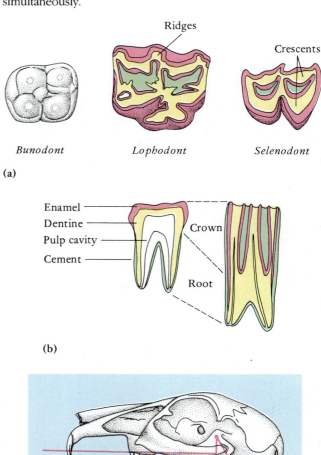

The hinging of the jaw changes so that all of the teeth are brought together at the same time (Fig. 40.19c). The configuration of the joint surfaces also permits more fore and aft and sideways movement of the lower jaw needed for grinding food. Parts of the digestive tract (stomach or cecum) are specialized to house a bacterial colony that digests the high cellulose content of their food.

Rodents, Lagomorphs, and Elephant Shrews

Cladistic analysis shows that rodents (order **Rodentia**), rabbits (order **Lagomorpha**), and elephant shrews (order **Macroscelidea**) are a closely related assemblage of herbivorous mammals (Fig. 40.12). Rodents gather their food by gnawing with an upper and lower pair of enlarged, chisel-like incisor teeth that grow out from the base as fast as they wear away at the tip. They grind their food with lophodont cheek teeth. The rodent mode of feeding has been very successful; there are more rodent species, and possibly more individuals, than of all other mammals combined. Rodents have undergone their own adaptive radiation and have evolved specializations for a variety of ecological niches (Fig. 40.20a and b). Rats, mice, and chipmunks live on the ground; gophers and woodchucks burrow; squirrels and porcupines are adept at climbing trees; the flying squirrel can stretch a skin fold located between its front and hind legs and glide to its destination; and muskrats and beavers are semiaquatic.

Rabbits and the related pika (Fig. 40.20c) of western North American mountains superficially resemble rodents for they have independently evolved a similar suite of adaptations for gnawing and grinding their food. One way by which lagomorphs differ from rodents is the presence of a reduced second pair of incisors hidden behind the large pair of upper incisors.

(a)

(b)

Figure 40.20
(a) The woodchuck, *Marmota monax*, and other rodents have enlarged incisor teeth for gnawing and gathering plant food. (b) A beaver, *Castor canadensis*, can use its teeth to fell a tree. (c) The pika, *Ochotoma princeps*, of the Western mountains of North America is a lagomorph that is closely related to rabbits and hares. (d) The rufus elephant shrew of Africa, *Elephantulus rufenscens*, has long legs and an elongated snout used to forage for invertebrates.

(c)

(d)

(a)

(b)

(c)

Figure 40.21
Contemporary ungulates include (a) the black rhinoceros, *Dioceros bicornis*; (b) the common zebra, *Equus burchelli*; (c) the hippopotamus, *Hippopotamus amphibius*; and (d) Grant's gazelle, *Gazella granti*. All of these species are African.

(d)

The elephant shrews of Africa are unusual creatures (Fig. 40.20d). They have a long, shrewlike snout used for probing in the ground litter for small invertebrates, but unlike the insectivore shrews they grind their food with lophodont cheek teeth that resemble those of rodents and lagomorphs.

Ungulates

Ungulates are generally large herbivores, such as horses, sheep, and cows. They crop their food with fleshy lips or with incisor teeth and grind it up with a battery of cheek teeth. These herbivores are the primary food supply of carnivores and protect themselves primarily by running away. Adaptations for speed have entailed a greater lengthening of the legs than seen in carnivores, especially of their distal portions. Their feet are very long, and the animals walk upon their toe tips, a gait termed **unguligrade** (Fig. 40.17c). Toes that do not reach the ground are vestigial or absent, and the primitive claws on the remaining toes are transformed into hooves, a characteristic that gives the name ungulate to these mammals (L. *ungula*, hoof). Capacity for limb rotation is lost. The limbs are in effect jointed pendulums that can swing rapidly back and forth in the vertical plane.

Contemporary ungulates are grouped into two orders. In the **Perissodactyla** (Fig. 40.21a and b), the axis of the foot passes through the third toe, and this is always the largest. Perissodactyls are characterized by having an odd

927

number of toes. Ancestral perissodactyls, including the primitive forest-dwelling horses of the early Tertiary, had three well-developed toes (the second, third, and fourth) and sometimes a trace of a fourth toe (the fifth). The tapir and rhinoceros retain the middle three toes as functional toes, but only the third is left in modern, plains-dwelling horses. The molar teeth, which are low-crowned and bunodont in primitive perissodactyls, have become high-crowned and lophodont in modern species.

In the order **Artiodactyla** (Fig. 40.21c and d) the axis of the foot passes between the third and fourth toes, which are equal in size and importance. Ancestral artiodactyls had four toes (the second, third, fourth, and fifth). Pigs and the aquatic hippopotamuses, which move across soft ground, retain these four toes, though the second and fifth are reduced in size. Vestiges of the second and fifth toes, the dew claws, are present in some deer, but camels, giraffes, antelope, sheep, and cattle retain only the third and fourth toes. Artiodactyls are even-toed ungulates. The molar teeth of pigs are low-crowned and bunodont, but those of cattle are high-crowned and selenodont. Antelope, sheep, and cattle have no upper incisor teeth, and the lower incisors crop against a pad of flesh. It is probable that perissodactyls and artiodactyls have evolved separately for a long time and owe their points of similarity to parallel evolution.

In addition to running away, ungulates protect themselves by kicking with their powerful legs and hooves. Many artiodactyls also have evolved weapons for defense or for combat among males: wild boars have **tusks** that are modified canine teeth, male deer have **antlers**, and sheep and cattle of both sexes have **horns** (Fig. 40.22). Both antlers and horns are bony outgrowths from the skull. Antlers branch, are covered by skin (the velvet) only during their growth period, and are shed annually; horns are nonbranching structures, are covered by heavily keratinized skin, and are not shed.

Porpoises and Whales

Porpoises, dolphins, and whales (order **Cetacea**) are highly specialized marine mammals that appear to have diverged from very primitive ungulates. Whales from early Tertiary deposits are already adapted for an aquatic life. One recently discovered Eocene specimen retains small hind legs, but only traces of the pelvic girdle remain in modern species. Tertiary species also retain the primitive, heterodont mammalian dentition. In more recent species, the teeth become more numerous and take on a simple conical shape well adapted to seize fish. Contemporary cetaceans have a fish-shaped body, pectoral flippers for steering and balancing, no pelvic flippers, and horizontal fins or **flukes** on a powerful tail that is moved up and down to propel the animal through the water (Fig. 40.23a). Some species have even evolved a fleshy dorsal fin. Despite these fishlike attributes, cetaceans are endothermic, air-breathing mammals that are viviparous and suckle their young. Some hair is present in the fetus, but it is vestigial or lost in the adult stage, in which its insulating function is performed by a thick layer of fat (blubber).

Whales are divided into two groups: the toothed whales and the baleen whales. Dolphins, porpoises, and sperm whales retain teeth and feed upon fishes, squid, and other relatively large marine animals. Although sight is important in finding food, the toothed whales have evolved high-frequency underwater echolocation systems that help them find their prey and avoid obstacles.

The baleen whales, which include the humpback whale and the blue whale, have lost their teeth and feed

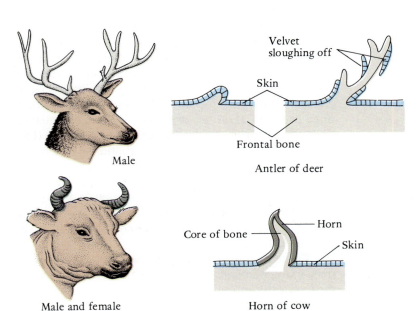

Male

Antler of deer

Velvet sloughing off

Skin

Frontal bone

Male and female

Horn of cow

Core of bone

Horn

Skin

Figure 40.22
Both antlers and horns are bony outgrowths from the skull, but the skin (velvet) covering antlers is shed as the antler matures, while the skin covering a horn becomes heavily keratinized and remains. Antlers are annual growths that are shed in winter; horns are permanent outgrowths.

(a)

(b)

Figure 40.23
Cetaceans. (a) A common porpoise, *Delphinus delphis*, is one of the toothed
cetaceans. Notice the fish-like dorsal and pectoral fins and the horizontal tail fluke.
(b) A young gray whale, *Eschrichtius robustus*. Notice the baleen plates on the roof
of its mouth.

upon plankton. With fringed horny plates (the **baleen**) that
hang down from the palate, a toothless whale strains small
shrimp-like crustaceans (krill) and other planktonic organisms from water passing through its mouth (Fig.
40.23b). The richness of the plankton together with the
buoyancy of the water has enabled these whales to attain
enormous size. The blue whale, which reaches a length of
27 m and a weight of 130 to 150 metric tons, is the largest
animal that has ever existed. Humpback whales make a
series of chirps, yups, groans, and other sounds ranging in
frequency from 40 Hz to 50 kHz. Some songs carry over
100 nautical miles. The songs are only emitted by males
and presumably function in finding mates and in other
sexual activities.

Subungulates

Elephants, hyraxes, and the manatees or sea cows appear
to be quite different kinds of animals, but they share many
distinctive characteristics indicative of a common origin.
Collectively they are often called **subungulates**. All arose in
Africa, but some have moved to other parts of the world.
None of them has a clavicle or collar bone. Elephants and
hyraxes have only three or four toes on each foot and these
end in hooflike nails. Canine teeth are absent in all subungulates, and the premolars and molars are lophodont in
elephants. New cheek teeth of elephants and manatees
develop at the back of each jaw and move forward. As they
move forward, they wear down and are finally lost.

Elephants (order **Proboscidea**) are the largest living
terrestrial vertebrates, and their adaptive features are associated with large size and the need to gather vast quantities
of coarse plant food (Fig. 40.24a). Their limbs are columnar. Most of the body weight is not supported by the toes
but by a large pad of elastic tissue behind them. A massive
head is needed to support the heavy trunk and dentition.
Their neck is short and elephants cannot reach the ground
with their mouths. The trunk or **proboscis**, which represents the drawn-out and fused upper lip and nose, is an
effective food-gathering organ. Elephants have one pair of
incisors, which are modified as tusks that are used to strip
bark from trees, to dig up roots, and for defense. A total of
six cheek teeth develop in each side of the upper and
lower jaws of an elephant during its lifetime, but they are
so large that there is room for only one tooth in each side of
a jaw at a time. As it is worn down, a new one moves in from
behind. By using up their premolars and molars one at a
time, total tooth life is prolonged. Large animals have a
proportionately smaller surface area relative to mass than
small animals; elephants compensate for this by having
enormous ears, through which heat can be lost.

Living elephants are restricted to Africa and tropical
Asia and are only a small remnant of a once worldwide and
varied proboscidean population. During the ice age, huge
species of proboscideans with thick coats of fur, known as
mammoths and mastodons, were abundant in North America.

The small hyraxes of Africa and the Middle East (order
Hyracoidea), though superficially resembling rodents,
show an affinity to the elephants in their foot structure and
in certain features of their dentition.

(a)

(b)

Figure 40.24
Subungulates. (a) The African elephant, *Loxodonta africana*. (b) A manatee,
Trichechus sp.

The manatees and dugongs (order **Sirenia**) live in warm coastal waters. They are unique among marine mammals in feeding upon seaweed, grinding it up with cheek teeth that are replaced from behind. Manatees have a powerful, dorsoventrally flattened tail and well-developed pectoral flippers (Fig. 40.24b). These features, together with a very mobile and expressive snout and a single pair of pectoral mammary glands, led mariners of long ago to regard them as mermaids.

Classification of Mammals

Class Mammalia (L. *mamma*, breast)   The mammals.
  Subclass Prototheria (Gr. *protos*, first + *therion*, wild beast)   Primitive mammals retaining many reptilian features, including the egg-laying habit and cloaca in living species. Teeth absent or variable, but molars not triangular.
    Order Triconodonta† (L. *tri*, three + Gr. *konos*, cone + *odous*, tooth)   Primitive Mesozoic mammals with molar teeth having three cones in a linear sequence.
    Order Monotremata (Gr. *monos*, single + *trema*, hole)   The platypus and echidnas.
  Subclass Theria (Gr. *therion*, wild beast)   Typical mammals. All living ones are viviparous.

Infraclass Trituberculata† (L. *tri*, three + *tuberculum*, small hump)   Two orders of Mesozoic mammals with molar teeth having three cusps arranged in a triangle; a talonid is sometimes present on lower molars. Ancestral to higher therians.
Infraclass Metatheria (Gr. *meta*, next to + *therion*, wild beast)   Marsupials. Young are born at an early stage of development and complete their development attached to teats that usually are located in a marsupium; typically three premolar teeth and four molars in each jaw.
    Order Marsupialia (L. *marsupium*, small purse)   Opossum, kangaroo, and other pouched mammals.

Infraclass Eutheria (Gr. *eu*, true + *therion*, wild beast) Placental mammals. Young develop to a relatively mature stage in the uterus; primitive dental formula is I 3/3, C 1/1, Pm 4/4, M 3/3. Living orders are listed below.

    Order Xenarthra (Gr. *xenos*, strange + *arthron*, joint) New World anteaters, sloths, and armadillos. Teeth reduced or lost; large claws on toes.

    Order Pholidota (Gr. *pholis*, scale) The pangolin, of Africa and southeastern Asia. Teeth lost; long tongue used to feed on insects; body covered with overlapping horny plates.

    Order Insectivora (L. *insectum*, insect + *vorare*, to devour) Insectivores, including tenrecs, shrews, moles, and the hedgehog. Small mammals, usually with long pointed snouts; sharp cusps on molar teeth adapted for insect eating; feet retain five toes and claws.

    Order Carnivora (L. *carno*, flesh + *vorare*, to devour) The carnivores or flesh-eating mammals: dogs, raccoons, bears, mink, cats, sea lions, seals. Large canines; certain premolars and molars often modified as shearing teeth; claws well developed in terrestrial species; appendages are flippers in marine species.

    Order Primates (L. *primus*, first) The primates: lemurs, tarsiers, monkeys, apes, and human beings. Primates are discussed in detail in Chapter 41.

    Order Chiroptera (Gr. *cheir*, hand + *pteron*, wing) The bats. Pectoral appendages modified as wings; hind legs small and included in wing membranes.

    Order Dermoptera (Gr. *derma*, skin + *pteron*, wing) The colugo of the East Indies and the Philippines. A gliding mammal with a lateral fold of skin.

    Order Macroscelidea (Gr. *makros*, large + *skelos*, leg) Elephant shrews. Shrewlike mammals with grinding molars.

    Order Rodentia (L. *rodere*, to gnaw) The rodents: squirrels, chipmunks, marmots, gophers, beavers, rats, mice, muskrats, lemmings, voles, porcupines, guinea pigs, capybaras, and chinchillas. Gnawing mammals with two pairs of chisel-like incisor teeth. The largest order of mammals.

    Order Lagomorpha (L. *lagos*, hare + *morphe*, form) Hares, rabbits, and pikas. Gnawing mammals with two pairs of chisel-like incisors and an extra pair of small upper incisors that lie behind the enlarged first pair.

    Order Tubulidentata (L. *tubulus*, small tube + *dens*, tooth) The aardvark of South Africa. Teeth without enamel; long tongue used to feed on insects.

    Order Artiodactyla (Gr. *artios*, even + *daktylos*, finger) Even-toed ungulates: pigs, hippopotamus, camels, deer, giraffes, antelopes, cattle, sheep, and goats. Axis of support passes between third and fourth toes; first toe lost; second and fifth toes reduced or lost.

    Order Perissodactyla (Gr. *perissos*, odd + *daktylos*, finger) Odd-toed ungulates: tapirs, rhinoceroses, and horses. Axis of support passes through third digit; lateral digits reduced or lost.

    Order Cetacea (L. *cetus*, whale) The whales and their allies. Large marine mammals; pectoral limbs reduced to flippers; pelvic limbs lost; large tail bears horizontal flukes, which are used in propulsion.

    Order Hyracoidea (Gr. *hyrax*, shrew mouse + *eidos*, form) Hyraxes. Small herbivores of the Middle East that superficially resemble guinea pigs; four toes on front foot, three on hind foot, each with a small hoof.

    Order Sirenia (Gr. *seiren*, a sea nymph that lured mariners to their death) Manatees and dugongs. Marine herbivores; pectoral limbs paddle-like; pelvic limbs lost; large horizontally flattened tail used in propulsion.

    Order Proboscidea (Gr. *proboskis*, trunk) Elephants and related extinct mammoths and mastodons. Massive ungulates; two upper incisors elongated as tusks; nose and upper lip modified as a proboscis.

†Extinct.

Summary

1. Endothermy probably evolved in ancestral mammals as they adapted to an insectivorous, nocturnal life style that was not available to the ectothermic reptiles. They probably maintained their body temperature close to the ambient nocturnal range, but those that later became diurnal maintained their temperature at a higher level. Mammals are able to maintain body temperature at a high level by controls on heat production and loss. Some have evolved special adaptations that enable them to survive in very cold or hot environments.

2. All of the organ systems have undergone changes as mammals became more active and agile. Their arched vertebral column includes three or more sacral vertebrae. Their limbs are usually rotated beneath the body. Their sense organs are highly developed, and the cerebrum and cerebellum of the brain are greatly expanded. Their jaw and tooth structures permit the mastication of food, the evolution of a secondary palate separates the food and air passages, and breathing can continue while food is in the mouth. Lung surface area is greatly expanded, and the lungs are ventilated by rib and diaphragm movements. Intestinal villi increase the digestive and absorptive area of the small intestine. Oxygen-depleted blood and oxygen-rich blood are completely separated in their passage through the heart. The kidneys can remove an increased volume of urea yet conserve salts and water.

3. Female mammals nourish their young on milk secreted by mammary glands. A delay in tooth eruption is correlated with nursing and is evidence that primitive, extinct species possessed mammary glands.

4. Mammalian characteristics gradually evolved among the synapsid, or mammal-like, reptiles, which diverged from other amniotes in the late Carboniferous. The change in the bones that bear the jaw joint and the correlated change in auditory ossicles occurred at the synapsid-mammal transition. Living monotremes are oviparous and have no teeth but are placed in the subclass Prototheria along with the ancestral triconodonts. All other mammals are viviparous and constitute the subclass Theria. Most have complex molar teeth.

5. Two different reproductive strategies have evolved among therian mammals. The trophoblast of the eutherian, or placental, embryo acts as an immunological barrier and prevents rejection of the embryo. The embryo develops an efficient chorioallantoic placenta. Eutherian mammals are born at a relatively advanced stage of development, but their maturity requires a considerable maternal investment. In contrast a shell membrane protects the early marsupial embryo from immunological rejection. After it has been absorbed, a placenta develops and functions for a short period. The embryos are born at a very immature stage, attach to nipples in the marsupium, and complete their development there. Few maternal resources are invested in the uterine young, and those in the marsupium can be aborted easily if food or other environmental resources fail.

6. With the extinction of many reptiles and the freeing up of environmental resources that occurred in the late Cretaceous period, marsupials and eutherian mammals underwent extensive adaptive radiations. An extensive marsupial radiation occurred in South America and Australia because these continents were isolated from other parts of the world for a long period of time.

7. Small, insect-eating ancestral eutherians rapidly diversified as they adapted to different diets and patterns of locomotion. Three orders (xenarthrans, pholidotans, and tubilidentates) specialized independently for a diet of ants and termites. Insectivores continued the ancestral mode of life. Carnivores specialized for feeding on larger prey organisms both on the land and in the sea. Primates, bats, and dermopterans became more arboreal, and bats evolved flapping flight. Elephant shrews, rodents, and lagomorphs are small herbivores with grinding, lophodont teeth. Artiodactyls and perissodactyls are larger herbivores with elongated legs and feet that allow them to run rapidly. Porpoises and whales became highly specialized for a marine life. Hyraxes, manatees, and elephants are a diverse group of herbivores that share several unusual dental characteristics.

References and Selected Readings

Many of the general references on vertebrates cited at the end of Chapter 35 also contain considerable information on the biology of mammals.

Eisenberg, J.F. *The Mammalian Radiations*. Chicago: University of Chicago Press, 1981. A scholarly analysis of evolutionary trends among mammalian groups and of mammalian adaptations and behavior.

Gingerich, P.D., et al. Hind limbs of Eocene *Basilosaurus*: Evidence of feet in whales. *Science* 249(1990): 154–157. The limbs are too small to have aided in locomotion and may have been used as copulatory guides.

Kanwisher, J.W., and S.H. Ridgeway. The physiological ecology of whales and porpoises. *Scientific American* 248 (June 1983):110–120. The adaptations of cetaceans for diving.

Kemp, T.S. *Mammal-like Reptiles and the Origin of Mammals*. London: Academic Press, 1982. A technical but very thorough account of synapsid evolution and mammalian origins.

Macdonald, D. (ed.). *The Encyclopedia of Mammals*. New York: Facts on File Publications, 1984. A beautifully illustrated account of the biology of all of the living mammalian families written by experts in the field.

Marshall, L.G. Land mammals and the great American interchange. *American Scientist* 76(1988):380–388. The merging of faunas following the late Tertiary connection of North and South America.

Novacek, M.J. Navigators of the night. *Natural History* (1988):67–70. A brief account of the echolocation system of bats.

————The skull of lepticid insectivorans and the higher-level classification of eutherian mammals. *Bulletin of the American Museum of Natural History* 183(1986): 1–112, 1986. A research paper on mammalian classification using cladistic techniques.

Schmidt-Nielsen, K. Countercurrent systems in animals. *Scientific American* 24(May 1981):118. A discussion of various countercurrent mechanisms, including the pattern of air flow in the camel's nasal passages, which reduces the amount of water lost in breathing.

Stonehouse, B., and D. Gilmore (eds.). *The Biology of Marsupials*. Baltimore: University Park Press, 1977. A collection of papers on the genetics, evolution, behavior, anatomy, and physiology, cell biology, and other aspects of marsupial biology.

Vaughan, T.A. *Mammalogy*, 3rd ed. Philadelphia: Saunders College Publishing, 1986. An excellent textbook with chapters on the origins of mammals, the various groups, ecology, zoogeography, behavior, and various aspects of mammalian physiology.

41

Primates and Human Evolution

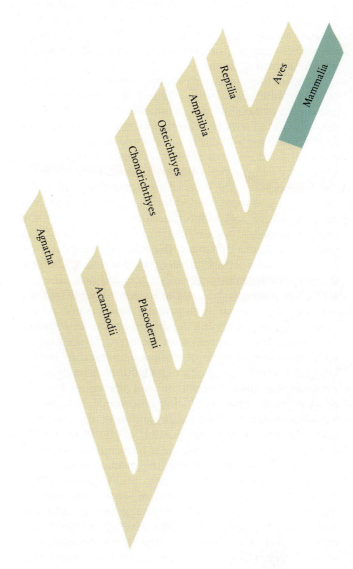

(Left) Their thick fur coats allow Japanese macaques, Macaca fuscata, to survive the snowy winters of Northern Japan. (Above) The Japanese macaque, or snow monkey.

The order **Primates** includes only about 180 species, but it is of great interest to us, for it includes human beings along with tree shrews, lemurs, monkeys, and apes. As we have seen (p. 920, Fig. 40.12), primates, chiropterans, and dermopterans are closely related to each other. They diverged in the late Cretaceous, some 60 million years ago, from a common primitive eutherian ancestor and quickly adapted to living in the trees of tropical forests.

PRIMATE ADAPTATIONS

A series of adaptations, allowing movement through the trees, evolved in the early primates. The limbs became increasingly flexible, hands and feet became increasingly prehensile, claws were transformed into fingernails and toenails, and the tail is retained in most species as a balancing organ.

Early primates fed on insects, cutting and crushing them with primitive triangular-shaped molar teeth. As primates became more omnivorous and included flowers, fruit, nuts, and other plant food in their diet, the molars became square and bunodont (see Fig. 40.19). Square molars now enable primates to crush food; the premolars remain as cutting teeth.

The early evolution of primates was profoundly influenced by their transition to arboreal life. Life in the trees required a major shift away from olfaction (smell) as the main source of sensory information, since olfactory trails were short-lived and easily interrupted in this new environment. In addition, movement through the trees—often high above the ground—placed a premium on vision and depth perception. The costs of misjudging the position of the branch you were headed for were great indeed. As a result of these and other pressures, the shape of the primate body and the architecture of the primate brain underwent dramatic changes. The olfactory organs and the olfactory centers of the brain decreased in size. In contrast, the eyes became far more important, and now face forward, rather than to the sides, as in many other mammals. The eyes cover overlapping visual fields (stereoscopic vision), greatly increasing depth perception. In addition, the visual cortex, the region of the brain devoted to the integration and analysis of images, increased in importance. The structure of the face changed accordingly, as the eyes became more prominent and the snout became proportionally smaller. Finally, movement through the trees demanded increased agility and muscular coordination. The cerebellum, the portion of the brain largely concerned with precise muscular movements, became a far more prominent feature in the primate brain.

The evolution of stereoscopic vision, increased agility, and the influx of a new sort of sensory information gained by handling objects with a grasping hand was accompanied by an extraordinary development of the cerebral hemispheres. To some extent the increased input of sensory information and the growth in size and complexity of the cerebrum were coupled. As more sensory information became available, increased cerebral size would have been favored. Conversely, as the cerebrum enlarged and became more complex, it in turn could deal with a greater sensory input. We do not know precisely at what point in our evolutionary past higher mental functions, such as symbolization and conceptual thought, first appeared. Such functions, however, represented a qualitative breakthrough in the abilities of the brain—a breakthrough that undoubtedly could not have taken place until the brain had attained a certain threshold of size and complexity under other selective influences. An increase in brain size and cranial capacity is one of the central themes and hallmarks of primate evolution.

Human beings have reverted to a completely terrestrial life, but bear the stamp of prior arboreal adaptations in the grasping hand, fingernails, tooth structure, stereoscopic vision, reduced sense of smell, large brain, and many other features. In a very real sense, we are a product of the trees.

THE GROUPS OF PRIMATES

Prosimians

Primates are divided into two suborders: the prosimians and the anthropoids (Fig. 41.1). The **Prosimii** include several groups of small, mostly arboreal and nocturnal primates of the Old World tropics, extending from Africa to the Philippine Islands. The tree shrews of India and southeastern Asia are small animals, superficially squirrel-like in appearance. They forage in the trees as well as in the ground litter (Fig. 41. 2a) and have acquired some arboreal adaptations, including large eyes and a first digit that diverges slightly from the others, giving it some rudimentary grasping ability. These are the sorts of specializations that we would expect to find in early primates, and thus we include the tree shrews in this order. However, their phylogenetic position is uncertain, and some authorities include them with the insectivores or place them in an order of their own. Other prosimians include the bush babies and pottos of Africa, the lorises of India and Sri Lanka, the tarsier (*Tarsius*) of the East Indies and Philippines, and the lemurs and aye-aye of Madagascar. All show the beginning of primate adaptations (Fig. 41.2b−d).

Although some lemurs retain rather long snouts, for olfaction is still important in their social structure, they have grasping feet, most of the claws are finger- and toenails, and the molar teeth, although still triangular, show indications of becoming bunodont.

The highest diversity of prosimians (including ten genera of lemurs and the aye-aye) is found on the island of Madagascar, an isolated fragment of the African Plate that separated from the African mainland some 80 million years ago. Prosimians should not be thought of as primitive monkeys, but rather as an early branch of the primates.

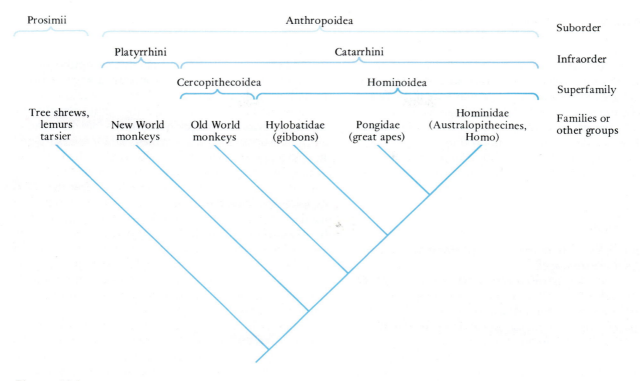

Figure 41.1
A cladogram showing the relationship between primate groups.

(a)

(b)

(c)

(d)

Figure 41.2
(a) The common tree shrew most resembles the ancient insectivores that gave rise to the primates. (b) A female black lemur, *Lemur macaco*, of Madagascar. (c) The large bat-like ears of the Senegal bush baby or galago, *Galago senegalensis*, enable it to track insect movements at night. (d) The tarsier, *Tarsius* sp.

Many evolved and diversified in the isolation of Madagascar. As a result of this chance isolation, prosimian lineages have not been replaced (as they have in the rest of Africa) by the more recent Old World monkeys and apes. Lemurs, however, may not survive the assault of the most recently evolved primate: human-produced deforestation of Madagascar has pushed many of the lemur species to the edge of extinction. The last living aye-aye was reported more than a half-century ago.

Anthropoids

Monkeys, apes, and human beings are placed in the suborder **Anthropoidea**. Anthropoids are primarily a diurnal group. They have a relatively flat face that is at least partly devoid of fur, well-developed stereoscopic vision, a complete bony wall forming the posterior wall of the orbit, the capacity to sit on their haunches and to examine objects with their hands, and an unusually large brain and globe-shaped braincase. They undoubtledly evolved from certain lemur-like prosimians, but the fossil record is not complete enough to be certain of the ancestral group.

New World monkeys and marmosets constitute the infraorder **Platyrrhini**. They differ from other anthropoids in having widespread nostrils in a rather flat nose and in retaining three premolar teeth. All are arboreal. Some have evolved a prehensile tail that serves as a fifth limb (Fig. 41.3a).

Old World monkeys, apes, and human beings are grouped in the infraorder **Catarrhini**. They are characterized by having downward-pointing nostrils in a narrower nose and only two premolar teeth. Contemporary Old World monkeys (superfamily **Cercopithecoidea**) constitute a large and diverse group, including the forest-dwelling langurs and green monkeys, the macaques (who are only partially arboreal), and the semiterrestrial baboons and mandrills. All are quadrupeds, but they tend to sit upright upon **ischial callosities**, which are hardened skin pads upon their buttocks (Fig. 41.3b). The bare skin around the callosities becomes swollen and brilliantly colored in many females during estrus. The thumb is completely opposable. Cusps of each molar tooth have fused to form two transverse ridges that help in grinding the plant food on which this group largely feeds. Many species have cheek pouches in which food can be stored temporarily. Social structure is highly developed in many species; baboons travel in troops led by a dominant male and cooperate in obtaining food and protecting the females and young.

Apes and human beings are placed together in the superfamily **Hominoidea** because they share many derived features (Fig. 41.1). They are larger than other anthropoids, lack a tail, and show more tendencies toward assuming an upright posture, at least some of the time. The smallest of the contemporary apes are the gibbon and siamang of Malaysia (family **Hylobatidae**). Their weight ranges from 5 to 13 kg. Ischial callosities develop late in life. Other, larger apes are called the great apes (**Pongidae**): the orangutan (*Pongo*) of Borneo and Sumatra, and the chimpanzee (*Pan*) and gorilla (*Gorilla*) of tropical Africa (Fig. 41.4). Some gorillas attain a weight of 270 kg.

Modern apes have limb specializations that enable them to **brachiate** (L. *brachium*, arm), i.e., to swing from branch to branch using their arms alternately. Their arms

Figure 41.3
(a) New World monkeys, such as this group of Costa Rican white-faced monkeys, *Cebus capuncius*, are characterized by their prehensile tails. (b) A rhesus monkey, *Macaca mulata*, showing the ischial callosities that characterize Old World monkeys.

(a)   (b)

(a)

(b)

(c)

Figure 41.4

Representative apes. (a) Gibbons, such as this white-handed gibbon, *Hylobates lar*, of Thailand and Malaysia, are the smallest of the apes and the best brachiators. (b) Orangutans, *Pongo pygmacus*. (c) An African gorilla, *Gorilla gorilla*, in its characteristic knuckle-walking gait.

are longer than their legs. They retain the grasping hind foot, but the thumb is short in relation to the elongated palm (Fig. 41.5a). The hand is used as a hook when brachiating; it is not as effective in grasping as is the hand of most monkeys and human beings, in which the thumb is relatively longer and can oppose the fingers (Fig. 41.5c). Gibbons and siamangs are the best brachiators and can

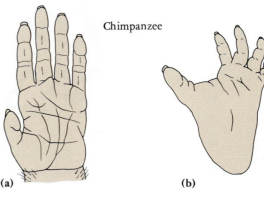

Chimpanzee

(a)   (b)

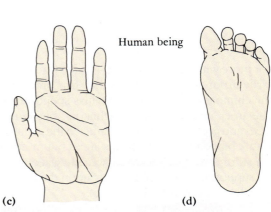

Human being

(c)   (d)

Figure 41.5

The hand of a chimpanzee (a) is adapted for brachiating and knuckle walking; its foot (b), for grasping. The human hand (c) is adapted for grasping; the foot (d), for bipedal walking.

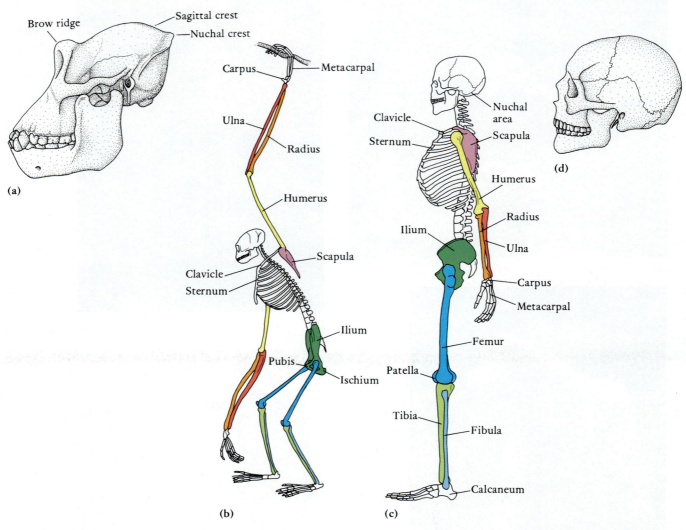

Figure 41.6

A comparison of the skeletons of an ape [(a) gorilla skull; (b) gibbon skeleton] and a human being (c and d).

clear 3 m or more with each swing (Fig. 41.4a). Orangutans are primarily arboreal climbers, but they can brachiate. The pygmy chimpanzee and young gorillas spend much time in the trees and are good brachiators, but larger chimpanzees and gorillas spend more time on the ground. They can walk bipedally over short distances but prefer a modified quadrupedal gait known as **knuckle walking**, in which they support the front of the body upon the knuckles of their elongated hands (Fig. 41.4c).

Apes are primarily herbivores, feeding upon fruit, young leaves, and other plant material. Occasionally they will eat meat. Their teeth are large and their jaws powerful, as is the case in many herbivorous mammals. Prominent bony brow ridges above the orbits help resist the stresses

set up in the skull by the powerful jaw mechanisms (Fig. 41.6a). The male gorilla's skull also has a large sagittal crest on the top of the head that increases the area available for the attachment of jaw muscles. The cusps of ape molars remain distinct and do not fuse to form transverse ridges, as they do in Old World monkeys. The tooth row has a somewhat squarish or **U**-shaped appearance, because the molars are in parallel rows, and the canines, which are used in displays of ritualized aggression, are large (Fig. 41.7a). Apes live in socially complex groups, organized in a hierarchy and led by a dominant male. They communicate with each other by primitive sounds and facial expressions, and can use sticks and other objects as tools to reach food (Fig. 41.8).

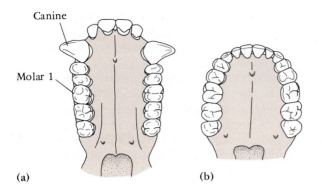

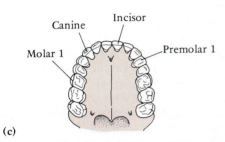

Figure 41.7
The palate and upper tooth row of (a) a gorilla, (b) *Austra-lopithecus*, and (c) a modern human. The large canine and parallel rows of cheek teeth give the tooth row of apes a nearly rectangular appearance. *Australopithecus* and humans have a more rounded tooth row.

Figure 41.8
The chimpanzee, *Pan troglodytes*, communicate effectively by facial expressions: (a) asserting dominance, (b) showing concern or fear, and (c) laughing.

HUMAN CHARACTERISTICS

Human beings differ morphologically from apes in so many ways that is is customary to place us in a separate hominoid family, the **Hominidae** (Fig. 41.1), but genetically we are very similar to the African apes. We differ from contemporary apes in being well adapted to a bipedal gait. Apes retain the primitive arched back of quadruped mammals, but we have a lumbar curve in our back that throws our shoulders back and places our center of gravity over the pelvis and hind legs (Fig. 41.6c). Our ilium is broad and flaring, providing a large surface for the attachment of gluteal and other muscles that hold us erect. Our legs are longer and stronger than our arms. The distal ends of our femur bones are brought close to the midline, giving us a knock-kneed appearance, but this places the foot under the projection of the body's center of gravity, thereby enabling us to balance easily on one foot when the other is carried forward off the ground (Fig. 41.9). Our foot has lost its primitive grasping ability, since the toes are short and parallel to each other (compare Fig. 41.5b and d). The heel bone (calcaneum) is large, the tarsals and metatarsals form strong supporting arches, and the great toe, with which we push off when walking, is enlarged. Our head is balanced upon the top of the vertebral column. The foramen magnum is far under the skull, and the nuchal area on the back of the skull, where neck muscles attach, is much reduced (Fig. 41.6). Body hair is very sparse in human beings, and a great increase in the number of sweat glands may be correlated with this.

Our thumb is longer and the metacarpal portion of our hand shorter than in apes (compare Fig. 41.5a and c), and the ends of our digits are broader. These features enable us to bring the ends of our fingers and thumb together and grip objects precisely. We use our hands not for locomotion, but for carrying things and for making and using tools.

We customarily include a great deal of meat in our diet as well as a wider variety of plant food. Although early humans retained powerful jaws, fire was used by later people to cook and soften food. Our teeth and jaws are less massive than those of apes, and our face does not protrude as much (Fig. 41.6). Our tooth row is more rounded, and the canines are small (compare Fig. 41.7a and c).

Our brain differs from that of apes not only in size but also in organization. The parietal, frontal, and temporal areas of the cerebral cortex are enlarged (see Fig. 17.21). These are regions that are important in sensory and motor integration and in association, memory, and speech.

Human patterns of reproduction and development also differ in important ways from those of apes. Adult female apes seldom copulate with males except during the

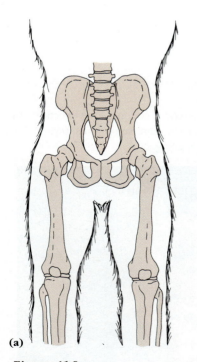

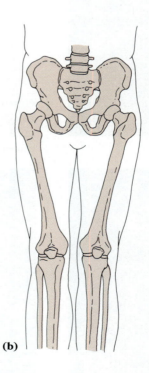

(a) **(b)**

Figure 41.9

(a) The femur bones of apes extend straight down from the pelvis, as they do in quadrupeds; (b) those of human beings slant inward, bringing the knees close to the projection of the body's center of gravity.

few days of estrus, which occurs near the middle of their monthly ovarian cycle. They do not copulate during pregnancy or lactation. Birth is relatively simple because the fetus's head is substantially smaller than the pelvic canal. The single young is at a relatively mature stage at birth and within a few days can cling to its mother's fur with its hands and feet as she moves about. Only one young is raised every few years. Correlated with increased pair bonding and a more complex social structure, human beings are sexually more active. Mature females are receptive to copulation throughout their ovarian cycles and during pregnancy and lactation. The human infant's head is relatively large, and birth sometimes is difficult. Newborn are quite immature and must be carried by one of the parents for months. During her reproductive years a human female can, and frequently does, give birth every year or two. Juvenile and adolescent life is prolonged for years as the children acquire the brain development and knowledge they will need to become integrated into a complex society.

EARLY EVOLUTION OF APES AND HOMINIDS

Humans differ, then, from contemporary apes in locomotion, the use of our hands, diet, brain size and capacity, reproduction and development, and complexity of our social organization. These differences become blurred as we trace lines of descent back in time. At some point, apes and hominids had a common ancestry, but when was this? A number of fossil primates have been found in Oligocene deposits in Egypt that go back 30 million years. Some of these had globe-shaped braincases and rather flat faces. They probably were early apes. One species shows some gibbon-like tendencies, so it is possible that gibbons and siamangs diverged from other apes very early. Other species may have been close to the ancestry of Old World monkeys.

The earliest pongids (great apes) are found in African Miocene deposits that date from 15 to 20 million years ago. Among the best known are *Proconsul* (Fig. 41.10) and several species of *Dryopithecus*. Remains of the limb skeleton indicate that they were not as specialized for brachiating as modern apes. They probably were agile creatures at home both in the trees and on the ground.

During the Miocene, Africa became more intimately connected to Europe and Asia (see Focus 9.3, p. 174), and primitive apes migrated into these areas. The first discovered and fragmentary remains of *Ramapithecus* from early to mid-Pliocene deposits in India showed certain humanlike tendencies, including small canines, thick enamel on the teeth, and what at first appeared to be a rounded tooth row. Many paleoanthropologists, students of human evolution, regarded it as an early ancestor of ours and believed that humans diverged from pongids 15 to 20 million years ago.

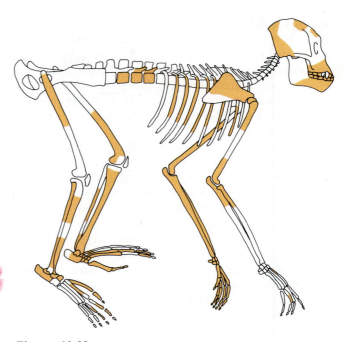

Figure 41.10

A reconstruction of the skeleton of *Proconsul*, a Miocene ape. Colored bones represent known parts of the skeleton.

A challenge to the phylogenetic position of *Ramapithecus* came from unlikely quarters: comparative genetics. In 1963, the biochemists Emil Zuckerkandl and Linus Pauling had proposed the concept of the molecular clock. They argued that the amino acid sequences of homologous proteins in different species could be used as a guide to the phylogenetic relationships of those species. Their argument rested on the assumption that the more closely related species on average share more similar proteins. They then proposed a bolder hypothesis: the differences between homologous proteins in two related lineages accumulated in roughly clocklike fashion. Proteins, in short, diverged at constant rates. If that were true, then the differences between, for example, hemoglobin in humans and that in house mice could be used to estimate the time elapsed since those two lineages last shared a common ancestor.

The notion of a molecular clock was profoundly appealing, since it offered the possibility of reconstructing evolutionary events even when fossil evidence was not available. Furthermore, it meant that the data of molecular biology could be used in evolutionary studies.

Beginning in 1967, two scientists, Allan Wilson and Vincent Sarich, began comparing the serum albumins of contemporary primates with the goal of establishing a molecular branching diagram of the primate lineages. Once the branching pattern was established, the rate at which the molecular clock was ticking could be calibrated. Given that the fossil record clearly indicated that monkeys and apes had separated some 30 million years ago, and that ape and monkey albumins differed, on average, by 24 amino acid residues, this suggested a rate of one amino acid substitution per 1.25 million years. Surprisingly, however, ape and human albumins differed at only four positions. This meant, if the model was correct, that apes and humans had last shared a common ancestor as recently as 5 million years ago.

This conclusion cast serious doubt on the status of *Ramapithecus*, 15 to 20 million years old, as an early hominid ancestor. Paleoanthropologists at first tried to refute the hypothesis of a molecular clock, but they were also forced to restudy the fossil material. More recent and complete discoveries of *Ramapithecus* from Pakistan showed it not to be a hominid. Instead, it now appears similar to *Sivapithecus*, an early ape from Turkey and Pakistan that is believed to be an ancestor of the orangutan. Fifteen million years ago, the split between the great apes and the hominid lineage that would eventually give rise to *Homo sapiens* had not yet taken place.

A large number of comparative studies seeking to establish the timing and order of branching within the Hominoidea have been carried out over the last decade. These studies have used a number of different proteins, as well as direct DNA sequence comparisons. A consensus is beginning to emerge, suggesting that the hominid branch split off from that of our closest relatives some 5 to 9 million years ago. Nevertheless, despite the enormous attention focused on this question, many of the details of our own recent past remain unclear.

Until about 10 million years ago, East Africa was heavily forested. Shortly thereafter, shifts in the earth's crust in East Africa, volcanic activity, and other geological disturbances led to the development of a more arid climate. Forests began to break up, and open woodlands interspersed with tropical grasslands, or savannahs, became more prevalent. It is likely that these changes provided the context in which ape and human lines of evolution separated. Apes remained forest animals; early hominids, although undoubtedly spending some time in the trees, adapted to the newly formed, more open savannah country.

THE AUSTRALOPITHECINES

In 1924, before many of the ape fossils we discussed above had been found, an endocranial cast and part of the skull of a child were discovered in cave deposits in South Africa. Anatomist Raymond Dart described it as *Australopithecus africanus* (L. *australis*, south + *pithecus*, ape). Subsequently, other specimens have been discovered in many parts of the Great Rift Valley extending north through East Africa into Ethiopia. Many of the fossils were assigned originally to different genera, but most paleoanthropologists now agree that they represent varieties of four or five species of the genus *Australopithecus*.

The best known of the species of *Australopithecus* were lightly built, slender or gracile creatures that stood 1.2 m tall. Most are assigned to *Australopithecus africanus*, but the oldest specimens, one discovered by Donald Johanson and Tim White in Pliocene deposits at Hadar in Ethiopia (a female, "Lucy") and one unearthed by Mary Leakey at Laetoli in Tanzania, are slightly different from later specimens. Johanson and White consider them to be a distinct species, which they named *A. afarensis*. (Fig. 41.11). The oldest individuals date to 3.7 million years ago. Many paleoanthropologists consider *A. afarensis* to be the ancestral species that gave rise to *A. africanus* between 2.5 and 3.0 million years ago; others regard these two as the same species (Fig. 41.12).

When first discovered, these gracile australopithecines were thought to be apes rather than hominids, but the configuration of the skull, pelvis, femur, and foot indicate that, like contemporary humans, they were bipeds. The small canines and rounded shape of the tooth row are human (Fig. 41.7b), but they had very large molar teeth with an unusually heavy coating of enamel. The jaws were massive and brow ridges well developed (Fig. 41.13a). Their cranial capacity ranged from about 430 to 500 cc, which is close to the brain size of the great apes (chimpanzee = 280 to 500 cc). In comparison, modern humans have an average cranial capacity of 1350 cc. The australopithecines are now regarded as the earliest members of the Hominidae, the family to which all humans belong.

Figure 41.11
The skeletal remains of "Lucy," *Australopithecus afarensis*, the oldest known hominid.

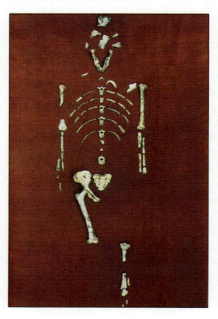

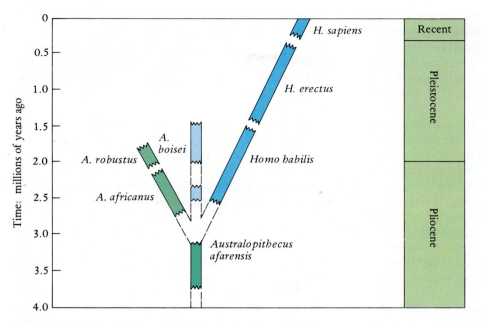

Figure 41.12
The hominid phylogenetic tree. The known geological distribution of the species
has been colored.

Figure 41.13
Lateral views of the skull and lower jaw of (a) *Australopithecus afarensis*; (b) *Homo
habilis* (lower jaw not known); (c) *Homo erectus*; (d) *Homo sapiens* (Neanderthal);
and (e) *Homo sapiens* (Cro-Magnon).

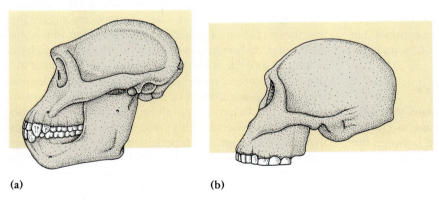

(a)   (b)

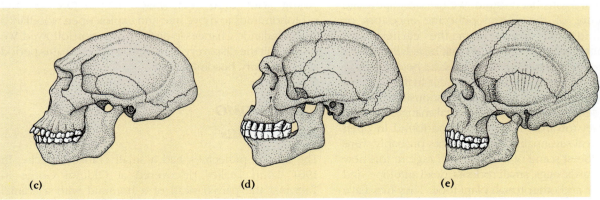

(c)   (d)   (e)

The remarkable increase in cranial capacity relative to body size that occurs throughout hominid evolution is the main theme of our own evolutionary past (see Focus 41.1). We, modern *Homo sapiens*, are unique among our fellow species in the dimension of our brains. With it comes the capacity for language, for symbolic thought, and for culture.

Tempting as it might be to see ourselves as the culmination of an inexorable evolutionary process, we are not. Rather than sitting at the top of an imaginary evolutionary ladder, we instead are the last surviving twig of what was once a bush containing several hominid species. The fossil remains of two species in the genus *Australopithecus* and a single species of the genus *Homo* (*Homo habilis*) have all been found in East African Pleistocene deposits of roughly the same age. The evidence from the Pleistocene is clear: approximately 1.5 to 2 million years ago, three (and possibly four) separate species of hominids coexisted on the African continent.

The lightly built *A. africanus* was eventually replaced in South Africa a little over 2 million years ago by a larger species, *A. robustus*, who stood about 1.5 m. tall (Fig. 41.12). Another large species, *A. boisei*, lived in East Africa and may have evolved independently from *A. afarensis*, since a *boisei*-like skull discovered in 1986 is about as old as the oldest *A. africanus* specimens. Both large species were adapted to eat tougher food, possibly even to crack open bones with their teeth, because the teeth and jaw apparatus are truly massive. As in male gorillas, a sagittal crest on the skull provided extra surface area for the attachment of large jaw muscles. *A. robustus* became extinct about 1.8 million years ago, but *A. boisei* continued in the early Pleistocene until about 1.2 million years ago.

Evolution of the Australopithecines

Although molecular data indicate that humans separated from apes between 5 and 9 million years ago, no fossils have yet been found for this crucial period between the apes of the Miocene and the first appearance of *Australopithecus* in the late Pliocene. We do not know just when the divergence occurred nor have any intermediate or transitional fossils been found. We can only speculate on the nature of the australopithecine ancestors or protohominids. We know from the earliest australopithecine fossils (*A. afarensis*) that bipedalism and a change in tooth structure are fully evolved before significant increases in brain size begin to occur in the hominid lineages. What selective forces are responsible for the changes in dentition and posture? Dental changes indicate a shift in diet towards the type of food found in open woodlands and savannahs. Our ancestors probably were leaving the forest some of the time to forage in this new habitat on lizards, eggs, small rodents, dead antelope, and seeds, tubers, and other tough plant food. They may have utilized sticks to help dig or catch food, in a manner similar

to that of contemporary chimpanzees and gorillas. They may have carried some of their food back to sheltered areas to eat. Many paleoanthropologists believe that bipedalism evolved as an efficient way to carry food back to the forest environment or to a central "camp." In addition, bipedalism makes it possible to carry infants in arms. In effect, a positive feedback loop is established: infants could no longer cling to their mothers in apelike fashion, because feet becoming adapted to bipedalism lose their ability to grasp. In response, the ability to carry infants in arms is favored, further increasing the selective advantages of bipedality. The ability to carry infants would also enable protohominids to forage and scavenge over a wider area.

Changes in reproductive habits leave no direct fossil record, but it is quite possible that changes in this important aspect of hominid behavior were taking place as well. A more dependent infant would require more parental attention. More frequent sexual access to the female, made possible by the evolution of continuous sexual receptivity, would act to strengthen the pair bond between males and females. Some indirect evidence for the evolution of pair bonding can be found in the fossil record of early *Australopithecus*. As a general rule, there is a close correlation between the degree of male-male competition for females and the differences in size between females and males of a given species. Thus, in species that are polygamous (males have access to many females), dimorphism is proportionately greater than in monogamous (pair-bonded) species. The degree of sexual dimorphism of early *Australopithecus* skulls is substantially smaller than that found in fossil or modern ape species, suggesting that the evolution of pair-bonding may be an important aspect of early hominid evolution. Pair-bonding may have helped ensure the survival of a greater proportion of the young. Our current models suggest that changes in habitat and diet, the acquisition of bipedalism, and changes in reproductive behavior may all be linked. A variant of this scenario, proposed in 1981 by C. Owen Lovejoy, argues that sexual changes and bipedalism evolved as a strategy that increased infant survival while the protohominids were still primarily forest dwellers. The acquisition of bipedalism in the context of providing the female and young with food then made it possible for the protohominids to move into and exploit open woodlands and grasslands. Changes in diet and dentition followed. We must await the discovery of more fossils before this period of our history becomes clear.

EARLY HOMO

Homo habilis

The australopithecines had a small cranial capacity. In 1961, Louis Leakey discovered at Olduvai Gorge in Tanzania the partial skull of a hominid with a cranial capacity of 700 cc, 200 cc larger than that of the australopithecines. Stone tools were associated with the remains, so Leakey named his discovery *Homo habilis* (L. *homo*, man + *habilis*, handy). The tools were little more than rounded stones sharpened by breaking off one end (Fig. 41.14a), yet cut marks on associated animal bones provide convincing evidence that they were used to cut meat from bones, possibly during scavenging. Subsequent discoveries, some made by Leakey's son, Richard, show that *H. habilis* had a cranial capacity ranging from a little over 500 to 750 cc (Fig. 41.13b). Fragments of arm and leg bones indicate a height of about 1 m. Body stature was no greater than in the gracile australopithecines, although the brain had begun to enlarge. Tool use may have been an ability first acquired by *Homo habilis*, but this view has recently been challenged by Randall L. Susman, who has discovered finger and thumb bones of *Australopithecus*

Figure 41.14
Paleolithic tool kits. (a) *Homo habilis* made cutting tools by knocking off large flakes from one end of a stone. (b) *Homo erectus* made more elaborate tools by first shaping a stone and then secondarily removing smaller flakes from the edges. (c) Neanderthals, archaic representatives of *Homo sapiens*, made a variety of more refined tools.

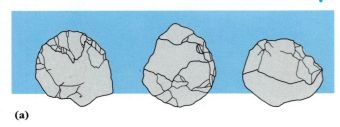

(a)

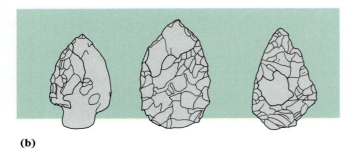

(b)

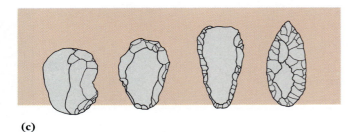

(c)

robustus that have the broad tips associated with the precision grip needed to make stone tools.

Increased cranial capacity and the use of tools indicate that hominid evolution was beginning to favor an increase of brain size and complexity and the development of hunting or scavenging skills. Growth in the size of the brain and improvement in tools—trends that dominate later human evolution—also have developmental and reproductive implications. For a brain to enlarge, its growth must accelerate relative to other parts of the body. Brain enlargement could pose problems at birth because head size would approach or exceed the size of the pelvic canal. One way large head size could be accommodated would be for the infant to be relatively immature at birth. Human infants are, in fact, less well developed at birth than those of apes and therefore far more dependent on parental care for longer periods. More parental care, in turn, requires greater cooperation between males and females, and increased pair bonding. More sophisticated tools also imply more complex social structures and more efficient ways of communicating. All of these changes must have been under way in *Homo habilis*. Most paleoanthropologists believe that *H. habilis* diverged from the australopithecines between 2.5 and 3 million years ago (Fig. 41.12).

Homo erectus

Late in the 19th century, when Charles Darwin's book prompted people to think about our evolutionary past, the German biologist Ernst Haeckel postulated the existence of a primate that would establish a connection between apes and modern human beings. This stirred the imagination of a Dutch physician, Eugene Dubois, who searched diligently for the "missing link" in Java, finally discovering in 1894 the fossils of a primitive hominid he called *Pithecanthropus erectus*. Shortly afterwards, Davidson Black discovered in caves near Beijing (Peking) the remains of similar creatures, which he named *Sinanthropus pekinensis*. Other fossils of similar hominids have been found in the East Indies, China, Africa, and Europe and have been given a variety of names (Algerian man, Heidelberg man), but paleoanthropologists now regard all of them as representing a single, widespread species called *Homo erectus*.

Homo erectus stood upright and, with a height of about 1.7 m, was considerably larger than *Homo habilis*. Brain size ranged from 730 to 1220 cc, which overlaps that of *Homo habilis* on one extreme and that of modern humans on the other. Frontal areas of the brain, however, were poorly developed, since *Homo erectus* had a low sloping forehead (Fig. 41.13c). The face was somewhat brutish, protruding slightly, with rather heavy brow ridges and no chin. Teeth, though large, were essentially modern in their configuration.

If we were to compare the brain sizes of a mouse lemur and a human being, there is little doubt as to which would be larger. By definition, larger organisms tend to have larger "parts." But we cannot be content with that superficial statement. Ideally, we want a technique that takes into account the differences in overall size that exist between mouse lemurs and humans, and therefore allows us to ask the following: is our brain *disproportionately* large for our size?

One simple technique involves plotting the cranial capacity of a variety of primates as a function of body weight (Fig. a). When the log of cranial capacity is plotted against body weight, the result is a straight line (Fig. b). This comes about because cranial capacity does not increase in direct proportion to body weight. The relationship instead is said to be *allometric*: Brain size increases as a power function of body size. The general relationship is of the form

$$\text{brain size} = C_1 (\text{body weight})^a$$

where C_1 is a constant.

For the primates as a whole, the value of a, the *allometric coefficient*, lies between 0.3 and 0.5, and corresponds to the slope on graph b. As you can see from the figure, however, three points lie far above the line: all 3 hominids depicted have disproportionately large brains for their size. Of these, the most extreme is *Homo sapiens*.

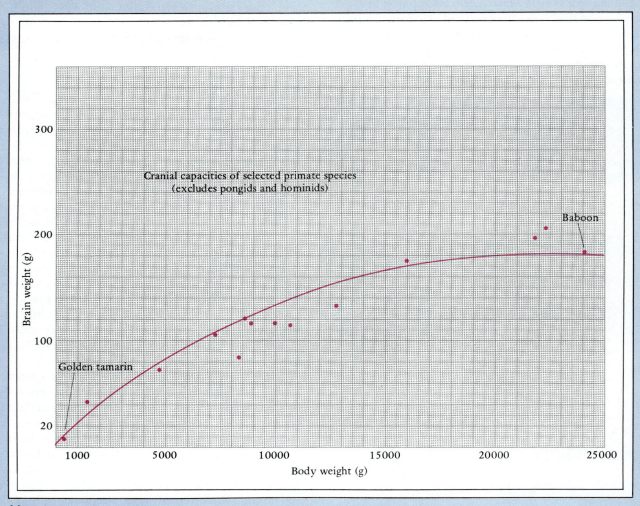

(a)

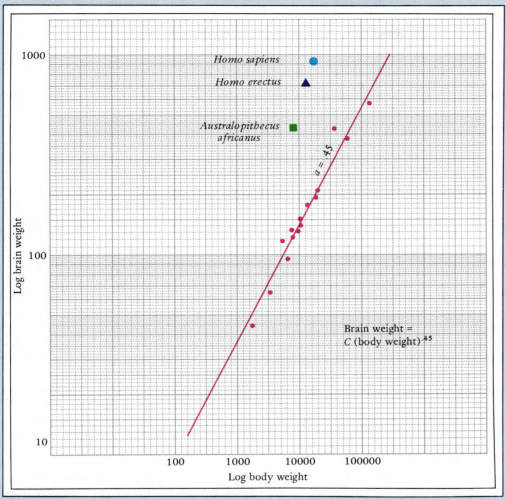

(b)

Compared to the rest of the primates, our brain size is roughly 2.5 times what we would expect on the basis of our body size.

The last graph (Fig. c) shows a similar analysis for fossil hominids. Hominids as a whole show a clear trend towards increased body size—Lucy, you will recall, stood roughly 1.3 meters tall. Superimposed on this trend, however, is an accelerated increase in cranial capacity, particularly in the members of the genus *Homo*. In contrast to Australopithecines, which do not show an increase in relative cranial capacity (and hence lie along a line of slope 0.33), the members of the genus *Homo* lie along a line of slope 1.73. Phrased differently, the brain size of *Homo* is increasing far faster than body size (nearly the *square* of body size). A larger brain and all the potential that comes with it were clearly of enormous selective advantage to our closest ancestors.

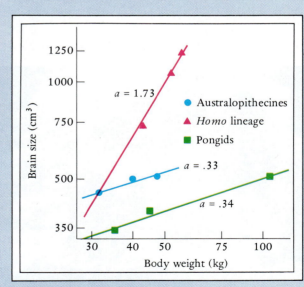

(c)

The oldest specimens of *Homo erectus* have been discovered in Olduvai Gorge in deposits lying above those containing *Homo habilis* and dating back in time 1.5 million years. The youngest remains so far discovered are about 300,000 years old (Fig. 41.12). During this period *Homo erectus* developed a sophisticated culture, including a variety of stone tools. Large, pear-shaped hand axes used for chopping and pounding were fashioned from a piece of flint or other fine-grain stone by secondarily chipping away the edges (Fig. 41.14b). A variety of smaller scraping and boring tools were made from flakes of stone chipped from the larger pieces. Charred bones indicated the use of fire. Fire, and perhaps crude clothing made of animal hides, would have been essential for this hominid to penetrate central Europe and Asia in the Pleistocene, about 410,000 years ago, when continental glaciers were advancing and the climate was quite cold. It is clear that *Homo erectus* was a hunter of large game, since their campsites contain the bones of bears, horses, and even elephants. Remains from a hunting site in Spain show that they set brush fires to drive game into ambush. These were intelligent people, living in socially complex groups, with an ability to communicate and teach their young tool making, hunting, and a knowledge of the seasons and habits of game (Fig. 41.15).

HOMO SAPIENS

People living in Africa and Eurasia 300,000 years ago had heavily built skulls with thick bones, traces of brow ridges, and large protruding jaws with receding chins, but all of their features are within the range of variation of modern people. The Steinheim skull from Germany and the Swanscombe skull from England are examples of this level of development. These people are assigned to our own species, *Homo sapiens* (L. *sapere*, to be wise), but are considered to be archaic representatives. Another archaic form, known as Neanderthals (Fig. 41.13d) because the fossils were first discovered in the Neander Valley of Germany (German *thal*, valley), occupied unglaciated parts of Europe during most of the last glacial period 70,000 to 35,000 years ago. Neanderthals are well known, for many specimens have been found. They often built their fires in hearths in cave floors, and their culture included a large tool kit of stone axes, scrapers, borers, knives, spear points, and saw-edged and notched tools probably used in making spear handles and other simple wooden implements. The stones show considerable secondary chipping to refine the shapes and sharpen the edges (Fig. 41.14c). Some paleoanthropologists believe the Neanderthals had developed a belief in an afterlife, because some fossilized skeletons have been found in depressions in cave floors that suggest intentional burial. But other paleoanthropologists argue that natural causes could have simulated burial. Another discovery at Shanidar in Iran suggested that the buried person wore a garland of flowers, for a circle of pollen was found on the chest.

Modern humans, *Homo sapiens*, have more delicate features, thinner skull bones, smaller jaws with well-developed chins, and a flatter face. Once again, keep in mind that we are not dealing with the transformation of a single lineage over time: Neanderthals and modern *Homo sapiens* were contemporaries for some time and may have been entirely separate species. Eventually, modern humans replaced Neanderthals in western Europe about 35,000 years ago, as the last continental glaciers retreated. Cro-Magnon people would have been indistinguishable from modern Europeans (Fig. 41.13e). Some 35,000 years ago this group developed a very sophisticated culture,

Figure 41.15
Reconstruction of the life of *Homo erectus* living in the Pleistocene at Peking.

including delicate stone, bone, and wooden tools. They are the artists responsible for the famous cave paintings of Lascaux in France and Altamira in Spain.

Although there is some controversy concerning the origins of modern *Homo sapiens*, they most likely radiated from their center of origin somewhere in Africa. They quickly occupied the African, European, and Asian continents, and eventually reached the American continent by crossing the Bering Strait. Just when modern humans first crossed the Bering Strait from Siberia to the New World is uncertain. The oldest reliable radiocarbon dates for human activity in the New World come from discoveries of a human encampment in Chile that may date from 13,000 years ago. To have reached Chile by this time, humans must have crossed from Siberia a thousand or more years earlier. Geological data indicate that a crossing could have been made between 14,000 and 25,000 years ago, near the end of the last glacial period, for sea level was about 100 m lower than today. A large plain would have connected Siberia and Alaska.

The worldwide spread of our ancestors placed them in a wide variety of different habitats and resulted in the evolution of certain differences between the various populations. Some of these appear to be direct responses to differences in the environment. Skin color, for example, appears to be closely linked to the levels of solar radiation. Body shape also appears to correlate with climate: shorter, stockier body shapes characterize peoples living in cold climates. Different human populations also exhibit differences in gene frequencies, including blood groups and certain serum proteins.

Despite the social importance placed on these differences between races, they are, from a biological standpoint, vastly overshadowed by differences between individuals. Furthermore, the modern history of human populations is one of constant gene flow between populations. Wherever people are brought together—by invasions, colonizations, migrations, or alliances—intermarriage will follow, reducing the differences between human populations.

Paleoanthropologists are not in agreement as to when populations of modern *Homo sapiens* diverged to form the races of humanity. Some believe that racial divergence can be traced back to already differentiated populations of *Homo erectus* living in different parts of the world, each of which gave rise independently to a population of modern humans. If this is the case, then *H. erectus* and *H. sapiens* are not distinct species in the normal biological sense. Most paleoanthropologists find the multiple independent origin of *Homo sapiens* implausible. They argue that *H. erectus* and *H. sapiens* are distinct species and that racial divergence occurred more recently, after *H. sapiens* arose in Africa, spread out throughout the Old World, and eventually replaced *H. erectus*. Important information concerning this question has come from comparative study of mitochondrial DNA (see Focus 41.2). The emerging consensus is that human races diverged very recently, perhaps as few as 40,000 years (2000 generations) ago.

HUMAN POTENTIAL AND RESPONSIBILITIES

The evolution of human beings was not predestined. We, like all other organisms, are the product of random mutation, natural selection, genetic drift, isolation, and other evolutionary forces. Evolution is a contingent process, and many chance events coincided to lead to our evolution. The outcome would have been different had certain small mammals not survived the Cretaceous extinction, or if the australopithecines had not adapted successfully to the climatic changes that occurred in East Africa 4 million years ago. We are but one of millions of contemporary species, each one the current representative of its particular evolutionary lineage. But of all animals, we alone are aware of our evolution and place in nature. Our genetic heritage has given us the ability to develop cultural and social structures that we can modify by experience and pass on through tradition and education to future generations. Our culture has enabled us to manipulate and control our environment and exploit the earth's resources to a greater extent than any other species. We have the potential of doing vast harm to the environment by overpopulation, pollution, and habitat destruction, or we have the potential to live in harmony with the rest of the natural world around us. To make rational judgments as to the effects of our actions, we must understand the interrelationships of organisms and the delicate balances of nature. It is to these topics that we turn in the concluding section of this book.

Classification of Primates

Order Primates (L. *primus*, the first)
Suborder Prosimii (Gr. *pro*, before + L. *simia*, ape) Lemurs, lorises, tarsiers.
Suborder Anthropoidea (Gr. *anthropos*, human being + *eidos*, resembling)

Infraorder Platyrrhini (Gr. *platys*, flat + *rhis*, nose) New World monkeys.
Infraorder Catarrhini (Gr. *kato*, downward + *rhis*, nose)

Two recurrent issues in primate evolution have been recently reinvigorated (though not necessarily resolved) by new molecular approaches to evolutionary reconstruction.

The first of these concerns the relationships and branching pattern of the Hominoidea. While it is clear on anatomical grounds that the gibbons, great apes, and humans belong together in a single monophyletic group, the exact relationships remain controversial. The most difficult pattern to resolve is the one closest to us: are chimps or gorillas our closest living relatives? The traditional answer places chimps and humans on the last fork of the branch, but recent DNA sequence analyses have not succeeded in actually resolving the trichotomy, and some workers suggest that gorillas and chimps may be equally distant from human beings.

The second controversy surrounds the origin of modern *Homo sapiens*. In an attempt to address that question, biologists have focused on mtDNA, the molecule of DNA carried by the mitochondria (see p. 78). Because during fertilization only the head of the sperm enters the egg, mitochondria are inherited only through the female line (see Fig. 21.8). Additionally, mutations accumulate more rapidly in mtDNA than in nuclear DNA, possibly because mtDNA lacks the repair mechanisms that tend to stabilize nuclear DNA. For these reasons, mtDNA is a fast molecular clock by which to date relatively recent evolutionary events. This "clock" has been calibrated by measuring the accumulated differences in the mtDNA of primates whose time of divergence can be dated from the fossil record. The rate has been estimated at 2% to 4% per million years, but it is not certain that the rate has remained uniform. The reasons underlying the study of human mtDNA diversity are clear: if all modern humans originate from a single original group of *Homo sapiens* somewhere in Africa, we all must have at some point carried similar or identical mtDNA molecules. Since then, as humans have spread over the globe, differences in mtDNA between isolated human populations must have accumulated. By quantifying the differences that have accumulated in the mitochondrial genomes of the human population, we in effect estimate the time when all humans shared a common mtDNA. Allan C. Wilson, Rebecca L. Cann, and Mark Stoneking have taken samples of mtDNA from the placentas of 147 Caucasians, Africans, Asians, native Australians, and New Guineans. The mtDNA from each individual is cut into fragments using 12 different restriction enzymes, each of which cleaves the molecule at unique groups of

bases. On average each molecule is cut at 370 sites. Mitochondrial DNA fragments from different individuals are separated electrophoretically (see Focus 3.1, pp. 50–51) and their positions subsequently visualized by staining or radioactive tagging. Mitochondrial DNA fragments from genetically different individuals form distinctive patterns or signatures on the gels.

The researchers found 134 different patterns or *haplotypes*. A "family tree" was then constructed by using a computer program that linked the different haplotypes by minimizing the number of steps needed to go from one haplotype to the next. An abridged version of this tree is shown in the Focus figure. The 133 different types fell into two major clusters, or trunks, one exclusively African and the other containing representatives of all human populations. The exclusively African branch of the tree contained the highest diversity of haplotypes and also the most divergent haplotypes (the deepest fork), confirming that African populations are the oldest populations of modern *Homo sapiens* and have therefore accumulated the most diversity. Secondly, the remaining clusters are all heterogeneous mixtures of the various races, further underscoring the unity of the human species. The scale in the lower part of the figure shows sequence divergence, an estimate of the extent to which two mtDNA sequences have diverged. We know that, on average, mtDNA molecules diverge at a rate of 2–3% per million years, allowing us to use the differences between mtDNA molecules as a "molecular clock." The two major branches of the tree converge near the base of the tree at a point estimated to be some 141,000 to 290,000 years ago. This last result, labeled the "mitochondrial Eve" hypothesis, suggested that all modern humans may have evolved from a single small band of females, all sharing the same mtDNA haplotype, that lived in Africa some 200,000 years ago (Focus Fig.). But the data support alternative interpretations. Modern *Homo sapiens* could have evolved much earlier, and then, about 200,000 years ago, the human population could have undergone a severe reduction in size, "crashing down" to a few individuals and eliminating the majority of haplotype diversity. Present populations would then have developed from the few survivors, creating the illusion of a mitochondrial "Eve." Alternatively, the common ancestral haplotype may have belonged to one or more females that were part of a large population 200,000 years ago. In the course of thousands of years, all other mtDNA lines might have died out by chance because they left no female descen-

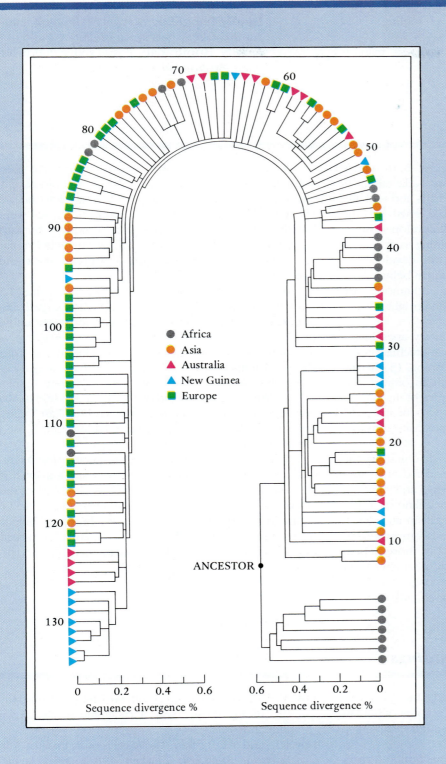

0   0.2   0.4   0.6
Sequence divergence %

0.6   0.4   0.2   0
Sequence divergence %

Africa
Asia
Australia
New Guinea
Europe

ANCESTOR

dants. Once again, the origin of modern *Homo sapiens* could be far older than the 200,000-year date. More genetic and fossil evidence is needed to make a choice among these alternative hypotheses.

Superfamily Cercopithecoidea (Gr. *kerkos*, tail + L. *pithekos*, ape + *eidos*, resembling) Old World monkeys.
Superfamily Hominoidea (L. *homo*, man + *eidos*, resembling)

Family Hylobatidae. Gibbon, siamang.
Family Pongidae. Orangutan (*Pongo*), chimpanzee, (*Pan*), gorilla (*Gorilla*).
Family Hominidae. Fossil and modern humans; *Australopithecus*, *Homo*.

Summary

1. Lemurs, tarsiers, monkeys, apes, and human beings all show indications of the arboreal adaptations of early primates: grasping hands, fingernails and toenails, crushing molar teeth, reduction of facial length and the sense of smell, keen vision, and a well-developed brain.

2. Primate features are just beginning to appear in lemurs and other prosimians and are well developed in the anthropoids: monkeys, apes, and human beings. Apes and human beings, collectively called hominoids, are larger, lack a tail, and assume more of an upright posture than monkeys.

3. Humans differ from contemporary apes in (1) being adapted for bipedalism rather than brachiation, (2) having a hand better adapted for grasping, (3) having dental and jaw features associated with an omnivorous diet, (4) having a larger and more complex brain, and (5) having sexual activity throughout the reproductive cycle, more dependent infants, and longer postnatal development.

4. Apes and monkeys probably diverged in the Oligocene, 30 million years ago. Molecular data suggest that humans diverged from apes between 5 and 9 million years ago, but fossils from this period are unknown. These ancestral protohominids evolved bipedalism and developed a more varied diet than apes. Bipedalism appears to be related to carrying food and, probably, more dependent infants. Changes in dentition indicate a shift from feeding on soft forest food to foraging on coarser and more varied food in open woodlands and savannahs.

5. The earliest known humans were the australopithecines of East and South Africa. The oldest species, *Australopithecus afarensis*, extends back into the late Pliocene, 3.7 million years ago. These hominids were bipeds and had a dentition similar to modern humans except for unusually large and thickly enameled molar teeth. The jaws were still massive and the brain size close to that of apes.

6. *Homo habilis*, who lived between 2.5 and 1.5 million years ago, probably evolved from *Australopithecus afarensis*. The cranial capacity was larger, and this hominid made primitive stone tools. The brain continued to evolve in *Homo erectus*, who replaced *Homo habilis* in East Africa about 1.5 million years ago. Cranial capacity ranged between that of *Homo habilis* and that of modern human beings. *Homo erectus* spread out of Africa, and by 410,000 years ago had reached Europe, Asia, and the East Indies. This species developed a sophisticated stone culture, used fire, and hunted large game.

7. Archaic representatives of our species, *Homo sapiens*, are first known from fossils in Africa and Europe that are about 300,000 years old. During the last part of the glacial period, a variant of this archaic type, classic Neanderthal, was isolated in western Europe. Modern *Homo sapiens* replaced this archaic type and then spread throughout the Old World as continental glaciers retreated.

References and Selected Readings

Campbell, B.G. *Human Evolution,* 3rd ed. Chicago: Aldine Publishing Co., 1985. An excellent account of human evolution with an emphasis on the unique anatomical adaptations of humans.

Cartmill, M., D. Pilbeam, and G. Isaac. One hundred years of paleoanthropology. *American Scientist* 74(1986): 410–420. A review of how changes and innovations in evolutionary biology, molecular biology, and earth sciences have affected our view of human evolution.

Day, M.H. *Guide to Fossil Man,* 4th ed. Chicago: University of Chicago Press, 1986. A catalogue of all known human fossils.

Hay, R.L., and M.D. Leakey. The fossil footprints of Laetoli. *Scientific American* 246(Feb. 1982):50–57. An account of the remarkable human footprints found in volcanic tuff 3.5 million years old.

Johanson, D., and J. Shreeve. *Lucy's Child: The Discovery of a Human Ancestor.* New York: William Morrow and

Co., Inc., 1989. A fascinating popular account of the discovery of human fossils and the interpretations that have been made about them.

Lewin, R. *Bones of Contention: Controversies in the Search for Human Origins.* New York: Simon and Schuster, 1987. A popular account by an editor of *Science* of the discoveries of the major hominoid fossils, the controversies they caused, and our changing interpretations of their significance.

———. The unmasking of mitochondrial Eve. *Science* 238(1987): 24–26. A review of the discovery of mitochondrial Eve and the different interpretations that can be made of her relation to the time of divergence of modern races.

Lovejoy, C.O. Evolution of human walking. *Scientific American* 259(Nov. 1988):118–125. An analysis of skeletal and muscular changes that occurred in the pelvis and hind limbs during the evolution of bipedalism.

———. The origin of man. *Science* 211(1981):341–350. An authoritative summary of human evolution with emphasis upon the reproductive and behavioral changes that may have led to the evolution of bipedalism.

Simons, E.L. Human origins. *Science* 245(1989):1343–1350. An excellent review of human evolution.

Susman, R.L. Hand of *Paranthropus robustus* from Member 1, Swartkrans: Fossil evidence for tool behavior. *Science* 241(1988):781–784. Evidence that the larger australopithecines may have used tools.

Tuttle, R.H. Apes of the world. *American Scientist* 78(1990):115–125. A review of the biology of living apes.

V

Interactions of Organisms and Their Environment

42   *Population Ecology*

43   *Communities and Ecosystems*

Ring-tailed lemurs, Lemur catta. *Destruction of Old World tropical rain forests is driving many lemur populations to extinction.*

42

Population Ecology

(Left) Black-browed albatross, Diomedea melanophis, *on a Falkland Island. Albatrosses usually nest in large colonies near their food supply and sheltered from predators. (Above) A pair of black-browed albatrosses.*

Seen from space, our planet shines white, blue, and brown. Clouds streak the surface, revealing the swirling movement of the atmosphere. Zoom in for a closer look and the blue oceans seem to come alive. Waves, currents, and upwellings mix seawater around the globe. Not even the land on this active planet stands still. Air and water reshape mountains and shorelines in relatively short periods of time. On a geological time scale—over millions of years—continents drift around the surface of the sphere. And with a closer look, the most dynamic aspect of the planet reveals itself—life. Earth's surface teems with living things swarming over and under nearly every nook and cranny on land, in the air, and in the sea. All of them call the earth home.

In 1866, the German biologist Ernst Haeckel conceived of studying animals' surroundings as well as the organisms themselves. He coined the word **ecology**, from the Greek *oikos*, meaning house or home, and *logos*, discourse. Since that time, biologists have expanded the notion of "house" to include the whole populated outer layer of our home planet—the **ecosphere**. Ecologists recognize four components of the ecosphere. To the three physical components—the air, the water, and the land—ecologists have added a fourth, living component, the **biosphere**. All organisms, from the smallest bacterium to the giant sequoia, are interconnected by a mesh of ecological interactions. That living web is the object of study for ecologists, who try to understand the relationships amongst species, and between organisms and their physical and chemical surroundings. Ecological studies may focus on the immediate interactions between species, on the annual cycles of an ecosystem, on the long-term patterns of climatic change brought about by human activity, and even on the interplay between organisms and the physical environment that has unfolded over geological time.

Ecology remains a young, rapidly growing science attempting to understand a complex set of phenomena. Faced with this vast challenge, ecologists take a variety of approaches. Some real world processes, for example, can be recreated on a small scale in the laboratory. Other processes may require a thorough understanding of a portion of the biological world, an understanding that can only be gained by painstaking work in the field. Finally, ecologists may choose to model the interactions they observe, using a mathematical approach to understand the dynamics of an ecosystem. Although mathematical models may seem somewhat removed from real organisms, such models can frequently distill the important ecological interactions, and arrive at certain generalizations about the living world. Naturally, ecological models are constantly tested and their predictions compared to the real world. Such tests help ecologists to reshape and refine their models, and occasionally lead them to discard a model completely.

Ecologists further simplify the study of the ecosphere by defining several levels of ecological organization. The smallest division, the **population**, comprises a group of individuals of the *same species* occupying a particular area at a particular time (p. 154). The aphids living on a potted plant or the deer mice in a field constitute populations. Several populations of *different species* living in the same general area at the same time form a **community**. Trees, squirrels, insects, birds, bacteria, and all the other organisms in a forest constitute a forest community. Adding the physical and chemical environment in which a community lives defines an **ecosystem**, a complete functional unit. The huge Serengeti plain of East Africa exemplifies an ecosystem. Its climatic conditions and water, minerals, and other resources support a characteristic flora of grasses, shrubs, and thorn trees that in turn provide shelter and food for a great diversity of animals and microorganisms. While Lake Victoria sharply defines the western boundary of this ecosystem, it blends into hills on the east and into woodlands on the south. In this respect, too, the Serengeti typifies ecosystems, which usually lack sharp boundaries. Terrain changes gradually from one habitat within an ecosystem to the next, and populations at the edge of one habitat interact with populations of adjacent habitats.

Certain attributes distinguish each level of ecological organization. Individuals reproduce and interact with the environment. Only populations have birth rates and death rates. Interactions between species must be studied in communities, which include a diversity of species.

In this chapter we examine population ecology, considering the distribution of populations, their growth, and factors that regulate their size. We ask why populations live in one area and not in others and what controls their spread, growth, and decline. We will take up community ecology, the study of species diversity, and interactions within communities in Chapter 43.

THE SPATIAL DISTRIBUTION OF POPULATIONS

Geographical Range

Each animal species occupies a characteristic geographical range determined by suitable biological, or **biotic**, and physical, or **abiotic**, aspects of the environment that allow populations to survive, grow, and spread. Plants and fungi may provide food and shelter; other animals may compete for food or space, or prey upon the population; bacteria and other organisms may parasitize an animal or degrade its wastes. Physical aspects of the environment also affect a population's range and in some cases can severely limit the spread of populations. Temperature extremes, rainfall, and terrain all determine the species that will inhabit a given area.

Range of Tolerance

The range of physical extremes within which populations will grow and spread is called the species' **range of tolerance**. Where some physical aspect of the environment consistently falls outside these limits, populations cannot survive. Generally, the factor for which a species has the narrowest range of tolerance exerts the strongest physical influence on its distribution—a concept formulated in 1913 by Victor E. Shelford and called the **law of tolerances**. Temperature and the availability of water commonly limit distribution in terrestrial environments. In soil or aquatic environments, oxygen availability may limit population distribution.

To better understand a species' distribution and abundance in different environments, ecologists first try to determine its tolerances for critical environmental factors, which is not always an easy task. Two species may have widely differing ranges of tolerance. In addition, a population may tolerate a wide range of conditions for one factor, but much narrower ranges for other factors. Moreover, a population's range of tolerance to one factor may be widened or narrowed by other environmental factors. Measuring a range of tolerance in the field presents so many variables that ecologists usually measure ranges of tolerance experimentally.

Most fish, for example, can gradually adjust to different water temperatures, but the ranges vary according to the species. The bullhead, *Ameiurus nebulosus*, can survive water temperatures as warm as 35° C, whereas the chum salmon, *Oncorhynchus keta,* cannot tolerate water warmer than 24° C. The temperature tolerance of other fish species is skewed by the pH of the water. Still other species tolerate a wide range of temperatures, a moderate range of pH, and a very narrow range of salinity.

The complex of biotic and abiotic requirements defines a species' **habitat** (L. *habitare,* to inhabit) or dwelling place. The habitat, a distinct area in which individuals normally live, may be as small as a termite's intestine for certain cellulose-digesting protozoans or as large as a forest for deer or an ocean for whales. Some physical character or plant type typically characterizes a habitat, for example, the running water of a stream habitat or the dominant vegetation in a grassland habitat.

Access

Even if an area meets all the physical and biological requirements of a species, individual animals may be unable to reach the area. A species normally spreads gradually from its point of evolutionary origin, often until it meets insurmountable geological barriers. But a species may find the proverbial grass greener on the other side of the barrier—if only it has some help crossing it.

Until the late 19th century, the range of the European starling, *Sturnus vulgaris,* was limited to Eurasia and northern Africa. Its further spread from its center of evolutionary origin had been halted by its intolerance to the cold polar regions and by its inability to fly across the North Atlantic. Then an American group bent on introducing every bird mentioned by Shakespeare to the United States came to the starling's aid. The group twice attempted to introduce the birds mentioned in *Henry IV* without success. But their 1890 release of 160 birds in New York's Central Park did the trick. The physical and biological conditions in North America at this time evidently met all the starling's needs. Americans were replacing forest and prairie with farmland, cities, and towns—perfect habitats for starlings. By 1950, starlings had reached the Rocky Mountains and established isolated populations on the West Coast. Today the species thrives throughout the United States, southern Canada, and northern Mexico, and has become a serious pest on farms and city parks (Fig. 42.1).

The starlings' unassisted access to North America was blocked by a **barrier**—a region biologically or physically inhospitable to the species. A barrier may halt the gradual generation-by-generation spread of a population. It may even prevent the passage of individuals if it presents severe environmental conditions. A land mass blocks the spread of an aquatic species. Water, a mountain range, or a desert can prevent the spread of a terrestrial species. Individuals or small groups of animals may be carried across barriers by another species, as in the starling example, or by wind or water-borne rafts of vegetation, as in island colonization. Barriers may appear or disappear over geological time as climates change and continents, riding on their crustal plates, drift over the Earth's surface.

One such barrier has profoundly shaped the fauna of the Americas. Some five million years ago, the rising Isthmus of Panama connected the North American continent to the previously isolated island continent of South America. The consequences of this geological event were dramatic. For marine organisms, the isthmus served as a barrier; several pairs of sister species of gastropods can be found respectively on the Atlantic and Pacific sides of the Isthmus. For terrestial organisms, the isthmus served not as a barrier, but as a bridge. The placental mammals that had evolved in North America poured across the isthmus, a migration that would eventually result in the extinction of most of the extraordinary marsupial fauna of South America. Only a few South American marsupials successfully migrated northwards; the opossum (*Didelphis*) now thrives in North America.

The **means of dispersal** used by a species determines to a large extent what sort of environment will constitute a barrier. Large organisms often face fewer barriers and spread more easily than smaller ones. Some mammals can travel great distances over land, and many birds can fly long distances over land or water. Small organisms, on the other hand, usually cannot travel far on their own. Some can, however, hitch a ride on the feet or fur of other

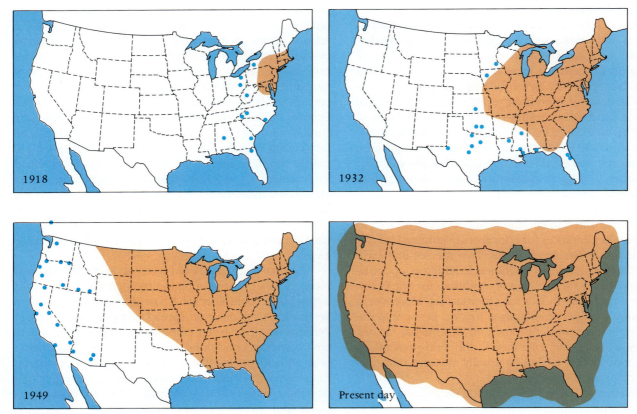

Figure 42.1

The westward expansion of the starling's range in the U. S. Orange areas indicate the breeding range; blue dots, isolated occurrences beyond its established range.

animals, as many arthropod parasites do, or on wind-blown leaves. Many spiders disperse using long silken "parachutes" to ride the wind. Rafts of vegetation carried out to sea by currents probably provide the transportation system that introduces new species to oceanic islands.

Normally, organisms crossing barriers fail to establish a viable population in the new territory. Their numbers may be too small, or the new area may not support growth of the population. Occasionally, however, a few members of a species succeed in crossing a barrier actively or passively and form a reproducing population well beyond the normal range of that species—a process known as **jump dispersal**. Species colonize islands by means of jump dispersal. Accidental or deliberate introduction of species by humans may also be called jump dispersal.

Species introduced by humans often devastate native populations. Early Galapagos Islands settlers brought goats, dogs, and pigs, and some escaped. When these domesticated animals reverted to the wild state (became feral) and spread across the islands, they decimated the native populations of giant tortoises. The agile goats browsed more efficiently than tortoises. Pigs and dogs dug up tortoise nests and ate the eggs and newly hatched young. Recent attempts to control feral animals and to breed tortoises in captivity have been only partly successful in reestablishing tortoise populations.

Dispersal Patterns

Both biotic and abiotic features occur in patches throughout the geographical range of a species—shelter, food, mates, and even local climate are not evenly distributed. Populations and individuals cluster around these resources (Fig. 42.2). Ecologists can gain clues about resource availability, variability of the environment, and interactions among individuals by studying the distribution of individuals and populations. In principle, if the environment were uniformly favorable, and the individuals were neither attracted to each other nor repelled by others, individual spacing would be determined by chance

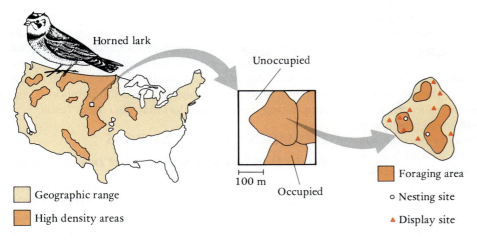

Horned lark

Unoccupied

Occupied

100 m

Foraging area
○ Nesting site
▲ Display site

Geographic range

High density areas

Figure 42.2
Hierarchy of distribution patterns of the horned lark (*Eremophila alpestris*) during
its breeding season. Although this species has a wide geographical range in the
U. S., its populations are most dense in the Plains states. Within dense regions the
birds occupy only locally favorable areas, and within these, they use particular sites
for foraging, nesting, and displaying.

and would appear **random** (Fig. 42.3a). But ecologists
rarely observe random distribution in the field. More
commonly, individuals form one or more **clumps**,
attracted to areas with the greatest availability of food or
shelter and avoiding less hospitable terrain. Individuals
may also be drawn together by some social interaction,

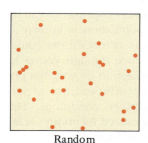

Random

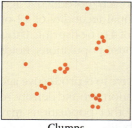

Clumps

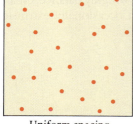

Uniform spacing

(a)

Figure 42.3
Individual dispersion. (a) Diagrams of random, clumping,
and uniform spacing of individuals in a population.
(b) Chicks of the black-browed albatross, *Diomedea
melanophis*, awaiting the return of their parents. Sea birds
commonly nest in colonies (clumps), but space their nests
uniformly.

(b)

such as mating or parental care. In other cases, antagonistic interactions between individuals or scarce food resources will lead to **uniform** spacing. Many sea birds, such as albatrosses, form large nesting colonies on parts of islands or coast lines offering nesting material, protection from predators, and access to food (Fig. 42.3b). The colony itself can be seen as a vast clump, but within the colony, each pair of nesting birds prevents neighbors from encroaching on territory near its nest. The nests tend to be uniformly spaced. Between breeding seasons, albatrosses live solitary lives at sea with no particular pattern of dispersal.

Density

To assess changes that occur in populations over time, ecologists must also know how many individuals live in an area—the population's **density**. They express density as the estimated average number of individuals per unit area, and select the spatial units according to the characteristics of the species and the purposes of the study. For barnacles on a wave-swept rock, square meters might be appropriate units, but for large mammals, numbers would be estimated per square kilometer.

Very few animal populations are small enough to allow the counting of every individual. Individual California condors and black-footed ferrets can be counted, for example, because all known individuals in these species live in captivity. Only 30 California condors, *Gymnogyps californianus*, survive. In 1985 as few as 10 black-footed ferrets, 6 of them captive, were known to exist. Later, ecologists captured all remaining wild individuals to establish a captive breeding program. The total population of these nearly extinct ferrets now numbers more than 100.

Because counting individuals becomes impractical in wild populations, ecologists resort to density estimates. In one technique, ecologists capture and count a small part of a population, then extrapolate to estimate the overall population density. To estimate the density of field mice, for example, an investigator might set live traps to cover 1000 m² of a field. The worker would then count, mark, and release the trapped mice. In a second trapping a day or two later, traps would contain some marked and some unmarked mice. The ratio of marked to unmarked mice in the second trapping indicates the efficiency of trapping and is proportional to the total number of animals in the entire population. From such a ratio, an ecologist can estimate the actual population density in an area:

Solving the equation for N gives an estimated total density of 364 mice/1000 m². The accuracy of such an estimate depends upon many factors, including the total number of animals trapped, the randomness of trap locations, mortality of marked animals, and behavioral factors such as trap shyness. Emigration and immigration occurring during the investigation may also affect results.

DYNAMICS OF POPULATIONS

Birth, Death, Emigration, and Immigration

Population distribution and density fluctuate from generation to generation as well as over longer periods of time. Within the population, individuals are born, reproduce, and die. Individuals also depart from the population or join the population under study. Looking at the population as a whole, a rate can be determined for each process, and those rates together with extrinsic factors—such as environmental fluctuations, natural disaster, and predators or parasites—affect a population's distribution and density.

Birth rate and death rate, or **natality** and **mortality**, are important factors. If natality increases, population size increases unless mortality rises as well. The adult members of every species produce a characteristic number of offspring. Ecologists call this measure **fecundity**. The maximum fecundity is limited in most species by the ability of females to produce eggs and to incubate or carry offspring. The theoretical maximum number of offspring that can be produced by the females of a population is known as **potential fecundity**. A single pair of organisms could, under ideal circumstances, produce enough young to quickly result in exponential population growth. But potential fecundity is practically never achieved, except perhaps under certain ideal laboratory conditions. The actual fecundity in the field, or **realized fecundity**, is far lower, because many individuals die from disease, predation, or accident long before they reach reproductive age. A female field mouse has a potential fecundity of 12 or more litters/year, each composed of 4 to 6 offspring. If this were the realized fecundity of mouse populations, fields would soon be awash with mice.

Those individuals that do survive to reproductive age may **emigrate**, reducing the population size and density. At the same time, individuals from other populations may **immigrate**, increasing both attributes. In most populations, emigration and immigration roughly balance, creating a population equilibrium, but occasionally extrinsic

First Trapping

$$\frac{260 \text{ mice trapped and marked}}{\text{Unknown total population size } (N)} = \frac{200 \text{ previously marked mice retrapped}}{280 \text{ mice trapped altogether}}$$

Second Trapping

$$\frac{260}{N} = \frac{200}{280}$$

factors upset this balance. A well-known human example unfolded near the middle of the 19th century, where blight repeatedly devastated potato crops in Ireland, leading to widespread famine. The famine's direct effect, starvation, reduced the Irish population, but its indirect effect was also profound, not only on Ireland, but also on the United States. Thousands of people emigrated from Ireland, decreasing its population, and many of them then immigrated to the northeastern United States.

Many of the same environmental effects that reduce fecundity increase mortality. Predators, disease, and natural disasters all take their toll. In addition to a potential fecundity, each species has an average physiological longevity, the life span of individuals that die of old age. But few animals in nature die of old age. A captive European robin that was protected and regularly fed lived for 11 years, for example, but robins in the wild rarely live longer than 2 years.

Figure 42.4

A Dall sheep, *Ovis dalli*, in Denali National Park, Alaska. Note the growth rings on its horns.

Survivorship Curves

Within a population, some individuals die at birth, others when young, and still others may live well beyond the average wild life span. Both species-specific and environmental factors affect age distribution. At any particular time, the number of individuals at each age can be graphed to produce an age distribution diagram.

Estimating age distribution in wild populations can be just as difficult as determining population size or density, and ecologists employ similar trapping and marking techniques. They may mark animals at birth, release them, and then live trap samples at regular intervals to estimate the number surviving to different ages. Or workers may find a measurable physical character from which they can determine the age of trapped or dead animals. In one such study, ecologists collected the skulls of 608 Dall sheep from Denali National Park. They determined the animals' age at death by measuring the skull and counting growth rings on the horns (Fig. 42.4). A **life table** showing the number of sheep surviving to different ages was constructed from these data (Table 42.1). From the table, investigators determined that 121 sheep died during their first year, leaving 487 at the beginning of the second year; 7 died the second year, leaving 480, and so on. The numbers of survivors at the beginning of each age group were then expressed as a decimal fraction of the initial population size.

By plotting survivorship fractions on a logarithmic scale, ecologists produce a **survivorship curve** (Fig. 42.5). On a logarithmic scale, a constant rate graphs as a straight line. In survivorship curves, a straight line slanting from the upper left corner to the lower right corner represents an exactly equal probability that an individual will die at any particular age and shows the initial population declining at a constant rate. This pattern is designated as a Type II

survivorship curve. It represents a theoretical baseline, a standard against which changes in real populations can be compared. The survivorship curves of most species lie above this baseline in a convex curve (Type I) or below it in a concave curve (Type III). The Type I pattern indicates that the majority of individuals survive to reproductive age and few die until old age. Many human populations, particularly in industrial countries, approximate this curve. Many newborns may die, but thereafter few people die until they are well beyond their reproductive years. Most other species have a concave survivorship curve, closer to Type III. Mortality is initially very high and most offspring die before they reach reproductive age. Thereafter, mortality decreases greatly.

Age and Sex Distribution

By determining the sex distribution as well as age distribution in a population, ecologists can better predict how a population may change over time. A population containing many prereproductive individuals and few of reproductive age has the potential for strong growth, but that potential will not be realized until the young mature. A population containing many reproductive females and at least some males of reproductive age has a high immediate growth potential. A population with few females or with most individuals past reproductive age has a much lower growth potential.

Consider, for example, the age and sex distribution in 1990 of humans in economically developed and less developed countries (Fig. 42.6). In developed countries, natality and mortality nearly balance, so that the population size grows only slowly. Natality is low and most people live 60 years or more, so the number of individuals in each age class remains about the same until old age. Knowing these

Table 42.1
A Life Table for Dall sheep in Denali National Park, Alaska

Age Interval (Years)	Number Dying During Age Interval	Number Surviving at Beginning of Age Interval*	Number Surviving as a Fraction of Newborn (Survivorship)
0–1	121	608	1.000
1–2	7	487	0.801
2–3	8	480	0.789
3–4	7	472	0.776
4–5	18	465	0.764
5–6	28	447	0.734
6–7	29	419	0.688
7–8	42	390	0.640
8–9	80	348	0.571
9–10	114	268	0.439
10–11	95	154	0.252
11–12	55	59	0.096
12–13	2	4	0.006
13–14	2	2	0.003
14–15	0	0	0.000

*Initial population = 608.
From R.E. Ricklefs. *Ecology*, 3rd ed. New York: W.H. Freeman and Co., 1990.

Figure 42.5
Survivorship curves plot the number of surviving individuals against their age. The Type II curve is a baseline that shows a steady rate of decline. Actual populations fall above or below this curve.

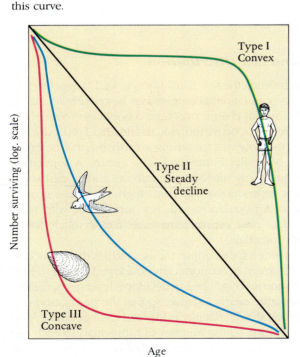

characters, demographers have predicted growth to the year 2025. They expect the population to grow slowly. Assuming that health care continues to improve, however, longevity will increase and the size of the older age classes will increase slightly. In the less developed countries, population size is larger and age distribution is skewed toward the younger and reproductive age classes. Natality is very high, but mortality increases with increasing age, so the age distribution graph has a pyramidal shape. Continued high natality and reduced mortality, due to advances in health care, would lead to a great population increase in these countries by 2025.

POPULATION GROWTH

By studying immigration, emigration, natality, and mortality in living populations, ecologists can estimate population growth rates and search for factors that affect population growth and decline. Such voluminous data are often condensed and understood through the use of mathematical models. Such models are not meant to represent any one population; their value instead resides in their generality. By collapsing a host of different forces and factors that affect fertility, mortality, and migration into a small number of variables, ecologists can simulate and test the effects of changes in one or more of these important variables. They can then return to the field to compare living populations with the model's predictions. In some cases, further

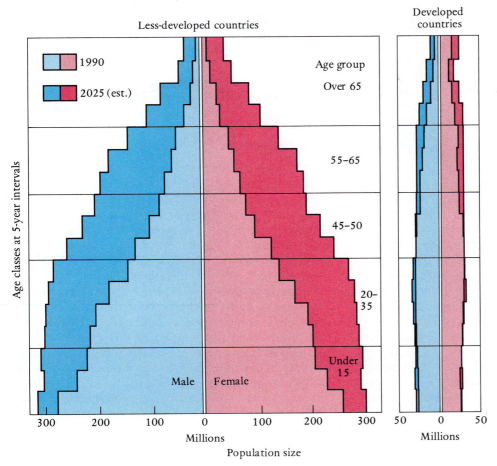

Less-developed countries

1990
2025 (est.)

Age group
Over 65
55–65
45–50
20–35
Under 15

Male Female

Age classes at 5-year intervals

300 200 100 0 100 200 300
Millions
Population size

Developed countries

Millions
50 0 50

Figure 42.6
Present and projected age and sex distribution of human populations in less developed and developed countries.

field work may show the models to be inaccurate or misleading, in which case theorists may modify the model or construct a new one.

Intrinsic Rate of Growth

To calculate the rate of growth of a population, one must first determine the contribution that each individual makes to the growth of that population. If 100 individuals produce a total of 20 individuals that themselves survive to reproduce, one could say that on average, each individual produces 0.2 offspring per generation. Of course, no individual produces a fractional offspring. The value we have derived is an abstraction called the **per capita rate of growth** or, more simply, the **intrinsic rate of growth**, and it is symbolized by r. Only surviving young who remain in the population contribute to r:

$$r = (\text{Natality} + \text{Immigration}) - (\text{Mortality} + \text{Emigration})$$

When r equals 0, a population maintains a stable size; when r is greater than 0, a population grows; and when r is less than 0, a population decreases in size. Many aspects of a species' life history affect the value of r, including litter size, the number of individuals surviving to reproductive age, the age of onset of reproduction, and the length of time an individual remains reproductive. Challenged by a sudden environmental change that drastically reduces population size, a species whose members have a high value for r has a greater potential to recover.

Exponential Growth

Ecologists build models one piece at a time, with each new piece intended to increase the reality and utility of a model. By multiplying the intrinsic rate of growth (r) by the number of individuals in a population (N), ecologists can create a mathematical model for population growth. To simplify modeling, r is assumed to remain constant, but it could and generally does vary in the real world. For any given value of r, a large population (N) will produce more surviving offspring than a small one for the simple reason that there are more breeding individuals.

Bacterial population growth illustrates this relationship. A common bacterium of the human intestine, *Escherichia coli*, can divide every 20 minutes under optimum conditions. A single cell, placed in a fresh flask of nutrient medium and kept at an appropriate temperature, will become a population of 512 cells in 3 hours. After 3 more hours, the population will total more than 250,000 cells. Nine hours after the bacteria were introduced into the hospitable territory of the flask, more than 100 million bacteria will swarm through the medium. This bacterial population is undergoing **exponential** growth made possible in a laboratory stiuation where food is abundant and there are no environmental limits. As the number of individuals increases, the number of new individuals added to the population also increases.

A simple model of such population growth can be shown graphically by plotting the number of individuals (N) against time (t) on an arithmetic scale (Fig. 42.7, upper curve). As long as population growth is unrestricted the population size increases exponentially and the curve becomes progressively steeper. The slope of the curve at any point shows the instantaneous growth rate of the population (the rate at which N is increasing, not to be confused with the intrinsic rate of growth, or r, which is considered constant). The change in numbers of individuals (dN) that occurs in an infinitesimally short period of time (dt)—the instantaneous rate of change—

can be expressed as (dN)/(dt). This term describes the rate of increase of the population per unit time and it is equal to rN:

$$(dN)/(dt) = rN$$

This differential equation describes the slope of the exponential growth curve at any point in time and for any value of N.

Logistic Growth

The exponential growth model demonstrates the extraordinarily high growth potential of populations if environmental conditions remain ideal. It assumes no limits on living space, food, water, and other resources. In wild populations, as population increases, the amount of space and food available to each individual decreases and population growth slows. The effect of these relatively declining resources can be thought of as an increasing **environmental resistance** to growth. When environmental resistance is considered in our model, the growth curve can become somewhat **S** shaped, or **logistic** (Fig. 42.7, lower curve). According to this model, when a species colonizes a new area, resource availability is high and environmental resistance to growth is low. While the number of individuals remains small, the population size increases slowly—graphically, the lower, nearly horizontal part of the **S**, just as it did in the exponential growth

Figure 42.7

Curves for exponential growth (*left*) and logistic growth (*right*). The difference between them is a measure of the environmental resistance.

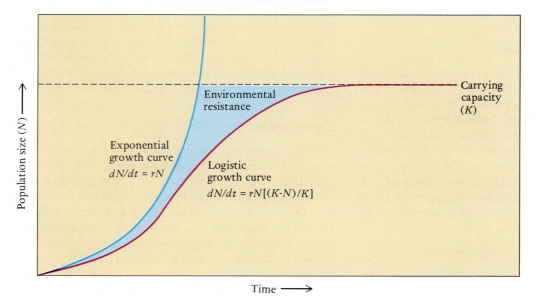

Population size (N) — Time →

Environmental resistance

Exponential growth curve
$dN/dt = rN$

Logistic growth curve
$dN/dt = rN[(K-N)/K]$

Carrying capacity (K)

Table 42.2
The Effect of the Decreasing Unutilized Opportunity for Growth on the Growth Rate of a Population

r	*Population Size (N)*	*Unutilized Opportunity for Population Growth [(K − N)/K]*	*Rate of Population Growth (dN/dt)*
1.0	1	99/100	0.99*
1.0	50	50/100	25.00
1.0	75	25/100	18.75
1.0	95	5/100	4.75
1.0	99	1/100	0.99
1.0	100	0/100	0.00

*The value for the intrinsic rate of growth (*r*) remains constant. The carrying capacity of the environment (*K*) is assumed to be 100 individuals.
From C.J. Krebs. *Ecology: The Experimental Analysis of Distribution and Abundance*, 3rd ed. New York: Harper & Row, 1985.

model. As *N* increases, the population growth rate increases, and *N* increases ever faster. This phase of growth shows as a nearly vertical portion of the graph. As the large number of individuals begins to put a strain on the resources available, population growth slows and the graph levels off. A population size is eventually reached where natality and mortality come into dynamic equilibrium, and growth rate averages 0. The population has reached the **carrying capacity (*K*)** of the environment.

The equation modeling exponential growth becomes an equation for logistic growth with the inclusion of a term representing the **unutilized opportunity for population growth**. This term, $(K−N)/K$, decreases rapidly as population size increases (Table 42.2). It is, therefore, the *inverse* of environmental resistance. When population size is small, the unutilized opportunity for growth is high and environmental resistance low. When population size is large, the unutilized opportunity for growth is low and environmental resistance is high.

With the addition of the unutilized opportunity for growth term, the exponential growth equation can be modified to give the equation for logistic growth:

$$dN/dt = rN \, [(K−N)/K]$$

Population Growth in Nature

The models for exponential and logistic population growth predict growth rates under certain idealized conditions. Exponential growth assumes no checks on growth, and logistic growth assumes a steadily increasing resistance to growth leading to a stable population size at the carrying capacity of the environment. Both assume

that individuals respond instantly to scarce resources, that all individuals are identical, and that the environment remains stable. These underlying assumptions rarely obtain in the real world.

Real populations change constantly. Population sizes and growth rates fluctuate up and down. To understand the wide variety of variables that can affect real population growth, ecologists return to field observations and gather information to improve their models or to modify them for specific situations. Both reproductive variables and chance (**stochastic**) environmental variables—such as seasonal and year-to-year weather fluctuations or changes in food supply—affect population growth.

Reproduction and individual growth introduce **time lags** into population growth. Lags occur between the time individuals are born and the time they start to reproduce. Young often become reproductive while their parents are still reproducing. Such lags and generational overlaps slow a population's reaction to resource scarcity or abundance. As a result, real populations frequently overshoot the carrying capacity of the environment. In such cases, resources become scarce, mortality may increase, fertility may be reduced, population growth slows, and population size eventually declines. When the population falls below the carrying capacity of the environment, resources may again become abundant, and the cycle is initiated again. In certain cases, steady oscillations develop around the carrying capacity; in other cases, the oscillations dampen, and the population eventually reaches a sustained equilibrium size very near the environmental carrying capacity. Finally, if the feedback between the environment and population growth is slow, the oscillations in population density increase.

Population Fluctuations

The effect of stochastic factors on populations can be illustrated by a study of random variability in bird fecundity. Pairs of swifts at Oxford, England, raise on average two young birds per brood, but some pairs have no reproductive success, while others hatch three or more eggs. Because each breeding pair's fecundity has no effect on the fecundity of other pairs, nothing prevents a widespread, random departure from the average population fecundity during a single season. Such a departure can have a large and lasting effect on population size.

Weather changes strongly affect population growth in species with short life spans. Populations of thrips, *Thrips imaginis*, an insect pest of roses, remain small in South Australia most of the year because the cool, dry climate does not support rapid population growth (Fig. 42.8a). But thrip populations bloom along with their host flowers in November and early December (summer in the southern hemisphere), when favorable climate and plant flowering coincide (Table 42.3). Thrip populations enter an exponential growth phase. The propitious weather does not last long enough for the population to reach the carrying capacity, however. Instead, the population **crashes** in late December and early January, when the cooler, drier weather cuts the population surge short. Bad weather does not kill every individual in the population, but caps the population at a low level. Some ecologists view this pattern not as a catastrophic disruption of logistic growth but as a normal population growth pattern for some species.

The growth of species with longer life cycles is less influenced by weather fluctuations. Sheep populations are not, of course, completely natural populations, for their growth is influenced by pastoral practices, which in turn react to markets for wool and meat. Yet sheep live outside all year, so their population growth is affected by normal environmental fluctuations. We have records for some sheep populations that extend over a far longer period of time than for any natural population. The data for sheep in Tasmania indicate that the growth of this population approaches the logistic model. Population size remains relatively constant, fluctuating slightly above and below a figure assumed to be the carrying capacity of the environment (Fig. 42.8b). Sheep were introduced into Tasmania about 1820. Environmental conditions were favorable for sheep husbandry and the population grew rapidly to slightly over 2,000,000 in 1850. Since then population size has fluctuated between 1,200,000 and 2,250,000 for more than a century.

Some populations show cyclic changes that extend over a period of years. The number of furs purchased from trappers by the Hudson Bay Company has been seen for decades as a reliable index of fur-bearing mammal population sizes in arctic Canada. Data accumulated since 1845 indicate that populations of lynx and hares peak approximately every 10 years, with the hares usually lead-ing the lynx by a short time. A crash follows each peak. (Fig 42.8c). The observation of these apparently coupled populations led ecologists to the conclusion that lynx, who feed upon hares, among other species, thrived when hare populations exploded. The heavy predation that followed checked the growth of the hare population. The lynx population declined in turn, as its food supply dwindled. But the question arose: Who follows whom? Subsequent studies questioned the seemingly obvious linkage. A first hint of suspicion grew out of the observation that in some years lynx population peaks coincided precisely with hare population peaks. Then ecologists found that populations of hares on islands with no lynxes show a similar oscillation. And predators were found to switch to other prey species before the dwindling hare population became a limiting resource. An alternative explanation emerged that the hare population exceeded the carrying capacity of the environment, overbrowsing its food plants. The population crashed when the winter food supply was inadequate to support a large population. If this hypothesis is correct, the lynx population does not drive the hare population to its crash, but follows it.

Regulation of Population Size

Natural curbs on population growth fall into two categories: those that change as population density changes, and those that remain independent of population density. The environmental influences that produce a decrease in the birth rate and an increase in the death rate as population density increases are called **density-dependent** curbs on the population (Fig. 42.9). Without a change in the environment, population growth stops when the birth rate equals the death rate (where the lines cross in Fig. 42.9). Density-dependent factors are generally biotic and include food supply, living space, and mates. They affect population growth because as populations approach the carrying capacity of their environment, members of the population must compete for these sorts of limiting resources.

If most of the individuals in a population have a similar chance to obtain food, as is the case with some browsing or grazing herbivores, the quantity and quality of the food supply will be relatively lower for each individual as the population density increases. Individuals must spend more time searching for food. As a byproduct of this search, they increase their exposure to predators. Inadequate food also can lead to lowered fecundity, often a result of smaller gonads in undernourished individuals. Even the frequency of breeding may drop, further lowering fecundity.

An increase in population size in some species increases the competition between individuals in the population for food and other resources that are becoming scarcer (**intraspecific competition**). Some individuals obtain an adequate supply, but others individuals are forced to marginal environments where resources are scarcer. If

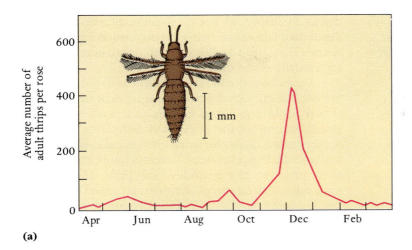

(a)

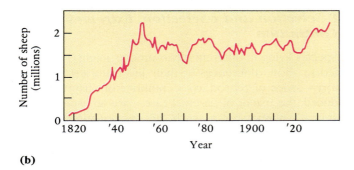

(b)

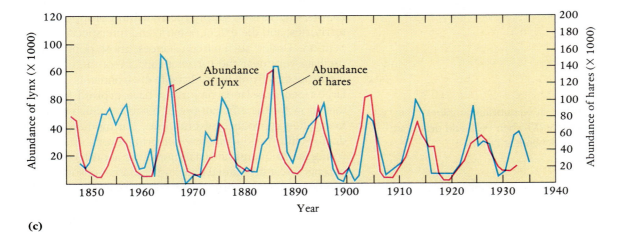

(c)

Figure 42.8

Examples of growth rates in natural populations. (a) The growth rate of thrips on roses in Adelaide, Australia, is exponential when conditions become favorable in November. The population crashes in late December and January. (b) The growth rate for sheep in Tasmania is close to logistic, and population size is maintained near the carrying capacity of the environment. (c) The growth rates of hares and lynx in Canada follow a 10-year cycle, increasing exponentially and then crashing.

Table 42.3
The Effect of Temperature on the Length of Development, Adult Life Span, and Fecundity of Thrips

	Temperature (°C)	
	8 to 12	23 to 25
Development period (days)	44	9
Length of adult life (days)	250	46
Total eggs laid per female	192	252
Daily egg production	1.4	5.6

From R.E. Ricklefs. *Economy of Nature*, 2nd ed. New York, Chiron Press, 1983.

population size becomes very large, even the successful individuals must devote more time and energy to defending their feeding ground and breeding sites. Aggressive encounters may become more frequent, and injuries and social stress may increase. Increased stress may upset delicate endocrine balances (p. 452), reducing fecundity. Stress may also adversely affect immune systems, allowing diseases to spread more rapidly and become more debilitating.

Interactions with other species may also put a damper on population growth. If two or more species utilize the same limited resource, **interspecific competition** can result. This could lower the growth rates of both species or it could, over evolutionary time, lead to a partitioning of the resources between them, as we will

discuss in Chapter 43. The combination of several density-dependent factors can bring the population size of some species close to an oscillating equilibrium near the carrying capacity of the environment.

Abiotic factors that have no relation to the density of the population, called **density-independent factors**, may also regulate population growth, independently of the size or density of the population. Storms or fires may reduce habitat quality for some species and check population growth. Drought or drastic temperature changes can affect longevity and fecundity (see Table 42.3). At the same time, such catastrophic events may create ideal habitats for other species. For example, many plants and associated animals thrive in areas recovering from recent fires (Focus 42.1). Density-independent factors may cause population size to fall or rise abruptly, as we saw in Australian thrip populations (Fig. 42.8a), but they do not usually lead to a stable equilibrium.

K-Selection and r-Selection: Extremes in a Continuum

Species living in different environments might be expected to show differences in their life histories. Life history—styles of reproduction, number of offspring produced, the extent of parental care—are an aspect of an animal's phenotype. As such, life history can be influenced by environment and shaped by natural selection.

Ecologists have sought to correlate the characteristics of particular environments with the life history "strategies" of the species that inhabit it. Some patterns have begun to emerge. As a general rule, the population sizes of species living in relatively constant and predictable environments appear to depend on **density-dependent factors**, such as food supply, the availability of appropriate territories, and the extent of intra- and interspecific competition. Under such circumstances, most populations oscillate very near the carrying capacity of the environment. Species living in predictable environments, such as large open grasslands or forests, have evolved features that allow for the efficient use of resources. Such species most often reproduce several times during their reproductive life, producing few offspring every time. The young are frequently altricial (immature at birth), and the parents invest substantial time and resources ensuring the survival of these offspring to reproductive age. In general, species occupying stable environments grow to a larger size than their counterparts in unpredictable environments, and are often longer-lived. Because the population hovers near carrying capacity, a premium is placed on the ability to successfully compete for resources with other members of the same species. Species exhibiting this package of life history and competitive strategies are sometimes said to be **K-selected**, a term meant to empha-

Figure 42.9
Population density reaches an equilibrium at the carrying capacity of the environment (*K*) when birth and death rates are controlled by density-dependent factors.

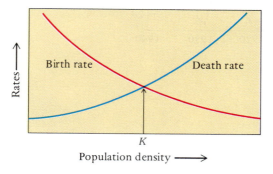

The great fires in Yellowstone National Park in the summer of 1988 devastated nearly 1.4 million acres. This was certainly a catastrophe for many species of plants and animals and for humans who use these forests for recreation, but fire is a natural, although unpredictable, phenomenon and a part of most ecosystems. Recognizing this, park managers have since the 1970s allowed fires caused by lightning to burn themselves out unless they threatened buildings or particularly scenic areas. Fires destroy, but they also clean out dead wood and debris that accumulate rapidly in this area, where climates are relatively dry, the winters long, and decay is slow. Fires open up parts of the forest, create meadows, and diversify the habitat.

The capacity of many types of forests to regenerate is great. The lodgepole pine is a dominant tree in Yel-lowstone and is quite resistant to fire. Its bark can withstand small fires, and one type of cone that it bears requires intense heat to open and release seeds. Shortly after the 1988 fires, one investigator counted released seeds and estimated that if 10% germinated, 5800 seedlings would start to grow on each acre. As the forest slowly regrows, grasses, shrubs, and other fast-growing plant species will develop and form habitats particularly favorable for many birds and other species that feed on their seeds and nest close to the ground. Meanwhile, lodgepole pine trees will continue to grow and eventually take over. Periodic fires are necessary to maintain the lodgepole pine forest because fires do not allow many hardwood trees, which are less resistant to fire, to become established.

(a) The extensive 1988 fires in Yellowstone National Park killed many trees, but
(b) created an environment favorable for the growth of other plants.

(a)

(b)

size the relevance of environmental carrying capacity. These species tend to have a relatively convex (Type I) survivorship curve (see Fig. 42.5). Most large herbivorous and carnivorous mammals fit the above description closely.

In contrast, species occupying highly variable, transient, or unstable environments are subjected to a very different set of selective pressures. Their population density is often determined by highly variable and often catastrophic **density-independent factors**, such as temperature, the disappearance of food resources, or disease. In such environments, we are likely to encounter many species favored for their invasive or colonizing ability. Weeds (such as dandelions) invading a newly tilled field are a good example of this scenario. Species that have high reproductive capacity, that are capable of producing many offspring when the opportunity arises, thrive in unstable environments. Although such species allocate a larger proportion of their resources to fast reproduction, they generally produce very small, rapidly developing offspring, in which they invest little or no parental care. These species are generally small, short lived, and may only reproduce once in their lifetime. Because the populations are most often far from the carrying capacity of the environment, there is little intraspecific competition for scarce resources. Mortality at all stages of the life cycle is high. Populations of such species generally increase exponentially during favorable periods, and will crash just as dramatically when conditions worsen. Resistant zygotes (or other developmental stages), or small populations survive unfavorable periods. This suite of life history characters is geared towards rapid increase in population size; species that exhibit them are said to be **r-selected**.

The black fly, a late spring pest in the northeastern United States, is a good example of such a strategy. For the few weeks when running water is abundant and temperatures remain cool, black flies thrive and populations skyrocket. Black flies survive the rest of the year as resistant zygotes. Many other invertebrates follow this strategy, and their survivorship curves approach the Type III pattern.

Although species were at one point classed as r-selected or K-selected, more recent ecological work emphasizes that these are only the extremes of a continuum of life history "strategies." More importantly, species exhibit considerable plasticity and flexibility in their life histories. Thus (within certain limits) populations of a given species occupying marginal or unstable environments may produce many offspring and invest little in their survivorship, while other populations of the same species living in predictable habitats produce far fewer offspring and expend considerable resources caring for them. Once again, it is the environmental context that determines the patterns we are most likely to encounter.

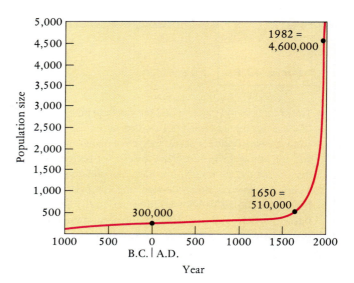

Figure 42.10
The growth of the human population has been exponential.

HUMAN POPULATIONS

World population in 1650 has been estimated at 0.5 billion. By 1950, 300 years later, it numbered 2.5 billion. In 1990 the population reached more than 5 billion (Fig 42.10). In 40 years the earth's human population increased as much as it did from the time of the evolutionary origin of our species through 1950. If this rate of increase continues, as appears likely, world population will reach 8.5 billion by the year 2025.

The rapid growth in world population is arguably the most serious issue, ecological or otherwise, humanity faces. Most of our problems with the environment and with social and political stability arise directly or indirectly from rapid population growth or large population size. We have used natural resources at an alarming rate, and the supplies of arable land, unpolluted fresh water, fossil fuels, and minerals are not infinite. We destroy natural habitats as we use more land for agriculture, home sites, factories, highways, and airports. Tropical rain forests are being decimated to provide hardwood for furniture or agricultural land that only produces crops or supports grazing for a few years. We are generating enormous volumes of wastes and are hard put to dispose of them without serious environmental damage—if not to our own backyard, then to someone else's. We are polluting our air, land, and water with toxic byproducts of an increasingly industrialized society. We may be doing ir-

The growth curve for the human population worldwide has been close to exponential, as shown in Figure 42.10. However, growth rates between countries are quite different. In forecasting the effect of different rates of growth it is sometimes useful to calculate the number of years (t) it would take the population to double, given different values for the intrinsic rate of growth (r) and assuming a continuation of exponential growth. To do so, we must use the natural base, $e = 2.71828. \ldots$ The formula for doubling time (2) is

$$2.0 = e^{rt}$$

This can be rewritten in logarithmic form as

$$\log_e (2.0) = rt$$

$$t = 0.69315/r$$

The value for r in recent years in the United Kingdom is 0.006, so doubling time would be 115.53 years. This is a slow rate of growth. On the other hand, in Algeria $r = 0.033$, and its doubling time is 21.01 years, a rapid rate of growth.

reparable damage to the atmospheric ozone layer that protects the earth's surface from dangerous ultraviolet radiation. And we may be altering our climate by rapidly increasing the carbon dioxide content of the atmosphere, thereby enhancing the greenhouse effect.

Some of these problems have been delayed because, unlike other species, we can increase certain of our resources as our population grows, although the growth of resources never is exponential. We have increased our food supply and have invented replacements for scarce materials. Much of our clothing is now made with synthetic fibers, and plastics have replaced metals in many applications. But we can increase agricultural production only so much, and even that increase carries a cost in fertilizer runoff and persistent pesticide and insecticide pollution. And we are daily discovering new problems created by synthetic materials, not the least of which is that these plastics are made from petroleum, a nonrenewable resource.

Population growth has slowed to near zero in some developed countries (Fig. 42.6), but much less so in developing parts of the world. As a result, 95% of the recent population increase has occurred in the less developed countries. The population growth rate in many less developed countries is high enough to double their populations in 22 years (see Focus 42.2). Clearly, world population growth cannot continue exponentially. The human species is not immune to the ecological principles that govern other species. We are in danger of exceeding the world's carrying capacity, if we have not done so already.

To wait for natural checks on growth is to invite dramatically lowered living standards, social and political upheaval, and unemployment. Disastrous environmental destruction and famine could soon follow.

Some ecologists optimistically believe that population growth can be slowed. Family planning is becoming more widespread, and the gradual industrialization of less developed countries may help check their growth rate, as it has for the industrialized nations. A decline in birth rate may result from an increase in standard of living, raised social status of women, and public support for the elderly.

Other ecologists believe that we may already have missed our opportunity. The bottom-heavy age distribution of populations in less developed countries (Fig. 42.6), coupled with improving health standards that lower mortality, point toward continued increases in population size. Development carries its own risks. Industrialization brings increased pollution. And because of the increased attention to nutrition and health care—especially for infants and the elderly—industrialization lowers the death rate before it lowers birth rate. Furthermore, a large increase in the industrial base of less developed countries may not be possible because of a shortage of resources.

The developed countries must recognize their responsibility to help less developed ones address their problems. They must also recognize the environmental damage that their own extravagant mode of life is causing. Corrective measures are imperative, and more careful long-range planning must be done for the development and management of renewable resources. It may be possible for the world to support a stable or slowly growing population with renewable resources, but only if we engage in planning on a global scale and help, rather than hinder, the less developed countries in their efforts toward family planning and industrialization. In addition, our own habits and our appetite for consumption must be tempered by the ecological realities of the time.

Summary

1. Ecology is the study of the relationship between organisms and their environment. In order to study these relationships, ecologists divide the ecosphere into several levels of organization: populations, communities, and ecosystems.

2. Each species has a characteristic geographical range determined by many biotic factors, the ranges of tolerance of the species to temperature changes and other abiotic factors, the point of evolutionary origin of the species, and its access to new areas. Population density is not uniform over a species range because individuals have characteristic dispersal patterns that are affected by the availability of resources and interactions between individuals.

3. Populations change constantly. Their size is controlled by birth rates (natality), death rates (mortality), and the immigration of individuals into a population or their emigration from it.

4. Survivorship curves can be calculated by determining the number of individuals in a population that die at different ages. In most species, mortality is high early in life and only a few individuals survive to old age. The human survivorship curve shows the opposite pattern, with a low mortality in early and middle life. The distribution of ages and sexes within a population can be used to predict a population's future growth.

5. By studying the dynamics of populations, ecologists can develop mathematical models that enable them to analyze the factors affecting growth. The exponential model assumes an unlimited supply of resources and predicts a steadily increasing rate of growth. The logistic model assumes steadily declining resources and predicts that populations will level off at the carrying capacity of the environment.

6. The size of natural populations is subject to so many additional variables (for example, time lags in reproduction, overlapping generations, and stochastic factors) that they usually fluctuate more than the models predict. Many populations grow exponentially for a while and then crash, others tend to level off and fluctuate above and below the carrying capacity, and still others display a pattern of cyclic rise and fall.

7. Growth is checked in all populations by density-dependent biotic factors such as intraspecific and interspecific competition for food and other resources, or by density-independent abiotic factors such as fires, floods, and drought.

8. *K*-selected species live in relatively stable and predictable environments, utilize resources efficiently over long periods of time, and maintain a population size close to the carrying capacity of the environment. At the other extreme, *r*-selected species live in unstable and unpredictable environments, reproducing rapidly when conditions are favorable, and their population crashes when conditions become unfavorable.

9. Human population growth has slowed in economically developed countries, but remains exponential in many of the less developed countries. The growth of world population is rapidly depleting resources and producing many environmental, social, and political problems.

References and Selected Readings

General references on ecology are cited at the end of Chapter 43. Several that emphasize distribution and populations are included here.

Ehrlich, A.H. The human population: Size and dynamics. *American Zoologist* 25(1985):395–406. A discussion of the population dilemma we face.

Hardin, G. Cultural carrying capacity: A biological approach to human problems. *Bioscience* 36(1986): 599–606. An essay on human population problems.

Keyfitz, N. The growing human population. *Scientific American* 261(Sept. 1989): 119–126. An excellent analysis of human population growth, its impact on the environment, and how it may be checked.

Krebs, C.J. *Ecology: The Experimental Analysis of Distribution and Abundance*, 3rd ed. New York: Harper & Row, 1985. A very good quantitative analysis of these topics.

43

Communities
and Ecosystems

(Left) Aerial view of a rain forest canopy in Madagascar. (Above) A kingfisher, Ceyx rufidorsus; *an inhabitant of a rain forest in Borneo.*

The Earth's biosphere consists of 30 million or more species, many of them as yet unstudied and unnamed. Each species, in turn, is organized into populations. No living population exists in a biological vacuum. Individuals interact with other members of their same population, with different populations, and with many different species. These interactions can take on many forms: predators and prey, parasites and hosts, beneficial mutualisms and outright competition. In attempting to understand these interactions, ecologists define units of biological organization above the population level. These include *communities*, made up of several species; *ecosystems* comprised of communities and their physical environments; and *biomes*, globe-spanning collections of ecosystems sharing overall geological and climatic features. In the following pages, we briefly review some of the best studied communities and ecosystems, and seek to describe certain laws and generalizations about the nature of ecological assemblages on Earth.

COMMUNITIES

Ecologists may define a community in different ways. All of the organisms living within an oak-hickory forest, for example, constitute a forest community. But an ecologist could also define a more restricted community living in the leaf mold on the forest floor or within the confines of a single decomposing log. Such a definition may seem arbitrary, but physically similar ecosystems often contain similar communities. Rotting logs in different forests frequently include the very same group of species, and even where different they may interact in similar ways.

Similar environments contain similar communities partly because the physical aspects of the environment constrain the variety of species that live there. Particular species live only where all physical parameters of the environment lie within their range of tolerance. Temperature, light, rainfall, salinity, and soils all have to be right. Furthermore, populations cannot survive adverse biological conditions, such as predators against which they have no defense, species with which they cannot successfully compete, parasites to which they succumb, or the lack of food organisms.

Geography also affects community composition. Some species may not occur in a particular habitat only because a geographical barrier has blocked immigration. But when transport transcends those barriers, they may successfully invade new territory. Many animals that humans have accidentally or deliberately introduced into new environments have readily joined existing communities, and some have even displaced native species. Introduced European earthworms, for example, have taken the place of native earthworms in the soil around large cities throughout the world.

While the physical environment constrains community composition, community composition also affects the environment. Each species modifies both physical and biological aspects of its environment. Forest trees modify the physical environment above and below ground. The leafy canopy dims light at lower levels. As leaves fall, they create a thick mat of humus on the forest floor, which is slowly buried by soil-dwelling organisms. The populations of birds, mammals, and insects living in the canopy of a tropical rain forest, for example, differ significantly from the populations that inhabit the dimly lighted, moist forest floor. Such **vertical stratification** of the environment appears in other communities as well. Water plants growing near the surface of a lake similarly limit light in the depths. The decreasing light intensity, combined with changes in water temperature and oxygen content, create a very different environment on the lake bottom.

Dominance

Some species affect the physical and biological aspects of an environment much more than other members of the community. These **dominants** set the character of the community. Large grazing mammals often dominate grasslands. Huge herds graze and trample much of the vegetation, determining which plant species thrive and often starting a chain of plant growth and consumption. Dominants affect other species directly and often modify physical conditions of the environment.

Nearshore marine communities off the coast of California resemble land forests in some respects. Huge, fast-growing algae called kelp play the dominant role here, producing food at a rate that rivals that of tropical rain forests. Kelp also shields the nearshore sea bed from sunlight, just as forest canopies limit light on the forest floor. Furthermore, kelp provides shelter, food, and substrate for communities of invertebrates, fish, marine mammals, and sea birds.

Unlike tropical rain forests, kelp changes dramatically from season to season. Kelp grows to its full 120-foot length in a single season. In the winter, the lush, brown forest turns to tons of rotting vegetation on the shore or scattered in the water, triggering a massive growth of decomposers. A similar change in a land forest sets off a different and much longer lasting series of events. If the dominant trees in a land forest die off due to disease or logging, or if the environment undergoes some other catastrophic change, the consequent changes in the community often follow a drawn-out sequence called succession.

Succession

A severely disturbed ecosystem undergoes a series of changes in which a new group of species may partially or completely replace the original biota. Over a period of years, decades, or even centuries, some species invade and others die out until a relatively stable group of populations occupies the site. This **ecological succession** of communities theoretically brings about a stable, long-lasting **climax community**. An observer might describe a different community on subsequent visits to a community undergoing succession. The entire series of these transitional communities leading to a climax community composes a **sere** and each transitional community is called a **seral stage**. In general, the number of species and the total biomass increase with successive seral stages as various populations approach the carrying capacity of the increasingly stable environment. Although ecologists often name seral stages after the dominant plant species, many of the species that make up the community are changing. For

example, an observer could describe the succession of spiders, birds, or mammals from one seral stage to the next in the succession of a forest.

Primary Succession

When wind-borne moss spores settle onto a rock tumbled over by spring floods, they initiate **primary succession** (Fig. 43.1). Any habitat as yet unchanged by the action of a biological community may support primary succession. On bare rock, on newly cooled lava or volcanic ash, even on a tree trunk felled by a storm, mosses and lichens often dominate the pioneer community. Arriving as hardy spores, these tiny plants and algal-fungal associations colonize the new surface. Lichens may degrade rock surfaces

Figure 43.1
Primary and secondary succession leading to an oak-hickory climax forest characteristic of many areas in the mid-Atlantic region of eastern U. S. All seral stages are not shown.

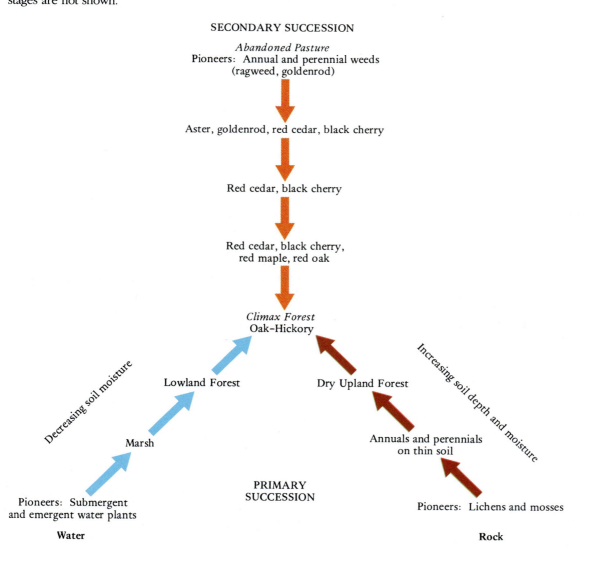

SECONDARY SUCCESSION

Abandoned Pasture
Pioneers: Annual and perennial weeds
(ragweed, goldenrod)

Aster, goldenrod, red cedar, black cherry

Red cedar, black cherry

Red cedar, black cherry,
red maple, red oak

Climax Forest
Oak–Hickory

Decreasing soil moisture

Lowland Forest

Dry Upland Forest

Increasing soil depth and moisture

Marsh

Annuals and perennials
on thin soil

PRIMARY
SUCCESSION

Pioneers: Submergent
and emergent water plants

Pioneers: Lichens and mosses

Water

Rock

chemically, and mosses trap dust and organic matter. Both contribute to the formation of small pockets of soil. The accumulation of soil permits seeds carried by the wind or by birds to germinate. These larger plants provide food, shelter, and habitat for a variety of animals.

While moss or fungi usually lead the colonization of bare terrain, other organisms can take precedence in some situations. When Washington State's Mount St. Helens erupted in 1980, huge parts of the mountain's flanks were buried in hot, flowing ash. Some of the first new permanent residents were spiders, which rode the wind into the new territory. On Mount St. Helens, their food literally fell out of the air. Insects, blown up the sides of the mountain, lost their ability to fly as they entered the cold air over bare or snow-covered patches of ash. The spiders only had to gather the bounty.

A shallow lake or the bend of a river that has become cut off from the parent stream may undergo a different succession, one involving the formation of new land. Pioneer water plants clog the edges of the pond, trapping an ever-increasing layer of sediment flowing into the pond along with rain runoff. The borders of the pond change from open water to marsh. This change in the edge environment prevents the water plants from thriving, but permits dry land plants to become established. A new community of birds, insects, and invertebrates comes to inhabit the marshy periphery of the pond. In this way, the pond slowly fills in from the edges toward the center, perhaps eventually becoming a forest overlying a sediment-filled basin. In these examples of primary succession, soil provides the basis for further development of the community. Because soil building proceeds slowly, pioneer seral stages can be long-lived. Eventually, however, increasing soil depth and changes in soil moisture permit the transition to new seral stages and, finally, to a climax community that will undergo its own slow changes.

Secondary Succession

Secondary succession follows major changes to an established ecosystem. Catastrophic weather events, fire, or human activities all disturb the environment (Fig. 43.1). After such an event on land, well-developed soil remains, giving pioneer species an easy foothold.

In newly disturbed land, the environment remains unstable. Pioneer species must be able to reproduce quickly, as relatively "r-selected" species do, but are unlikely to survive in such a rapidly changing environment. As the ecosystem stabilizes, with major changes gradually slowing, longer-lived populations invade, and they may be able to approach carrying capacity. These "K-selected" species, which tend to produce relatively fewer but hardier seeds, tend to dominate the dynamic equilibrium of the climax community.

In the eastern United States, the successional pioneers typically belong to a weed community that includes annual and perennial herbaceous plants, such as horseweed, ragweed, and aster. Conspicuous dominants in later seral stages include evergreen trees, such as pine in the south and cedar in the north. The mix of animal species in the community changes along with the dominant plants. Early inhabitants of disturbed areas include soil-dwelling ants, for example. These species eventually give way to ants that nest in logs or on trees.

Secondary succession may take place without destroying or disturbing all species in a community. In the mid-Atlantic U.S., the American chestnut once held the dominant place in the typical climax forest community. An epidemic of chestnut blight, caused by an accidentally introduced Asian fungus, swept through the eastern U.S. beginning around 1900. Within 50 years, 3.5 billion chestnut trees died. The chestnut climax community changed rapidly into a oak-hickory climax community, with a consequent change in makeup of the community. For example, the size of squirrel populations plummeted when chestnuts—which made up the bulk of their diet—disappeared.

Prior to the advent of widespread agriculture in North America, secondary succession occurred mostly following fire set by lightning or by humans. As we clear-cut forests, bulldoze land for housing complexes, and change the physical conditions of some environments with pollutants, we continue to foster secondary succession.

When agricultural fields are abandoned, a particular sequence of seres ensues. Old fields abound in the United States, and ecologists can study succession easily. But because old-field succession follows human activities—logging, ploughing, fertilizer or herbicide use, and continuous depletion of the soil by crops—lessons learned at abandoned farms must be evaluated with caution. Southeastern U.S. old fields show a classical example of one species changing the physical environment and thereby excluding the previously dominant species. A grass called broomsedge dominates one sere, providing a favorable environment for the germination of pine seeds blown in from adjacent forests. As the young pines grow they gradually shade the ground, forcing out the light-requiring broomsedge.

Climax Communities

In the view of classical ecologists, succession resulted in a stable, self-perpetuating climax community. Dominant plants in the climax community would not inhibit the germination and growth of their own seeds. Animal species would not deplete the region of food. According to this view, only a shift in climate would change the character of a climax community. Most ecologists today, however, see climax communities as dynamic equilibria, subjected to continual disturbances, such as fires, storms, flood, and drought. Climatic and geological changes on a time scale of decades and very long-term geological changes, such as continental drift, also prevent stability in climax communities.

Along with early assumptions about the stability of climax communities, ecologists holding to the monoclimax hypothesis predicted that within a given geographical area, a single type of climax community would come to occupy the whole area. This model has fallen before observations that climax communities actually vary throughout a geographical area. The local climax community may be determined by soil factors, elevation, slope, frequency of fires, or water availability. Each factor affects the climax community and contributes to the overall pattern of climax vegetation throughout the geographical area.

Some communities, such as southern California's chaparral community, appear to depend on occasional major disturbances. This community reaches climax in 30 to 50 years. If the ecosystem experiences no fires in 60 years, species diversity declines and the chaparral plants themselves stop growing. Occasional fires appear to rejuvenate this ecosystem, allowing cyclic succession.

Microseres

Continual and rapid succession also takes place on the "micro scale," within small parts of ecosystems. When a tree crashes to the forest floor, insects, fungi, and bacteria attack its trunk and branches. These pioneers begin the slow physical and chemical decomposition of the log, and the various stages of decomposition are marked by correlated changes in the populations of organisms inhabiting it. These **microseres** also occur in decomposing vegetation, animal carcasses, and feces.

Cow droppings (pats), for example, support a rich flora of bacteria, fungi, and other microorganisms, which in turn support a food web including rotifers, roundworms, mites, flies, beetles, and other animals. Some animals, such as dung beetles and flies, eat the dung itself or use it as food for their larvae. Like a newly emerged volcanic island, a fresh cow pat serves as new territory, ripe for colonization by immigrants from the soil and neighboring cow pats. The microenvironment of the cow pat changes steadily. As its original bovine gut flora dies off or is eaten, immigrant organisms tunnel into and digest the pat. Exposure to the air dries the surface, maintaining moisture within.

Among the early immigrants to cow pats are small roundworms, or nematodes. Some nematodes invade within the first few days of the cow pat's deposition. They hitchhike on the legs, beneath the wings, or between the genital segments of flies and dung beetles feeding on the pat. Later arriving nematodes tunnel up from the underlying soil. The changing microenvironment controls a succession of inhabitants until the cow pat eventually disappears, distributed into the upper layer of the soil.

Walter Sudhaus studied nematodes in cow pats by taking 3.3-cm-diameter core samples from dated cow pats over a 46-day period. Sudhaus separated the nematodes from the cored samples and examined them under a dissecting microscope.

Over the 46 days, the zoologist found a complex microsere of nematode species, identifying 34 species in all. The total nematode count peaked twice, jumping from an average of 750 nematodes per sample to more than 1500. Populations of the most numerous species peaked at different times, and no one species persisted for the full 46 days (Fig. 43.2). Cow pat nematodes feed on bacteria, microorganisms, and fungi, with more than one species exploiting each resource. In general, the first nematode populations to mature in the cow pat eat bacteria. Preda-

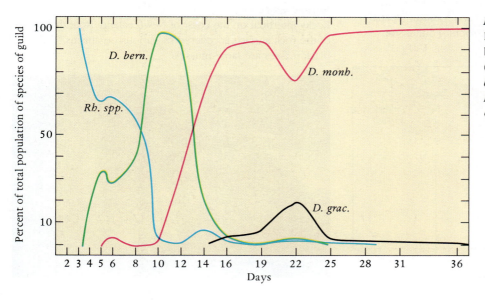

Figure 43.2

Population changes of species of bacteria-feeding nematodes (*Rhabditis* spp., *Diplogaster bernensis, D. monbysteroides,* and *D. gracilis*) during the life span of a cow pat.

tory populations peak later. At the end of the 46 days, nematode populations consisted entirely of immigrants from the soil below the pat.

Niche

Each animal species in each seral stage of every community exhibits a certain life style. Some forage at night and some in the day. Some only become active during certain parts of the year. Each species occupies a certain amount of space, prefers a particular range of temperature and humidity, eats particular foods, and may in turn be the preferred food of another species. Collectively, all of the interactions between a species, its community, and the physical environment in which it lives define the **ecological niche** that the species occupies. Unless all resources remain abundant, no two species can coexist in the same niche in any community, because one or the other would have a slight competitive edge in exploiting any limited resource. As a result, one species may die out or begin to utilize slightly different resources, essentially defining a new niche.

The niche concept is multidimensional and defies easy definition, but remains a useful concept to ecologists. The usual three dimensions—length, width, and height—describe a theoretical niche in space, but that description would leave out much information about the species. Ecologists describing niches include a trophic dimension (for food), a temporal dimension (to describe when individuals are active), and so on (Fig. 43.3). The biological dimensions describe what the species does in the ecosystem. Physical dimensions such as humidity and light determine where those activities take place.

Zoologists commonly say that a certain animal is "filling" a particular niche, but it is important to realize that there are no pre-existing niches. The niche is defined only when a species is present to define it. The expression "filling a niche" is derived, in part, from the fact that we would expect similar kinds of communities in different parts of the world to have similar kinds of species occupying similar types of niches.

A marine ecologist familiar with the rocky shores of the Pacific Northwest coast would recognize the members of similar communities on the coasts of Chile or New Zealand. Several species of barnacles and mussels, along with certain algae, form conspicuous bands in the intertidal zone, the region between low and high tide (Fig. 43.4). These sessile populations are usually the most conspicuous members of rocky coastline communities. Grazing species, such as chitons, snails, and limpets, feed on the fine algal mat growing on the rocks. Species occupying similar niches in widely separated environments are called **ecological equivalents**.

Diversity: Why so Many Species?

The species composition of communities varies widely. Some communities have few species, others have many. Tropical ecosystems comprise very large numbers of species—they show high **species diversity**. On the other hand, population density, the number of individuals of a species per unit area, often remains low in tropical communities. These features of community composition change with latitude. The number of species per unit area decreases as latitude increases, while average population density increases. This latitudinal gradient holds true in the marine environment as well. Between the tropics and the Arctic, for example, the number of copepods col-

Figure 43.3

Three hypothetical niches diagrammed as occupying three-dimensional space. Note that the niches of species A and B overlap in the temporal (activity time) and location (spatial) dimensions but not in their feeding (trophic) dimensions.

Figure 43.4

Rocky intertidal on the coast of New Zealand. Note that much of the rock surface is covered by various organisms. The black masses are large mussels.

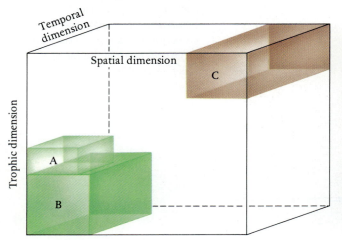

lected per cubic meter of water rises steadily, but the number of species declines (Fig. 43.5).

Biologists have long recognized this puzzling latitudinal pattern of species richness, with a peak in the tropics and paucity in the arctic. Many have attempted to solve the puzzle, but as yet no one explanation has proved universally acceptable. One hypothesis relates relative stability of the ecosystem to species diversity. Compared to the tropics, high latitudes undergo great seasonal stresses. The harsh climate in the north restricts primary production to a brief period of warm weather. For much of the rest of the year the environment remains inhospitable. In addition, high latitudes have been free of Ice Age glaciers for only a relatively short time; they are geologically younger. Such conditions might theoretically favor *r*-selected species—those that reproduce rapidly, produce large numbers of young (most of which die), and have short life spans. In these conditions, a species' long-term competitiveness is less important than its speed in exploiting fleeting resources. The tropics, which are subjected to much less seasonal stress, might be expected to favor species with longer life spans and with smaller numbers of young (many of which survive to reproduce). Ability to compete within this stable environment would be of great importance. But harsh Arctic ecosystems also contain long-lived, relatively *K*-selected species such as the polar bear, bowhead whale, and ring seal, and tropical rain forests house some species of animals, which typify the *r*-selected model in many respects.

Recently ecologist George Stevens has added another dimension to this hypothesis, arguing that the more seasonally stressful the environment, the wider should be a resident species' observed range of tolerance. Species with wide tolerance ranges can thrive under a relatively wide range of conditions. They should also be relatively widely distributed, and, because of drastic seasonal variation, must occupy large niches. The reverse situation would be true in the tropics: where seasons vary little, species should show a much narrower tolerance to varying environmental conditions. Thus modest distances, such as from the side of a ravine to its upper edge, would pose variations in conditions extreme enough to favor

Figure 43.5
Relationship of number of species (*blue bars*) and individuals (*red bars*) of copepods to latitude (temperature) in the north Pacific. All samples were taken from the upper 50 m of the water column.

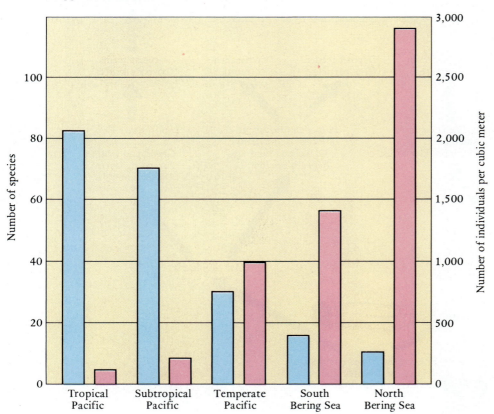

different species. The tropics, according to Stevens, becomes a mosaic of many narrowly distributed species occupying small niches.

Other ecologists attempt to explain the difference in species diversity by citing the high productivity of the tropics. A longer, highly productive growing season may elevate the carrying capacity of the tropical environment for many species; increase the total biomass, thereby increasing the number of niches available; and increase the structural complexity. That structural complexity itself could lead to greater species diversity in a positive feedback effect. Still another hypothesis predicts that predation should be more intense in the tropics, limiting populations of each prey species, but therefore fostering the coexistence of many different prey species.

Finding the solution to the species diversity puzzle—whether it turns out to be any or all of these explanations or an entirely different one—will have to wait for further field research and theoretical modeling.

Resource Partitioning

When many species coexist in a particular habitat, such as a tropical rain forest, more than one species may exploit the same resource in slightly different ways. One species may eat in the daylight while another eats the same food at night, or two species may forage at different heights in the same tree. Such **resource partitioning** may come about when a species extends its range and must compete for food with another species already resident in the community. Over time, the two species may evolve differing feeding times or locations, or they may subdivide the food resource in other ways.

In the waters surrounding Aldabra Atoll in the Indian Ocean, for example, five species of intertidal sea cucumbers live beneath boulders near the shore. The fact that most sea cucumbers move little and use their tentacles to eat detritus led Canadian zoologist Norman Sloan to ask how these species partitioned resources.

Figure 43.6

Resource partitioning of intertidal sea cucumbers on Aldabra Atoll in the Seychelles.

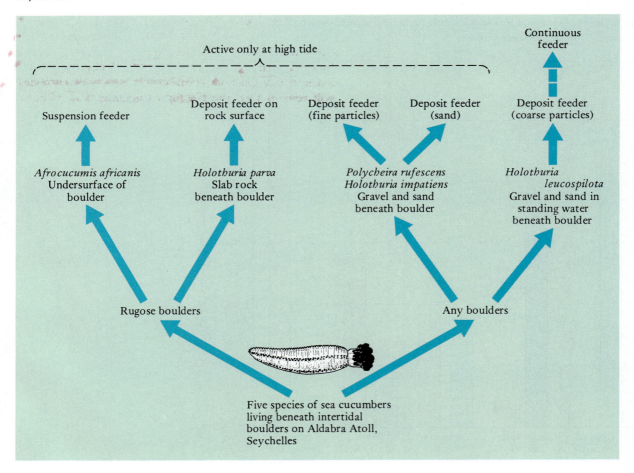

Sloan found that two of the five species lived only beneath rough, ridged boulders: one attached to the boulder and the other to the smooth, rocky substrate on which the boulders rest (Fig. 43.6). The rock dweller held its tentacles out into the surrounding water to catch falling detritus, while the species attached to the substrate skimmed fallen detritus from the flat rock surface. All three remaining species gleaned detritus from ingested sand-gravel. But each utilized different-sized particles.

The sea cucumber species differed in other ways as well. Four fed only at high tide, but one lived in tidal pools beneath boulders and thus fed continuously. Altogether, three niche dimensions separated these five sea cucumbers and allowed them to share a single food resource: the microhabitat in which they lived (the spatial dimension), the way they fed (the trophic dimension), and the time they fed (the temporal dimension).

Even distantly related species compete and may therefore partition resources. On the plains of the Serengeti in Africa the grazing mammalian herbivores interact to share food effectively (Fig. 43.7). During the rainy season in March and April the herds feed on rapidly growing, short grasses in the south Serengeti. In May and June, with the short grass cropped and the rains ended, these large mammals migrate to the western and then northern parts of the Serengeti, where tall, coarse grasses predominate. The coarse, close-growing stems restrict initial grazers to the upper parts of the grass plants, protecting low, leafy herbs from grazers. Herds of zebras feed on the stems of the grass. Wildebeests, an antelope, follow zebras, feeding primarily on grass leaves and sheathes. These grazers not only crop but also trample the plain. In doing so, they expose low-growing herbs. With the increased sunlight, the low plants burgeon, transforming the trampled sod into a green carpet. About a month later, Thomson's gazelles follow wildebeest, exploiting this growth brought about by the previous grazers. Each of these species exploits the grassland in a different way, allowing all to share a potentially limited resource.

These same grasslands also support large numbers of herbivorous rodents and insects. Ecologists know little about the biology of these animals, but can predict that resource partitioning has taken place among these smaller herbivores just as it has for the large mammals.

The study of resource partitioning remains an active and controversial research field. Some ecologists question partitioning studies, asking how observed patterns of physical traits or behavior rely on the researcher's choice of traits to study and their inclusion or exclusion of particular species. A researcher's classification of foods different species eat as a single resource may also be open to question. And lastly, ecologists are still unsure if the evolution of resource partitioning can be inferred from studies of present-day feeding patterns.

THE EARTH'S MAJOR ECOSYSTEMS

Different parts of the Earth's surface support similar ecosystems. On land similar ecosystems constitute a **biome**. Each biome consists of the climax communities living within a particular geographical area characterized by similar climate and terrain. A particular biome, such as a desert, may exist on several continents and in each case will remain recognizable for its geographical, climatic,

Figure 43.7
Wildebeest grazing in the east African grasslands.

and community features. Similarly, an aquatic ecosystem, such as a freshwater lake, on any continent will resemble in many particulars a similar aquatic ecosystem on another continent.

Biomes

Many of the biomes, such as the tropical rain forest, temperate deciduous and boreal coniferous forests, and tundra, span the globe like a belt. Thus in traveling from the equator to the Arctic, you could pass through many of the same biomes whether you started in South America or Africa. But you may have a hard time saying just when you traveled from one biome to the next. The line of demarcation between adjacent biomes remains vague. Biome margins are typically transitional and fingers of one biome can extend far into another.

An additional phenomenon blurs biome definitions. Because climate becomes colder with increasing altitude, as it does with increasing latitude, high mountains exhibit altitudinal zones of vegetation that parallel some of the latitudinal shifts in biomes. In the Rocky Mountains, for example, a low-altitude woodland zone contains deciduous trees—those that drop their leaves each fall. At higher altitudes, coniferous or needle-bearing evergreen trees replace the deciduous trees. The timberline marks the separation of the subalpine forest and the alpine tundra, composed primarily of low herbaceous and woody plants. These zones roughly correspond to part of the sequence of biomes encountered while traveling from the equator toward the poles.

Tropical Rain Forests

A great band of forest encircles much of Earth's equatorial land area. Interrupted only by mountain ranges, small areas of desert, and savanna, the tropical forest band includes both rain forests and seasonally hot, dry forests (Fig. 43.8).

The greatest existing expanse of tropical rain forest lies in the Amazon basin of Brazil, Peru, and Ecuador. Temperatures there remain relatively constant around 27° C, and as much as 1000 cm of rain may fall in a year. Farther from the equator, rain forest seasons become more pronounced and both temperature and rainfall variations increase.

Tall, broad-leaved evergreens dominate tropical rain forest communities, forming a dense canopy about 50 m above the ground. A few taller trees jut through the canopy 60 m or more above ground. Shorter trees fill the air with dense foliage below the canopy. These several leafy layers intercept most sunlight before it reaches the forest floor, which remains dimly lighted and moist.

With their several leafy layers, constant warm temperatures, and abundant water, tropical rain forests support a greater diversity of species than any other biome.

Whereas a temperate climax forest may have as many as 10 tree species, a tropical rain forest can contain 400. With so many tree species, it may be difficult to find two individuals of the same species in a small area. That difficulty illustrates the flip side of the great species diversity of the tropics. Populations of most species tend to be small and scattered, and seldom occur at high densities.

The great diversity of trees in turn supports a correspondingly great diversity of animals. On one preserve in Peru ornithologists counted nearly 600 species of birds in 300 km². Only a few more species nest in all of North America, an area 80,000 times as large. Insect species also abound in the tropics. Harvard entomologist E.O. Wilson once collected 43 species of ants from a single tree in Peru—about the same number of ant species found in all of the British Isles.

Many tropical trees support large numbers of epiphytic plants on their limbs and trunks (Fig. 43.9). The photosynthetic epiphytes anchor onto the bark of the host tree, but do not parasitize it. Epiphytes such as bromeliads, orchids, and ferns all have adaptations for water capture or storage. The small pools of water surrounding epiphyte leaf bases, for example, function as tiny ponds housing many animal species, including frogs (Connections 32.1, p. 752).

Each year logging crews clearcut an area of tropical rain forest about the size of Kentucky for timber, highways, ranching, or agriculture. Small abandoned clearings may be quickly reclaimed by forest, but large cleared tracts become long-lasting shrub savanna containing relatively few species. E.O. Wilson conservatively estimates that 4000 to 6000 plant and animal species a year become extinct as a result of deforestation. Some ecologists predict that by the year 2025, only a few remnants of the world's tropical rain forests will remain, chiefly in parks and on mountainous land with little economic value. The worldwide loss of species diversity will be irreparable.

Ironically, land stripped of rain forest and converted to agriculture usually supports crops for only a few years and then must be abandoned. The dense, rapidly growing vegetation in the rain forest sucks nutrients out of the soil as soon as decomposers release them, leaving nutrient-poor soil. Land cleared for ranching similarly supports cattle for only a short time. And to further compound the exploitation of the land, many cattle ranches are foreign owned and managed, supplying beef for foreign markets.

Deforestation also may affect rainfall in the tropics. A considerable part of the precipitation that characterizes

Figure 43.8 ▶

The world's terrestrial biomes. Precipitation and temperature are the principal factors governing their distribution.

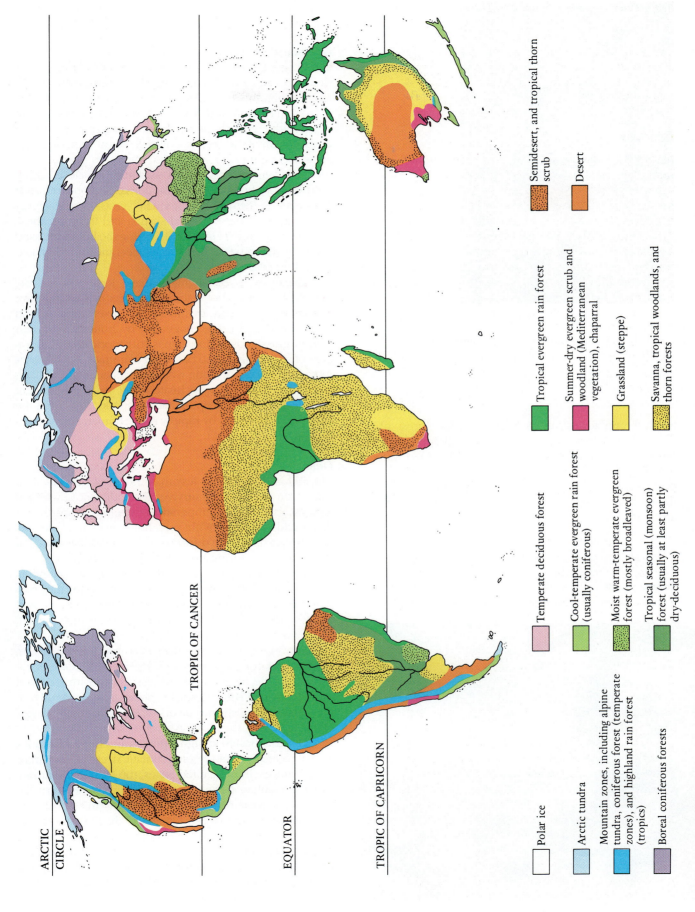

ARCTIC CIRCLE

TROPIC OF CANCER

EQUATOR

TROPIC OF CAPRICORN

Polar ice

Arctic tundra

Mountain zones, including alpine tundra, coniferous forest (temperate zones), and highland rain forest (tropics)

Boreal coniferous forests

Temperate deciduous forest

Cool-temperate evergreen rain forest (usually coniferous)

Moist warm-temperate evergreen forest (mostly broadleaved)

Tropical seasonal (monsoon) forest (usually at least partly dry-deciduous)

Tropical evergreen rain forest

Summer-dry evergreen scrub and woodland (Mediterranean vegetation), chaparral

Grassland (steppe)

Savanna, tropical woodlands, and thorn forests

Semidesert, and tropical thorn scrub

Desert

Figure 43.9
Epiphytic bromeliads growing on the trunk of a tree in the Caribbean National Forest of Puerto Rico.

tropical rain forests evaporates from the forest plants themselves. When this vegetative cover disappears over a large geographical area, the volume of recycled water is greatly reduced. Moreover, without the forest cover, run off increases. The rain water may eventually escape from the former forest entirely, breaking the local evaporation and rainfall cycle.

Figure 43.10
The east African savanna near Mt. Kilimanjaro in Kenya.

Grasslands and Savanna

Temperate grasslands—North American prairies, Soviet steppes, and Argentinian pampas—cover vast areas of many continents (Fig. 43.8). A hot, dry summer with little or no rain typically follows cool spring rains too sparse to support the growth of trees. In addition to perennial grasses, many other herbaceous species also thrive in the grassland biome.

Some grasslands, called savannas, lie close to the equator, and these too enjoy moderate, seasonal rain, but temperature changes little annually. Grass dominates savannas as well, but scattered shrubs and trees dot these African landscapes (Fig. 43.10). During the hot, dry season many of these trees and shrubs become dormant and shed their leaves.

Great herds of ungulates have long roamed the world's grasslands, but in temperate regions humans have eliminated most large mammals by converting grassland to farmland and cattle range and by hunting. The East African grasslands and savannas still support large numbers of such grazing and browsing mammals, but diseases introduced to Africa along with cattle, poaching, and habitat destruction have decimated the once-magnificent herds everywhere except in national parks. Now even the parks are threatened by rapidly rising human populations (p. 974).

Deserts

In mid-latitudes, where less than 25 cm of rain falls in a year, even grass cannot survive. These dry areas, with the greatest day-to-night temperature extremes on Earth, are the world's deserts. With very little vegetation, deserts bake under direct sun during the day and radiate much of that heat away during the clear, chilly nights.

Deserts can be found in the interiors of continents (Australian, Gobi, and Saharan deserts), in the rain shadows of mountain ranges (Great Basin Desert), and along coasts with cold ocean currents (Chilean, Peruvian, and Namibian deserts) (Fig. 43.8).

Some deserts, such as the Sahara, receive virtually no rainfall and consequently support little life. Others receive sporadic, brief downpours associated with localized thunderstorms. These deserts support greater overall biomass and greater species diversity, including many desert plants.

Deserts located at high altitudes and higher latitudes are subjected to the additional climatic stress of low winter temperatures. In the United States, the Mohave, Sonoran, and Chihuahuan deserts of the Southwest remain relatively warm year round, but snow sometimes dusts the Great Basin Desert of Nevada and Utah (Fig. 43.11).

Desert vegetation consists of two types: annuals with very brief life cycles immediately following rain, and perennials that are adapted for tapping deep soil water and retaining the water they absorb. Many of the adaptations of desert animals were described on p. 365.

Figure 43.11
Southwestern desert showing fishhook cacti in bloom in the foreground. The shrubby trees in the background include mesquite.

Dry Shrub Biomes

Between deserts and temperate forests lie dry shrub biomes. Bridging a climatic and often a latitudinal gap between desert and forest, tall, drought-resistant shrubs and widely spaced small trees thrive with short growing seasons, mild winters, and long, hot, dry summers (Figs. 43.8 and 43.12). Dominants in the U.S. shrub biome include chaparral, California scrub oak, chamise, buckthorns, and California lilac.

Figure 43.12
Chaparral in Napa Co., California. The vegetative cover of small trees and shrubs is composed largely of manzanita and chamise.

Summer wildfires often ravage shrub lands. In some cases, destruction may spark renewal in a complicated community life cycle. Following fires, many shrubs sprout from the unburned root stocks, rejuvenating the community. Periodic fires caused by lightning leveled shrub lands even before the presence of human populations, removing old, moribund shrubs.

In other parts of the world—Australia, South Africa, and some regions around the Mediterranean—similar shrub land communities occur. The dominant species there fill similar niches but are unrelated to U.S. chaparral.

Temperate Deciduous Forests

Closer to the poles, rainfall increases again, supporting large numbers of taller trees and creating the temperate deciduous forests. These forests cover much of eastern North and South America and Europe, but the dominant species differ on different continents (Fig. 43.8). Warm summers and freezing winters characterize most parts of the biome, but the length of the growing season varies with latitude.

The dominant species and the number of species forming the canopy vary not only from continent to continent but also from one part of a biome to another. Beech, maple, and birch commonly dominate northern deciduous forests in the U.S., while oak and hickory occur more often in the south.

Like tropical forests, temperate forests show clear vertical stratification. The canopy and lower layers of leaves provide homes to insects and birds. The leaves also intercept much light, creating a dim forest floor.

Boreal Coniferous Forests

A climate of very cold winters and cool summers, harsher than that of the temperate deciduous forest biome, characterizes the boreal coniferous forest, or taiga (Fig. 43.8). Here the growing season averages 120 days or less, with only moderate rainfall, but considerable winter snowpack. Conifers, such as spruce, fir, and larch, dominate these forests.

Tundra

Between boreal coniferous forest to the south and Arctic ice to the north lies the cold, treeless tundra. A land of permanently frozen subsoil, or **permafrost**, tundra forms a belt across the far north of Eurasia and North America, and also occurs at high altitudes (Fig. 43.8). A short, cool growing season and a long, freezing winter characterize its harsh climate. Because sunlight strikes the tundra at a low angle, passing through a large volume of atmosphere, light levels in the tundra remain quite low. Little precipitation falls, depositing a winter snowpack of only 10 to 20 cm, but what little water reaches the ground evaporates slowly. Mosses, lichens, low sedges, grasses, and dwarf woody plants dominate the landscape, and the

Figure 43.13
Tundra in northern Alaska in autumn.

species diversity remains low (Fig. 43.13). Mammals common in the tundra include the herbivorous voles, lemmings, arctic hare, and caribou, and carnivores such as arctic foxes and wolves. Large numbers of migratory birds make their summer home in the tundra, but few reside there year round.

Aquatic Ecosystems

For all biomes, including the driest deserts, water is a crucial environmental factor. Water moderates temperatures, modifies soil during freeze-thaw cycles, and takes part in all biochemical processes. Some ecosystems, however, exist entirely within water, and these environments have a unique set of physical and biotic features.

Marine Environments

The oceans cover 70% of the Earth's surface, forming a continuous globe-spanning body of water. But life in the ocean is not evenly distributed. Most oceanic species live near coasts and in certain surface waters where nutrients well up from the bottom. Light, limited by water depth, and temperature are two major factors affecting the distribution of marine plants and animals. Salinity varies little in the open ocean, averaging 34 to 35 parts per thousand. Estuaries, where rivers and oceans meet, contain water with highly variable salinity.

Ocean water moves constantly. Winds and the Earth's rotation swirl surface waters into great clockwise currents (gyres) in the northern hemisphere and counterclockwise gyres in the southern hemispheres (Fig. 43.14). The north-flowing Gulf Stream off the southeastern coast

of the U.S. and the north-flowing Humboldt Current off the southwestern coast of Chile form parts of those gyres (Fig. 43.14). In the deep ocean, salinity and temperature changes generate slow-moving currents by modifying water density. Surface freezing of Antarctic waters, for example, can create oceanwide currents. When seawater freezes, at around $-2°$ C, it leaves much of its salt behind in the water. The extra salt increases the density—and therefore the weight—of this very cold water. It sinks and creeps slowly northward beyond the equator, eventually mixing with the bottommost layer of water in the deep ocean basins.

Ocean floors compare geologically with land masses. Slopes, ridges, mountains, deep canyons, and trenches abound. A gently sloping **continental shelf** usually extends some distance offshore; beyond this, the **continental slope** drops steeply to the ocean floor (Fig. 43.15). Communities differ as the ocean floor changes. The **littoral**, or **intertidal**, **zone** lies between high and low tide lines, supporting a tremendous variety of algae and marine invertebrates, as well as myriad single-celled organisms. The continental slope itself, or **bathyl zone**, drops to about 4000 m, blending into the deep plains of the ocean bottom, the **abyssal zone**. Marine scientists call the deepest ocean trenches, some as deep as 11,500 m below the surface, the **hadal zone**, after Hades, Greek god of the nether world. In areas of rifting crustal plates on the ocean floor research submarines find the remarkable hydrothermal vent ecosystems, where food chains begin not with photosynthesis but with chemosynthesis, the bacterial utilization of sulfur compounds brought up in hot seawater circulating through the vent system.

Ecologists also divide the water itself into zones. Over the continental shelf lies a region of relatively shallow water, the **neritic zone**, with water as deep as 200 m. Beyond the shelf lies the open ocean, or **oceanic zone**. Light penetrates these waters and powers photosynthesis only down to about 100 m (**euphotic zone**). In the dark waters below (**aphotic zone**), most life depends on the gentle rain of organic matter from above.

Ecologists make two broad generalizations about the distribution of biomass in the world's oceans. First, marine biomass decreases from the shoreline out into the open ocean: neritic waters generally support a richer flora and fauna than oceanic waters (Fig. 43.16). Second, unlike biomes on land, oceanic biomass decreases toward the tropics. In other words, most cold, temperate oceans waters are biologically richer than most parts of tropical seas (Fig. 43.16).

Phytoplankton carries on much of the photosynthetic production in the ocean, called **primary production**, although some nearshore kelp communities are phenomenally productive. The supply of inorganic nutrients such as phosphates and nitrates proves crucial in determining levels of primary production in the ocean. A continual

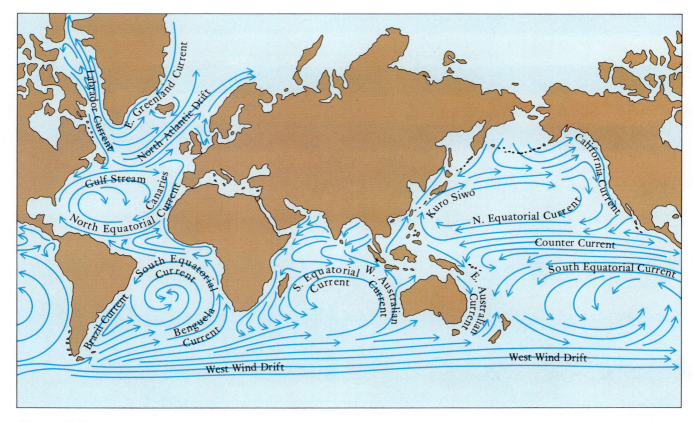

Figure 43.14
Pattern of the world's ocean currents.

==supply of nutrients pours into the oceans from rivers and from nutrients cycled through food webs as organisms die, settle to the bottom, and decay. Thus shallow coastal waters support a greater biomass than deeper oceanic waters because sunlight penetrates much or all of the water column and nutrient levels remain high.==

In some parts of the world, surface currents carry water away from shore. Deeper, nutrient-rich water rises to replace the surface water in a process called **upwelling**. Upwelling provides nutrients to boost primary production, which in turn supports high species diversity and density. Some of the world's greatest fisheries center on

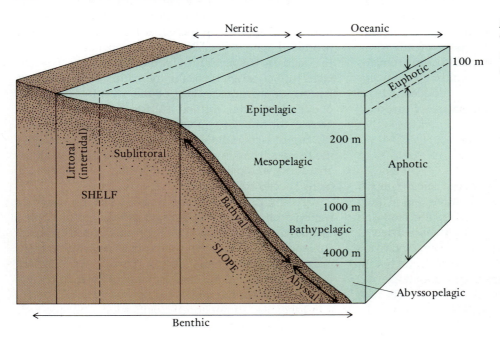

Figure 43.15
The continental margin of a marine basin, showing the different environmental zones.

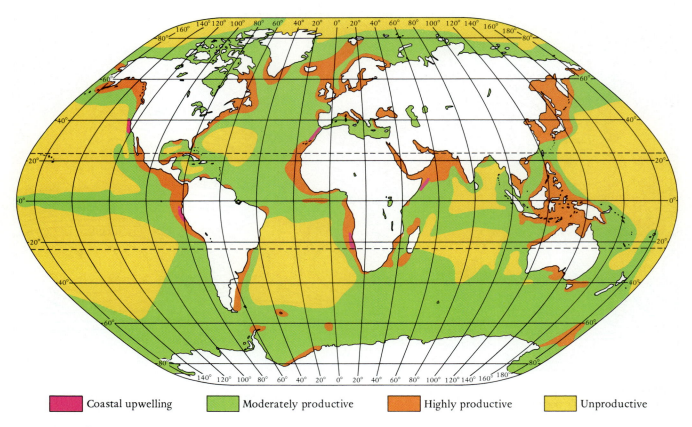

| Coastal upwelling | | Moderately productive | | Highly productive | | Unproductive |

Figure 43.16
Variation in primary production across the world's oceans.

areas of upwelling—along the coasts of Chile and California, and the Antarctic, for example. Mixing may also occur as surface water in temperate latitudes cools during the winter (Focus 43.1). In contrast, the permanently warm and less dense surface water in tropical oceans remains on the surface and the nutrient-rich cold water remains below. An exception to this is a narrow production band along the equator where countercurrents produce upwelling.

Freshwater Ecosystems

When seawater evaporates from the ocean, it leaves its salt behind, entering the atmosphere as water vapor, then condensing as fresh water. As it falls on the land and gradually flows back toward the sea, it creates a variety of freshwater ecosystems: rapidly moving streams and rivers and relatively still marshes, ponds, and lakes, each with characteristic ecosystems. The science of **limnology** (Gr. *limne*, marsh or pond + *logos*, discourse) deals with these freshwater environments. The following brief discussion will focus on lakes.

The great glacial ice sheets that ploughed across North America in the Pleistocene scoured basins in the northern U.S., forming the Great Lakes. Similar glaciers produced many other lake basins throughout the world. In contrast to these relatively young **glacial** lakes, **tectonic** lakes formed when blocks of the Earth's crust slid apart, leaving depressions. Some of the world's largest and deepest lakes were formed this way, including the great lakes of the Rift Valley in Africa and Lake Baikal in Siberia, the largest lake in the world in terms of volume. Many of the species living in tectonic lakes have a long evolutionary history within the lake. The Rift Valley lakes all contain many species found nowhere else on Earth.

Freshwater lakes exhibit zonation like that in oceans but on a miniature scale. Around the margin of a lake a **littoral zone** of shallow water contains various types of submerged and emergent water plants and myriad other organisms. The central, deeper part of the lake, or **limnetic zone**, and the bottom, or **profundal zone**, support different communities.

Lakes also show vertical stratification. Light adequate for photosynthesis reaches no deeper than the shallow

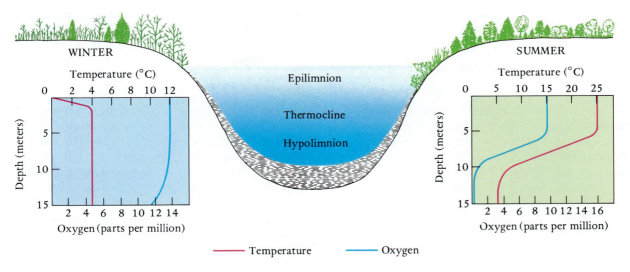

Figure 43.17
Thermal stratification in a north temperate lake (Linsley Pond, Connecticut). Summer conditions are shown on the right, winter conditions on the left. Note that in summer the oxygen-rich circulating layer of water, the epilimnion, is separated from the cold oxygen-poor hypolimnion waters by a broad zone, called the thermocline, that is characterized by a rapid change in temperature and oxygen with increasing depth.

photic zone. Furthermore, lake water lies in layers differing in temperature during most of the year. In summer, temperatures decrease from the surface to the bottom (Fig. 43.17). Wind circulates the upper layer, but because surface water remains warmer and therefore lighter than deeper water it mixes little with deep water. During spring and autumn in temperate regions, changes in air temperature alter the surface water temperature. Differences between surface and deep water disappear, and winds can stir the whole lake, causing spring and autumn **overturns**.

The onset of winter, with its colder air temperatures, reestablishes thermal stratification. As surface water cools, circulation slows, and when the surface freezes, shielding the lake from the wind, winter stratification sets in. Fresh water near freezing is slightly less dense than warmer water—up to about 4° C. So the lake water attains a winter temperature gradient just the reverse of summer stratification: 0° C at the surface and gradually warming toward the bottom, where deep lake water never falls below about 4° C (Fig. 43.17). Spring and autumn overturns circulate not only nutrients but also oxygen, which can become depleted in deeper water where photosynthesis does not take place.

Biologically, lakes also resemble oceans in some respects. Phytoplankton carry on primary production,

many zooplankton graze on the phytoplankton, and fish and other animals prey upon plankton.

In addition to inhabiting the several underwater zones in lakes and ponds, some animals live primarily at the air-water interface. Because of water's extremely high surface tension, a thin film separates air from water. Water striders, some beetles, many insect larvae, and even hydras live on, or attached to the underside of, this film between two worlds.

ENERGY FLOW THROUGH FOOD WEBS

Trophic Levels

Within a community, most populations depend on some other population for food. All end up being consumed by other organisms, if only decomposers. Ecologists call these links **trophic** relationships and call a trophic dependency through a series of species a **food chain**. Nearly every food chain begins with some photosynthetic organism storing the sun's energy in the chemical bonds of complex sugar molecules. For example, mesquite bushes of the southwestern U.S. deserts store energy in seeds. Kangaroo rats eat the seeds. The rats in turn become prey of snakes. The snakes may be the prey of certain birds, such as roadrunners.

In temperate regions, oceanic phytoplankton typically undergo two seasonal population explosions, or blooms—one in the spring, the other in late summer or fall (Focus Fig.). In winter, low temperatures and reduced light restrict photosynthesis. Spring brings higher temperatures and more light that, combined with the nutrient-rich waters, produces a bloom of phytoplankton. Soon, however, the huge phytoplankton populations exhaust the nutrients. Replacement from lower layers no longer occurs because warming of the surface water (which lowers its density) keeps it on top and prevents mixing. Whereas temperature and light were the limiting factors during the winter, the nutrient level limits population growth during the summer. As fall approaches, the upper layers of water begin to cool again and thermal stratification disappears. Mixing is no longer impeded and water rich in phosphates and nitrates rises from below, firing a second phytoplankton bloom. The bloom is halted by low winter temperatures and reduced light. The animals grazing on the phytoplankton show similar population changes, lagging somewhat behind those of the phytoplankton.

Cycle of primary production in a temperate sea.

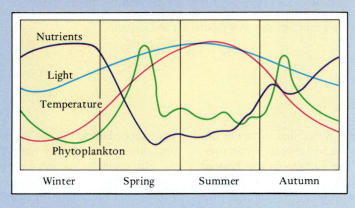

In this food chain each organismal link represents a step up to a new **trophic level**. The relationship can be diagrammed as a linear chain of energy flow.

Sun ⇒ Mesquite → Rat → Snake → Roadrunner

But the trophic relationships of a community actually appear more weblike than linear. Many species eat mesquite seeds, and the kangaroo rat has other food sources. Similarly, a number of carnivores, including roadrunners, eat kangaroo rats, and the rat represents only one of the prey items of each carnivore. Furthermore, organisms at every level are host to parasites and bacteria, and all feed a wide variety of decomposers when they die. Food webs usually turn out to be complex, and most remain only partially known (Fig. 43.18).

A food web can be viewed as a map of energy flow through the community. Energetically, all communities are open systems. An initial energy input from outside the community (the sun in most ecosystems) fuels primary production. Energy enters most sea bottom, or benthic, communities as living or dead organic matter drifting down from the heavily populated photosynthetic zone far above. In a cave, organic material—fecal droppings, leaves, and wood—washes or blows in. As energy flows through the system, some continuously escapes in the form of heat and export. And whatever the source, food webs only delay the flow of energy, all of which eventually escapes as heat.

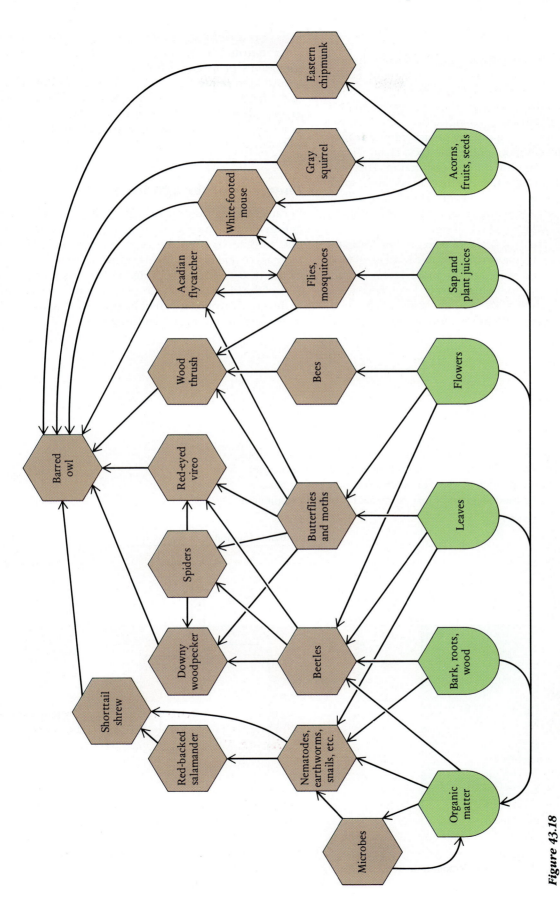

Figure 43.18
Diagram of the food web of a temperate deciduous forest community. Many species and species groups have been omitted. Hexagons are consumers; round-bottom shields depict parts of primary producers.

Production

Plants and other photosynthetic organisms form the first trophic level for communities receiving sunlight. Called **producers**, they convert about 1% of the solar radiation that impinges on the Earth into chemical bonds in ATP, sugars, cellulose, and other compounds. Not all of the chemical bond energy trapped by photosynthesis, called **gross primary production**, enters the food chain. Photosynthesizers use some for cellular respiration. This fraction eventually leaves the plant as heat. In addition, some energy escapes as heat during every chemical reaction. The amount of energy trapped, or fixed, in new biomass is called the **net primary production**.

Organisms that feed on producers, or **primary consumers**, constitute the second trophic level of all food chains. In every food web, one or more levels of consumers eat organisms in the next lower level. Secondary consumers eat primary consumers, and tertiary consumers eat secondary consumers. If a consumer eats ani-

mals, it is called a carnivore. At every trophic level, parasites detour some energy through their bodies, sometimes providing a link to a different food web. A special kind of consumer, active at every trophic level in every food web, devours feces and dead tissue. These **decomposers**, primarily bacteria and fungi, in turn serve as a food source for many animals, linking the first food web to yet another.

Energy Transfer Efficiency

Because biomass stores energy, ecologists can estimate energy flow by measuring the transfer of biomass from one trophic level to the next. Only a fraction of the biomass of one trophic level becomes part of the biomass of the next trophic level (Fig. 43.19). On average, only 10% of the biomass in one trophic level becomes biomass in the next, carrying much of its energy along with it. The 90% biomass loss occurs in three ways.

Figure 43.19

Energy flow through four trophic levels within a community. At the first trophic level, part of the absorbed light energy is lost as heat and part of the primary production is used in respiration. At the consumer levels only part of that consumed can be assimilated and part of that assimilated is used in respiration.

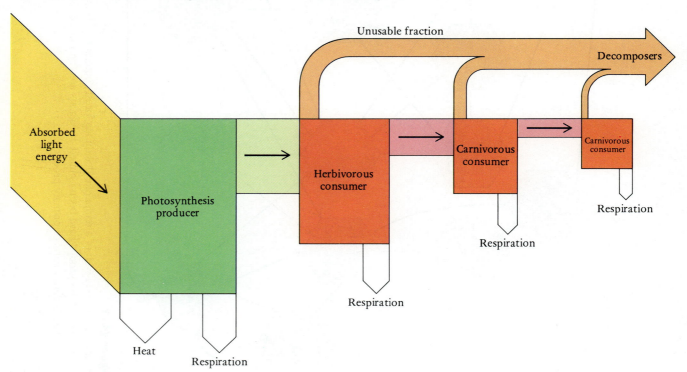

First, some biomass leaves the community. Wind, water, and parasites all aid this export. Gravity carries some biomass down from forest canopy communities. Running water carries leaves, twigs, and the bodies of small animals downstream. In marine ecosystems, the bodies and feces of surface dwellers fall through the water column, eventually reaching the sea floor, where this detritus feeds the benthic community. Some parasites spend part of their life cycle in one host and a subsequent part in a different host, which may live in a different community.

Second, of the biomass ingested by any animal, a fraction leaves the body as undigested food or feces. Much of this biomass feeds decomposers, sometimes in a separate food web. For example, insect-eating birds cannot digest the chitinous exoskeleton of their prey, and most mammals pass cellulose through their gut undigested.

Third, of the total amount of food biomass that animals digest and assimilate, a considerable fraction fuels cell respiration. This biomass leaves the food web as exhaled carbon dioxide (CO_2) and water (H_2O). Some of

Figure 43.20

Annual production in a North Sea food web. Figures represent biomass equivalents in kcal/m²/year. A number in the middle of an arrow indicates the part of the lower trophic level in kcal consumed by the particular group of animals of the higher trophic level. Percent conversion efficiencies are shown for three steps.

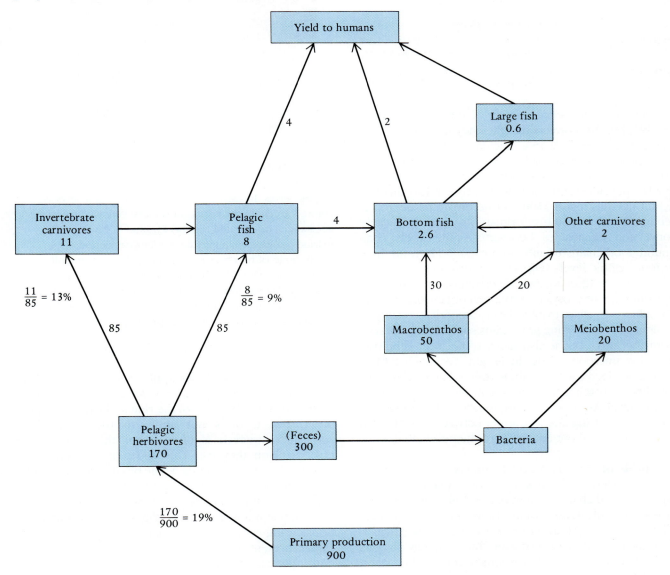

the energy once contained in the chemical bonds of carbon-containing compounds powers movement and growth, but much escapes as heat.

Two factors affect the overall efficiency of energy transfer. First, animals differ in how efficiently they assimilate food. Plants contain much indigestible cellulose, reducing the herbivore's **assimilation efficiency**. Carnivores, on the other hand, have an average higher assimilation efficiency because more of their food is digested and assimilated. The second factor measures how organisms use the food they assimilate. Because warm-blooded animals, or endotherms, use so much energy to maintain body heat—which is constantly lost to the environment—they convert food less efficiently into tissue. Cold-blooded animals, or ectotherms, achieve higher **tissue growth efficiency**. Combining the two rates of efficiency yields the **ecological growth efficiency**, expressed as new biomass/gross energy intake:

Ecological Growth Efficiency

Herbivores	Endotherms	0.3% – 1.5%
	Ectotherms	9.0% – 25.0%
Carnivores	Endotherms	0.6% – 1.8%
	Ectotherms	12.0% – 35.0%

An energy transfer efficiency of 15% may seem reasonable, but this **conversion efficiency** only applies to a single trophic level. At 15% efficiency, 100 grams of biomass produced by a plant yields 15 g of tissue in an ectothermic primary consumer. To gain 15 g of tissue, a third trophic level endothermic carnivore at 1% efficiency would need to ingest 1500 g of herbivore tissue, representing nearly 10,000 g of primary production. Because conversion efficiencies between trophic levels compound, food chains usually do not extend beyond three or four trophic levels (Fig. 43.20; Focus 43.2).

A trophic level analysis yields important general information about ecosystems. But real organisms often do not adhere to strict trophic level categories created by ecologists. A cow eating grass gains some energy directly from the grass, but it also gains some from cellulose-digesting microorganisms in its gut. Seed-eating birds sometimes feed insects to their young. Some carnivores eat almost anything they can catch, ignoring the prey's trophic level. As a result, ecologists often take a pragmatic approach, studying energy transfer between organisms in a community without attaching trophic level labels.

Pyramids of Numbers and Biomass

With each step up the trophic ladder, much energy escapes the food chain. A graph of the total energy trapped in each trophic level would always appear pyramidal, because no energy transfer process can be 100% efficient. If we measure biomass, however, the seemingly obvious conclusion that each higher trophic level should contain

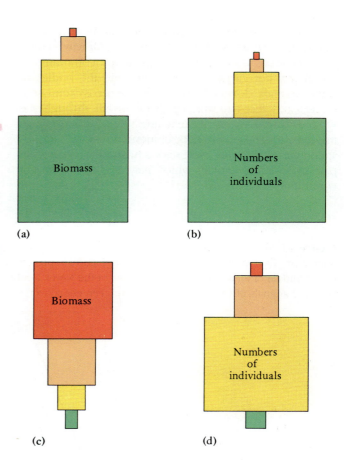

Figure 43.21
Hypothetical pyramids of numbers and biomass of four trophic levels. In each case the producer trophic level is colored green; herbivorous consumers, yellow; carnivorous consumers, orange and red. (a) Pyramid of biomass in a marine pelagic community over a ten-year period. (b) Pyramid of numbers in a marine pelagic community. (c) Inverted pyramid of biomass of the standing crop in a marine pelagic community. (d) Partially inverted pyramid of numbers in a forest community, reflecting the large size and smaller numbers of producer organisms.

less mass does not always apply. It does apply to the biomass of polar bears in the Arctic, which is much smaller than the biomass of the ring seals the bears prey upon. But it does not apply to the biomass of zooplankton, which is often larger than that of the phytoplankton they eat. This apparent paradox stems from the fact that ecologists define biomass as an instantaneous measure. A pyramid of biomass will be evident when the total production of all individuals in a trophic level over a number of years (Fig. 43.21a) is compared. Considering only **standing crops**—the number of individuals present at any particular point in time—the pyramids often appear inverted.

Marine food chains are commonly graphed as a pyramid of numbers—with fewer individuals in each higher trophic level—and an inverted pyramid of biomass, if only standing crops are considered (Fig. 43.21c). For example, there may be more diatoms in the standing crop than grazing copepods and more copepods than menhaden, a fish that eats copepods, but the menhaden biomass is greater than that of copepods, which is greater than that of diatoms. Menhaden achieve large individual biomass by consuming large numbers of copepods over many copepod generations. Copepods in turn are grazing on diatoms, which produce many generations during a copepod's one-year life span.

Standing crops in forest communities exhibit a different pattern: the pyramid of numbers stands on a small base. Trees form the photosynthetic base of a forest community, and there are few trees compared to the number of individual herbivores feeding on them. Considering biomass instead of numbers, a forest community pyramid appears normal and sits on a wide base. The biomass of the trees outstrips that of the herbivores (Fig. 43.21d).

SOLAR RADIATION: WHERE THE CHAIN BEGINS

Nearly all ecosystems on Earth owe their existence to solar radiation that bathes the planet. Sunlight provides heat and light, allowing animals to be active and see their environment. Sunlight also powers most food chains. Photosynthesizing organisms capture less than 1% of the solar energy striking the Earth and convert it into high-energy bonds between carbon atoms, a process called carbon fixation. That energy can then be used as cellular fuel for both the producer and consumers. Almost all other organisms are dependent indirectly on that fixation.

A large part of the total radiation striking the Earth bounces right back into space. The remaining energy—nearly all absorbed as heat by land masses, water, and vegetation—only remains on Earth a short time. Radiation absorbed as heat slowly radiates back into space. Earth and all of its life only delays the inevitable spread of heat away from the sun. Even the energy captured in photosynthesis eventually radiates away as heat, delayed only a short while on its one-way journey. The solar energy striking the Earth equals the energy reradiated into space. This balanced exchange is referred to as the Earth's **heat budget** (Fig. 43.22).

The Ozone Layer

The Earth's atmosphere modifies the heat budget in important ways, in the process making our planet habitable. A layer of ozone (O_3) in the outermost part of the atmosphere (the stratosphere) absorbs a small fraction of solar radiation, about 3% (Fig. 43.22). But a part of that fraction is high-energy ultraviolet light, which is mutagenic and damaging to living tissues. Ozone blocks 99% of all the high-energy ultraviolet radiation reaching the stratosphere, shielding organisms from excessive exposure. Few organisms could survive on the Earth's surface without this shield.

Unfortunately, the ozone layer has turned out to be rather fragile. Chlorofluorocarbons—chemicals used in refrigerators and air conditioners, as a propellant in some spray cans, and in some industries—break down ozone when released into the atmosphere. The alarming appearance of a hole in the ozone layer over the Antarctic, first noticed in 1986, dramatically focused attention on the problem. Atmospheric scientists remain cautious about predicting the current rate and ultimate extent of ozone depletion. Manufacturers of products in spray cans took a small step by discontinuing their use of chlorofluorocarbons as a propellant.

While we worry about preserving the high-altitude ozone layer, we also worry about low-altitude ozone produced by automobiles and power plants. Ozone in the lower atmosphere acts as a serious pollutant.

The Greenhouse Effect

Much solar energy reaching the Earth consists of short-wavelength radiation, which passes easily through the atmosphere. About 35% bounces back as short-wavelength light and about 65% warms clouds, land, and water (Fig. 43.22). All of the energy retained on Earth eventually radiates back into space as longer-wavelength radiation, or heat waves. Atmospheric CO_2 and water vapor absorb some of this outgoing long-wavelength radiation, reradiating the energy in all directions, including back toward Earth. The resulting cycle traps heat within the shell of the atmosphere.

A car sitting in the sun with its windows rolled up illustrates this greenhouse effect. Incoming short-wavelength radiation passes easily through glass, which absorbs and reradiates the long-wavelength, outgoing radiation. These heatwaves pass through the glass and car body slowly. As a result, the interior temperature of the car soars.

Without the atmospheric greenhouse effect, the temperature on the Earth's land surface would be about 34°C colder than it is today, too cold to support most life. But human activities—including widespread removal of trees, which absorb CO_2 from the atmosphere, and fossil fuel burning that spews large amounts of CO_2 into the air—continue to increase atmospheric CO_2. Altering the gaseous content of the atmosphere concerns ecologists and atmospheric scientists alike. The potential results of increasing the greenhouse effect have proved difficult to gauge. Some scientists predict increased productivity and milder climate, but most worry that even a slight world-wide temperature increase of 3° to 5° C could begin melting the polar ice caps and raise the sea level signifi-

The principles of food chains and energy conversion efficiencies have considerable application to human welfare. A much larger human population could be supported by eating grain than by feeding the grain to cattle and then eating the cattle. The food energy available when we become carnivores represents only about 5% of what was available had we eaten the grain itself. Even if we feed the grain to pigs—which are twice as efficient at converting grain to meat as cows—the apparent value of reducing human meat consumption even slightly staggers the imagination. It has been estimated that a 10% reduction of annual meat consumption in the U.S. would provide enough grain to support 60 million additional people.

Any proposal to convert humans to a vegetarian diet, however, faces several pragmatic challenges:

1. Transporting grain worldwide to the populations that need it most has proved to be a major problem.

2. Much of the world's rangeland cannot support grain suitable for human consumption.

3. Humans usually require some animal protein, if only dairy products, to supplement a vegetarian diet—and milk cows require high-quality grain.

4. Finally, if human populations have reached the Earth's carrying capacity, as some ecologists contend, only halting worldwide population growth will have a lasting effect.

Food chains also account for a quite different human problem. Some toxic substances cannot be degraded or excreted by organisms. Small amounts eaten or absorbed by primary consumers remain in their tissues. If a carnivore at the next trophic level eats a large number of primary consumers, its tissues may retain the toxin from each meal. At yet higher trophic levels, the toxin may become sufficiently concentrated to have direct deleterious effects.

Many pesticides degrade slowly in the environment and may enter food chains. DDT and related pesticides are insoluble in water. They persist in soil and on plants, but if eaten by herbivores, they dissolve in fat, becoming sequestered in fatty tissues. When predatory birds consume contaminated herbivores, the DDT can cause the birds to lay thin, weak eggs that often break before hatching (Focus Fig.) Because of these effects on birds, the U.S. government has banned DDT use in this country, even though it still allows manufacturers to export the pesticide. Many other pesticides and other industrial chemicals undergo similar biological magnification, causing death or low viability in animals.

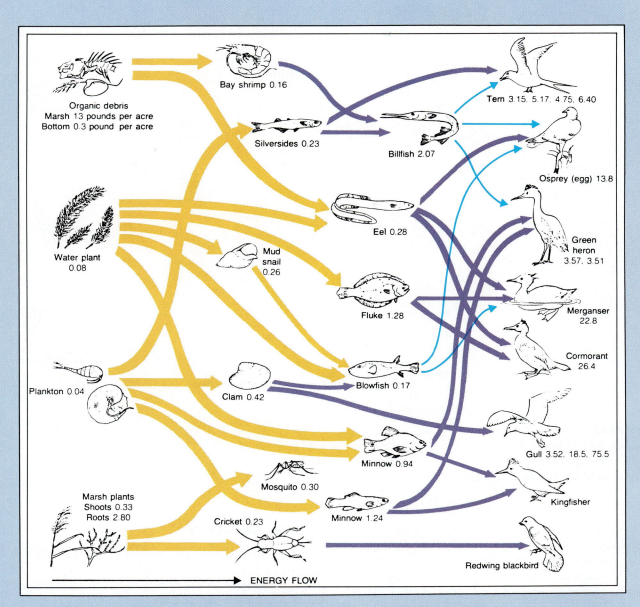

ENERGY FLOW

Part of the food web in the estuaries of Long Island. The thickest arrows are between the first and second trophic levels; the thinnest arrows are between the second and third levels. This figure was taken from a study of DDT accumulation in the food chain, and the numbers after each organism indicate the DDT content in parts per million. The rising numbers thus reflect the energy flow through the food web and the accumulation of pesticides in biomass at higher levels.

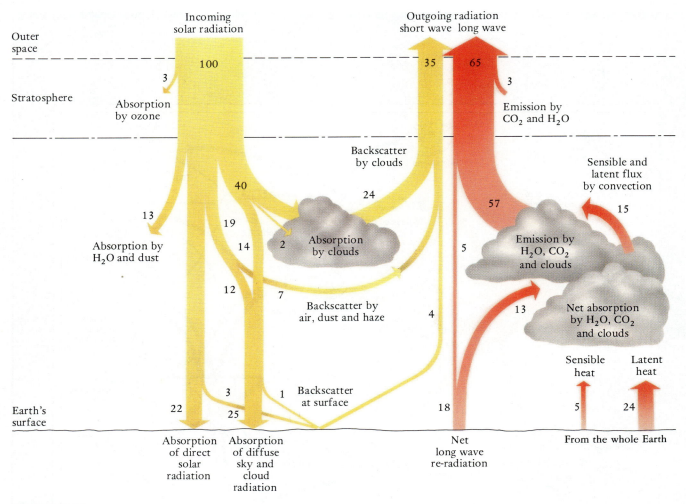

Figure 43.22
The annual heat budget of the Earth, showing the balance between the heat gained
through solar radiation and that lost from the Earth. The numbers indicate the
average percent loss or gain from various pathways.

cantly. A rise of only 10 meters would flood 10 million
square kilometers, including many of the world's major
cities. In the short term, slight global warming could affect
regional precipitation patterns and increase evaporation,
leading to drier soils and profoundly affecting agriculture.

BIOGEOCHEMICAL CYCLES

Life on Earth only delays solar energy on its one-way trip
through space. But almost all matter remains on Earth,
cycling constantly. Respiration, excretion, and death all
release matter from food webs. Some of this matter enters
other webs, and some enters chemical cycles. Because
matter cycles thorough living and nonliving systems, and
its movement depends on geological processes, these
cycles are called **biogeochemical cycles**. Ecologists find
several cycles especially relevant.

Hydrological Cycle

All water takes part in the **hydrological cycle**, the continual movement of water molecules from oceans to the
atmosphere, soil, bodies of fresh water, living things, and
eventually back to the oceans.

  The sea contains 97% of the Earth's water, providing
the largest source of evaporation. Five-sixths of the water
evaporating into the atmosphere comes from the ocean
surface. The remaining one-sixth evaporates from soil
and plants. Water returns from the atmosphere to the land
and sea as precipitation. Some of the water evaporating
from the ocean falls back into the ocean, but a significant
part falls on land. That fraction falling on land, and eventually returning to the sea in streams and rivers, makes
possible a richer terrestrial biota. Water also acts as a
carrier, taking many chemicals along with it as it flows

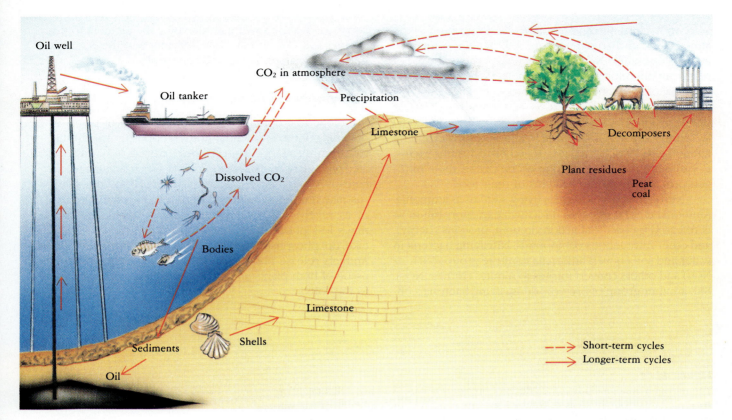

Figure 43.23
Simplified diagram of the carbon cycle. The dashed arrows indicate rapid processes; the solid arrows denote relatively slow ones, for example the geological uplift that carries limestone to the surface where its carbonate dissolves in water in the form of bicarbonate ion and is utilized by aquatic plants as a carbon source.

over and through the land. Eventually any chemical dissolved in water arrives in the sea. If the chemical involved acts as a poison or a pollutant, the carrying action of water can cause pollution problems far from the original site.

Carbon Cycle

Making up the backbone of all organic molecules, carbon is one of the four major elements composing the bodies of organisms. Carbon enters and leaves ecosystems as part of a wide variety of biomolecules—carbohydrates, fats, proteins, and nucleic acids. Ultimately, however, carbon enters nearly all living systems as CO_2 absorbed by plants and other photosynthetic organisms (Fig. 43.23). Even hydrothermal vent bacteria—which extract energy chemosynthetically—use CO_2 dissolved in seawater as a carbon source. Photosynthesizers release some CO_2 during their own respiration and convert the remainder to sugars and tissue. Consumers and decomposers ultimately recombine the carbon with oxygen during respiration, releasing CO_2 back into the atmosphere or water.

There are a number of points in the carbon cycle, called sinks, where carbon can become unavailable to organisms for long periods of time. The wood of trunks and branches of trees immobilizes large amounts of organic carbon for years or decades. When organic material becomes coal or oil, the carbon disappears from the cycle for millennia. Atmospheric CO_2 dissolves in water, producing bicarbonate and carbonate ions. Some of this dissolved CO_2 diffuses back into the atmosphere, but marine organisms use some to build calcium carbonate shells and skeletons, which eventually fall to the bottom, contributing to the formation of great limestone deposits. This carbon may stay out of circulation for millennia. Many such deposits laid down in the past have become subsequently uplifted and exposed. Their slow dissolution recycles some of that carbon back into the water.

Nitrogen Cycle

Nitrogen, another essential element in proteins and nucleic acids, makes up 78% of the atmosphere. Like CO_2, molecular nitrogen gas (N_2) can enter organisms. But since molecular nitrogen remains inert, returning to the atmosphere unchanged and unused, neither plants nor animals can utilize N_2. Certain bacteria come to the aid of plants and therefore all consumers. These **nitrogen-fixing bacteria**, often symbionts on the roots of legumes such as peas and beans, can convert N_2 into ammonia (NH_3) or nitrate groups (NO_3). Plants then use nitrate or ammonia to build nitrogen-containing biomolecules, including amino acids, which they join together to form proteins. Consumers, including decomposers, obtain their nitrogen primarily from the proteins of their food.

Nitrogen is released from living systems as ammonia. Ammonia may be released directly in excretion, but most is released by bacterial action on urea, proteins (decay), and other nitrogenous compounds. **Nitrifying bacteria** in water and soil convert ammonia to nitrite groups ($NH_3 \rightarrow NO_2$) and others convert nitrites to nitrates ($NO_2 \rightarrow NO_3$). Still another group of bacteria, **denitrifying bacteria**, con-

vert nitrites to nitrogen gas, returning the element to the atmosphere (Fig. 43.24). However, much of the recycled nitrogen bypasses the atmospheric sink.

Phosphorus and Sulfur Cycles

Symbiotic microorganisms come to the aid of plants in obtaining another essential nutrient, phosphorus. In this case the microorganisms are mycorrhizae, fungi associated with plant roots. A component of such organic molecules as ATP and nucleic acids, and also of bone, phosphorus may limit individual and population growth because of its relative unavailability. Soil may contain a million times less phosphorus than plants growing in it. Plants take up phosphates ($H_2PO_4^-$ or HPO_4^{2-}) from the soil. Animals release phosphates in wastes, and the decaying tissue of both plants and animals also returns phosphate to the soil.

Figure 43.24

The nitrogen cycle, including the effects of human activities. We have deliberately *not* simplified this diagram so as to convey some impression of the multitude of pathways whereby this vital substance travels throughout the biosphere.

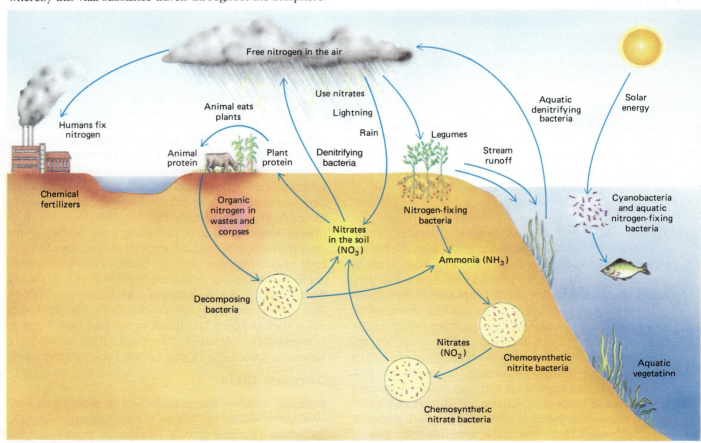

In the sea, phosphates can become incorporated in large sedimentary deposits, which act as long-term sinks. Over geological time such deposits become elevated and eroded, allowing phosphates to dissolve back into solution.

Because of the element's relative scarcity, farmers often use phosphate-rich fertilizers to increase yield. Runoff from fertilized fields and phosphates in sewage find their way into lakes. The element promotes growth in lake ecosystems as well as on farms, and when lake algae are well fertilized, their populations bloom, beginning a process called eutrophication. When the algae bloom dies and sinks to the bottom of the lake, it in turn feeds massive growths of bacteria. That bacterial bloom can seriously deplete the oxygen in the deep water, often killing many species of fish and causing other ecological changes.

Unlike phosphate, sulfur, a component of some amino acids and other organic molecules, abounds in soil. Plants usually absorb ionic sulfur (e.g. SO_4^{2-}) and animals recycle the element, releasing it back into the soil in urine and feces. Bacterial decomposition of wastes releases hydrogen sulfide (H_2S). H_2S may in turn enter the atmosphere, where it combines with oxygen to form sulfur dioxide (SO_2) (Focus 43.3), or it may enter the soil or water, where various groups of bacteria convert it to sulfates.

EPILOGUE: GLOBAL CHANGE

The explosive increase in human populations and the resultant massive pollution, rampant resource depletion, and wholesale habitat destruction are four powerful and interrelated forces driving global change. There have been great swings in global temperatures and fluctuations in levels of atmospheric gases in past geological ages. At the end of the Permian and Cretaceous periods species diversity plummeted because of widespread extinctions. But never in the history of our planet has one species brought about so much change so rapidly.

From the human point of view, planetary changes seem slow. Each generation, knowing little of the past, witnesses only a small part of the change and accepts the world as it appears. Given this temporal shortsightedness, it is not surprising that short-term economic demands, social issues, and national self-interests take precedence. But considered on a geological or evolutionary time scale, massive changes in the atmosphere, widespread destruction of tropical rain forests, and poisoning of the oceans with toxic waste appear rapid indeed. Concerted international efforts for long-term accommodation of human existence with the ecological realities of planet Earth may have to wait for a more enlightened time, but some action can be taken now.

Individuals can make a difference by politically supporting family planning, conservation, recycling, strict control on automobile and factory emissions, and development of clean alternative energy sources. Making such political decisions, and related decisions within our own lives, requires knowledge about the state of the environment and the ecological processes modifying it. Viewed from this perspective, the study of ecology is not an option, but a necessity.

Focus 43.3 *Biogeochemical Cycles and Acid Rain*

Cars and coal-burning power plants discharge large amounts of sulfur dioxide (SO_2) and nitrogen oxides (NO_x) into the atmosphere. Fossil fuel burning in the United States and the industrialized nations of Europe releases about ten times the amount of SO_2 released by natural decay. In the atmosphere, SO_2 and NO_x undergo various photochemical reactions and dissolve in water droplets to form sulfuric and nitric acid, producing acidic precipitation with pH readings far below the neutral pH 7.

Under natural conditions, rainwater can approach pH 5.6 because water vapor reacts with atmospheric CO_2 to form carbonic acid. However, the rain in most of eastern North America now reads below pH 5, and levels below pH 2 have been recorded, which are more acid than lemon juice. The detrimental effects of acid rain are clear. Acidified lakes and streams no longer support the growth of many kinds of aquatic plants, most fish, and many other organisms. Acid rain has contributed to high tree mortality in Europe and the U.S. Ecologists still do not completely understand how acidification causes these forest declines, but interference with nutrient uptake appears to be a major mechanism by which soil acidification produces stress in forests.

Summary

1. A community is an association of interacting populations sharing a particular set of environmental conditions. A community and its physical environment make up an ecosystem.

2. Dominant species in a community interact with the environment in ways that modify environmental conditions. Such changes determine in large part the character of the community and the mix of other species in the community. One example of environmental modification is the vertical stratification in tropical rain forests produced by dominant tree species of the canopy.

3. A succession of communities, called a sere, occupies vacant environments. The seral stages eventually develop into a relatively stable, self-perpetuating climax community characteristic of the climatic region in which it occurs. The dominants of each seral stage prior to climax create an environment that favors the invasion of dominants by the next stage. Seres in microcosm, called microseres, may be described within communities.

4. Each species within a community occupies a niche defined by the particular environmental requirements and interactions of that species. An animal's diet and mode of feeding constitute an important trophic dimension of its niche. Other dimensions may also be described. Competition may lead to resource partitioning, thereby allowing large numbers of species to coexist.

5. Biomes comprise many communities sharing similar climate and terrain. Four types of terrestrial biomes—tropical rain forests, temperate deciduous forests, boreal coniferous forests, and tundra—form more or less latitudinal bands over much of the globe. Grassland, desert, and chaparral biomes show a more patchy global distribution. Available moisture and temperature extremes largely determine the character of terrestrial biomes.

6. The distribution of marine organisms is largely determined by temperature, water depth, and nutrient concentration. In general, coastal waters are richer than oceanic waters, and cold seas are richer than tropical seas. Freshwater ecosystems comprise streams, rivers, and both glacial and tectonic lakes. The seasonal pattern of primary production in temperate lakes is similar to that of temperate seas. Nutrients and oxygen are mixed by spring and fall overturns in lakes.

7. Food webs map the energy flow through a community and show trophic relationships between species. Photosynthetic producers form the base of most food chains, converting light energy to the chemical bond energy of organic compounds. In its passage from one trophic level to the next, energy leaves the food chain as heat and waste. The quantitative relationships of trophic levels are reflected in pyramids of population size and biomass.

8. All of the solar radiation received by the Earth is reflected or eventually radiated back into space. Atmospheric gases absorb long-wavelength radiation leaving the surface of the Earth. Some of this heat is reradiated toward Earth and trapped within the atmospheric envelope. This greenhouse effect makes life possible on our planet, but the excessive amounts of CO_2 and other gases released into the atmosphere by human activities may increase the greenhouse effect, leading to global warming and serious climatic changes.

9. The elements composing the molecules of living systems take part in biogeochemical cycles. Carbon enters and leaves living systems as CO_2 via photosynthesis and cell respiration. Major carbon sinks include wood, the atmosphere, and limestone deposits.

10. Nitrogen and sulfur enter living systems as ammonia, nitrates, and sulfate ions. Plants incorporate these elements into amino acids and other organic compounds. The elements escape food chains as ammonia and sulfate ions. Microorganisms play important roles in the cycling of nitrogen, sulfur, and phosphorus.

References and Selected Readings

Brewer, R. *The Science of Ecology.* Philadelphia: Saunders College Publishing, 1988. A textbook of ecology.

Colinvaux, P. *Ecology.* New York: John Wiley & Sons, 1986. A textbook of ecology.

Editors of Scientific American. *Managing Planet Earth. Scientific American* 261(Sept. 1989) This special issue contains many articles on aspects of global change and strategies for coping with that change.

Ehrlich, P. R., and J. Roughgarden. *The Science of Ecology.* New York: Macmillan Publishing Co., 1987. A balanced treatment stressing the interplay of theoretical models, laboratory experiments, and field observations in modern ecology.

Gross, M. G. *Oceanography: A View of the World,* 5th ed. Englewood Cliffs, N.J.: Prentice-Hall, 1990. A textbook of oceanography.

Kemp, D. *Global Environmental Issues, A Climatological Approach.* London and New York: Routledge, 1990. A multidisciplinary analysis of the environment and society, including such topics as acid rain, ozone depletion, the greenhouse effect, and nuclear winter.

McConnaughey, B. H., and R. Zottoli. *Introduction to Marine Biology,* 5th ed. Prospect Heights, Ill.: Waveland Press, 1989. A textbook of marine biology.

Miller, G. T. *Environmental Science: An Introduction,* 2nd ed. Belmont, Calif.: Wadsworth Publishing Co., 1988. A textbook of environmental science concerned primarily with human impact on the environment.

Nebel, B. J. *Environmental Science: The Way the World Works,* 2nd ed. Englewood Cliffs, N.J.: Prentice-Hall, Inc., 1987. A textbook of environmental science dealing primarily with human impact on the environment.

Ricklefs, R. E. *Ecology,* 3rd ed. San Francisco: W.H. Freeman, 1988. A textbook of ecology.

Stevens, G. C. The latitudinal gradient in geographical ranges: How so many species coexist in the tropics. *American Naturalist* 133(1989):240–256.

A P P E N D I X   A

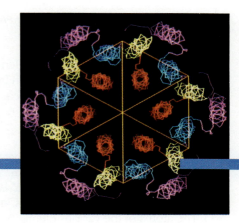

The Chemical and Physical Basis for Zoology

Zoologists utilize the tools and concepts of mathematics, chemistry, and physics in seeking to understand the nature of the living world and the biochemical and physiological processes that underlie it. Most of you will have received a background in these subjects as part of your secondary education or in other science courses in college. The material in this appendix will be familiar to you if your background is good, if not, it will introduce you to material that you will need in order to understand the chemical and physical processes discussed in this book.

Matter and Energy

The entire universe consists only of matter and energy in their various forms. **Matter** occupies space and has a mass. A lump of sugar and the particles into which it can be subdivided are examples of matter. The *quantity* of matter is known as its **mass**. We often equate mass with **weight**, which is the force that gravity exerts upon the mass, but technically there is a difference. The quantity or mass of an object does not change regardless of its location, but its weight will vary according to the pull of gravity. A rock of a given mass would weigh about $1/6$ less on the moon than on earth because of the moon's lower gravitational force.

    Energy is the capacity of any system, including living organisms, to do work, i.e., to produce a change of some sort in matter.* Gravity, light, electricity, heat, and motion are examples of energy. Matter and energy are interrelated by Einstein's famous equation

$$E = mc^2$$

where E is energy, m is mass, and c is the speed of light. The speed of light is a constant that in a vacuum is approxi-

*See Appendix B for definitions of the basic units of measurement and of the quantities, such as energy, derived from them.

mately 299,744 km/second. Nothing moves faster than light. Einstein's equation can, in principal, go in either direction. Energy can be transformed into matter; matter, into energy.

Liquids, Solids, and Gases

Because of the heat energy in the environment, the units, or molecules, into which a substance can be divided are in constant motion, that is, they have a certain **kinetic energy** (Gr. *kinein*, to move). The random movement of water molecules can be seen indirectly by observing a drop of water containing minute particles of India ink under a microscope. The water molecules themselves are, of course, not seen, but as they impinge upon the particles of ink these are set in random motion.

    Molecular motion would cease only at absolute zero, which is the lowest temperature theoretically possible, i.e., $-273°$ C or zero on the Kelvin scale. Many kinds of matter exist in three physical forms (a solid, liquid, or gas) depending on the amount of kinetic energy in the molecules and their freedom of movement. We are all familiar with the three forms of water: ice, water, and water vapor. The molecules of a **solid** are closely packed, and strong intermolecular forces limit their motions to vibrations. As temperatures increase, the molecules in a substance become increasingly dispersed, intermolecular attractive forces decrease, and the random movement of the molecules increases. A solid turns into a **liquid**, and a liquid into a **gas**.

Atomic Structure

Innumerable kinds of matter are known, but all can be reduced into smaller units. The smallest units into which a substance can be broken down *chemically* are the **elements**, each of which has different and unique properties. Ninety-two naturally occurring elements are known; but only about a dozen are common in biological systems (Table A.1). Oxygen, carbon, hydrogen, and nitrogen, in that sequence, are the most abundant.

Atomic Nuclei

The smallest unit of an element that still retains the properties of the element is an **atom**. The atoms of all elements can be broken down *physically* into the same subatomic particles: protons, neutrons, and electrons (Fig. A.1). The atoms of the various elements differ only in the numbers of these particles that they contain. (Atomic physicists have split protons and neutrons into yet smaller components, but these need not concern us.) **Protons**, which carry a positive electrical charge, and **neutrons**, which are electrically neutral, are held tightly together in the **nucleus** of the atom by nuclear forces. Each type of element has a characteristic number of protons, which determines the number of negatively charged electrons around the nucleus and hence the chemical properties of the element. The number of protons in an atom is its **atomic number** (Table A.1). It is indicated as a subscript in front of the symbol for the atom, e.g., $_1H$ for hydrogen and $_8O$ for oxygen. The nucleus of a typical hydrogen atom consists only of a single proton and has no neutron. Nearly all of the mass of an element is in its protons and neutrons, so their *combined* number is a measure of the **atomic mass** of the element. An element's mass is indicated by a superscript in front of the symbol, e.g., $_1^1H$ hydrogen. Atomic masses range from 1 for hydrogen to 238 for uranium.

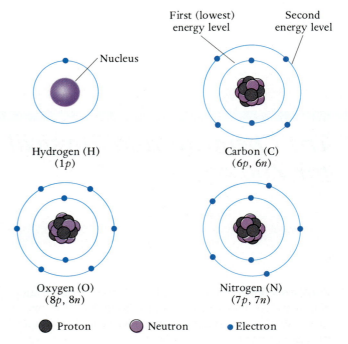

Figure A.1

Diagrams of the four most common elements in biological molecules. Notice that the number of electrons equals the number of protons in the nucleus. This type of two-dimensional diagram, devised by Niels Bohr, is an oversimplification of atomic structure, but it is useful to understand the main features of an atom and chemical reactions.

Table A.1
Common Elements in Animals*

Name	Symbol	% by Mass in Human Body	Atomic Number	Atomic Mass
Oxygen	O	65	8	16
Carbon	C	18	6	12
Hydrogen	H	10	1	1
Nitrogen	N	3	7	14
Calcium	Ca	1.5	20	40
Phosphorus	P	1	15	31
Potassium	K	0.4	19	39
Sulfur	S	0.3	16	32
Sodium	Na	0.2	11	23
Magnesium	Mn	0.1	12	24
Chlorine	Cl	0.1	17	35

*The 11 most common elements in biological molecules are listed from the most abundant to the least. Of these, oxygen, carbon, hydrogen, and nitrogen comprise 96% of the mass of a human. A few other elements also are needed by all animals, but in trace amounts; among them are iron (Fe), iodine (I), manganese (Mn), copper (Cu), zinc (Zn), cobalt (Co), fluorine (F), and selenium (Se).

Isotopes

All of the atoms of a given element contain the same number of protons and electrons, and hence have the same chemical properties, but in some elements the number of neutrons in the nucleus, and hence the atomic mass, varies. Most hydrogen, for example, is $_1^1H$. There are variants of hydrogen that contain, in addition to one proton, one neutron, thus giving hydrogen a mass of 2 ($_1^2H$ = deuterium) or two neutrons, giving it a mass of 3 ($_1^3H$ = tritium). Atoms that have the same atomic number but differ in the number of neutrons and hence in their mass are known as **isotopes** of an element. Some isotopes are unstable and emit gamma rays (a type of electromagnetic radiation without mass), beta rays (electrons), or alpha rays (a helium nucleus, i.e., two protons and two neutrons). As they do so, they change or decay slowly into other elements.

Some unstable isotopes are very useful in tracing the sequence of chemical reactions within the body because they can be identified by their differences in atomic mass or by the radiations they emit. They can, therefore, be used as **tracers**. Normal carbon has an atomic mass of 12; radioactive carbon, a mass of 14. A simple sugar can be prepared using $_6^{14}C$ instead of $_6^{12}C$, and its biochemical breakdown in the body can be followed by identifying the presence of $_6^{14}C$ in intermediate compounds.

Some other radioactive isotopes with much longer decay times are used to date the geological past. A radioactive isotope of potassium occurs in small amounts in many rocks that have crystallized from lavas or other igneous rocks that originally occurred in a molten state. Radioactive potassium decays slowly into a particular isotope of argon. Its half-life is 1.3×10^9 years, by which we mean that half of the radioactive potassium originally present at the time of the cooling and crystallization of the rock will have become this isotope of argon in this period of time. It will take an equal time for one-half of the remainder to decay, and so on. By measuring the amount of radioactive potassium left in the rock and the amount of argon formed, one can calculate when the rock crystallized. Such dating is important to zoologists when a fossil-bearing sedimentary rock that was formed by the compaction of water-borne or wind-blown sediments occurs just beneath the igneous rock layer. In this case we would know that the fossil must be somewhat older than the igneous rock we have dated, and older than the absolute age of the igneous deposit.

Electron Orbitals

One or more **electrons**, each of which carries a single negative charge and has a mass of only about 1/8000 of either a proton or neutron, whirl around the nucleus of an atom. The electrons do not escape from an atom for they are held by the positive charges of the protons in the nucleus. As they whirl around, at times close to the nucleus and at other times farther away, they form sort of an electron cloud (Fig. A.2). The exact position of an electron at any time cannot be known with certainty, but its average position forms an electron **orbital** around the nucleus. Some electrons have spherical orbitals; some, dumbbell-shaped ones. Some electrons move in orbitals close to the nucleus; others, in orbitals farther away. Since electrons moving in orbits farther away from the nucleus have more energy than those closer to the nucleus, the orbitals also can be thought of as **energy levels**. Electrons normally stay at their energy level, but by gaining or losing energy, they can jump from one energy level to another.

Although orbitals vary in shape, electrons at each energy level can be depicted diagrammatically as moving in concentric and circular orbits or shells around the nucleus (Fig. A.1). A maximum of two electrons can be held in the innermost shell. The second and third shell can hold up to eight each; the fourth, up to 18. The total

Figure A.2

Electron orbitals and energy levels of a carbon atom.
(a) Two electrons lie in the innermost energy level and have spherical orbitals (1*s*). They may lie anywhere in the dotted area, but are most likely to be where the dots are most dense. (b) and (c) Four electrons occupy the second energy level. One has a spherical orbital (2*s*), and three have dumbbell-shaped orbitals (2*p*).

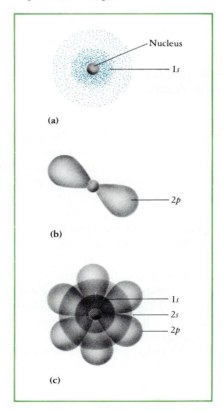

number of electrons equals the number of protons in the nucleus, resulting in an atom that is electrically neutral. A hydrogen atom contains only one electron. Oxygen, nitrogen, and carbon have two electrons in their inner shells, but differ in the number in the second shell. Oxygen has six electrons in this shell; nitrogen, five, and carbon, four. In none of these elements does the number of electrons in the second shell reach the maximum of eight.

Molecules, Compounds, and Mixtures

Most elements are not present in nature as free atoms, but are combined with other atoms of the same or different elements to form molecules and compounds. The oxygen we breathe is not made up of single atoms of oxygen, but of molecular oxygen consisting of two atoms of oxygen bound together and symbolized O_2. A molecule of water contains one atom of oxygen bound to two of hydrogen, H_2O. A molecule of a simple sugar, glucose, consists of 24 atoms (6 of carbon, 12 of hydrogen, and 6 of oxygen): $C_6H_{12}O_6$. Statements of this type that indicate in symbols the type and number of atoms in a molecule are known as **empirical chemical formulas**. A **molecule** is an aggregate of at least two atoms of the same or different elements that are present in precise proportions reflecting the ways the atoms bind with one another. A **compound** is somewhat similar to a molecule for it too consists of atoms held together in a fixed proportion. The difference is that a compound always contains two or more *different* types of atoms, whereas a molecule can contain one or more than one type of atom. Thus O_2 is a molecule and not a compound, but H_2O can be thought of as a molecule or a compound. Molecules and compounds have chemical properties quite different from their constituent atoms. The properties of water are distinct from those of hydrogen and oxygen.

We have seen that the mass of an atom equals the number of protons and neutrons present in its nucleus (Table A.1). Similarly, the **molecular mass** of a molecule of glucose is calculated by adding up the atomic masses of 6 carbons, 12 hydrogens, and 6 oxygens: $(6 \times 12) + (12 \times 1) + (6 \times 16) = 180$. The very small units used to measure atomic and molecular masses are called **atomic mass units** (**amus**), but they are much too small to measure in a laboratory. They are translated into grams. The amount of glucose, or any other compound, whose weight in grams is equivalent to its molecular mass is known as one gram molecule, or one **mole**. One mole of glucose weighs 180 grams. The number of elemental particles in a mole of any substance (atoms, molecules) is known as **Avogadro's number** and equals 6.022×10^{23}.

Many substances in biological systems are present in **solution** in water, i.e., the substance has dissolved and its individual molecules or atoms have separated. It is useful to have a measure of the concentration of molecules in solution. One mole of a substance dissolved in enough water to make one liter is known as a one **molar solution** (1M). Zoologists usually work with smaller concentrations, millimoles (1 mM $= 10^{-3}$M).

A **mixture** in the chemical sense is a combination of two or more types of atoms, molecules, or compounds that are not bound together and therefore retain their identity and may be present in any proportions. The properties of a mixture depend on the proportions of its constituents. Dry atmospheric air is about 20% oxygen (O_2), 79% nitrogen (N_2), 1% carbon dioxide (CO_2), and trace gases. Air within our lungs has different properties because it contains proportionally less O_2, more CO_2, and considerable water vapor.

Chemical Bonds

The chemical properties of an element, and the way it reacts with other elements, depend on the number of electrons in its outermost shell. If this shell is full, the element is inert. Neon is an inert gas because its second and outer shell contains the maximum of eight electrons. It does not react with other elements. Atoms whose outer shell is not full move towards a more stable configuration by losing, gaining, or sharing electrons with other atoms. The atoms become bound together by attractive forces known as **chemical bonds** that represent potential **chemical energy**. Many types of bonds are known, but the ones of greatest importance in living systems are ionic, covalent, and hydrogen bonds.

Ionic Bonds

When different types of atoms come together, those with three or fewer electrons in their outer shell lose electrons, and atoms with five to seven in the outer shell generally gain additional electrons. Sodium (Na) has a single electron in its outer shell and chlorine (Cl) has seven. When these atoms are together, sodium loses its electron to chlorine and becomes electrically positive (Na^+), for it now has more protons in its nucleus than electrons in its shells (Fig. A.3). Chlorine, on the other hand, gains an extra electron and becomes negative (Cl^-). Charged particles of this type are known as **ions** (Gr., moving particle). Positive ions are called **cations** because they move when in a solution to which an electric current is applied toward the cathode, or negative electrode; negative ions are **anions**, for they move toward the positive electrode, or anode. Sodium and chloride ions remain as distinct ions in solution, but when precipitated from solution they are held together by weak **ionic bonds** in an orderly pattern and form crystals of sodium chloride, table salt:

$$Na^+ + Cl^- \rightarrow Na^+Cl^+$$

Solution   Solid

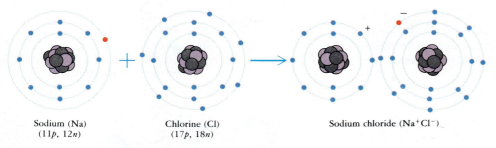

Sodium (Na)
(11*p*, 12*n*)

Chlorine (Cl)
(17*p*, 18*n*)

Sodium chloride (Na⁺Cl⁻)

Figure A.3

When sodium and chlorine come together, the single electron (shown in red) in the outer orbit of sodium is transferred to the outer orbit of chlorine. Both atoms become charged ions. The ions remain separate in an aqueous solution, but in crystals of sodium chloride (table salt) they are held together by weak ionic bonds.

Covalent Bonds

In many cases the stability of the outer electron shell is achieved not by transferring electrons from one atom to another but by the sharing of electrons between two atoms. Hydrogen may lose its single electron to some electron acceptor and become a hydrogen ion (H^+). Alternatively, two hydrogen atoms may come close enough together to share their two electrons in a **covalent bond** (Fig. A.4a and b). Covalent bonds are very strong. The shared electrons are attracted to both nuclei and whirl

Figure A.4

The formation of covalent bonds in a molecule of hydrogen (a), oxygen (b), and water (c).

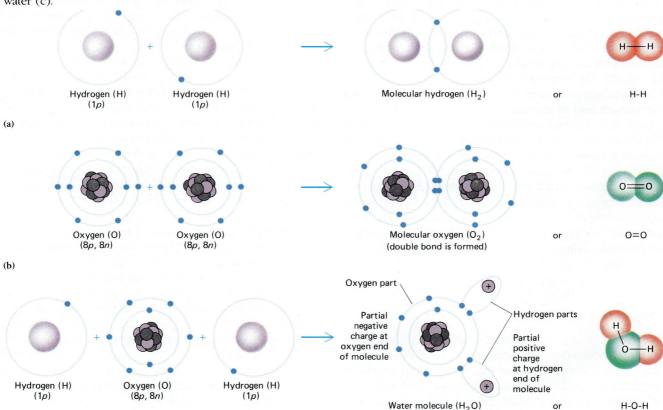

Hydrogen (H)
(1*p*)

Hydrogen (H)
(1*p*)

Molecular hydrogen (H₂)

or

H-H

(a)

Oxygen (O)
(8*p*, 8*n*)

Oxygen (O)
(8*p*, 8*n*)

Molecular oxygen (O₂)
(double bond is formed)

or

O=O

(b)

Oxygen part

Partial negative charge at oxygen end of molecule

Hydrogen parts

Partial positive charge at hydrogen end of molecule

Hydrogen (H)
(1*p*)

Oxygen (O)
(8*p*, 8*n*)

Hydrogen (H)
(1*p*)

Water molecule (H₂O)

or

H-O-H

(c)

around both. The resulting molecule (H_2) can be depicted by a **dot diagram** showing the electrons in the outer energy level and how they are shared or by a **structural formula** in which a single line indicates that a pair of electrons are shared:

$$H:H \quad \text{or} \quad H—H$$

When two oxygen atoms join to form molecular oxygen (O_2), they share four electrons in a double covalent bond:

$$:\overset{..}{O}::\overset{..}{O}: \quad \text{or} \quad O{=}O$$

Each oxygen alone has six electrons in its outer shell. By sharing four they have eight and complete their outer shells. Methane (CH_4) is formed when four hydrogens, each with a single electron, bind covalently with the four electrons in the outer shell of carbon:

$$H:\overset{\overset{\textstyle H}{..}}{\underset{..}{C}}:H \quad \text{or} \quad H—\overset{\overset{\textstyle H}{|}}{\underset{\underset{\textstyle H}{|}}{C}}—H$$

The major atoms of organisms (oxygen, hydrogen, carbon, and nitrogen) easily form covalent bonds, giving great stability to biological compounds.

Hydrogen Bonds

When two hydrogen atoms bond covalently to form water (H_2O), they do not bond on opposite sides of the oxygen atom but toward one side with an angle of 105° between them (Fig. A.4c). As a consequence the water molecule, although electrically neutral as a whole, is asymmetrical electrically. The negative electrons of hydrogen are closer to the nucleus of oxygen than to that of hydrogen, giving the oxygen end of the water molecule a slight negative charge and the hydrogen end a slight positive charge. Electrically asymmetrical molecules of this type are described as **polar**. Polar molecules attract each other and are held together by weak electrostatic forces known as **hydrogen bonds** that form between the hydrogen of one molecule and the oxygen of another (Fig. A.5). Hydrogen bonds are weak, but when many are present they can have a considerable collective strength. They play an important role in maintaining the shape of protein and nucleic acid molecules.

Properties of Water

The dielectric or polar nature of water molecules, and the hydrogen bonds that they form, gives water properties that are of great importance for life. Water is an excellent **solvent** because the charges on the water molecules enable them to surround other charged molecules. Crystals of sodium chloride, for example, quickly dissolve in water

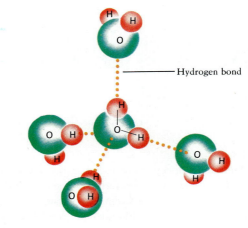

Figure A.5
Hydrogen bonding between molecules of water.

because clusters of water molecules surround the individual ions. The slight negative charge at the oxygen end of the water molecules is attracted to the positive charge of the sodium ions, and the positive charge at the hydrogen end of the water molecules is attracted to the negative charge of the chlorine ions. **Hydration shells** of water molecules surround the ions (Fig. A.6). Many organic molecules that have polar qualities, including alcohols, sugars, and proteins, dissolve in water for the same reason. Much of the water in cells is in the form of hydration shells around other substances.

The hydrogen bonding between water molecules gives water other very important properties. Although the hydrogen bonds in water are very transitory, lasting in any individual bond only 10^{-10} seconds, there are so many water molecules that a certain fraction are bound together at all times. Raising the temperature of a substance requires adding heat energy so that its molecules move faster. As a consequence of the hydrogen bonding between some of the water molecules, more heat must be added to break these bonds than would otherwise be the case. Water has considerable thermal stability, or a high **specific heat**, and does not change temperature as easily as many other substances. The temperature at which water boils and vaporizes is also quite high, 100° C. This is far above the temperatures that most organisms encounter.

As water temperature falls and approaches 0° C, the hydrogen bonds between water molecules become more numerous and evenly spaced. The molecules are held farther apart, and water begins to expand and becomes less dense. As a consequence, ice is lighter than water and floats. Ponds, streams, and large bodies of water do not freeze solid, and life continues beneath a sheet of ice.

Hydrogen bonding also gives water a greater cohesiveness than other solutions. Water has a high surface

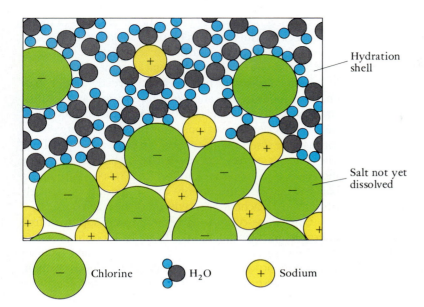

Chlorine H_2O Sodium

Figure A.6
Molecules of water form hydration shells around sodium and chloride ions as a crystal of salt is dissolved.

tension that supports small organisms on the surface of ponds and streams. Because of its high cohesiveness and a high adhesion (tendency to stick to surfaces), water is also drawn into small spaces by **capillary action**.

The Nature of Energy and the Laws of Thermodynamics

The movements of atoms and molecules, the formation of chemical bonds, and all of the activities of life require an input of energy. Energy, as pointed out at the outset of this appendix, is the capacity of any system to do work. We must examine energy more closely before considering the nature of chemical reactions.

Energy takes many forms including light, chemical, mechanical, electrical, and thermal energy. One form can be changed into others, but energy is neither lost nor created during the conversion. For example, the chemical energy present in the covalent bonds of certain molecules that are abundant in muscle can be released when the muscle is activated. Some of this chemical energy causes a

contraction of the muscle, some is dissipated as heat. Together, the movement and heat equal the chemical energy released. This is an example of the principle of **the conservation of energy**, or the **first law of thermodynamics**, which states that the total amount of energy in a closed system, that is, an object and its surroundings, always remains the same. Energy is transformed, but in the process no new energy is created nor is any lost from the system.

Although energy may take many forms, its capacity to do work derives from either the position or state of a body, or from the motion of the body (Fig. A.7). The former, known as **potential energy**, can be exemplified by a boulder at rest on the top of a hill. When the boulder is given a shove, it begins to roll down the hill and its potential energy is converted into the **kinetic energy** of motion. Similarly, some of the potential energy in chemical bonds can be converted to the kinetic energy of muscle contraction.

As energy flows through any closed system and undergoes various transformations, some energy is neces-

Figure A.7
A boulder perched on top of a hill has potential energy. Most of the potential energy is converted to kinetic energy as the boulder rolls down the hill, but some is lost as heat by friction with the ground.

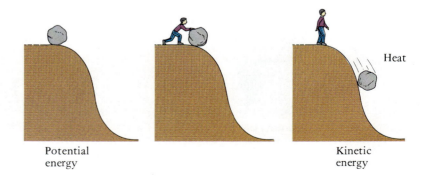

Potential energy   Heat   Kinetic energy

sarily lost at each step as heat, which is dissipated into the surroundings and manifested as an increase in the random movement of molecules. In our solar system, energy moves from a highly organized state (the atoms within the sun) toward increasing disorder and randomness (heat) (Fig. A.8). In biological systems, some of the light from the sun is trapped by green plants as potential energy: some light is used in photosynthesis to form the chemical bonds of sugars and other molecules. Animals derive their energy directly or indirectly from plants and use it to drive their numerous biochemical processes and to cause their movements, but all energy is ultimately lost as heat, the least useful form of energy. Heat can be used as a source of energy in heat engines where there is a pronounced thermal gradient, but heat cannot be used to do work in biological systems, the parts of which operate at nearly the same temperature. Physicists term the increasing disorder and randomness of energy **entropy** (Gr. *en*, in + *trope*, change). The **second law of thermodynamics** states that over time energy becomes increasingly disordered or that entropy increases. Although there may be temporary reversals, the flow of energy is ultimately unidirectional, flowing from an ordered to an increasingly disordered state.

The potential energy in the chemical bonds of molecules of organisms that can be used to do useful work or change the system at a particular temperature and pressure is called **free energy**. The total amount of free energy in a molecule cannot be measured directly; rather, biochemists measure the change in free energy as the initial reactant molecules are converted to their products and heat. As a series of reactions takes place and energy flows through the system, some energy is lost at each step as heat. The amount of free energy left decreases and entropy increases. Eventually an equilibrium is reached in which a balance between the energy in the reactants and their products is reached. Free energy is then zero.

Organisms continually receive a new supply of free energy from solar radiation or from the food they eat. Life is a highly organized form of energy and as an organism grows it incorporates more free energy. Indeed, throughout the course of evolution species have become more numerous and diverse, and most have become more highly organized. One may at first think that organisms violate the second law of thermodynamics, but on further reflection it should be clear that living organisms by themselves are not a closed thermodynamic system. If we take a broader perspective on organisms, the environment

Figure A.8
Energy flows from the sun through organisms, but eventually all of it is lost as heat. Free energy decreases at each step and entropy increases.

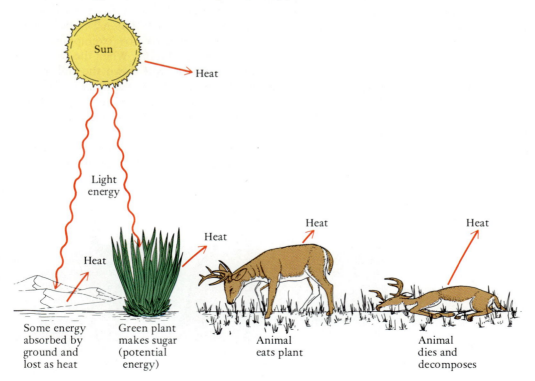

in which they live, and the entire solar system, entropy is increasing. Life exists only by the constant input of solar energy, and this ultimately is dissipated into the environment as heat. Entropy is increasing, the solar system is running down, and when the nuclear fuel in the sun is consumed and dispersed into space as heat, life in the solar system will cease to exist.

Chemical Reactions

A chemical reaction is a change in the molecular structure of one or more substances, and it is governed by the laws of thermodynamics. Two molecules may combine to form a new substance with different chemical properties, or a large molecule may split into different ones. Chemical reactions occur when atoms and molecules, as a result of their kinetic energy, collide with each other and form bonds, or pull existing bonds apart. Reaction rates are dependent on the amount of kinetic energy in the system. By applying heat, kinetic energy and reaction rates increase. Conversely, cooling a system reduces kinetic energy and reaction rates. Since molecular motion ceases at absolute zero, chemical reactions cannot take place at this temperature.

An example of a chemical reaction is the breakdown of glucose in the presence of oxygen to form carbon dioxide and water. In Chapter 13, we examine the many intermediate steps in this reaction, but its overall empirical formula, which simply states the types and number of molecules involved, is

$$C_6H_{12}O_6 + 6O_2 \rightarrow 6CO_2 + 6H_2O + Energy$$

Reactants   Products

In words, one molecule of glucose and six of molecular oxygen, which are called the **reactants**, form six molecules of carbon dioxide and six of water (the **products**), with the release of the potential energy that was locked in the bonds of the glucose. You will notice that the atoms have changed positions and have bound to different ones, but no matter is lost. The same number of hydrogen atoms (12) is present in the six molecules of water as was present in the one molecule of glucose. The number of oxygen and carbon atoms on each side of the equation is also equal. This is an example of another fundamental principle known as the **conservation of matter,** Matter is not lost or destroyed during chemical reactions, but only changed.

Some of the energy that is released in the body during this reaction is used to synthesize other compounds, but some is dissipated as heat. Energy was not lost in this reaction either, but was changed. The amount of free energy (in chemical bonds) decreased, and entropy (heat) increased. Reactions of this type that release energy are described as **exergonic** (Gr. *ex*, out + *ergon*, work).

Chemical reactions in principle are reversible, and carbon dioxide and water can be combined to form glu-

cose and oxygen. However, a considerable input of free energy is needed in this case. The reaction consumes energy, or is **endergonic** (Gr. *endon*, within). This reaction occurs during photosynthesis, when green plants use the energy of sunlight to synthesize sugar and oxygen from carbon dioxide and water.

Many exergonic reactions can start spontaneously when the reactants are brought together, but sugar is a very stable compound and will not spontaneously combine with oxygen in a test tube or pan unless heat is applied. In many reactions sufficient kinetic energy is not present to increase collision rates of molecules, or strain their bonds, to the point that a reaction occurs. The input of energy required to overcome this barrier is known as the **activation energy** (see Fig. 3.16, p. 45). The initial shove needed to cause a boulder that is resting on the top of a hill to roll down a hill is another example of activation energy. Once started, many exergonic reactions proceed spontaneously to completion, at which time all of the reactants are used up and free energy is zero. In other exergonic reactions, the increasing concentration of products will start to move the reaction in the opposite direction.

$$Reactants \rightleftharpoons Products$$

At some point a dynamic equilibrium is reached where the forward and backward reactions balance each other. At this time energy is evenly distributed between the two sides and free energy is zero. In contrast to exergonic reactions, endergonic reactions do not go far in the direction of completion unless there is a continual supply of free energy.

The amount of activation energy that is needed can be reduced if a catalyst is present (see Fig. 3.16). **Catalysts** are chemical substances that mediate reactions and enable them to proceed at lower temperatures and faster rates than would be the case in their absence. They do not make a reaction possible that otherwise could not occur. Catalysts bind temporarily with the reacting molecules and take part in initiating the reaction, but they are released and can be used again. Therefore, only a small amount of a catalyst is needed. Fine powders of many metals, including ion, nickel, and platinum, are used to catalyze industrial reactions such as cracking petroleum to make gasoline. Enzymes in biological systems are organic catalysts, as we discuss in Chapter 3.

Oxidation-Reduction Reactions

Many types of reactions occur in living systems. **Oxidation-reduction reactions** are a common type that require further explanation. A familiar example of this type of reaction is the rusting of iron. When four atoms of iron (4Fe) combine with three molecules of molecular oxygen ($3O_2$), each iron atom loses three electrons, creating four iron cations ($4Fe^{3+}$). A total of 12 electrons have been released.

Each of the six atoms of oxygen in three molecules of molecular oxygen accepts two of the released electrons and forms six oxide anions ($6O^{2-}$). The cations and anions combine to form two molecules of ferric oxide or rust ($2Fe_2O_3$):

$$4Fe + 3O_2 \rightarrow 4Fe^{3+} + 6O^{2-} \rightarrow 2Fe_2O_3$$

The iron is said to be **oxidized** because it loses electrons; the oxygen is **reduced** because it accepts them. Oxidation-reduction reactions are always coupled because electrons cannot be lost from one atom unless they can be accepted by another one.

Covalently bonded organic compounds do not give up individual electrons, but rather release hydrogen atoms, which in turn lose their single electrons to some electron acceptors and become hydrogen ions (H^+). The electron acceptor often binds with the hydrogen ion too, and becomes reduced. The loss of electrons (oxidation) in biological systems almost always involves a **dehydrogenation**. Reduction often involves **hydrogenation**.

$$\text{AH} + \text{B} \rightarrow \text{A} + \text{BH}$$

| Electron donor | Electron acceptor | Oxidized | Reduced |

The breakdown of glucose into carbon dioxide and water with the release of energy involves a series of oxidation-reduction reactions (Chap. 13).

Electrolytes: Acids, Bases, Salts

Electrolytes

Acids, bases, and salts are important compounds in biological systems. An **acid** may be defined as a compound that releases hydrogen ions (H^+) when dissolved in water. Since a hydrogen ion consists of only the nucleus or proton of a hydrogen atom, acids can also be defined as proton donors. A strong acid, such as hydrochloric acid (HCl), will dissociate completely into its constituent ions:

$$HCl \rightarrow H^+ + Cl^-$$

A weak acid, such as carbonic acid (H_2CO_3), only partially dissociates and hence releases fewer hydrogen ions:

$$H_2CO_3 \leftrightharpoons H^+ + HCO_3^-$$

(HCO_3^- is known as the **bicarbonate ion** and it is very common in biological systems.)

A **base** is an alkaline compound that releases **hydroxyl ions** (OH^-) in aqueous solution. Bases are proton acceptors, for the hydroxyl ion combines with the hydrogen ion to form water. Sodium hydroxide (NaOH) is a strong base, for it completely dissociates:

$$NaOH \rightarrow Na^+ + OH^-$$

Weaker bases do not dissociate as much.

When an acid and a base are mixed, water is formed and the rest of the acid and base bind ionically to form a **salt**. An example is the formation of water and table salt (Na^+Cl^-) from hydrochloric acid and sodium hydroxide:

$$HCl + NaOH \rightarrow H_2O + Na^+Cl^-$$

An electric current can flow through a solution containing acids, bases, and salts because the molecules in solution dissociate to form ions. These substances are called **electrolytes** in contrast to **nonelectrolytes**, such as sugars, that do not form ions in solution and cannot conduct a current.

Hydrogen Ion Concentration or pH

It is important to have a measure for the degree of acidity or alkalinity of a solution. Our reference is pure water, which partially dissociates into hydrogen and hydroxyl ions:

$$H_2O \leftrightharpoons H^+ + OH^-$$

Since relatively few molecules dissociate, the hydrogen ion concentration at equilibrium is only 0.0000001 gram per liter, or 10^{-7} molar. The negative logarithm is used as the measure of the amount of hydrogen ions present, or the **pH**. In pure water it is 7, and this is taken as neutrality. Since we are dealing with a negative logarithm, a *decrease* in the number represents an *increase* in the hydrogen ion concentration. A solution with a pH of 6 is ten times more acid than one with a pH of 7; one with pH 8 is alkaline, for it is ten times less acid than pure water.

Buffers

Buffers are substances that prevent the pH of a solution from changing greatly by sopping up excess hydrogen or hydroxyl ions. They are very important in biological systems because the pH of body fluids must be held within narrow limits. Wide departures can disrupt many biochemical processes and are incompatible with life. The pH of human blood is held close to 7.35, for example.

Carbonic acid (H_2CO_3), which is formed by the union of carbon dioxide with water, partially dissociates into hydrogen and hydroxyl ions:

$$CO_2 + H_2O \leftrightharpoons H_2CO_3 \leftrightharpoons H^+ + HCO_3^-$$

If an acid that is stronger than carbonic acid is added to the solution, its hydrogen ions will drive the above reactions to the left and H^+ will be taken out of solution. If a strong base is added, its hydroxyl ions will combine with hydrogen ions released by carbonic acid to form water. The pH of the solution remains remarkably stable. Many proteins are also important buffers.

A P P E N D I X   B

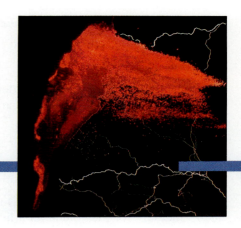

Scientific Units and Quantities

Basic SI Units

Scientists must be able to measure quantities such as distance, time, force, and so on. The units for measuring these quantities have varied. We often measure distance in inches, feet, and miles, but people in most other countries use centimeters, meters, and kilometers. A uniform standard was developed in 1960 at a General Conference of Weights and Measures, and these standard units, known as **SI Units** (from the French, Système International d'Unites), have gradually been adopted by scientists throughout the world.

Seven quantities have been selected as basic units, and all others can be expressed in terms of them. The seven basic units are length, mass, amount of a substance, time, temperature, electric current, and luminous intensity. These are listed in Table B.1. Each can be reduced or amplified by powers of ten by applying the prefixes shown in Table B.2.

Table B.1
Basic SI Units

Quantity	Name of Unit	Symbol
Length	Meter	m
Mass	Gram	g
Amount of substance	Mole	mol
Time	Second	s
Temperature	Degree:	
	Celsius	°C
	Kelvin	K
Electric current	Ampere	A
Luminous intensity	Candela	cd

Table B.2
The Most Common Multiples and Submultiples of Units*

Prefix	Abbreviation	Factor and Example
Multiples of basic unit		
giga-	G	1×10^9 m = 1 gigameter (Gm)
mega-	M	1×10^6 m = 1 megameter (Mm)
kilo-	k	1×10^3 m = 1 kilometer (km)
Basic unit		
Subunits of basic unit		
deci-	d	1×10^{-1} m = 1 decimeter (dm)
centi-	c	1×10^{-2} m = 1 centimeter (cm)
milli-	m	1×10^{-3} m = 1 millimeter (mm)
micro-	μ	1×10^{-6} m = 1 micrometer (μm)
nano-	n	1×10^{-9} m = 1 nanometer (nm)

*The factors are applied to meters as an example, but they could be applied to most of the units. Multiples of seconds, however, are usually stated as minutes and hours.

Derived SI Units

All other units of quantity can be derived from these standard units. The ones that you are most likely to encounter are listed below.

Area = m^2, or some multiple or submultiple of a meter squared.

Volume = m^3

Although not an SI Unit, **liter** (L) is a widely used measure for the volume of a liquid. One liter is the volume occupied by one cubic decimeter (dm^3), or 1000 cm^3. Liters are commonly subdivided into milliliters (mL). One mL is the volume occupied by one cubic centimeter (cm^3).

Velocity is the change in distance with time:

$$Velocity = change\ in\ distance/time$$

Its SI Units are meters and seconds, so V = m/s.

Acceleration is the change in velocity with time:

$$Acceleration = change\ in\ velocity/time$$
$$A = (m/s)/s$$
$$A = m/s^2$$

Force = mass × acceleration:

$$F = kg \times m/s^2$$

The unit of force is the Newton (N). 1 N = 1 $kg\ m/s^2$.

Pressure = force per unit area. The SI Unit for pressure is the Pascal (Pa):

$$1\ Pa = 1\ Newton\ per\ square\ meter$$
$$1\ Pa = (1\ kg\ m/s^2)/m^2$$

Pressures are frequently expressed in **atmospheres.** One atmosphere is the weight of a column of dry air at sea level and at a temperature of 0° C. One atmosphere raises a column of mercury (Hg) 760 mm; 1 mm Hg = 0.1333 K Pa.

Energy is the capacity to do work or produce a change. In physics, energy = force × distance through which the force acts:

$$E = kg\ m/s^2 \times m$$
$$E = kg\ m^2/s^2$$

The SI Unit of energy is the joule (J)

$$1\ J = 1\ kg\ m^2/s^2$$

Energy can also be expressed as **calories.** One calorie is the energy needed to raise 1 g water (at 16° C) 1° C.

Density is the mass of an object divided by its volume.

$$D = g/cm^3$$

Conversion Factors

It is sometimes helpful to convert standard units to other units in common usage. The factors listed below can be used.

1 inch = 2.54 cm
1 mile = 1.609 km
1 km = 0.622 miles
1 pound = 453.6 g
Temperature (°Fahrenheit) = (9/5 × temperature °C) + 32
Temperature (°Celsius) = 5/9 × (temperature °F − 32)
1 calorie = 4.184 joules

GLOSSARY

This glossary is restricted to terms that reoccur to some degree in the text. The names of animal taxa are not included. Where a term does not form the plural with an "s," the plural is indicated in parentheses. Other definitions can be found in specialized dictionaries such as Lawrence, E., ed., Henderson's Dictionary of Biological Terms, 10th ed. New York: John Wiley and Sons, 1989.

A

Aboral　Opposite the mouth, e.g., the aboral surface of a sea star.

Absorption　The taking up of a substance by diffusion or active transport through the lining of the digestive tract, tubule walls, or other surfaces.

Accommodation　Adjustments of the eye that bring an image into sharp focus on the retina, usually accomplished by moving the lens or changing its curvature; adaptation of receptors to variations in the stimulus.

Acetylcholine　The acetic acid ester of choline. A neurotransmitter released at many synapses and myoneural junctions.

Acid　A substance whose molecules or ions release hydrogen ions (protons) in water. Acids unite with bases (*q.v.*) to form salts.

Acoelomate body plan　The condition of having no body cavity between the body wall and internal organs as in flatworms.

Acousticolateral system　A sensory system in vertebrates consisting of groups of hair cells that detect low frequency vibrations or movements in the water (the lateral line system of fishes and larval amphibians), movement and changes in position of the body (part of the ear), and high frequency vibrations and pressure changes in the external medium (part of the ear).

Acrosome　A caplike structure covering the head of the spermatozoan of many species of animals and releasing enzymes that enable the sperm to penetrate the membrane covering the egg.

Actin　Minute protein filaments found in most contractile cells. In striated muscle cells, thin filaments of actin overlap and alternate with thicker filaments of myosin to form myofilaments. The interactions between actin and myosin lead to muscle contractions.

Action potential　The rapidly moving, progressive depolarization of nerve and other cell membranes produced by the inflow of Na^+ ions. The depolarization is responsible for the conductance of the nerve impulse in neurons.

Active transport　The transfer of a substance into or out of a cell across the cell membrane against a concentration gradient by a process that requires the expenditure of energy.

Adaptation　An evolutionary process by which an organism becomes fitted to the environment in which it lives; a structure fitted for a particular environment; the physiological habituation of a receptor to a particular stimulus such that the receptor ceases to respond unless the stimulus becomes more intense.

Adaptive radiation　The evolution from a single ancestral species of a variety of species that come to occupy different habitats and exhibit different life styles.

Adaptive value　A measure of the success of a given genotype in a population relative to the most successful genotype in the population (which is given the adaptive value of 1). The complement of selection coefficient (*q.v.*).

Adenosine triphosphate (ATP)　A compound composed of adenine, ribose sugar, and three phosphate groups; the principal energy source for cellular reactions.

Aerobic　Growing or metabolizing only in the presence of molecular oxygen.

Aestivation　The dormant state of decreased metabolism in which certain animals endure hot, dry seasons.

Afferent　Conveying toward a center; designating vessels or neurons that transmit blood or impulses toward a point of reference; afferent neurons are sensory neurons conducting impulses toward the central nervous system; *cf.* efferent.

Agonistic behavior　Aggressive behavior between animals; usually used in reference to such behavior between individuals of the same species.

Allantois　One of the extraembryonic membranes of reptiles, birds, and mammals; a pouch growing out of the posterior part of the digestive system and serving as an embryonic urinary bladder or as a source of blood vessels to and from the chorion or placenta.

Allele　One of a group of alternative forms of a gene that may occur at a given site (locus) on a chromosome.

Allopatric　Pertaining to species or populations occupying different geographic areas; *cf.* sympatric.

Alveolus(i)　A small sac-like dilation or cavity; the terminal chamber of air passages in the mammalian lung.

Ambulacrum　In echinoderms, pertaining to the area containing the tube feet.

Amino acid　An organic compound containing an amino group ($-NH_2$) and a carboxyl group ($-COOH$) joined to the same carbon; amino acids may be linked together to form the peptide chains of protein molecules. Twenty different amino acids form the building blocks of all proteins.

Amnion One of the extraembryonic membranes of reptiles, birds, and mammals; a fluid-filled sac around the embryo.

Amniote A vertebrate characterized by having an amnion during its development; a reptile, bird, or mammal.

Amoebocyte An amoeboid cell, often phagocytic, found in the blood, coelom, and interstitial spaces of many animals.

Amoeboid movement The movement of a cell by means of the slow flowing of the cellular contents, which forms pseudopodia.

Ampulla Any membranous, enlarged vesicle such as the ones on the semicircular ducts of the inner ear and those associated with the tube feet of echinoderms.

Amylase An enzyme that catalyzes the hydrolysis of starches by cleaving at the alpha 1, 4 linkages in starch.

Anabolism Chemical reactions in which simpler substances are combined to form more complex substances, resulting in the storage of energy, the production of new cellular materials, and growth; *cf.* catabolism.

Anaerobic Growth or metabolism carried out in the absence of molecular oxygen.

Analogy A similarity in structure or function among biological features that does not arise from common ancestry. Analogous structures perform similar functions, and may or may not resemble each other superficially; *cf.* homology.

Anamniote A vertebrate characterized by the absence of the amnion during embryonic development; a fish or amphibian.

Anaphase Stage in mitosis or meiosis, following the metaphase, in which the chromosomes move apart toward the poles of the spindle.

Androgen Any hormone (*q.v.*) that possesses masculinizing activities, such as testosterone or one of the other male sex hormones.

Anion An ion carrying a negative charge and moving toward the anode, or positive electrode; *cf.* cation.

Antenna(e) A projecting, usually filamentous organ equipped with sensory receptors.

Anterior The front end or toward the front end of the body.

Antibody A protein produced by a subset of cells in the immune system, in response to the presence of some foreign substance (antigen, *q.v.*). These proteins, or immunoglobulins, are capable of binding selectively with the antigen, thus aiding in eliminating the antigen.

Anticodon A sequence of three nucleotides in transfer RNA that is complementary to, and pairs with, the three nucleotide codon on messenger RNA, thereby binding the amino acid-transfer RNA combination to the mRNA.

Antigen A foreign substance, usually a protein or protein-polysaccharide complex, that elicits the formation of specific antibodies (*q.v.*) within an organism.

Aorta One of the primary arteries of the body, e.g., the ventral aorta of fishes that distributes venous blood to the gills or the dorsal aorta that distributes blood to the body.

Aposematic coloration A conspicuous color pattern that helps to protect an animal by advertising that it is toxic or dangerous; also found in some harmless species that mimic dangerous ones.

Appendicular skeleton The skeleton of the appendages, usually referring to vertebrates.

Archenteron The central cavity of the gastrula, lined with endotherm, which forms the rudiment of the digestive system.

Archinephric duct The primitive kidney duct; drains the pronephros, mesonephros, and opisthonephros.

Arteriole A minute arterial branch, especially one just proximal to a capillary.

Artery A vessel through which the blood passes away from the heart to the various parts of the body; typically has thick elastic walls.

Assimilation The conversion of absorbed simple molecules into complex macromolecules.

Atom The smallest quantity of an element that can retain the chemical properties of the element, composed of an atomic nucleus containing protons and neutrons together with electrons that circle the nucleus in specific orbitals.

ATP (See adenosine triphosphate)

Atrium An entrance chamber to another structure or organ; a chamber of the heart receiving blood from a vein and pumping it to a ventricle.

Auditory ossicle One or more small bones (malleus, incus, and stapes in mammals) that transmit vibrations across the middle ear cavity from the tympanic membrane to the inner ear.

Auricle The ear-shaped portion of the mammalian atrium (*q.v.*); sometimes used as a synonym for atrium. The external ear flap of mammals.

Autosome Any chromosome occurring in similar form in males and females of a species, and therefore able to pair fully at meiosis and mitosis. Distinguished from the sex chromosomes, which differ chromatically in the two sexes.

Autotroph An organism, including plants, that manufactures organic nutrients from inorganic raw materials; *cf.* heterotroph.

Axial skeleton That portion of the somatic (*q.v.*) skeleton that lies in the longitudinal axis of the body, e.g., the skull, vertebral column, ribs, and sternum.

Axon The part of a neuron (*q.v.*) that conducts electrical impulses, often over long distances through the progressive depolarization of the membrane (*cf.* action potential). The nuclear region of a neuron may be at one end, in the middle, or to one side of an axon.

B

B cell A type of lymphocyte that matures in the bursa of Fabricius in birds, and probably in the bone marrow of mammals, circulates in the blood, and plays a key role in the humoral immune response.

Baroreceptor A receptor that detects changes in pressure.

Basal body The organelle from which a cilium or flagellum arises; also called a basal granule. Its ultrastructure is the same as that of a centriole (*q.v.*).

Basal lamina Fibrous membrane located at one side, usually the basal side, of many epithelial cells; sometimes called the basement membrane.

Base A compound that releases hydroxyl ions (OH^-) when dissolved in water.

Benthic Living on the bottom of oceans or lakes.

Biological clock An internal timing mechanism of animals that is involved in controlling circadian rhythms (*q.v.*) and other cycles, and also plays a role in some types of navigation.

Biological oxidation The removal of electrons from an atom or molecule, accompanied in biological systems by dehydrogenation. The removed electrons are accepted by other molecules in the electron transport system of the mitochondria. Molecules accepting electrons are said to be reduced.

Biomass The total weight of all the organisms in a particular habitat.

Biome Large, ecological unit composed of similar types of climax communities arising as a result of complex interactions of climate, other physical factors, and biotic factors, e.g., a rainforest, temperate grassland.

Biosphere The earth's surface and envelope, including all living organisms.

Biotic potential Highest possible rate of population increase for a given species, assuming that all environmental conditions are optimal.

Biramous Two branched.

Blastocoel The fluid-filled cavity of the blastula (*q.v.*).

Blastocyst The modified blastula stage of embryonic mammals; it consists of an inner cell mass, which develops into the embryo, and a peripheral layer of cells, the trophoblast, which contributes to the placenta.

Blastomere An embryonic cell resulting from the cleavage division of the egg.

Blastopore The opening, in the gastrula stage of development, from the archenteron to the surface.

Blastula Usually a spherical structure produced by cleavage of a fertilized ovum, consisting of one or more layers of cells surrounding a fluid-filled cavity.

Bowman's capsule Double-walled, hollow sac of cells that surrounds the glomerulus at the end of each kidney tubule in vertebrates.

Brachial Pertaining to the arm.

Branchial Pertaining to the gills or gill region.

Budding Asexual reproduction in which a small part of the parent's body separates from the rest and develops into a new individual, eventually either taking up an independent existence or becoming a more or less independent member of a colony.

Buffer Substance in a solution that tends to lessen the change in hydrogen ion concentration (*pH, q.v.*) which otherwise would be produced by adding acids or bases.

Bulk flow The transport of materials by the movement of the gas or liquid in which they are contained.

C

Capillary Microscopic thin-walled vessel located in the tissues connecting an artery and vein. Ions, gases, and organic molecules pass to the interstitial fluid through the capillary walls.

Carapace A bony or chitinous shield covering the back of an animal.

Carbonic anhydrase An enzyme that catalyzes the reaction carbon dioxide + water to carbonic acid; abundant in erythrocytes.

Cardiac Pertaining to the heart.

Carnivore A flesh-eating animal, or one whose diet includes a large component of animal protein.

Carotid artery One of the two large arteries in the neck carrying blood to the head.

Cartilage replacement bone Bone that develops in and around a cartilaginous rudiment that it replaces.

Catalyst A substance that accelerates a chemical reaction, usually by temporarily combining with the reagents and lowering the activation energy, and is recovered unchanged at the end of the reaction, e.g., an enzyme (*q.v.*).

Cation An ion bearing a positive charge and moving toward the cathode, or negative electrode; *cf.* anion.

Cecum(a) A blind pouch associated with a part of the intestine in which digestion and absorption may occur; located at the junction of the ileum and colon in mammals.

Cell constancy An extreme example of mosaic development that results in all individuals in a species having the same number of cells in comparable tissues performing similar functions.

Cell theory The generalization that all living things are composed of cells and cell products, and that new cells are formed by the division of preexisting cells.

Centriole One of a pair of small, barrel-shaped, dark-staining organelles, usually seen near the nucleus in the cytoplasm of animal cells, that form the spindle during mitosis and meiosis.

Centrolecithal egg Type of arthropod egg having the yolk arranged as a large sphere around the cental nucleus.

Centromere The chromosomal region at which sister chromatids are attached during early stages of cell division; it divides in late metaphase of mitosis and during metaphase II of meiosis. The centromere contains the kinetichore to which certain spindle fibers attach during cell division.

Cephalization Head formation, concentration of nervous tissue and sense organs at the anterior end of the body.

Cerebellum The part of the vertebrate hindbrain that coordinates posture and muscular movement.

Cerebrum A major portion of the vertebrate forebrain; the two cerebral hemispheres, united by the corpus callosum, form the largest part of the central nervous system in mammals and integrate most sensory impulses and motor responses.

Character displacement The morphologic divergence that occurs between related species as they compete for limited resources, each eventually specializing on the utilization of a subset of these resources.

Chelate appendage Having the terminal parts of an appendage in the form of pincers, as in crabs and scorpions.

Chelicera(ae) A pincer-like head appendage found in spiders, scorpions, and other arachnids, typically paired.

Chemosynthesis The synthesis of complex organic compounds utilizing energy derived from the oxidation of inorganic compounds; occurs in some bacteria.

Chitin An insoluble horny polysaccharide consisting of modified glucose molecules containing an N-acetyl group and often bound to protein to form a glycoprotein. Abundant in the exoskeleton of arthropods.

Choanocyte A unique cell type with a flagellum surrounded by a thin cytoplasmic collar of microvilli; characteristic of sponges and one group of protozoa.

Chondrocranium Cartilage or cartilage replacement bone that contributes to the vertebrate brain case and encapsulates the nose and inner ear.

Chorion An extraembryonic membrane in reptiles, birds, and mammals that forms an outer cover around the embryo and in mammals contributes to the formation of the placenta.

Chromatin The readily stainable granular portion of the cell nucleus; composed of nucleoprotein and condensing to form chromosomes at cell divison.

Chromatophore A pigment-containing cell capable of producing color change in the skin.

Chromosome Minute rod-shaped bodies that become evident in the nucleus during cell division; composed of DNA and associated proteins. Genes are physically located along the length of the chromosomes.

Ciliary movement Movement of a cell, or of material across the surface of a cell, by the beating of microscopic cilia.

Cilium(a) One of numerous cytoplasmic processes on the exposed surface of many cells. Cilia typically are shorter and more numerous than flagella (*q.v.*) and have a more oarlike action.

Circadian rhythm Repeated sequence of biological or biochemical events that occur regularly and periodically at about 24-hour intervals.

Cladistics The classification of organisms based entirely on derived characters shared by two or more species. Cladistic classifications are composed entirely of monophyletic groups that include all descendants from a single common ancestor.

Class A higher taxonomic rank between a phylum and order.

Cleavage The early mitotic divisions of a zygote that leads to the blastula.

Cleidoic egg The yolky, terrestrial eggs of insects, reptiles, and birds which are enclosed by a shell and are largely self-contained systems.

Clitellum(a) Glandular segments in earthworms and leeches that secrete the cocoon.

Cloaca A common chamber receiving the discharge of the digestive, excretory, and reproductive systems in most of the lower vertebrates.

Clone A population of genetically identical cells or individuals descended by mitotic division from a single ancestral cell.

Closed circulatory system A circulatory system in which the blood is always confined to vessels.

Cnidocyte Cell of cnidarians containing an explosive stinging structure, called a nematocyst (*q.v.*).

Cochlea The elongated and often coiled portion of the inner ear of crocodiles, birds, and mammals containing the sound receptive cells.

Codon A sequence of three adjacent nucleotides in DNA or mRNA encoding a single amino acid. Sixty-one of the sixty-four possible codons specify a single amino acid to be added to the growing peptide chain; the three remaining codons terminate protein synthesis.

Coelom Body cavity of triploblastic animals lying within the mesoderm and lined by it.

Coenzyme A nonprotein molecule that is required for some particular enzymatic reaction to occur; participates in the reaction by donating or accepting some reactant; loosely bound to an enzyme.

Coevolution The evolution of two species in close relationship to each other such that modifications in one species affect the evolution of the other, e.g., the structure of certain flowers and the insects or animals that pollinate them.

Colony An association of unicellular or multicellular organisms of the same species; individuals may be separate from or connected to other members of the colony and there may be some division of labor among them.

Commensalism A symbiotic (*q.v.*) relationship between members of two species in which one member benefits and the other is neither benefited or harmed.

Community An assemblage of populations of different organisms that live in a defined area or habitat. The organisms constituting the community interact in various ways with one another.

Complement system A system of plasma enzymes that, when activated by antigen-antibody complexes, leads to the lysis and destruction of invading microorganisms.

Connective tissue Type of animal tissue in which the cells are widely separated and the intercellular space is filled with a matrix of various substances, e.g., fibrous connective tissue, cartilage, and bone.

Conservation of energy, law of The first law of thermodynamics, which states that in any given system the amount of energy is constant; energy is neither created nor destroyed, but only transformed from one form to another.

Consumer organism A species within an ecosystem, plant or animal, that eats other plants or animals.

Contractile vacuole Osmoregulatory organelle of protozoa and sponges.

Convergent evolution The evolution of similarities between two or more different species (or organs or molecules) resulting from their independent adaptation to similar circumstances rather than to their descent from a common ancestor.

Copulation Sexual union; act of physical joining of two animals during which sperm cells are transferred from one to the other.

Corpus luteum A yellow glandular mass in the vertebrate ovary formed by the cells of an ovarian follicle that has matured and discharged its ovum; secretes progesterone.

Cortex The outer layer of an organ.

Countercurrent exchange The exchange of gases, heat, solutes, or water between opposite flowing streams as occurs in some gills, blood vessels of extremities, and kidneys.

Covalent bond Bond between atoms formed by shared electrons.

Cranium The part of the skull that surrounds the brain; the braincase.

Crop A sac in the digestive tract in which food is stored before digestion begins.

Crossing over The exchange of chromosome segments during meiosis between homologous but nonsister chromatids; leads to allelic combinations that differ from those of either parent.

Cryptic coloration A color pattern that conceals an animal by helping it visually blend with its surroundings, often found in prey species and sometimes in predators that stalk or lie in wait for their prey.

Cutaneous Pertaining to the integument or skin.

Cuticle A noncellular protective layer on the surface of an organism.

Cytochrome An iron-containing heme protein of the electron transport system that is alternately oxidized and reduced in biological oxidation (*q.v.*).

Cytokinesis The division of the cytoplasm at the conclusion of mitosis or meiosis; results in cell division.

Cytopharynx Gullet-like organelle of ciliates and certain other protozoa.

Cytoplasm The living substance of a cell surrounding the nucleus and enclosed by the cell membrane.

Cytostome The mouth opening of ciliates and certain other protozoans.

D

Deamination Removal of an amino group (—NH₂) from an amino acid or other organic compound.

Definitive host Host of the sexually mature parasite.

Dehydrogenation A form of oxidation in which hydrogen atoms are removed from a molecule.

Denaturation Alteration of physical properties and three-dimensional structure of a protein, nucleic acid, or other macromolecule by mild treatment that does not disrupt the primary structure (*q.v.*), but generally results in the inactivation of the molecule.

Dendrite The process of a neuron (*q.v.*) that is stimulated by a receptor or another neuron and initiates a nerve impulse.

Deoxyribonucleic acid, DNA The helical molecule in the nucleus that forms the chromosomes and carries the genetic code for protein synthesis; consists of two complementary strands of linked nucleotides.

Deposit feeding Utilization as a food source of organic detritus that has become mixed with mineral particles in aquatic and terrestrial habitats.

Derived character A recently acquired feature of a group that differentiates it from a closely related group, e.g., hair is a derived character of mammals that differentiates them from non-mammalian vertebrates (and all other organisms).

Dermal bone Bone that develops directly in connective tissue and is not preceded by a cartilaginous rudiment.

Dermis The deeper layer of the skin of vertebrates and some invertebrates.

Determinate cleavage A cleavage pattern in which the fate of embryonic cells is fixed very early in development (during cleavage).

Detritus, organic Decomposing fragments of plants and animals used as a food source by many organisms.

Deuterostomes A major branch of the animal kingdom containing animals in which the mouth forms as a new opening at the opposite end of the body from the blastopore; *cf.* protostomes.

Diastole Relaxation of the heart muscle, especially of the ventricles, during which the lumen becomes filled with blood; *cf.* systole.

Diffusion The movement of molecules from a region of high concentration to one of low concentration, brought about by their kinetic energy.

Digestion The process whereby large food molecules are broken down by hydrolytic enzymes into smaller units that can be absorbed.

Dioecious Organisms in which male and female sexes are distinct individuals; *cf.* monoecious.

Diploid Containing two sets of chromosomes, usually one from each parent; a chromosome number twice that found in gametes; symbolized as 2n; *cf.* haploid.

Direct development A developmental pattern in which the young at hatching have the form of the adult; larval stages are absent; *cf.* indirect development.

Divergent evolution Evolutionary changes that lead to the division of one group into two or more lineages; the evolution in different directions of previously similar organisms, genes, or proteins that shared a common ancestor.

DNA (See deoxyribonucleic acid)

Dorsal The upper surface of a bilateral animal.

E

Ecdysis The shedding, or molting, of the arthropd exoskeleton.

Ecology The study of the interrelations between living things and their environment, both physical and biotic.

Ecosystem All of the organisms of a given area and the encompassing physical environment.

Ectoderm The outermost of the three germ layers (*q.v.*) of the early embryo; gives rise to the epidermis of the skin and nervous system.

Ectothermy Condition in which the internal temperature of an animal is dependent upon the temperature of its environment; cold blooded; *cf.* endothermy.

Edema An excessive accumulation of liquid in the tissues.

Effector Structures of the body by which an organism acts; the means by which it reacts to stimuli, e.g., muscles, glands, cilia.

Efferent Pertaining to a structure, such as a nerve or blood vessel, that leads away from a point of reference. Efferent neurons are motor neurons that carry impulses from the central nervous sytem to the effectors; *cf.* afferent.

Egestion The elimination of undigestible waste material; defecation.

Endocrine Secreting internally; applied to ductless glands that secrete into the blood or lymph a substance that has a specific effect on another target organ or tissue.

Endocytosis A process whereby cells take in material by invaginations of the cell membrane to form vesicles enclosing the material; *cf.* phagocytosis.

Endoderm The innermost of the three germ layers of the early embryo, lining the archenteron; forms the epithelial cells of the digestive tract and its outgrowths—the liver, lungs, and pancreas of vertebrates.

Endopeptidase A proteolytic enzyme that cleaves peptide bonds within a peptide chain; *cf.* exopeptidase.

Endoskeleton Internal skeleton; bony and cartilaginous supporting structures within the body of vertebrates.

Endostyle A longitudinal groove in the floor of the pharynx of certain protochordates and larval cyclostomes; its glandular and ciliated cells secrete mucus that is moved by ciliated cells through the pharynx and entraps minute food particles.

Endothermy Condition in which the internal temperature of an animal is dependent upon its metabolic processes and is held at a relatively high and constant level; warm blooded; *cf.* ectothermy.

Enterocoel A coelomic cavity formed by outpocketing from the primitive gut; *cf.* schizocoel.

Enzyme A protein catalyst (*q.v.*) produced within a living organism that accelerates specific chemical reactions.

Epiboly One of the gastrulation processes by which the smaller blastomeres at the animal pole of the embryo grow over and enclose the cells of the vegetal hemisphere.

Epidermis The outermost layer of cells of an organism.

Epiglottis The lidlike structure that deflects food around the glottis.

Epinephrine Hormone secreted by the adrenal medulla of mammals that mobilizes glucose in the tissues, among other functions; derived from the methylation of norepinephrine (*q.v.*).

Epithelial tissue Tissue composed of cells that fit together tightly, leaving little intercellular space; typically covering body surfaces, lining cavities, and forming glands.

Epitheliomuscle cell A cell type having both a contractile and an epithelial component, as in the epidermis of hydra.

Erythrocyte Red blood cell; it contains hemoglobin and transports gases.

Esophagus A region of the animal gut that transports food from the mouth or pharynx.

Estrogen One of a group of female steroid sex hormones, of which estradiol is the most common, that are synthesized primarily by the ovary and placenta. They are responsible for the development of female secondary sex characteristics and for the maturation and functioning of the female reproductive organs.

Estrus The recurrent, restricted period of sexual receptivity in many female mammals, marked by intense sexual urge.

Ethology The study of animal behavior.

Eukaryotes Those organisms having cells that contain a membrane-bound nucleus and in which many cellular functions are performed by membrane-bound cytoplasmic organelles, such as the endoplasmic reticulum, mitochondria, and Golgi apparatus; *cf.* prokaryotes.

Euryhaline animal An animal that can tolerate a relatively wide range of environmental salt concentrations (salinities).

Eutely Condition in which the organs of an animal are composed of fixed numbers of cells, the numbers characteristic of the species; cell constancy; found in rotifers and nematodes.

Excretion The removal of metabolic wastes by an organism.

Exopeptidase A proteolytic enzyme that cleaves only the bond joining a terminal amino acid to a peptide chain; *cf.* endopeptidase.

Exoskeleton External, calcareous, chitinous, or other hard material covering the body surface and providing protection or support.

Extracellular digestion Digestion that takes place outside of cells within a gut cavity or in some animals outside of the body.

F

Facilitation The promotion or hastening of any natural process; the reverse of inhibition.

Fat An organic compound composed of fatty acids and glycerol, generally solid at room temperature (20°C); one of the lipids (*q.v.*).

Fatty acid An organic compound composed of a chain of carbon atoms having a methyl group at one end and a carboxyl group at the other.

Fertilization The fusion of a spermatozoon with an ovum to form a zygote.

Fetus The unborn offspring after its major features have developed and it is recognizable as the species it will become; the human embryo after 7 weeks.

Filtration The passage of a liquid through a filter following a pressure gradient; occurs in capillary beds, including the glomeruli of the kidney.

Fission Process of asexual reproduction in which an organism divides into two or more parts.

Flagellum(a) A whiplike cytoplasmic process on the exposed surface of many cells. Structurally similar to a cilium (*q.v.*), but typically flagella are longer, are fewer in number, and have a more undulating movement.

Fluke Common name of the parasitic flatworms belonging to the class Trematoda.

Food Organic compounds used in the synthesis of new biomolecules and as fuels in the production of cellular energy.

Food chain A sequence of organisms through which energy captured from sunlight by photosynthesis is transferred from one consumer (or trophic level) to the next; each organism in the chain eats the preceding and is eaten by the following member of the sequence.

Foramen(mina) A small opening in a body structure.

Fouling organism A sessile organism, such as many cnidarians and barnacles, that attaches to the bottom of ships or other man-made objects and impedes their progress or function.

G

Gamete A reproductive cell; an egg or sperm whose union, in sexual reproduction, forms a zygote and initiates the development of a new individual.

Gametogenesis Formation of gametes.

Ganglion(ia) A knotlike mass of the cell bodies of neurons located outside the central nervous system; in invertebrates includes the swellings of the central nervous system.

Gap junction A tight union between cells that contains channels facilitating the passage of ions or action potentials from one cell to the next.

Gastrodermis The epithelial tissue lining the gut cavity in cnidarians that is responsible for digestion and absorption.

Gastrula Early embryonic stage that follows the blastula; consists initially of two layers, the ectoderm and the endoderm, and contains an internal cavity, the archenteron, that opens to the exterior through the blastopore.

Gastrulation The process by which the blastula becomes a gastrula.

Gene The biologic unit of genetic information (DNA) by which a hereditary feature is transmitted from parent to offspring; self-reproducing and located in a definite position (locus) on a particular chromosome.

Gene flow The spread of alleles through a population, or between populations, that result from migration or sexual reproduction.

Gene pool The totality of all of the genes and their alleles of all of the individuals of a population.

Genetic drift Random changes in gene frequencies occurring primarily in small isolated populations. The changes result from factors other than natural selection, such as chance.

Genetic equilibrium The situation in which allele frequencies in a population remain constant in successive generations as a consequence of the operation of the Hardy-Weinberg-Castle law (*q.v.*), or of balancing selective forces.

Genome A complete set of hereditary factors; technically those contained in the haploid assortment of chromosomes, but sometimes used for the total genetic content of the nucleus.

Genotype The genetic composition, or assortment of genes, of any given organism that together with environmental influences determines the appearance or phenotype (*q.v.*) of an organism.

Genus(era) The lowest taxon embracing a number of species; the first name of the binomial species name.

Germ layer One of the three primary embryonic layers—ectoderm, mesoderm, and endoderm—laid down in gastrulation and from which adult tissues derive.

Gill slit An opening to the outside from the pharynx that arises during development. Water taken in at the mouth passes out through the gill slits, aiding gas exchange and sometimes the filtering of food.

Gizzard A region of the digestive tract that mechanically breaks up food.

Globulin One of a family of compact proteins in blood plasma, some of which (gamma globulins) function as antibodies.

Glomerulus(i) A tuft of minute blood vessels or nerve fibers; specifically, the knot of capillaries at the proximal end of a kidney tubule.

Glottis The opening between the pharynx and larynx; it is bounded by the vocal cords in mammals.

Glycolysis The anaerobic breakdown within cells of glucose to pyruvate, with the accompanying release of energy, some of which is stored as ATP.

Glycoprotein A protein combined with a carbohydrate; component of cellular secretions and cell membranes.

Gonad A gamete-producing organ; an ovary or testis.

Gonoduct General term for the reproductive duct of any animal.

Gonopore External opening of any reproductive system.

Green gland Excretory organ of certain crustaceans, such as shrimp and crabs; also called antennal gland.

H

Habitat The natural abode of an animal or plant species; the physical area in which it may be found.

Habituation A gradual decrease in response to successive stimulation due to changes in the central nervous system.

Haploid Containing one set of chromosomes, as in a gamete; symbolized by *n*; *cf.* diploid.

Hardy-Weinberg-Castle Law Allele frequencies in a population remain the same from generation to generation if the population is large, mating is random, and mutation, migration, and natural selection are not occurring.

Helix A spiral having a uniform diameter and a periodic spacing between the coils; a common secondary structure of proteins and DNA.

Hemocoel The spaces between the cells and tissues of many invertebrates, which are filled with blood. Part of an open blood vascular system.

Hemocyanin A blue copper-containing respiratory pigment found in the blood of many molluscs and arthropods.

Hemoglobin The red, iron-containing respiratory pigment found in the blood of vertebrates and some invertebrates; transports gases and aids in the regulation of pH.

Hemolymph The bloodlike fluid of animals with open circulatory systems; combines the properties of blood and lymphlike interstitial fluid.

Herbivore A plant-eating animal.

Hermaphroditism A state characterized by the presence of both male and female sex organs in the same organism.

Heterotroph An organism that cannot synthesize its own food from inorganic materials and therefore must obtain organic compounds by consuming the bodies or products of other organisms; *cf.* autotroph.

Heterozygous Possessing two different alleles at a given locus on homologous chromosomes.

Hibernation The dormant state of decreased body temperature and metabolism in which certain animals pass the winter.

Histological Pertaining to tissues.

Homeostasis The tendency to maintain uniformity or stability in the internal environment of the organism.

Homeothermic Constant-temperature animals, e.g., birds and mammals, that maintain a constant body temperature despite variations in environmental temperature; endothermic (*q.v.*); *cf.* poikilothermic.

Homology Basic similarity in structure between organs in different animals that results from inheritance from a common ancestor; homologous organs have a common basic plan and mode of development.

Homozygous Possessing two copies of the same allele for a given locus of homologous chromosomes.

Hormone Substance produced in cells in one part of the body that diffuses or is transported by the blood stream to cells in other parts of the body, where it regulates and coordinates their activities.

Hybrid Any cross-bred organism combining different characters from parents of the same species or different species; macromolecules (especially DNA) formed by combining portions of different origin.

Hydrogen bond A weak bond between two molecules formed when a hydrogen atom is shared between two atoms, one of which is usually oxygen; an important bond in stabilizing the secondary and tertiary structure of nucleic acids and proteins.

Hydrolysis The splitting of a compound into parts by the addition of a molecule of water between certain of its bonds, the hydroxyl group being incorporated in one fragment and the hydrogen atom in the other; occurs in digestion.

Hydroskeleton A turgid column of liquid within one of the body spaces that provides support or rigidity to an organism or one of its parts, e.g., the coelom and coelomic fluid of an annelid worm.

Hydrostatic skeleton Same as hydroskeleton.

Hyperosmotic Having a higher osmotic concentration of solutes than that of another solution with which it is being compared.

Hypo-osmotic Having a lower osmotic concentration of solutes than that of another solution with which it is being compared.

Hypothalamus A region of the forebrain, the floor of the diencephalon, that contains centers controlling visceral activities, water balance, temperature, sleep, and so on.

I

Immune response A response by the immune system in which antibodies and/or cytotoxic cells are produced in response to cells or other materials that are recognized as foreign.

Immunity A resistance to disease that may be due to nonspecific factors, such as phagocytosis of invading bacteria, or specific factors (antibodies, cytotoxic cells) produced by the immune system as a result of a previous infection, or induced by vaccination with treated microorganisms or their products.

Immunoglobulin An antibody (*q.v.*), a highly variable protein synthesized by plasma cells derived from B cells which have responded to a foreign substance (antigen). Antibodies recognize and bind with the specific antigen that led to their formation and initiate responses that lead to the destruction or neutralization of the antigen.

Incomplete cleavage A cleavage pattern in which the cleavage planes do not completely divide the zygote.

Indeterminate cleavage A cleavage pattern in which the fate of embryonic cells is not fixed early in development.

Indirect development A developmental pattern containing a larval stage; *cf.* direct development.

Induction The production of a specific morphogenetic effect in one tissue of a developing embryo through the influence of an organizer or another tissue.

Ingestion The act of taking food into the body by mouth.

Instar The stage of an insect or arthropod between successive molts.

Integration The coordination of nerve impulses from many different sources in a neuron, or group of neurons, that combines them to produce a new signal.

Integument Skin, the covering of the body.

Interneuron A neuron that receives impulses from another neuron, often a sensory neuron, and transfers them to other interneurons or to a motor neuron.

Interstitial Referring to the small spaces, fluid, or animals that lie between cells or sand grains.

Intracellular digestion Digestion that occurs in a food vacuole within a cell. Characteristic of protozoa and some metazoa.

Invagination The infolding of one part within another, specifically a process of gastrulation in which one region infolds to form a double-layered cup.

Ion An atom or a group of atoms bearing an electric charge, either positive (cation) or negative (anion).

Isolecithal Pertaining to eggs containing a small amount of yolk nearly evenly distributed throughout the cytoplasm.

Isosmotic Pertaining to a solution having an osmotic concentration equal to that of another solution with which it is being compared.

Isozyme Different molecular form of a protein, the result of small differences in amino acid sequence, exhibiting the same enzymatic activity but differing in other properties such as electrophoretic mobility or optimal pH.

K

Keratin A horny, water-insoluble protein found in the epidermis of terrestrial vertebrates and in nails, feathers, and horn.

Kinetosome Cytoplasmic organelle from which a cilium or flagellum arises; same as basal body.

L

Lacuna(ae) A cavity; the small cavities in bone and cartilage in which the cells of these tissues are located.

Larva(ae) A motile and sometimes feeding stage in the early development of many animals.

Larynx The cartilaginous structure located at the entrance of the trachea; houses the vocal cords in many vertebrates.

Lateral Toward the side.

Leukocyte White blood cell; colorless cell capable of phagocytosis and amoeboid movement. Important in inflammatory and immune reactions.

Linkage The tendency for a group of genes located on the same chromosome to be inherited together in successive generations.

Lipase An enzyme that catalyzes the hydrolysis of fats; it cleaves the ester bonds joining fatty acids to glycerol.

Lipid Any of a diverse group of organic compounds that are insoluble in water but soluble in organic acids; includes waxes, fats, oil, fatty acids, and steroids.

Littoral The region of shallow water near the shore between the high and low tide marks.

Locus(i) The particular physical location on the chromosome at which the gene for a given trait occurs.

Lymph The fluid in the lymphatic system; similar in composition to interstitial fluid. It contains many white blood cells, especially lymphocytes.

Lymph node One of the nodules of lymphatic tissue that occur in groups along the course of the lymph vessels; produces and contains lymphocytes and phagocytic cells.

Lymphatic system System of vessels in vertebrates draining excess interstitial fluid back into the venous system.

Lymphocyte A type of vertebrate blood cell present in lymphoid tissues and circulating in the blood and lymph; consists of two populations, B cells (*q.v.*) and T cells (*q.v.*), that participate in different ways in immune responses.

Lysis The process of disintegration of a cell or some other structure.

Lysosome A membrane-bound intracellular organelle present in eukaryotic cells, it contains hydrolytic enzymes that are released when the lysosome ruptures.

M

Macroevolution Evolutionary processes responsible for large-scale patterns in the history of life, extending over long periods of time and leading to the formation of genera and higher taxa; *cf.* microevolution.

Macromere The larger cell resulting from unequal cleavage division; *cf.* micromere.

Macrophage Large phagocytic cells in connective tissues, hepatic sinusoids (Kupffer cells), the alveoli of the lungs, and other organs that differentiate from certain leukocytes (monocytes); their numbers increase at the site of an infection.

Malpighian tubule Fine, thin-walled excretory tubule of many arthropods that discharges into the gut.

Mantle Integument of molluscs that is covered by and secretes the shell.

Marsupium(a) Pouch or chamber in which young are brooded. Characteristic of certain mammals and amphipod and isopod crustaceans.

Mast cell Cells of the immune response that become active in inflammatory reactions and release heparin and histamine.

Matrix A network of fibers, secreted by and surrounding the connective tissue cells; the inner region of a mitochondrion enclosed by the inner membrane.

Medulla The inner part of an organ, e.g., the medulla of the kidney; the most posterior part of the brain, lying next to the spinal cord.

Meiosis Kind of nuclear division, involving two successive cell divisions, that results in daughter cells with the haploid number of chromosomes, one-half the number of chromosomes in the original diploid cell.

Melanin A dark-brown or black pigment common in the integument of many animals and sometimes found in other organs; usually occurs within chromatophores.

Mesenchyme A meshwork of loosely associated, often stellate cells, found in the embryos of many animals.

Mesentery(ies) One of the membranes in vertebrates that extend from the body wall to the visceral organs or from one organ to another; consists of two layers of coelomic epithelium and enclosed connective tissue, vessels, and nerves; a septum extending inward from the body wall of some cnidarians.

Mesoderm The middle layer of the three primary germ layers of the embryo, lying between the ectoderm and the endoderm; gives rise to muscles, connective tissue, blood, the kidney, and the dermis of the skin.

Mesoglea A membrane or gelatinous matrix located between the epidermis and gastrodermis of cnidarians.

Messenger RNA (mRNA) A particular kind of ribonucleic acid that is transcribed in the nucleus and passes to the ribosomes in the cytoplasm, where it serves as a template guiding the synthesis of a protein.

Metabolic rate Organic fuel utilization per unit of time.

Metabolic waste The waste product of a metabolic process, largely CO_2, ammonia, and uric acid.

Metabolism The sum of all the physical and chemical processes by which living substance is produced and maintained; the biochemical reactions by which energy and matter are made available for the uses of the organism.

Metamerism The division of the body into a linear series of similar parts or segments, as in annelids and chordates.

Metamorphosis An abrupt transition from one developmental stage to another, e.g., from a larva to an adult.

Metanephridium(a) An excretory tubule of invertebrates in which the inner end opens into the coelom by way of a ciliated funnel; *cf.* protonephridium.

Metanephros The adult kidney of reptiles, birds, and mammals.

Metazoa Motile, multicellular, heterotrophic organisms; animals.

Microevolution The genotypic and phenotypic evolutionary changes that occur within populations and often lead to the formation of subspecies and species; *cf.* macroevolution.

Micromere The smaller cell, following unequal cleavage division; *cf.* macromere.

Microvillus(i) Ultrastructural finger-like projections of the free surfaces of some epithelial cells, such as the intestinal epithelium; appear as a brush border with the light microscope.

Mimicry An adaptation whereby an animal becomes camouflaged by taking on the appearance of some other living or nonliving object.

Mitochondrion(ia) Spherical or elongate cytoplasmic organelle having a double membrane and containing the electron transport system and certain other enzymes; site of oxidative phosphorylation, and principal site of ATP synthesis in aerobic organisms.

Mitosis A form of cell or nuclear division in which each of the two daughter nuclei receives exactly the same diploid complement of chromosomes present in the nucleus of the parent cell.

Mole The amount of a chemical compound whose mass in grams is equivalent to its molecular weight, the sum of the atomic weights of its constituent atoms.

Molecule An aggregation of at least two atoms of the same or different elements held together by special forces (covalent bonds) and present in precise proportions, e.g., O_2, H_2O.

Monoecious Organisms in which male and female reproductive organs are contained by the same individuals; these organs may or may not function at the same time; *cf.* hermaphroditic.

Monophyletic Pertaining to a taxon whose members have a common ancestry; when defined cladistically (*q.v.*), taxon includes all descendants of common ancestor; see also paraphyletic.

Motor neuron A neuron that carries a nerve impulse from an interneuron or sensory neuron to an effector such as a muscle; often called an efferent neuron.

Motor unit All the muscle fibers supplied by a single motor neuron.

Mucus A slimy secretion rich in glycoproteins (*q.v.*); mucous is the adjective.

Multiple alleles Three or more alternate conditions of a gene at a single locus that may produce different phenotypes.

Mutualism A symbiotic relationship (*q.v.*) between two species that is beneficial to both (host and symbiont).

Mutation A change in the chemical structure, organization, or amount of DNA. When occurring in the coding region of a gene, mutations often result in a change in the characteristics of the cell or organism. Mutations include base substitution, insertions or deletions of bases at particular loci, rearrangement of material in and between chromosomes, and changes in the normal number of chromosomes or chromosome sets.

Myelin The fatty material that forms a sheath around the axons of nerve cells in the white matter of the central nervous system and in most peripheral nerves.

Myofibril One of many contractile fibrils visible in muscle tissue with light microscopy. Composed of groups of myofilaments of actin and myosin.

Myoglobin A type of hemoglobin found in some muscles; facilitates the transfer of oxygen from the blood to the muscles.

Myomere The adult muscle segment of segmented animals.

Myosin The thick protein filaments found in muscles and some other contractile cells; usually associated with actin (*q.v.*).

N

Nares The external opening of a nasal cavity; singular *naris*.

Natural selection The hypothesis proposed by Charles Darwin that argues that individuals in a population best suited to the existing environment will survive and reproduce to a greater extent than other individuals. To the extent that the features that led to the survival of these individuals have a genetic basis, a differential transmission from generation to generation of traits will occur.

Nauplius(i) The earliest hatching crustacean larva, having three pairs of appendages—future head limbs.

Nematocyst A minute explosive structure found in cnidarians and used for defense, for the capture of prey, and for anchorage; located within a cnidocyte.

Neoteny A type of paedomorphosis (*q.v.*) in which the retention of certain larval features into the adult stage results from the retardation of somatic development relative to the development of reproductive organs; common in salamanders.

Nephridium(a) The excretory tubules of many invertebrate animals; may be a protonephridium (*q.v.*) or a metanephridium (*q.v.*).

Nephron The anatomical and functional unit of the vertebrate kidney; consists of Bowman's capsule and a tubule.

Nephrostome The cilated funnel-like opening into the coelom of the inner end of a metanephridium.

Nerve A bundle of neuron axons connecting the central nervous system with peripheral receptors and effectors.

Nerve net A diffuse net of nerve cells with both unidirectional and bidirectional transmission of impulses.

Neuroglia Connecting and supporting cells in the central nervous system; certain ones (oligodendrocytes) surround and myelinate central axons.

Neuron A nerve cell with its processes, collaterals, and terminations; the structural and functional unit of the nervous system.

Neurosecretion The production of hormone-like substances by nerve cells.

Neurotransmitter A chemical released by an axon terminal in response to a nerve impulse that mediates the transmission of that impulse across a synapse (*q.v.*) to a postsynaptic neuron or across a myoneural junction to a muscle cell.

Niche The role of an organism within its community, including the resources it utilizes, its period of activity, and its effect on other members of the community; often conceived as a multidimensional space.

Nondisjunction The failure of a pair of homologous chromosomes to separate normally during the reduction division at meiosis; both members of the pair are carried to the same daughter nucleus and the other daughter cell is lacking in that particular chromosome.

Norepinephrine A neurosecretion produced by the axon terminals of post-ganglionic motor neurons of the sympathetic nervous system that mobilizes glucose in the tissues; *cf.* epinephrine.

Notochord A rod of turgid cells located along the back of chordate embryos ventral to the nerve cord. Acts as a hydroskeleton; usually replaced by the vertebral column in adult vertebrates.

Nucleic acid Very large molecules composed of one or two long chains of nucleotides (*q.v.*); deoxyribonucleic acid (DNA) and ribonucleic acid (RNA).

Nucleotide A molecule composed of a phosphate group, a 5-carbon sugar—ribose or deoxyribose, and a nitrogenous base—a purine or a pyrimidine; the subunits of nucleic acids. Nucleotide triphosphates, such as adenosine triphospate (ATP), store chemical energy within cells.

Nucleus(i) The organelle of a eukaryotic cell containing the hereditary material; a group of nerve cell bodies in the central nervous system.

O

Ocellus(i) A simple light receptor found in many different invertebrate animals.

Olfaction The sense of smell.

Omnivore An animal that feeds on both plants and animals.

Oogenesis The origin, development, and maturation of the ovum or egg; *cf.* spermatogenesis.

Open circulatory system A circulating system in which the blood is not always confined to vessels, but flows through interstitial tissue spaces some of the time; the site of exchange is the interstitial spaces.

Opisthonephros The adult kidney of most fishes and amphibians; its tubules extend from the mesonephric region to the posterior end of the coelom; drained by the archinephric duct and sometimes also by accessory urinary ducts.

Order A taxon above the level of family and below that of class.

Osmoconformer A marine animal in which the salt concentration of the extracellular body fluids conforms to, or is the same as, that of the environment.

Osmoregulator An animal in which the salt content of the body is maintained at a different level from that of the environment.

Osmosis The diffusion (*q.v.*) of solvent molecules (water in biological systems) from the lesser to the greater concentration of solute when two solutions are separated by a membrane that selectively prevents the passage of solute molecules but is permeable to the solvent.

Osmotic pressure The osmotic activity of a solution; correlates directly with the concentration of solutes in a solution; measured as the minimum pressure that must be applied to a solution to prevent the passage of water by osmosis into it from another solution when the two solutions are separated by a membrane permeable to water but not to the solute.

Osphradium(a) A chemical sense organ in the mantle cavity of many molluscs.

Ossicle A small skeletal element; a small bone in vertebrates or a calcareous element in echinoderms.

Oviduct The duct that carries eggs from the ovary or coelom to the outside of the body; differentiates in mammals into part of the vagina, the uterus, and the uterine tube.

Oviparity The deposition of eggs outside the body of the female.

Ovoviviparity Brooding of eggs within the body of the female, but the developing young derive food primarily from egg yolk and not directly from the mother; sometimes called aplacental viviparity.

Ovulation The discharge of an ovum from the ovary.

Ovum The female reproductive cell, which after fertilization by a sperm develops into a new member of the same species.

Oxidation (See biological oxidation)

Oxygen debt The amount of oxygen required to oxidize completely the lactic acid that accumulates during vigorous muscular exercise. (See glycolysis).

P

Paedomorphosis The retention of certain larval or juvenile features by sexually mature individuals; may result from a retardation of somatic development relative to sexual development (neoteny, *q.v.*), or less commonly from an acceleration of sexual development relative to somatic development (progenesis).

Palp An external appendage-like process, usually located near the mouth.

Papilla A small, often conical protuberance.

Parallel evolution The independent evolution of similarities between two or more closely related species; *cf.* convergent evolution.

Paraphyletic Pertaining to a taxon, such as reptiles, that has evolved from a common ancestor but does not include all of the descendants of that common ancestor; *cf.* monophyletic.

Parapodium(a) One of a pair of appendages extending laterally from each segment of polychaete worms and bearing setae.

Parasitism A type of symbiotic relationship in which one member, the parasite, lives in or on the living body of a plant or animal (host) and obtains its nourishment at the expense of the host.

Parasympathetic system A division of the autonomic nervous system in which fibers leave the central nervous system with certain cranial and lumbar nerves and go to visceral organs; has an effect that is antagonistic to sympathetic (*q.v.*) stimulation.

Parthenogenesis The development of an unfertilized egg into an adult organism; occurs naturally in a several groups of animals.

Partial pressure The pressure of a particular gas when in a mixture of gases, as in atmospheric gases.

Pelagic Living in open water either at the surface or at intermediate depths.

Penis The copulatory organ of the male, found in most of those species of animals in which fertilization is internal.

Peptide bond A covalent bond joining the amino group of one amino acid with the carboxyl group of another amino acid.

Pericardial cavity The chamber of the coelom containing the heart.

Pericardial sinus An enlarged portion of the hemocoel surrounding the heart in many invertebrates with open circulatory systems.

Peristalsis Powerful, rhythmic waves of muscular contraction and relaxation in the walls of hollow tubular organs such as the ureter or the parts of the digestive tract; serve to move the contents through the tube.

Peritoneum Coelomic epithelium and supporting connective tissue that lines a coelomic cavity; in mammals, limited to the abdominal cavity.

pH A measure of the acidity of a solution; it represents the negative logarithm of the hydrogen ion concentration, thus acidity increases as pH falls. Neutrality is pH 7.0, which is the pH of pure water.

Phagocytosis The engulfing by endocytosis (*q.v.*) of microorganisms, other cells, and foreign particles by a cell such as a white blood cell or an amoeba; *cf.* pinocytosis.

Pharynx Anterior region of the gut, generally muscular and adapted for ingestion. In vertebrates, that part of the digestive tract from which the gill pouches or slits develop; in higher vertebrates it is bounded anteriorly by the mouth and nasal cavities and posteriorly by the esophagus and larynx.

Phenotype The appearance of an individual; results from the interaction of the genotype (*q.v.*) and its environment.

Pheromone A substance secreted by one individual into the external environment that influences the behavior or development of another individual of the same species.

Photoperiodism The physiologic response of animals and plants to the relative duration of day and night.

Phylogeny The evolutionary history or relationships of a group of organisms.

Phylum(a) The highest subdivision of a kingdom in the biological system of classification of animals.

Phytoplankton Microscopic floating autotrophic organisms that are distributed throughout the ocean or a lake.

Pinocytosis The uptake by cells of droplets of solution by endocytosis (*q.v.*); *cf.* phagocytosis.

Pituitary A small gland that lies just below the hypothalamus of the vertebrate brain, to which it is attached by a narrow stalk; the anterior lobe (adenohypophysis) forms in the embryo as an outgrowth of the roof of the mouth and the posterior lobe (neurohypophysis) grows down from the floor of the brain; often called the hypophysis cerebri.

Placenta(ae) A structure formed partly from embryonic tissues and partly from the mother's uterine lining by means of which the embryo receives nutrients and oxygen and eliminates wastes.

Plankton Minute algae and animals suspended in fresh or salt water.

Planula(ae) Larval stage of cnidarians (hydras, jellyfish, sea anemones, and corals).

Plasma The liquid portion of the blood in which the corpuscles are suspended; differs from serum in containing fibrinogen.

Plasma cell Antibody-producing cell of the immune response that has differentiated from an antigen-activated cell.

Platelet Small non-nucleated and colorless cell fragment present in the blood of mammals involved in blood clotting.

Pleura Coelomic epithelium and supporting connective tissue that lines the cavities containing the lungs of mammals and extends over the surface of the lungs.

Podocyte Specialized cell with interdigitating processes between which filtration occurs.

Poikilothermic Animals whose body temperature fluctuates with that of the environment; ectothermic (*q.v.*); *cf.* homeothermic.

Polar body The smaller of the cells formed by meiotic divisions during oogenesis (*q.v.*); consists of little more than a nucleus.

Polymorphism (morphological) Differences in form among the members of a species; occurrence of several distinct phenotypes in a population.

Polyploidy The condition of having more than two full sets of homologous chromosomes.

Polyp Hydra-like animal; the sessile form of many cnidarians; protruding growth from a mucous membrane.

Portal vein A vein, or group of veins, that drain one region and lead to a capillary bed in another organ rather than directly to the heart, e.g., the hepatic portal vein and renal portal vein.

Posterior Pertaining to the rear end of an animal.

Preadaptation The acquisition by an ancestral group of certain characteristics that usually are adaptive to the ancestral mode of life yet at the same time enable a shift to a new mode of life, e.g., lungs in the fish lineages ancestral to tetrapods.

Primitive character A character that appeared early in the evolution of a lineage; an ancestral character. Primitive characters are usually less complex than advanced ones.

Proboscis Any tubular process of the head or anterior gut region of an animal, usually used in feeding.

Prokaryotes Organisms that lack membrane-bound nuclei and specialized cytoplasmic organelles; the bacteria and blue-green algae; *cf.* eukaryotes.

Protandry A hermaphroditic condition found in various animals in which the individual first produces sperm and then produces eggs.

Protease An enzyme that catalyzes the digestion or cleavage of proteins.

Protein A crucial category of biological macromolecules composed of a chain of covalently joined amino acids, capable of adopting a wide variety of three-dimensional structures (see protein structure) and of carrying out a wide range of biological activities including binding, catalysis, and transport.

Protein structure The sequence and three-dimensional configuration adopted by a protein. Includes the linear sequence of amino acids (primary structure), the local configuration of short stretches into helices or pleated sheets (secondary structure), the overall three-dimensional structure of the protein in biological conditions (tertiary structure), and in the case of multimeric proteins, the relationship between the monomeric subunits (quaternary structure).

Protoctista A kingdom of unicellular and multicellular organisms that includes the algae, protozoans, and slime molds.

Protogyny An hermaphroditic condition found in a few animals, such as certain fish, in which the individual is first female and then changes to male.

Protonephridium(a) A type of nephridial tubule in which the inner end is blind and bears one or more cilia; *cf.* metanephridium.

Protostomes Members of an evolutionary line of the animal kingdom characterized by having the mouth derive from the blastopore; *cf.* deuterostomes.

Pseudocoel A body cavity between the mesoderm and endoderm; a persistent blastocoel.

Pseudopodium(a) Temporary cytoplasmic protrusion of an amoeboid cell; functions in amoeboid locomotion and feeding.

Pulmonary Pertaining to the lungs.

Punctuated equilibria The hypothesis proposing that the morphological features of species remain unchanged for long periods of time (stasis) and that change is concentrated in geologically short periods that punctuate the history of a lineage; frequently associated with episodes of speciation.

Pupa(ae) A stage in the development of an insect, between the larva and the adult; a pupa neither moves nor feeds, but tissues are reorganized within it.

Pygidium(a) The nonsegmental posterior part of a metameric animal that usually bears the anus.

R

Radial cleavage Type of cleavage pattern characteristic of echinoderms and vertebrates in which the spindle axes are parallel or at right angles to the polar axis; *cf.* spiral cleavage.

Radial symmetry Type of body symmetry characteristic of cnidarians and echinoderms in which similar parts are arranged around a central axis.

Radula(ae) A rasplike structure in the buccal cavity of chitons, snails, squids, and certain other molluscs.

Range The geographical area over which a given species is found.

Receptor A sensory cell or a free nerve ending that responds to a given type of stimulus.

Recessive allele An allele that does not express its phenotype in the presence of an alternative "dominant" allele at the same locus, i.e., an allele that produces its effect only when homozygous, when present in "double dose."

Reflex An automatic, involuntary response to a given stimulus, typically involving only a small number of neurons.

Refractory period The period of time that elapses after the response of a neuron or muscle fiber to one stimulus before it can respond again.

Regeneration Regrowth of a lost or injured tissue or part of an organism.

Releasing hormone One of a number of short peptides synthesized in the hypothalamus, secreted into the hypothalamo-hypophyseal portal system, and carried to the anterior lobe of the pituitary, where it initiates the synthesis and release of another specific hormone.

Renal Pertaining to the kidney.

Respiration, cellular The cellular degrading of organic compounds during which energy is transferred to ATP.

Respiratory pigment Metal-containing protein that binds reversibly with oxygen and makes possible the transport of more oxygen in the blood than could be carried in simple solution.

Ribonucleic acid (RNA) Nucleic acid (*q.v.*) containing the sugar ribose; present in both nucleus and cytoplasm and of prime importance in the synthesis of proteins.

Ribosome Minute organelle composed of protein and ribonucleic acid (rRNA) that is either free in the cytoplasm or attached to the membranes of the endoplasmic reticulum of a cell; the site of protein synthesis.

S

Salinity The concentration of salts dissolved in a solution; the salt concentration of sea water.

Schizocoel A body cavity formed by the splitting of embryonic mesoderm into two layers; *cf.* enterocoel.

Secretion A substance released by a gland or cell onto a surface or into a body fluid.

Segmentation Division of a body or structure into more or less similar parts; *cf.* metamerism.

Selection See Natural Selection

Selective reabsorption The removal of certain substances from the fluid passing through an excretory tubule.

Seminal receptacle A portion of the female reproductive tract in which sperm are stored after mating.

Seminal vesicle A portion of the male reproductive tract in which sperm are stored before mating; in mammals produces part of the seminal fluid.

Sensory neuron A neuron that carries a nerve impulse from a receptor to an interneuron or motor neuron; often called an afferent neuron.

Serum(a) The clear portion of a biological fluid separated from its particulate elements; light yellow liquid left after clotting of blood has occurred.

Sessile Living attached to the substratum as is true of sea anemones, sponges, and barnacles.

Seta(ae) Bristle-like projection of the body surface composed of some skeletal material, most commonly chitin.

Shell A relatively heavy exoskeleton of calcium carbonate or chitin.

Skeletal muscle In vertebrates the striated muscle that is largely attached to the bony and cartilaginous skeleton.

Sinus A cavity or channel within a tissue or organ, not completely lined by epithelium.

Sinusoid A small, blood-filled sinus (*q.v.*), as in the liver or spleen.

Smooth muscle Type of muscle tissue in which the actin-myosin protein fibrils are not aligned; composed of spindle-shaped uninucleated cells; muscle cells that are not striated; *cf.* striated muscle.

Social animal A group of individuals of the same species whose members cooperate to varying degrees in performing necessary activities such as feeding, rearing young, and defense; members are interdependent but not physically united. In some social insects (termites), the individuals are physiologically and morphologically specialized for the performance of their particular tasks, but this is not always the case.

Solute The substance present in smaller amounts in a solution (*q.v.*); often the substance dissolved in a solvent.

Solution A mixture of two or more substances in which their individual molecules have separated and are homogeneously dispersed.

Solvent The substance present in larger amounts in a solution (*q.v.*); often the liquid in which a solute is dissolved; a liquid that is capable of dissolving a solvent.

Somatic Pertaining to the body or body cells as opposed to the germ or gamete-forming cells; pertaining to structures in the body wall as opposed to those in visceral organs.

Somatic skeleton In vertebrates, that portion of the skeleton associated with the body wall; includes the axial skeleton (*q.v.*) and appendicular skeleton.

Somite One of a paired, blocklike mass of mesoderm, arranged in a longitudinal series alongside the neural tube of the embryo, forming the vertebral column and most of the muscles of the body wall and appendages.

Speciation The evolutionary process by which one species arises from another.

Species A population of similar individuals that are capable of interbreeding with each other but are reproductively isolated in nature from other interbreeding groups; a taxonomic unit having two names, the generic name and the specific name or epithet, e.g., *Homo sapiens* for modern human beings.

Spermatogenesis The origin, development, and maturation of sperm; *cf.* oogenesis.

Spermatophore A package secreted by part of the male reproductive tract, enclosing a mass of sperm.

Spicule Microscopic mineral needle-like or rodlike skeletal piece found in sponges and certain other animals.

Spiracle An external opening associated with breathing such as the opening into the tracheal system or book lungs of certain arthropods, or the opening (a modified gill slit) into the pharynx in certain fishes.

Spiral cleavage A cleavage pattern characteristic of a number of invertebrate phyla, such as annelids and molluscs, in which the cleavage planes are oriented obliquely to the polar axis of the egg; *cf.* radial cleavage.

Splicing The process, taking place in the nucleus, whereby the non-coding portions of the mRNA (introns) are removed, and the coding portions (exons) are stitched together prior to the export of mRNA to the cytoplasm.

Statocyst A chambered sense organ containing one or more granules that is used in a variety of animals to sense the direction of gravity.

Stenohaline Referring to animals restricted to a narrow range of environmental salt concentrations.

Steroid A complex molecule containing carbon atoms arranged in four interlocking rings, three of which contain six carbon atoms each and the fourth of which contains five; the male and female sex hormones and the adrenal cortical hormones.

Stomodeum An ectodermal invagination at the front of an embryo; contributes to the mouth cavity.

Striated muscle Type of muscle tissue in which the repeating actin-myosin protein fibrils are aligned to give the appearance of cross striations, composed of long multinucleated cells; *cf.* smooth muscle.

Stylet A hard structure shaped like a needle or dagger, usually functioning in feeding or sperm transfer.

Substrate The substance on which an enzyme acts.

Substratum(a) The surface on which an organism lives or to which it is attached.

Summation The adding of one response to another, as in types of muscle contraction in which stimuli follow one another before the effects of previous ones have dissipated (temporal summation), or additional motor units are activated (spacial summation).

Superficial cleavage Type of incomplete cleavage characteristic of many arthropod eggs in which the nuclei and peripheral cytoplasm divide but not the large central mass of yolk.

Suspension feeding The utilization of plankton and other particles suspended in the water column as a food source.

Swim bladder A gas-filled chamber, homologous to a lung, situated in the dorsal part of the abdominal cavity of most bony fishes. By controlling its gas content, a fish can attain neutral buoyancy.

Symbiosis(es) The living together of two organisms of different species; association may be mutualistic, commensal, or parasitic.

Sympathetic system A division of the autonomic nervous system in which fibers leave the central nervous-system with certain thoracic and lumbar nerves and go to the sweat glands, blood vessels and visceral organs; has an effect on most organs that is antagonistic to parasympathetic (*q.v.*) stimulation.

Sympatric   Pertaining to species occupying the same geographic area or whose ranges are partially overlapping; *cf.* allopatric.

Synapse   The point of communication between neurons, usually occurs between the axon terminals of one neuron and the dendrites or cell body of another; sometimes used for the junction between a neuron and muscle cell (such points are more properly called myoneural junctions).

Synapsis   The pairing and union side by side of homologous chromosomes from the male and female pronuclei early in meiosis.

Syncytium(a)   Tissue in which nuclei and adjacent cytoplasm are not separated by cell membrane; a multinucleate cell or tissue.

Systole   The contraction of the heart, especially its ventricles; *cf.* diastole.

T

T cell   A type of lymphocyte that matures in the thymus, circulates in the blood, and plays a key role in the cell-mediated immune response.

Taxon(a)   One of the units in the classification of organisms. The major taxa from the largest, or most inclusive, smallest are kingdom, phylum, class, order, family, genus, and species.

Taxonomy   The science of naming, describing, and classifying organisms.

Telolecithal egg   Type of egg having the yolk material concentrated toward the vegetal pole.

Territoriality   Behavior pattern in which an animal (usually a male) delineates a territory of his own and defends it against intrusion by other members of the same species and sex; territoriality is most pronounced shortly before and during the breeding season.

Tetrad   A bundle of four homologous chromatids produced at the end of the first meiotic prophase.

Tetrapod   The terrestrial vertebrates, most of which have four limbs; amphibians, reptiles, birds, and mammals.

Thorax   The anterior part of the trunk of many animals, especially crustaceans, insects, and mammals.

Tissue   An organized association of similar cells that perform a common function, e.g., epithelial tissue, muscle tissue, etc.

Torsion   Twisting, generally referring to the embryonic or evolutionary twisting of the molluscan visceral mass on the head-foot.

Trachea(ae)   An air-conducting tube. In terrestrial vertebrates it is the main trunk of the system of tubes through which air passes to and from the lungs; in terrestrial arthropods it is one of a system of minute tubules that permeate the body and deliver air to the tissues.

Transcription   The biochemical process that occurs within the nucleus of eukaryotic cells whereby a DNA strand is copied to produce a complementary RNA strand that, after editing, is exported from the nucleus as messenger RNA (mRNA) (*q.v.*).

Transducer   Device receiving energy from one system in one form and supplying it to a second system in a different form; e.g., converting radiant energy to chemical energy and stimulus energy to a wave of membrane depolarization (neuron impulse) in a receptor.

Transfer RNA (tRNA)   Small RNA molecules found in the cytoplasm that carry specific amino acids and bring them to the ribosomes during protein synthesis. Each type of amino acid is recognized by and bound to a specific tRNA molecule; a triplet in the tRNA (the anticodon) recognizes each of the triplets in the messenger RNA.

Translation   The biochemical process whereby genetic information encoded in mRNA directs the synthesis of specific proteins. Translation occurs on the ribosomes (*q.v.*) where mRNA is used as a template for the formation of polypeptide chains. Amino acids are lined up and joined in the appropriate sequence by base pairing between the anticodon on each tRNA and the codons on mRNA.

Triplet code   The sequences of three nucleotides that compose the codons (*q.v.*), the units of genetic information in DNA or RNA that specify the order of amino acids in a peptide chain.

Trochophore   The top-shaped larva of marine molluscs, polychaete annelids and some other marine invertebrates that bears an apical tuft and girdles of cilia.

Tube foot   The external tube-like appendage of the water vascular system of echinoderms used in locomotion, gas exchange, and feeding.

U

Urea   One of the end products of protein metabolism, less toxic than ammonia from which it is formed; excreted by many animals in which water must be conserved, including cartilaginous fishes and mammals.

Ureter   The fibromuscular tube that conveys urine from a metanephric kidney to the cloaca or to the bladder.

Urethra   The membranous canal conveying urine from the bladder to the exterior of the body.

Uric acid   An end product of nucleic acid and protein metabolism with a low solubility in water; the form of the nitrogenous waste excreted by insects, land snails, birds, and most reptiles.

Urine   The fluid eliminated from the excretory organ. Composed of water, salt, nitrogenous wastes, and some other substances.

Uterus(i)   The womb; the hollow, muscular organ of the female reproductive tract in which the embryo undergoes development.

V

Vacuole   A spherical intracellular vesicle bounded by a membrane, e.g. food vacuole, contractile vacuole.

Vascular   Pertaining to blood supply or vessels.

Vein   Vessel through which blood passes from the tissues toward the heart, typically has thin walls and valves that prevent a reverse flow of blood; extension of the tracheal system that supports the wings of insects.

Veliger   Larval stage of many marine clams and snails; often is a second larval stage that develops from the trochophore (*q.v.*).

Ventral Pertaining to the lower surface of an animal.

Ventricle A cavity of an organ such as one of the chambers of the heart that receive blood from the atria, or one of the several chambers of the brain.

Vertebrates The subphylum of chordates characterized by the presence of a vertebral column; includes fishes, amphibians, reptiles, birds, and mammals.

Vestigial organ An organ present in one animal that is a remnant of a well-developed homologous organ in an ancestral animal; vestigial organs are without function, or their original functions are greatly reduced.

Villus(i) A small, finger-like process or protrusion, especially a protrusion from the free surface of a membrane such as the lining of the intestine.

Visceral Pertaining to the internal organs, or viscera, as opposed to structures in the body wall; *cf.* somatic.

Visceral arch or skeleton An arch of cartilage or bone that develops in the wall of the pharynx between the gill slits; they support the gills and jaws in fishes, but certain ones become incorporated into the skull, hyoid apparatus, and larynx of higher vertebrates.

Vitamin An organic substance necessary in small amounts for the normal metabolic functioning of a given organism; must be present in the diet because the organism cannot synthesize an adequate amount of it.

Vitelline Pertaining to yolk.

Viviparity Brooding of eggs within the body of the female, and the developing young derive food directly from the mother.

W

Warning coloration See Aposematic coloration

Y

Yolk Stored food reserves within an egg or, in some species, within yolk or nurse cells.

Yolk sac A pouchlike outgrowth of the digestive tract of certain vertebrate embryos that grows around the yolk and makes it available to the rest of the organism.

Z

Zoochlorellae The green algal symbionts (*Chlorella*) of some animals, such as certain sponges and hydras.

Zooplankton The animals or animal-like organisms (protozoa) composing plankton.

Zooxanthellae Dinoflagellates living symbiotically with certain marine animals, especially corals.

Zygote The cell formed by the union of two gametes; a fertilized egg.

CREDITS

Preface

Page v: Frans Lanting/Minden Pictures; **vi**: (right) M. Kazmers/M. L. Dembinsky, Jr., Photography Associates

Contents Overview

Page viii: Skip Moody/M. L. Dembinsky, Jr., Photography Associates; (right) Gerard Lacz/Peter Arnold, Inc.; **ix**: Frans Lanting/Minden Pictures

Contents

Page xi: (left) ScienceVU/Visuals Unlimited; (right) © Robert Carlyle Day 1986/Photo Researchers, Inc.; **xii**: (left) Lennart Nilsson, in *The Incredible Machine*, National Geographic Society, pp. 118–119. © Boehringer Ingelheim International GmbH; **xiii**: Frans Lanting/Minden Picture; **xiv**: (left) Skip Moody/M. L. Dembinsky, Jr., Photography Associates; (right) Frans Lanting/Minden Pictures; **xv**: Frans Lanting/Minden Pictures; **xvii**: (left) SIU/Visuals Unlimited; (right) Frans Lanting/Minden Pictures; **xviii**: M. L. Dembinsky, Jr., Photography Associates; **xix**: Lennart Nilsson, from *A Child Is Born* (New York: Dell Publishing, 1990):109; **xxi**: (left) A. J. Copley/Visuals Unlimited; (right) Eric Gravé/Photo Researchers, Inc.; **xxii**: Frans Lanting/Minden Pictures; **xxiii**: Fred McConnaughey/Photo Researchers, Inc.; **xxiv**: Grant Heilman; **xxv**: Doug Wechsler/VIREO, Academy of Natural Sciences, Philadelphia

Part Opening Photos

I: Skip Moody/Marvin L. Dembinsky, Jr., Photography Associates; **III**: Gerard Lacz/Peter Arnold, Inc.; **IV**: George E. Stuart/Marvin L. Dembinsky, Jr., Photography Associates; **V**: Frans Lanting/Minden Pictures

Chapter 1

Openers: M. P. Kahl/VIREO; **Figure 1.1**: Bettmann Archive; **1.2a**: Bettmann Archive; **1.2b**: Bettmann Archive; **1.3**: Bettmann Archive; **1.4**: Bettmann Archive; **1.5**: Science VU-NCI/Visuals Unlimited; **1.6**: Bettmann Archive; **1.7**: NASA; **1.8**: Bettmann Archive; **1.9**: Lennart Nilsson, from *A Child Is Born* (New York: Dell Publishing Co., Inc.); **1.10a**: Frans Lanting/Minden Pictures; **1.11a**: Oxford Molecular Biophysics Laboratory/Science Photo Library/Photo Researchers, Inc.; **1.11b**: C. Fox/Photo Researchers, Inc.; **1.11c**: Frans Lanting/Minden Pictures; **1.11d**: Frans Lanting/Minden Pictures; **1.11e**: Frans Lanting/Minden Pictures

Chapter 2

Openers: (left) Stan Osolinski/Marvin L. Dembinsky, Jr., Photography Associates; (above) William E. Ferguson; **Figure 2.1a**: G. R. Roberts; **2.1b**: Biophoto Associates; **2.3**: James L. Castner; **2.4**: Y. Arthus-Bertrand/Peter Arnold, Inc.; **2.5**: Visuals Unlimited/Steve McCutcheon; **2.6**: Sinclair Stammers/Science Photo Library/Photo Researchers, Inc.; **2.7**: *The Structure of Marine Ecosystems* by John H. Steele, Cambridge, Mass.: Harvard Univ. Press, Copyright © 1974 by the President and fellows of Harvard College; **2.9**: Biophoto Associates, National Portait Gallery, London; **2.10**: William E. Ferguson; **2.11**: Bettmann Archive; **2.12b**: Christopher Ralling; **2.13a**: David Cavagnaro; **2.13b**: Frans Lanting/Minden Pictures; **2.13c**: Frans Lanting/Minden Pictures; **2.14a**: Carl R. Sams, II/Marvin L. Dembinsky, Jr., Photography Associates; **2.14b**: Animals Animals © 1991 Zig Leszczynski; **2.15**: From *Recherches sur les ossements fossiles*, 1836, by G. Cuvier. Reprinted from Alan Moorehead, *Darwin and the Beagle,* (New York: Harper and Row, 1969): 84; **2.16**: John D. Cunningham/Visuals Unlimited; **2.18**: Frans Lanting/Minden Pictures; **2.19**: Bettmann Archive; **2.20**: Biophoto Associates; **Focus 2.1**: Biophoto Associates, Linnaean Society, London

Chapter 3

Openers: Biophoto Associates

Chapter 4

Openers: E. R. Degginger; **Figure 4.2**: William E. Ferguson; **4.3**: Courtesy of Andrew Knoll, Harvard University; **4.4**: William E. Ferguson; **4.5**: Pearson/Milon/Photo Researchers, Inc.; **4.6a**: Courtesy of Dr. Stanley Miller; **4.7a**: David M. Phillips/Visuals Unlimited; **4.7b**: Dr. Lyle C. Dearden; **4.8**: Dr. Sidney Fox

Chapter 5

Openers: (left) Biophoto Associates; (above) Tom Adams; **Figure 5.3**: Dr. Susumu Ito, Harvard Medical School; **5.5d**: Omikron/Photo Researchers, Inc.; **5.10b**: Charles Sanders/Biological Photo Service; **5.13a**: Don Fawcett/Photo

Researchers, Inc.; **5.13b**: Dr. E. Anderson; **5.14a**: After a model by J. Kephart; **5.14b**: Don Fawcett/Photo Researchers, Inc.; **5.16a**: Courtesy of K. R. Porter; **5.18a**: Dr. Susanne M. Gollin and Wayne Wray, Kleberg Cytogenetics Laboratory, Baylor College of Medicine and Biology Department, The Johns Hopkins School of Medicine; **5.19**: M. Schliwa/Visuals Unlimited; **Focus 5.1a**: E. R. Degginger

Chapter 6

Openers: (left) Animals Animals © 1991 Leonard Lee Rue, III; (above) Robert Carlyle Day/Photo Researchers, Inc.; **Figure 6.1**: William E. Ferguson; **6.8**: David M. Phillips/Visuals Unlimited; **6.9a**: Dr. Dorothy Warburton/Peter Arnold, Inc.; **6.9b**: Dr. Jamie Lee/Peter Arnold, Inc.; **6.11**: Animals Animals © 1991 E. R. Degginger; **6.14**: David M. Phillips/Visuals Unlimited; **6.15**: C. A. Hasenkampf, Univ. of Toronto/Biological Photo Service

Chapter 7

Openers: (left) CNRI/Science Photo Library/Photo Researchers, Inc.; (above) Science VU/Visuals Unlimited; **Figure 7.1**: After Hartseeker's drawing from "Essay de Dioptrique," Paris, 1694; **7.2a**: Dennis D. Kunkel/Biological Photo Service; **7.5a**: Courtesy of Dr. Lee D. Simon, Institute for Cancer Research, Philadelphia; **7.9**: Courtesy of Biophysics Research Unit, Medical Research Council, King's College, London; **7.17b**: K. G. Murti/Visuals Unlimited; **7.20**: Dennis D. Kunkel/Biological Photo Service

Chapter 8

Openers: Dwight R. Kuhn; **Figure 8.2c**: David M. Phillips/Visuals Unlimited; **8.3a**: Charles C. Brinton, Jr. and Judith Carnahan; **Focus 8.1**: R. W. Van Norman/Visuals Unlimited; **8.7a**: Medslide, Inc.; **8.7b**: Richard Hutchings/Photo Researchers, Inc.

Chapter 9

Openers: John D. Cunningham/Visuals Unlimited; **Figure 9.1**: From Ayala, F. J. et al., *Modern Genetics*, 2nd ed., Benjamin Cummings, 1984; **9.3**: Redrawn from Cavalli-Sforza, L. L. and W. F. Bodmer, *The Genetics of Human Populations*, W. H. Freeman, 1971; **9.5**: From Ayala, F. J. et al., *Modern Genetics*, 2nd ed., Benjamin Cummings, 1984; **9.6b**: After D. L. Harte, *Principles of Population Genetics*, Sinauer, 1980; **9.7**: Modified from Cavalli-Sforza, L. L. and W. F. Bodmer, *The Genetics of Human Populations*, W. H. Freeman, 1971; **9.8**: Michael Tweedie/Photo Researchers, Inc.; **9.9**: From Levine, L., *Biology of the Gene*, C. V. Mosby, 1973; **9.10**: Data from T. Dobzhansky and B. Spassky, *Genetics*, Vol. 29(270–290), May, 1944; **9.12**: Modified from Harte, D. L., *Principles of Population Genetics*, Sinauer, 1980; **9.15**: After Dudley, 1977; **9.16**: Hope Entymology Collection, Oxford/Biological Photo Service; **9.17**: After Geological Survey publication; **9.18**: Redrawn from Raup, D. and S. Stanley, *Fundamentals of Paleontology*, W. H. Freeman, 1978; **9.19**: Courtesy of the University of Michigan Museum of Zoology; **9.20**: From Keeton, W. and J. Gould, *Biological Science*, 4th ed., W. W.

Norton, 1986; **9.21**: From Keeton, W. and J. Gould, *Biological Science*, 4th ed., W. W. Norton, 1986; **9.22**: Modified from Ayala, F. J. et al., *Modern Genetics*, 2nd ed., Benjamin Cummings, 1984; **9.23a**: From the U.S. Geological Survey pamphlet, *The Great Ice Age*, U.S. Government Printing Office, 0-357-128, 1969; **9.23b**: From Remington, C. L., "Suture zones of hybrid interactions between recently joined biota," in *Evolutionary Biology*, 2:321–428, 1968; **Focus 9.3a–d**: From K. Norstog and R. W. Long; **9.24**: From Moore, Lalicker, and Fischer, *Invertebrate Paleontology*, McGraw-Hill, 1952; **9.27**: From B. H. Erwin et al., *Evolution* 41, 1177 (1988)

Chapter 10

Openers: (left) Frans Lanting/Minden Pictures; (above) G. R. Roberts; **Figure 10.1b**: William S. Ormerod/Visuals Unlimited; **10.8a**: Larry Tackett/Tom Stack & Associates; **10.8b**: Animals Animals © 1991 Tim Rock; **10.8c**: Hans Pfletschinger/Peter Arnold, Inc.; **10.8d**: M. P. Kahl/Photo Researchers, Inc.

Chapter 11

Openers: (left) G. Ziesler/Peter Arnold, Inc.; (above) R. S. Virdee, from Grant Heilman; **Figure 11.4**: Modified from Pillsbury, Shelley, and Kligman, *Manual of Cutaneous Medicine*, W. B. Saunders, 1961; **11.6**: Based on a photograph by McNamara, 1981; **11.7a**: After Weber, from Kaestner; **11.7b**: After Weber, from Vandel; **11.10a**: Fred Hossler/Visuals Unlimited; **11.17**: Modified after Neal and Rand; **11.18a**: Modified after Gregory; **11.18b**: After Romer; **11.21**: Modified from Bloom, W. and D. W. Fawcett, *A Textbook of Histology*, 10th ed., W. B. Saunders, 1975; **11.22**: From Guyton, A. C., *Textbook of Medical Physiology*, 5th ed., W. B. Saunders, 1976; **11.25**: Modified from Schmidt-Nielsen, K., *Animal Physiology*, Cambridge University Press, 1975; **11.27**: Modified after Howell; **11.28**: From Walker, W. F., *Vertebrate Dissection*, 7th ed., Saunders College Publishing, 1986; **11.29**: Modified from Allen, R. D., "Amoeboid Movement," in Brachet, J. and A. E. Mirsky (eds.), *The Cell*, Academic Press, 1961; **11.30**: (a) and (b) from Satir, P., "How Cilia Move," *Scientific American*, 231(Oct.):44–52, 1974; (c) and (d) modified from the work of I. R. Gibbons; **11.31**: From Schmidt-Nielsen, K., *Animal Physiology*, 2nd ed., Cambridge University Press, 1979; photograph by R. L. Hammersmith; **11.33**: After Gray and Lissmann; **11.34**: Modified from Marshall and Hughes; **11.36c**: Modified after Norberg

Chapter 12

Openers: (left) Stan Osolinski/M. L. Dembinsky, Jr., Photography Associates; (above) Stan Osolinski/M. L. Dembinsky, Jr., Photography Associates; **Figure 12.1a**: Animals Animals © 1991 William D. Griffin; **12.4b**: After P.A.X.; **12.5**: Richard C. Johnson/Visuals Unlimited; **12.7a**: St. Bartholomew's Hospital/Science Photo Library/Photo Researchers, Inc.; **12.7b**: Runk/Schoenberger, from Grant Heilman; **12.8**: Reproduced by permission of the Royal Society of Edinburgh and Nichol, E. A. T., "The feeding mechanism, formation of the tube, and physiology of digestion in Sabella pavonina." *Trans. R. Soc.*, Edinburgh,

56(23):537–596, 1930; **12.10b**: From Jacob, Francone, and Lossow, *Structure and Function in Man*, 4th ed., W. B. Saunders, 1978; **12.11**: Runk/Schoenberger, from Grant Heilman; **12.17**: Photograph by E. Strauss. From Bloom, W. and D. W. Fawcett, *A Textbook of Histology*, 11th ed., W. B. Saunders, 1986; **12.18**: E. R. Degginger; **12.19**: Food and Agricultural Organization of the United Nations, Photo by P. Pittet

Chapter 13

Openers: Jerome Wexler/Photo Researchers, Inc.; **13.3b**: Ross: *A Textbook of Entomology, 3/e.* Reprinted by permission of John Wiley and Sons, Inc. **Figure 13.8**: Modified in part from Lehninger; **13.10**: Modified after Hughes; **13.11b**: Modified from Calman; **13.13**: (a) and (d) after Snodgrass; (b) after Ross; **13.15**: (b–e) after Gans, Dejongh, and Farber; **13.18**: Ed Reschke; **13.20a&b**: From Kessel, R. G. and R. H. Kardon, *Tissues and Organs: A Text-Atlas of Scanning Electron Microscopy*, W. H. Freeman, 1979; **13.27**: From Schmidt-Nielsen, K., *Animal Physiology: Adaptation and Environment*. Cambridge University Press, 1975; **13.28**: After Hemmingsen, from Schmidt-Nielsen, K., *Animal Physiology: Adaptation and Environment*, Cambridge University Press, 1975

Chapter 14

Openers: Lennart Nilsson, in *The Incredible Machine*, National Geographic Society, pp. 118–119, © Boehringer Ingelheim International GmbH; **Figure 14.1a–c** from Prosser, C. L., *Comparative Animal Physiology*, 3rd ed., Saunders College Publishing, 1973; **Focus 14.1**: From Harvey, *Exercitatio Anatomica de Motu Cordis et Sanguinis in Animalibus*. Translation by C. D. Leake, *Anatomical Studies of the Motion of the Heart and Blood in Animals*, Charles C. Thomas, 1931; **14.4**: From *Science* 173(3993): 1971, Cover picture, photograph by Emil Bernstein, the Gillette Company Research Institute; **14.7b**: Modified from Nicoll; **14.21**: Modified from Williams, P. L. and R. Warwick (eds.), *Gray's Anatomy*, 36th British edition, W. B. Saunders, 1980

Chapter 15

Openers: Lennart Nilsson, in *The Incredible Machine*, National Geographic Society, pp. 170–171, © Boehringer Ingelheim International GmbH; **Figure 15.1**: UPI/Bettmann Archive; **15.2**: After Tizard, I. R., *Immunology: An Introduction*, 2nd ed., Saunders College Publishing, 1988; **15.5**: Bettmann Archive; **15.9a&b**: R. Rodewald, Univ. of Virginia/Biological Photo Service; **15.12a**: Lennart Nilsson © Boehringer Ingelheim International GmbH, from *The Incredible Machine*, p. 170; **15.12b**: Lennart Nilsson © Boehringer Ingelheim International GmbH, from *The Incredible Machine*, p. 171; **15.13**: After Tizard, I. R., *Immunology: An Introduction*, 2nd ed., Saunders College Publishing, 1988; **15.14**: Don Fawcett/Science Source/Photo Researchers, Inc.; **15.16**: From Podack, E. R. and Dennert, G., 1983, *Nature* 301:442, used with permission; **15.17a–c**: Lennart Nilsson © Boehringer Ingelheim International GmbH; **15.18a**: From Fawcett, D. W. and W. Bloom, *Textbook of Histology*, 11th ed., W. B. Saunders, 1986; **15.18b**: Lennart Nilsson © Boehringer

Ingelheim International GmbH, from *The Incredible Machine*, p. 211; **15.20**: From Bellanti, J. A., *Immunology II*, W. B. Saunders, 1979; **15.21**: From Kimball, J. W., *Introduction to Immunology*, Macmillan, 1983; **15.22**: After Tizard, I. R., *Immunology: An Introduction*, 2nd ed., Saunders College Publishing, 1988

Chapter 16

Openers: Stan Osolinski/M. L. Dembinsky, Jr., Photography Associates; **16.4**: (a) after Reisinger; (b) and (c) from Kummel, G., *Z. Zellforsch* 57:172–201, 1962; **16.11**: From Kessel, R. G. and R. H. Kardon, *Tissues and Organs: A Text-Atlas of Scanning Electron Microscopy*, W. H. Freeman, 1979, p. 231; **16.12**: Modified from Guyton, A. C., *Human Physiology and Mechanisms of Disease*, 3rd ed., W. B. Saunders, 1982; **Focus 16.1**: Werner H. Muller/Peter Arnold, Inc.; **16.13**: Adapted from Schmidt-Nielsen; **16.15**: From Schmidt-Nielsen, K., "Salt Glands," *Scientific American* 200(1959):109–116; **16.17**: John D. Cunningham/Visuals Unlimited

Chapter 17

Openers: (left) Reinhard Kunkel/Peter Arnold, Inc.; (above) Stan Osolinski/M. L. Dembinsky, Jr., Photography Associates; **Figure 17.3d**: C. S. Raine/Visuals Unlimited; **17.9a**: Courtesy of E. R. Lewis; **17.13**: Modified from Williams, P. L. and R. Warwick, *Gray's Anatomy*, 37th British edition, W. B. Saunders Co., 1989; **17.14**: Modified from Williams, P. L. and R. Warwick, *Gray's Anatomy*, 37th British edition, W. B. Saunders Co., 1989; **17.17**: After Howell; **17.18**: After Patten; **17.22**: From Penfield and Rasmussen, *The Cerebral Cortex*, Macmillan

Chapter 18

Openers: (left) Bruce Iverson/Visuals Unlimited; (above) Animals Animals © 1991 Ashod Frances; **Figure 18.3b**: Runk/Schoenberger, from Grant Heilman; **18.4**: Modified from Williams, P. L. and R. Warwick, *Gray's Anatomy*, 36th British edition, W. B. Saunders, 1980; **18.5**: John R. MacGregor/Peter Arnold, Inc.; **18.6a**: Modified from Williams, P. L. and R. Warwick, *Gray's Anatomy*, 36th British edition, W. B. Saunders, 1980; **18.6b**: Biophoto Associates/Photo Researchers, Inc.; **18.8c**: Ed Reschke; **Focus 18.1a–f**: Modified from A. J. Kalmijn, "The electric sense of sharks and rays," *Journal of Experimental Biology* 55(1971):371–377; **18.14a**: Bettmann Archive; **18.14b**: IFA/Peter Arnold, Inc.; **18.21a**: Christian Petron/Seaphot, Ltd.; **18.22a**: David Scharf; **18.22d**: Thomas Eisner; **18.23**: E. R. Degginger; **18.25a**: E. R. Lewis, Y. Y. Zeevi, T. E. Everhart/Biological Photo Service; **18.28**: Gregory G. Dimijian, M.D./Photo Researchers, Inc.

Chapter 19

Openers: SIU/Visuals Unlimited; **Focus 19.1**: R. Calentine/Visuals Unlimited; **Figure 19.6a**: Animals Animals © 1991 Breck P. Kent; **19.6b**: Skip Moody/M. L. Dembinsky, Jr., Photography Associates; **19.6c**: E. R. Degginger; **19.9**: Courtesy of Dr. Gordon Williams; **19.13b**: Astrid & Frieder Michler/ Peter Arnold, Inc.

Chapter 20

Openers: Frans Lanting/Minden Pictures; **Figure 20.1a**: Kjell B. Sandved/Visuals Unlimited; **20.1b**: Modified from Sargent, R. C. and G. E. Wagenbach, "Cleaning behavior of the shrimp Peridimenes anthopilus," *Bulletin of Marine Science* 25(4): 466–472, 1975; **20.2**: From Lorenz, K. Z. and N. Tinbergen, *Zeitschrift für Tierpsychologie* 2, 1, 1938; **20.3**: From Willows, A. O. D., "Giant brain cells in mollusks," *Scientific American*, 224(Feb. 1971):69–75; **20.4**: From Drickamer, L. C. and S. H. Vessey, *Animal Behavior* © 1982 by PWS Publishers. All rights reserved. Used by permission of Willard Grant Press; **20.6**: From Alcock, J., *Animal Behavior: An Evolutionary Approach*, 2nd ed., Sinauer, 1979; **20.7**: From Levi, H. W., "Orb-weaving spiders and their webs," *American Scientist* 66:734–742, 1978; **20.8**: From Alcock, J., *Animal Behavior: An Evolutionary Approach*, 2nd ed., Sinauer, 1979; **20.9**: Spectrogram photographs by M. Konishi, from Alcock, J., *Animal Behavior: An Evolutionary Approach*, 2nd ed., Sinauer, 1979; **20.10a**: R. F. Ashley/Visuals Unlimited; **20.10b**: Doug Wechsler; **20.10c**: Animals Animals © 1991 Michael Fogden; **20.10d**: John D. Cunningham/Visuals Unlimited; **20.11a**: Courtesy of E. S. Ross, California Academy of Sciences; **20.11b**: Thomas Eisner; **20.12a**: Doug Wechsler; **20.12b**: Doug Wechsler; **20.13**: L. E. Gilbert, Univ. of Texas at Austin/Biological Photo Service; **20.14**: Fritz Polking/M. L. Dembinsky, Jr., Photography Associates; **20.15**: VIREO-Academy of Natural Sciences of Philadelphia; **20.16**: Animals Animals © 1991 Patti Murray; **20.17**: (a) and (b) from Moynihan, M. and Rodaniche, A. F., *The Behavior and Natural History of the Caribbean Reef Squid (Sepioetenthis sepioidea)*, Paul Parey Publishers, 1982; (c) and (d) Reprinted by permission of the publishers from *The Insect Societies* by Edward O. Wilson, Cambridge, MA: The Belknap Press of Harvard University Press, Copyright © 1971 by the President and Fellows of Harvard College; **20.18**: Modified after Tinbergen, N., *The Study of Instinct*, Oxford University Press, 1951; **20.19**: Arthus Bertrand/Peter Arnold, Inc. By permission of Oxford University Press.

Chapter 21

Openers: (left) Glenn Oliver/Visuals Unlimited; (above) Marvin L. Dembinsky, Jr., Photography Associates; **Figure 21.1a**: M. I. Walker/Science Source/Photo Researchers, Inc.; **21.1b**: After Child; **21.1c**: R. D. Campbell, Univ. of Calif., Irvine/Biological Photo Service; **21.1d**: Fred Bavendam/Peter Arnold, Inc.; **21.3**: Modified after Bloom, W. and D. W. Fawcett, *A Textbook of Histology*, 11th ed., W. B. Saunders, 1986; **21.5**: (a–c) modified after Walbot, V. and H. Holder, *Developmental Biology*, Random House, 1987, with permission of McGraw-Hill, Inc.; (d) modified after Bloom, W. and D. W. Fawcett, *A Textbook of Histology*, 11th ed., W. B. Saunders, 1986; **21.7**: Biophoto Associates; **21.8a**: Modified after Walbot, V. and H. Holder, *Developmental Biology*, Random House, 1987, with permission of McGraw-Hill, Inc.; **21.8b–c**: G. Schatten, Univ. of Wisconsin/Biological Photo Service; **21.9a**: Lynn Funkhouser/Peter Arnold, Inc.; **21.9b**: After Girdholm; **21.9c**: David C. Fritts © 1991 Animals Animals;

21.10: (a) after Pavlovsky; (b) after Watase; **21.11**: H. Wes Pratt/Biological Photo Service; **21.12a**: Doug Wechsler; **21.17**: From Walker, W., Jr., *Anatomy and Dissection of the Fetal Pig*, 4th ed., W. H. Freeman, 1988

Chapter 22

Openers: (left) Lennart Nilsson, from *A Child Is Born* (New York: Dell Publishing Co., 1990):134; (above) Lennart Nilsson, from *A Child Is Born* (New York: Dell Publishing Co., 1990): 190; **Figure 22.7**: Modified from Balinsky, B. I., *An Introduction to Embryology*, 5th ed., Saunders College Publishing, 1981; **22.10**: Modified from Balinsky, B. I., *An Introduction to Embryology*, 5th ed., Saunders College Publishing, 1981; **22.13b**: E. R. Degginger; **22.15**: Animals Animals © 1991 Oxford Scientific Films; **22.18**: (a) modified after Patten; (b) modified after Williams, P. L. and R. Warwick, *Gray's Anatomy*, 36th British edition, W. B. Saunders, 1980; **22.19**: After Williams, P. L. and R. Warwick, *Gray's Anatomy*, 36th British edition, W. B. Saunders, 1980; **Focus 22.1a–c**: After Arey; **22.20**: After Gilbert, S., *Developmental Biology*, 2nd ed., Sinauer, 1988; **22.21**: After Gilbert, S., *Developmental Biology*, 2nd ed., Sinauer, 1988; **22.22**: Modified from Jamrich et al., 1985, based on a photograph by I. Dawid and M. Sargent; **22.23**: After Gilbert, S., *Developmental Biology*, 2nd ed., Sinauer, 1988; **22.24**: After Gilbert, S., *Developmental Biology*, 2nd ed., Sinauer, 1988; **22.25a**: Runk/Schoenberger, from Grant Heilman; **22.25b**: David Scharf/Peter Arnold, Inc.; **22.26**: After Gilbert, S., *Developmental Biology*, 2nd ed., Sinauer, 1988

Chapter 23

Openers: (left) Richard H. Gross; (above) A. J. Copley/Visuals Unlimited; **Figure 23.6a**: Courtesy of Mark McMenamin, from *Trends in Ecology and Evolution*, Vol. 3, No. 8, p. 207; **23.6b**: William E. Ferguson; **23.6c**: Courtesy of S. Conway Morris, from Boardman et al., *Fossil Invertebrates*, Blackwell Scientific, 1987, reprinted from *Philosophical Transactions of the Royal Society of London*, Vol. 285 (1979):277–284; **23.6d&e**: From Boardman et al., *Fossil Invertebrates*, 1987. Reprinted by permission of Blackwell Scientific Publications, Inc.; **23.6f**: From McMenamin, M. A. S., "The emergence of animals," *Scientific American* 256(4):94–102, 1987; **23.6g&h**: Redrawn from Gould, S. J., *Wonderful Life: The Burgess Shale and the Nature of History*, W. W. Norton, 1989; **23.6i&j**: Redrawn from Moore, R. C. (ed.), *Treatise on Invertebrate Paleontology*, Vols. 0, 1, and 5, 1(2). Geol. Soc. America and Univ. of Kansas Press, 1959 and 1967; **23.7**: Modified from Erwin, D. H. et al., "A comparative study of diversification events: the early Paleozoic versus the Mesozoic," *Evolution* 41(6):1177–1186; **23.8**: Modified and redrawn from Leadbeater, B. S. C., "Life history and ultrastructure of a new marine species of Proterospongia," *Journ. Mar. Biol. Assoc. U.K.*, 63:135–160

Chapter 24

Openers: (left) M. I. Walker/Photo Researchers, Inc.; (above) Eric Gravé/Photo Researchers, Inc.; **Figure 24.1**: (a) from Biophoto Associates; (b–e) from Jahn, T. L. and E. C. Bovee,

"Motile behavior of protozoa," in Chen, T. (ed.), *Research in Protozoology*, Pergamon Press, 1967; **24.2**: Klebs, G. "Über die Organisation einiger Flagellaten-Gruppe und ihre Beziehungen zu Algen and Infusorien," Untersuchungen aus dem botanisches Institut zu Tübingen, 1:233–362, 1883. **24.3a**: Modified after Chen; **24.3b**: After Pennack; **24.3c**: Biophoto Associates; **24.3d**: M. I. Walker/Photo Researchers, Inc.; **24.4**: (a) from Farmer, J. N., *The Protozoa*, C. V. Mosby, 1980; (b) from Sleigh, M. A., *The Biology of Protozoa*, 5–12g p. 115. 1989. By permission of Edward Arnold Publishers; **24.5a**: Courtesy of S. S. Hendrix; **24.5c**: From Sleigh, M. A., *The Biology of Protozoa*, 5–15c p. 119. 1989. By permission of Edward Arnold Publishers; **24.6**: Redrawn from Ratcliffe, 1927; **24.7a**: Biophoto Associates; **24.7b**: Modified from Deflandre; **24.7c**: Dr. C. G. Ogden, Natural History Museum, London; **24.8**: (a–c) based on micrographs by J. W. Murray (d) from Moore, R. C. (ed.), *Treatise on Invertebrate Paleontology* Vol. C(2), Geological Society of America and Univ. of Kansas Press, 1964; **24.9a**: After Doflein, from Tregonboff; **24.9b**: L. E. Roth, Univ. of Tennessee/Biological Photo Service [518-017#4]; **24.10**: (a) from Farmer, J. N., *The Protozoa*, C. V. Mosby, 1980; (b) after Haekel; **24.11**: From Farmer, J. N., *The Protozoa*, C. V. Mosby, 1980; **24.12**: Redrawn and modified from Blacklock and Southwell; **24.13**: From Sleigh, M. A., *Protozoa and Other Protists*, 1–5c p. 9. 1989. By permission of Edward Arnold Publishers; **24.14**: (a) and (c) after Fauré-Fremiet, from Corliss, J. O., *The Ciliated Protozoa*, Pergamon Press, 1961; **24.15**: After Ehret and Powers, from Corliss, J. O., *The Ciliated Protozoa*, Pergamon Press, 1961; **24.16**: (b) after Pierson, from Kudo; (c) after Stein, from Grell, K. G., *Protozoology*, Springer-Verlag, 1973; (d) after Sleigh, M. A., *The Biology of Protozoa*, 7–16a p. 211, 1989. By permission of Edward Arnold Publishers; (e) after Corliss; **Connections 24.1**: Courtesy of Tom Fenchel; **24.17**: Adapted from Fenchel, T., "Protozoa filter feeding," *Prog. Protistology*, 1986; **24.20**: Adapted from Laybourn-Parry, J., *A Functional Biology of Free-Living Protozoa*, Univ. of California Press, 1984; **24.21**: Modified from Curds, C. R., "An ecological study of ciliated protozoa in activated sludge," *Oikos* 15(1986); **24.22**: From Sleigh, M. A., *Protozoa and Other Protists*, 2–4b p. 21, 1989. By permission of Edward Arnold Publishers, 1989; **24.23**: Modified after Clakins, from Wichterman

Chapter 25

Openers: (left) J. Mathias/Peter Arnold, Inc.; (above) Jeff Rotman/Peter Arnold, Inc.; **Figure 25.1a**: Modified from Buchsbaum; **25.2**: (a) redrawn from Hyman and other sources; (b) after Minchin, from Jones; (c) after Borojevic, from Bergquist, P. R., *Sponges*, Hutchinson, 1978; **25.3**: Courtesy of the General Biology Supply House, Inc.; **25.4a**: Modified after Kilian; **25.4b**: From Rutzler, K. and Rieger, G., 1973, "Sponge burrowing: fine structure of Cliona lampa penetrating calcareous substrata," *Mar. Biol.*, 21(2):144–162; **25.6**: From Weissenfels, N., 1982: Bau and Funktion des Susswasserschwamms Ephydatia fluviatilis IX. Rasterelektronen-mikroskopische Histologie und Cytologie. *Zoomorphology* 100:75–87; **25.8**: After Evans; **Connections 25.1a&b**: From Rutzler, K. and Rieger, G., 1973, "Sponge burrowing: fine structure of Cliona lampa

penetrating calcareous substrata," *Mar. Biol.*, 21:144–162; **25.10**: Courtesy of the American Museum of Natural History; **25.11**: Rassat, J. and A. Ruthmann "*Trichoplax adhaerens* F. E. Schulze (Placozoa) in the Scanning Electron Microscope," *Zoomorphologie* 93:59–72, Springer-Verlag, 1979

Chapter 26

Openers: (left) Gary Milburn/Tom Stack & Associates; (above) William Curtsinger/Photo Researchers, Inc.; **Figure 26.2**: After Gelei; **26.3c**: Westfall, J., S. Yamataka, and P. Enos, *20th Annual Proc. Electron Micro., Soc. AM.*, 1971; **26.4a**: Reprinted with permission from: Zoologica Scripta, 11(4): 227–41, Ostman, Nematocysts and Taxonomy in Laomedea, Gonothyrea, and Obelia, 1982; **26.4b**: From Tardent, P., and T. Holstein, "Morphology and morphodynamics of the stenotele nematocysts of *Hydra attenuata*." *Cell Tissue Res.* 224:269–290, 1982; **26.6**: From Mackie, G. O. and L. M. Passano, "Epithelial conduction in hydromedusae," *Journal of General Physiology* 52:600, 1968; **26.7**: Based on a figure from Hyman, L. H., *The Invertebrates*, Vol. I, McGraw-Hill, 1940; **26.8a**: R. DeGoursey/Visuals Unlimited; **26.8b**: Phil Degginger; **26.10**: (a) modified after Naumov; (b) after Hyman, L. H., *The Invertebrates*, Vol. I, McGraw-Hill, 1940; **26.11a**: Runk/Schoenberger, from Grant Heilman; **26.11b**: After Lane; **26.12a**: Redrawn from Mayer; **26.12b**: C. E. Mills, Univ. of Washington/Biological Photo Service; **26.13**: From Bayer, F. and Owre, H. B., *The Free-Living Lower Invertebrates*, Macmillan, 1968; **26.14a**: Animals Animals © 1991 Zig Leszczynski; **26.14b**: Fred Bavendam/Peter Arnold, Inc.; **26.14c**: After Mayer; **26.15a**: By J. H. Walker. In Rees, W. J. (ed.), *The Cnidaria and Their Evolution*, Zoological Society of London, 1966; **26.16a**: Fred Bavendam/Peter Arnold, Inc.; **26.20a&b**: Neil G. McDaniel/Photo Researchers, Inc.; **26.21a**: After Hyman, L. H., *The Invertebrates*, Vol. I, McGraw-Hill, 1940; **26.22b**: Betty M. Barnes; **26.22c**: W. Ober/Visuals Unlimited; **26.22d**: Charles Seaborn; **26.23**: Modified and redrawn from Bayer, F. M., 1956, *Octocorallia*. In Moore, R. C., *Treatise on Invertebrate Paleontology*, Courtesy of Geological Society of America and Univ. of Kansas; **Connections 26.1a&b**: Courtesy of Clay Cook; **26.24a**: Pamela Roe; **26.24c**: Fred Bavendam/Peter Arnold, Inc.; **26.25b**: Douglas Faulkner/Photo Researchers, Inc.; **26.25c**: David Hall/ Photo Researchers, Inc.; **26.28**: C. E. Mills, Univ. of Washington/Biological Photo Service

Chapter 27

Openers: (left) Dwight R. Kuhn; (above) CBS ©/Visuals Unlimited; **Figure 27.1a**: Stan Elems/Visuals Unlimited; **27.1d**: Drawn from a photograph by D. P. Wilson; **27.1e**: From Rieger, R. and J. Ott, *Vie et Milieu*, Supplement 22, 1971; **27.1f**: After Steinmann and Bresslau; **27.3**: L. S. Dembo, *Confucian modes of Ezra Pound*. Copyright © 1963 The Regents of the University of California; **27.6b**: After Burr; **27.8**: (a) from Doe, D. A. *Zoomorphology* 101:39–59, 1984; (b) redrawn from Kepner, Carter, and Hess; **27.9**: After Threadgold, C., from Smyth, J. D., *Introduction to Animal Parisitology*, p. 154, figure 107, greatly modified, by permission of Hodder and Stoughton Ltd; **27.10**: Modified from Noble, E. R. and G. A. Noble, *Parasitology*, 5th ed., Lea and Febiger, 1982; **27.12**:

After Bychowsky, from Schmidt, G. D. and L. S. Roberts, *Foundations of Parasitology*, 4th ed., Times Mirror/Mosby College Publishing, 1989; **27.14:** (a) based on figures from Coe and Gibson; (b) redrawn from Pennack, R. W., *Fresh-Water Invertebrates of the United States*, 2nd ed., reprinted by permission of John Wiley and Sons, Inc, 1978; **27.15:** After Atkins

Chapter 28

Openers: (left) James Bell/Photo Researchers, Inc.; (above) Tom Branch/Photo Researchers, Inc.; **Figure 28.2:** Redrawn from Hyman; **28.3:** T. E. Adams/Visuals Unlimited; **28.4:** (a) redrawn from Meyers; (b) from Nogrady, T., *Synopsis and Classification of Living Organisms*, Vol. 1, McGraw-Hill, 1982; (c) after Hudson; **28.5a&b:** After Beauchamp; **Connections 28.1:** (a) E. R. Degginger; (b) from Greven, H., 1971: On the morphology of tardigrades: A stereoscan study of Macrobiotus hufelandi and Echiniscus testudo. *Forma et Functio*, 4:283–302; **28.6:** Modified after Zelinka; **28.7:** From Tyler, S. and Rieger, G. E., 1980: Adhesive organs of the Gastrotricha.I Dvo-gland organs. *Zoomorphologie*, 951:15; **28.8a:** L. S. Stepanowicz/Photo Researchers, Inc.; **28.8b:** After Hope; **28.9:** (a) after de Coninck; (b) and (c) after Thorne; **28.12:** From Schmidt, G. D. and Roberts, L. S., *Foundations of Parasitology*, 3rd ed., Times Mirror/Mosby College Publishing, St. Louis, 1985, p. 491; **28.13:** ScienceVU/Visuals Unlimited; **28.14:** After Chandler, A. C. and C. P. Read, *Introduction to Parasitology*, reprinted by permission of John Wiley & Sons, 1961; **28.15:** Based on Conway Morris, S., 1982; Kristensen, R. M., 1983; and others; **28.16:** From Sterrer, W., 1972, *Syst. Zool.* 21(2):151–173; **28.17:** R. Calentine/Visuals Unlimited; **28.18:** After Yamaguti, from Cheng, T. C., *The Biology of Animal Parasites*, W. B. Saunders, 1964; **28.19:** Redrawn after R. P. Higgins; **28.20:** From Kristensen, R. M., "Loricifera, a new phylum with Aschelminthes characters from the meiobenthos," *Z. f. Zool. Systematik U. Evolutionsforachung*, 21(3):163–180, 1983; **28.21:** After Theel

Chapter 29

Openers: (left) Ed Robinson/Tom Stack & Associates; (above) Richard Herrmann; **Figure 29.2d:** From Solem, A., *The Shell Makers: Introducing Mollusks*. Reprinted by permission of John Wiley & Sons, 1974; **29.3:** After Patten; **29.4:** (a), (b), and (d) adapted from Lemche and Wingstrand; **29.6:** Ray Coleman/Photo Researchers, Inc.; **29.7a:** After Borradaile and others; **29.9:** E. R. Degginger; **29.10:** Modified from Graham; **29.11:** Modified from Yonge; **29.14b:** After Yonge; **29.14c:** Brian Parker/Tom Stack & Associates; **29.15:** (a) and (b) from Miller, S. L., 1974, "The classification, taxonomic distribution, and evolution of locomotor types among prosobranch gastropods." *Proc. Malac. Soc. London*, 41:233–272; (c) modified after Denny, M. W., and after Trueman, E. R., and H. D. Jones; **Connections 29.1:** Adapted from Hartnoll, R. G. and J. R. Wright, "Foraging movements and homing in the Pimpet *Patella vulgata*," *Animal Behavior* 25:808, 1977; **29.16:** Adapted from Graham, Morton, and Yonge; **29.18a:** Modified from Hyman; **29.18b:** Fred Bavendam/Peter Arnold, Inc.; **29.18c:** E. R. Degginger; **29.18d:** Neil G. McDaniel/Photo

Researchers, Inc.; **29.18e:** Daniel W. Gotshall/Visuals Unlimited; **29.19a:** Modified from Hyman; **29.19b:** Animals Animals © 1991 Roger Archibald; **29.19c:** William J. Weber/Visuals Unlimited; **29.20:** Courtesy of David T. Barnes; **29.21a:** From Carriker, M. R., 1969, "Excavation of boreholes by the gastropod Urosalpinx" *AM 2001*, 9:917–933; **29.21b:** From Solem, A., 1974, *The Shell Makers: Introducing Mollusks*. Reprinted by permission of John Wiley and Sons, p. 163; **29.22a:** From Nielsen, C., "Observation on *Buccinum undatum* attacking bivalves and on prey responses, with short review on attack methods of other prosobranchs." *Ophelia* 13(1975):87–108; **29.22b:** George Whiteley/Photo Researchers, Inc.; **29.22c:** From Kohn, A. J. et al., 1972, *Science*, 176:49–51. Copyright 1972 by the American Association for the Advancement of Science; **29.22d:** Based on a photograph by Robert F. Sisson and Paul Zahr; **29.24a:** George Lower/National Audubon Society/Photo Researchers, Inc.; **29.24b:** Barbara J. Miller/Biological Photo Service; **29.25:** After Werner, from Raven; **29.28b:** After Kennedy, W. J., et al., "Environmental and biological controls on bivalve shell minerology." *Biological Review* 44:499–530, 1969; **Connections 29.2:** Adapted from Evans, J. W., "Tidal growth increments in the cockle *Clinocardium nuttalli*," *Science* 176: 416–417, 1972; **29.30a–c:** After Yonge; **29.32:** (a) and (b) after Atkins; (c) after Yonge; **29.33:** After Morton, J. E. *Molluscs*, Hutchinson University Library, 1967; **29.34a:** Modified after Trueman, E. R., "Bivalve Mollusks: Fluid Dynamics of Burrowing." *Science* 152:523. © 1966 by American Association for the Advancement of Science; **29.36:** Fred Bavendam/Peter Arnold, Inc.; **29.37a:** After Galtsoff; **29.38a&b:** Reprinted from "The bearing of the late Cambrian monoplacophoran genus Knightoconus upon the origin of cephalopoda," by Yochelson, E. L., R. H. Flower and G. F. Webers from LETHAIA 6:275–310 1973, by permission of Universitetsforlaget AS; **29.40a:** Animals Animals © 1991 W. Gregory Brown; **29.40b:** After Stenzel; **29.41a:** Animals Animals © 1991 G. I. Bernard/Oxford Scientific Films; **29.41b:** From Roper, C. E. F. and Brundage, W. L., 1972: Cirrate octopods with associated deep sea organisms: New biological data based on deep benthic photographs. *Smithsonian Contrib. Zool.*, 21:1–46; **Connections 29.3:** From Herring, P. J., "Luminescence in Cephalopods and fish," in Nixon, M. and Messenger, J. B. (eds.), *The Biology of Cephalopods*, Symp. Zool. Soc. London, 38, Academic Press, London, pp. 127–159, 1977; **29.43a:** Based on a photograph by Robert F. Sisson; **29.43b:** Robert F. Myers/Visuals Unlimited

Chapter 30

Openers: (left) Carl Roessler/Tom Stack & Associates; (above) Dave Woodward/Tom Stack & Associates; **Figure 30.1a:** After Brown, F. A., *Selected Invertebrate Types*, reprinted by permission of John Wiley and Sons, Inc., 1950; **30.2:** Based on a figure by A. Kaestner in *Invertebrate Zoology*, Vol. I, New York, *Interscience Publ.*, 1967. Reprinted by permission of John Wiley and Sons, Inc.; **30.3:** From Renaud, J. C., "A report on some polychaetous annelids from the Miami-Bimini area," *Am. Mus. Novit.* 1812:1–40, 1956. Courtesy of the American Museum of Natural History, New York, NY; **30.4a:** Larry Lipsky/Tom Stack & Associates; **30.4b:** After Mettam; **30.5a:**

E. R. Degginger; **30.5b**: After Oekelmann and Vahl; **30.6**: (a) modified after Brown, F. A., *Selected Invertebrate Types*, reprinted by permission of John Wiley and Sons, Inc., 1950; (c) from Wells, G. P., "The behavior of arenicola marina in sand and the role of spontaneous activity cycles," *Journal of Marine Biology Association* (U.K.) 28:465–478, 1949. By permission of Cambridge University Press; **30.10**: Greatly modified after MacGinitie; **30.11**: After Woodsworth, from Fauvel; **30.12**: All after Cazaux, C., 1967, *Vie et Milieu*, 18:559–571; **30.13a**: Courtesy of Globe Photos; **30.13b**: Betty M. Barnes; **30.15**: After Hesse; **30.16a**: After Grove, A. J. and L. F. Cowley, "On the reproductive processes of the brandling worm *Eisenia foetida*," *Quarterly Journal of Micro. Sci.* 70: 559–581, 1926, with permission of The Company of Biologists Ltd.; **30.17**: (a) and (b) from Sawyer, R. J., *North America Freshwater Leeches*, Univ. of Illinois Press, 1972; (c) adapted from Keegan et al.; **30.18**: (a) after Harding and Moore; (b) modified after Pfurtscheller; **Connections 30.1**: D. Foster/ WHOI/Visuals Unlimited; **30.19b**: Animals Animals © 1991 Kathie Atkinson/Oxford Scientific Films; **30.20**: After Yamaguchi

Chapter 31

Openers: (left) Daniel W. Gotshall/Visuals Unlimited; (above) William S. Ormerod, Jr./Visuals Unlimited; **Figure 31.1**: After Weber; **31.3**: After Welcott and Raymond; **31.4a**: Kenneth Lucas/Biological Photo Service; **31.5**: Barbara J. Miller/ Biological Photo Service; **31.6**: Frans Lanting/Minden Pictures; **31.7**: R. J. Erwin/Photo Researchers, Inc.; **31.8**: After Comstock; **31.10a**: Stephen Dalton/Photo Researchers, Inc.; **31.11a**: Charles W. Mann/Photo Researchers, Inc.; **31.11b&c**: After Baker and Wharton; **31.11d**: Biophoto Associates; **31.11e**: John Walsh/Science Photo Library/Photo Researchers, Inc.; **31.12**: (a) from Wilson, R. S., *American Zoologist*, 9:103–111, 1969; (b) based on a figure from Levi, H. W. and L. R. Levi, *A Guide to Spiders and Their Kin*, A Golden Nature Guide, Golden Press, 1968; **31.13a**: E. R. Degginger; **31.13b**: Larry Miller/Photo Researchers, Inc.; **31.13c**: Ed Reschke; **31.13d**: Rod Planck/Tom Stack & Associates; **31.14**: After Angermmann, H., *Zeitschrift für Tierpsychologie* 14:276–302, 1957; **31.15**: After Gerhardt; **31.16**: J. Alcock/Visuals Unlimited; **31.17**: Animals Animals © 1991 Doug Allan/ Oxford Scientific Films; **31.18a&b**: After Calman; **31.19a**: Minden Pictures/Frans Lanting; **31.19b**: Brian Parker/Tom Stack & Associates; **31.19c**: A. J. Copley/Visuals Unlimited; **31.20**: After Green; **31.21**: After Howes; **31.22b**: Barbara J. Miller/Biological Photo Service; **31.23**: From Glaessner, M. F., *Treatise on Invertebrate Paleontology* Part R4(2), courtesy of the Geological Society of America and Univ. of Kansas Press, 1969; **31.24b**: Joe McDonald/Tom Stack & Associates; **31.25a**: Daniel W. Gotshall/Visuals Unlimited; **31.25b**: From Costlow, J. D., and C. G. Bookhout, 1971, *Fourth European Marine Biology Symposium*, Cambridge University Press, p. 214; **31.26**: After Sars; **31.27a**: After VanNance, W. G., 1936, "The American land and freshwater isopod crustacea," *Bull. Amer. Mus. Nat. Hist.* 71–77; **31.27b**: After Gruner; **31.27c**: Richard Walters/Visuals Unlimited; **31.28a**: E. R. Degginger; **31.28b**: Dwight R. Kuhn; **31.29**: (a) modified from Coldwell, R. L., and H. Dingle, "Stomatopods," *Scientific American* 234(1):80–89,

1976; (b) after Calman; **31.30a**: Dr. William J. Jahoda/Photo Researchers, Inc.; **31.30b**: After Sars; **31.31a**: After Matthes; **31.32a**: From Gerritsen, J. and Porter, K. G., 1982: The role of surface chemistry in filter feeding by zooplankton, *Sci.* 216: 1225–1227; **31.32b**: From Gerritsen, J. and Porter, K. G., 1982: The role of surface chemistry in filter feeding by zooplankton. *Sci.* 216:1225–1227; **31.33**: (a) after Matthes; (b) after Griesbrecht; **31.34b**: Courtesy of J. E. Hazel and T. M. Cronin, from Boardman et al., *Fossil Invertebrates*, 1987; **31.34c**: After Claus; **31.35b**: After Broch, from Kaestner; **31.35c**: Francois Gohier/Photo Researchers, Inc.; **31.36a**: After Gruvel; **31.36b**: William S. Ormerod, Jr./Visuals Unlimited; **31.37**: F. Gohier/Photo Researchers, Inc.; **31.38**: From Walley, L. J., *Philosophical Transactions of the Royal Society of London* (Biol.), 256, 807:237–280, 1970; **31.39**: Dennis W. Williams

Chapter 32

Openers: (left) Skip Moody/M. L. Dembinsky, Jr., Photography Associates; (above) Skip Moody/M. L. Dembinsky, Jr., Photography Associates; **Figure 32.1a**: Dwight R. Kuhn; **32.1b**: William E. Ferguson; **32.1d**: Greatly modified from Snodgrass; **32.2a**: After Snodgrass; **32.2b**: John D. Cunningham/Visuals Unlimited; **32.2c**: After Snodgrass; **Connections 32.1**: Raphael Gaillarde/Gamma-Liaison; **32.3b**: Modified from Romoser; **32.4**: (a) after Rolleston; (b) after Snodgrass; **32.5**: Based on text and diagrams of Evans, H. E., *Insect Biology: A Textbook of Entomology*, Addison-Wesley, 1984; **32.6**: Scott Camazine/ Photo Researchers, Inc.; **32.7**: After Ross, H. H. *A Textbook of Entomology*, 3rd ed., reprinted by permission of John Wiley and Sons, Inc., 1965; **32.8a**: John Gerlach/M. L. Dembinsky, Jr., Photography Associates; **32.8b**: Animals Animals © 1991 Oxford Scientific Films/G. I. Bernard; **32.8c**: Animals Animals © 1991 Jack Wilburn; **32.9a**: Peter J. Bryant/Biological Photo Service; **32.9b**: E. R. Degginger; **32.10**: After Ross, H. H., *A Textbook of Entomology*, 3rd ed., reprinted by permission of John Wiley and Sons, Inc., 1965; **32.11**: After Ross, H. H., *A Textbook of Entomology*, 3rd ed., reprinted by permission of John Wiley and Sons, Inc., 1965; **32.12**: (a) and (g) after Snodgrass; (b) and (c) after Waldbauer; (d) after Hickmann; (e) after Poisson; (f) after Kullenberg; **32.13c**: William E. Ferguson; **32.13d**: Rod Planck/M. L. Dembinsky, Jr., Photography Associates; **32.14**: After Lutz; **32.15**: After Von Frisch; **32.16**: After Snodgrass; **32.17a**: Animals Animals © 1991 Michael Fogden; **32.17b**: After Marcus

Chapter 33

Openers: (left) Daniel W. Gotshall/Visuals Unlimited; (above) M. I. Walker/Photo Researchers, Inc.; **Figure 33.1**: Modified from Boardman et al., *Fossil Invertebrates*, Blackwell Scientific, 1987; **33.2a**: Modified from Marcus; **33.2b**: Fred Bavendam/Peter Arnold, Inc.; **33.3a**: Fred Bavendam/Peter Arnold, Inc.; **33.3b**: Kjell B. Sandved/Visuals Unlimited; **33.3c**: After Maturo; **33.4**: From Cook, P., "Settlement and early colony development in some cheilostomata," in Larwood, G. P. (ed.), *Living and Fossil Bryozoa*, Academic Press, 1973; **33.5**: Gary R. Robinson/Visuals Unlimited; **33.6**: (a) from Williams, A. and A. J. Rowell, in Moore, R. C. (ed.), *Treatise on*

Invertebrate Paleontology, Geol. Soc. of Amer. and Univ. of Kansas Press, 1965; (b) modified from Francois; (c) from Rudwick, M. J. R., *Living and Fossil Brachiopods*, Hutchinson Univ. Library, 1970; **33.7**: After Ehlers

Chapter 34

Openers: (left) Phil Degginger; (above) Minden Pictures/Frans Lanting; **Figure 34.1**: From Ubaghs, G., in Moore, R. C. (ed.), *"Treatise on Invertebrate Paleontology."* Pt. S, Vol. 1, Courtesy of the Geological Society and the University of Kansas Press, 1967; **34.2a**: Chuck Davis; **34.2b**: Animals Animals © 1991 F. Roessler; **34.4**: (a) after Fisher; (c) after Hyman, L. H., *The Invertebrates*, Vol. IV, McGraw-Hill, 1955; **34.7a**: Charles Seaborn; **34.7b**: H. Wes Pratt/Biological Photo Service; **34.9**: After Streklov; **34.10a**: Norbert Wu/Peter Arnold, Inc.; **34.10b**: Brian Parker/Tom Stack & Associates; **34.10c**: E. R. Degginger; **34.10d**: Animals Animals © 1991 Steve Earley; **34.11**: After Ried, W. M., in Brown, F. A., *Selected Invertebrate Types*, reprinted by permission of John Wiley and Sons, Inc., 1950; **34.12**: From DeRidder, C. and J. M. Lawrence, "Food and feeding mechanisms: Echinoides." In Jangoux, M. and Lawrence, J. M. (eds.), *Echinoderm Nutrition*, A. A. Balkema, Old Post Road, Brookfield, VT 05036, 1982; **34.13**: After Nichols, in part; **34.14**: Modified after Petrunkevitch, from Reid; **34.19**: Ed Robinson/Tom Stack & Associates; **34.20**: (a) after Clark; (b) greatly modified from Carpenter and Hyman; **34.21**: Jeff Rotman/Peter Arnold, Inc.; **34.22**: From Byrne, M. and A. R. Fontaine, 1980, "The feeding behavior of Florometra serratissima," *Can. Jour. Zool.*, 59:11–18; **34.23**: From Baker, A. N., V. E. Rowe, and H. E. S. Clark, "A new class of echinodermata from New Zealand," *Nature* 321, 1986; **34.24**: (a) from Kesling, R. V., in Moore, R. C. (ed.), *"Treatise on Invertebrate Paleontology,"* Part S., Vol. 1, Courtesy of the Geological Society and the University of Kansas Press, 1967; (b) after Hyman, L. H., *The Invertebrates*, Vol. IV, McGraw-Hill, 1955; **34.25a**: Animals Animals © 1991 Peter Parks/Oxford Scientific Films; **34.25b**: After Ritter-Zahoney

Chapter 35

Openers: Dave Woodward/Tom Stack & Associates; **Figure 35.1a**: After Stiasny; **35.3**: From Lester, S. M.: "Cephalodiscus sp.: Observations of functional morphology, behavior and occurrence in shallow water around Bermuda" *Marine Biology* 85:263–268, 1985; **35.5a**: F. Stuart Westmorland/Tom Stack & Associates; **35.5b**: C. R. Wyttenbach, Univ. of Kansas/Biological Photo Service; **35.7**: (a) after Yonge; (b) after Delage and Hervuard; **35.8**: (a) modified after Uljanin and Barrios; (b) after Alldredge, A. *Sci. Am.* 235(1)95–102, 1976; **35.9a**: Runk/Schoenberger, from Grant Heilman; **35.12**: After Romer; **35.13**: Redrawn from Gould, S. J., *Wonderful Life: The Burgess Shale and the Nature of History*, W. W. Norton, 1989

Chapter 36

Openers: (left) David Hall/Photo Researchers, Inc.; (above) Fred McConnaughey/Photo Researchers, Inc.; **Figure 36.2**: (a) and (b) from McFarland, W. N., H. Dough, T. J. Code, and J. B.

Heiser, *Vertebrate Life*, 3rd ed., Macmillan, 1985; (c) after Alexander; **36.3**: After Jordan; **36.5a&b**: Tom Stack/Tom Stack & Associates; **36.8**: After Jarvik; **36.9**: After Watson; **36.11**: After Dean; **36.13a**: Charles Seaborn; **36.13b**: Alan Desbonnet/Visuals Unlimited; **36.13c**: Kenneth Lucas/ Biological Photo Service; **36.13d**: Animals Animals © 1991 Zig Leszczynski; **36.13e**: Tom McHugh, Steinhart Aquarium/Photo Researchers, Inc.; **36.13f**: Rod Allin/Tom Stack & Associates; **36.15**: After Dean; **36.16**: Science VU/Visuals Unlimited; **36.17**: (a) from Romer, A. S. and T. S. Parsons, *The Vertebrate Body*, 6th ed., Saunders College Publishing, 1986; (b) from Carroll, R. L., *Vertebrate Paleontology and Evolution*, W. H. Freeman, 1988; **36.18a**: Photo Researchers, Inc./Tom McHugh; **36.18b**: Kenneth Lucas/Biological Photo Service; **36.19a**: E. R. Degginger; **36.19b**: Carlton Ray/Photo Researchers, Inc.; **36.21**: (a) after Alexander; (b) after Lum; **36.22a–c**: After Norman; **36.22d**: Runk/Schoenberger, from Grant Heilman; **36.23a**: Alan Pitcairn, from Grant Heilman; **36.23b&c**: Modified after Carey; **36.24a**: Animals Animals © 1991 Zig Leszczynski; **36.24b**: Chuck Davis; **36.24c**: Runk/ Schoenberger, from Grant Heilman; **36.25a**: Peter Scoones/ Seaphot, Ltd.; **36.25b**: Tom McHugh, Steinhart Aquarium/ Photo Researchers, Inc.; **36.25c**: Patrice/Visuals Unlimited

Chapter 37

Openers: Stephen G. Maka; **Figure 37.3**: (a) after Marshall and Hughes; (b) after Romer; **37.6a**: Hans Pfletschinger/Peter Arnold, Inc.; **37.6b**: Animals Animals © 1991 G. I. Bernard/ Oxford Scientific Films; **37.6c–f**: Hans Pfletschinger/Peter Arnold, Inc.; **37.7**: (a) from Carroll, R. L., *Vertebrate Paleontology and Evolution*, W. H. Freeman, 1988; (b) Neg. no. 322872 (photo by Logan): Courtesy Department of Library Services, American Museum of Natural History; **37.8**: Courtesy of the Ohio Department of Natural Resources; **37.9a**: Jane Burton/Bruce Coleman, Inc.; **37.9b**: E. R. Degginger; **37.9c**: Courtesy of the Ohio Department of Natural Resources; **37.9d**: David M. Dennis/Tom Stack & Associates; **37.10a**: Doug Wechsler; **37.10b**: Courtesy of Ohio Department of Natural Resources; **37.11**: David M. Dennis/Tom Stack & Associates; **37.13**: Courtesy of Mr. Earle R. Edminston; **37.14a**: Courtesy of the Ohio Department of Natural Resources; **37.14b**: E. R. Degginger; **37.14c**: Stephen G. Maka; **37.14d**: Courtesy of the Ohio Department of Natural Resources; **37.15a&b**: Animals Animals © 1991 Michael Fogden; **37.15c**: E. S. Ross, California Academy of Sciences; **37.15d**: Science VU/Visuals Unlimited; **37.15e**: Courtesy of B. R. Zug, Smithsonian Institution; photograph by K. Miyata, National Museum of Natural History; **37.16a**: Kenneth Lucas/Biological Photo Service; **37.16b**: Animals Animals © 1991 Michael Fogden

Chapter 38

Openers: Francois Gohier/Photo Researchers, Inc.; **Figure 38.3**: Modified from Halliday, T. and K. Adler, *The Encyclopedia of Reptiles and Amphibians*, Facts on File, Inc., by permission of Andromeda Oxford Limited, 1987; **38.4**: John Cancalosi/Peter Arnold, Inc.; **38.5**: From Carroll, R. L., *Vertebrate Paleontology and Evolution*, W. H. Freeman, 1988, after Clark and Carroll; **38.6**: (a–c) after Romer; (d) modified from Carroll, R. L., *Vertebrate Paleontology and Evolution*, W. H. Freeman, 1988; **38.7**: After Gregory; **38.8a**: Tom McHugh/Photo Researchers, Inc.; **38.8b**: Runk/Schoenberger, from Grant Heilman; **38.8c**: Frans Lanting/Minden Pictures; **38.8d**: Frans Lanting/Minden Pictures; **38.9**: Courtesy of R. Goellner, St. Louis Zoological Park; **38.10a**: E. R. Degginger; **38.10b**: Stephen Dalton/Photo Researchers, Inc.; **38.10c**: Courtesy of C. Gans, University of Michigan, Ann Arbor; **38.10d**: Kenneth Lucas/Biological Photo Service; **38.10e**: Stan Osolinski/M. L. Dembinsky, Jr., Photography Associates; **38.10f**: Stephen J. Kraseman/Photo Researchers, Inc.; **38.10g**: Courtesy of E. R. Ross, California Academy of Science; **38.11**: Jany Sauvanet/Photo Researchers, Inc.; **38.12**: Modified from Bellairs, A., *The Life of Reptiles*, Vol. 1, Weidenfeld & Nicolson, 1969; **38.13**: Tom McHugh/Photo Researchers, Inc.; **38.15**: Animals Animals © 1991 Stephen Dalton/Oxford Scientific Films; **38.16**: (a) after Fenton and Fenton; (b) after Spinar and Burian; **38.17a**: John Kaprielian/Photo Researchers, Inc.; **38.17b**: Stan Osolinski/M. L. Dembinsky, Jr., Photography Associates; **38.18**: (a) modified after Romer, A. S., *Osteology of Reptiles*, Univ. of Chicago Press, 1956; (b–d) from Bakker, R. T., *The Dinosaur Heresies*, William Morrow, 1986; **38.19**: Trans. no. 1000 (Painting by Constantin Astori) Courtesy Dept. of Library Services, American Museum of Natural History

Chapter 39

Openers: A. Carey/VIREO; **Figure 39.2**: From Hertel, H. *Structure, Form and Function*, Reinhold Publishing Co., 1963; **39.3b**: Fritz Polking/M. L. Dembinsky, Jr., Photography Associates; **39.3c**: Stan Osolinski/M. L. Dembinsky, Jr., Photography Associates; **39.3d**: From "The Soaring Flight of Birds" by Cone, C. D., Jr., © by Scientific American, Inc. All rights reserved; **39.4a**: Modified from Rayner, J. M. V., in Elder, H. Y., and E. R. Trueman (eds.), *Aspects of Animal Locomotion*, Cambridge Univ. Press, 1980; **39.4b&c**: Animals Animals © 1991 Stephen Dalton/Oxford Scientific Films; **Focus 39.1**: Modified from Rayner, J. M. V., "Voracity and Animal Flight," in Elden, H. Y., and E. R. Trueman (eds.), *Aspects of Animal Movement*, Cambridge, England: Cambridge Univ. Press, 1980; **39.5b**: Steve Maslowski/Photo Researchers, Inc.; **39.6a**: Modified after Young; **39.6b**: From Welty, J. C., *The Life of Birds*, 4th ed., Saunders College Publishing, 1988; **39.6c**: From Welty; **39.7**: (a) modified after Heilman; (b) from D'Arcy Thompson; **39.8**: After Storer; **39.9**: After Walls; **39.10**: From Welty, J. C., *The Life of Birds*, 4th ed., Saunders College Publishing, 1988; **39.12b**: After Schmidt-Nielsen; **39.12c**: Photography by H. R. Duncker, from Schmidt-Nielsen, K., *How Birds Breathe*. Copyright by Scientific American, Inc. All Rights Reserved; **39.13a**: G. R. Roberts; **39.13b**: P. Davey/VIREO; **39.14**: After Lincoln; **39.15**: After Lincoln; **39.17**: "Bird Flight: How Did It Begin?" by John H. Ostrum. American Scientist 67:46–56, 1979; **39.18**: From Feduccia, A., *The Age of Birds*, Cambridge, Harvard Univ. Press; **39.19a**: P. Davey/VIREO; **39.19b**: B. Chudleigh/VIREO; **39.20a**: D. & M. Zimmerman/VIREO; **39.20b**: Carl R. Sams, II/ M. L. Dembinsky, Jr., Photography Associates; **39.20c**: Gunter Ziesler/Peter Arnold, Inc.; **39.21a**: Animals Animals © 1991 Robert A. Lubeck; **39.21b**: A. Forbes-Watson/VIREO; **39.21c**: A. Morris/VIREO; **39.22a**: J. H. Dick/VIREO; **39.22b**: M. P. Kahl/ VIREO; **39.22c**: From Carroll, R. L., *Vertebrate Paleontology and Evolution*, W. H. Freeman, 1988; **39.23**: Skip Moody/M. L. Dembinsky, Jr., Photography Associates; **39.24**: A. Cruickshank/VIREO; **39.25**: Frans Lanting/Minden Pictures; **39.26**: T. Fitzharris/VIREO; **39.27**: O. S. Pettingill, Jr./VIREO

Chapter 40

Openers: Grant Heilman; **Figure 40.1**: Frans Lanting/Minden Pictures; **40.2**: After Schmidt-Nielsen, K., *Animal Physiology*, 2nd ed., Cambridge Univ. Press, 1979, after C. R. Taylor; **40.3**: (d) and (e) from Stahl, B. J., *Vertebrate History: Problems in Evolution*, McGraw-Hill, 1974, after G. G. Simpson; **40.4**: After Sisson and Grossman; **40.5a**: Trans. no. 2420 (Painting by C. R. Knight): Courtesy of Dept. of Library Services, American Museum of Natural History; **40.5b**: Trans. no. 203 (Painting by John C. Germann): Courtesy Dept. of Library Services, American Museum of Natural History; **40.6**: From Jenkins, F. A., Jr. and F. R. Parrington, "The Postcranial skeleton of the Triassic mammals Eozostrodon, Magazostrodon and Erythrotherium," *Philosophical Transactions of the Royal Society of London* B273:387–431; **40.8a**: Dave Watts/Tom Stack & Associates; **40.8b**: Tom McHugh/Photo Researchers, Inc.; **40.9a**: Animals Animals © 1991 Joe and Carol McDonald; **40.9b**: Animals Animals © 1991 Breck P. Kent; **40.10**: Tom McHugh/Photo Researchers, Inc.; **40.11a**: Chip Isenhart/Tom Stack & Associates; **40.11b**: E. R. Degginger; **40.12**: After M. J. Novacek; **40.13a**: Warren Garst/Tom Stack & Associates; **40.13b**: Jany Sauvanet/Photo Researchers, Inc.; **40.13c**: E. P. I. Nancy Adams/Tom Stack & Associates; **40.13d**: Gary Milburn/ Tom Stack & Associates; **40.14a**: Dwight R. Kuhn; **40.14b**: Rod Planck/Tom Stack & Associates; **40.15a**: Steve Kaufman/Peter Arnold, Inc.; **40.15b**: Carl R. Sams II/M. L. Dembinsky, Jr., Photography Associates; **40.15c**: Carleton Ray/Photo Researchers, Inc.; **40.15d**: George H. Harrison, from Grant Heilman; **40.17**: From Vaughan, T. A., *Mammalogy*, 2nd ed., Saunders College Publishing, 1978; **40.18a**: Animals Animals © 1991 Stephen Dalton; **40.18b**: Gerald Cubitt/Bruce Coleman, Ltd.; **40.19a&b**: After Vaughan, T. A., *Mammalogy*, 2nd ed., Saunders College Publishing, 1978; **40.20a**: Stephen J. Lang/Visuals Unlimited; **40.20b**: John D. Cunningham/Visuals Unlimited; **40.20c**: Thomas Kitchin/Tom Stack & Associates; **40.20d**: Tom McHugh/Photo Researchers, Inc.; **40.21a**: Fritz Polking/M. L. Dembinsky, Jr., Photography Associates; **40.21b**: Frans Lanting/Minden Pictures; **40.21d**: Fritz Polking/M. L. Dembinsky, Jr., Photography Associates; **40.23a**: E. R. Degginger; **40.23b**: C. Allan Morgan/Peter Arnold, Inc.; **40.24a**: Stan Osolinski/M. L. Dembinsky, Jr., Photography Associates; **40.24b**: Larry Lipsky/Tom Stack & Associates

Chapter 41

Openers: (left) Francois Gohier/Photo Researchers, Inc.; (above) Steve C. Kaufmaan/Peter Arnold, Inc.; **Figure 41.2a**: Warren Garst/Tom Stack & Associates; **41.2b**: Frans Lanting/ Minden Pictures; **41.2c**: Gary Milburn/Tom Stack & Associates; **41.2d**: Gary Milburn/Tom Stack & Associates; **41.3a**: Jack Swenson/Tom Stack & Associates; **41.3b**: E. Hanumantha Rao A.F.I.A.P./Photo Researchers, Inc.; **41.4a**: Tom McHugh/Photo Researchers, Inc.; **41.4b**: Brian Parker/Tom Stack & Associates;

INDEX

Page numbers in *italics* indicate illustrations.

A

Aardvark, 920, 931
Abalones, 665, 666, *666*, 688
Abomasum, 244
Absorption, 239–241, *240*, 254
Abyla, 617
Abyssal zone, 992, *993*
Acantharea, 581
Acanthaster, 779, *779*
Acanthocephala, 653, *653*
Acanthodii, 823, 824, 838
Acanthometra, *571*, 581
Acanthopterygii, 839
Acari, 721, 745
Acariformes, 745
Acceleration, 413
Accommodation, 426
Acetabularia, experiments in nuclear
    control, 110, *111*
Acetate, 266
Acetyl-glucosamine, 39
Acetylcholinase, muscle contraction, 218
Acetylcholine, 379, 397
   muscle contraction, 218
Acetylcholinesterase, 379
Acetylcoenzyme, 266
Aciculum, 695
Acid rain, 1007
Acmaea, 688
Acoela, 637
Acoelomate body, structure, 624
Acontium, 610
Acorn barnacles, 741
Acorn worms, 798, *799*, 811
Acquired Immune Deficiency Syndrome
    (AIDS), 324
Acromegaly, *446*
Acron, 212
Acropora, 618
Acrosomal reaction, 491
Acrosome, 486
Actin, amoeboid movement 223
Actin filaments, 79
Actiniaria, 618
Actinopoda, 581

Actinopods, 569
Actinopterygii, 828, 830–837, 838
   mouth evolution, *833*
Actinula larva, 603, *604*
Action potential, 374, *375*, *376*
   muscle contraction, 218
Activation energy, 45
Active site, 44
Active transport, 71–72
Adaptation, 17
Adaptive diversity, concept, 554
Adaptive radiation, 178
Adenine, 46
Adenohypophysis, *444*, 445
Adenosine monophosphate (AMP),
    49, 436
Adenosine triphosphate (ATP), 49
   in cellular respiration, 262
Adenylate cyclase, 436
Adhesive glands, Aschelminths, 642
Adipose tissue, 202
Adrenal gland, 450–452, *451*
Aedeagus, 763
Aelosoma, 711
Aeolidia, 688
Aerobic respiration, 265
Aestivation, 366
African sleeping sickness, 566, *567*
After discharge, 381, *383*
Agalma, 617
Aggression, 471
Aggressive mimicry, 469
Aglaura, *604*, 617
Agnatha, 819–822, 838
   characteristics, 819
   ostracoderms, 819–820
Agonistic behavior, *471*
   and dominance, 471
Agriolimax, *670*
AIDS. *See* Acquired Immune Deficiency
    Syndrome
Air breathing, evolution in vertebrates,
    828
Air sac, 278
Airfoils, 229

Albatrosses, 903, *903*, 905
Alciopidae, 701
Alcyonacea, 618
Aldosterone, 361, 451
Algal symbiosis, 612
Alkaptonuria, 115, *115*, *116*
Allantois, 524
Allele frequencies, 155
Alleles, 91
   multiple, 150
Allelic exclusion, 330
Allergies, 338
Allochronic isolation, *171*
Allometric growth rates, 191, *192*
Allopatric speciation, 172
Altruistic behavior, 477
ALU sequences, 139
Alula, 882
Alveolus
   ciliates, 576
   lung, 282, *282*, *283*
Ambulacral groove, 779
Amino acids, 41, *42*, *43*
   as cellular fuel, 271
   essential, 255
   origin of life, 57
Ammocoetes larva, 822
Ammonia, 349, *349*
Ammonites, *18*
Ammonoidea, 689
Amnion, 524
Amniotes, 524, 862
   phylogeny, *862*
Amoeba, 581
Amoebas, 567–568, *569*
Amoebic dysentery, 566, 568
Amoeboid movement, 223, *223*
AMP. *See* Adenosine monophosphate
Amphibia, 843–859
   adaptations, 845–849
   characterisitcs, 843
   classification, 858
   evolution, 849
   gas exchange, 279
   phylogeny, *844*

Amphibia *(continued)*
    transition from water to land, 844
Amphiblastula larva, *592*
Amphioxus, 804–807, *806*, 811
Amphipoda, 735, *735*, 745
Amphisbaenia, 870, *870*, 877
Amphitrite, 697, 711
Ampulla, 780
Amylase, 237
Anaerobic glycolysis, 265
Anagenesis, 169
Analogy, 544
Anaphase
    meiosis, 101–102
    mitosis, 100
Anaphylaxis, 338
Anapsida, 866, 877
Anatomical planes, 186
Ancylostoma, 650
Andracantha, 654
Androctonus, 745
Androgens, 451, 501
Angiotensin, 361
Angler fish, *836*
Animal kingdom, origin, 547–552
Animal phyla, synopsis, 555–557
Animalia, 548
Annelida, 693–612
    characteristics, 693
    evolutionary relationships, 709
    Hirudinea, 707–709
    locomotion, 226
    metamerism and locomotion, 694
    Oligochaeta, 702–707
    Polychaeta, 695–702
Anomalocaris, *551*
Anseriformes, 905
Antagonistic muscles, 215
Anteaters, *921*, 931
Antedon, 794
Antennae
    crustaceans, 729
    uniramians, 750
Antennal glands, 351
Anterior end, 186
Anthozoa, 608–614, 618
    octocorals, 612–614
    sea anemones, 609–611
Anthropoidea, 938, 951
Anti-Mullerian Duct Factor, 535
Antibody, 325, *326*
    diversity, 328–333
Antibody titer, 325
Anticodon, 128
Antigen, 325, *326*
Antigen-antibody complex, 325, *333*
Antigenic drift, 343
Antigenic shift, 343
Antigenicity, 326
Antipathes, 618

Antlers, 928, *928*
Antlions, 766
Ants, 766
    alarm pattern, *473*
    leaf-cutting, 469
Anura, 849, 854–858, 859
    adaptations, 850–853
    reproduction, 857
Anus, 251
Aorta
    dorsal, 522
    ventral, 522
Aortic arches, 304, 522
    evolution, *306*
Aortic bodies, 284
Apes, 931
    evolution, 943, *943*
    femur, *942*
    skeleton, *940*
    teeth, *941*
Aphids, 763, 765
Aphotic zone, 992, *993*
Apicomplexa, 571, *572*, 581
Apis, 761
Aplacophora, 688
Aplysia, *669*, 688
Apnea, 279
Apocrine sweat glands, 204
Apodiformes, 905
Aposematic coloration, 470, *471*
Appendicular propulsion, 228
Appendicular skeleton, 214
Appendicularians, 804
Apterygiformes, 904
Apterygota, 755, 765
Aquatic ecosystems, 992–995
Arachnida, 720–728
Araneae, 745
Araneus, 745
Arbacia, 787, 788, 794
Arcella, *569*, 581
Archaeopteryx, 25, *25*, 897
Archaeornithes, 897, 904
Archenteron, 514, 522
Archeocyathan, *551*
Archeocytes, 590
Archeognatha, 755, 765
Archinephric duct, 353
Archipallium, 393
Architeuthis, 684
Archosauria, 877
Archosaurs, 874
Archsauromorpha, 877
Arenicola, 697, *699*, 711
    hemoglobin, 285
Arginine phosphate, 218
Argiope, 460
Aristotle, concept of adaptation, 19
Armadillos, 931

limb biomechanics, 231
Arms, cephalopods, 685
Arrowworm, *794*
Artemia, 739
Arteries, 297, 304, *317*
    coronary, 312, *313*
Arterioles, 316
Arthropoda, 715–764
    characteristics, 715
    Chelicerata, 719–729
    Crustacea, 729–746
    evolutionary relationships, 764
    exoskeleton, 205–206, *205*
    ground plan, 716–719
    neuromuscular system, 220
    Uniramians, 750–764
Articular membranes, 206
Artiodactyla, 928, 931
Ascaris, 650, 651, *651*
Ascaroid nematodes, 650
Ascetospora, 572
Aschelminths, 641–655
    characteristics, 641
    classification, 653–654
    evolutionary relationships, 652
    Gastrotricha, 646
    Nematoda, 646–652
    rotifers, 643–645
Ascidiacea, 802–804, *802*, *803*, 811, *811*
Ascidian larva, *801*
Asconoid sponge, *586*, *588*, *589*
Ascorbic acid, 256, 257
Aselius, 736
Asellus, 745
Asexual reproduction, 484
    advantages, 485
Aspect ratio, 883
Asplanchna, *643*, *644*
Assimilation efficiency, 1000
Association areas, 395
Asterias, 780, 782, 794
Asteroidea, 779–783, 794
Asterozoa, 794
Astraea, *664*
Astropecten, 794
Astrophyton, 794
Atlas, 212
Atmospheric gases, 272
Atolls, 614, *615*, *616*
ATP. *See* Adenosine triphosphate
Atrioventricular (AV) node, 314
Atrium, 304, 307, 586, 802
Attack angle, 230
Auditory ossicles, evolution, 416
Auditory tube, 248, 414
Aulacantha, 581
Aurelia, 607, 618
Auricularia larva, 790
Australopithecines, 944–946

Autogamy, 581
Autoimmune disease, 337
Autonomic nervous system, 388, *389*
Autosomes, 104
Autotomy, 300
    brittle stars, 784
Auxotrophic mutations, 149
Aves, 881–907
Avicularia, 772, *773*
Axiothella, 698
Axis, 212
Axon, 372
Axopods, 569

B
B lymphocytes, 329, *331, 335*
Babesia, 572
Bacterial transposons, 139
Bacteriophage, experiments in role of
    DNA, 114, *114*
Balantidium, 566
Balanus, 745
Baleen, 928
Bamboo worms, 698, *700*
Barbulanympha, 565
Barnacles, 741, *743*
Baroceptors, 315
Barrier reefs, 614
Barriers, 961
    in reproductive isolation, 171
Basal lamina, 201
Base-pairing rules, 119
Basement membrane, 201
Basket stars, 784, *784*, 794
Basophils, 300, 302
Batesian mimicry, 470, *470*
    disruptive selection, *167*
Bathyal zone, 992, *993*
Batoidea, 826, 838
Bats, 924, *924*, 931
    echolocation, 924
Bdelloura, 622, 637
Beach fleas, 735
Beagle, voyage, 22–23, *22*
Bear, *922*
    winter sleep, 290
Beaver, *926*
Beef tapeworm, 621, 634, *635*
Bees, 766
    colony, *193*
    cuckoo, 544
    dance, 762
    pheromones, 474
    stings, 340
    workers, 762
Beetle mites, 722, *724*, 745
Beetles, *757*, 766
Behavior, 458–481
    causation, 459–464
    development, 464–468

ecology, 468–479
physiology, 460–462
sexual, 475–476
social, 476
Behaviorism, school, 459
Bends, 281
Beriberi, 256
Beroe, 617
Bicarbonate, gas transport, 285
Biceps, 223
Bichirs, 830, *831*
Bilateral symmetry, 186
Bilaterality, 622
Bilayers, phospholipids, 67
Bile, 249
Bile duct, 249
Bile pigments, 250
Bilirubin, 299
Bills, birds, 890, *891*
Biogeochemical cycles, 1004
Biological diversity, history of life, *169*
Biological rhythms and clocks, 462–464
Biology, 4
Biomes, 980, 988–995
Biosphere, 960, 980
Bipinnaria larva, 782, *783*
Biramous appendages, 730
Birds, 881–907
    adaptations, 885–895
    adaptive radiation, 899–904
    characteristics, 881
    classification, 904
    evolutionary origin, 897, *898, 899*
    flight, 882–885
    gas exchange, 892, *893*
    internal organs, *891*
    land, 899
    migration and navigation, 895–897
    reproduction and development,
        894–895
    shore and water, 902–904
    skeleton, 888, *889*
    song learning, 465, *467*
    toothed, 897, *899*
    water balance, 367
Birth, 526
    multiple, 527
Birth weights, *166*
Biston betularia, 161, *161*
Bithorax genes, 534, *534*
Bivalvia, 673, 689
    mantle and shell, 673
    nutrition, 675
    reproduction, 682
Black coral, 618
Black widow spider, 724, *726*
Bladder, urinary, 353
Bladder worm, 634
Blastaea, 552
Blastocoel, 512

Blastoderm, 513, 516
Blastoids, 794
Blastomeres, 510
Blastopore, 514, *530*
Blastula, 510
Bleaching, coral, 612
Blood, 297–300
    circulation, 298
    pressure, *316*
    velocity, *316*
Blood-brain barrier, 397
Blood cells, *299*
Blood clot, *302*
Blood clotting. *See* Hemostasis
Blood flow, 315
    regulation, 316
Blood groups, 150, 340
    MN, 157
Blood islands, 521
Blood pressure, *452*
Blood sucking, *244*
Blood sugar level, regulation, *450*
Bloodworms, 697, *698*
Blooms, phytoplankton, 996
Blowflies, 462
Blue crab, *732*
    osmoregulation, 362
Blue-gill sunfish, behavior, 476
Blue heron, behavior, 468
Blue-tailed skink, *869*
Body cavity, 190
    grouping of phyla, 553
Body covering, 200–204
Body size, metabolic rates, 290
Body wall, 189
Bohr effect, 286
Bombardier beetle, 470, *470*
Bombykol, 474
Bone, 208, *208*
Bone marrow, 208
Bonnets, 688
Bony fishes, 828–838
Boobies, *23*, 903
Book gills, 720
Book lungs, 278, 727
Boreal coniferous forests, 991
Borers, marine, 591
Boring
    bivalves, 680
    gastropods, 671
    sponges, 591, *591*
Boundary layer, 226
Bowerbirds, *472*
    lek, 472
Bowfin, 830, *832*
Bowman's capsule, 355
Box jellies, 606, *606*
Brachiation, 938
Brachiolaria larva, 783, *783*
Brachioles, 792, *793*

Brachiopoda, 774, *774*
Brachioria, Gonopods, 171
Brachyuran crab, 732
Brain, *382*, 384
    development, *390*
    evolution, 390–393, *392*
    nutrition, 397
Brain coral, 611
Brain regions, 391
Branchial arches, 211
Branchial chamber, 274
Branchial muscles, 222, 223
Branchiobdella, 711, *711*
Branchiobdellida, 711
Branchiopoda, 737–739, 745
Breathing, 282–284, *283*
Brine shrimp, 737
Bristletails, 765
Brittle stars, 783, *784*, *785*, 794
Bronchiole, 281
Bronchus, 281, *282*
Brooding, 524
    arachnids, 728
    birds, 895
Brown recluse spider, 725, *726*
Brugia, 650
Bryozoa, 770–774
    characteristics, 770
    colonies, 771–773
    evolutionary relationships, 775
    reproduction, 773–774
    zooid structure, 770–771
Bubble shells, 668, *669*, 688
Bubulcus ibis, 539
Buccal cavity, 242
    ciliates, 577
Buccinum, 671
Buccopharyngeal respiration, 847
Bucerotiformes, 905
Budding, *484*, 485
    cnidarians, 599
Bugs, 765
Bugula, 772, *773*
Bulk flow, 273
Bullfrog, *843*, 856
Bunodont teeth, 925
Buoyancy, 207
    cephalopods, 685
Burgess Shale, 548–549
Burgessochaeta, *551*
Burnet, 331
Burrowing, bivalves, 679
Bursae, 785
Bush baby, *837*
Busycon, *668*, 688
Butterfly, 756, 766
    checkerspot, *482*
    monarch, life cycle, *442*
Byssal secretion, 680

C
Caddis flies, 766
Caecilians, 858, *858*
Caenorhabditis, 649
Calanus, 740, *740*, 745
Calcarea, 592
Calciferous glands, 704
Calcitonin, 453
Calcium, 255
Calcium carbonate, skeletons, 205
Calcium homeostasis, 453, *453*
Callinectes, *732*, 746
Calyces, 356
Cambarus, 746
Camber, 229
Cambrian diversification, 549
Camels, 928
    water balance, 367
Canada goose, *896*
Canaliculi, 208
Canals, sponges, 588
Cancellous bone, 208
Cancer, 746
Candona, 745
Capillaries, *294*, 297, *317*
    fenestrated, 319
    structure, *319*
Capillary exchange, 318, *318*
Capitulum, 741
Caprella, 745
Caprellid amphipod, *736*
Caprimulgiformes, 905
Captorhinida, 865, *865*, 877
Carapace
    chelicerates, 719
    crustaceans, 729
    turtles, 866
Carapus, 790
Carbohydrases, 237
Carbohydrates, 37–39 *38*, *39*
    digestion, 237, *238*
Carbon cycle, 1005, *1005*
Carbon dioxide
    carbon cycle, 1005
    greenhouse effect, 1001
    product of cellular respiration, 263, 266, 267
Carbon fixation, 1001
Carbonic anhydrase, 285
Carcharodon, 827
Carcinus, osmoregulation, 362
Cardiac control, 315
Cardiac muscle, 311, *312*
Cardiac output, 315
Cardiac sphincter, 249
Cardiac stomach, 734
    Echinoderms, 782
Carnassial teeth, 922, *923*
Carnivora, 922–923, 931
Carnivores, 236

Carotid bodies, 284
Carpals, 214
Carrying capacity, 969
Cartilage, 207, *207*
Cartilage replacement bone, 209
Cartilaginous fishes, 824–828
Cassiopeia, 612, 618
Cassowaries, 898, 904
Castes, 761
Castle, 155
Casuariiformes, 904
Catabolism, 262
Catarrhini, 938, 951
Catenulida, 637
Caterpillars, 757
Cats, 931
Cattle, 928, 931
Caudal vertebrae, 212
Caudofoveata, 688
Causation, behavior, 459–464
Ceca, 242
Cell
    eukaryotic, 64
    structure, 66–80
Cell cycle, 98
Cell-mediated immune response, 335, *336*
Cell membrane, 67–68
Cell theory, 65
Cellular respiration, 262–271
    aerobic respiration, 265
    ATP yield, 269
    controls, 269
    fermentation, 263
    fuels other than glucose, 270
    glycolysis, 263
Cellulase, 237
Cellulose, 38
Cellulose digestion, *243*
Cenocrinus, 794
Centimorgans, 145
Centipedes, 751, *751*, 765
Central dogma, Protein synthesis, 124, *124*
Central nervous system, 372, 383, *385*
Central neuronal oscillators, 385
Centrioles, 100
Centrolecithal egg, 719
Centromere, 100
Centrum, 211
Centruroides, 745
Cephalization, 186
Cephalochordata, 804–807, 811
Cephalopoda, 682–687, 689
    locomotion, 685
    nutrition, 685
    reproduction, 686
    shell, 682–685

Cephalothorax
chelicerates, 719
crustaceans, 732
Cerata, 668
Ceratium, 563
Cercaria larva, 631
Cercopithecoidea, 938, 954
Cerebellum, *390, 391*
Cerebral cortex, 381, *382,* 393
Cerebral ganglia, 384
Cerebral hemisphere, *390, 391, 393, 394*
Cerebratulus, 636
Cerebrospinal fluid, *396,* 397
Cerebrum, 391, 394–397
Ceriantharia, 618
Cerianthus, 618
Cervical vertebrae, 212
Cestoda, 634, 638
Cetacea, 928, 931
Ceyx, 979
Chaetognatha, 794, *794*
Chaetopleura, 688
Chaetopteridae, 700
Chaetopterus, 700, *701,* 711
Chagas' disease, 565, 566, *567*
Chalmys, 681
Chameleon, *869*
Chapparal, 991
Charadriiformes, 905
Chelate appendages, 720
Chelicerata, 719–729, 745
Arachnida, 720–728
classification, 745
Merostomata, 719–720
Pyenogonida, 728
Chelifer, 745
Chelipeds, 731
Chelonia, 866, 877
Chemical reactions, *38*
Chemiosmosis, 268, *269*
Chemoreceptors, 404–406
Chemosynthetic bacteria, 710
Chiasmata, 101
Chickens, 901
Chiggers, 727, 745
Chilopoda, 751, 765
Chimpanzee, 938, 954
facial expressions, *941*
Chinese liver fluke, 631, *632*
Chiracanthium, 728
Chironex, 606, 618
Chiroptera, 924, 931
Chitin
arthropod exoskeleton, 205
chemical structure, 38
Chitinase, 237
Chitinous exoskeleton, arthropods, 716
Chiton, 688
Chitons, 662, *662, 663*

Chlamydomonas, 564
Chloragogen cells, 704
Chloride cells, 362
Chlorine, 255
Chlorofluorocarbons, 1001
Choanae, 248
Choanocytes, 586, *588*
Choanoflagellates, 552, 565, *565, 565,* 581
Cholecalciferol, 453
Cholecystokinin, 448
Cholesterol, 501
chemical structure, 41, *41*
Chondrichthyes, 824–828, 838
adaptive radiation, 826
basic features, 825
characteristics, 824
Chondritic meteorites, 56
Chondrocranium, 211
Chondrocytes, 207
Chondrostei, 830, 838
Chordamesoderm, 514
Chordata, 800–811, 811
characteristics, 797
evolutionary relationships, 809, *810*
Chorella, 612
Chorion, 524
Chorionic gonadotropin, 505
Choroid, 425
Choroid plexuses, 398
Christmas tree worm, *700*
Chromaffin cells, 450
Chromatid, 100
Chromatin, 78
Chromatographic technique, 50
Chromatophores, 204, *204*
cephalopods, 686
Chromosomal abnormalities, 142
Chromosomal changes, 140
Chromosome theory of inheritance, 104
Chromosomes, *79*
behavior in mitosis, 98
mapping, 145
as part of nucleus, 78
recombination, 143
sex, 104
Chrysaora, 605, 607, 618
Chyme, 249
Chymotrypsin, 237
Cicadas, sound production, 761, 765
Ciconiiformes, 905
Ciguatoxin, 565
Cilia, 224–225, 573
Ciliary body, 425
Ciliary movement, 224, *224*
Ciliature
buccal, 577
somatic, 575
Ciliophora, 573–581, 582
locomotion, 576
nutrition, 577–579

reproduction, 580
water balance, 580
Circadian rhythms, 463
Circulation
adult mammal, *310*
fetal, 308, *310*
neonatal, 310, *310*
Circulatory systems
amphibians and reptiles, 306
annelid worms, *304*
birds and mammals, 307
clam, *302*
closed, 297
crayfish, *303*
fishes, 304–305, *305*
formation, 521, *521*
human, *307*
invertebrate, 302–304
open, 297
vertebrate, 304–311
Cirri, *575*
barnacles, 742
crinoids, 791, *791*
Cirripedia, 741–743, 745
Cirroteuthis, 685
Cisternae, 218
Citric acid cycle, 267, *267*
Cladistics, 547
Cladocera, 737, *738*
Cladograms, 547
Clam shrimp, 745
Claspers, sharks, 825
Clavicle, 214
Clawed frog, *856*
Cleaning shrimp, 458, *458*
Clearance rate, kidney, 357
Cleavage, 510–513
Cleavage patterns, 510
Cleidomastoid, 223
Cleiodoic eggs, 523, *523*
insects, 753
Climax community, 981, 982
Cliona, 592
Clione, 669
Clionidae, 591
Clitellata, 709
Clitellum, 706
Clitoris, 499
Cloaca, 251, 353
Clocks, biological, 462–464
Clonal selection theory, 330, *332*
Clone, 581
Closed instincts, 465
Clown fish, 611
Clupeomorpha, 839
Clymenella, 698
Cnidaria, 595–619
Anthozoa, 608–614
characteristics, 595
classification, 617–618

Cnidaria *(continued)*
  coral reefs, 614–616
  Cubozoa, 605–608
  evolutionary relationships, 616
  Hydrozoa, 599–605
  Scyphozoa, 605–608
  structure and function, 596–599
Cnidocil, 597
Cnidocyte, 596, *598*
Coccidians, 572
Coccyx, 212, *213*
Cochlea, 416, *418*
Cockroaches, 765
Cocoon, oligochaetes, 706
Code, genetic, 122–124
Codominants, 91, *92*
Codons, 122
Codosiga, 565
Coelacanthiformes, 838, 839
Coelenterata, 596
Coelom, 190, 658
Coelom formation, *515*
Coenenchyme, 614
Coenzyme Q, 268
Coenzymes, 45
Cofactor, 45
Coleoidea, 689
Coleoptera, 766
Colies, 905
Coliiformes, 905
Collagen, 201
Collar cells, 586
Collateral ganglia, 390
Collecting tubule, 356, 359
Collembola, 766
Collotheca, 644
Colon, 251
Colonial organization, 192
Colonial origin of the animal kingdom, *551*
Color blindness, 104
Coloration
  aposematic, 470, *470*
  cryptic, 470
  skin, 204
  warning, 470, *470*
Colugos, 924, *924*, 931
Columbiformes, 905
Columella, 665
  amphibians, 846
Columnar epithelium, *200*, 201
Comb jellies, 616
Commensalism, definition, 193
Commissures, 384, 395
Communication, behavior, 473–475
Compact bone, 208, *208*, *209*
Comparative anatomy, evidence for animal evolution, 549
Competition, interspecific, 972
Complement, DNA, 119

Complement cascade, 333, 335, *335*
Complete metamorphosis, 757, *758*
Compound ciliary organelles, 577
Concentricycloidea, 792, 794
Conchae, 280
Conchiolin, 205
Conchs, 688
Condensation reaction, 37
Conditional mutants, 148
Conditioning, 468
Cone shells, 670, *671*, 688
Cones, 426, *426*
Coniferous forests, 991
Conjugation, 580
Conjugation bridge, bacteria, 139
Connective tissue, 201, *201*
Consumers, 998
Continental drift, 174
Continental shelf and slope, 992, *993*
Contractile vacuoles, ciliates, 580, *580*
Contraction, 218
Conus, 688
Conus arteriosus, 304
Convergent evolution, 544
Convoluta, 637
Convoluted tubule, 356
Cooling mechanisms, 289
Coots, 902
Copepoda, 739, *740*, 745
Copulation, 494
Coraciiformes, 905
Coracoid process, 214
Coral reefs, 614–616
Corallimorpharia, 618
Corals
  gorgonian, 614
  scleractinian, 611–612, *611*
  soft, 614
Cormorants, 235, 903
Cornea, 425
Corona, 643
Coronary arteries, 312, *313*
Coronatae, 618
Corpus callosum, 395
Corpus luteum, 502, 505
Corpus striatum, 393
Cortex, kidney, 356
Cortical areas, 394
Cortical reaction, 491
Cortisol, 451
Corynactis, 618
Cosmine, 830
Cotransport, 241
Cotransport system, 71
Countercurrent, gas exchange, 274
Countercurrent multiplier mechanism, kidney, *358*, 359
Cowpox, 327
Cowries, 688
Crab body form, 732, *733*

Crab spiders, 725
Crabs, 745
  brachyuran, 732
  gills, 274, *275*
  hermit, 732
  land, 733
Cranes, 902, *902*, 905
Cranial nerves, 387, 388
Cranium, 213
Craspedacusta, 603, *604*, 617
Crayfish, *731*, 745
  gills, 274, *275*
  green gland, *352*
Creatine phosphate, 218
Creation-science, 4–6, 9
Crematogaster, alarm pattern, *473*
Cretinism, 447
Cri du chat syndrome, *141*, 142
Cribrinopsis, 609
Crick, DNA, 118
Crickets, 765
  escape behavior, 460
  sound production, 761
Crinoidea, 790–792, 794
Crinoids, 541
Crinozoa, 794
Cristatella, 769
Cro-Magnon people, 950
Crocodiles, 874, *874*, 877
  phylogeny, 545
Crocodylia, 874, 877
Crop, 242
  birds, 890
Cross-fiber patterning, 403
Cross-reactivity, 327
Crossing-over, 101, *103*, 144, *147*
Crossopterygii, 837, 839
Crossopterygian fish, appendicular skeleton, *214*
Crown, 643
  crinoids, 791, *791*
Crown-of-thorns sea star, 779, *779*
Crustacea, 729–746, 745
  Branchiopoda, 737–739
  Cirripedia, 741–743
  classification, 745–746
  Malacostraca, 731–737
  Ostracoda, 740
  Remipedia, 743
Cryptic coloration, 470
Cryptobiosis, 645
Crypts of Lieberkuhn, 253
Crystalline style, 679, *679*
Ctenoid scale, *832*, 833
Ctenophora, 616–617
Cuboidal epithelium, *200*, 201
Cubozoa, 605–608, 618
Cuckoo bees, 544
Cuckoos, 901, 904
  parasitic, 465

Cucuilformes, 904
Cucumaria, 790, 794
Cuspidaria, 689
Cutaneous respiration, amphibians, 847
Cuticle
   arthropod exoskeleton, 205
   Aschelminths, 642
   nematodes, 648
Cuttlefish, *683*, 685, 689
Cyanea, 606
Cyanobacteria, in sponges, 588
Cycloid scale, *832*, 833
Cyclops, 745
Cypris, 745
Cypris larva, 741, *744*
Cystic duct, 249
Cystic fibrosis, 116
Cysticercus, 634
Cystoids, *793*, 794
Cysts, ciliates, 580
Cytochromes, 268
Cytokinesis, 100
Cytopharynx, 577
Cytoplasmic organelles, 73
Cytoproct, 579
Cytosine, 46
Cytoskeleton, 79, 223
Cytostome, 563
Cytotoxic T cells, 335, *338*

D

Dactylogyrus, 633, *633*, 637
Daddy longlegs, 721, 745
Dall sheep, *965*
Damselflies, 756, 765
Daphnia, *738*, 745
Darwin
   development of evolutionary biology, 19–29
   fossil record, 175–176
   problem of diversity, 169
Darwinian fitness, 162
Data, in science, 9
Deamination, in use of amino acids as fuel, 271
Decapoda, 731–735, 745
Decarboxylation, 263
Decomposers, 998
Decussation, 381
Deer, 928, 931
   agonistic behavior, 471
Deer mice, behavior, 467
Defense, behavior, 470
Definitive host, 628
Dehydration reaction, 37
Dehydrogenation, 263
Delamination, 516
Deletions, 141, *141*
Demospongiae, 592
Dendraster, *786*, 794

Dendrites, 372
Dendronephthya, *614*
Dense connective tissue, 201, *201*
Density-dependent factors, 970
Dental formula, 922
Dentalium, 689
Dentine, 246
Deoxyribonucleic acid (DNA), 46, *46*, *47*, *48*, *49*
   as part of chromosome, 78
   repair, 136
   replication, 119, *120*
   structure, 117–119, *118*
Deoxyribonucleic acid ligase, 119
Deoxyribonucleic acid polymerase, 119
Deoxyribonucleic acid topoisomerase, 119
Dermacentor, *724*, 745
Dermal bone, 209
Dermal papilla, 203
Dermaptera, 765, 931
Dermasterias, 777
Dermatophagoides, *724*, 745
Dermis, vertebrate, 203
Dermoptera, 924
Deserts, 990
Desmodus, 478
Determinate development, 529
Detorsion, 668
Detritus, 236
Deuterostomes, 554, 778
Development
   behavior, 464–468
   embryonic, 510–537
   embryonic control, 529–535
   human embryonic, 526–529
Developmental patterns, grouping of phyla, 553
Diabetes mellitus, 448
Diadema, 785
Dialysis, 360
Diapause, 758
Diaphragm, 282
Diapsida, 866, 877
Diastole, 311
Dictyoptera, 765
Didinium, *575*, 582
Diencephalon, *390*, 391, 392
Diets, kinds of, 236
Difflugia, *569*, 581
Diffusion, 69–70
   facilitated, 241
Digenetic flukes, 631
Digestion, 237, 249, 253, 254
   carbohydrates, 237, *238*
   cellulose, *243*
   extracellular, 242
   intracellular, 241
   lipids, 237, *238*
   proteins, 237–239, *238*

Digestive enzymes, 239
Digestive secretions, regulation, 254
Digestive tract
   evolution, 241–243, *241*
   human, *248*
   vertebrate, 246–254
Digitigrade posture, 922
Dihybrid crosses, 94, *94*
Dinoflagellates, *563*, 564, *564*, 581
Dinornithiformes, 904
Dinosaurs, 874–877, *875*, *877*, *878*
   thermoregulation, 876
Diopatra, *699*, 711
Diophrys, 576
Diphyllobothrium, 630
Diploblastic condition, 596
Diploid condition, 91
Diplopoda, 751, 765
Diplorhina, 819, 838
Diploria, *611*, 618
Diplosegments, *750*, 751
Diplura, 766
Dipnoi, 837, 839
Diptera, 766
Direct development, *522*
Direct fitration theory, 83
Directional selection, 167
Dirofilaria, 652
Dirt daubers, 760
Disaccharides, 37
Discoidal cleavage, 512, *513*
Dispersal mechanisms, 961
Dispersal patterns, 962, *963*
Disruptive selection, 167
Distal convoluted tubule, 356
Distal end, 186
Divergent evolution, 544
Diversity
   history of life, *169*
   in the living world, 16
Diving adaptations, 281
DNA. *See* Deoxyribonucleic acid
DNA hybridization, 545
DNA repair, 486
Dobsonflies, 766
Dobzhansky, 170
Dog tick, *724*, 726
Dogfish, *827*
   muscular system, *222*
   skeleton, 211, *211*
   swimming, *228*
   visceral organs, *826*
Dogs, 931
   evolution, 24
Dominance, ecological, 980
Dominant alleles, 90
Dopamine, 397
Doris, 688
Dormant eggs
   rotifers, 645

Dormant eggs *(continued)*
  water fleas, 739
Dorsal aorta, 304
Dorsal lip, *530*
Dorsal surface, 186
Double circulation, evolution, 305
Double helix, *48, 49*
  DNA, 117, *119*
Dove, behavior, 461, *461*
Down's syndrome, 142
Dracunculus, 650, 652
Drag, 226
Dragline, 723
Dragonflies, 755, *756*, 765
Drilling
  bivalves, 680
  gastropods, 671
Drilonereis, *698*
Drones, 761
Drosophila
  endogenous rhythms, 464
  fitness, *163*
  gene linkage, 144
  inheritance of eye color, 104, *105*
  population genetics, 155
  populations, 154, *154*
  sex recognition, 468
Duchenne muscular dystrophy, 104, 116
Ducks, 902, *903*, 905
Duct
  bile, 249
  cystic, 249
Ductus arteriosus, 308
Dugesia, *624, 625, 626, 627*, 637
Dugongs, 930, 931
Dunkleosteus, *824*
Duodenum, 251
Duogland systems, 625
Duplications, 141, *141*
Dusky salamander, *854*
Dust cells, 282
Dynein, 225

E
Eagle ray, *827*
Eagles, 901
Ear
  auricle, *415*
  external, 414, *415*
  human, *415*
  inner, 414, *415*
  middle, 414, *415*
Earthworms, 702, *704, 705*, 711
  body wall, *189*
  locomotion, 226, *227*
  metanephridia, 352
  water balance, 366
Earwigs, 765
Eccrine sweat glands, 204
Ecdysiotropin, 442

Ecdysis, 717
  arthropods, 206, *206*
Ecdysone, 441
Echidna, 917, *917*
Echinocardium, 794
Echinococcus, 630, 634, 638
Echinoderes, 654
Echinodermata, 777–795
  characteristics, 777
  classification, 794
  Concentricycloidea, 792
  Crinoidea, 790–792
  Echinoidea, 785–788
  evolution of symmetry, *188*
  evolutionary relationships, 792
  Holothuroidea, 788–790
Echinoidea, 785–788, 794
Echinopluteus larva, 788
Echinosphaerium, *570*, 581
Echinozoa, 794
Echiura, 709
Echiurus, *710*
Ecological equivalents, 984
Ecological growth efficiency, 1000
Ecological isolation, 171
Ecological succession, 980–984
Ecology
  behavioral, 468–479
  communications and ecosystems,
    979–1009
  populations, 960–976
Ecosphere, 960
Ecosystem, 980, 987–995
Ectoderm, 514
Ectoderm differentiation, 517
Ectolecithal condition, 628
Ectoparasites, 245
Ectoparasitism, 628, *629*
Ectoplasm, amoeboid movement, 223
Ectoprocta, 770
Ectothermic animals, 288
Ediacaran fauna, 549
Eels, 833, *834*
Effectors, 372
Eft, red, *470*, 471
Egestion, 237
Egg, 485
Egg deposition, 496
Egg types, 510
Eijkman, 256
Eimeria, 572
Elasmobranchii, 826, 838
Elastic cartilage, 207
Elastic fibers, 201
Eldredge, 176
Electra, *772*
Electrocardiagram, *314*
Electron microscope, 82
Electron transport, 267, *268*
Electrophoretic technique, 50

Electroreception, 411
Eledone, *685*
Elephant shrews, *926, 927*, 931
Elephantiasis, 650, 651, *651*
Elephants, 929, *930*, 931
  agonistic behavior, *471*
  social behavior, 476
Elkhorn coral, 615
Elopomorpha, 838
Embolus, 302
Embryo, frog, *519*
Embryology, 510
Embryonic development, 510–537
  cleavage, 510–513
  control, 529–535
  gastrulation, 513–517
  human, 526–529
  organogenesis, 517–522
  sex determination, 535
Embryonic diapause, 918
Embryonic disc, 516
Emerita, 733, 746
Emigration, 964
Emus, 898, 904
Enamel, 246
Endemism, in oceanic islands, 175
Endocrine, integration, 435
Endocrine glands, 434–454, *434*
Endocuticle, 206, 716
Endocytosis, 72–73, *72*
Endoderm, 514
Endoderm differentiation, 522
Endogenous rhythms, 463
Endometrium, 504
Endoparasites, 245
Endoparasitism, 628
Endopeptidases, 237
Endoplasm, amoeboid movement, 223
Endoplasmic reticulum, 73–75, *74*
Endopodite, 730
Endopterygota, 757, 766
Endorphins, 397
Endoskeletons, 206
Endostyle, 800
Endosymbiotic theory, 81–83
Endothermic animals, 289
Endothermy, birds, 885
Energy flow, ecological, 995–1001
Energy needs, 255
  source of calories, 256
Eniwetok, 616
Enkephalins, 397
Entamoeba, 566, 568
Enterobius, 650, 652
Enterocoel, 515
Enterocoely, 554
Enterokinase, 239
Enteropneusta, 798, 811
Entoprocta, 775
Entropy, 64

Environmental resistance, 968
Enzymes, 41, 44–45, *45*
   digestive, 239
Eosinophils, 300
Ephemeroptera, 765
Ephydatia, 589, 592
Ephyra, 607
Epiboly, 516
Epicuticle, 206, 716
Epidermis, 202, 596
Epididymis, 496
Epigamic selection, 475
Epiglottis, 280
Epilimnion, *995*
Epimysium, 215
Epinephrine, 450
Epiphanes, 642
Epiphragm, 670
Epiphyseal plate, *209*
Epiphysis, 208
Epiphytic plants, 988, *989*
Epistatic interactions, 97
Epithelial tissue, *200*, 201
Epitheliomuscle cells, 596, *597, 599*
Epitopes, 325, *326*
Equational division, 102
Equilibrium, 408
Ermine, *922*
Erythroblastosis fetalis, 341, *341*
Erythrocytes, 299
Erythropoietin, 300, 448
Esophagus, 242, 248
Essential amino acids, 41, 255
Essential fatty acids, 255
Estivation, gastropods, 670
Estradiol, 501
Estrogens, 501
Estrus cycles, 504
Ethogram, 459
Ethological isolation, 171
Ethology, 458–481
Euchlanis, 644
Euglena, 563, 563, 581
   fission, *568*
Euglenids, 562–563
Eukaryotes, 66
Eukaryotic cell, 66, *74*
   origin, 80–83
Eunice, 702, 711
Euphausia, 746
Euphausiacea, 737, 746
Euphotic zone, 992, *993*
Euplectella, 592, *592*
Euplotes, 575, 582
Eupolymnia, 700
Euryhaline animals, 362
Eurypterids, 720, *720, 820, 820*
Eustachian tube, 248, 414
Eutely, 642
Eutheria, 917, 931

adaptive radiation, 919–930
   cladogram, *920*
Eutrophication, 1007
Evolution
   convergent, 544
   and creation-science, 4
   Darwin and history of evolutionary
      thought, 16
   divergent, 544
   macroevolution, 175–180
   microevolution, 168–175
   origin of variation, 136–150
   populations, 154–160
   rates, 178
Evolutionary relationships, 544
Evolutionary systematics, 545
Excision repair system, 136, *137*
Excitation, 379
Excretion, 348–361
   excretory organs, 350–361
   urine formation, 356–361
Excretory organs, 350–361
   green glands, 351
   malpighian tubules, 351
   Metanephridia, 352
   vertebrate kidney, 353–361
Exocrine glands, 242
Exocuticle, 206, 716
Exocytosis, 73, *73*
Exogenous rhythms, 463
Exons, 125
Exopeptidases, 237
Exopodite, 730
Exopterygota, 757, 765
Exoskeletons, 205–206
   arthropods, *716*
Extension, 215
External oblique, 223
Extinctions, macroevolution, 178, *180*
Extracellular digestion, 242
Extraembryonic membranes, 523, *523*.
   See also names of individual
   membranes
Extrinsic ocular muscles, 424
Eye worm, 650
Eyes
   compound, 422–424, *422*
   development, *520*
   everted, 420, *421*
   image-forming, 420, *422*
      of octopus, *422*
   inverted, 420, *420*
   light-orienting, 420
   mammalian, *425*
   optics, *421*
   pigment cup, 420, *420*
   retinal cup, 420, *421*
   vertebrate, 424–429

F
Facial muscles, 223
Facial pit, *429*
Facial skeleton, 213
Facilitated diffusion, 70, 241
FAD. *See* Flavin adenine dinucleotide
Fairy shrimp, 737, *738*, 745
Falconiformes, 904
Falcons, 901, 904
Fallopian tube, 499
Falsifiability, in science, 7
Fangs, snakes, 871
Fanworm, *245*, 698
Fasciola, 631, 632
Fat
   as a cellular fuel, 270
   as a storage compound, 271
Fat cell, 202
Fatty acids, 39, *40*
   essential, 255
Feather duster worms, 698
Feather mites, 726
Feather stars, 791, *792*, 794
Feathers, 886–888, *887*
   origin of, 897
Feces, 237, 254
Fecundity, 964
Feeding, *236*
   deposit, 246, *246*
   filter, 245
   raptorial, 244
   suspension, 245
Feeding mechanisms, 243
Femur, vertebrates, 214
Fermentation, 263, *265*
Fertilization, 490–496, *493, 494*, 505
   external, 493
   internal, 494
   mammalian, 499
Fertilization cone, 491
Fertilization envelope, 491
Fetal hemoglobin, 288
Fetus, 526
Fever, 289
Fibrin, 301
Fibrinogen, 301
Fibroblast, 201, *201*
Fibrous cartilage, 207
Fibrous tunic, 424
Fibula, 214
Fiddler crabs, 732, 733, *733*
Fig tree formation, 55, 56
Fight-or-flight response, 388
Filaments, actin, 79
Filariasis, 651
Filaroid nematodes, 651
File shells, 680
Filtration, glomerula, 357
Finches, Galapagos, *26*, 27, *27*
Fire, ecological effect, 973

Fire coral, *600*, 601, 617
Fire sponge, *586*
Fireflies, 760
   reproductive isolation, *170*
Fireworm, *697*
Fish leeches, *707*
Fishes, 816–841
   adaptations, 816–819
   Agnatha, 819
   Chondrichthyes, 824–828
   classification, 838
   early jawed, 822–824
   evolution, *817*
   gills, 275
   muscular system, *222*
   osmoregulation
      in freshwater, 364
      in the sea, 362
   Osteichthyes, 828–838
   skeleton, 211
Fission, 484, *484*
Fissurella, 688
Fitness, 161–163
Fixed action patterns, 459
Flagella, 224–225
Flagellar movement, 224, *224*, 562
Flagellated chamber, *588*
Flagellates, 562–567
   phytoflagellates, 562–565
   reproduction, 566–567
   zooflagellates, 565–566
Flagellum, flagellates, 562
Flame cell, 350
Flamingos, 905
Flat-backed millipedes, *750*, 751
Flatworms, 621–639
   body wall, *189*
   characteristics, 621
   classification, 637–638
   evolutionary relationships, 634
   parasitic classes, 628–634
   structure, 624
   Turbellaria, 623–628
Flavin adenine dinucleotide (FAD), 267
Fleas, 759, *760*, 766
Flexible learning, 467
Flexion, 215
Flies, 766
Flight, 229, *229*
   birds, 882–885
   flapping, 884, *884*
   hovering, 884, *886*
   muscles, 888, *890*
   vortex theory, 885
Florometra, *792*, 794
Flounder, *836*
Flukes, 630, 638, 928
Flying lemur, *924*
Flying squid, 685
Follicle, *503*

Follicle mites, 745
Follicle-stimulating hormone, 501
Following response, 465
Food, storage, 271
Food chain, 995, 1002
Food compounds, interconversion, 271
Food supply, balancing with cellular
   demand, 271
Food vacuole, 241
   ciliates, 577
Food webs, 996, *997*
Foot
   molluscs, 658
   rotifers, 643
Foraging strategies, 468–470
Foramen magnum, 213
Foramen ovale, 308
Foramina, bone, 213
Foraminiferans, 568–569, *570*, 581
Forebrain, 391
Foregut, 718
Form, animal, 186–194
Fossa ovalis, 310
Fossil record, 548
   early life, 54
Fouling organisms, 680, 742
Founder effect, 175
Fovea, 428
Frame-shift mutations, *122*, 137
Freshwater ecosystems, 994
Frogs, *171*, 859
   adaptations, 854–858
   circulation, *848*
   life cycle, *850*
   reproduction, 857
   skeleton, *854*
   tongue, *247*
   water balance, 366
Frontal plane, 186
Frontal surface, bryozoans, 770
Frontonia, 576
Fructose-1, 6–diphosphate, 263
Fruit flies
   endogenous rhythms, 464
   sex recognition, 468
Fulcrum, 230
Functional groups, 36, *37*
Fungi, 547
Fungia, 618
Funiculus, 772

G
GABA. *See* Gamma-aminobutyric acid
Galapagos finches, *26*, 27
Galapagos islands, Darwin, 25–27
Galapagos rift fauna, 710
Galapagos tortoise, *867*
Galen, 298
Gallbladder, 249

Gallery dwellers, Polychaetes, 697
Galliformes, 904
Galls, 759, *760*
Gametes, 485
   in genetics, 91
Gametocytes, *Plasmodium*, 572
Gametogenesis, 486–490
Gamma-aminobutyric acid (GABA), 397
Gammarus, *735*, 745
Ganglia, 385
   cerebral, 384
   collateral, 390
   sympathetic, 390
Ganglion, spinal, 380
Gannets, *894*
Ganoid scale, 830
Ganoine, 830
Gap junction, *200*
Gap period, cell cycle, 98
Gar, 830, *832*
Garter snakes, *861*
Gas exchange, 272–288
   in aquatic animals, 273–276
   steps, 273
   terrestrial animals, 276–284
Gas transport, 284–288
Gastric ceca, 758
Gastric glands, 249, *250*
Gastric mill, 734
Gastrin, 448
Gastrodermis, 596
Gastropoda, 662–673, 688
   evolution, 666
   locomotion, 666
   nutrition, 670
   reproduction, 672
   shell, 665
   water circulation, 666–669
Gastrotricha, 646, *646*, 653
Gastrovascular cavity, 596
Gastrozooids, 601
Gastrula, 514
Gastrulation, 513–517
   amphioxus, *514*
   bird, *517*
   frog, *516*
Gaviiformes, 905
Gazelle, *927*
   thermoregulation, 911
Gecko, *869*
Geese
   egg rolling, 459, *459*
   following response, 465
Gemmules, 590, *591*
Gene, nature of, 110–119
Gene expression, 530, *531*
Gene flow, 159, *160*
Gene frequencies, 155
Gene interaction, 97
Gene pool, 155

Generic name, 542
Genes, 91
  bithorax, 534
  homeotic, *533*, 534
  hormonal regulation, 439
  maternal effect, 533
  segmentation, 533
  ZFY, 535
Genetic code, 122–124, *123*
Genetic drift, 158, *159*
Genetic isolation, 172
Genetics, 88–133
  basic vocabulary, 91
  dihybrid crosses, 94–95
  material basis, 95–98
  Mendel, 88–91
  monohybrid crosses, 91–93
  nature of gene, 110–119
Genome, human, 149
Genotype, 91
Geographical isolation, 172
Geographical range, 960–962, *962*
Geologic timetable, *168. See also inside
    front cover*
Germ layers, 514
Germ-line mutations, 136
Gharial, *874*
Ghost crabs, 732
Giant clams, 676
Giardia, 566
Gibbon, 938, *939*, 954
Gila monster, *869*
Gill bailer, 274
Gill bladder, 732
Gill rakers, 276
Gill slits, 522
Gills, 273
  bivalves, 677
  fishes, *276*
  generalized molluscs, 658
  invertebrates, 274–275
  vertebrates, 275–276
Giraffes, *371*, 931
Gizzard, birds, 890
Gizzards, 242
  earthworms, 704
Glacial lakes, 994
Gland
  adrenal, 450–452, *451*
  gastric, 249, *250*
  lacrimal, 424
  parathyroid, 453
  pituitary, 442–445
  salivary, 247
  thyroid, 446–448
Gland cells, Cnidarians, 596
Glands, in vertebrate integument, 203
Glass sponges, *592*
Glenodinium, *564*
Glider, *919*

Gliding, 882
Global change, 1007
Globigerina, 568, *570*, 581
Glochidium, 682, *682*
Glomerula filtration, 357
Glomerulus, 355
Glossiphonia, 707, *708*, 711
Glottis, 248, 280
Glucagon, 448
  role in food storage, 271
Glucocorticoids, 451
Glucose metabolism, 448
Glycera, 697, *698*, *703*, 711
  oxygen conformer, 272
Glycerol, 39, *40*
  as a cellular fuel, 270
Glycogen, 37
  as a storage compound, 271
Glycolipids, 41
Glycolysis, 263
Glycolytic fibers, 220
Gnathostomes, 822, *823*
Gnathostomula, *652*
Gnathostomulida, 653
Gnats, 766
Goats, 931
Goblet cells, 253
Golden plover, *896*
Golgi apparatus, 75–76
Golgi tendon organs, 419, *419*
Gonadotropin-releasing hormone, 440,
    501
Gondwana, 174
Gonionemus, *603*, *604*, 617
Gonodactylus, 745
Gonoducts, 491
Gonopods, millipedes, 171
Gonozooids, 601
Gonyaulax, *564*
Gordius, 653
Gorgonacea, 618
Gorgonia, 618
Gorgonian corals, 614, *614*
Gorgonocephalus, *784*
Gorilla, 938, *939*, 954
Gould, 176
Gradual metamorphosis, 757, *757*
Graft rejection, 335
Granuloreticulosea, 581
Grasshoppers, 765
  sound production, 760
Grasslands, 990
Gray matter, 380
Great white shark, *827*
Grebes, 903, *903*, 905
Green crab, osmoregulation, 362
Green glands, 351, 734
  crayfish, *352*
Greenhouse effect, 1001
Greenland, oldest rocks, 55

Grouse, *900*, 901, 904
  lek, 472
Growth
  logistic, 968
  rates, 967, *968*, *969*, *971*
Growth factors, 448
Grubs, 757
Gruiformes, 905
GTP. *See* Guanosine triphosphate
Guanine, 46, 350
Guanosine triphosphate (GTP), 49
Gueese, 902, 905
Guinea worm, 650, 652
Gulf Stream, 992
Gulls
  behavior, 465
  chick recognition, 475
Gunflint formation, 56
Gut, 241
Gymnophiona, 849, 858, 859
Gypsy moths, pheromones, 474

H
Habitat, 961
Habituation, 465
Hadal zone, 992
Haemadipsa, 707
Hagfishes, *821*, 822
  immune response, 342
Hair, in vertebrate integument, 203
Hair cells, 408
Hair mites, 726
Haliclona, 592
Haliclystus, 618
Haliotis, 688
Halteres, 756, *757*
Haploid condition, 91
Hardy, 155
Hardy-Weinberg-Castle Law, 154–160,
    156
Hares, 931, *971*
Harvestmen, 721, *723*, 745
Harvey, 298
Haversian system, 208, *208*
Hawks, *882*, 901
Hay fever, 339
Heart
  conducting system, *314*
  evolution, *306*
  human, *313*
  mammalian, 311
  types, 311, *311*
  valves, 312
Heart urchins, 786, *786*, 794
Heartbeat, integration, 312
Heartworm, 652
Heat budget, 1001, *1004*
Hedgehogs, 920, 931
Heliaster, 794
Heliopora, 618

Helioporacea, 618
Heliozoea, 569, *570*, 581
Helix, 689
  water balance, 366
Helmet shells, 688
Helper T cells, 335
Hemagglutinin, 343, *343*
Hemal system, 782
Heme, 285
Hemichordata, 797–800
  characteristics, 797
Hemimetabolous development, 757,
    757
Hemiptera, 765
Hemocoel, 297, 718
Hemocyanin, 285, 718
Hemoglobin, 285
  sickle-cell anemia, 164
  structure, *44*
Hemolymph, 297
Hemophilia, 104, *106*, 302
Hemostasis, 300–302, *301*
Heparin, 300, 302
Hepatic portal system, 305
Hepatopancreas, 734
Herbivores, 236, 243
Heredity, 113. *See also* Genetics
Hermaphroditism, *495*, 495
Hermit crab, *730*, 732
Herons, behavior, 468
Herring gull, behavior, 465
Heterocentrotus, 786
Heterogametic sex, 104
Heterotherms, 289
Heterozygotes, 91
Hexactinellida, 592
Hexose, 37
Hibernation, 290
  amphibians, 848
  mammals, 911
Hierarchy, in biological explanation, 11
High-crowned teeth, 925
Hindbrain, 391
Hindgut, 718
Hinge, bivalves, 673, *675*
Hippocampus, 393
Hippodisplosia, 769
Hippopotamus, *927*, 928
Hirudinea, 707–709, 711
Hirudo, 707, *708*, 711
Histamine, 300
Historicity, in biological explanation, 10
Holoblastic cleavage, 511
Holocephali, 826, 838
Holometabolous development, 757, *758*
Holonephros, 353
Holothuria, 789, *791*, 794
Holothuroidea, 788–790, 794
Homarus, 746
Home ranges, 473

Homeobox, *534*, 535
Homeothermic animals, 289
Homeotic genes, *533*, 534
Homeotic mutants, 534
Hominid
  phylogeny, *945*
  skulls, *945*
Hominidae, 954
Hominoidea, 938, 954
Homo erectus, 947, *950*
Homo habilis, 946
Homo sapiens, 950
Homogametic sex, 104
Homogentisic acid, 115
Homologous chromosomes, 101
Homology, 544
Homozygotes, 91
Honeybees
  dance, 762
  endogenous rhythms, 463
  social organization, 761
Hooke, 65
Hookworms, 649, *649*
Hoopes, 905
Hormones, 434. *See also names of
    individual hormones*
  action, 436, *437*
  adrenal, 450
  adrenocorticotropic, 445
  antidiuretic, 445
  behavior, 461
  corticotropin-releasing, 452
  discovery, 436
  follicle-stimulating, 445
  gastrointestinal, 448, *449*
  gonadotropic, 440
  gonadotropin-releasing, 440
  growth, 445, 446
  inhibiting, 445, 446
  juvenile, 441
  levels, regulation, 440
  luteinizing, 445
  male reproductive, 502, *502*
  mammalian, 438
  melanophore-stimulating, 445
  menstrual cycle, *503*
  molting, 440
  ovarian cycle, *503*
  parathyroid, 453
  receptors, *451*
  releasing, 445, 446
  reproductive, 501–505, *501*
  thyrotropic, 445
  thyrotropin-releasing, 446
  types, 436
Hornbills, 905
Horned lark, *963*
Horns, 928
Horse nematode, 647
Horseflies, 756

Horsehair worms, 653, *653*
Horses, 928, 931
  evolution, 178, *179*
  limb biomechanics, 231
  urine concentration, 361
Horseshoe crabs, 719, *719*, 745
Hotspots, genetic, 150
Human, 954
  characteristics, 942
  evolution, 942–955
  femur, *942*
  reproduction, 942
  skeleton, *940*
  teeth, *941*
Human development, 526–529, *527*
Human embryo, early development,
    525
Human immunodeficiency virus, 324
Humbolt current, 992
Humerus, 214
Hummingbird moths, 756
Hummingbirds, *886*, 901, 905
Humoral immune response, 325–335,
    332
Humus, 236
Huntington's chorea, 116
Hutton, 55
Hyaline cartilage, 207, *207*
Hyalospongia, 592
Hybrid mortality, 171
Hybridization, DNA, 545
Hydatid cyst, 634
Hydra, *604*
Hydractinia, *601*, 617
Hydras, 597, 599, 617
Hydrocorallina, 617
Hydrocorals, 601
Hydroid colony, 599, *600*, *601*
Hydroida, 617
Hydrological cycle, 1004
Hydrolysis, 237
Hydromedusae, 602, *603*
Hydroskeletons, 207
Hydrostatic skeletons, 207
Hydrothermal vents, 710
Hydrozoa, 599–605, 617
  colony formation, 599
  life cycles, 603
  Medusae, 602
  polymorphism, 601
  skeleton formation, 600
Hyena, *18*
Hylobatidae, 938, 954
Hymenolepis, 630
Hymenoptera, 766
Hyoid arch, 211
Hyoid muscles, 223
Hyomandibular, 211
Hypersensitivity, 338
Hypertension, 452

Hypodermic impregnation, 495, 628
  leeches, 709
  rotifers, 645
Hypolimnion, *995*
Hypopharynx, 754
Hypophyseal portal system, 445
Hypophyseal sac, 821
Hypophysis cerebri, 442
Hypothalamus, 392, 442
  circadian rhythms, 464
Hypotheses, in science, 8
Hypotrichia, 576
Hyracoidea, 929, 931
Hyraxes, 929, 931

I

Ichthyopterygia, 873, 877
Ichthyosauria, 873, *873*
Ileum, 251
Ilium, 214
Imaginal discs, 757
Immigration, 964
Immune responses, 324–341
  cell-mediated, 335
  evasion of, 343–344
  evolution, 341–342
  humoral, 325–335
  invertebrate, 341–342
  non-specific mechanism, 324–325
  vertebrate, 342
Immune system, 324–344
Immunoglobulin fold, 342
Immunoglobulins, 328, *328*
Immunological barriers, in reproductive
    isolation, 171
Immunological memory, 325
Immunosuppression, 344
Implantation, 524
Imprinting, 465
Inbreeding, 158, *158*
Incomplete metamorphosis, 757
Incus, 214, 414
Independent assortment, law of, 95
Independent effectors, 381
Indeterminate development, 530
Indirect development, 522, *522*
Induction, 517, 530
Inertial drag, 226
Inferior colliculus, *390*, 392
Influenza viruses, 343
Infraciliature, 573
Ingestion, 236
Ingression, 515, 517
Inhibition, 379
Ink, cephalopods, 686
Insecta, 752–764
  communication, 760
  ground plan, 753
  immune response, 342
  insect-plant interactions, 758

malpighian tubules, 353
  nutrition, 758
  parasitism, 759
  radiation, 754
  reproduction, 763
  social organization, 761
  water balance, 366
Insectivora, 920–921, 931
Insemination reactions, in reproductive
    isolation, 171
Instars, 717
Instinctive behavior, 464–465
Insulating barriers, 289
Insulin, *433*, 448
  role in food storage, 271
Integration
  endocrine, 435
  neurons, 435
Integument, 189, 200–204, 203
  invertebrates, 202
  vertebrate, *202*, 203
Intercalated discs, 311
Interconversion of food compounds,
    271
Intercostal muscles, 283
Interleukins, 448
Intermediate host, 628
Internal transport, 296–320
Interneurons, 380
Interphase, 98
Interspecific competition, 972
Interstitial animals, 622
Interstitial cells, 598
Interstitial fluid, 296
Interstitial lamellae, 208
Intertidal zone, 992
  rocky, 984–986
Intervertebral disc, 211
Intervertebral foramina, 386
Intestine, 242, 251, *252*
  large, 251
  small, 251
Intracellular digestion, 241
Introns, 125, *127*
Inulin, 357
Invagination, 514
Inversions, 140, *140*
Involution, 516
Ioxdes, 745
Iris, 425
Irritability, 372
Ischial callosities, 938
Ischium, 214
Islets of Langerhans, 448
Isogametes, 571
Isolating mechanisms, 170–172
Isolation
  geographical, 172
  reproductive, 170–172
Isometric contraction, *219*, 220

Isopoda, 735, 745
Isoptera, 765
Isotonic contraction, 220
Itch mite, 727

J

Jacana, *902*
Jacobson's organ, 406, 870
Jaws, vertebrate evolution, 823
Jejunum, 251
Jellyfish, 596
  cannon ball, *605*
  Hydrozoans, 602
  Scyphozoa, 605
Jenner, 327
Jet lag, 464
Joint, 208, *209*
Julus, 765
Jumping, frogs, 854, *855*
Jumping genes, 139
Jumping spiders, 725
Juxtaglomerular apparatus, *356*, 361

K

K-selection, 972
Kala-azar, 565, 566
Kangaroo, *918*, 930
Kangaroo rats, water balance, 367, *367*
Kaposi's sarcoma, 324
Karyotype, *96*
Katydids, 765
Keratin, 203
Ketoglutarate, 267
Keyhole limpet, 666, *666*
Kidney, vertebrates, 353–361, *354,
    355*
Killer T cells, 335
Kilocalories, 255
Kinase, 436
Kinetochore, fiber, 100
Kinetofragminophorea, 582
Kinetosomes, 225, 573
Kinety, 573
King penguin, *12*
Kingfishers, 901, 905
Kinorhyncha, *654*
Kite, *900*
Kittiwakes, nesting, 475
Kiwis, 898, *900*, 904
Klinefelter's syndrome, 142
Knuckle walking, 940
Koala bears, *919*
Komoda dragon, *869*
Kori bustard, *11*
Krebs cycle, 267
Krill, 737, 746

L

Labia majora, 499
Labia minora, 499

Labial palps, 675
Labium, 754
Labrum, 754
Labyrinthodontia, 849, *851*, 858
Lacewings, 766
Lacrimal glands, 424
Lactase, 237
Lactation, 505
Lactic acid, 265
Lacunae, 207
Lagomorpha, 926, 931
Lake Baikal, 994
Lamellae, bone, 208
Lamellibranchia, 676, 689
Lamp shells, 775
Lampreys, 821, *821*
    immune response, 342
Lancelets, 804–807, *806*, 811
Land crabs, 733
Land snails, 669
    water balance, 366
Langur monkeys, infanticide, 475
Laomedea, 598
Large intestine, 251
Larva, 522
    snail, *523*
Larvacea, 804, *805*, 811
Larvae, echinoderms, *778*
Larynx, 280, *281*
    cartilage, 214
Lasionectes, 744, 745
Latent learning, 467
Latent period, muscle contraction,
    219
Lateral body surface, 186
Lateral line system, 408, *410*, 414
Lateral plate, 521
Latimeria, 837
Latissimus dorsi, 223
Latouchella, 551
Latrodectus, 724, *726*, 745
Laveran, 572
Law of priority, 542
Law of tolerances, 961
Leaf-cutting ants, 469
Leaf hoppers, 765
Leaf mold fauna, 722
Learned behavior, 464–468
Learning, 395
Leeches, *244,* 707–709, 711
Leeuwenhoek, 65
Leishmania, 565, 566
Leishmaniasis, 566
Lek, 472
Lemurs, 931, 936, *937,* 951
Lens, 425
Lens vesicle, 520
Lepas, 745
Lepidodermella, 646, *646*
Lepidoptera, 766

Lepidosauria, 868, 877
Lepidosauriomorpha, 877
Lepospondyli, 849, 858
Leptocephalus larva, 833
Leptogorgia, 618
Leptoplana, 637
Leuconoid sponge, 588, *589*
Leucosolenia, 592
Leukocytes, 300
Leydig cells, *487*
Lice, 759, *760,* 765
Lichenoides, 551
Life, characteristics, 59
Lift, 229–230
Ligaments, 208
Ligamentum arteriosum, 311
Limax, 689
    water balance, 366
Limbic system, 393
Limnetic, zone, 994
Limpets, 665, 688
    keyhole, 666, *666*
Limulus, 719, 745
Linckia, 779
Lineus, 636
Lingula, 774, 775
Linkage, in genetics, 105
Linked genes, 144
Linnaeus, 20, *20,* 542
Lion fish, 401
Lions, cooperative hunting, 476
Lipase, 237
Lipids, 39–41
    digestion, 237, *238*
Lipoprotein lipase, 271
Liposomes, 67, *68*
Lissamphibia, 849, 858
Lithobius, 765
Littoral zone, 992, *993*
Littorina, 688
Liver, 249, 251, 522
    role in food storage, 271
Lizards, 868–869, 877
    viscera, *863*
Loa, 650
Lobosea, 581
Lobster, slipper, *730*
Lobsters, 745
Locomotion, 225–231
    snakes, 872
Locus, 91
Loligo, 657, 680, *683, 686,*
    688
Loons, 903, 905
Loop of Henle, 356, 359
Loose connective tissue, 201, *201*
Lophodont teeth, 925
Lophophore, Bryozoans, 770
Lorenz, 459, 465
Lorica, rotifers, 643

Loricifera, 654, *654*
Lorises, 951
Lovenia, 786
Loxosceles, 725, 745
Lucy, *944*
Lugworms, 697, *699*
Lumbar vertebrae, 212
Lumbricus, 703, *705,* 706, *706,* 711
Lunar cycles, 463
Lungfish, 279, 837
    cocoon, *830*
Lungs, 278–284, 522
    bony fishes, 828
Lunules, 788
Luteinizing hormone, 501
Lycosa, 745
Lygia, 745
Lyme disease, 727
Lymph nodes, 308
Lymphatic return, 319
Lymphatic system, 308, *309*
Lymphocytes, 300
Lymphokines, 337
Lynx, *971*
Lysosomes, 76
Lytechinus, 794
Lytic complex, 334

M

Macaques, *935*
Machilids, 755, 765
Macrocyclops, 740
Macroevolution, 175–180
Macronucleus, 580
Macrophages, 202, 323, *334*
    immune response, 337
    in immune system, 333
    in lungs, 282
    phagocytosis, 73
Macroscelidea, 926, 931
Macrostomida, 637
Macrostomum, 629, 637
Macula densa, 361
Madreporite, 779
Maggots, 757
Magnesium, 255
Major histocompatibility complex, 337
Malacostraca, 731–737, 745
Malaria, 566, 572
    sickle-cell anemia, 164
Malleus, 214, 414
Malpighian tubules, 351, 718
    insects, *353,* 754
Maltase, 237
Malthus, 28
Mammal-like reptiles, 878, *914*
Mammalia, 909–933
    adaptations, 910–914
    characteristics, 909
    classification, 930

gas exchange, 280–284
herbivory, 925, *925*
hind leg, *923*
origin, 914–915
phylogeny, *916*
reproductive strategies, 917
teeth, 912, *913*
thermoregulation, 910
Mammary glands, 914
Manatees, 930, *930*, 931
Mandibles
crustaceans, 729
uniramians, 750
Mandibular arch, 211
Mandibular cartilage, 213
Mandibular gland, 247
Mandibular muscles, 223
Mange mites, 745
Manta ray, *827*
Mantis shrimp, 737, *737*, 745
Mantle
Bivalvia, 673
molluscs, 658
Manubrium, 602
Mapping, chromosome, 145
Marine environments, 992
Marmots, foraging, 468
Marrella, *551*
Marrow cavity, 208
Marsupalia, 930
Marsupial tree frog, *857*
Marsupials, adaptive radiation, 918–919
Marsupium, 917, *918*
crustaceans, 735
Masseter, 223
Mast cells, 302, 339, *339*
Mastax, 643, *644*
Mastigophora, 562–567, 581
Materialism, in science, 7
Maternal effect genes, 533
Mating, Hardy-Weinberg-Castle Law, 157
Mating systems, 475
Mating types, 581
Matrix, connective tissue, 201
Mawsonites, *551*
Maxillae
crustaceans, 729
uniramians, 750
Maxillipeds, 731
Mayflies, 756, *756*, 765
Mayr, 170
McClintock, transposable elements, 138
Meadowlark, *901*
Mechanoreceptors, 406–419
Medulla, kidney, 356
Medulla oblongata, *390*
Medusae
hydrozoans, 602
Schyphozoa and Cubozoa, 605
Medusoid body form, 596, *596*

Megakaryocytes, 300
Megaloptera, 766
Megascolides, 702
Meiocytes, *145*
Meiosis, 101–103, *103*
Meissner's corpuscles, 407
Melanin, in skin, 204
Melanism, industrial, 161
Melanophore-stimulating hormone, 445
Melatonin, 445
Mellita, *789*, 794
Membranelles, 577
Membranous labyrinth, 412, *412*
Memory, 395
Memory B cells, 333
Mendel, *89*
pea chromosomes, 146
Meninges, *396*, 398
Menstrual cycle, 502, *503*, 504
Mercenaria, *674*
Merkel's discs, 408
Meroblastic cleavage, 512
Merostomata, 719–720, 745
Merozoites, 572
Meselson-Stahl experiment, *120*
Mesencephalon, *390*, 391
Mesenchyme, 518
Mesenterial filament, 610
Mesenteries, Anthozoans, 608
Mesoderm, 514
Mesoderm differentiation, 520, *520*
Mesoderm formation, *515*
Mesoglea, 596
Mesohyl, 586
Mesonephros, 353
Mesostoma, 637
protonephridium, *351*
Mesozoa, 637
Messenger RNA, 124
Metabolic rate, 288–290
body size, 290
temperature, 288–290
Metabolic wastes, 348
Metabolism, 262
glucose, 448
Metacarpals, 214
Metacercaria, 631
Metachronal ciliary waves, 225, *225*
Metamerism, 190, *190*
Annelids, 694, *695*
Metamorphosis
complete, 757
gradual, 757
incomplete, 757
Metanephridia
earthworm, *352*
molluscs, 659
Metanephridium, 351
Metanephros, 353

Metaphase
meiosis, 101–102
mitosis, 100
Metatarsals, 214
Metatheria, 917, 930
Metazoa, 548
Metencephalon, *390*, 391
Meteorites, origin of life, 57
Metridium, 618
Micelles, 67, *68*
Microciona, 592
Microevolution, 168–175
forces, 160–168
Microfilariae, 651
Micronuclei, 580
Microscope, 82
Microseres, 983
Microspora, 572, 581
Microstomum, 626, 637
Microtubules, 79
in centriole, *100*
of spindle, 100
Microvilli, 253, *253*
Midbrain, 391
Midges, 766
Midgut, 718
Midsagittal plane, 186
Migration
affecting Hardy-Weinberg-Castle Law, 159
birds, 895, *896*
Milk secretion, 506
Millepora, *600*, 601, 617
Miller, 57
Millipedes, *750*, 751, 765
reproductive isolation, *171*
Mimicry
aggressive, 469
Batesian, 470, *470*
Mullerian, 471
Mineral balances, *452*
Mineralocorticoids, 451
Minerals, 254, 255
Miniaturization, 191
Minks, 931
Miracidium larvae, 631
Mites, 721
Mitochondria, 76–78, 77
Mitochondrial DNA, 78
Mitochondrial Eve, 952
Mitochondrion
in cellular respiration, 266, *266*
in chemiosmosis, *269*
Mitosis, 98–101, *99*
Mitotic spindle, 100
MN blood groups, frequencies, 157
Moas, 898, 904
Mockingbirds, Galapagos, 27
Mole crabs, 733, *733*
Moles, 544, 920, *921*, 931

Mollusca, 657–691
  Bivalvia, 673
  body plan, 658–661, *659*
  Cephalopoda, 682–687
  characteristics, 657
  classification, 688
  evolutionary relationships, 687
  Gastropoda, 662–673
  Monoplacophora, 661
  Polyplacophora, 661, 662
Moloch, *869*
Molting
  arthropods, 206, *206*
  crayfish, *443*
  insect, *443*
Molting-inhibiting hormone, 441
Monera, 547
Monkeys, 931, *938*, 951
  New World, 938
  Old World, 938
Monoclimax hypothesis, 983
Monocytes, 300
Monogamy, 476
Monogenea, 633
Monohybrid crosses, 90, *92*
Monophyletic taxon, 543
Monoplacophora, 661, 688
Monorhina, 819, 838
Monosaccharides, 37
Monotremata, 916, 930
Monounsaturated fatty acids, 39
Montastrea, *611*, 618
Moon shells, 688
Morgan, gene linkage, 144
Morphogenetic movements, 513,
  532
Morphogens, 533
Mortality, 964
Mosaic development, 529, *529*
Mosquitos, role in malaria, 572
Moths, 760, 766
Motility, and symmetry, 186
Motor cortex, 395, *395*
Motor end plate, 218
Motor program, behavior, 459
Motor unit, 218
Mount St. Helens, 982
Mouth, 246
Mouth cavity, 242
Movement, 214–231
Mucosa, 253
Mucous membrane, *252*
Mucus, in movement, 225
Mud puppy, *852*
Mudskipper, *836*
Muggiaea, *602*
Mules, hybrid sterility, 171
Mullerian mimicry, 471, *471*
Multiple alleles, 150
Multiunit smooth muscle, 218, *219*

Muscle, 214–223, *215*
  cardiac, 311, *312*
  physiology, 218–221
  structure, 214–216
Muscle spindles, 419, *419*
Muscularis, 253
Mussel shrimp, 745
Mutagens, 136
Mutants, homeotic, 534
Mutation, 136–144
  consequences, 147–150
  direction and rate, 149
  point, 136–137
  random nature, 150
Mutations, affecting Hardy-Weinberg-
  Castle Law, 157
Mutualism, definition, 192
Myelencephalon, *390*, 391
Myelin sheath, 372, *374*
Myoepithelial cells, 214
Myofibrils, 214, *217*
Myofilaments, 216
Myoglobin, 220
Myomeres, 804
  fish muscles, 222
Myonemes, 576
Myosin, 216
Myotome, 521
Myriapods, 750–752
Myxiniformes, 822, 838
Myxozoa, 572

N

Nacre, 673
NAD. *See* Nicotinamide adenine
  dinucleotide
Nanaloricus, *654*, 655
Nannoplanktron, 565
Natality, 964
Natural selection, development of
  Darwinian concept, 27–29
Natural selection, 160–161
  affecting Hardy-Weinberg-Castle Law,
  159
Natural theology, 21
Naupliar eye, 730
Naupliar larva, *730*, 731
Nautiloidea, 689
Nautilus, 682, *684*, 685, 689
Navigation, birds, 895–897
Neanderthals, 950
Necator, 650
Nectarina, *472*
Nectocaris, *551*
Negative feedback system, 440
Negative work contraction, 220
Nematocyst, 597, *598*
Nematoda, 646–652, 653
  succession in dung, 983
Nematomorpha, 653, *653*

Nematoplana, *622*
Nemertea, 636
Neognathae, 898, 904
Neopallium, 393
Neopilina, *661*, 688
Neoptera, 765
Neopterygii, 830, 838
Neornithes, 898, 904
Neoteny, 447
  salamanders, 852, *852*
Nephric ridge, 521
Nephridia, 350
  molluscs, 659
Nephridiopore, 350
Nephrogenic mesoderm, 353
Nephron, 353, 355, *356*
Nephrostome, 351
Nereis, *694*, *696*, 711
Neritic zone, 992, *993*
Nerve cord, Chordates, 802
Nerve impulse, 374–377, 376, *376*, 377
  velocity, 377
Nerve net, 382
Nerve nets, Cnidarians, 598, *599*
Nerves, 380
  cranial, 387, 388
  spinal, 386, *387*
  splanchnic, 388
Nervous integration, 435
Nervous system, 371–399
  autonomic, 388, *389*
  control, 372
  evolution, 381–385
  invertebrate, *386*
  organization, 379–381
  parasympathetic, 388, *389*
  peripheral, 286, 372
  sympathetic, 388, *389*
  vertebrate, 385
Neural crest, 518, 808
Neural tube, 517, *518*
Neuraminidase, 343
Neurilemma cells, 372
Neuroendocrine reflexes, 506
Neuroglial cells, 372
Neurohypophysis, *444*, 445
Neuromasts, 408
Neuron pools, 381, *383*
Neurons, postganglionic sympathetic,
  372–374, *373*
  motor, 380, 386
  sensory, 380, 386
  visceral motor, 386
Neuropodium, 695
Neuroptera, 766
Neurosecretory cells, *435*, 436
Neurotransmitter, 379, 397
Neurulation, 517
Neutrophils, 300
Newt, *853*

red-spotted, *470*, 471
Niche, ecological, 984, *984*
Nicotinamide adenine dinucleotide
    (NAD), 263
Nictitating membrane, 424
Nighthawks, 905
Nitrogen cycle, 1005, *1006*
Nitrogen excretion, 849
Nitrogen-fixing bacteria, 1006
Nitrogenous wastes, 348
Noctiluca, 564
Nodes of Ranvier, 372
Nonion, *570*, 581
Noradrenalin. *See* Norepinephrine
Norepinephrine, 397, 450
Nosema, 582
Nostrils, 280
Notochord, 802
Notophthalmus, *470*
Notopodium, 695
Nuchal organs, 701
Nucleases, 239
Nucleic acid, 46–49, *46*
    as hereditary material, 113
    origin of life, 57
Nucleolus, 79
Nucleosome, 78
Nucleotides, 46
Nucleus, 78–79, *78*
    and genes, 110
Nucula, 689
Nudibranchs, 668
Numerical taxonomy, 545
Nummulites, *570*, 581
Nutrition, 254–257
Nutritive muscle cells, 596
Nymphon, *728*, 745

O

Obelia, *600*, 603, 617
Oceanic zone, 992, *993*
Ocelli, 419
    hydrozoans, 602
    insects, 754
Octocorallia, 618
Octocorals, 612–614, *613*
Octopods, 685, *688*, 689
Octopus, 657, 689
Oculina, 618
Ocypode, 732
Odonata, 765
Odontognathae, 898, 904
Odontophore, 658
Olenellus, *551*
Olfactory bulb, *390*, 393
Olfactory epithelium, *405*
Oligochaeta, 702–707, 711
    nutrition, 703
    physiology, 704
Oligodendrocytes, 372

Oligohymenophorea, 582
Omasum, 244
Ommatidia, 422
Omnivore, 236
Onchocerca, 650
Oncogenes, 136
Oncosphere larva, 634
Oniscus, 745
Onychophora, 764, *764*
Onycoteuthidae, 685
Oocytes, 490
Oogenesis, *488*, 489–490
Oogonia, 489
Ootid, 490
Oozes, marine, 571
Opalinata, 581
Oparin, 57
Open instincts, 465
Opercular apparatus, 847
Operculum
    cartilaginous fishes, 826
    fishes, 275
    gastropods, 665
Ophidia, 870
Ophioderma, 794
Ophiopluteus larva, 785
Ophiotrix, 794
Ophiuroidea, 783–785, 794
Opilioacariformes, 745
Opiliones, 745
Opisthaptor, 633
Opisthobranchia, 668, *669*, 688
Opisthonephros, 353
Opisthorchis, 630, 631, *632*
Opossums, *918*, 930
    behavior, 470
Opsonization, 333
Optic chiasma, 391, 429
Optic cup, 520
Optic disc, 427
Optic lobes, 391
Opuntia, *23*
Oral arms, 606
Orangutan, 938, *939*, 954
Orb-weaving spiders, *466*
    behavior, 460
Orb web, 725
Organ of Corti, 416
Organelles, of eukaryotic cell, 73
Organic compounds, 35, *35*, 36, *36*
Organizer, 530
Organogenesis, 517–522
Oribatid mites, 722
Origin, 215
Origin of life, 54
    complex molecules, 56
    monophyletic, 56
Origin of new species, 168–175
Origin of Species, On the, Darwin, 21
Orionina, *741*

Ornithischia, 876, 878
Orthoptera, 765
Osculum, 586
Osmoconformers, 361
Osmoregulation, 363
Osmoregulators, 362–363
Osmosis, *70*, 70–71
Osmotic regulation
    freshwater animals, 363–364
    marine animals, 361–363
    terrestrial animals, 364–367
Osphradium, 660
Ossicles, Echinoderms, 779, *781*
Ostariophysi, 839
Osteichthyes, 828–838, *838*
    basic features, 828
    characteristics, 828
Osteocyte, 208
Osteoglossomorpha, 838
Ostium, 499
Ostracoda, 740, *741*, 745
Ostracoderms, 819–820, *820*
Ostriches, *11*, *900*, 904
Otic capsule, 211
Otoliths, 412
Ovarian cycle, 502, *503*
    hormonal control, *504*
Ovarian follicle, 502
Ovary, 489, *490*, *503*
Overturns, lakes, 995
Ovicells, 772, 773, *773*
Oviducts, 491
Ovigerous legs, 728
Oviparous, 524
Ovipositor, 763
Ovisacs, 499
Ovovivparous, 524
Ovulation, 490
    cyclic, 505
    induced, 505
Owls, 901, *901*, 905
    eye, 890
Oxaloacetate, 267
Oxidation-reduction reaction, in cellular
    respiration, 262
Oxygen
    availability in environment, 272
    role in cellular respiration, 267–268
Oxygen conformers, 272
Oxygen debt, 265
Oxygen dissociation curves, 286, *287*
Oxygen regulators, 273
Oxyhemoglobin, 285
Oxytocin, 445
Ozone layer, 1001

P

P-elements, 139
Pacinian corpuscles, 408
Paddlefish, 830

Pagurus, 746
Painted turtle, *867*
Palate, 248
Palatoquadrate cartilage, 213
Paleognathae, 898, 904
Paleopallium, 393
Paleoptera, 765
Pallial line, 673
Palolo worm, 702, *702*
Palps, Annelids, 695
Palythoa, 618
Pampas, 990
Pancreas, 249, 251, 522
Pangaea, 174
Pangolins, 919, *921*, 931
Papilio, disruptive selection, *167*
Papulae, 779, *781*
Paracanthopterygii, 839
Paragonimus, 630
Paramecium, *577*, 582
    buccal ciliature, 577
    conjugation, 580
    food vacuole, 578
    movement, 225
    pellicle, 576
    swimming, 576
Paramylon, 563
Paramyosin muscle, 805
Paraphyletic taxon, 547
Parapodia, 695, *696*
Parasagittal plane, 186
Parascaris, *647*
Parasitiformes, 745
Parasitism, 245, 628–629
    definition, 193
    insects, 759
Parasympathetic nervous system, 388
Parathyroid glands, 453
Parenchyma, 624
Parenchymula larva, *592*
Parietal eye, 445, *864*
Parotid gland, 247
Parrots, 905
Parthenogenesis, 485
    insects, 763
    rotifers, 645
Partial pressure, 272
Parturition, 506
Passeriformes, 905
Passive immunity, 325
Pasteur, 65
Patiria, 794
Pattern formation, 532
Pauropoda, 765
Pauropus, 765
Paxillae, *781*
Peanut worms, 709
Pearl oyster, 673
Pearlfish, 790
Pearls, 673

Peas, Mendel, 90
Pecking order, 471
Pectoral girdle, 211
Pectoralis, 223
Pedal disc, 610
Pedicel, 775
Pedicellaria, 780, *781*
Pedipalps, 719
Peduncle, 741
Pelecaniformes, 905
Pelecosauria, 914
Pelecypoda, 673
Pelicans, 903, *903*, 905
    diving, 281
Pennaeus, 746
Pennatulacea, 618
Pentamerous radial symmetry, 778
Pentose, 37
Peppered moth, 161, *161*
Pepsin, 237
Pepsinogen, 239
Peptidases, 237
Peptide bond, 41
Peptidyl transferase, 129
Peranema, 563, *564*, 581
Perch, internal anatomy, *829*
Perching mechanism, *901*
Perforins, 337
Pericardial cavity, 659
Pericardial sinus, 718
Perichondrium, 207
Periclimenes, 458, *458*
Perimysium, 215
Periosteum, 208
Periostracum, 658
Peripatric speciation, 175
Peripheral nervous system, 372, 386
Periphylla, 618
Periproct, 785
Perissodactyla, 927, 931
Peristalic locomotion, 226
Peristalsis, 249
Peristomium, 695
Peritoneum, 190, 253
Peritrophic membrane, 758
Peritubular capillaries, 357, 360
Periwinkles, 688
Permafrost, 991
Peromyscus, 467
Peroxisomes, 76
Petaloid, 788
Petrels, 903
Petromyzontiformes, 821
PGAL. *See* Phosphoglyceraldehyde
Phaedarea, 581
Phaenicopteriformes, 905

Phagocata, 637
Phagocytosis, 73, 299
Phalanges, 214
Phanocrinus, 541
Pharyngeal pouches, 522
    hemichordates, 799
Pharynx, 242, 248, 522
    anthozoans, 608
Phasic muscle, 220
Phenetics, 545
Phenogram, 547
Phenotype, 91
Phenotypic distribution, 164–165
Phenylketonuria, 116, 149
Pheromones, 406, 474
    insects, 760
Phidippus, 745
Philodina, *643*
Pholidota, 919, 931
Phonation, 280
Phoronida, 774
Phoronis, 774
Phosphagens, 218
Phosphate homeostasis, 453
Phosphoglyceraldehyde (PGAL), 263
Phospholipids, 41
    cell membrane, 67, *68*
Phosphorus, 255
Phosphorus cycle, 1006
Phosphorylation, 263
Photinus, 760
Photochemistry of vision, 426, *427*
Photophores, 686
Photoreceptors, 419–429
Phthiraptera, 765
Phyletic gradualism, 178
Phylogenetic systematics, 547
Phylogenetic tree, 545
Phylogeny, 542–547
Physa, 689
Physalia, *602*, 617
Physiological adaptation, 403, *404*
Phytoflagellates, 562–565
Phytomastigophorea, 581
Phytoplankton, 992
Piciformes, 905
Pigeons, 905
Pigs, 931
Pikas, 926, *926*, 931
Pill bugs, 735
    water balance, 366
Pinacocytes, 586
Pinacoderm, 586
Pinctada, 673
Pineal eye, 445
Pineal gland, 445
    circadian rhythms, 464
Pinnules, 791
Pinworms, 652
Pisaster, 794

Piscicola, 707, 711
Pituitary gland, 442–445, *444*
Placenta, 525, *526*, 917
Placental lactogen, 506
Placentation, 524
Placodermi, 824, *824*, 838
Placodes, 520
Placoid scale, *825*
Placozoa, 593
Planarians, 622, *624, 625, 626, 627,*
    637
Plankton, 245
Planospiral shells, 663
Plantae, 548
Plantigrade posture, 912
Planula larva, 599, *599*
Plasma, 297–299
Plasma cells, 333
Plasma membrane, *69*
Plasmids, 139
Plasmodium, 566, 572, *573*, 581
    sickle-cell anemia, 164
Plastron, 866
Platelets, 300
Platyhelminthes, 621–639
    characteristics, 621
    evolutionary relationships, 634
    parasitic classes, 628–634
    Turbellaria, 623–628
Platyias, *641*
Platypus, 916, *917*
Platyrrhini, 938, 951
Plecoptera, 765
Pleistocene, allopatric speciation, *173*
Pleopods, 731, 732
Plesiosaur, *873*
Plesiosauria, 877
Pleura, 281
Pleurobrachia, 617, *617*
Pleuroperitoneal cavity, 278
Plovers, 905
Pneumatic bones, 888
Pneumocystis carinii, 324
Pocket mouse, urine concentration, 361
Podia, 779, 782
Podicipediformes, 905
Podocyte, 351, 357
Pogonophora, 709, 710
Poikilothermic animals, 288
Point mutations, 136
Poison, arachnids, 724
Poison arrow frog, *857*
Poison claws, *751*
Polar body, 490
Polar filament, 572, *573*
Polarization, sky, 763
Polycelis, *622*, 637
Polychaeta, 695–702, 711
    adaptive groups, 695–700
    physiology, 701

reproduction, 701
Polychoerus, 637
Polycladida, 637
Polycystic kidney disease, 116
Polycystinea, 581
Polydesmus, 765
Polygenic inheritance, 163–165, *165*
Polygenic traits, 97
Polygyny, 476
Polyhymenophorea, 582
Polymorphism
    alleles, 164
    morphological, 192
        hydrozoans, 601
        insects, 761
Polypeptides, 41
Polyphyletic taxon, 544
Polyplacophora, 688
Polypoid body form, 596, *596*
Polyribosome, 129
Polysaccharidases, 237
Polysaccharide, 37
Polytomella, *563*
Polyunsaturated fatty acids, 39
Pongidae, 938, 954
Pons, *390*, 391
Population
    age and sex distribution, 965, *967*
    density, 964
    dynamics, 964–966
    evolution, 154–160
    fluctuations, 970
    growth, 966–974
    human, *974*, 974–976
    regulation, 970
    survivorship curves, 965
Population ecology, 960–976
Population genetics, 154–160
Porcupines, 926
Porifera, 585–593
    characteristics, 585
    evolutionary relationships, 592
    regeneration and reproduction,
        590–591
    structure and function, 586–590
Pork tapeworm, 634
Porocytes, 586
Porpoises, 928
    osmoregulation, 363
Portuguese man-of-war, 601, *602*, 617
Portunid crabs, 732
Posterior end, 186
Postganglionic fibers, 388
Postnatal development, 528
Postsynaptic inhibition, 379
Postzygotic isolating mechanisms, 171
Potassium, 255
Potassium channels, 374
Prairies, 990
Precapillary sphincters, 316

Preganglionic fibers, 388
Pregnancy, 505
Pressure drag, 226, *226*
Preying mantids, 765
Prezygotic isolating mechanisms,
    170
Priapulida, 655
Priapulus, 655
Primary features, 187
Primary host, 628
Primary production, 998
    oceanic, 992, *994*
Primary structure, 41
Primary succession, 981–982
Primates, 924, 931, 935–955
    adaptations, 936
    brain evolution, 948
    cladogram, *937*
    classification, 951
    groups, 936
    hands and feet, *939*
Primer, DNA replication, 119
Primitive species, 554
Primitive streak, 516
Proboscidea, 929, 931
Proboscis
    elephants, 929
    insects, 758, *759*
Procellariiformes, 905
Proctodeum, 718
Procuticle, 206
Producers, 998
Production, ecological, 998
Proenzyme, 239
Profundal zone, 994
Progesterone, 501
Proglottids, 634
Programmed learning, 465
Prokaryota, 547
Prokaryotic cell, 66
Prolactin, 445
Promoter, 125
Pronephros, 353
Pronucleus, 491
Propagule, speciation, 172
Prophase
    meiosis, 101–102
    mitosis, 98
Proprioceptors, 417–419
Prorodon, *574*, 582
Prosimii, 936, 951
Prosobranchia, 666, 688
Prostaglandins, 454
Prostheceraeus, *623*, 637
Prostoma, *636*
Prostomium, 694, 695
Protandry, 495
Proteases, 237
Protein deficiency, 255, *256*
Proteins, 41–45, *44*

Proteins *(continued)*
  deficiency, 255, *256*
  digestion, 237–239, *238*
  enzymes, *45*
  membrane, 68
  metabolism, 348
  structure, 41–45, *43, 44*
  synthesis, 124–131, *131*
  turnover in tissues, 271
Proteoglycans, 201
Prothrombin, 300
Protoacanthopterygii, 839
Protobranchia, 675, *677*, 689
Protoctista, 547, 561
Protogyny, 495
Protonephridia, 350, *351*
  turbellarians, 626
Protoplasm, 67
Protopodite, 730
Protostomes, 554, 658
Prototheria, 915–917, 930
Prototroch, 660
Protozoa, 561–582
  characteristics, 561
  Ciliophora, 573–581
  Sarcomastigophora, 562–571
  sporozoans, 571–573
Protura, 766
Proventriculus, 758
Proximal convoluted tubule, 356
Proximal end, 186
Pseudocella, 647
Pseudocoel
  aschelminths, 642
  embryonic origin, *642*
Pseudopodia, 223
Pseudoscorpionida, 745
Pseudoscorpions, 721, *723*
Pseudotriton, *470*
Psittaciformes, 905
Pteridinium, *551*
Pterobranchia, 800, *801*, 811
Pterosauria, 877, 878
Pterygota, 755, 765
Ptilocrinus, *791*
Ptilosarcus, *614*
Pubis, 214
Pulmonata, 669, 689
Pumpkin-seed sunfish, behavior, 476
Punctuated equilibrium, 176, *177*
Punnett square, *92*
Pupil, 425
Purine, 46
Purkinje cell, *373*
Purkinje fibers, 314
Pycnogonida, 728, *728*, 745
Pycnopodia, *794*
Pygidium, 694
Pygostyle, 888
Pyloric ceca, 830

Pyloric sphincter, 249
Pyloric stomach, 734
  echinoderms, 782
Pyramids of numbers and biomass, 1000, *1000*
Pyrimidine, 46
Pyruvate, 263
Python, *871*

Q
Q_{10} value, 288
Quahog, *674*

R
r-selection, 974
Rabbits, 926, 931
Racoons, 931
Radial cleavage, *511*
Radial symmetry, 187
Radiations, 148, 178
Radiolarians, 569–571, *571*, 581
Radioles, 698
Radius, 214
Radula, 658, *660*
  gastropods, 670, *671*
Rails, 902, *902*
Ram ventilation, 276, 835
Range, geographical, 960–962, *962*
Ratfishes, *827*
Rattlesnake, *873*
Rays, 828
Razor clam, allometric growth, *192*
Reabsorption, tubule, 358–359
Reaction kinetics, 64
Receptor mechanisms, 402–404
Receptor potential, 402
Receptors, 372, 402
  equilibrium, *410*
  gravitational, 408–413, *410*
  gustatory, 406
  motion, 408
  olfactory, 404
  phasic, 402
  pressure, 407, *409*
  sound, 413–417
  tonic, 402
  touch, 407, *409*
  types, 402
Recessive alleles, 90
Recessive lethals, 148
Recombination, 486
  chromosomal, 144–147
  chromosome, *103*
Rectal ceca, 782
Rectal gland, sharks, 363, 825
Rectum, 242, 251
Red blood cells, 299
Red bone marrow, 208
Red eft, *470*, 471
Red salamander, *470*, 471

Red-spotted newt, *470*
Red tides, 565
Red-water fever, 572
Redbugs, 727
Rediae, 631
Reduction division, 101
Reef flat, 615
Reefs, 614
Reflex, neuroendocrine, 440, *441*
Reflexes, 381
Refraction of light, 420
Regeneration, sponges, 590
Regulative development, 529
Relaxin, 506
Releaser, 459
Releasing mechanism, 459
Remipedia, 743, *744*, 745
Renal corpuscle, 355
Renal pelvis, 356
Renal portal system, 305
Renal pyramids, 356
Renal threshold, 359
Renal tubule, *356*
Renilla, 618
Renin, 361
Rennin, 249
Replication, DNA, 119–121
Reproduction, 484–507
  asexual, 484
  fertilization, 490–496
  gametogenesis, 486–490
  hormones, 501–505
  modes, 484–486
  sexual, 485
  vertebrate pattern, 496–501
Reproductive isolation, 170–172
Reproductive strategies, 476
Reproductive synchrony, 491
Reproductive system
  invertebrate, *497*
  vertebrate female, *500*
  vertebrate male, *498*
Reproductive tract design, 496
Reptilia, 961–879
  adaptations, 863–865
  characteristics, 861
  classification, 877
  evolution, 865
  skull types, 866, *866*
  water balance, 366
Residual air, 284
Resource partitioning, 986–987
Respiration
  cellular, 262–271
    aerobic respiration, 265
    ATP yield, 269
    controls, 269
    fuels other than glucose, 270
  gas exchange, 272–288
Respiratory pigments, 184–185

Respiratory trees, 789, *790*
Resting potential, 374, *375*
Restricted learning, 465
Reticulopods, 568
Reticulum, 243
Retina, 425–429, *426, 428*
Retinal, 427
Retinol, 427
Retinular cells, 422
Retroperitional condition, 190
Rh antigen, 339
Rhabdites, 625
Rhabdocoela, 637
Rhabdome, 422
Rheas, 904
Rheiformes, 904
Rhesus monkey, *938*
Rheumatoid arthritis, 338
Rhinoceros, *927,* 928, 931
Rhipidistia, 837, 839
Rhizopoda, 567, 581
Rhizostomae, 618
Rhodopsin, 419, 427
Rhopalia, 607
Rhopalura, 637
Rhynchocephalia, 868
Rhynchocoela, 636, 636
Rib, 211
Ribbon worms, 636
Ribonucleic acid
    origin of life, 59
    structure, 125
Ribonucleic acid polymerase, 125
Ribonucleic acid (RNA), 46, *46*
Ribosomal RNA, 127
Ribosomes, *127*
    endoplasmic reticulum, 75
    translation, 127–128
Ribozymes, 59
Richards, 357
Rift Valley lakes, 994
Rigor mortis, 218
Ring dove, behavior, 461, *461*
Ritualized combat, 471
RNA. *See* Ribonucleic acid
Roach, wood, cellulose digestion, 566
Roadrunner, *900,* 901, 904
Robberflies, *757*
Rocky Mountain spotted fever, 727
Rodentia, 926, 931
Rodents, 931
Rods, 426, *426*
Roly-polies, 735
Ross, 572
Rostroconchia, 673
Rotifera, 643–645, 653
Roundworms, 646–652
Ruffini's end organs, 408
Rugosa, 618
Rugose corals, 616

Rumen, 243
Running, mechanics, 228

S
Sacculus, 412
Sacral vertebrae, 212
Sacrum, 212, *213*
Sagittal plane, 186
Salamander, 859
    adaptations, 850–853, *851, 852*
    red, *470,* 471
    reproduction, 853
Salinity, oceanic, 992
Salivary glands, 247
Salps, 804, *805*
Salt glands, 363
Salt lick, *255*
Saltatory conduction, *377*
Sand dollars, 786, *786, 789,* 794
Sarcodina, 567, 581
    amoebas, 567–568
Sarcolemma, 214, *217*
Sarcomastigophora, 562–571, 581
    flagellates, 562–567
Sarcomere, 214, 216
Sarcoplasmic reticulum, *217*
Sarcopterygii, 828, 830, 837, 839
Sarcoptes, 727, 745
Saturated fatty acids, 39
Saturation deficit, 365, *365*
Saurischia, 874, 877
Sauropterygia, 873, 877
Savannas, 990
Sawfish, *827*
Scabies, 727
Scales
    ctenoid, *832,* 833
    cycloid, *832,* 833
    in vertebrate integument, 203, *203*
Scales ganoid, 830
Scallops, 680, *681*
    adductor muscle, 220
Scanning electron microscope, 82
Scaphopoda, 689
Scapula, 214
Scarcopterygii, *831*
Scars, bivalve shells, 673, *674*
Scent glands, in vertebrate integument,
    203
Schistosoma, 630, 632, *633*
    camouflage strategy, 344
Schistosomiasis, 630, 632
Schizocoel, 515
Schizocoely, 554
Schizogamy, 572
Schleiden, 65
Schwann, 65
Schwann cells, 372, *374*
Science, characteristics, 6–9
Sclera, 425

Scleractinia, 618
Scleractinian corals, *611,* 611–612
Sclerospongiae, 592
Sclerotization, *716*
Scolex, 634
Scolopendra, 765
Scoloplos, 696, 711
Scopelomorpha, 839
Scopes Trial, 5–6
Scorpionida, 745
Scorpions, 720, *721, 727,* 745
Screw-worm fly, 760
Scrotum, 496
Scurvy, 256
Scutigera, 751, 752, 765
Scutigerella, 765
Scyphistoma, 607
Scyphozoa, 605–608, 617
Sea anemones, *609,* 609–611, 618
    immune response, 342
Sea bass, 815
Sea butterflies, 668, 669, *669,* 688
Sea cucumbers, 788–790, *789, 790,* 794
    resource partitioning, 986, *986*
Sea dollars, 794
Sea fans, 614, 618
Sea hare, 668, *669,* 688
Sea lilies, 791, *791,* 794
Sea lions, *922*
Sea nettles, *605,* 606
Sea pansies, 614, 618
Sea pens, 614, *614,* 618
Sea rods, 614, *614,* 618
Sea slug, 668, *669,* 672
    behavior, 460, *460*
Sea spiders, 728, *728,* 745
Sea squirt colony, *193*
Sea stars, 779–783, 794
Sea turtle, *867*
Sea urchins, 785–786, *786, 787, 788,*
    794
Sea walnuts, 616
Sea wasps, *606,* 607, 618
Sea whips, 614, 618
Seal, elephant, dominance, 472
Sea, lions, 923
Seals, 923
    diving, 281
    osmoregulation, 363
Sebaceous glands, 204
Second messenger, 436
Secondary features, 188
Secondary palate, 913, *913*
Secondary succession, 982
Secondary sympatry, 173, 175
Secretin, 436, 448
Secretion
    tubular, 359
    wastes, 350
Sedgewick, and Darwin, 23

Seed shrimp, 740, 745
Segment, metamerism, 190
Segmentation, 253
Segmentation genes, 533
Segments, annelids, 694, *695*
Segregation, Law of, 93
Selection
   on complex traits, 163
   directional, 167
   disruptive, 167, *167*
   epigamic, 475
   and fitness, 162
   natural, 160–161
   stabilizing, 166
Selective coefficient, in relation to
      fitness, 162
Selective reabsorption, 350
Selenocytes, 807
Selenodont teeth, 925
Self and nonself recognition, 337
Semaeostomae, 618
Semicircular ducts, 412
Seminal receptacle, 495
Seminal vesicle, 494
Seminiferous tubules, 486, *487*
Sense organs, 402
Sensilla, 404, *405*, 719
Sensory coding, 403, *403*
Sensory cortex, *395*
Sensory placodes, 808
Sepia, *683*, 689
Sepioteuthis, *473*
Septa, annelids, 694
Septibranchia, 689
Serengeti, resource partioning, 987
Serial homology, 717
Serotonin, 397
Serpent stars, 783
Serpentes, 870
Sertoli cells, 486, *487*
Serum, 301
Sessility, 186
Setae
   annelids, 695
   function, 228
Severe Combined Immunodeficiency
      Disease, 324
Sex chromosomes, 104
Sex determination, 535
Sex differentiation, 535
Sex linkage, 104
Sex-linked traits, 104, *105*
Sexual behavior, 475–476
Sexual competition, 475
Sexual reproduction, 485
   advantages, 486
Sharks, 826
   muscular system, *222*
   osmoregulation, 362
   skeleton, 211, *211*

Sheep, 928, 931, *971*
Sheep liver fluke, *631*
Shell
   Bivalvia, 673
   molluscs, 658
Shields, 784
Shipworms, 681
Shivering, 289
Shrew, 920, *921*, 931
   following response, *467*
Shrimp, *730*, 745
Shrub biomes, 991
Siamang, 954
Sickle-cell anemia, 116
   and genetic variation, 164
Sickle-cell hemoglobin, *138*
Sideneck turtle, *867*
Sign stimulus, 459
Siliceous compounds, skeletons, 205
Silk, 722
Silkworm moths, pheromones, 474
Silverfish, 755, *755*, 765
Sinoatrial (SA) node, 312
Sinus venosus, 304
Siphon, gastropods, 668, *668*
Siphonaptera, *760*, 766
Siphonoglyph, 610
Siphonophora, 601, 617
Siphons, bivalves, 679
Siphuncle, 684
Sipuncula, 709, *710*
Sirenia, 930, 931
Size
   as a factor in animal form, 190–191
   metabolic rates, 290
Skates, 828
Skeletal muscle, 214, *215*
Skeleton shrimp, 735, *736*
Skeletons, 204–214
   exoskeleton, 205
   fish, 211
   human, *212*
   Tetrapod, 211
   vertebrates, 210–214
Skin
   defense, 324
   vertebrate, *202*, 203
Skin friction, 226
Skinner, 459
Skogsbergia, *741*
Skull, 213, *213*
Sleep-wake cycle, 464
Sleeping sickness, antigenic variation,
      343
Sliding filament hypothesis, 216
Slipper shells, *495*, 672
Slit shells, 666
Sloth, *921*, 931
Slugs
   land, 669, *670*

sea, 668, *669*, 672
   water balance, 366
Small intestine, 251, *252*
Smallpox, 327
Smooth endoplasmic reticulum, 73
Smooth muscle, 214, *215*, 218
   innervation, *219*
Snails, 662
Snake, skull, *871*
Snakeflies, 766
Snakes, 870–873, 877
   water balance, 366
Snow buntings, monogamy, 476
Soaring, 883, *883*
Social behavior, 476
Sociobiology, 478
Sodium, 255
Sodium channels, 374
Sodium-potassium pump, 241, 376
   membrane transport, 71, *71*
Soft corals, 614, *614*
Solar radiation, 1000
Solemya, 689
Solenogastres, 688
Somatic muscles, 221
Somatic mutations, 136
Somatic recombination, 329
Somatic skeleton, 210
Somatomedins, 446
Song learning, birds, 465, *467*
Songbirds, 901, 905
Sound waves, *413*
Sow bugs, 735
Sparrow, behavior, 462, *463*
Spatangus, 794
Speciation, allopatric, 172
Species, origin, 168–175
Species concept, 170
Species diversity, 984
Species selection, 178
Specific name, 542
Sperm, 485, *490*
Sperm ducts, 491
Sperm web, *728*
Spermatids, 486, *489*
Spermatocytes, 486
Spermatogenesis, 486–489, *488*
Spermatogonia, 486
Spermatophores, 494, *495*
   arachnids, 727
   cephalopods, 686
   insects, 763
Spermatozoon, 486, *489*
Sphaeroma, *736*, 745
Sphenisciformes, 905
Sphenodonta, 868, 877
Spicules, sponges, 586, *587*
Spider crab, osmoconformity, 362
Spider mites, 723, 745
Spider wasps, 760

Spiders, 721, *721*, 745
  orb-weaving, behavior, *466*
  water balance, 366
Spinal cord, *380*, 381, 382
Spinal ganglion, 380
Spinal nerves, 386, *387*
Spindle, 100
Spines, Echinoids, 785
Spinnerets, 722
Spiracle, 277, *277*
  sharks, 825
Spiral cleavage, 512, *512*
Spiral valve, 825, *826*
Spirostomum, 575, 582
Splanchnic nerves, 388
Spleen, 308
Spliceosome, 125
Splicing, transcription, 125
Sponges, 585–593
  characteristics, 585
  evolutionary relationships, 592
  immune response, 342
  regeneration and reproduction,
    590–591
  structure and function, 586–590
Spongilla, 592
Spongin, 586, *587*
Spongocoel, 586
Spongy bone, 208, *209*
Sporocyst, 631
Sporozoans, 581
Sporozoite, 572
Sporozonas, 571–573
Spot desmosome, *200*
Squalomorpha, 826, 838
Squamata, 877
Squamous epithelium, *200*, 201
Squids, *683*, 685, *686*, 689
  alarm pattern, *473*
Squilla, 745
Squirrel fish, 401
Stabilizing selection, 166
Staghorn coral, 612
Standard metabolic rate, 288
Stapes, 214, 414
  amphibians, 846
Starch, 37, *39*
Starling, *962*
Starling's law of the heart, 315
Stasis, evolution, 176
Statocysts, 408
  hydrozoans, 602
Stauromedusae, *605*, 618
Stelleroidea, 779–785, 794
  Asteroidea, 779–783
  Ophiuroidea, 783–785
Steno, fossils, 20
Stenohaline animals, 361
Stenostomum, 622, 626, 637
  protonephridium, *351*

Stentor, 574, 576, 582
  buccal ciliature, 577
Stephanodrilus, 711
Steppes, 990
Sternomastoid, 223
Sternum, 212
  birds, 888
Steroids, chemical structure, 41, *41*
Sticklebacks, polyandry, 476, *477*
Stigma, 563
Stinging coral, 617
Stomach, 242, 249
  cow, *243*
  human, *250*
Stomatiiformes, 839
Stomatopoda, 737, 745
Stomodeum, 718
Stomolophus, *605*
Stone canal, 780
Stone tools, *947*
Stoneflies, *756*, 765
Stony corals, 611
Storage, food compounds, 271
Storage mites, 745
Storks, 905
Strain energy, 231
Stratified epithelium, 201
Stratum corneum, 203
Stratum germinativum, 203
Striated muscle, 214, *215*, *217*
Stridulation, 413
Strigiformes, 905
Strobila, 607, 634
Stromatolites, *56*
Strombus, 688
Strongylocentrotus, 786, 794
Strongyloides, 650
Struthioniformes, 904
Sturgeon, 830, *831*
Stylaria, *704*, 711
Style sac, 659
Stylet, *629*
  nematodes, *648*, 649
Sublingual gland, 247
Submucosa, 253
Subungulates, 929
Succession
  ecological, 980–984
  primary, 981–982
  secondary, 982
Succinate, 267
Sucrase, 237
Sugars, origin of life, 58
Sulfur, 255
Sulfur cycle, 1006
Summation, 379
  muscle contraction, *219*, 220
Sun star, *777*
Sunbird, winged, territoriality, 472, *472*
Sunfish, behavior, 476

Superficial cleavage, 513, *513*
Superior colliculus, *390*, 392
Support, 204–214
  exoskeleton, 205
Suppressor T cells, 336
Suprabranchial cavity, 678
Surface area to volume ratio, 190–191,
    191
Surface-volume relationship, 242
Surfactants, 282
Surinam toad, *857*
Survivorship curves, *966*
Swallow, nesting, 476
Swallowing, 247, *249*
Swallows, bank, nesting, 476
Swallowtail butterfly
  disruptive selection, *167*
  toxic secretions, 470, *470*
Swans, 905
Swarming
  bees, 761
  polychaetes, 702
Sweat glands
  cooling, 289
  in vertebrate skin, 204
Swifts, 901, 905
Swim bladder, 207
  bony fishes, 828, *829*
Swimmerets, 732
Sycon, 592
Syconoic sponge, 588, *589*
Symbiodinium, 612
Symbiosis
  algal, 612
  definition, 192
Symmetry
  in animals, 186
  grouping of phyla, 552
Sympathetic cords, 388
Sympathetic ganglia, 390
Sympathetic nervous system, 388, *389*
Symphyla, 765
Synapsida, 866, 878
Synapsis, 101
Synaptic transmission, *378*, 378–379
Synaptic vesicles, 379
Synaptid sea cucumbers, 789, *789*
Synovial fluid, 208
Synsacrum, 888
Systema Naturae, 542
Systematic zoology, 544
Systematics, 542–547
Systole, 311

T
T lymphocyte, 335, *335*
Tabulata, 618
Tabulate corals, 616
Tadpole shrimp, 745
Taenia, 630, 634, 638

Taeniarhynchus, 630, 634, *635*, 638
Taenidia, 278
Talitrus, 745
Tanning
    arthropod skeleton, 206
    vertebrate skin, 203
Tapetum lucidum, 425
Tapeworms, 634, 638
Tapirs, 931
Tarantulas, 725
Tardigrada, *764*
    cryptobiosis, 645
Tarsals, 214
Tarsier, 936, *937*, 951
Taste buds, 406, *407*
Taxa, 542
Taxonomic characters, 545
Taxonomy, 542
Tay-Sachs disease, 116
Tectonic lakes, 994
Tedania ignis, *586*
Teeth, human, *247*
Tegument, 630
Telencephalon, *390*, 393
Teleostei, 830, 838
    adaptive radiation, 833
Telestacea, 618
Telophase
    meiosis, 101–102
    mitosis, 100
Telson, horseshoe crabs, 720
Temperate deciduous forests, 991
Temperature, metabolic rates, 288
Temporalis, 223
Tendons, 208
Tenrecs, *910*, 931
Tergum, insects, 755
Termination factor, 129
Termites, 765
    cellulose digestion, 566
Territoriality, 472
Test, echinoids, 785
Testability, in science, 7
Testis, 486, *487*
Testudinata, 866, 877
Tetanus, *219*, 220
Tetrad, 101
Tetrahymena, *575*, 582
Tetrapods, skeleton, 211
Thalamus, 381, 392
Thalassicola, 581
Thaliacea, 804, *805*, 811
Theory, in science, 8
Therapsida, 878, 915
Theria, 917, 930
Thermal stratification, *995*
Thermocline, *995*
Thermodynamics, second law, 64
Thermoreceptors, 429

Thermoregulation, reptiles, 864, *865*
Thomson's gazelle, resource partioning,
    987
Thoracic duct, 308
Thoracic vertebrae, 212
Thorax, crustaceans, 731
Thorny corals, 618
Thrips, 765, *971*, *972*
Thrombin, 300
Thrombocytes, 300
Thrombus, 302
Thymine, 46
Thymus, 308
Thyone, *790*, 794
Thyroglobin, 446
Thyroid gland, 446–448
Thyroid secretion, *447*
Thyroxine, 446
Thysanoptera, 765
Thysanura, 755, 765
Tibia, 214
Ticks, 721, 726, 745
Tidal air, 284
Tidal cycles, 463
Tiedemann's bodies, 780
Tiger salamanders, *852*
Tinamiformes, 904
Tinamou, 904
Tinbergen, 459
Tintinnopsis, *574*
Tissue growth efficiency, 1000
Toads, *856*
    adaptation, 854–858
    reproduction, 857
    water balance, 366
Tongue, 247
    sensory areas, *408*
    snake, 870, *871*
Tongue bar, 799
Tonic muscle, 220
Tonus, 220
Tooth replacement, 246
Tooth shells, 689
Tooth structure, 246
Tooth types, 246
Tornaria larva, 799, 800
Torsion, 664
Tortoise
    Saddleback, *23*
    Galapagos, *27*
Toxocara, 651
Toxoplasma, 566
Trace elements, 255
Trachea, 280, 718
    insects, 754
Tracheal systems, 277–278
Tracheloraphis, 576
Tracheole, 277, *277*
Trachylina, 617
Tracts, 381

Transamination, 271
Transcription, 124–127, *125*, 436
Transfer RNA, 124, 128, *129*
Transforming principle, 112
Transitions, mutations, 136
Translation, 124, 127–131, *131*
Translocations, 141
Transport
    active, 71
    internal, 296–320
Transport systems, *296*
    function, 296
    types, 296
Transposable elements, 138–140
Transposons, 139
Transverse fission, *484*
Transverse plane, 186
Transversions, mutations, 136
Trapezius, 223
Tree frogs, *857*
Tree shrew, *937*
Trematoda, 630, 638
Trends, macroevolution, 178
Triacylglycerol, 39, *40*
Triarthrus, *718*
Tricarboxylic acid cycle, 267
Triceps, 223
Trichinella, 650, 651, *652*
Trichinellid nematodes, 651
Trichinosis, 651
Trichocysts, 576
Trichomonas, 566
Trichoplax, 593, *593*
Trichoptera, 766
Trichurus, 650
Triconodonta, *915*, 916, 930
Tridacna, 676
Trilobite, *551*, *718*
Trilobite larva, 720
Trimethylamine oxide, 362
Triple X females, 142
Triploblastic condition, 596
Tripsinogen, 239
Trisomy, 142, 144
Tritonia, 460, *460*
Trituberculata, 917, 930
Trochophore larva, 660, *660*
    polychaetes, *703*
Trombicula, 745
Trophic levels, 995
Trophoblast, 524
Tropical rain forests, 988
Troponin, 216
Tropoyosin, 216
Trypanosoma, *563*
    antigenic variation, 343
Trypanosoma brucei, 566, *567*
Trypanosoma cruzi, 565, 566
Trypanosomes, 565, 581
Trypsin, 237

Tsetse fly, 566, *567*
Tuatara, *868*
Tube-dwelling polychaetes, 698
Tube feet, 779
Tubular reabsorption, *358–359*
Tubular secretion, kidney, 359
Tubularia, *600*
Tubules of Cuvier, 790, *791*
Tubulidentata, 920, 931
Tubulin, 225
Tularemia, 727
Tullgren funnel, 722
Tuna, 835, *835*
Tundra, 991, *992*
Tunicate colony, *193*
Tunicates, 802–804
Tunicin, 802
Turban shell, *664*
Turbanella, *647*
Turbellaria, 623–628, 637
Turkeys, altrusitic behavior, 477
Turner's syndrome, 142
Turtles, 866–867, 877
   sea, osmoregulation, 362
   skeleton, *867*
Tusk shells, 689
Tusks, 924, 928
Twinning, 528
Twitch, *219*, 220
Tydeus, *724*
Tympanic cavity, 414
Tympanic membrane, 414
Typhlosole, 704

U

Uca, 732, *733*
Ulna, 214
Undulating membranes, 577
Ungulates, 927–928
Unguligrade gait, 927
Uniramia, 750–764
   classification, 765
   Insecta, 752–764
   myriapods, 750–752
Upwelling, 993
Uracil, 46
Urea, 349
Ureter, 353
Urethra, 353
Urey, 57
Uric acid, 349, *349*
Urinary bladder, 353
Urine, 350
   concentration, 359
   formation, 356–361
Urobilinogen, 357
Urochordata, 802–804, 811
Urodela, 849, 859
Uropods, 731, 732
Uropygeal gland, 885

Urosalpinx, *671*
Urostyle, 855
Urticina, *609*
Uterine tube, 499
Uterus, 499
Utriculus, 412

V

Vaccination, 325
Vacuole
   digestion, 578
   food, 241
Vagina, 495, 499
Vampire bat, altrusitic behavior, 478
Vasa efferentia, 496
Vascular tunic, 425
Vasopressin, 445
Vasopressin antiduretic hormone, 361
Vasotectin, 445
Veins, 297, 304, *317*
Vejovis, *721*
Veliger, 673
Veliger larva, *682*
Velum, 602, 673
Vena cava, 308
Venom, 871
Venous return, 319
Ventilation
   gas exchange, 273
   mammalian lung, 282–284
Ventral aorta, 304
Ventral body surface, 186
Ventricle, 304, 307
Ventricles, brain, 391, 397
Ventriculus, 758
Venus' flower basket, 592
Vermetus, 688
Vermiform appendix, 251
Verongia, 592
Vertebrae, 211, 212, *213*
Vertebral column, 211
Vertebral ossicles, 784
Vertebrata, 811
Vertebrates
   body wall, *189*
   characteristics, 807, *808*
   cladogram, *809*
   muscular system, 221
   origin, 807
   skeleton, 210–214
Viability, in reproductive isolation, 171
Vibracula, 772
Vicariant speciation, 173
Villi, intestinal, 242
Vintacrinus, 541
Virchow, 65
Visceral arches, 211
Visceral mass, 658
Visceral muscles, 221
Visceral skeleton, 201

Visceral smooth muscle, 218, *219*
Viscous drag, 226
Vital capacity, lungs, 284
Vitamin A, 256, 257
Vitamin C, 256, 257
Vitamin D, 256, 257, 453
Vitamins, 256, 257
Vitellaria larva, 792
Vitelline arteries, 522
Vitelline envelope, 491
Vitelline veins, 521
Viviparity
   aplacental, 524
   placental, 524
Viviparous, 524
Vocal cords, 280
Voltage-gated channels, 376
Volvocida, 563
Volvox, 564, *564*, 581
   reproduction, 566
Vomeronasal organ, 406, *406*
von Frisch, 762
Vorticella, *575*, 576, 582
   buccal ciliature, 577
Voyage of the *Beagle*, 22–23
Vultures, *900*, 901

W

Walking, mechanics, 228
Walking sticks, 765
Wallace, concept of evolution, 30
Walruses, *922*, 923
Warning coloration, 470, *470*
Wasps, 766
   parsitism, 759
Water, 254
Water bears, *764*
Water fleas, 737, 745
Water mites, 721, *724*, 745
Water vascular system, 778, 780, *781*
Watson, 459
   DNA, 118
Wax, 41
Weasel, *922*
Weaver bird, *894*
Web-building spiders, 725
Weberian ossicles, 414, *414*
Weinberg, 155
Whale, 928, 931
   Humpbacked, *18*
   osmoregulation, 363
   Sperm, diving, 281
Wheel organ, 805
Whelks, *668*, 670, *671*, 688
Whippoorwills, 901
Whipworm, 650
White blood cells, 300
White-crowned sparrow
   behavior, 462, *463*
   song learning, 465, *467*

White matter, 380
Wildebeests, resource partioning, 987, *987*
Wilson, 478
Wing, birds, 882, *882*
Wing loading, 883
Wings
   flight, 229
   insects, 755, *755*
Wobble, DNA code, 123
Wolf spider, *726*
Wolves, evolution, 24
Wood lice, 735
Wood roaches, cellulose digestion, 566
Woodchuck, *926*
Woodpeckers, 901, 905
Worm shells, 672, *672*, 688

Wuchereria, 650

X
X chromosome, 104
Xenarthra, 919, 931
Xeroderma pigmentosum, 147
Xyloplax, 792

Y
Y chromosome, 104
Yeast, fermentation, *265*
Yellow bone marrow, 208
Yolk sac, 524

Z
Zebra, *20*, 347, *927*
   resource partioning, 987

Zebra clam, 680
Zeitgeber, 464
ZFY genes, 535
Zoantharia, 618
Zoanthidea, 618
Zoanthus, 618
Zoea larva, 734, *734*
Zoochlorellae, 612, *613*
Zooflagellates, 565–566
Zooid, bryozoan, 770
Zoology, 4
Zoomastigophorea, 581
Zooxanthellae, 612, *613*
Zygote, 491